TREATISE ON INORGANIC CHEMISTRY

SOLE DISTRIBUTORS FOR THE U.S.A. AND CANADA:
ELSEVIER PRESS, INC., 2330 HOLCOMBE BOULEVARD,
HOUSTON 25, TEXAS. — FOR THE BRITISH COMMON-
WEALTH EXCEPT CANADA: CLEAVER-HUME PRESS, LTD.,
31 WRIGHT'S LANE, KENSINGTON, LONDON, W.8

TREATISE ON
INORGANIC CHEMISTRY

by

H. REMY

PROFESSOR OF INORGANIC CHEMISTRY,
UNIVERSITY OF HAMBURG (GERMANY)

Translated by

J. S. ANDERSON, F. R. S.

PROFESSOR OF INORGANIC CHEMISTRY,
UNIVERSITY OF MELBOURNE (AUSTRALIA),
FORMERLY SENIOR DEPUTY OFFICER, ATOMIC ENERGY RESEARCH
ESTABLISHMENT, HARWELL (ENGLAND)

Edited by

J. KLEINBERG

PROFESSOR OF INORGANIC CHEMISTRY,
UNIVERSITY OF KANSAS, LAWRENCE,
KANSAS (U.S.A.)

VOLUME I

INTRODUCTION AND MAIN GROUPS
OF THE PERIODIC TABLE

ELSEVIER PUBLISHING COMPANY

AMSTERDAM   HOUSTON   LONDON   NEW YORK

1956

Library of Congress Catalog Card Number: 55–11852

PRINTED IN THE NETHERLANDS BY N.V. DRUKKERIJ G. J. THIEME, NIJMEGEN

PREFACE

Since its first German edition in 1931, this book has been repeatedly revised and supplemented in accordance with the progress of chemistry; but its original plan has proved entirely satisfactory as a framework for new knowledge. In particular the systematic treatment of inorganic substances on the basis of the Periodic System has so far proved its worth as to have found its way into many other textbooks. Equally successful, when dealing with particular classes in each group, has been found the method of giving first a general introduction and then, before the detailed discussion of each substance, a survey of such topics of wider significance as may present themselves in its connection.

It is likely that future textbooks will give increasing prominence to the behaviour of groups of substances, as opposed to the discussion of individual ones; and in successive editions of this book, such chapters have become more numerous and fully developed. Each starts from some special property which for its further explanation needs the general theory, or else points the way to it. Thus is taken into account the line of thought of the chemist, progressing from particulars to generalities; and such knowledge then throws further light on subsequent problems.

The present translation is based essentially on the 7th and 8th German editions, but with further revision, improvement and supplementation before translation; notably in the sections on the chemical bonds, radioactivity, nuclear chemistry and the transuranic elements. I am very grateful to Professor J. S. Anderson, F.R.S., for the translation, carried out with great keenness and insight into the author's mind, and for suggesting many valuable improvements and additions to meet the needs of the wider circle of readers to whom the book is now addressed. And to the Editor, Professor Kleinberg, for his great interest my thanks are equally due.

Hamburg, October 1955   H. REMY

TRANSLATOR'S FOREWORD

During the past ten or fifteen years there has been a marked resurgence of interest in inorganic chemistry in Great Britain and the United States. The need for a detailed understanding of the chemical relationships between the elements in many pure and applied research problems has become apparent, and increased attention has been given to the chemistry of the elements in university courses. In consequence, many have felt the need for a comprehensive text in the English language setting out the subject in its present state of development. Such a work needs to cover the whole of the factual material and to bring it into proper relationship with the relevant thermodynamic, kinetic and structural data.

Amongst German texts, successive editions of Remy's *Lehrbuch der anorganischen Chemie* have gone far to meet this need, and it is to be hoped that this translation will prove valuable to a wide circle of readers. The translation has been based upon Professor Remy's latest revision of this standard work, and has been brought up to date as far as is practicable in a rapidly changing field of activity. With the author's permission, a few passages have been modified by the translator. Thermodynamic conventions have been changed where necessary, to follow American usage.

The author is indebted to Mr. J. D. M. McConnell B.Sc. for undertaking the task of proof correction, and for his care in the final preparation of the text.

Melbourne, September 1955   J. S. ANDERSON

CONTENTS

CONTENTS

Chapter 15. Sixth Main Group of the Periodic System: Oxygen-Sulfur Group

Chapter 16. Oxidation and Reduction

ABBREVIATED CONTENTS OF VOLUME II

SOME IMPORTANT PHYSICAL CONSTANTS

Absolute temperature of the melting point of ice, $T_{0°C} = 273.16°K$.

Atmosphere (normal pressure), 1 atm. $= 1.013250 \cdot 10^6$ dyne cm^{-2}.

Avogadro's number, $N_A = 6.0238 \cdot 10^{23}$.

Boltzmann's constant, $k = 1.38026 \cdot 10^{16}$.

Elementary quantum of electricity, $e = 4.8022 \cdot 10^{-10}$ e.s.u. $= 1.60186 \cdot 10^{-20}$ e.m.u. $= 1.60186 \cdot 10^{-19}$ coulomb.

Specific charge on the electron $= \dfrac{e}{m} = 1.7591 \cdot 10^8$ coulomb g^{-1}.

Units of energy

1 erg $= 10^{-7}$ joules (watt-seconds) $= 2.777778 \cdot 10^{-14}$ kWh $= 0.239006 \cdot 10^{-7}$ cal.

1 liter-atmosphere $= 1.013278 \cdot 10^9$ erg $= 101.3278$ joules $= 2.81466 \cdot 10^{-5}$ kWh $= 24.2180$ cal.

1 cm^{3}atm. $= \dfrac{1 \text{ l-atm.}}{1000.028} = 0.101325$ joules $= 2.81458 \cdot 10^{-8}$ kWh $= 2.42177 \cdot 10^{-2}$ cal.

1 cal (thermochemical gram-calorie) $= 4.1840$ joule $= 1.16222$ kWh $= 0.041292$ l-atm. $= 41.293$ cm^{3}atm.

1 cal$_{15°}$ ($15°$ calorie) $= 4.1855$ joules $= 1.00036$ cal.

1 ev (electron-volt) $= 1.6020 \cdot 10^{-19}$ joules $= 3.829 \cdot 10^{-20}$ cal.
   1 ev per molecule is equivalent to 23.064 kcal per mol.

Energy of a light quantum (photon) of wave length λ cm, $\dfrac{hc}{\lambda} = \dfrac{1}{\lambda} \cdot 1.98574 \cdot 10^{-16}$ erg $= \dfrac{1}{\lambda} \cdot 1.23954 \cdot 10^{-4}$ ev.

This corresponds to $\dfrac{1}{\lambda} \cdot 11.9617$ joules per mol or $\dfrac{1}{\lambda} \cdot 2.85892$ cal per mol.

Faraday (electrical charge per gram equivalent), $1 \, \mathfrak{F} = 96493$ coulombs. This represents 26.804 ampere-hours.

Gas constant, $R = 0.082054$ l-atm. $= 8.3144$ joules $= 1.9872$ cal.

Velocity of light, $c = 2.9979 \cdot 10^{10}$ cm sec^{-1}.

Liter (volume of 1 kg of air free water at its temperature of maximum density), $1 \, l = 1000.028$ cm^3.

Molar volume of ideal gas (at $0°$, 760 mm pressure) $= 22.414 \, l = 22414.5$ cm^3.

Acceleration of gravity (normal value, at sea level and $45°$ latitude), $g_0 = 980.665$ cm sec^{-2}.

Quantum of action (Planck's constant), $h = 6.6238 \cdot 10^{-27}$ erg sec.

Smythe factor $\left(= \dfrac{\text{physical atomic weight}}{\text{chemical atomic weight}} \right) = 1.000279$.

INTRODUCTION

Chemistry is concerned with the occurrence, isolation and artificial preparation of the different sorts of matter; with the study of their composition, properties, and reactions; and with the systematics and the reasons for the phenomena observed.

From analysis of the various sorts of matter presented to us in nature, it has been concluded that the whole known material world is formed from a limited number of fundamental sorts of matter—the *chemical elements*. A chemical element was formerly defined as a substance which could not be resolved by chemical means. However, it has been found that it is not always possible to delimit the concept of an element with sufficient precision by this criterion of resolvability. A chemical element is now defined as a species of matter of which all the atoms bear *the same nuclear charge* (cf. p. 1). When it became possible, by processes directly affecting the nucleus of the atom, to alter the charge on the nucleus, and thereby to transform atoms of one element into atoms of another element (cf. Vol. II, Chap. 13), it became possible to prepare elements which do not occur in Nature. Including these artificially produced elements (the products of nuclear transformations), 101 different species of atoms, differing in their nuclear charge, are now known with certainty—i.e., 101 chemical elements are now known (cf. Table II, Appendix). Only 91 of these have been detected in Nature.

One of the chemical elements plays a special role in the living world—namely, *carbon*. All living organisms consist of carbon compounds, some of great complexity. The part of chemistry which is particularly concerned with the compounds of carbon is therefore called *organic chemistry*. As distinct from this field, the study of the composition, properties, reactions, etc., of the whole material world, in so far as it does not properly fall within the subject of organic chemistry, is called *inorganic chemistry*.

Even among the compounds of carbon, it is usual to class as *organic compounds* only those which, in composition, structure and properties are related to the carbon compounds of importance for living organisms. This includes, admittedly, the great majority of the compounds of carbon, but there is a considerable number of carbon compounds which are quite different from these in composition and structure, and therefore in properties—for example, the metallic carbides. Such carbon compounds are regarded as *inorganic compounds*. 'Carbon compounds' and 'organic compounds' are thus not identical concepts. Elementary carbon also falls within the scope of inorganic chemistry, while its simplest compounds (especially its oxides and the acids derived therefrom, and the simplest hydrocarbons) are closely related both to organic and inorganic compounds. It is therefore usual to discuss these compounds in both inorganic and organic chemistry.

The aim and content of *inorganic chemistry* are concerned with the extraction and utilization of the different elements, and with a knowledge of their chemical individuality (i.e., with the occurrence, preparation and properties of the elements and their compounds). It treats also of the general and systematic laws deduced from a comparison of the elements with one another, and of the causes which, in the ultimate analysis, underlie the observed properties and behavior.

There are other branches of chemistry which overlap both organic and inorganic chemistry. The particular field of *analytical chemistry* is the study and application of those properties and reactions of matter which make it possible to decompose a substance into its constituents. It is then possible to determine whether a substance is simple or compound in nature, and to determine the proportions of its constituents. *Synthetic chemistry* is sometimes set in contrast with this, as the science which builds up substances from their constituents. It is now more usual to take *preparative chemistry* as the antithesis of analytical chemistry. Preparative chemistry, of which synthetic chemistry is a part, comprises the preparation of substances, with the techniques involved. It includes not only the preparation of compounds from simple substances, but also the preparation of elements from compounds—a by no means unimportant aspect of preparative inorganic chemistry. The methods and apparatus which are used in industry for chemical operations, form part of *chemical technology*, which deals also with the applications and methods of testing the substances manufactured. *Industrial chemistry* merges into chemical technology, in so far as it treats of the organization and operation of the chemical industry; it is also concerned with the economic importance of raw materials and the products derived from them by chemical processes.

There are also other important special branches of chemistry. *Electrochemistry* deals with the use of the electric current in carrying out chemical reactions. *Photochemistry* deals with the effects of light on chemical change. *Colloid chemistry* is the study of the properties of matter in a state characterized by a certain range of particle size (cf. Vol. II). *Radiochemistry* deals with the chemistry of radioactive substances, and such of their properties as are significant for chemistry. *Nuclear chemistry* is the study of the transformations of atomic nuclei, and of the properties and behavior of artificially produced species of atoms. *Crystal chemistry* is the study of the relation between crystal structure and chemical constitution. *Geochemistry* is concerned with the composition of the earth, and the principles underlying the terrestrial distribution of the elements. *Metallography* employs special experimental methods which have found application in other parts of chemistry. Mention should also be made of the *history of chemistry* as an important and instructive branch of the science.

In so far as they are concerned with inorganic substances, these branches of chemistry form part of the wider subject of inorganic chemistry. They will be dealt with in such scope as seems appropriate in a book covering the whole field of inorganic chemistry.

Mathematics and physics are the sciences which are most valuable subsidiary subjects in the study of chemistry. Some knowledge of the fundamental ideas of *crystallography* is also indispensable, and for an understanding of certain parts of chemistry an acquaintance with mineralogy and geology, or with the biological sciences, is also profitable.

During the last century, the application of physical methods and of mathematical reasoning created a new branch of science, *physical chemistry*. The principal objects of this subject lie in the quantitative application of physical methods to substances and reactions, with a view to discovering general laws, and deducing the behavior of matter from simple fundamental principles (e.g., from the laws of thermodynamics). It is not possible to draw a sharp line of distinction between the proper scope of physical chemistry, as compared with physics or with general chemistry, the more so as, during the last few decades, physical methods have been applied increasingly in all parts of chemistry. This is especially true of inorganic chemistry, so that it is no longer possible to present the subject from a modern and scientific standpoint without making extensive use of the results of physico-chemical investigations.

The field of study of inorganic chemistry is very wide. It comprises more than 100 elements, no two of which are exactly identical in behavior, and the enormous number of compounds which these elements form with one another. A survey of this wide field, and a grasp of those facts which are significant, can be made easier by organizing the multiplicity of facts in a suitable way. Such a synoptic viewpoint is provided by the Periodic System. This is based on the observed periodicity of chemical properties among the elements and—as will shortly be shown—it can be regarded as an inherently natural classification of the elements. Before considering the individual elements and their most important compounds, we shall therefore discuss the Periodic System. A consideration of the regularities which it reveals furnishes, at the same time, a brief survey of some of the most important properties of the elements and their compounds.

CHAPTER 1

THE PERIODIC SYSTEM

1. General

(a) The Natural Order of the Elements

In the Periodic System of the chemical elements [1]*, which was first formulated by Lothar Meyer and Mendeléeff, the elements are arranged in natural groups; the elements of each group are placed together not merely on the basis of superficial resemblances, but because they are necessarily related by a fundamental regularity. For this reason it is also called the *natural system of the chemical elements*. It is called the *Periodic System* because it is based on a recognition of the fact that all the chemical properties of the elements, and most of their physical properties, are a *periodic function* of some fundamental quantity which is characteristic of, and uniquely defined for, each of the elements, and which changes regularly from element to element. If the elements are arranged in the order fixed by this property, it is found that elements with almost the same (or at least very similar) properties recur at definite intervals.

An element is *uniquely characterized* by the nuclear charge of its atoms. Each element corresponds to a *definite nuclear charge number*. As will be shown later, this determines the number of electrons in the electrically neutral atoms and, as a consequence, the chemical properties of the element. The basis of the Periodic System is thus the fact that the chemical (and most physical) properties of the elements are a periodic function of the nuclear charges.

At the time of the discovery of the Periodic System, nothing was known of the structure of the atom, and therefore of nuclear charge numbers. The elements were therefore put in the order of their atomic weights, although in course of time it became obvious that in a few places this principle must be abandoned, and certain transpositions made in the order of the elements. It was therefore necessary to express the order in which the elements fit into the Periodic System by some special numbers. These were called the *atomic numbers* of the elements. It was later found that the atomic numbers could be defined independently of the Periodic System, and determined. Finally, it emerged that the atomic numbers of the elements are identical with their nuclear charge numbers.

When the elements (listed alphabetically in Table I, Appendix) are arranged in the order of their nuclear charges, Z (also given in the table), beginning with hydrogen (nuclear charge number $Z = 1$) up to californium, with the nuclear

* Numbers in square brackets refer to the correspondingly numbered recommendations for supplementary reading, at the end of each Chapter.

charge number $Z = 98$, it is found that roughly the same properties recur periodically in the resulting series. Thus in Table II (Appendix) the elements are arranged in horizontal series, in order of increasing nuclear charge. The series have been interrupted at certain places, and the sections so obtained have been written one below another. The periodicity of the properties is expressed in the fact that the elements which display similarity in behavior are found vertically below one another throughout. Such elements are said to be *homologous* elements.

Elements with particularly far reaching similarities in properties, *strictly homologous elements*, follow each other initially after every eight, later after every eighteen places. There are thus two sorts of Periods: short Periods with eight, and long Periods with eighteen members. The latter can, however, be sub-divided, as has been done in Table II (Appendix), in such a way that all the elements are fitted into eight vertical columns (the families of the Periodic System). Although, indeed, the elements of the long periods are not strictly homologous with the other elements in the same vertical columns, they are nevertheless related in certain respects, particularly as regards their valence states. They also display fairly close analogies in their chemical behavior, even if these are not as clear as the resemblances between the elements in the short periods. It is thus customary to sub-divide each of the eight families of the Periodic System into two groups, which are distinguished as the *Main Groups* (A-groups) and *Sub-Groups* (B-groups) respectively.

If each long period is sub-divided into two series in this way, it is necessary in one of the series to accommodate 10 elements in 8 places. This can be achieved by putting the last 3 of the 10 elements together into a single Sub-group each time, namely the eighth. Justification for this can be found in the fact that at this point the three *consecutive* elements display particularly far reaching similarities, whereas elsewhere the similarities between *vertically* related elements tend to predominate.

The *inert gases* constitute Group VIII. Being zero-valent, however, they are very often separated from Group VIII, and put into a 'Group o' at the beginning of the Table.

This grouping will be used in the following chapters for the systematic discussion of the elements and their compounds. Hydrogen will be treated first, then the inert gases, and after them the remaining elements. It is, however, better not to separate the inert gases from Group VIII of the Periodic System in drawing up a tabular presentation, since to do so obscures the relationship between the elements of Groups IVA to VIIA and the inert gases that follow them, which is of great importance in connection with valence theory. From the point of view of *atomic structure* it is also inappropriate to place the inert gases at the beginning of the Periodic System. The principles of atomic structure (which will be discussed in later chapters) show that every Period is characterized by the completion of a fresh electron shell. The formation of a new shell commences in each case with the elements of Group IA, and finds its completion in the inert gases.

At one point in the series of the elements, as arranged in order of their nuclear charges, there occur fifteen immediately consecutive elements which are peculiar in that their chemical properties are extraordinarily similar. In particular, these consecutive elements do not display the characteristic change of valence on going from one element to another. These fifteen elements are those having nuclear charges from $Z = 57$ to $Z = 71$; they cannot be accommodated in the Periodic System as written in its ordinary form. It is only a makeshift to put them together in a single place, namely that belonging to lanthanum (nuclear charge $Z = 57$) (as was formerly done). It is now usual to take the fourteen elements following lanthanum right out of the Periodic System and to put them as a special group, known commonly as the *lanthanide series*.

The elements following uranium, the *transuranic elements*, constitute a similar special series.

From the point of view of atomic structure, as will be seen later, it is probable that, following actinium ($Z = 89$), a special group of elements begins, of such a kind that it can be compared with the lanthanides. It has therefore been called the *actinide series*. The first elements of the actinide series (thorium, protactinium and uranium), however, have such a close chemical similarity to the elements of Groups IVB, VB and VIB that it is better, in considering their chemical behavior, to leave these elements in the Sub-groups. The elements following uranium, however, are completely different in their chemical properties from the elements of Groups VIIB and VIIIB, and fall therefore into a special group on chemical grounds.

Hydrogen also occupies a unique position. This is connected with the fact that it is the element with the smallest nuclear charge, and is therefore the first member of the series. Since it immediately precedes an inert gas (helium), it is appropriate in some ways to put it in Group VIIA, all the members of which come immediately before inert gases. This is the position given to hydrogen in Table II of the appendix. Although there are many parallels between hydrogen and the typical elements of Group VIIA, the halogens, it is also very different from them—again because of the consequences of its position as the first of the elements. It is better, in discussing its properties and chemical behavior, to treat hydrogen as an element apart.

The difficulties of fitting the elements cited into the Periodic System, as ordinarily written, come from the fact that there are really not two, but four sorts of periods in the natural system of the elements. The first period comprises only 2 elements, hydrogen and helium; the next two each contain 8; the two following each 18; the next Period however comprises in all 32 elements, namely all the elements from cesium to radon inclusive. The lanthanide series constitutes a Sub-group of this especially long Period.

With the elements following radon, another long Period begins, but the series of elements breaks off before its end is reached. Uranium was formerly the heaviest known element. Still heavier elements, known as *transuranic elements* are now known (cf. Table II, Appendix, p. 838).

The number of elements in the various long periods mentioned are in the ratios $2 : 8 : 18 : 32$ i.e., as $2 \times 1^2 : 2 \times 2^2 : 2 \times 3^2 : 2 \times 4^2$. This series of numbers makes it appear as if the progression of the number of elements in the periods follows some comparatively simple law. This is indeed the case, and we shall later discern the basis for the increasing lengths of the Periods, in terms of the concepts which have developed from Bohr's theory of the structure of the atom.

The customary numbering of the Periods differs from what would logically follow from the foregoing. The regular increase in the length of the Periods has been recognized only relatively lately (or at least has only lately received attention). For this reason, it has become customary in numbering the Periods to leave out of account the Period consisting only of the two elements hydrogen and helium, and to begin the numbering with the Period lithium to neon. This is therefore designated the first short Period, and so on, as given in Table II (Appendix.)

It is usual to denote the families in the Periodic System by Roman numbers, the Periods or series by Arabic numbers. The Main and Sub-groups into which the individual families are broken up will be represented by Roman figures throughout the book.

TABLE 1

LONG-PERIOD FORM OF THE NATURAL SYSTEM OF THE ELEMENTS

Period	IA	IIA	IIIB	IVB	VB	VIB	VIIB	VIIIB	VIIIB	VIIIB	IB	IIB	IIIA	IVA	VA	VIA	VIIA	VIIIA
1																	1.0080 H 1	4.003 He 2
2	6.940 Li 3	9.013 Be 4											10.82 B 5	12.011 C 6	14.008 N 7	16.0000 O 8	19.00 F 9	20.183 Ne 10
3	22.991 Na 11	24.32 Mg 12											26.978 Al 13	28.09 Si 14	30.975 P 15	32.066 S 16	35.457 Cl 17	39.944 A 18
4	39.100 K 19	40.08 Ca 20	44.96 Sc 21	47.90 Ti 22	50.95 V 23	52.01 Cr 24	54.94 Mn 25	55.85 Fe 26	58.94 Co 27	58.69 Ni 28	63.54 Cu 29	65.38 Zn 30	69.72 Ga 31	72.60 Ge 32	74.91 As 33	78.96 Se 34	79.916 Br 35	83.80 Kr 36
5	85.48 Rb 37	87.63 Sr 38	88.92 Y 39	91.22 Zr 40	92.91 Nb 41	95.95 Mo 42	[99] Tc 43	101.6(z) Ru 44	102.91 Rh 45	106.7 Pd 46	107.880 Ag 47	112.41 Cd 48	114.76 In 49	118.70 Sn 50	121.76 Sb 51	127.61 Te 52	126.91 I 53	131.3 Xe 54
6	132.91 Cs 55	137.36 Ba 56	138.92 La 57	178.6 Hf 72	180.95 Ta 73	183.92 W 74	186.31 Re 75	190.2 Os 76	192.2 Ir 77	195.23 Pt 78	197.0 Au 79	200.60 Hg 80	204.39 Tl 81	207.21 Pb 82	209.00 Bi 83	210 Po 84	210 At 85	222 Rn 86
7	223 Fr 87	226.05 Ra 88	227 Ac 89	232.05 Th 90	231 Pa 91	238.07 U 92	93 etc. Transuranic elements, see below											

(Lanthanides)

Lanthanides

Period	IVB	VB	VIB	VIIB	VIIIB	VIIIB	VIIIB	IB	IIB	IIIA	IVA	VA	VIA	VIIA
6	140.13 Ce 58	140.92 Pr 59	144.27 Nd 60	[145] Pm 61	150.43 Sm 62	152.0 Eu 63	156.9 Gd 64	158.93 Tb 65	162.46 Dy 66	164.94 Ho 67	167.2 Er 68	168.94 Tm 69	173.04 Yb 70	174.99 Lu 71

Transuranic elements

Period	VIB	VIIB	VIIIB	VIIIB	IB	IIB	IIIA	IVA	VA
7	237 Np* 93	239 Pu** 94	[241] Am** 95	[242] Cm** 96	[243] Bk** 97	[246] Cf** 98	[255] E* 99	[255] Fm** 100	[256] Mv* 101

* Longest lived and most accessible isotope.

The form of presentation used in Table II (Appendix), in which the elements of the Long Periods are divided into two series, so that the short Periods determine the horizontal breadth of the system, is known as the *Short Period presentation*, as contrasted with the *Long Period presentation* in which the elements of the Long Periods are each time included in a single series. The Short Periods must then be broken up accordingly. Mendeléeff had already used both the short-periodic and the long-periodic mode of tabulation. The adjacent Table I sets it out in a form which is based directly on that already used by Mendeléeff, but completed by the insertion of the elements discovered subsequently.

If the arrangement is so chosen as to group the elements around the inert gases (as has been done in Table 24, p. 125), the resulting long-periodic presentation avoids the need for splitting up the Short Periods into two portions, one placed at the beginning, and the other at the end of the System. A combination of Long and Short Periodic form, obtained by taking the elements of the Sub-groups, as well as the lanthanides, out of the main system, has been proposed by E. Wiberg (*Z. angew. Chem.*, 49 (1936) 480).

A form of the Periodic Table which makes provision for the lanthanides, and into which the transuranic elements can also be fitted, was devised by Bohr (1923), following the proposals of Thomsen (1895) (see Table 2). In this form, homologous elements are linked by lines, and the relationships which follow from considerations of atomic structure are emphasized. The so-called 'transition' elements—i.e., the elements in which electronic *d*-levels (see pp. 117–120) are filled, and also the elements in which *f*-levels are filled, are shown by surrounding them with a frame.

The chemical behavior of the elements of the Main Groups of the Periodic System is determined principally by the position of these elements with respect to the inert gases, as will be shown in Chap. 5. On this basis, the classification of the elements into those of the Main groups and those of the Sub-groups can be made unambiguously as follows: Elements (other than hydrogen, which takes a special place) with atomic numbers which are *1 to 2 units larger*, or *5 to 1 units smaller* than the atomic number of an inert gas are assigned to the Main Groups. The remaining elements (apart from the lanthanides and the transuranic elements, which constitute special series) belong to the Sub-groups of the Periodic System. The whole of the elements fall thus into the following 5 Groups: [i] hydrogen, [ii] Main Groups of the System, [iii] Sub-groups of the System, [iv] lanthanide series, [v] transuranic elements.

Similarities in chemical behavior and physical properties are not restricted to elements belonging to the same Main Group or the same Sub-group. They can be discerned, to a greater or less degree, also between elements belonging to the same *family*, although not to the same Main group or same Sub-group. This shows itself most distinctly in the IIIrd and IVth families of the Periodic System, and is, indeed, so marked that for a long time there was uncertainty as to which elements should be regarded as the closer homologues of boron and aluminum (or of carbon and silicon respectively)—i.e., which elements should be put in the Main Group, and which in the Sub-group. Defining the Main groups as has been done above, the same subdivision of the elements is obtained as can be reached on grounds of atomic structure (cf. pp. 118–120). This allocation of the elements has now gained general acceptance, as being the most appropriate and most useful.

(b) The Three Origins of the Periodic System

The Periodic System, in the form in which we now possess it, has grown from three roots. These were:

(i) the attempt to arrive at a natural classification of the chemical elements;

(ii) the recognition that there is an inner connection between some fundamental quantity, characteristic of each element, and certain other properties of the elements;

TABLE 2

PERIODIC SYSTEM ACCORDING TO THOMSEN AND BOHR

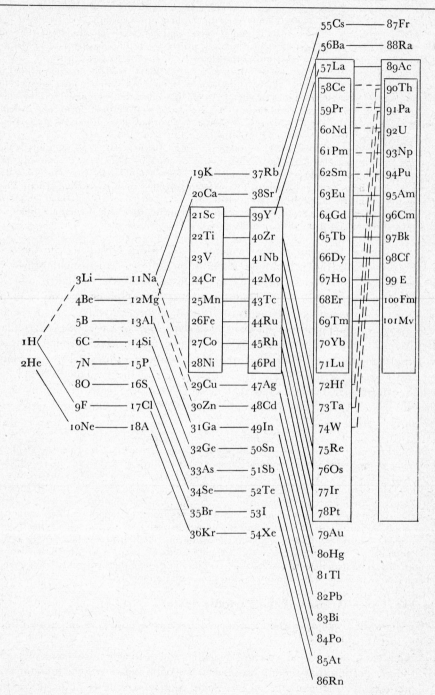

(iii) recognition of periodicity in properties, and in chemical behavior, as being dependent on this fundamental quantity.

These concepts, fundamental to the Periodic System, developed to some extent independently of each other. They were then summarized in the form of the Periodic System at almost the same time (1869) by two workers, Lothar Meyer and Mendeléeff.

(c) Historical

The fact that certain elements (the alkalis, alkaline earths, halogens) constitute natural groups was already known long before the formulation of the Periodic System. Attempts to arrange the remaining elements in groups in corresponding manner did not at first prosper, for the reason that no criterion was known by which it was possible to make a clear assignment in the case of the less clearly related elements. The Rule of Triads, published by Döbereiner in 1829, brought an important advance in this direction.

Döbereiner [2] noticed that when closely related elements were arranged in groups of three (triads), the atomic weight of the middle one appeared as the arithmetic mean of the highest and lowest atomic weight. For example, the atomic weight of selenium is roughly equal to the mean of the atomic weights of sulfur and tellurium.

On the basis of the Rule of Triads, Lenssen in 1857 was able to arrange all the elements then known into one system, from a consistent standpoint. The system proposed by Odling in the same year, also based on Döbereiner's Rule of Triads, already contained many elements grouped in the same way as in our present Periodic System. The Rule of Triads, however, left much room for chance, since it is possible to group elements of quite dissimilar character together in triads, on the basis of atomic weights alone. Nevertheless, it brought the important advance that relationships which could be determined quantitatively were made the basis of 'similarity'. It was, moreover, important that the Rule of Triads brought out, for the first time, the idea that there is some connection between the properties and the atomic weights. At that time, however, there was no procedure known whereby atomic weights could be established with certainty. For this reason, Cannizzaro's proposal (1860), that the molecular weights of gases and vapors based upon Avogadro's hypothesis provide a basis for fixing atomic weights, was of crucial importance for the discovery of the Periodic Law, since it created a reliable basis for the determination of atomic weights.

In 1862 Béguyer de Chancourtois arranged the elements, in the order of Cannizzaro's atomic weights, on a spiral line, wound around a cylinder in such a way that elements with similar properties stood one above another. Here, for the first time, the concept of periodicity found its expression; nevertheless the work did not influence the development of chemistry, since it received little attention and was soon completely forgotten.

The law of periodicity was discovered for the second time in 1864 by John Newlands, who again arranged the elements according to the atomic weights based on Cannizzaro's system, and found that elements with similar properties recurred each time after every seven elements (the inert gases were not known at that time). He called this relation that he had discovered *the Law of Octaves*, but found at first no appreciation of it. Only in 1869, when Mendeléeff brought out his table, and in the following year when that of Lothar Meyer was published (already sketched out in 1868, and agreeing substantially with Mendeléeff's), was the conception of an inner relationship between atomic weights and properties able to achieve acceptance. It must be remembered that the systems of de Chancourtois and Newlands still contained considerable arbitrariness and discrepancies which must have detracted from the plausibility of the underlying ideas. The systems of Mendeléeff and Lothar Meyer [3] were largely free of such errors; these could not be completely excluded, however, because of the inadequacy of reliable knowledge of all the atomic weights.

At the time the Periodic System was propounded, the number of the known elements was considerably smaller than it is today. Quite a number of gaps were left in the series of the elements when these were set out so that homologous elements were everywhere placed one below another. In particular, the places for the homologues of boron, aluminum and silicon with the (present day) atomic numbers 21, 31, and 32, were at that time unoccupied. From the need for these gaps, Mendeléeff deduced the existence of as yet unknown elements belonging to these places —eka-boron*, eka-aluminum, eka-silicon—and he predicted their properties on the basis of the regular relationships of the Periodic System. The discovery of these elements, which followed shortly afterwards, (cf. under scandium, gallium, and germanium) then constituted a striking confirmation of the correctness of the views underlying the Periodic arrangement.

A few gaps still remained in the Periodic Table, even after the number of the known elements had been considerably enlarged by the discovery of the inert gases (1894–99) and of new elements of the rare earth group, and after the discovery of the radio-elements polonium, radium, actinium and protactinium. This became very apparent with the discovery of Moseley's law (p. 225), since it then became possible to assign the atomic number of every element unambiguously, by measurement of its 'characteristic X-ray spectrum' (see pp. 223–226). It was thereby found that the places belonging to the elements of atomic numbers 72 and 75 were still unoccupied. The elements concerned were discovered in 1922 and 1925, and were named *hafnium* and *rhenium*. The possibility of identifying the elements by their X-ray spectra, without having first separated them from the elements accompanying them in Nature, was the decisive factor in their discovery.

The search for the elements of atomic numbers 85 and 87 (given the names eka-iodine and eka-cesium before their discovery) was for a long time fruitless, although their existence had long been suspected on the basis of the Periodic Table. As may be seen from Table II, (Appendix), both elements (now given the names of *astatine* [At] and *francium* [Fr]) come in a region of the Periodic System in which all the elements are *unstable*—i.e., undergo radioactive decay. From this it may be concluded that, in all probability, *no stable nuclear species* of elements 85 and 87 exist. It can, moreover, be predicted that their unstable nuclei will be short-lived, since experience shows that elements of odd atomic number tend to have shorter half-lives than neighboring elements of even atomic number. Radioactive elements of rather short life can only exist in Nature in appreciable amounts if they are continuously generated by the disintegration of elements of long life. Only thorium and uranium (U I and AcU) fulfil this condition. Hence the only hope of finding elements 85 and 87 in Nature lay in the possibility that they were formed in the radioactive disintegration of thorium or uranium. Obviously all attempts to discover these elements in non-radioactive minerals must be fruitless. The elements sought for were, indeed, discovered among the decay products of thorium and uranium (actinouranium)—francium in 1939 by Miss Perey, and astatine in 1943 by B. Karlik and T. Bernert (see Chaps. 6 and 17).

Astatine had already been prepared *artificially* (by E. Segré, in 1940), by nuclear reactions before it was discovered in Nature. Two other gaps in the Periodic Table have also been filled by the artificial preparation of the elements concerned—namely elements 43 and 61. The rules governing nuclear disintegration (Vol. II, Chap. 11) indicate that these elements must also be unstable, and this conclusion is in agreement with experiment. The artifically prepared elements 43 and 61 have been given the names of *technetium* (Tc) and *promethium* (Pm). Technetium and promethium do not form part of the natural radioactive decay series. Although the longest-lived isotopes of these elements decay much more slowly than those of astatine and francium, their disintegration is rapid enough for no detectable quantity of either element to remain in the earth's crust, even if it had originally been formed. These two elements therefore do not occur in Nature, and can only be obtained artificially.

* Eka, Sanskrit = one.

In the series of elements from hydrogen to uranium, i.e., from atomic number 1 to atomic number 92, there are thus now no remaining gaps. It has, in fact, been possible (by atomic transmutation), to prepare elements of *higher atomic number* than uranium. These are the transuranic elements already mentioned, with atomic numbers 93, etc. (at present 93–101). They will be considered in Vol. II.

2. Periodicity of Chemical Properties

The periodicity of chemical properties shows itself particularly clearly in the composition of the chemical compounds. The composition of a compound is determined by the valences of its constituent elements.

(a) Valence

The valence of a chemical element is a number which states how many atoms of hydrogen can combine with, or be replaced in some other compound by, one atom of the element. The valences so defined are sometimes called *formal* or *stoichiometric valences*, to distinguish them from the electrochemical valence or electrovalence.*

The *electrochemical valence* is equal to the number of unit electric charges borne by each atom of the element in the state considered.** According as the atom of an element in any compound is positively or negatively charged, we distinguish between positive and negative electrovalence. In the elementary state, the elements are electrochemically zero-valent.

In a classical *structural formula****, the stoichiometric valence of each atom represents the number of 'valence bonds' going out from that atom. Today, the 'valence bond' in a structural formula is usually employed as a symbol for a chemical bond effected by an electron pair (see Chap. 5). The number of electron pairs which one atom shares with others is now often called the *bond number* (cf. p. 130). In organic compounds, the bond numbers of the atoms are usually the same as their stoichiometric valences. This is often not the case among inorganic compounds. Numerous examples of this will be encountered later.

In many cases, the formal valence can be strictly stated, whereas the electrovalence is not known with certainty. The definition of electrochemical valence is inapplicable to many compounds. Thus it is obviously pointless to enquire what is the electrovalence of carbon in chloroform, $CHCl_3$, for example, since this is a covalent compound—i.e., is not built up from oppositely charged atoms. On the other hand there are cases in which the electrovalence is well defined but the formal valence not uniquely determined. This often

---

* The expression *valence* is also often used for the force which binds the atoms together. In order to avoid confusion springing from this usage, Biltz has suggested that the two concepts should be distinguished by the terms *valence number* and *valence force*.
** The charge on the hydrogen ion which is equal in magnitude, but opposite in sign, to the electronic charge is used as the unit of charge both here and later, unless otherwise stated.
*** The chemical structural formula, in the classical sense, relates to the structure of the *molecule* of a compound. For compounds which, in the crystalline state, are not built up from molecules (i.e. from atomic or ionic aggregates of finite size), classical structural formulas can be strictly valid only for the gaseous state. In so far as they are not constructed from molecules, but are built up directly from atoms or ions in (in principle) infinite numbers, the crystalline compounds can be classed structurally with the coordination compounds (cf. Chap. 11). The concept of valence (stoichiometric or structural valence) derived from the substitution principle, cannot be applied to these (cf. p. 392 and 399).

applies to the valence of the central atoms in coordination compounds (cf. Chap. 11). The electrovalence always refers to the element in a particular state. The stoichiometric valence can also be applied to the element as such. Thus one says not only 'barium can function as bivalent' but also, in full accordance with the definition of valence, 'barium *is* bivalent', 'sulfur *is* bivalent, quadrivalent, and sexavalent'. According to the definition of stoichiometric valence the statement means that one atom of barium is able to combine with two, and one atom of sulfur with two, four, or six atoms of an element equivalent to hydrogen. The example last mentioned shows, at the same time, that the valence may be variable.

Coordinative valence or covalence is also to be distinguished from stoichiometric valence and electrochemical valence; on this point see p. 394.

A concept which can be applied to any compound is that of the *oxidation state* or *oxidation number*. The oxidation numbers of the various elements making up a compound can be found by dividing up the charges on the atoms in such a way that, wherever two dissimilar elements are linked, the valence electrons are assigned to the more electronegative element. Between bonded atoms of the same kind, the valence electrons are equally shared. By virtue of its definition, the oxidation number makes no statement about the *actual* charge distribution in the compound. It can therefore be applied also to covalent compounds. Thus carbon in CCl_4 has the oxidation number $+4$, in CH_4, the oxidation number -4, and in $CHCl_3$ the oxidation number $+2$. It is often very convenient to make use of the idea of the oxidation number when considering oxidation-reduction reactions.

If the electrons in every covalent bond are distributed uniformly between the linked atoms (irrespective of their electrochemical character), the so called *formal charges* borne by the atoms are obtained.

For the elements of the Main Groups of the Periodic System, the electropositive maximum valence number of each element is identical with *its group number*.

Oxygen, fluorine and the inert gases do not come into this rule, for they never function as electropositive, since they are not able to give up electrons to other elements (cf. p. 129). Considering hydrogen a special case, the only exception to this rule is presented by bromine which, as far as is known, exhibits a maximum oxidation number of five, although it belongs to Group VIIA.

In so far as the elements are able to function as electropositive*, they exert their highest valence towards oxygen, although they also frequently display lower valences towards oxygen as well. The elements frequently exert the same valence towards fluorine as towards oxygen, and often also towards chlorine, sulfur, and other non-metals. Many elements display a lower valence towards hydrogen. This rises from one to four in Families I to IV, and then decreases from four to zero in Families IV to VIII. In the Main Groups IV to VII, the sum of the group number and the valence towards hydrogen is 8.

It is characteristic of the hydrogen compounds of the elements of the IVth and following Main Groups that they are volatile. The valences which these elements exert in their volatile hydrides can be regarded, in the electrochemical sense, as their electronegative valences.** The same applies to the compounds of these elements with strongly electropositive metals. These metal compounds can be included within the above rule if it is stated as follows: the sum of the Group Number and the maximum electronegative valences is equal to 8, for the elements of Groups IVA to VIIA.

* The justification for considering the non-metals as the electropositive constituents in their compounds with oxygen is considered in Chap. 5.
** For a reservation which must be made in using this expression see, for example, p. 412.

The composition of the most important compounds, as dependent on position in the Periodic System, can be represented by the following Table.*

TABLE 3

MOST IMPORTANT TYPES OF COMPOUND IN THE MAIN GROUPS
OF THE PERIODIC SYSTEM

Main group	I	II	III	IV	V	VI	VII	VIII
Highest normal oxide	R_2O	RO	R_2O_3	RO_2	R_2O_5	RO_3	R_2O_7	—
Simplest hydrogen compound	RH	RH_2	$\begin{cases} RH_3 \text{ or} \\ R_2H_6 \end{cases}$	RH_4	RH_3	RH_2	RH	—

On the whole, the elements of the Sub-groups of the Periodic System display towards oxygen and the non-metals the same valences as do the elements of the Main Groups. They do, however, tend to exhibit a much greater variability of valence than the elements of the Main Groups. In the Ist Sub-group, univalence is quite subordinate in the case of copper and gold. In the VIIIth Sub-group, the octavalence corresponding to the group number has been observed only with ruthenium and osmium.

Hydrogen compounds, similar in type to those which appear in the Main Groups of the Periodic System, and obtainable in weighable amounts, are not ordinarily formed by the elements of the Sub-groups.

Hydrogen compounds of different type are formed by a number of Sub-group elements, however. Many of the elements of the Sub-groups—especially those of Groups IIIB and IVB—take up considerable amounts of hydrogen into solid solution. The IIIB and IVB elements also form compounds with hydrogen. These are variable in composition, and it would appear that they are related in type to the intermetallic compounds (cf. p. 45).

The existence of a considerable number of volatile diatomic hydrides has been detected by band spectroscopic investigations, although they are not isolable in weighable quantities. They are formed not only by many elements of the B Sub-groups, but also by elements of the Main Groups, which do not often otherwise function as univalent. In a few cases it has been possible to confirm their existence by other methods also (cf. Vol. II). All these cases apparently involve compounds of which the molecules are only stable as long as they do not undergo any collisions. In this connection, it may be pointed out that aggregates of atoms which cannot be isolated because of their short lifetime are customarily regarded by physicists as the molecules of compounds. Among these belong, for example, many which the chemist is accustomed to consider as *radicals*, incapable of existence in the free state.

(b) Electrochemical Character

Elements which can readily exist as electropositive elementary ions are said to be *electropositive* elements; those which generally form only negative elementary ions

* In this Table, R represents any chemical element belonging to the Group in question. The letter R in chemical formulas will subsequently be used to represent any element or radical quite generally.

are *electronegative* elements. The metals are electropositive, the non-metals more or less distinctly electronegative. The distinction is not sharp, however; one speaks of a more or less strongly developed electropositive or electronegative character, depending on how clearly the tendency to form the corresponding ions is manifested. The tendency to form positive or negative ions is also referred to as *electroaffinity*. With many elements, the electroaffinity is so small that it is not possible to detect the formation of any free ions. It is often possible, however, to show (or at least to obtain indication) that they bear an electrical charge in their compounds.

Compounds made up from constituents which are obviously oppositely charged are termed *ionic* or *heteropolar* compounds; those which are built up from constituents not obviously bearing charges of opposite sign are called *covalent* or *homopolar* compounds. From the foregoing it follows that the boundary between hetero- and homopolar compounds is not sharp.

The terms polar and nonpolar compounds are also used instead of 'heteropolar' and 'homopolar'. However, in accordance with a proposal of Debye, we shall consider as *polar* such compounds as behave in a polar manner, i.e., which act on their environment as dipoles.*

This may also happen on the part of molecules of homopolar structure. It does not follow that every molecule which is heteropolar in structure is dipolar in properties. If the charges are symmetrically distributed, in such a way that their effects on the surroundings mutually compensate, the resultant dipole moment is zero.

In the English literature, molecules built up from uncharged atoms are generally called *covalent* molecules, since in them the atoms are bound together by quantum mechanical resonance forces or covalences (see later).

In the Main Groups of the Periodic System the electropositive character decreases from group to group, going from left to right (i.e., with increasing group number), whereas the electronegative character rises correspondingly. In each series, therefore, the most strongly electropositive element stands in the Ist and the most strongly electronegative in the VIIth Main Group (the inert gases may be omitted from consideration because of their inability to form chemical compounds).

Within each Main Group, the electropositive character increases with rising atomic number, i.e., descending each column in the Periodic System; the electronegative character diminishes in the same sense. The most strongly electropositive elements accordingly stand in the Periodic System at the bottom on the left (cesium and francium), the most strongly electronegative at the right at the top (fluorine). Because of this regularity, the non-metals are all congregated towards the upper right hand corner of the Periodic System. The metals, insofar as they belong to the Main Groups, stretch upwards and sideways from the lefthand lower corner. The Sub-groups consist exclusively of metals as do the lanthanides and the transuranic series. The boundary between the regions occupied by metals and non-metals in the Main Groups of the Periodic System is marked by the elements boron–silicon–arsenic–tellurium–astatine.

* By a *dipole* is meant a system of electric charges which are so distributed that the centers of gravity of the positive and negative charges do not coincide. If the dipole is produced only under the influence of external forces, one speaks of an *induced dipole*. The process of formation of dipole is then termed *polarization*. A single neutral atom may also become a dipole as a consequence of polarization (cf. p. 48).

The more two elements differ in their electrochemical character, the greater, in general, is their tendency to form compounds with each other. In view of what has been said, it follows that, in general, the tendency of the elements to combine with each other is the greater, the further apart they stand in the Periodic System.

Tammann pointed out that, as a rule, the elements of any one Group (Main Group or Sub-group) generally do not combine with one another. If one excludes the elements of the first two Short Periods, this rule is followed by most elements both of the Main and the Sub-groups of the Periodic System, and especially by the metals.

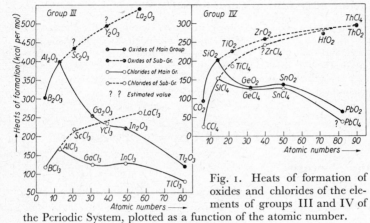

Fig. 1. Heats of formation of oxides and chlorides of the elements of groups III and IV of the Periodic System, plotted as a function of the atomic number.

(c) Basic and Acidic Character

In the Main Groups of the Periodic System, the basic character of the hydroxides increases concurrently with the electropositive nature of the elements forming them. The strongest bases are derived from the most strongly electropositive elements.

In the Main Groups, the acid character of the volatile binary hydrogen compounds increases from left to right in each series of the Periodic System, with the electronegative character of the elements combined with hydrogen. Within each group, on the other hand, it increases from top to bottom, that is in a direction opposite to the trend in electronegative character. The reason for this is discussed later.

The strength of the oxy-acids is not quite so simply related to position in the Periodic System, being dependent upon several different factors. If oxy-acids of analogous composition and constitution are compared with one another, the rule applies here also, that their strength increases from left to right within one and the same series of the Periodic System—e.g., in the series H_4SiO_4, H_3PO_4, H_2SO_4, $HClO_4$ —, but on the other hand, it decreases in going from top to bottom within any one Group.

These rules do not hold for the Sub-groups of the Periodic System.

(d) Heats of Formation of the Oxides and Chlorides

An approximate measure of the ability of the chemical elements to combine is often provided by the heats of formation of their oxides, chlorides, and similar simple compounds.

These heats of formation are clearly seen to depend on the position of the elements in the Periodic System.

As an example of the trend of heats of formation within a family of the Periodic System, the heats of formation of the oxides and chlorides of the elements of Groups III and IV are represented in Fig. 1, as they vary with the atomic number of the elements. The curves bring out very clearly that in both families, both the Main Group elements and the Sub-group elements display a close and systematic relationship, especially to the second Main Group elements (aluminum and silicon respectively). A similar trend in the curves is obtained also for the other Groups of the Periodic System (W. A. Roth, *Z. physikal. Chem.*, A, 159 (1932) 1).

(e) Tendency of Oxides to Combine with Water

The tendency of the oxides to form compounds of definite stoichiometric composition with water decreases strongly in the Main Groups of the Periodic System from Group I to Group IV, and then rapidly increases again on going further towards the right. In and around Group IV, the oxide hydrates often appear in the form of jelly-like products, of very variable composition, which mostly contain the water combined, not so much chemically, as by capillarity and osmotic forces (see Chap. 12).

3. Periodicity of Physical Properties

(a) Atomic Volumes

The product of the atomic weight and the specific volume of an element, or the quotient of the atomic weight and density*, is known as the *atomic volume*.

$$\text{Atomic volume} = \text{Atomic weight} \times \text{Specific vol.} = \frac{\text{Atomic weight}}{\text{Density}}$$

This expression is called the atomic volume because it gives a rough relative measure for the space occupied by the atoms.

The specific volume of any substance is the volume occupied by 1 g; hence the atomic volume is the volume taken up by 1 g atom of the element in question. Since 1 g atom of all elements contains the same number of atoms, the atomic volumes of different elements would stand in the same relation to one another as the volume actually occupied by their individual atoms, provided that the atoms of all elements were similarly arranged, and filled the same fraction of the space available to them. This is not strictly true, so that the atomic volumes only give these ratios approximately.

The atomic volume is a periodic function of the atomic number. The validity of this rule, first formulated by Lothar Meyer in 1870 (with reference to atomic weight, not atomic number), can be seen from the graph shown in Fig. 2. The atomic numbers of the elements are taken here as the abscissae, their atomic volumes as ordinates. It can be seen that, starting from the alkali metals, the atomic volumes on each occasion first decrease, and then increase again, till a peak and a sharp maximum of the curve is reached at the position of the next alkali metal. Not only the alkali metals however, but all elements in the Periodic System belonging to the same Group, lie at corresponding positions on the curve. The most strongly electropositive alkali metals lie on the peaks, the less strongly electropositive alkaline earth metals all on the falling branches of the curve, but closer to the alkali metals than the still more weakly electropositive aluminum and its homologues. The electronegative elements lie on the sections of the curve which rise sharply once more, homologous elements of the Periodic System again being in corresponding positions. The most important sets of homologous elements are joined by dotted lines in the figure.

* The density is numerically equal to the specific gravity, referred to water at 4° C. Since the atomic weight is a pure number, the atomic volume has the dimensions $cm^3 \ g^{-1}$.

The *compressibilities* of the elements appear to depend on their Atomic Volumes, and may also be represented as a periodic function of the atomic number, accordingly.

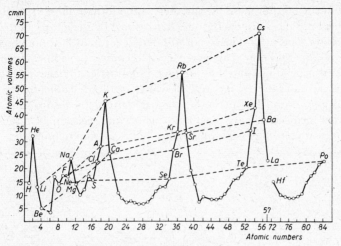

Fig. 2. Curve of atomic volumes.

(b) Apparent Atomic and Ionic Radii

The space taken up by the atoms in the elements, in their crystalline state, can now be represented much more precisely than by the function called the atomic volume. It is now possible to determine the distance of the centers of the atoms from one another in crystals with considerable accuracy, by means of measurements which will be discussed later. If we suppose these centers to be surrounded by spherical shells, so as just to touch one another, the radius of these spheres can be defined, in the manner of W. L. Bragg and V. M. Goldschmidt, as the *apparent atomic radius*. This holds for crystals built up from uncharged atoms—as, for example, the crystals of the elements. With substances built up from electrically charged atoms (ions) we arrive in a similar manner at the apparent *ionic radii*. In the latter case, however, the distance between the mid-points of the atoms gives directly only the sum of the radii. However, if the ionic radius of one partner has been determined in some way*, that of the other can at once be obtained by subtracting the known radius from the total distance. This can be used in turn to determine the radius of another element combining with the first, and so on. V. M. Goldschmidt showed in 1926 that for any one ion, in crystals of widely varying compounds, the radius is constant, with only small deviations. However, comparison may only be made between compounds which crystallize in certain structural types which Goldschmidt called commensurable (this is further treated in Vol. II). Subject to this restriction, the apparent radius of a particular ion is always approximately the same, irrespective of the nature of the compound. Thus the values obtained for the apparent radii of the alkaline earth ions are substantially independent of whether they are calculated from the crystal dimensions of the fluorides, the chlorides, or the oxides.

The atomic and ionic radii calculated from the dimensions of crystals are termed

* Radii of some important ionic species, useful in crystal chemistry, were deduced in 1923 by Wasastjerna from optical data, e.g., $O^{2-} = 1.33$Å, $F^- = 1.32$Å.

apparent radii because they are not claimed to represent the 'real extension' of the atoms and ions, but rather represent only to what distances the mid-points of the atoms and ions can approach one another in the formation of crystallized compounds. These distances are magnitudes determined by the nature of the atoms or ions in question, and are, indeed, not absolutely constant,* although experience nevertheless shows that they may be regarded as constant to a first approximation. In other words, the atoms and ions are found in practice to behave in this respect as more or less solid spheres, the radii of which are represented by the apparent atomic and ionic radii. H. Jensen (1938) has derived theoretical values for the ionic radii, on the basis of wave mechanics, which are very similar to those derived purely empirically, from crystal dimensions, by V. M. Goldschmidt.

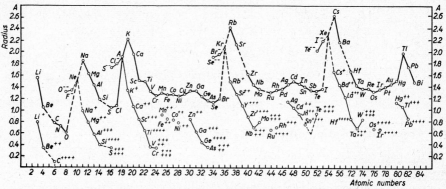

Fig. 3. Periodicity of apparent atomic and ionic radii.

In Fig. 3. Goldschmidt's apparent radii of the atoms and ions in crystals are plotted as a function of the atomic number. The curve for the atomic radii corresponds closely to the atomic volume curve, as may be seen, but reproduces the relationships more exactly since it eliminates the irregularities arising from differences of structure. The ionic radii form a similar curve. It may be seen that the radii of *positive* ions are *smaller*, those of *negative* ions *larger* than the radii of the neutral atoms. With the elements of the Main Groups of the Periodic System, the atomic and ionic radii run largely parallel to one another, as shown by Fig. 3. The radii of positive ions of these elements are in general about 0.7 to 0.9 Å smaller than the atomic radii, those of the negative ions about the same amount larger.

(c) Optical Spectra

It is well known that many elements impart characteristic colorations to the non-luminous flame of the Bunsen burner. This property is associated with certain Groups of the Periodic System. Similarities in respect of the ability to emit light, and of the mechanism by which light is emitted, become more apparent when the light dispersed by a spectroscope is considered. The spectra of elements in any one Group of the Periodic System are extremely closely related in their fine structure. We shall see later that this fact originates from the periodicity of atomic structure, and goes back to the same causes which underlie the periodic character of the chemical properties.

* For the influence of the charge and the number of the other ions surrounding an ion upon its radius, see Chap. 7.

(d) Ionization Potentials

The periodicity of the ionization potentials is also closely connected with the periodicity of chemical properties.

The *ionization potential* of the atoms of any element signifies that potential, measured in volts, through which an electron must be accelerated, in order that the energy acquired shall just suffice to bring about ionization (i.e., conversion into the electrically charged state) on collision with the atoms of the element in question.

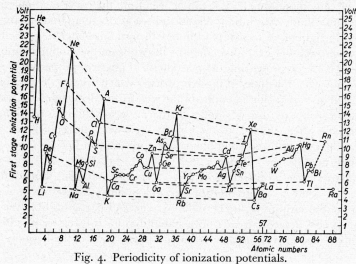

Fig. 4. Periodicity of ionization potentials.

In Fig. 4, the first ionization potentials of the most important elements are plotted against the atomic numbers,—i.e., the potentials which refer to the conversion of the free atoms of the elements into singly positively charged free ions. Their periodicity is very clearly displayed. The alkali metals lie every time at the lowest points, the inert gases at the summit points of the curve. Homologous elements are joined to one another by dotted lines as in Fig. 2 (see further p. 112).

(e) Other Physical Properties

Many other physical properties may also be represented as periodic functions of the atomic number,—for example the melting points and boiling points of the elements, their crystal structures, and their magnetic susceptibilities. [4]

With very few exceptions (e.g., alkali metal superoxides), all the elements which are paramagnetic in the form of electrolytic ions or in heteropolar compounds, are found in the Sub-groups of the Periodic System or in the lanthanide and transuranic groups.

The property of forming colored ions also displays a periodic character. All the elements which are able to form colored elementary electrolytic ions belong in the Sub-groups, the lanthanide group or the transuranic group.

4. Change of Properties in the Horizontal Series
of the Periodic System

Close relationships exist not only between a certain element and its neighbors within the same Group (i.e., immediately above and below it), but also between it and its lateral neighbors. Examples of this have already been provided by the properties already discussed. Thus it can be seen from the curve of atomic volumes that the atomic volume of sodium is related at least as closely to that of magnesium, its right hand neighbor, as to that of lithium which stands above it, in the Periodic System. Numerous other examples could be drawn from the atomic volume curve and from the curves representing the periodicity of physical properties. The same is also true, however, for many of those properties which are usually considered to be chemical. This is true, for example, of the electrochemical character, as well as of the basic or acidic nature. It applies furthermore to solubility relationships of compounds, and often, indeed, to the ability of elements to replace one another mutually in their crystallized compounds (isomorphism), a property otherwise so closely bound up with the valence.

The similarity of adjacent elements (i.e., elements following immediately one after the other in the same series of the Periodic System) is found to be especially marked in the Sub-groups of the Periodic System, and particularly in the VIIIth Sub-group. It has, indeed, already been stated that for this reason it is justifiable to assign three consecutively adjacent elements to one place of the Periodic System, in this Group.

The similarity between consecutively adjacent elements is manifest to quite an exceptional extent in the lanthanide series, i.e. in the series of 14 elements with atomic numbers 58–71 inclusive, following after lanthanum.

It is here so far reaching that it includes even the valence. The same holds true for the transuranic elements. Just why, at these places in the Periodic System, there should be such a very close similarity between adjacent elements can be understood from the atomic structure of the elements. This will be discussed at the beginning of Volume II of this book.

In general, however, it may be said that apart from the lanthanides and transuranic elements, similarities in a vertical direction are the most striking, both chemically and in other respects. The elements of the first period depart more or less strongly from this rule. Most of them show the greatest similarity not with their higher homologues within the same Group, nor with their immediate neighbors, but rather each with the element standing to its right in the following series, i.e., lithium with magnesium, beryllium with aluminum, boron with silicon, oxygen with chlorine. The similarities in chemical behavior between beryllium and aluminum are so close that, for a long time, doubt was cast upon the bivalence of beryllium and its homology with magnesium and the alkaline earth metals.

The elements of the second short period (sodium, etc.) show another peculiarity. Of all the elements of the Main Groups, they are the ones which in each case are most closely related to the Sub-groups bearing the same number—often indeed so closely that in many respects they are more like the elements of the Sub-group than those of the Main Group. This appears most markedly perhaps in the center of the Periodic System, mainly in the IIIrd and IVth families.

5. **Periodicity of Atomic Structure**

During the last forty years it has become possible to understand how the properties of the elements are conditioned by the structure of the atoms. Although this field, in which chemistry and physics come intimately together, is in no sense a completed chapter in Science, much has been achieved and a great deal can be regarded as complete and well established. We shall discuss in the following chapters those concepts which are most important for the chemist. It can no longer be doubted that many of the properties which differentiate the various elements from one another can be interpreted as the *immediate consequences of the atomic structure* [1]. In all cases in which experimental evidence (especially from spectroscopy) has made it possible to draw more or less direct conclusions about atomic structure, it has emerged that the atoms of elements which are closely analogous in physical and chemical properties (e.g., of elements standing one above another in the Periodic System) are also in large measure similar in structure. The fact that similarities between the elements recur periodically is a direct consequence of periodically recurring homologies in the structure of their atoms.

In the Periodic Table at the end of this book, symbols characterizing their atomic structure are given under the names of each of the elements. The meaning of these symbols is explained on p. 120. It is at once apparent that everywhere in the Periodic Table, elements in the same vertical column correspond in their atomic structures. Certain typical differences between the Main Group elements and the Sub-group elements also stand out immediately.* The special feature of the structure of the lanthanides (filling of a $4f$ shell), which makes these form a group on their own, can also be clearly seen from the symbols given in the Table. A corresponding circumstance (filling of a $5f$ shell) affects the elements following actinium. It has already been stated that the elements following actinium constitute a special group, like the lanthanides (the actinide group), although the first three elements on practical grounds are classed with the elements of Groups IVB, VB and VIB. Both this similarity, and its disappearance in the later actinides (transuranic elements) can also be understood from their atomic structure (cf. Vol. II). We shall later see from numerous examples how the chemical properties of the elements can be deduced from their structure, and how the rules based on the Periodic System are determined by the systematics of the structure of the atom.

A detailed knowledge of atomic structure enables us to perceive more relationships than can be deduced merely from the positions of the elements in the Periodic System. What appears to be an exception to the rule when based only on the Periodic System, turns out to be a necessary consequence when it is derived from the laws governing the structure of the atom. Thus we shall discover, as consequences of the systematic change of atomic structure in progressing from element

---

* These differences show up most obviously if the elements of the Main groups are compared with the so-called 'transition' elements—i.e., with the elements of groups IIIB to VIIIB. That the difference also exists between the A and B groups of the Ist and IInd families, can be recognized by reference to the data in the first column of the Table. From this it may be seen that in potassium, for example, the outermost electron shells are represented by $3s^2 3p^6 4s$, whereas those of copper are represented as $3d^{10} 4s$. The influence of d-electrons on the chemical properties is considered in Vol. II.

to element, not only the regularities discussed, such as the change of valence from group to group, and the systematic change in electrochemical character, but also apparent irregularities, such as the special position of hydrogen, the peculiar division of the Periodic System into periods of various lengths, and the particularly close relationship between the elements of the rare earths.

References

1 E. RABINOWITSCH and E. THILO, *Periodisches System, Geschichte und Theorie*, Stuttgart 1930, 302 pp.

2 LOTHAR MEYER (Editor), *Die Anfänge des natürlichen Systems der chemischen Elemente [(Arbeiten von J. W. Döbereiner (1829) und M. Pettenkofer (1850)] nebst einer geschichtlichen Übersicht der Weiterentwicklung der Lehre von den Triaden der Elemente*, Sammlung Ostwalds Klassiker, No. 66, 34 pp.

3 K. SEUBERT (Editor), *Abhandlungen über das natürliche System der chemischen Elemente von L. Meyer und von D. Mendelejeff (1864–1869 und 1869–1871)*, Sammlung Ostwalds Klassiker, No. 68, 134 pp.

4 A. VON ANTROPOFF and M. VON STACKELBERG, *Atlas der physikalischen und anorganischen Chemie*, Berlin 1929, 64 pp, 29 Plates; *Supplement*, Berlin 1932, 10 pp.

CHAPTER 2

HYDROGEN

Atomic number 1.   Atomic weight 1.0080.
Atomic volume (at the b.p.) 14.4 cc.   Valence 1.

1. Introduction

(a) General

Hydrogen is the lightest of all the elements. Its atomic weight and atomic number place it at the beginning of the series of the chemical elements. It cannot be strictly assigned to any definite family of the Periodic System. There is, however, a noteworthy difference between its relationships to the elements of Group IA and to those of Group VIIA. The properties in which it resembles the alkali metals are, except for its valence, due to quite different causes from those operative in the alkali metals. On the other hand, those properties of hydrogen in which it resembles the halogens arise from the same causes as with the elements of Group VIIA. In brief, hydrogen can be regarded as a halogen which, because of its position as the first in the series of the elements, also has superficial similarities in chemical behavior to the alkali metals.

In considering hydrogen by itself, the numerous instructive relationships between hydrogen and the halogens should not be overlooked. In spite of all the differences, it shares a number of their typical properties. Thus it is a non-metal, and like the halogens, it forms a diatomic molecule in the elementary state. In these molecules, the atoms (of hydrogen or of the halogens respectively) are linked by single bonds. The energy necessary to dissociate a molecule into atoms decreases in the sequence H_2—Cl_2—Br_2—F_2—I_2. Hydrogen resembles the halogens in that the atom can acquire a negative charge, and has a positive electron affinity—i.e., energy is liberated in the process of adding an electron to a neutral hydrogen atom. The halogens are invariably univalent, like hydrogen, in compounds in which they are negatively charged. The metal compounds in which hydrogen is the negative component have the same structure and bond character as the corresponding halogen compounds: they are *salts*, so that hydrogen can be regarded as a halogen ('salt former') in the literal sense of the word. The energy to be expended in forming a positively charged hydrogen atom—i.e., in removing an electron from a hydrogen atom—is no smaller than that required for any of the halogen atoms other than fluorine, as can be seen by comparing the ionization potentials (see p. 115).

It is only the fact that, as the first of the elements, the atom of hydrogen contains only one electron, that makes hydrogen differ from the halogens in many of its properties. These properties are very prominent, however, in the behavior of hydrogen as a whole. If the electron is expelled from the hydrogen atom, there remains only the excessively minute nucleus of the atom, the *proton*. In this respect, hydrogen is completely different from all other elements*. The amount of energy liberated when the very minute hydrogen nucleus

---

* For all other elements, the diameter of the core of the atom, remaining after expulsion of the valence electrons, is between 0.2 and 3.3 Å. The diameter of the hydrogen nucleus is only about 0.00001 Å.

adds itself on to an electronegatively charged atom is particularly great. The same applies to the formation of the hydronium ion $[H_3O]^+$, in aqueous solution, by addition of a proton to a water molecule, and also for other similar reactions. This is why, in spite of its high ionization potential, hydrogen behaves as a relatively strongly electropositive element. Hydrogen has a strong tendency to form covalent, and therefore volatile, compounds.* The volatility of these compounds leads to a displacement of the equilibria in reactions in which they are involved, favoring their formation. In consequence, hydrogen has strongly reducing properties.

Its univalence, and its apparently strongly electropositive character, as shown by its reducing power, have often led to a relationship being traced between hydrogen and the alkali metals. Some have therefore assigned hydrogen to Group IA of the Periodic Table. Except for its valence, however, there is little in common between hydrogen and the alkali metals. It is a non-metal, and its ability to become negatively charged is in direct contrast with the alkalis. It also differs greatly in respect of its electropositive character. The alkali metals are the most electropositive of the elements, they have the lowest ionization potentials, and stand at the extreme left of the electrochemical series (p. 30). Hydrogen comes fairly well to the right in this series, and its ionization potential is higher even than those of the noble metals. It stands to the left of these in the electrochemical series solely because of its very high hydration energy—i.e., for quite a different reason from that which places the alkali metals at the head of the series. The extremely small radius of its positive (unhydrated) ion is in marked contrast to the alkalis, which give rise to the largest of all the positive ions. In the light of our present knowledge it is therefore no longer justifiable to include hydrogen in the group of the alkali metals.

The peculiar position of hydrogen arises from the fact that the real first period of the natural system of the elements comprises only two elements, hydrogen and helium, and not, like the following periods, 8 or more elements. It thus happens that hydrogen combines in itself the characteristics of the Ist and the penultimate (VIIth) Main Groups of the Periodic System.

(b) Occurrence

As a constituent of water and other compounds, hydrogen is very widely distributed in nature. Its proportion in the make-up of the earth's crust (including the oceans and the atmosphere) has been calculated as 0.87 % by weight, or 15.4 atom%. In the neighborhood of the earth's surface it occurs with great rarity in the free state.** It is occasionally found mixed with other gases in the exhalations from volcanoes and fumaroles, and also, in small amounts, occluded in potassium salts. It is believed that above a height of several hundred kilometres the earth's atmosphere consists predominantly of hydrogen and helium. It occurs, moreover, in large quantities in the sun and in most stars, as has been found by means of spectroscopic analysis.

(c) History

Hydrogen was discovered in 1766 by Cavendish. He found that when metals were dissolved in dilute acid, a combustible gas was evolved. He was also the first to recognize, in 1781, that water is the product of the combination of hydrogen and oxygen. Lavoisier, in 1783, succeeded in decomposing water by means of red hot iron (by passing steam through a gun barrel heated to red heat). The decom-

---

* See Chap. 9 on the relation between bond type and volatility.

** The average hydrogen content of the atmosphere in the neighborhood of the earth's surface amounts to $5 \cdot 10^{-5}$ %. (F. A. Paneth, 1937). For the mean deuterium and tritium content of the atmosphere, see Vol. II.

position of water by the electric current was observed for the first time in 1789, by Nicholson and Carlisle.

(d) Formation, Preparation and Uses

In Nature, hydrogen is formed for the most part during the decomposition of organic substances (e.g., of cellulose and albumen) by certain bacteria. Large quantities of hydrogen are set free in the coking of coal. For this reason, coal gas and coke oven gas contain on the average, about 50 % by volume of elementary hydrogen. Use has been made of this in recent times for the technical preparation of hydrogen. By a liquefaction process, based on the Linde process for the manufacture of liquid air, the other constituents of coke oven gas are separated from hydrogen, which is condensable only with difficulty. Hydrogen obtained in this way finds particular application for the hydrogenation of coal, and the constituents of the coke oven gas separated from it are used for the supply of towns' gas.

Apart from this, hydrogen is almost invariably prepared from water. When other compounds are used they are themselves mostly prepared from water.

Hydrogen gas is produced industrially today on an enormous scale. [1] Very large quantities are used for the synthesis of ammonia, also for the hardening of fats and for the hydrogenation of coal, oils, and hydrocarbons. In addition, hydrogen gas is used for the synthesis of hydrogen chloride, methanol, and hydrocyanic acid, for the welding and forging of metals, in the manufacture of incandescent lamps, and in the production of synthetic gems. In aeronautics, it serves to fill captive and free balloons. Hydrogen is transported and handled commercially in steel cylinders, under a pressure of 120–150 atmospheres.

Of the methods of preparation for hydrogen mentioned below, those of industrial importance are: from *water-gas* by conversion of CO (water gas catalytic process, steam-water gas process); from natural gas or coke-oven gas by methane decomposition (steam-hydrocarbon process); from coke-oven gas or water gas by fractional condensation; from the electrolysis of water; from the steam-iron process. Hydrogen is obtained as an important by-product in the electrolysis of alkali chlorides in aqueous solution, and in the manufacture of acetylene by the arc process. The reaction of steam with phosphorus (Liljenroth process) also finds a certain limited industrial application, as also does the thermal decomposition of hydrocarbons

$$(CH_4 \xrightarrow{1000\,°C} C + 2H_2).$$

Hydrogen is also occasionally made by the catalytic decomposition of methanol with steam

$$(CH_3OH + H_2O \xrightarrow{250\,°C} CO_2 + 3H_2)$$

or by the catalyzed thermal decomposition of ammonia

$$(2\,NH_3 \xrightarrow{950\,°C} N_2 + 3H_2).$$

It is true that these compounds are produced on the large scale from hydrogen, but the preparation of hydrogen from them is very simple, and may therefore be advantageous in installations which require relatively small amounts (less than 500 m³ per day).

2. Most Important Methods of Preparation of Hydrogen

(1) Dissolution of zinc in dilute hydrochloric acid

$$Zn + 2HCl = ZnCl_2 + H_2$$

This process is mostly used in the laboratory.

Instead of hydrochloric acid, dilute sulfuric acid may also be used. However if this is too concentrated, the hydrogen evolved from it may easily be contaminated with SO_2 and

H_2S. If the zinc used is not quite pure, the hydrogen also contains other impurities, e.g., AsH_3 and PH_3; the unpleasant smell of the hydrogen evolved in this way is due to the impurities.

To *purify* it, the hydrogen is passed through an acidified solution of potassium permanganate or potassium dichromate, and then through caustic potash, and possibly also through concentrated sulfuric acid or over silica gel (cf. p. 496) to dry it. The very fine spray of liquid droplets carried over by the gas from the generating vessel is not removed by the wash liquids, since the droplets remain suspended within the gas bubbles; it is best removed by a filter of tightly compressed cotton wool or glass wool.

If absolutely pure zinc is used, it is necessary to add a few drops of chloroplatinic acid or copper sulfate solution, as the zinc is otherwise not attacked (see Vol. II).

The hydrogen used to fill balloons was formerly prepared by dissolving iron filings in dilute sulfuric acid, as for example by Charles in 1783. The process was still used until shortly before the First World War.

(2) Dissolution of aluminum or silicon in caustic alkalis

$$Al + NaOH + 3H_2O = Na[Al(OH)_4] + \tfrac{3}{2}H_2$$

$$Si + 4\,NaOH = Na_4SiO_4 + 2H_2$$

These reactions have found considerable application in more recent times (roughly since the Russo-Japanese war) for the generation of hydrogen in the field, for captive balloons. For the generation of 1 cu.m. of hydrogen (measured at 0° and 760 mm pressure), only 0.81 kg of aluminum or 0.63 kg of silicon are required, as compared with 2.9 kg of zinc or 2.5 kg of iron.

In place of silicon, ferrosilicon is also used (silicol process). A mixture of ferrosilicon and soda lime, known as *hydrogenite*, introduced by the French army shortly before the First World War, has the property of burning with the vigorous evolution of hydrogen after it has once been ignited. The following reaction occurs

$$Si + Ca(OH)_2 + 2NaOH = Na_2SiO_3 + CaO + 2H_2$$

(3) Action of sodium on water

$$Na + HOH = NaOH + \tfrac{1}{2}H_2$$

Since pure sodium reacts very violently, sodium amalgam is generally used in its place, especially for the production of hydrogen intended to be used for reduction in the nascent state (cf. p. 41). The other alkali metals and the alkaline earth metals react with water in the same manner as sodium.

(4) Action of calcium hydride on water

$$CaH_2 + 2H_2O = Ca(OH)_2 + 2H_2$$

This process provides a convenient means of generating hydrogen in the field. For 1 cubic meter of hydrogen, 0.94 kg of CaH_2 is theoretically required, and no other reagents except water. The high price of calcium hydride is a disadvantage.

(5) Passage of steam over red hot iron

$$4H_2O + 3Fe = Fe_3O_4 + 4H_2$$

It was by means of this reaction that Lavoisier, in 1783, first demonstrated analytically the compound nature of water.

The iron oxide formed in the reaction can readily be reduced again to metallic iron, by passing producer gas (see p. 445) over it, so that the passage of steam can be performed repeatedly. In this form, the process was of considerable technical importance for a while. It is still employed to some extent.

(6) Passage of steam over heated coke

Above 1000°, the reaction proceeds for the most part according to the equation

$$H_2O + C = CO + H_2$$

Water gas is first obtained, i.e., a mixture of hydrogen with carbon monoxide, containing some carbon dioxide and nitrogen (cf. p. 446). The gas can be freed from carbon dioxide by washing it with water under pressure. In the Frank-Caro-Linde process, the removal of carbon monoxide and nitrogen is effected by liquefying them, by cooling with liquid air boiling at —200°. The last traces of carbon monoxide can be removed by passing the gas over heated soda lime

$$CO + NaOH = NaCO_2H \text{ (sodium formate)}$$

This process furnishes very pure hydrogen, such as is used, for example, in the hardening of fats. [2]

A more usual procedure is to pass the water-gas, mixed with steam, over a suitable catalyst (e.g., iron- or cobalt oxide) at about 400° ('catalytic water gas process' or steam-water gas process). The CO then undergoes reaction with the H_2O: $CO + H_2O_{vapor} \rightleftarrows CO_2 + H_2$ ('CO-conversion').

The CO_2 thus formed is removed by washing with water, under pressure. The remainder of the CO (about 1 % by volume) is usually washed out with ammoniacal copper(I)-chloride solution. The water gas used in this process is produced not only by passing steam over red hot coke, but also, to an increasing extent, by the reaction of steam with powdered coal (powdered coal gasification). Water gas obtained in this way has a higher H_2 content than that from the ordinary method. Hydrogen from the catalytic water gas process, which contains nitrogen, is used chiefly for ammonia synthesis and for the hydrogenation of coal.

(7) Fractional liquefaction of coke oven gas

Hydrogen may be obtained from coke oven gas, of which it is the principal constituent, by liquefaction, as from water gas (cf. p. 432).

Substantially similar processes for this were developed independently in Germany by the Linde-Gesellschaft and in France by G. Claude. Carbon dioxide is first removed from the desulfurized coke oven gas, by washing with water under pressure and subsequent treatment with sodium hydroxide. The remaining constituents are then removed in turn by progressive condensation, leaving hydrogen, which can be freed from the last impurities, for example, by washing with chilled liquid nitrogen. This process also is chiefly used to generate hydrogen for ammonia synthesis.

(8) Reaction of methane with steam (Methane decomposition, steam-hydrocarbon process)

In the presence of suitable catalysts, methane reacts with steam at high temperatures (e.g., at 1100°) according to the equation

$$CH_4 + H_2O_{vapor} + 48.9 \text{ kcal} = CO + 3H_2 \text{ (at const. press.)}$$

The heat absorbed in this reaction must either be supplied by external heating, or produced by 'internal combustion'—i.e., air or oxygen is added, so that a portion of the methane burns to carbon dioxide:

$$CH_4 + 2O_2 = CO_2 + 2H_2O_{vapor} + 191.8 \text{ kcal} \quad \text{(at const. pressure).}$$

The composition is therefore chosen in such a way that the reaction as a whole is weakly exothermic. E.g.,

$$12CH_4 + 5H_2O_{vapor} + 5O_2 = 29H_2 + 9CO + 3CO_2 + 20.4 \text{ kcal.}$$

Hydrogen is also produced from the carbon monoxide, by the CO-conversion process. Carbon dioxide is removed at that stage, either by washing it out with water under pressure, or by the Girbotol process (cf. p. 442). Natural gas is used for this process in the United States and coke oven gas in most other countries. Hydrogen prepared by the steam-hydrocarbon process is used chiefly for ammonia synthesis and the hydrogenation of coal.

(9) Reaction of steam with phosphorus (Liljenrot)

$$P + 4H_2O = H_3PO_4 + 2\tfrac{1}{2}H_2$$

The process is generally so operated that phosphorus vapor, obtained by the reduction of calcium phosphate in the electric furnace (cf. p. 625), is brought into contact with steam at about 400°–600° in the presence of a catalyst. With increase of temperature, the equilibrium of the above equation shifts from right to left. To avoid the decomposition of H_3PO_4 primarily formed, into H_3PO_3 and PH_3, the products of the reaction are rapidly cooled. The process is used to make hydrogen for synthetic ammonia, when this in turn is to be worked up into ammonium phosphate, which is important as a ballast-free fertilizer.

(10) Electrolytic decomposition of water

Since water by itself is practically a non-conductor of the electric current, an electrolyte (generally KOH) is added. The hydrogen is set free at the cathode. At the anode, an equivalent quantity of oxygen is evolved, and is thus obtained as a by-product of the process.

Apart from a minute oxygen content, which can readily be removed by passing the gas over a catalyst (e.g., gently heated palladium asbestos), the electrolytically evolved hydrogen is very pure. It therefore finds application for the hardening of fats and for other catalytic hydrogenations. The energy consumed in the electrolytic decomposition of water is, however, rather considerable. Hydrogen is also obtained as a by-product in the electrolysis of alkali chlorides, when this is carried out in aqueous solution (see p. 182).

For the discharge of 2 g ions of H^+,—i.e., for the evolution of 1 g molecule of hydrogen—, $2 \times 96,500$ coulombs $= 53.6$ ampere hours are necessary. The decomposition potential is 1.23 volts (see p. 32). Hence, to generate 1 cubic meter of hydrogen (18° and 760 mm) the requisite energy is, theoretically,

$$\frac{53.6 \cdot 1.23 \cdot 1000}{22.43 \cdot 1.06} \text{ Watt-hours, } \sim 2.8 \text{ kWh.}$$

In practice, considerably more is required (4.2–5 kWh), since it is generally necessary to allow for a terminal voltage greater than 2 volts per cell, because of the overvoltage at the electrodes (cf. p. 31), and the ohmic resistance of the electrolyte and the diaphragms. It is necessary to separate the anode and cathode spaces by a diaphragm, generally made of asbestos, to prevent the mixing of the gases liberated at the two electrodes, especially as the electrodes are brought as close together as possible, in order to diminish the voltage losses in the electrolyte.

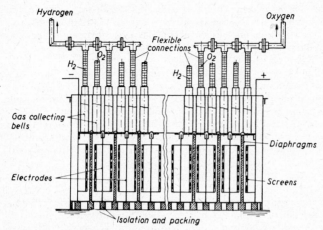

Fig. 5. Technical electrolysis of water with bi-polar cells.

A water electrolytic cell is represented in section in Fig. 5. This is of the type in which the cells are so connected in series that the anode of one cell functions at the same time as cathode in the next cell (bi-polar cells). By this arrangement (due to the Swiss chemist O. Schmidt), the voltage loss in the passage of the current from one cell to the next is reduced to the minimum. The electrodes of iron sheet, coated with nickel on the anode side, which are also the partitions between compartments, are provided with perforated front sheets on which the evolution of gas takes place preferentially. By this means the gases can be removed more rapidly from the path of the current, and the voltage losses in the electrolyte are thereby diminished.

The reduction of the conductivity by reason of the gas bubbles suspended in the electrolyte—which can reduce the volume of the latter by more than a quarter—causes a considerable extra consumption of electrical energy. Considerable interest therefore attaches to experiments to diminish the bubble volume by using high pressures, of 100 to 200 atm. (high pressure electrolysis). If the evolved gases are to be filled into steel cylinders, the work of compression is saved at the same time. The small space occupied by the cells for high pressure electrolysis is also of importance.

With bi-polar cells, of which up to 250 may be connected in series, operating currents up to 10,000 amperes are used, and with uni-polar cells up to as much as 20,000 amperes. This method furnishes hydrogen with a purity of 99.9–99.95 %, together with oxygen of 99.6–99.8 % purity.

3. Electrolysis of Water;
Electrochemical Potential Series [22, 23, 25, 26]

The electrolytic decomposition of water depends on the discharge of *hydronium ions*, $[H_3O]^+$, at the cathode. Hydrogen atoms thus liberated at once combine in pairs to form H_2 molecules:

$$[H_3O]^+ + e = H_2O + H; \quad 2H \rightarrow H_2.$$

[The symbol e represents the free negative electric charge (i.e., the *electron*)].

At the anode, the amount of positive electricity entering the solution must, of course, be the same as the negative electricity entering at the cathode.

The hydronium ion is often referred to simply as the '*hydrogen ion*', and represented by the symbol H+. The hydrogen ion in this sense (i.e., the hydrogen ion in aqueous solution) must not be confused with the free, positively charged H-atom (the hydrogen nucleus, or proton). The free hydrogen nucleus, or proton, has a very different energy content from that which is combined with a water molecule (cf. p. 75), and is also quite different in its chemical reactivity (cf. Vol. II, Chap. 18).

Instead of the statement that 2 positive electric charges are given up to the solution by the positive electrode (anode), we can also say that 2 negative electric charges are given up by the solution to the positive electrode. In the case of an alkaline solution, the negative charges given up to the positive electrode originate, for the most part, from the discharge of hydroxyl ions, OH-. The neutral residues OH then at once react in pairs, reforming a molecule of water, and eliminating an atom of oxygen. The O atoms at once combine in pairs to form O_2 molecules. The process at the anode can thus be represented by the following equation:

$$2OH^- - 2e = H_2O + \tfrac{1}{2}O_2$$

If the solution contains salts or acids, other anions may in some circumstances also be discharged at the anode in place of hydroxyl ions. Similarly, hydrogen ions are discharged exclusively at the cathode only if the solution contains no positive ions which can be more easily discharged than the hydrogen ions.

(a) Electrode Potential and Decomposition Voltage

The work which must be expended in order to discharge the ions of the various elements can be precisely expressed by the difference in potential set up between the element in question and a solution in contact with it, containing the element in ionic form. This potential difference E, the *electrode potential* of the element in question depends on the *concentration* of the solution and on the *temperature*. The important quantity is the '*effective concentration*' or '*activity*' of the ions in the solution. This is obtained from the analytical molar concentration c, in mols or g-ions per liter, by multiplying it by a coefficient which makes allowance for the deviations from the laws of 'ideal solutions'—the so-called '*activity coefficient*', f_a (cf. p. 68). The activity a is thus given by $a = c \cdot f_a$. [11, 12] The variation of electrode potential with temperature and concentration is then described by the Nernst equation

$$E = E_0 - \frac{RT}{n\mathfrak{F}} \ln a \qquad (1)$$

R is the gas constant, T the absolute temperature (°K), a the activity of the ions of the substance in question in the solution, n = number of faradays of electric charge, borne by 1 g ion of the substance.* Measuring E in volts, and substituting ordinary logarithms in place of natural logarithms, then at 18 °C ($T = 291.16°$),

$$E = E_0 - \frac{0.05777}{n} \log a \qquad (1a)$$

* 1 Faraday = 96493 coulombs. R is also to be expressed in electrical units (= 8.3144 Joules). There is thus obtained for

$$\frac{RT}{\mathfrak{F}} = \frac{2.30259 \times 8.3144 \times 291.16}{96493} = 0.057767.$$

The numerical value for 20° is 0.058164, for 25° 0.059156.

Since the second term in the equation becomes zero for $a = 1$, E_0 represents the potential difference between the element and a solution of its ions, when the activity of the latter is unity. E_0 is called the *normal potential* or *standard potential* of the element in question. It is a quantity which is temperature-dependent, but is otherwise characteristic of each element.

The contact potential difference between an electrode and its solution—i.e., the absolute value of the electrode potential—is difficult to measure exactly. It is, however, easy to measure the difference in potential between two electrodes, each of which dips into a solution of the corresponding electrolyte. If the two solutions are in contact with one another they are at the same potential, apart from a small potential difference at the junction (the diffusion potential). If the activities of the ions concerned $= 1$ in both of the solutions, the potential difference between the electrodes thus measures directly the difference of the normal potentials.

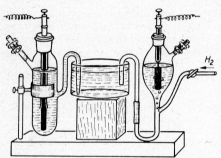

Fig. 6. Zinc electrode (left) and hydrogen electrode (right) combined to a galvanic element.

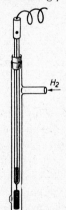

Fig. 7. Simpler form of hydrogen electrode.

The experimental arrangement represented in Fig. 6 may serve as an example. In the one vessel (left), there is a zinc rod in a solution of zinc sulfate, in the other, a platinum foil covered with platinum sponge, which dips into dilute sulfuric acid, and over which hydrogen gas is bubbled. Such an electrode assumes a well defined potential, which is characteristic of hydrogen. It is accordingly called a *hydrogen electrode*; a simpler form of hydrogen electrode, adequate for many purposes is illustrated in Fig. 7. The potential difference between the zinc and the hydrogen electrode can readily be measured. Neglecting the diffusion potential, which can generally be reduced to a practically vanishingly small value by suitable experimental means,* the measured potential is equal to the difference $E(Zn) - E(H)$, where $E(Zn)$ represents the potential difference between the zinc rod and the zinc sulfate solution, and $E(H)$ that between the hydrogen electrode and the sulfuric acid solution, since no detectable differences of potential occur except at the three points of contact $Zn \mid ZnSO_4$ solution $\mid H_2SO_4$ solution $\mid H_2(Pt)$. When experimental conditions are chosen so that the activity of zinc ions in the zinc sulfate solution, and of hydrogen ions (i.e., of $[H_3O]^+$ ions) in the sulfuric acid are both equal to 1, it is found that the zinc electrode is 0.76 volt more positive than the hydrogen electrode; this means that the normal potential of zinc lies 0.76 volt higher than that of hydrogen, i.e.,

$$E_0(Zn) - E_0(H) = + 0.76 \text{ volts.}$$

This is usually summarised by the statement: the *normal* (or *standard*) *potential of zinc, relative to the normal hydrogen electrode, is* $+ 0.76$ *volts.*

The normal potential of any other metals may be compared with that of hydrogen in a corresponding manner. Indeed, as has already been illustrated by the example of hydrogen, measurements need not be restricted to pure metallic electrodes. The essential is that the electrode shall be a good conductor of the electric current, and that the process which takes place at the electrode shall be strictly defined and reversible.

* On the exact calculation of diffusion potentials cf. J. J. HERMANS, *Z. physik. Chem.*, A, 176 (1935) 55 and 131.

In Table 4 are collected the normal potentials for a number of elements at 18°, referred to the normal hydrogen electrode at the same temperature. Where the values refer to gas electrodes, they are valid for a pressure of 1 atm. of the gas in question. The normal potentials of elements not listed in Table 4 are given elsewhere in this book, when the elements are described.

TABLE 4

NORMAL POTENTIALS, RELATIVE TO THE NORMAL HYDROGEN ELECTRODE

Cations		Anions
K | K$^+$ +2.9 Volt	Ni | Ni^{++} +0.25 Volt	F$^-$ | F$_2$ −2.8 Volt
Ca | Ca^{++} +2.8 ,,	Sn | Sn^{++} +0.16 ,,	Cl$^-$ | Cl$_2$ −1.36 ,,
Al | Al^{+++} +1.69 ,,	Pb | Pb^{++} +0.13 ,,	Br$^-$ | Br$_2$ −1.08 ,,
Mn | Mn^{++} +1.1 ,,	H$_2$ | H$^+$ 0.00 ,,	I$^-$ | I$_2$ −0.58 ,,
Zn | Zn^{++} +0.76 ,,	Cu | Cu^{++} −0.35 ,,	OH$^-$ | O$_2$ −0.41 ,,*
Fe | Fe^{++} +0.51 ,,	Hg | Hg$_2$$^{++}$ −0.793 ,,	S$^-$ | S +0.51 ,,
Cd | Cd^{++} +0.40 ,,	Ag | Ag$^+$ −0.808 ,,	
Co | Co^{++} +0.29 ,,	Au | Au^{+++} −1.38 ,,	

* It is to be noticed that the normal potential of oxygen refers to a solution in which the OH$^-$ activity = 1. In neutral solution, the electrode potential of oxygen referred to the normal hydrogen electrode = − 0.81 volts, as follows from equation (1a).

If the metals, including hydrogen, are arranged in a horizontal series, in the order of decreasing normal potentials, the most strongly electropositive metals stand on the extreme left, the most weakly electropositive on the right (*electrochemical potential series*, Table 5).

TABLE 5

ELECTROCHEMICAL SERIES OF THE ELEMENTS

Decreasing normal potentials

———————————————————————————————————————→

K,   Ca,   Al,   Mn,   Zn,   Fe,   Cd,   Co,   Ni,   Sn,   Pb,   H$_2$,   Cu,   Hg,   Ag,   Au

←———————————————————————————————————————

Increase in electropositive character

The normal potentials could formally be referred to some other combination, instead of to hydrogen in a solution in which the [H$_3$O]$^+$ activity is unity. The proposal to use the normal hydrogen electrode as a basis originated with Nernst, and is today generally accepted. Although other comparison electrodes, and especially the calomel electrode, are generally used for convenience in measurement, the values so found are nevertheless generally converted to the normal hydrogen electrode scale.

In older work, the electrolytic potentials are referred to a hydrogen electrode in which hydrogen (at atmospheric pressure) is in contact with a solution having an apparent [H$_3$O]$^+$ ion concentration (calculated from the conductivity) equal to 1. The potential of such an electrode is about 2.3 millivolts lower than that of the normal hydrogen electrode, as defined above*—i.e., a hydrogen electrode in contact with a solution in which the [H$_3$O]$^+$ ion *activity* = 1.

* See CLARK [*18*], also KOLTHOFF, *Rec. Trav. chim.*, 49 (1930) 401. Cf. also F. MÜLLER, *Z. Elektrochem.*, 48 (1942) 288; G. KORTÜM [*15*], and *ibid.*, 48 (1942) 145; H. REUTHER, *ibid.*, 49 (1943) 176.

If the ionic concentration in the solution is not unity, the normal potential of an element M, referred to the normal hydrogen electrode, can be calculated from equation (1) if the ionic concentration c is known. It is then given by

$$_0\varepsilon_h = \underbrace{E^M_0 - E^H_0}_{\text{measured}} - \frac{RT}{n\mathfrak{F}} \ln a$$

If the two electrodes of the cell shown in Fig. 6 are joined by a wire, an electric current flows continuously from the hydrogen electrode to the zinc; a voltage difference is perpetually maintained between the two poles, because the zinc goes into solution, taking up positive charges, and at the other electrode hydrogen ions are discharged and deposited as hydrogen. If the left hand vessel in Fig. 6 were replaced by one having a copper electrode in copper sulfate solution, the current would then flow through the wire in the reverse direction, from the copper to the hydrogen electrode, since according to Table 4, copper has a lower normal potential than hydrogen and consequently the voltage difference is in this case reversed, compared to the case zinc-hydrogen mentioned before.

Corresponding relations apply if an electric current is passed through a solution by an externally applied voltage (potential difference). That element is always deposited which involves the smallest expenditure of potential for its deposition. It is clear that the minimum value of the applied voltage necessary to bring about deposition (*deposition voltage* or *discharge voltage*) must be equal and opposite in sign to the potential difference between the electrode and the solution. The discharge potentials—which are, of course, also referred to the normal hydrogen electrode as zero—are therefore equal and opposite to the electrode potentials, relative to the normal hydrogen electrode. The electrode potentials referred to the normal hydrogen electrode are also briefly designated the *normal* or *standard potentials* of the corresponding elements.

(b) Overvoltage

Hydrogen and other gases do not always actually deposit at the potentials theoretically required. Depending upon the nature of the electrodes provided for their liberation, considerably higher voltages are frequently necessary. The magnitude of the voltage in excess of the theoretical requirement is termed the *overvoltage*. For example, for the deposition of hydrogen on polished platinum (as contrasted with platinum coated with platinum black), this amounts to 0.08 volts; on a lead electrode to as much as 0.78 volts; and on mercury 0.8 volts. For the evolution of oxygen, the overvoltage on nickel amounts to about 0.1 volts, on iron 0.24 volts, and on smooth platinum 0.44 volts. These values hold only for low current densities.

The phenomenon of overvoltage arises from the fact that the discharge of ions on the electrode surface—e.g., $H_3O^+ + e \rightarrow H_2O + H$ —proceeds at a finite rate (Smits 1919, Volmer and Erdey-Gruz 1930). The ions are first deposited at the electrode surface, forming a double layer which behaves like a condenser, and increases the potential drop relative to the solution.* The greater the current density, the more undischarged ions collect at the

---

* It is not impossible that there may also be cases in which slow recombination of the atoms to molecules ($H + H \rightarrow H_2$) causes the overvoltage (C. A. Knorr 1936). For the great majority of cases so far investigated, however, this interpretation, originally due to Tafel, has been disproved by Volmer.

surface of the electrode. The increase of overvoltage (η) with increasing current density (I), first observed by Tafel in 1900, is explained in this way. It is represented by the equation

$$\eta = a + b \log I$$

The constant a depends on the nature of the electrode, b only on the nature of the gas deposited; on the basis of Volmer's theory this should be 0.118 for hydrogen at 25°. This value is found with most metals. Neutral salts, the ions of which displace H_3O^+ ions from the surface, diminish the overvoltage (Heyrovsky); it is also influenced by other additions, e.g., by colloids. The overvoltage falls with increasing temperature, and to a small extent also if the pressure is raised. It is also reduced if an alternating current is superimposed on the direct current. This is connected with the fact that the overvoltage does not instantaneously reach its maximum value at the beginning of the passage of the current, but rises gradually. The gradual increase, according to G. Masing (1936), is to be attributed to active centers, initially present on the surface of the electrode, becoming ineffective. The overvoltage varies in a systematic way with the curvature of the surface. It is smaller on a flat surface than on a convex, and greater than on a concave surface (Sederholm 1932). This explains why it is a minimum on metals with a high degree of roughness, such as platinum sponge, and quite vanishes in this case at small enough current densities.

(c) Decomposition Voltage

The absolute minimum voltage necessary for the electrolytic decomposition of any substance (decomposition voltage) is obtained by the addition of the deposition potentials of the cations and the anions in question. The uniformly negative values of the decomposition potentials are in accord with the fact that the corresponding chemical reactions are not spontaneous but require the application of electrical energy. It should also be noticed that in equations (1), (1a) and (1b), n is negative for anions.

The numerical value of the decomposition potential of an electrolyte is thus obtained by subtracting the single potential of the element forming the cations from that of the element forming the anions. The single potential E_e for an n-valent ion, having the concentration c, is obtained from the normal potential E_0 of the same ion (Table 4) by means of the equation

$$E_c = E_0 - \frac{0.05777}{n} \log c.f_a \tag{1b}$$

where f_a represents the activity coefficient of the ion concerned at the concentration c. When no great accuracy is required in the calculations, f_a may be taken as $= 1$, i.e. the difference between the analytical concentration c and the activity a may be neglected.

For a caustic potash solution, 1 normal in hydroxyl ions, the concentration c_{H^+} of hydrogen ions in the solution at room temperature is $0.74 \cdot 10^{-14} = 1 \cdot 10^{-14}$.[13] Applying equation (1b), and taking $f_a = 1$, the decomposition voltage Z_1 may be calculated as

$$Z_1 = E^O{}_0 - (E^H{}_0 - 0.058 \log c_{H^+}) = -0.41 - 0 - 14.13 \cdot 0.058 = -1.23 \text{ volts}$$

For a sulfuric acid solution 1 normal in hydrogen ions, the decomposition potential Z_2, calculated from the single potentials, (since $c_{OH^-} = 10^{-14 \cdot 13}$) is

$$Z_2 = (E^O{}_0 + 0.058 \log c_{OH^-}) - E^H{}_0 = -0.41 - 0.058 \cdot 14.13 - 0 = -1.23 \text{ volts}$$

Thus the decomposition potential should be the same for the acid as for the alkaline solution. Decomposition does, indeed, take place at this potential, if one combines a small platinum point as cathode with a large platinized platinum sheet as anode. The decomposition is recognizable by the lively evolution of hydrogen from the cathode. If, however, one combines a large cathode with a small anode, the decomposition becomes perceptible through the formation of bubbles of oxygen only at a considerably higher voltage. This is because the normal potential given in Table 4 relates to the process of the discharge of O^{2-} ions*. However, the concentration of these ions, in acid solution, is so extraordinarily small that it cannot lead to a perceptible evolution of oxygen under the conditions given. If the voltage is increased, the discharge potential of the hydroxyl ions is reached, being 1.68 volts in a solution 1-normal in hydrogen ions. Since, however, the hydroxyl ions are also present only at a low concentration in the acid solution, the evolution of oxygen only becomes vigorous when the discharge potential of the SO_4^{2-} ions, (1.9 volts) is attained.

To calculate the *consumption of electrical energy* we must add to the decomposition potential the sum of the overvoltages on both electrodes, and must also bear in mind that a certain potential is required to overcome the ohmic resistance in the electrolysis vessel and in the leads. The total consumption of electrical energy is given by the product of the voltage, current strength, and time. The product of current and time gives the quantity of electricity passed through the solution. In order to decompose 1 gram-equivalent of an electrolyte, a quantity of electricity equal to 96,493 coulombs (= 1 faraday), or 26.804 ampere-hours, must be passed through the electrolyte.

(d) Conditions under which Hydrogen is Evolved from Solutions

According to the foregoing, hydrogen will always be liberated from a solution by electrolysis if the solution contains no other substance which is deposited at a lower potential than hydrogen. In applying this rule, it is to be noticed that, as required by equation (1), account must be taken of the activity of the $[H_3O]^+$ ions in the solution. For each power of 10 by which the molar concentration or activity of the $[H_3O]^+$ ions is decreased, the potential necessary for the deposition of hydrogen ions is raised by about 0.058 volts, as shown by equations (1) and (1b). Since the hydrogen ion concentration in pure water at room temperature is about 10^{-7}, and in 1-normal alkali hydroxide solutions about 10^{-14}, the requisite potential for deposition of hydrogen from these solutions is about 0.4 and 0.8 volts respectively higher than that for deposition from solutions normal with respect to hydrogen ions.

When any element is brought into contact with a solution which contains hydrogen ions (as do all aqueous solutions), it passes into solution if its electrode potential lies above that of hydrogen in the solution in question, and hydrogen is deposited accordingly, i.e. escapes as gas.

This is the basis of the processes 1 to 3 for the preparation of hydrogen. It may be seen that an explanation is found here, not only for the evolution of hydrogen from hydrochloric acid by means of zinc, or from water by means of sodium, but also for the decomposition of alkali hydroxide solutions by aluminum. For, in spite of the very low hydrogen ion concentrations in these solutions, the discharge potential of the hydrogen in them is still numerically less than that of aluminum. The process is, in any case, substantially favored by the fact that aluminum does not remain as Al^{+++} ions in the alkaline solution, but is transformed mostly into the hydroxo-aluminate ion $[Al(OH)_4]^-$, whereby the magnitude of discharge potential is still further raised. The same is true for zinc. The insolubility of metals like zinc and aluminum in pure water is conditioned by a secondary process, namely, by

---

* One can, nevertheless, calculate the single potential of the oxygen electrode according to equation (1b) just as if it involved the discharge of OH^- ions, because the factor 2, which should be inserted for n in equation (1b), since oxygen ions are doubly charged, is cancelled out because the O^{2-} concentration is proportional to the square of the OH^- concentration.

the deposition of insoluble hydroxide in a very thin layer on the surface of the metals, whereby these are protected from further attack by the water.

4. Thermal Decomposition of Water. Law of Mass Action

Of the processes listed on pp. 23–26 for the preparation of hydrogen, several (in particular the decomposition of steam by red hot coke or red hot iron) depend ultimately on the *thermal decomposition* of water. This takes place according to the equation

$$2H_2O \rightleftharpoons 2H_2 + O_2 * \tag{2}$$

Even though the decomposition is extremely minute at the temperatures which can be employed practically, the reaction nevertheless proceeds substantially from left to right because the oxygen formed by the decomposition is at once removed from the mixture by reaction with the carbon or iron. That this must be so follows from the *chemical Law of Mass Action*.

(a) Law of Mass Action

The chemical law of mass action, in its general form (that is, applied to reactions between any specified number of substances) states that:
for a reaction between the substances A, B, C, etc., which proceeds reversibly** to form the substances M, N, O, etc. according to the equation

$$aA + bB + cC + \ldots \rightleftharpoons mM + nN + oO + \ldots \ldots,$$

at equilibrium the relation holds that

$$\frac{[A]^a \cdot [B]^b \cdot [C]^c \cdot \ldots}{[M]^m \cdot [N]^n \cdot [O]^o \cdot \ldots} = K_c \tag{3}$$

In this expression the symbols in square brackets represent the molecular concentrations of the substances, i.e., their concentration expressed in moles per liter. K_c is a constant (the *equilibrium constant*) which depends on the temperature. The mass action law applies only to stable homogeneous (one phase) systems.

A homogeneous system is one in which (on a scale greater than molecular dimensions) there are no interfaces. Gases in the state of complete admixture, liquids (liquid *mixtures* only when one substance is completely dissolved by the other), solutions, and solid bodies of uniform composition, including mixed crystals, are homogeneous. Examples of inhomogeneous systems are; a gas in contact with a liquid or with a solid, immiscible or partly miscible liquids in contact with each other, liquids in contact with solid substances, two or more solid substances of different composition present together. The portions of a mixture which are of the same kind are called the *phases* of the mixture. Homogeneous systems are thus one-phase systems.

---

* In addition, some thermal dissociation of water takes place according to the equation $2H_2O \rightleftharpoons H_2 + 2OH$, as Bonhoeffer showed in 1928. This does not, however, affect the present discussion.

** In principle, every reaction can be considered to be reversible.

In applying the law of mass action to the reactions of gaseous systems, the *partial pressures* of the reactants can be inserted in place of their concentrations, since for gases the pressures are proportional to the concentrations. It must be noticed however, that the numerical value of the equilibrium constant will be different if the pressures are inserted in place of the concentrations, except in the case where the total number of molecules (and therefore the total pressure) remains unaltered by the reaction.

If the equilibrium constant, defined with respect to the pressures, is represented by K_p, and that related to the concentrations is K_c, then

$$K_p = (RT)^{\Sigma\gamma} \cdot K_c, \quad \text{or} \quad \ln K_p = \ln K_c + \Sigma\gamma \cdot \ln RT \tag{4}$$

In (4), the symbol $\Sigma\gamma$ represents the arithmetical sum of the numbers of molecules of each kind in the equation for the reaction. The numbers of molecules of substances entering into the left-hand side of the equation are thereby taken as positive, and those of substances on the right-hand side as negative. Example: for the reaction $2H_2 + O_2 = 2H_2O$, $\Sigma\gamma$ is $3 - 2 = 1$. Thus the equilibrium constants of equations (7) and (8), p. 37, are connected by the relation $K_p = RT \cdot K_c$.

Equation (4) is obtained if the molar concentration $\dfrac{n}{v}$ of each gas is replaced in equation (3) by the expression $\dfrac{p}{RT}$, derived from the gas equation $p \cdot v = n \cdot RT$, where p is the partial pressure of the gas considered.

Establishment of the equilibrium of equation (3) naturally presupposes that the corresponding reactions take place sufficiently fast. This is frequently not the case at low temperatures. Such systems, which are not in equilibrium, but which nevertheless are not undergoing transformation because the reaction velocity is too small, are said to be *metastable*.

The Law of Mass Action was originally discovered empirically, by Guldberg and Waage in 1864; it received its name because the molar concentrations which enter into it were formerly called the 'active masses'. It was shown later that the law follows as a necessary consequence from the *kinetic theory of gases and liquids*. It can also be derived *thermodynamically*—i.e., it is a consequence of the first and second laws of thermodynamics.

In the thermodynamic derivation of the law of mass action, it is assumed that the reactants obey the laws which are valid for 'ideal gases' and 'ideal dilute solutions'. Since real gases and solutions depart from these laws, to an extent which increases with the concentration, the mass action law in the form given above— i.e., with molar concentrations inserted—is only a limiting law, applying exactly only at infinite dilution. The law of mass action may be made to hold exactly for any substance, and at high concentrations, by multiplying the analytical molar concentrations by certain coefficients, known as *activity coefficients*. The products of analytical molar concentrations and activity coefficients are called *activities*. The use of activities was introduced by G. N. Lewis, in 1907. As defined, activities are quantities which must be substituted in the law of mass action, in place of the analytical molar concentrations, in order that the law may be truly valid for real substances at finite concentrations, and not merely at infinite dilution.

The activity can be defined thermodynamically, independently of the law of mass action.*
Activity coefficients can accordingly be determined experimentally, without having to
carry out chemical transformations on the substances concerned. Table 6 gives values of

TABLE 6

ACTIVITY COEFFICIENTS OF VARIOUS SUBSTANCES
IN AQUEOUS SOLUTION AT 25°

m = mols of dissolved substance in 1000 g solvent

	m =	1	0.5	0.2	0.1	0.05	0.01	0.005	0.001
Non-electrolytes	Methanol	0.80	0.82	0.84	0.87	0.90	0.95	—	—
	Ethanol	0.85	0.87	0.91	0.94	—	—	—	—
	Acetone	0.935	0.968	0.988	0.994	—	—	—	—
	Cane sugar	1.24	1.11	1.05	1.03	1.01	—	—	—
	Acetic Acid	0.900	0.960	—	—	—	—	—	—
Strong electrolytes	HCl	0.802	0.750	0.760	0.791	0.825	0.902	0.927	0.965
	H_2SO_4	0.130	0.154	0.209	0.265	0.304	0.544	0.639	0.830
	$NaNO_3$	0.565	0.633	0.714	0.771	0.820	0.905	0.929	0.966
	$BaCl_2$	0.388	0.392	0.436	0.499	0.564	0.716	0.774	0.881
	$Ca(NO_3)_2$	0.35	0.38	0.42	0.48	0.545	0.71	0.77	0.88
	$MnSO_4$	0.080	0.110	0.176	0.247	0.333	0.536	0.621	0.780
	$CuSO_4$	0.041	0.061	0.098	0.149	0.209	0.41	0.53	0.74

activity coefficients for a few substances at various concentrations. The figures show that
the activity coefficients—which may be either greater or less than unity— all tend to the
value 1 with increasing dilution. This must be so since, by definition, the difference between
the activity and the analytical concentration must vanish. The table also shows that, at
moderate concentrations (0.1 to 1 molar), the activity coefficients of non-electrolytes and
weak electrolytes differ but little from 1, whereas for strong electrolytes the differences are
considerable.

* Thus, for a dissolved substance which does not dissociate, the activity a is defined by
$a = e^{-A_c/RT}$ or $A_c = -RT \ln a$, where A_c is the work which must be expended in order to
bring 1 mol of the dissolved substance from its standard state, in which its activity is taken
as $a_0 = 1$, by an isothermal, reversible process, to the concentration c. For the standard
state, a state is selected in which the substance obeys the laws of dilute solutions. Strictly
speaking, this would be true only at infinite dilution ($c = 0$) — i.e., $\lim_{c=0} a_0 = c_0$. However,
it is always possible to take some finite concentration c_0, at which a_0 and c_0 are identical
within experimental error. The activity may be similarly defined on thermodynamic
grounds for a gas, vapor or solvent. For definition of activity of strong electrolytes, see
p. 69.

It follows from the foregoing that for *gaseous* substances, under conditions such that they behave practically as ideal gases (i.e., at normal pressures and not too low temperature) the analytical molar concentrations or partial pressures can be introduced into the law of mass action. If these conditions are not fulfilled, it is necessary to replace concentrations by activities for gases or vapors also.

The equilibrium constant K of the mass action law is temperature dependent, its variation being given by the following equation which, following the work of Gibbs (1879), was put forward by Van 't Hoff

$$\frac{d \ln K}{dT} = \frac{W}{RT^2} \tag{5}$$

or, assuming the constancy of W, in the integrated form

$$\ln K_2 - \ln K_1 = \frac{W}{R}\left(\frac{1}{T_1} - \frac{1}{T_2}\right) \tag{6}$$

W = heat evolved by the reaction; R = gas constant (expressed in heat units = 1.987 calories); T = absolute temperature; K_1 and K_2 values of the equilibrium constant at the temperatures T_1 and T_2.

According as K is the equilibrium constant K_c or K_p, the quantity substituted for W must be the heat of reaction at constant volume, W_v, or that at constant pressure, W_p.

Since the equation of the mass action law (equation 4) holds at constant temperature, it was termed by Nernst the *reaction isotherm*.

(b) Application of the Mass Action Law to the Thermal Decomposition of Water

If we apply the chemical law of mass action to the thermal dissociation of water (equation 2, p. 34), we have

$$\frac{[H_2]^2 \cdot [O_2]}{[H_2O]^2} = K_c \tag{7}$$

or, substituting the partial pressures of the gases p_{H_2}, p_{O_2}, p_{H_2O}, in place of the concentrations,

$$\frac{p_{H_2}^2 \cdot p_{O_2}}{p_{H_2O}^2} = K_p \tag{8}$$

If, now, to this gas mixture, a substance is added having an oxide with a smaller oxygen pressure than that of water at any particular temperature, it will combine with oxygen, forming its oxide, until the oxygen pressure has fallen to the value corresponding to that oxide. In order that K_c or K_p respectively may remain constant the ratio $\dfrac{p_{H_2}^2}{p_{H_2O}^2}$ must increase in the same proportion as the oxygen pressure p_{O_2} has diminished. If, for example, the oxygen pressure over the added substance (e.g., iron) is one millionth part of the oxygen pressure of water at the temperature in question, then the hydrogen pressure in contact with it must increase one thousandfold as compared with the pressure of undissociated water vapor.

According to Emmet (1933) and Kapustinsky (1934), the following equilibrium pressures are established over a mixture of Fe and FeO

Temperature °C	600	700	800	900	1000	1200
p_{H_2}/p_{H_2O}	3.0	2.4	2.0	1.6	1.4	1.1

With decreasing temperature, the equilibrium is displaced in favor of the formation of hydrogen. This is because the oxygen pressure of FeO diminishes more rapidly with decrease of temperature than does that of water vapor. In general, however, it is not possible to go to temperatures below about 600 to 800°, as the reaction otherwise proceeds too slowly, even in the presence of added substances which accelerate the reaction. For decomposition by coke, even higher temperatures (over 1000°) must be used. In this case, the effective O_2 pressure is determined not only by the equilibrium $C + \frac{1}{2}O_2 \rightleftharpoons CO$, but the equilibrium $CO + \frac{1}{2}O_2 \rightleftharpoons CO_2$ is also involved (cf. p. 444).

If the numerical values given in the foregoing table are substituted in eqn. (8), the oxygen pressure of FeO at the given temperatures can be calculated, since K_p is known from other measurements. Thus, at 1000° K (= 727 °C), $K_p = 4.5 \cdot 10^{-21}$.

Hence, $p_{O_2} = K_p \left(\dfrac{p_{H_2O}}{p_{H_2}} \right)^2 = 4.5 \cdot 10^{-21} \left(\dfrac{1}{2.3} \right)^2 = 0.85 \cdot 10^{-21}.$

At this temperature, then, FeO has an oxygen pressure of $0.85 \cdot 10^{-21}$ atm. The example shows how, from the equilibrium attained in reversible reactions, it is possible to determine such dissociation pressures, which are far too small to be measured directly. Below 570°, Fe_3O_4 has a still smaller oxygen pressure than FeO. Iron therefore reacts with steam below 570°, to form Fe_3O_4.

5. **Properties of Hydrogen**

Hydrogen is a colorless gas, without taste or smell, which burns with a weak, bluish, very hot flame when ignited in air. It is 14.38 times lighter than air, and has a thermal conductivity 7 times that of air.

As regards its thermal expansion and the dependence of its volume upon pressure, hydrogen at not too low temperatures and not too high pressures behaves practically as an ideal gas, that is, it very closely obeys the Boyle-Mariotte and the Gay-Lussac laws, i.e., the gas equation $pv = nRT$ (in which $R = 0.082054$ liter atm.).

Since hydrogen does not show the Joule-Thomson effect (i.e., the phenomenon in which a gas is cooled down when it expands without performing external work— as for example when it flows into a vacuum —) above about —80°, the liquefaction of the gas at first presented great difficulty. Dewar was the first to succeed in liquefying the gas in 1898, by allowing highly compressed hydrogen, first cooled to —205° by means of liquid air, to flow through a narrow aperture into a vacuum. Liquid hydrogen is a very light, colorless fluid, which does not conduct electricity, and which solidifies to a solid mass of density 0.08 when it is boiled under reduced pressure.

The most important physical constants of hydrogen are collected in Table 7. On account of its low molecular weight, hydrogen has the greatest rate of effusion and the greatest rate of diffusion of all gases, in accordance with the laws of Graham and Bunsen.

Hydrogen is very sparingly soluble in water and other solvents, and also in india

TABLE 7

PHYSICAL CONSTANTS OF HYDROGEN*

Liter weight at 0° and 760 mm pressure, at 45° latitude:	=	0.089870 g
Molecular volume	=	22.428 l
Gas density (relative to oxygen = 1)	=	0.062893
(relative to air = 1)	=	0.06952
Boiling point	=	−252.8 °C = 20.4 °K
Density at the boiling point	=	0.0700
Melting point	=	−257.3 °C = 15.9 °K
Density at −260°	=	0.0763
Critical temperature	=	−239.9°
critical pressure	=	12.8 atm.
critical density	=	0.0310
Specific heats c_p (at 20°)	=	3.396 cal.
c_v (at 18°)	=	2.404 cal.
Ratio of specific heats: c_p/c_v (at 17°)	=	1.407
Molecular heats: C_p	=	6.847 cal.
C_v	=	4.847 cal.
Specific heat of liquid hydrogen at −256°, c_p	=	1.93 cal.
of solid, at −259.8°, c_p	=	0.630 cal.
Heat of evaporation of liquid hydrogen	=	108–114 cal per g (at 20.4 °K)
Heat of fusion of solid hydrogen	=	14 cal. per g
Crystal structure: hex. close packing (of H₂ molecules); a	=	3.75 Å
c	=	6.12 Å
Thermal conductivity of hydrogen at 0°: λ (i.e., the quantity of heat λ flows through unit cross section each second, for a temperature gradient of 1° per cm).	=	0.000412 cal.cm.$^{-1}$sec.$^{-1}$degrees^{-1}
Solubility in water: at 0°	=	0.0215 volumes
at 18°	=	0.0185 volumes (reduced to 0°) of hydrogen in 1 volume of water
H₂ *molecule*. Heat of formation** from 2H	=	102.72 kcal/mol H₂ (at 0° K)
Internuclear distance	=	0.75 · 10^{-8} cm
Moment of inertia	=	0.467 · 10^{-40}
Dipole moment		0
Mean free path at 0° and 1 atm.	=	1.123 · 10^{-5} cm
H-*atom*		
Ionization potential	=	13.54 volts
Electron affinity	=	0.715 e-volts

* Except for the properties of the H₂ molecule, the data relate to ordinary hydrogen, i.e., for hydrogen with 0.02 % D₂ (cf. p. 43).
** Refers to p-H₂ (cf. p. 43). For the mixture p-H₂ : o-H₂ = 1 : 3, the value is 102.47 kcal per mol.

rubber—a property of importance when it is used as a filling gas for balloons.* It is taken up by many metals in fairly considerable quantities, and diffuses through

* Since hydrogen does not dissolve in india rubber it cannot diffuse through this material in the strict sense, but can only effuse through pores. Hydrogen gas is therefore retained considerably longer by a rubber membrane than are gases such as hydrocarbons, etc., (coal gas), which are soluble in rubber, even though the latter have considerably smaller effusivities than hydrogen.

them at a red heat, as also through quartz. Palladium has the greatest solvent power for hydrogen.

Molecular hydrogen is chemically rather inert at ordinary temperature. It generally combines directly only with fluorine,* with chlorine when it is illuminated, and with oxygen only after ignition. When heated, it combines with numerous elements—especially with non-metals, but also with the strongly electropositive metals (alkali and alkaline earth metals).

The combination of hydrogen with oxygen to form water:

$$H_2 + \tfrac{1}{2}O_2 = H_2O **$$ (9)

takes place with a very great evolution of heat. The heat of formation of water from H_2 and O_2, at 25° and constant pressure, amounts to 68.35 kcal per mole of *liquid* H_2O, or 57.85 kcal per mole of *gaseous* H_2O.

According to Lewis and Randall, the temperature-dependence of the heat of formation of gaseous H_2O, at constant pressure, can be represented over the temperature range 0 — 2,000°, by the empirical equation.

$$W_p = 57.44 + 9.4 \cdot 10^{-4}T + 1.65 \cdot 10^{-6}T^2 - 7.4 \cdot 10^{-10}T^3,$$

where T represents the absolute temperature and W_p the heat of formation in kcal per mole H_2O. At 0°, for example, this is 57.80; at 100°, 57.98 kcal.

A mixture containing hydrogen and oxygen in exactly the proportions in which they react according to equation (9) (*detonating gas*) explodes with exceptional violence when ignited.

In a wider sense, the name detonating gas is often applied also to other explosive mixtures of hydrogen with oxygen or air. The values given for the limits of explosibility of hydrogen-air mixtures vary. In general, it can probably be stated that such mixtures no longer explode if they contain less than 6 or more than 67 volume per cent of hydrogen. One volume of hydrogen requires 2.39 parts by volume of air for complete combustion. The ignition temperature of a detonating gas mixture depends to a considerable degree upon the vessel in which the gas is contained; with pure detonating gas, explosions have been observed from about 500° upwards. In the presence of suitable catalysts, combination already sets in at considerably lower temperatures, or even at room temperature in the presence of platinum metals. In some circumstances explosion does not occur, but even under these conditions ignition and explosion generally occur after a short time, because of the heat evolved in the reaction.

In the blowpipe burner, an oxy-hydrogen flame can be produced without danger. The temperature attainable with the oxy-hydrogen blowpipe flame is said to be about 3100°. The oxy-hydrogen flame is often used for autogenous welding. It gives the highest temperature when it burns most quietly; with an excess of oxygen it hisses, with a hydrogen excess it rustles.

Hydrogen also combines directly with bromine and sulfur, and with iodine, selenium, and tellurium. With the last three, the reaction leads to a distinct equilibrium state. However, whereas hydrogen iodide decomposes to an increasing ex-

---

* The velocity of reaction of H_2 with F_2 depends very much upon the material of the containing vessel. Practically no reaction takes place at room temperature in a vessel made of magnesium (Bodenstein 1937).

** Contrary to older assumptions, the reaction does not necessarily proceed by way of H_2O_2 as an intermediate (see Vol. II).

tent with rising temperature, rise of temperature shifts the equilibrium with hydrogen selenide and telluride in favor of the formation of the compounds. These are thus endothermic, as may be deduced by means of the Le Chatelier principle. Hydrogen combines with nitrogen to any considerable extent only in the presence of a catalyst; the reaction is exothermic (see further, under ammonia synthesis). The combination of a substance with hydrogen is called hydrogenation.*

Because of the great affinity of hydrogen for oxygen, hydrogen is very suitable for removing oxygen from combination with other substances: it is a strong *reducing agent*. The oxides of many heavy metals are transformed into the metals when heated in hydrogen.

A few substances are reduced by gaseous hydrogen even at the ordinary temperature,—in particular palladium(II) chloride in aqueous solution, and also, to a much smaller degree, silver nitrate. Numerous substances are reduced or hydrogenated by hydrogen at the ordinary temperature in the presence of substances which 'activate' the hydrogen, e.g., platinum sponge or palladium. In the presence of the latter, hydrogen adds on especially easily to unsaturated organic compounds (hydrogenation by the methods of Paal and Skita). At higher temperatures, nickel also acts as a hydrogen carrier (hydrogenation method of Sabatier and Senderens).

Catalytic hydrogenation processes have become very important technically in recent times. The most important of these are the hydrogenation of nitrogen to ammonia, the hydrogenation of coal, petroleum, and tar to gasoline, the hydrogenation of carbon monoxide to methanol and higher alcohols, or to gasoline, and the hydrogenation of unsaturated fatty acids (hardening of fats).

In the laboratory, and especially in organic chemistry, hydrogenations are often carried out in solution, by shaking with a suspension of platinum black or palladium. According to Biesalski (1930), it is better to introduce the hydrogen through a fine pored filter disc, into a solution containing a foam-forming agent.

Hydrogen is also particularly reactive at the instant of its liberation from its compounds ('*in statu nascendi*'). This is attributed to the fact that when, for example, it is evolved electrolytically, or liberated from sulfuric acid by means of zinc, it is initially formed in the atomic state, whereas in the ordinary gaseous state the atoms are combined in pairs to form molecular H_2. Use is often made of the great reactivity of 'nascent' hydrogen for energetic reductions. For this purpose it is generally evolved by dissolving zinc in sulfuric acid or in caustic alkali, by the action of sodium amalgam on water, or electrolytically. In the last case, the reducing power is particularly great if a cathode (e.g., of lead) is used on which the hydrogen displays a considerable overvoltage. Fichter (1933) has achieved similar effects by the dissolution of lead-sodium alloys.

Under high pressures (100 to 200 atm.) hydrogen can precipitate metals from warm solutions of their salts even if no activating substances are present (e.g., at 200°); this occurs the more readily, the nobler the metal displaced. Even active metals such as zinc can be displaced from their salts in this way, although they are more readily deposited as their hydroxides, through reduction of the acid radicals combined with them—e.g., of the NO_3^- ion (Ipatieff).

* In organic chemistry combination with hydrogen is also frequently termed reduction. See Chap. 16.

6. Special Forms of Hydrogen

(a) Atomic Hydrogen

If a wire of tungsten, platinum, or palladium is heated in hydrogen at low pressure (p less than 0.01 mm), it is possible to detect the formation of a gaseous, active modification of hydrogen. This is also formed when a glow discharge is passed through hydrogen at a pressure of a few tenths of a millimeter, as was found by R. W. Wood (1922). Wood's process is particularly convenient for preparing active hydrogen; the apparatus used by Bonhoeffer is sketched in Fig. 8. This enables the active hydrogen, which has been shown to consist of free H atoms, to be obtained under suitable conditions in a concentration of up to 95 %. Atomic hydrogen is stable, however, only for a very short time (about one-third to one-half second at the pressure mentioned). This time suffices to remove it from the discharge tube by a rapid stream of gas, and to pass it over materials with which it may react. Atomic hydrogen is able to reduce a whole series of metallic oxides (e.g., CuO, PbO, HgO, Ag_2O, Bi_2O_3), sulfides and halides, even at ordinary temperature. It combines with sulfur, arsenic, and phosphorus, forming the corresponding hydrides. It combines with molecular oxygen, the principal product being hydrogen peroxide H_2O_2 when the reaction occurs at low temperatures; water is formed otherwise. The first product at low temperatures is an isomer of H_2O_2, and this is transformed into normal hydrogen peroxide at $-115°$, with partial decomposition, (Harteck 1932). Atomic hydrogen reacts with mercury, forming a solid hydride stable only at low temperatures. It hy-
drogenates oleic acid immediately, and also poly-
merizes it in such a way that for one molecule
which is hydrogenated to stearic acid by addition
of 2H, two molecules are linked together with the
addition of 1H per molecule. Atomic hydrogen
can also exert a dehydrogenating action on
unsaturated organic compounds, by splitting out
H from them with the formation of H_2.

Bonhoeffer's apparatus (Fig. 8) consists essentially of the electrolytic hydrogen generator H,
and the discharge tube E. The latter is a glass
tube 2 cm wide and 2 m long, bent for compact-
ness into an S shape, and furnished with cylindrical sheet aluminium electrodes. It withstands
a loading of 0.4 amp. D.C. at 3,000 to 4,000
volts. A metal regulating valve (shaded in the
drawing) is inserted between H and E; this is

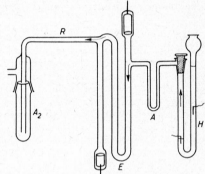

Fig. 8 Apparatus for preparing active
(atomic) hydrogen.

adjusted so that by pumping hydrogen off rapidly at the end of the apparatus (preferably by means of a diffusion pump), the pressure in the discharge tube is between 0.1 and 1 mm. The hydrogen can be freed from water vapor by immersion of the U-tube A in liquid air. Substances may be introduced into the tube at R when their behavior towards active hydrogen is to be examined, and the products of reaction are condensed by means of liquid air in A_2.

That active hydrogen, characterized by the reactions mentioned, consists of hydrogen atoms, follows in particular from the fact that the Balmer spectrum emitted by the hydrogen *atom* H increases in intensity as compared with the band spectrum emitted by the *molecule* H_2, under the same conditions that favor the formation of active hydrogen. In view of this fact, the lifetime of the active hydrogen given above appears surprisingly large, since it implies that on the average a million collisions must take place between pairs of hydrogen atoms, before recombination occurs to form the molecule H_2, even though the heat of formation of the H_2 molecule from the atoms is quite large (cf. Table 7, p. 39). The explanation is that, in order for recombination to take place, a third partner must take part in the collision, to carry off part of the energy of recombination (see further in Vol. II under mechanism of reactions). The recombination of H to H_2 is particularly strongly accelerated on the surface of many solids. The catalytic efficiency can be measured by placing the solid in question on the bulb of a thermometer, which is suspended in the gas stream. The greater the catalytic effect, the higher is the temperature to which the thermo-

meter is raised by the liberated heat of recombination. A tungsten wire introduced near the discharge tube is heated to incandescence. Small particles of metal, such as Pt, Pd, W, glow even at a considerable distance from the discharge tube, so that their incandescence, together with the reaction with sulfur (formation of H_2S), can be used as a means for the detection of atomic hydrogen.

The high heat of recombination of atomic hydrogen finds technical application for the welding of especially high-melting metals, by means of the Langmuir blowpipe. This consists of two tungsten rods, over which hydrogen is blown, and between which an electric arc is struck. A narrow jet of hydrogen is blown through the arc and as it passes through the arc, the hydrogen is partially dissociated into atoms, which recombine to form molecules, at a sharply defined spot on the surface of the metal on which the jet impinges. Extremely high local temperatures—up to 4000°—are thereby set up, so that it is possible, with the atomic hydrogen blowpipe, to melt even the most infusible metals, e.g., tungsten and tantalum. The hydrogen atmosphere in which melting is carried out protects the metal from oxidation.

Modern work has not confirmed the existence of the triatomic hydrogen, H_3, formerly postulated by some workers. However, the existence of the positively charged species H_3^+ in canal rays was discovered by J. J. Thomson and finally established by Aston, who determined its mass accurately.

(b) Heavy Hydrogen. Deuterium

An isotope of atomic weight 2, which has been called *heavy hydrogen* or *deuterium* (chemical symbol D; see further Vol. II) is present in ordinary hydrogen to the extent of 0.02 volume per cent. Because of its low concentration, this affects the properties of hydrogen only to a slight extent. When accurate numerical values are involved, however, it is necessary to notice whether they refer to ordinary hydrogen—i.e., the mixture of isotopes—or to deuterium-free hydrogen.

(c) Tritium

A third isotope of hydrogen, with mass 3, was discovered in 1934 by Oliphant, Harteck and Rutherford. It is known as *tritium* (chemical symbol ^3H or T). This is an *unstable* isotope of hydrogen: it is radioactive, and emits β-rays, decaying with a half-life of 12.5 years (see Vol. II, Chap. 11). For this reason it is not present in any appreciable amount in naturally occurring hydrogen compounds or the hydrogen prepared from them, but has been obtained in the laboratory by artificially induced nuclear reactions. The formation and properties of tritium are accordingly discussed in Volume II, Chap. 13.

(d) Ortho- and Para-hydrogen

The specific heat of hydrogen gas at low temperatures (e.g., at the boiling point of liquid air) is considerably smaller than it should be for a diatomic gas, according to the kinetic theory.* This is partly explained in that the rotation of the hydrogen molecules ceases at low temperatures, so that hydrogen then behaves as a monatomic gas. However, the course of the curve for the specific heat of hydrogen at low temperatures was first completely explained by Heisenberg and Hund (1927), who deduced theoretically that ordinary hydrogen consists of two modifications of H_2 molecules with different rotational heats.** These

---

* According to the kinetic theory of gases the molecular heat of a monatomic gas should be $\frac{3}{2}R$, i.e., about 3 calories, independent of the temperature. The molecular heat of a diatomic gas should also be independent of temperature at not too high temperatures, and not less than $\frac{5}{2}R$ or about 5 calories. For hydrogen the molecular heat at low temperatures approaches the value 3 calories.

** For the significance of the difference between the modifications of H_2 on the basis of the theory of atomic structure see Footnote, p. 171.

are distinguished as *ortho-* and *para-hydrogen*. They are identical in their chemical behavior, but differ not only in their different specific heats, but in other physical properties also. K. F. Bonhoeffer and P. Harteck succeeded, in 1929, in preparing pure para-hydrogen, by the adsorption of ordinary hydrogen on wood charcoal at about 20° absolute. Pure para-hydrogen is stable, in the strict sense, only in the neighborhood of absolute zero. In the absence of catalysts however,—even the walls of the vessel can admittedly act as such—it does not change into ortho-hydrogen with perceptible speed even at room temperature. *Ordinary hydrogen* consists of *3 parts of ortho-hydrogen* with *one part para-hydrogen*. Mixtures richer in para-hydrogen are stable at low temperatures. However, even in this case, the change of one modification into the other does not generally occur at a measurable rate. It can be accelerated, however,—e.g., by high pressure,—and the enrichment of para-hydrogen at low temperatures was successfully carried out by this means almost simultaneously by Bon-hoeffer and Harteck, and by A. Eucken. The latter had already noticed the anomalous temperature dependence of the specific heat of hydrogen at low temperatures in 1912.

Para-hydrogen (p-H_2) has a rather higher vapor pressure than ordinary hydrogen (n-H_2), and boils at $-252.871°$ (ordinary hydrogen boils at $-252.754°$). The heat of transformation n-$H_2 \rightarrow p$-H_2 is 0.337 kcal per mole at absolute zero.

7. Hydrogen Compounds. Hydrides

The binary compounds of hydrogen are called hydrides, and may be divided into four classes:(1) gaseous or volatile hydrides, (2) polymeric hydrides (solid hydrides), which are neither salt-like nor metallic in character, (3) salt-like hydrides, (4) pseudo-metallic hydrides.

Gaseous or volatile hydrides are formed by boron, by gallium, and by all the elements of the fourth to seventh Main Groups of the Periodic System. In so far as one can speak of the polarity of hydrogen in these molecules, it is positive. Thus the halogen hydrides, when brought into water split off the hydrogen in the form of a positive univalent ion.

Salt-like hydrides are formed by hydrogen with most of the elements of the Ist and IInd Main Groups. Hydrogen is the *electronegative* component in these hydrides, and behaves in them like a halogen (cf. lithium hydride p. 173).

All the elements of the Main Groups of the Periodic System (except for the inert gases, which form no compounds at all) are now known to form hydrides; the gaps which remained until recently have now been filled through the researches of E. Wiberg and H. I. Schlesinger (1940 etc.). It may be seen from the survey given in Table 8 that the valence of the elements towards hydrogen increases from 1 to 4 in Groups IA to IVA, and decreases from 4 to 1 in Groups IVA to VIIA.

In Groups IIA and IIIA, there are some solid hydrides which take an intermediate position between the salt-like hydrides and the volatile hydrides. They differ quite characteristically from the salt-like hydrides, however, and constitute a distinct class—hydrogen compounds of high-polymer type. Their involatility—or the underlying association of their simple molecules such as BeH_2 into aggregates of indefinite complexity—is probably the result of the formation of hydrogen-bridge compounds (cf. p. 217), in the same way as the dimerization of the compounds BH_3 and GaH_3 to B_2H_6 and Ga_2H_6.

Only the simplest of the volatile hydrogen compounds are listed in Table 8. In addition to these, volatile hydrides of higher molecular weight are also formed in some cases. These arise from the union of simpler radicals to form chains. This occurs especially with carbon, which forms a very large number of higher hydrocarbons, as well as CH_4. It is true to a lesser degree of boron and silicon also. In the volatile hydrides of other elements, the capacity for chain formation, if found at all, is generally limited to the union of a pair of radicals. Only in the case of germanium is the compound Ge_3H_8 known as well as GeH_4 and Ge_2H_6, whereas sulfur forms hydrides with sulfur chains of enormous length (see hydrogen polysulfides).

TABLE 8

HYDROGEN COMPOUNDS FORMED BY ELEMENTS OF THE MAIN GROUPS
OF THE PERIODIC SYSTEM

Volatile hydrogen compounds are indicated by *circles*, and salt-like hydrides by *rectangles*.

Increasing salt character   ←   Increasing acid character →

I	II	III	IV	V	VI	VII
LiH	$[BeH_2]_x$	$(BH_3)_2$	CH_4	NH_3	OH_2	FH
NaH	$[MgH_2]_x$	$[AlH_3]_x$*	SiH_4	PH_3	SH_2	ClH
KH	CaH_2	$(GaH_3)_2$	GeH_4	AsH_3	SeH_2	BrH
RbH	SrH_2	$[InH_3]_{\dot{x}}$	SnH_4	SbH_3	TeH_2	IH
CsH	BaH_2	TlH_3**	PbH_4	BiH_3	PoH_2	AtH

* AlH_3 is present in monomeric form in ethereal solution.
** TlH_3 is only known in the form of double compounds such as $TlH_3 \cdot 3GaH_3$.

Many of the elements which form volatile hydrides can also form peculiar non-volatile hydrogen compounds (e.g., polysilene, $[SiH_2]_x$, solid phosphorus hydride $[P_2H]_x$ or $P_{12}H_6$, solid arsenic hydride, $[AsH]_x$. The nature of these compounds is not well established.

In so far as the elements of the Sub-groups of the Periodic Table can combine with hydrogen at all, they generally form '*pseudo-metallic*' hydrides. These very often do not have compositions fixed by stoichiometric ratios. The metals that form them often first incorporate hydrogen purely in solid solution, without change of crystal structure, although there is a dimensional increase in the crystal lattice. The amount of hydrogen so taken up is usually proportional to the square root of the hydrogen pressure (Sieverts). With many metals, however, especially those of groups IVB and VB, a structural transformation sets in when the amount of hydrogen taken up exceeds some limiting concentration. Since this is associated with a discontinuous change in properties, it is considered that a *compound* is formed (cf. Vol. II, Chap. 1). Hydrogen compounds of this type are called 'pseudo-metallic' hydrides. The hydrides of the Group IIIB elements (especially lanthanum hydride) occupy an intermediate position between the pseudometallic and the salt-like hydrides.

It is also possible to obtain compounds with the character of salt-like hydrides by the action of atomic hydrogen on the metals of the Sub-groups (Pietsch 1933), but nothing is known of their stoichiometric composition. Numerous diatomic hydrides exist in flames and in the electric arc, as is shown by band spectroscopic evidence. Most of these however,— such as CH, NH, OH, MgH, ZnH, AlH, AuH,—are either unstable under normal conditions, or at least not obtainable in weighable amounts.

The most important compound of hydrogen is *water*, H_2O. Besides this, there exists a second oxygen compound of hydrogen, namely hydrogen peroxide, H_2O_2. Only these two compounds of hydrogen will be treated in detail in this chapter, compounds with the other elements being described thereunder. However, a general discussion of two particularly important classes of hydrogen compounds must be given at this stage—namely the aqueous *acids* and *bases*. In connection with these, we may consider the nature of electrolytes, to which class the acids and bases belong. Particular importance attaches to these in the field of inorganic chemistry.

8. Water

(a) Occurrence and Purification

Water, H_2O, in the form of the oceans, is the principal constituent of the earth's surface. It is also overwhelmingly important in the structure of animal and vegetable organisms. The water content of the human body accounts for more than half of its total weight. Water is contained in the atmosphere in the form of vapor, and is precipitated in the form of mist, cloud, rain, frost, snow or hail, by a fall in temperature.

Natural water is never chemically pure. According to its origin, it contains various substances in solution, or in some circumstances also in suspension. Sea water, as is well known, is strongly saline. Rain water always contains small amounts of ammonium nitrate, and often traces of other salts as well. Spring water and ground water contain in solution the constituents of the soil from which they come. In addition to iron and manganese salts, which make the water unsuitable for many purposes, such as for laundering, if they are present in considerable amount, the hydrogen carbonates and sulfates of the alkaline earth metals and magnesium are of practical significance. The *hardness* of the water is measured by the concentration of these. The gas content of water is also important, and especially the amounts of oxygen and carbon dioxide, which are of decisive importance for the formation of rust on iron in contact with water.

The requirements set upon the purity of water vary according to whether it is wanted for drinking water or for technical purposes (boiler feed water, laundering, etc.). [3–6] *Spring* water is in general most suitable for drinking water; if this is not available, ground water is used. The latter is usually clarified in waterworks by filtration through gravel and sand. Where water is taken from river courses, it is often necessary to carry out chemical purification as well. Sterilization, which is particularly important in this case, is performed by treatment with chlorine or ozone or, more recently, by irradiation with ultra-violet light also. The purification of water for technical purposes involves principally the removal of hardness. This is done either by addition of chemicals ($Ca(OH)_2$, Na_2CO_3) which precipitate the substances giving rise to the hardness, or by treatment with a base-exchanger such as Permutit (cf. p. 513) or ion-exchange resins. Hard water is unsuitable for use in laundering, not only because part of the soap is precipitated through the formation of calcium soaps (that is of insoluble calcium salts of the fatty acids contained in the soap), so that there is an increased consumption of soap, but also because the calcium soaps are precipitated to some extent within the fiber of fabrics, giving a greasy feel and often an unpleasant smell.

Distilled water is used in the chemical laboratory and also in many technical installations. It is practically free from salts; if it leaves a residue on evaporation, it must be redistilled. For many more industrial purposes water with a similar degree of purity is prepared by the so-called 'electro-osmotic' process. This is, in reality, an electro-dialytic removal of salts, based on the fact that the ions of the impurities (salts) are concentrated by means of an applied electric potential in anodic and cathodic compartments, enclosed by diaphragms, and are washed away from these. This method does not, however, remove colloidally dispersed substances (e.g., silica). With the ion-exchange resins which have come recently into use, it is possible to achieve complete removal of dissolved salts. For this purpose, a cation-exchanging resin is combined with an anion-exchange resin.

$$\left. \begin{array}{lll} CaSO_4 + 2H^+_{resin} & = H_2SO_4 + Ca^{++}_{resin} \\ NaCl \ + \ H^+_{resin} & = HCl \quad + Na^+_{resin} \end{array} \right\} \ cation \ exchange$$

$$\left. \begin{array}{lll} H_2SO_4 + 2OH^-_{resin} & = 2H_2O + SO_4^{2-}_{resin} \\ HCl \ + \ OH^-_{resin} & = \ H_2O + Cl^-_{resin} \end{array} \right\} \ anion \ exchange$$

In the chemical laboratory, and especially for analytical purposes, the use of distilled water is generally preferable to that purified by ion-exchange techniques.

Especially pure distilled water, such as is used for conductivity measurements, is termed *'conductivity water'*. Even this, however, still contains traces of impurities which come from the air, especially carbonic acid. The effect of these is that the specific conductivity, which for absolutely pure water is $0.04 \cdot 10^{-6}$ at $18°$, is raised to about $1 \cdot 10^{-6}$. Large amounts of so-called ultra-pure water with a specific conductivity $= 0.05$ to $0.07 \cdot 10^{-6}$ were prepared for the first time by Washburn and Wieland (1918), by the prolonged passage of CO_2-free air and distillation from a quartz apparatus. Various arrangements for the preparation of ultra-pure water have been described since then, a particularly simple one having been recently given by Thiessen (1937).

(b) Properties

Pure water is tasteless, odorless, and practically colorless, although it has a bluish tinge in thick layers.* It solidifies to ice when sufficiently strongly cooled. The temperature at which ice and water are stable in presence of one another, under normal pressure (760 mm Hg), is taken as the zero point of the Centigrade (Celsius) scale. The temperature of $100°C$ is defined by the point at which water boils under normal pressure. Other standard physical units are also referred to water. The unit of *mass*, the kilogram, is based on the mass of a cubic decimeter of water at $4°C$, the temperature at which the density of water is a maximum. The quantity of heat which is necessary in order to heat one gram of water from $14.5°$ to $15.5°C$ was formerly used as a unit of heat (the small calorie or gram-calorie). The unit of heat is now defined in terms of the equivalent quantity of electrical energy. For thermochemical measurements the calorie is defined as the quantity of heat equivalent to 4.1840 joules (or watt-seconds). Thus 1 *calorie* $= 4.1840$ *joules*. The quantity of heat defined by the specific heat of water at $15°$ is called the '$15°$ calorie' (1 $cal_{15°} = 4.1855$ joules), to distinguish it from the 'thermochemical calorie' (cal). Heats of chemical reactions are generally expressed in *kilocalories* (1 kcal $= 1000$ cal).

Water has a perceptible vapor pressure at room temperature, and even at $0°$. This must be taken into account when the volumes of gases which are confined over water are corrected to normal pressure (760 mm). Before doing so, the pressure of the water vapor present must be deducted from the total pressure under which the gas stands. The vapor pressure of water at various temperatures is shown in Table 9. In Table 10 are found the boiling

TABLE 9

VAPOR PRESSURE p OF WATER BETWEEN —60° AND +350° C

t in °C	—60	—40	—20	—15	—10	—5	—4	—3	—2	—1
Ice: p in mm Hg	0.007	0.09	0.77	1.24	1.95	3.01	3.28	3.57	3.88	4.22
Water: p in mm Hg	—	—	—	1.43	2.14	3.16	3.40	3.67	3.95	4.25
t in °C	0	2	4	6	8	10	12	14	16	18
p in mm Hg	4.58	5.29	6.10	7.01	8.04	9.21	10.52	11.99	13.63	15.48
t in °C	20	22	24	26	28	30	32	34	36	38
p in mm Hg	17.53	19.83	22.38	25.21	28.35	31.82	35.66	39.90	44.56	49.69
t in C°	40	42	44	46	48	50	60	70	80	90
p in mm Hg	55.32	61.50	68.26	75.65	83.71	92.51	149.38	233.7	355.1	525.8
t in °C	100	110	120	130	140	150	200	250	300	350
p in atm.	1.00	1.41	1.96	2.67	3.57	4.70	15.35	39.26	84.79	163.20

* For the manner in which the optical absorption curve depends upon the degree of purity, see R. LANGE, *Z. physik. Chem.*, A, 159 (1932) 303; 160 (1932) 468.

TABLE 10

BOILING POINT OF WATER UNDER VARIOUS BAROMETRIC PRESSURES
(REDUCED TO 0° C)

p in mm Hg	600	650	700	750	755	756	757	758	759	76
t in °C	93.48	95.66	97.71	99.63	99.81	99.85	99.89	99.93	99.96	100.0
p in mm Hg	761	762	763	764	765	770	780	790	800	82
t in °C	100.04	100.07	100.11	100.15	100.18	100.39	100.73	101.09	101.45	102.1

points in the most important range of pressure. The most important other physical constants of water are collected in Table 11. The data in Tables 9 to 11 apply to water with the normal content of D_2O. (For 'heavy water', see Vol. II).

Physical measurements indicate that in the H_2O molecule, the two hydrogen nuclei and the midpoint of the oxygen atom together form an angle of about 106°. Because of its unsymmetrical structure, the H_2O molecule has a strongly dipolar character. Polar molecules, or dipole molecules, are those in which the centers of gravity of the positive and negative electric charges do not coincide. Upon very close approach, such molecules must exert attractive forces upon one another, for as soon as the distances between the centers of gravity of opposite charges are substantially less than the distances between centers of gravity of like charges (cf. fig. 9), the attraction between opposite charges outweighs the repulsion between like charges. The electrical charges in non-polar molecules may, in some circumstances, be so displaced by the operation of the electric fields of other molecules, that the molecules become polar. This process is known as *polarization*, and the resulting dipole is an *induced dipole*. Absence of dipole character implies a symmetrical structure. The strongly dipolar character of the H_2O molecule is responsible for the particular readiness with which water forms addition compounds. Such compounds are formed chiefly with substances built up from ions ('salt-like' substances) or with those which, although not ionic in structure, have a dipolar character like water itself.

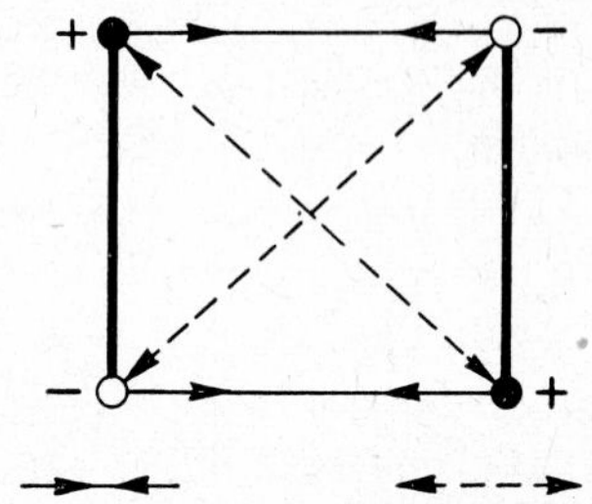

Fig. 9. Mutual attractions of dipole molecules

From the weight of a liter of water vapor at 100°, (Table 11), the molecular weight at 100° is given as 18.31, which is very little in excess of that required by the formula H_2O (18.02). Water is thus associated only to a very slight extent in the vapor state at ordinary pressure. The extent of association increases if the pressure is raised; at 4 atm. pressure the density has been found to correspond to the molecular weight 19.06. In organic solvents, water consists entirely of double molecules $(H_2O)_2$. It is even more highly associated in the liquid state. The phenomenon that water has a maximum density at 4°, expanding again when it is cooled further, arises from the increase in volume which accompanies association. Since the degree of association increases with a fall of temperature, the dilatation brought about by this eventually outweighs the usual contraction associated with cooling.

Tammann, in 1926, was the first to calculate the degree of association approximately, from the change in volume. It can also be inferred, from a comparison of the infra-red spectra of water and ice that some, at least, of the associated molecules in liquid water have a structure similar to that of (ordinary) ice. Redlich (1929) showed that the number of 'ice molecules' decreased rapidly with rising temperature. They also begin to disappear at high pressures, or when electrolytes are added (increase of the 'internal pressure', Tammann). Even at 100°, however, liquid water is still associated to a considerable degree, as follows from the fact that the boiling point of water is abnormally high compared with the hydrogen compounds of the homologues of oxygen (cf. Chap. 15). Eucken (1948–49) has shown that the association of water leads to the formation of two-fold, four-fold (possibly also three- and five-fold) and, in the neighborhood of 0°, especially of eight-fold aggregates. It is possible roughly to assess how their proportions vary with temperature. The eight-fold aggregates (which decrease very rapidly in abundance as the temperature is raised) are the ones with the 'ice-like' structure. They are much more open in structure than the others, and

TABLE II

PHYSICAL CONSTANTS OF WATER

Freezing point 0.00° C (fixed point)	= 273.16° K ⎫ at 760 mm
Boiling point 100.00° C (fixed point)	⎬ pressure
Density of ice at 0° C	= 0.9168
Density of water at 0° C	= 0.999868
at 4°	= 1.000000
at 10°	= 0.999727
at 15°	= 0.999126
at 20°	= 0.998230
at 25°	= 0.997071*
Density of water vapor, relative to air (at the same pressure) = 1, at 0° C	= 0.624
Weight of 1 liter of saturated steam at 100° C and 760 mm pressure	= 0.5974 g

Critical data

Critical temperature	= 374.1° C
Critical pressure	= 218.5 atm
Critical density	= 0.324
Specific heat of ice at 0°	= 0.487 cal
molecular heat	= 8.78 cal
Specific heat of water at 15° C = 1.00000 cal$_{15°}$	= 1.00036 cal (cf. p. 47)
Specific heat of steam at constant pressure c_p	= 0.462 cal at 100° C
molecular heat C_p	= 8.32 cal
ratio of specific heats c_p/c_v	= 1.28
Latent heat of fusion of ice	= 79.40 cal per g or
	1.430 kcal per mole
Latent heat of evaporation of water in cal per g at 100°	= 539.1
at 25°	= 583.0
at 0°	= 597.3
Surface tension (at 20° C, in contact with moist air)	= 72.7 dynes cm^{-1}
Viscosity: at 0°	= 1.789 centipoise
at 20°	= 1.002 centipoise
Molar elevation of the boiling point	= 0.513
Dielectric constant at 0° C	= 87.8
at 18°	= 80.1
at 25°	= 78.3
Specific electrical conductivity $\varkappa$ of absolutely pure water	
at 0° C	= 0.01 · 10^{-6} ohm^{-1}
at 18° C	= 0.04 · 10^{-6} ohm^{-1}
at 50° C	= 0.17 · 10^{-6} ohm^{-1}

H_2O *molecule*

Dipole moment μ	= 0.84 · 10^{-18}
Internuclear distance O–H	= 1.013 Å
H–H	= 1.63 Å
Apparent radius in crystals	= 1.7 Å
Energy of dissociation $H_2O \rightarrow H + OH$	= 117.6 kcal per mol
$OH \rightarrow O + H$	= 100.6 kcal per mol

* These values relate to air-free water. For water containing dissolved air, below 20° the last decimal place is lower by 2–3 units.

the thermal anomalies of water are largely bound up with their formation. X-ray investigations have shown that ordinary ice has a structure like tridymite (cf. p.489), so that in it each H_2O molecule is surrounded tetrahedrally by four others. On melting, the tetrahedra lose the regularity of their mutual orientation, but on the whole the configuration is at first substantially maintained (Bernal, 1933).

In the act of fusion, aggregates each consisting of eight H_2O molecules are broken out of the crystal lattice. Regularity of orientation is therefore lost over regions which are larger in size than these aggregates, but within the individual aggregates the arrangement of the H_2O molecules is at first substantially the same as it was in the crystal (cf. Fig. 10).

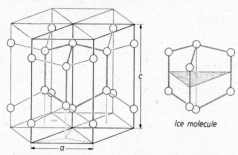

Fig. 10. Crystal lattice of ice and structure of an $(H_2O)_8$ molecule ('ice molecule').

$$a = 4.5 \text{ Å}, \qquad c = 7.3 \text{ Å}$$

The unit cell is denoted by heavier lines.

The expansion of water when it freezes is due to the fact that the H_2O molecules occupy less volume when they are irregularly oriented (or ordered only within small regions) than when they are completely ordered, to form the tridymite structure. Because water expands when it freezes, the freezing point is lowered by increase of pressure, in accordance with the Le Chatelier principle. If, however, the pressure is increased above a certain value before solidification occurs, other forms of ice are produced, which are denser than ordinary ice and mostly denser, indeed, than liquid water. For this reason the bursting action which water generally exerts on freezing, when it is confined in iron vessels, or the crevices of stones, is not found if it already stands under a very high pressure before it freezes.

(c) Equilibrium Diagram of Water and the Phase Rule

The equilibrium diagram of water is reproduced in Fig. 11. An equilibrium diagram is a figure in which are represented the ranges of existence of the various phases, as dependent

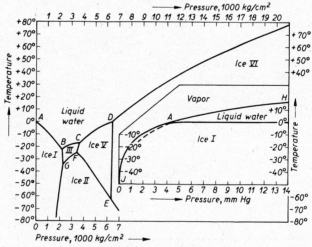

Fig. 11. Equilibrium diagram of water

upon external conditions (e.g., upon pressure and temperature). According to Gibbs' phase rule, [7, 8, 24] any system in a state of equilibrium obeys the relation:

$$P + F = C + 2, \tag{10}$$

in which P is the number of phases, F the number of degrees of freedom, and C the number of components of the system. Since, in the present case, the system contains only one component, H_2O, the maximum number of phases which can simultaneously exist in equilibrium with one another is here *three*. The number of degrees of freedom is then o, i.e., neither the pressure nor the temperature can be altered without the disappearance of one of the phases. Thus the three types of ice I, II, and III, can co-exist only at $t = -34.7°$ and $p = 2170$ kg per sq. cm (point G of the diagram), the ice types I and III together in contact with water only at $t = -22.0°$ and $p = 2115$ kg per sq. cm (point B). Ice I (ordinary ice), liquid water, and water vapor, are likewise in equilibrium with one another only at one pressure and at one temperature, namely at $p = 4.58$ mm, and $t = +0.0075$ °C.* The regions of the phase diagram in the neighborhood of this point are represented on a larger scale on the lower righthand side of Fig. 11. The points at which three phases co-exist (i.e. the points A, B, C, D, E, F, and G, of Fig. 11) are called *triple points*. The curves intersecting in the triple points give the pairs of pressures and temperatures at which two phases are co-existent—e.g., ice I and ice II together, or ice I with liquid water, or liquid water with water vapor. In accordance with equation 10, the two-phase systems possess one degree of freedom. Within these systems, one of the external conditions can be chosen at will; the second is then fixed by the curve. Thus liquid water and water vapor can be in equilibrium at various temperatures (curve AH); to each of these temperatures, however, there corresponds one quite definite value of the vapor pressure. If this is exceeded, at constant temperature, one enters the region in which water is stable only in the liquid state; if a lower pressure is maintained, one enters the range of existence of the vapor phase. Within the one-phase regions, however, the temperature and pressure can be altered independently of each other at will, since equation 10 shows that the system in this case has two degrees of freedom. The concentration is then once more defined by both pressure and temperature. In the present case this means that the volume occupied by a definite amount of water is determined, in the one-phase regions, by the temperature and pressure.

If the vapor pressure curve of liquid water—i.e., the curve HA which delimits the range of existence of the liquid phase from that of the vapor phase—is extended beyond its intersection with the vapor pressure curve of ice (curve $\mathcal{J}A$), it passes into a range in which the only stable phase is ice. This continuation of the vapor pressure curve of liquid water is shown dotted in the righthand lower portion of Fig. 11. Although liquid water is unstable in this region, it can be obtained temporarily in this state because of the phenomenon of *supercooling*. J. Meyer has recently observed supercooling of water to $-33°$. Vapor pressure measurements have confirmed that supercooled water has a higher vapor pressure than ice at the same temperature (cf. Table 9). It may be deduced thermodynamically that the stable form of a substance is always that form in which it possesses the lowest vapor pressure under the given conditions. The instability of supercooled water is clearly shown in that, immediately the inhibition of crystallization is removed (e.g., by shaking, or by touching with a crystal of ice), it is transformed spontaneously into ice, with the evolution of heat.

The relations discussed for the example of water apply, in an analogous way, to the equilibrium diagrams of any systems with one component. For systems with several components, see p. 72 *et seq.*

For the chemist, water is particularly important as a *solvent*. Furthermore, water reacts chemically with many substances, and also takes part as an intermediary in numerous reactions. This reveals itself by the fact that in the complete absence of water many reactions do not proceed at all, or only extremely slowly.

(d) Electrolytic Dissociation of Water

The intrinsic conductivity of absolutely pure water, mentioned above, arises from the fact that it does dissociate to form hydronium ions and hydroxyl ions, even although only to the most minute extent:

$$2H_2O \rightleftharpoons [H_3O]^+ + OH^- \tag{11}$$

* o° C is the temperature at which water and ice are at equilibrium under 1 atm. *pressure*. A rise in pressure of 760 mm lowers the melting point of ice by 0.00753°. Reduction of the pressure by (760— 4.6) mm thus raises the melting point by

$$0.00753(760 - 4.6)/760 = 0.00748°.$$

The dissociation constant of water at various temperatures is shown in Table 12. At 22°, 1 g molecule is present in dissociated form in ten million liters of water, and at 18°, 0.86 g molecules.

According to the mass action law, at any particular temperature the following relation holds:

$$\frac{[H_3O^+] \cdot [OH^-]}{[H_2O]^2} = K$$

This relation states that with increase of the H_3O^+ ion concentration (e.g., by addition of an acid), the concentration of the OH^- ions decreases correspondingly, and vice versa. The amount of undissociated water remains practically unaltered; it makes no practical difference if one additional gram mole is added to 560,000,000 (1 liter contains 56 gram moles H_2O). The concentration of undissociated water can therefore be included in the constant:

$$[H_3O^+] \cdot [OH^-] = [H_2O]^2 \cdot \text{const.} = K_w \qquad (12)$$

It is generally the constant K_w, defined by equation (12), which is referred to as the dissociation constant of water.

Since equal numbers of H_3O^+ and OH^- ions are formed by the dissociation of pure water, $[H_3O^+]$ for pure water $= [OH^-] = 0.86 \cdot 10^{-7}$ at 18°. For $[H_3O^+] = 1$, $[OH^-] = 0.74 \cdot 10^{-14}$ at 18°; conversely $[H_3O^+] = 0.74 \cdot 10^{-14}$ for $[OH^-] = 1$. Since it is often useful to calculate with the negative logarithm of K_w, these are given in Table 12 for the most important temperature range.

TABLE 12
DISSOCIATION CONSTANT OF WATER
$$K_w = [H_3O^+] \cdot [OH^-]$$

$t =$	16	17	18	19	20	21	22	23	24	25	°C
$K_w =$	0.63	0.68	0.74	0.79	0.86	0.93	1.01	1.10	1.19	1.27 · 10^{-14}	
$-\log K_w =$	14.200	14.165	14.130	14.100	14.065	14.030	13.995	13.960	13.925	13.895	

$t =$	30	35	40	50	60	70	80	90	99	°C
$K_w =$	1.89	2.71	3.80	5.95	12.6	21.2	35	53	72 · 10^{-14}	

(e) Thermal Dissociation of Water

Since the reaction of equation (9), p. 40, is exothermic, it must be to some extent reversed at high temperatures, in accordance with Le Chatelier's principle. Indeed, as has already been mentioned, water is split up to a measurable extent into O_2 and $2H_2$ by strong heating. According to Nernst and von Wartenberg, the degree of thermal dissociation a of water vapor at 1 atm. total pressure, is:

$T =$	1000°	1200°	1400°	1600°	1800°	K
$a =$	0.00003	0.00081	0.00861	0.051	0.199	%
$K_c =$	5.49 · 10^{-23}	2.73 · 10^{-18}	2.78 · 10^{-15}	4.96 · 10^{-13}	2.66 · 10^{-11}	
$K_p =$	4.50 · 10^{-21}	2.69 · 10^{-16}	3.19 · 10^{-13}	6.51 · 10^{-11}	3.93 · 10^{-9}	

$T =$	2000°	2200°	2400°	K
$a =$	0.588	1.42	2.92	%
$K_c =$	6.21 · 10^{-10}	8.10 · 10^{-9}	6.62 · 10^{-8}	
$K_p =$	1.02 · 10^{-7}	1.46 · 10^{-6}	1.30 · 10^{-5}	

The numerical values for the equilibrium constants K_c and K_p at different temperatures are given in the lower lines, for the case of the mass action law applied to the equilibrium between H_2, O_2, and H_2O. To calculate the equilibrium constant K_c at any particular temperature, from the experimentally determined degree of dissociation a, the following equation (13) applies.

$$K_c = \frac{p}{RT} \frac{a^3}{(2+a)(1-a)^2} \tag{13}$$

p = pressure in atm., a = fractional degree of dissociation.

This equation is obtained in the following way from (2) and (7). It is assumed that we start from pure water, the amount of this before dissociation occurs being $2a$ moles. After heating to the temperature at which the dissociation is measured, the amount of water remaining undissociated is $2a(1-a)$ moles, $2aa$ moles having dissociated. Since $2H_2$ and $1\ O_2$ molecules appear, for every $2\ H_2O$ molecules thermally dissociated, the amount of hydrogen present at equilibrium is $2aa$, and that of oxygen is aa. The total number n of moles present is now

$$n = 2a(1-a) + 2aa + aa = a(2+a).$$

The molar concentration of each molecular species is obtained by dividing its amount by the volume v taken up by the whole gas. Thus

$$[H_2O] = \frac{2a(1-a)}{v}, \quad [H_2] = \frac{2aa}{v}, \quad [O_2] = \frac{aa}{v}$$

If we substitute this in equation (7) we first obtain (14)

$$\frac{a \cdot a^3}{v(1-a)^2} = K_c \tag{14}$$

We can use this equation to calculate K_c if the quantities a and v are known, as well as a. It is more convenient to utilize equation (13), in which the total pressure p of the gas is introduced in place of the volume v. This is obtained from equation (14), noting that according to the gas equation,

$$v = \frac{a(2+a)RT}{p},$$

since in the present case $n = a(2+a)$. As follows from eqn (4), $K_p = .0821\ T \cdot K_c$.

9. Hydrogen Peroxide and Its Derivatives

(a) Hydrogen Peroxide [9, 10]

Hydrogen peroxide is always formed when hydrogen in an active state (e.g., dissolved in palladium, 'nascent', or evolved electrolytically) reacts with molecular oxygen dissolved in water. It is also formed by the slow oxidation of various organic and inorganic substances by atmosphere oxygen. It is generally so formed only in extremely minute concentrations, however.

At the present time, the most important technical method of preparing hydrogen peroxide or its aqueous solutions is by the decomposition of peroxysulfuric acid or its salts, prepared by electrolytic methods. It is most convenient first to prepare ammonium peroxysulfate, which is obtained in better yield than the free acid. With suitable conditions of operation, H_2O_2 formed by hydrolysis can be directly distilled at reduced pressure from the warm solution of this salt, acidified with sulfuric acid, or from a solution of $H_2S_2O_8$.

Anode		Cathode
$2SO_4^{2-} - 2e = S_2O_8^{2-}$		$2H_3O^+ + 2e = 2H_2O + H_2$

$$S_2O_8^{2-} + 4H_2O = 2SO_4^{2-} + 2H_3O^+ + H_2O_2$$

Since the SO_4^{2-} ions are regenerated by the hydrolysis, no by-products, or waste products other than hydrogen, appear in the process.

Some use is also made of the old process, in which hydrogen peroxide was obtained by adding moist barium peroxide (for preparation see Chap. 8) to approximately 20 % sulfuric acid or concentrated phosphoric acid solution:

$$BaO_2 + H_2SO_4 = BaSO_4 + H_2O_2; \qquad BaO_2 + H_3PO_4 = BaHPO_4 + H_2O_2$$

Relatively concentrated solutions of hydrogen peroxide can readily be obtained by fractional distillation under reduced pressure. Hydrogen peroxide generally comes into commerce in the form of aqueous solutions, containing 3 or 30 % by weight H_2O_2, the latter under the name 'perhydrol'. For industrial purposes, it is transported exclusively in the form of highly concentrated solutions (30 to 60 %, or even 90 %), which are more stable than the dilute solutions, and are not liable to danger from frost. 30 % H_2O_2 freezes only at —30°, 60 % at —53°.

Pure anhydrous hydrogen peroxide, which is only of scientific interest at present, is a syrupy liquid, generally colorless, but blue in very thick layers. Density $d_0^{20} = 1.448$, m.p. —0.46° C, refractive index $n_D^{25} = 1.4067$, surface tension 75.94 dyne/cm, viscosity 1.307 centipoises at 18°. Liquid hydrogen peroxide supercools very readily. The freezing point is very greatly lowered by water, which forms a hydrate melting at —51°.

It can be distilled without decomposition at low pressure (boiling point at 28 mm pressure = 69.7° C).

Hydrogen peroxide is to be regarded as an acid, although an extremely weak one. When anhydrous, it does not, indeed, redden litmus paper, but the aqueous solution has an acid reaction. The dissociation constant of hydrogen peroxide, i.e., the constant K in the equilibrium

$$H_2O_2 + H_2O \rightleftharpoons H_3O^+ + HO_2^- \qquad \frac{[H_3O^+] \cdot [HO_2^-]}{[H_2O_2]} = [H_2O] \cdot const. = K *$$

is $1.5 \cdot 10^{-12}$ at 20° (V. A. Kargin (1929)). This corresponds roughly to the 3rd dissociation constant of orthophosphoric acid (cf. p. 631), but is very small as compared, e.g., with the 1st dissociation constant of carbonic acid. Nevertheless, the H_3O^+ ion concentration of H_2O_2 in 1 molar aqueous solution is considerably greater than that of pure water. By increasing the OH^- concentration, the dissociation of H_2O_2 is increased in accordance with the mass action law, and is practically complete when $NaOH : H_2O_2 > 2$ (Bredig, 1934).

Sodium hydrogen peroxide (hydroperoxide) $NaOOH$, which can be obtained pure as a white crystalline powder from non-aqueous solutions, is to be regarded as a sodium salt of hydrogen peroxide. It can also be obtained from aqueous solution, in the form of its addition compound with H_2O and H_2O_2.

The dielectric constant of H_2O_2 is high, like that of H_2O, being 84.9 at 0°. It increases upon admixture with water to a maximum value 121, at a content of about 35 % H_2O_2.

* See p. 52, regarding the inclusion of $[H_2O]$ within the equilibrium constant.

According to Sutherland (1934) and A. Simon (1935), the ordinary H_2O_2 molecule is built up of two OH groups joined through the O atoms, and making an angle of about $110°$ with the O——O axis, but twisted through $90°$ relative to each other. The distance $O \leftrightarrow O = 1.47$ Å, $O \leftrightarrow H = 0.97$ Å (Pauling, Giguère and Schomaker, 1943).

Like water, hydrogen peroxide is highly associated in the liquid state, although the vapor is monomolecular. Solid H_2O_2 has a tetragonal structure ($a = 4.02$Å, $c = 8.02$ Å).

Hydrogen peroxide can be extracted by ether from its concentrated aqueous solution. It is miscible with water in all proportions. It adds on to numerous substances, forming loose addition compounds—in particular to various organic substances, and also to a series of salts. The H_2O_2 loosely bound by these is called hydrogen peroxide of crystallization.

Hydrogen peroxide is a strongly endothermic compound. Nevertheless, in practice it is stable in the pure state, and also in aqueous solution. This is because its rate of decomposition at ordinary temperature is practically zero. Decomposition is considerably accelerated, however, by the presence of many substances, particularly by platinum, silver, or manganese dioxide, also by materials with rough surfaces, and by alkalis. The decomposition follows the equation $H_2O_2 = H_2O + \frac{1}{2}O_2 + 23.47$ kcal. Traces of finely divided platinum introduced into aqueous solution of hydrogen peroxide bring about a vigorous evolution of oxygen. Highly concentrated hydrogen peroxide solutions can be made to explode violently by dust particles falling into them. Slow decomposition of hydrogen peroxide is brought about even by traces of alkali given up by glass. For this reason, pure hydrogen peroxide solutions (i.e., without added stabilizers) are kept in paraffin wax coated vessels. Many organic substances also bring about decomposition of hydrogen peroxide. Substances derived from living organisms, and characterized by a specific action in decomposing hydrogen peroxide are called *catalases*. Such catalases are found, for example, in blood and in the mucous membrane.

The catalases belong to a class of so-called ferments or *enzymes*. These rather unstable compounds, occurring only in minimal concentrations, play in organisms a role analogous to that of catalysts in the rest of chemistry. It was formerly usual to consider it a peculiarity of the ferments that they were very specific in their action, that is, acted only with certain definite compounds and in a particular way. This rule can however be maintained only to a restricted extent.

In addition to catalase* there is another ferment with a specific action on hydrogen peroxide: *peroxydase*. This has the property of activating hydrogen peroxide, so that in the presence of oxidizable materials it gives up 1 oxygen atom to the latter. In the absence of such substances hydrogen peroxide is not decomposed by peroxidase.

On the other hand, there are also substances which slow down the catalytic and fermentative decomposition of hydrogen peroxide, i.e., which more or less completely inhibit the action of catalysts or enzymes. Such substances include phosphoric acid (which serves especially to remove manganese and iron salts which bring about decomposition) and other acids, especially barbituric acid and uric acid. One gram of the latter is sufficient to stabilize 30 liters of concentrated hydrogen peroxide solution. For most purposes, such small additions are quite unimportant as contaminants. For this reason, hydrogen peroxide solutions, including those of high concentration, are now almost all stabilized by added substances of this kind, and sold commercially in ordinary glass bottles.

* It is not known whether there is only one catalase, or several catalases.

Hydrogen peroxide can act both as an oxidizing agent and as a reducing agent. It oxidizes iron(II) sulfate to iron(III) sulfate, sulfurous acid to sulfuric acid, nitrous acid to nitric acid, arsenious acid to arsenic acid, lead sulfide to lead sulfate. It deposits iodine from hydriodic acid, and decolorizes indigo solution. It reduces substances which readily give up oxygen, such as potassium permanganate, and bleaching powder. Compounds of noble metals are also reduced by hydrogen peroxide; thus it deposits gold from gold salt solutions, reduces silver oxide to silver, and mercury oxide to mercury.

The excellent disinfectant action of hydrogen peroxide depends on its ability to give up oxygen readily. For this reason it finds extensive uses in medicine. For these purposes it is also marketed in the form of a solid loose addition compound with urea, under the name 'ortizone'. It is also used as a bleaching agent, e.g., for cosmetic purposes, (bleaching hair). It finds technical application on a large scale today, especially as a bleaching agent.

Hydrogen peroxide was originally used chiefly for the bleaching of straw, feathers, ivory, starch, glue, leather and skins. Recently it has been applied increasingly in the bleaching of textiles, (wool, cotton, silk, and artificial fibers). In this field it is likely to displace chlorine bleaching to an increasing extent, since it is faster and more lasting in action, and attacks the fibers less. Large quantities of hydrogen peroxide are also used to bleach fats and oils. For this purpose, the most concentrated solution possible (up to 60 %) are used, in order to introduce as little water as possible into the substances treated.

Considerable quantities of hydrogen peroxide are used in the production of other bleaching agents, particularly sodium borate peroxyhydrate, and for the manufacture of organic peroxy compounds. Mixtures of hydrogen peroxide with hydrazine hydrate, methanol or hydrocarbons have found application as propellants for rocket aircraft and similar purposes. The world production is at present about 30,000 tons of 30 % H_2O_2 per annum.

In analysis, hydrogen peroxide is easily detected by the intense yellow coloration which it confers to a titanium sulfate solution. It is determined quantitatively by means of its reaction with potassium permanganate, which takes place in sulfuric acid solution, according to the equation

$$5H_2O_2 + 2KMnO_4 + 3H_2SO_4 = 2MnSO_4 + K_2SO_4 + 8H_2O + 5O_2$$

The velocity of this reaction at first increases with rising H_2O_2 concentration and then markedly decreases again (Riesenfeld, 1934). In testing hydrogen peroxide solutions for H_3PO_4, it must be remembered that H_2O_2 oxidizes the colorless $MoO_4^=$ ion to the intensely orange-yellow $MoO_5^=$ ion. H_2O_2 must therefore be removed from the solution by evaporation, before testing for H_3PO_4.

(b) Derivatives of Hydrogen Peroxide

Replacement of the H atoms of hydrogen peroxide by other atoms or positively charged radicals yields *peroxides*; exchange of one or both H atoms for acid radicals, gives *peroxy-acids* and *peroxy-salts* (see for example p. 340). Hydrogen peroxide also forms addition compounds, (peroxyhydrates) with many substances, as has been mentioned. For the differentiation of peroxyhydrates from peroxy-salts see p. 340.

An isomer of H_2O_2, stable only at low temperatures, and formed by the action of atomic H on O_2 was mentioned on p. 42. From its mode of formation it is likely that in this substance both H atoms are linked to the same O atom.

10. **Acids, Bases, and Salts**

(a) **Acids**

Acids in aqueous medium are substances which can form hydronium ions, H_3O^+, by proton transfer to water.

$$RH + H_2O \rightleftharpoons R^- + H_3O^+ \quad (R = \text{acid racical}) \tag{15}$$

Substances which are able to give up protons are called *proton donors*; those which can add on protons are called *proton acceptors*. These terms may be applied equally to chemical compounds in the ordinary sense or to ions (such as NH_4^+). Many substances can act either as proton donors or proton acceptors, according to the other substances with which they are brought in contact.*

The process of proton transfer is said to be the *electrolytic dissociation* of acids. The process comes to an equilibrium, which is displaced to the right by dilution of the solution. The extent to which different acids are dissociated in solutions of the same equivalent concentration is a measure of their *strength*. Some acids are dissociated to an even smaller extent than is water. Acids which are more highly dissociated than water will redden litmus paper.

In applying the Mass Action law to electrolytic dissociation, the values substituted for the activities of the species present in solution (ions and undissociated molecules) are those obtained when the activity of the solvent is taken as $= 1$, irrespective of the concentration. Thus, for the dissociation equilibrium of acids

$$\frac{[R^-] \cdot [H_3O^+]}{[RH] \cdot [H_2O]} = \frac{[R^-] \cdot [H_3O^+]}{[RH]} = K_a \tag{16}$$

K_a is known as the *dissociation constant* of the acid.

If the symbol H^+ is used instead of H_3O^+, equations (15) and (16) become

$$RH \rightleftharpoons R^- + H^+, \quad \text{and} \quad \frac{[R^-] \cdot [H^+]}{[RH]} = K_a$$

The numerical value of K_a is quite unaffected by the different way in which the equilibrium is represented.

An atom, such as the H atom of an acid RH, which may be split off under suitable conditions, to form an electrolytic ion, is said to be ionogenically bound. This concept says nothing on the subject of whether the ion formed by electrolytic dissociation already existed in an electrically charged state in the undissociated molecule, or whether it was an electrically neutral atom which acquired a charge in the process of dissociaton. Water has strongly dissociating properties as a solvent. This is, in part, a consequence of its high dielectric constant.

Such acids as contain only one element other than hydrogen (e.g., HCl) are called *hydro-acids*, those which also contain oxygen (e.g., $HClO_4$, H_2SO_4) are called *oxy-acids*. For *thio-acids* see p. 737, for *halogeno-acids* see p. 798.

---

* Water, for example, acts as a proton acceptor towards HCl:
$$H_2O + HCl = H_3O^+ + Cl^-,$$
but as proton donor towards ammonia:
$$H_2O + NH_3 = OH^- + NH_4^+.$$

(b) General Methods of Preparation of Acids

Oxy-acids are obtainable;

(1) By the *action of water* on certain oxides.

Examples: Sulfur trioxide combines with water forming sulfuric acid. Phosphorus pentoxide absorbs water and forms metaphosphoric acid, on standing in the air.

$$SO_3 + H_2O = H_2SO_4; \qquad P_2O_5 + H_2O = 2\,HPO_3$$

Those oxides which combine with water to form acids are called *acid anhydrides*.

(2) By the *hydrolysis* of compounds of which both constituents are acid-forming.

Two acids always result from this, and the method can be used preparatively, provided that the two acids can be separated from each other. The hydrolysis of covalent chlorides is frequently used for the preparation of acids.
Examples:

$$PCl_5 + 4H_2O = H_3PO_4 + 5HCl; \qquad PCl_3 + 3H_2O = H_3PO_3 + 3HCl$$

Hydrochloric acid, being volatile, can easily be removed by evaporation. Decomposition of acid chlorides is more frequently used for the preparation of acids in organic chemistry than in inorganic chemistry.

(3) Oxy-acids result, further, from the *oxidation of acid-forming elements in aqueous solution*, in the absence of bases.

Example. Preparation of orthophosphoric acid. by the oxidation of phosphorus with nitric acid.

$$3\,P + 5HNO_3 + 2H_2O = 3H_3PO_4 + 5NO$$

(4) The free acids are always formed at the anode when solutions of salts of oxy-acids are *electrolyzed*.

The following method is applicable to the preparation of hydro- and oxy-acids.

(5) Preparation of the acids from their salts by treatment with other acids (double decomposition):

$$MA + HB \rightarrow HA + MB$$
$$\text{salt} \quad \text{acid} \quad \text{acid} \quad \text{salt}$$

As a rule the process is distinctly reversible, and therefore takes place only incompletely. For the reaction to go to completion, at least one of the substances resulting from the reaction must be removed from the reaction mixture. The following possibilities are open for this:

(a) The resulting acid is volatile, often only at high temperature, or it decomposes into volatile anhydride and water.

The best known methods of preparing acids are of this kind—e.g., the preparation of HCl from NaCl and H_2SO_4 (cf. p. 790), of HNO_3 from KNO_3 or $NaNO_3$ and H_2SO_4 (p. 598), of $HClO_4$ from $KClO_4$ and H_2SO_4 (p. 808), of H_2S from FeS and HCl (p. 732); cf. also the preparation of CO_2 from $CaCO_3$ and HCl.

(b) The resulting acid is sparingly soluble.

Thus 'silicic acid' (or gel of its anhydride) is deposited from acidified solutions of alkali silicates. Metatitanic acid is precipitated when solutions of titanates are boiled.

(c) The resulting salt is sparingly soluble.

This is also a widely used method. It is especially useful for preparing acids which will not withstand strong heating.

Examples:

H_2O_2 from BaO_2 and H_2SO_4. HIO_3 from $Ba(IO_3)_2$ and H_2SO_4.

$HClO_4$ from $KClO_4$ and H_2SiF_6 H_5IO_6 from $Ag_2H_3IO_6$ and HCl

H_2SeO_4 from $BaSeO_4$ and H_2SO_4

(d) The acid, or the salt formed simultaneously, can be extracted by another solvent. *Example* for this mode of preparation: hydroferrocyanic acid (cf. Vol. II).

(6) Preparation of acids from their salts by treatment with cation exchangers. 'Exchangers' are solid substances which have the property of taking up certain substances from solutions, but putting others into the solution in their place. The term is used particularly for materials which will replace one ion in the solution (either cation or anion) by another. The property was first observed with zeolites and permutits (cf. p. 513), and is more or less highly developed in many adsorbents —e.g., surface active aluminum oxide (p. 352). *Synthetic resins* have become the most important class of ion exchangers during recent years.

Ion-exchanging resins were first prepared in England by Adams and Holmes in 1935. They have the advantage that, by modification of their structure and composition, their ion exchange properties can be adapted to the most varied requirements. Thus some resins will exchange only anions, and others only cations. With ion-exchange resins. the quantities of ions exchanged are *equivalent;* this is generally not the case in ion-exchange on adsorbents. Compounds capable of salt formation—e.g., organic sulfonic acids—are built into the ion exchange resins, and their action is not one of adsorption, but involves chemical double decomposition reactions.

R. Klement (1949) has shown that the cation exchanging properties of such resins can be used for the *preparation of acids from their alkali salts.* For this purpose, a suitable resin (e.g., Dowex 50, Amberlite IR-100, Wofatite KS, Zeocarb 215) is placed in a vertical glass tube, drawn out and provided with a tap at the bottom. The resin bed, swelled with water, which rests on a glass frit or a porcelain filter plate, is first enriched in H^+ ions by treatment with $5N$-hydrochloric acid. It is then washed with distilled water, and filled up with the alkali salt solution. The acid is thereby formed from the alkali salt, by cation exchange—e.g., $NaH_2PO_2 + H^+_{resin} = H_3PO_2 + Na^+_{resin}$. When reaction is complete (about 15 min, depending on the particle size of the resin), the tap is opened and the acid solution is run out. The resin bed is regenerated—i.e., the Na^+ ions taken up are removed again—by passage of $5N$ hydrochloric acid slowly through the bed.

Preparation of acids by the ion-exchange process is particularly advantageous when the acid can otherwise be prepared only by troublesome methods—e.g., by roundabout methods involving heavy metal salts, which must usually first be prepared from the alkali metal salts.

Similar ion exchange processes can be used for *analytical* purposes—e.g., for the determination of alkali salts by acidimetric titration. [O. SAMUELSON, 1939; E. WIESENBERGER, 1942; cf. KLEMENT, *Z. anal. Chem.*, 127 (1944) 2, 128 (1948) 106. For a general review see J. F. DUNCAN and B. A. J. LISTER, *Chem. Soc. Quart. Rev.*, 4 (1948) 307.]

Ion-exchange resins have found technical application especially for water purification and similar purposes (cf. p. 46), but also in preparative technical chemistry—e.g., for the manufacture of sodium nitrate from lime-saltpeter, and for other salt conversions; for the elimination of electrolytes from colloidal dispersions; and for removal of traces of heavy metals from organic substances. Thus it has been shown that it is possible to free wine and other drinks completely from traces of copper, lead and arsenic.

(c) Bases

Bases are substances which give rise to hydroxyl ions in aqueous solution. The aqueous solutions of bases which are more strongly dissociated than water turn red litmus paper blue.

The alkali and alkaline earth hydroxides are the most important strong bases. The most important weak base is the hydrate of ammonia, $NH_3 \cdot H_2O$ (cf. p. 611), which dissociates to a small extent in aqueous solution, according to the equation

$$NH_3 \cdot H_2O = [NH_4]^+ + OH^-$$

There are numerous hydroxides which can function both as acids and as bases; these are said to be *amphoteric*.

The amphoteric behavior of a hydroxide ROH may arise from either of two quite distinct causes. The hydroxide may either be capable of giving up protons to water molecules to a perceptible extent, as well as of splitting off hydroxyl ions. Alternatively, it may be able not only to split off hydroxyl ions, but also to add on hydroxyl ions. Depending on which of these explanations is valid, the following equilibria will be established.

Either $\quad ROH + H_2O \rightleftharpoons RO^- + H_3O^+ \quad$ and $\quad ROH \rightleftharpoons R^+ + OH^-$

or $\qquad\qquad\qquad ROH + OH^- \rightleftharpoons R(OH)_2^- \rightleftharpoons R^+ + 2OH^-$

Application of the law of mass action to these gives either

$$\frac{[RO^-] \cdot [H_3O^+]}{[R^+] \cdot [OH^-]} = const. \qquad or \qquad \frac{[R(OH)_2^-]}{[R^+] \cdot [OH^-]^2} = const.$$

From eqn. (12), $[H_3O^+] = \dfrac{K_w}{[OH^-]}$. Hence the equilibrium in both cases depends on the H_3O^+ or OH^- ion concentration in the same way. If a strong acid is added to the solution, the H_3O^+ concentration is raised (or the OH^- concentration lowered), and the ratio $\dfrac{[RO^-]}{[R^+]}$ or $\dfrac{[R(OH)_2^-]}{[R^+]}$ must decrease accordingly. Hence the process

$$RO^- + H_3O^+ \rightleftharpoons R^+ + OH^- + H_2O, \qquad or \qquad R(OH)_2^- \rightleftharpoons R^+ + 2OH^-$$

must proceed from left to right. The converse must occur if the OH^- ion concentration is raised by the addition of a strong base. It follows from this that amphoteric substances must function as bases towards strong acids, but as acids towards strong bases. This is in fact the case.

(d) General Methods of Preparation of Bases

Methods available for the preparation of bases are formally similar to those used for oxy-acids. There are, however, typical differences in carrying them out, and in the importance of the individual processes. We may therefore consider how the methods given apply to the preparation of the bases.

(1) Preparation from *oxides* (in this case the basic anhydrides) and water.

The majority of metallic oxides, especially those containing metals in their lower valence states, are the anhydrides of bases. This method however, is not nearly so important for the preparation of bases as for that of acids, since most metallic oxides take up water only very slowly if at all.

The preparation of $Ca(OH)_2$ by this method is important (slaking of lime p. 261). With elements whose hydroxides are not themselves stable, the existence of the equilibrium $M_2O + H_2O \rightleftharpoons 2MOH \rightleftharpoons 2M^+ + 2OH^-$ makes it possible to use suspensions of the oxides in water instead, e.g., silver oxide suspended in water, in place of silver hydroxide, which can itself not be obtained.

In addition to the oxides, which function as base anhydrides, there are also other substances which form bases by the addition of water, e.g., ammonia, which forms the base ammonia hydrate, $NH_3 \cdot H_2O$. Following Werner's proposal, the name *anhydro-base* is applied to a substance of this kind (cf. p. 62).

(2) Hydrolysis of *salts*.

This method has analytical, rather than preparative, importance. It is often used in analytical chemistry for the deposition of metal hydroxides in the pure state, or for the separation of trivalent metals from bivalent metals, by reason of the greater hydrolysis of the former. Examples of this will be found under aluminum, chromium, and iron.

Only the salts of very weak bases with very weak acids, such as aluminum sulfide, are completely decomposed on treatment with water. More generally the hydrolysis leads to an equilibrium, e.g., with aluminum chloride:

$$AlCl_3 + 5H_2O \rightleftharpoons AlO(OH) + 3H_3O^+ + 3Cl^-$$

which is, however, easily displaced, so that the process proceeds completely from left to right. This can be achieved by adding suitable substances which, without themselves furnishing OH^- ions, react with the H^+ ions and so hold their concentration at a low value. Thus barium carbonate, sodium thiosulfate, and sodium acetate are used for this purpose. A mixture of potassium iodide and potassium iodate acts similarly (cf. p. 813).

With many salts, decomposition takes place on treatment with superheated steam, e.g., with barium carbonate

$$BaCO_3 + H_2O = Ba(OH)_2 + CO_2$$

(3) Oxidation of *metals* in the presence of water.

Examples: Iron forms hydrous ferric oxide in moist air.

$$2Fe + \tfrac{3}{2}O_2 + xH_2O = Fe_2O_3, xH_2O$$

Water alone has an oxidizing action on alkali and alkaline earth metals.

$$Na + H_2O = NaOH + \tfrac{1}{2}H_2$$

(4) The hydroxide is formed at the cathode, when neutral salt solutions of strongly electropositive metals are *electrolyzed*. This method finds extensive application in the technical preparation of alkali hydroxides.

(5) Preparation of bases from salts by *double decomposition*.

$$\underset{\text{salt}}{MX} + \underset{\text{base}}{BOH} = \underset{\text{base}}{MOH} + \underset{\text{salt}}{BX}$$

This is the most important method for preparing all bases except sodium, potassium, and calcium hydroxides.

The three most important cases to be considered here are the following:

(a) The resulting base or its anhydride is volatile. In this case the base can readily be removed from the reaction mixture by raising the temperature. The preparation of ammonia and many of its organic derivatives is based on this.

(b) The resulting base is insoluble in water. Since the hydroxides of all metals, except for the alkali hydroxides, are either insoluble in water or sparingly soluble (alkaline earth hydroxides), this method is capable of the widest application. Every precipitation of a hydroxide is an application of it.

(c) In many cases use can be made of the formation of a sparingly soluble salt. A technically important application of this method is the causticizing of soda (cf. p. 181.) One process for preparing strontium hydroxide is also based on it

$$SrS \text{ (from celestine) } + CuO + H_2O = Sr(OH)_2 + CuS$$

Barium hydroxide can be prepared in an analogous manner.

(e) Summary of the Preparation of Bases of Various Classes of Metals

Alkali hydroxides. Most important preparation: electrolysis of salt solutions (method 4). Also, for sodium hydroxide, by causticizing soda solution with quick lime (method 5c). Purest sodium hydroxide: from sodium and water (method 3).

Ammonia hydrate. Aqueous solution obtained by passing ammonia into water. Ammonia is prepared in the laboratory from ammonium salts and caustic soda or quick lime (method 5a). For technical processes for preparing ammonia, see p. 607 *et seq.*

Calcium hydroxide. Slaking of quicklime (method 1).

Strontium and barium hydroxides. From concentrated solutions of the chlorides by treatment with caustic soda (method 5b), or by methods 5c or 1.

Other bases. By precipitation from solutions of their salts, either by addition of alkali hydroxide or ammonia (method 5b), or by addition of a salt which favors hydrolysis (method 2).

(f) Salts

If the solution of a base is added to that of an acid, the hydrogen ions of the one unite with the hydroxyl ions of the other, forming water: $H_3O^+ + OH^- = 2H_2O$. If so much of the base is added that practically all H_3O^+ ions disappear (i.e., their concentration is reduced roughly to the small value that it has in pure water), then the negative ions of the acid (acid anions) and the positive ions of the base (mostly metal cations) remain in the solution. This process is known as *neutralization*. The oppositely charged ions can combine with one another because of the electrostatic attraction between them. This generally occurs only when the solution is sufficiently concentrated. The products resulting thereby are called *salts*. In brief, salts result from the combination of acids with bases, by the elimination of water.

Salts are formed from acid anhydrides and basic anhydrides without the formation of water, e.g., $CO_2 + CaO = CaCO_3$. Anhydro-bases also combine with acids forming salts without elimination of water, e.g., $NH_3 + HCl = NH_4Cl$. There are also substances which can combine with bases to form salts without the elimination of water; these are called anhydro-acids. Like the electrolytic dissociation of acids, salt formation by anhydro-acids depends on proton transfer. Thus

$$H_2O + HCl = H_3O^+ + Cl^-; \qquad NH_3 + HCl = NH_4^+ + Cl^-$$

There is a corresponding proton transfer between the anhydro-base and H_2O; to this is due the basic reaction exhibited by aqueous solutions of anhydro-bases.

$$NH_3 + H_2O \rightleftharpoons NH_4^+ + OH^-$$

Thus anhydro-bases are compounds which give rise to OH⁻ ions because they can abstract a proton from the H_2O molecule.

Just as anydro-bases do not split off OH⁻ ions, but add on protons instead, so anhydro-acids do not split off protons, but add on OH⁻ ions. In particular, there is a series of amphoteric metal hydroxides which function as anhydro-acids towards strong bases. Thus:

$$Al(OH)_3 + HOH = [Al(OH)_4]^- + H^+, \quad \text{or,}$$
$$Al(OH)_3 + OH^- = [Al(OH)_4]^-$$

The salts formed by the addition of bases to anhydro-acids are termed hydroxo-salts, e.g. $K[Al(OH)_4]$, potassium hydroxoaluminate.

(g) Weak and Strong Electrolytes

Acids, bases, and salts are electrolytes, i.e., their solutions in solvents of high dielectric constant, and especially in water, conduct the electric current, and simultaneously undergo decomposition (electrolytic decomposition). Many of the substances mentioned also display electrolytic conductivity in the molten state, and some even in the solid state.

Electrical conductivity may be of two kinds: *metallic* (which is not however necessarily restricted to metals), and *electrolytic*. These differ fundamentally, in that metallic conduction does not involve any material change in the conductor, whereas electrolytic conduction is accompanied by a chemical change. They also display many other differences, e.g., metallic conductivity decreases with rising temperature, but electrolytic conductivity generally increases.

To explain the phenomena of electrolytic conduction, it is familiar that we assume that electrically charged particles—the electrolytic ions— are present in solutions of electrolytes, and that the transport of electricity through the electrolyte results from the migration of these from one electrode to the other. [11-13] The migration of electrolytic ions can be directly detected, by the changes of concentration which occur whenever the two ions of opposite sign which constitute an electrolyte differ in their velocities of migration (mobilities), as is almost always the case. In addition to the concentration changes, the discharge of ions at the electrodes brings about chemical transformations there. This process has already been discussed in the electrolysis of water.

The ratio of the specific conductivity of an electrolyte solution to its concentration in mols per cubic centimeter is termed the *molar conductivity* of the electrolyte.* Whereas the specific conductivity generally diminishes rapidly, with decrease in the concentration of electrolyte, the molar conductivity usually increases with decreasing concentration, up to a constant limiting value, which is termed the *molar conductivity at infinite dilution.*

Electrolytes can be divided into two classes, according to the manner in which the molar conductivity varies with the concentration of electrolyte. Substances are said to be *strong electrolytes* if their molar conductivity is large at high concentrations, and increases relatively little with the dilution. *Weak electrolytes* are those for which the molar conductivity is small at high concentrations, but increases greatly with dilution. Hydrochloric acid and potassium chloride are examples of strong electrolytes; formic acid and acetic acid are examples of weak electrolytes (cf. Fig. 12).

It is at once apparent that the larger the number of ions in any solution, and the greater the mobility of these ions, the higher must its conductivity be.

* See footnote p. 64.

It is found experimentally that the molar conductivity μ of the electrolyte (that is the specific conductivity, always referred to the same quantity of electrolyte* *increases* with dilution. If we assume for the moment that the mobility of any given free ion is independent of its concentration, we can explain this observation in terms of an increase in dissociation with dilution, until eventually the electrolyte is completely dissociated into its ions. When this is the case, the molar conductivity must have reached its maximum value, and will remain unaltered on further dilution. If μ_c denotes the molar conductivity at the concentration c, and μ_0 its limiting value for infinite dilution (concentration = 0), then the degree of dissociation a of the electrolyte, that is the fraction of the electrolyte which is split up into ions, is given by

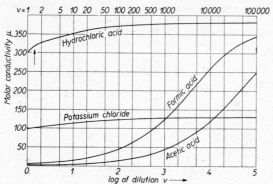

Fig. 12. Increase of molar conductivity with solution, for strong and weak electrolytes respectively.

$$a = \frac{\mu_c}{\mu_0} \left(\text{or also } a = \frac{\Lambda_c}{\Lambda_0} \right), \tag{17}$$

as follows at once if the conductivity of any substance is taken as being proportional to the quantity of its ions. The assumption that the degree of dissociation is given by the ratio of the molar conductivities was first made by Arrhenius in 1887. He found it to be confirmed by the results of measurements of the lowering of freezing point by electrolytes. As Van't Hoff had shown in 1886, the depression of the freezing point and elevation of the boiling point are proportional to the osmotic pressure of a dissolved substance, and this in turn to the number of molecules contained in a definite volume, just as is the gaseous pressure of gases. Thus the same equation holds for the osmotic pressure as for gases

$$P = \frac{n}{v} RT$$

P = osmotic pressure, g = number of gram moles contained in the dissolved state in the volume v, R = gas constant, T = absolute temperature.

For non-electrolytes, the number of gram molecules, n, contained in a given volume of the solution is equal in general to the number of gram molecules which

* The molar conductivity μ is given by $\mu = \dfrac{\varkappa}{m}$ or $\mu = \varkappa \cdot 1000\, v$, where $\varkappa$ = the specific conductivity, m = concentration in gram moles per cc or v = the volume in liters containing 1 gram mole of electrolyte. The equivalent conductivity Λ is often given, instead of the molar conductivity μ. This is defined analogously by $\Lambda = \dfrac{\varkappa}{\eta}$, or $\Lambda = \varkappa \cdot 1000\, \varphi$, where η = concentration in gram-equivalent of dissolved electrolyte, or φ = volume in liters containing 1 g equivalent. The conductivity is often referred to units of gram-equivalents per 1000 g solvent, instead of gram-equivalents per liter. This method of expressing concentrations has the advantage of being independent of temperature.

were dissolved therein. Association between the dissolved molecules or with the solvent can bring about deviations from this. For electrolytes, however, since dissolution is accompanied by dissociation, the number of molecules undergoes an increase. The osmotic pressure (and therefore the elevation of boiling point and depression of freezing point) of electrolyte solutions is accordingly greater than would be calculated from the formal molecular weight. For solutions of electrolytes in dissociating media, such as water, it is given by the formula

$$P = \frac{i \cdot n}{v} RT$$

According to the classical theory of electrolytic dissociation (cf. later), the factor i in this formula gives the proportion by which the number of molecules has been increased through dissociation. It can thus be defined as the ratio of the number of molecules, as found by cryoscopic or similar measurements, to the number of molecules brought into solution, or as the ratio of the depression of the freezing point (or elevation of the boiling point) actually found, to that calculated from Raoult's law on the assumption that no change in the number of molecules occurred upon dissolution.

Raoult's law states that, for any solvent, the depression of the freezing point, or elevation of the boiling point, in not too concentrated solutions, is proportional to the amount of a substance dissolved in a given quantity of solvent, and is inversely proportional to its molecular weight.

$$\Delta t = \frac{p}{M} G$$

Δt = depression of freezing point or elevation of boiling point, p = weight of dissolved substance in grams, M = molecular weight of the solute, G is a constant, characteristic of each solvent, which is termed the *molar freezing point depression* or *molar boiling point elevation*. If Δt_{calc} is the freezing point depression calculated by this formula, and Δt_{obs} is that found, then i is defined as

$$\frac{\Delta t_{obs}}{\Delta t_{calc}} = i \tag{18}$$

According to the 'classical' theory, i in the case of complete dissociation represents the number of ions into which each dissolved molecule dissociates. For a uni-univalent electrolyte (e.g., of the type of sodium chloride, which forms two univalent ions by dissociation), $i = 2$ for complete dissociation. For incomplete dissociation of a uni-univalent electrolyte,

$$i = 1 + a \tag{19}$$

If n molecules are introduced into a solution, and a denotes the fractional degree of dissociation, the number of molecules remaining undissociated = $n(1 - a)$. The number of molecules resulting from the dissociation is $2na$ for the case of the uni-univalent electrolyte. The total number of particles is thus $n(1 - a) + 2na$, and the ratio of (number of molecules after dissociation) to (initial number) = $1 + a$.

Quite generally, for an electrolyte dissociating into x ions,

$$i = 1 + (x - 1)a \tag{19a}$$

Arrhenius showed from abundant experimental evidence that practically the same values were obtained for a, the degree of dissociation, if this were calculated either from equations (18) and (19) or (19a), or from the change of molecular conductivity with dilution, according to equation (17). He concluded from this that values of a, calculated by the two different methods, really represented the degree of dissociation of electrolytes. This 'classical' theory of electrolytic dissociation does hold for *weak electrolytes*. It does not apply at all to *strong electrolytes*.

For weak electrolytes, the degree of dissociation can, indeed be calculated from the change of conductivity with dilution, or from cryoscopic data. By applying the law of mass action to the dissociation of a uni-univalent electrolyte, we obtain

$$\frac{a^2}{v(1-a)} = \text{const.}, \qquad \text{or} \qquad \frac{a^2 \cdot c}{1-a} = \text{const.} \tag{20}$$

in which concentrations are expressed in terms of a, the degree of dissociation, or, utilizing equation (17),

$$\frac{\Lambda^2_c \cdot c}{\Lambda_0(\Lambda_0 - \Lambda_c)} = \text{const.}$$

where $v \left(= \dfrac{1}{c} \right)$ represents the volume of solution per gram equivalent (*Ostwald's dilution law*).

Weak electrolytes conform to the Ostwald dilution law. It is quite inapplicable, however, to strong electrolytes, as may be seen from the figures in the last two columns of Table 13.

TABLE 13

VALIDITY OF OSTWALD'S DILUTION LAW FOR WEAK ELECTROLYTES AND ITS INAPPLICABILITY TO STRONG ELECTROLYTES

The Table gives values at 18°

Equivalent concentration c	Weak Electrolyte		Strong Electrolyte	
	Acetic Acid	Ammonia-water	Potassium chloride	Magnesium sulfate
1	$1.40 \cdot 10^{-5}$	$1.40 \cdot 10^{-5}$	2.430	0.0879
0.1	$1.70 \cdot 10^{-5}$	$1.92 \cdot 10^{-5}$	0.535	0.0333
0.01	$1.70 \cdot 10^{-5}$	$1.68 \cdot 10^{-5}$	0.151	0.0133
0.001	$1.50 \cdot 10^{-5}$	$1.56 \cdot 10^{-5}$	0.046	0.0060
0.0001	$1.31 \cdot 10^{-5}$	$1.06 \cdot 10^{-5}$	0.013	0.0023
	Mean $1.52 \cdot 10^{-5}$	Mean $1.51 \cdot 10^{-5}$		

The Ostwald dilution law fails with strong electrolytes, because the value of a calculated from equation (17) does not, in this case, represent the degree of dissociation. The calculation is based on the assumption that the mobility of electrolytic ions is independent of their concentration. This is approximately true only when the number of ions per unit volume is very small. For weak electrolytes, this condition is true to some extent.* In solutions of strong electrolytes, however, the concentration of the ions is so high that their mobility is considerably lowered as a result of the forces arising between the ions, due to their electrical charge. As the

---

* Even for weak electrolytes, the effects of inter-ionic forces must be allowed for in accurate calculations.

solutions are progressively diluted, these forces diminish, and the mobility of the ions increases accordingly. This increase of ionic mobility is the explanation of the increase in the equivalent conductivity of strong electrolytes with increasing dilution. It is therefore not possible to deduce that there is any increase in the degree of dissociation. Since the ions of strong electrolytes behave in a different manner from uncharged particles, it is also impossible to use the formula for osmotic pressure as a means of calculating the degree of dissociation.

It is assumed today that the typical strong electrolytes, and especially the majority of the salts, are completely dissociated at high dilution (below 0.01 molar) and dissociated almost completely at higher concentrations, or at least to a considerably greater extent than would be deduced, according to the classical theory, from conductivity and cryoscopic data. This conception was principally developed by Sutherland, Milner, Bjerrum, and Ghosh, and formulated exactly for very dilute solutions by Debye and Hückel in 1923.* [*14–16*]

It was first shown by Milner (1913) that the ions are not distributed completely at random in the solution of an electrolyte, in which case the electrostatic attractions and repulsions which they exert on each other would mutually cancel out. Instead, because of the electrostatic forces, more negative ions than positive are present in the environment of each positive ion, and near each negative ion more positive ions than negative are present. Quite apart from its effect on the conductivity, this also leads to the ions in solution being more strongly held than are uncharged molecules. In calculating the osmotic pressure, one can accordingly not consider ions and uncharged molecules as precisely equivalent.

Even at high electrolyte concentrations, the ionic concentrations in solutions of weak electrolytes are small, and the distances between the ions are correspondingly large. The influence of the inter-ionic forces is therefore of subordinate importance for weak electrolytes, and need be taken into consideration only in very precise calculations.

The conditions obtaining in solutions of strong electrolytes are complicated yet further, in that account should also be taken of the forces of interaction between the solvent and the ions. In general, these lead to the formation of *solvates*, or in the case of water to *hydrates* (cf. p. 75).

In addition to the electrolytic conductivity and the osmotic pressure, other properties depending on the concentration of the ions are also influenced by interionic forces—e.g. discharge potentials or electrode potentials and, in fact, all concentrations involved in the solution equilibria of dissolved substances. As already mentioned, these are known as *activities*. With increasing dilution, the activities of ions, like the activities of non-electrolytes, approach and eventually become equal to the true concentrations. However, even in relatively very dilute solutions of strong electrolytes ($c = 0.001$ mol per liter) the ionic activities deviate quite significantly from the true concentrations. On the other hand, in strong electrolytes the true concentrations of the ions differ only slightly from those that would be calculated on the assumption of complete dissociation. In practice, it is possible to consider that *strong electrolytes are completely dissociated*, especially in their relatively dilute solutions ($c < 0.1$ mol per liter).

Following Bjerrum's proposal, the factors by which the true concentrations of the ions must be multiplied, in order to obtain the apparent concentrations found from activity measurements (by equation (1)), from conductivity measurements

* See also D. A. McInnes, *J. Franklin Inst.*, 225 (1938) 661 for a clear and brief presentation of the Debye-Hückel theory. For the exact calculation of dissociation constants, see p. 824.

(by equation (17)), or from osmotic methods (by equation (18)), neglecting inter-ionic forces in the first instance, are termed the *activity coefficients* f_a, the *conductivity coefficients* f_μ, or the *osmotic coefficients* f_0, or generally as *deviation coefficients*.

The deviation coefficients are defined by the following equations:

$$\text{Activity coefficient} \quad f_a = \frac{a}{c_w}, \tag{21}$$

where a represents the activity, i.e., the effective concentration, and c_w the true concentration of the corresponding ion.

$$\text{Osmotic coefficient} \quad f_0 = \frac{P}{P_0}, \tag{22}$$

where P = measured osmotic pressure, P_0 = osmotic pressure which would be exerted if the ions behaved as neutral molecules.

$$\text{Conductivity coefficient} \quad f_\mu = \frac{1}{a_w} \cdot \frac{\mu}{\mu_0}, \tag{23}$$

in which a_w represents the true (not the measured) degree of dissociation.

The coefficients f_a, f_0, f_μ, depend upon the nature of the electrolyte, the dilution, and the temperature. They are generally not equal to one another, but all three attain the limiting value 1 at infinite dilution. With weak electrolytes, they differ from 1 by a negligible amount, even in less dilute solutions. For weak electrolytes, therefore, the ionic concentrations found by different methods can be taken as equal to one another, without introducing serious errors. In practice, therefore, for weak electrolytes and especially at great dilutions,

$$f_a = f_0 = f_\mu = 1; \quad \text{and} \quad \frac{a}{c_w} = \frac{P}{P_0} = \frac{1}{a_w} \cdot \frac{\mu}{\mu_0} = 1$$

For the typical strong electrolytes (most salts) however, at sufficient dilution

$$a_w = 1$$

and thus if x denotes the number of ions into which each dissolved molecule dissociates

$$f_0 = \frac{i}{x} \quad \text{and} \quad f_\mu = \frac{\mu}{\mu_0}$$

In Table 14 are shown the deviation coefficients for potassium chloride solutions of various concentrations. In the last column of the Table are given the values of $i - 1 \ (= 2 f_0 - 1)$. These are the values which, on the classical theory, would be regarded as the degree of dissociation, as also would the f_μ values listed in the preceding column. It

TABLE 14

VALUES OF THE DEVIATION CONSTANTS FOR KCl

Molar concen.	f_a	f_0	f_μ	$i-1$
1.0	0.558	0.854	0.755	0.708
0.1	0.762	0.932	0.861	0.864
0.01	0.882	0.969	0.941	0.938
0.001	0.943	0.985	0.979	0.970

can be seen that the values separated in the Table by the double ruling agree quite well, over the concentration range from 0.1 to 0.001. In the present state of the theory it is not easy to see why this should be, since no simple relation can be derived theoretically between f_0 and f_μ. A relation between f_a and f_0 may be deduced thermodynamically. It can be used to calculate f_a, if f_0 and its dependence upon concentration are known.

Equation (21) serves to define the activity coefficient for any given *species of ion* (e.g., for the Cl⁻ ion). The activity coefficient of a strong *electrolyte* (e.g., KCl) is understood as signifying the geometric mean of the *activity coefficients of its ions*. Thus the activity coefficient f_a^{AX} of an electrolyte which dissociates into p ions having an activity coefficient f_1, and into q ions with an activity coefficient f_2 is

$$f_a^{AX} = (f_1{}^p \cdot f_2{}^q)^{\frac{1}{p+q}}$$

Thus, for $BaCl_2$ ($p = 1$, $q = 2$), the activity coefficient is

$$f_a^{BaCl_2} = [f_a^{Ba++} \cdot (f_a^{Cl-})^2]^{\frac{1}{3}}$$

The *activity of a strong electrolyte* is the *product of the activities of its individual ions*. Thus the activity of $BaCl_2$ (at an analytical molar concentration c) is:

$$a = c \cdot f_a^{Ba++} \cdot (2\,c \cdot f_a^{Cl-})^2 = 4\,c^3 \cdot f_a{}^{Ba++} \cdot (f_a^{Cl-})^2$$

At infinite dilution this has the limiting value $\lim\limits_{c=0} a = 4\,c^3$.

Almost all salts are strong electrolytes. On the basis of the newer theory, it is possible to understand the striking fact that, for salts of the same type, e.g., for all uni-univalent salts irrespective of the elements composing them, the ratios of molecular conductivities, μ_c/μ_0 (or of equivalent conductivities Λ_c/Λ_0) are almost the same—i.e., that salts of the same type have approximately the same apparent degree of dissociation. According to the theory, all salts, apart from a few exceptions, are almost completely dissociated into their ions upon dissolution in water. The forces exerted upon one another by these ions, present in the solution at relatively great distances from one another, are determined in dilute solutions almost entirely by the number of charges borne by the ions, and are thus but little dependent on differences in their chemical nature. In contrast with this, some of the acids display strongly individual differences in their conductivity quotients. The monobasic strong acids are, however, again substantially similar among themselves. However, these can also be regarded as almost completely dissociated. For

TABLE 15

CONDUCTIVITY QUOTIENTS Λ_c/Λ_0 OF SALTS AND OF BASES (AT 18°)

For *salts* and *strong bases*, the quotients Λ_c/Λ_0 give the *apparent*,
and for *weak bases* the *true* degree of dissociation

Type	Examples	1-normal	0.1-normal
Uni-univalent salts	KCl	0.75	0.86
Uni-bivalent or bi-univalent salts {	K_2SO_4	0.54	0.72
	$CaCl_2$	0.58	0.75
Bi-bivalent salts	$MgSO_4$	0.25	0.43
Monacidic strong bases	KOH	0.77	0.89
Diacidic strong bases	$Ba(OH)_2$	—	0.77
Weak bases	$NH_3 \cdot H_2O$	0.004	0.014

weak acids, the conductivity quotient gives the degree of dissociation to a close approximation. Sulfuric acid, oxalic acid, phosphoric acid and similar acids take an intermediate position, and are said to be moderately strong acids. The conductivity quotients of strong bases correspond to those of the salts.

In Tables 15 and 16 are given the conductivity quotients Λ_c/Λ_0 for the most important types of salts and for a number of bases and acids. The values shown in the Tables represent

TABLE 16

CONDUCTIVITY QUOTIENTS Λ_c/Λ_0 OF ACIDS (AT 18°)

For *weak* acids the quotients Λ_c/Λ_0 give the *true*-, and for
strong acids the *apparent* degree of dissociation.

Type	Examples		1-normal	0.1-normal
Strong acids	{	HNO_3	0.82	0.93
	{	HCl	0.79	0.92
	{	H_2SO_4	0.51	0.59
Moderately strong acids	{	$H_2C_2O_4$ oxalic acid	0.16	0.31
	{	H_3PO_4	0.06	0.12
Weak acids	{	$CH_3 \cdot COOH$ acetic acid	0.004	0.013
	{	H_2CO_3	—	0.001

in each case the ratios of the equivalent conductivity Λ_c for a given concentration, to Λ_0, the limiting value of the equivalent conductivity at infinite dilution.

(h) Dissociation in Stages

Polybasic acids, such as sulfuric acid, oxalic acid, and phosphoric acid, do not dissociate off all the replaceable hydrogen atoms in one stage. The dissociation takes place rather in steps, i.e., one hydrogen ion is lost first, and the dissociation of the second begins only at considerably greater dilution, the loss of a third hydrogen ion, etc., taking place only at a yet higher dilution, or only on the addition of alkali hydroxide. In so far as the mass action law can be applied to these processes, a separate dissociation constant applies to each stage in the dissociation. This dissociation in stages is the reason for the ability of the polybasic acids to form acid salts, i.e., salts formed by the incomplete neutralization of the acids by bases, in which hydrogen atoms replaceable by metals are therefore still present.

(i) Range of Applicability of Mass Action Law to Strong Electrolytes

The failure of the Ostwald dilution law with strong electrolytes shows that, in this form, the law of mass action cannot be applied to strong electrolytes; the newer theory of strong electrolytes has shown why this is. According to our present knowledge it is not merely inexact, but fundamentally wrong, to regard the conductivity quotient μ/μ_0 (of a strong electrolyte) as a true degree of dissociation, and to apply the law of mass action to it. The activities of the ions must also be inserted in other mass action equations into which the ionic concentrations of strong electrolytes enter (as for example in the calculation of solubility products). In considering the influence of added electrolytes upon ionic equilibria, it must be borne in mind that, according to the newer theory, the activities of the ions of an electrolyte are affected by *all electrolytes*, and not only by those which possess a common ion. The rule that applies is that, in dilute solutions, the activity coefficient of any ion depends solely on the *total ionic strength* of the solution.* The ionic strength I signifies the sum

$$I = \tfrac{1}{2}(c_1 z_1{}^2 + c_2 z_2{}^2 + \ldots.),$$

in which c_1, c_2, etc. are the real molar concentrations, and z_1, z_2 etc. the charges of the various ions. Thus in a pure 0.01 molar $BaCl_2$ solution, I has the value

$$\tfrac{1}{2}(0.01 \cdot 2^2 + 0.02 \cdot 1^2) = 0.03.$$

In this solution, the activity coefficients of the Ba^{++} and the Cl^- ions have the values $f_a{}^{Ba++} = 0.50$ and $f_a{}^{Cl-} = 0.86$. If 0.1 mole of a uni-univalent strong electrolyte is added to the solution, then I increases by the amount $\tfrac{1}{2}(0.1 + 0.1) = 0.1$. The total ionic strength is then $I = 0.13$, and in a solution of this ionic strength the activity coefficients have the values $f_a{}^{Ba++} = 0.10$ and $f_a{}^{Cl-} = 0.79$, irrespective of whether the added electrolyte were, for example, KCl or KNO_3**. A 0.0433 molar solution of pure $BaCl_2$ would have this same ionic strength ($I = 0.13$); $f_a{}^{Ba++}$ and $f_a{}^{Cl-}$ in such a solution also have the values given above. Since the activities of the individual ions are obtained by multiplying their true concentrations (practically the same as the analytical concentrations in the case of strong electrolytes) by the relevant activity coefficients, the activity of the Ba^{++} ions $= 0.01 \cdot 0.50 = 0.0050$, and that of the Cl^- ions $= 0.02 \cdot 0.86 = 0.0172$ in a pure 0.01 molar $BaCl_2$ solution. The activity (effective concentration) of the Cl^- ions therein is thus not even approximately equal to twice that of the Ba^{++} ions, as would be presupposed by the classical dissociation theory.

For more or less rough approximations, it is often possible to use the analytically determined, true concentrations for strong electrolytes (i.e., the concentrations given by the content of electrolyte, assuming complete dissociation), without taking into account the activity coefficients. This can be done, for example, when making approximate calculations of equilibria in mixtures of strong and weak electrolytes, or when it is required to estimate the effect of additions of small amounts of an equi-ionic strong electrolyte upon the solubility of a sparingly soluble electrolyte. If greater accuracy is required, it is quite essential to work with 'activities'. The use of apparent ionic concentrations obtained from conductivity or osmotic data, instead of activities, is not justifiable from the theoretical standpoint, for strong electrolytes. Moreover, such data rarely afford, in practice, any greater accuracy than would be obtained by using the analytical concentrations directly.

In theory, the mass action law should be applicable to the dissociation process of the strong electrolyte itself. There must, in fact, always be at least a few ions of opposite charge so close together in the solution that one can regard them as having combined to form undissociated molecules. For strong electrolytes, the number of undissociated molecules in this sense is not closely connected in any way with that given by the apparent degree of dissociation.

11. Hydrates

Water has the property of combining with many substances to form chemical compounds. The products resulting therefrom are called *hydrates*, and the process *hydration*. It is not now usual to speak of the resulting compounds as hydrates when water is not combined as such, i.e., when the process of combination demonstrably involves splitting up the molecule of water, so that the elements of water are added on separately. This is the case, for example, in the formation of a hydroxide from an oxide:

$$CaO + HOH = Ca(OH)_2$$

* The rule of the dependence of activity coefficients upon the ionic strength was discovered empirically in 1921 by Lewis and Randall. Its basis was later established by the Debye theory.

** For the determination of ionic activity coefficients see for example LEWIS and RANDALL, *Thermodynamics*, McGraw Hill Book Co., New York (1923).

In such a case, water is said to be 'constitutively' combined.

The designation 'hydrate' has been retained in archaic nomenclature, corresponding to the former outlook whereby compounds such as $Ca(OH)_2$ were regarded as addition compounds,—for example in the older name 'calcium oxide hydrate' for calcium hydroxide. At the present time, hydrates are understood to be compounds containing water, which is shown by their behavior to be combined as such.

There are substances in which it is uncertain whether water is present in a chemically combined state, or is only retained mechanically. This is the case, in particular, with certain metallic oxides when they are obtained in the form of *hydrogels* (cf. p. 352). As long as it is open to question whether these are hydroxides, or whether they are oxides containing firmly but physically bound water, it is advantageous to refer to them as 'oxide hydrates'.

(a) Solubility Diagram of Hydrates and the Phase Rule

Many substances crystallize from aqueous solution with a varying water content, according to their concentration. Thus $CaCl_2$ crystallizes with 6, 4, 2, and 1 H_2O. The conditions under which the various hydrates are stable in the presence of the solution can be inferred from the *solubility diagram*. Fig. 13 represents the solubility of calcium chloride in

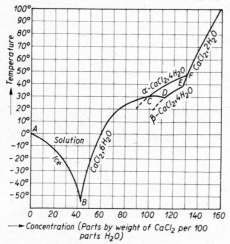

Fig. 13. Solubility diagram of calcium chloride in water.

water, as a function of the temperature (solubility curves). From solutions which contain less than 42.5 parts by weight of $CaCl_2$ per 100 parts by weight of H_2O, *ice* separates out on freezing. The concentration of the solution is increased thereby, and the freezing temperature decreases simultaneously, along the curve AB. At B, i.e., at a content of 42.5 parts of $CaCl_2$ to 100 of H_2O and at $-55°$, the remainder solidifies as a *cryohydrate*. From solutions with more than 42.5, and less than 100.6, parts of $CaCl_2$ per 100 of H_2O, calcium chloride hexahydrate separates out first upon cooling. The concentration of the solution, and the temperature of deposition of the crystals, thereby alter along the curve CB. At $-55°$, the remainder once more solidifies as a cryohydrate. At each value of the concentration and the temperature represented by the curves AB and CB, three phases are present in equilibrium with one another, namely solution, ice, and water vapor; or solution, hexahydrate and water vapor, respectively. Since we are dealing with a three-phase system of two components (H_2O and $CaCl_2$), this has 1 degree of freedom, in accordance with the phase rule (eqn. (10)). In this case therefore, in contrast to the example shown in Fig. 10, three phases can be in equilibrium with one another over a range of temperatures. At every temperature, however, there is a quite definite concentration, and a definite vapor pressure, of the solution in equilibrium with a given solid phase. In the same way, for each concentration of the solution, there is a definite temperature and a definite vapor pressure at which it is in equilibrium with the appropriate solid phase. Point B, being the intersection of the two equilibrium curves for ice and the hexahydrate, gives the concentration and the temperature at which both solid phases are in equilibrium with the solution and its vapor. We have here 4 phases in all, and accordingly no degree of freedom according to equation (10). Such a point as B, at which 4 phases co-exist, is called a *quadruple point*. The mixture of the two solid phases—the cryohydrate in this instance—thus separates from the solution at a perfectly definite temperature $(-55°)$*. It has a fixed composition, accordingly (42.5 parts of $CaCl_2$ per

---

* The name cryohydrate is actually misleading, since the term *hydrate* is otherwise applied only to compounds and not to mere mixtures. It is in general use, however.

100 of H_2O), and a definite vapor pressure (0.015 mm Hg). The cryohydrate has the same composition, and at the temperature of its deposition the same vapor pressure, as the solution. Although it has a constant composition, it is a mixture, and not a chemical compound.

The α modification of the tetrahydrate ought to crystallize from solutions having a $CaCl_2$ content between the values on the abscissa corresponding to C and F (100.6 to 130.2 parts of $CaCl_2$). Because of delayed crystallization of this form, however, the 6-hydrate generally still crystallizes out between C and D (100.6—112.8 parts $CaCl_2$), and the unstable β-modification of the 4-hydrate between D and E*. Although it is unstable in this region, the 6-hydrate usually crystallizes from a solution containing 102.7 parts of $CaCl_2$ per 100 of H_2O, i.e., having the actual composition of the 6-hydrate. The 6-hydrate therefore usually melts without decomposition (at 30.2°) on warming. Since seeding with the 6-hydrate leads to this phase crystallizing out first, from solutions having $CaCl_2$ concentrations between 102.7 and 112.8 parts of $CaCl_2$ per 100 parts of H_2O, the concentration of the solution increases on cooling until the point D is reached. At the concentration (112.8 parts of $CaCl_2$ per 100 of H_2O) and temperature (29.2°) corresponding to this point, a mixture of crystals of the 6-hydrate, and of the β-4-hydrate separates out without further change in the concentration of the solution. At point D therefore, as at B, 4 phases are in equilibrium with one another; solution, vapor, and the two solid phases (both being unstable in this case). The mixture of the two latter, crystallizing at 29.2°, also has the same composition as the solution. It is noteworthy that, as the diagram shows, the 6-hydrate can be in equilibrium at one and the same temperature with two solutions of different concentration, e.g., at 29.2°, with a solution containing 112.8 parts of $CaCl_2$ and with one containing 94.5 parts of $CaCl_2$ per 100 of H_2O. At the point where the salt hydrate co-exists with a solution of the same concentration (102.7 parts $CaCl_2$) the equilibrium temperature is a maximum (30.2°).

The unstable β-modification of the tetrahydrate, in contact with the saturated solution, is at once transformed into the stable α modification if it is rubbed with a glass rod. The transformation takes place immediately if it is seeded with an α crystal.

The dihydrate is the first solid to separate from solutions containing more than 127.5 parts of $CaCl_2$ (or, if they are seeded with the α-4-hydrate, more than 130.2 parts of $CaCl_2$) and after the concentration of the solution has been hereby reduced to the value corresponding to the points E or F respectively, the remainder separates as a mixture of 2-hydrate and 4-hydrate. In the range from 130.2—297 parts of $CaCl_2$ per 100 of H_2O, the dihydrate is stable in contact with the solution, and at still higher concentrations, the monohydrate is stable.

(b) Vapor Pressure Curves. Thermal Degradation of Hydrates

The vapor pressure curves of the $CaCl_2$ hydrates are represented in Fig. 14.** If crystals of the 6-hydrate are placed in an evacuated vessel they give up water vapor, forming the tetrahydrate until the water vapor pressure in the vessel has reached the value corresponding to equilibrium. There is then present a system of two components ($CaCl_2$ and H_2O), with three phases (6-hydrate, 4-hydrate, and vapor), for which $F = 1$ according to equation (10).

This can accordingly exist in equilibrium at various temperatures, but to each temperature there corresponds a definite vapor pressure. The associated values of temperature and pressure make up the *vapor pressure curve*.

If, at some definite temperature (e.g., 25°), water vapor is pumped out of the vessel, the pressure remains constant ($p = 5.1$ mm) as long as 6-hydrate is still present, provided that the rate of decomposition of the hydrate suffices to maintain the equilibrium. As soon as

---

* This corresponds to Ostwald's rule of successive stages, cf. p. 489. The β-tetrahydrate generally crystallizes from solutions having a concentration a little below D (112.8 parts of $CaCl_2$ per 100 of H_2O). If, however, the solutions are seeded with the 6-hydrate, this separates out. The form which is actually stable in this range of concentrations, the α-tetrahydrate, likewise only crystallizes if it is used for seeding.

** Of the two modifications of the 4-hydrate, only the stable one is included. The vapor pressure curve of the monohydrate is also omitted, since in the temperature range represented in Fig. 14, no accurate measurements for this hydrate have been recorded.

the 6-hydrate has disappeared, the 4-hydrate begins to decompose. The vapor pressure consequently drops at this moment to the value which corresponds to the equilibrium

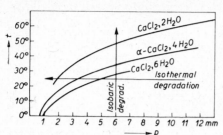

pressure of the 4-hydrate at $25°$ ($p = 3.4$ mm cf. Fig. 14). This pressure in turn remains constant until the 4-hydrate has been completely degraded to the 2-hydrate. As soon as the 4-hydrate disappears, the vapor pressure falls suddenly to the equilibrium pressure of the dihydrate ($p = 2$ mm) and so on. If the vapor is pumped off in portions, and the composition of the solid being degraded is determined each time (either from the loss in weight, or by measuring the amount of vapor pumped off), we obtain the stepped curve represented in Fig. 15 by plotting the pressure against the composition. Each step

Fig. 14. Vapor pressure curves of calcium chloride hydrates.

represents the discontinuous change in pressure taking place at that moment when one hydrate has disappeared, and the phase of lower water content resulting from it is present practically undecomposed. The composition of the latter can therefore be read off from the degradation curve.

Instead of carrying out the degradation at constant temperature (*isothermal degradation*) it can also be done by heating (*isobaric degradation*). One then observes that at a given pressure each individual hydrate decomposes at a definite temperature. Thus, in an atmosphere with a water vapor pressure of 6 mm, the 6-hydrate of calcium chloride passes over into the 4-hydrate at $28°$ with the loss of water vapor, the 4-hydrate at $34°$ into the 2-hydrate, and the latter at $49°$ into the 1-hydrate, as can be read off from Fig. 14. Decomposition takes place discontinuously in this case also, i.e., the loss of water continues at constant temperature until the hydrate in question has disappeared, and the next hydrate only begins to dissociate when its decomposition temperature is reached. In the range of temperature lying between these points, the composition of the hydrate does not alter.

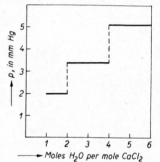

Fig. 15. Degradation curve of $CaCl_2.6H_2O$ (isothermal degradation).

It may be seen from Fig. 13 that the unstable (or metastable) β-modification of $CaCl_2.4\,H_2O$ has a higher solubility than that of the stable α-modification at the same temperature. It also has a higher water vapor pressure than the stable form at the same temperature. It is a general principle that when a substance exists in several modifications, the *form which is stable under given conditions is that with the lowest solubility and the lowest vapor pressure or dissociation pressure.* This is a consequence of the second law of thermodynamics. It is also possible to deduce thermodynamically another rule, the validity of which was first pointed out by O. Sackur (1908), *the vapor pressure of a hydrate depends upon the nature of the product poorer in water, formed by its dissociation.* Thus the 6-hydrate of calcium chloride has a higher vapor pressure if it loses water to form the stable α-4-hydrate than when it dissociates to give the unstable β-4-hydrate. This rule of Sackur applies generally to the thermal decomposition of all substances.

The relations illustrated here for the case of hydrates apply correspondingly to other solid compounds which dissociate to form a gaseous product, e.g., for ammoniates and also for hydroxides, oxides, carbonates, etc. The essential condition, however, is that a second solid phase shall be formed by the dissociation; otherwise the system has two degrees of freedom instead of one. A definite temperature would then no longer correspond to a definite decomposition pressure, but the latter would be dependent also on the composition of the solid phase. Among hydrates, this phenomenon was first observed for certain minerals, the *zeolites*, of which the water content can vary to a considerable extent, according to the pressure and the temperature, without the crystal becoming inhomogeneous. In such a case the water is said to be *zeolitically* combined. The vapor pressure can also be influenced by the degree of subdivision of a substance, whether by reason of its great surface area, or because it contains the water occluded in fine capillaries. In either case, the vapor pressure of water may be reduced to such an extent that it is only given off along with the crystal water, or even with constitutively bound water. The steps in the dehydration curves may be smeared out by the simultaneous loss of adsorbed water or water bound in capillaries and the discontinuous loss of combined water, or discontinuities may even be simulated at false values (cf. pp. 494 and 514).

(c) **Hydration of Ions**

As numerous substances already exist as hydrates in the solid state, such addition compounds must certainly already be present in aqueous solutions, and indeed the latter will in some circumstances be still richer in water than the solid hydrates. In the case of ions, the hydration can most simply be determined, in not too dilute solutions, by measuring the electrolytic transference of water, that is the transport of water from anode to cathode in virtue of the different degree of hydration of anion and of cation.* Relatively concordant values for the electrolytic transference of water have been obtained by various observers (Buchböck, Washburn, Remy, Baborowsky), and lead approximately to the quantities shown in Table 17 for the hydration of the ions.

TABLE 17

HYDRATION OF IONS IN 1-NORMAL SOLUTION (H. REMY)

Ion	H^+	Li^+	Na^+	NH_4^+	K^+	Rb^+	Cs^+	Mg^{++}	Ca^{++}	Sr^{++}	Ba^{++}	Cl^-	Br^-	I^-
Number of water molecules bound by each ion	1	13.0	8.6	4.4	4.2	4.0	3.5	14	10–12	8	4	3	2	3–4

Many of the properties of ions are considerably modified as a result of hydration—for example, their electrolytic mobility, and the deposition potentials. The effect of hydration on the *hydrogen ion* is especially great. The behavior of the *hydronium ion* H_3O^+, formed by addition of a proton to a water molecule, is quite different from that which would be exhibited by a free proton. It has already been pointed out that the proton is extraordinarily firmly bound to a H_2O molecule. By comparison, the H_2O molecules are only loosely bound in other hydrated ions. Their binding energies are even smaller than the values obtained by dividing the heats of hydration of ions (p. 156) by the number of bound water molecules. This is because the energy of hydration contains not only the binding energy of the water molecules immediately linked to the ion, but also the energy liberated through the orientation effect exerted by the ion on the more remote water molecules, in virtue of their dipolar nature. The hydronium ions also exert this orienting effect on the water molecules around them, but the magnitude of this energy of orientation is hardly significant in comparison with the large energy of binding.

According to Wicke (1954), the $[H_3O]^+$ ion is additionally hydrated, i.e., there are attached to it further H_2O molecules whose binding strength corresponds almost to that of the H_2O molecules in other hydrates, e.g., the alkali metal ions. On this basis, the hydronium ion $[H_3O]^+$ is characterized by a particularly high binding energy of the water molecule to the proton. Wicke claims that by addition in aqueous solution of three further H_2O molecules, the $[H_3O]^+$ ion (whose occurrence in crystalline compounds has long been

* With increasing dilution, the transport of the water chemically bound by the ions is overlaid, to an increasing extent, by the hydrodynamic streaming which the wandering ion produces around itself (Ulich, 1933). Remy had already pointed out, in 1915, that it was possible that a portion of the water was carried over hydrodynamically.

recognised, cf., p. 809) gives rise to a pyramidal $[H_9O_4]^+$ complex, in which the excess proton is probably almost freely mobile, so that after formation of this complex it can no longer be assigned to any definite H_2O molecule. To the 'inner' hydration shell so formed, further H_2O molecules can be similarly added in an 'outer' hydration shell.

As in aqueous solutions, the dissolved substances in other kinds of solutions frequently combine with the solvent to form more or less loose addition compounds. These, including the hydrates present in aqueous solution, are known by the collective name *solvates*. The phenomenon is called *solvation*.

12. The Hydrogen Ion (Hydronium Ion)

(a) Properties

References to the hydrogen ion in aqueous solution are always to be understood as signifying the hydrated hydrogen ion (hydronium ion $[H_3O]^+$ or $[H_9O_4]^+$). This is the most important of all electrolytic ions. Numerous processes which take place in aqueous solution are markedly dependent upon the concentration of hydrogen ions present. Many processes are catalytically accelerated by hydrogen (or rather, hydronium), ions, as for example the saponification of esters and the inversion of cane sugar. In other reactions, hydrogen ions have a retarding effect—e.g., on the decomposition of hydrogen peroxide, as has already been mentioned. Vital processes are extremely sensitive to the hydrogen ion concentrations in the solutions involved; in particular the function of enzymes is influenced to a far reaching extent by even small alterations of the hydrogen ion concentration in the fluids in which they are contained. The determination of hydrogen ion concentrations has therefore become of great importance in biology. Hydrogen ions act as precipitants of many colloidal materials. This is also of importance in analytical chemistry; it is the basis of the rule that substances which readily form colloidal dispersion should be precipitated, if possible, in acid solution.

The electrolytic mobility of the hydrogen (hydronium) ion far exceeds that of all other ions. This is the reason for the high conductivity of the strongly dissociated acids. At 18° the mobility of the hydrogen ion at infinite dilution is 315; the mobility of other ions lies in general between 40 and 70, only the hydroxyl ion OH^- having a considerably greater mobility, namely 174. The very high mobility of hydrogen ions is probably due to the fact that they do not migrate over the whole path as hydrated ions ($[H_3O]^+$ or $[H_9O_4]^+$ ions), but that an exchange of H^+ nuclei takes place as well, in such a way that one $[H_3O]^+$ ion gives up a H^+ nucleus to a neighboring water molecule, and so on.*

The number of free H^+ nuclei is infinitesimally small, even in strongly acid solutions. According to Kolthoff (1930), they would be present in the ratio of roughly $1 : 10^{130}$ as compared with the $[H_3O]^+$ ions in a solution normal with respect to the latter.

(b) Methods of Determining Hydrogen Ion Concentrations

Since the determination of the hydrogen (hydronium) ion concentration [17–21] is of great importance for many purposes, the most important methods available

---

* For an explanation of the very high mobility of the 'hydrogen' ion see P. WULFF, *Z. Elektrochem.*, 47 (1941) 858 and G. KORTÜM and EUCKEN, *ibid.*, 52 (1948) 268. See also E. WICKE ET AL., *Z. physikal. Chem., (New Series)*, 1 (1954) 340.

for its determination will be shortly enumerated here. All these methods give the apparent concentration and not the true concentration, insofar as these differ significantly. In accordance with what has already been stated (p. 67 *et seq.*), the values obtained for the apparent concentration are not quite independent of the methods by which they are determined.

Methods for determining the hydrogen ion concentration must not be confused with the methods for determining the content of 'free acid' in a solution. This is found by titration. If the solution contains only the free acid, its concentration naturally gives the hydrogen ion concentration also, provided that the degree of dissociation is known. If, in addition to the free acid, its salts are also present, then in simpler cases the hydrogen ion concentration (or hydrogen ion activity) can again be calculated by correct application of the law of mass action. In this case, however, direct determination of the hydrogen ion concentration is generally to be preferred. This is certainly true for complicated mixtures, such as are present for example in the body fluids of living organisms.

The following methods find the most general application for determination of hydrogen ion concentrations.

The methods listed under 1 to 3 give the *activity* of the hydrogen ion (strictly hydronium ion). They are thus applicable to any aqueous solutions. The apparent hydrogen ion concentrations found by methods 4 and 5, on the other hand, cannot be immediately taken as equivalent either to the activities or to each other. However, in fairly dilute solutions, or in those containing no strong electrolytes, the deviations are so small that they can often be neglected in practice.

1. *Potentiometric determination of hydrogen ion concentration.*

This depends on the measurement of the potential difference between a comparison electrode and a platinized platinum electrode, over which hydrogen is bubbled, and which dips into the solution of unknown hydrogen ion concentration. The hydrogen ion concentration (more precisely, the activity) of the solution is found from equation (1), p. 28. This method is now generally employed for accurate measurements.

2. *Colorimetric determination of hydrogen ion concentration* (indicator method).

The indicator method is based on the fact that the color of certain organic dyestuffs depends on the hydrogen ion concentration. If such an 'indicator' is added to the solution in question, the hydrogen ion concentration can be found as in other colorimetric determinations, by comparing the color of the solution containing the indicator, with the color of comparison solutions of known hydrogen ion concentration, to which indicator has also been added. See also Chap. 18.

3. *Determination of hydrogen ion concentration from the catalytic action of hydrogen ions.*

The concentration of hydrogen ions in aqueous solutions can also be determined from their catalytic effect upon many reactions. Suitable reactions are:

a) *Hydrolysis of esters* to free acids and alcohol. The progress of the reaction is followed by titrating the acid liberated. Other factors being equal, the rate of hydrolysis is proportional to the hydrogen ion concentration of the solution.

b) *The inversion of cane sugar*. The rate of this reaction is also proportional to the hydrogen ion concentration. The course of the inversion—i.e., the hydrolysis of sucrose into glucose and fructose—can be followed by simple polarimetric measurements. The strength of many acids has been determined by this means.

c) Decomposition of diazoacetic ester. The rate of this reaction is also, in general, proportional to the hydrogen ion concentration. The velocity can conveniently be determined by measuring the rate of evolution of nitrogen.

The three reactions mentioned are affected by various factors, such as the presence of neutral salts, in a way which is not, as yet, fully explained. The results must therefore be evaluated with some care. It seems that marked effects of neutral stalts, etc., are not found when account is taken of the fact that the activity of the hydrogen ions determines their catalytic efficiency, and that, in such solutions, this may differ considerably from the apparent concentration derived from the classical theory, unless proper account is taken of the ionic strength of the solutions (cf. the work of Duboux, 1924 and later).

4. *Determination of hydrogen ion concentrations from conductivity measurements.*

The specific conductivity $\varkappa$ of an electrolyte solution is given by

$$\varkappa = \eta_1 u_1 + \eta_2 u_2 + \eta_3 u_3 + \ldots\ldots, \tag{24}$$

where η_1, η_2 etc. represent the concentrations of the individual ionic species (in gram equivalents per cc.) and u_1, u_2, etc., their mobilities. This relation (with the limitations indicated by the modern theory of electrolytes) can be used quite generally to determine the concentration of any species of ion of known mobility. It is especially suitable for determining hydrogen ion concentrations, however, since the concentration of hydrogen ions is of particularly great influence on the conductivity because of their great mobility. The mobilities of the other ions must, of course, also be known. The concentrations of all but one of them must be known, since the requirement that the total equivalent concentrations of positive and negative ions must be equal, furnishes a second equation. If, in addition to hydrogen ions, the solution contains only one anion, a determination of the conductivity suffices to determine the hydrogen ion concentration, provided that the mobility of the anion is known. Conductivity measurements are more frequently used to determine the degree of dissociation of acids than for this purpose. For weak acids, the degree of dissociation a is given from the specific conductivity $\varkappa_c$ at the equivalent concentration c ($= 1000\,\eta$) by means of

$$a = \frac{\Lambda_c}{\Lambda_0} = \frac{1000\ \varkappa_c}{c \cdot \Lambda_0} \tag{25}$$

In place of Λ_0 one can substitute the sum of the mobilities of anion and cation at infinite dilution. The hydrogen ion concentration of such an acid solution is equal to ca, i.e., is found directly from the specific conductivity of the solution, being equal to $1000\,\varkappa_c/\Lambda_0$, as follows from equation (24).

5. *Determination of hydrogen ion concentration from freezing point or boiling point determination.*

For weak electrolytes, the degree of dissociation a is given by the deviation of the freezing point depressions or boiling point elevations from the Raoult-Van't Hoff law, in accordance with equations (19) or (19a). For weak acids, the hydrogen ion concentration is then obtained from the Van't Hoff factor and the known total concentration.

References

1 H. Pincass, *Die industrielle Herstellung von Wasserstoff*, Dresden 1933, 82 pp.
2 H. Schönfeld, *Die Hydrierung der Fette*, Berlin 1932, 152 pp.
3 J. Leick, *Das Wasser in der Industrie und im Haushalt*, Leipzig 1935, 118 pp.
4 J. Holluta, *Die Chemie und chemische Technologie des Wassers*, Stuttgart 1937, 219 pp.
5 F. J. Matthews, *Boiler Feed Water Treatment*, London 1936, 256 pp.
6 H. Klut, *Untersuchung des Wassers an Ort und Stelle*, 8th Ed., Berlin 1943, 260 pp.
7 A. Findlay, *The Phase Rule and its Applications*, 7th Ed., London 1931, 326 pp.
8 F. F. Purdon and V. W. Slater, *Aqueous Solutions and the Phase Diagram*, New York 1946, 167 pp.

9 W. Machu, *Das Wasserstoffperoxyd und die Perverbindungen*, Vienna 1937, 408 pp.

10 O. Kausch, *Das Wasserstoffperoxyd, Eigenschaften, Herstellung und Verwendung*, Halle 1938, 254 pp.

11 F. Foerster, *Elektrochemie wässeriger Lösungen*, 4th Ed., Leipzig 1923, 900 pp.

12 S. Glasstone, *The Electrochemistry of Solutions*, London 1937, 551 pp.

13 P. Walden, *Das Leitvermögen der Lösungen* (Vol. IV of the *Handbuch der allgemeinen Chemie*, edited by Wi. Ostwald and C. Drucker), Leipzig 1924, 383+346+397 pp.

14 H. Falkenhagen, *Elektrolyte*, Leipzig 1932, 346 pp.

15 G. Kortüm, *Elektrolytlösungen*, Leipzig 1941, 483 pp.

16 C. W. Davies, *The Conductivity of Solutions and the Modern Dissociation Theory*, 2nd Ed., London 1933, 281 pp.

17 H. T. S. Britton, *Hydrogen Ions, Their Determination and Importance in Pure and Industrial Chemistry*, 3rd. Ed., London 1942, Vol. I, 420 pp; Vol. II, 443 pp.

18 W. M. Clark, *The Determination of Hydrogen Ions*, 3rd Ed., Baltimore 1928, 717 pp.

19 I. M. Kolthoff and H. A. Laitinen, *pH and Electro Titrations, The Colorimetric and Potentiometric Determination of pH*, 2nd Ed., New York 1941, 190 pp.

20 L. Michaelis, *Hydrogen Ion Concentration, Its Significance in the Biological Sciences and Methods for Its Determinations.* Vol I, *Principles of the Theory* (transl. by W. A. Perlzweig), London 1926, 295 pp.; Vol. II, *Oxydation-Reduction Potentials* (transl. by L. B. Flexner), Philadelphia 1930, 199 pp.

21 M. Huybrechts. *Le pH et sa Mesure*, 4th Ed., Paris 1946, 474 pp.

22 B. E. Conway, *Electrochemical Data*, Amsterdam 1952, 374 pp.

23 G. Milazzo, *Elektrochemie, Theoretische Grundlagen und Anwendungen*, Vienna 1952, 474 pp.

24 J. E. Ricci, *The Phase Rule and Heterogeneous Equilibria*, New York 1951, 503 pp.

25 H. S. Harned and B. B. Owen, *The Physical Chemistry of Electrolytic Solutions*, 2nd Ed., New York 1950, 675 pp.

26 G. Kortüm and J. O'M. Bockris, *Textbook of Electrochemistry*, Amsterdam 1951, Vol. I, 352 pp.; Vol. II, 530 pp.

CHAPTER 3

THE HYDROGEN SPECTRUM AND THE STRUCTURE OF THE HYDROGEN ATOM

1. The Hydrogen Spectrum

(a) Examination and Interpretation

Hydrogen can readily be made to emit light by introducing it under reduced pressure into a glass tube provided with electrodes (Geissler tube), through which

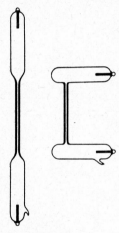

a high voltage electric current is passed e.g., by means of an induction coil. Tubes of the form shown in Fig. 16, which were first constructed by Plücker, are generally used; the luminosity is particularly intense in the constricted portion of the tube. If the light emitted by such a hydrogen tube is examined through a spectroscope, it can be seen that the spectrum consists of only a few sharp lines, namely one in the red, one in the greenish blue and two violet lines. These are designated by the symbols H_α, H_β, H_γ, and H_δ. The founders of spectroscopic analysis, Kirchhoff and Bunsen, first recognized them in 1860 as lines peculiar to hydrogen. They had been detected as absorption lines in the sun's spectrum by Fraunhofer as early as 1814.

With a quartz, or better a grating spectrograph, a whole series of other lines peculiar to hydrogen, lying in the ultra-violet, can be detected photographically. The spectrum reproduced in Fig. 17 is thereby obtained, from which it may be seen that in going from light of long wave length to light of shorter wave lengths, the individual lines crowd continually closer together. At the same time they become progressively weaker until they completely disappear.

Fig. 16. Plücker tubes.

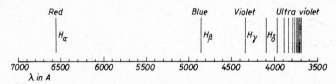

Fig. 17. Hydrogen spectrum (Balmer series).

Even superficial consideration of the spectrum suggests that the sequence of the lines must be given by a comparatively simple relationship. Balmer succeeded in

1885 in formulating this relation mathematically, namely that the wave lengths λ of all the lines can be reproduced by the simple formula

$$\frac{1}{\lambda} = R_\mathrm{H} \left(\frac{1}{2^2} - \frac{1}{n^2} \right), n > 2 \qquad (1)$$

in which R_H denotes a constant, and n can be in turn any one of the series of whole numbers from 3 upwards. For $n = 3$ one obtains the first line of the visible hydrogen spectrum H_α, for $n = 4$, H_β, for $n = 5$ H_γ, etc. As n increases, λ becomes progressively smaller. The smallest value which λ can attain according to the formula, namely that for $n = \infty$ is

$$\lambda = \frac{4}{R_\mathrm{H}} \qquad (2)$$

This gives the limit of the sequence of lines on the ultra-violet side. It cannot, however, be determined directly from the emission spectrum since the lines become weaker with diminishing wave length, and already disappear before reaching the limit.

The Balmer formula holds with extraordinary accuracy as may be seen from Table 18. In this, the wave lengths λ are given in Ångstrom units (Å). One Ångstrom unit = 0.1 mμ = 10^{-8} cm. It has been possible to measure the emission spectrum of hydrogen in vacuum tubes up to the 25th, and in plates of nebulae up to the 33rd. line.

TABLE 18

THE FIRST LINES OF THE HYDROGEN SPECTRUM

	H_α $n = 3$	H_β $n = 4$	H_γ $n = 5$	H_δ $n = 6$
observed =	6562.80	4861.33	4340.47	4101.74 Å
calculated =	6562.80	4861.38	4340.51	4101.78 Å

The constant R_H, the so-called Rydberg constant, can be found from spectroscopic measurements with an accuracy which has hardly ever been obtained in any other field. Its value is

$$R_\mathrm{H} = 109677.581 \pm 0.007 \text{ cm}^{-1}$$

(b) Series and Their Terms

A spectrum which, like the hydrogen spectrum just considered, consists of a sequence of lines belonging together is called a *series spectrum* or, briefly, a *series*. The spectrum just considered is called the Balmer Series.

In addition to the Balmer series, hydrogen possesses several other series spectra, namely one (the Lyman series) lying entirely in the ultra-violet region, and three (the Paschen series, the Brackett series, and the Pfund series) situated in the infrared. These series can be represented by formulas strictly analogous to that of the Balmer series.

Lyman Series $\qquad\qquad \frac{1}{\lambda} = R_\mathrm{H} \left(\frac{1}{1^2} - \frac{1}{n^2} \right), n = 2, 3, 4 \ldots \ldots$

Paschen Series
$$\frac{1}{\lambda} = R_H \left(\frac{1}{3^2} - \frac{1}{n^2} \right), n = 4, 5, 6 \ldots \ldots$$

Brackett Series
$$\frac{1}{\lambda} = R_H \left(\frac{1}{4^2} - \frac{1}{n^2} \right), n = 5, 6 \ldots \ldots$$

Pfund Series
$$\frac{1}{\lambda} = R_P \left(\frac{1}{5^2} - \frac{1}{n^2} \right), n = 6, 7 \ldots$$

The constant R_H is the same as for the Balmer series. All four series can be represented by the single formula

$$\frac{1}{\lambda} = R_H \left(\frac{1}{n_2^2} - \frac{1}{n_1^2} \right) \quad (n_1, n_2 = \text{whole numbers } n_1 > n_2) \tag{3}$$

According to this, every hydrogen line can be represented as the difference between two expressions of the form $\frac{R_H}{n^2}$ ($n =$ a whole number). Expressions of this form are known in spectroscopy as *terms*. The wave lengths of the lines of a *series* are obtained by keeping n_2 in equation (3) constant, and allowing n_1 to follow the series of whole numbers. Accordingly in the difference

$$\frac{1}{\lambda} = \frac{R_H}{n_2{}^2} - \frac{R_H}{n_1{}^2}$$

the first member is known as the constant term, the second member as the variable term. The constant term is the *series limit*.

(c) Line Spectra and Band Spectra

In addition to its line spectrum, hydrogen also possesses a band spectrum. This is obtained with small voltages on the Geissler tube. Band spectra differ from line spectra in that when they are observed through a spectroscope of small resolving power they appear to consist of coherent bands. In reality however, these bands also consist of individual lines following one another in close sequence; their arrangement however is typically different from that in the line spectra. In general, we attribute the band spectra to diatomic or polyatomic molecules, the line spectra to the free atoms. Thus in the case of iodine, one can observe how the characteristic band spectrum of the vapor, seen in absorption, disappears with rising temperature in the same measure as the I_2 molecules progressively dissociate into I atoms. In the same way, the band spectrum of hydrogen is attributed to the hydrogen molecule, whereas the line spectra considered are those of the hydrogen atom.

2. Bohr's Theory of the Structure of the Hydrogen Atom [2–4, 16]

(a) Rutherford's Theory of Atomic Structure

It had long been suspected that the light emitted by the atom must give us some information about its structure. Electromagnetic waves, to which light belongs, always result, as is well known, from the oscillation of electrical charges. This must at once suggest that the cause of emission of light should be sought in the oscillation of electrical charges, or in related processes. Like matter, electricity is atomistic in nature. The smallest particles of free negative electricity are known as *electrons*. The mass of an electron, compared with the mass of the lightest atom, the hydrogen atom, is as $1 : 1838$ (cf. p. 110). A neutral atom must contain equal numbers of electrons and positive charges. It was formerly thought that the positive electricity was distributed over the whole volume of the atom and that the electrons were, so to speak embedded in it (J. J. Thomson's atom model). However, this model did not provide a satisfactory explanation for the motion of the electrons, which manifests itself in the emission of light. A completely different basis for the problem was provided when Rutherford, in 1911, from the results of investigations by Geiger and Marsden on the scattering of α-rays in their passage through matter*, drew the conclusion that the positive charge of every atom must be concentrated in quite a tiny region in the interior of the atom. This region is known as the *nucleus* of the atom. The nucleus of the hydrogen atom has a diameter of at most $2 \cdot 10^{-13}$ cm, and that of gold, at most $3 \cdot 10^{-12}$ cm, whereas the diameter of the atom, as can be calculated, for example, from the density of crystalline solids, (cf. Chap. 7) is of the order of magnitude of 10^{-8} cm.

Rutherford's atomic theory is based on the observation that α-rays are generally deflected only to a small extent in their passage through matter, but that in rare instances extremely large deflections are produced. This can only be interpreted as meaning that, in individual cases, the particles come during their trajectory into the immediate neighborhood of what must be considered almost point-centers of electric force, containing in themselves the entire positive charge of the atoms. Being themselves positively charged, the α-particles are repelled by this positive charge, according to Coulomb's law. The enormously strong repulsion which is necessary to impose a drastic deflection on the fast moving and relatively heavy α-particle can be achieved only if the available positive charge of the atom is concentrated almost into a single point. Exact numerical evaluation of the results shows that the values given above represent the maximum space over which the positive charges can be distributed, if the observed large deflections are to be accounted for by the Coulomb repulsion.

Coulomb's law, which is the basis of the calculation of the interaction of electrical forces, states that the force exerted upon one another by two electrically charged bodies is proportional to the product of their electric charges, and inversely proportional to the square of the distance between them. If the units of the electrical charges are so chosen that the proportionality factor is unity, the Coulomb law can be expressed by the formula

$$K = \frac{e_1 \cdot e_2}{r^2} \tag{4}$$

where K is the force between the electrical charges e_1 and e_2, and r is the distance between the charges. The law was discovered experimentally by Coulomb in 1785. It can readily be proved that an electric charge, uniformly distributed over the surface of a sphere, acts on an electrically charged particle outside the sphere as if the charge on the sphere were concentrated at its center. The force exerted by electric charges upon one another is a repulsion if the charges have the same sign, and an attraction if they are of opposite sign. K thus denotes an attractive force if its value is negative. Equation (4) holds for the case that there is empty space between the charges. If there is a medium of dielectric constant D between them, the force must be divided by D.

The work A which must be expended to bring the electrical charges e_1 and e_2 from an infinite distance to the distance r from one another, in empty space, is given by

$$A = \frac{e_1 \cdot e_2}{r} \tag{5}$$

---

* Lenard had already concluded in 1903, from his experiments on the passage of cathode rays through matter, that the positive charges could not be continuously distributed over the whole region of the atom.

as follows from Coulomb's law if one constructs the path-integral of the force. If the charges differ in sign, A is negative; this means therefore that work is gained by bringing them together. In every case, A represents the potential energy (ability to do work) of the system composed of the two charges. The work which must be expended in order to bring a unit charge up to a certain distance from the charge e is called the potential energy or *potential* of the charge e with respect to this point.

(b) The Bohr Theory

Since hydrogen, in ionic form, always bears one unit of electric charge, it is natural to assume that its nuclear charge is unity. The neutral hydrogen atom, which has the mass 1.008146, referred to oxygen* = 16.0000, then consists of one unit of positive charge, together with one electron (i.e., the unit of negative charge)

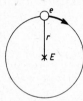

Fig. 18.
Electron e, re-
volving around
the nucleus E.

with the mass 0.000549. The electron is attracted by the nucleus according to Coulomb's law, and as it does not fall into the nucleus, another force must counteract this attraction. We will assume with Bohr (1913) [1] that the electron revolves around the nucleus, and is thereby kept in an equilibrium by the opposed action of the Coulomb attraction and centrifugal force (Fig. 18).

If we denote the charge of the nucleus by E, and that of the electron by $-e$ (in this case $E = e$, but for the present this will not be taken into account), then since the centrifugal force is equal to the product of the mass m, radius r, and the square of the angular velocity ω, the condition for equilibrium between centrifugal force and Coulomb attraction is given by

$$mr\omega^2 - \frac{Ee}{r^2} = 0 \qquad \text{or} \qquad Ee = mr^3\omega^2 \tag{6}$$

As it stands, this condition would permit of any value of r, according to the velocity of rotation of the electron. Such a system could not be stable, however, according to the Maxwell theory, whereby every alteration of the velocity or direction of motion of an electrically charged particle is associated with the radiation of electromagnetic energy. If the electron were to radiate energy, this could happen only at the expense of its kinetic and potential energy, the first being given by the square of the velocity, the latter by the distance from the nucleus. The electron would accordingly follow a contracting spiral orbit and must ultimately fall into the nucleus. Bohr accordingly rejected this consequence of the Maxwell theory, and replaced it by the postulate that there are certain orbits in which the electron may revolve without loss of energy (and in consequence without emission of light), and that these orbits should be such that the *orbital angular momentum* is a whole number multiple of $\dfrac{h}{2\pi}$.

By the *angular momentum* is meant the product of the momentum of a rotating body and its distance from the center of rotation. h is a constant, known from the quantum theory by the name of *Planck's quantum of action*. It has the value 6.6238 · 10^{-27} erg.sec.

* More accurately referred to the oxygen isotope ^{16}O = 16.0000.

The postulate stated above is known as *Bohr's first quantum condition*. It may be formulated

$$p = n \frac{h}{2\pi}$$

where p represents the orbital angular momentum, and n any whole number, the *quantum number* of the orbit in which the electron circulates. Remembering that

$$p = m \cdot v \cdot r = m \cdot r^2 \cdot \omega$$

we obtain

$$\frac{nh}{2\pi} = m \cdot r^2 \cdot \omega \tag{7}$$

and, dividing equation (6) by equation (7),

$$\frac{2\pi Ee}{nh} = r\omega \tag{8}$$

When substituted in (7), this gives

$$r = \frac{n^2 h^2}{4\pi^2 meE} \tag{9}$$

This equation implies that only *discrete orbits* are possible. Their radii are related as the squares of whole numbers which, as already indicated, are called the *quantum numbers* assigned to the orbits. Thus, for an orbit with the quantum number 3, the radius is given by substituting $n = 3$ in equation (9). As long as the electron moves in this orbit, according to Bohr, it emits no light. The radiation of light occurs when the electron 'jumps' from one orbit to another with a lower quantum number. The frequency of vibration of this light is fixed by *Bohr's second quantum condition*, the so-called frequency condition, which states that the energy difference between initial and final orbits is equal to the product of the frequency of the radiation and Planck's quantum of action:

$$\varepsilon_1 - \varepsilon_2 = h\nu \tag{10}$$

$\varepsilon_1 = $ energy of the atom in the initial state,
$\varepsilon_2 = $ energy of the atom in the finalstate,
$\nu \;= $ frequency of the radiation, $= c/\lambda$, where λ is the wave length and c the velocity of light.

According as the energy of the atom in the initial state is greater or less than that in the final state, equation (10) gives the frequency of light emitted or absorbed. The validity of this equation has been demonstrated by direct measurement of the energy which is necessary in order to excite the atom to particular radiations.

The energy ε of the electron circulating around the nucleus of the atom is made up of the kinetic energy ε_{kin} and the potential energy ε_{pot}:

$$\varepsilon = \varepsilon_{\text{kin}} + \varepsilon_{\text{pot}}$$

The kinetic energy is related to the mass m and velocity v of the electron by

$$\varepsilon_{\text{kin}} = \frac{1}{2} m \cdot v^2 = \frac{1}{2} m \cdot r^2 \omega^2 = \frac{n \cdot h}{4\pi} \omega \qquad \text{(from eq. (7))}$$

By substituting (9) in (8), we get

$$\omega = \frac{8\pi^3 m e^2 E^2}{n^3 h^3}$$

that is

$$\varepsilon_{\text{kin}} = \frac{n \cdot h}{4\pi} \cdot \frac{8\pi^3 m e^2 E^2}{n^3 h^3} = \frac{2\pi^2 m e^2 E^2}{n^2 h^2}$$

The potential energy ε_{pot} is

$$\varepsilon_{\text{pot}} = \frac{-eE}{r} = -\frac{4\pi^2 m e^2 E^2}{n^2 h^2} = -2\varepsilon_{\text{kin}}$$

The total energy is therefore

$$\varepsilon = \varepsilon_{\text{kin}} + \varepsilon_{\text{pot}} = \varepsilon_{\text{kin}} - 2\varepsilon_{\text{kin}} = -\frac{2\pi^2 m e^2 E^2}{n^2 h^2} \tag{11}$$

There is thus found

$$\nu = \frac{\varepsilon_1 - \varepsilon_2}{h} = \frac{2\pi^2 m e^2 E^2}{h^3} \left(\frac{1}{n_2^2} - \frac{1}{n_1^2} \right) \tag{12}$$

In this equation n_1 denotes the quantum number assigned to the atom in the initial state, n_2 that in the final state.

If we introduce the nuclear charge number $Z = E/e$, and also, for simplification, a quantity R_∞ defined by the equation

$$\frac{2\pi^2 m e^4}{h^3} = c \cdot R_\infty \qquad (c = \text{velocity of light})$$

then, since $c = \lambda \cdot \nu$,

$$\frac{1}{\lambda} = R_\infty \cdot Z^2 \left(\frac{1}{n_2^2} - \frac{1}{n_1^2} \right) \tag{13}$$

In the case of hydrogen, $E = e$, i.e., $Z = 1$. If R_∞ is computed*, we obtain exactly the same formula which reproduced the hydrogen line spectrum (equation (3), p. 82), namely

$$\frac{1}{\lambda} = 1.097 \cdot 10^5 \left(\frac{1}{n_2^2} - \frac{1}{n_1^2} \right) \tag{14}$$

Not only does equation (14) agree formally with (3), but within the limits of accuracy with which e, e/m and h are known, the correct value for the Rydberg constant has been thus obtained from the Bohr theory.

Strictly speaking, allowance should be made for the motion of the nucleus as the electron circulates about it. Just as the moon does not strictly revolve about a stationary earth, but the earth and moon both move about their common center of gravity, so also the nucleus and electron, in the case of the atom, revolve about their common center of gravity. The distances x of the nucleus, and y of the electron, from the centroid are in the ratio of the masses of the electron (m) and nucleus (M): $y : x = M : m$. To allow for the motion of the

---

* The value of h has already been given. e is the elementary quantum of electricity = $= 4.8029 \cdot 10^{-10}$ e.s.u.; the ratio $e/m = 1.7589 \cdot 10^7 \cdot c$, where c ($= 2.99793 \cdot 10^{10}$ cm per sec) is the velocity of light.

nucleus, the expression $Mm/(M + m)$ should replace m in equation (11), and correspondingly in (13), so that the latter in more exact form is

$$\frac{1}{\lambda} = R_\infty \frac{M}{M + m} Z^2 \left(\frac{1}{n_2^2} - \frac{1}{n_1^2} \right) \tag{13a}$$

The Rydberg constant for hydrogen is thus represented exactly by the expression $R_\infty \cdot M_H/(M_H + m)$, where M_H stands for the mass of the hydrogen nucleus. The greater M is in comparison with m, the less is the difference made in the Rydberg constant by allowance for the motion of the nucleus. For the hydrogen atom $M_H/(M_H + m) = 0.999455$.

The origin of the various hydrogen series spectra may now be explained as follows, on the basis of the Bohr concept just discussed.

The electron can move around the nucleus, without emitting radiation, in orbits having radii in the ratios $1^2 : 2^2 : 3^2$, etc. Normally it will be found in the innermost orbit ($n = 1$), since in this case the atom has the minimum total energy, according to equation (11). From equation (9), when $n = 1$, the radius of the innermost orbit is found to be $0.530 \cdot 10^{-8}$ cm. If the atom acquires energy—from inelastic collision or from absorption of radiation—the electron can be raised to a higher orbit. Suppose it were raised to the third orbit. If the energy is supplied by radiation, it must absorb a quantum of light of which the frequency (or wave length) is given by

$$\nu = \frac{E_2 - E_1}{h}, \qquad \frac{1}{\lambda} = R_H \left(\frac{1}{1^2} - \frac{1}{3^2} \right)$$

The atoms spend only a short time in states of higher energy ('excited states'). After 10^{-8} to 10^{-9} sec, on the average, the electron drops back to a lower orbit, with the emission of a quantum of light. In the case considered, it can jump back directly to the first level, and thereby emit light corresponding to the second line of the Lyman series, of a wave length given by

$$\frac{1}{\lambda} = R_H \left(\frac{1}{1^2} - \frac{1}{3^2} \right)$$

The electron might, however, fall back first to the second orbit and then to the ground state. In this case it would emit, one after the other, the first line of the Balmer series

$$\frac{1}{\lambda} = R_H \left(\frac{1}{2^2} - \frac{1}{3^2} \right)$$

and the first line of the Lyman series

$$\frac{1}{\lambda} = R_H \left(\frac{1}{1^2} - \frac{1}{2^2} \right)$$

It may be seen that electron transitions which correspond to lines belonging to the same series always have the same lower state (end state). For the lines of the Balmer series, for example, this is the second quantum state, and for those of the Lyman series the first, or ground state. Fig. 19 represents the transition of the electron from one orbit to another, and gives the emission lines corresponding to the transitions. The converse transitions are, of course, to be ascribed to the corresponding absorption lines.

The number n in each term is the *quantum number* of the orbit corresponding to

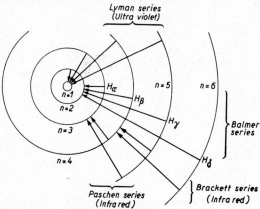

Lyman series
(Ultra violet)

Balmer series

Paschen series
(Infra red)

Brackett series
(Infra red)

Fig. 19. Origin of the various series in the hydrogen spectrum, according to Bohr's theory.

each term. According to the magnitude of n, one distinguishes between one quantum, two quantum, three quantum orbits, etc.

(c) Principal and Subsidiary Quantum Numbers

For the sake of simplicity we have assumed, initially, that the electron moves about the nucleus in circular orbits. We already know, however, from planetary orbits, that motion about a center of attraction can also take place in elliptical orbits.

The theory of this kind of motion about the atomic nucleus was developed by Sommerfeld in 1915. Although only *one* parameter need be given for the construction of a circle (e.g., the radius), *two* parameters are required to specify an ellipse—e.g., the major and the minor axes.

Correspondingly, in place of the one quantum number of the original Bohr theory, two quantum numbers n and k, which are termed the principal and the subsidiary quantum number, are needed in Sommerfeld's theory. The principal quantum number n gives the semi-major axis a of the ellipse (Fig. 20) in the same way as the radius does for a circular orbit (cf. equation (9)). It is thus

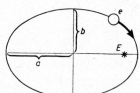

Fig. 20. Electron moving in an elliptical orbit (3_2 orbit).

$$a = \frac{n^2 h^2}{4\pi^2 meE} \tag{15}$$

The semi-minor axis b of the ellipse is given by the relation

$$\frac{b}{a} = \frac{k}{n} \; (k \leqq n) \tag{16}$$

For $k = n$, $b = a$, i.e., the electron moves in a circular orbit. For $k < n$, the electron moves in an elliptical orbit, the eccentricity being greater, the smaller k is as compared with n. It is usual to denote orbits characterized by the principal quantum number n and the subsidiary quantum number k as n_k orbits. Thus an orbit with principal quantum number = 3, subsidiary quantum number = 2, as represented in Fig. 20, is a 3_2 orbit.

In the absence of any external field, such as a strong magnetic field, and when no account is taken of the variation of the mass of the electron with its velocity, as required by the Theory of Relativity, the energy of the hydrogen atom according to the Sommerfeld theory is determined solely by the principal quantum number. Just the same spectral line is emitted—e.g., the red H_α line ($\lambda = 6562.8$ Å)— irrespective of whether the electron jumps from the orbit 3_2 to the orbit 2_2, from the orbit 3_2 to the orbit 2_1, or from the orbit 3_1 to the orbit 2_2.

The *quantum numbers* give not only the radii of the orbits in which the electrons move, according to the Bohr model (equations (15) and (16)), but also, in accordance with equation (11), the energy states which the atom can assume. These energy states are related, as shown by equation (11), as the reciprocals of the squares of the corresponding quantum numbers. The production of the various lines of the hydrogen spectrum can thus be described in terms of transitions from one energy state (*energy level*) to another, according to Fig. 21, instead of by transitions from one orbit to another, as represented in Fig. 19.

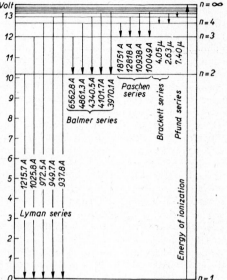

Fig. 21. Energy levels of the hydrogen atom.

[10–13, 15] The description in terms of energy states will be used exclusively in the subsequent pages, since according to wave mechanics (see below), the concept of definite electronic

[The energies of the H atom corresponding to the various quantum numbers ($n = 1, 2, 3$, etc.) are given in *e*-volts. For the significance of this measurement of energy, and the relation of the quantum number $n = \infty$ to the ionization energy, see p. 112 *et seq*.]

orbits must be given up, whereas the energy levels defined as in equation (11) retain their meaning in the newer theory.

An increase in the accuracy of spectroscopic measurement to the highest possible degree has shown that the lines of the Balmer series are made up in reality of several (at least two) lines, differing in wave length from each other only by an extremely minute extent. For the H_α line the difference is only about 0.05 Å, or less than 0.001 % of the wavelength. As Sommerfeld showed, the Bohr theory is able to explain, in principle, even these fine points of the spectrum. If elliptical orbits are permitted, and allowance is made for the variation of the mass of the electron with its velocity, as required by the Theory of Relativity, an expression for the energy E is obtained into which the subsidiary quantum number also enters. The energy levels of Fig. 21 are thereby split into sub-levels. Although these sub-levels are, indeed, extraordinarily close together, the transitions $3_3 \rightarrow 2_2$, $3_2 \rightarrow 2_1$, $3_1 \rightarrow 2_2$, no longer give rise to identically the same frequencies (wave lengths) on the basis of equation (10). In this respect, the Bohr-Sommerfeld theory admittedly no longer fits the experimental observations completely and quantitatively, and its inadequacy becomes obvious when it is applied to heavier atoms.

(d) Inadequacy of the Bohr Theory. Wave Mechanics and Quantum Mechanics

Although the agreement between the Bohr theory of the hydrogen atom and the spectroscopic data for the Balmer series and other series spectra of hydrogen is so excellent, it has nevertheless been found that the range of application of the Bohr theory is limited. Thus it does not apply at all to the behavior of hydrogen atoms in a magnetic field, nor has a satisfactory interpretation of the hydrogen *molecule* been possible on the basis of the Bohr theory.

These difficulties appear to be fundamental in nature, rather as it is impossible to derive the limits of resolving power of the microscope from geometrical optics alone (i.e., without taking account of the phenomena of diffraction). A way out of the difficulties indicated can be found, on the basis of a theory of wider scope, based on the theory of material waves— namely *wave mechanics* or *quantum mechanics*. [5–9, 14] Wave mechanics had its origin in the work of de Broglie (1924), and was shortly afterwards (1926) applied by Schrödinger to the theory of the hydrogen atom. It describes motion of material particles, such as electrons, in terms of the differential equations of wave motion, just as the properties of rays of light are described in the undulatory theory of light.* Bohr's first quantum postulate, which appeared, in the setting of the Bohr theory, to be introduced arbitrarily, turns out, when considered from the standpoint of quantum mechanics, to be a necessary consequence of the laws applying generally to wave systems. The second postulate of Bohr can be deduced from wave mechanics in a similar way. The same equations are obtained for the spectrum as on the basis of Bohr's assumptions. In addition to this, however, wave mechanics confers on the hydrogen atom a symmetry corresponding to its behavior in a magnetic field; it leads to a satisfactory understanding of the hydrogen molecule; and it gives, for the heavier atoms, values of the ionization potential which are in better agreement with measured values than are those derived from the Bohr theory.

The *quantum mechanics* developed by Heisenberg, Born and Jordan (1925) leads to the same results as those which are obtained on the basis of wave mechanics. This approach dispenses with readily visualized physical concepts, and derives purely mathematical relations between the directly observed quantities of atomic physics—in particular, the frequencies and intensities of the spectral lines characteristic of the atoms. It has been possible to show that quantum mechanics and wave mechanics are equivalent to one another mathematically—that is to say, the formulas of the one theory can be derived from the other theory by purely mathematical transformations. Wave mechanics, however, starts out from rather more concrete ideas than does quantum mechanics. The results derived from it, regarding the structure of atoms and molecules, are therefore more easily translated into physical concepts.

The wave mechanical atom model does not, however, have quite the same direct physical significance as the Bohr-Sommerfeld theory of the atom. In the following pages, we shall therefore make use of the more readily grasped Bohr model wherever possible. Provided the limits of applicability of this model are remembered, this can be done with the same justification as the conceptions of geometrical optics are customarily employed in interpreting the action of lens systems, such as the telescope or microscope. In considering the action of these optical instruments, it is usual to use the expressive and very simple model of 'rays of light', as infinite thin lines, even though we know that we cannot explain all phenomena by this model, and that as soon as diffraction phenomena enter, for example, we must revert to the undulatory theory, which it is indeed also possible to visualize, though less simply than geometrical optics. In the same way, we shall make use of the instructive models of the structure of the atom, which the Bohr theory and its later developments have provided, wherever this leads to results which do not contradict those derived from quantum mechanics.

Chemistry, however, poses a number of very important problems which cannot be treated on the basis of the Bohr-Sommerfeld theory. In such cases, we cannot dispense with the results obtained by the use of wave mechanics. This is true, in particular, of the problems connected with the *chemical bond*. It is convenient at this point to summarize briefly the fundamental principles of wave mechanics, and their application to the theory of atomic structure, so that it is possible to understand these problems and the way in which they have been solved.

---

* The conflict between the corpuscular and the undulatory theory of light loses its meaning when viewed from the quantum-mechanical standpoint.

3. The Hydrogen Atom in Wave Mechanics

The same lines are found for the spectrum of the hydrogen atom according to wave mechanics as from the Bohr-Sommerfeld theory. Nevertheless, the conception of the structure of the hydrogen atom derived from wave mechanics differs very considerably from that of the Bohr-Sommerfeld theory. In Bohr's model, the electron moves in a circular orbit, in the ground state of the atom, so that the atom possesses circular (or disc) symmetry. According to wave mechanics, it possesses spherical symmetry, not only in the ground state, but also in those excited states in which, according to Sommerfeld, the electron moves in highly eccentric orbits (namely, those with the auxiliary quantum number $k = 1$). The conception of the structure of the H-atom based on wave mechanics has been found to accord with experiment. In the following section we consider how these conceptions follow as a necessary consequence of the theory, and what meaning must be attached to them.

(a) Basic Ideas of Wave Mechanics

According to wave mechanics, a wave motion is associated with every moving electron (and, indeed, with every moving massive particle). The wave length λ of this wave motion is given by

$$\lambda = \frac{h}{mv} \tag{17}$$

where h is Planck's quantum of action, m the mass and v the velocity of the massive particle. This assumption, basic for wave mechanics, was originally put forward as a hypothesis by de Broglie. It was confirmed by experiment, when it was shown (first by Davisson and Germer in 1927) that cathode rays were diffracted when they passed through matter, just like X-rays (cf. Chap. 7). Analogous diffraction phenomena can be observed also for atom or molecular rays, as was first proved experimentally by O. Stern (1929).

The 'material wave' propagates itself in space with a velocity u, which is given by

$$uv = c^2 \tag{18}$$

(v = velocity of particle, c = velocity of light)

As with all waves, the frequency ν of the material wave is related to its velocity of propagation u and wave length λ by the expression $u = \nu\lambda$.

(b) Equation for Vibration, and the Wave Equation

In the simplest case, a vibration can be represented by the equation

$$y = a \sin (2\pi\nu t) \tag{19}$$

In this expression, a is the *amplitude* of the vibration—i.e., if we apply the equation to the easily visualized case of the vibration of a massive particle, the maximum displacement from the rest position* attained by the particle in the course of its vibrations, and ν is the *frequency* (number of vibrations per second).

---

* The 'rest position' is the position which the particle would take up if it were not excited to vibration. As it vibrates, the particle possesses its maximum velocity each time it passes through the rest position.

Equation (19) applies, for example, to the vibration of a pendulum. It also represents the vibration of a string, the periodic fluctuations of density of the air due to a sound, or an electromagnetic vibration—always provided that these are cases of *simple* vibrations—i.e., vibrations with a single frequency and constant amplitude.

Equation (19) is true quite generally for the case of simple harmonic motion— i.e., when the restoring force on the particle is proportional at every moment to the displacement. In this case, the acceleration of the particle at any moment is proportional to y, or—denoting the factor of proportionality by ω^2,

$$\frac{d^2y}{dt^2} = -\omega^2 y \qquad (20)$$

The negative sign on the right hand side of the equation comes from the fact that the force acts in the opposite direction to y. A vibration described by (20) is a *harmonic vibration.*

If we take as time zero some instant at which the vibrating particle passes through its position of rest (i.e., if we substitute $t = 0$ for $y = 0$), the foregoing differential equation has the solution

$$y = a \sin \omega t. \qquad (21)$$

a is here a constant, and its physical significance may be seen from the following considerations.

It can be shown that eqn. (21) is a solution of eqn. (20), by double differentiation of the latter, whereby eqn. (21) is obtained:

$$\frac{dy}{dt} = a\frac{d\,(\sin \omega t)}{dt} = a\omega.\cos \omega t;$$

$$\frac{d^2y}{dt^2} = a\omega\frac{d\,(\cos \omega t)}{dt} = -a\omega^2.\sin \omega t = -\omega^2 y.$$

The relation between y and t expressed in eqn. (21), is shown graphically in Fig. 22. It may be seen that the vibrating point passes through its position of rest ($y = 0$) at definite, equal intervals of time, and does so alternately from both directions (upwards and downwards). The interval of time between one transit through the position of rest and the next transit in the same direction is said to be the *period of vibration.* If this is denoted by τ, it is at once apparent from Fig. 22 that $y = \sin \omega t$ assumes the value $y = 0$ for $t = 0$, $t = \dfrac{\tau}{2}$, $t = \tau$, $t = \dfrac{3}{2}\tau$, etc. Since $\sin \omega t = 0$ for $\omega t = 0$, $\omega t = \pi$, $\omega t = 2\pi$, etc, expressing ωt in angular measure, we have that $\omega \tau = 2\pi$. If, in place of τ we substitute its reciprocal, the frequency ν, then $\omega = 2\pi\nu$. If this is substituted in eqn (21), we obtain eqn. (19). This shows that eqn. (19) represents the vibration of a massive particle about its position of rest.

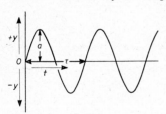

Fig. 22. Graphical representation of the function $y = a \sin \omega t$ (periodical vibration).

$$\omega = \frac{2\pi}{\tau}$$

(τ = period of vibration)

If we differentiate eqn. (19) twice, or substitute $2\pi\nu$ for ω directly in eqn. (20), we obtain

$$\frac{d^2y}{dt^2} + 4\pi^2\nu^2 y = 0 \qquad (22)$$

This equation may be called a *vibration equation*, as distinct from the *wave equation* considered later.

The maximum value of y is attained for $t = \dfrac{\tau}{4}$, $t = \tau + \dfrac{\tau}{4}$, etc., as can be seen directly from Fig. 21. The result can be obtained analytically from eqn. (19) by substituting the maximum value for sin $2\pi\nu t$ $\left(\text{i.e., sin } 2\pi\nu t = \sin\dfrac{\pi}{2} = 1, \text{ or } t = \dfrac{1}{4\nu} = \dfrac{\tau}{4}\right)$. Substituted in eqn. (19), this gives $y = a$ as the maximum value of y, or as the amplitude of the vibration.

If the conditions are such that the vibratory motion can be propagated through space, a *wave* is set up. The wave length—i.e., the shortest distance between two points which are in the same state of vibration, is given by $\lambda = \dfrac{u}{\nu}$, where u is the velocity of propagation of the wave (wave velocity) and ν the frequency (number of vibrations per second). The wave velocity along a certain direction—e.g., in the x-direction—is given by $u = \dfrac{x}{t}$. This holds on the assumption that u is constant. If $t = \dfrac{x}{u}$ is substituted in eqn. (19), we obtain

$$y = a \sin\left(\frac{2\pi\nu}{u}x\right) \tag{23}$$

This equation describing the propagation of a vibration along one axis of a coordinate system, which can be written in the alternative form $y = \sin\dfrac{2\pi}{\lambda}x$, is in every way analogous to the equation for the periodic course of a vibration. In the graphical representation of the function (Fig. 23), the wave is shown on both sides of the origin of coordinates.* The significance of this is that, in general, a vibration does not spread in only one direction in space but is propagated simultaneously in opposite directions. In eqn. (23), a denotes the amplitude, as it does in eqn. (19).

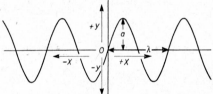

Fig. 23. Graphical representation of the function

$$y = a \sin\frac{2\pi x}{\lambda} \qquad (a \; wave)$$

By double differentiation of eqn (23), we obtain

$$\frac{d^2y}{dx^2} = -a\left(\frac{2\pi\nu}{u}\right)^2 \sin\left(\frac{2\pi\nu}{u}x\right) = -\left(\frac{2\pi\nu}{u}\right)^2 y, \qquad \text{or}$$

$$\frac{d^2y}{dx^2} + \frac{4\pi^2\nu^2}{u^2}y = 0 \tag{24}$$

This equation describes the linear propagation of a vibration. It is therefore known as the *linear wave equation*.

It may be shown that at every moment the *total energy* of a vibrating massive particle (i.e., the sum of the kinetic and potential energy) is proportional to the square of the

---

* This could also have been done in Fig. 22, since equation (19) contains nothing to specify at what instant the vibration commenced. Just as either positive or negative values of x can be inserted in eqn. (23), either positive or negative values of t could be substituted in eqn. (19).

amplitude of its vibrations. In other types of vibration—e.g., light and sound waves— the square of the amplitude is a measure of its magnitude, and is known as the *intensity* of the wave.

(c) General Solution of the Vibration Equation

Equation (21) is not the only solution of the differential equation (20). By changing the 'boundary conditions'—namely by taking $y = a$ at $t = 0$ (instead of $y = 0$), the solution found is $y = a \cos \omega t$. We obtain a *general solution* (i.e., one which does not depend on the boundary conditions) if we substitute $y = a\,e^{at}$, where a is a constant which is for the moment not specified. We then obtain from eqn. (20).

$$a a^2 e^{at} + a\omega^2 e^{at} = 0, \qquad \text{or} \qquad a e^{at}(a^2 + \omega^2) = 0$$

In order that this equation may hold,

$$a^2 + \omega^2 = 0, \qquad \text{or} \qquad a = \pm i\omega \qquad (i = \sqrt{-1})$$

The corresponding solutions of (20) are thus

$$y_1 = a_1\, e^{+i\omega t} \qquad\qquad y_2 = a_2\, e^{-i\omega t}$$

According to de Moivre's theorem, $e^{ix} = \cos x + i \sin x$, so that we may write $y_1 = a_1(\cos \omega t + i \sin \omega t)$; $y_2 = a_2(\cos \omega t - i \sin \omega t)$. y_1 and y_2 are independent solutions of the differential equation (20). The expression obtained by adding these solutions—

$$y = (a_1 + a_2) \cos \omega t + (a_1 - a_2)\, i \sin \omega t$$

is also a solution of (20). Writing $a_1 + a_2 = A_1$ and $(a_1 - a_2)i = A_2$, and replacing ω by $2\pi v$, we obtain

$$y = A_1 \cos 2\pi v t + A_2 \sin 2\pi v t \tag{25}$$

This is the general solution of the vibration equation (22). A_1 and A_2 are, in general, complex factors defining the amplitude. Their numerical values are defined by the boundary conditions of the physical problem.

(d) Three Dimensional Vibration Equation

In eqn. (22), y is a variable which depends only on the time t. The form of the equation remains unaltered if y is replaced by a function Ψ, which depends on the three spatial coordinates, x, y, z, as well as on the time, except that the ordinary differential is replaced by a partial differential expression. The 3-dimensional vibration equation is therefore

$$\frac{\partial^2 \Psi}{\partial t^2} + 4\pi^2 v^2 \Psi = 0 \tag{26}$$

Its general solution (cf. eqn. (25)) is

$$\Psi = A_1 \cos 2\pi v t + A_2 \sin 2\pi v t \tag{27}$$

A_1 and A_2 are again the amplitudes of the vibrations, but are in this case functions of x, y and z, instead of being constants.

The *intensity* of the vibration (cf. above) is given in this case by

$$|\Psi|^2 = \Psi . \Psi^* \tag{28}$$

In (28), $|\Psi|$ is the 'absolute' magnitude of Ψ,—i.e., the size of Ψ without reference to its sign. Ψ^* is the complex conjugate* of Ψ.

(e) Three Dimensional Wave Equation

We obtain the three dimensional wave equation from eqn. (24), by replacement of the variable y (which depends on x only) by a function ψ which is a variable of all three spatial coordinates, at the same time writing the sum $\dfrac{\partial^2\psi}{\partial x^2} + \dfrac{\partial^2\psi}{\partial y^2} + \dfrac{\partial^2\psi}{\partial z^2}$ in place of $\dfrac{d^2y}{dx^2}$.

The three dimensional wave equation is therefore

$$\frac{\partial^2\psi}{\partial x^2} + \frac{\partial^2\psi}{\partial y^2} + \frac{\partial^2\psi}{\partial z^2} + \frac{4\pi^2 v^2}{u^2}\psi = 0 \tag{29}$$

An equation representing the variation of the fluctuating quantity with both time and spatial coordinates is obtained if ψ in the wave equation is replaced by Ψ. By combination with the wave equation we then obtain

$$\frac{\partial^2\Psi}{\partial x^2} + \frac{\partial^2\Psi}{\partial y^2} + \frac{\partial^2\Psi}{\partial z^2} - \frac{1}{u^2}\frac{\partial^2\Psi}{\partial t^2} = 0 \tag{30}$$

It is shown in theoretical physics that every quantity which can be represented as a function of time and space by the foregoing equation can spread put as a wave motion, u being the velocity with which the waves travel. Conversely, any quantity which travels as a wave can be represented by the above equation. It may be noted that Huygens' principle, which is the basis of the explanation of diffraction phenomena and wave optics can be derived directly from eqn. (30). The point of immediate importance, however, is that eqn. (30) can be applied to material waves. The physical significance of the constant u in eqns. (29) and (30) can be seen from the equations. It represents the velocity of propagation of the material wave; this is not a constant quantity (in empty space), like the velocity of light, but depends upon the velocity v of the material particle with which the wave motion is associated, as described by eqn. (18). The significance to be attached to the quantity Ψ will be discussed later.

(f) The Schrödinger Equation

The energy ε of any moving particle is equal to the sum of its kinetic energy $(= \frac{1}{2} mv^2)$ and its potential energy ε_{pot}. Thus $\varepsilon = \frac{1}{2} mv^2 + \varepsilon_{pot}$. From eqn. (17) and $u = v\lambda$, it follows that $\dfrac{v^2}{u^2} = \dfrac{m^2v^2}{h^2}$. Substituting $mv^2 = 2(\varepsilon - \varepsilon_{pot})$, it follows that $\dfrac{v^2}{u^2} = \dfrac{2m(\varepsilon - \varepsilon_{pot})}{h^2}$. This may be substituted in the wave equation (29). The resulting equation is the Schrödinger equation, which is of fundamental importance in the application of wave mechanics to atomic structure:

$$\frac{\partial^2\psi}{\partial x^2} + \frac{\partial^2\psi}{\partial y^2} + \frac{\partial^2\psi}{\partial z^2} + \frac{8\pi^2 m}{h^2}(\varepsilon - \varepsilon_{pot})\psi = 0 \tag{31}$$

(g) The Hydrogen Atom in Wave Mechanics

We start from the fact that the hydrogen atom consists of a nucleus, with a charge E, and one electron with the charge $-e$. As on p. 84, we shall at first take

---

* The quantity $a^* = x - iy$ is the *complex conjugate* of the quantity $a = x + iy$. The square root of the product $a \cdot a^*[= \sqrt{(x^2 + y^2)}]$ has the same absolute magnitude as either a or a^*.

no cognizance of the fact that $E = e$, in order to make the results directly applicable to the general case of atoms with a higher nuclear charge.

If the electron is at a distance r from the nucleus of the atom, its potential energy is $\varepsilon_{pot} = -\dfrac{eE}{r}$. If this value is inserted in eqn. (31), we obtain

$$\frac{\partial^2\psi}{\partial x^2} + \frac{\partial^2\psi}{\partial y^2} + \frac{\partial^2\psi}{\partial z^2} + \frac{8\pi^2 m}{h^2}\left(\varepsilon + \frac{eE}{r}\right)\psi = 0 \tag{32}$$

It has long been known that certain differential equations can be solved only if the parameters occurring in the equation (that is, the arbitrary constants occurring in it) have certain definite values. The particular values of the parameters for which it is possible to find a solution of the equation which is everywhere finite, and which converges sufficiently rapidly to zero at infinity*, are known as the *eigenvalues* of the differential equation, and the corresponding solutions are called *eigenfunctions*.

Thus the differential equation $\dfrac{d^2y}{dx^2} + (a - x^2)y = 0$ can be given a definite solution only if a is an odd integer. The parameter of the equation is a, and the eigenvalues of the parameter are $1, 3, 5, \ldots$ etc. If $a = 1$, the solution obtained is $y = e^{-\frac{x^2}{2}}$. The expression $e^{-\frac{x^2}{2}}$ is thus the eigenfunction corresponding to the eigenvalue $a = 1$. The eigenvalues 3 and 5 correspond to the eigenfunctions $2x \cdot e^{-\frac{x^2}{2}}$ and $(4x^2 - 2) \cdot e^{-\frac{x^2}{2}}$. It may readily be seen, by double differentiation, that these expressions are, in fact, solutions of the above differential equation, corresponding to the eigenvalues $1, 3, 5$ respectively. A strict mathematical proof that this differential equation has no solutions which satisfy the given criteria of uniqueness, except when a is an odd integer would be quite complicated.

Schrödinger demonstrated that eqn. (32) is uniquely soluble for all positive values of ε, but only for certain discrete negative values of ε. These are the values given by the expression

$$\varepsilon = -\frac{2\pi^2 m e^2 E^2}{n^2 h^2} \tag{33}$$

where n is any whole number. This means that *the energy of an electron bound to an atomic nucleus can have only certain discrete values.* This is so, since the energy ε of such an electron is *negative***. The states that the atom can assume without the electron being split off are known as the *stationary states* of the atom.

The same discrete energy values are obtained for the stationary states of the

---

* These are the conditions for definite solutions.

** That this is so follows from the expression $\varepsilon = \frac{1}{2}mv^2 - \dfrac{eE}{r}$. The electron is only bound firmly by the nucleus if its kinetic energy is smaller than the work involved in removing the electron altogether—i.e., if $\frac{1}{2}mv^2 < \dfrac{eE}{r}$ i.e., $\varepsilon < 0$. A *free* electron possesses only *kinetic* energy (since $r = \infty$). This is always positive and a free electron can therefore acquire any velocity, without restriction (up to the velocity of light).

atom on the basis of the Schrödinger equation as from the Bohr model.* However, whereas the Bohr theory derived these values only by introducing a quite unproven postulate (that the electron can circulate round the atom only in certain discrete orbits without radiating energy), they follow as a necessary consequence of the basic ideas of wave mechanics, as solutions of the Schrödinger equation, without any auxiliary hypotheses.

In wave mechanics, the stationary states of the atom are *those states in which the electron waves in the Coulomb field of the atom are not everywhere extinguished by interference.* An electron moving in the field of force of a nucleus, and not possessing sufficient kinetic energy to get out of that field, must necessarily stay close to the nucleus. Its energy must then have a value conforming to the conditions laid down by eqn. (33).

The Bohr frequency condition $(\varepsilon_1 - \varepsilon_2 = h\nu)$ can also be deduced directly, without additional postulates, from wave mechanics or quantum mechanics. In conjunction with eqn. (33) this gives the frequencies of the lines of the hydrogen spectrum, as was considered on p. 86).

(h) Principal and Subsidiary Quantum Numbers in Wave Mechanics

The values found for ε from eqn. (32) are the *eigenvalues* of the Schrödinger equation, as it applies to the hydrogen atom. To each one of these eigenvalues, characterized by the principal quantum number n, there correspond, in general, *several eigenfunctions.* These are distinguished from one another by the values of two additional quantum numbers, l and m, which are also integers. These are called *subsidiary quantum numbers.*

The subsidiary quantum number l in wave mechanics corresponds to the Bohr subsidiary quantum number k; however, it does not have the same values, but $l = k - 1$. l differs from k, and can have a minimum value of zero, because it serves as a measure of the *angular momentum* of the atom. l is therefore known as the *angular momentum* (or simply as the *angular*) *quantum number.* The angular momentum of the atom is given by wave mechanics as $\sqrt{l(l+1)}\,\dfrac{h}{2\pi}$. If the angular momentum of the atom is zero, $l = 0$.

The subsidiary quantum number m takes account of the fact that the spectral lines emitted by the excited atom are split up into several components under the influence of a magnetic field (Zeeman effect). It is therefore known as the *magnetic quantum number.* If the atom is not subject to the action of any magnetic field, its energy state is independent of the value of m.

The influence of the subsidiary quantum numbers on the energy of the atom in its stationary states does not appear in eqn. (33). Indeed, in so far as the assumptions underlying the derivation of eqn. (32) are valid, the energy of the atom is independent of l and m. The effect of a magnetic field is not taken into consideration in deriving eqn. (32). Hence the energy of an atom is not dependent upon the subsidiary quantum number m *as long as the atom is not situated in an external electromagnetic field.* Furthermore, the mass of the electron m is assumed to be constant in deriving eqn. (32). It may be shown that this can not be exactly true, when account is taken of the theory of relativity. A relativistic treatment leads to the conclusion that there are small differences of energy between states with the same principal quantum number but different values of l; these lead to the very slight splitting of the

---

* For an accurate calculation, the reduced mass $\mu = \dfrac{mM}{m+M}$ must be substituted in eqns. (32) and (33) in place of m, just as was done in eqn. (11).

spectral lines of hydrogen, to which reference was made on p. 89.* The effect of the angular momentum quantum number l on the magnitude of ε shows itself more clearly when the relation $\varepsilon_{\text{pot}} = -\dfrac{Ee}{r}$ is no longer strictly valid, because of the perturbation of the Coulomb field between electron and atomic nucleus. This perturbation can be produced by the action of an external electrical field (Stark effect). It must also occur if the atom contains more than one electron. Hence in all atoms of higher nuclear charge than hydrogen—except in states where all except one electron has been stripped off—the subsidiary quantum number l exerts a distinct, and often very great effect on the energy levels of the atom, and therefore on the spectrum lines.

(i) Significance of the ψ Function

The function ψ which appears in the Schrödinger eqn. (31) or (32) can be interpreted as follows. $\psi \cdot \psi^* \cdot dv = |\psi|^2 dv$ is a *measure of the probability of finding the electron in some element of volume dv close to the atom.* This probability can be considered

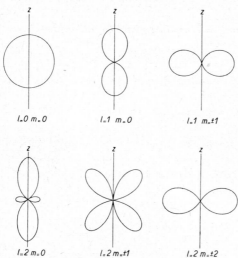

l₌0 m₌0   *l*₌1 m₌0   *l*₌1 m₌±1

l₌2 m₌0   *l*₌2 m₌±1   *l*₌2 m₌±2

Fig. 24. Probability (density) distribution function for the electron in various stationary states of the H atom, plotted as a function of direction (*angular distribution function*). (After H. E. WHITE, *Introduction to Atomic Spectra*, 1934). [The distribution function has rotational symmetry about the vertical axis of each figure, in the plane of the drawing.]

as a function of the distance of the electron from the nucleus, or as a function of direction in space. The latter is represented in Fig. 24 by an angular distribution function, for various states characterized by the quantum numbers as shown. The significance of the curves reproduced in Fig. 24 is that if a straight line is drawn from the origin to some point on one of these curves, the length of the line is a measure of the probability of finding the electron along the direction of the line. The curves apply to hydrogen and to 'hydrogen-like' atoms—i.e., all atoms which are ionized so far that they contain only one bound electron. The curves give the probability distribution in the plane of the paper; the probability distribution in space is obtained by rotating the curves about their z-axis, (represented as vertical and in the plane of the paper). The hydrogen atom thus has rotational symmetry in all its stationary states. In all states characterized by $l = 0$ it is, indeed, spherically symmetrical, whatever the value of the principal quantum number n. The fact that different values of l and m correspond to different probability distributions for the electron accords with the statement that the same values for the eigenvalues in eqn. (32) lead, in general, to several eigenfunctions. It may also be seen that the case $n = 1$ is the only one which leads to a single eigenfunction.

* The spin quantum number considered in the next chapter also enters into this 'fine structure'.

Fig. 25 represents the probability of finding the electron as a function of distance r from the nucleus (radial distribution function), for those states in which the atom has spherical symmetry (i.e., for $l = 0$). Distance from the nucleus (in Å) is plotted as abscissa, and the corresponding values of $r^2\psi\psi^*$ as ordinate. This expression

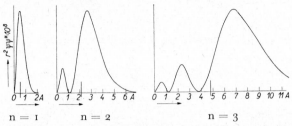

Fig. 25. Probability (density) distribution function for the electron in stationary states of the H atom, as a function of distance r from the nucleus (*radial distribution function*).
[The probability of finding the electron within a spherical shell, of radius r from the nucleus, and thickness dr, is given by $4\pi r^2\psi\psi^* dr$. The scale of ordinates is not the same in the various figures, and is arbitrarily chosen to make the principal maximum the same height in all the curves.]

(multiplied by $4\pi dr$) gives the probability of finding the electron at a distance r (more exactly, within a thin spherical shell of thickness dr) from the nucleus. The figure shows that in every stationary state the electron is not restricted in any way to a single orbit of definite radius, as is postulated in the Bohr model. It does not move over a single spherical surface, but may occur at any distance from the nucleus. There is, however, a certain spherical shell within which the probability function has a maximum value. Outwards from this, the probability falls off rapidly to zero. Inwards, there may be additional, but appreciably smaller, maxima in the probability function, separated by spherical shells in which the probability function has the value zero. These are 'nodal surfaces' of the stationary wave system set up in the neighborhood of the nucleus. They correspond to the nodal points of a (one dimensional) vibrating string. In some circumstances, there may be nodal surfaces which pass through the mid-point of the system; these, of course, do not have spherical symmetry, but only rotational symmetry. The number of such nodal surfaces is l, and the total number of nodal surfaces is $(n-1)$. Thus in wave mechanics the principal quantum number n and the angular quantum number l serve to define the number and kind of nodal surfaces in the electron density distribution.

It is permissible to think of the spatial distribution of the electron density probability function as a sort of cloud surrounding the nucleus. This 'electron cloud' may have rotational symmetry or spherical symmetry, depending on the value of l. One can also think of a 'boundary' (albeit very much smeared out) to the electron cloud, since the density falls off very rapidly after attaining its maximum value. The symmetry of the electron cloud defines, in a sense, the 'shape' of the atom. Since the electron cloud will tend to have its greatest extension in that direction in which the probability of finding the electron is greatest, the solids of revolution obtained by rotating the angular distribution functions of Fig. 24 about their z axes give a rough picture of the electron clouds. It is only a very schematic picture, since the clouds do not have a 'boundary' like the surface swept out by the curves, and because—except in the case of spherically symmetrical electron density distributions—the density distribution within the cloud is not the same in all radial directions. The concept of an electron cloud surrounding the nucleus can therefore be looked upon as a purely pictorial way of looking at the electron density probability function. As has already been stated, the electron moves about the nucleus in closed orbits, in wave mechanics as in classical mechanics. At any instant, therefore, its location is definite. It is inherently impossible to determine this position exactly (Heisenberg uncertainty principle), so that it is not possible to specify more than the probability that the electron is in a certain position, or the relative frequency with which the electron would be encountered in that element of volume.

(j) Application of the Schrödinger Equation to Atoms with Several Electrons

Equations corresponding to (32) could, in principle, be set up for atoms with several or many electrons. However, the mathematical treatment of such problems becomes much more difficult even when only 2 electrons are involved. In such a case, the second partial derivatives for each particle, for each of the three spatial coordinates, must be substituted in eqn. (31)—i.e., for 6 spatial coordinates in all. ψ then becomes a function in 6-dimensional space. With a further increase in the number of electrons, the mathematical difficulties increase correspondingly, so that it becomes impossible to find an exact solution for the Schrödinger equation, as applied to the many-electron problem. It is, however, often possible to make certain simplifying assumptions, and many results of considerable importance in chemistry have obtained from wave mechanics or quantum mechanics, by the introduction of suitable simplifying assumptions.

References

1 N. Bohr, *The Theory of Spectra and Antomic Constitution*, 2nd Ed., Cambridge 1925. 148 pp.
2 A. Haas, *Atomtheorie*, 3rd Ed., Berlin 1936, 292 pp.
3 G. Herzberg, *Atomic Spectra and Atomic Structure* (translated by J. W. T. Spinks), New York 1937, 257 pp.
4 E. N. da C. Andrade, *The Structure of the Atom*, London 1934, 750 pp.
5 A. Haas, *Materiewellen und Quantenmechanik*, 5th Ed., Leipzig 1934, 299 pp.
6 L. de Broglie, *An Introduction to the Study of Wave Mechanics* (translated by H. T. Flint), London 1930, 250 pp.
7 S. Flügge and A. Krebs, *Experimentelle Grundlagen der Wellenmechanik*, Dresden 1936, 238 pp.
8 C. Schaefer, *Quantentheorie (Einführung in die theoretische Physik*, Vol. III, Part 2), Berlin 1937, 510 pp.
9 A. E. Ruark and H. C. Urey, *Atoms, Molecules and Quanta*, London 1930, 780 pp.
10 F. Hund, *Linienspektren und Periodisches System der Elemente*, Berlin 1927, 221 pp.
11 W. Grotrian, *Graphische Darstellung der Spektren von Atomen und Ionen mit ein, zwei und drei Valenzelektronen*, Berlin 1928, Vol. I, 245 pp.; Vol. II, 168 pp.
12 R. F. Bacher and S. Goudsmit, *Atomic Energy States, as Derived from the Analyses of Optical Spectra*, London 1933, 562 pp.
13 *Die Spektren, Entstehung und Zusammenhang mit der Struktur der Materie* (Vol. 9 of the *Hand- und Jahrbuch der chemischen Physik*, edited by A. Eucken), Leipzig 1934–37, 266+408+ 305 pp.
14 L. C. Pauling and E. B. Wilson, *Introduction to Quantum Mechanics*, New York 1935, 468 pp.
15 G. R. Harrison, R. C. Lord and J. R. Loofbourow, *Practical Spectroscopy*, New York 1948, 604 pp.
16 F. O. Rice and E. Teller, *The Structure of Matter*, New York 1949, 361 pp.

CHAPTER 4

GROUP o OF THE PERIODIC SYSTEM: THE INERT GASES

A-tomic numbers	Names of the elements	Symbols	Atomic weights*	Liter-weight, grams	relative to oxygen = 1	Melting points	Boiling points	Valences
2	Helium	He	4.003	0.17848	0.12490	—272.1°**	—268.98°	0
10	Neon	Ne	20.183	0.8990	0.6280	—248.6°	—246.03°	0
18	Argon	A	39.944	1.7837	1.2483	—189.4°	—185.87°	0
36	Krypton	Kr	83.80	3.736	2.614	—156.6°	—152.9°	0
54	Xenon	Xe	131.3	5.891	4.123	—111.5°	—107.1°	0
86	Radon	Rn	222	9.96	6.97	— 71°	— 65°	0

* Except for isotopically 'pure' elements, the atomic weights listed in this and similar tables apply to the ordinary, naturally occurring mixture of isotopes, and are on the scale O (ordinary natural oxygen) = 16.0000. Atomic weights of the individual isotopic species are given in Volume II. Other data also apply to the natural mixture of isotopes.

** At 25 atm pressure.

1. Introduction

(a) General

The inert gases ('noble gases') namely helium, neon, argon, krypton, xenon and radon constitute a group of elements which, under ordinary conditions, form no chemical compounds at all in the strict sense, (that is, no 'valence compounds'— see below). [1, 2]

The inert gases are thus zero-valent. To bring the group number into agreement with the maximum positive valence of the inert gases, as with the other main groups of the Periodic System, the term 'Group o' has been introduced, and as such the inert gases have been placed at the beginning of the Periodic Table. When the elements are written down in an unbroken series, as in Table II (Appendix), the inert gases are found to fall into the eighth family, constituting its main group. This is in agreement with the laws of the Periodic System, since zero valence might be expected in the eighth group from the way in which negative valence decreases with increase of group number, from group IV onwards. The dual position of the inert gases accords with their special chemical character, as compared with the elements of the remaining main groups. This is dealt with in the next chapter.

The inability of the inert gases to enter into combination also finds expression in the fact that—unlike the other elements that are gaseous at the ordinary temperature—they are monatomic.* Their molecular weights are thus the same as their atomic weights. That the inert gases are monatomic is shown by the ratio of the specific heats at constant volume and at constant pressure, c_p/c_v; for the inert gases this is $5/3 = 1.667$, as is required by the kinetic theory of gases for a monatomic gas. The inert gases can, indeed, be liquefied, and then transformed into the solid state by a comparatively small further diminution of temperature. However, their low boiling points show that the 'secondary valences', or Van der Waals forces, which the atoms exert upon each other are only weak. This is especially true of helium, the boiling point of which is even lower than that of hydrogen—i.e., helium in the *atomic* state is more 'saturated' than is hydrogen in the *molecular* state. The boiling points and melting points of the inert gases rise progressively with their atomic weights.

The Van der Waals forces are the forces which are responsible for the deviation of gases from the ideal gas laws, and ultimately for their liquefaction. F. London (1931) has shown that in many cases they can be attributed to the operation of wave mechanical resonance forces between the electrons of the different atoms (cf. p. 131 and p. 228). In other cases, they arise chiefly from polarization effects (cf. p. 48).

The inert gases have zero dipole moment, and also have low polarizabilities. Nevertheless, some of them are able to form *loose addition compounds* with certain substances, and especially with H_2O. The *hydrates* of the inert gases increase in stability with the atomic weight of the inert gas (Villard, 1896; de Forcrand 1923 and 1925). No hydrates of He and Ne could be obtained at 0°, even under 260 atm. pressure. At 0° the hydrate of A has a dissociation pressure of 105, that of Kr 14.5 and that of Xe 1.45 atm. Crystals of the hydrate of radon, the existence of which was proved indirectly by Nikitin (1936), can be obtained simply by passing radon through water at 0°, since its dissociation pressure is less than 1 atm. at 0°. The crystalline hydrates of the inert gases contain $5\frac{3}{4}$ H_2O, as may be deduced from their crystal structure (found, for krypton hydrate 5.7 H_2O; cf. p. 220), and they form mixed crystals with the corresponding hydrate of SO_2. The variation in stability of the inert gas hydrates makes it feasible to use the formation of such mixed crystals as a means of separating the inert gases from one another. Contrary to earlier statements, it appears from recent work (Wiberg, 1948) that BF_3 does not form addition compounds with A, Kr and Xe, and is immiscible with them in the liquid state. Liquid Xe is also immiscible with SO_2, H_2S, CH_3OH and $(CH_3)_2O$, and it is improbable that it forms addition compounds with any of these substances.

The inert gases form definite solid addition compounds with a number of organic substances, such as quinol, $C_6H_4(OH)_2$. Quinol crystallizes from solution in a structure such that the molecules in the crystal are linked together in a rigid arrangement by means of secondary valence forces ('hydrogen bonds'). The orientation of the molecules of the organic compound is of such a nature that the framework encloses certain cavities which are just of the appropriate size to contain a small gas molecule without bringing any atoms closer together than corresponds to the Van der Waals radii of the atoms. Without being bound by any specific chemical forces, a molecule which was enclosed within one of these 'cages' could not diffuse out without overcoming strong repulsive forces. There is one of these cavities in the crystal structure for each three molecules of quinol. When quinol is crystallized from solution in the presence of an inert gas (argon, krypton or xenon, under high pressure), or of some other substance of suitable molecular dimensions (e.g., H_2S, HBr, CO_2, C_2H_2), the small molecules are incorporated in the cage-like structure. With argon, the compound obtained has the idealized formula $A \cdot 3C_6H_6O_2$, but may contain less argon if not all the available crystallographic positions are occupied. The compound is a solid, with no perceptible dissociation pressure of argon, although the concentration of argon atoms in the crystal corresponds to the distances between argon atoms in argon gas under 70 atmos-

* As regards the extremely unstable poly-atomic molecules of inert gases observed in discharge tubes, see Footnote 3, p. 148.

pheres pressure. When the crystalline cage structure is broken down, by heating the crystals or by dissolving them in alcohol, the gas is at once liberated.

Compounds of this kind do not owe their stability to chemical forces *between* the components, but to the inclusion of the smaller molecule within a continuous framework from which it cannot escape because of the high repulsive forces which would arise from the closer approach of the gas molecule to the atoms of the organic component. The gas molecule within the cage is in a position of minimum potential energy. Its size, rather than its chemical nature or even its polarizability, is the property of importance. The nature and structure of these compounds were first elucidated by H. M. Powell (1950), who has termed them 'clathrate' compounds.

The ideal formula given above for the inert gas hydrates (W. F. Claussen, 1951; von Stackelberg and Müller, 1951) is based on the view that they are of the same type. The water molecules in ice are linked by secondary valence forces (hydrogen bonds) into a continuous structure. With a slight rearrangement of the molecular packing, a modified structure can be envisaged theoretically, which would enclose one 'cage' of appropriate size for every $5\frac{3}{4}$ water molecules in the continuous framework. If this hypothesis is valid, the gas hydrates owe their stability to the forces between the water molecules, and not primarily to dipole interactions between the water molecules and the gas molecules or inert gas atoms.

The inert gases display an extremely close chemical resemblance to each other, in the sense that under normal conditions they form absolutely no valence compounds—that is, compounds of composition determined by ordinary (formal or electrochemical) valence considerations. It would, however, be too sweeping to say that the inert gases are absolutely identical chemically. There are other instances in which elements of the same valence form compounds which are completely analogous, and capable of differentiation only by reference to their physical properties (solubilities, etc.). The inert gases manifest these same characteristic differences—different volatility, solubility, adsorption, etc.—in the uncombined state.

(b) History [3]

In 1892, during an investigation of the density of the common gases (oxygen, nitrogen, etc.), Lord Rayleigh discovered that nitrogen, obtained from the air by removal of the oxygen, had a higher density than that prepared from chemical compounds, such as ammonia or nitrates. The idea that the difference might be attributable to the presence of a hitherto undiscovered, heavier gas in the air, was first proposed by Ramsay. After removal of the oxygen by means of red hot copper, he was able to absorb the nitrogen by means of heated magnesium. The residual gas proved to be a new chemical element, with a characteristic spectrum. It was isolated simultaneously by Lord Rayleigh, who removed the nitrogen by the old method of Priestley and Cavendish (cf. p. 589). Both workers gave to the newly discovered element the name of *argon*, because of its chemical inertness (from ἀργός or ἀ-εργός, inert). It was known that treatment of the mineral cleveite with sulfuric acid evolved a gas resembling nitrogen. When Ramsay investigated the gas from cleveite, suspecting that it might be argon, he found that its spectrum displayed a new line, situated very close to the yellow sodium line, but clearly distinguished from the latter. This yellow spectral line had already been observed in the chromosphere of the sun by various astronomers—first in 1868, on the occasion of a total eclipse. When it was established that this solar line was, in fact, not identical with the yellow sodium line, the conclusion was first drawn by Secchi that some element must exist in the sun, which was as yet unknown on the earth;

this was named *helium* (ἥλιος = sun) by Lockyer and Frankland. This name was accordingly applied to the element discovered by Ramsay, since the evidence of its spectrum showed that it was identical with the solar element. A few days after Ramsay's discovery, and without knowledge of it, the occurrence of helium in cleveite was also discovered by Langlet, in Cleve's laboratory.

Because of the exceptionally close similarity in their properties, and their complete difference from all other elements known up to that time, it could hardly be doubted that the two newly discovered gases belonged to the same group of the Periodic System, and in fact to a new group. Their atomic weights would, in that case, imply that between them there must be a vacant place in the Periodic Table. On this reasoning, Ramsay in 1887 foretold the existence of another inert gas. The publication, at that time, of the Linde and Hampson process for the liquefaction of air, which made possible the liquefaction and fractional distillation of argon, came at an appropriate moment for the discovery of this element. By this means, within a few days, Ramsay discovered *two* new inert gases—first *krypton* (κρυπτός, hidden), relatively heavier than argon, and recognizable by its two characteristic lines in the yellow and green part of the spectrum, and then the predicted lighter element, *neon*. The latter was named in accordance with a suggestion from Ramsay's twelve-year-old son, who looked with fascination at the crimson light emitted from the discharge tube, and suggested that the element so characteristically identified should be called 'the new one' (τὸ νέον, new). The residue from the fractional distillation of argon proved to contain yet another new gas, *xenon* (τὸ ξένον, the stranger). Rutherford and Soddy later found that the radioactive gases, the *emanations*, formed from certain radioactive substances, were also members of the group of inert gases. Ramsay characterized radium emanation by its spectrum, and also determined its atomic weight, from its gas density, in 1910. He suggested that it should be designated *niton* (the luminous, from *nitere*), as an allusion to the phosphorescent glow of the liquefied gas in a glass tube. The name has subsequently been repeatedly altered, but agreement has ultimately been reached on the name *radon*.

(c) Occurrence

The atmosphere contains the inert gases to the extent of about 1 volume per cent. Argon is the principal constituent thereof, with the other gases present only in quite small quantities. The inert gas content of the atmosphere is unexpectedly low, as compared with the cosmic abundance of their atoms. The reason for this is considered in Vol. II, Chap. 15.

100 liters of air contain 934 cc. of argon, 1.82 cc. of neon, 0.524 cc. of helium, 0.114 cc. of krypton, and 0.0087 cc. of xenon (Glueckauf, 1951). The inert gas content of air, like its content of oxygen and nitrogen, is practically constant. Argon can be detected wherever air enters—e.g., dissolved in the blood.

Gases rich in argon, and especially in helium, are given off by many mineral waters. Neon has occasionally also been found therein in large amounts. Helium is formed by all radioactive transformations in which α-rays are emitted, and it therefore occurs in all radioactive waters and rocks. It escapes from the latter when they are dissolved in acids or are strongly heated. In a few places in North America, natural gases with relatively high helium content escape from the earth's surface in enormous quantities.

The natural gas wells containing helium extend in the United States over a region from Texas, where the richest sources are, through Oklahoma, Kansas and Ohio to New York. Profitable gas wells have also been discovered in Canada (at Inglewood near Toronto). The helium content of gases in some cases exceeds 1 %; the principal constituent in some wells is nitrogen, but more often methane. The European helium sources have but a small yield. Many mineral spring gases do, indeed, contain relatively important amounts of helium (e.g., Dürkheim 1.8 %, Baden-Baden 0.85 %, Wildbad 0.71 %), but at most these could furnish only a few hundred cubic metres of He per annum. Helium has also been found in volcanic gases (0.26 % in the soffioni of Tuscany). Small amounts of helium are also found in many places in the gases of coal mines. Enormous quantities of helium are present in the sun's atmosphere, in the gaseous nebulae, and in certain stars, as is shown by their spectra.

The occurrence of radon and the other radioactive emanations is associated with the radioelements from which they are produced, since the emanations themselves rapidly disintegrate. The amount of radon in equilibrium with 1 g of radium (*i.e.*, under conditions that the amount formed at each instant equals that which disintegrates) is 0.59 cu.mm. at 0° and 1 atm. pressure. This quantity is called 1 curie. It serves as a unit for measuring the emanation content of radioactive water.

2. Isolation and Applications

Helium has been produced in North America since 1917 from the rich natural gases there, and has been used as a filling gas for lighter-than-air craft, since it is not inflammable.

The United States Navy used 200,000 cu.m. for this purpose in 1929. The amount which could be recovered from the North American natural gases has been estimated as over 1 million cu.m. annually, of which about 30 % is due to Canadian gas wells. Helium is, indeed, twice as heavy as hydrogen, but since the lifting force of an airship is determined by the difference between its own weight and that of the same volume of air, the replacement of hydrogen by helium only diminishes the load capacity of an airship by the factor $\frac{28.8 - 2}{28.8 - 4} = \frac{108}{100}$.

The process used in North America for recovering helium is based essentially on the fact that helium, unlike other gases, is adsorbed only to a small extent on active charcoal, cooled with liquid air. This principle had already been employed by the Gesellschaft für Lindes Eismaschinen, for the preparation of neon and helium from the gas residues obtained in the liquefaction and rectification of air. Practically pure neon can be obtained by fractional distillation of the mixture of helium and neon, with the aid of solid hydrogen. However, for most technical purposes (though not as a filling gas for balloons) the helium-neon mixture is quite applicable.

According to Sieverts (1912), helium can most conveniently be prepared on a small scale by heating minerals that contain the element—e.g., cleveite or thorianite—to 1000–1200° in a porcelain tube, closed at one end. The gas evolved is passed over heated copper oxide, contained in the same tube, and over a boat containing solid caustic potash, to absorb H_2, H_2O and CO_2. It is freed from nitrogen by repeated passage over a strongly heated mixture of lime, magnesium and sodium, and from argon by means of well out-gassed adsorption charcoal.

Helium [4] is used as a filling for gas thermometers, and for the production of the lowest attainable temperatures. Neon (usually mixed with helium) [5] is used chiefly for illuminated advertising signs and similar purposes (e.g., neon arc lamps and neon glow lamps). Neon tubes are also used, however, in various fields of electro-technology— for rectifiers, voltage regulators, indicators, etc. Argon finds applic-ation chiefly in the production of completely inert atmospheres—e.g., for metal-lurgical purposes. As such it is used (mixed with 15 % of nitrogen) in place of nitrogen as the filling for gas-filled electric filament lamps.

An electric arc is easily struck in an atmosphere of *pure* argon, but this is inhibited by the presence of nitrogen. The low thermal conductivity of argon, as well as its chemical inertness, is important for its use in gas-filled lamps. Krypton and xenon, having a still lower thermal conductivity, have recently also been used for this purpose. It may be mentioned that it has been suggested that krypton and xenon might be used in the radiological examination of the respiratory organs, because of their high absorption for X-rays.

Argon (more accurately 'crude argon') can be prepared in the laboratory by the chemical absorption of oxygen and nitrogen from air, previously freed from water vapor and carbon dioxide by the usual methods.

Oxygen is generally removed by passage over heated copper. Nitrogen is combined with magnesium or calcium, or with a mixture of magnesium, lime and sodium. Oxygen and nitrogen can also be absorbed at the same time, by means of heated calcium carbide. It is better to start with the oxygen, still containing some nitrogen, obtained by fractionally distilling liquid air. Argon is considerably enriched in this (up to about 3 %), since, like oxygen, it is less volatile than nitrogen.

In the Linde-Gesellschaft process, argon is first enriched to at least 60 % by repeated fractional distillation of liquid air. The oxygen (making up about 30 %) is then removed chemically, and the argon-nitrogen mixture which remains is suitable for lamp filling without further treatment.

As prepared, argon obtained from the air contains the other inert gases of the atmosphere. The small amount of impurity (about 0.25 volume-%) is quite un-important for the usual uses of argon. The 'crude argon' can be freed from the other inert gases by repeated fractional distillation. Krypton and xenon may also be obtained pure by fractional distillation.

In scientific laboratories, fractional adsorption on active charcoal is now generally used, as well as fractional distillation, for the separation of the inert gases from one another. The adsorbability of the inert gases increases considerably with increase in their atomic weight. According to the pressure, either He and Ne, or Ne and A may be separated from one another by fractional adsorption on charcoal, followed by fractional desorption (K. Peters, 1930). Rn can be completely adsorbed from its solution in liquid air, by means of silica gel.

For technical purposes, methods have also been worked out depending on the various diffusion rates of the inert gases in some other gas, which can subsequently be easily removed again, or employing the different rates of effusion through a porous wall for the separation. Considerable quantities of krypton and xenon are obtained by means of the Claude process, which depends on the observation that liquid air will wash out these two gases from air which has been cooled down approximately to their liquefaction temperature.

To obtain radon, a solution of a radium salt is allowed to stand for about 4 weeks in a closed flask. This time is necessary for attainment of the equilibrium mentioned above, between radium and its emanation. The radon can then be boiled off from the solution, or pumped off by evacuation.

3. **Properties**

The inert gases are all colorless and odorless. They are relatively difficult to liquefy, though the ease of liquefaction increases with their atomic weights. The boiling point of radon at atmospheric pressure is only —62— —65°; its melting point, at which it has a vapor pressure of 500 mm, is —71°. Radon can thus be condensed even when in great dilution. If it is passed through a U-tube cooled with liquid air, it is deposited on the walls, as can be detected from the vivid green fluorescence of the glass produced by the radon. When the radon is under very low pressure, condensation takes place between about —152° and —154°. The melting points of the inert gases lie, in general but little below the boiling points. In the case of helium, conversion to the solid state is possible only under pressure.

The melting point of helium is 1.13° K at 25.3 atm. and 4.21° K at 140 atm. Helium forms two liquid phases, which transform, one to the other, at 2.3° K, with evolution (or absorption) of heat, and which show themselves in a discontinuous change in the properties.

The latent heats of melting and vaporization of the inert gases are extremely small; their magnitude, and also the critical data, are shown in Table 19. The crystal structure of the solidified inert gases is discussed on p. 212.

TABLE 19

HEATS OF FUSION, HEATS OF EVAPORATION AND CRITICAL DATA FOR THE INERT GASES

	Helium	Neon	Argon	Krypton	Xenon	Radon
Latent heat of evaporation	0.022	0.44	1.50	2.31	3.27	43 kcal/mol
at $t =$	—270°	—246°	—186°	—151°	—109°	—62 °C
Latent heat of fusion	0.0033	0.078	0.265	0.36	0.49	0.8 kcal/mol
at $t =$	—270.7°	—249°	—189°	—157°	—140°	—71° C
Critical temperature	—267.9°	—228.7°	—120°	—62.5°	+16.6°	+104.5° C
Critical pressure	2.26	26.9	50	54.3	58.2	62.4 atm
Critical density	0.069	0.4	0.4	0.7	0.9	1.2

The inert gases have a relatively high solubility in water. According to Lannung (1930), 1 liter of water at 20° can dissolve 8.8 cc. of He, 10.4 cc of Ne and 33.6 cc of A (volumes reduced to 0°). The solubility of argon in water is thus rather greater than that of oxygen. The solubility diminishes with rise in temperature; it increases with the atomic weight of the inert gas, and amounts to about 51 vol% in the case of radon, at 0°. It has already been mentioned that the inert gases also form crystalline hydrates at high pressures. In organic solvents, the solubility is, in some cases, even higher than in water. Except for helium, the inert gases are more or less avidly adsorbed at low temperatures by active wood charcoal (cf. p. 106). Unlike hydrogen, helium does not diffuse through red hot platinum. It does, however, diffuse through quartz glass at high temperatures (as also does hydrogen). This property can be used for the separation of helium from neon (Paneth 1925).

The inert gases are characterized by a fairly good conductivity of the electric current, in which respect they considerably surpass such gases as hydrogen, nitrogen, etc. The low resistance that they oppose to the passage of current, and the low voltages at which luminous discharges may be started in them, are of great importance for their use in illuminating technology.

The spectra of the inert gases are very characteristic, and the analytical detection of the inert gases is always based thereon. The capillary of a Plücker tube emits an intense yellow glow with a helium filling, and a brilliant crimson with neon. With the other gases, the color of the glow depends somewhat upon the experimental conditions. Argon generally gives a red discharge, krypton greenish to lilac, xenon violet, and radon a bright white.

4. The Spectra of the Inert Gases

The most prominent lines in the visible region of the inert gas spectra are listed in Table 20. The lines of the *helium spectrum* have been determined with especial accuracy, as they are frequently used for the calibration or adjustment of spectrographic apparatus. Where the wave lengths are given in the table to the second decimal place, the last figure may be taken as accurate to within a few units.

The capillary of a Plücker tube filled with argon at about 3 mm pressure gives a red glow. At lower pressures, and when the voltage of the discharge is raised by placing a Leyden jar condenser across the tube, the color changes to steel-blue and finally to white. In the 'red spectrum', the strongest lines are in the red and infra-red; in the 'blue spectrum' they are in the violet and ultra-violet. For the other inert gases, the position is much as with argon; only the strongest lines of the ordinary spectra are included in Table 20.

The spectrum of helium is particularly interesting from the theoretical standpoint. Even before the formulation of the Bohr theory, it was known that, when excited in a discharge tube, the spectral lines of helium, (which are far more numerous than the list given in Table 20, especially as there are many lines in the ultra-violet as well), could be fitted into *six series*. These six series were found to form two quite independent groups, the 'terms' of which could not be combined with one another, as is otherwise usual with the lines of different series of one and the same element. For this reason, it was doubted for a time whether helium was a single element.

The *combination principle*, formulated by Ritz in 1908, states that the combination (i.e., the addition or subtraction) of terms of different series, gives the wave numbers (reciprocal wave-lengths) of lines which should also be found in the spectrum of the element concerned. Thus, in the hydrogen spectrum, if the ground term of the Balmer series is subtracted from the ground term of the Lyman series, we obtain the *wave number* of the first line of the Lyman series. This rule has its origin in the fact that, as was seen in the previous chapter, the spectral terms define the energy levels of the atom, corresponding to the various *stationary states* (cf. p. 96). The meaning of the combination principle is thus that the atom can undergo a transition (by absorption or emission of light) from any one stationary state to any other stationary state. It may be noted here, however, that the combination principle needs some restrictions when the energy levels are defined not merely by the principal quantum number, but by the subsidiary quantum number also (cf. p. 114).

Spectroscopists formerly believed that the two groups of series displayed by helium must be ascribed to two different elements, which they named orthohelium and parahelium (or asterium). It has, however, been shown that helium is definitely a simple element. The differences in the spectra arise from the fact that, in the case of helium, combinations between the two groups of terms are in general forbidden by the restrictions placed on the combination principle. In the end state of the electron transitions responsible for the emis-

TABLE 20

PRINCIPAL LINES OF THE INERT GAS SPECTRA, IN THE VISIBLE REGION

Wave lengths in international Ångstrom units (Å), $1 \text{ Å} = 10^{-8}$ cm

Spectral color	Helium	Neon	Argon 'red spectrum'	Argon 'blue spectrum'	Krypton	Xenon	Radon
			8115.3		8112.9	8231.6	
			8006.2		7854.8		
Extreme					7694.5		
red					7601.5	7642.0	
		7544.0			7587.4		
		7535.8					
	7281.35	7245.2					
	7065.19	7173.9					
		6929.5	6965.4				
	6678.15	6678.3					
Red		6599.0					
		6506.5				6469.7	
		6402.2					
Orange		6266.5					
		6163.6				6182.4	
		5975.5					
	5875.62	5881.9			5870.9		
Yellow		5852.5					
		5400.6			5570.2		5582
	5047.74	5037.7					5085
Green	5015.68	4957.0					4979
	4921.93	4827.3		4806.0			4817
Blue	4713.14	4715.3			4739.0	4734.1	4766
					4671.2	4671.2	4681
							4644
					4624.3	4614.3	4626
							4605
		4540.4					
		4537.8	4510.7		4502.4	4501.0	
Indigo	4471.48				4475.0		
					4463.7		
	4437.55			4426.0	4453.9		
					4376.1		
	4387.9		4300.1	4348.0	4355.5		4350
					4319.6		4307
			4272.2	4277.5	4274.0		
			4266.3	4266.3			
			4259.4				
			4200.7				4203
			4198.3			4193.5	
	4143.8		4158.6				4167
	4120.8			4104.0	4088.4	4078.8	
Violet	4026.2		4044.4		4057.0		
			4057.0				3982
	3964.7		3949.0	3928.6	3920.1	3951.0	3972

sion of the parahelium spectrum, the two electrons of the neutral helium atom are bound in a different manner from that of the end state reached by the emission of the orthohelium spectrum. Although, in this latter state, the helium atom is in a state of higher energy than in the former, it cannot undergo a direct radiative transition to this state. Transitions from other orthohelium states to states of parahelium (or vice versa), with emission of radiation, are also practically incapable of occurrence, except by way of a complete removal of one electron (ionization).

In 1922, Lyman discovered yet another series, belonging to parahelium, in the farthest ultra-violet, so that four important series are now known. A few other series of parahelium and orthohelium, represented by only a very few lines, are of interest only to the spectroscopist and the atomic physicist.

Under certain conditions—e.g., when a powerful condensed discharge is passed through helium at a pressure of 0.25–1 mm.—helium emits lines belonging to neither of the series discussed. The same lines have also been found in stellar spectra. This spectrum, produced in discharge tubes only by strong sparks, is called the *spark spectrum* of helium, as distinct from the usual *arc spectrum*. Its lines correspond exactly to those which must be expected, on the basis both of the Bohr theory and of quantum mechanics, for the *singly ionized helium atom*, on the assumption that the latter differs from the hydrogen atom in having a nucleus with twice the charge on a proton. The lines of the helium spark spectrum are thus given by equation (13a) (p. 87), by substituting $Z = 2$ and $M = 4$ (atomic weight of helium). In agreement with the theory, the Rydberg constant for helium has a slightly higher value than for hydrogen, the spectroscopic measurements giving

$$R_{He} = 109722.26 \pm 0.01 \text{ cm}^{-1}$$

This value can be used to calculate m/M_H, the ratio of the mass of the electron to that of the proton (hydrogen nucleus), very accurately, since

$$\frac{R_{He}}{R_H} = \frac{M_{He}(M_H + m)}{M_H(M_{He} + m)}, \quad \text{or} \quad \frac{m}{M_H} = \frac{R_{He} - R_H}{R_H - R_{He}\left(\dfrac{M_H}{M_{He}}\right)}$$

Using the values given in Chap. 13 of Volume II for the mass numbers of the H and He nuclei, one thus obtains the value 0.00054448. Multiplying this by the ratio of the mass of the proton to the mass of the hydrogen atom, we obtain for the ratio of the mass of the electron to the mass of the hydrogen atom 0.00054418, or approximately 1/1838.

Table 21 shows how accurately the wave lengths λ, calculated for the lines of the helium spark spectrum from the formula

$$\frac{1}{\lambda} = 4R_{He}\left(\frac{1}{n_2^2} - \frac{1}{n_1^2}\right) \tag{1}$$

agree with those observed experimentally.

TABLE 21

WAVE LENGTHS OF THE SPARK SPECTRUM OF HELIUM

Series	n_2	for $n_1 =$ 2	3	4	5	6	7	8	9
1st series	1	λ obs. 303.6	256.3	—	—	—	—	—	—
		λ calc. 303.7	256.25	—	—	—	—	—	—
2nd series	2	λ obs.	1640.4	1215.2	1085.2	1025.6	—	—	—
		λ calc.	1640.5	1215.2	1085.0	1025.3	—	—	—
3rd series	3	λ obs.		4685.8	3203.1	2733.3	2511.2	2385.4	2306.2
(Fowler series)		λ calc.		4685.9	3203.2	2733.4	2511.3	2385.5	2306.25
4th series	4	λ obs.			—	6560.1	5411.6	4859.3	4541.6
(Pickering series)		λ calc.			10123.8	6560.2	5411.6	4859.4	4541.7

The spark spectrum which is assigned with certainty to ionized helium He⁺ thus corresponds in structure to the spectrum of the neutral hydrogen atom. This exemplifies a general rule, that *the spark spectrum of each element* (i.e., the spectrum emitted by its [singly] ionized atoms) *is similar in character to the arc spectrum* (that of the neutral atom) *of the element preceding it in the Periodic System.* This is Sommerfeld's spectroscopic displacement law. The spectra emitted by neutral atoms are called the *arc spectra* because, for most elements, they are excited by means of the electric arc.

The spectra of the other inert gases are much richer in lines even than the spectrum of helium—e.g., nearly 900 lines have been measured for neon. Their structure is much more complex than that of the helium spectrum. This indicates that there are more electrons present in the outer shells of these inert gas atoms than in the case of helium. We shall learn a means of determining this number in a later section.

5. Ionization and Radiation

As will be seen later, it is of considerable interest in assessing chemical reactivity, to know the energy required (the amount of work to be expended) to transform an atom of any element into the electrically charged state, by removing an electron. The energy expended in transforming an element from the normal state (electron in the innermost available orbit) to the ionized state (electron taken right out of the atom) is known as the *ionization energy* of the atom in question.* The ionization energy can be calculated accurately from the spectrum, even when no exact knowledge of the atomic structure is available, provided that it is certain that the largest of the ground terms found from the spectroscopic data really corresponds to the normal state of the atom. In many cases, however, the ionization energy can also be measured directly, by bringing about ionization by means of electron collisions.

Equation (11) of the previous chapter gives directly the energy required to ionize the hydrogen atom, by substitution of $n = 1$. The same energy is obtained from the generalized Balmer formula [equation (3) of previous chapter] by substituting $n_2 = 1$ and $n_1 = \infty$— i.e., from the ground term or series limit of the hydrogen spectrum of shortest wave length (Lyman series), multiplied by $c \cdot h$. In the same way, the energy necessary to remove the second electron from the singly ionized helium atom—i.e., the energy for the process

$$He^+ \rightarrow He^{++} + e,$$

can be derived from the spark spectrum of helium. From equation (1), p. 110, this comes to $\varepsilon = 4R_{He} \cdot c \cdot h.$

The spectroscopic measurements do not, however, always lead to a ground state with $n = 1$ as the normal state of the atom. The shortest wave length series of the orthohelium spectrum leads to a ground term with quantum number $n = 2$. No lines can be found capable of combination with the other lines of the orthohelium spectrum, but belonging to a ground state with $n = 1$, although they would lie in the region of the spectrum which is accessible to measurement. It must be concluded from this that the ground orbit of the second electron in orthohelium is a 2-quantum orbit. For parahelium, by contrast, the series limit leads to a 1-quantum orbit as the lowest state. We can thus state more precisely the difference between ortho- and parahelium: that in the orthohelium atoms the lowest orbit of the second electron is a 2-quantum orbit, whereas in parahelium the second electron, like the first, has a 1-quantum lowest orbit. The absence of combination lines implies that

---

* More strictly, this is the energy for *single* stage ionization. The energy for multiple ionization can be defined in a corresponding manner.

the helium atom can pass from one ground state to the other only by way of complete ionization.

(a) Direct Measurement of Ionization Energy

The method of direct measurement of ionization energies by means of electron impact was worked out by Franck and Hertz in 1913. It is based on the following principle: electrons of well defined velocity are produced within a tube containing the gas under investigation—e.g., helium. This is achieved by utilizing the continuous emission of electrons of very low energy from an incandescent wire; these are allowed to pass through an electric field, set up between the incandescent wire, as negative electrode, and a positively charged wire gauze electrode, as in the thermionic valves of radio technology. They thereby acquire a velocity v and a corresponding kinetic energy defined exactly by the potential difference between filament and electrode ($\frac{1}{2} mv^2 = eV$, where e is the charge of the electron, and V the accelerating potential). It is usual to define the velocity of the electrons in terms of the potential which must be applied in order to produce it—i.e., in volts—and the energy, corresponding to the velocity, in *electron volts* (eV). To convert the latter units to ergs, it is necessary to multiply by $1.602 \cdot 10^{-12}$, or by $3.8285 \cdot 10^{-20}$ to convert electron volts to calories, i.e., $1 \text{ eV} = 1.602 \cdot 10^{-12} \text{ erg} = 3.8285 \cdot 10^{-20} \text{ cal.}$ (equation (2)).

If, starting from zero, the applied voltage is gradually increased, it is found that the strength of the current at first rises gradually. Suddenly—at a potential of 19.75 volts, in experiments with helium—the current falls again practically to zero. The current strength is determined by the number and the velocity of the electrons. As long as the temperature of the filament is held constant, the number of electrons emitted is constant, so that the initial increase in the current is to be explained by the increase in velocity of the electrons as the potential is raised. It follows from this that the electrons at first undergo elastic collisions with the helium atoms, without suffering any appreciable change in velocity. This holds only until their velocity has reached a certain value, characterized in this case by the accelerating field of 19.75 volts. As soon as this is reached, the electrons completely lose their kinetic energy in a single act of collision—i.e., they give up their entire energy to the helium atoms with which they collide. If the potential is increased further, the current once more rises, corresponding to the excess of the voltage over the critical value 19.75 volts. The electrons thus give up to the helium atoms only as much energy as corresponds to the potential 19.75 volts. With further increase of voltage —at 20.55 volts in the case of helium—a second break appears in the current-voltage curve. Two further breaks have been also observed with helium, at 21.2 and 22.9 volts.

It is generally possible to observe that on the occasion of each break in the current-voltage curve, one or more spectral lines is simultaneously excited, and continues to be emitted on a further increase of the potential. It has been shown that a very simple relation exists between the energy $\varepsilon = eV$ of the colliding electrons at the moment of the first appearance of a given line or lines and the spectral terms (i.e., energy levels) assigned to them. Since the colliding electron has given up its energy ε to the atom struck, ε also represents the increment of energy acquired by the atom through the collision, that is $\varepsilon = \varepsilon_2 - \varepsilon_1$, where ε_1 is the energy of the atom before, and ε_2 that after the collision. The atom raised by the collision into the state with the energy ε_2 can now undergo transitions to states with an energy smaller than ε_2, in so far as the transitions are not forbidden by the 'selection rules'. If a transition to the ground state is possible, the line of shortest wave length which is emitted is given by

$$\varepsilon_2 - \varepsilon_1 = \frac{hc}{\lambda} = h\nu \tag{3}$$

This equation is nothing other than the second Bohr postulate (cf. Chap. 3, eqn. (10)), the validity of which is thus not only confirmed by wave mechanics, but is demonstrated experimentally.

We must suppose, accordingly, that in the process of excitation of the atom by electronic impact one electron of the atom struck is lifted out of its normal orbit (in the sense of the Bohr model) into a higher orbit by the colliding electron. In jumping back to the original orbit, it emits in the form of light the energy transferred to it by the collision. As long as the energy of the colliding electron is insufficient to lift one of the electrons of the atom into a higher orbit, the electron is reflected according to the laws of elastic collision. The voltage V which is necessary in order to excite an atom in the manner indicated, so that a certain spectral line is emitted, is called the *excitation potential* of the line in question. The potential which is required, in order to impart to the colliding electron so great a velocity that it completely removes an electron from the atom, and so ionizes it, is called the *ionization potential*, V_I. The work of ionization A_I is accordingly given by the product of the ionization potential and the electronic charge.

$$A_I = eV_I \tag{4}$$

Equation (4) gives the energy in electron volts to be expended in ionizing a free atom or molecule. The energy required to ionize one gram atom or one gram molecule, in calories, is obtained by multiplying the ionization energy, converted to calories, by the Avogadro number, $N_A = 6.0238 \cdot 10^{23}$.* Thus: Ionization Energy in kilocal. per mole = 23.062 × *ionization energy in electron volts per molecule*.

The terms first, second, third, etc. ionization potentials, are used to refer to the removal of the first electron from the neutral atom (i.e., conversion of the atom into a singly charged positive ion) or to the removal of a second electron (conversion of the singly charged into a doubly charged ion) or of a third electron, etc. The total energy to be expended for multiple ionization is obtained by addition of the individual energies to be expended in ionization at each stage. Thus the first ionization potential of helium is 24.48 volts, the second 54.14 volts. The total energy which must therefore be expended to convert an He atom into an He²⁺ ion is 24.48 + 54.14 = 78.62 electron volts. To produce one gram ion of helium nuclei from one mole of helium gas would accordingly require the expenditure of 78.62 · 23.064 = 1813.3 kilocal. of energy.

---

* Avogadro's number gives the number of molecules in one gram molecule of any substance. The name Loschmidt's number has recently been used for this in the physical literature whereas a distinction was formerly drawn between Avogadro's and Loschmidt's number. If this distinction is made, one should strictly use Loschmidt's number for the number of molecules per cubic millimeter of a gas at 0° and atmospheric pressure ($= 27 \cdot 10^{15}$) as was first deduced by Loschmidt in 1865. The customary meaning is the number of gas molecules per cubic cm ($= 27 \cdot 10^{18}$).

Direct spectroscopic observation of the emission of light accompanying excitation by electronic impact is made difficult, in the case of helium, by the fact that the lines in question all lie in the farthest ultra-violet. If, however, the lines of the series of helium observed by Lyman in the far ultra-violet are

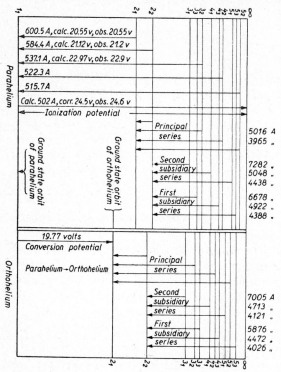

compared with the results obtained by the method of electronic impact, it is found that the observed potential differences correspond exactly to the energy differences calculated from the wave-lengths. Fig. 26 illustrates this point. It represents the energy levels corresponding to the excitation potentials and the spectrum. The relative energies are not represented strictly to scale, and energy levels with principal quantum number higher than $n = 5$ are omitted. So also are those energy levels which are of importance only for the series in the infra-red, which is not shown and of which only a few lines are known. The arrows represent the electron jumps corresponding to the emission of the wave-lengths shown. The differences between the energy levels obtained according to eqn (3) are shown as 'calculated'; the corresponding potentials measured directly by the electron impact method are shown as 'observed' values. All energies are calculated in electron volts. The wave lengths of the series limit of the shortest wave length series, $\lambda = 502$Å, is

Fig. 26. Energy levels in parahelium and orthohelium

not directly measured but has been calculated from the series formula. The first break in the current-voltage curve (at 19.75 volts) does not correspond to any spectral line, but to the energy required for conversion of the stable parahelium atom into the metastable orthohelium atom. The energy level of the latter is obtained also from the ground term of the principal series of orthohelium. In this way a level 19.77 volts above the normal state of parahelium is found, in very good agreement with the observed conversion potential.

The scheme of Fig. 26 shows how the lines of the most important series of helium come about, by electron transitions from one level to another, in the direction of the arrows shown. According to this, the impossibility of combining the terms of the parahelium and the orthohelium spectra is due to the fact that an electron never jumps spontaneously from an energy level of parahelium to a level of orthohelium, or vice versa. The Figure also shows how the restrictions of the combination principle, already mentioned on page 108 apply. It can be seen that the transition always takes place in such a way that the subsidiary quantum number k either changes by one unit or remains unaltered. The restriction of the combination principle by the selection rules cannot be more closely discussed here. The example of helium shows how, even in complicated cases, the combination of the results of spectroscopic measurements with measurements by the electron impact method can provide a complete and exact determination of the energy levels in the atom, even without requiring that the model of the atom shall be known.

In Table 22 are collected the first ionization potentials of the elements arranged according to the Periodic System. The indices preceding the symbols represent the atomic numbers. For the inert gases, the values calculated from spectroscopic data

TABLE 22

FIRST IONIZATION POTENTIALS OF THE ELEMENTS, IN VOLTS

The values given in brackets have not been measured, but are calculated from the regular trend in the screening of the nuclear charge by the inner electrons (W. Finkelnburg, Z. Naturforsch., 2a (1947) 16).

MAIN GROUPS OF THE PERIODIC SYSTEM

Main group: I	II	III	IV	V	VI	VII	VIII	spectroscopic	24.6 obs.
						$_1$H 13.54	$_2$He 24.48		24.6 obs.
$_3$Li 5.37	$_4$Be 9.30	$_5$B 8.28	$_6$C 11.24	$_7$N 14.46	$_8$O 13.57	$_9$F 17.46	$_{10}$Ne 21.47		21.5 ,,
$_{11}$Na 5.09	$_{12}$Mg 7.63	$_{13}$Al 5.94	$_{14}$Si 8.14	$_{15}$P 10.43	$_{16}$S 10.42	$_{17}$Cl 13.01	$_{18}$A 15.68		15.4 ,,
$_{19}$K 4.32	$_{20}$Ca 6.09	$_{31}$Ga 5.97	$_{32}$Ge 8.10	$_{33}$As 10.05	$_{34}$Se 9.75	$_{35}$Br 11.82	$_{36}$Kr 13.94		13.3 ,,
$_{37}$Rb 4.19	$_{38}$Sr 5.68	$_{49}$In 5.75	$_{50}$Sn 7.34	$_{51}$Sb 8.35?	$_{52}$Te 8.89	$_{53}$I 10.43	$_{54}$Xe 12.08		11.5 ,,
$_{55}$Cs 3.86	$_{56}$Ba 5.21	$_{81}$Tl 6.08	$_{82}$Pb 7.37	$_{83}$Bi 7.25?	$_{84}$Po (8.3)	$_{85}$At (9.4)	$_{86}$Rn 10.69		— ,,
$_{87}$Fr (3.97)	$_{88}$Ra 5.21								

SUB-GROUPS OF THE PERIODIC SYSTEM

Sub-group: I	II	III	IV	V	VI	VII	VIII		
$_{29}$Cu 7.67	$_{30}$Zn 9.37	$_{21}$Sc 6.7	$_{22}$Ti 6.81	$_{23}$V 6.74	$_{24}$Cr 6.74	$_{25}$Mn 7.41	$_{26}$Fe 7.83	$_{27}$Co 7.8	$_{28}$Ni 7.61
$_{47}$Ag 7.58	$_{48}$Cd 8.94	$_{39}$Y 6.6	$_{40}$Zr 6.92	$_{41}$Nb (6.9)	$_{42}$Mo 7.2	$_{43}$Tc (7.1)	$_{44}$Ru 7.5	$_{45}$Rh 7.7	$_{46}$Pd 8.3?
$_{79}$Au 9.20	$_{80}$Hg 10.41	$_{57}$La 5.59	$_{72}$Hf (7.6)	$_{73}$Ta (7.6)	$_{74}$W (7.6)	$_{75}$Re 7.8	$_{76}$Os 8.7	$_{77}$Ir (8.7)	$_{78}$Pt 8.8
		$_{89}$Ac (5.5)	$_{90}$Th (5.7)	$_{91}$Pa (5.7)	$_{92}$U (5.7)				

Lanthanides:	I	III	IV	V	VI	VII	VIII
	$_{58}$Ce 6.54	$_{59}$Pr 5.76	$_{60}$Nd 6.31	$_{61}$Pm (6.3)	$_{62}$Sm 6.55	$_{63}$Eu 5.64	$_{64}$Gd 6.65
	$_{65}$Tb 6.74	$_{66}$Dy 6.82	$_{67}$Ho (6.9)	$_{68}$Er (6.9)	$_{69}$Tm (6.9)	$_{70}$Yb 7.06	$_{71}$Lu (7.3)

(Spectr.) and those found by the electron impact method (Obs.) are set out for comparison. Values calculated from spectroscopic data are numerically the more accurate. It can be seen that, with the inert gases, the energies needed for removal of one electron from the atom are particularly high (cf. also Fig. 4, p. 17). The Table shows, further, that in the main groups of the Periodic System the energy required to remove an electron from the neutral atom increases in general in each series from left to right, and decreases in each group from top to botton. It is characteristic of the elements of the Sub-groups that the ionization energies vary but little, and follow a quite irregular course as compared with the Main groups.

As will be shown later, the energy which must be expended in order to remove the electrons is of decisive importance for the chemical behavior of the elements. The inability of the inert gases to form true valence compounds is very closely connected with their high ionization potentials.

(b) Significance of the Subsidiary Quantum Numbers in the Helium Atom

For the hydrogen atom, and for the singly ionized helium atom, the levels corresponding to different subsidiary quantum numbers differ in energy only by extremely small amounts, so that they make themselves evident only in the fine structure of the spectral lines. With the neutral helium atom, however, the differences are very much larger. This is because, in the neutral helium atom, and indeed in all atoms possessing more than one electron, the energy of the orbit is dependent to a considerable degree upon the subsidiary quantum number, even apart from the variation of electronic mass with velocity which is required by the relativity theorem. It may readily be seen from the Bohr-Sommerfeld theory of atomic structure that this dependence must exist, by considering the orbits represented in Fig. 27*.

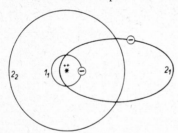

Fig. 27. Screening of nuclear charge by an electron circulating in the 1_1 orbit, according to the theory of Bohr-Sommerfeld.

This may perhaps apply to orthohelium. Whether the orbits lie in one plane or not is unimportant. In any case, the electron which is rotating in the circular orbit characterized by the quantum numbers $n = 1$, $k = 1$ exerts a repulsive force upon any electron which may be further out. In consequence the full attraction of the positively charged nucleus (doubly charged in the present example) for the outer electron does not come into play. The nuclear charge is *screened* by the electrons circulating in the inner orbit. The amount of screening is dependent on the distance. When it is far from the nucleus an electron circulating in an elliptical orbit, such as might be characterised by the quantum numbers $n = 2$, $k = 1$, will be attracted in substantially the same way by a doubly charged nucleus with one electron circulating round it as by a singly charged nucleus. In one part of its orbit however, it is closer to the nucleus than the electron 1_1, so that in this part of the orbit practically the full nuclear charge is operative. On the other hand, for an electron moving in the circular orbit 2_2, the screening acts to the same degree throughout its path. It is accordingly plausible that, in the example cited, the energy of an electron in the elliptical

* The wave mechanical theory of atomic structure leads to substantially the same final result, the effect of screening on the total energy of the atom depending on the ratio between the principal and subsidiary quantum numbers. Since, however, wave mechanics abandons the concept of the movement of the electrons in definite orbits, the interpretation which it provides cannot be so clearly derived as on the basis of the Bohr-Sommerfeld model, though it is more exact in its ultimate conclusions. The discussion is given here in terms of the Bohr-Sommerfeld model. See p. 98, however, for the symmetry of the electron density distribution in the wave mechanical model, for states with $l = 0$, $l = 1$ etc.

orbit $n = 2$, $k = 1$ would be different from that in the circular orbit $n = 2$, $k = 2$ and more exact calculation shows that this must indeed be the case. Because of the inconstancy of the force acting from the focal point, the elliptical orbit suffers a distortion, such that the movement of the electron can be roughly described as circulation round an ellipse, of which the major axis itself rotates around the focus. The energy difference between the circular and elliptical orbits may indeed be emitted as a spectral line, as can be seen from the scheme of Fig. 26. The first line of the principle series both of ortho- and of parahelium originates from the transition of electrons from circular orbits $n = 2$, $l = 2$ to elliptical orbits $n = 2$, $l = 1$.

Closer examination of Fig. 26 will show that, in every case, lines belonging to any one series are characterized not only by the same ground level, but also by the same subsidiary quantum numbers of their various initial levels. This is true also for the series of other heavier elements.

With helium, and also with other elements, it is usual to recognize the following four series: the *principal* series, *sharp subordinate* series, *diffuse subordinate* series, and *fundamental* or *Bergmann* series*. The initial orbits for these series are known quite generally (i.e., not only for helium but for all other atoms also) as p-, s-, d-, and f-orbits respectively and the corresponding energy levels are p-, s-, d-, and f-levels. In so far as it has been possible to assign quantum numbers to the terms it has been shown that the p-orbits correspond to the subsidiary quantum number $k = 2$, the s-orbits to $k = 1$, the d-orbits to $k = 3$, and the f-orbits to $k = 4$. As has already been stated, the subsidiary quantum numbers k of the Bohr-Sommerfeld theory correspond in wave mechanics to subsidiary quantum numbers l which are one unit smaller than the Bohr subsidiary quantum numbers ($l = k - 1$). We thus have

Name of series	Symbol denoting initial level of each line	Subsidiary quantum number assigned to initial level	
		In Bohr theory	In quantum mechanics
Principal	p-levels	$k = 2$	$l = 1$
Sharp subsidiary	s-levels	$k = 1$	$l = 0$
Diffuse subsidiary	d-levels	$k = 3$	$l = 2$
Fundamental (Bergmann)	f-levels	$k = 4$	$l = 3$

The different kinds of series are distinguished from one another in characteristic ways, so that as a rule the spectroscopist can deduce at once from the general character of the series to which initial orbits they should be assigned.

(c) The Neutral Helium Atom

It follows from the spark spectrum of helium, in accordance with what was said earlier, that helium has the nuclear charge 2. The neutral helium atom thus has 2 electrons. In the normal state, i.e., the state of lowest energy, both of these are equivalent in binding (apart from their anti-parallel spins, cf. p. 120). The ground term of the shortest wave length series of parahelium shows, namely, that the second electron is to be assigned the principal quantum number $n = 1$ in its ground level—i.e., the lowest possible principal quantum number. The same must accordingly apply to the first, which can naturally not be more loosely bound than the second. That the system comprised by these two electrons is of quite peculiar stability, is shown, for one thing, by the extremely high ionization potential of helium. The especial stability of the electronic system in normal helium is shown

* The term fundamental series has been used chiefly in English and American literature because of the particularly simple structure of this series.

even more clearly if one compares the energies which are necessary, on the one hand to raise the electron from the $1s$ level to the next higher ($2s$) level, and on the other hand to raise it from the $2s$ to (e.g.) the $3p$ level (cf. Fig. 26). The first amounts to 20.55 eV, the second only to 2.42 eV. The extremely large energy difference between the energies of normal helium (parahelium) and the metastable form (orthohelium) also brings out the exceptional stability of the electronic system present in the former. This peculiar stability is of great importance not only for the chemical behavior of helium itself, but also for the behavior of the elements following helium, as will be shown in the following chapter. What applies to helium is true, in this respect, of the other inert gases also; the electronic systems of these are also marked by special stability, as their high ionization potentials show (cf. Table 22).

6. Number of Electrons in the Inert Gas Shells

Neutral helium possesses 2 electrons, as has been shown, both being normally bound in '1-quantum' levels. The element following helium, lithium, has 3 electrons in all, as will be shown in Chap. 6. Of these, one is quite loosely bound, as may be seen from the ionization potentials set out in Table 23, while the other two are very tightly bound. Beryllium, the next element, has 4 electrons, of which 2 are relatively loosely and the other two again very firmly bound. The next element has 5 electrons, of which 3 are considerably less tightly bound that the other two (cf. Table 23). With each one of the elements following helium, as far as the VIIth Group, as many electrons as are represented by the Group number can be removed by ionization, with far greater ease than can the remaining two. Thus in each of these elements there are two electrons which are particularly tightly bound, as in helium.

TABLE 23

IONIZATION POTENTIALS OF THE ELEMENTS H TO Ne

Columns I to X give the ionization potential (in volts) for the removal of the 1st, 2nd, 3rd,......10th electron

Element	I	II	III	IV	V	VI	VII	VIII	IX	X
$_1$H	13.54									
$_2$He	24.48	54.14								
$_3$Li	5.37	75.26	121.84							
$_4$Be	9.30	18.12	153.11	216.63						
$_5$B	8.28	25.00	37.74	258.03	338.53					
$_6$C	11.24	24.26	47.64	64.17	390.02	487.55				
$_7$N	14.46	29.44	47.20	72.04	97.40	549.08	663.73			
$_8$O	13.57	34.94	54.63	77.03	113.30	137.42	735.22	867.09		
$_9$F	17.46	34.81	62.35	86.72	—	156.37	184.26	—	1097.7	
$_{10}$Ne	21.47	40.91	63.30	ca. 97	—	—	—	—	—	1355.5

It appears likely, therefore, that the electronic system present in helium, which is distinguished by particular stability, recurs in the atoms of each of the succeeding elements. The electrons additional to this would then not be assigned to this system, but bound outside it 'in a second shell', as it is customarily and graphically expressed. The results of X-ray spectroscopy accord with this hypothesis (Chap. 7); they show that in all the succeeding elements, a one-quantum system with two electrons is present in the interior of the atom, as in helium.

When we proceed further in the Periodic System to the element of atomic number 10, we come again to an inert gas, neon. This possesses 10 electrons in all, of which 8 are in the 'second shell'. Since the electronic system of neon is again characterized by particular stability, we conclude that an especial stability is to be ascribed to this 'octet', as it is to the helium system. Proceeding further from neon, we come again to elements with 1, 2, 3 etc. more loosely bound electrons. We may assign these in turn to a 'third shell', as is again substantiated by the results of X-ray spectroscopy. The element with atomic number 18 is again an inert gas, argon. This means that the third shell is also completed when its electron number has reached the total of 8.

The next inert gas, krypton, has 18 electrons more than argon. One might suppose from this that after two 'octets' have been built up, a shell with 18 electrons is next formed. The actual situation is somewhat different. In this case also, i.e., in the fourth shell, an octet is once more completed, but in order to attain this a total of 18 electrons is required, since during the process 10 of these are built into an 'intermediate shell', between the third and fourth shells. The evidence for this will be brought forward in discussing the Sub-groups of the Periodic System, since the origin of the Sub-groups is very closely bound up with the filling up of the intermediate shell by 10 electrons. The same applies with the following inert gas, xenon. This, in its turn, has 18 electrons more than krypton; of these, only 8 are once more in the outermost—now the fifth—shell, and 10 in an intermediate shell between the fourth and the fifth shells. Finally, radon possesses 32 electrons more than the inert gas preceding it. Once more, only 8 are contained in the outermost shell; all the rest are in intermediate shells, whereby among other things, the existence of the subgroup of the *lanthanides* can be explained as will be shown later. We thus come to the result that each of the inert gases contains an outer electron shell marked by particular stability, containing 2 electrons in the case of helium, but built up from 8 electrons in all the other inert gases.

The kind of statement that the atoms are 'built up from successive electron shells' should be understood in a schematic sense only. The real meaning of the 'shells' is that all electrons with the same principal quantum number are assigned to the same 'shell'. Such a statement as 'the electron belongs to the third shell' therefore merely means that the ground level of the electron is a level with the principal quantum number $n = 3$. The shells in the heavier atoms (krypton etc) were originally numbered without including the 'intermediate' shells in the numbering.* The 'shells' can be identified with the energy levels in the atom, which are indeed given by the quantum numbers (cf. Fig. 55, p. 229, which represents a section through the energy level surfaces of a heavy atom); they possess the same reality as any other equipotential surfaces in the neighborhood of an electrically charged body.

The similarities between the structures of the inert gas shells is still more evident if one considers the quantum numbers of the electrons, and especially the subsidiary quantum numbers, as well as the number of electrons. It also becomes clear why the inert gas configu-

* On the assignment of quantum numbers to the intermediate shells, see p. 230.

ration is attained, in the case of helium, with 2 electrons instead of only at 8 electrons. For a complete description of the mode of binding of an electron in an atom, it is necessary to specify not only the *principal quantum number n*, the *angular momentum quantum number l*, and the *magnetic quantum number m*, but also a fourth quantum number, the so-called *spin quantum number s*. This defines the spin of the electron (in the Bohr-Sommerfeld model)—i.e., its rotation about its own axis (cf. p. 171). For any one electron it can assume only two values $+\frac{1}{2}$ or $-\frac{1}{2}$, according to the direction of rotation. The numerical values which may be possessed by the quantum numbers named can be briefly summarized by the following relations:

$$0 \leq l \leq n-1 \qquad |m| \leq l \qquad s = \pm\frac{1}{2} \qquad (n, l, m \text{ all integral})$$

It has been found that in their outer shells, in the ground state, the inert gases never contain an electron for which l is greater than 1. If $l = 1$, then according to the above relations, m can assume the values -1, 0 or $+1$. For $l = 0$, only $m = 0$ is permitted. To each of the possible pairs of parameters l, m thus permitted, there may belong two values of s, namely $+\frac{1}{2}$ or $-\frac{1}{2}$. There are, accordingly, $2 \times 4 = 8$ different combinations of quantum numbers possible in this case, and exactly this number are present in the inert gas shell. Only in the case of helium, since $n = 1$, and consequently $l = 0$, are only two different combinations possible, namely $l = 0$, $m = 0$, $s = +\frac{1}{2}$ and $l = 0$, $m = 0$, $s = -\frac{1}{2}$. The inert gas shells therefore contain just as many electrons as can be accommodated in different quantum states provided that l does not exceed the value 1.

7. The Pauli Principle

Going beyond the inert gases, to the elements with higher nuclear charge, the additional electrons taken up are built into other shells, as has already been mentioned. It may be inferred that a shell is completed when each of the quantum states possible in it is occupied by an electron. Pauli, on this basis put forward the principle that in one and the same atom there cannot be two electrons which have all four quantum numbers, n, l, m, and s the same.

The varying length of the Periods in the natural system of the elements can at once be derived from the Pauli principle, and the occurrence of the Sub-groups and the groups of lanthanides can thereby be explained. It has also become of great importance for the explanation of the homopolar (covalent) chemical bond, as will be shown in the next chapter.

8. Symbols for Atomic Structure

Although *four* quantum numbers must be specified in order to define precisely the binding energy of an electron, the relation between *atomic structure* and *chemical behavior* can usually be expressed by giving *two* quantum numbers—the principal quantum number n and the angular quantum number l. The following abbreviated symbolism has come into use for stating these quantum numbers. An electron occupying an energy level characterized by the angular quantum number $l = 0$ is said to be an s electron, one in a level with angular quantum number $l = 1$ is a p electron, one in a level with $l = 2$ is a d electron, and one in a level with $l = 3$ is an f electron.

The basis of this nomenclature is that, as in the case of helium, the lines in the atomic spectra of the elements of higher atomic number than hydrogen can mostly be grouped into the four kinds of series referred to on p. 117. The initial levels giving rise to the various lines in these series are then denoted by the initial letters of the names of the series.

Energy levels with angular momentum quantum numbers higher than $l = 3$ are never met with in the ground states of the atoms of even the heaviest elements. The symbols s, p, d, f thus enable us to write down the binding states of the electrons in the atoms of all the elements in a compact but precise manner. The *principal quantum number* specifying the energy level is written down immediately preceding the symbol for the angular quantum number. The *number of electrons* in the atom having the particular principal and angular quantum numbers thus described (if more than one) is written as a superscript (or exponent) after the symbol for the angular quantum number. If electrons are present in several different energy states their symbols are written down one after another.

Thus a '$2s$ electron' is an electron in an energy state with principal quantum number $n = 2$, and angular quantum number $l = 0$. If there are four electrons in p-states, with principal quantum number $n = 2$, this is represented by the symbol $2p^4$*. Similarly, the symbol $2s^2 2p^6$ expresses the fact that in the atom referred to there are two s-electrons with principal quantum number $n = 2$, and six p-electrons with principal quantum number $n = 2$.

The *electronic configuration* (i.e., the mode of binding of all the electrons) of the atoms of hydrogen, helium and neon can therefore be represented in the following way.

H	He	Ne
$1s$	$1s^2$	$1s^2 2s^2 2p^6$

The characteristic feature of the atomic structure of all the inert gases is that their 'outer shells' are completely filled with s and p electrons. A completely filled shell is one in which every possible quantum state is occupied by an electron. The Pauli principle implies that an s level can accomodate not more than 2 electrons, and a p level not more than 6 electrons. Hence in all the inert gas atoms, the outermost shell has the configuration $ns^2 np^6$—except in the case of helium for which, since $n = 1$, l necessarily $= 0$, so that there can be no p electrons.

In Table II (Appendix) the electronic configurations for all the elements are represented by the symbolism explained above. It may be seen that the special place of the inert gases among the elements is connected with the fact that the inert gas atoms mark, on each occasion, the completion of an outer shell, and that with the element following the inert gas, a fresh electron shell is begun.

9. The Emanations

The *emanations* (from Latin *emanare*, to stream out) are gaseous substances which are continuously given off by certain radioactive elements, and which in their turn are themselves radioactive. Chiefly three different emanations are known, from the naturally occurring radioactive elements**; they are named after

---

* Note that the figure 4 is not a *power*. The symbol is therefore spoken as 'two-p-four'.

** There are *four* natural emanations known. In the radium series there occurs, besides that arising from radium, another emanation from one of the radium decomposition products (astatine emanation). Its concentration is, however, extremely small (less than 10^{-12} of the

the radioelements from which they are directly produced. They are therefore distinguished as radium emanation (RaEm), actinium emanation (AcEm) and thorium emanation (ThEm). The first emanation discovered was that of thorium, by Rutherford in 1900. In the following year Dorn observed the formation of radium emanation, and shortly afterwards actinium emanation was discovered by Giesel and by Debierne. Radium emanation is the inert gas *radon*, already discussed.

It has been shown that not only radium emanation, but also the others, have the character of inert gases, and accordingly all belong in the same group of the Periodic System. It will be shown in the chapter on Radioactivity and Isotopy (Vol. II) that it is possible to deduce the atomic weights of disintegration products from the theory of radioactive disintegration. In every case where experimental verification has been possible, atomic weights deduced from this theory have been found to agree with those found experimentally. This is so, for example, in the case of radon, for which Ramsay and Whytlaw Gray found approximately 223 from density determinations, as the mean of several experiments, whereas the disintegration theory leads to the value 222. When one considers the experimental errors unavoidably associated with density determinations carried out on such minimal quantities of gas as were used, the agreement is quite satisfactory. The theoretical value can, however, be regarded as numerically the more trustworthy. The disintegration theory gives 220 for the atomic weight of thorium emanation (*thoron*) and 219 for actinium emanation (*actinon*). These values of the atomic weight exclude any possibility of fitting the three emanations into different series of the Periodic System. They must in fact all three belong both in the same series and in the same group—that is, they all have to be accommodated *in the same place* in the Periodic System, so that the other two emanations are *isotopic* with radium emanation.

The same conclusion is reached more directly—though the argument follows less closely from the experimentally observable facts—from the fact that the place appropriate to each element in the Periodic Table is determined unequivocally by its atomic number. The same atomic number, 86, is derived from the theory of radioactive decay for all three emanations, so that all three should be assigned to the place characteristic of the atomic number 86.

Being isotopic substances, the three emanations do not differ from one another in their chemical properties. They do, however, differ as regards their radioactive properties, which will be discussed in Vol. II.

10. Isotopy of the Inert Gases

Although, once they have been mixed together, radon and the other emanations cannot be separated again by the processes usual in chemistry, they can readily be obtained pure from the radioelements that form them by their disintegration. The

---

radium emanation). There are also three other nuclear species, isotopic with the naturally occurring emanations, which do not occur in Nature, but are formed by the radioactive disintegration of artificially prepared nuclear species of atomic number 88. See Vol. II.

other inert gases, however, occur in Nature as isotopic mixtures. The number of isotopes in these mixtures increases considerably from helium to neon. He consist-of only 2 isotopes, of which the lighter (atomic weight 3) is mixed only in vanishs ingly small amount (1 in 10 million) with the heavier (atomic weight 4). Ne ha. 3 isotopes, as has argon. Kr consists of 6 and Xe of 9 isotopes (cf. Vol. II, Chap. 12).

References

1 W. RAMSAY and H. RUDOLF, *Die Edelgase* (Vol. II of Ostwald-Drucker, *Handbuch der allgemeinen Chemie*), Leipzig 1918, 416 pp.
2 W. RAMSAY, *The Gases of the Atmosphere*, 4th Ed., London 1915, 306 pp.
3 M. W. TRAVERS, *The Discovery of the Rare Gases*, London 1928, 128 pp.
4 W. H. KEESOM, *Helium*, Amsterdam 1942, 494 pp.
5 S. GOLD, *Neon*, London 1934, 178 pp. (deals with the technical basis of neon tube illumination).

CHAPTER 5

VALENCE AND AFFINITY

1. Valence

(a) Grouping of the Elements about the Inert Gases

One arrangement of the Periodic Table which brings out the properties of the elements and their compounds with particular clarity is obtained by grouping the elements about the inert gases, as in Table 24.

This form of the Periodic System differs from the usual one, in that the group of inert gases is placed neither at the beginning nor the end of the System, but is so placed that the Main Groups are arranged without a break, even in the long periods. In this way the inert gases come roughly into the middle of the Main Groups, and form an axis about which the other elements of the Main Groups are balanced. The mutual relations of the elements in a horizontal direction are thereby particularly clearly shown.

As set out in Table 24, the *valences* of the elements increase quite regularly by a unit at a time in going from one group to another, both in the Main Groups and in the Sub-groups. In the latter, this is subject to the restrictions that the elements can generally alter their valence (or ionic charge) relatively easily, and that the valence state corresponding to the Group Number is often unimportant as compared with other valence states. In the Main Groups, the maximum positive valence state at first increases in steps of unity from 3 to 7. The maximum negative valence state decreases by the same amount and passes through 0 at the inert gases; the valence then rises to positive values. In all, the valence stages from —4 to +2 are run through, in regular unit steps, as one goes from left to right, and this trend continues right up to the value +8 where the periods continue into the Sub-groups. A zone of elements with a preference for moderately high valences follows, in the VIIIth Group, whereupon the regular rise of positive valences from +1 onwards begins once more. This apparently finds its natural continuation in the rise of the positive valence from 3 to 7 in the elements of the Main Groups immediately following, as has been already mentioned.

It may be seen, further, that elements which precede the inert gases form volatile hydrides, those which immediately follow the inert gases form solid, salt-like hydrides. Moreover the great majority of the elements preceding the inert gases have a *non-metallic character*; the higher in the Table, and the closer to the inert gases, that any element stands, the more marked this is. All the elements standing to the right of the inert gases are *metals*. The closer an element is to the next-following inert gas, the more acidic is the nature of the hydride. The closer any element is to the inert gas next preceding it, the stronger is its base-forming character.

Colored elementary ions, or paramagnetic ions, are formed by none of the elements preceding the inert gases in Table 24, and by none of the elements immediately following; (i.e. standing not more than 3 places from an inert gas).

TABLE 24

GROUPING OF THE ELEMENTS AROUND THE INERT GASES

Main Groups of the Periodic System

	III	IV	V	VI	VII	VIII / O	I	II
Principal valence states	+III	+IV / −IV	+V / −III	+VI / −II	+VII / −I	O	+I / −	+II / −
Simplest hydrides		Volatile				—		Solid, salt-like
					[$_1$H]	$_2$He	$_3$Li	$_4$Be
	$_5$B	$_6$C	$_7$N	$_8$O	$_9$F	$_{10}$Ne	$_{11}$Na	$_{12}$Mg
	$_{13}$Al	$_{14}$Si	$_{15}$P	$_{16}$S	$_{17}$Cl	$_{18}$A	$_{19}$K	$_{20}$Ca
	$_{31}$Ga	$_{32}$Ge	$_{33}$As	$_{34}$Se	$_{35}$Br	$_{36}$Kr	$_{37}$Rb	$_{38}$Sr
	$_{49}$In	$_{50}$Sn	$_{51}$Sb	$_{52}$Te	$_{53}$I	$_{54}$Xe	$_{55}$Cs	$_{56}$Ba
	$_{81}$Tl	$_{82}$Pb	$_{83}$Bi	$_{84}$Po	$_{85}$At	$_{86}$Rn	$_{87}$Fr	$_{88}$Ra
Magnetism	All elementary ions are diamagnetic							
Color	Form *colorless* elementary ions exclusively							

The properties of the last element of each series are related in each case to those of the first element in the next series. B forms volatile hydrides, like C.

Lanthanides
$_{58}$Ce, $_{59}$Pr, $_{60}$Nd, $_{61}$Pm, $_{62}$Sm, $_{63}$Eu, $_{64}$Gd, $_{65}$Tb, $_{66}$Dy, $_{67}$Ho, $_{68}$Er, $_{69}$Tm, $_{70}$Yb, $_{71}$Lu

Sub-groups of the Periodic System

	III	IV	V	VI	VII	VIII			I	II
Principal valence states	+III	+IV	+V	+VI	+VII	+VIII	Readily variable valence		(+I)	+II
	…	…								
Simplest hydrides	Hydrides quasi-metallic									
	$_{21}$Sc	$_{22}$Ti	$_{23}$V	$_{24}$Cr	$_{25}$Mn	$_{26}$Fe	$_{27}$Co	$_{28}$Ni	$_{29}$Cu	$_{30}$Zn
	$_{39}$Y	$_{40}$Zr	$_{41}$Nb	$_{42}$Mo	$_{43}$Tc	$_{44}$Ru	$_{45}$Rh	$_{46}$Pd	$_{47}$Ag	$_{48}$Cd
	$_{57}$La	$_{72}$Hf	$_{73}$Ta	$_{74}$W	$_{75}$Re	$_{76}$Os	$_{77}$Ir	$_{78}$Pt	$_{79}$Au	$_{80}$Hg
	$_{89}$Ac	$_{90}$Th	$_{91}$Pa	$_{92}$U	$_{93}$Np	$_{94}$Pu	$_{95}$Am	$_{96}$Cm	$_{97}$Bk	$_{98}$Cf
									$_{99}$E $_{100}$Fm $_{101}$Mv	
Magnetism	Some elementary ions paramagnetic									
Color	Form *colored* elementary ions in many cases									

Transuranics: $_{93}$Np $_{94}$Pu $_{95}$Am $_{96}$Cm $_{97}$Bk $_{98}$Cf $_{99}$E $_{100}$Fm $_{101}$Mv

The relationships mentioned—to which several others might be added—are sufficient to lead to the supposition that many properties of the elements, and especially their chemical properties, must be connected in some way with their position in the Periodic System, in relation to the inert gases. In other words, the properties of any element, and especially its valence, must depend to a considerable extent on the number of units by which its atomic number differs from that of an inert gas. An important advance was made simultaneously and independently, during the year 1915–1916, by G. N. Lewis in the United States, and by A. Kossel in Germany, when this relationship was brought into the forefront of all considerations of valence theory.

(b) Kossel's Theory

If, among the elements adjacent to the inert gases, we first consider those which are demonstrably able to form elementary ions*—i.e., the alkali and alkaline earth metals, aluminum and its homologues, and on the other hand the halogens, sulfur and its higher homologues—it is striking that the total number of electrons possessed by each of these elements in its elementary ions is *in every case equal to the number of electrons in the nearest inert gas*.

Thus the $S^=$ ion possesses $16 + 2$ electrons, the Cl^- ion $17 + 1$ the K^+ ion $19 - 1$, the Ca^{++} ion $20 - 2$, the Sc^{+++} ion $21 - 3$, etc., so that each has in all 18 electrons—the same number as has argon, about which these elements are grouped. In the same way, the F^-, Na^+, Mg^{++} and Al^{+++} ions all have 10 electrons, as has neon.

If, now, we compare the chemical compounds which dissociate to furnish these ions with the compounds of other elements of the Main Groups, which are analogous in constitution and function, it is reasonable to suppose that the elements are present in the charged state in these compounds also, even in those cases where the existence of ions in the free state cannot be proved. It has been possible to show by other means (e.g., by X-ray methods, (Chap. 7)) that some compounds are built up of oppositely charged atoms, even in the solid state. This is true, for example, of magnesium oxide MgO (cf. p. 213), which can be shown to be built up from positive bivalent magnesium ions Mg^{++} and negative bivalent oxygen ions $O^=$. Generalizing from these facts, Kossel concluded that the oxides of other metals are also composed of positive metal ions and negative oxygen ions, and indeed that the compounds of metals with non-metals always contain the metal in the positively charged, and the non-metal in the negatively charged state. Kossel made the further assumption that the atoms are also electrically charged in those compounds of non-metals with one another which are analogous in composition to the demonstrably heteropolar (electrovalent or ionic) compounds; it was assumed that the more strongly electronegative atom would always be negatively charged in the resulting compound. Weakly electropositive metals would also bear a negative charge, in their compounds with strongly electropositive metals or with hydrogen. In every case, where simple molecules are involved, the number of charges would correspond with the formal valence.

* By elementary ions are meant ions in solution (usually aqueous solution); for ions in crystals, see further below. In canal rays, etc., the atoms may occur in quite anomalous states, which do not permit one to draw any direct conclusions about their nature in chemical compounds.

Abegg (1904) had already advanced numerous facts to support the thesis that the majority of compounds are heteropolar (ionic) in structure. Except for the cases discussed, however, strict proof is lacking. We must therefore bear in mind that these generalizations are hypothetical in nature: they are tentative hypotheses, leading to a general theory which enables us to understand how various properties, and especially the valence, are related to the inert gases. When we once have this, and have learned how to support it by other evidence, we can better judge, retrospectively, how far these generalizations are really essential to the theory, how far they are subsequently confirmed and how far they must be altered or queried.

Kossel [1] considered that compounds in which the elements exert the valence typical of their position in the Periodic System, are generally heteropolar (ionic)

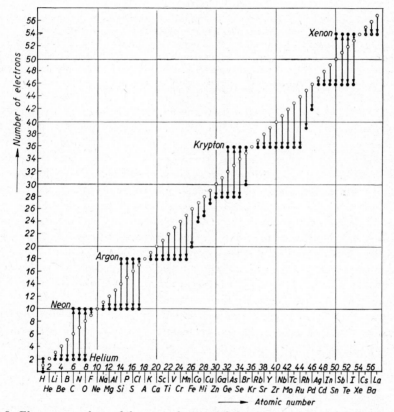

Fig. 28. Electron numbers of the neutral atoms of elements 1–56 (open circles) and their maximally charged forms (black points), after Kossel.

in structure. He worked out the effective atomic numbers of the first 57 elements (i.e., up to the Sub-group of lanthanides) in those compounds in which they exert their highest positive and negative valences. In Fig. 28 the elements are set out in the sequence of their atomic numbers, and the electron number (Sidgwick's 'effective atomic number') is marked as an ordinate. For elements which can occur both positively and negatively charged, there are two points marked, which are, of course, vertically above one another, and 8 units apart, corresponding to the fact that the sum of the highest positive and negative valences is 8 as pointed out by

Abegg. The open circles in the diagram represent the electron numbers of the neutral atoms of the elements. Whereas these last naturally increase regularly from one element to the next, and lie on a line inclined at $45°$ to the axes, the black points for elements adjacent to the inert gases all lie on lines parallel to the abscissa, and passing through the points representing the neutral inert gas atoms. Hence, in their typical compounds the atoms of the elements of the Main Groups have as many electrons as correspond in number to the electron number of the adjacent inert gas. It follows from this that when two elements, such as sodium and fluorine, combine to form a chemical compound, the one gives up to the other as many electrons as will enable both to acquire the electron number of the nearest inert gas.

It may be seen that other electron numbers than the inert gas number are found at some places in Fig. 28, although only with the elements of the Sub-groups of the Periodic System, which are far from the preceding inert gas. Electron systems which are not of inert gas like type are formed only through the loss of electrons, never through acquisition of electrons The electron systems with 28 and with 46 electrons which are formed only through the los. of electrons, seem to be marked by a certain stability.

Kossel deduced from these rules that the electronic systems of the inert gases possess a particular stability, such that the atoms of the elements immediately preceding the inert gases readily form an electronic system corresponding to that of the inert gases, by taking up extra electrons. On the other hand, the elements following the inert gases readily split off as many electrons as they possess over and above the number in the atom of the preceding inert gas. This conclusion, reached by Kossel from chemical evidence, as to the special stability of the electron system present in the inert gases, has since been substantiated in general terms by purely physical means (ionization potentials) as was shown in the previous chapter.

According to Kossel, then, the heteropolar chemical bond (or *electrovalence*) arises from the attraction between oppositely charged atoms, according to Coulomb's Law. In doing so they behave, to a first approximation, as spheres, with their charge uniformly distributed over their surface.

Thus, in Kossel's view, the formation of potassium chloride from the elements chlorine and potassium, depends on the transfer of one electron per atom from potassium to chlorine. The ions Cl^- and K^+, so formed, associate themselves into the potassium chloride crystal, in accordance with Coulomb's law. The same applies to the formation of MgO from magnesium and oxygen, etc.

It is, of course, not necessary that both atoms should acquire the electronic configuration of the *same* inert gas. In place of potassium and chlorine, we could have chosen sodium and chlorine, or lithium and bromine as examples. Further, the number of electrons per atom accepted by the one element need not be the same as that given up by each atom of the other elements: only the total numbers of electrons given up and accepted respectively must agree. Hypothetically, one silicon atom gives up 4 electrons to oxygen. Two oxygen atoms thereby become negatively charged, so that by the union of Si^{4+} with $2O^{2-}$ the molecule SiO_2 is formed. The same reasoning may be extended to the formation of any more complicated compounds.

The Kossel theory of compound formation could be summarized as follows: the combining atoms become oppositely charged, through the exchange of electrons, and then cohere because of the electrostatic attraction produced by their charges.

(c) Applications of the Kossel Theory

Not only does the Kossel theory attribute the forces of chemical combination to a very simple principle, but it enables this same principle to be applied at once to the interpretation of the coordination compounds which, in the valence theories based on individual valence bonds, had presented considerable difficulties (see Chap. 11). It was also far superior to the older inorganic valence theories in that it furnished not only the *composition* of the chemical compounds but also,—at least in the simpler cases—their most important properties without auxiliary hypotheses. Thus, as will be shown, it is possible on the basis of the Kossel theory, to predict whether the most important hydroxides should be acidic or basic in behavior. Many other examples of the application of the Kossel theory to particular problems will be cited later. In this place it will be shown by means of a few examples how the theory shows up the natural relationships and leads much further than the older valence theories.

In the first place, it provides an explanation for the zero-valence of the inert gases—or more correctly, for their inability to form heteropolar compounds. If the ability of the elements to enter into combination with one another depends on acquiring an ionic charge through their tendency to achieve the inert gas configuration, there is, of course, no reason for the inert gases to form compounds. They already possess the most stable electronic configuration, so that there is no question of acquiring electrons (becoming negatively charged). There is also no reason for them to become positively charged—except of course, in electrical discharge tubes, or at extremely high temperatures: it is hardly likely that any other element will be able to abstract their outermost electrons in order to build up for itself the configuration already present in them. This conclusion is completely borne out by quantitative calculations of the free energy of the reactions involved (cf. p. 145 *et seq.*).

We can also now explain why oxygen and fluorine never appear in the positive 6- or 7-valent state, as do their higher homologues. In oxygen the electrons are particularly tightly bound (cf. Tables 22 and 23): there is no element able to abstract them entirely.*

The exceptional place of hydrogen in the Periodic System follows at once from its relation to helium. It resembles the alkali metals, which also give up only 1 electron, in that, in the neutral state, it has only 1 electron to give up. It differs from the alkali metals, however—and indeed from all metals—in that this electron is relatively tightly bound (cf. Table 22). In spite of the fact that it generally functions as positively charged, it proves to be a non-metal. Its position preceding helium, the inert gas with 2 electrons, leads to the possibility of appearing as electronegative also; it need acquire only one electron to attain the inert gas configuration. It may, perhaps, be said that in this respect, and in view of its preponderably non-metallic physical properties, it stands in closer relation to the halogens than to the alkali metals.

(d) The Valence Theory of Lewis [3]

The valence theory published by G. N. Lewis in January 1916 shared with Kossel's theory, published at the end of December 1915, the fundamental conception that the motivating principle in the formation of chemical compounds was the especial stability of the electronic configuration present in the inert gases. In contrast with Kossel, Lewis placed the emphasis in his considerations rather on the *homopolar (covalent)* compounds. He assumed, accordingly, that the electronic configurations distinguished by special stability could frequently arise through the *sharing* of electrons between the atoms concerned as well as by the complete transfer

* The reason that O and F have maximum valences of 2 and 1 respectively in *homopolar (covalent)* compounds will be explained in Chap. 15.

of an electron from one atom to another. His original presentation of the theory was based by Lewis on certain views of atomic structure which have been super-seded, but which are not essential to the theory. They are therefore not elaborated here.

The Lewis theory was extended by Langmuir in 1919, who showed that its application led, in many cases, to striking results. Thus it provides a probable explanation for the con-siderable resemblance in physical properties between substances of very different chemical character, such as CO and N_2, or CO_2 and N_2O. According to the Lewis-Langmuir theory, these pairs of molecules would possess the same outer electronic systems. Langmuir called such substances *isosteric* (cf. J. Goubeau, *Naturwiss.*, 35 (1948) 246).

Lewis and Langmuir assumed that the inert gases each contain 8 electrons in their outer-most shell, with the exception of helium, with 2 electrons. This assumption is in accord with the conceptions of atomic structure which have grown out of the Bohr theory. Although Kossel made the same assumption, it did not possess any fundamental significance in his theory. The assumption of a shell of 8 electrons is essential, however, in the theory of Lewis, who attributed the exercise of valence to the attempt to complete as many groups of eight ('octets'–Langmuir) as possible. Langmuir accordingly termed the theory the *Octet Theory* of valence.

In addition to the completion of octets, Lewis and Langmuir attach considerable impor-tance to the tendency to form electron pairs—formation of *doublets*. This conception of the tendency of electrons to form pairs has become a prominent feature of the modern develop-ment of valence theory (see below).

Lewis introduced a simple mode of representing the electrons in chemical formulas. He indicated the electrons of the outermost shells—which alone are involved in binding, in his sense—by dots, placed around the corresponding chemical symbol. Those electrons which are shared by two atoms are represented by dots placed between them, and put closer to the chemical symbol representing the more electronegative atom. As a rule, each pair of dots written between the symbols represents a valence bond as written in the usual way. Examples

				H
Ordinary representation	Cl—Cl	O=O	H—Cl	H—N—H

As written by Lewis	: Cl : Cl :	: O : : O :	H : Cl :	H : N : H
				H

The characteristic feature of the Lewis-Langmuir theory is that it is concerned primarily with homopolar compounds (*covalent* compounds), whereas that of Kos-sel, conversely, starts from the heteropolar compounds. There is no doubt that the Kossel concept is substantially true for typical heteropolar (ionic) compounds, but can be extended only with reservations to include compounds with less clearly marked heteropolar character. [2] For pure homopolar combination, Kossel devel-oped views which are related to some extent to those of Lewis and Langmuir. The latter theory contains some features which are very fruitful for the understanding of the homopolar bond. [4–12, 27–29] The essential core of truth in its underlying ideas has been confirmed through the *wave mechanical* treatment of the chemical bond. It must be remembered, however, that formulas based on the Lewis-Lang-muir theory are in most cases largely hypothetical, so that caution must be exer-cised in making deductions from them.

(e) The Wave Mechanical Treatment of the Covalent Bond

The formation of heteropolar compounds can at once be understood on the basis of the Kossel theory, since the attractive force of the oppositely charged atoms for one another is given directly by Coulomb's law. It is, however, not at once obvious how atoms which are not oppositely charged should be firmly bound to each other through the mutual sharing of electrons, as is postulated in the Lewis theory. A satisfactory explanation of how this comes about, through the operation of Coulomb forces, has been furnished only by *wave mechanics*.

For a system consisting of two hydrogen atoms, the Schrödinger equation takes the form

$$\frac{\partial^2 \psi}{\partial x_1^2} + \frac{\partial^2 \psi}{\partial y_1^2} + \frac{\partial^2 \psi}{\partial z_1^2} + \frac{\partial^2 \psi}{\partial x_2^2} + \frac{\partial^2 \psi}{\partial y_2^2} + \frac{\partial^2 \psi}{\partial z_2^2} +$$

$$\frac{8\pi^2 m}{h^2} \left\{ \varepsilon - e^2 \left[\frac{1}{r_{ab}} + \frac{1}{r_{12}} - \frac{1}{r_{a1}} - \frac{1}{r_{a2}} - \frac{1}{r_{b1}} - \frac{1}{r_{b2}} \right] \right\} \psi = 0$$

In this equation, the subscripts $_1$ and $_2$ refer to the two electrons and the subscripts $_a$ and $_b$ refer to the two atomic nuclei. Thus r_{ab} denotes the distance of the two nuclei from one another, r_{12} the mutual distance between the two electrons, r_{a1} the distance of electron I from nucleus a, etc. Heitler and London (1927), on the basis of this equation, worked out the interaction between two neutral hydrogen atoms in their ground state. It emerged that two different solutions could be obtained for the energy of interaction. From the first solution, there is no finite probability that an electron belonging to the one atom will be found in close proximity to the second nucleus. According to the second solution, the probability of such an event is real and finite. The first solution leads to *positive* values of the interaction energy at all interatomic distances—i.e., to a state of *repulsion* between the atoms. The second solution leads to negative values for the energy of interaction for a certain range of values of the interatomic distance—i.e., to *attraction* between the atoms. The former solution is obtained if the two electrons in the system have parallel directions of spin (↑ ↑); the second solution is obtained if the electronic spins are antiparallel (↑↓). The result means that as the two atoms approach one another, the potential energy of the system either increases of decreases, according as the two electrons have the same or opposite spins.

The forces of interaction between the two H-atoms which are thus bound together to form a H$_2$ molecule are known as *exchange forces*. They are an example of quantum mechanical *resonance*, and they correspond, in a mechanical analogy, to the well known manner in which two equal and (more or less loosely) coupled pendulums interact with one another. If one pendulum is made to vibrate while the other is at rest, the motion of the first pendulum is gradually transferred to the second. Ultimately this is vibrating with the maximum amplitude while the first pendulum is at rest, whereupon the whole process starts again, but in the opposite sense. There is thus a continuous exchange of energy between the two coupled pendulums. Similar considerations apply to a pair of H-atoms coupled through one or two electrons. They can be regarded as continuously exchanging the one electron (in the case of the H$_2^+$ ion) or both electrons (in the H$_2$ molecule).

$$\mathrm{H}_a^e \;\; \mathrm{H}_b \qquad \text{and} \qquad \mathrm{H}_a \;\; {}^e\mathrm{H}_b ; \qquad \mathrm{H}_a^1 \;\; {}^{e2}\mathrm{H}_b \qquad \text{and} \qquad \mathrm{H}_a^{e2} \;\; {}^{e1}\mathrm{H}_b .$$

This is the origin of the term 'exchange forces'. In the H$_2^+$ ion, the resonance leads to a looser binding between the nuclei than it does in the H$_2$ molecule. The resulting attraction is about twice as strong in the latter case. In the case of the H$_2$ molecule, calculation shows that the potential energy has a minimum value when the two nuclei have approached to $0.74 \cdot 10^{-8}$ cm (Wang, 1928)—i.e., to the distance which must be ascribed to the H$_2$ molecule on the basis of physical measurements (cf. Table 5). In this way the formation of the H$_2$ molecule from two H atoms is explained by wave mechanics, which also furnishes the correct value for the energy of formation. H. M. James (1935), who has made the most exact calculations, found for this the value 4.454 ± 0.013 eV, in excellent agreement with the most reliable experimental determinations (4.455 ± 0.001 eV).

It must be emphasized that the so-called *exchange forces*, to which the mutual attraction of the two H atoms is attributed in wave mechanics are electrostatic in origin, and are not magnetic forces, even though the spin of the electrons—i.e., their circular rotation about their own axis—is important. The possibility that two bodies can exert electrostatic forces on each other without bearing opposite electrical charges, through the action of stationary electric waves may perhaps be understood from a model. If we suppose that stationary electric waves are set up (by induction, for instance), in two parallel wires, of equal length, lying close together, each wire remains uncharged as a whole, but electric charges oscillate to and fro. If at the beginning of a cycle the charges are distributed as shown by the full lines in Fig. 29, after one half cycle the charges will have changed their signs (dotted wave lines, Fig. 29) and so forth. If the charges oscillate in phase in the two wires, as in the left

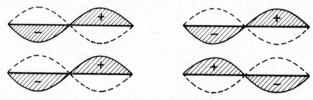

Fig. 29. Electrostatic attraction or repulsion originated by stationary electric waves.

hand diagram, the two wires must repel each other, as similar charges are always adjacent. If the charges are oscillating in opposite phase (as in the right hand diagram, Fig. 29), opposite charges are always adjacent to each other, and in this case the wires must attract each other.

The Heitler-London calculations show that in many cases the number of so-called *unpaired* electrons determines the homopolar valence, or *covalence* of an atom. By 'paired' electrons are meant two electrons of the same atom, which have all their quantum numbers identical except the spin quantum number. Thus in the normal state of helium the two electrons are paired (cf. p. 120). Since helium has no unpaired electrons, it is just as incapable of forming covalent compounds as (by reason of its high ionization potential) it is of forming electrovalent compounds. Li, however, like H, has one unpaired electron. Hence Li can form covalent compounds, as well as electrovalent compounds, so that Li atoms can, for example, combine forming the molecule Li_2. Calculation shows, however, in this case, in accord with experience, that the firmness of binding is much less than for the case of the H_2 molecule (1.14 eV, as compared with 4.45 eV).

The importance of the Heitler-London theory lies in that it has shown that, in principle, covalent binding can be attributed to known forces. On account of the mathematical difficulties arising in the treatment of more complicated systems, its range of rigorous application is limited. Generalizations that are not supported by exact calculation should be regarded with caution: thus it is certainly not permissible to deduce from the theory that covalent binding invariably depends on the pairing of electrons which were initially present, unpaired, in the component atoms of the compound.

Thus, rigorous quantum mechanical calculations (E. Teller, 1930) have shown, in agreement with experimental observations, that the molecule H_2^+, which contains only *one* electron, is a free ion of greater stability (dissociation energy 2.7 eV) than the molecule Li_2, in which the union is brought about by two electrons. Further, spectroscopic and other observations (cf. p. 690) show that the molecule O_2 contains 2 unpaired electrons. Its formation is therefore not bound up with the pairing of all the electrons, and although it possesses unpaired electrons, two O_2 molecules do not unite to form a valence compound O_4 (see below). Furthermore, the molecule NO possesses one unpaired electron, as its spectrum shows, but, as far as is known, two NO molecules combine to form a compound N_2O_2 only at very low temperatures. These few examples suffice to show that the assumption that unpaired electrons form pairs (so-called 'spin coupling'), when atoms unite to form covalent compounds, is to be regarded as a working hypothesis, to be applied with due care. Consequences deduced from the assumption remain open to doubt unless it is proved that the formulas based upon it are in agreement with exact experimental results (e.g., spectroscopic data). It is not superfluous to emphasize this, since electronic formulas based on the hypothesis are finding increasing use in place of the older structural formulas.

Any rigorous extension of the procedure used by Heitler and London, to the calculation of bond energies and interatomic distances in systems more complex than the H_2 molecule, is practically excluded by the mathematical difficulties that arise. There is therefore no exact theoretical treatment of the covalent compounds of the heavier atoms, or of more complex molecules. Nevertheless, Slater (1931) and Pauling (1931 and later[†]) succeeded in extending the Heitler-London method in principle, in such a manner as to afford an insight into covalent bond formation by multivalent atoms. Pauling, in particular, showed how the known stereochemistry of the elements could be accounted for by wave mechanics.

The *maximum number of covalent bonds* that can be formed by an atom, on the Heitler-London model, is equal to the number of unpaired electrons. It therefore varies in a systematic way with the electronic configuration of the elements. In the first period of the Periodic System, the ground state configurations and number of unpaired electron spins of the elements are as follows.

	B	C	N	O	F	Ne
Configuration	s^2p	s^2p^2	s^2p^3	s^2p^4	s^2p^5	s^2p^6
Unpaired spins	1	2	3	2	1	0

It would therefore appear that boron and carbon should be univalent and bivalent respectively, if their covalent compounds are derived from the ground state of their atoms. Since boron is trivalent and carbon quadrivalent, it is clear that bonds must be formed from excited states of the atom, with three unpaired spins (sp^2) and four unpaired spins (sp^3), respectively. The energy required to promote one s electron to a p state is more than compensated by the exchange energy of an additional pair of covalent bonds.

(f) Bond Formation and Wave Functions

It has been seen that, in the Heitler-London treatment of the H_2 molecule, the binding forces arise from the resonance exchange of electrons between the atomic nuclei. There is, accordingly, a high probability of finding the electrons in some element of volume *between a pair of bonded atoms*. In more complex cases, where the Schrödinger equation cannot be exactly solved to give molecular wave functions, it is permissible to argue that in the immediate neighborhood of each nucleus the motion of the electrons (or their associated stationary wave pattern) is approximately defined by atomic wave functions. Between the bonded atoms, the electron density function $\psi \cdot \psi^*$ must have a relatively high value with respect to wave functions centered about both nuclei. The condition for establishing a bond between two atoms can therefore be expressed in a rather different form: *wave functions of the two atoms* (each occupied, in the separated atoms, by a single electron) *must overlap* so as to give a maximum electron density between the nuclei. In Pauling's

† See especially L. Pauling, *The Nature of the Chemical Bond*, Cornell University Press, Ithaca, 1939.

view, the greater the overlap (i.e., the greater the value of each of the wave functions in the space between the nuclei), the stronger is the resulting bond.

It has been seen that s wave functions are spherically symmetrical, whereas the p wave functions p_x, p_y, p_z have cylindrical symmetry about three mutually orthogonal axes, their angular distribution function being such that the electron density falls off rapidly along all directions deviating much from the axis of the wave function. The p electrons thus spend most of their time near this axial direction; they are more concentrated in space than are s electrons, and therefore form stronger bonds. Moreover, the maximum overlap of wave functions is achieved when the bound atom lies along the axis of the p wave function.

Viewed along the internuclear direction, the shared s–s, s–p or p–p wave functions have the same (circular) symmetry as an s atomic wave function. Bonds formed by overlap of this kind are therefore sometimes denoted as σ-bonds, to distinguish them from other bonds (π-bonds) of different symmetry, which are important for the understanding of unsaturated and aromatic carbon compounds and coordination compounds of some transition metals (cf. Chap. 11). π-bonds will not be considered in this chapter.

It follows from the foregoing that if an atom forms two or three σ-bonds by the use of p wave functions (e.g., the H-O bonds in H_2O, or H-N bonds in NH_3), the bonds should, to a first approximation, be at right angles to one another, so that the linked atoms can each lie along the axis of the appropriate p wave function. In fact, the valence angles in simple molecules such as H_2O and NH_3 are somewhat greater than $90°$, as a result of the repulsive forces between hydrogen nuclei and other factors. The displacement is relatively small, however, and it is clear that *bonds formed by p wave functions are directed in space*.

(g) Hybridization

From the considerations advanced up to this point, it would appear that the quadrivalent (sp^3) carbon atom should form three strong bonds ($p\sigma$-bonds), directed approximately at right angles to one another, and one weaker bond ($s\sigma$ bond). The latter would not be spatially directed, and the atom bound by it would be located in whatever position conformed with the volume or polar repulsion of the groups bound by the three $p\sigma$ bonds. The bond angles would be inherently indeterminate. This conclusion is at variance with firmly established fact. The evidence of organic chemistry, and of accurate molecular structure determinations, makes it certain that the carbon atom forms four equivalent bonds, at equal angles to one another ($109° \ 28'$)—i.e., directed towards the vertices of a regular tetrahedron.

Pauling (1931) introduced the idea that, in some cases, the strongest bonds could be formed, not from pure s or pure p wave functions, but from certain combinations of these, which result in a greater concentration of electron density along certain spatial directions. Such wave functions are termed *hybrid* wave functions.

Justification for the procedure can be obtained as follows. The total wave function of the atom defines the total distribution of electron density. It may be represented as a combination of s and p single electron wave functions, or as a combination of any other set of wave functions which, together, lead to the same electron density throughout space. In constructing the newly defined set of wave functions, certain restrictions are imposed. In particular, Pauling assumes that the radial part of the wave functions ψ_s and ψ_p are closely similar, so that it is necessary to derive only a new set of composite angular functions. This restriction is obeyed by the $2s$ and $2p$ wave functions sufficiently closely. If ψ_s, ψ_{px}, ψ_{py}, ψ_{pz} represent the set of s and p wave functions of the carbon atom, it is necessary to derive a new set of four wave functions f_1, f_2, f_3, f_4, such that

$$f = a_i.\psi_s + b_i.\psi_{px} + c_i.\psi_{py} + d_i.\psi_{pz}$$

where $a_i{}^2 + b_i{}^2 + c_i{}^2 + d_i{}^2 = 1$. ($i = 1, 2, 3, 4$.) The coefficients a, b, c, d can be so chosen (i.e., the independent s and p functions mixed in such proportions) that f_1, f_2, f_3, f_4 are equal in magnitude, and have the greatest possible value. When this is carried out, it is found that a set of four wave functions is obtained, directed towards the vertices of a regular tetrahedron. Their angular distribution function is such that the electron density is concentrated very greatly along the axis of the wave function. These 'tetrahedral' hybrid wave functions give the maximum possible overlap with wave functions of other atoms, and form stronger bonds than pure p wave functions. The four bonds formed by a carbon atom (e.g., in CH_4 or diamond) are of this [sp^3] hybrid type.

It must be emphasised that hybridization is a purely formal mathematical device for arriving at the steric requirements and the form of the overlap integrals involved in constructing the strongest possible set of bonds. It involves no change in the electron density distribution, and it is possible to arrive at the same conclusions from different formal postulates.

If three electrons are available for bonding, as in the sp^2 configuration of boron, the strongest possible bonds are again a set of three equivalent hybridized [sp^2] bonds; these are coplanar and at $120°$ to one another, in conformity with the known stereo-chemistry of boron (Chap. 10). The elements of Sub-group II have the ground state s^2, but must form two covalent bonds (as in $Zn(CH_3)_2$ or $HgBr_2$) from the configuration sp, with two unpaired electrons. The strongest possible bonds are here also formed from hybrid [sp] wave functions, and are collinear. The resulting molecular configuration is therefore completely different from that which results from bond formation by two p electrons, giving a valency angle of about $90°$.

The subsequent chapters will make it clear how far these considerations account for the spatial arrangement of atoms in molecules, as revealed by experiment. The importance of hybrid wave functions will be again considered (Vol. II) in relation to the stereochemistry of the metallic elements.

Another method for the wave mechanical treatment of the covalent bond which can be applied also to the heavier atoms, the results of which are in good qualitative agreement with observations, is the method of Hund (1928) and Mulliken (1928), which was later developed further by Herzberg. The essential feature of this is that the energy levels or quantum states of the individual electrons in the free atoms are compared with those which can be deduced on certain assumptions for the same electrons in the molecules formed from the atoms. According as any particular electron enters a higher level or not when the atoms combine with each other, we may distinguish between *bonding* and *antibonding* electrons. The simple rule that stable molecules are formed when the number of bonding electrons is greater than that of antibonding electrons is roughly true. The difference between the numbers of bonding and antibonding electron pairs gives the multiplicity of the bond. Thus it follows from the quantum states calculated by Hund that two N atoms contain in their outer shells 8 bonding and 2 antibonding electrons in all*; in this case there are, accordingly, $8 - 2 = 6$

* The interaction of the inner shells (helium shells) can be neglected.

bonding electrons, or 3 bonding electron pairs, i.e., a triple bond for the N_2 molecule. The O_2 molecule contains two electrons more than the N_2 molecule, both being antibonding electrons. Since these compensate one bonding electron pair, two bonding electron pairs remain and there results a double bond between the two O atoms, which is weaker than the triple bond between the N atoms. In the same way, there is found for F_2 a still weaker single bond, since two further antibonding electrons have entered. Two Ne atoms are completely unable to combine, since the introduction of two additional antibonding electrons means that all the bonding electron pairs are now compensated. As has already been mentioned, the O_2 molecule contains two unpaired electrons. If the formation of electron pairs were the sole principle governing the formation of covalent compounds, it might be expected that two O_2 would unite by the pairing of electrons to form a valence compound O_4, which is contrary to experience.* In the Hund-Mulliken theory, the formation of a valence compound O_4 from $2O_2$ is excluded, for it follows from their quantum state that the two unpaired electrons of the O_2 molecule are antibonding electrons. In consequence of this, no combination can take place with a second O_2 molecule, which would likewise contribute two antibonding electrons. Combination can, however, occur with an O atom, which would furnish two bonding electrons (formation of O_3). In the same way, it follows that NO can not combine with NO, but can combine with O (formation of NO_2).

It follows from the above that the only electrons involved in ionic valence forces are those which are arranged outside the inert gas core, the so-called 'outer electrons' of the atoms. The same is substantially true for the exercise of covalence. Because of their significance for the exercise of valence forces, the outer electrons are also known as the *valence electrons*.

(h) Transition between Electrovalence and Covalence

The assumption that oppositely charged atoms or ions behave like uniformly charged spheres is only approximately true. The greater the *polarization* which the ions exert on one another when they approach to atomic distances, the less valid this assumption is. This polarization can be pictured as arising from deformation of the electron clouds of the atoms.

The electron clouds of the negative ions, in particular, are deformed by the forces exerted by the positive ions, and the smaller the radius, and the larger the charge of the positive ions, the greater the deformation. Furthermore, the larger the negative ions are, the more strongly they are deformed, so that for example the S^{2-} ion is more deformable that the O^{2-} ion. Ions with the inert gas configuration have a less strong deforming action than the others. For this reason the effect of polarization is exhibited in very marked degree by the compounds of the elements of the Sub-groups.

In every case the first element of a main group of the Periodic System resembles in its properties the second element of the following main group (e.g., Li like Mg, Be like Al, B like Si). This phenomenon is largely due to the fact that the ions of these elements are similar in their deforming action, since the two factors determining polarizing action—the increase in charge and the increase in ionic radius—here roughly compensate each other.

With increasing deformation of the anion, the properties of ionic compounds (heteropolar compounds) approximate increasingly to those of covalent compounds (homopolar compounds). This can be explained in that the more strongly the negative ion is deformed, the more its electronic system is attracted to that of the cation, or even merged with it. In this way the forces of interaction between the electrons (resonance forces or exchange forces), to which the covalent bond is

* It is true that in highly compressed or liquid oxygen $(O_2)_2$ molecules are present. It has been shown, however, that their formation depends on Van der Waals forces, and not on the exercise of true valences.

attributed in wave mechanics, must be of increasing importance in the binding. This accounts for the fact that no sharp demarcation can be drawn between ionic and covalent compounds, but rather that all possible transitional forms are observed, even though the two kinds of bonding are fundamentally different.

In principle, the forces to which covalent bonding is ascribed in wave mechanics are *always* operative between the two partners in chemical combination. With ideal ionic compounds (pure heteropolar compounds), these forces are insignificant in practice as compared with the Coulomb attraction of the ions, in that the repulsion deduced from wave mechanical theory for the approach of the atoms to less than a certain distance can be replaced by the concept of almost rigid spheres, whose radii and compressibilities may be determined empirically.* The covalent binding forces can also be neglected for the case of moderately strong deformation.** For pure covalent compounds these forces alone operate. There is, however, a wide intermediate range within which account must be taken of both types of force for any complete description of the bond, especially if this is to be quantitative.

If it is true, in principle, that the quantum mechanical exchange forces (which can produce either attraction or repulsion) are operative for every bond, and that superposed on these in certain cases there may be attractions between oppositely charged ions, it is clear that the additional effect of the ionic bond must always increase the strength of a bond, and can never decrease it. Pauling has shown that on the basis of these considerations it is possible to calculate the proportion of the bond energy arising from ionic binding for cases intermediate between the two extreme types (i.e., for so-called mixed covalent and electrovalent compounds). He makes the assumption (which is admittedly not strictly valid) that purely covalent bond energies are additive, i.e., that in the case of pure covalent compounds we have $A \leftrightarrow B = \frac{1}{2}[A \leftrightarrow A + B \leftrightarrow B]$, where $A \leftrightarrow A$, $B \leftrightarrow B$, $A \leftrightarrow B$ represent the bond energies for the molecules A_2, B_2 and AB respectively. Thus the bond energies for H_2, I_2, and F_2 are 4.45, 1.54 and 2.80 eV per molecule, respectively. From these, the contribution of the covalent bonding can be calculated on the above assumptions to be 3.00 eV in HI, and 3.63 eV in HF. The bond energies found experimentally are 3.07 and 6.39 eV. For HI the calculated and experimental values differ but little; in this case the bond is almost a pure covalent one, as may also be inferred from many properties of the compound. In the case of HF, however, the calculated bond energy is much smaller than that found experimentally. Pauling concludes from this that a considerable proportion of the bond energy (2.76 eV or about 43 %) is due in this case to the ionic contribution to the bonding. This accords well with the properties of the compound, which point to a strongly polar character. Pauling has made similar comparisons for many compounds, and has shown that the more the various elements differ from one another in electrochemical character (electronegativity), the greater is the contribution made by the ionic bond energy to the total energy of the bond between them.

TABLE 25

ELECTRONEGATIVITIES OF THE ELEMENTS, ACCORDING TO PAULING

Cs	0.7	Ca	1.0	Ti	1.6	Te	2.1	Br	2.8
Rb	0.8	Mg	1.2	Sn	1.7	P	2.1	Cl	3.0
K	0.8	Y	1.3	Ge	1.7	H	2.1	N	3.0
Na	0.9	Sc	1.3	Sb	1.8	Se	2.4	O	3.5
Ba	0.9	Be	1.5	Si	1.8	I	2.4	F	4.0
Li	1.0	Al	1.5	As	2.0	S	2.5		
Sr	1.0	Zr	1.6	B	2.0	C	2.5		

* H. Jensen, 1936, showed that the forces of repulsion calculated from wave mechanics for the close approach of the ions provide a justification for the concept of almost incompressible spheres.

** The polarizabilities of the ions can also be determined empirically. In principle, the wave-mechanical treatment of the bond should enable the polarizabilities of ions to be predicted.

On this basis it is possible to arrange the elements in a series of decreasing electropositive character, or increasing electronegative character, as displayed in their compounds. Such an electronegativity sequence is given in Table 25.

(i) The Semipolar (Coordinate Covalent) Bond. Electrovalence and Covalence

A particular kind of superposition of covalent and ionic bonding is found in the *semipolar* (*coordinate covalent*) *bond*. Carbon monoxide may be cited as a simple example of this, even though its constitution cannot be regarded as definitely established.* From the very close agreement in the physical properties of CO and N_2, Langmuir concluded in 1919 that there must be an exactly equivalent electron configuration in the molecules of the two substances, i.e., $: C :: O :$ corresponding to $: N :: N :$. This is only possible, however, if the O atom has given up one electron to the C atom, so that the C is formally negative and the O positively charged. In effect, of the three electron pairs shared between carbon and oxygen, two are contributed to equally by each atom, and one pair is supplied by the oxygen.

In contrast to *electrovalence*, arising from the opposite charges on atoms or ions, the valences arising from exchange forces are frequently called *covalences*, following Sidgwick. In compounds having semipolar bonds, the distribution of charge in the resulting molecule is often represented by the charge signs $+$ and $-$. Thus, on the view developed above for the constitution of CO, the molecule may be represented by the formulas

$$\overset{-}{C}\!\!\equiv\!\!\overset{+}{O} \qquad \text{or} \qquad : \overset{-}{C} :: \overset{+}{O} :$$

The formation of semipolar (coordinate covalent) bonds or *dative bonds* is of great importance in molecular compounds, or compounds of higher order, to be considered in a later chapter (Chap. 11). The oxygen atom in the molecules of water or the alkyl ethers,

$$\begin{matrix} H \\ \\ H \end{matrix} : \overset{..}{\underset{..}{O}} : \qquad\qquad \begin{matrix} H_3C \\ \\ H_3C \end{matrix} : \overset{..}{\underset{..}{O}} :$$

or the nitrogen atom in the molecules of ammonia and the amines,

$$\begin{matrix} H \\ .. \\ H : N : H \\ .. \end{matrix} \qquad\qquad \begin{matrix} CH_3 \\ .. \\ H_3C : N : CH_3 \\ .. \end{matrix}$$

have complete octets and are not able to form additional covalences in the way first postulated by G. N. Lewis. They possess, however, pairs of electrons (two such pairs in the O atom of R_2O molecules, one pair in the N atom of R_3N molecules) which are not employed in the formation of covalent bonds. Such unused electron pairs were termed 'lone pairs' of electrons by N.V. Sidgwick. They can be used for the formation of additional bonds if they are shared with atoms which, although 'saturated' in the usual sense, can accommodate more electrons in the valence group. Thus the boron atom in boron trimethyl is 'saturated'; it is forming three covalent bonds by sharing all its valence electrons with the carbon atoms of the methyl groups. It does not, however, possess a complete octet (i.e., a completed, stable electron configuration), but has 6 electrons in the valence shell. It can complete the octet of electrons, and form an additional bond, by sharing the lone pair of electrons on, e.g., the nitrogen atom of a molecule of trimethylamine:

$$\begin{matrix} CH_3 \\ .. \\ H_3C : B \\ .. \\ CH_3 \\ \text{I} \end{matrix} \qquad\qquad \begin{matrix} CH_3 \;\; CH_3 \\ .. \;\;\;\; .. \\ H_3C : B : N : CH_3 \\ .. \;\;\;\; .. \\ CH_3 \;\; CH_3 \\ \text{II} \end{matrix} \qquad\qquad \begin{matrix} CH_3 \;\; CH_3 \\ | \qquad | \\ H_3C\!-\!B\!\leftarrow\!N\!-\!CH_3 \\ | \qquad | \\ CH_3 \;\; CH_3 \\ \text{III} \end{matrix}$$

* The very small dipole moment is evidence against the semipolar bond in CO. The structure given below for this molecule may be regarded as being merely a contributing structure to the actual state of the molecule.

The resultant bond is equivalent to that considered above, in the CO molecule. It is commonly represented by an arrow, to distinguish it from a normal covalent bond.

Semipolar bonds are involved not only in addition compounds of the type considered, and in coordination compounds of the metals (Chap. 11), but also in the oxyacids of the elements and similar compounds. The series of anions $[XO_4]$ formed by the elements from silicon to chlorine would be represented on the Kossel approximation as built up from a central cation X^{n+} (where $n =$ the number of the Group to which X belongs), and four O^{2-} ions. Thus the perchlorate anion would be (I). The formulation with 7-covalent chlorine (IIa) would imply that, in forming three double bonds and one single bond with oxygen, the valence shell of the chlorine was extended to accomodate 14 electrons (IIb). In this formulation, not all the oxygen atoms of the $[PO_4]^{3-}$, $[SO_4]^{2-}$ and $[ClO_4]^-$ anions would seem to be equivalent, although other evidence shows that these anions are symmetrical, tetrahedral atomic groups. An alternative formulation is that shown in (IIIa) and (IIIb), in which the central atom and the oxygen atoms are linked by semipolar bonds.

$$
\begin{array}{ccccc}
\text{O}^{2-} & \text{O} & :\text{O}: & \text{O}^- & :\overset{\cdot\cdot}{\text{O}}: \\
 & \| & & \uparrow & \\
\text{O}^{2-}\quad \text{Cl}^{7+}\quad \text{O}^{2-} & \text{--O--Cl}=\text{O} & :\text{O}:\text{Cl}::\text{O} & \text{--O--Cl}^{3+}\!\!\rightarrow\!\text{O}^- & :\text{O}:\text{Cl}:\text{O}: \\
 & \| & & \downarrow & \\
\text{O}^{2-} & \text{O} & :\text{O}: & \text{--O} & :\overset{\cdot\cdot}{\text{O}}: \\
\text{I} & \text{IIa} & \text{IIb} & \text{IIIa} & \text{IIIb}
\end{array}
$$

As will be seen later, the actual bonds between the central atom and oxygen in these oxyacid anions and similar structures is not exactly represented by any of these formulations, but the structures (II) and (III) are both partially true.

2. Chemical Affinity

In order to make predictions about the stability of a compound, or the conditions under which it may be obtained in practice it is not, in general, necessary to know the nature of the valence forces involved. For this, it is sufficient to know the affinities or *free energies* of the reactions involved. The free energy may be determined by various methods, which are based on the measurement of directly observable quantities, and which do not presuppose any detailed knowledge of binding forces in order to evaluate them. [13–20]

Since the work of Van't Hoff (1883), the free energy of a chemical reaction has been understood to mean the maximum work (more strictly, the maximum useful work) which can be extracted in the transformation of unit quantity of substance: the unit taken is the gram-molecule, and the free energy of a reaction is usually referred to the number of gram-molecules entering into the reaction concerned. The free energy thus has the dimensions of *work per unit quantity of material*, and is usually expressed in heat units.

(a) Maximum Work and Useful Work

According to thermodynamics, any process (such as a chemical reaction) can take place spontaneously—i.e., without absorbing energy from its surroundings—only if it is capable of performing external work. The process furnishes the maximum amount of external work (the *maximum work*) when it takes place *isothermally* and *reversibly*. If the process occurs under the constant pressure p of the atmosphere, and is attended with a volume change ΔV of the system considered, the external work W performed may be resolved into two portions:

$$W = W_u + p \cdot \Delta V \tag{1}$$

The quantity of work $p \cdot \varDelta V$ must, as it were, be performed as an extra if the process is to be carried out at atmospheric pressure, and cannot be utilized. Only the portion W_u can be utilized, and is therefore termed the *useful work*. In the case of chemical processes, the maximum useful work, referred to the unit quantity of reacting substance, is taken as a measure of the affinity, or free energy change $-\varDelta F$, i.e., $-\varDelta F = W_u$.

The concept of the change in free energy has been introduced as a measure of the tendency of the elements to combine with one another. It is evident that one can attach a meaning to such a tendency only if the elements concerned can combine with one another spontaneously, i.e., without being constrained to do so by external action (supply of added energy).* It is logical to measure this tendency by the magnitude of the opposing forces which would be necessary to prevent the elements from combining. For a reversible process, these forces are at every instant equal to the forces exerted by the system under consideration. The work which can be performed when the reaction takes place reversibly, i.e., the maximum work, is thus found to be a measure of the affinity of the reaction. Since we are generally concerned with reactions occurring at constant (usually atmospheric) pressure, it is useful to take, as a measure of the combining tendency of the elements, not the total external work, but the portion of it which has been defined as 'useful work' in equation (1) above.

(b) Energy of Formation and Affinity of Formation

The maximum useful work which could be obtained from the formation of one gram-molecule of a compound from its component elements in some standard state is called the *free energy of formation*, $\varDelta F°$, of the compound in question. If work is *performed* by the process—i.e., if the affinity of the process is positive— the free energy is written with the *negative sign***. The affinity and the free energy of formation of a compound, referred to the same quantity of material, are thus equal and opposite. It is customary in English and American chemical literature to employ the quantity which we have defined as the free energy.

Thus the free energy change of the (exothermic) reaction $H_2 + Cl_2 = 2HCl$ is -45.4 kcal. at $25°$. The free energy of formation of hydrogen chloride, is therefore half as great, $\varDelta F°_{HCl} = -22.7$ kcal. at $25°$ C. The *heat of reaction* Q_r $(= -\varDelta H)$ often provides an approximate measure of the affinity of a reaction. In calculating the free energy changes of chemical reactions, care must be exercised to avoid confusion in the signs of the energy quantities, since the chemical usage (heat of reaction written as positive for exothermic reactions) and the thermodynamic convention (energy of reaction written as negative if external work is performed) are opposite.

When reference is made simply to the free energy, without specially specifying its dependence on the concentrations of the substances undergoing reaction, this means the free energy of the reaction when the molar concentrations of all the substances involved (their pressures, in the case of gaseous reactants) are equal to unity (the so-called *standard free energy*, cf. p. 144).

* This does not, of course, include external conditions which affect merely the *speed* of the reaction; even in the absence of these, the process would go to completion, though in some circumstances an enormously long time might be taken for reaction to take place to a measurable degree.

** It is the convention in thermodynamics always to regard quantities representing energy *imparted to* the system as *positive*, those representing energy *given up by* the system as *negative*. Thus A represents the work *performed on*, $-A$ the work performed *by* the system. In the same way, $\varDelta H$ represents the heat *added* to the system; and $\varDelta H = -Q$, where Q stands for the heat of reaction as ordinarily defined in chemistry, is the heat *liberated* in a reaction. In the same way, $W_u = -\varDelta F$.

(c) Temperature Variation of Free Energy

The variation of the free energy with temperature, at constant pressure, is shown by thermodynamics to be given by

$$\left(\frac{\mathrm{d}\Delta F}{\mathrm{d}T}\right)_p = \frac{\Delta F - \Delta H}{T} = -\Delta S \tag{2}$$

where $-\Delta H$ represents the heat of reaction (at constant pressure). The expression

$$\frac{\Delta F - \Delta H}{T} = -\Delta S$$

is called the change in *entropy* of the system (i.e., of the reacting substances) in the reaction. [26]

If $\Delta H > \Delta F$ (i.e., if the 'affinity' $W_u > Q_r$, the heat of reaction), the entropy of the system increases; i.e., the system absorbs heat from its environment when the reaction takes place isothermally, and this cannot be transformed into work without lowering the temperature of the surroundings; if the process is carried out reversibly, this energy can be returned to the surroundings in the form of heat when the reaction is allowed to proceed in the reverse direction. If $\Delta H < \Delta F$ (i.e., $W_u < Q_r$), the entropy of the system decreases, and if the process takes place isothermally heat is given up to the surroundings. This, likewise, cannot be converted into work without lowering the temperature. At 18°, for example, the free energy of formation of water is (numerically) 11.7 kcal. per mole less than its heat of formation (see below). If 18 g of detonating gas are allowed to combine reversibly to form water (as can be approximately achieved by allowing the process to take place in a galvanic cell consisting of a hydrogen electrode and an oxygen electrode), only the amount $\Delta F_{H_2O} = -56.7$ kcal. can be utilized in the form of work (as electrical energy in the example quoted), although the heat of formation Q_f ($= -\Delta H_{H_2O}$) is 68.4 kcal. The difference is given up to the surroundings in the form of heat. The entropy of the water at the temperature concerned ($T = 291°$ K) is less than that of the detonating gas by $11.7/291 = 0.039$ kcal. per degree. Conversely, to decompose 18 g of water, electrical energy corresponding to only 56.7 kcal. must be expended. The difference, 11.7 kcal., is absorbed from the surroundings in the form of heat if the process is carried out isothermally and reversibly.

(d) Determination of Free Energy

The following are the most important methods of determining free energy changes.

(1) *Determination of free energy from electromotive force.*

The reaction taking place in a galvanic cell furnishes electrical energy, and the amount of this is given by the product of the cell voltage and the quantity of electricity transferred from one pole to the other. The maximum work is obtained from the reaction when no energy is lost through transformation of electrical energy into heat. The cell voltage accordingly attains its maximum value when the passage of current is prevented by an exactly compensating opposed e.m.f. Provided that no irreversible side reactions take place, the process is then reversible. The maximum work, and therefore the free energy, for a reaction taking place in a galvanic cell, operating reversibly, is thus given by the product of the voltage of the cell when no current is passing (open circuit voltage) and the quantity of electricity which would be transported from one electrode to the other if the reaction were not prevented by the opposed electromotive force. This quantity of electricity is determined by the change in the charge of the ions concerned. Thus

$$-\Delta F = n.\mathfrak{F}.E \tag{3}$$

where $n =$ the change in valence of the ions, $\mathfrak{F}$ is the quantity of electricity associated with 1 g ion of hydrogen ($=$ 1 Faraday), and E the open circuit potential of the cell. This is numerically equal to the decomposition potential defined on p. 32. If E is measured in volts and $\mathfrak{F}$ is expressed in coulombs ($1\mathfrak{F} = 96493$ coulombs), eqn (3) gives the free energy in watt-seconds, or joules: this may be converted to calories per g ion by multiplying it by 0.2390.

If the free energy referred to 1 *ion* (instead of to 1 g ion) is represented by Δg, eqn (3) becomes

$$-\Delta g = n \cdot e \cdot E \tag{3a}$$

—i.e., if the measured e.m.f. is multiplied by the change of valence, we obtain directly the free energy in electron-volts per ion. The free energy in kcal per g ion is obtained by multiplying this by 23.062. Conversely, we can obtain the electromotive force of a galvanic cell in which the reaction might be made to furnish useful work by dividing the measured free energy (expressed in kcals per g ion) by 23.062 × the change of valence.

Since the change in energy when a system passes from one state into another is independent of the process by which the change is brought about, the maximum work obtainable from a process occurring in a galvanic cell is also that obtainable if the process takes place outside the cell, provided that it is carried out isothermally and reversibly. The free energy calculated from the electromotive force according to eqn (3) therefore applies to the reaction as such, and is not bound up with the particular process used for the calculation.

Example. The free energy of the reaction

$$Zn + CuSO_4 = ZnSO_4 + Cu \tag{4}$$

can be calculated from the e.m.f. of a galvanic cell consisting of a zinc plate dipping in a $ZnSO_4$ solution, and a copper plate dipping in a $CuSO_4$ solution. If the solutions are 1-molar* with respect to the ions in question, the electromotive force, which could, of course, be measured directly, is given, from the data of Table 4, p. 30 as $E = 0.35 + 0.76 = = 1.11$ volts (at 18°). Since the change in charge of the ions is 2 ($Zn = Zn^{2+} + 2e$, $Cu^{2+} + 2e = Cu$), the free energy of the reaction is found to be $-\Delta F = 2 \cdot 96493 \cdot 1.11 = = 214214$ joules, or 51.20 kcal. The heat evolved in reaction (4) $-\Delta H$ is found to be 50.11 kcal. The heat of reaction is thus found in this case to be a little less than the free energy change**.

In the same way, for the free energy of the reaction $2H_2 + O_2 = 2H_2O$, (using the data contained in pp. 30 and 32) we find

$$-\Delta F = 4 \cdot 96493 \cdot 1.23 \cdot 0.2390 = 113464 \text{ cal (at 18°).}$$

This value applies to the formation of liquid water from the gases at atmospheric pressure. The heat of formation Q_f ($= -\Delta H$) of 2 moles H_2O_l at 18° $= 136740$ cal. In this case the

* Unless otherwise specified, the free energy is always referred to unit concentration.
** It follows from these results and eqn (2). that the magnitude of the free energy change, and the e.m.f. of the cell considered, must accordingly *increase* with temperature, though only to a small extent, since $\Delta F - \Delta H$ is small. This is in accord with experience.

heat of formation is considerably greater than the free energy of formation. Much more energy, in the form of heat, could thus be obtained from the combination of the constituents of 1 g of detonating gas than must be expended, as electrical energy, to decompose 1 g of water. In the latter process the difference, amounting to 17 % of the heat of formation, is abstracted in the form of heat from the surroundings.* According to the second law of thermodynamics, it is impossible to convert this portion into mechanical work. It is thus inherent that not more than 83 % of the heat of formation of water can be made available for the performance of work. In the case of carbon, on the other hand, the heat of combustion and the free energy of formation of CO_2 at ordinary temperature are roughly of the same magnitude. It would therefore be possible to utilize almost the entire heat of combustion of carbon for the performance of work if the union of C and O_2 could be made to proceed reversibly at ordinary temperature, although by the indirect course, with the use of heat engines, a substantial proportion of the energy is inevitably lost.**

(2) *Determination of free energy from chemical equilibrium.*

For any reaction which proceeds according to the equation

$$a\text{A} + b\text{B} + \ldots = m\text{M} + n\text{N} + \ldots,$$

the free energy change is given by

$$\Delta F = RT \left(\ln \frac{[M_2]^m \cdot [N_2]^n \cdots}{[A_1]^a \cdot [B_1]^b \cdots} - \ln K_c \right) \tag{5}$$

Here R is the gas constant (1.986 cal per degree), T the absolute temperature, and K_c the equilibrium constant for the reaction, found by application of the law of Mass Action. The quantities in square brackets are the molar concentrations (more accurately the activities) of the reactants *before* the reaction ($[A_1]$, etc.), and of the reaction products *after* the reaction ($[M_1]$, etc.). Note that the concentrations of the substances disappearing in the reaction must appear in the denominator, and those formed in the reaction, in the numerator, both in K_c and in the other term in the bracket.*** The use of partial pressures in place of concentrations or activities, in the case of gas reactions, applies in the same way as for the Mass Action Law.

Equation (5) shows that for a system in chemical equilibrium, the free energy of the reaction is zero. This follows, indeed from the definition of the free energy, for a system in chemical equilibrium cannot spontaneously alter its state, and so can perform no work.

It may be seen from (5) that the free energy of a reaction depends on the concentrations of the substances involved. If the concentrations of the substances on the left hand side of the equation are greater than the equilibrium concentrations, the progress of the reaction from left to right is accompanied by a decrease in free energy (ΔF is negative), and vice versa. The more the concentrations of the reactants differ from the equilibrium concentrations, the greater is the magnitude of the free energy change. If the concentrations of disappearing and resulting substances are all taken as unity, then

$$-\Delta F^\circ = +2.3026 \, RT \log_{10} K_c$$

---

* In practice it is true that no cooling is usually observed but the electrolyte heats up instead, since the effect is more than compensated by the Joule heating of the current. This appears as the result of an irreversible process, the overcoming of the ohmic resistance.

** Even in the most modern power plants, 28 % at most of the energy of combustion of carbon is converted to electrical energy. The possibility of directly converting the chemical energy of carbon or other fuels into electrical energy ('fuel cells') therefore has considerable technical interest (cf. on this subject *Z. Elektrochem.* 39 (1933) 169; 41 (1935) 93 and 405). W. Ostwald said that to solve this problem was the most important task of electrochemistry.

*** If the converse convention is used, the sign of the corresponding term inside the bracket must be changed.

This quantity, $\Delta F°$, is known as the *standard free energy* of the reaction.

Example. The equilibrium constant of the reaction $H_2 + I_2 = 2HI$ at $441°$ ($= 714°$ K) has the value $K_c = 50.4$. The standard free energy of the reaction at the given temperature is, accordingly

$$-\Delta F° = +2.303 \cdot 1.986 \cdot 714 \cdot \log_{10} 50.4 = +5556 \text{ cal}$$

Work equivalent to this amount of heat can therefore be obtained from the combination of 1 mole of I_2 and 1 mole of H_2, at unit concentration, to form 2 moles of HI at the same concentration. If one starts with a mixture in which the H_2 concentration is 10 times as great as the concentration of I_2 before the reaction, and of HI after the reaction, eqn (5) leads to the considerably larger value

$$-\Delta F = -2.303 \cdot 1.986 \cdot 714 \cdot (\log_{10} 0.1 - \log_{10} 50.4) = +8820 \text{ cal}$$

for the free energy change accompanying the formation of 2HI. In the example chosen, the free energy change for the reaction is considerably greater than the heat of reaction, which amounts to about 2800 cal for the formation of 2 moles of HI from the gaseous components at $441°$.

(3) Calculation of free energy from heat data.

By use of equation (2), p. 141, the free energy change at any temperature can be calculated from calorimetric data for a reaction, provided that the free energy is known at some one temperature. For example, the free energy of conversion of one modification of a substance into another at some temperature T can be calculated from the transition temperature (i.e., that temperature T_0 at which the two modifications are in equilibrium with each other, and at which the free energy change is therefore zero), and from the heat of transformation at the temperature T. In order to integrate eqn. (2), however, it is necessary to know how the heat of transformation W varies with temperature.

If, as is generally the case in the temperature ranges concerned in practice, it is possible to use an equation of the form (6) to represent the difference ΔC_p between the molar heats of the two modifications at constant pressure, as a function of temperature

$$\Delta C_p = a + \beta T + \gamma T^2 \tag{6}$$

(where a, β and γ are empirical constants), the variation of the heat of transformation with temperature at constant pressure, is given by

$$W = W_0 - aT - \tfrac{1}{2}\beta T^2 - \tfrac{1}{3}\gamma T^3 \tag{7}$$

Here W_0 represents a constant, which can be evaluated by means of eqn. (7), together with (6), if W is known for some particular temperature (within the range of applicability of eqn. (6)—i.e., at temperatures not too close to absolute zero). The integration of eqn. (2) then furnishes the following equation for the calculation of the free energy change

$$-\Delta F = W_0 + aT \ln T + \tfrac{1}{2}\beta T^2 + \tfrac{1}{6}\gamma T^3 + CT$$

The constant C in this equation can be determined by substituting for T the transition temperature T_0, since ΔF then becomes 0.

Example. It is required to calculate the free energy of conversion of monoclinic into rhombic sulfur at $25°$. The molecular heats of the two modifications between $0°$ and $100°$ can be represented, from the data available, by $C_p(\text{S monocl.}) = 3.62 + 0.0072\,T$, $C_p(\text{S rhomb.}) = 4.12 + 0.0047\,T$. ΔC_p in eqn (6) represents the *increase* in molar heat (or atomic heat) of the system when it changes from state 1 to state 2. We have, therefore,

$$\Delta C_p = (4.12 - 3.62) + (0.0047 - 0.0072)\,T = 0.50 - 0.0025\,T;$$
$$W = W_0 - 0.50\,T + 0.00125\,T^2.$$

According to Lewis and Randall, the heat of transformation is 77.0 cal per mole at $0°$ ($= 273°$ K). From this we obtain $W_0 = 77.0 + 0.50 \cdot 273 - 0.00125 \cdot 273^2 = 120.3$. Using this value, and the experimentally determined transition temperature $T_0 = 368°$ K, the value of C is found from eqn. (8), by substituting $\Delta F = 0$. We thus have

$$C = \frac{-120.3}{368} + 0.5 \cdot 2.303 \cdot \log 368 - 0.00125 \cdot 368 = -2.821.$$

The equation for calculating the free energy of transformation of sulfur for any temperature (within the range of validity of the equation used for the temperature dependence of the specific heats) thus becomes

$$-\Delta F = 120.3 + 0.50 \cdot 2.303 \, T \log T - 0.00125 \, T^2 - 2.821 \, T \tag{9}$$

This equation yields the value $\Delta F = -17.6$ cal per g atom of S for the free energy of conversion of monoclinic into rhombic sulfur at $25°$. The negative value of the free energy change shows that the change takes place spontaneously, as may readily be observed experimentally. Thus the rhombic modification of sulfur is stable at ordinary temperature, but above $95°$ the monoclinic modification is stable. For example, at $100°$ C ($= 373°$ K), eqn. (9) gives the value $\Delta F = +1.2$ cal, so that in this case the reaction proceeding in the opposite direction leads to a decrease in free energy.

The calculation of free energies by means of eqn. (8) is possible only if the free energy change is already known at some particular temperature, or if the temperature is known at which the free energy change of the process is zero (transition temperature, melting point, etc). By use of the Nernst heat theorem (cf. p. 174), however, it is possible to derive equations which enable the free energy to be calculated when all that is known is the heat effect ΔH of the process, and its variation with temperature.*

(4) Calculation of free energy from spectroscopic data.

The calculation of free energies, and of chemical equilibria, from spectroscopic data has recently acquired considerable importance. In particular, such data provide information as to energies of ionization and dissociation. The *ionization energy* is derived, according to eqn. (3), p. 113, from the wave-length of the appropriate series limit. Its value, in electron-volts per atom or per molecule, is obtained by multiplying the reciprocal of the wave length (in cm^{-1}) by $1.2395 \cdot 10^{-4}$; or, in cal per g molecule, by multiplying by 2.858.

Just as the ionization energy may be derived from atomic spectra (line spectra), so it is possible to calculate from molecular spectra (band spectra) the dissociation energy of molecules. In this case it is necessary to take into account whether the molecules dissociate into free atoms in their ground state, into excited atoms, or into ions.**

Spectroscopic data yield directly the free energy changes at absolute zero. The temperature-dependence of the free energy can, however, be calculated from these data by using the quantum mechanical formulas of the kinetic theory of gases.

The use of ionization and dissociation energies to calculate the free energy changes in chemical reactions, or the free energy of formation of chemical compounds, may be illustrated here by a few simple examples.

---

* This is discussed more fully in textbooks of thermodynamics and of physical chemistry. For the derivation of the formulas cited above see, in particular, the references [13] and [14] in the bibliography at the end of this chapter.

** More concerning the calculation of dissociation energies from band spectra will be found in the works [21–24] cited on p. 151.

1st Example. Calculation of the free energy of formation of IBr. The energies required to dissociate the molecules Br_2, I_2 and IBr have been determined from the band spectra;

$$Br_2 = 2Br - 45.23 \text{ kcal} \qquad \text{(Gordon and Barnes 1933)}$$
$$I_2 = 2I - 35.40 \text{ kcal} \qquad \text{(Brown 1931)}$$
$$IBr = I + Br - 41.68 \text{ kcal} \quad \text{(Brown 1932)}$$

Adding the first two equations, and subtracting twice the third,

$$Br_2 + I_2 = 2IBr + 2.73 \text{ kcal}$$

The free energy of formation $-\Delta F°$, from spectroscopic data, is thus found to be 1365 cal per mole (at $T = 0°$ K), whereas Yost (1931) found, from equilibrium measurements $-\Delta F°_{IBr} = 1270$ cal per mole, extrapolated to $0°$ K.

The dissociation energies found spectroscopically refer to the free molecules. The free energy of formation of IBr calculated from them therefore apply to the hypothetical case that the substance remains in the gaseous state at absolute zero; the same is true of the value extrapolated from Yost's measurements. To relate the data to other states of aggregation, the energies of fusion and vaporization must be allowed for.

2nd Example. Calculation of the energy of formation of potassium iodide. We may suppose that the formation of potassium iodide from potassium and iodine could be brought about in two ways: first, by the direct combination of the constituents; and second, by a method involving the formation of the gaseous ions, which, when they came together to form the crystal lattice, would liberate an amount of energy that could be calculated on the basis of Coulomb's law. On the assumption that all the processes take place isothermally and reversibly, the two ways could be combined in one *cyclic process*. With the use of a Born-Haber cycle, this can be symbolized in the following way:

$$
\begin{array}{ccc}
& +\,S_K + S_I + \tfrac{1}{2}D_{I_2} & \\
[K] + [I] & \xrightarrow{\hspace{3cm}} & (K) + (I) \\
{\scriptstyle -\Delta F°_{IK}}\Big\uparrow & & \Big\downarrow {\scriptstyle +\,I_K - E_I} \\
[KI] & \xleftarrow[\;\; -G_{KI} \;\;]{} & (K^+) + (I^-)
\end{array}
$$

We start with 1 g atom each of solid potassium, [K], and solid iodine, [I].* Both substances are first transformed into the free atomic state. For potassium, this involves expenditure of the heat of sublimation, S_K; for iodine, in addition to the heat of sublimation S_I, the heat of dissociation referred to half a gram molecule of I_2, $\tfrac{1}{2}D_{I_2}$, is required. If we suppose these processes to be performed at a temperature close to absolute zero, we may take the heat effects as equivalent to the work which must be expended, in accordance with the Nernst heat theorem. We next convert the K atoms into free K^+ ions; for this the ionization energy I_K must be expended. By conversion of the I atoms to I^- ions, however, the energy E_I is liberated. The energy released by the conversion of the atom of an element into a negative ion (i.e., by the combination of an atom with an electron) is known as the *electron affinity* of the element, E. The free ions K^+ and I^- are now allowed to come together, to form the crystal lattice of KI. The energy G set free thereby is known as the *lattice* energy of the compound in question. In order to decompose potassium iodide into its elements—i.e., in order to reproduce the initial state of the system—the quantity of energy which must be *expended* is equal to the maximum useful work which could be *obtained* by the formation of 1 g mol of KI from its elements—i.e., $-\Delta F°_{KI}$, the free energy of formation of KI. In the scheme as set down above, amounts of energy expended *on* the system (i.e., energy added to the system from without) are written as positive, energy set free *by* the system, with the negative sign. Since we have carried out a cyclic process, the total energy change is zero, i.e., we have that

$$S_K + S_I + \tfrac{1}{2}D_{I_2} + I_K - E_I - G_{KI} - \Delta F°_{KI} = 0 \tag{10}$$

* For the use of square brackets to denote the solid state of aggregation, and of round brackets to denote gaseous substances, see p. 174.

We can use this equation in order to compute any one of the energy quantities contained therein, provided all the others are known. In the present instance, we shall use it to calculate $\Delta F°_{KI}$. The individual energy quantities needed for this purpose are known from experimental data, with the exception of the lattice energy G_{KI}. This, however, can be calculated for an ionic lattice, from Coulomb's law.

If we consider the ions as solid spheres, with the radii r_1 and r_2, and the charges $+e$ and $-e$, then the work which must be expended to bring them into contact with each other is given by

$$A = -\frac{e^2}{r_1 + r_2} \qquad \text{(cf. p. 83)}$$

The negative value of A shows that energy will be set free in forming the molecule KI from the free ions K^+ and I^-. If we substitute the appropriate values for the ionic radii of K^+ and I^- (see pp. 156 and 773) we find

$$A = -\frac{(4.8022 \cdot 10^{-10})^2}{(1.33 + 2.20) \cdot 10^{-8}} = -6.534 \cdot 10^{-12} \text{ erg}$$

for the energy of formation of a KI molecule from the gaseous ions. Multiplying by Avogadro's number ($N_A = 6.0238 \cdot 10^{23}$) we obtain the free energy of formation in ergs per mole, which can be converted into cal per mole by multiplying again by $2.390 \cdot 10^{-8}$ i.e., $-6.534 \cdot 10^{-12} \cdot 6.0238 \cdot 10^{23} \cdot 2.390 \cdot 10^{-8} = -9.40 \cdot 10^4$ cal/mole or -94.0 kcal per mole. If we correct this value, which applies to gaseous KI, for the heat of sublimation of KI ($= 48.9$ kcal per mole), we obtain -142.9 kcal per mole as the approximate free energy of formation of crystalline KI from the gaseous ions, at absolute zero. Since polarization was neglected, this calculation must furnish too low a value for the free energy of formation from the ions, and for the lattice energy.

The lattice energy may be obtained more accurately by the application of Coulomb's law, not to the formation of the molecule, but directly to the formation of the crystal lattice (cf. Chap. 7) from the gaseous ions. In such crystal lattices as that of KI (rock salt structure, cf. Chap. 7), in which all the ions are surrounded uniformly by ions of opposite sign, (so-called *coordination* lattices), the polarizing action of the surrounding ions almost cancels out. In molecules, because of the one-sided attraction, the polarization (deformation of the electron clouds) is far greater. A direct calculation of the lattice energy on the basis of Coulomb's law, taking into account the mutual interaction of all the ions, was first successfully carried out by E. Madelung (1918). For binary compounds, this can be done by means of the formula

$$G = \frac{n-1}{n} \cdot \frac{Z^2 e^2}{r} \cdot A \cdot N_A \qquad (11)*$$

Here r is the shortest distance between the centers of ions in the crystal lattice, Z is the valence of the ions, e the elementary quantum of electricity, N_A Avogadro's number, and A a constant (the Madelung constant), the magnitude of which depends on the crystal structure. (For rock salt crystals, $A = 1.748$, cesium chloride type $A = 1.763$, zinc blende type $A = 1.639$. For fluorspar type lattices, $A = 5.039$, taking $Z = 1$ for compounds of the general formula $M^{ii}R_2$, and $Z = 2$ for those of the type $M^{iv}R^{ii}_2$). The so-called 'repulsion exponent' n takes account of the repulsion between the ions when they approach each other very closely, and its approximate value can be determined from the compressibility of crystals. Since the repulsion arises from the interaction of the electron clouds of the individual ions, the value of n may also be calculated theoretically. For inert-gas like ions, according to Pauling, n has the following values, according to their configuration.

Electron configuration	He	Ne	A	Kr	Xe	Rn
$n =$	5	7	9	10	12	14

* For the derivation of this formula see, for example, [2]

For compounds formed from ions of different electron configuration (e.g., from K^+ (argon structure) and I^- (Xe configuration)), the average value is to be substituted for n (e.g., $n = 10.5$ for KI). The lattice energy is reduced by the fraction $1/n$ through the mutual repulsion of the electronic systems, as shown by equation (11).*

For the lattice energy of KI (which crystallizes with the rock salt structure), equation (11) gives the value

$$G_{KI} = \frac{9.5}{10.5} \cdot \frac{(4.8022 \cdot 10^{-10})^2}{3.525 \cdot 10^{-8}} \cdot 1.748 \cdot 6.0238 \cdot 10^{23} = 6.233 \cdot 10^{12} \text{ ergs per mole or}$$

148.9 kcal per mole. As would be expected, the value is higher than that calculated by the first method. If we substitute this calculated value in equation (10), with the other energy quantities determined experimentally**, we obtain

$$-\Delta F^\circ_{KI} = -21.85 - 8.03 - 17.70 - 99.65 + 79.3 + 148.9 = 81.0 \text{ kcal per mole.}$$

This value for the free energy of formation of KI (at absolute zero) agrees satisfactorily with that found experimentally by Ishikawa (1934), from electromotive force measurements ($\Delta F^\circ_{KI} = -77.5$ kcal per mole), when one considers that the latter value applies to ordinary temperature.

(e) Importance of Free Energy Determinations

Just as the free energies may be determined from electromotive force, chemical equilibria, etc., so, conversely, may these be calculated from free energy data. Thus the equilibrium constants and heat effects of reactions can be calculated with the aid of free energies deduced from spectroscopic data. Furthermore, it is possible from free energy calculations to make predictions as to the possibility of existence of chemical compounds. If a *small* positive free energy change is found for the formation reaction of some hypothetical compound, this may perhaps, in some circumstances be changed to a negative value by appropriate changes in the temperature and concentration conditions. However, if a *large* positive free energy change is found for the corresponding reaction, it may be said with certainty that this cannot be carried out directly.

For example, the impossibility of obtaining ionic compounds of the inert gases can be shown by calculating the free energies of the reactions leading to them. Thus, for NeF and NeCl, on the assumption that they have ionic structures, the Born-Haber cycle indicates free energies of formation of $+241$ and $+254$ kcal per mole respectively. Such compounds cannot possibly be obtained by chemical reactions in the ordinary sense.*** The non-existence or instability of salt-like compounds of the inert gases is not so much the direct conse-

---

* A more complicated expression for the repulsion is furnished by quantum mechanics. However, so long as polarization effects are not too great, equation (11) generally provides a good approximation for lattice energies. The methods to be used for exact calculations are rather tedious (cf. p. 156).

** The value calculated by Bichowski and Rossini [25] from the measurements of J. E. Mayer (1930) has been used for E_1. Other values are taken from Landolt-Börnstein.

*** The possibility is, of course, not excluded that excited inert gas atoms, or inert gas ions produced by electron impact or by radiation, may unite with each other or with other atoms to form unstable, short-lived polyatomic molecules; nor that such transient molecules might be formed by union of normal inert gas atoms with ionized atoms of other elements (by the action of Van der Waals forces). Such polyatomic molecules, formed by inert gases —e.g., He_2^+, HgA—have, indeed, been detected spectroscopically. These are, however, not chemical compounds in the usual sense. The addition compounds of the inert gases with highly polar molecules, such as H_2O, discussed in Chap. 4, are also not affected by the above argument.

quence of their high ionization energies, but arises rather because the large amount of energy to be expended on ionization is not compensated by a correspondingly high lattice energy of the compounds in question. For example, the energy needed to remove two electrons from a Be atom (632 kcal per g atom) is considerably greater than that required to remove one electron from a Ne atom (495 kcal per g atom), but in the case of the Be compounds the lattice energy is also correspondingly high. In the case of BeF_2 it amounts to about 826 kcal per mole, whereas for NeF it is calculated as only about 181 kcal per mole. It should be noted that what determines the possibility of forming compounds is not the *absolute magnitude of the ionization energies, but their magnitude as compared with the amount of energy liberated in the association of the oppositely charged ions to form molecules or crystals* (and in some cases also with energy liberated from other factors, such as the further addition of H_2O molecules, etc.).

Since we shall frequently make use later of the Born-Haber cycle for calculating free energies of formation to assess the stability of compounds, the cycle may be represented by a more general scheme:

$$
\begin{array}{ccc}
\text{Initial substances} & \dfrac{+\ \Sigma Q_f^i + \Sigma Q_v^i + \Sigma Q_D^i}{+\ \Sigma Q_s^a} & \text{Free gaseous atoms} \\
\text{solid, liquid or gaseous} & \xrightarrow{\hspace{3cm}} & \text{of reactants} \\[2mm]
-\Delta F \uparrow & & \downarrow \begin{array}{l}+\Sigma A_I \\ -\Sigma E\end{array} \\[2mm]
\text{Reaction products} & \dfrac{-\Sigma B_{Mx} - \Sigma Q_v^e - \Sigma Q_f^e}{-\Sigma G} & \text{Gaseous ions} \\
\text{solid, liquid or gaseous} & \xleftarrow{\hspace{3cm}} &
\end{array}
$$

Q_f here represents the (molar) latent heat of fusion, Q_v the heat of evaporation, Q_s $(= Q_f + Q_v)$ the heat of sublimation, all corrected to $T = 0°$ K. The quantities marked with the superscript i refer to the initial reactants, those marked with e to the end products. Q_D represents the molar dissociation energy for the dissociation of a gas, consisting of polyatomic molecules, into atoms. A_I is the ionization energy (per g atom) for the formation of positive ions, E the electron affinity for formation of negative ions. G is the lattice energy per mole; for a gas this is replaced by the energy of formation of 1 g molecule of the gas from the ions. ΔF is the free energy change in the reaction *initial reactants → end products*. The summation signs imply that the sum of the corresponding quantities for all the individual substances should be substituted, each being multiplied by the number of such molecules taking part in the reaction. Since, for the cyclic process

$$\Sigma Q_S + \Sigma Q_D + \Sigma A_I - \Sigma E - \Sigma G - \Delta F = 0$$

we finally obtain, as the equation for calculating the free energy

$$-\Delta F = \Sigma G + \Sigma E - (\Sigma A_I + \Sigma Q_D + \Sigma Q_S) \tag{12}$$

For gases, B_{MX} is substituted in place of G, and for liquids G is replaced by $(B_{MX} + Q_v^e)$, and Q_S by Q_v^i.

The influence which the various energy quantities of eqn. (12) exert upon the magnitude of the free energy change may be seen from a few examples in Table 26. All the quantities in this table relate to the formation of 1 gram equivalent of the compound; lattice energies have been calculated from eqn. (11), and the free energies listed under $\Delta F°_{calc}$ from eqn. (12). For comparison with these, some free energies determined by purely experimental means are listed under $\Delta F°_{obs}$, and experimentally found heats of formation (at ordinary temperature only) under W_f. The accuracy of free energy calculations based on the Born-Haber cycle is, in general, not very great at present. Nevertheless, the cycle does provide an excellent survey of the factors which are important for the stability of a chemical compound. It may be seen from Table 26 that two chief factors influence the magnitude of the free energy decisively: the *ionization potential* of the electropositive constituent of the compound, and the *lattice energy* (or, for gases, the energy of formation of the molecule from the ions). The electron affinity of the electronegative constituent generally exerts a smaller influence.

According to eqn. (11), the lattice energy depends chiefly on the valence of the ions, and the interionic distance in the crystal,—i.e., on the sum of the apparent ionic radii (cf. Chap. 7). The sum of the ionic radii of cation and anion, $r = r_c + r_a$ is shown in the last column of Table 26. Because of the exceedingly small radius of the hydrogen nucleus, hydrogen is able to form stable compounds with negative ions, even though *its ionization energy is greater than that of xenon* and, moreover, an additional 51.3 kcal per g atom must be expended, in the formation of H⁺, to dissociate the H_2 molecule into atoms. Because of its small size, the H⁺ nucleus also exerts a particularly strong polarizing action. The energy of formation of the HF molecule from the ions, for example, is increased by about 50 % by the polarization effect, as compared with that calculated neglecting polarization. For the same reason, the hydrides of the electronegative elements are all volatile, since the energy of formation of free molecules is greater, by reason of polarization, than that for an ionic lattice.

TABLE 26

EXAMPLES OF DATA INVOLVED IN THE CALCULATION OF AFFINITIES
BY MEANS OF THE BORN-HABER CYCLE

Compound	G	B_{MX}	E	A_I	Q_D	Q_S	$+\Delta F°_{calc}$	$\Delta F°_{obs}$	W_f	r
HF	—	364.2	95.3	312.2	82.7	—	—	−64.6	64.4	—
LiH	218.6	—	16.3	123.8	51.3	39.0	− 23.7	—	21.6	2.04
NaF	214	—	95.3	118.0	31.3	26.2	−134	—	136.4	2.31
½CaF₂	308	—	95.3	206.4	31.3	21.4	−144	—	144.7	2.36
½CaO	420	—	−86.5	206.4	29.4	21.4	− 76	−72.7	76.1	2.40
½CaS	360	—	−42	206.4	25.7	28.7	− 57	−54.9	55.6	2.84
XeF	ca. 150	—	95.3	278.2	31.3	0	ca. + 64	—	—	ca. 3.4
½NaF₂	ca. 335	—	95.3	601.3	31.3	12.9	ca. +215	—	—	ca. 2.2
CaF	ca. 183	—	95.3	140.3	31.3	47.8	ca. − 59	—	—	ca. 2.8

In the lower part of Table 26 are shown the lattice energies for a few unknown compounds. Following Grimm and Herzfeld (1923), these have been calculated on the assumption that if the compounds existed, they would form crystal lattices similar to those of analogous, known compounds. The reason for the non-existence of compounds like XeF and NaF₂ may be seen in the *positive* free energies of formation found in this way. The compound CaF has, indeed, a negative free energy of formation; nevertheless, as may be deduced from the Table, far more energy is liberated when 1 g atom of fluorine combines with ½ g atom of Ca to form ½CaF₂, than if it combined with 1 g atom of Ca to form CaF. In other words, even although CaF is stable with respect to the free components, it is unstable as compared with CaF₂, so that CaF would spontaneously decompose into Ca + CaF₂. It must, however, be remembered that with rising temperature these equilibria may be markedly displaced (cf. p. 239).

It emerges, as a generalization, that the lattice energies are not sufficient to provide the energy of ionization of an electron bound in an inert-gas shell (in the inert gases themselves, or in the elements following them). This is why the valences of the elements of the Main Groups are determined by their position with respect to the inert gases. The lattice energy decreases with increase in the radius of both cation and anion. The stability of compounds of analogous composition and similar structure therefore generally decreases in descending each of the Main Groups of the Periodic System—i.e., with increasing radius of the homologous ions. If one compares a series of compounds of different valence states of the same cation, a sharp maximum is obtained, for most elements, for the free energy of formation of that compound in which the cation has given up all electrons external to the inert-gas core. This is most strikingly so in the case of the elements of the first two Main Groups and for boron and aluminum. The fact that these elements, in their ionic compounds, exercise almost exclusively the valences corresponding to their group numbers is to be explained in this way.

References

1 W. KOSSEL, *Valenzkräfte und Röntgenspektren*, 2nd Ed., Berlin 1924, 89 pp.

2 A. E. VAN ARKEL and J. H. DE BOER, *Chemische Bindung als elektrostatische Erscheinung*, Leipzig 1931, 320 pp.

3 G. N. LEWIS, *Valence and the Structure of Atoms and Molecules*, New York 1923, 172 pp.

4 N. V. SIDGWICK, *The Electronic Theory of Valency*, Oxford 1927, 310 pp.

5 N. V. SIDGWICK, *Some Physical Properties of the Covalent Link in Chemistry*, Ithaca 1933, 249 pp.

6 L. C. PAULING, *The Nature of the Chemical Bond and the Structure of Molecules and Crystals; An Introduction to Modern Structural Chemistry*, 2nd Ed., Ithaca 1940, 450 pp.

7 G. BRIEGLEB, *Atome, Ionen und Molekeln*, Berlin 1940, 404 pp.

8 R. DE L. KRONIG, *The Optical Basis of the Theory of Valency*, Cambridge 1935, 246 pp.

9 C. A. COULSON, *Valence*, Oxford 1952, 340 pp.

10 G. I. BROWN, *Simple Guide to Modern Valency Theory*, London 1953, 174 pp.

11 L. C. PAULING and E. B. WILSON, *Introduction to Quantum Mechanics*, New York 1935, 468 pp.

12 Y. K. SYRKIN and M. E. DYATKINA, *The Structure of Molecules and the Chemical Bond* (translated by M. A. Partridge and D. O. Jordan), London 1950, 509 pp.

13 O. SACKUR, *Die chemische Affinität und ihre Messung*, Braunschweig 1908, 129 pp.

14 G. N. LEWIS and M. RANDALL, *Thermodynamics and the Free Energy of Chemical Substances*, New York 1924, 653 pp.

15 H. ULICH, *Chemische Thermodynamik*, Dresden 1930, 353 pp.

16 J. J. VAN LAAR, *Die Thermodynamik einheitlicher Stoffe und binärer Gemische*, Groningen 1935, 379 pp.

17 F. HABER, *Thermodynamics of Technical Gas Reactions* (translated by A. B. Lamb), London 1908, 296 pp.

18 I. PRIGOGINE and R. DEFAY, *Chemical Thermodynamics* (translated by D. H. Everett), London 1954, 542 pp.

19 T. DE DONDER and P. J. VAN RYSSELBERGHE, *The Thermodynamic Theory of Affinity*, Stanford 1936, 142 pp.

20 E. A. GUGGENHEIM, *Modern Thermodynamics by the Methods of J. Willard Gibbs*, London 1933, 206 pp.

21 R. RUEDY, *Bandenspektren auf experimenteller Grundlage*, Braunschweig 1930, 124 pp.

22 R. MECKE, *Bandenspektren und ihre Bedeutung für die Chemie*, Berlin 1929, 87 pp.

23 H. SPONER, *Molekülspektren und ihre Anwendung auf chemische Probleme*, Berlin 1935–36, Vol. I, 154 pp.; Vol. II, 506 pp.

24 R. DE L. KRONIG, *Band Spectra and Molecular Structure*, Cambridge 1930, 163 pp.

25 F. R. BICHOWSKY and F. D. ROSSINI, *Thermochemistry of the Chemical Substances*, New York 1936, 460 pp.

26 W. BÜTTNER, *Die Entropie. Physikalische Grundlagen und technische Anwendungen*, Berlin 1939, 131 pp.

27 J. C. SPEAKMAN, *An Introduction to the Modern Theory of Valence*, 2nd Ed., London 1943, 160 pp.

28 J. A. A. KETELAAR, *Chemical Constitution. An Introduction to the Theory of the Chemical Bond*, (translated by L. C. Jackson), Amsterdam 1953, 398 pp.

29 O. K. RICE, *Electronic Structure and Chemical Binding*, London 1940, 511 pp.

CHAPTER 6

FIRST MAIN GROUP OF THE PERIODIC SYSTEM: THE ALKALI METALS

Atomic numbers	Names of the elements	Symbols	Atomic weights	Densities	Melting points	Boiling points	Specific heats	Valence States
3	Lithium	Li	6.940	0.534	179°	1340°	0.837	I
11	Sodium	Na	22.991	0.97	97.8°	883°	0.295	I
19	Potassium	K	39.100	0.86	63.5°	760°	0.177	I
37	Rubidium	Rb	85.48	1.52	39.0°	696°	0.080	I
55	Cesium	Cs	132.91	1.87	28.45°	708°	0.048	I
87	Francium	Fr	223	—	—	—	—	I

1. Introduction

(a) General

Group IA includes the elements lithium, sodium, potassium, rubidium, cesium, and francium, which occurs in minimal quantities as a product of radioactive decay. These are grouped together under the common name of the *alkali metals*, because the hydroxides of the principal members have long been known under the name of 'alkalis'. The alkali metals are very soft, and have very low densities for metals. Their extreme fusibility and their low boiling points are noteworthy.

They are extremely reactive chemically, and decompose not only water but also alcohol, with the evolution of hydrogen:

$$Na + HOH = NaOH + \tfrac{1}{2}H_2$$
$$Na + C_2H_5OH = NaOC_2H_5 + \tfrac{1}{2}H_2$$

In the first case, solutions are obtained with the characteristic behavior, known as 'alkaline', which is due to the high hydroxyl ion content of the solutions. Solutions with a high hydroxyl ion concentration are, accordingly, quite generally spoken of as alkaline solutions. Their burning taste, and especially their property of affecting the color of certain organic dyestuffs—e.g., of turning tincture of litmus blue,—are typical properties. Concentrated solutions of the alkali hydroxides have a strongly corrosive action on the skin; the compounds are therefore known as *caustic alkalis*.

The decomposition of water by the alkali metals is a necessary consequence of the fact that their normal potentials lie considerably above the potential taken up by a hydrogen electrode in a solution having the hydrogen ion concentration of pure water, or even in a solution normal with respect to hydroxyl ions. They react, accordingly, with hydrogen ions even at that minute concentration which is present in strongly alkaline solutions. The essentials of the reaction can be represented by the equation

$$M + H^+ = M^+ + \tfrac{1}{2}H_2,$$

where M is an atom of alkali metal.

The loss of an electron from the alkali metal atoms takes place so easily that it can even be given up to uncharged hydrogen, which thereby acquires a negative charge

$$M + \tfrac{1}{2}H_2 = M^+H^-$$

This reaction takes place when dry hydrogen is passed over gently heated alkali metal. The alkali metals react very readily with those elements which have a marked tendency to acquire a negative ionic charge. Even lithium, the least reactive, ignites immediately in chlorine, and burns therein with a brilliant white light.

The alkali metals tarnish immediately in moist air, and are soon covered with a thick crust of hydroxide; they must therefore be stored under petroleum or paraffin oil. They ignite when moderately heated in air or oxygen, the products of combustion being dependent on the size of the alkali metal, *e.g.* $Li \rightarrow Li_2O_3$; $Na \rightarrow Na_2O_2$; $K \rightarrow KO_2$. The valence is invariably I in all compounds and the alkali metals are always electropositively univalent. The alkali metals are among the most strongly electropositive of the elements. The rule that the electropositive character within any one group increases with increasing atomic weight is distinctly exemplified by the alkali metals, when their whole chemical behavior is taken into account.

In particular cases this can be clearly demonstrated. For example, the stronger electropositive character of potassium, as compared with sodium, is manifested in the differing behavior of the metals towards water. Sodium reacts vigorously, indeed, so that it is melted by the heat of reaction, but provided the globule of sodium can move freely about on the water, ignition does not take place, even in the presence of air. Potassium, however, reacts so violently with water that, in the presence of air, the hydrogen evolved burns immediately. The two metals differ even more markedly in their behavior towards bromine and iodine. Sodium reacts only superficially with bromine at ordinary temperature, and it can even be cautiously melted with iodine without any reaction becoming noticeable. If, however, a small fragment of potassium is thrown on to liquid bromine, a violent detonation takes place; the same occurs if potassium is heated with iodine.

The strong electropositive character of the alkali metals is due, essentially, to the small expenditure of energy necessary to remove one electron from the atoms of the elements. This finds direct expression in their low ionization potentials (see Table 22). Their good electrical conductivity, and the fact that they exhibit the photoelectric effect to an especially marked extent (i.e., the emission of electrons when their clean surfaces are irradiated with ultraviolet light), also depend on the loose binding of the electrons .The ease with which the alkali metals are transformed into positively charged ions is of decisive importance for their chemical reactivity. On the other hand, the heats of formation of their compounds (cf. Table 27) do not provide any direct measure of the differing electroaffinities of the several elements, since they depend also on a number of other factors, and especially on the lattice energies.

In Table 27 are collected the heats of formation of the normal oxides M_2O and other salt-like compounds of the alkali metals; these may be taken as an approximate measure of the free energies of formation of the compounds concerned. It might be thought that the increasing electropositive character in the series lithium to cesium must show itself in an increase in the free energies of formation (or heats of formation) of the salt-like compounds. Table 27 shows that this is not the case. This is because, as was shown in the previous Chapter, the free energies of formation of the oxides, etc., are made up of several partial contributions, some of which change in opposite directions in going from the lightest to the heaviest element of the group.

TABLE 27
HEATS OF FORMATION OF ALKALI COMPOUNDS

Heats of formation shown in the Table are in kcal per g atom of alkali metal (at ordinary temperature). For non-metals, the gaseous state is taken as standard, except for sulfur, for which the rhombic modification is the standard state.

Compounds	Normal oxides	Hy- droxides	Hy- drides	Fluo- rides	Chlorides	Bromides	Iodides	Sulfides
Lithium	71.5	116	21.6	144.7	98.7	87.1	71.2	—
Sodium	51.5	102.3	12.8	136.6	97.8	90.4	76.7	44.9
Potassium	43.4	101.4	9.8	134.5	104.2	97.5	86.4	43.6
Rubidium	41.7	101.2	—	132.8	105.0	99.2	87.4	43.5
Cesium	41.3	100.3	—	131.5	106.4	101.2	90.4	—

If one compares compounds with the same anion, then according to eqn. (12) p. 149, the difference $G - (A_I + Q_S)$ is the factor which determines the trend of the free energy of formation. From the data collected in Table 28 (p. 156) it may be seen that the quantity $(A_I + Q_S)$ decreases from 159.8 kcal to 108.1 kcal—i.e., in the ratio 3 : 2 as one goes from lithium to cesium. If the lattice energy diminishes in the same, or even greater, proportion, the difference between G and $(A_I + Q_S)$ must also decrease. The free energy of formation, $-\Delta F°$, will decrease, in that case, in passing from lithium compounds to cesium compounds. According to eqn (11) (p. 147), the lattice energy is inversely proportional to the interionic distance r in the crystal (i.e., the sum of the radii of anion and cation). In the fluorides, the interionic distance increases from 2.01 Å in LiF to 3.01 Å in CsF (cf. Table 39, p. 210). The lattice energy therefore decreases roughly in the ratio of 3 : 2.* The free energy of formation must therefore decrease from LiF to CsF. The same is true for the oxides. If G decreases to a relatively smaller extent than does $(A_I + Q_S)$ the net change in $[G - (A_I + Q_S)]$ can be given only when the magnitude of the individual terms is considered. For LiI and CsI, G may be calculated from eqn. (11) (p. 147) as 170.6 and 130.7 kcal respectively, so that in this case G decreases only in the proportion 4 : 3. The difference $[G - (A_I + Q_S)]$ is found from the data of Table 28 as 10.8 kcal for LiI, and 22.6 kcal for CsI. For the iodides therefore, the free energy of formation increases along with the electropositive character of the alkali metals; this is true for the bromides also.

The trend of the heats of formation of the compounds cited in Table 27 can thus be explained as being due to the superposition of two factors—the increase in electropositive character in the direction from lithium to cesium, and a decrease in lattice energy, in the same direction, because of the increase in ionic radii. The smaller the charge and the larger the ionic radius of the anion (for any one cation), the smaller is the influence exerted by the second factor. It is thus already of subordinate importance in the chlorides, and even more so in the bromides and iodides, for which the heats of formation consequently run parallel to the electropositive character.

The radii of the alkali ions are shown in Table 28; those of the halogen ions are given on p. 773. For the calculation of lattice energies it is better to use the interionic distances obtained directly from cell dimensions, since the rule of the constancy of ionic radii is only approximately true. The distance between centers of the ions in alkali halide crystals may be taken from Tables 39 and 40, p. 210 et seq., noting that for the rock salt type, $r = a_0/2$, and for the cesium iodide type $r = \sqrt{3} \cdot a_0/2$.

---

* The lattice energy does not diminish in exactly the same ratio as the interionic distance increases, since the value of $(n - 1)/n$ increases by 8 % in the series in question. The Madelung constant A has the same value for all the alkali fluorides since they all have crystal structures of the same type.

TABLE 28

ATOMIC AND IONIC RADII, HEATS OF FUSION AND EVAPORATION, IONIZATION POTENTIALS AND STANDARD POTENTIALS OF ALKALI METALS, AND HEATS OF HYDRATION OF ALKALI IONS

Latent heats of fusion and evaporation refer to the melting point and boiling point of the metals. Hydration energies are based on the assumption that the heat of hydration of the hydrogen ion is 250 kcal.

Elements	Lithium	Sodium	Potassium	Rubidium	Cesium
Atomic radius in Å	1.56	1.86	2.33	2.43	2.62
Ionic radius in Å*	0.60	0.95	1.33	1.48	1.69
Heat of fusion in kcal/g atom	0.76	0.63	0.57	0.53	0.50
Heat of vaporization in kcal/g atom	32.25	23.12	18.92	18.06	16.32
Heat of sublimation at 0° K	35.96	26.20	21.85	20.58	18.74
Ionization energy $M \rightarrow M^+ + e$ in kcal/g atom	123.80	118.00	99.65	95.90	89.40
$M^+ \rightarrow M^{++} + e$	1736.86	1084.5	730.59	631.56	538.7
Normal potential in volts** obs.	+2.96	+2.71	+2.92	+2.92	—
calc.	+2.95	+2.71	+2.90	+2.99	+2.92
Heat of hydration in kcal/g ion	112	99	82	77	71

* Values from Pauling, *The Nature of the Chemical Bond*.
** The signs of the potentials have been changed from those given in the original text, to conform with those commonly used by American chemists.

Table 29 gives the lattice energies of the halides of Li, Na, K and Rb, based on calculations by De Boer (1936), and those of the halides of Cs and NH_4, after J. E. Mayer (1932 onwards). These values are more accurate than those obtained above, by means of eqn (11), p. 147, since in calculating them more exact allowance has been made for the effects of the forces superposed on the Coulomb attraction. The calculations are thereby made much more complicated. Calculations made by means of eqn (11) suffice, however, if less importance attaches to great accuracy of the individual values than to an approximate determination for the purpose of comparing the stability of compounds, or for surveying the effect of ionic radii and interionic distances in the crystal lattice upon the lattice energy, and consequently upon the free energy of formation of compounds.

TABLE 29

LATTICE ENERGIES OF HALIDES OF THE ALKALI METALS AND OF AMMONIUM

(in kcal per mol at 0° K)

	Li	Na	K	Rb	Cs	NH_4
Fluorides	237	213	188	180	174	191
Chlorides	195	180[1]	164	158	152	162
Bromides	185	172	157	152[2]	146	154
Iodides	171	160	149[3]	143	139[4]	145

[1] Obs. 181. [2] Obs. 151. [3] Obs. 154. [4] Obs. 141. These values are calculated by means of eqn. 12 (p. 149) from measured ΔF values, ionization energies etc.

The small heat of formation of the normal oxides of the alkali metals is noteworthy, in comparison with the heats of formation of the halides. This is not because the lattice energy of the oxides is less, but because energy must be expended for the process $O + 2e \rightarrow O^{2-}$

whereas energy is liberated in the union of an electron with the free atom in the case of the halogens.

The somewhat irregular sequence of the *normal potentials* (Table 28) is also due to the superposition of several factors. The normal potential is a measure of the difference in energy between the metal and a 1-normal solution of its ions. This energy difference is not given just by the sum of the heat of sublimation and the ionization energy, but it is neces-sary to take account of the rather considerable amount of energy liberated because the alkali ion surrounds itself with a *hydration sheath*. (Hydration, cf. p. 75). The high normal potential of lithium, lying out of sequence, arises from the particularly strong hydration of the lithium ion (cf. Table 17). The extent of hydration decreases in the series from lithium to cesium, and the electrolytic *mobilities* increase in the same order (see Table 30). Although the lithium ion is the smallest of the alkali ions, it has the lowest mobility in aqueous solution because it is the most strongly hydrated.

TABLE 30

ELECTROLYTIC MOBILITIES OF ALKALI IONS
AT INFINITE DILUTION, AT 18°

Li^+	Na^+	K^+	Rb^+	Cs^+
33.5	43.5	64.6	67.5	68

The energies set free through the hydration of ions are measured by the *heats of hydration*. The sums and differences of these can be calculated from heats of solution.* If the heats of hydration given in Table 28 are converted to *e*-volts, and added to the normal potentials given in the same Table, we obtain the normal potentials which would be found for the alkali metals if their ions were not hydrated: Li^+, —1.9 volts, Na^+, —1.6 volts, K^+, —0.6 volts, Rb^+, —0.4 volts, Cs^+, —0.2 volts. These potentials do, indeed, increase regu-larly from lithium to cesium, as would be expected from the increase of electropositivity of the metals, since the greater the tendency of an element to pass into the ionic state, the higher must its solution potential lie.

If allowance is made for the heats of hydration, it is possible to calculate the normal potentials from the heats of sublimation of the metals and the ionization potentials of the free atoms, by means of the cyclic process described on p. 149. The starting materials for this are the metals, and the products of reaction the hydrated ions; the energy of hydration thus enters in place of the lattice energy. For exact calculations, allowance must be made for the difference between the heat effects and the free energy changes, and for the variation with temperature. Makishima (*Z. Elektrochem.* 41 (1935) 697) carried out accurate calcu-lations of this kind for a large number of metals, and obtained values agreeing well with the direct measurements. Table 28 shows the normal potentials of the alkali metals calculated by Makishima, as well as the measured values.

The *salts* of the alkali metals are colorless (unless derived from colored anions), and are almost all readily soluble; only lithium forms a fair number of moderately sparingly soluble salts. The aqueous solutions of the salts contain colorless, posi-tively univalent alkali ions, which are all hydrated to a more or less considerable extent in dilute solutions. The salts of the lighter alkali metals often contain con-siderable amounts of water in the crystalline state. Many lithium salts, in particu-lar, can also crystallize with alcohol in place of water, but except for the lithium compounds the salts of the alkali metals are insoluble, or only slightly soluble, in alcohol. The alkali salts are almost completely dissociated in aqueous solutions. The same is true of the hydroxides, which are accordingly the strongest bases.

---

* Cf. footnote on p. 786. The sums and differences of the ionic heats of hydration are more accurately known than the absolute values. The absolute heat of hydration of the hydrogen ion according to Fajans (1921) is 252, according to Latimer (1922) 257, and according to Webb (1926) 249.6 kcal per g ion.

Not only are the alkali salts and hydroxides completely dissociated in solution, but the forces which the free ions exert on one another in solution are relatively small. This is in part due to the small charge, and in part to the large radii of the ions. The influence of the latter makes itself noticeable especially in concentrated solutions, and is responsible for the increasing strength of the bases (i.e., the increasing 'apparent degree of dissociation', or more correctly the increase in the conductivity coefficient f_μ of the hydroxides), in the series from lithium to cesium hydroxide. According to Hlasko (1935) the following are the values of f_μ in $0.031 N$ solutions at $25°$:

for	LiOH	NaOH	KOH	RbOH	CsOH
f_μ	0.918	0.935	0.938	0.944	0.955

Many other properties of the alkali compounds such as the relatively great volatility of the chlorides, for example, are due to the large radii of the alkali ions.

The salts of the alkali metals with weak acids have a strongly basic reaction in consequence of hydrolytic decomposition.

Among the characteristic properties of the alkali metals is the ease with which they can be excited to the emission of light. If any not too involatile compounds of the alkali metals are introduced into the flame of the Bunsen burner, this becomes colored, and if examined spectroscopically a few characteristic lines may be seen in the visible region. The ease with which the emission of the lines is excited, and the simple structures of the spectra are both closely related to the strongly electropositive character of the alkali metals, as will be shown in the section on the *spectra of the alkali metals.*

In terms of the Kossel theory, the characteristic properties of the alkali metals follow from their position in the Periodic System. They stand in the group immediately following the inert gases, and their atoms accordingly contain one electron more than the atoms of the preceding inert gases. As shown by the spectroscopically determined ionization energies, this electron can easily be removed, whereas for the ionization of a second electron a quite disproportionate amount of energy must be expended (Table 28), for which the energy of formation of the crystal lattice is not nearly sufficient. This is why the alkali metals always behave as positively *uni*valent. The cores remaining after loss of one electron—i.e., the univalent ions of the alkali metals—have not merely the same number of electrons, but also the same electronic configuration as the immediately preceding inert gas, in each case; i.e., the electrons are in the same quantum states as in the inert gas atoms. The impossibility of obtaining compounds in which the alkalis are negatively charged is to be associated with the wide separation between each alkali metal and the next following inert gas.

From the fact that the ground terms in the absorption spectra of the alkali metals are s terms (cf. p. 170), it may be concluded that each has one electron which, in the normal state of the atom, moves in an orbit with the azimuthal quantum number $l = 0$. This electron is bound, in each case, outside the electron shell of the preceding inert gas, so that the principal quantum number of the corresponding orbits are one unit greater, each time, than those of the inert gases. The ground states for the outermost electrons of the alkali metals are accordingly characterized by the following quantum numbers

	Li	Na	K	Rb	Cs
$n_l =$	2_0	3_0	4_0	5_0	6_0

According to the concept of the Bohr theory, the outer electrons thus move in highly eccentric orbits. According to the theory, this eccentricity of the orbit and the small effective charge in the outer sphere of the atoms are to be associated with the large atomic radii of the alkali metals, and the great difference between their atomic and ionic radii. The latter refer, in fact, to the core which remains after the outermost electron has been lost. The atomic and ionic radii increase considerably within the group, as the atomic weight increases.

According to the Heitler-London theory (p. 131 *et seq.*) the alkali metals should also be capable of forming covalent compounds, though only of low stability, in which they are also univalent. This is confirmed by experiment. Vapor density measurements show that diatomic molecules as well as monatomic molecules are present in the vapors of the alkali metals just above the boiling point. The *alkyl* compounds of the alkalis, e.g. $Na.CH_3$, sodium methyl, may possibly be regarded as covalent compounds, although their solutions in other metal alkyls—for example in zinc methyl—display electrolytic conductivity, as was found by Hein (1922). The alkali alkyls were first isolated by Schlenk in 1917; they are colorless powders, insoluble in most inert solvents; they decompose on heating, without melting, and inflame in air.

The first element in the series, *lithium*, occupies a special position in many respects, as compared with the other alkali metals. Thus it forms various compounds which are formed by none of the others, and in various ways its behavior constitutes a transition between the alkalis and the group of alkaline earth metals. It shares with the latter the sparing solubility of many salts—e.g., the phosphate, carbonate, and fluoride. It also resembles the elements of the second Main Group in its ability to form double salts, which is much more marked than for the other alkalis.

Sodium also deviates in many respects from the behavior which is typical for the remaining members of the group, although to a much smaller extent than lithium. The fact that the third element, potassium, first provides the really typical representative of the alkali group exemplifies a rule which also applies elsewhere in the Periodic Table, that the group character is fully developed only with the second or third elements of the Main Groups, whereas the first element (and to a smaller extent the second also) shows departures from the typical character. The behavior of the first element is frequently rather like that of the next higher Main Group; the second element often forms the link with the Subgroup belonging to the same family.

Sodium and cesium are pure elements, i.e., atoms of only one mass number are known in each case. The other elements have been shown, on examination by means of Aston's mass spectrograph, to be mixed elements.

As was discovered by N. R. Campbell in 1907. potassium is radioactive, though only very feebly; its activity is only about one thousandth that of uranium. Rubidium displays the same phenomenon. Potassium and rubidium are both β-ray emitters. It has been deduced from their penetrating power that the β-rays of potassium have an initial velocity of about $2.5 \cdot 10^{10}$ cm per sec., and those of rubidium about $1.8 \cdot 10^{10}$ cm per sec.

The mass spectrograph has shown that potassium is a mixture of 3 isotopes, with the mass numbers 39, 40, and 41, present in the ratio 93.41 : 0.012 : 6.59 respectively.* Of these three isotopes, only that with the mass number 40 is radioactive. From the number of β-particles emitted in unit time, the half-life of ^{40}K has been found to be about $4.5 \cdot 10^8$ years.

By partially separating the mixture of isotopes by 'ideal distillation' (see Vol. II), v. Hevesy was able to show, in 1927, that the phenomenon of radioactivity was connected

---

* This isotopic ratio is strictly correct only for the potassium salts in sea water. Values have been found for the ratio ^{39}K : ^{41}K in minerals which vary by as much as 3% from one another, while variations up to 15% have been observed in plant ashes. (Cf. A. K. Brewer, *J. Am. Chem. Soc.*, 58 (1936) 370).

with one of the heavier isotopes, since the fraction in which they were enriched possessed the higher radioactivity. The radioactivity, however, was increased by only 4.4 % as compared with ordinary potassium, whereas the isotope ^{41}K was enriched by more than 10 % above its original concentration, as shown by the determination of the atomic weight. This indicated that ^{40}K, not ^{41}K, possessed the radioactive properties, since a smaller degree of enrichment of ^{40}K is to be expected in the ideal distillation. That ^{41}K is not responsible for the radioactive properties of potassium also follows from another argument. According to the radioactive displacement law, ^{41}Ca would be formed from the β-disintegration of ^{41}K and should be detectable in considerable amounts in potassium minerals of great geological age, if ^{41}K were radioactive. Tests with the mass spectrograph, however, showed the complete absence of ^{41}Ca (Aston and v. Hevesy, 1935). More recently, Smythe (1937) has been able to prove directly that ^{40}K is the bearer of the radioactive properties, by measuring directly the activity of ^{40}K separated by means of the mass spectrograph. By the emission of a β-particle, the nucleus of ^{40}K is converted into that of ^{40}Ca. About 12 per cent of the atoms of ^{40}K disintegrate by a second process, however, known as K-electron capture. In this process, an electron from the most deep-seated, or K, electron shell enters the nucleus of the atom, which is thereby converted to ^{40}A. At the same time, the vacant K-electron level is filled by an electron which makes a transition from some level of higher energy. As follows from considerations discussed later (Chapter 7), such transitions in the extra-nuclear structure result in the emission of the characteristic K series X-ray spectrum of the daughter element. The emission of the X-ray spectrum (and in some cases of γ-rays also, from nuclear transitions) but not of massive charged particles is characteristic of K-capture processes. The production of argon from potassium amounts to $3.6 \cdot 10^{-12}$ cm^3 per g potassium per year.

The radioactivity of rubidium is due to the disintegration of the isotope ^{87}Rb, which is present to the extent of 27.2 %, along with ordinary ^{85}Rb; by emission of a β-particle (with a maximum energy of 0.132 Mev) it is transformed into the stable strontium isotope, ^{87}Sr. The half-life of ^{87}Rb is $4.3 \cdot 10^{10}$ years (Huster, 1954). The 'strontium method' of determining the age of minerals (cf. Vol. II) is based on these facts. No radioactive isotope of cesium occurs in Nature (Hahn and Mattauch, 1942).

On the basis of the Periodic System, the existence of an alkali metal (eka-cesium), with the atomic number 87 had been suspected, but for a long time all attempts to discover it were fruitless. The conclusions drawn by Allison (1930 and later) from magneto-optic measurements are not convincing, and the findings of Papish (1931), who believed he had discovered the X-ray lines of element 87, were later shown by I. Noddack (1934) to be incorrect. Recently, however, the existence of eka-cesium in nature has been proved by Mlle. Perrier. She detected that this element is formed as the product of the (branching) α-decay of actinium, and named it first *Actinium K*, and later *Francium, Fr*. It is unstable (radioactive), as had already been suspected from its position in the Periodic System, and from the rules governing nuclear stability. See further Vol. II, Chap. 13.

(b) Occurrence

Because of the excessive ease with which they are oxidized, the alkali metals never occur in the free state, but invariably in the form of their compounds. In this form, however, *sodium* and *potassium* are among the most widely distributed of the elements, constituting 2.6 % and 2.4 %, respectively, of the earth's crust. They are thus about equally abundant, on the whole; nevertheless the compounds of potassium are less generally dispersed than those of sodium. *Felspar*, $K[AlSi_3O_8]$, and *mica* $KAl_2[AlSi_3O_{10}](OH,F)_2$ occur as constituents of many widely distributed rocks, and particularly of *granite*.* The most widespread sodium minerals are *oligoclase* (sodium potash felspar) and *albite* (soda felspar) $Na[AlSi_3O_8]$, which make up similar rocks. In the form of *rock salt*, sodium chloride, $NaCl$, occurs in enormous

---

* Granite is a rock, composed of the minerals *orthoclase*, *mica*, and *quartz*, which occur in it as a coarse-grained mixture. A rock of the same composition, but with a different structure (a slaty texture) is *gneiss*, which is also widespread.

deposits, which owe their origin to the drying up of former seas, or of great lakes. Rock salt is frequently over-laid by potassium and magnesium salts, the so-called 'abraum' salts, as for example at Stassfurt, in the north western plain of Germany. Considerable quantities of sodium chloride are contained in solution in sea water. The content varies but little in the great oceans, and in them, and also in the North Sea, amounts to 2.5 %; the Baltic Sea, on the other hand, contains only 0.6–1.7 % NaCl, whereas 3 % has been found in the Mediterranean Sea, and as much as 3.5 % in the Red Sea. In inland seas, from which there is no outlet, the salt content may rise to much higher values. Thus the Dead Sea contains about 20 % sodium chloride, with large amounts of other salts. By comparison with sodium chloride, the potassium chloride content of sea water is small, being only about $1/_{40}$ of the sodium chloride content. This is probably because, unlike sodium compounds, potassium compounds are strongly adsorbed by the soil, so that they cannot be transported into the sea. From the soil, potassium enters into plants, and is contained in plant ashes, in the form of the carbonate (potash). The sodium content of the ashes of land plants, on the other hand, is small; only coastal and marine plants contain larger quantities of sodium, chiefly in the form of organic compounds. A few plants—e.g., many varieties of tobacco—also contain quite significant amounts of lithium.

Potassium chloride occurs in very small amounts in human and animal tissues and body fluids, along with large quantities of sodium chloride. Ringer's solution, which is used in experiments on animal physiology, being similar in composition to blood serum and able, for example, to maintain the activity of the excised heart, contains 1 part each of a 1 % solution of $NaHCO_3$, a 1 % solution of $CaCl_2$, and a 0.75 % solution of KCl, to each 100 parts of 0.6 % NaCl solution. Considerable amounts of potassium are contained in *suint*, the wool sweat of sheep.

Among technically important sodium compounds must be mentioned *cryolite*, Na_3AlF_6, which is found in Greenland, and *Chile saltpeter*, $NaNO_3$, found in vast deposits on the Chilean coast. These minerals are not important for their sodium content, however. Chile saltpeter is used as a fertilizer, because of the combined nitrogen it contains, and cryolite is important primarily because it is used in the extraction of aluminum.

Lithium occurs in small amounts along with sodium and potassium especially in rocks and in many medicinal springs. There are also minerals, however, in which it occurs in considerably greater amounts—e.g., in *amblygonite* $LiAl[PO_4]F$, *triphyline* (Li, Na) (Fe,Mn) $[PO_4]$, *spodumene* (triphane) $LiAl[Si_2O_6]$, *lepidolite* (lithium mica)

$$K(Al_2[AlSi_3O_{10}],Li_2[AlSi_3O_6(OH,F)_4](OH,F)_2,$$

and *petalite* (Li,Na)$AlSi_4O_{10}$. A cesium mineral of similar composition, $Cs_4Al_4Si_9O_{26}H_2O$, occurs along with petalite in druses or inclusions in the granite of the island of Elba; the two minerals are therefore known also as Castor and Pollux (*pollucite*). Rubidium and cesium occur otherwise only in minute amounts accompanying the other alkali metals, but are widely distributed in low concentrations. A relatively large amount of rubidium (up to more than 1 %) occurs in lepidolite. Much smaller amounts of rubidium (about 0.015 %), with traces of cesium, are contained in carnallite, and a process for their extraction from that source was worked out by G. Jander (1929–30).

A review of the minerals of the alkali metals is provided by Table 31.

(c) History

As early as in the books of the Old Testament, mention is made of a substance which served as a washing agent, and was referred to as *neter*. The same substance, which was also known to the ancient Egyptians, is found in the writings of the

TABLE 31

ALKALI MINERALS

HALIDES, SULFATES AND NITRATES

Simple salts	Double salts and mixed salts	
Rock salt NaCl	*Carnallite* $KCl \cdot MgCl_2 \cdot 6H_2O$	Douglasite $2KCl \cdot FeCl_2 \cdot 2H_2O$
Sylvine KCl	*Kainite* $KCl \cdot MgSO_4 \cdot 3H_2O$	Erythrosiderite $2KCl \cdot FeCl_3 \cdot H_2O$
Thenardite Na_2SO_4	*Glauberite* $Na_2SO_4 \cdot CaSO_4$	
Arcanite K_2SO_4	Syngenite $K_2SO_4 \cdot CaSO_4 \cdot H_2O$	
Glauber salt $Na_2SO_4 \cdot 10H_2O$	Glaserite $Na_2SO_4 \cdot 3K_2SO_4$	
	Blödite (Astrakanite) $Na_2SO_4 \cdot MgSO_4 \cdot 4H_2O$	
	Löweite $Na_2SO_4 \cdot MgSO_4 \cdot 2\frac{1}{2}H_2O$	
	Vanthoffite $3Na_2SO_4 \cdot MgSO_4$	
	Polyhalite $K_2SO_4 \cdot MgSO_4 \cdot 2CaSO_4 \cdot 2H_2O$	
	Schönite $K_2SO_4 \cdot MgSO_4 \cdot 6H_2O$	
	Langbeinite $K_2SO_4 \cdot 2MgSO_4$	
	Alunite $K(AlO)_3(SO_4)_2 \cdot 3H_2O$	Jarosite $\overset{III}{K}(FeO)_3(SO_4)_2 \cdot 3H_2O$
	Sodium alum (Solfatarite) $NaAl(SO_4)_2 \cdot 12H_2O$	Natrojarosite $\overset{III}{Na}(FeO)_3(SO_4)_2 \cdot 3H_2O$
	Potash alum (Kalinite) $KAl(SO_4)_2 \cdot 12H_2O$	
Chile saltpeter (Sodium-salpeter) NaNO3	Darapskite $NaNO_3 \cdot Na_2SO_4 \cdot H_2O$	
Saltpeter KNO3		
	Cryolite Na_3AlF_6	Chiolite $5NaF \cdot 3AlF_3$
	Hieratite K_2SiF_6	Pachnolite (Thomsenolite) $NaCaAlF_6 \cdot H_2O$
		Sulfohalite $NaF \cdot NaCl \cdot 2Na_2SO_4$

CARBONATES (a) Simple	
Soda $Na_2CO_3 \cdot 10H_2O$	(b) Double carbonates
Trona $Na_2CO_3 \cdot NaHCO_3 \cdot 2H_2O$	Pirssonite $Na_2CO_3 \cdot CaCO_3 \cdot 2H_2O$
	Natrocalcite $Na_2CO_3 \cdot CaCO_3 \cdot 5H_2O$
	Dawsonite $Na_2CO_3 \cdot [Al(OH)_2]_2CO_3$

BORATES	
Borax (Tincal) $Na_2B_4O_7 \cdot 10H_2O$	Boronatrocalcite (Tincalcite) $NaCaB_5O_9 \cdot 6H_2O$
	Franklandite $Na_2CaB_6O_{11} \cdot 7H_2O$

PHOSPHATES and ARSENATES	
Triphyline $LiFe[PO_4]$	Amblygonite $LiAl[PO_4]F$
	Durangite $NaAl[AsO_4]F$

SILICATES (For structures of silicates see p. 498 *et seq.*)

Felspars
Soda felspar (Albite) $Na[AlSi_3O_8]$
Potash felspar (Orthoclase)
$\quad K[AlSi_3O_8]$
Soda orthoclase (mixed crystals of
$\quad$ orthoclase and albite)
Potash-soda felspars (mixed crystals of
$\quad$ albite and anorthite, cf. p. 242).

Nepheline $Na[AlSiO_4]$
Leucite $K[AlSi_2O_6]$
Sodalite $Na_3[Al_3Si_3O_{12}] \cdot NaCl$
Nosean $Na_3[Al_3Si_3O_{12}] \cdot Na_2[SO_4]$
Hauyne
$\quad Na_3[Al_3Si_3O_{12}] \cdot (Na_2,Ca)[SO_4]$
Lazurite (Ultramarine)
$\quad Na_3[Al_3Si_3O_{12}] \cdot Na_2S_2$
Scapolite, mixed crystals of
$\quad Na_3[Al_3Si_9O_{24}] \cdot NaCl$ and
$\quad Ca_3[Al_6Si_6O_{24}] \cdot Ca[SO_4,CO_3]$

Micas
Muscovite (Potash mica)
$\quad KAl_2[AlSi_3O_{10}](OH,F)_2$

Biotite (Magnesia mica)
$\quad K(Mg,Fe)_3[AlSi_3O_{10}](OH,F)_2$
Phlogopite (Magnesia mica)
$\quad KMg_3[AlSi_3O_{10}](OH,F)_2$
Lepidolite (Lithium mica), mixed crystals of
$\quad KAl_2[AlSi_3O_{10}](OH,F)_2$ and
$\quad KLi_2[AlSi_3O_6(OH,F)_4](OH,F)_2$
$\quad$ often containing Rb and Cs.
Zinnwaldite (Lithium-iron mica), mixed
$\quad$ crystals of
$\quad K(Mg,Fe)_3[AlSi_3O_{10}](OH,F)_2$ and
$\quad KLi_2[AlSi_3O_6(OH,F)_4](OH,F)_2$
Paragonite (soda mica)
$\quad NaAl_2[AlSi_3O_{10}](OH,F)_2$

Alkali-Augites
Spodumene (Triphane) $LiAl[Si_2O_6]$
Jadeit $NaAl[Si_2O_6]$

Agirine (Akmite) $NaFe[Si_2O_6]$
Alkali-Hornblendes
e.g.,
Riebeckite
$\quad Na_2Fe_3^{II}Fe_2^{III}[Si_8O_{22}](OH)_2$
Glaucophane
$\quad Na_2(Mg,Fe)_3(Al,Fe)_2[Si_8O_{22}](OH)_2$
Arfvedsonite
$\quad (Na_2,K_2,Ca)(Mg,Fe)_3(Al,Fe)_2[Si_8O_{22}](OH)_2$

Petalite (Castor) $(Li,Na)AlSi_4O_{10}$
Pollucite (Pollux) $Cs_4Al_4Si_9O_{26} \cdot H_2O$

Zeolites
Natrolite $Na_2[Al_2Si_3O_{10}] \cdot 2H_2O$
Analcime $Na[AlSi_2O_6] \cdot H_2O$
Apophyllite $KCa_4[Si_8O_{20}]F \cdot 8H_2O$
Chabasite $(Na_2,Ca)[Al_2Si_4O_{12}] \cdot 6H_2O$
Thomsonite (Comptonite)
$\quad NaCa_2[Al_5Si_5O_{20}] \cdot 6H_2O$
Desmine $(Na_2,Ca)[Al_2Si_6O_{16}] \cdot 6H_2O$
Phillipsite $(K_2,Ca)[Al_2Si_4O_{12}] \cdot 4H_2O$
Harmatome $(K_2,Ba)[Al_2Si_5O_{14}] \cdot 5H_2O$
The water content of zeolites is subject to
$\quad$ wide variations.

Greeks (Aristotle, Dioscorides) under the name of νίτρον, and in the Latin authors (Pliny) as *nitrum*. We must take this to refer to soda, and on occasion to potash, which could not at that time be distinguished from soda.* The name *natron* then gradually came into use among the Arabic alchemists, from the name nitrum. In addition, in the writings ascribed to the alchemist Geber (dating to the 14th and 15th centuries) the name alkali occurs with the same meaning**, together also with the name *soda*, there used for the first time. Terms referring to the origin of the substances concerned were, however, in more common use among the alchemists; thus potash obtained from wine lees (tartar) was called *sal tartari*, and that obtained from plant ashes *sal vegetabile*. After about 1600, the name *sal lixiviosum* was used for alkali carbonate.

* The designation *nitrum* for *saltpeter* first arose towards the end of the 16th. century.
** Probably from the arabic *qualjan*, plant ashes.

The distinction between *natron* (that is, the base giving rise to common salt) and *kali* or *potash* (then generally obtained in the form of its carbonate, from wood ashes) was first stated in 1702 by the phlogistonist chemist Stahl (1660–1734), but was first proved experimentally in 1736 by Duhamel de Monceau (1700–1781). Marggraf, in 1758, discovered the possibility of distinguishing the two elements by their flame coloration, and Klaproth (1797) first showed that, contrary to its usual designation as *alcali vegetabile*, potassium also occurs in the mineral kingdom. Davy first succeeded in isolating the free metals, in 1807, by electrolysis of a piece of moistened caustic soda or caustic potash, which lay in a platinum dish that served at the same time as the cathode.

Lithium was discovered in 1817 by Arfvedson, a pupil of Berzelius, first in petalite and soon afterwards in spodumene and lepidolite. It was named on account of its occurrence in rocks (λίθος = stone). The similarity between the lithium compounds and those of the alkalis was recognized by the discoverer; the red flame coloration characteristic of lithium was observed by Gmelin (1818). Bunsen and Matthiessen (1855) first isolated the metal, by electrolysis of the molten chloride.

Rubidium and *cesium* were discovered by Bunsen in the mineral water of Dürkheim by means of their characteristic spectra, from which they also received their names; cesium, discovered in 1860, after the two blue lines peculiar to its spectrum (*caesius*, blue-grey), and rubidium, discovered in 1861, after the two characteristic lines in the red (*rubidus*, dark red). The preparation of cesium metal was accomplished by Setterberg (1882) by electrolysis of a molten mixture of cesium and barium cyanides, after Bunsen had already prepared the amalgam. Rubidium was prepared by Bunsen in the metallic state, by electrolysis of the molten chloride.

2. Preparation

The technically important alkali metals are almost invariably prepared by electrolytic methods. The alkali metals can, indeed, also be liberated from their compounds by purely chemical means, by means of strong reducing agents such as carbon, calcium carbide, or iron carbide. For the lighter alkali metals, however, these processes present greater technical difficulties than does electrolysis. Lithium and sodium, especially, are produced on a large scale almost exclusively by electrolysis. Sodium was formerly prepared by electrolysis of the molten hydroxide. Since the hydroxide must first be prepared by electrolysis of the chloride, this method of extraction requires in all a greater expenditure of electrical energy than the direct electrolysis of the chloride itself. Originally, however, it presented simpler technical problems to carry out, since the hydroxide has a lower melting point than the chloride. After it had been found that the melting point of the chloride could be sufficiently lowered by means of suitable additions, the electrolysis of the chloride became increasingly of importance as compared with the electrolysis of caustic soda.

The older process depending on the electrolysis of molten caustic soda was developed by Castner (1890) who used the apparatus represented in Fig. 30. This consists of an iron vessel heated in its upper part and containing the melt, into which an iron rod is introduced from below as the cathode, the anode being an iron cylinder surrounding the cathode. Between these is a shorter, narrower iron cylinder, which is joined to a wire gauze cylinder surrounding the upper thickened portion of the cathode. The purpose of this is to hinder metallic sodium, produced at the cathode, from reaching the anode. Because of its low specific gravity, the sodium rises to the top and collects above the melt, and can readily be removed. Oxygen and water are evolved at the anode by the discharge of hydroxyl ions, according to the equation

$$2OH^- = \tfrac{1}{2}O_2 + H_2O + 2e$$

The water evaporates for the most part, but some is decomposed by the current, so that hydrogen is always evolved at the cathode along with the sodium, and burns at the lid of the inner cylinder. The heavier impurities in the melt collect in the lower narrowed part of the electrolysis vessel, which is closed at the bottom by means of solidified caustic soda. It is important, as was pointed out by Castner, that the temperature of the bath should rise as little as possible above the temperature of fusion, as otherwise the sodium metal is dispersed in the melt, and is then oxidized again by the oxygen, which also dissolves in the melt to a considerable extent at high temperatures.

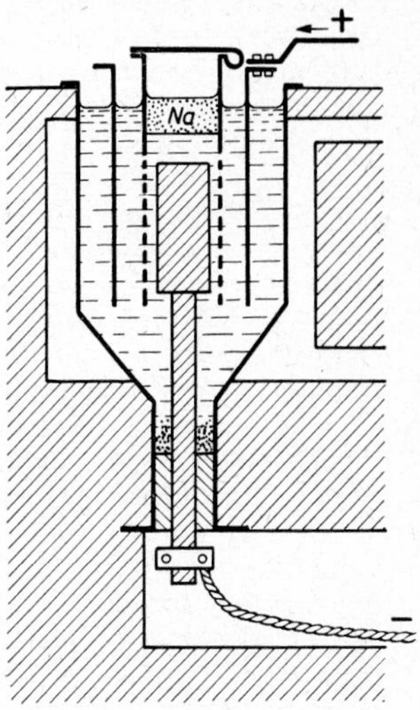

The attempts to prepare sodium by the direct electrolysis of molten sodium chloride originally failed because of the dispersion of sodium metal in the melt at high temperatures. It proved possible, however, to lower the melting point considerably by the addition of calcium chloride, so that the electrolysis of such a mixture can be carried out a little above 600°. On this is based the process of Ciba, which set up the large scale preparation of sodium by electrolysis of the chloride in 1910. A very convenient cell for this process was constructed by Downs. The Downs cell (see Fig. 31) consists of a steel vessel, lined with acid-resisting brick into which are introduced an Acheson graphite anode A from below, and iron cathodes K from the side. The anode is covered by a sheet iron hood H, from which is suspended a wire gauze D to separate the anode and cathode spaces. The chloride mixture, which is maintained in the molten state by the heating effect of the current, is covered on its surface by a crust of solid salt which protects the melt both from the entry of air and from excessive losses of heat. The sodium liberated at the cathode rises upwards and flows along the edge of the hood H, which is bent into a runnel R, to an iron exit pipe F, from which it overflows into the collecting vessel G. By this means the liquid sodium is completely protected from access of air. At the same time, the edge of the hood, which almost closes off the lower part of the vessel, prevents the salt which is added from above from getting into the path of the current before it has been completely dehydrated and melted.

Fig. 30. Electrolysis of molten caustic soda

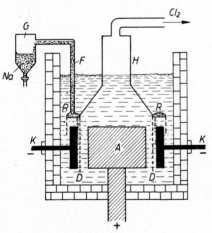

Fig. 31. Downs cell for the electrolysis of molten sodium chloride

From the heats of formation of NaCl and $CaCl_2$ given in Tables 27 and 48, one might be inclined to the view that calcium, and not sodium, must be deposited in the electrolysis of a mixture of NaCl and $CaCl_2$, since the heat of formation of $\tfrac{1}{2}CaCl_2$ is smaller than that of NaCl. This conclusion relates, however, to the heats of formation at ordinary temperature. With increasing temperature, the heat of formation of NaCl decreases more than does that of $CaCl_2$, and at 620° the heats of formation are: for NaCl 90.5, and for $\tfrac{1}{2}CaCl_2$ 90.9 kcal. At this temperature NaCl thus has the smaller heat of formation. Insofar as the heats of formation provide any basis for conclusions about the free energies, one must therefore expect that at 620° NaCl will have a smaller decomposition potential than $CaCl_2$.* In agreement with this, experience shows that the sodium metal obtained by electrolysis of the fused mixture is contaminated with calcium only to a small extent (generally less than 1 %). If the metal is allowed to cool slowly, the calcium separates out almost completely (to a few hundredths per cent), since the solubility of calcium in molten sodium diminishes greatly with falling temperature. The Ciba or Downs process thus yields directly sodium of the degree of purity required for technical use.

Potassium is also prepared to some extent by electrolysis of melts,—recently by the electrolytic decomposition of oxyacid salts of potassium dissolved in fused potassium halides, whereby CO_2 is set free at the graphite anode. Purely chemical processes also possess considerable importance for the preparation of potassium, however, especially the carbide process (reaction of KF with CaC_2). In the United States, potassium is produced primarily by reaction between potassium chloride and sodium at elevated temperatures. This process also gives Na-K alloys, which are becoming useful as heat exchangers. Although it is possible to prepare metallic lithium by electrolysis of the fused pure chloride, since it is considerably more readily fusible than sodium chloride, more fusible mixtures are used for the purpose technically,—generally a mixture of LiCl and KCl. The electrolysis of lithium chloride dissolved in pyridine can also be employed in the laboratory.

For rubidium, and especially for cesium, the electrolytic preparation is less advantageous. The metals are better prepared by chemical methods, e.g., by heating the hydroxides with metallic magnesium in a current of hydrogen, or with metallic calcium in a vacuum. According to de Boer, zirconium is a particularly suitable reducing agent. Small amounts of Rb and Cs can be prepared by heating their chlorides, mixed with barium azide, in a high vacuum (De Boer, 1927). The Ba formed by decomposition of the azide liberates the alkali metals from their chlorides; these evaporate and are deposited on the walls of the unheated portions of the container.

Metallic cesium can be prepared directly from pollucite by heating the mineral with calcium at 900° (Hackspill, 1950). It can be purified by redistillation in vacuum at 350°.

3. Properties

The alkali metals display a silver-white metallic luster on clean surfaces, except for cesium which (according to Costeanu), is golden yellow in the pure state. The lustrous colors of these metals can be observed, however, only on preparations inside evacuated vessels, or at least on freshly cut surfaces. In air, the alkali metals at once tarnish and become dull. All the alkali metals are extremely soft and compressible. Lithium, the hardest of them, has a hardness of 0.6 on Mohs' scale, and is thus appreciably softer than talc (the hardness of which defines the first degree of Mohs' scale). With increasing atomic weight the hardness diminishes yet further, so that sodium is about of the hardness of white phosphorus, and cesium is as soft as wax.

* The data given in Tables 37 and 47, based on measurements by B. Neumann, show directly that the decomposition potential of NaCl is lower at high temperatures, and higher at ordinary temperature than that of $CaCl_2$. For 25°, using the temperature coefficients given, one finds $\zeta_{NaCl} = 3.77$ and $\zeta_{CaCl_2} = 3.26$ volts. However, according to these measurements the decomposition voltage of $CaCl_2$ would exceed that of NaCl only above 650°, whereas technical experience indicates that this must already be the case below 620°. If one calculates the difference in the deposition potentials of Na and Ca at 620° on the assumption that the difference in the heats of formation is equal to the difference in free energies according to eqn (3), page 142, it is found that this amounts only to 0.02 volts. This appears to lie below the limits of the values given in Tables 37 and 47. It is also possible that the addition of NaCl to the melt may raise the decomposition voltage of $CaCl_2$ as a result of complex formation.

The increasing compressibility from lithium to cesium may be seen from Table 32, which gives the compressibilities measured at room temperature. The figures in the Table give the decrease in volume, as a fraction of the total volume, for a pressure of 1 bar, i.e., 10^6 dynes per sq. cm or 75 cm of mercury.

TABLE 32

COMPRESSIBILITIES OF ALKALI METALS

Metal	Lithium	Sodium	Potassium	Rubidium	Cesium
Compressibility per 1 Bar	$8.8 \cdot 10^{-6}$	$15.4 \cdot 10^{-6}$	$31.5 \cdot 10^{-6}$	$40 \cdot 10^{-6}$	$61 \cdot 10^{-6}$

The alkali metals have a high electrical conductivity. At $0°$, the specific electrical conductivity of lithium is 10.9 times, of sodium 22 times, of potassium 15 times, of rubidium 8 times, and of cesium 5.2 times that of mercury at the same temperature. They are, however, inferior in conductivity to silver, the best conducting metal; sodium, the best conductor of the alkali metals, has a resistivity 3 times as great as that of silver.

The alkali metals are extremely light. Potassium and sodium are lighter than water; lithium floats on petroleum, and is the specifically lightest of all the elements solid at ordinary temperature. The melting points and boiling points of the alkali metals are very low (cf. summary table).

The alkali metals are substantially monatomic in the vapor state, as was first proved by determination of the ratio of the specific heats $c_p : c_v$. More recent vapor density determinations have shown, however, that a by no means inconsiderable association of the atoms into diatomic molecules takes place, especially with the lighter alkali metals; the existence of diatomic molecules was first proved spectroscopically. Thus Rodebush (1930) found from the vapor density that sodium vapor at 13.5 mm pressure was associated to the extent of 13 % at $570°$, in agreement with the value obtained from the energy of dissociation of the Na_2 molecule, as determined spectroscopically. For sodium vapor at atmospheric pressure, the degree of association is 16 % at the boiling point of the metal (Ladenburg and Thiele, 1930). Energies of dissociation determined spectroscopically are

for	Li_2	Na_2	K_2	Rb_2	Cs_2
	26.3	17.5	11.8	10.8	10.4 kcal per mol.

The vapors of the metals are intensely colored. Sodium vapor is purple, like a dilute permanganate solution; potassium vapor is blue-green; and rubidium greenish-blue. The colors of colloidal dispersions, made by electrical dispersion of the metals in inert solvents, such as ethyl ether, are similar. The colors of these sols are

Metal	Li	Na	K	Rb	Cs
Color of colloidal dispersion	Brown	Purple-violet to blue	Blue to blue-green	Greenish blue to greenish	Blue-green to greenish blue

In the brown color of its ether sol, lithium resembles magnesium, which also forms a brown ethereal colloidal dispersion.

The alkali metals dissolve in liquid ammonia. Dilute solutions have a deep blue color and concentrated solutions are golden in appearance.

All the alkali metals decompose water. Lithium does so without melting, sodium melts in the process, and potassium catches fire. Rubidium and cesium react much more vigorously still. These latter ignite on exposure to oxygen, even in the absence of water, whereas the other alkali metals ignite in dry air or oxygen only when moderately heated. Lithium burns to the oxide Li_2O, contaminated only with traces of the peroxide. The other metals form higher oxides when burned (cf. p. 176). The reaction with alcohol, forming alcoholates, has already been mentioned. No reaction takes place with anhydrous ether, paraffin, or petroleum; the yellowish or bluish crusts with which sodium or potassium gradually become covered, when stored under petroleum, arise from oxygen compounds present in the petroleum.

Lithium, sodium and potassium form *hydrides* when they are gently warmed in an atmosphere of hydrogen. Only lithium combines with molecular nitrogen, forming the nitride Li_3N; the reaction takes place particularly vigorously at a dull red heat, and lithium reacts slowly with moist nitrogen in the cold. The heat effect of the reaction

$$3Li + \tfrac{1}{2}N_2 = Li_3N + 47.2 \text{ kcal}$$

is, however, less than that for formation of the hydride, when referred to the same amount of lithium.

The other alkalis can also unite with nitrogen excited in the electric glow discharge, and either the nitrides M_3N or the azides MN_3 may then be formed according to the experimental conditions (Wattenberg 1930). The nitrides of Na and K can also be obtained by the thermal decomposition of the azides. They are intensely colored (red at room temperature), stable in dry air, but decomposed at once by water or alcohol, forming ammonia. Na_3N reacts with hydrogen at 120°:

$$Na_3N + 3H_2 = 3NaH + NH_3$$

If ammonia gas is passed over the molten alkali metals, they react, forming the amides: e.g., $Na + NH_3 = NaNH_2 + \tfrac{1}{2}H_2$. This reaction, like the others cited, is based on the facility with which the alkali metals assume a positive charge. The alkali amides are shown by the high conductivity of their melts and their solutions in liquid ammonia to be ionic—i.e., to have a structure expressed by the formula $M^+[NH_2^-]$.

Lithium alone combines directly with carbon, forming the carbide Li_2C_2, which decomposes again into its elements at higher temperatures. The carbides of the other alkalis can be obtained only by indirect means. Lithium is also the only alkali metal to combine with silicon; it unites directly when heated, forming the silicide Li_6Si_2, in dark violet, hygroscopic and very reactive crystals.

4. The Spectra of the Alkali Metals

If the alkali metals, or any of their compounds which are not too involatile, are introduced into the flame of the Bunsen burner, they impart characteristic colorations to the flame; carmine red by lithium, yellow by sodium, and violet by potassium, rubidium and cesium. When the flame is examined in a spectroscope, individual sharp lines are observed, having such characteristic positions that they provide an easy means of identifying the alkali metals.

The lines characteristic of the alkali metals are shown in Fig. 32, as they are seen when the colored flames are observed through a small hand spectroscope. The

yellow sodium line marked with the letter D appears even when vanishingly small
traces of sodium are present, and is contained in every flame spectrum produced

in the chemical laboratory unless special precautions are taken. Since it serves as a useful means of orientation for
the position of the other lines, it is shown
dotted in the other spectra. A rough
measure of the relative intensity of the
lines is given by the varied thickness of
the strokes in the diagram. As a rule,
the most intense of them suffice for purposes of identification, and the weaker
are not visible in a hand spectroscope if
the metal in question is present only in
small amounts. The violet potassium
line, lying on the extreme right is also
often hard to see.

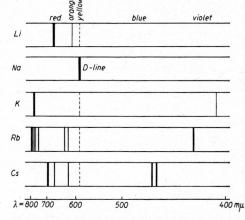

Fig. 32. The spectra of the alkali metals

Table 33 gives the wave lengths of the
lines of the alkali metal spectra for the
visible region, in mμ (millimicrons, $= 10^{-7}$ cm). This unit of wave length remains
the most suitable for ordinary spectral analysis, since the spectroscopes ordinarily
used for the purpose do not enable the tenth-mμ (0.1 mμ = 1 Å) to be measured
accurately. It must be noticed that, in passing through a prism, light of short wave
length undergoes much greater dispersion than does light of long wavelength.
Since Fig. 32 represents the appearance of the lines as formed by a prism spectroscope, the lines in the red appear to be strongly crowded together, as compared
with those in the violet. For the sake of completeness, Table 33 also contains some
lines which are not ordinarily observed when emission is excited by introducing
the compounds concerned into a flame. They are readily observed only when a
rather higher temperature is employed, such as is attainable by striking a small arc
between a solution of the salt concerned and an iridium wire point placed close
above it. To ensure that the arc burns steadily, the positive current should flow in

TABLE 33
LINES OF THE ALKALI SPECTRA IN THE VISIBLE REGION
Wave lengths in mμ. Lines bracketed together cannot be resolved in
spectroscopes with a small dispersion

Lithium	Sodium	Potassium	Rubidium	Cesium
670.8	616.1	**769.9** ⎱	794.8 ⎱	794.4
610.4	**589.3**	**766.5** ⎰	**780.0** ⎱	760.9
460.3	568.5	693.1	775.8 ⎰	**697.3**
413.2	498.1	691.1	761.9	672.3
391.5		583.2	740.8	621.3
		581.2	629.8	**455.5**
		580.2	620.6	**459.3**
		578.3	421.6	
		404.5	420.2	

the direction from the solution to the iridium wire electrode. Under these conditions one obtains, for example, a second orange sodium line ($\lambda = 616$ mμ), as well as the yellow D line. In contrast with the latter, however, this line is observable only when sodium is present in weighable amounts. In Table 33 the main gradations of intensity are shown by the thickness of the type. At low concentrations only the lines shown in heavy type can be observed.

Next to the hydrogen spectrum and the spark spectrum of helium, the spectra of the alkali metals are the simplest of all in structure. Their study has therefore been of fundamental importance for the discovery of the general laws of the structure of spectra. Fig. 33

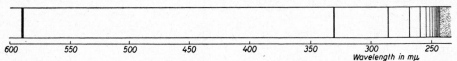

| 600 | 550 | 500 | 450 | 400 | 350 | 300 | 250 |

Wavelength in mμ

Fig. 33. Absorption spectrum of sodium vapor

shows the absorption spectrum of sodium vapor; the similarity to the Balmer series of the hydrogen spectrum is unmistakable (Fig. 17). It can, indeed, be represented by a series formula similar in appearance to the Balmer formula. The same series is also emitted by glowing sodium vapor as the strongest, in addition to several others which do not appear in absorption. It is accordingly known as the *principal series* of sodium, and no less than 57 lines belonging to it have been measured. The continuous spectrum abutting on the short wavelength end of the series is not visible in emission. In addition to the principal series, sodium possesses two *subsidiary series* and a so-called Bergmann ('Fundamental') series. Thus the orange sodium line ($\lambda = 616$ mμ) mentioned above belongs to the so-called '*sharp subsidiary series*' of sodium. Similar considerations apply to the other alkali metals. In each case, the principal series is the strongest, and was therefore discovered first. The Bergmann series are named after Bergmann, who discovered the first series of this type (actually for the alkali metals) in 1907. The types of series mentioned differ not only in the intensities of their lines (which fall off strongly towards the end in all series), but in other characteristics also. These features have interest, however, only for spectroscopists, who are thereby assisted in the assignment of the lines into series. We may take a series to be sufficiently well defined as all the lines which can be formally represented by a mathematical expression as a single regular sequence. They are of importance to us because they enable us to make more precise statements about the nature of the binding of the outer electrons in the alkali metal atoms.

From what has been said (p. 87) about the absorption of energy during light absorption, it can readily be seen that the initial orbit for all the lines of the absorption series is the *ground state* of the outermost electron. The lines of the Principal series in emission coincide with those of the absorption spectrum, so that the ground orbit (level) is the final orbit for all lines of the Principal series. Its energy, determined from the series limit, agrees well with measurements of the ionization potential; these have been carried out for Na, K, Rb, and Cs. The principal quantum number to be assigned to the ground orbit follows from the fact that the levels corresponding to the inert gases are fully occupied. In helium, the orbits with principal quantum number $n = 1$ are occupied, so that for lithium we must have $n = 2$. The subsidiary quantum number l is obtained, on the basis of the Bohr theory, from the following considerations. The constant term of the series is given by the energy of the ground orbit. If, in the normal state of the atom, the electron were found in a $2p$ orbit ($l = 1$, or $k = 2$ in the older Bohr representation of quantum states), i.e., in a circular orbit enclosing the $1s$ orbit ($l = 0$ or $k = 1$) and at a fairly great distance therefrom (cf. Fig. 27, p. 116, making the nuclear charge $= 3$ instead of 2, and placing two electrons in the $1s$ orbit), it might be expected that the value $Z = 1$ could be inserted in the constant term for the 'effective nuclear charge', since of the 3 charges borne by the lithium nucleus, 2 would be permanently screened. Such a term may be briefly described as 'hydrogen-like'. It emerges, however, that the effective nuclear charge which determines the constant term of the Principal series is substantially greater than 1. It may be inferred from the discussion of Fig. 27, Chap. 4, that the orbit involved is one which brings the electron, for part of its course, into the immediate proximity of the nucleus—i.e., a highly eccentric orbit according to the Bohr-Sommerfeld theory. This implies, in the case of lithium, that a $2s$ orbit is the only possible ground orbit. It may be shown similarly for the other alkali metals that their ground orbits are also s orbits (with the subsidiary quantum number $l = 0$). Their eccentricity naturally increases with increasing principal quantum number.

The ground terms of the subsidiary series are more hydrogen-like than those of the Principal series, and this applies even more to the Bergmann or Fundamental series. It was because of the specially hydrogen-like nature of their constant terms that the latter were also called the Fundamental series. The more hydrogen-like the ground term of a series, the less eccentric is the ground orbit corresponding to that series according to the Bohr-Sommerfeld theory.

The familiar yellow sodium line, the so-called D line, is the first line of the Principal series of sodium. Its reciprocal wave length is given by the difference between the ground term of the Principal series and the ground term of the 1st Subsidiary (Sharp) series. The line originates in emission in the following way: the electron which, in the ground state of the atom, circulates in the $3s$ orbit, is first raised to the next higher orbit, the $3p$ orbit (e.g., by the energetic collisions occurring at the temperature of the flame), and then drops back again to the original $3s$ orbit. It is apparent that this process occurs particularly frequently, and thus explains why the D line is the brightest line in the sodium spectrum. The whole sequence of lines in the Principal series of sodium arises from the transitions:

$$3p \rightarrow 3s, \ 4p \rightarrow 3s, \ 5p \rightarrow 3s, \ 6p \rightarrow 3s, \ 7p \rightarrow 3s, \text{ etc.}$$

The converse transitions correspond to the absorption lines. The spectra of the other alkali metals arise in exactly analogous manner.

Under high spectral resolution, the D line appears doubled; it forms a so-called *doublet*. This is explained by the assumption that yet another quantum number j, in addition to the subsidiary quantum number l, determines the energy of the p-levels. The values assumed by this quantum number, which are obtained from the term analysis of the spectra, can be understood on the assumption that the electron behaves as if it rotates about its own axis, simultaneously with its motion around the nucleus. This rotation can occur in the same direction, or in the opposite direction to its orbital motion, and in this way the splitting of the energy level into two levels can be explained naturally. The rotation of the electron about its own axis is referred to as *spin*, to distinguish it from the orbital rotation about the nucleus of the atom.* If the concept of electron spin (which is due to Uhlenbeck and Goudsmit) is correct, all the terms of the alkali spectra must be doubled, with the exception of the s terms which (for reasons which cannot be discussed here) are always single. It follows that all the lines in the alkali spectra form at least doublets. This is, indeed, the case. The separation of the lines in the doublets increases markedly with increasing atomic weight; for the yellow sodium line it is so small (wavelength difference $= 5.97$ Å) that quite a good spectroscope is needed to separate the two components, while for cesium it has so far increased that the two characteristic blue cesium lines can be distinguished even with quite a small dispersion (wavelength difference here 37.94 Å). For the doublet of the first member of the Principal series of cesium, which lies in the infrared, the difference is as great as 422.4 Å.

It should be mentioned that in addition to the arc spectra (the spectra excited in the flame appear in the electric arc also), ascribed to the neutral atoms as already discussed, the alkali metals also possess *spark spectra*. These latter are difficult to excite and are very rich in lines (of the character of the inert gas spectra); they are therefore attributed to the singly ionized atoms M^+. Their occurrence corresponds to the 'spectroscopic displacement law' cited on p. 111; it provides further confirmation for the view that the core of each alkali metal atom, remaining after the loss of one electron, exactly corresponds in structure to the atom of the preceding inert gas.

---

* In polyatomic molecules one can also speak of the atomic nuclei as possessing spin. Thus it is assumed that in orthohydrogen both of the nuclei in the molecule rotate in the same direction, and in parahydrogen they rotate in the opposite direction about their axes.

5. Uses of the Alkali Metals and Their Compounds

Since metallic sodium can be prepared at low cost, it finds extensive technical applications. It is used as the starting material for the manufacture of sodium peroxide (for bleaching materials), and also for sodium amide and sodium cyanide. It is also much used for organic syntheses (e.g., in the dye industry). It is applied, in illuminating engineering, for sodium vapor discharge lamps. In the laboratory, sodium is employed as a reducing agent, as well as in synthetic work; for this purpose the amalgam, which is more easily manipulated, is commonly used in place of the pure metal. Potassium metal is also occasionally used in the laboratory. Potassium (and especially cesium) is also used in photoelectric cells; rubidium and cesium otherwise find little application as free metals. Lithium, on the other hand, has attained considerable technical importance, being used to an increasing extent in alloys, since small additions of this metal improve the properties of many alloys to a notable degree. Lithium (with sodium and calcium) is used especially in lead bearing metals (see p. 542), and for the preparation of Sklerone (see p. 348); also as a deoxidant for copper and for the improvement of sulfur-containing nickel.

Among the compounds of the alkali metals, sodium peroxide is used in large amounts in the manufacture of washing materials. In the laboratory, it is used as a strong oxidizing agent, both in aqueous solutions and in melts. The hydroxides of sodium and potassium are extensively used as strong bases, both in the laboratory and in industry. Because they are not hygroscopic, lithium oxide and lithium hydroxide are used for the production of photographic developers in powdered form.

The salts of the alkali metals, and especially those of sodium and potassium, find the most extensive applications. When any acid radical is required in a reaction, it is generally used in the form of the corresponding alkali (sodium or potassium) salt. The choice of alkali salts for reactions in this way is based partly on their ready solubility and fusibility, and in part because, since almost all their salts are soluble, the alkali salts can bring about no interfering side reactions. For technical purposes, however, compounds other than the corresponding alkali salts are frequently preferred as being cheaper. More about the application of the various individual alkali salts will be found in the discussion of the compounds in question. As regards lithium salts, however, it may be mentioned here that many of them, such as the carbonate, salicylate and the citrate, find application as medicines for the treatment of gout; lithium salts are also the effective constituent of the natural waters used to treat this ailment.

The world production of sodium chloride in 1948 was about 42 million metric tons, of which 15 million tons was produced in the United States, and 3 million tons in Britain. Potassium salts extracted in the same year were equivalent to about 3.5 million tons of K_2O. About two thirds of this potash came from the European continental deposits (1.34 million tons of K_2O from the Stassfurt area, and 0.77 million tons from Alsace). United State production amounted to about 1 million tons of K_2O.

6. Compounds of the Alkali Metals

The compounds of the alkali metals correspond to the general type MX, where M signifies the alkali metal and X a univalent atom or radical, or the equivalent amount of a multivalent radical. The following summary gives a survey of the most important alkali metal compounds. The hydrides, oxides and hydroxides, higher oxides and classes of salts listed below are considered in more detail in this section. Data regarding other compounds will be found under the elements from which they are derived.

SURVEY OF THE MOST IMPORTANT TYPES OF ALKALI COMPOUNDS

Hydrides MH

Normal oxides M_2O, Higher oxides M_2O_2 and MO_2

Hydroxides MOH

Nitrides M_3N (cf. p. 168)

Carbides (acetylides) M_2C_2 (pp. 437, 466)

Metal compounds
(cf. Tables 85–87, pp. 574–576 and
92, p. 585)

Salts e.g., Chlorides MCl
Nitrates MNO_3
Carbonates M_2CO_3
Sulfates M_2SO_4

Double salts
e.g., $Na_2SO_4 \cdot 3K_2SO_4$, glaserite

Sparingly soluble alkali salts
e.g., $KClO_4$ and $Na[Sb(OH)_6]$

(a) Hydrides

The most stable of the alkali hydrides is lithium hydride LiH. In this it resembles the alkaline earth hydrides, which are considerably more stable than the alkali hydrides. It is fairly easily obtained, by passage of hydrogen over gently heated lithium, contained in an iron boat, and forms a hard white mass, consisting of regular crystals, melting at 680°.

Although, like the free alkali metals, lithium hydride is extremely reactive at high temperatures, it is remarkably stable at ordinary temperature, and is not attacked by dry gases such as O_2, Cl_2 and HCl. It is vigorously decomposed by water, however: $LiH + HOH = LiOH + H_2$.

Evolution of hydrogen through thermal dissociation, $LiH = Li + \frac{1}{2}H_2$, begins to be appreciable in a vacuum at 450°. As was first proved by Nernst and Moers (1920), molten lithium hydride conducts the electric current, being decomposed into lithium and hydrogen. The latter is evolved at the anode, and is therefore the electronegative constituent of the compound.

The best criterion of the electrolytic character of the decomposition produced by the passage of the electric current through any material is furnished by Faraday's law; this states that equivalent quantities of material are deposited at the electrodes by the same quantity of electricity. Half a millimole (= 11.2 cc) H_2 must accordingly be evolved by the passage of a quantity of electricity sufficient to deposit 107.88 mg of silver. That this is, indeed, the case in the electrolysis of lithium hydride was shown by K. Peters in 1923, at the suggestion of Nernst. To avoid complications from convection, the experiment was actually carried out not on molten lithium hydride, but on the solid, at temperatures rather below the melting point, where the conductivity is already quite good.

In nature and properties, the other alkali hydrides correspond to lithium hydride, but are considerably less stable. According to Ephraim (1921), their dissociation temperatures are much lower than those of the hydrides of lithium and the alkaline earths, and lie close together. It appears that the stability increases somewhat in the direction from cesium to sodium. The heats of formation of the alkali hydrides are shown in Table 27.

The compounds crystallize with the rock salt structure (cf. p. 209). The edge length a of the unit cubic cells and the densities (determined directly and by X-rays), as found by Zintl (1931), are

	LiH	NaH	KH	RbH	CsH
a	4.085	4.880	5.700	6.037	6.376 Å
$d_{x\text{-ray}}$	0.77	1.36	1.43	2.59	3.41
$d_{obs.}$	0.82	1.38	1.47	2.60	3.42

The last line gives the values for the densities found directly by various investigators. From the data above, the ionic radius of H^- is found to be 1.54 Å.

(i) Indirect Determination of Heats of Reaction

For compounds such as the alkali hydrides, for which no direct determination is possible, it is none the less possible to find the heats of formation indirectly, on the basis of Hess' *law of constant heat summation* and the law of Lavoisier and Laplace of the *equality of heats of formation and decomposition*. Thus Sieverts (1930) found the following heat effects for the reactions of Na and NaH with water:

$$[Na] + H_2O + aq = NaOH.aq + \tfrac{1}{2}(H_2) + 44.4 \text{ kcal} \qquad (1)$$

$$[NaH] + H_2O + aq = NaOH.aq + (H_2) + 31.6 \text{ kcal} \qquad (2)$$

$$[Na] + \tfrac{1}{2}(H_2) = [NaH] + 12.8 \text{ kcal} \qquad (3)$$

Equation (3) is obtained by subtraction of eqn (2) from eqn (1). According to the laws cited above, eqn (3) furnishes the heat set free when one g atom of solid Na combines directly with $\tfrac{1}{2}$ mol H_2 to form solid NaH.

In the above equations, NaOH.aq represents 1 mole of sodium hydroxide, dissolved in (much) water. In such equations, the symbol 'aq' generally denotes an indeterminate amount of solvent water, usually so much water that a further increase in its amount is without influence on the heat liberated. The heat of solution should not be neglected in thermochemical formulas; it amounts, for example, to nearly 10 kcal for sodium hydroxide. Since changes in the state of aggregation, and the transformation of one modification into another are also associated with heat effects, the states of aggregation, or modifications, must be specified in thermochemical equations if they are not self-evident. The solid state is denoted by square brackets [ ], and the gaseous state by round brackets ( ), while the formulas of reactants which are in the liquid or dissolved state are written without brackets. It should be stated, further, that the heat effects in thermochemical equations are generally to be taken as applying to *constant pressure*. If the observations are made at *constant volume*, they are recalculated to constant pressure, by subtraction of the quantity of heat corresponding to the work which must be performed in expansion against the pressure.

(ii) Calculation of Heats of Reaction from Equilibrium Data

The *Nernst approximation formula* is often used to calculate heats of reaction from measured temperature-pressure data for dissociation equilibria. For the frequently occurring case of a solid substance giving rise to one gaseous dissociation product, this takes the form

$$Q = 4.571 \; T \cdot (1.75 \log T + C - \log p) \qquad (4)$$

Here Q represents the number of gram calories which must be expended at ordinary temperature to dissociate 1 gram molecule of the gaseous dissociation product; p is the measured dissociation pressure, in atmospheres, at the temperature T° K; and C is the *conventional chemical constant* of the corresponding gas. Its value, for example, for H_2 is 1.6, O_2 2.8, N_2 2.6, Cl_2 3.1, Br_2 3.2, I_2 3.9, CO 3.5, CO_2 3.2, NH_3 3.3, H_2O 3.6.

Equation (4) can be deduced from the reaction isochore on the basis of *Nernst's heat theorem* with the use of simplifications which are approximately valid. If the heat effect is known it can of course also be used to compute approximately the temperature-pressure curve; this was indeed formerly its chief field of application.* Nernst's heat theorem states that the temperature coefficient of the maximum work which can be performed by a reaction tends to the limiting value zero as absolute zero of temperature is approached

$$\text{Limit } \frac{dA}{dT} \to 0 \text{ for } T \to 0 \qquad (A = \text{maximum work})$$

Thus for the dissociation pressure of sodium hydride, Hüttig and Brodkorb found 10 mm (= 0.013 atm.) at 270° C (= 543° K). Substituted in eqn (4), these values give $Q = 20{,}500$ cal, or for 1 g atom of Na (i.e., for removal of $\frac{1}{2}H_2$) 10.25 kcal. This agrees approximately with the values found by Sieverts, given above.

The importance of the Nernst approximation formula for the calculation of heats of reaction lies in the fact that *only one* pair of temperature-pressure values need be known in order to employ it. If two or more not too widely separated points on the pressure curve are known one obtains more exact values by using the integrated reaction isochore (Chap. 2 eqn (5)). Reference should be made to text books of physical chemistry for more detailed treatment of its use for this purpose.

(b) Oxides

The normal oxides M_2O of the alkali metals are relatively unimportant. In this case also, the lithium compound is the most stable, and is the only one to be formed as the chief product when the metal burns in air. When sufficient air is present, the other alkali metals burn forming higher oxides.

Even with lithium, the oxide Li_2O obtained by combustion contains the peroxide Li_2O_2 as impurity. The normal oxide can be obtained pure, however, by heating lithium hydroxide, nitrate, or carbonate, to about 800° in hydrogen. It is reduced neither by hydrogen nor by carbon or carbon monoxide. Lithium oxide forms a porous white mass of density 2.02. It combines with water only slowly, forming the hydroxide, whereas the other alkali oxides react vigorously with water, and in some cases with great violence (incandescence).

Sodium oxide Na_2O is usually contained as an impurity in the sodium peroxide Na_2O_2 formed by burning sodium. It is obtained in the pure state when sodium peroxide or sodium hydroxide are heated with metallic sodium

$$Na_2O_2 + 2Na = 2Na_2O \qquad\qquad (5)$$

$$NaOH + Na = Na_2O + \tfrac{1}{2}H_2 \qquad\qquad (6)$$

It is also obtained when sodium is burned with an insufficient supply of oxygen; apparently in this case the excess of sodium reacts according to eqn (5) with the peroxide primarily formed. A better method for its preparation, however, is the reaction of sodium (or of sodium azide according to Zintl, 1931) with sodium nitrite.

$$NaNO_2 + 3Na = 2Na_2O + \tfrac{1}{2}N_2 \qquad \text{or}$$

$$3NaN_3 + NaNO_2 = 2Na_2O + 5N_2$$

The normal oxides of the other alkali metals can be obtained by methods like those for sodium oxide. With increasing atomic weight they display an increasing yellowish coloration. The oxide of sodium, like that of lithium, is pure white; potassium oxide however is yellowish white, rubidium oxide bright yellow, and cesium oxide orange. The volatility increases in the same sequence. Rubidium and cesium oxides react with hydrogen when gently heated, forming the hydroxide and hydride: $Rb_2O + H_2 = RbOH + RbH$.

The alkali oxides have crystal lattices of the fluorspar type (cf. p. 265), except for Cs_2O,

---

* Today whenever possible the Ulich approximation formula is used for this purpose (cf. H. ULICH, *Kurzes Lehrbuch der physikal. Chemie*, 4th Ed. 1942, p. 96; J. R. PARTINGTON, *Advanced Treatise or Physical Chemistry*, 1949, Vol. I, p. 231). This gives considerably more accurate values but the entropies and the specific heats of the substances involved in the reaction must be known, as well as the heat of reaction in order to apply it.

which has the cadmium chloride type structure (cf. Vol.II). Most of the sulfides, selenides and tellurides of the alkali metals also have the fluorspar structure, although Cs_2S, Rb_2Se, Cs_2Se, Rb_2Te and Cs_2Te have complicated structures. Table 34 brings together the unit cell dimensions of the alkali chalcogenides with fluorspar structure (a = cell edge of unit cube, in Å). The densities of the compounds are also listed.

TABLE 34

LATTICE CONSTANTS AND DENSITIES OF ALKALI CHALCOGENIDES

	a	Density X-ray	Density Pycnometric		a	Density X-ray	Density Pycnometric
Li_2O	4.62	2.01	2.01	Li_2Se	6.00	2.83	2.91
Na_2O	5.55	2.39	2.27	Na_2Se	6.81	2.61	2.58
K_2O	6.44	2.33	—	K_2Se	7.68	2.29	—
Rb_2O	6.74	4.05	—	Rb_2Se	—	—	3.57
Cs_2O	—	—	4.69	Cs_2Se	—	—	4.39
Li_2S	5.71	1.63	1.63	Li_2Te	6.50	3.39	3.24
Na_2S	6.53	1.85	1.86	Na_2Te	7.31	2.93	2.90
K_2S	7.39	1.80	1.81	K_2Te	8.15	2.51	2.52
Rb_2S	7.65	2.99	—	Rb_2Te	—	—	3.62
Cs_2S	—	—	4.15	Cs_2Te	—	—	4.30

A series of *suboxides* of cesium is known—Cs_7O, Cs_4O (?), Cs_7O_2, Cs_3O, and Cs_2O. The existence of these peculiar compounds was first demonstrated by E. Rengade (1909), and is has been later confirmed by the work of G. Brauer [*Z. anorg. Chem.*, 255 (1948) 101].

(c) Higher Oxides

The following higher oxides are formed when the alkali metals are burned in oxygen.

Na_2O_2	KO_2	RbO_2	CsO_2
pale yellow	orange	dark brown	orange

The higher oxides of the higher alkali metals are fairly readily fusible. On strong heating, cesium hyperoxide (CsO_2) loses oxygen and is transformed into black Cs_2O_3. On the whole, however, the alkali higher oxides are remarkably stable towards heat, provided that no oxidizable substances are present. They readily give up their oxygen to oxidizable substances when heated, including substances that are otherwise not readily attacked, and they therefore find application (especially sodium peroxide) in the laboratory, for attacking oxidizable substances that are insoluble in acids.

Lithium peroxide, Li_2O_2, is formed only to a small extent by burning the metal in oxygen. It can be prepared by the addition of hydrogen peroxide to a solution of lithium hydroxide; the crystalline precipitate, of the composition $Li_2O_2.H_2O_2.3H_2O$, which is thrown down by means of alcohol, is dehydrated by prolonged drying over phosphorus pentoxide.

Essentially the same higher oxides are obtained by the action of oxygen on solutions of the alkali metals in liquid ammonia. In addition, other oxides are formed as well—e.g., $NaO_{1.67}$, K_2O_2, K_2O_3 (?). With cesium, under these experimental conditions, Cs_2O_2, Cs_2O_3 and CsO_2 are formed successively. The last is the same oxide as is formed at higher temperatures.

The oxides of the type M_2O_2—e.g., sodium peroxide Na_2O_2— can be regarded as direct derivatives of hydrogen peroxide H_2O_2. They react quantitatively with acids or with water, forming hydrogen peroxide:

$$M_2O_2 + H_2SO_4 = M_2SO_4 + H_2O_2; \qquad M_2O_2 + 2H_2O = 2MOH + H_2O_2$$

The double decomposition with water is quite analogous to the hydrolysis of a salt. Oxides of this type can therefore be regarded as salts of hydrogen peroxide. Hydrolysis of the oxides of the type MO_2 (hyperoxides), however, furnishes 1 molecule of O_2 as well as 1 molecule of H_2O_2:

$$2MO_2 + H_2SO_4 = M_2SO_4 + H_2O_2 + O_2; \qquad 2MO_2 + 2H_2O = 2MOH + H_2O_2 + O_2$$

The alkali dioxides MO_2 (hyperoxides) were formerly formulated as tetroxides, M_2O_4. However, Kassatoschkin (1936) and Klemm (1939) showed by X-ray studies that they have the same structure as the alkaline earth peroxides, of the type of BaO_2 (see p. 264), and as the acetylides of the type of BaC_2 (see p. 466). The KO_2 crystal lattice can be derived from that of KCl, by replacement of every Cl atom by the mid-point of an O_2 group, with the interatomic distance $O—O = 2.08$ Å. The lines joining every pair of O atoms forming such groups all lie parallel to the c-axis, which is elongated in comparison with the other two axes ($c/a = 1.178$), so that a tetragonal lattice results instead of the cubic lattice (cf. Fig. 84, p. 466).

(d) Sodium Peroxide

Sodium peroxide, Na_2O_2, is manufactured by burning sodium metal in aluminum vessels. It forms a pale yellow powder, which can be melted almost without decomposition, and is not in itself explosive. The heat of formation of Na_2O_2 is 124 kcal per mole (W. A. Roth, 1947). In contact with easily oxidized materials such as cotton wool, sawdust, straw, charcoal, or with aluminum powder, however, it reacts with such vigor that violent explosions can occur. If ether, glacial acetic acid, nitrobenzene, or moist glycerol is poured over a little sodium peroxide, a most vigorous deflagration occurs. Sodium peroxide also reacts with sulfur with incandescence. Reaction with carbon monoxide is gentler, and forms sodium carbonate (7). Carbonate is also formed by carbon dioxide, oxygen being liberated (8)

$$Na_2O_2 + CO = Na_2CO_3 \qquad (7)$$

$$Na_2O_2 + CO_2 = Na_2CO_3 + \tfrac{1}{2}O_2 \qquad (8)$$

The latter reaction is important for the use of sodium peroxide in breathing apparatus, such as is used by divers and firemen, and for the regeneration of the air in enclosed spaces, as for example in submarines. Preparations containing sodium peroxide for these purposes have been introduced commercially under the name of *oxone*. Potassium hyperoxide, KO_2, is used in the United States in breathing apparatus.

Sodium peroxide also reacts vigorously with water and a considerable amount of heat is evolved. This arises from the formation of the hydrate $Na_2O_2.8H_2O$, which can be obtained from the solution, in the form of fusible crystal leaflets or plates. The dihydrate and monohydrate are also known. Sodium peroxide decomposes in dilute solution, forming sodium hydroxide and hydrogen peroxide:

$$Na_2O_2 + 2H_2O = 2Na^+ + 2OH^- + H_2O_2$$

If such a solution, which behaves exactly like a mixture of sodium hydroxide and hydrogen peroxide, is to be prepared from the anhydrous peroxide, efficient cooling is necessary, as the heat of solution otherwise gives rise to a vigorous evolution of oxygen.

Sodium peroxide finds widespread applications in the preparation of bleaching baths for silk, wool, artificial silk, straw, hair, bristles, sponges, wood, horn, bone and ivory.

For this purpose, the sodium peroxide is dissolved in cold water, to which has previously been added just sufficient sulfuric acid to neutralize the sodium hydroxide formed. A small excess of sulfuric acid remaining after the reaction is neutralized with ammonia. Alternatively, magnesium sulfate is previously added to the water, so that the hydroxyl ions liberated on dissolution of the sodium peroxide are precipitated as magnesium hydroxide. Washing powders containing sodium peroxide as well as soap powder are also manufactured commercially.

(e) Hydroxides

The hydroxides of the alkali metals, also known as caustic alkalis, are colorless, strongly corrosive, rather fusible masses. It should be noted that the molten caustic alkalis not only attack glass and porcelain vessels, but also attack platinum quite considerably in the presence of air. They should therefore not be fused in platinum crucibles; silver crucibles are generally used for this purpose in the laboratory, and iron vessels are used on a technical scale.

The alkali hydroxides are very soluble in water, and dissolve with the evolution of a considerable amount of heat. Thus the heat of solution for NaOH is about 10 kcal per mole. In aqueous solutions of the usual concentrations, the alkali hydroxides are almost completely dissociated into ions: $MOH = M^+ + OH^-$, although in solutions of moderate concentrations the conductivity coefficients $f_\mu = \mu/\mu_0$ fall quite appreciably below the value 1, because of the action of interionic forces.

Thus:

for potassium hydroxide in 0.01   0.1   1 molar solution
$\qquad$ f_μ (at 18°) = 0.96 $\qquad$ 0.89 $\qquad$ 0.77

Similar figures have been found for the other alkali hydroxides.

The effective hydroxyl ion concentrations in solutions of the alkali hydroxides are greater than those in solutions of all other metal hydroxides at the same equivalent concentration. The alkali hydroxides are therefore the strongest bases of all the metal hydroxides.

The increase in basic character of the alkali hydroxides from lithium to cesium hydroxide is connected with the increasingly loose binding of the hydroxyl ions in this series.* As a measure for the strength of this binding, we can consider the energy necessary to separate the hydroxyl ion from the alkali ion in a 'molecule' of alkali hydroxide. If we consider the ions as rigid spheres, and neglect the influence of polarization, this work is found on the basis of Coulomb's law as

$$A = \frac{e^2}{D} \left(\frac{2}{r + r_0} - \frac{1}{r + 2r_0} \right)$$

* According to the theory of strong electrolytes the ionic radii play a part in the interionic forces in moderately dilute solutions, even if complete dissociation is assumed, since the more concentrated the solutions, the more frequently do the ions collide because of their thermal motion. The work which must be done to separate them again depends on the ionic radii, in accordance with Coulomb's law.

where D is the dielectric constant, r the radius of the alkali ion, and r_0 that of the oxygen ion. Substituting Goldschmidt's values of the radii, we obtain for the dissociation energies the values joined by the lowest curve in Fig. 34. In Fig. 34 the alkali hydroxides are re-

presented at equal intervals along the abscissa, and the corresponding values of A (in units of $(e^2/D) \cdot 10^8$), as ordinates. The points so obtained are joined by the lowest curve. It can be seen that the calculated dissociation energies decrease from lithium to cesium hydroxide. Represented in the figure, on the same scale, are also the energies A' calculated for the dissociation of an H^+ ion, i.e., for dissociation according to MOH = MO' + H$^+$. These are calculated as

$$ A' = \frac{e^2}{D} \left(\frac{2}{r_0} - \frac{1}{r + 2r_0} \right) $$

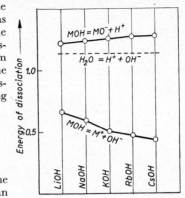

Fig. 34. Dissociation energies of the alkali hydroxides

These energies, the values of which are joined by the uppermost curve, are very considerably greater than those necessary for the dissociation of an OH$^-$ ion.* In practice, therefore, the hydroxides in question only dissociate to OH$^-$ ions. For comparison, the energy necessary for the dissociation of a water molecule is represented by the ordinate of the dotted line.

The melting points of the alkali hydroxides are

for	LiOH	NaOH	KOH	RbOH	CsOH
m.p.	445°	328°	360.5°	301°	272.3° C

The melting points of the alkali amides show a trend similar to that of the melting points of the alkali hydroxides (R. Juza, 1937).

The heats of formation of the alkali hydroxides from the elements are given in Table 27, p. 155. The crystal structure of lithium hydroxide is given on p. 216. The other hydroxides undergo a phase change at elevated temperatures, the transition temperatures being:

NaOH 229.6°, KOH 248°, RbOH 245°, CsOH 223°

The high temperature forms of NaOH and KOH (and probably of RbOH) have the sodium chloride structure ($a = 5.00$ Å and 5.48 Å resp. at 300°) (West 1935; Klemm 1939). The detailed structure of the low temperature forms is not known. The effective radius of the hydroxyl ion decreases from $1.53 \rightarrow 1.50 \rightarrow 1.45$ Å in the sequence KOH, NaOH, LiOH, as the effect of polarization increases with diminishing radius of the cation.

The hydrogen sulfides MSH and hydrogen selenides MSeH are related structurally to the hydroxides, and are also dimorphous. Their structures are summarized in the following Table 35, in which the a-form is that stable at ordinary temperature, and the β-form is the high temperature modification

The alkali hydroxides dissolve very readily in alcohol as well as in water. Only lithium hydroxide is moderately soluble both in water and in alcohol, and so forms a transition to the alkaline earth hydroxides. It is also less hygroscopic than the other alkali hydroxides, but crystallizes from aqueous solution as the hydrate LiOH · H$_2$O. The alkali hydroxides are perceptibly volatile at red heat, the volatility increasing in the series from lithium to cesium hydroxide.

* If the influence of polarization is considered, still larger energies are found for the dissociation of the H$^+$ nucleus. On the other hand, a considerable amount of energy is liberated through the formation of [H$_3$O]$^+$ from H$^+$ and H$_2$O, which has not been taken into account. The errors introduced by neglecting these two effects roughly compensate each other.

TABLE 35

CRYSTAL STRUCTURES OF ALKALI HYDROGEN SULFIDES AND SELENIDES

M =	Li	Na	K	Rb	Cs
MSH	not yet investigated	α-form rhombodehral $a = 3.99$ Å $a = 96°\,21'$	α-form rhombohedral $a = 6.61$ Å $a = 97°\,09'$	α-form rhombohedral $a = 6.89$ Å $a = 97°\,13'$	CsI-type $a = 4.30$ Å (at 20°)
		β-form: NaCl-type $a = 6.07$ Å (at 200°)	β-form: NaCl-type $a = 6.67$ Å (at 200°)	β-form: NaCl-type $a = 6.97$ Å (at 200°)	
MSeH	not known	α-form rhombohedral $a = 6.24$ Å $a = 96°\,27'$	α-form rhombohedral $a = 6.83$ Å $a = 97°\,21'$	α-form rhombohedral $a = 7.12$ Å $a = 98°\,07'$	CsI-type $a = 4.43$ Å (at 20°)
		β-form: NaCl-type $a = 6.30$ Å (at 150°)	β-form: NaCl-type $a = 6.92$ Å (at 180°)	β-form: NaCl-type $a = 7.21$ Å (at 180°)	

All the methods for the preparation of bases discussed on p. 60 *et seq.* can, in principle, be used for the preparation of the alkali hydroxides. In practice, however, only the 4th and 5th methods are important: electrolysis of aqueous solutions of the chlorides, and double decomposition between an alkali salt and an alkaline earth hydroxide. The causticization of soda by means of quick lime and electrolysis are both used for the technical preparation of sodium hydroxide. The former method is also suitable for the preparation of rubidium and cesium hydroxides. In this case, one starts from the sulfates, which are treated with barium hydroxide: $Rb_2SO_4 + Ba(OH)_2 = 2RbOH + BaSO_4$. Use is occasionally made of the reaction of the alkali metals with water vapor (method 3) for the preparation of absolutely pure hydroxides.

(i) *Sodium hydroxide*, caustic soda, NaOH, forms a white, opaque, crystalline, brittle, very hygroscopic mass of density 2.13, which is generally sold commercially in the form of sticks or pellets. It melts below red heat, and vaporizes at higher temperatures. The compound crystallizes from the strongly alkaline aqueous solution in hydrated form; various hydrates have been detected, with a water content between 1 and 7 H_2O. These can be dehydrated by fusion. The dissolution of sodium hydroxide in water is very exothermic because of the strong hydration of the sodium ion.

The solubility of sodium hydroxide is:

at	0°	20°	100°
	42	109	342 g of NaOH in 100 g of water

It is also quite soluble in methyl and ethyl alcohols (alcoholic caustic soda). For the density and viscosity of caustic soda at temperatures up to 70° see W. KRINGS, *Z. anorg. Chem.*, 255 (1948) 294. Absolutely pure sodium hydroxide melts at 328°.

The melting point of caustic soda is generally about 10° lower than that given above, because of the presence of water and of carbonate.

As already stated, sodium hydroxide is manufactured either electrolytically, by the process described for potassium hydroxide or by the double decomposition of soda solution with slaked lime:

$$Na_2CO_3 + Ca(OH)_2 = 2NaOH + CaCO_3$$

The soda, which is only weakly alkaline in reaction, is made corrosive or 'caustic' (from καυστικός = burning) by this treatment. Sodium hydroxide obtained in this way is therefore described as 'causticized' soda. Pliny describes how in Egypt the soda was made more effective by treatment with burned lime. The preparation of 'caustic kali' was common practice amongst the Arabic and western alchemists.

Caustic soda finds technical application on a very large scale. Very large quantities are used in the different branches of the chemical industry—e.g., in the soap industry and dyestuff industry, where it is used, in oxidizing melts—e.g., in the preparation of alizarin dyes from anthraquinone sulfonic acid derivatives. Increasing quantities (about 400,000 tons in 1945 in the United States) are consumed by the rayon (viscose) industry, and for the production of paper pulp. Cellulose is obtained by heating straw or wood with caustic soda solution. The adsorptive properties of cotton fibers towards dyestuffs are improved by treatment with concentrated sodium hydroxide solution (mercerization). If stretched cotton fabrics are subjected to this treatment, they acquire a silky luster, and at the same time their tensile strength is increased. Caustic soda is also employed in the purification of fats, and in the refining of petroleum. It is also a reagent much employed in analytical chemistry. The caustic soda melt is an excellent flux for analytical purposes (Brunck and Höltje, *Angew. Chem.*, 45 (1932) 331).

(ii) *Potassium hydroxide*, caustic potash, KOH, a hard, white brittle mass, with a fibrous texture, melting at a red heat, closely resembles sodium hydroxide. Its density is 2.04, and its heat of solution 12.5 kcal per mole. Like sodium hydroxide, it is very hygroscopic and forms hydrates with water; mono-, di-, and tetrahydrates are known in the solid state. The dihydrate is stable at ordinary temperature; this hydrate has a negative heat of solution (—0.03 kcal. per mole), and its solubility accordingly increases with increasing temperature, amounting to

at	0°	20°	100°	
	97	112	178	g of KOH in 100 g of water.

Potassium hydroxide is also readily soluble in methyl and ethyl alcohols. The alcoholic solutions are known as (methyl- or ethyl-) alcoholic potash, as distinct from ordinary (aqueous) potash.

Although potassium hydroxide can be obtained by the double decomposition of potassium carbonate with lime (as for sodium hydroxide), it is prepared technically almost exclusively by the electrolysis of potassium chloride solutions. Most potassium hydroxide is transported in the form of concentrated solutions (with about 50 % KOH); solid caustic potash can be obtained by evaporation. This, however, loses the last traces of water less readily than does caustic soda. Caustic potash is used largely in the manufacture of soaps. Because of its hygroscopic nature, fused caustic potash is used as a drying agent, and also as an absorbent for carbon dioxide and for alkaline fusions. It is used in surgery as a caustic.

7. Electrolysis of Alkali Metal Chlorides [1]

When an electric current is passed through an aqueous solution of an alkali metal chloride, chlorine is evolved at the anode:

$$Cl^- - e = \tfrac{1}{2}Cl_2 \tag{9}$$

Hydrogen is evolved at the cathode, since the deposition potential for hydrogen ions, in spite of their excessively small concentration is still much lower than that of the alkali ions. Hydrogen ions are therefore discharged, and the corresponding number of hydroxyl ions are set free:

$$H^+ + e = \tfrac{1}{2}H_2 \tag{10}$$
$$H_2O = H^+ + OH^- \tag{11}$$
$$\overline{H_2O + e = \tfrac{1}{2}H_2 + OH^-} \tag{12}$$

so that the total cathodic process is represented by eqn (12). Formation of alkali hydroxide, as well as the evolution of hydrogen, accordingly takes place at the cathode. A quantity of alkali ions equivalent to the hydroxyl ions is available, since an equivalent amount of Cl^- ions has disappeared from the solution through discharge at the anode. The alkali which accumulates in the neighborhood of the cathode must be prevented from coming into contact with the chlorine evolved at the anode, since reaction would occur, according to

$$2OH^- + Cl_2 = Cl^- + ClO^- + H_2O$$

—i.e., hypochlorite would be obtained instead of the desired hydroxide, chloride being simultaneously reformed. Various procedures have been worked out to hinder this: the diaphragm process, the older diaphragm-less 'bell jar' process, and the amalgam process. Of these, the diaphragm process is now the one most commonly employed, the Siemens-Billiter cell being principally used in Europe.

In the *diaphragm process*, the anode and cathode solutions are kept apart by means of a porous partition or diaphragm. Such a diaphragm cell is shown schematically in Fig. 35. The current enters the cell by means of the anode A, made of gas carbon, Acheson graphite, or fused magnetite, and chlorine is there evolved, escaping at a. The current passes through the diaphragm D into the cathode space, where it liberates hydrogen at the negative electrode K, which generally consists of an iron wire gauze. The hydrogen escapes by b, while the potassium hydroxide solution produced simultaneously flows out through c. In practice, it is usual to work with the cathode placed so close to the diaphragm that the cathode space contains but a small volume of solution. A suitable diaphragm was first successfully prepared by A. Breuer, in 1884, by using cement mixed with finely ground common salt which was subsequently leached out after the cement had set. Disadvantages of the diaphragm process are the high electrical resistance which the diaphragm offers to the passage of current, and its limited durability. These disadvantages have, however, been largely overcome in the modern forms of the process (see below).

The *bell jar process* operates without any diaphragm. Separation of the two solutions is achieved through the fact that the alkaline cathode solution is specifically heavier than the anode solution. The bell jar G (Fig. 36) consists of a long, narrow inverted earthenware vessel, shown in cross section in the diagram, through which is introduced the carbon or graphite anode A. The liquid is placed in an earthenware trough T, and the denser caustic potash solution formed at the sheet iron cathodes K collects on the bottom. This eventually fills the entire cathode space, and is allowed to flow out from the top of this, while fresh

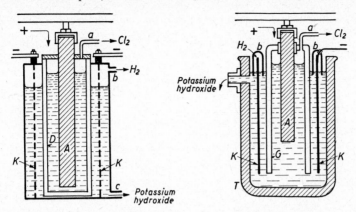

Fig. 35. Diaphragm process   Fig. 36. Bell jar process

saturated potassium chloride solution is continually run into the anode compartment. This hinders the boundary between the alkali and the anodic chloride solution from being displaced towards the anode, through the migration of the OH^- ions in the direction of the negative current. The chloride solution is allowed to flow into the anode compartment at a rate which must be so regulated that migration of the OH^- ions towards the anode is exactly compensated by the downward flow of the electrolyte solution. The bell jar type of cell requires a higher voltage, and is not suited to high efficiency operation. It has therefore been almost entirely replaced by the diaphragm and amalgam types of cell.

The mercury process is based on the fact that the deposition potential of the alkali metals can be very considerably reduced by employing a mercury cathode, because of the great tendency of the alkali metals to alloy with mercury, while at the same time the potential needed for the liberation of hydrogen is considerably raised because of the considerable overvoltage for the deposition of hydrogen on mercury. For these reasons, when a concentrated alkali chloride solution is electrolyzed in a cell using a mercury cathode, hydrogen is not evolved but alkali ions are discharged instead. The alloy formed by the alkali metal with the mercury is decomposed by water in a special cell, forming alkali hydroxide. The decomposition of the very dilute amalgam is hindered by the overvoltage of the hydrogen when this has to be evolved on a mercury surface. The mercury amalgam is therefore made the anode, in a second electrolytic cell, with an iron cathode at which hydrogen is liberated almost without overvoltage in alkaline solutions. The experimental arrangement is shown schematically in Fig. 37. The trough T is divided in the middle by a partition (which is not porous), reaching nearly to the bottom of the cell. The path taken by the current is indicated by the short arrows. In the left hand side, the saturated alkali chloride solution is electrolyzed between the graphite anode A and the mercury layer on the bottom of the cell, which acts as the cathode in this section of the cell. The eccentric E oscillates the cell, and thereby brings the amalgam so formed into the

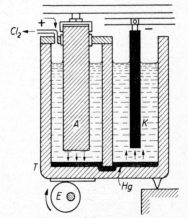

Fig. 37. Mercury process

right-hand section of the cell, into which dips the iron electrode K, connected with the negative current lead. In this compartment, the mercury (or amalgam) constitutes the anode, in that the alkali metal that it contains passes into solution again, in ionic form, while the equivalent quantity of hydrogen is set free at the cathode K. Caustic alkali is thus formed in the right hand compartment of the cell, and is notable for its high purity (freedom from chlorides and sulfates) when prepared by this process.

The mercury process not only furnishes very pure caustic alkali, but can also be operated so as to furnish solutions of considerably greater concentration than the other processes. By heating the amalgam decomposition cell, solutions of 50 % or even of 85 % concentration can be obtained directly. As against this, not only are the costs of the installation higher, because of the cost of the mercury, but there is the further disadvantage that the voltage required is higher than in the modern forms of the diaphragm process. The bell jar process which, in theory, involves the smallest voltage losses, is but little used, since it operates without trouble only when given the most careful attention. The diaphragm process is the one most commonly operated today. This was improved considerably by Billiter, and in the Siemens-Billiter cell the diaphragm is arranged horizontally. The electrolyte solution drops continuously through the diaphragm, and through the wire gauze lying immediately against it, into the bottom section of the cell, which is filled only with hydrogen and from which it flows as alkali hydroxide solution. The electrodes (shaped iron rods, surrounded by asbestos tubes) are also placed horizontally in the Siemens-Pestalozza cell. Processes operating with vertical diaphragms have been developed chiefly in the United States (e.g., the Gibbs-Vorce, Nelson and Hooker cells). Good diaphragm cells today work with a current efficiency of 95 % at a voltage of 4–5 volts. This corresponds to a yield of 41–52 %, calculated on the electrical energy which must theoretically be expended for the decomposition.* The present world's consumption of electrical energy for the electrolysis of alkali chlorides is estimated as $2\frac{1}{2}$–3 thousand million kWh annually.

8. Alkali Metal Salts

Table 36 provides a summary of the solubility relations of the most important alkali metal salts. The naturally occurring chlorides of sodium and potassium are the usual starting materials for the preparation of salts of these two metals. With lithium, the carbonate is generally prepared from the naturally occurring compounds, being relatively easy to obtain pure because of its sparing solubility. Most of the other salts can readily be obtained from it, by decomposing it with acids. For rubidium and cesium, the main problem lies in effecting a separation from potassium, since the mother liquors from working up potassium salts are the principal sources of the salts of these elements. The purification is generally carried out by means of the alums after rubidium and cesium have been enriched in the mother liquors from the crystallization of carnallite (cf. p. 189). Rubidium may be separated from cesium by means of the differing solubilities of their carbonates in alcohol.

In many respects the lithium salts occupy a special position as compared with the salts of the other alkalis, not only in that many of them are sparingly soluble in water, but also because of the relatively high solubility of various lithium salts in non-aqueous solvents. They are mostly dissociated ionically to a considerable extent in these non-aqueous solvents also. The freezing point depressions and the boiling point elevations of aqueous solutions of lithium salts not infrequently exceed the values which would be calculated on the assumption of complete dissociation. This can be explained by the considerable hydration of the lithium ion, in consequence of which the water available as actual solvent water is markedly reduced in amount.

The fact that nearly all lithium salts crystallize in hydrated form may also be considered to be a consequence of the strong tendency of lithium ions to be hydrated. Thus the chloride usually crystallizes with 1–, the chlorate and also the nitrate with $\frac{1}{2}$–, the perchlorate, nitrate, thiosulfate and dithionate with $3H_2O$; the bromate, sulfate, sulfite, selenate and selenite of lithium also crystallize with $1H_2O$.

* The decomposition voltage, calculated from the normal potentials, is 2.2 volts.

TABLE 36

SOLUBILITIES OF THE MOST IMPORTANT ALKALI (AND AMMONIUM) SALTS IN WATER

g = grams of anhydrous salt in 100 g of solution.   t = temperature, °C

	General Formula	M = Lithium		M = Sodium		M = Potassium		M = Rubidium		M = Cesium		M = Ammonium	
		g	t	g	t	g	t	g	t	g	t	g	t
Fluorides	MF	0.26	18°	4.22	18°	48.0	18°	v. sol	—	v. sol	—	deliquesc.	—
Chlorides	MCl	44.7	20°	26.4	20°	25.5	20°	47.66	20°	65.0	20°	27.1	20°
Chlorates	M[ClO$_3$]	75.4	18°	49.7	20°	6.78	20°	5.09	19.8°	5.91	19.8°	v. sol	—
Perchlorates	M[ClO$_4$]	37.85	25°	65.6	15°	1.52	15°	0.76	14°	1.17	14°	20.0	25°
Permanganates	M[MnO$_4$]	29	16°	deliquesc.	—	5.96	19.8°	1.06	19°	0.23	19°	—	16.9°
Nitrates	M[NO$_3$]	41.2	20°	46.8	20°	24.1	20°	34.8	20°	18.7	20°	64.6	—
Carbonates	M$_2$[CO$_3$]	1.31	20°	17.6	20°	53.2	25°	v. sol	—	v. sol	—	v. sol	20°
Sulfates	M$_2$[SO$_4$]	25.7	20°	16.0	20°	10.07	20°	32.5	20°	64.1	20°	43.0	20°
Chromates	M$_2$[CrO$_4$]	52.6	18°	44.6	21.2°	38.6	20°	42.4	20°	v. sol	—	v. sol	—
Dichromates	M$_2$[Cr$_2$O$_7$]	56.6	30°	64.3	20°	11.6	20°	5.0	18°	moderately insoluble	16°	23.9	16°
Hexachloro-platinates	M$_2$[PtCl$_6$]	v. sol	—	v. sol	—	1.09	20°	0.028	20°	0.009	20°	0.67	20°
Phosphates	M$_3$[PO$_4$]	0.039	18°	11	20°	v. sol	—	v. sol	—	vl sol	—	v. sol	20°
Alums	MAl(SO$_4$)$_2$·12H$_2$O	—	—	16.8	13°	4.80	15°	1.25	15°	0.35	15°	6.19	20°
Hydrogen tartrates	MH[C$_4$H$_4$O$_6$]	v. sol	—	6.34	18°	0.49	20°	1.17	25°	8.9	25°	2.14	15°

Lithium chlorate, $LiClO_3$, is notable for its quite extreme solubility, being probably the most soluble inorganic salt. At 18°, 100 g of water dissolve 313.5 g of $LiClO_3$.

Lithium salts do not as a rule form mixed crystals with other alkali salts. They, do however, form mixed salts and double salts, examples of which are given on p. 200.

The properties of the *ammonium salts*—i.e., the salts derived from the univalent positive radical $[NH_4]^+$, ammonium, display a great similarity to the alkali salts. They are very similar to the potassium salts, especially. The solubilities of pure ammonium salts $[NH_4]X$ are therefore included in Table 36 for comparison.

The similarity in behavior between the ammonium ion and the alkali ions is due to the fact that, like the latter, the ammonium ion is positively univalent, and has an ionic volume similar to that of the potassium ion. Although ammonium compounds and alkali salts are alike in many of their properties, there are also a number of characteristic differences, arising from the close relation of the ammonium radical to ammonia, from which it is derived by the addition of a H^+ ion. In spite of the similarity between alkali and ammonium salts it therefore appears better to defer consideration of the latter until it can be done in immediate conjunction with ammonia. In addition to the ammonium radical, other positive univalent radicals are known which also behave as analogues of the alkali ions.

The salts of potassium, rubidium, and cesium, the three alkali metals most closely related among themselves, are often isomorphous with one another, and the isomorphism usually extends to the ammonium salts also.

(a) Chlorides

The chlorides of sodium and potassium, which occur in nature in vast amounts, are rarely quite pure. Not only is potassium chloride generally contaminated mechanically, but it occurs chiefly in the form of double compounds with magnesium salts. These decompose, however, when dissolved in water, so that irrespective of whether the extraction involves separation from mechanically admixed impurities, or preparation from the double salts, the method of purification is the same, namely fractional crystallization.

The chlorides of rubidium, cesium, and lithium are usually prepared by decomposition of the carbonates with hydrochloric acid.

The melting points and boiling points of the alkali chlorides are collected in Table 37, together with the decomposition potentials for the fused alkali chlorides (according to measurements by Neumann, 1925). It may be seen from this that the deposition potentials of the alkali metals from molten salts increase in magnitude regularly from Li to Cs, as is to be expected from the increase of electropositive character in the series. The irregularity in the deposition potentials from aqueous solutions, arising from the effect of hydration (cf. p. 157), has here disappeared.

TABLE 37

MELTING POINTS AND BOILING POINTS OF ALKALI CHLORIDES
DECOMPOSITION POTENTIALS OF MOLTEN ALKALI CHLORIDES.

Compound	LiCl	NaCl	KCl	RbCl	CsCl
Melting point	606°	801°	768°	717°	638° C
Boiling point	1350°	1440°	1411°	1385°	1300° C
Decomposition potential at 800°	—2.38	—2.62	—2.77	—2.91	—3.04 Volts [3]
Temp. coefficient*	1.35	1.49	1.51	1.50	$1.545 \cdot 10^-$

* Decrease in decomposition potential per degree rise in temperature.

(i) *Sodium chloride*, common salt, NaCl, occurs in nature both in sea water, which contains on the average about 2.7 % NaCl, and as rock salt, in great deposits which may be 1000 meters or more in thickness. These are widely distributed—e.g., in Cheshire, England, in the United States in New York, Louisiana and Kansas, and other parts of the world. In Europe there are deposits of especial importance in the North German plain, and in Galicia (Wieliczka). [3–5]

The salt deposits of North Germany originated from the drying up of a great inland sea, which stretched in former times (towards the end of the Middle Magnesian Limestone* period) from the Urals far into the present France, and which included the present North Sea which at that time was temporarily cut off from the Atlantic Ocean. Towards the South it extended about to the present Danube Plain. In the hot climate then existing, especially in summer, evaporation rapidly occurred. This resulted in the deposition of the dissolved constituents of the sea water, in accordance with their concentrations and solubilities, and with the different temperature of the water in summer and in winter. The first to be deposited was calcium carbonate, the least soluble in water, which therefore now lies (as the magnesian limestone) underneath the actual salt deposits. On this the other salts were subsequently deposited,—in summer principally gypsum, anhydrite, and polyhalite, while in winter, rock salt was deposited ('annual rings'). The deposition of potassium salts occurred last of all. Dried out portions of the sea were next covered over by sand masses, and later in part inundated once more, so that at many places several deposits now occur on top of one another. The uppermost layers, containing the potassium salts, were mostly later washed away again. Only in a few places have they been preserved, through having been covered over, at a sufficiently early stage, with water-impervious clay, and they now have great importance as potassium salt deposits.

The extraction of common salt is carried out chiefly by three methods. (1) By mining rock salt; (2) By dissolution of the rock salt underground, and evaporation of the *brine* so obtained, (in part also from natural brines); (3) By allowing sea water to evaporate in so-called 'saltans' or, in very cold regions, by freezing it.

On the continent of Europe, the rock salt used for industrial purposes is obtained chiefly by mining, in some places as an auxiliary to the extraction of potassium salts. Mining is economic only for high grade rock salt, with 98–99 % NaCl. Salt of lower purity is indeed not raised at present, but is left in the mine to serve as a filling for worked out galleries in the potassium salt deposits.

The chief impurities in rock salt are calcium and magnesium sulfates. A certain purification can be achieved by handpicking of lumps of anhydrite and gypsum, after coarsely crushing the salt. A more thorough purification is effected by melting or by extracting the material with pure brine.

The preparation of *edible salt*, which makes the highest demands upon purity, is carried out chiefly by evaporating natural or artificially prepared aqueous salt solutions, so-called brines. The salt so obtained is called evaporated salt. Before evaporation, which is carried out in flat pans, the brines are brought to the saturation concentration. This was formerly done by evaporation in air, by allowing the brine to trickle over a high stack of brushwood. It is now usual to operate in such a manner that the brine dissolves rock salt to the point of saturation.

The brine is purified before evaporation, by adding calcium chloride (to remove sulfates), and lime (to remove magnesium). The precipitated magnesium hydroxide simultaneously carries organic impurities down with it, since it acts as an adsorbent.

* The Magnesian Limestone period in geology signifies the second portion of the so-called Permian, i.e., the epoch following immediately after the Carboniferous age.

Pure sodium chloride is not hygroscopic. That cooking salt becomes moist in damp air, as is well known, is due to the impurities remaining in it. Sodium chloride crystallizes in colorless regular cubes of density 2.17.* At its melting point (801°) it is already perceptibly volatile, though less so than potassium chloride. According to Horiba the vapor pressure of NaCl at 800° is 1 mm, whereas that of KCl is 4.5 mm at 800°, and 1.5 mm at 700°. The vapor density corresponds to the formula NaCl.

Rock salt occasionally occurs in nature with a blue color. A similar coloration can also be produced artificially, by the action of sodium vapor or by exposure to cathode rays, or the rays of radium, with subsequent heating. Siedentopf suggested that the blue color of natural rock salt is due to sodium metal dissolved in colloidal form, since this was taken to be the origin of the artificial coloration. However, it appears from more modern work that the natural blue rock salt does not contain colloidal sodium. The change produced in alkali chlorides by irradiation is not identical with that brought about by exposure to sodium vapor; although the colorations have similar absorption spectra in the visible region, they differ in the ultraviolet. The coloration produced by irradiation (which is the same as that of natural blue rock salt) arises from the incorporation of *free electrons* in the crystal lattice ('F centres'). These occupy definite positions, in that they are trapped at positions in the halogen ion lattice which are unoccupied, in consequence of 'lattice disorder' (cf. Vol. II, p. 26; F. SEITZ, *Rev. Mod. Physics*, 19 (1946) 384).

The solubility of sodium chloride changes but little with temperature (see Table 38, p. 189); the (negative) heat of solution of sodium chloride is correspondingly small (—1.2 kcal per mole). Crystals obtained by evaporation of solutions decrepitate on heating, since inclusions of mother liquor vaporize and break up the crystals. The decrepitation of rock salt from Wieliczka, which takes place when the salt is dissolved in water, has a different origin, and is due to inclusions of compressed gases occurring in the salt (chiefly N_2 and O_2, according to Tammann), which rupture the crystal walls as soon as these have grown thin enough through dissolution.

Common salt is indispensable as a foodstuff, especially with a vegetable diet. It is therefore also fed to cattle ('salt licks'). It is also used for the preservation of meat and fish (salting and pickling). However, the use of common salt in the foodstuff industries is far exceeded by its other industrial uses. In industry, sodium chloride constitutes the starting material for the manufacture of almost every other sodium compound.

Salt forms the basis of the hydrochloric acid and sulfate industries, and of the manufacture of soda, chlorine, and caustic soda. In addition, it has many other industrial and technical uses, e.g., for salting out soaps and organic dyestuffs; for 'chloridizing roasting' in many metallurgical processes; for the preservation of hides in tanning; for glazing common stone ware; for melting snow and the preparation of freezing mixtures, etc.

Sodium chloride crystallizes from aqueous solutions at low temperatures, as hexagonal plates of a hydrate with the composition $NaCl \cdot 2H_2O$. At 0.15°, both the dihydrate and the anhydrous salt are stable in contact with the saturated solution. A solution saturated with sodium chloride boils at 109.7°, and contains 40.4 g of NaCl to 100 g of water. The solubility of NaCl in water is greatly reduced by the addition of HCl. At 18°, a 1-normal HCl solution is saturated with sodium chloride when it contains 20.6 g of NaCl in 100 g of solution, a 3-normal HCl solution when it contains 10.6 g of NaCl per 100 g of solution. If sodium chloride is prepared in anhydrous benzene solution, by the reaction of organic sodium compounds with organic compounds containing chlorine, it is obtained in colloidal

* In the presence of certain substances, sodium chloride occasionally crystallizes in regular octohedra—e.g., from evaporated urine, as a result of the action of urea.

form. It forms a yellow to yellowish-red sol with benzene, which is fairly stable in the absence of water. Sodium bromide, but not sodium iodide, can form a similar, though less stable, organosol.

(ii) *Potassium chloride*, KCl, occurs native in the potassium salt deposits, as *sylvine*. This is one of the most valuable potassium minerals since, when ground up, it can be used directly as a fertilizer. It is however, often strongly contaminated with admixed sodium chloride. If the latter forms the main constituent the product is known as *sylvinite*.

The most important deposit of potassium salts occurs in the North German Plain, especially near Stassfurt. Deposits also occur in Alsace. The latter however, are not direct marine deposits but are of secondary origin, potassium salts having been leached out of original saltbeds and once more deposited there. Deposits of potassium salts have also recently been found in Siberia east of the Urals, and in Yorkshire, England.

The chief constituent of the potassium salt beds, and therefore the most important starting material for the preparation of potassium chloride, is the double salt of potassium chloride with magnesium chloride, *carnallite*, $KCl \cdot MgCl_2 \cdot 6H_2O$.

Other compounds of importance include the so-called 'hard salt' (a mixture of 35–70 % NaCl, 10–48 % kieserite, $MgSO_4 \cdot H_2O$, 12–23 % KCl, and varying amounts of other potassium salts); *kainite*, $KCl \cdot MgSO_4 \cdot 3H_2O$; and *sylvinite*, mentioned above. These salts are known as '*Abraumsalze*' (from German 'Abraum', refuse) since they were originally excavated in order to get at the underlying rock salt, and rejected as valueless. In 1865 Liebig drew attention to their great value as fertilizers. At the present time most of the potassium salts extracted (more than 95 % of the production) are used for fertilizers. The salts can often be used for this purpose without any processing except coarse grinding.

In the industrial preparation of potassium chloride from carnallite the latter is dissolved in water. It breaks up thereby into its components, KCl and $MgCl_2$, of which the former crystallizes out first when the solution is evaporated because of its smaller solubility.

In technical practice, a moderately concentrated magnesium chloride solution is used for dissolving, in place of pure water, since by use of such a solution the impurities contained in crude carnallite ($MgSO_4$ and NaCl) mostly remain undissolved.

After so much potassium chloride has crystallized out that the solution contains a very large excess of magnesium chloride, the double salt $KCl \cdot MgCl_2 \cdot 6H_2O$, (artificial carnallite) crystallizes from the mother liquors in the cold. A further quantity of potassium chloride can be separated from this by treatment with water.

Hard salt and sylvinite are worked up for potassium chloride in a similar manner. The process depends on the fact that potassium chloride is more soluble than sodium chloride at temperatures near the boiling point of water, but is less soluble in the cold (see Table 38). It is therefore possible to dissolve potassium chloride out of its mixture with sodium chloride by means of hot water or, better, by means of hot sodium chloride solution, such as the mother liquor from the crystallization of KCl. The separation of the constituents of the Stassfurt salt deposits has been the subject of many phase rule studies by Van't Hoff and his pupils, especially in recent years by J. d'Ans.

TABLE 38

SOLUBILITIES OF NaCl AND KCl

Temperature	0°	10°	20°	30°	50°	70°	90°	100°
Grams of NaCl in 100 g of water	35.6	35.7	35.8	36.0	36.7	37.5	38.5	39.1
Grams of KCl in 100 g of water	28.5	32.0	34.7	37.4	42.8	48.3	53.8	56.6

Pure potassium chloride is colorless, and crystallizes in the regular (plagihedral-hemihedral) system in cubes, often combined with the octohedron. It begins to volatilize noticeably even below its melting point (768°), and at 2000° the vapor density corresponds to the formula KCl.

(iii) *Rubidium and cesium chlorides* are obtained by neutralization of the carbonates, or by ignition of the chloroplatinates. Both chlorides crystallize in cubes. Cesium chloride is markedly poisonous. Rubidium *iodide* is used in medicine since it is less harmful than potassium iodide. It may be mentioned here that rubidium and cesium have a marked tendency to form *polyhalides*.

Rubidium and cesium chlorides are frequently used in analytical and preparative chemistry for the preparation of pure double chlorides with the chlorides of heavy metals. The double salts so formed are mostly noteworthy for their sparing solubility and good power of crystallization.

(iv) *Lithium chloride*, LiCl, a colorless salt, has a considerable volatility even at a red heat, and can be completely evaporated at white heat in a current of hydrogen chloride. Unlike the chlorides of the other alkali metals, it is deliquescent, and is soluble in alcohol and in a mixture of alcohol and ether. It dissolves not only in ethyl alcohol, but also in methanol, amyl alcohol, and other alcohols; also in polyhydric alcohols such as glycerine, and in other organic solvents containing oxygen such as aldehydes, acetone, formic acid, etc. Well defined compounds with these solvents may frequently be isolated. Lithium chloride crystallizes from its aqueous solution above 98° as the anhydrous salt, and as hydrates at lower temperatures (with 1, 2, or $3H_2O$, according to the temperature).

Lithium chloride (and also the other lithium halides) absorbs ammonia, both in solution and in the dry state; the dry salt takes up 1 to 4 molecules NH_3, according to the temperature. Numerous addition compounds of organic amines with lithium chloride have been prepared, containing as a rule not more than 3 and often only 1 molecule of the organic amine.

Lithium chloride is generally prepared by the action of hydrochloric acid on the carbonate, which is the lithium salt most readily obtained in the pure state.

$$Li_2CO_3 + 2HCl = 2LiCl + CO_2 + H_2O$$

Water must be evaporated by heating in a stream of hydrogen chloride, as hydrolytic decomposition otherwise occurs.

Lithium bromide and lithium iodide are very similar to the chloride. They also form hydrates in the solid state, likewise with 1, 2, or 3, H_2O.

(v) *Lithium fluoride*, LiF, is of interest because of its sparing solubility. It is obtained as a granular powder by evaporation of a solution of lithium carbonate in hydrofluoric acid. It forms regular octohedra, of density about 2.6, when recrystallized from fused potassium chloride or potassium hydrogen fluoride. Its small solubility in water (100 g of water dissolves 0.27 g of LiF at 18°) is reduced yet further by the addition of alcohol. It dissolves more easily in hydrofluoric acid, because of the formation of the hydrogen fluoride $LiHF_2$ which can also be obtained crystalline. Stockbarger (1936) has described the preparation of large single crystals of LiF (up to 7 cm diameter). Since these exceed fluorspar in transparency in the short-wave ultraviolet region, vacuum spectrographs for investigations in the very short wave region ('Schumann Region'), down to 110 mμ have since then been fitted with lenses and prisms of lithium fluoride in place of fluorspar.

(b) Nitrates

(i) *Sodium nitrate*, $NaNO_3$, is a colorless salt, crystallizing in rhombohedra, which were formerly mistaken for cubes. For this reason, sodium nitrate was sometimes called 'cubic saltpeter' in the older literature. The density is 2.26, the melting point 311°; decomposition commences at 380°. Sodium nitrate has a cold, bitter taste.

It is very soluble in water, and brings about a marked fall of temperature on dis-solution, in accord with the fact that its solubility increases considerably with rise of temperature: at 0°, 73 g and at 100°, 175.5 g of $NaNO_3$ dissolve in 100 g of water.

Sodium nitrate occurs native in a few places. It occurs in very large quantities on the rainless Chilean coasts of the Pacific Ocean, and is known as 'Chile salt-peter'. Smaller deposits have been found in Egypt, Trans-Caspia and Colombia. The origin of the Chilean deposits is not yet settled. They are generally attributed to the bacterial decomposition of animal or—more probably—vegetable materials, and especially sea weeds. Modern authors, however (e.g., Perroni, Stoklasa), prefer the assumption that Chile saltpeter is of volcanic origin. Large quantities of am-monia are often liberated in volcanic eruptions, and these may subsequently be transformed into nitrates.

Crude chile saltpeter, called *caliche*, is usually very impure, being mixed with sand and clay, as well as with other salts; it may contain on the average 30–50 % $NaNO_3$. It is purified by dissolution in hot water and recrystallization by cooling. Potassium perchlorate, $KClO_4$, which acts as a plant poison, is not removed thereby, and must be removed by special methods if it is present to the extent of more than 0.5 %. Sodium iodate, $NaIO_3$, is also present in Chile saltpeter. This is, however, not poisonous but rather promotes the growth of plants.

Sodium nitrate is used directly as a nitrogenous fertilizer, and also as a source of potassium nitrate (for gunpowder) and nitric acid. Sodium nitrate fertilizer is manufactured, to some extent, from the reaction of soda with synthetic nitric acid, but other, synthetically prepared nitrogenous fertilizers have displaced it in-creasingly, and it is no longer the principal source of nitric acid.

(ii) *Potassium nitrate*, saltpeter, KNO_3, is a colorless, dimorphous salt—i.e., it crystal-lizes in two different froms. It crystallizes from acid solutions in rhombohedral crystals, but in other conditions as rhombic crystals of density 2.105. The rhombo-hedral crystals are metastable at ordinary temperature, but are stable above the transition temperature of 128°

$$KNO_{3\text{rhombic}} \overset{128°}{\rightleftharpoons} KNO_{3\text{rhombohedral}}$$

The rhombohedral crystals are not strictly isomorphous with sodium nitrate, since the two salts form mixed crystals only to a very restricted extent at ordinary temperature (miscibility gap between 0.5 and 99.9 weight-% $NaNO_3$). Above 175°, the salts are completely miscible. Mixtures of the two salts have a marked minimum melting point for the composition 45 wt.% $NaNO_3$. Unmixing in the solid state has frequently been found in binary mixed crystal systems with such a minimum melting point.

Potassium nitrate melts at 339°, and at higher temperatures is transformed, with loss of oxygen, into the nitrite. Scheele first prepared pure oxygen in this way. Saltpeter has a cool, bitter taste. It is easily soluble in water, and the cooling attending its dissolution is even more marked than for sodium nitrate, corre-sponding to its still larger positive temperature coefficient of solubility. The solu-bility is:

at	0°	20°	100°
	13.27	31.59	246 g of KNO_3 in 100 g of water.

Saltpeter is formed whenever animal excrements decay in the presence of potassium hydroxide or carbonate. It is formed naturally in this way (e.g., in cattle
yards and stables), and was prepared artificially by this method for hundreds of
years ('niter plantations'). Today it is manufactured either by double decomposition of sodium nitrate with potassium chloride ('conversion saltpeter') or by
neutralization of nitric acid with potassium hydroxide or potassium carbonate.

The conversion process depends on the fact that when KCl is added to a roughly equivalent amount of a hot, saturated solution of $NaNO_3$, NaCl is thrown out of solution. After
filtration of NaCl, KNO_3 crystallizes out on cooling. Thus the process

$$NaNO_3 + KCl \rightleftharpoons NaCl + KNO_3$$

proceeds from left to right in hot solution, because NaCl is thrown down. In the cold it
proceeds yet further in the same direction because the KNO_3 now separates. Two pairs of
salts such as these which can be mutually transformed into one another, are called *reciprocal
salt pairs*. The solubility of the salts are

	$NaNO_3$	KCl	$NaCl$	KNO_3
at 100°	175	56.6	39.1	246 g of salt in 100 g of water
at 0°	73	28.5	35.6	13 g of salt in 100 g of water

Potassium nitrate is used in the manufacture of gunpowder, for which it is suitable since, unlike sodium nitrate, it is not hygroscopic. It is also used to prepare
potassium nitrite which, like sodium nitrite (see p. 602), is used in the dyestuff
industry for diazotizations.

Potassium nitrate (and also sodium nitrate) always crystallize in the anhydrous
condition. It can, however, combine with nitric acid to form 'acid nitrates',
$KNO_3 \cdot HNO_3$ and $KNO_3 \cdot 2HNO_3$.

(iii) *Rubidium and cesium nitrates*, $RbNO_3$ and $CsNO_3$, are isomorphous with potassium
nitrate, which they closely resemble in all respects. The solubility of rubidium nitrate also
increases very strongly with rise of temperature, but this is not true to the same extent of
cesium nitrate. They form addition compounds with nitric acid, similar in composition to
those formed by potassium nitrate.

(iv) *Lithium nitrate*, $LiNO_3$, is generally made by decomposition of the carbonate with
nitric acid. It is a colorless deliquescent salt, which crystallizes from its aqueous solution at
ordinary temperature with $3H_2O$. Although the solubility of lithium nitrate increases
strongly with rise of temperature, its heat of solution is not correspondingly large and negative, but is actually weakly positive, because of the considerable heat of hydration produced
by addition of water to the lithium ion.

(c) Carbonates

Two series of alkali carbonates are derived from carbonic acid, H_2CO_3. The
neutral or secondary carbonates M_2CO_3 result from the complete neutralization of
the carbonic acid, whereas if only one of the two hydrogen atoms is replaced, *acid*
or primary carbonates, $MHCO_3$ are formed. The latter are also known as *hydrogen
carbonates*, since they still contain hydrogen which is replaceable by metals. This
terminology is in accordance with the international rules concerning nomenclature; in older nomenclature they were termed *bicarbonates* since they arise from
the neutralization of one equivalent of alkali with 2 equivalents of carbonic acid.
Both the 'neutral' and the 'acid' alkali carbonates have a *basic* reaction in aqueous

solution, since they are partially hydrolysed because of the weakness of carbonic acid:

$$M_2CO_3 + H_2O \rightleftharpoons MOH + MHCO_3$$

and $$MHCO_3 + H_2O \rightleftharpoons MOH + H_2CO_3$$

The hydrogen carbonates have, however, a much weaker basic reaction than the 'neutral' carbonates. Whereas the latter react basic towards phenolphthalein, as well as towards methyl orange and litmus, the hydrolysis of the hydrogen carbonates is not indicated by phenolphthalein, which is less sensitive towards hydroxyl ions.

With the exception of lithium carbonate, the neutral carbonates of the alkali metals are all readily soluble in water. Sodium hydrogen carbonate has a relatively low solubility, but the other hydrogen carbonates are all easily soluble. The low solubility of sodium hydrogen carbonate is of great technical importance, as will be seen.

The strength of the binding of the CO_2 increases considerably from Li to K, but then decreases again to Cs. This is shown by Fig. 38, in which the dissociation pressures corresponding to the equilibrium $M_2CO_3 \rightleftharpoons M_2O + CO_2$, as measured by Lebeau, are plotted as a function of temperature.

The carbonates of K and Rb form congruently melting double compounds with the corresponding fluorides, with the composition $K_2CO_3 \cdot KF$ and $Rb_2CO_3 \cdot RbF$, respectively. Neither double compounds nor mixed crystals are formed, however, in the systems $Li_2CO_3 + LiF$, $Na_2CO_3 + NaF$ and $Cs_2CO_3 + CsF$. The influence of cation radius on the energy of formation of such compounds has been discussed by SCHMITZ-DUMONT, Z. anorg. Chem., 260 (1949) 49.

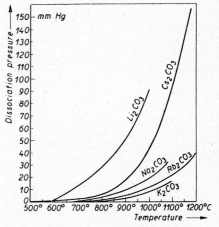

Fig. 38. Dissociation pressures of the alkali carbonates

(i) *Sodium carbonate*, soda, Na_2CO_3, forms a white powder when anhydrous, with the density 2.4–2.5, and melting at about 850°. It is readily soluble in water, and evolves heat because of the formation of hydrates. The most important of the hydrates obtainable in the solid state, *washing soda*, $Na_2CO_3 \cdot 10H_2O$, crystallizes from aqueous solution below 32°, in the form of large, water-clear monoclinic crystals, density 1.45, which melt at 32° in their water of crystallization. Aqueous solutions of soda have a decidedly basic reaction, the salt being extensively hydrolyzed since carbonic acid is very weak.

In addition to the 10-hydrate, a rhombic 7-hydrate exists, stable in contact with the solution between 32.017° and 35.3°. There is also a rhombic monohydrate which is transformed into the anhydrous salt in contact with the solution, at 112.5° and 1.27 atm. pressure (Waldeck 1932). The 7-hydrate also exists in a second modification which is, however, not stable in contact with the aqueous solution at any temperature.

Soda is occasionally found in Nature in lakes, notably in Owens Lake, California, which is estimated to contain nearly 100 million tons of soda. Soda is obtained from it in an impure state by evaporation of the water by the sun's heat. The soda lakes frequently contain the hydrogen carbonate, as well as the neutral carbonate, and a double compound of the two salts, $Na_2CO_3 \cdot NaHCO_3$, called *trona* or *urao* crystallizes out in places. Na_2CO_3 and $NaHCO_3$ also occur together in the waters of alkaline medicinal springs, as for example at Carlsbad.

Sodium carbonate occurs in the ashes of many marine plants in place of the potassium carbonate of land plants. Up to about 150 years ago, soda was chiefly obtained from the ashes of such plants.

Soda is manufactured today [2] almost entirely by the Solvay, or ammonia-soda process. The Leblanc process, devised in 1794 and until recently the most widely used, is now almost completely obsolete. The manufacture of soda by carbonating caustic soda, prepared electrolytically, has attained only limited importance, in contrast with the analogous preparation of potassium carbonate. Caustic soda is, in fact, often manufactured by the converse process, the causticization of sodium carbonate. The cryolite-soda process is used to some extent in the United States.

In the *Leblanc process*, salt was first heated with concentrated sulfuric acid, whereby sodium sulfate (known industrially, as *salt cake*) was obtained, with hydrochloric acid as an important by-product

$$2NaCl + H_2SO_4 = Na_2SO_4 + 2HCl \tag{1}$$

To convert salt-cake to soda, it was mixed with calcium carbonate (limestone) and coal, and heated (to 700–1000°) in a rotary reverberatory furnace. Reactions (2) and (3) thereby went to completion.

$$Na_2SO_4 + 2C = Na_2S + 2CO_2 \tag{2}$$

$$Na_2S + CaCO_3 = Na_2CO_3 + CaS \tag{3}$$

The Na_2CO_3 was leached out of the product (known as *black ash*), while the insoluble CaS remained as a waste product of little value. The process was worked out in 1791 by Nicholas Leblanc, in response to a prize of 100,000 francs offered by the French Academy, but the inventor died in poverty, by his own hand, in 1806. Soon afterwards the soda industry, which until 1870 was based exclusively on the Leblanc process, developed to full fruition, first in England and later in Germany and France. Only in recent times has the Leblanc process been displaced by the Solvay process, which is economically superior to it.

The *Solvay* or *ammonia-soda* process is based on the formation of the relatively sparingly soluble sodium hydrogen carbonate, $NaHCO_3$, by the double decomposition of sodium chloride with ammonium hydrogen carbonate in aqueous solution:

$$NaCl + NH_4HCO_3 = NaHCO_3 + NH_4Cl$$

This is carried out technically by passage first of ammonia and subsequently of carbon dioxide into a nearly saturated brine. The $NaHCO_3$ formed is filtered off and converted by calcination into Na_2CO_3 (soda ash):

$$2NaHCO_3 = Na_2CO_3 + CO_2 + H_2O$$

Half of the CO_2 originally used is thereby recovered, and is led back into the process. To recover the ammonia, more ammonia and steam are first blown into the mother liquors from which sodium hydrogen carbonate has separated. The ammonium hydrogen carbonate still contained therein is thereby first converted to the neutral carbonate (4), which decomposes above 58° into carbon dioxide, ammonia and water (5)

$$NH_4HCO_3 + NH_3 = (NH_4)_2CO_3 \tag{4}$$

$$(NH_4)_2CO_3 = CO_2 + 2NH_3 + H_2O \tag{5}$$

The ammonia contained in the mother liquors in the form of ammonium chloride (about 75 % of the whole) is then driven out by the addition of milk of lime:

$$2NH_4Cl + Ca(OH)_2 = CaCl_2 + 2H_2O + 2NH_3$$

Apart from unreacted sodium chloride, the only waste product obtained is the calcium chloride liquor, which is commonly discharged into rivers.

The ammonia-soda process, although devised and even operated for a time by others (e.g., by Dyar and Hemming, 1838) was worked out technically by the Belgian, E. Solvay, in 1863. It furnishes soda of high purity.

The *cryolite soda* process is carried out to some extent in the United States, although only on a small scale. The process depends on the fact that *cryolite* Na_3AlF_6 is decomposed when heated with chalk: thus, at a red heat

$$Na_3AlF_6 + 3CaCO_3 = Na_3AlO_3 + 3CaF_2 + 3CO_2$$

The sodium aluminate thereby formed is subsequently decomposed by means of water and carbon dioxide

$$2Na_3AlO_3 + 3H_2O + 3CO_2 = 3Na_2CO_3 + 2Al(OH)_3$$

Cryolite soda is distinguished by its high purity.

Soda is one of the most important products of the heavy chemical industry. It is consumed in very large quantities in the manufacture of glass and soap. It is also the starting material for the production of many other important sodium compounds, such as caustic soda, borax, sodium phosphate, water glass, etc. Large amounts are also used in laundering, paper making and the paint industry, as well as for the softening of boiler feed water. It is much used domestically as a cleaning agent. The world production of soda ash amounted to about 8 million tons in 1944.

(ii) *Sodium hydrogen carbonate*, sodium bicarbonate, $NaHCO_3$, is a white powder, (density 2.2), with an alkaline taste, and stable in dry air. It is obtained by passage of carbon dioxide into a cold saturated solution of Na_2CO_3:

$$Na_2CO_3 + CO_2 + H_2O = 2NaHCO_3 \tag{6}$$

Sodium hydrogen carbonate is obtained as an intermediate in the Solvay process. To obtain pure sodium hydrogen carbonate from this material, which is contaminated with ammonium hydrogen carbonate, it is dissolved in warm water. Pure sodium hydrogen carbonate separates on cooling. The preparation by the reaction of equation (6) is carried out in the laboratory as a means of purifying sodium carbonate, since $NaHCO_3$ is readily reconverted into sodium carbonate by heating at about 300°:

$$2NaHCO_3 = Na_2CO_3 + CO_2 + H_2O$$

In aqueous solution, or in the moist state, sodium hydrogen carbonate slowly loses CO_2, even at ordinary temperature. Above 65°, the evolution of CO_2 becomes vigorous. According to Fedotieff, the solubility in water, for solutions saturated with CO_2 at one atmosphere pressure, is

at	0°	15°	30°	45°	
	6.9	8.8	11.02	13.86	g of $NaHCO_3$ in 100 g of water

The solution has a very weakly basic reaction, by reason of hydrolysis, being alkaline towards litmus and methyl orange but not (at 0°) towards phenolphthalein. The principal use of sodium hydrogen carbonate is in baking powders; also as a pharmaceutical, for the neutralization of stomach acidity, and for the preparation of effervescent powders.

(iii) *Potassium carbonate, potash,* K_2CO_3, is a white hygroscopic powder of density 2.30 and m.p. 894° which is very soluble in water with the evolution of heat (9.5 kcal per mol). The solubility is: at 0°, 105; at 25°, 113.5; and at 100°, 156 g of K_2CO_3 in 100 g of water. Of the solid hydrates, the dihydrate, $K_2CO_3 \cdot 2H_2O$, is stable at ordinary temperature.

Potash is prepared technically for the most part by two methods: (1) by carbonation of potassium hydroxide, (2) directly from potassium salts, either by the *Stassfurt process* or, recently by the formate process. In addition, potash is extracted from calcined beet sugar residues, from the ash of the wool sweat or 'suint' of sheep (wool washings) and, in countries where wood is very plentiful, from wood ashes.

The cleansing properties of wood ashes, due to potassium carbonate, have been known from earliest times. The ancient Greeks and Romans knew how to prepare potassium carbonate from wood ashes, but did not distinguish between the salt obtained from this source, and soda, which occurs native. The alchemists prepared pure potassium carbonate by igniting tartar, $KHC_4H_4O_6$; at a later date tartar was deflagrated with potassium nitrate.

Leblanc worked out for the manufacture of potash a process exactly analogous to his well known method for the production of soda. For a time this process had considerable technical importance.

The carbonation of potassium hydroxide, by leading in carbon dioxide, is carried out in direct conjunction with the electrolytic production of the base.

$$2KOH + CO_2 = K_2CO_3 + H_2O$$

Hargreaves suggested that the carbon dioxide should be led directly into the electrolysis cell, so that K_2CO_3 is formed at once. However, it is difficult by this process to obtain the potash free from potassium chloride.

The Stassfurt (or magnesia) process, of Engel and Precht, is based on the sparing solubility of the double salt of potassium hydrogen carbonate with magnesium carbonate. It is carried out by passage of carbon dioxide into a suspension of $MgCO_3 \cdot 3H_2O$ in a solution of potassium chloride at 20°:

$$2KCl + 3[MgCO_3 \cdot 3H_2O] + CO_2 = 2[MgCO_3 \cdot KHCO_3 \cdot 4H_2O] + MgCl_2$$

The sparingly soluble double salt is precipitated, and is decomposed by warming it with water at 60°

$$2[MgCO_3 \cdot KHCO_3 \cdot 4H_2O] = 2[MgCO_3 \cdot 3H_2O] + K_2CO_3 + CO_2 + 3H_2O$$

or by treating it with magnesium oxide.

In the formate process (Goldschmidt process), potash is prepared by way of potassium formate, HCOOK. A solution of potassium sulfate is mixed with milk of lime in equimolecular proportions and carbon monoxide is passed into the mixture at an elevated temperature (about 220°) under a pressure of 30 atm. A solution of potassium formate is so obtained, according to the equation

$$K_2SO_4 + Ca(OH)_2 + 2CO = CaSO_4 + 2HCOOK + 7 \text{ kcal}$$

This is evaporated to dryness after separation of the calcium sulfate. The carbonate is obtained by calcining the potassium formate under oxidizing conditions:

$$2HCOOK + O_2 = K_2CO_3 + CO_2 + H_2O + 106.2 \text{ kcal}$$

Potassium carbonate finds application in the manufacture of soap and glass, also in dyeing, bleaching and wool scouring. It is also used in the manufacture of potassium cyanide, and is frequently employed in preparative chemistry as a desiccant.

(iv) *Potassium hydrogen carbonate*, $KHCO_3$, is rather less soluble than the neutral potassium carbonate, but much more soluble than sodium hydrogen carbonate (36.1 g of $KHCO_3$ in 100 g of water at 26°). It forms colorless monoclinic crystals of density 2.17. When heated it decomposes similarly to sodium hydrogen carbonate

$$2KHCO_3 = K_2CO_3 + CO_2 + H_2O$$

(v) *Rubidium and cesium carbonates*, Rb_2CO_3 and Cs_2CO_3, are best prepared by treatment of the sulfates with barium hydroxide, and evaporation with ammonium carbonate. The salts, thoroughly dehydrated by ignition, are deliquescent, and dissolve in water with considerable evolution of heat (8.75 and 11.84 kcal, respectively). Rubidium carbonate like the other alkali carbonates, is only slightly soluble in alcohol, but cesium carbonate dissolves readily. 100 g of alcohol at 19° dissolves 0.74 g of Rb_2CO_3 and 11.1 g of Cs_2CO_3. This property is utilized for the separation of rubidium from cesium. The hydrogen carbonates of rubidium and cesium are more soluble in water than potassium hydrogen carbonate.

(vi) *Lithium carbonate*, Li_2CO_3, is precipitated as a white crystalline salt, sparingly soluble in water and insoluble in alcohol, when solutions of lithium salts are treated with $CO_3^=$ ions. Density 2.11, m.p. 618°.

Lithium carbonate is generally prepared technically from *amblygonite*, $LiAl[PO_4]F$, which occurs in large deposits in the United States. This is digested with hot concentrated sulfuric acid. After the addition of ammonia and ammonium sulfide to throw down aluminum and iron (present as impurity), the aqueous extract is treated with sodium carbonate. Lithium carbonate is thereby precipitated. It is purified by dissolution in hydrochloric acid, and removal of admixed sulfuric acid by precipitation with barium chloride. Lithium carbonate is finally re-precipitated. The preparation from other lithium minerals is carried out in a similar manner, except that in some cases they must be opened up by different methods— e.g., by fusion with soda, in the case of spodumene, $LiAl[Si_2O_6]$. The silicates can also be attacked by ignition with $CaCO_3$ or CaO. Concentrated sulfuric acid, to which nitric acid has been added, is best for triphylline. Lithium can be separated from the other alkali metals through the insolubility of the phosphate in water, or through the solubility of the chloride in a mixture of alcohol and ether.

Lithium carbonate is more soluble in cold water than in hot, the solubility diminishing from 1.54 g per 100 g of water at 0°, to 0.73 g at 100°. The solubility is considerably greater in water containing CO_2. Use is occasionally made of this fact for the purification of the carbonate. It is assumed that the hydrogen carbonate is formed in water containing CO_2, as with the alkaline earths, even though lithium hydrogen carbonate cannot be isolated in the solid state.

Lithium carbonate constitutes the starting point for the preparation of most other lithium salts. It formerly found application in medicine for the treatment of gout; the use of organic lithium salts, such as lithium citrate or lithium salicylate, is now preferred for this purpose.

(d) Sulfates

Two series of alkali sulfates are derived from sulfuric acid, as a dibasic acid. The *neutral sulfates*, M_2SO_4, dissolve in water with a neutral reaction, and the *acid sulfates* or *hydrogen sulfates* $MHSO_4$ with an acid reaction. All the alkali sulfates are freely soluble in water.

All the alkali sulfates, except Li_2SO_4, form congruently melting double compounds, $M_2SO_4.MF$, with the corresponding fluorides. (Schmitz-Dumont, 1949).

(i) *Sodium sulfate*, Na_2SO_4, is prepared industrially in large quantities as a by-product ('salt cake') of the manufacture of hydrochloric acid from sodium chloride and sulfuric acid. It is also obtained as a by-product from the leaching residues in the extraction of potassium chloride. These residues contain $NaCl$ and $MgSO_4$, and from the solution of this mixture the hydrated salt $Na_2SO_4.10H_2O$ crystallizes at low temperatures (e.g., winter cold), as a result of the double decomposition

$$2NaCl + MgSO_4 \rightleftharpoons MgCl_2 + Na_2SO_4$$

Glauber obtained sodium sulfate by the action of sulfuric acid on sodium chloride, as long ago as 1658. Sodium sulfate occurs in nature in many mineral waters; the anhydrous salt occurs as *thenardite*. It is present in the potassium salt deposits chiefly in the form of double salts, such as *glauberite*, $Na_2SO_4 \cdot CaSO_4$; *astrakanite*, $Na_2SO_4 \cdot MgSO_4 \cdot 4H_2O$; *löweite*, $Na_2SO_4 \cdot MgSO_4 \cdot 2\frac{1}{2}H_2O$; *van't-hoffite* $MgSO_4 \cdot 3Na_2SO_4$; *glaserite* $Na_2SO_4 \cdot 3K_2SO_4$.

From cold aqueous solutions (below 32.383°), sodium sulfate crystallizes as the hydrate, $Na_2SO_4 \cdot 10H_2O$, in large colorless monoclinic prisms, which gradually effloresce on exposure to air. When heated, the crystals melt in their own water of crystallization above 32°, and deposit the anhydrous salt. Above 32.383°, only the anhydrous salt is stable in contact with the solution.

The solubility of sodium sulfate reaches a maximum at 32.383°, the values being

at	0°	10°	20°	30°	32.38°	35°	40°	50°	100°	
	4.5	8.24	16.1	28.9	33.2	33.1	32.5	31.8	29.8	g of Na_2SO_4 in 100 g of solution.

The temperature (32.383°) at which the anhydrous salt is in equilibrium with the 10-hydrate, the saturated solution and its vapor, is suitable for use as a thermometric fixed point (T. W. Richards, 1903), and it can be maintained with practically as little trouble as the melting point of ice. Variations in the small traces of D_2O which ordinarily are present in water cannot affect the fixed point to any significant extent, since even if the H_2O is replaced completely by D_2O, the transition point is raised only by 2.10° (H. S. Taylor, 1934).

Sodium sulfate very readily forms supersaturated solutions. The anhydrous sulfate, obtained from aqueous solution, forms rhombic bipyramidal crystals, of density 2.68 and m.p. 884°. It dissolves in water with a mild evolution of heat. The crystalline hydrated sodium sulfate (commonly known as *Glauber's salt*) dissolves in water with strong absorption of heat (—18.76 kcal per mole). It is therefore used occasionally for the production of cold. It is used in medicine as a purgative, and industrially in dyeing and for the finishing of cotton fabrics. The anhydrous sulfate, which is generally known in industry as salt cake, is used in large quantities in glass making, and also in the manufacture of ultramarine.

A mixture of equimolar amounts of Glauber's salt and common salt displays a transition point at 17.9°. The transition (dehydration of the Glauber's salt) takes place so slowly that the temperature remains constant for hours on end. The mixture is therefore very suitable for producing an exactly defined 'room temperature'. It was shown by R. Löwenherz (1895) that the temperature of transition of Glauber's salt into anhydrous sodium sulfate is depressed by substances dissolved in the melt, in the same way as the melting point of a pure solvent. If the concentration of the added substance is not too great, the depression of the transition point follows the same law (Raoult-Van't Hoff law) as the depression of the freezing point of a solution. The molar depression of the transition point of Glauber's salt is 3.25°. 'Salt cryoscopy' can often be a very useful technique, especially in the determination of ionic weights (cf. K. F. JAHR, *Angew. Chem.*, 63, [1951] 220).

(ii) *Sodium hydrogen sulfate*, $NaHSO_4$, a colorless soluble salt, is formed when common salt is moderately heated with concentrated sulfuric acid:

$$NaCl + H_2SO_4 = NaHSO_4 + HCl$$

On stronger heating with salt it is transformed to the neutral sulfate:

$$NaCl + NaHSO_4 = Na_2SO_4 + HCl$$

A fused mixture of the acid sulfate and normal sulfate (or possibly a compound, $Na_2SO_4 \cdot NaHSO_4$) is obtained in the manufacture of nitric acid from chile saltpeter, and is known as *niter cake*.

Heated by itself, sodium hydrogen sulfate loses water and forms sodium pyrosulfate, $Na_2S_2O_7$. When more strongly heated, the latter decomposes also, giving up sulfur trioxide:

$$2NaHSO_4 = Na_2S_2O_7 + H_2O; \qquad Na_2S_2O_7 = Na_2SO_4 + SO_3$$

Sodium hydrogen sulfate and sodium pyrosulfate are used in analysis to attack and open up insoluble compounds.

(iii) *Potassium sulfate*, K_2SO_4, crystallizes only in the anhydrous state, in the form of rhombic crystals of hexagonal habit. It is transformed at 587° into a hexagonal modification. The density is 2.67 at 18°, and the melting point is 1074°. The solubility in water is:

at	0°	10°	20°	30°	40°	50°	100°	170°	
	7.35	9.22	11.11	12.97	14.76	16.50	24.1	32.9	g of K_2SO_4 in 100 g of water

Potassium sulfate is insoluble in alcohol.

Potassium sulfate is used industrially for the manufacture of glass and of alum. It is also a valuable fertilizer. Potassium chloride and the magnesium sulfate (kieserite) present in the potassium salt deposits, together with their double salts, are the principal starting materials for its technical preparation.

In aqueous solution, magnesium sulfate first reacts partially with potassium chloride, according to the equation

$$MgSO_4 + 2KCl = K_2SO_4 + MgCl_2$$

The K_2SO_4 so formed combines with $MgSO_4$ to form a sparingly soluble double salt, schönite, $K_2SO_4 \cdot MgSO_4 \cdot 6H_2O$. This is filtered off and washed, and treated once more with potassium chloride solution, whereupon double decomposition takes place to a considerable extent, according to the equation

$$K_2SO_4 \cdot MgSO_4 \cdot 6H_2O + 2KCl = 2K_2SO_4 + MgCl_2 \cdot 6H_2O$$

Potassium sulfate was formerly prepared by the action of concentrated sulfuric acid on potassium chloride or nitrate. Glauber prepared it in this way, and as long ago as the 14th century its formation by the action of heated iron vitriol (which gives off sulfuric acid) on saltpeter was known. Potassium sulfate ranks amongst the first chemical compounds of which the composition was discovered (Glauber, Tachenius, Boyle).

(iv) *Potassium hydrogen sulfate*, potassium acid sulfate, $KHSO_4$, was formerly obtained as a by-product of the preparation of nitric acid by the action of sulfuric acid on saltpeter (cf. p. 598). It is now prepared by the dissolution of neutral potassium sulfate in an excess of dilute sulfuric acid.

$$K_2SO_4 + H_2SO_4 = 2KHSO_4$$

It crystallizes as a hydrate from aqueous solution.

Pure potassium hydrogen sulfate is not hygroscopic. The compound is most simply obtained in the pure state by dissolution of potassium pyrosulfate, $K_2S_2O_7$, in water, evaporation to dryness, and heating of the residue to 120° to constant weight.

The anhydrous salt melts at about 200°, and, like the sodium compound, on further heating it passes first into the pyrosulfate, by loss of water, and then further into the neutral sulfate. It finds the same uses as sodium hydrogen sulfate.

(v) *Pure potassium pyrosulfate* is conveniently prepared in the laboratory by the thermal decomposition of potassium peroxydisulfate, $K_2S_2O_8$. This loses exactly 1 atom of oxygen when it is heated to about 290° for about $\frac{1}{2}$ hour.

(vi) *Rubidium and cesium sulfates*, Rb_2SO_4 and Cs_2SO_4, form rhombic crystals, isomorphous with K_2SO_4. At higher temperatures (above 657°) Rb_2SO_4 forms a second modification. The densities at 20° are 3.61 and 4.24, and the melting points 1074° and 1019°, respectively. Both compounds are notable for the ease with which they form excellently crystallized double salts (or mixed salts) with the sulfates of aluminum, ferric iron and the bivalent metals. The formation of cesium alum, $CsAl(SO_4)_2 \cdot 12H_2O$, finds application in the detection of aluminum.

(vii) *Lithium sulfate*, Li_2SO_4, crystallizes from hot solutions as the anhydrous salt, in the form of (apparently rhombic) needles. At ordinary temperature it crystallizes as the hydrate $Li_2SO_4 \cdot H_2O$ in thin monoclinic plates.

The anhydrous salt (density 2.21) melts at 859°. The solubility is

at	0°	20°	100°
	35.5	34.5	29.5 g of Li_2SO_4 in 100 g of water,

and is thus quite considerable, although not so great as that of the chloride and nitrate. Lithium sulfate is much less volatile than the chloride.

Like other lithium salts, lithium sulfate does not form mixed crystals with the corresponding salts of the other alkali metals. A whole range of *mixed salts* and *double salts* exists, however, e.g.,

$NaLiSO_4$	$KLiSO_4$	$(NH_4)LiSO_4$
$Na_3Li(SO_4)_2 \cdot 6H_2O$	—	—
$Na_4Li_2(SO_4)_3 \cdot 9H_2O$	$K_4Li_2(SO_4)_3$	—
$Na_2Li_8(SO_4)_5 \cdot 5H_2O$	$K_2Li_8(SO_4)_5 \cdot 5H_2O$	—

Lithium sulfite also tends to form such double salts—

$Na_2Li_{12}(SO_3)_7 \cdot 6H_2O$	$K_2Li_2(SO_3)_2 \cdot H_2O$

Other lithium salts do not display nearly such a marked tendency to form double salts with the other alkali metals as does the sulfate.

(e) Sparingly Soluble Alkali Salts

A few sparingly soluble alkali salts are discussed. These are of importance chiefly in analytical chemistry.

Of the simple sodium salts, *sodium hydroxoantimonate*, $Na[Sb(OH)_6]$, (formerly referred to as 'acid sodium pyroantimonate') is notable for its sparing solubility. It is precipitated from not too dilute neutral, or weakly basic, solutions of sodium salts, on the addition of the corresponding potassium salt. It forms a white granular precipitate, consisting of microscopic crystals of characteristic form.

$$K[Sb(OH)_6] + NaCl = KCl + Na[Sb(OH)_6].$$

The reaction serves for the detection of sodium in qualitative analysis. The potassium hydroxoantimonate used as precipitant (for preparation cf. p. 667) is, indeed, itself sparingly soluble in cold water, but freely soluble at 40–50°, whereas 1 part of the sodium salt needs 350 parts of boiling water for dissolution.

Among sparingly soluble *potassium salts*, the *acid tartrate* (potassium hydrogen tartrate, $KHC_4H_4O_6$) has long been known as *cream of tartar*. This is soluble to the extent of 0.57 g of salt in 100 g of water at 20°, 6.9 g at 100°; this solubility is markedly diminished by the addition of alcohol. The *neutral potassium tartrate*, crystallizing as the hydrate $K_2C_4H_4O_6 \cdot \frac{1}{2}H_2O$, on the other hand, is very soluble in water (1 part in 0.66 parts water at 20°), as also is the mixed salt $KNa[C_4H_4O_6] \cdot 4H_2O$ (Seignette's salt).

The potassium salts of perchloric acid and chloroplatinic acid are moderately sparingly soluble in water (for data see Table 36), and practically insoluble in the presence of much alcohol. *Potassium chloroplatinate*, K_2PtCl_6 forms small, yellow, regular octohedra. When heated, a residue of potassium chloride and metallic platinum is obtained. The solubility of *potassium perchlorate* is markedly dependent on temperature. It increases from 0.70 g in 100 g of water at 0° to 18.7 g at 100°. The corresponding rubidium and cesium salts are even less soluble than the potassium salts.

Among insoluble lithium salts, the fluoride, LiF, and the carbonate, Li_2CO_3, have already been discussed. *Lithium phosphate*, Li_3PO_4, is obtained as an insoluble, white, crystalline precipitate when phosphate ions are introduced into lithium salt solutions. If acid phosphates are used as precipitants, NaOH must be added to make precipitation complete, as otherwise the lithium acid phosphate, LiH_2PO_4, which is much more soluble, is formed. The latter salt forms an addition compound with phosphoric acid, $LiH_2PO_4 \cdot H_3PO_4 \cdot H_2O$, in large deliquescent crystals.

9. Analytical

The alkali metals are most simply detected by their characteristic spectra, their flame colorations giving the simplest indication of their presence. It must be borne in mind that even unweighably small amounts of sodium can give rise to the yellow sodium flame. The violet potassium flame is masked by the presence of sodium, but can be rendered visible if the flame is observed through a light-filter of cobalt glass (i.e., glass colored blue by cobalt oxide) or of chrome alum solution, which, if thick enough, is not transparent to the yellow sodium light.

Among precipitation reactions, the formation of sodium hydroxoantimonate $Na[Sb(OH)_6]$ is used as a test for sodium. The complex salt

$$NaC_2H_3O_2 \cdot [UO_2](C_2H_3O_2)_2$$

is also very suitable as a means of identification since, although only moderately insoluble it crystallizes in very characteristic forms (yellow microscopic tetrahedra). It is obtained by evaporation of a drop of the solution nearly to dryness on a watch glass or microscope slide, followed by the addition of a drop of a saturated solution of uranyl acetate $[UO_2](C_2H_3O_2)_2$ in dilute acetic acid.

Potassium can be detected by precipitation of the perchlorate $KClO_4$; the other sparingly soluble salts mentioned above can also be used. More sensitive is the reaction with sodium hexanitrocobaltate ('sodium cobaltinitrite'), $Na_3[Co(NO_2)_6]$ (cf. Vol. II), which produces a yellow crystalline precipitate of potassium hexanitrocobaltate, $K_3[Co(NO_2)_6]$, soluble in water to the extent of 1 part in 1000. Most of the reactions for the detection of potassium—but not the reaction with perchloric acid—are also given by ammonium salts. Cognizance must be taken of this in analytical work.

Rubidium and cesium salts are hard to distinguish from potassium salts on the basis of precipitation reactions, and use is always made of the spectroscope in their detection. The detection of lithium is also most simply and certainly achieved by spectroscopic means.

For their gravimetric analytical determination, sodium and potassium are usually converted into the chlorides or sulfates, and weighed in that form. Care must be exercised in ignition, because of the volatility of potassium chloride in particular. Potassium can be separated from sodium by precipitating the former as the perchlorate. Sodium can then advantageously be precipitated from the filtrate in the form of sodium magnesium (or zinc) uranyl acetate, and weighed as such (Kögler, 1935) Alkali salts of volatile acids can. be converted to the fluosilicates, and weighed in that form, by heating them with SiO_2 and hydrofluoric acid (Treadwell, 1933), whereas a mixture of two alkali salts can be determined by weighing as fluosilicate and titration of the F^- content. A method for determining potassium colorimetrically, as the picrate, has been given by Caley (1931).

Lithium is separated from the other alkali metals by treatment of the mixture of the chlorides either with amyl alcohol or with a mixture of ether and absolute alcohol (in equal parts), saturated with hydrogen chloride. Lithium chloride is dissolved out of the mixture by this means, and may be converted to the sulfate for weighing.

References

1 J. BILLITER, Die technische Chloralkali-Elektrolyse, Dresden 1924, 80 pp.

2 G. LUNGE, Handbuch der Soda-Industrie, 3rd Ed., Braunschweig 1909, Vol. II, 868 pp.; Vol. III, 641 pp.

3 E. JÄNECKE, Die Entstehung der deutschen Kalisalzlager, 2nd Ed., Braunschweig 1923, 111 pp.

4 J. D'ANS, Die Lösungsgleichgewichte der Systeme der Salze ozeanischer Salzablagerungen, Berlin 1933, 254 pp.

5 J. H. VAN'T HOFF, Untersuchungen über die Bildungsverhältnisse der ozeanischen Salzablagerungen, Leipzig 1912, 374 pp.

CHAPTER 7

CRYSTAL STRUCTURE AND X-RAYS

1. X-ray Diffraction

(a) Molecule and Crystal

When a hydrogen ion and a chloride ion come together, a *molecule* of hydrogen chloride is formed. Vapor density determinations, based on Avogadro's law, show that gaseous hydrogen chloride consists of individual molecules of HCl. If the gas is cooled down, so that it liquefies and ultimately solidifies, the obvious assumption is that the crystals formed by the frozen material are also built up from HCl molecules. In this instance, such an assumption is valid.

The situation is different, however, when alkali metal ions and chloride ions come together. In the first place, it is not possible to detect the formation of undissociated molecules in the aqueous solutions to any appreciable extent; it is possible to explain the decrease in molecular conductivity on other grounds, as has been seen. If the water is allowed to evaporate no *gas* is evolved, as was the case with hydrogen chloride, but a salt remains, in the *crystalline state*, and this is converted into the gaseous state only at relatively high temperatures. In this respect, crystals such as those of sodium chloride differ characteristically from those of hydrogen chloride, or of other substances which are either gaseous at ordinary temperature, or at least are readily volatile. Nevertheless, it was formerly assumed that crystals of the former type, like those of the latter, were built up from *molecules* of the compounds in question. This was not a satisfactory assumption, in that it provided no clear reason why the two classes of compound should differ characteristically in their ease of volatilization and also in their solubilities in non-polar solvents. Determinations of the structures of typical crystals, by means of X-rays, have proved the assumption to be wrong, since it has been shown that crystals of the alkali metal chlorides, and of other typical salts, are built up *directly from the ions* of which the compound is composed, and not from molecules.

(b) Diffraction of X-rays by Crystals

As is well known, light is diffracted when it passes through a narrow slit; the diffracted rays interfere with one another and—for the case of monochromatic light—produce a system of light and dark bands. Corresponding interference patterns result from the passage of light through a glass plate ruled with regularly and closely spaced lines (a so-called grating), or from the reflection of light from a similar metal grating. There is a simple relation between the *wave length* of the light, the corresponding *angle of diffraction*, and the *grating constant* (the distance between neighboring rulings), and this is the basis of the production of grating spectra and the determination of wave lengths. In order to produce interference, however, the grating constant must not be less than *one-half* the wave length of the light.

The regular forms of crystals [*13–19*] indicate that their structural units, whether they be atoms or molecules, must be arranged in ordered fashion. A regular arrangement of points is called a *point lattice*. The wave theory of light teaches that interference must also occur on transmission through such a point lattice, or by reflection from those planes in the lattice which are densely occupied by points (*lattice planes*), provided that the lattice constant—in this case the distance between adjacent lattice planes—and the wave length of the incident light stand in an appropriate relation to each other.

It can be shown that ordinary light cannot be diffracted by crystals. Thus, from the Avogadro number $\mathcal{N}_A$ we can calculate the number of atoms in a cubic centimeter of a substance of known density, and can estimate the order of magnitude of lattice constants on that basis.

For common salt, NaCl, the density is 2.17 and the formula weight 58.45. The value of Avogadro's number is $\mathcal{N}_A = 6.023 \cdot 10^{23}$. The crystal of salt accordingly contains

$$\frac{6.023 \cdot 10^{23} \cdot 2.17}{58.45} = 2.24 \cdot 10^{22} \text{ 'molecules' per cc,}$$

or twice that number of atoms. If we think of these as being uniformly arranged, there will be $3.55 \cdot 10^7$ atoms per centimeter length, so that the distance between adjacent atoms would accordingly be about 2.8 Å. This is less than one thousandth of the shortest wave length present in visible light, and much too small even to produce diffraction in the region of the furthest vacuum ultraviolet (about 100 Å).

On the other hand, von Laue was able to show, in 1912, that when Röntgen rays passed through a crystal, an interference pattern was obtained, as a result of the diffraction of the rays. Evidence was thereby provided that crystals have a fine structure in the form of a space lattice, as pictured in the model just indicated. At the same time, it was rigorously proved that X-rays are of the same nature as light waves, and differ from them only by reason of their much shorter wave length.

The wave lengths of the X-rays ordinarily used are of the magnitude 0.1 to 1 Å. Röntgen radiation containing rays with wave lengths distributed over a continuous range is called 'white' X-radiation, by analogy with sunlight. *Homogeneous* X-radiation, on the other hand, contains X-radiation of one definite wave length only. The method of producing this is discussed below.

Laue allowed white X-radiation to pass through a crystal of zinc blende,

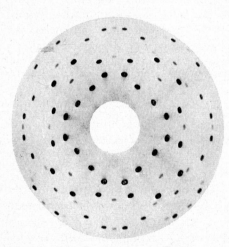

Fig. 39. Laue diffraction pattern (zinc blende, ZnS, irradiated perpendicularly to a cube face).

perpendicular to a cube face—i.e., along the direction of a fourfold axis,—and obtained, on a photographic plate placed behind the crystal, a diffraction pattern as reproduced in Fig. 39. The symmetry of the crystal is clearly reflected in this pattern. Each spot on the plate corresponds to some plane in the crystal which is especially thickly populated by atoms. These are essentially the same planes which commonly occur as bounding surfaces on crystals of the same species.

It can be shown that the diffraction of X-rays by *transmission* through a crystal lattice occurs in just the same way as when X-rays are *reflected* from the lattice planes. The theory of X-ray interference is most simply developed for this case.

(c) Theory of X-ray Interference

As with light rays, interference must take place with X-rays when neighboring wave fronts, after reflection from lattice planes, either reinforce of cancel each other out, according to the angle of reflection. The condition that the X-rays reflected from a set of equidistant lattice planes, should reinforce each other, can easily be deduced from Fig. 40.

Consider first the two parallel rays S_1, S_2, reflected from adjacent planes. These rays are incident at the grazing angle ϑ upon the planes considered as giving rise to the reflection, and according to the law of reflection, they are reflected at the same angle. After leaving the crystal, the rays

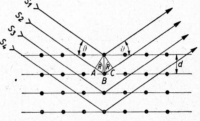

Fig. 40. Origin of X-ray interferences.

differ in path by an amount equal to AB + BC, which the ray S_2 has to travel further than S_1. When they fall on a screen, the two rays can only mutually reinforce each other if the path difference AB + BC is equal to a whole number multiple of the wave length λ of the X-radiation used. We thus have the condition for reinforcement of the rays, that

$$AB + BC = n\lambda, \quad \text{where } n = \text{whole number}$$

$$\text{or} \qquad 2d \sin \vartheta = n\lambda; \quad \lambda = \frac{2d \sin \vartheta}{n}, \tag{1}$$

since AB = BC = $d \sin \vartheta$, where d = the distance between lattice planes. Reinforcement by the rays S_3, S_4 also occurs under the same conditions. If, however, $\frac{2d \sin \vartheta}{\lambda}$ is not a whole number, the rays can always be combined in pairs in such a way that mutual extinction takes place. Thus if $\frac{2d \sin \vartheta}{\lambda} = \frac{1}{2}$ (or $\frac{3}{2}$, $\frac{5}{2}$ etc), the rays reflected from neighboring planes—S_1 and S_2, etc.—cancel out. If $\frac{2d \sin \vartheta}{\lambda} = \frac{1}{4}$ (or $\frac{3}{4}$, $\frac{5}{4}$ etc.), rays S_1 and S_3, S_2 and S_4, etc. extinguish each other. If the rays are allowed to fall on a photographic plate or film, they produce blackening only at those points where the condition (1) is satisfied.

If, as Laue did, we allow white X-radiation to fall on a crystal, there will always be present some wave lengths which fulfill condition (1) for the spacings corresponding to various lattice planes—e.g., in the case of a cubic lattice, for planes parallel to the cube faces, for those parallel to octahedral faces, rhombic dode-cahedron faces, etc. Since reflection can take place only from the atoms*, the more densely any plane is occupied by atoms, the stronger is the intensity of the corre-sponding reflected beam. The only planes in which the density of atoms is great enough to lead to a perceptible blackening are the planes which are most important crystallographically. The plate therefore displays spots corresponding only to the most important crystallographic planes, and it is possible to deduce from their relative intensities which planes are concerned.

(d) Methods of Producing X-ray Diffraction

The most important methods of producing X-ray diffraction at present in use are:

  (i) the Laue method

  (ii) the Bragg method, and its variants (*rotating crystal* method of Seemann and Schiebold),

 (iii) the Debye-Scherrer-Hull method.

The Laue method has already been considered. The experimental arrangement is shown schematically in Fig. 41. The basic form of the rotating crystal method was worked out in 1913 by W. H. Bragg and his son, W. L. Bragg. In this method

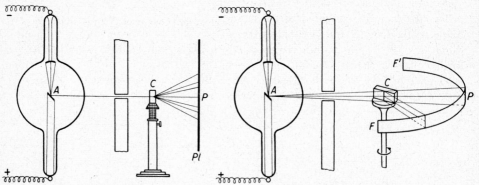

Fig. 41. The Laue method.
A anticathode; C crystal; Pl photo-graphic plate; P point of impact of the primary (undiffracted) X-ray on the plate.

Fig. 42. The Bragg method.
A anticathode; C crystal; FF' film; P point of impact of the primary radiation beam.

monochromatic, or homogeneous, X-rays are used (cf. Fig. 42), and are allowed to fall on a crystal which may, at first, have any orientation. The condition of eqn (1) will, in this case, generally not be satisfied, so that no diffraction pattern is at first produced. The crystal is now slowly rotated—usually about some selected crystallographic direction. At some instant, the angle of incidence on a particular set of planes will attain a value satisfying eqn (1), and at that instant deflection of the X-rays will take place, as indicated in Fig. 40. On further rotation, the re-flected beam vanishes again, until once again some lattice plane is encountered at

  * Strictly speaking, reflection takes place from the electrons of the atoms—see below.

the appropriate grazing angle. In this method also, every possible plane in the crystal records itself by corresponding reflection of the beam of X-rays; the planes do not, of course, need to be present as bounding surfaces of the crystal. The preferred lattice planes which reflect the X-rays flash out during the rotation of the crystal in the same way as a mirror rotated in a beam of sunlight.

To observe the directions of reflection, Bragg originally utilized the ability of X-rays to ionize gases, and make them electrically conducting. Although the Bragg ionization chamber is still of use for special purposes (absolute measurements of intensity), later workers have replaced it by a strip of film, wrapped around so as to be concentric with the crystal. The places struck by the reflected X-radiation appear blackened on the film after development. Recent advances in the experimental methods for detecting and measuring ionizing radiations enable the original Bragg method to be modified, by the use of Geiger-Müller tubes and counting circuits. In this form it is as convenient as photographic registration of the diffraction pattern, and may be applied to all types of problems involving X-ray diffraction either by single crystals or by powders (Debye-Scherrer method, see below).

A combination of the Bragg and the Laue process, originally worked out independently by Seemann and by Schiebold is the 'rotating crystal' or 'complete diagram' method. This has been elaborated by a number of workers, especially in the laboratories of W. H. and W. L. Bragg (by J. D. Bernal, J. M. Robertson, and others). This is the most powerful of all experimental methods for the investigation of crystal structures.

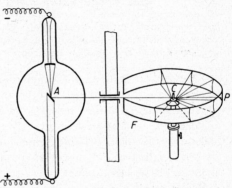

Fig. 43. The Debye- Scherrer method.
A anticathode; C crystal powder; F film;
P point of impact of the primary X-ray.

The Debye-Scherrer-Hull method (Fig. 43) is based on the same principle as the Bragg method. In place of the rotating crystal, Debye and Scherrer (1916) employed a fine crystal powder. Since the various particles of such a powder are orientated at random in all directions, they behave, taken all together, in much the same way as a single particle which is brought successively into all possible positions.

(e) Application of X-ray Diffraction [1–5, 12]

X-ray diffraction has found extensive application during recent years, in the determination of crystal structures (*structural analysis*). The *lattice constant* of a crystal —i.e., the distances between the principal lattice planes—can readily be determined from the measurement of a few diffraction angles in the case of crystals of simple structure. The calculation is based on eqn (1). Even with substances of very complex structure, however, it is possible, from the occurrence of X-ray diffraction, to state at least whether or not they are crystalline, i.e., whether they have a completely regular structure. As follows from the theory, only in this case can sharp interference spots or interference rings be produced. *Very small crystallites*, however, furnish by the Debye-Scherrer method only diffuse, and more or less broadened interference rings; the more these latter are broadened, the smaller must the crystals be, so that it is possible to draw conclusions concerning the size of the crystals from the breadth of the interference rings. If the diameter of the crystals is appreciably less than 10 mμ, distinct interference rings can in general no longer be observed. Cognizance must be taken of this fact when judging whether or not a crystalline structure is present.

A perceptible broadening of the X-ray diffraction lines in a Debye-Scherrer pattern can be observed when the particle size* is smaller than 0.5 to 0.2 μ. The broadening increases in a regular manner as the particle size becomes smaller still. This is true so long as the atomic arrangement in the crystallite is perfectly ordered. If the regularity of the atomic arrangement is not perfect (i.e., if there are 'lattice imperfections', as discussed in Vol. II, Chap. I), broadening of the diffraction lines may be found even with considerably larger particles; whether this is so depends upon the nature of the lattice imperfections. Certain types of imperfection may give rise not to broadening, but to an abnormal decrease in intensity of the diffraction lines. For further discussion see R. FRICKE, Z. *Elektrochem.*, 44 (1938) 291.

Even in amorphous solids, and in liquid and gaseous substances, the atoms are not completely disordered, in the sense that certain interatomic distances occur more frequently than others. More or less marked maxima and minima are observed in the blackening of a photographic film with these substances also, and it is possible to draw certain inferences about the structure of the molecules from the position (angular diameter) of these maxima. To investigate the molecular structure of such substances, and especially of gases, use has been widely made in recent years of the *diffraction of electrons*, as well as of X-rays. A beam of electrons is diffracted like a beam of X-rays on passage through matter—e.g., through a gas—and gives rise to similar interference phenomena. [6, 7]

Equation (1) shows that if X-radiation is passed through a crystal of known lattice constant, the *wave length* λ of the X-rays can be determined by measuring the grazing angle ϑ at which reflection takes place. This use of X-ray diffraction for the determination of wave lengths (*X-ray spectroscopy*) has also found important applications. One of its especially important results has been the relation, discovered by Moseley, between the wave length of the characteristic X-radiation of any element and its atomic number.

Before entering more closely into this we shall survey the results of X-ray crystal structure determinations, in so far as they relate to substances already discussed.

2. X-ray Structure Analysis

(a) Crystal Structures

The first determinations of crystal structures were carried out by the Laue method in 1913, by W. L. Bragg, then a student in Cambridge; namely those of rock salt, NaCl, sylvine, KCl, and potassium bromide, KBr. From the distribution of intensities, Bragg was able to deduce at one and the same time the structure of the crystal and the wave length of the X-rays giving rise to each spot on the diffraction pattern. The structure of such a crystal can today be determined much

---

* 'Particle size' here implies the dimensions of those regions in the solid which give rise to coherent scattering of X-rays; it is also referred to as 'crystallite size'. An actual physical particle of the solid may contain a large number of such crystallites, clumped together at random, or forming a dendritic structure. Hence the 'particle size' as found by X-ray studies may be very much smaller than the particle size determined by other methods (sedimentation, surface area measurements, etc.), but cannot be greater than this value.

more simply by using X-rays of known wave length, since the wave lengths of X-rays can now be readily measured by using crystals of known interplanar spacing. Bragg found that the crystal lattice common to rock salt, sylvine and potassium bromide is built up in the manner shown in Fig. 44. The individual lattice points are occupied alternately by alkali metal atoms and halogen atoms (or more accurately, *ions*, since the atoms are charged, as will be seen later). Although the three compounds are isomorphous and are of similar structure as shown by the X-ray analysis, the diffraction patterns of NaCl and of KBr differ from that of KCl in quite a characteristic manner when the intensities are considered. As Bragg

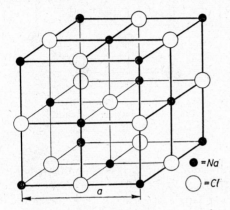

Fig. 44. Rock salt lattice ($a = 5.628$ Å).

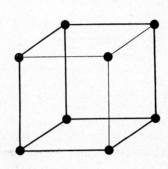

Fig. 45. Simple cubic lattice.

showed, the intensity of the reflected ray depends not only on the number of lattice points, but also upon the atomic weight of the scattering atoms, being proportional to the square of the atomic weight. Since the atomic weights of potassium and chlorine are close to one another, a crystal of KCl behaves practically as though its crystal lattice were built up entirely of one kind of atom, and it gives the diffraction pattern of a simple cubic lattice (Fig. 45). It may be seen that the lattice represented in Fig. 44 no longer corresponds to this when account is taken of the difference between the two kinds of atoms making up the structure. The diffraction patterns given by NaCl and KBr accordingly show the interferences not of a simple cubic lattice, but those belonging to two interpenetrating face center cubic lattices.

The smallest portion of a space lattice which possesses all the symmetry properties of the whole is known as the elementary parallelopiped, or *unit cell*. If this is a cube, the crystal lattice derived from it is cubic. A lattice of which only the corner points are occupied is called a *primitive* lattice (cf. Fig. 45). If the center of the faces, as well as the corners, are occupied by atoms, it is said to be a *face-centered* lattice (Fig. 46); if the central point of the unit cell, as well as the corners, is occupied, the lattice is said to be *body-centered* (Fig. 47). A crystal lattice which can be built up by the interpenetration of two or more simple lattices is described as a *compound* lattice; it is not necessary that the individual space lattices should be built up from different sorts of atoms. The rock salt structure, however, would pass over into a primitive lattice if all the lattice points were occupied by atoms of the same sort. The face-centered lattice can also be regarded as formed from four interpenetrating primitive lattices. It may be mentioned that there is no substance known which forms a simple cubic lattice.

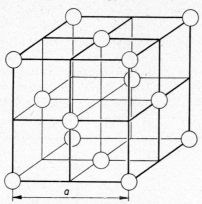

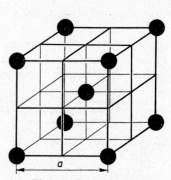

Fig. 46. Face-centered lattice (cubic closest packing). Example: Argon ($a = 5.42$ Å).

Fig. 47. Body-centered cubic lattice. Example: Sodium ($a = 4.30$ Å).

(i) Rock Salt Type

The rock salt type comprises most of the alkali halides. It is noteworthy that the alkali hydrides also possess this structure (cf. p. 173).

The alkali halides which form a crystal lattice of rock salt type are shown in Table 39, which gives for each substance the edge length a of the cubic unit cell. The very accurately determined edge length of the unit cell of rock salt itself is

TABLE 39

ALKALI HALIDES OF ROCK SALT TYPE

Fluorides	a	Chlorides	a	Bromides	a	Iodides	a
LiF	4.02 Å	LiCl	5.14 Å	LiBr	5.49 Å	LiI	6.00 Å
NaF	4.62 ,,	NaCl	5.628 ,,	NaBr	5.96 ,,	NaI	6.46 ,,
KF	5.33 ,,	KCl	6.277 ,,	KBr	6.58 ,,	KI	7.05 ,,
RbF	5.63 ,,	RbCl	6.54 ,,	RbBr	6.85 ,,	RbI	7.32 ,,
CsF	6.01 ,,	β-CsCl	7.02 ,,	—	—	—	—
—	—	β-NH$_4$Cl	6.53 ,,	β-NH$_4$Br	6.90 ,,	β-NH$_4$I	7.24 ,,

frequently used as a standard of reference for other X-ray measurements; the value given in the table is that of native rock salt. Since this contains traces of potassium chloride as an impurity, built into the crystal structure, its cell constant is greater by about 0.001 Å than that of absolutely pure sodium chloride; allowance must be made for this in precision measurements. The distance between the centers of ions in the rock salt structure is equal to half the length of the edge of the unit cube, i.e., $\frac{1}{2}a$, as may be seen from Fig. 44.

(ii) Cesium Iodide Type

The higher halides of cesium, especially CsBr and CsI, and ordinarily CsCl also, crystallize with a structure different from that of rock salt. They form a crystal lattice formed from two simple cubic lattices, in such a way that every atom is in the center of the primitive cube formed by 8 atoms of the other kind (cesium iodide type, Fig. 48). If all the lattice points in a structure of cesium iodide type were to be occupied by atoms of the same kind, it would pass over into a body-centered cubic lattice (Fig. 47).

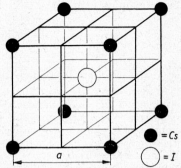

The unit cubic cell of the cesium iodide lattice contains equal numbers of atoms of Cs atoms and I atoms, as is required by the chemical formula of the compound, although this may not at once appear obvious. The cube in Fig. 48 contains one I atom. Each of the Cs atoms at the cube corners belongs, simultaneously, to 8 unit cells—i.e., contributes only one eighth of a Cs atom to the cell under considera-tion. The total number of Cs atoms in the cube shown in Fig. 48 is therefore $8 \cdot \frac{1}{8} = 1$, i.e., equal to the num-ber of I atoms. The origin of coordinates could equal-ly well have been taken at the position of an I atom, instead of at a Cs atom. The unit cell of the rock salt structure contains $1 + \frac{12}{4} = 4$ Cl atoms and $6 + \frac{3}{8} = 4$ Na atoms. The distance between centers of atoms or ions in crystal lattices of cesium iodide type is equal to half the body diagonal of the unit cube, i.e., to $\frac{1}{2}a\sqrt{3}$.

Fig. 48. Cesium iodide lattice
$(a = 4.562 \text{ Å})$.

The halides of ammonium can also crystallize according to the cesium iodide type. They do this at low temperatures, and on heating them the cesium iodide structure is transformed into the rock salt structure. Ammonium iodide, which crystallizes with the rock salt structure at the ordinary temperature, is transformed to the cesium iodide type when it is strongly cooled. A similar transition has been established in the case of CsCl (cf. Tables 39, 40). G. Wagner has also obtained RbCl crystallized with the rock salt structure at $-190°$ $(a = 3.74\text{Å}$ at $-190°)$. In Tables 39, 40, the modifications stable at low temperatures are

TABLE 40

ALKALI HALIDES OF CESIUM IODIDE TYPE

Chlorides	a	Transition point	Bromides	a	Transition point	Iodides	a	Transition point
a-CsCl	4.11 Å	445°	CsBr	4.29 Å	—	CsI	4.56 Å	—
a-NH$_4$Cl	3.86 Å	184.3°	a-NH$_4$Br	4.05 Å	137.8°	a-NH$_4$I	4.37 Å	$-17.6°$

designated as a-, those stable at higher temperatures by β-. For the modifications which are unstable at ordinary temperature, the lattice constant a relates to temperatures close to the transition point.

Ammonium fluoride NH_4F crystallizes with quite a different structure, namely according to the wurtzite type (Fig. 60, p. 258) $a = 4.39$, $c = 7.02$ Å.

(iii) Crystal Structure of the Alkali Metals

The alkali metals all crystallize with body-centered cubic crystal lattices (Fig. 47). The cell dimensions are

	Li	Na	K	Rb	Cs
a, in Å =	3.50	4.29	5.31	5.66 at $-10°$	6.133 at $-10°$

In the case of K, Rb and Cs this holds at low temperatures. At ordinary temper-ature, potassium forms tetragonal single crystals.

(iv) *Crystal Structure of the Inert Gases*

In so far as they have been investigated, the inert gases crystallize in face-centered cubic structures, with the cell dimensions

	Ne	A	Kr	Xe
$a =$	4.52	5.42	5.59	6.18 Å

(v) *Crystal Structure and Coordination Number*

As may be seen, there is no fundamental difference between the crystal structure of the alkali metals and that of the alkali halides; in both cases the lattice points are occupied by *single atoms.** This is not surprising in the case of the metals, since these also occur in the monatomic state in the gaseous condition. For the salts, however, a chemist of the older school might have expected a different sort of structure, in view of their chemical formulas. The formerly used formula Na—Cl states that 1 atom of sodium combines with exactly 1 atom of chlorine, and in terms of the older valence theories the combining forces of the sodium are thereby saturated. It cannot bind further chlorine atoms with equal strength. We find, nevertheless, from the crystal structure, that every sodium atom is surrounded by *six equidistant* chlorine atoms. Each chlorine atom is similarly surrounded by six sodium atoms at equal distances. If the bonding between any pair of atoms were in any way unique, this would inevitably find expression in the interatomic distances. Since this is not the case, we must conclude that it has no real meaning to assign the atoms in the sodium chloride crystal to one another in pairs. Each sodium atom exerts the same forces on *six* chlorine atoms. In this respect the newer theory of electrovalence shows its superiority over the older outlook. In terms of the theory of electrovalence, the valence in ionic compounds denotes only the *ionic charge*. The statement that sodium chloride consists of electrochemically univalent, oppositely charged atoms or ions** tells nothing of the *structure* of the compound formed when the ions come together. The structure results from the operation of the general laws of force between electrically charged bodies. These do not require that oppositely charged particles should always come together in pairs. It may be shown that, provided the attraction is due practically entirely to Coulomb forces, the energy liberated through the formation of an ionic lattice is greater than that which would be liberated by the formation of a molecular lattice consisting of atoms associated in pairs. Molecular lattices are formed when their energy of formation is greater than that of the corresponding ionic lattice. This can be the case, for example, when the association of atoms in pairs or groups is favored by the special circumstances of strong polarization.

A concept which has materially facilitated our understanding of the principles on which crystal lattices are built up is that of the *coordination number*, transferred to crystal chemistry from the chemistry of the complex compounds. In the latter field, coordination number signifies the number of atoms or groups which are directly bound to the central atom of a complex (cf. chap. 11). Following Pfeiffer, we may regard the crystal as a coordination compound in a wider sense. The coordination number is then that number which states how many atoms or groups any one atom or ion has as nearest neighbors. In the sodium chloride crystal, the chlorine has the coordination number 6, as may be seen at once from Fig. 44; it is surrounded by 6 sodium atoms at the same distance, $a/2 = 2.814$ Å. The sodium has the same coordination number. This holds for all compounds of the sodium chloride type. In cesium iodide, the iodine at the cube center is surrounded by 8 cesium atoms, at the cube corners. It thus has the coordination number 8, as also, for reasons of symmetry, has the cesium. In the elementary state all the alkali metals exhibit coordination number 8, since they form body center cubic structures (Fig. 47). Argon has the still higher coordination number of 12 (Fig. 46). Goldschmidt (1927) has shown that the types of crystal structure can very usefully be classified on the basis of the coordination numbers.

---

* We neglect, for the moment, the distinction that in the one case these atoms are uncharged and in the other are charged.

** See the next section for detection of the opposite ionic charges in crystalline compounds.

(b) Electrons as the Origin of the Scattering of X-rays

(i) *Ionic Lattices*

The result established by Bragg, and mentioned above that the intensity of X-rays, reflected from a lattice plane occupied by atoms of a particular kind, is proportional to the square of the atomic weight, holds only for small grazing angles of incidence, and only approximately even then. It stands in contradiction to a rule discovered earlier by Barkla, that the intensity of the X-radiation emitted by an element is *directly* proportional to its atomic weight. This contradiction was resolved by Debye (1918). The deviation of X-rays by their passage through crystals, or by their reflection at the lattice planes of crystals depends, like the diffraction (and also the reflection) of ordinary light, on the scattering of electromagnetic radiation when it falls on a single small particle. The particle on which the light is incident behaves as a self-luminous point*, from which a spherical wave is propagated. It is clear, therefore, that the greater the scattering power of the individual particles occupying any lattice plane, the greater will be the intensity with which that plane reflects X-rays. On the basis of classical electrodynamics, Debye deduced that the intensity of scattering, and therefore of the reflection, of X-rays must be proportional to the number of scattering electrons. It is the *electrons* that are responsible for the scattering of X-radiation. The distribution of intensity of the scattered radiation therefore gives direct evidence of the number and location of the electrons. Since, in neutral atoms, the number of electrons is equal to the atomic number of the atoms, and this is roughly proportional to the atomic weight**, we obtain as the general case, approximate proportionality between scattered radiation and atomic weight—i.e., Barkla's law. For small angles of incidence, however, as Debye showed, the theory leads to proportionality between intensity and the square of the number of electrons, in approximate agreement with Bragg's rule.

In the case of potassium chloride, investigated by Bragg, the numbers of electrons would be *equal* if one assumes *ions*, but *unequal* if one assumes *atoms* to occupy the lattice points. However, the accuracy attained by Bragg was not sufficient, in view of the small difference involved in this case. The difference should be much more evident in the case of sodium fluoride, since as neutral atoms sodium will have 11 electrons and fluorine 9, whereas if they are ions, each has 10 electrons. Consideration of Fig. 44 will show that in a crystal of sodium chloride (or fluoride), lattice planes parallel to the octahedron faces must each be occupied only by atoms of one kind. Suppose for the moment that only half these planes of Fig. 44 are occupied—e.g., by sodium ions,—and consider some point at which the X-rays are not extinguished on a fluorescent screen, placed to receive the X-rays. The relation expressed by eqn (1), p. 205, must hold, d being the distance apart of planes occupied by Na^+ ions. Next suppose the planes of F^- ions to be inserted between them. Then the spot on the screen considered previously can remain bright only if n in eqn (1) happened to be an even number, since if we halve d, we must also halve n if the other quantities are to remain the same. If n were previously odd, its half is not an integer, and the condition for reflection is no longer fulfilled. In all, therefore, half of the reflections from the lattice planes under consideration must be extinguished. The case is different if the planes are occupied alternately by *atoms* of fluorine and sodium. In this situation, they will reflect the X-rays with different intensities, and there can be no complete *extinction*, but only a *diminution* of intensity at the point in question. Debye found that one line which appeared strongly in the diffraction patterns of other lattices of this type was completely absent from the diffraction pattern of sodium fluoride, and he concluded that the lattice points were occupied by *ions*. Gerlach and Pauli found the same for magnesium oxide. We may assume, quite generally, that atoms are present in the electrically charged state in crystal lattices of this kind, and especially in those formed by substances which dissociate into ions in aqueous solution.*** Since the scattering of X-rays is due to the electrons, the hydrogen ion (the proton) makes absolutely no contribution to the interferences, so that its position cannot be directly deter-

---

* As is well known, this is the basis of ultramicroscopy.

** See the final section of this chapter.

*** To indicate the charges of atoms and radicals, the signs $^+$ and $^-$ will subsequently be generally used.

mined in this way[*]. Moreover, it is not possible to detect elements of low atomic number in the presence of those with many electrons in the atom, since the interferences due to the former are far too weak by comparison. For example, in the crystal structure of lithium iodide, only the position of the iodide ions can be found directly by scattering of X-rays, and the position of the lithium ions must be deduced by indirect means.

(ii) *Dependence of Crystal Type upon Ionic Radius*

In the case of ionic structures, as Goldschmidt showed, the structural type can in many cases be predicted from the principle that every ion attempts to attract as many oppositely charged ions as possible into the closest possible proximity, as a consequence of electrostatic attraction. In this respect, one may consider the ions to be almost rigid spheres, with the apparent radii determined by Goldschmidt (cf. p. 15). They do not behave as absolutely rigid spheres, in that their apparent radii do depend to some extent on the crystal type. Thus the transition from the CsI type (coordination number 8) to the NaCl type (coordination number 6) is attended with a decrease of 3 % in the interionic distance; the transition from NaCl type to structures with coordination number 4 brings a decrease of 5–7 %. The—admittedly small—effects of temperature and pressure are superimposed on this. On the whole, however, the ions behave as almost rigid spheres, so that the Cl^- ion, for example, has practically the same apparent radius whether it is associated with Li^+, Na^+, K^+, or Rb^+ or—subject to the limitation indicated above—is combined with Cs^+ or even Sr^{++}, for example. The interionic distances in crystals can therefore be approximately predicted by adding together the apparent, or Goldschmidt, radii of the ions concerned. The values calculated in this way are not very accurate, however, since no account is then taken of the variation of apparent radii with the coordination number. Moreover, it neglects the extent to which the apparent radius of any particular ion is affected by forces exerted by the other ion with which it is combined. It can be shown theoretically that there must be such an effect (Pauling 1927). Thus the radius of the Na^+ ion must be larger when it is surrounded by univalent ions than when it is surrounded by bivalent ions.

(iii) *Reduced ('Univalent') Ionic Radii*

Interionic distances in crystals of binary compounds can be more accurately calculated by use of the values for ionic radii *reduced to the univalent state and the coordination number 6*, as proposed by Pauling and Zachariasen (*Z. Kristallog.*, 80(1931) 137), and listed in Table 41. The distance r between ionic centers, in Å, is then found to be

$$r = \frac{(r_1^\circ + r_2^\circ) \cdot f_{CN}}{\sqrt[n-1]{z_1 \cdot z_2}} \tag{2}$$

TABLE 41

REDUCED ('UNIVALENT') IONIC RADII (IN Å), AFTER ZACHARIASEN

			H^{1-}	Li^{1+}	Be^{2+}	B^{3+}	C^{4+}	N^{5+}	
			1.36	0.68	0.55	0.42	0.38	0.35	
C^{4-}	N^{3-}	O^{2-}	F^{1-}	Na^{1+}	Mg^{2+}	Al^{3+}	Si^{4+}	P^{5+}	S^{6+}	Cl^{7+}
2.49	2.02	1.76	1.33	0.98	0.89	0.79	0.69	0.66	0.64	0.63
Si^{4-}	P^{3-}	S^{2-}	Cl^{1-}	K^{1+}	Ca^{2+}	Sc^{3+}	Ti^{4+}	V^{5+}	Cr^{6+}	Mn^{7+}
2.97	2.56	2.20	1.81	1.33	1.17	1.03	0.88	0.82	0.70	0.68
	As^{3-}	Se^{2-}	Br^{1-}	Rb^{1+}	Sr^{2+}	Y^{3+}	Zr^{4+}	Nb^{5+}	Mo^{6+}	
	2.62	2.29	1.96	1.48	1.34	1.19	1.07	0.98	0.90	
	Sb^{3-}	Te^{2-}	I^{1-}	Cs^{1+}	Ba^{2+}	La^{3+}	Th^{4+}			
	2.77	2.47	2.19	1.67	1.49	1.30	1.24			

Ce^{4+}

1.14

---

[*] The same is true for the use of electron beams. On the other hand, the position of protons can be directly determined by the use of *neutron* beams (cf. Vol. II), since these are scattered by atomic nuclei and not by the extra-nuclear electrons.

In (2), r_1° and r_2° are the radii given in Table 41, z_1 and z_2 are the valences of cation and anion, n and f_{CN} are coefficients which make allowance for the repulsion manifested when the ions approach very closely, and the consequent variation of the interionic distance with the coordination number. n is the *repulsion exponent* (cf. p. 147), which depends on the electronic configuration. f_{CN} depends upon n and the coordination number; its values may be taken from Table 42. For the case of C.N. = 6, f_{CN} is unity, by definition.

TABLE 42

VALUES OF f_{CN}, AFTER ZACHARIASEN

$n =$		5	6	7	8	9	10	11	12
C.N. =	2	0.793	0.834	0.857	0.877	0.890	0.902	0.911	0.919
,,	3	0.862	0.889	0.906	0.919	0.928	0.936	0.942	0.947
,,	4	0.920	0.936	0.946	0.954	0.959	0.963	0.967	0.970
,,	8	1.067	1.053	1.044	1.038	1.033	1.029	1.026	1.024
,,	12	1.160	1.126	1.104	1.088	1.077	1.068	1.061	1.056

The coordination number is determined essentially by the ratio of the reduced ionic radii of cation and anion (i.e., by the radius ratio). In general, for

$\dfrac{r^\circ_{cation}}{r^\circ_{anion}} =$	0.15–0.23	0.23–0.41	0.41–0.73	0.73–1.00	> 1.00
C.N.	3	4	6	8	12

The interionic distances are also somewhat dependent upon the radius ratio. This effect can be neglected in practice, however, except for the case where the anions quite, or almost, touch each other. This leads to a distinct expansion of the crystal lattice. It is not possible to use eqn (2) to calculate the interionic distance in such cases.

The reduced ('univalent') radii do not directly furnish a measure of the space occupied by the individual ions in crystals, but only do so when allowance is made for the conversion factors given in eqn (2). The use of eqn (2) to calculate interionic distances may be illustrated by the example of CaF_2. From Table 41 we find $r_1^\circ = 1.17$, $r_2^\circ = 1.33$ Å; $z_1 = 2, z_2 = 1$. For Ca^{2+} (argon configuration), $n = 9$ (cf. p. 147), for F^- (neon configuration), $n = 7$; mean $n = 8$. For the radius ratio $r_1^\circ : r_2^\circ = 1.17 : 1.33 = 0.88$, we have C.N. = 8. For C.N. = 8 and $n = 8$, $f_{CN} = 1.038$ (Table 42). The interionic distance Ca—F is accordingly found as

$$r = \frac{(1.17 + 1.33) \cdot 1.038}{(1 \cdot 2)^{1/7}} = 2.35 \text{ Å}$$

This agrees well with the value found experimentally, $r = \dfrac{a}{4}\sqrt{3} = 2.36$ Å (cf. p. 266). The value calculated from Goldschmidt's apparent ionic radii is: $r = 1.06 + 1.33 = 2.39$ Å, in distinctly poorer agreement. The Goldschmidt radii are still important, however, since they give a direct indication of the space occupied by ions in crystals, whereas the reduced or 'univalent' radii of Zachariasen are merely parameters for calculations, except for the case of univalent ions in compounds with the coordination number 6.

(iv) *Polarization*

Polarization can produce considerable alterations in apparent ionic radii, and these can not be readily calculated. The cases considered hitherto refer only to ions of small polarizing power—i.e., to ions with inert gas configuration.

Polarization, as has already been indicated, signifies a mutual displacement of the positive core and electron cloud, brought about by an external electric field, such as that of an adjacent ion. *Dipoles* may thus be produced or, if such are already present, they can be strengthened (cf. p. 48).

According to Goldschmidt, polarization leads to the formation of radicals such as $[NH_4]^+$, $[NO_3]^-$, $[SO_4]^{2-}$, etc., which appear in crystals as distinct units, or structural groups. In these radicals the atoms are brought particularly close together. However, if the radicals may be supposed to arise from the association of inert-gas like ions, the interatomic distances can generally be calculated with satisfactory accuracy from the 'univalent' ionic radii.

In other cases, polarization leads to the formation of *layer lattices*, and with still stronger polarization, *molecular lattices* result. Highly symmetrical structures such as those discussed above (called by Hund *coordination lattices*) are formed when polarization effects are of subordinate importance. In general, it may be said that the polarization effects lead to a reduction of symmetry.

(c) Layer Lattices

A layer lattice is a structure in which the atoms or ions are arranged in sheets, often in such a manner that the individual layers contain only atoms or ions of one kind—e.g., are built up only from positive or only from negative ions. A typical layer lattice is formed by lithium hydroxide; the unit cell is illustrated in Fig. 49a. To bring out the layer structure more clearly, the lattice points of the lower half of the unit cell are represented in Fig. 49b,

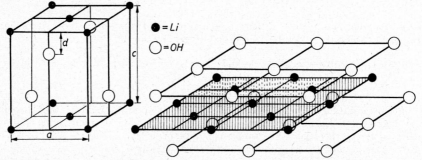

Fig. 49a and b. Lattice structure of lithium hydroxide (typical layer lattice)
$$a = 3.55 \text{ Å}; \quad c = 4.33 \text{ Å}; \quad d = 0.867 \text{ Å}$$

with the addition of further lattice points belonging to neighboring unit cells. It may be seen that every sheet of Li^+ ions is sandwiched between two sheets of OH^- ions. Since the latter are half as densely occupied as the Li^+ sheets, there is 1 OH in all for each Li atom, as required by the formula. The oxygen atoms are so strongly polarized in LiOH that not only do they form sheets without incorporation of any lithium ions, but such sheets of oxygen—more strictly, of hydroxyl ions—succeed one another in pairs throughout the structure.

Since the occurrence of layer lattices is associated with strong polarization, they are formed chiefly by compounds in which the cation does not possess the inert gas configuration—i.e., predominantly in the Subgroups of the Periodic Table. They are also found in the main groups, however, as the example of LiOH shows, and occur especially when the anions are strongly polarized by cations of especially small radius. In the example cited, the polarization may be due chiefly to the hydrogen of the OH^- ions. Many other hydroxides form layer lattice structures—e.g., $Mg(OH)_2$, $Ca(OH)_2$, α-$Zn(OH)_2$, $Cd(OH)_2$, $Ni(OH)_2$, $Co(OH)_2$, $Fe(OH)_2$ and $Mn(OH)_2$, all of which have the brucite structure (cf. p. 261), and also hydrargillite or gibbsite, $Al(OH)_3$. There are, however, a number of hydroxides which form definite coordination lattices and not layer lattices—e.g., the high temperature modifications of NaOH and KOH (NaCl type), ε-$Zn(OH)_2$ (Vol. II, Chap. 9), and $Y(OH)_3$, $La(OH)_3$ and the hydroxides of the rare earths, in which the metal ions exert the coordination number 9 towards the OH^- ions.

Table 43 gives the 'effective radius' of the hydroxyl ion in a series of crystalline hydroxides of the bivalent and trivalent elements. The 'effective radius' is here defined as half the distance between the centers of hydroxyl ions, which are adjacent, but not coordinated to the same metal atom. In general, this radius decreases as the basic character of the hydroxide diminishes, and its acidic character increases. This decrease in 'effective radius' is partly due to the polarizing action

TABLE 43

EFFECTIVE RADIUS OF THE HYDROXYL ION IN COMPOUNDS OF THE TYPES $M(OH)_2$, $M(OH)_3$ AND $MO(OH)$

Hydroxide	Effective radius of the OH$^-$ion (Å)
$Ca(OH)_2$	1.68 Å
$Mg(OH)_2$	1.61
$Ni(OH)_2$	1.55
$Cd(OH)_2$	1.49
$Y(OH)_3$	1.45
$La(OH)_3$	1.45
$Nd(OH)_3$	1.45
β-$Be(OH)_2$	1.43
ε-$Zn(OH)_2$	1.41
γ-$Al(OH)_3$	1.39
$B(OH)_3$	1.35
γ-$FeO(OH)$	1.35
a-$FeO(OH)$	1.34
a-$AlO(OH)$	1.33

of the cation, but is also probably to be ascribed in part to the effect of hydrogen bond formation. This view is supported by Glemser's observation [*Naturwissenschaften*, 40, (1953) 199] that the absorption maximum of the OH vibration, in the infrared, is displaced progressively towards lower frequencies as the effective radius of the hydroxyl ion decreases.

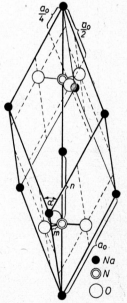

Fig. 50. Lattice of sodium nitrate (calcite type) $a_0 = 6.065$ Å

$$a = 47° \ 14'$$
$$n = 3.24 \text{Å} \quad m = 1.31 \text{Å}$$

Substances with layer lattice structures generally possess a good cleavage along one plane, since the cohesive forces in the direction perpendicular to the layers are weaker than the forces within the layers. In lithium hydroxide, sheets of similarly charged hydroxide ions succeed each other and, if it were not for the strong polarization, would actually repel each other. In other cases, such as the lattice of graphite (p. 423), it may be concluded that the attractive forces perpendicular to the sheets are only weak, since the interatomic distances in this direction are considerably larger than those in the plane of the sheets.

(d) Structure of Sodium Nitrate

The crystal structure of sodium nitrate may be cited as an example of how radicals function as special structural groups in crystals. $NaNO_3$ forms trigonal rhombohedral crystals, like calcite. The structure is represented in Fig. 50. It may be seen that the oxygen atoms are closely grouped around the nitrogen atoms, to form a structural unit. This typical structure was first determined for calcspar, and is therefore generally called the *calcite* type.

If they have a sufficiently low moment of inertia, radicals which act as discrete structural groups, like the NO_3^- ion, may undergo rotation about their axis, within the crystal, far below the melting point. This has been shown by X-ray investigations and by measurements of the variation of the specific heat with temperature, and has been observed both for radicals and for molecules which exist as closed units in a crystal lattice. In $NaNO_3$, the NO_3^- ion begins to vibrate about its axis (the vertical axis of Fig. 50) at 150°. At 215°, the vibratory movement has become so strong that a definite position can no longer be assigned to the oxygen atoms, on the basis of the X-ray diagram, and at 275° all the NO_3^- ions are in full rotation. Radicals with very small moments of inertia, such as the NH_4^+ ion, can enter into rotation even at quite low temperatures. For example, the NH_4^+ ions in the ammonium halides rotate at temperatures far below 0°.

Potassium nitrate, KNO_3, forms a layer lattice of *aragonite* type. In this structure, sheets of K^+ ions are interposed between pairs of sheets made up of NO_3^- ions (half as densely occupied), in such a way that on a NO_3^- sheet there is superimposed, at a short distance, a K^+ sheet and another NO_3^- sheet close to it; at a considerably greater distance there is the next NO_3^- sheet, which in its turn is close to the next K^+ sheet and so on. The NO_3^- radicals stand out clearly as structural units in this crystal lattice.

(e) Double Salts

If two salts crystallize from a melt or from a solution in simple stoichiometric proportions, forming a distinct crystal lattice, they are said to form a double salt.

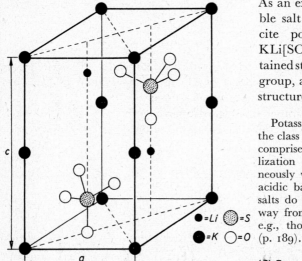

As an example of how such a double salt may be formed, we may cite potassium lithium sulfate $KLi[SO_4]$ (Fig. 51). A self-contained structural group, the $[SO_4]^{2-}$ group, again clearly appears in the structure.

Potassium lithium sulfate belongs to the class of so-called 'mixed salts'. This comprises the salts formed by neutralization of a polybasic acid simultaneously with several bases, or a polyacidic base with several acids. Such salts do not differ structurally in any way from the ordinary double salts—e.g., those of the type of carnallite (p. 189).

Fig. 51. Lattice of potassium lithium sulfate.
$a = 5.13$ Å; $c = 8.60$ Å

$\bullet = Li$ $\circledcirc = S$
$\bullet = K$ $\circ = O$

(f) Isomorphism and the Formation of Mixed Crystals

Mixed crystals are said to be formed when two substances crystallize from a solution or a melt, to form a common crystal structure in which, within certain limits at least, the proportions of the components may vary continuously. The phenomenon of mixed crystal formation is restricted, as a rule, to substances which crystallize in the same form. It is therefore spoken of as *isomorphism* (ἴσος same, μόρφη form). The essential feature of isomorphism and mixed crystal formation is that the isomorphous substances are capable of replacing one another within the crystal lattice. In a crystal of potassium chloride, for example, the chlorine may be replaced practically completely, ion for ion, by bromine, without the crystal lattice losing its stability. The properties of potassium chloride are transformed continuously into those of potassium bromide in the process. Absolutely unlimited miscibility, which represents the highest degree of isomorphism, is not so frequently encountered, however. More usually, there are wider or narrower *miscibility gaps*. This is more particularly the case when the two substances themselves form crystal lattices which either differ in structure, or have substantially different dimensions.

We may therefore formulate the chief requirements for the formation of a wider range of mixed crystals between two substances as

(1) chemical similarity

(2) equality in valence

(3) approximate equality in atomic or ionic radius.

If these conditions are fulfilled, it is also possible for elementary atoms to be replaced by radicals.

Isomorphism and the formation of mixed crystals are not precisely equivalent terms. The concept of isomorphism is often given a broader meaning than that of mixed crystal formation, so as to include as isomorphous crystals that are sufficiently analogous in structure for the crystals of the one to continue to grow when placed in the saturated solution of the other substance, even although they are incapable of forming mixed crystals. This is the case, for example, with calcite and sodium nitrate. On the other hand, substances are frequently not spoken of as isomorphous when, as with the pair NaCl and KCl, their ability to form mixed crystals is but very limited, because of the difference in their cell dimensions, even although they have identical crystal structure. Still less is the term applied to substances, such as NaCl and MgO, which are alike only with regard to the fine structure of their crystals.

In contrast with the double salts, mixed crystals do not, as a rule, have a distinct crystal structure of their own, differing from that of their components. Examples are known, however, in which such is the case.

(g) Isotypy

It has been suggested by Rinne that where two crytsalline substances, such as NaCl and MgO, are crystallographically equivalent in structure (i.e., form crystal lattices of the same type), without it being possible to regard them as *isomorphous* in the usual sense, the phenomenon should be termed *isotypy*. Isotypic substances are such as have analogous chemical composition and the same type of crystal structure, without any ability to form mixed crystals.

Substances which have the same type of crystal lattice, but with the cations and anions interchanged therein, are called *anti-isotypic*. Thus the alkali oxides M_2O are anti-isotypic with fluorspar, CaF_2; they form a crystal lattice of fluorite type, but with positive calcium ions replaced by a face-centred cubic lattice of negative oxygen ions, and the negative fluoride ions replaced by positive alkali metal ions. Isotypy and anti-isotypy are of very frequent occurrence among binary compounds.

(h) Structure of the Gas Hydrates

The name 'gas hydrates' is used to describe the crystalline addition compounds formed by water with chemically saturated molecules (of normally gaseous substances)—e.g., with A, Kr, Xe, Rn, CH_4, CH_3Cl, CO_2, N_2O, SO_2, H_2S, Cl_2 and Br_2. These are all loosely bound compounds, arising from the action of Van der Waals' forces. Although such forces cannot in themselves cause the aggregation of molecules in any definite ratios, the gas hydrates are substances of definite stoichiometric composition. This is conditioned by the volume relationships between the components in the crystal lattice. The structure of the gas hydrates has been largely explained through the work of von Stackelberg (1949–52). In their crystal lattices, the H_2O molecules are arranged in such a way that vacant spaces are created between them in a regularly ordered way: the molecules of the other component—e.g., A or Kr—are incorporated in these holes.

Two types of structure may be formed, depending on the molecular size of the interstitial component. In the first type of structure, the cubic unit cell (a = about 12 Å) contains 46 H_2O-molecules, which surround 6 larger and 2 smaller cavities (of diameters 5.9 and 5.2 Å, respectively). The unit cell of the other type (a = 17.2 Å) contains 136 molecules of H_2O, with 8 larger and 16 considerably smaller cavities (diameters 6.8 and 4.8 Å). The compositions which result for the gas hydrates, depending on whether all the holes or only the larger ones are occupied, are exemplified by the following examples: $6Br_2 \cdot 46H_2O$ or $Br_2 \cdot 7\frac{2}{3}H_2O$; $8Kr \cdot 46H_2O$ or $Kr \cdot 5\frac{3}{4}H_2O$; $8CHCl_3 \cdot 136H_2O$ or $CHCl_3 \cdot 17H_2O$; $8CHCl_3 \cdot 16H_2S \cdot 136H_2O$ or $CHCl_3 \cdot 2H_2S \cdot 17H_2O$ (a 'double hydrate'). The formulas which had previoulsy been assigned to the gas hydrates, from analytical data, generally agree (within the experimental uncertainty) with those based on the structure determination.

The hydrates of the *smaller* molecules crystallize with structures of the first type (*gas hydrates in the stricter sense*), and hydrates of *larger* molecules according to the second type ('liquid hydrates', since the second component is itself generally a liquid under normal conditions if its molecules are large). In both structures, each molecule enclosed in the void spaces is surrounded by a very large number of H_2O molecules (28, 24 or 20, according to the size of the void space). The fact that the gas molecules are coordinated with such a large number of water molecules is connected with the observation that hydrates of this kind are formed only by substances the molecules of which have *no polarizing action*, or *very weak polarizing power* at most. *Strongly* polarizing molecules, such as NH_3 and HCl, form hydrates of a completely different type (cf. pp. 611, 791), and their composition depends upon the individual nature of the substance concerned; the composition of the gas hydrates corresponding to the structural types mentioned above depends solely on the molecular size and the volatility of the substance forming the hydrate, and is the same for substances which are completely different in chemical nature [for the conditions of formation of gas hydrates, see M. von Stackelberg, *Naturwissenschaften*, 36, (1949) 327].

Most of the gas hydrates are stable only at low temperatures, or under high pressures (cf. Table 44). It is curious that their stability is enhanced by the presence of an indifferent gas, such as O_2 or N_2, under high pressure. This observation had already been made in 1897 by Villard, who made use of it for the preparation of iodine hydrate. If no indifferent gas is present, it is impossible to form this hydrate at all, but it is stable up to 8° in the presence of O_2 at 330 atm. Von Stackelberg has shown that gas hydrates, such as bromine hydrate, are able to take up perceptible amounts of O_2 and N_2 into the structures; less H_2 can be incorporated. These observations may have some bearing on the stability of bromine hydrate, which decomposes at 6.2° under its own vapor pressure, is stable up to 20° under 150 atm of O_2, but only stable up to 9° under 200 atm pressure of hydrogen.

TABLE 44

HEATS OF FORMATION, DECOMPOSITION TEMPERATURES AND DECOMPOSITION
PRESSURES OF GAS HYDRATES

Heats of formation are for 0° C, and apply to the reation of 1 mole of the hydrate-forming
component (in gaseous form) with liquid water

Hydrate-forming substance	B.p. °C	Structure type	Hydrate				
			No. of H$_2$O molecules		Heat of formation Kcal/mole	Decompn. temp. (at 1 atm)	Decompn. press. at 0° C
			Theor.	Obs.			
	—186°	I	5¾	~6	13.3	—42.8°	105 atm
H$_4$	—164°	,,	,,	~6	14.5	—29.0°	26 ,,
r	—153°	,,	,,	5.7	13.9	—27.8°	14.5 ,,
F$_4$	—128°	,,	,,		—	—	
e	—107°	,,	,,	~6	16.7	— 3.4°	1.5 ,,
$_2$H$_4$	—104°	,,	,,	~6	15.3	—13.4°	5.5 ,,
$_2$O	— 89°	,,	,,	~6	14.8	—19.3°	10.0 ,,
$_2$H$_6$	— 88°	,,	,,	5.8 ±0.5	15.0	—15.8°	5.2 ,,
H$_3$	— 87°	,,	,,	5.9	16.4	— 6.4°	1.6 ,,
$_2$H$_2$	— 84°	,,	,,	5.7	15.2	—15.4°	5.7 ,,
O$_2$	— 79°	,,	,,	~6	14.6	—24°	12.3 ,,
H$_3$F	— 78°	,,	,,	~6	—	—	2.1 ,,
$_2$S	— 61°	,,	,,	5.7	16.3	+ 0.35°	731 mm
sH$_3$	— 55°	,,	,,	—	—	+ 1.8°	613 ,,
$_3$H$_8$	— 45°	,,	,,	—	—	0°	760 ,,
$_2$Se	— 42°	,,	,,	5.9	16.8	+ 8.0°	346 ,,
I$_2$	— 34°	,,	,,	5.9 ±0.3	16.2	+ 9.6°	252 ,,
$_2$H$_5$F	— 32°	,,	,,	~6	20.1	+ 3.7°	530 ,,
H$_3$Cl	— 24°	,,	,,	~6	18.1	+ 7.5°	311 ,,
O$_2$	— 10°	,,	,,	6.1 ±0.6	16.6	+ 7.0°	297 ,,
H$_3$Br	+ 4°	,,	7⅔	~8	19.5	+11.1°	187 ,,
O$_2$	+ 10°	,,	,,	~8	—	+15°	~160 ,,
H$_5$Cl	+ 13°	II	17	~16	31.9	—	201 ,,
H$_5$Br	+ 38°	,,	,,	—	—	—	~155 ,,
H$_2$Cl$_2$	+ 42°	,,	,,	—	29	—	116 ,,
H$_3$I	+ 43°	,,	,,	17	31.4	—	74 ,,
H$_3$CHCl$_2$	+ 57°	,,	,,	—	—	—	~56 ,,
$_2$	+ 59°	I	7⅔	7.9 ±0.5	19.6	—	45 ,,
HCl$_3$	+ 61°	II	17	—	30	—	~45 ,,

(i) Volume Chemistry of Substances of Non-polar Structure

The distances between the ions, in crystals built up from ions of inert gas configuration,
can generally be calculated with satisfactory accuracy with the reduced ('univalent') ionic
radii of Zachariasen; polarization effects are generally not large enough, in this case, to
alter the interatomic distances materially. The greater the importance of polarization
effects, however, the less successful is this mode of calculation. The interatomic distances in
layer lattices* which Goldschmidt considered to be incommensurable with the coordination
lattices, (cf. p. 15) can often be calculated to a good approximation with the use of the
atomic radii derived from the atomic lattices of the elements themselves. Since this empirical

* In the case of layer lattices made up of inert-gas-like ions, calculation of the distances
using reduced ionic radii seems to furnish satisfactory results—e.g., for the distance Be-O
in beryllium oxide (cf. p. 258), calc. 1.64 Å, obs. 1.65 Å.

result lacks any theoretical basis, however, it is not possible to discern how far it is valid. Still less is it at present possible to devise any general formulas relating the interatomic distances in pure homopolar (covalent) compounds to the radii which their component elements possess in atomic or ionic lattices. For many purposes, however, it is important to be able to compute in advance the specific volumes or molecular volumes possessed by such compounds in the solid state.

The molecular volume of a crystalline compound of known structure can be computed from the volume of the unit cell, v, the number of molecules (or of formula units made up of ions) of the compound contained in the unit cell, z, and Avogadro's number N_A.* Thus molecular volume $= N_A \cdot v/z$. In so far as the sizes and the number of atoms in the unit cell can be predicted,** this method provides the most accurate values for the calculation of molecular volumes. Its applicability is limited, however, and it fails in just those fields for which the theoretical calculation of specific volumes of compounds and mixtures would be of particular interest to the chemist—e.g., with alloys and glasses. It is, however, in this field that volumes can be calculated to a very good approximation from the law of *additivity of volumes*, formulated by Biltz. In particular, this permits the calculation with very fair agreement of the molecular volumes (and from this the specific volumes) of *atomic and molecular aggregates*. These terms are applied to substances built up from atoms (or from positive ions plus electrons) or from molecules, and not (directly, at least) from oppositely charged ions. The metals and their alloys provide examples of atomic aggregates in this sense. To the class of molecular aggregates belong liquids, melts, supercooled liquids and melts, and glasses, as well as crystals built up from molecules.

The researches of Biltz and his co-workers [11] have shown that the molecular volumes of atomic and molecular aggregates at absolute zero can generally be calculated to a fair approximation by adding together quantities, called *volume increments* by Biltz, which are representative of the spatial requirements of the individual atoms. Kopp (1855) had already shown that the molecular volumes of liquid carbon compounds at their boiling point could be computed by adding volume values characteristic of the individual constituents. In many cases, these volume values agreed with the atomic volumes derived from the density of the corresponding elements, but this was not always the case, since the spatial requirement of any atom or group is not uninfluenced by different types of chemical bonds (e.g., double bonds, aromatic systems, etc.).

Biltz proved that the volumes of solid substances, both inorganic and organic, referred to a temperature of absolute zero, could be similarly calculated, by adding together his volume increments. The volume increments are dependent upon the mode of combination or the type of compound concerned. Thus, for some particular metal, they are different in its compounds with other metals and in its compounds with non-metals. They are, therefore, in general not the same as the atomic volumes obtained from the densities of the elements. They are highly dependent upon the charge borne by the atom in question. For example, the volume increment of the K atom is 43.4, but that of the K^+ ion is 16. By allowing for the influence of the charge upon the magnitude of the volume increment, it is also possible to calculate approximately the molecular volume of compounds forming typical ionic structures, as Biltz has shown. Such calculated values usually differ from the observed values to a greater extent, however, than the volumes calculated for atomic or molecular aggregates. It is therefore generally preferable to use ionic radii for the calculation of the volumes of ionic compounds.

---

* For example, the unit cell of NaCl, which has the volume $a^3 = 5.628^3 \cdot 10^{-24}$ cm^3, contains $4 + 4$ ions. The molecular volume is the volume containing 1 g ion of each sort,

i.e.,
$$\frac{5.628^3 \cdot 10^{-24} \cdot 2 \cdot 6.023 \cdot 10^{23}}{8} = 26.85 \text{ cm}^3$$

** The coordination number, and therefore the lattice type, can often be predicted from the radius ratio r_1°/r_2°. The significance of the prediction of interionic distances lies not so much in the possibility of estimating molecular volumes (which, in this field, are mostly known already from measured densities), but rather in assisting in the determination of structures by the X-ray method, by calculation in advance of the interplanar distances, or lattice parameters, which are to be expected.

If the ions behaved as absolutely rigid spheres, the law of additivity of ionic radii, in the sense of Goldschmidt, would be strictly true. On the other hand, the law of additivity of volumes would hold if the atoms or ions formed completely 'plastic' entities, i.e., if they fused together over their whole surface, without any gaps. The first case would necessarily exclude the validity of the law of additivity of volumes,* for if the volumes were proportional to $(r_1 + r_2)^3$ they could not at the same time be proportional to $(r_1^3 + r_2^3)$. Since, however, the inert-gas-like ions do not behave as absolutely rigid spheres, their real behavior is 'bracketed' by the two laws—i.e., the real molecular volumes usually fall between that calculated from the law of additivity of radii, and that derived from additivity of volumes. For compounds with crystal lattices built up from only slightly polarized ions, the observed volumes approximate to those calculated from additivity of the radii. The greater the role of polarization, the greater is the discrepancy between the values so calculated and the observed values, and it is found experimentally that they approximate increasingly closely to the volumes obtained by summation of the volume increments.

The calculation of molecular volumes from the volume increments may be illustrated by a few examples. A tabular comparison of the volume increments of the elements is given by Biltz [11]. (See also Z. anorg. Chem., 223 (1935) 321 and 234 (1937) 253).

Example 1. Molecular volume of intermetallic compounds. Two classes of intermetallic compound may be distinguished on volume-chemical grounds: those in which the volume increments of the metals are equal, or almost equal to their atomic volumes, and those whose formation is attended by a considerable contraction. In the latter, the volume increments to be substituted for the components are considerably smaller than their atomic volumes. The compound Mg_2Pb belongs to the former class. The atomic volume of Mg (at $0°$ K) is 13.8, that of Pb is 17.9. Since, in this instance, the volume increments may be taken as equal to the atomic volumes, the zero-point molecular volume is calculated as $2 \cdot 13.8 + 17.9 = 45.5$ (found 46.5).

The second class comprises chiefly the compounds of the alkali and alkaline earth metals—e.g., $NaPb_3$. In compounds of this class, Pb has the volume increment 17.0, and Na also has the volume increment 17.0. The zero-point molecular volume of $NaPb_3$ is thus $17.0 + 3 \cdot 17.0 = 68.0$ (found 68.3).

Example 2. Molecular volume of a silicate. Biltz gives 6.5 as the average volume increment of Na^+, 11 for that of O^{2-}, and 0 for Si^{4+}. The zero-point volume of $Na_2Si_2O_5$ is thereby reckoned as $2 \cdot 6.5 + 5 \cdot 11 = 68$ (found 72).

Example 3. Specific volume of a glass. A glass may have the analytical composition $6Na_2O \cdot 13SiO_2$. According to Biltz, O^{2-} in SiO_2 has a considerably higher volume increment than in all other compounds including the silicates, namely 13.6. In calculating the numerical molecular volume of a silicate glass, we must consider how much excess silica it contains. The well defined compound of highest silica content formed between SiO_2 and Na_2O is $Na_2Si_2O_5$. In the glass cited, if $6Na_2Si_2O_5$ are deducted, 1 SiO_2 remains over. We then obtain for the molecular volume calculated $2 \cdot 13.6 + 12 \cdot 6.5 + 30 \cdot 11 = 435$. Dividing this by the formal value of the 'molecular weight' (= 1152.8) gives the value 0.377 for the specific volume at $0°$ K. On adjustment to room temperature we obtain 0.383 (found 0.406). Much better agreement is obtained if, instead of going back to the volume increments of the individual ions, the partial volumes of vitreous $Na_2Si_2O_5$ and vitreous SiO_2 are added directly, with the use of the empirical molecular volumes (73.1 and 27.2). In this way, for the specific volume of the glass in question we have

$$\frac{(6 \cdot 73.1 + 27.2)}{1152.8} = 0.404 \quad \text{(found 0.406)}.$$

3. X-ray Spectroscopy

(a) Characteristic Radiation and 'Bremsstrahlung'

As is well-known, X-rays are produced when fast cathode rays (i.e., electrons moving with high velocity) are suddenly brought to rest by collision with solid

---

* Though it would be roughly true if the radii differed but little among themselves.

substances, such as the anticathode (target) of an X-ray tube. The maximum energy of the radiation emitted by each individual electron is given by the relation

$$E = e.V = h\nu \tag{3}$$

(E = energy, e = charge on the particle, V = discharge potential across the X-ray tube, h = Planck's constant, ν = frequency of X-radiation).

Since, in general, a cathode ray particle does not give up all its energy in a single act*, the frequency of most of the energy radiated is lower than this value.

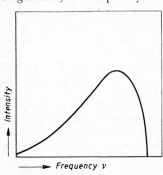

Fig. 52. Distribution of intensity in the spectrum of 'white' X-rays

Hence, for the radiation as a whole, the maximum intensity lies rather below the short-wave length limit given by (3), and the distribution of intensity may be represented by a curve such as that of Fig. 52. It somewhat resembles the curve for the distribution of energy in the spectrum of light emitted by an incandescent body, but differs in that the short wave length (high frequency) end displays a sharp limit. It follows from (3) that this limit, and the intensity maximum with it, shifts towards higher frequencies (shorter wave lengths) as the discharge voltage is increased. The distribution of intensity in the so-called 'bremsstrahlung' is practically independent of the material of the target, and is determined by the voltage applied across the tube.

If the discharge voltage is gradually increased, so that the limit shifts more and more towards higher frequencies, then at a definite stage, depending on the nature of the target material, the emission of an extremely intense radiation will be observed. Its intensity falls right outside the normal intensity curve, and remains unaffected even long after the maximum of intensity has moved away from the wave length in question. This radiation is known as *characteristic radiation*. As the name indicates, the wave length of this radiation is determined by the material of the target (or at least of its surface).

The phenomenon was observed by Barkla even before the discovery of X-ray diffraction. He found, in 1905, that with increasing discharge potential the characteristic radiation did not change in frequency. He deduced the latter from the *penetrating* power or *hardness* of the radiation, which he assumed to be proportional to the frequency. The posibility of making accurate wave-length measurements by means of X-ray diffraction at once brought with it an important result—namely the simple relation between the frequency of the characteristic radiation of an element and its atomic number, discovered by Moseley in 1913.

(b) Moseley's Law

Spectral resolution of the characteristic X-radiation by means of a crystal [8] yielded a *series*, consisting of quite a few lines, of different intensity. These lines are denoted (substantially in the sequence of their intensity) by the Greek letters α, β, γ etc. The strongest line, the α-line, is that of longest wave length. Strictly, the α-line consists of two lines, α_1 and α_2; the other lines may, in some cases, also be further resolved. Most elements possess several series, which lie, however, in widely sepa-

---

* Moreover, most of the energy of the cathode rays is degraded to *heat*.

rated spectral regions. They are denoted by the letters K, L, M, etc; the individual lines are correspondingly indicated by $K_\alpha, K_\beta, L_\alpha, L_\beta$ etc. The various series differ in their structure or arrangement of lines, but the same series, for different elements, has completely analogous structure (see Fig. 53). With most elements, only one or two series lie in the region of wave lengths accessible to measurement by use of a crystal lattice as a diffraction grating—namely the K-series for the lighter elements from sodium onwards, and the L-series also for the heavier elements. There are also several other series (M-, N-, O- and P-series) for the elements of very high atomic weight. For any one element, the K series always lies in a region of higher frequency than the L series, and this in turn than the M series, etc.

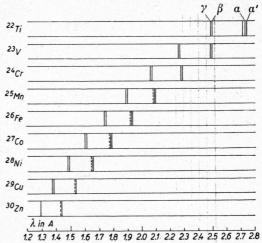

Fig. 53. X-ray spectra of the elements of atomic numbers 22–30 (lines of the K-series)

In 1913, Moseley discovered that for corresponding lines—e.g., for the K_a line—the frequency increased quite uniformly from one element to another as the atomic number increased.

For corresponding lines in the characteristic X-ray spectrum the frequency is proportional to the square of the atomic number—*Moseley's law*. This may be written

$$\frac{\nu}{(Z-a)^2} = \text{const.} \qquad \text{or} \qquad \frac{\sqrt{\nu}}{(Z-a)} = \text{const.} \qquad (4)$$

ν = frequency of the characteristic radiation of an element of atomic number Z, a = a constant which is nearly equal for the corresponding lines of different elements.

Fig. 54 shows how well this relation holds for the K-series, and indeed, as may be seen, for each individual line in the series. The atomic numbers Z are here plotted as abscissas, the square roots of the reciprocals of the wave lengths, in place of the square roots of the frequencies, as ordinates. The reciprocal of the wave length is known as the *wave number, ν'*. The wave number of a line with the wave length λ or frequency ν is therefore defined by $\nu' = \dfrac{1}{\lambda} = \dfrac{\nu}{c}$

(c = velocity of light). Equation (4) requires that all the points plotted in Fig. 54 should lie on a straight line, which is very closely true. The slight curvature of the lines arises from the fact that the quantity a in the Moseley equation is not exactly a constant.

Moseley's law furnished, in the first place, a striking confirmation that the atomic numbers deduced from the Periodic System are correct. Moreover, in cases of uncertainty, it made it possible to decide whether (as for example in the lanthanide group) elements which were difficult to separate had been assigned to the correct order, or whether in fact they were correctly to be regarded as distinct elements.

An element may be identified with certainty, and relatively easily, by means of its characteristic X-ray spectrum, even in cases where the difficulties attending quantitative separation processes, and the complexity of the optical spectra, result in the failure of chemical methods of analysis. This is the basis of X-ray spectrum analysis, which will first be briefly discussed. We shall return in the last section to the relation between *atomic weight* and *ordinal number* or *atomic number*, which was revealed in principle by Moseley's law, and which proved so important for the

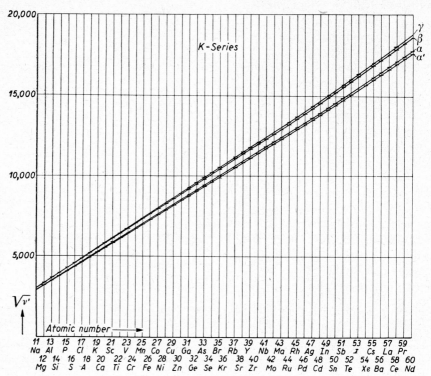

Fig. 54. Relation between frequency of characteristic radiation and atomic number (Moseley's law)

history of the Periodic System. From the X-ray spectra we may first reach certain conclusions concerning atomic structure, which form the basis for an understanding of the relation between atomic weight and atomic number.

(c) X-ray Spectrum Analysis

It is of fundamental importance for X-ray spectroscopy that the characteristic X-radiation is almost independent of the form in which an element is present— i.e., whether as the free element or in some one of its compounds. In ordinary spectroscopic analysis this is true only in so far as compounds are *decomposed* at the temperature of the flame or arc. In itself, the optical spectrum of an element is completely different from the spectra of its compounds.

In order to excite the characteristic radiation, it is only necessary to rub a small quantity of the material under investigation on a water-cooled target. This is mounted in a specially constructed X-ray tube, and the spectrum is recorded photographically when the high tension voltage is applied. X-ray spectroscopy gives not only a qualitative detection of the elements present, but, by comparison of intensities, it is possible to make an approximate quantitative determination.

It has been possible to calculate the X-ray spectra of hitherto undiscovered elements by means of Moseley's law. As a result of this, the formulation of the law soon led to the discovery of new chemical elements. The discoveries of *hafnium* in 1922 by von Hevesy and Coster, and of *rhenium* in 1925 by Noddack and Tacke (see Vol. II) were based on Moseley's law. The most valuable contribution of X-ray spectroscopy, however, has been to our knowledge of atomic structure and the energy levels in the atom.

(d) Characteristic Radiation and Atomic Structure

The wave lengths of the characteristic radiations of the elements in the X-ray region are far shorter than those of the ordinary optical spectra. They are therefore conveniently measured in special units, known as X-units; 1 X-unit (X.U.) = 0.001 Å = 10^{-11} cm. For the line of longest wave length in the K spectrum of potassium, K_{a_2}, $\lambda = 3737.1$ X.U., i.e., more than a thousand times shorter than the wave length of the potassium violet line. With increasing atomic number, the wave lengths of the X-ray lines rapidly diminish, in accordance with the Moseley relation.

It can be taken that the source of X-ray emission lies in the transition of electrons, from higher levels to lower levels, in accordance with the Bohr theory. If we consider the Moseley formula more closely, as applied to a single line, it is at once apparent how the origin of the characteristic radiations should be interpreted in terms of Bohr's principles. This was, indeed, perceived by Moseley himself. For the K_a line, according to Moseley,

$$\nu = R_\infty \, (Z - 1)^2 \, \frac{3}{4} \qquad (R_\infty = \text{Rydberg constant})$$

or
$$\nu = R_\infty \, (Z - 1)^2 \left(\frac{1}{1^2} - \frac{1}{2^2} \right) \tag{5}$$

This corresponds exactly to the Bohr formula (eqn (13) of Chap. 3*), except that Z of that equation is replaced by $Z - 1$. This can be taken as meaning that the total nuclear charge Z does not act on the electron, but only a portion of it, the remainder being screened by another bound electron. The constant a in Moseley's equation (= 1 in the case considered) is therefore called the *screening constant*.

Since *all* the elements, so far as the corresponding spectral region is accessible to investigation, have the K_a line in their spectrum, it follows that the electron concerned is bound in a similar manner in all the elements, i.e., that the innermost structure corresponding to this electron, is the same in all elements.

From the small magnitude of the screening constant a in the Moseley equation (eqn (4)) for the K_a line, it follows that the electron circulates immediately around the nucleus in its innermost orbit. The emission of the K_a line results, as shown by eqn (5), from a transition from a 2-quantum orbit to a 1-quantum orbit.

For the L_a line, Moseley found the relation

$$\nu = R_\infty \, (Z - a)^2 \left(\frac{1}{2^2} - \frac{1}{3^2} \right) \qquad (\text{where } a = \text{about } 7.4).$$

It may be inferred in an analogous way that the line is emitted by an electron which jumps from a 3-quantum to a 2-quantum orbit, and is acted upon by the

---

* Since $\nu = c/\lambda$, R_∞ in the above formula $= c \times R_\infty$ of eqn (13), Chap. 3.

nuclear charge screened by about 7.4 units. This electron, and the sphere corresponding to it, is again common to all the elements, with the exception of the very lightest.

The characteristic X-radiation thus shows that the atoms of the heavier elements possess a common sequence of energy levels. These are known as the K-, L-, etc. levels, according to the line (K_a, L_a, etc.) which is radiated when an electron falls back into them from the next higher level. The remaining (higher frequency) lines of the various series originate when the electron falls back from some higher lying level, and not from the next higher level. In agreement with this, these other lines are of lower intensity, since the corresponding transitions occur more infrequently.

It is natural to correlate the energy levels recurring constantly in all the atoms with the 'shell' structure of the atoms, deduced from the optical spectra and the relations with the inert gases. The spectra and the chemical behavior of the alkali metals indicate that in each of them a new 'shell', or as we may now say a new *energy level* is begun. The preceding energy level is thus completed each time with the completion of the inert gas configuration. From this standpoint we must, accordingly, assign a definite energy level to each inert gas configuration: the lowest, or K level, to the configuration attained in helium; the L level to that reached in neon; the M level to the argon configuration, etc. If we represent the X-ray lines by 'terms' as in eqn (5), the following principal quantum numbers are obtained for the various levels:

Helium shell	Neon shell	Argon shell	Krypton shell	Xenon shell	Radon shell
K-level	L-level	M-level	N-level	O-level	P-level
$n = 1$	$n = 2$	$n = 3$	$n = 4$	$n = 5$	$n = 6$

In Fig. 55, the energy levels are represented by circles, and the electron transitions by arrows, as in Fig. 19. The difference between Fig. 55 and Fig. 19 lies only in that in Fig. 19 the circles represented possible levels or orbits, to which the electron, usually revolving about the nucleus in the innermost orbit, *might* be raised, but which are usually vacant. In Fig. 55, however, the higher levels are generally occupied by electrons; the circles do not represent the orbits in which the electrons move, but rather the energy levels corresponding to these orbits. Their distances are quite arbitrarily represented.

The emission of the characteristic radiation takes place in the following way: an electron is first knocked out of an inner orbit—e.g., by a cathode ray. It is thereby expelled completely from the atom, since the higher orbits are all occupied. The energy which must be expended is given by the relation $E = h\nu$ where E represents the energy of the level from which the electron was removed. The resulting vacancy in the level from which the electron was removed is at once filled, through an electron jumping in from some outer-lying shell. If the first electron was expelled from the K-level, the vacancy will as a rule, be filled by an electron from the adjacent L-level, and the K_a line results. An electron may, alternatively, drop down from the M level (K_β line) or one from the N-level (K_γ line). The latter cannot happen, however, with those atoms in which the N-level is as yet empty— e.g., certainly not before potassium, if our identification of the X-ray levels with the inert gas shells is correct. In general, however, such transitions do not occur with appreciable probability when there are only a few electrons in the higher

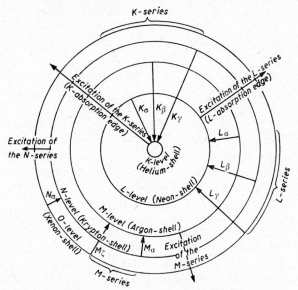

Fig. 55. Energy levels in the atom and the origin of X-ray spectra (schematic).

level. In fact, the K line is first detectable with certainty for vanadium, which has 5 electrons in the N-level.

If the transition did not occur from the highest level, but from one of the deeper-lying levels, then a vacancy is produced in this level, in turn; to fill this, an electron jumps in from a higher level. Thus the production of the K_a line is necessarily associated not only with the appearance of the other lines of the K-spectrum, but also with the simultaneous appearance of the entire L-, M-, N-spectra etc., according to the atomic number of the element. However, whereas the lines K_β, K_γ, etc. always appear only with the K_a line, the L spectrum can also be excited without the K-spectrum, the M-spectrum without the K- and L- spectra, etc., but not the L-spectrum without the M-spectrum, etc. These regularities were observed soon after the discovery of X-ray spectroscopy, and interpreted by Kossel in 1914 in the manner set out here.

It follows from this theory, in agreement with experiment, that in order to excite an X-ray line it is not sufficient to impart the energy given by the frequency of the line in question, as is the case with the optical spectra originating in the periphery of the atom, but a *higher energy* is essential. For example, the energy $E = h \cdot \nu_a$ does not suffice to excite the K_a line with the frequency ν_a. This would be enough to raise the electron only into the L-level, whereas it must be excited at least into the periphery of the atom. If we call the energy needed for this E_K, and the voltage which must be applied to the tube to give a cathode ray particle this energy is V_K, we can assign a corresponding frequency by means of the equation

$$eV_K = E_K = h\nu_K \tag{6}$$

We can also impart energy to the atom by means of X-rays, as well as by cathode rays. If we irradiate the atom with 'white' X-radiation, it will absorb only those rays having an energy sufficient to raise an electron from one of the inner orbits, again to the periphery of the atom. For these rays, $E \geq E_K = h\nu_K$. Radiation corresponding to energy lower than E_K will not be absorbed. We thus find an *absorption limit*, at a frequency ν_K given by eqn (6). Equation (6) thus states that the *excitation potential for the K_a line*—i.e., for the whole *K*-series—represents the energy of the corresponding *K-series absorption edge*, and vice versa. We may therefore calculate the energy of the ground state orbit of the *K*-series both from the excitation potential and from the position of the absorption edge. The same holds for the other series.

As has already been mentioned, the K_a-line is shown by more exact resolution to consist of two lines. This is to be explained in that, strictly speaking, there is not a single *L*-level, but there are *two L-levels*, differing somewhat in energy. According as the electron which drops down to fill the vacancy in the *K*-shell originates in one or other of these, the K_{a_1} or K_{a_2} line results. Just as for the levels deduced from the optical spectra, each of these levels can again be assigned a subsidiary quantum number k (or l) in addition to the principal quantum number n: indeed, we require two subsidiary quantum numbers k_1 and k_2 in order to represent by means of terms the sum total of all the X-ray lines. Exact analysis of the X-ray spectra has shown that the *L*-level may be resolved into not more than 3 sub-levels, the *M*-level into 5, the *N*-level into 7, the *O*-level into 5, and the *P*-level into 3 sub-levels. In the atoms following radon, the X-ray spectra afford evidence for the following energy states, characterized by the Bohr principal and subsidiary quantum numbers $n_{k_1,\,k_2}$ given in Table 45. In the lighter atoms, all the levels are,

TABLE 45
ENERGY LEVELS IN THE ATOMS FOLLOWING RADON

K-level	L-level			M-level				
$1_{1.1}$	$2_{1.1}$	$2_{2.1}$	$2_{2.2}$	$3_{1.1}$	$3_{2.1}$	$3_{2.2}$	$3_{3.2}$	$3_{3.3}$

N-level						
$4_{1.1}$	$4_{2.1}$	$4_{2.2}$	$4_{3.2}$	$4_{3.3}$	$4_{4.3}$	$4_{4.4}$

O-level					P-level		
$5_{1.1}$	$5_{2.1}$	$5_{2.2}$	$5_{3.2}$	$5_{3.3}$	$6_{1..}$	$6_{2.1}$	$6_{2.2}$

of course, not yet occupied. We shall see later how the occupation of levels proceeds in detail.

The electron shells, or energy levels, of the atom as revealed by the X-ray spectra, are important not only for an understanding of the Main Groups, as indicated above; the subdivision of the shells, revealed by the fine structure of the X-ray spectra, has made it possible to understand for the first time the origin of the Sub-groups of the Periodic System, including the Lanthanide Series and the Actinide Series.

(e) Atomic Number and Nuclear Charge

Moseley's eqn (4) defines for each element a number Z which, as has already been seen, is identical with its *atomic number*, as given by the sequence of the elements in the Periodic Table. The significance of the atomic number so determined lies, in the first place, in that it furnishes an unambiguous order of the elements, based on their X-ray spectra. This order agrees with that given by the Periodic Table *in every case*, where the elements can be unequivocally arranged on the basis of their chemical properties. A determination of the atomic number by means of Moseley's law therefore makes it possible to assign with certainty to their places in the Periodic System those elements which could not formerly be unambiguously placed by their atomic weight and other properties.

The definition of the atomic number according to Moseley's law requires some restriction only in that it has not yet been possible to measure the characteristic radiation of the inert gases or of the elements lighter than sodium. In the case of the inert gases, the reasons for this are purely experimental: the material of which the radiation is to be excited must either be solid itself, or must be brought on to the target in the form of a solid, non-volatile compound. The latter is impossible for the inert gases, and the strong heating of the target by the impact of the cathode ray beam is an obstacle to their investigation as frozen solids. With the elements having atomic weights less than that of sodium (or neon), the difficulty lies in the fact that their characteristic frequencies are so low (since Moseley's law indicates that these decrease very rapidly with atomic number in this region) that none of the available crystal lattices are suitable for the determination of wave lengths. It has, however, been possible to measure the *excitation potentials* of the K-series for the elements from Li to O, and from these the frequency of the K-absorption edge is obtained by means of eqn (5).

Holtsmark (1923) has observed an interesting deviation from Moseley's law in the case of these elements. Only when the frequencies of their absorption edges have been measured for compounds in which the elements concerned can be considered as being negatively charged (e.g., for the oxygen atom in copper oxide) has it been found that the measured values lie on the extrapolation of the straight line joining the corresponding absorption edges of the elements, as required by Moseley's law. The neutral elements below neon show distinctly lower excitation potentials for the K-limit. This is because the condition which always holds for the K-series in the higher elements—that the electron orbits of the next higher energy level shall be fully occupied—applies here in general only if the atom is negatively charged. If the next higher energy level (the 2-quantum level) is not fully occupied, it is of course possible for one of the electrons circulating directly about the nucleus to be raised into a 2-quantum orbit, and Moseley's law can no longer be applied to the necessary excitation energy, and to the corresponding frequency for this process. An apparent exception to the rule thus confirms the true law and the theory based upon it.

Whereas, as far down as lithium, the excitation potentials correspond only to *absorption limits*—i.e., to frequencies of radiation which the atoms can only absorb, but cannot emit—the analogous absorption limits of hydrogen and helium, which lie in the ultra-violet, represent frequencies of radiation which may be *both absorbed and emitted*. This means that the electrons immediately revolving about the nucleus are not, in these atoms, in an *inner sphere*: they are no longer surrounded by electrons in higher energy levels. For these two atoms Moseley's law— the applicability of which is very striking in this case—holds in the field of optical spectra.

We have seen in the preceding section (p. 227) that the number Z in the Moseley equation is to be interpreted as the *nuclear charge* number. It represents the number of unit positive charges borne by the atomic nucleus. Moseley's law thus shows that the nuclear charge increases regularly, by steps of one unit, from element to element in the Periodic System.

In the neutral atom, the positive charge of the nucleus must be compensated by an equal number of negative charges—i.e., electrons—outside the nucleus. Thus for every neutral atom

Atomic number or ordinal number = nuclear charge number = number of electrons
in extra-nuclear structure

The extra-nuclear structure comprises all the electrons which circulate around the nucleus. This should not be confused with the electrons of the so-called 'outer sphere' of the atom, commonly called the valence electrons because of the part they play in chemical combination.

(f) Atomic Number and Atomic Weight

The atomic weight increases, on the whole, in rough proportionality to the atomic number.

This statement provides the basis for the fact that the properties of the elements, which should actually be represented as periodic functions of the atomic number appear also to be approximately periodic functions of the atomic weight. Fig. 56 shows how far the atomic weight is proportional to the atomic number. It may be seen that for the lighter elements, up to about calcium, the atomic weight is roughly twice the atomic number (except for hydrogen, of which the atomic weight and atomic number are equal). From calcium onwards, the atomic weights increase rather more rapidly than the quantity $2Z$, twice the atomic number. The increase in atomic weight, unlike the increase in atomic number, is irregular from element to element. This is why in certain cases the atomic weights come out of sequence, as compared with the atomic numbers, and is the origin of the irregularities met with in attempting to arrange the elements in the Periodic System according to their atomic weights.

The atomic numbers or nuclear charges, and also the atomic weights, are atomic constants which have *aperiodic* character. They are determined only by the *nucleus of the atom*. The properties which alter in *periodic* fashion have their origin in the *extranuclear structure*, and especially in the outermost sphere of the atom—i.e., in the valence electrons. For this reason, the periodic character of the elements is most strikingly manifested in their *valence properties*.

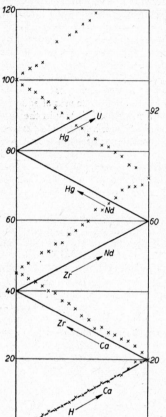

Fig. 56. Atomic numbers and atomic weights. The atomic numbers are plotted as abscissas and change from left to right or vice versa at multiples of 20. The half atomic weights of the elements are plotted as ordinates and marked with crosses. If the half atomic weights were exactly the same as the atomic numbers, one would obtain the continuous line.

References

1 P. P. Ewald, *Kristalle und Röntgenstrahlen*, Berlin 1923, 327 pp.
2 F. Halla and H. Mark, *Leitfaden für die Röntgenographische Untersuchung von Kristallen*, Leipzig 1937, 354 pp.
3 *Strukturberichte*, Vols. I to IV, Leipzig 1931–37.
4 R. W. G. Wyckoff, *The Structure of Crystals*, 2nd Ed., New York 1931, 497 pp., *Supplement*, New York 1935, 256 pp.
5 W. L. Bragg, *Atomic Structure of Minerals*, Ithaca 1937, 292 pp.

6 M. VON LAUE, *Die Interferenzen von Röntgen- und Elektronenstrahlen*, Berlin 1935, 46 pp.

7 J. T. RANDALL, *Diffraction of X-Rays and Electrons by Amorphous Solids, Liquids and Gases*, London 1934, 290 pp.

8 M. SIEGBAHN, *The Spectroscopy of X-Rays* (translated by G. A. Lindsay), Oxford 1925, 299 pp.

9 G. L. CLARK, *Applied X-Rays*, 3rd Ed., London 1940, 674 pp.

10 E. BRANDENBERGER, *Röntgenographisch-analytische Chemie*, Basel 1946, 287 pp.

11 W. BILTZ, *Raumchemie der festen Stoffe*, Leipzig 1934, 338 pp.

12 W. P. DAVEY, *Study of Crystal Structure and its Applications*, New York 1934, 695 pp.

13 R. C. EVANS, *Introduction to Crystal Chemistry*, Cambridge 1939, 388 pp.

14 C. W. BUNN, *Chemical Crystallography*, Oxford 1948, 422 pp.

15 W. H. and W. L. BRAGG, *X-Rays and Crystal Structure*, 4th Ed., London 1924, 338 pp.

16 W. L. BRAGG, *The Crystalline State*, Vol. I, London 1949, 352 pp.; R. W. JAMES, *idem*, Vol. II, London 1948, 623 pp.

17 M. J. BUERGER, *X-Ray Crystallography*, New York 1942, 531 pp.

18 N. F. M. HENRY, H. LIPSON and W. A. WOOSTER, *Interpretation of X-Ray Diffraction Photographs*, London and New York 1951, 258 pp.

19 K. LONSDALE, *Crystals and X-Rays*, London 1948, 199 pp.

CHAPTER 8

SECOND MAIN GROUP OF THE PERIODIC SYSTEM: THE ALKALINE EARTH GROUP

Atomic numbers	Names of the elements	Sym- bols	Atomic weights	Densities	Melting points	Boiling points	Specific heats	Normal valences
4	Beryllium	Be	9.013	1.86	1285°	2970°	0.540	II
12	Magnesium	Mg	24.32	1.74	650°	1100°	0.242	II
20	Calcium	Ca	40.08	1.54	845°	1439°	0.168	II
38	Strontium	Sr	87.63	2.60	757°	1366°	—	II
56	Barium	Ba	137.36	3.74	710°	1696°	0.068?	II
88	Radium .	Ra	226.05	ca. 6	ca. 700°	—	—	II

1. Introduction

(a) General

The second Main Group of the Periodic System comprises the elements beryllium, magnesium, calcium, strontium, barium and radium. It is known as the Alkaline Earth Group, after its principal members, calcium, strontium and barium, which are known as the metals of the alkaline earths.

They were so called—magnesium being frequently included with them—because their oxides, being intermediate in properties between the *alkalis* (i.e., the oxides and hydroxides of the alkali metals)—and the *'earths'*, (the infusible oxides of the elements, typified by alumina), were given the name of alkaline earths.

Except for its valence, beryllium, the first element of the group, is more closely related in behavior to aluminum than to the higher members of its own group. The second element of the group, magnesium, also differs in many respects from the alkaline earth metals proper, although to a lesser extent. It shows many resemblances to the second Sub-group, and especially to zinc. Thus the sulfates of magnesium and zinc, in contrast to the alkaline earth sulfates, are readily soluble, are isomorphous, and form double salts of analogous composition. The alkaline earth group illustrates particularly clearly the rule cited in Chapter 1—that the first element is apt to constitute a transition to the next *Main Group*, the second element to the *Sub-group* belonging to the same family, whereas the group character is fully developed for the first time in the third element.

The highest member of the group, radium, corresponds exactly to the alkaline earth metals in its chemical properties. Nevertheless, it is not usual to include it in the group of the alkaline earth metals proper, because from its occurrence and its most important property—the radioactivity from which it derives its name—it deserves a special place. In considering general properties, we shall not at first include radium in the discussion since but little quantitative knowledge is extant.

Except for radium, the elements of the alkaline earths are all to be numbered among the 'light metals', this term being applied to those metals which have a density lower than 5. They are much harder than the alkali metals; barium, which is the softest of them and which in all its properties stands closest to the alkalis, is about as hard as lead. The melting points are all much higher than those of the alkali metals.

The apparent atomic and ionic radii of the elements of the alkaline earth group in crystals are set out in Table 46. In each case, the atomic or ionic radius is a little less than that of the preceding alkali metal. This contraction is attributed to the increase of nuclear charge by one unit in each case.

TABLE 46

ATOMIC AND IONIC RADII; LATENT HEATS OF FUSION AND EVAPORATION;
IONIZATION ENERGIES AND NORMAL POTENTIALS FOR ELEMENTS OF GROUP IIA
HEATS OF HYDRATION OF IONS; LATTICE ENERGIES OF COMPOUNDS

[Heats of fusion and evaporation given at the m.p. and b.p., respectively. Heats of hydration calculated assuming 250 kcal for heat of hydration of the hydrogen ion. Lattice energies after Sherman (1932) and De Boer (1936)]

Element		Beryl-lium	Mag-nesium	Calci-um	Stron-tium	Barium
Atomic radius		1.05	1.62	1.97	2.13	2.17 Å
Ionic radius*		0.31	0.65	0.99	1.13	1.35 ,,
Latent heat of fusion, kcal/g atom		2.5	1.70	3.14	—	—
Heat of evaporation, kcal/g atom		53.5	32.52	36.58	33.61	35.66
Heat of sublimation at 0° K		—	36.56	42.82	49.2	—
Ionization energy,	(M → M^+ + e	213.7	175.4	140.3	130.7	119.6
kcal/g atom	(M^+ → M^{2+} + e	417.6	344.8	272.4	253.0	229.3
	(M^{2+} → M^{3+} + e	3532.7	1837.7	1173	986	818
Normal potential in volts { obs.		+1.69	+2.40	+2.76	—	—
{ cal.		+1.73	+2.51	+2.83	+2.87	+2.92
Heat of hydration in kcal/g ion		570	437	361	324	297
Lattice energy in { Chlorides		713	595	537	506	488
kcal/mole { Oxides		1080	936	830	784	740

* Values from Pauling '*The Nature of the Chemical Bond*'.

The elements of the second Main Group are alike in being positively bivalent in their compounds. Only in extremely rare and exceptional cases do they exhibit positive *uni*valence as well as bivalence. This typical bivalence is the property which, with their atomic numbers and atomic weights, most unambiguously gives these elements their place in the second Main Group. They share also a strongly electropositive character, which shows itself in their position near the alkali metals in the electrochemical series of the elements and in their great affinity to electronegative elements.

In accordance with the values of their normal potentials, all the Group IIA metals decompose water. Beryllium and magnesium do so only slowly; this is because their hydroxides, which result from the reaction—e.g., according to the equation

$$Mg + 2HOH = Mg(OH)_2 + H_2$$

are insoluble, and when they have formed on the surface they hinder any further attack on the metal. For this reason, in the case of magnesium, it takes days at ordinary temperature before fine shavings in contact with water are completely converted to the hydroxide. The other alkaline earth metals react much more quickly with water, a fact corresponding to the greater solubility of their hydroxides. Barium, which has the most soluble hydroxide and the highest normal potential, reacts quite vigorously not only with water but with alcohol. The stability of the alkaline earth metals in air decreases from magnesium to barium in a corresponding manner. They displace the heavy metals from their solutions, in accordance with their place in the electrochemical series.

Direct measurements of the normal potentials, which determine their place in the electrochemical series, have not been made for all the alkaline earth metals. In Table 46, the normal potentials calculated by Makishima (cf. p. 157) are given, as well as the measured values. It may be seen that the alkaline earth metals closely rival the alkali metals in the strength of their electropositive character. The sequence of normal potentials is the same as that of ionization energies in this Group. The effect of the hydration of the ions upon the normal potentials, which is prominent in the alkali metal series, is thus less important for the alkaline earths. This is because the difference between the heats of hydration and the ionization energies is much greater in the alkaline earth group than it is for the alkali metals. In both Groups, the heats of hydration of the ions are smaller than the ionization energies, and very much smaller than the sum of these and the heats of sublimation. The reason that these metals liberate hydrogen from water or acids, or that they assume a positive potential with respect to the hydrogen electrode, is thus not really 'the tendency of these metals to form positive ions in solution', as is commonly stated, but rather the tendency for hydrogen ions to discharge themselves, forming H_2 molecules. More accurately, it may be said that the free energy change in the reaction

$$H^+ aq + e = \tfrac{1}{2}(H_2) + aq \tag{1}$$

is greater than that of the reactions

$$M^+ aq + e = [M] + aq \tag{2a}$$

or

$$\tfrac{1}{2}M^{++} aq + e = \tfrac{1}{2}[M] + aq \tag{2b}$$

The normal potential, referred to the normal hydrogen electrode, is a direct measure of the difference in standard free energies between the reactions represented by eqn (1) and (2a) or (2b), in accordance with eqn (3), Chap. 5.

From the normal potentials given in Table 46, and the normal potential of chlorine, there may be derived the values given in Table 47 for the decomposition potentials of the chlorides of the alkaline earth group in molar aqueous solutions. They fall into the same sequence as the decomposition potentials of the molten chlorides, measured by B. Neumann in 1925. These latter, with their temperature coefficients, are also given in Table 47. The measurements were made on melts containing KCl and LiCl, with solidification temperatures below 750°.

TABLE 47

MELTING POINTS AND DECOMPOSITION POTENTIALS OF ALKALINE EARTH CHLORIDES

Compound	$BeCl_2$	$MgCl_2$	$CaCl_2$	$SrCl_2$	$BaCl_2$
Melting point	$405°$	$715°$	$780°$	$872°$	$960°$ C
Decomposition potential in aqueous solution $18°$ C	-3.05	-3.96	-4.12	-4.23	-4.28 volt
Decomposition potential of melt, $750°$	-1.46	-2.25	-2.74	-2.90	-2.99 ,,
Temperature coefficient of the latter	$0.965 \cdot 10^{-3}$	$0.712 \cdot 10^{-3}$	$0.714 \cdot 10^{-3}$	$0.714 \cdot 10^{-3}$	$0.714 \cdot 10^{-3}$

The products of combustion of the alkaline earth metals are the *normal oxides* MO in all cases. Where peroxides are formed at all, they are much less stable than those of the alkali series.

The *heats of formation* of the oxides and other compounds of the Group IIA metals are collected together in Table 48. The oxides of this group are among those with the highest heats of formation. The relation between electropositive character and heats of formation of oxides was discussed in connection with Table 27, Chap. 6. The other heats of formation given in Table 48 follow the same general trends as for the corresponding columns of Table 27, except that, in this case, heats of formation for the chlorides, as well as for the bromides and iodides, increase with increasing electropositive character of the metals throughout the group.

TABLE 48

HEATS OF FORMATION OF COMPOUNDS OF BERYLLIUM, MAGNESIUM AND THE ALKALINE EARTHS

[Values in kcal per g equivalent of metal]

Elements	Oxides	Hydrides	Fluorides	Chlorides	Bromides	Iodides	Sulfides	Nitrides
Beryllium	68.7	—	—	56.3	—	—	28.1	22.3
Magnesium	72.9	—	130.7	75.5	60.8	42.4	41.9	19.2
Calcium	76.0	23.0	144.7	95.5	81.0	64.0	57.1	17.1
Strontium	70.6	21.1	144.6	98.8	85.4	68.0	55.1	15.2
Barium	66.7	20.4	143.1	102.5	90.1	77.2	51.2	15.0

The oxides of the alkaline earths combine with water, forming hydroxides, with an energy which increases considerably from beryllium to barium. The *solubility* of the hydroxides increases greatly from beryllium hydroxide to barium hydroxide: nevertheless, barium hydroxide is only moderately soluble at ordinary temperature. The basic character increases in the same order from beryllium hydroxide, which is amphoteric, to barium hydroxide, which is strongly basic.

The ability of the elements of Group IIA to combine directly with *nitrogen* is noteworthy. This ability also increases with increase in atomic weight (although the heats of formation of the nitrides decrease), and for the alkaline earth metals proper it is so marked that they combine slowly with nitrogen at ordinary temperature.

The alkaline earth metals in the narrower sense (*i.e.*, Ca, Sr, Ba), like the alkali metals, combine with hydrogen forming hydrides—e.g.,

$$Ca + H_2 = CaH_2$$

These are salt-like in character, and hydrogen is the electronegative component in them, as it is in the alkali hydrides. The preparation of MgH_2 directly from the elements is effected with greater difficulty, and BeH_2 cannot be obtained in this way at all. The compounds MgH_2 and BeH_2 are solid and non-volatile, like the alkaline earth hydrides, but they do not have a well defined salt-like character.

All the elements of Group IIA form colorless, doubly charged ions: Be^{++}, Mg^{++}, Ca^{++}, Sr^{++}, Ba^{++}, Ra^{++}. Beryllium also forms the anions $[BeO_2]^=$ and $[Be(OH)_4]^=$, which are also colorless. All the salts MX_2 of these elements are colorless, unless they are derived from colored anions.

The salts of radium are also intrinsically colorless. Under the influence of the rays emitted from the radium that they contain, however, many of them—e.g., the chloride and bromide—have the property of becoming gradually colored, so that they ultimately turn brown or black. They become pure white once more, however, when they are recrystallized.

Many of the salts of the alkaline earth metals are sparingly soluble in water. The solubilities often show a regular gradation e.g., in the case of the sulfates, the solubilities of which fall off steeply as the atomic weight of the alkali metal increases. The trend in the solubilities of the chromates is similar. Most of the salts of the alkaline earth metals with weak or moderately strong acids are sparingly soluble— e.g., the phosphates, oxalates and carbonates. Many are quite soluble, however, especially the sulfides, cyanides, thiocyanates and acetates. The extent to which the hydroxides differ in the degree of their basic character shows itself clearly in the carbonates, which are hydrolyzed to an increasing extent in going from barium to beryllium. Their ease of decomposition changes in the same direction. Whereas barium carbonate is incompletely decomposed, even at a white heat, calcium carbonate can easily and completely be decomposed by ignition, into CaO and CO_2, and magnesium carbonate still more readily, giving up all its CO_2 even on gentle heating.

The bivalence of the elements of the alkaline earth group is explained, according to the electronic theory, in that each follows two places after an inert gas in the Periodic System. Thus each possesses two electrons more than the preceding inert gas. Because the inert gas configuration is stable, the atoms readily give up these two electrons, but not more, since to do so would involve breaking into the inert gas configuration.

This view is strengthened by consideration of the spectroscopic data. The spectra (cf. p. 250 *et seq.*) show that in each of these elements *two* electrons are relatively weakly bound, as compared with the rest. In terms of the Bohr theory, these electrons, in the ground state of the atom, revolve around the core in strongly eccentric orbits, as in the alkali metals. The principal and subsidiary quantum numbers of these orbits are the same as in the neighboring alkali metal. If one electron is first ionized off, the remaining electron gives rise to a spectrum which stands in the same relation to the spectrum of the preceding alkali metal as does the spectrum of singly ionized helium (cf. p. 110) to that of the hydrogen atom; corresponding to the high eccentricity of the orbit (higher principal quantum number), the binding is not nearly as strong as in helium, however. The strongly electropositive nature of the elements of Group IIA, like that of the alkali metals, is explained in terms of the older Bohr atom model by the high eccentricity of the ground state orbits. As the eccentricity increases with increasingly difference between the principal and the subsidiary quantum number (cf. Chap. 6) the ionization energy diminishes and the electropositive character of the atoms increases. That the electrons are more tightly bound than in the alkali metals, explains why the electropositive character of the alkaline earth metals is weaker, on the whole, than that of the alkali metals.

It may be seen from the ionization energies given in Table 46, which are derived from spectroscopic data, that to ionize off a third electron from the elements of the alkaline earth group—i.e., an electron from the inert gas core—would require the expenditure of an enormously greater amount of energy than is needed to remove both electrons from outside the inert gas core. This energy is far greater than could be compensated by the formation of a crystal lattice. The possibility of obtaining, by chemical means, any ionic compounds in which the metals of the alkaline earth group display a valence greater than 2 is thus absolutely excluded.

In order to remove 2 electrons it is necessary in the case of these elements to expend considerably more energy than for the ionization of one electron. However, the extra energy requirement is not so great that it cannot be met by the energy liberated when the ions come together to form a crystal lattice. As the example of CaF_2, given in Table 26 shows, the ionization energy is in fact considerably *over-compensated* by the lattice energy. As has already been discussed, this is really why the metals of the alkaline earth group do not in general behave as univalent, but are always bivalent in their stable compounds at ordinary temperature.

It is not impossible, theoretically, that circumstances may be different at high temperatures. The data of Table 26 show that CaF does not occur under ordinary conditions, simply because it is less stable than CaF_2 at low temperatures. Theory does not exclude the possibility that, with rising temperature, we may reach a range of temperature in which CaF is roughly as stable as CaF_2, and so can exist in equilibrium with the latter, provided that free calcium is present. Molecules such as CaF, CaCl, etc. have, in fact, been detected by band spectroscopy (Mecke, R. C. Johnson). They therefore can exist in the gaseous state (in equilibrium with the dihalides). It is probable, however, that their range of existence lies far above the melting points of the metals and the dihalides. This follows both from theoretical estimates of the way in which their free energy of formation changes with temperature, and from recent experiments (cf. p. 271). The possibility is also not to be excluded that alkaline earth monohalides (or suboxides) might be present in molten reaction mixtures. If this were so, it might be possible, by chilling such a reaction mixture, to hinder the decomposition, which would otherwise occur when the temperature was lowered. The monohalide or suboxide could then be obtained as a metastable compound at ordinary temperature.

Application of the Heitler-London theory to the metals of the alkaline earths leads, in the first place, to the conclusion that in their *normal state* the atoms are *unable* to form covalent compounds, since they contain no unpaired electrons. Since, however, only a small expenditure of energy is required to raise one of the two outer electrons to the next higher level ($l = 0 \rightarrow l = 1$) (62.2 kcal per g atom for Mg, 43.1 kcal for Ca), the theoretical possibility is still open that covalent compounds might be formed through a transition of the atom to the 'excited' state. It is, however, not at present possible to carry out exact calculations of the stability of covalent bonds formed by electrons with the azimuthal quantum number $l = 1$ (Bohr subsidiary quantum number $k = 2$). Experience shows that purely covalent compounds of metals of the alkaline earth group are not very stable—e.g., the compounds formed with organic radicals. The particular ability of magnesium to form compounds of the type MgXR (where X = halogen atom, R = organic radical) is very striking, however. Because of the great practical importance of these compounds, an explanation of the bonding mechanism in them, and of the energetics of their formation, would be of considerable interest to chemists.

(b) Occurrence

With the exceptions of beryllium and radium, the elements of the IInd Main Group occur in nature in very large amounts, in the form of their compounds. Calcium and magnesium are among the most widely distributed of the elements, calcium contributing 3.5 % and magnesium 2.5 % of the mass of the earth's crust. As accords with their great chemical reactivity, the alkaline earth elements *never occur native*, but always in the combined state.

The carbonates of calcium and magnesium, in the form of *limestone* and *chalk* ($CaCO_3$), and also of *dolomite* ($CaCO_3 \cdot MgCO_3$), make up whole mountain ranges. The simple carbonate of magnesium, *magnesite*, $MgCO_3$, also known as *bitter spar* or *talc spar*, also occurs in rich deposits in many places, although not in such vast quantities as the double carbonate. It is found especially in Styria (Austria), on Euboea, in Canada, California, and Chinese Manchuria.

Marble is another mineral found in many places in large amounts, and consists of practically pure calcium carbonate. The carbonates of the heavier alkaline earth metals—*strontianite*, $SrCO_3$, and *witherite*, $BaCO_3$—are not so widely distributed.

Gypsum, $CaSO_4 \cdot 2H_2O$, also known as *selenite*, forms huge deposits. *Alabaster* is a variety of gypsum. Anhydrous calcium sulfate, *anhydrite*, $CaSO_4$, is found, together with *kieserite* $MgSO_4 \cdot H_2O$, as an almost invariable accompaniment of rock salt: it is also commonly found, however, in stratified deposits. Mention has already been made of the occurrence of anhydrite together with various magnesium double salts, in the potassium salt deposits (pp. 187, 189). The sulfates of strontium and barium form *celestine*, $SrSO_4$ and *barytes* or *heavy spar*, $BaSO_4$: the latter, in particular, occurs widely.

Of the *silicates* of magnesium, *olivine*, $(Mg,Fe)_2[SiO_4]$, *enstatite* $Mg_2[Si_2O_6]$ and the hydrous silicates *serpentine*, *asbestos*, *talc* and *meerschaum* may be mentioned. The *double silicates* of magnesium, and especially of calcium, are extremely numerous (cf. Table 49). Among other calcium compounds, *phosphorite* (cf. p. 624), *apatite*, $3Ca_3(PO_4)_2 \cdot Ca(F,Cl)_2$ and *fluorite* or *fluorspar*, CaF_2, may be named as minerals of special importance. *Spinel*, $MgO \cdot Al_2O_3$, should also be mentioned as a magnesium mineral which is fairly widely distributed, although in small amounts, various varieties of it being prized as gem stones.

As products of the weathering of minerals, calcium and magnesium compounds are always found in soils, and also in most natural waters, the 'hardness' of which is due to their content of these salts.

TABLE 49

MINERALS OF THE ALKALINE EARTHS

CHLORIDES	NITRATES and IODATES	OXIDES and HYDROXIDES
Bischofite $MgCl_2 \cdot 6H_2O$	Magnesia saltpeter (Nitromag-	Brommelite BeO
Carnallite $KCl \cdot MgCl_2 \cdot 6H_2O$	nesite) $Mg(NO_3)_2 \cdot H_2O$	Periklase MgO
Tachhydrite	Lime saltpeter (nitrocalcite)	Brucite $Mg(OH)_2$
$CaCl_2 \cdot 2MgCl_2 \cdot 12H_2O$	$Ca(NO_3)_2 \cdot H_2O$	
	Baryta saltpeter $Ba(NO_3)_2$	
	Lautarite $Ca(IO_3)_2$	

TABLE 49 *(continued)*

SULFATES

ieserite $MgSO_4 \cdot H_2O$

itter salt (Epsomite; Epsom salts) $MgSO_4 \cdot 7H_2O$

nhydrite $CaSO_4$

ypsum (Selenite) $CaSO_4 \cdot 2H_2O$

elestine $SrSO_4$

eavy spar (barytes) $BaSO_4$

Kainite $KCl \cdot MgSO_4 \cdot 3H_2O$

Schönite $K_2SO_4 \cdot MgSO_4 \cdot 6H_2O$

Langbeinite $K_2SO_4 \cdot 2MgSO_4$

Leonite $K_2SO_4 \cdot MgSO_4 \cdot 4H_2O$

Astrakanite (Blödite) $Na_2SO_4 \cdot MgSO_4 \cdot 4H_2O$

Loeweite $Na_2SO_4 \cdot MgSO_4 \cdot 2H_2O$

Polyhalite $K_2SO_4 \cdot MgSO_4 \cdot 2CaSO_4 \cdot 2H_2O$

Syngenite $K_2SO_4 \cdot CaSO_4 \cdot H_2O$

Glauberite $Na_2SO_4 \cdot CaSO_4$

Fauserite $(Mg,Mn)SO_4 \cdot 7H_2O$

Magnesium alum (Bosjesmanite) $Mg[Al(SO_4)_2]_2 \cdot 24H_2O$

Quetenite $MgO \cdot Fe_2O_3 \cdot 3SO_3 \cdot 12H_2O$

Botryogen $2MgO \cdot Fe_2O_3 \cdot 4SO_3 \cdot 15H_2O$

MOLYBDATES and TUNGSTATES

Belonosite $MgMoO_4$ (very rare)

Pavellite $CaMoO_4$ (very rare)

Scheelite $CaWO_4$

CARBONATES

Magnesite (Talc spar, bitter spar, giobertite) $MgCO_3$

Dolomite (Pearl spar, brown spar) $MgCO_3 \cdot CaCO_3$

Calcite (Iceland spar) ⎫
ragonite   ⎬ $CaCO_3$

trontianite $SrCO_3$

lstonite $(Ca,Ba)CO_3$

Witherite $BaCO_3$

arnowitzite $(Ca,Pb)CO_3$

Hydromagnesite $Mg(OH)_2 \cdot 3MgCO_3 \cdot 3H_2O$

Natrocalcite $Na_2CO_3 \cdot CaCO_2 \cdot 5H_2O$

Pirssonite $Na_2CO_3 \cdot CaCO_3 \cdot 2H_2O$

Uranothallite $UO_2(CO_3) \cdot 2CaCO_3 \cdot 10H_2O$

PHOSPHATES, ARSENATES, etc.

truvite $NH_4MgPO_4 \cdot 6H_2O$

Wagnerite $Mg_3(PO_4)_2 \cdot MgF_2$

apatite $3Ca_3(PO_4)_2 \cdot Ca(F,Cl)_2$

vabite $3Ca_3(PO_4)_2 \cdot Ca(F,Cl,OH)_2$

Phosphorite (see p. 624)

Herderite $CaBe(F,OH)PO_4$ (rare)

wanbergite – a mixed phosphate and sulphate of calcium and aluminum

obierite $Mg_3(PO_4)_2 \cdot 8H_2O$

Monite $Ca_3(PO_4)_2 \cdot H_2O$

Monetite $CaHPO_4$

rushite $CaHPO_4 \cdot 1\frac{1}{2}H_2O$

Newbergite $MgHPO_4 \cdot 3H_2O$

soclase $Ca_2PO_4(OH) \cdot 2H_2O$

irrolite $3CaO \cdot Al_2O_3 \cdot P_2O_5 \cdot 1\frac{1}{2}H_2O$

avistockite $3CaO \cdot Al_2O_3 \cdot P_2O_5 \cdot 3H_2O$

oyazite $3CaO \cdot Al_2O_3 \cdot P_2O_5 \cdot 9H_2O$

Berzelite $(Mg,Ca)_3(AsO_4)_2$

Atopite $Ca_2Sb_2O_7$

Lewisite $5CaO \cdot 3Sb_2O_5 \cdot 3TiO_2$

Hörnesite $Mg_3(AsO_4)_2 \cdot 8H_2O$

Haidingerite $CaHAsO_4 \cdot H_2O$

Pharmakolite $CaHAsO_4 \cdot 2H_2O$

Wapplerite $(Ca,Mg)HAsO_4 \cdot 3\frac{1}{2}H_2O$

Rösslerite $MgHAsO_4 \cdot \frac{1}{2}H_2O$

Microlite, essentially $Ca_2Ta_2O_7$

Koppite, essentially $Ca_2Nb_2O_7$

Samarskite, essentially a niobate and tantalate of Ca, Fe, Y and Er (containing uranium)

Hjelmite, essentially Ca-, Fe-, Mn-tantalate

Calcium phosphates containing Fe and Mn are messellite, tamanite, (anapaite), reddingite, fillowite, dickinsonite, and fairfieldite.

Calcouranite is a calcium uranyl double phosphate, barium uranite (uranocirite) a barium uranyl double phosphate

Arseniosiderite $6CaO \cdot 3As_2O_3 \cdot 4Fe_2O_3 \cdot 9H_2O$

TABLE 49 *(continued)*

BORATES

Pinnoite $Mg(BO_2)_2 \cdot 3H_2O$
Ascharite $3Mg_2B_2O_5 \cdot 2H_2O$
Boromagnesite $Mg_5B_4O_{11} \cdot 2\frac{1}{2}H_2O$
Pandermite, Priceite $Ca_2B_6O_{11} \cdot 3H_2O$
Hydroboracite $CaMgB_6O_{11} \cdot 6H_2O$
Colemanite $Ca_2B_6O_{11} \cdot 5H_2O$
Borocalcite, Bechilite $CaB_4O_7 \cdot 4H_2O$
Franklandite $Na_2CaB_6O_{11} \cdot 7H_2O$
Boronatrocalcite (Ulexite, Tinkalcite) $NaCaB_5O_9 \cdot 6H_2O$
Sussexite $(Mg,Mn)_2B_2O_5 \cdot H_2O$

Sulfoborite $Mg_2B_2O_5 \cdot MgSO_4 \cdot 4\frac{1}{2}H_2O$
Lüneburgite $Mg(BO_2)_2 \cdot 2MgHPO_4 \cdot 7H_2O$
Ludwigite $3MgO \cdot Fe_3O_4 \cdot B_2O_3$
Boracite $2Mg_3B_8O_{15} \cdot MgCl_2$

SPINELS

Spinel $MgO \cdot Al_2O_3$
Magnesioferrite $MgO \cdot Fe_2O_3$
Iron spinel (Pleonast, Ceylonite, black spinel) $(Mg,Fe)O \cdot (Al,Fe)_2O_3$
Chrome spinel (Picotite) $(Mg,Fe)O \cdot (Al,Cr,Fe)_2O_3$

SILICATES

Olivine, Peridote
  $(Mg,Fe)_2[SiO_4]$
Enstatite $Mg_2[Si_2O_6]$
Bronzite $(Mg,Fe)_2[Si_2O_6]$
Hypersthene $MgFe[Si_2O_6]$
Chrysotile (fibrous serpentine)
  $Mg_3[Si_4O_{11}] \cdot 3Mg(OH)_2 \cdot H_2O$
Antigorite (leafy serpentine)
  $Mg_3[Si_4O_{10}](OH)_2 \cdot 3Mg(OH)_2$
Meerschaum
  $Mg_3[Si_6O_{15}] \cdot Mg(OH)_2 \cdot 3H_2O$
Talc, Steatite
  $Mg_2[Si_4O_{10}] \cdot Mg(OH)_2$
Asbestos p. 254
Strahlstein
  $Ca_2(Mg,Fe)_5[Si_8O_{22}](OH)_2$
Cordierite $Mg_2Al_3[AlSi_5O_{18}]$
Anorthite (lime felspar)
  $Ca[Al_2Si_2O_8]$
Humite $3Mg_2[SiO_4] \cdot Mg(OH,F)_2$

Lime soda felspars, many zeolites, micas, chlorites, and similar minerals also belong to this group.

Wollastonite $Ca_2[Si_2O_6]$
Diopside $CaMg[Si_2O_6]$
Hedenbergite
  $CaFe[Si_2O_6]Ca_2Al_3[SiO_4]_3[OH]$
  rhomb: Zoisite
  monocl: Clinozoisite
Epidote
  $Ca_2(Al,Fe)_3[SiO_4]_3[OH]$
Monticellite $CaMg[SiO_4]$

Datolite $Ca(OH)BSiO_4$
Homolite $FeCa_2B_2Si_2O_{10}$
Danburite $CaB_2Si_2O_8$
Axinite $Ca_3Al_2BSi_4O_{15}(OH)$
Barylite $Ba_4Al_4Si_7O_{24}$
Tourmaline p. 323
Beryl $Be_3Al_2[Si_6O_{18}]$
Euclase $BeAlSiO_4(OH)$
Phenacite $Be_2[SiO_4]$
Leucophane
  $(Ca,Na)_2[BeSi_2(O,F)_7]$
Helvine
  $(Fe,Mn)_4[Be_3Si_3O_{12}]S$

GARNETS

Vesuvianite $Ca_{10}Mg_2Al_4[SiO_4]_5[Si_2O_7]_2(OH)$
Pyrope $Mg_3Al_2Si_3O_{12}$
Grossular $Ca_3Al_2Si_3O_{12}$
Uvarovite $Ca_3Cr_2Si_3O_{12}$
Andradite $Ca_3Fe_2Si_3O_{12}$

Calcium and magnesium are also almost always encountered in vegetable and animal material. Magnesium forms a necessary constituent of the *chlorophyll* of the green leaf. Calcium forms hydroxy-apatite, the hard substance of the bones and teeth. Egg shells, mussel and oyster shells, and corals are built up from calcium carbonate.

Sea water contains appreciable amounts of magnesium and calcium salts. On the average, about 0.30 % $MgCl_2$, 0.04 % $MgBr_2$, 0.18 % $MgSO_4$ and about 0.16 % $CaSO_4$, in addition to about 2.9 % alkali metal chlorides. Many mineral waters, namely the 'bitter springs', contain considerable quantities of magnesium in the form of the sulfate.

Beryllium occurs in nature in a few minerals which are not very widely distributed. *Beryl*, $Be_3Al_2[Si_6O_{18}]$, is the most abundant, fairly large deposits being known in North America, Brazil, Norway, Spain, India, and in the Urals. In addition, there are *euclase*, $BeAlSiO_4(OH)$ (very rare), *gadolinite* $Be_2Y_2FeSi_2O_{10}$, *chrysoberyl* $BeO \cdot Al_2O_3$ or $Al_2[BeO_4]$ (which is not isomorphous with spinel, but crystallizes in the rhombic system), and the rare *phenacite* $Be_2[SiO_4]$.

Other rare minerals which contain beryllium, although only in subordinate amounts, are *helvine* $(Fe,Mn)_4[Be_3Si_3O_{12}]S$, *danalite* $(Fe,Zn)_4[Be_3Si_3O_{12}]S$, *leucophane* $(Ca,Na)_2[BeSi_2(O,F)_7]$, *melinophane* $(Ca,Na)_2[Be(Si,Al)_2(O,F)_7]$, *trimerite* $MnBeSiO_4$, and *bertrandite* $Be_8[Si_2O_6][SiO_4]_2(OH)_4$.

Certain beryllium minerals are frequently found in very beautiful pieces, which are used for gems. Thus *emerald* and *aquamarine* are varieties of beryl, and *alexandrite* is a variety of chrysoberyl. Euclase and phenacite are also used for gems.

Radium is contained in extraordinarily small quantities in the ores of uranium, from which it is formed by radioactive disintegration. The ratio of radium to uranium is almost constant, amounting to 1 : 3,000,000. The principal uranium mineral, *pitchblende*, contains an average of 0.14 g of radium per ton (1000 kg). The radium content of other uranium ores, such as autunite, carnotite, etc. is even less.

Carnotite is found chiefly in Colorado and Utah in the United States. The principal source of pitchblende was formerly St. Joachimsthal (in the Bohemian Erzgebirge). Much larger deposits were later found at Katanga (Belgian Congo), so that Belgium was for long practically the only producer of radium compounds, the Belgian output (several grams per month) being adequate for the needs of the world. Other important deposits of uranium minerals were opened up about 1938 at Great Bear Lake, in the far North-West of Canada.

(c) History

Quicklime, CaO, obtained by burning limestone or marble, was slaked and used for the preparation of builder's mortar even in very ancient times. Gypsum was also used as a mortar. Dioscorides, who lived in Asia Minor in the first century of our present era, used for calcium oxide the term 'unslaked lime' still current in building practice. The caustic nature of quick lime was also known in antiquity. It later became customary to call calcium oxide lime-earth, and indeed to designate all metallic oxides as 'earths'. Thus magnesium oxide, which became known at the beginning of the 18th century, received the name of 'bitter earth'. By the end of the 17th century, magnesium sulfate (bitter salts) was used as a medicine, first of all in England, where it was obtained from the waters of the mineral springs of Epsom. It therefore received the old name of sal anglicum or Epsom salts. Lime earth and bitter earth were first clearly distinguished from one another by Joseph Black, who demonstrated the different solubilities of the two oxides and the sulfates derived from them.

Heavy spar, $BaSO_4$, first became known through the discovery, made by a shoemaker of Bologna (1602), that it acquired the property of phosphorescing when it was ignited with organic matter ('Bolognese phosphorus'). Scheele, in 1774, discovered that barium oxide was a new 'earth', differing from the various earths hitherto known, but did not immediately perceive that it was the earth giving rise to heavy spar. This was first recognized by Gahn, and was then confirmed by Scheele. The oxide then received the name *baryta* (from βαρύς, heavy). Barium carbonate was discovered in nature by Withering in 1782, and was later named *witherite* after him. Soon afterwards a similar mineral was found at Strontian, in Scotland, and called strontianite. Klaproth, in 1793, showed that this was derived from a new 'earth'. Beryllium oxide was discovered soon afterwards, in 1797, by Vauquelin; in the course of an analysis of beryl and emerald he encountered an earth which was in many respects similar to alumina, but which—unlike the latter—was unable to form alums. It received the name of beryl-earth. The metal from which it is derived, beryllium, has been known in French literature as glucinum, a reference to the sweet taste of beryllium salts.

Radium was discovered in 1898 by Prof. P. and Mme. M. Curie, in conjunction with G. Bémont. The discovery followed from the observation that whereas, in general, the radioactivity of uranium compounds is proportional to their uranium content, there are some uranium minerals which have a higher activity than corresponds with their uranium content. This led Mme Curie to the hypothesis that the abnormally high radioactivity of these minerals must be attributed to the fact that they contained one or more other radioactive elements, in addition to uranium. She then succeeded, together with her husband, in discovering in pitchblende first the new element *polonium* (cf. pp. 753 *et seq.*) and soon afterwards a second new element, *radium*.

The free alkaline earth metals were isolated for the first time by Davy, in 1808, from the amalgams obtained when the slightly moistened hydroxides were electrolyzed, with mercury serving as the cathode. The amalgams were first obtained by Berzelius, following the discovery of the alkali metals by Davy. The first preparation of metallic radium, by M. Curie and A. Debierne (1910) was also carried out by way of the amalgam. Metallic beryllium was first prepared by Wöhler (1828), by reduction of beryllium chloride with metallic potassium, but is was obtained for the first time in the pure state by Lebeau (1898), by electrolysis of sodium beryllium fluoride.

The preparation of metallic magnesium was also first effected by electrolysis (1808, by Davy). It was first obtained by a chemical process (action of potassium vapor on anhydrous magnesium chloride) by Bussy (1928). Its preparation by electrolysis of molten salts dates back to Bunsen (1852). The technical use of alloys with a high percentage of magnesium was introduced by Pistor (1908).

2. Preparation of the Metals

Magnesium is the most important of the elements of the IInd Main Group, in the metallic state. It is usually prepared by the electrolysis of fused, pure, anhydrous carnallite, $KCl \cdot MgCl_2$, or similar salt mixtures, at temperatures above the melting point of magnesium. Acheson graphite is employed for the anode, and iron as the cathode. Molten magnesium rises to the surface, and can be removed by means of perforated ladles.

Since only the $MgCl_2$ is decomposed during the electrolysis, this is the only constituent which needs to be replenished. It can be obtained from magnesite or burned dolomite by the reaction

$$MgO + CO + Cl_2 = MgCl_2 + CO_2$$

which utilizes the chlorine liberated at the anode.

Reference has been made to the presence of magnesium salts in seawater. Although the concentration of magnesium salts is low, the oceans represent an enormous reserve of readily accessible raw material. Various processes have therefore been developed for the isolation of pure anhydrous magnesium chloride from sea water, although not all have been economically statisfactory. As operated at the very large plant at Velasco River, Texas, on the Gulf of Mexico, the first stage consists in precipitating magnesium from the sea water as its hydroxide, by means of the very pure lime made by calcining oyster shells: $MgCl_2 + Ca(OH)_2 = = Mg(OH)_2 + CaCl_2$. The supernatant liquor from the precipitation stage is pumped back to the sea at a point remote from the intake, in order to ensure that the magnesium content of the feed solution does not become lowered through contamination with effluent solution. The magnesium hydroxide is converted to magnesium chloride, which crystallizes as $MgCl_2 \cdot 6H_2O$. This can be dehydrated as far as the composition $MgCl_2 \cdot 1.5H_2O$ without undergoing hydrolysis, and the incompletely dehydrated chloride is added directly to the fused chloride electrolytic bath. In order to obtain an output of approximately 100 tons of metallic magnesium per day from the plant, it is necessary to treat approximately 3 million cubic feet of sea water each day.

In recent years the so called direct *electrothermal* processes have become increasingly important for the production of magnesium. The most important of these are (a) reduction of MgO with calcium carbide or carbon (*carbothermal process*), and (b) reduction of MgO with silicon (*silicothermal process*).

In the carbothermal process, MgO is reduced by carbon in the electric arc furnace. It was shown by Slade (1909) that the reaction $MgO + C \rightleftharpoons Mg + CO$ can proceed from left to right only above 2000°. If the reaction is not to be reversed during cooling and condensation, with reformation of MgO, the escaping mixture of Mg vapor and CO must be cooled with extreme rapidity. 'Shock cooling' was originally effected by injecting hydrogen (F. Hansgirg, 1931). In the United States, where the carbothermal process is chiefly employed, natural gas is now used for quenching the reaction mixture.

Reduction of MgO is more easily carried out by the use of other reducing agents, with a higher heat of oxidation than that of carbon. In this way it is possible to effect reduction at a much lower temperature, whereby the back reaction can be partially or completely obviated. CaC_2 may be used for this purpose (Matignon and Thierry, 1915). Reduction is carried out in externally heated, evacuated retorts, and follows the equation $MgO + CaC_2 = Mg + CaO + 2C$. This process—used principally in England (by Murex) and in France—is economically practicable only where carbide is cheap.

The silicothermal (or Pidgeon) process depends in principle on the reversal of the exothermic reaction $SiO_2 + 2Mg = Si + 2MgO$ at high temperatures. This is carried out, not with pure magnesium oxide, but with calcined dolomite, which is heated with silicon (usually as 75 % ferrosilicon) in cast steel retorts or electrically heated high vacuum furnaces to 1200–1300°. The equilibrium pressure of magnesium vapor in the system $2MgO + 2CaO + Si(Fe) = Ca_2SiO_4 + 2Mg$ would be about 18 mm at 1150°. The reaction therefore goes to completion, and the magnesium can be condensed (98–99 % pure) in air cooled receivers.

The world production of magnesium in 1948 amounted to about 20,000 tons (9,100 tons in the United States, 3,500 tons in Britain), but during the war of 1939–45, the consumption of light alloys (see below) was such that the annual production of magnesium metal exceeded 200,000 tons.

Calcium is prepared technically by electrolysis of molten calcium chloride, mixed with calcium fluoride or potassium chloride. It has been found advantageous to arrange the cathode so that its surface just touches the melt, and then to raise it gradually as the calcium is deposited upon it. The temperature of the melt is not raised higher than is necessary to maintain it in the liquid state; if it is heated above the melting point of the calcium, considerable amounts of the metal diffuse into the melt, and are lost by re-oxidation. On the other hand, if the metal is deposited below its melting point, it separates in a spongy, very impure form. The arrangement of the cathode indicated, however, leads to the deposition of the calcium on the cathode in the molten state, because of the strong local heating brought about by the high current density there, without any significant amounts of calcium diffusing into the colder melt. As the electrode is raised, the calcium solidifies, so that a continually growing rod of metal is formed, and this remains covered by a crust of the melt, which protects it from oxidation.

Very pure calcium can be prepared on a technical scale by heating pure calcium chloride with metallic aluminum. The volatile aluminum chloride thereby distils out of the system (Hackspill, 1951). A very high degree of purity can be achieved by redistillation of calcium in vacuum.

Although strontium can be obtained in a manner similar to calcium, the electrolytic preparation of barium presents difficulties. According to Guntz (1906) and Matignon (1913), however, the reduction of barium oxide (in vacuum) with aluminum or silicon, at 1200°, is very suitable for the isolation of this metal:

$$3BaO + 2Al = Al_2O_3 + 3Ba; \qquad 3BaO + Si = BaSiO_3 + 2Ba$$

On a very small scale, barium is most readily obtained by the thermal decomposition of barium azide, $Ba(N_3)_2$:

$$Ba(N_3)_2 = 3N_2 + Ba \qquad\qquad (cf. \ p.\ 605).$$

Metallic beryllium, according to Fichter, can best be obtained, with the use of ordinary laboratory resources, by electrolysis of a fused mixture of BeF_2 and NaF in the molar ratio 2 : 1. The melt is contained in a nickel crucible, which serves also as the cathode, and a stout carbon rod, dipping into the melt, is used for the anode. The temperature must not be too high, or the beryllium alloys with the nickel. The technical preparation of beryllium, which was worked out by A. Stock and H. Goldschmidt (1921 and 1925), is based upon the electrolysis of a mixture of $NaBeF_3$ and $Ba(BeF_3)_2$ or some other suitable mixture, with a composition so chosen as to combine the highest possible conductivity with the lowest possible volatility, at temperatures above the melting point of beryllium. A cooled contact cathode is employed, as in the electrodeposition of calcium, but at a far higher temperature (about 1400°).

a Beryllium of the highest purity can be obtained by vacuum distillation, as described by Sloman and Kroll (1932–34–35). The crude metal is heated in a BeO crucible, by means of high frequency induction furnace.

3. Properties

The elements of Group IIA are grey or white metals. A freshly cut or broken surface is lustrous, but tarnishes more or less rapidly in the air. Their hardness and tensile strength decrease from beryllium, through calcium, to barium. In so far as moderately accurate measurements are available, the data, together with figures for electrical conductivity, are collected in Table 50.

In Table 50 is given the hardness determined by scratching (measured on Mohs' scale). The tensile strength gives the weight (in kg per mm² cross section) which a wire or rod can just support without breaking.

TABLE 50

HARDNESS, TENSILE STRENGTH AND ELECTRICAL CONDUCTIVITY OF METALS OF GROUP IIA

Metals	Beryllium	Magnesium	Calcium
Scratch hardness	Between 6 and 7	2.6	2.2–2.5
Tensile strength in kg per sq. mm	—	20	5
Electrical conductivity in ohms^{-1} per cm^3	$5.41 \cdot 10^4$ at 20°	$23.2 \cdot 10^4$ at 0°	$21.8 \cdot 10^4$ at 20°

The boiling points of the alkaline earth metals, given in the summary table at the beginning of this chapter, show a quite irregular course, in contrast with those of the alkali metals, which decrease regularly from the lightest element to the heaviest. The same applies to the melting points of the elements, again in contrast with those of the alkali metals. The irregular order of the melting points is probably connected with the fact that there is a change in crystal structure in passing from magnesium to calcium, and again from strontium to barium, (see below).

Magnesium is a silver white metal, which soon becomes a dull white in the air. It is of moderate hardness, and fairly ductile, so that it can be rolled into thin sheet and drawn into wire. These properties are strongly affected, however, even by traces of impurities. The electrical conductivity is about $\frac{9}{25}$ that of copper or $1\frac{2}{19}$ that of aluminum: the thermal conductivity is about $\frac{5}{13}$ that of copper or $\frac{7}{9}$ that of aluminum. Cold water acts only slowly upon magnesium, but the evolution of hydrogen is made quite apparent by the rising gas bubbles, especially if magnesium shavings are used, freed from grease by washing them with ether. The reaction occurs much faster with boiling water.

Magnesium also reacts with alcohol when heated. If it has been previously etched with iodine, it reacts with alcohol almost as quickly as with water. Magnesium reacts with dilute acids with strong evolution of hydrogen. The *amalgam* reacts very vigorously with water, even at ordinary temperature. Magnesium in the form of ribbon or powder burns with a dazzling white light when ignited in air, forming a dense white smoke of the oxide; the light is rich in actinic, or photochemically efficient, rays and use is made of this in photography, for flashlight exposures. Magnesium ignites spontaneously in moist chlorine, and burns with a vigorous flame. It has a strong affinity for other non-metals also. Thus, when heated, it combines readily with nitrogen, forming the nitride Mg_3N_2. This is also formed in appreciable amounts, in addition to the oxide, when the metal burns in an insufficient supply of air. Magnesium is able to abstract the more electronegative elements from many other compounds—e.g., it reacts with explosive violence when it is heated with many alkali oxides or hydroxides. It forms alloys with numerous metals, but only a few of these alloys have any technical significance, since they are mostly brittle and too easily oxidized. It was discovered by Grignard that magnesium reacts with organic iodine compounds in ethereal solution, forming alkyl magnesium iodides—e.g.,

$$Mg + CH_3I = Mg{<}^{I}_{CH_3} \qquad \text{(methyl magnesium iodide).}$$

Magnesium reacts in an analogous way with many organic bromine and chlorine compounds, especially those of the aliphatic series. In these cases it is of advantage to have some iodine present also, or to 'activate' the magnesium by prior treatment with iodine. These alkyl magnesium halides play an important part in preparative organic chemistry (Grignard reactions). [5]

The true alkaline earth metals, calcium, strontium and barium, are much more closely related in properties to the alkali metals than to magnesium. They are considerably softer than the latter, though their melting points are higher. They do not corrode in the air so quickly as do the alkali metals, though, like these they are best stored under petroleum. They burn readily—thus barium sometimes ignites when it is merely crushed up in the air. In addition to oxide, nitride is present in the products of combustion, and is also formed slowly even at ordinary temperature. The formation of the nitrides can readily be carried to completion by heating the metals to a red heat in a stream of nitrogen.* Vigorous reactions occur with other non-metals also. The alkaline earth metals abstract oxygen from metallic oxides. Calcium reacts sluggishly with water at ordinary temperature, strontium more vigorously, and barium quite energetically; they also react with alcohol, liberating hydrogen. They are dissolved by dilute mineral acids, with rapid evolution of hydrogen—e.g., $Ca + 2H^+ = Ca^{++} + H_2$.

The alkaline earth metals dissolve in liquid ammonia, forming deep blue-black solutions, and coppery or gold-like lustrous residues are obtained when the ammonia is evaporated. These are *ammoniates*, addition compounds of ammonia with the metals concerned, having a definite composition—namely $[Ca(NH_3)_6]$, $[Sr(NH_3)_6]$ and $[Ba(NH_3)_6]$. Biltz (1930) showed that only these hexammines exist; there are no other ammoniates of these metals.

In the presence of catalysts, the ammoniates gradually decompose, forming the amides— e.g.,

$$[Ca(NH_3)_6] = Ca(NH_2)_2 + 4NH_3 + H_2$$

If the amides are heated under reduced pressure, they may be transformed into the yellow *imides*, which crystallize with face-centered cubic structures having the cube edge-lengths a and densities d

	CaNH	SrNH	BaNH
a	5.16 Å	5.45 Å	5.84 Å
d	2.66	4.18	5.07

By further heating of the imides in a high vacuum, *pernitrides* can be obtained—e.g., Sr_3N_4, strontium pernitride, a reddish-brown crystalline powder which is decomposed by dilute acids, with the evolution of nitrogen:

$$Sr_3N_4 + 8HCl = 3SrCl_2 + 2NH_4Cl + N_2 \quad \text{(Hartmann, Fröhlich and Ebert, 1934)}.$$

When heated in ammonia, the alkaline earth metals react to form nitride and hydride. Their affinity for hydrogen is so great that calcium inflames when it is heated in hydrogen, forming the hydride—$Ca + H_2 = CaH_2$. The alkaline earth metals also form compounds with many other metals. A great number of such

---

* In the case of Ca, the formation of nitride proceeds most rapidly around 450°, since the allotropic change occurring at this temperature (cf. p. 249) causes a 'loosening' of the crystal lattice. Sr and Ba do not show this phenomenon. Small traces of sodium in the surface have an activating effect, since they hinder the formation of a coherent skin of nitride.

intermetallic compounds is known, at least in the case of calcium, which has been most thoroughly investigated from this standpoint.

Metallic *radium* is very similar to barium, but considerably more volatile, and extremely unstable in air.

Metallic *beryllium* is steel-grey in color, and hard enough to scratch glass. It is brittle at ordinary temperature, so that it is broken by hammering: it is said to be ductile at red heat. Its electrical conductivity is about $\frac{1}{12}$ that of copper. It remains bright in dry air. In contact with water, it becomes covered with a thin film of oxide, and is not further attacked, but it dissolves vigorously in dilute acids. It is not noticeably attacked by cold concentrated nitric acid, and the reaction with dilute (2 normal) nitric acid also soon comes to a standstill in the cold. It is rapidly dissolved by these acids when heated, however. Beryllium differs sharply from the other metals of Group IIA by its solubility in aqueous alkalis. Although dilute caustic potash dissolves it only when heated, beryllium reacts with 50 per cent caustic potash even at ordinary temperature.

4. Crystal Structure

Beryllium and magnesium, in the elementary state, crystallize with a hexagonal crystal lattice of the type shown in Fig. 57. The unit cell of this structure is formed by the black circles, joined by strongly drawn lines. To make it clear how the hexagonal symmetry comes about, a few other atoms (circles filled with dots) are shown in the diagram, lying outside the unit cell marked. These are linked by thinner lines. The unit cell represented in Fig. 57 can be divided into two trigonal prisms, and the center of one of these two prisms, as well the eight corners of the unit cell, is occupied by an atom. In all, the unit cell contains 2 atoms, since each of the corners is common to 8 unit cells. The structure represents a *hexagonal closest packing* of spheres: if we suppose the atoms to be rigid spheres, and a great number of them to be so packed together that they fill the available space as completely as possible—i.e., so that as little interstitial space as possible remains between them—we arrive at the arrangement shown in Fig. 57. This represents a space filling of 74.05 %. There is, however, a second possible arrangement of spheres in which 74.05 % of the available space is also filled up. This is the structure shown in Fig. 46, p. 210, the *face-centered cubic structure*. Metallic calcium crystallizes with this structure (below 300°). These two lattice types, both representing the closest pack-

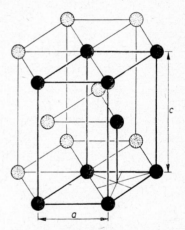

Fig. 57. Hexagonal closest pa-
cking of spheres.
Example: Mg ($a = 3.22$ Å,
$c = 5.23$ Å)

ing of spheres, are the ones most commonly met with among the elements, although it is true that there are often small deviations from hexagonal closest packing in the strictest sense, in that the ratio of the height c of the unit cell to the length a of the two equal edges of the base does not correspond exactly to the ideal value $c/a = 1.633$.

For beryllium, $a = 2.268$ Å, $c = 3.594$ Å, $c/a = 1.585$. Each Be atom has 12 neighbors, but not at exactly equal distances from each other; 6 in the basal plane, at the distance a, and three in each of the neighboring planes above and below at a distance 2.232 Å. For Mg, $a = 3.22$ Å, $c = 5.23$ Å, $c/a = 1.624$. Each Mg atom has 12 neighbors at practically the same distance, 3.22 Å.

For the face-centered cubic form of calcium (a-Ca), $a = 5.56$ Å. Each Ca atom has 12 neighbors at equal distances, $\sqrt{2}\,a/2 = 3.94$ Å.

At 300°, Ca undergoes an allotropic change, forming a crystal lattice of lower symmetry (β-Ca), and this in turn changes at 450° into a third form (γ-Ca) with a hexagonal lattice, like that of magnesium. Calcium containing some nitride crystallizes above 450° in the body-centered cubic form (like Ba), but this structure is not found for *pure* calcium.

Strontium, like calcium, crystallizes face-centered cubic, $a = 6.05$ Å, distance Sr—Sr = = 4.27 Å. Barium crystallizes cubic body-centered (Fig. 47, p. 210), $a = 5.01$ Å. This structure gives a space-filling of 68.02 %. Each Ba atom has 8 neighbors at equal distances,

$$\frac{a}{2}\sqrt{3} = 4.34 \text{ Å}.$$

Calcium and strontium form a continuous series of mixed crystals with each other. Ca and Ba, and Sr and Ba, are also able to form mixed crystals with an extensive range. In these systems, however, corresponding to the change in crystal type, there are gaps of miscibility, having a breadth of about 4–6 atoms % (W. Klemm, 1941). There is no formation of mixed crystals between Ca and Mg.

5. Flame Colorations and Spectra

When brought into the Bunsen flame in the form of volatile compounds,—e.g., the chlorides—calcium, strontium and barium give *characteristic flame colorations*. Calcium salts color the flame brick red, strontium salts carmine, and barium salts

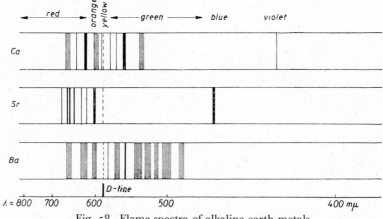

Fig. 58. Flame spectra of alkaline earth metals

yellow green. Radium salts also color the flame a carmine red. When viewed through a spectroscope, the lines or groups of lines sketched in Fig. 58 may be observed.

Unlike those of the alkali metals, the flame spectra of the alkaline earths are not due to the free atoms. Under high resolution it may be seen that many of the 'lines' of their flame spectra—which are indeed striking for their breadth and unsharp edges—are in reality dense groupings of many lines, so-called 'bands'. It has already been stated that band spectra are attributed to molecules. Quite different bands are obtained, in consequencef depending on whether the chloride, fluoride or oxide of an alkaline earth is investigated. I, a drop of the solution of an alkaline earth chloride is introduced into the flame on a platinum wire, one obtains in the first moment the chloride spectrum. This changes very soon into the oxide spectrum, and at the same time the lines of the free metal appear. If the platinum wire is moistened with hydrochloric acid, the chloride spectrum is obtained once again, and so on. In spite of their varying appearance, the alkaline earth flame spectra can be used for detecting the elements if those lines or bands which appear regularly, and with especial clarity, are used for identification. These are collected in Table 51; in so far as the wave lengths given refer not to sharp lines but to bands, they represent the positions of the centers of the bands.

TABLE 51

LINES OF THE ALKALINE EARTH FLAME SPECTRA
SUITABLE FOR ANALYTICAL DETECTION

Calcium		Strontium		Barium	
red	622.0 mμ	orange	605.0 mμ	green	524.2 mμ
green	553.5 mμ	blue	460.8 mμ	green	513.7 mμ

If the temperature is raised considerably higher—e.g., if one generates the arc spectra, as described on p. 169, one obtains exclusively the sharp lines which are to be ascribed to the free atoms. The lines of beryllium and magnesium are also obtained in this case; these produce no coloration when introduced into the ordinary Bunsen flame. The lines of the *arc spectra* suitable for the analytical detection of the elements concerned are shown in Table 52.

TABLE 52

CHARACTERISTIC LINES IN THE ARC SPECTRA OF THE GROUP IIA METALS

Beryllium	Magnesium	Calcium	Strontium	Barium	Radium
467.5 mμ	516.7 mμ	612.2 mμ	460.7 mμ	493.4 mμ	482.6 mμ
436.2 mμ	518.4 mμ	422.7 mμ	421.6 mμ	455.4 mμ	468.2 mμ

Table 52 contains only a very limited selection of the lines which may be observed in an intense arc if this is struck, not as described on p. 169, between a salt solution and an iridium wire electrode but, e.g., between carbon rods impregnated with salts of the elements. Under these conditions, for example, at least 35 lines of calcium may be observed in the visible region, all arising from the free atoms of the element. In addition there are numerous lines in the ultraviolet and infrared regions. Quite generally, the spectra of the elements of the second Main Group prove to be considerably more complex than the spectra of the alkali metals. Nevertheless, it has been possible in this case also to analyze them into *series*. In so doing it has been found that, whereas the series of the alkali metals consist entirely of double lines (doublets), which pass into single lines only when the differences in wave length between them become so small that they can no longer be resolved, three different sorts of series occur with the alkaline earth metals. These are series consisting of nothing but single lines (*singlet* series), those consisting entirely of *doublets*, and those consisting entirely of threefold lines (*triplet* series). Of these, the singlet and triplet series belong together, since they combine among themselves, but not with the doublet series.

The doublet series of the alkaline earth group show considerable analogies with the series of the alkali metals. They may be represented by exactly analogous series formulas, except that for the constant depending on the effective nuclear charge (analogous here to the expression $R_\infty Z^2$ for hydrogen and singly ionized helium) it is necessary to insert a value four times that for the alkali metals. The constant in question is proportional to the square of the 'effective nuclear charge' i.e., essentially to the square of the charge on the core of the atom. The value of the charge on the core of the atom is thus found to be 2 in this case. That implies, however, that the doublet series of the alkaline earth group originates from transitions of an electron revolving around a doubly charged atomic core. They are thus related not to the neutral atom, but to the singly charged atom, and are thus 'spark spectra' according to the definition given on p. 111. Paschen has, indeed, found that the doublet systems appear with relatively greater intensity through spark excitation than in the arc. That these doublet series are spark spectrum series—i.e., that their carriers are the *singly ionized atoms*—is further confirmed in that the ionization energies, for the removal of an electron, as calculated from the limits of the principal doublet series, do not correspond with the ionization potentials for the process $M = M^+ + e$, but with the considerably higher value for the process $M^+ = M^{++} + e$. Thus, for magnesium, the ionization potential corresponding to the formation of the doubly charged atom was determined by Foote, Meggers and Mohler, by the electron impact method, and was found to be 15 volts, in agreement with that calculated from the series limit, whereas the ionization potential for the formation of the singly ionized atom is 7.6 volts. The doublet systems of the alkaline earth elements thus furnish a confirmation of Sommerfeld's spectroscopic displacement law (cf. p. 111).

Fig. 59 shows how complete is the analogy between the arc spectra of the alkali metals and the spark spectra of the alkaline earth group. In this figure, the energy levels corre-

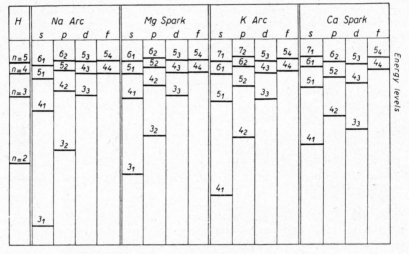

Fig. 59. Comparison of energy levels derived from arc spectra of the alkali metals and the spark spectra of the alkaline earth group

sponding to the individual spectral terms are represented by horizontal lines, as in Fig. 21. The energy content of the atom when the electron in question is completely removed is taken as the zero of energy (upper line). The step-by-step decrease in energy, as the electron approaches progressively closer to the nucleus, is measured downwards from this. Farthest on the left are the energy levels of the hydrogen atom, for the principal quantum numbers $n = 2, 3, 4$, etc.; the energy level for $n = 1$ lies deep, and outside the boundaries of the diagram. The quantum numbers n_k assigned to the individual energy levels are marked above each. At the top are given the designations of the corresponding orbits, as used in spectroscopy (s-orbit, p-orbits, etc.). The energy levels of the singly ionized alkaline earth atoms are represented on a scale one-fourth that used for the neutral atoms of the alkalis. Apart from the factor 4, attributable to the double charge of the atomic core (and eliminated in the diagram by a change of scale), the energy levels in the singly ionized alkaline earth atoms correspond throughout to those in the neutral alkali atoms. Closer inspection, however, reveals a typical difference in one place. Whereas for Mg^+, as for Na, the 3d level lies higher than the 3p level, it is *lower* than the 4p level in Ca^+, but *higher* than the 4p level in K. We shall see later that this difference, at first sight quite unimportant, is in fact of fundamental significance. The same effect, that manifests itself in calcium only in the displacement of a single spectral term, becomes of such profound influence on the chemical behavior, on progressing only one place further along the sequence of the elements, that it is responsible for the branching off of the Sub-groups of the Periodic System that begins with the IIIrd Group.

In contrast to the doublet series, the *singlet* and *triplet* series belong to the neutral atoms. This can be inferred from the much smaller numerical value of the effective nuclear charge in these series. For the Principal singlet series, this conclusion follows most directly, however, from the value of the series convergence limit, which corresponds to the energy of ionization of an electron from the neutral atom, and which has been confirmed in the case of magnesium, for example, by direct measurement of the ionization potential by the electron impact method (V_I, found 7.75 volts, calculated from the series limit 7.61 volts). The appearance of the 'spark lines' of the alkaline earth elements even in the arc—though less intensely than in the spark—is due to the fact that the singly ionized atoms of the metals of this group give up yet another electron relatively easily. As has already been discussed, it is because of this fact that these elements are almost exclusively bivalent in their chemical behavior.

6. Uses

The only element of Group IIA which finds application in large quantities as the pure metal is magnesium. Because of the strong emission of light when it is burned, it is extensively used in pyrotechnics and in photography. The 'flashlight powders' which are widely used consist of magnesium powder mixed with substances which readily give off oxygen when they are heated, such as potassium chlorate, potassium permanganate, manganese dioxide, or nitrates of the rare earth metals. Flash powders deflagrate instantaneously when they are ignited by a fuse of saltpeter paper, or by special firing arrangements. Alternatively, pure magnesium powder may be used, by blowing it through a flame ('flash light'). This use of magnesium depends on the fact that the light it emits when it burns is especially rich in just those rays towards which the photographic plate is most sensitive. In particular, of the energy set free in the combustion, the fraction which appears in the form of light is especially high in the case of magnesium, being estimated at about 50 times that of the ordinary gas light. It would need about 140 times as much stearin as of magnesium, by weight, to produce the same amount of light by burning stearin candles as by burning magnesium wire. The yield of light from magnesium is exceeded in the combustion of zirconium or tantalum. According to van Liempt (1934), the yield of light, in lumens per watt, when these metals are burned in pure oxygen is: for Zr 36, Ta 32, Mg 28, Al 26, Ce 9.3, and C 1.9.

Magnesium is used in preparative organic chemistry for Grignard reactions. It is used in metallurgy to free metals from dissolved oxides and sulfides, the magnesium abstracting the oxygen or sulfur and forming compounds which are insoluble in the melt.

Alloys of magnesium are used today on a very extensive scale. [*1–3*, *11*] They are characterized by particularly low densities, and are used especially in the aircraft and automobile industries, but also in other aspects of engineering and for optical instruments, etc.

Thus *magnalium* (a magnesium-aluminum alloy with 10–30 % Mg) has the density 2.5–2, and the density of *elektron metal* (alloys containing about 90 % or more magnesium, with small quantities of other metals such as Al, Zn, Cu, Mn, and for special purposes Si) is but little greater than that of magnesium itself (1.74). In spite of the strongly electropositive character of magnesium, these alloys are stable in air once they have covered themselves with a film of oxide. They cannot be used, however, in prolonged contact with water. More recently, however, it has been possible to produce magnesium alloys which are practically stable towards water (e.g., from 2 % Mn and 98 % Mg).

Unlike aluminum, elektron metal is not sensitive towards alkaline solutions or hydrofluoric acid. It can be protected against atmospheric corrosion if the surface is etched with a solution of alkali dichromate in nitric acid. It is very suitable for castings, and can also be rolled or forged at high temperatures. The compressive strength is about 30, the tensile strength 10–20 kg/mm², according to the composition of the alloy. By peening the castings, the tensile strength may be raised to 25–40 kg/mm². The use of elektron metal makes it possible to achieve a saving of weight of more than 80 % as compared with iron, 20–40 % over duralumin, and more than 40 % over wood.

Among the compounds of magnesium, the carbonate, oxide, and sulfate, in particular, find application. Further mention is made of these uses later, when the compounds are described. In addition many of the naturally occurring *silicates* are useful, especially *asbestos* which is an incombustible, soft, fibrous material. The name of asbestos is given to fibrous, felted varieties of serpentine (chrysotile) and hornblende. Dense dark-green or brownish serpentine is carved into ornaments, plates and vessels. Steatite, which stands close to serpentine in composition, is used for objects such as burner jets (e.g., for acetylene or gas burners), and for tailor's chalk. *Talc* is also used for the latter, but principally as a polishing powder, as a dressing, and for ointments. Another hydrous magnesium silicate, *meerschaum*, is used especially for smokers' pipes and similar utensils.

The compounds of calcium find very extensive uses. Limestone is [6,7] particularly important in the building industry. The quicklime manufactured from it, or its hydrate, calcium hydroxide, is the alkali most used in industry. Gypsum [8] is used for plaster work, moulds and plaster-casts. The chloride, fluoride and nitrate also have great technical importance. Calcium carbide and cyanamide will be discussed as carbon compounds in Chapter 12. The glasses, discussed in the same chapter, are also, for the most part, calcium compounds. Metallic calcium is used in the production of certain lead alloys—e.g., an alloy containing 0.7 % calcium, 0.6 % sodium, and 0.04 % lithium has been extensively used as a bearing metal.

Of the strontium compounds, the nitrate is chiefly important in pyrotechnics. The carbonate and oxide are also used in the sugar industry.

The oxide, hydroxide, peroxide and sulfate are technically the most important compounds of barium. The chloride is much used as a chemical reagent, and the nitrate and chlorate in pyrotechnics. The poisonous character of barium compounds must be borne in mind.

The most widely used compound of beryllium is the nitrate; it is employed chiefly in the gas mantle industry. The oxide is an excellent refractory, suitable for the manufacture of crucibles, etc. for use at high temperatures. Beryllium compounds are exceptionally toxic when inhaled as dust.

Metallic beryllium is used to make windows for X-ray tubes, the metal absorbing X-rays only $\frac{1}{17}$ as strongly as does aluminum. Heavy metals are greatly hardened by alloying them with beryllium—e.g., a copper-beryllium alloy with about 6.3 % beryllium is as hard as steel and is also extremely resistant both mechanically and chemically. Quite small additions of beryllium to cast copper are technically important, because of their deoxidizing action. The 'conductivity copper' made with the addition of 0.01–0.02 % beryllium is superior to the phosphor bronzes—copper-tin alloys to which phosphorus has been added as a deoxidant in the process of casting—because of its considerably better conductivity.

The uses of radium compounds arise solely from their radioactive properties. [*9*, *10*] They are employed chiefly for medical and scientific purposes, the chloride, bromide, carbonate and sulfate being most commonly used.

7. Compounds of the Elements of the Second Main Group

The composition of the compounds discussed here is summarized in the following table:

Hydrides MH_2	Salts:
Oxides MO, Hydroxides $M(OH)_2$	Halides MX_2 (X = F, Cl, Br or I)
Peroxides MO_2, Hyperoxides $M(O_2)_2$	
Nitrides M_3N_2	Nitrates $M(NO_3)_2$
Carbides or acetylides MC_2	Carbonates MCO_3
	Sulfates MSO_4
Metallic compounds – see Table 85 (p. 574),	Double salts. For numerous
Table 87 (p. 576) and Table 92 (p. 585)	examples, see Table 49

Compounds of magnesium and calcium are obtained as waste by-products in very large amounts in several branches of industry. They are otherwise prepared chiefly from the naturally occurring carbonates and sulfates. The chief starting material for the preparation of barium compounds is barytes, and for strontium compounds, celestine. Beryllium compounds are generally prepared technically from beryl.

In order to open up beryl, it is fused with potassium carbonate, and silica is precipitated by dissolution of the melt in dilute sulfuric acid and heating. Most of the aluminum is separated from the filtrate in the form of potassium alum, and the rest, together with the iron that is present as an impurity, is precipitated with ammonium carbonate. After acidifying the filtrate with hydrochloric acid, and driving off the carbon dioxide, the beryllium is precipitated as the hydroxide by means of ammonia. Several repetitions of the ammonium carbonate treatment are necessary in order to remove the admixed aluminum completely. Recrystallization of the basic acetate from chloroform can also be used to purify beryllium; according to Stock (1925), the distillation of the basic acetate is still more suitable.

The compounds of radium [*9*, *10*] will not be described separately in the following sections. So far as they are known, they are very similar to the compounds of barium.

(a) Hydrides

All the metals of Group IIA can form compounds with hydrogen; in these, they are bivalent, as in their normal compounds.* The stability of the hydrides increases from Be to Ca, and then falls off again from Ca to Ba. The hydrides of the alkaline earth metals proper are definitely salt-like, resembling the alkali hydrides. This is not equally true of the hydrides of Be and Mg, although their hydrolysis

---

* The existence of molecules such as BeH, MgH, CaH has been proved by their band spectra. These, however, exist only in the gaseous state, and in equilibrium with free atoms of the elements, and cannot be isolated in weighable amounts.

shows that in them the metal is the electropositive constituent, as compared with the hydrogen. All the hydrides MH_2 are colorless, non-volatile solids.

(i) *Beryllium hydride*, BeH_2, was obtained by Wiberg (1951) by the action of LiH on $BeCl_2$ in ether solution, and by Schlesinger (1951), by the reaction of $Be(CH_3)_2$ with $LiAlH_4$ or $Al(CH_3)_2H$:

$$BeCl_2 + 2LiH = BeH_2 + 2LiCl; \qquad 2Be(CH_3)_2 + LiAlH_4 = 2BeH_2 + LiAl(CH_3)_4$$

$$Be(CH_3)_2 + 2Al(CH_3)_2H = BeH_2 + 2Al(CH_3)_3.$$

Beryllium hydride is a white, non-volatile solid, which decomposes into its elements at 125°. It is insoluble in ether, and is decomposed by water or methanol, with evolution of hydrogen.

$$BeH_2 + 2HOH = Be(OH)_2 + 2H_2; \qquad BeH_2 + 2CH_3OH = Be(OCH_3)_2 + 2H_2.$$

(ii) *Magnesium hydride*, MgH_2, can be obtained by direct synthesis from the elements (by heating under high pressure in presence of MgI_2; Wiberg, 1951). It was first obtained by Wiberg (1950) by the thermal decomposition of magnesium diethyl in high vacuum: $Mg(C_2H_5)_2 = MgH_2 + 2C_2H_4$, (at 175°). It can also be prepared by the reaction of magnesium alkyls, in ether solution, with B_2H_6 or with $LiAlH_4$ (Schlesinger, 1951). It is a white, non-volatile solid, and is slightly soluble in ether. In compact pieces it is not spontaneously inflammable. It is stable in vacuum up to 300°. Water decomposes it vigorously: $MgH_2 + 2HOH = Mg(OH)_2 + 2H_2$, as also does methanol.

(iii) *Alkaline earth hydrides*. In appearance and properties, the hydrides of Ca, Sr and Ba closely resemble the alkali metal hydrides. According to Ephraim and Michel, their dissociation pressures become appreciable at higher temperatures than those of the hydrides of the typical alkali metals, but lower than for lithium hydride. Their stability decreases from calcium hydride to barium hydride. The metal resulting from the decomposition is held in solid solution by the hydride still remaining, and reduces the dissociation pressure of the latter considerably. This phenomenon does not occur with the alkali metal hydrides.

Zintl (1935) found that the metal ions in the crystal lattice of the alkaline earth hydrides are arranged approximately in hexagonal closest packing (Fig. 57). The H^- ions are probably inserted in the largest interstices between the metal ions, in such a way that each metal ion is surrounded by 7 H^- at slightly different distances. The densities are: for CaH_2 1.90; SrH_2 3.27; BaH_2 4.15.

(b) Oxides and Hydroxides

Although they are formed as the products of combustion of the metals, the normal oxides of the alkaline earth elements are most generally obtained by the thermal decomposition of the salts of oxyacids—e.g., the carbonates and nitrates. They form white, very infusible masses, having a loose texture if they are not formed at too high a temperature. They combine with water to form *hydroxides*: the heats of formation and solubilities of these may be seen from Table 53, which (except for $Ca(OH)_2$) gives the heats of formation, at constant pressure, of the stable crystalline hydroxides from the stable, crystalline oxides and liquid water. In some cases, the hydroxides also exist in a metastable crystalline form, having a lower heat of formation than the stable form. The degree of subdivision of the oxide also has a considerable influence on the heat of formation of the hydroxide.

TABLE 53

HEATS OF FORMATION OF HYDROXIDES FROM OXIDES; SOLUBILITY OF HYDROXIDES

Hydroxide	$Be(OH)_2$	$Mg(OH)_2$	$Ca(OH)_2$	$Sr(OH)_2$	$Ba(OH)_2$	
Heat of formation	2.7	7.3	15.2	14.7	17.3	kcal per mol
Solubility	$2 \cdot 10^{-5}$ at 20°	$2 \cdot 10^{-3}$ at 18°	0.1 at 20°	0.7 at 20°	3.4 at 20°	g of oxide in 100 g of solution

The hydroxides of the alkaline earth group provide a typical example for the rule that the basic character of the hydroxides increases with the atomic weight inside any one Main Group of the Periodic System. Beryllium hydroxide is amphoteric: it can function both as an acid and as a base. Magnesium hydroxide has only a weakly basic character; calcium and strontium hydroxides are moderately strong to strong bases, and barium hydroxide is a base which approaches the alkali hydroxides in strength.

The increase in basic character in the direction from $Mg(OH)_2$ to $Ba(OH)_2$ is reflected also in the conductivity coefficients f_μ. Hlasko (1935) gives for these the values, at 25°:

for	$Mg(OH)_2$	$Ca(OH)_2$	$Sr(OH)_2$	$Ba(OH)_2$
in 0.031 N solution f_μ =	—	0.703	0.737	0.831
0.00006 N solution f_μ =	0.882	0.962	0.967	0.979

The regular change in properties of the hydroxides can be understood in terms of the concepts developed by Kossel, whereby electrostatic attractions, following Coulomb's law, are primarily responsible for the cohesion of the oppositely charged atoms in the molecule or crystal. If, as was done for the alkali hydroxides, we calculate the energy necessary to split off on the one hand an OH^- ion, and on the other hand a H^+ ion from the hydroxides of the alkaline earth group, and plot the values graphically, as was done in Fig. 34, p. 179, we obtain two curves. These resemble the curves in Fig. 34, but are closer together, and are steeper, so that they intersect in the neighborhood of beryllium hydroxide. The difference between the energy of dissociation of the OH^- and the H^+ ions is no longer large in this case, so that it may be expected that beryllium hydroxide could dissociate not only as a *base*

$$Be(OH)_2 = Be^{++} + 2OH^- \tag{1}$$

but also as an *acid*

$$Be(OH)_2 = BeO_2^= + 2H^+ \tag{2}$$

The mode of calculation quoted is far too crude to enable us to predict which of the two processes preponderates. The behavior of beryllium hydroxide shows that process (1) is the dominant one. In the equilibrium

$$Be^{++} + 2OH^- \rightleftharpoons BeO_2^= + 2H^+ \tag{3}$$

the ions on the left hand side of the equation predominate in pure aqueous solutions. If, however, the hydrogen ion concentration is depressed to a very small value—or, in other words, if the solution is made strongly alkaline—then the reaction must proceed from left to right, in accordance with the Mass Action law. Thus beryllate ions $[BeO_2]^=$ are formed in strongly alkaline solutions. The corresponding salts, the *beryllates* $M^I_2[BeO_2]$, in which beryllium oxide is the acidic component, can be isolated from alcoholic solutions, though not indeed, from aqueous solutions.

Hydroxides can also display acidic properties because they are able to add on OH⁻ ions. It can be shown, from Coulomb's law, that energy can also be liberated by this process (cf. Chap. 11). It appears that with most metallic hydroxides in aqueous solution, the *addition* of OH⁻ ions takes place more readily than the removal of H⁺ ions, so that the amphoteric metal hydroxides behave towards strong bases in aqueous solution not as true acids, but as *anhydro-acids*. It is therefore probable that $Be(OH)_2$ in alkaline aqueous solutions does not dissociate according to eqn (2), but rather forms *hydroxo-ions*, by adding on OH⁻ ions according to the equations

$$Be(OH)_2 + OH^- = [Be(OH)_3]^- \text{ trihydroxoberyllate ion}$$
$$Be(OH)_2 + 2OH^- = [Be(OH)_4]^= \text{ tetrahydroxoberyllate ion}$$

Application of the mass action law to the equation

$$Be^{++} + 4OH^- \rightleftharpoons [Be(OH)_4]^= \tag{4}$$

leads to the same expression as does the process represented in eqn (3) as regards dependence of the beryllium and beryllate (or hydroxoberyllate) ions upon the hydrogen ion concentration.

It would follow from the foregoing that even relatively strong bases could assume the functions of acids, or of anhydro-acids, provided one could displace the equilibria of equations (3) or (4) sufficiently to the right, by raising the hydroxyl ion concentration. It has, indeed, been shown by Scholder (1935) that even the quite strongly basic *barium hydroxide* may behave towards very concentrated sodium hydroxide like an amphoteric substance. Whereas the solubility of $Ba(OH)_2$ in water is strongly diminished by small amounts of NaOH, it increases rapidly once again if the concentration of the NaOH rises above 10 mols per liter. The behavior of barium hydroxide with concentrated caustic soda is essentially the same as that shown by the typical amphoteric hydroxides, such as $Be(OH)_2$ and $Al(OH)_3$, even towards dilute sodium hydroxide.

The crystal structure of the oxides of Mg, Ca, Sr, and Ba is the same as that of sodium chloride (Fig. 44, p. 209), except that the lattice points are occupied by bivalent ions in place of univalent ions. These oxides are therefore *isotypic* with NaCl and most of the other alkali halides. Gerlach and Pauli were able to prove in the case of magnesium oxide, from the X-ray interferences, that these oxides also had *ionic* lattices, and not *atomic* lattices. The cubic unit cells have the dimensions

	MgO	CaO	SrO	BaO
$a =$	4.20	4.80	5.15	5.53 Å

The crystal structure of BeO bears to that of MgO, CaO, etc. the same relation that the structure of elementary Be and Mg bears to that of elementary Ca. In the rock salt structure, each kind of ion in itself makes up a face-centered cubic structure—i.e., a lattice of the type of cubic closest packing. In the same way in the structure of beryllium oxide (Fig. 60), each kind of atom is itself arranged as in hexagonal closest packing. By the interpenetration of the two partial lattices the structure shown in Fig. 60, a type of layer lattice, is obtained. This is generally known as the *wurtzite* type, since it was first discovered in wurtzite, the hexagonal modification of ZnS. For BeO, $a = 2.69$ Å, $c = 4.37$ Å. Each Be atom is surrounded by 4 O atoms, and every O atom by 4 Be atoms, arranged tetrahedrally. The distance Be—O = $p = 1.65$ Å.

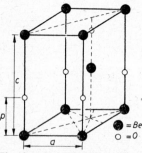

Fig. 60. Crystal lattice of the wurtzite type.
Example: BeO ($a = 2.69$ Å, $c = 4.37$ Å, $p = \frac{3}{8} c = 1.65$ Å)

● = Be
O = O

As corresponds with the difference in crystal structure, BeO forms no mixed crystals with MgO and CaO. MgO also forms no mixed crystals with CaO, however, because of the great difference in the ionic radii. The eutectic temperatures of the melts are: for BeO+MgO, 1955°; for BeO+CaO, 1475°; and for

MgO+CaO, 2360° (Ruff 1933). These are notably low, in view of the very high melting points of the pure oxides.

	BeO	MgO	CaO	SrO	BaO
Melting point	2530°	2800°	2576°	2430°	1923° C

(i) *Beryllium oxide*, BeO, (beryllia, glucina) results from the ignition of the hydroxide, or of beryllium salts such as the carbonate or sulfate. It is a loose white powder, which can be melted, and at the same time sublimes, only at the temperature of the electric furnace. It can be converted into the crystalline form by heating it with *mineralizers*—substances which promote crystallization by lowering the melting point, or in some other way. Although its melting point lies above 2500°, beryllia is distinctly volatile in the presence of steam above 1200° (Hutchison and Malm, 1950). Its solubility in water is extremely low, and it is the least soluble oxide of the alkaline earth group (cf. Table 53). It is not reduced by alkali metals; when it is heated with carbon in the electric furnace, it is converted into the carbide, Be_2C, which resembles aluminum carbide.

(ii) *Beryllium hydroxide* is precipitated by means of hydroxyl ions from solutions of beryllium salts as a white gelatinous precipitate, soluble both in acids and in alkalis. The alkaline solutions gradually decompose, however, or more rapidly on heating, and deposit a less soluble modification of the hydroxide. When more strongly heated, it loses water and changes to the oxide. Unlike aluminum hydroxide, freshly precipitated beryllium hydroxide dissolves easily in aqueous ammonium carbonate, as well as in acids and strong alkalis. It is also distinguished from aluminum hydroxide by its insolubility in ethylamine. The solubility in alkalis depends on the formation of beryllates, and the sodium and potassium salts, Na_2BeO_2 and K_2BeO_2 have been prepared from alcoholic solutions. The salts are hydrolyzed in water—e.g.,

$$K_2BeO_2 + H_2O = Be(OH)_2 + 2K^+ + 2OH^-.$$

For this reason they cannot be prepared from aqueous solution. Beryllium hydroxide prepared from beryl forms the starting material for the preparation of other beryllium compounds.

(iii) *Magnesium oxide*, MgO, (magnesia, bitter earth), is formed when magnesium burns in air, or by the ignition of the hydroxide, carbonate, nitrate and other oxyacid salts of magnesium; the sulfate is completely decomposed on strong ignition. It is a loose, white, very infusible powder (magnesia usta, burned magnesia, etc.). In the electric furnace it sublimes, and condenses in the crystalline form. This latter may more readily be obtained by heating it with mineralizing agents— e.g., with calcium borate, or by strongly heating it in a current of hydrogen chloride. Crystalline magnesium oxide is found in nature, in the form of very small regular octohedra and cubes, as *periclase* (hardness 6, density 3.7). This is colored greenish grey to dark green by its iron content. Whereas crystallized magnesium oxide is hardly attacked by water, and attacked only with difficulty by acids, the finely divided material dissolves readily in acids and is slowly converted by water into the hydroxide.

The technical uses of magnesium oxide depend principally on its infusibility, and it is used especially for the manufacture of refractory materials. It is also used to prepare magnesia cement and xylolite, and in medicine as an antidote for poisoning by acids. Magnesia is manufactured technically by ignition of either magnesite or magnesium hydroxide, which is obtained from the end-liquors of potassium salt extraction, containing magnesium chloride. For this purpose, the end liquors are first treated with milk of lime; iron is thereby precipitated as hydrous oxide, and sulfate ion as calcium sulfate. The magnesium is then precipitated by the further addition of milk of lime. Magnesium oxide can also be obtained by decomposition of magnesium chloride with superheated steam

$$MgCl_2 + H_2O = MgO + 2HCl$$

Hydrochloric acid is obtained as a by-product of this process, but since complete decomposition takes place only around $500°$, the consumption of fuel is high.

Reference has been made (p. 245) to the preparation of magnesium oxide and hydroxide from sea water, as a part of the process of the extraction of magnesium from the magnesium salts dissolved therein. The usual procedure is first to precipitate the HCO_3^- ions in the form of $CaCO_3$ (by the addition of milk of lime), and after filtration of the $CaCO_3$, to precipitate Mg^{++} ions as $Mg(OH)_2$ by further addition of milk of lime.

The principal difficulty in this process is presented by the poor sedimentation of the magnesium hydroxide precipitated from sea water, and the difficulty experienced in filtering it. This can be overcome by treatment of the washed slurry of $Mg(OH)_2$ with CO_2, thereby converting it into a hydrate of the carbonate, which filters more readily. This is subsequently converted to MgO by ignition.

The mother liquors or 'bitterns' of salt marshes, which were formerly rejected, are also now worked up for magnesium oxide. After removal of bromine, and precipitation of $SO_4^=$ as $CaSO_4$, the magnesium is precipitated as hydroxide by the addition of milk of lime.

(iv) *Magnesium hydroxide*, $Mg(OH)_2$, is formed by the action of water on powdered magnesium oxide, or on magnesium turnings; also by the action of hydroxyl ions on magnesium ions.

$$Mg^{++} + 2OH^- = Mg(OH)_2 \qquad (5)$$

Its properties depend to some extent upon the method by which it was prepared. Since the solubility product $K_{sp} = [Mg^{++}] \cdot [OH^-]^2$ is relatively high, magnesium hydroxide is incompletely precipitated by aqueous ammonia, and not at all if considerable amounts of ammonium salt are present in the solution, since these repress the dissociation of the $NH_3 \cdot H_2O$.* Magnesium hydroxide can be dried at $100°$ without decomposition, but is converted to the oxide at a dull red heat.

Crystallized magnesium hydroxide is found in nature as *brucite*. This forms thin, flexible colorless to green leaflets of the hexagonal system, with a glassy luster and mother-of-pearl shimmer on broken surfaces. The hardness is 2 and the density 2.3–2.4.

---

* As Fredholm (1934) has shown, the solubility of sparingly soluble magnesium *salts* is considerably raised by the addition of ammonia—a sign that the Mg^{++} ion adds on NH_3 forming a loose complex. Complex formation in this way does not markedly affect the equilibrium of eqn (5), except at very high concentrations of ammonia.

The unit cell of brucite is represented in Fig. 61. This lattice type (a layer lattice) is frequently found for the hydroxides of bivalent metals, for iodides, such as CdI_2, MgI_2, CaI_2, also for $MgBr_2$ and for sulfides such as SnS_2.

(v) *Calcium oxide*, quick lime, burned lime, CaO. Formation—as for magnesium oxide. It is always made by ignition of the carbonate. For the industrial preparation, limestone is generally used. This is heated to about 800° ('burned') in shaft furnaces or, in larger works, in rotary kilns. To carry the decomposition to completion at the temperature cited, the carbon dioxide must be carried away by suction, as the CO_2-pressure exceeds the atmospheric pressure only at temperatures considerably

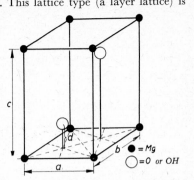

Fig. 61. Brucite lattice
($a = b = 3.12$ Å, $c = 4.73$ Å, $d = 1.05$ Å)

● = Mg
○ = O or OH

above 800° (cf. Table 57, p. 277). The requisite draught is often provided by the ascent of the hot gases. Burning the lime at higher temperatures leads not only to a wasteful consumption of fuel, but, if limestone with clay as an impurity is used, the product reacts only sluggishly with water because of sintering ('dead-burned' lime). Pure calcium oxide is extremely infusible. Oxide formed by the ignition of quite pure carbonate (marble) therefore forms a loose amorphous powder which react fairly readily with water, even when it has been strongly heated. The reaction of quick lime with water is known as the 'slaking' of lime; it is accompanied by the evolution of a good deal of heat (cf. Table 53, p. 257). Calcium oxide emits a very brilliant light when it is heated in the oxy-coal gas flame (Drummond's lime light). Calcium oxide melted in an electric furnace solidifies in crystalline form, and has the density 3.40. When heated to a high temperature with carbon, it forms calcium carbide—a reaction of considerable technical importance. In addition to being used for the production of slaked lime, calcium oxide finds application as a furnace lining, as a basic addition in the smelting of metals, in glass making, and as a fertilizer. It is also used occasionally as a drying agent which simultaneously absorbs CO_2. Pure calcium oxide, prepared from marble, finds some medical application, e.g., as a caustic for destroying warts and moles.

(vi) *Calcium hydroxide*, slaked lime, $Ca(OH)_2$, the product of combination of calcium oxide and water, forms a white, dusty amorphous powder, of density 2.08. It gives up its water only at temperatures considerably above 100°, and the H_2O-pressure reaches 1 atmosphere at 450°. It is sparingly soluble in water, with slight evolution of heat (2.8 kcal.). The solubility is somewhat increased by the presence of alkali metal salts and especially of ammonium chloride. Its aqueous solution (lime water) has a strongly alkaline reaction, though weaker than an equi-normal solution of potassium hydroxide. Calcium hydroxide combines with cane sugar, forming saccharates which may contain 1–6 CaO per molecule of sucrose. Calcium hydroxide therefore dissolves much more readily in cane sugar solutions than in pure water—e.g., 100 cc of a solution containing 8 % of sugar will dissolve 22.4 g of CaO. In pure water the solubility is

at	$0°$	$20°$	$50°$	$100°$ C
	0.131	0.126	0.098	0.060 g of CaO in 100 g of H_2O

Calcium hydroxide is used principally for the preparation of builders' mortar. It is used in the sugar industry for desugaring molasses—i.e., the residual mother liquors from the crystallization of sugar, from which it deposits sugar in the form of calcium saccharate when it is added in excess. Calcium hydroxide is also used in the manufacture of bleaching powder and on a large scale industrially as a cheap base—e.g., for the liberation of ammonia, the causticizing of soda—and as a caustic, e.g., for the removal of hair from the hides in tanning, and to destroy rotting organic matter. As lime water it is used medicinally—as a dressing for burns, and as an antidote for poisoning by sulfuric or oxalic acid. The *milk of lime* used for many industrial purposes—e.g., as a ceiling paint—is a suspension of calcium hydroxide in lime water.

Crystalline calcium hydroxide, which may be obtained by evaporation of the aqueous solution in hexagonal platelets, (crystal structure as for brucite, $a = 3.58$ Å, $c = 4.90$ Å), is much less reactive than the calcium hydroxide prepared in the ordinary way by slaking quick lime. This has the character of a *hydrogel*, and contains accordingly more water than corresponds to the formula $Ca(OH)_2$. Even under strongly reduced pressure this water is not removed completely. The colloidal properties are of material importance for the reactivity of slaked lime, and so for the production of mortar.

(vii) *Strontium oxide and hydroxide.* Strontium oxide, strontia, SrO, is generally prepared, like calcium oxide, by ignition of the carbonate. Since however, the dissociation pressure of strontium carbonate is smaller than that of calcium carbonate—it exceeds 1 atm. only above 1100°—a much higher temperature is necessary. It forms a white amorphous mass, of density 3.93–4.61, but can also be obtained crystalline. Amorphous strontium oxide combines with water with vigorous evolution of heat, first forming the hydroxide $Sr(OH)_2$. With excess water this forms an 8-hydrate, $Sr(OH)_2 \cdot 8H_2O$, which can be degraded to the monohydrate. The 8-hydrate has a negative heat of solution in water (—14.6 kcal.), whereas the anhydrous hydroxide dissolves with evolution of heat (+11.6 kcal.). The solubility in water increases with rise of temperature (unlike that of calcium hydroxide) and is quite considerable at 100°. 100 g of a saturated solution contains

at	$0°$	$20°$	$50°$	$100°$ C
	0.35	0.68	2.13	18.60 g of SrO

Strontium hydroxide is a strong base. It combines with cane sugar forming saccharates. Strontium saccharate, $C_{12}H_{22}O_{11}.2SrO$ is less soluble than calcium saccharate, and therefore more suitable for de-sugaring molasses. Apart from this, strontium oxide is used as the starting material for the preparation of other strontium compounds.

(viii) *Barium oxide*, BaO (caustic baryta, baria), is best obtained in the pure state by strong ignition of barium nitrate (or barium iodate). If the salt is insufficiently ignited, the preparation is contaminated with peroxide. The oxide is generally prepared technically by ignition of barium carbonate mixed with carbon. Barium carbide can be employed in place of carbon. The purpose of the admixture is to remove CO_2 by converting it to CO, so that, in spite of the low CO_2 pressure, the decomposition of the barium carbonate proceeds to completion. The dissociation pressure of CO_2 reaches 1 atmosphere only at a full white heat, at which the barium oxide is strongly sintered, or begins to melt. Pure barium oxide is a white powder, of density 4.7–5.8; that prepared industrially is generally colored grey by traces of carbon. It is more fusible than strontium or calcium oxides, and solidifies in crystalline form. It unites with water with the evolution of much heat ($+17.3$ kcal.). When it is gently heated in air (at about 500°), it is converted to the peroxide,

$$BaO + \tfrac{1}{2}O_2 = BaO_2 + 16.2 \text{ kcal.}$$

It is used chiefly for the preparation of the hydroxide and the peroxide.

(ix) *Barium hydroxide*, $Ba(OH)_2$, the product of union of barium oxide with water, is an amorphous white powder when anhydrous, and melts without decomposition. It is considerably more soluble in water than the hydroxides of the other alkaline earth metals, the solubility being

at	0°	20°	50°	80° C
	1.5	3.84	11.75	90.8 g of BaO in 100 g of water.

The solubility thus increases considerably with temperature, and hot water dissolves considerable amounts of barium hydroxide. The solubility is increased somewhat by the addition of neutral salts, but greatly diminished by alkali hydroxides. The aqueous solution is strongly basic in reaction. It is a sensitive reagent for CO_2 (white precipitate of $BaCO_3$). Hydrated barium hydroxide crystallizes when the solution is evaporated, the 8-hydrate, $Ba(OH)_2 \cdot 8H_2O$ being stable in contact with the solution at room temperature and up to 100° or above; this may be dehydrated to the monohydrate. The octohydrate forms colorless tetragonal crystals, of density 1.66, which begin to melt in their water of crystallization at 78°. The heat of solution of the 8-hydrate is negative (-15.2 kcal.), that of the anhydrous hydroxide is positive ($+12.3$ kcal.). Barium hydroxide can be largely thrown out of aqueous solution by the addition of alcohol. A solution of barium hydroxide saturated at ordinary temperature is extensively used in the laboratory (baryta water)

(c) Peroxides

The metals of the alkaline earths form white *peroxides* of the type MO_2. In addition, there is evidence that calcium and barium can give yellow *hyperoxides* of the formula $M(O_2)_2$. The peroxides give only hydrogen peroxide, when cautiously treated with acids, whereas the hyperoxides form molecular oxygen as well:

$$CaO_2 + H_2SO_4 = CaSO_4 + H_2O_2; \quad Ca(O_2)_2 + H_2SO_4 = CaSO_4 + H_2O_2 + O_2$$

The peroxides, like hydrogen peroxide and the alkali metal peroxides, contain the group O_2^{2-}, whereas the hyperoxides contain the group O_2^{-} (P. Ehrlich, 1944). The peroxides are sparingly soluble in water; the solutions have an alkaline reaction, because of hydrolysis:

$$MO_2 + 2H_2O \rightleftharpoons M(OH)_2 + H_2O_2.$$

The alkaline earth peroxides can be obtained as hydrates in the wet way by the action of hydrogen peroxide on the hydroxides, e.g.,

$$Ca(OH)_2 + H_2O_2 = CaO_2 + 2H_2O.$$

This mode of formation corresponds exactly to the neutralization of a base by a dibasic acid, and accords with the view, developed on p. 177, of the peroxides as salts of hydrogen peroxide.

The stability of the peroxides increases with increase in electropositive character of the metals. No peroxide of beryllium is known, and only peroxide hydrates of magnesium. The anhydrous peroxide of calcium, CaO_2, can be obtained only by dehydrating the hydrate $CaO_2 \cdot 8H_2O$; strontium peroxide SrO_2 may be obtained by the direct action of oxygen on strontium oxide though only under high pressure; barium peroxide, however, can easily be obtained by passing air over heated barium oxide. It was the first of the peroxides to be discovered.

The alkaline earth hyperoxides have not yet been obtained in the pure state, but only admixed with the ordinary peroxide. These mixtures, with about 8–9 % of hyperoxide, are lentil-yellow powders, which can withstand gentle heating without decomposition, but decompose to form the peroxides when more strongly heated. If the temperature is raised still further, the peroxides also decompose. In the case of barium peroxide, the decomposition pressure of 1 atmosphere is reached only at about 800°.

(i) *Barium peroxide*, BaO_2, is a white powder, fairly sparingly soluble in water and insoluble in alcohol and ether. It combines with water, forming the hydrate $BaO_2 \cdot 8H_2O$. The aqueous solution of barium peroxide acts as an oxidant for iron(II) salts, but as a reducing agent towards potassium hexacyanoferrate(III) ($K_3[Fe(CN)_6]$) and many other heavy metal salts. Thus it reacts with mercuric chloride:

$$HgCl_2 + BaO_2 = Hg + BaCl_2 + O_2$$

Dilute acids liberate hydrogen peroxide from barium peroxide. As a very weak acid, hydrogen peroxide is displaced from its salt by stronger acids.

$$BaO_2 + 2H^+ = Ba^{++} + H_2O_2$$

Barium peroxide begins to lose oxygen above 500° in vacuum or above 700° in air. The dissociation pressure p at various temperatures is

$t =$	525°	720°	790°
$p =$	20	210	570 mm Hg

It is prepared technically by heating porous barium oxide to about 500° in a stream of air. At this temperature, the dissociation pressure of barium peroxide is still so small that the simple oxide is completely converted to the peroxide by the air.

Barium peroxide is used in bleaching silk, vegetable fibers and straw, and—on a large scale—for the preparation of hydrogen peroxide. It is used also for preparing barium peroxycarbonate, as a decolorizing agent for lead glass, and as a disinfectant. The igniter pellets used in aluminothermy consist of a mixture of barium peroxide and magnesium powder.

Brin's process for the manufacture of oxygen, which has been superseded by the liquid air process (cf. p. 689), was based upon the formation of barium peroxide when barium oxide was gently heated in air, and the converse liberation of oxygen by lowering the pressure and raising the temperature.

The crystal structure of BaO_2 can be derived from that of BaO by replacement of the O^{2-}

ions by O_2^{2-} ions (Bernal 1935). The lines joining the pairs of O-atoms in all the O^2_2- groups (1.37 Å between them) all lie parallel to the c-axis, which is lengthened with respect to the other two axes, to make a tetragonal cell ($a = 5.34$ Å, $c = 6.77$ Å, $c/a = 1.27$) in place of the cubic cell. SrO_2 has a corresponding structure with $a = 5.02$ Å, $c = 6.55$ Å.

Most of the acetylides have the same crystal structure (barium carbide type, Fig. 84, p. 466).

(ii) *Strontium peroxide*, SrO_2, and *calcium peroxide*, CaO_2, are much less important than barium peroxide. These are obtained in the form of their hydrates, $SrO_2 \cdot 8H_2O$ and $CaO_2 \cdot 8H_2O$, which are isomorphous with the corresponding barium compound, by addition of hydrogen peroxide or sodium peroxide to a solution of strontium hydroxide or to lime water. The anhydrous compounds are obtained from these by gentle heating (to 100–130°) or by direct precipitation from hot solutions. Oxygen is evolved only on much stronger heating—only above the melting point (a red heat) in the case of strontium peroxide. Both compounds are sparingly soluble in water, and are decomposed by acids, as is barium peroxide.

The heats of formation of the peroxides from the oxides are: for CaO_2 5.4; SrO_2 13.5; and BaO_2 16.2 kcal per mol.

(iii) *Magnesium peroxide* is manufactured in the impure state (containing water and magnesium oxide) by various methods—e.g., by the action of H_2O_2 on MgO. It is used in pharmacy.

8. Salts

(a) Halides

The *fluorides* occupy a special place among the halides of the Group IIA elements, since (with the exception of beryllium fluoride) they are insoluble, whereas the remaining halides of this Group are all readily soluble, and in some cases deliquescent. The halides of beryllium are hydrolyzed to a considerable extent in aqueous solution. With the soluble halides of magnesium and the alkaline earth metals, the tendency for hydrolysis becomes apparent only at higher temperatures. The halides of beryllium and magnesium readily form double salts with the alkali metal halides.

A number of the anhydrous halides of the alkaline earths crystallize in the fluorite type (Fig. 62) structure. In fluorspar, the Ca^{++} ions make up a face-centered cubic lattice. The

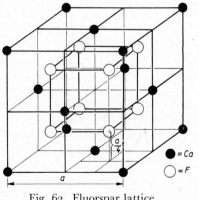

Fig. 62. Fluorspar lattice
(For CaF_2, $a = 5.45$ Å)

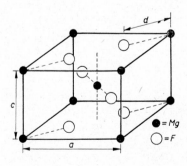

Fig. 63. Lattice of the rutile type
Example: MgF_2 ($a = 4.62$ Å, $c = 3.06$ Å, $d = 2.05$ Å)

F^- ions occupy the centers of the 8 smaller cubes into which the unit cell may be subdivided. Cell dimensions of a number of the halides are given below:

	CaF_2	SrF_2	BaF_2	RaF_2	$SrCl_2$
$a =$	5.45	5.78	6.19	6.37	6.97 Å

The fluorite structure is widely found among compounds of the general formula AB_2. The interionic distances in structures of this type are equal to $\dfrac{a}{4}\sqrt{3}$.

MgF_2 has the *rutile* structure (Fig. 63). The rutile structure is frequently encountered also among the *dioxides*. It was first observed in rutile, TiO_2. The unit cell of this structure consists of a tetragonal prism, the edges and mid-point of which are occupied by ions or atoms of one kind—e.g., Mg^{++} ions—whereas the ions or atoms of the other kind—e.g., F^- ions—lie on the diagonals through the corners and mid-point, at a distance d, as represented in Fig. 63.

$CaCl_2$ forms a crystal lattice like that of MgF_2, but rhombically deformed; each Ca^{++} is surrounded by $6Cl^-$, lying on the vertices of a deformed octohedron, 2 being at a distance of 2.70 Å and 4 at 2.76 Å. $MgCl_2$ has a layer lattice similar to $Mg(OH)_2$; distance Mg—$Cl = 2.54$ Å, Cl—Cl, between adjacent layers, 3.59 Å.

$MgBr_2$, MgI_2 and CaI_2 are isotypic with $Mg(OH)_2$ (Brucite structure, cf. Fig. 61). $MgBr_2$: $a = 3.81$ Å, $c = 6.26$ Å, $d = c/4$; MgI_2: $a = 4.14$ Å, $c = 6.88$ Å, $d = c/4$; CaI_2: $a = 4.48$ Å, $c = 6.96$ Å, $d = c/4$.

Fluorides

(i) *Beryllium fluoride*, BeF_2, is obtained in the hydrated state by evaporation of a solution of beryllium hydroxide in hydrofluoric acid. It must be heated in a stream of hydrogen fluoride in order to remove the water, as hydrolysis occurs otherwise. The anhydrous compound is more simply obtained by heating ammonium fluoberyllate, $(NH_4)_2[BeF_4]$ in a stream of CO_2.

It is a glassy, hygroscopic mass, easily soluble in water. With the alkali fluorides, it forms well crystallized complex salts, fluoroberyllates, of the types $M^I[BeF_3]$ and $M^I_2[BeF_4]$. These are readily obtained by evaporation of mixed solutions of the components. Beryllium fluoride is decomposed when heated in air, forming the oxyfluoride. It is reduced to the metal by magnesium or the alkali metals.

BeF_2, like SiO_2, generally solidifies in the form of a glass. It is also isotypic with SiO_2 (in the β-cristobalite modification) in the crystallized state. Venturello (1941) stated that BeF_2 formed an unbroken mixed crystal series with MgF_2, but this report needs confirmation since it would imply that BeF_2 can crystallize with the rutile type of structure. This is rather improbable, and has not been found experimentally. The hexagonal modification of BeF_2 described by Venturello may be a form with a tridymite-like structure.

(ii) *Magnesium fluoride*, MgF_2, is formed as a precipitate on the addition of F^- ions to magnesium salt solutions. It is sparingly soluble in water (87 mg per liter at 18°), melts at 1265° (b.p. 2260°), and crystallizes in the tetragonal system on cooling. The rare mineral *sellaite* (colorless, vitreous, translucent, hardness 5, density 2.97) is natural, crystallized MgF_2.

With the alkali fluorides, MgF_2 forms double salts of the composition $M^IF \cdot MgF_2$ and $2M^IF \cdot MgF_2$. These are obtained by the addition of MgF_2 or MgO to fused alkali fluorides.

(iii) *Calcium fluoride*, CaF_2, may be obtained crystalline by neutralization of calcium carbonate with dilute hydrofluoric acid. It forms as a gelatinous precipitate when F^- ions are added to a calcium salt solution. It is very sparingly soluble in water (16 mg per liter at 18°), but readily forms colloidal dispersions. It unites with hydrofluoric acid, forming a more soluble acid salt, $CaF_2 \cdot 2HF \cdot 6H_2O$, and many other substances also increase the solubility. Dehydrated calcium fluoride forms a white powder, melting at 1403° without decomposition (b.p. 2500°). When it is heated with concentrated sulfuric acid, hydrogen fluoride is evolved:

$$CaF_2 + H_2SO_4 = CaSO_4 + 2HF$$

Calcium fluoride is practically insoluble in dilute strong acids. It occurs native in large quantities, as *fluorite* or *fluorspar*. Apart from the preparation of hydrofluoric acid and the etching of glass, calcium fluoride is used as an opacifier in the enamel industry, and as an antiseptic. It is administered in small amounts, together with other calcium salts, as a medicament for certain bone diseases.

(iv) *Strontium fluoride*, SrF_2, is obtained as a white precipitate by neutralisation of the hydroxide or carbonate with hydrofluoric acid, or by metathetic reaction between the ions. Its solubility in water is 117 mg per liter at 18°. When obtained at high temperatures, it is crystalline (regular octohedra). M.p. about 902°, b.p. 2460°.

(v) *Barium fluoride*, BaF_2, prepared like SrF_2, is also sparingly soluble in water (1.63 g per liter at 18°). It forms small transparent crystals, density 4.83, m.p. 1353°, b.p. 2260°.

Chlorides, Bromides and Iodides

(vi) *Beryllium chloride*, $BeCl_2$, can be obtained anhydrous by heating metallic beryllium in dry chlorine or hydrogen chloride, as a snow-white, fusible mass (m.p. 405°), easily sublimed (b.p. 488°). It dissolves in water with considerable evolution of heat. The solution has a strongly acid reaction, since the chloride is partially hydrolyzed (about 2 % in 0.1-normal solution at 40°). On evaporation, the hydrate $BeCl_2.4H_2O$ crystallizes in deliquescent, monoclinic plates, provided that hydrolysis is repressed by addition of hydrochloric acid; otherwise, basic salts are precipitated. The hydrated salt loses HCl when it is heated. Anhydrous beryllium chloride dissolves readily in alcohol and ether, and forms an addition compound $BeCl_2.2(C_2H_5)_2O$ with the latter. It is, indeed, very prone to the formation of addition compounds with organic compounds.

(vii) *Beryllium bromide*, $BeBr_2$. Preparation and properties as for the chloride. It is still more volatile than the chloride (b.p. 473°).

(viii) *Beryllium iodide*, BeI_2, is best obtained by heating Be_2C in hydrogen iodide. It forms small hygroscopic crystals, m.p. 480°, b.p. 488°. It is converted to the oxide if it is heated in air, and is reduced by the alkali metals. It unites with ammonia to form $BeI_2.3NH_3$, and is highly reactive towards many organic compounds.

(ix) *Magnesium chloride*, $MgCl_2$, occurs in vast amounts dissolved in sea water. As a mineral, it occurs as *bischofite*, $MgCl_2·6H_2O$ in the potassium salt deposits, but in much greater quantities as the double salt *carnallite*, $MgCl_2·KCl·6H_2O$. In the course of working up carnallite for KCl, magnesium chloride is manufactured from the end liquors. [2] Depending upon the temperature, it crystallizes out when these are evaporated, either as the 6-hydrate, $MgCl_2.6H_2O$, stable at ordinary temperature, or, if evaporation is carried further, as products poorer in water. Only a relatively small proportion of the end-liquors is worked up for magnesium chloride, most being discharged as waste, into the rivers (for the preparation of magnesium chloride from sea water, see p. 245). In addition to the 6-hydrate, magnesium chloride also forms hydrates with 2, 4, 8 and $12H_2O$. The 6-hydrate, however, has the greatest range of stability, namely from —3.4° to 116.7°. It forms deliquescent monoclinic crystals, with a bitter taste and density 1.56. The water cannot be removed completely without decomposition of the salt, since hydrogn chloride is lost on heating, and *basic chlorides* (oxychlorides) of variable composition are formed – e.g.,

$$2MgCl_2 + H_2O = Mg_2OCl_2 + 2HCl$$

Magnesium chloride is very soluble in water. The solution contains.

at	—33.6°*	0°	20°	100°	186° C
	26	53	54.5	73	128 g of $MgCl_2$ to 100 g of water

* Cryohydric point

The aqueous solution of magnesium chloride has a neutral reaction. If strongly ignited magnesium oxide is added to a concentrated solution of magnesium chloride, the slurry so obtained solidifies in the course of a few hours to a solid mass,— the so-called *magnesia cement* or Sorel cement— because the oxide combines with the chloride, forming basic chlorides.

Sorel cement is best prepared by mixing 10 parts by weight of MgO with 5 parts by weight of anhydrous $MgCl_2$. A reaction similar to that between MgO and $MgCl_2$ takes place also between MgO and other magnesium salts, such as $MgSO_4, Mg(NO_3)_2$, $Mg(C_2H_3O_2)_2$, etc. The cements so obtained are inferior in strength and hardness to Sorel cement. By mixing Sorel cement with saw-dust, cork waste, etc., woody, fairly weatherproof masses are obtained, which are used as coverings for floors, laboratory tables, etc. (xylolite).

Magnesium chloride forms double salts with alkali chlorides, the naturally-occurring carnallite $MgCl_2 \cdot KCl \cdot 6H_2O$ being one of these. *Tachhydrite*, $2MgCl_2 \cdot CaCl_2 \cdot 12H_2O$, also found in potash deposits, is of the same type. The corresponding ammonium chloride double salt, $MgCl_2 \cdot NH_4Cl \cdot 6H_2O$ (obtained from the components) decomposes without any hydrolysis when it is heated, and can therefore be used for the preparation of anhydrous magnesium chloride.

In addition to the preparation of magnesia cement and xylolite, magnesium chloride is used as a textile dressing to keep fabrics pliable, for the impregnation of wood, for spraying dusty roads, and for the preparation of magnesium oxide and other magnesium compounds. Chlorine and hydrochloric acid can also be obtained from it. In medicine, it finds application as a purgative and for enemas. Considerable quantities of magnesium chloride, waste end-liquors from the extraction of potash, are returned unused to the workings in order to avoid making the rivers too strongly saline.

(x) *Magnesium bromide*, $MgBr_2$, melting point $711°$, is largely similar to magnesium chloride. Like the latter, it crystallizes from solution at ordinary temperature as the hygroscopic 6-hydrate, which is isomorphous with the 6-hydrate of the chloride, and is present in small amounts in the minerals bischofite and carnallite. Magnesium bromide forms a 10-hydrate as well as the 6-hydrate. The double salt $MgBr_2 \cdot KBr \cdot 6H_2O$ corresponds to carnallite. When a concentrated magnesium bromide solution is boiled with magnesium oxide, a double compound of magnesium bromide and magnesium hydroxide,

$$MgBr_2 \cdot 3Mg(OH)_2 \cdot 9H_2O,$$

is obtained, crystallizing in colorless needles.

(xi) *Magnesium iodide*, MgI_2, crystallizes from aqueous solution as the 8-hydrate (at ordinary temperature) and as the 10-hydrate. A white, hygroscopic powder, very soluble in water, it finds some medicinal applications (against syphilis, scrofula, rheumatism).

(xii) *Calcium chloride*, $CaCl_2$, when anhydrous, forms a white, extraordinarily hygroscopic mass, which melts at $780°$ and is volatile at a white heat. The density of calcium chloride, solidified from the melt, is 2.2. It is prepared from hydrated calcium chloride by heating above $260°$. This dehydration must be carried out carefully, however, as partial hydrolysis and elimination of HCl occurs if the salt is heated too quickly. The hydrolysis of $CaCl_2$ by superheated steam affords a possible technical method for the preparation of hydrochloric acid (cf. E. Briner, *Helv. Chim. Acta*, 31 (1948) 556). Anhydrous calcium chloride dissolves in water with the evolution of much heat (17.41 kcal.), which is attributable to hydration. The best known hydrate, the hexahydrate $CaCl_2 \cdot 6H_2O$, crystalizes out when a solution of calcium chloride is evaporated at ordinary temperature. It forms hexagonal prisms, of density 1.65, and dissolves in water with strong absorption of heat (—4.31 kcal.). It is therefore very suited to the preparation of freezing mixtures, and by mixing the hexahydrate with ice in the ratio 1.44 : 1 by weight, a temperature of —54.9° can be attained, corresponding to the cryohydric point. The solubility of calcium chloride in water is:

at —54.9°   0°   +10°   20°   40°   60°   100°   260° C
    42.5   60.0   65.0   74.5   115   137   159   347 g of $CaCl_2$ in 100 g of H_2O

It readily forms supersaturated solutions. In addition to the hexahydrate, there are two tetrahydrates, a dihydrate and a monohydrate (cf. p. 72).

Solutions of calcium chloride are obtained as waste products in a number of chemical processes, especially in the Solvay soda process. These are the materials from which technical calcium chloride is manufactured, the solutions being purified by boiling them with quick lime. Pure calcium chloride is prepared by dissolution of pure calcium carbonate (marble) in hydrochloric acid.

Calcium chloride solutions are often used as liquid heating baths, as liquids for transmitting hydraulic pressure, and as coolants. Tables 54 and 55 indicate the boiling points and freezing points of such solutions. Calcium chloride solutions are also used to impregnate wood, textiles, etc., to make them incombustible. They have also been proposed for fire extinguishing. Show case windows can be preserved from deposits of dirt by wiping them with calcium chloride solution.

TABLE 54

BOILING POINTS OF CALCIUM CHLORIDE SOLUTIONS
(Gerlach and Schlamp)

Grams $CaCl_2$ per 100 g water	6.0	11.5	16.5	21.0	25.0	41.5	69.0
Boiling point	101°	102°	103°	104°	105°	110°	120°
Grams $CaCl_2$ per 100 g water	101	137.5	178	222	268	292	305
Boiling point	130°	140°	150°	160°	170°	175°	178°

TABLE 55

FREEZING POINTS OF CALCIUM CHLORIDE SOLUTIONS
(Pickering, Loomis and Rodebush)

am $CaCl_2$ in g solution	0.1	1	5	10	15	20	25	30	32.5
eezing point	—0.051°	—0.46°	—2.44°	—5.89°	—10.96°	—18.6°	—29.9°	—48.0°	—51.0°

Calcium chloride dissolves readily in anhydrous alcohols, as well as in water, (including the higher alcohols, such as propyl, isobutyl and amyl alcohol; it forms solid addition compounds with the two last-named). From solutions containing calcium hydroxide, as well as calcium chloride, – e.g., the residual solutions from the preparation of ammonia from NH_4Cl and $Ca(OH)_2$ – a double compound $CaCl_2 \cdot 3Ca(OH)_2 \cdot 12H_2O$ crystallizes in long, fine needles. According to Werner, this should be regarded not as an ordinary basic salt, but as a 'hexol' salt, belonging to the class of coordination compounds (on this point see Vol. II).

Anhydrous calcium chloride also combines with ammonia; this must be remembered when calcium chloride is employed as a drying agent. This use of calcium chloride depends on its great hygroscopicity. For this purpose, it is generally used in the form of porous lumps which are obtained by not quite complete dehydration, under conditions which avoid fusion. Pure calcium chloride is used in medicine.

(xiii) *Calcium bromide*, $CaBr_2$, a white mass melting at 760° and boiling at 806–812° when anhydrous, is very similar to the chloride in properties. Only one hydrate is definitely known, however, the hexahydrate $CaBr_2 \cdot 6H_2O$ (m.p. 38.2)°, which forms silky needles, but crystallizes poorly. Calcium bromide is readily soluble in water and in alcohol. It is used in the photographic industry and in medicine.

(xiv) *Calcium iodide*, CaI_2, is very similar to the bromide; density 4.9, melting point 740°, boiling point 708–719°. It crystallizes from solution as the hexahydrate $CaI_2, 6H_2O$. A hydrated compound of CaI_2 with 3 moles of $Ca(OH)_2$, analogous in constitution to the corresponding chloride, has been described by Tassily.

(xv) *Strontium chloride*, $SrCl_2$, is most simply prepared by dissolution of strontium carbonate in hydrochloric acid. It crystallizes from solution below 60° as the hexahydrate, $SrCl_2 \cdot 6H_2O$—deliquescent hexagonal needles, isomorphous with the corresponding calcium salt. Above 60°, the dihydrate crystallizes in rectangular plates. By heating the hydrated compound above 100°, the anhydrous salt is obtained. This melts at 872°. If moisture is present when it is melted, traces of hydrolysis occur, as with calcium chloride. Strontium chloride is very soluble in water. The solution has a sharp, bitter taste. Solubility

at	0°	20°	100° C
	44.2	53.9	101.9 g of $SrCl_2$ in 100 g of H_2O

The saturated solution boils at 117°. Unlike calcium chloride, $SrCl_2 \cdot 6H_2O$ is sparingly soluble in ethyl alcohol, and less soluble hot than cold. Like $CaCl_2$, $SrCl_2$ combines with ammonia.

(xvi) *Strontium bromide*, (m.p. 643°) and *strontium iodide* (m.p. 507°) differ from strontium chloride in being readily soluble in alcohol. Both are used in medicine. The molecular solubilities of the strontium halides in water increase in the order $SrCl_2$, $SrBr_2$ and SrI_2.

(xvii) *Barium chloride*, $BaCl_2$, when anhydrous, forms a white mass, m.p. 878°, density 3.86. It is easily soluble in water, sparingly soluble in absolute ethyl alcohol, and moderately soluble in methyl alcohol (2.18 parts of $BaCl_2$ in 100 parts at 15.5°). The solubility in water is:

at	0°	10°	20°	50°	100°	104.1°	
	31.5	33	36	43.5	59	60	g of $BaCl_2$ in 100 g of H_2O

At ordinary temperature, the salt crystallizes from solution as the dihydrate, $BaCl_2 \cdot 2H_2O$, in flat, four-sided colorless rhombic plates of density 3.06. The crystals are stable in air, when heated they lose first one molecule of water, and on stronger heating the second.

Barium chloride is used industrially to soften boiler feed water containing gypsum, and for the preparation of other barium compounds, especially Permanent White. It is employed as a reagent in the laboratory, especially for the detection and determination of sulfuric acid. It is also used in medicine for various purposes. It has an unpleasant taste and is *very poisonous*. Antidote: after evacuating the stomach, wash out with 1% Glauber salt solution.

The chloride is prepared commercially either from witherite or from heavy spar. The first is dissolved in hydrochloric acid. The latter is converted into chloride by heating with carbon and calcium chloride (waste from Solvay process). The carbon reduces barium sulfate to sulfide, and this reacts with calcium chloride

$$BaSO_4 + 4C = BaS + 4CO; \qquad BaS + CaCl_2 = BaCl_2 + CaS$$

Barium chloride is leached out of the resulting melt by means of hot water. Small amounts of unreacted barium sulfide that have been dissolved are decomposed by passage of carbon dioxide into the solution; the carbonate so formed is converted to the chloride by means of hydrochloric acid, and the solution is evaporated and crystallized.

(xviii) *Barium bromide*, $BaBr_2$, most readily prepared by neutralization of $Ba(OH)_2$ or $BaCO_3$ by means of hydrobromic acid, is a white substance, melting point 847°, density 4.79. It is very soluble in water (100 g of water dissolve 98 g of $BaBr_2$ at 0°, 104 g at 20°, 149 g at 100°). The dihydrate, $BaBr_2 \cdot 2H_2O$, isomorphous with the corresponding chloride, crystallizes from the solution. This loses water rather less readily than does the chloride; the first molecule is lost at 75°, the second only above 100°. Barium bromide is very soluble in methyl alcohol, and markedly soluble in pure ethyl alcohol (about 3.1 g of $BaBr_2$ in 100 g of alcohol). In the presence of oxygen from the air, carbon dioxide decomposes aqueous solutions of barium bromide, forming the carbonate:

$$BaBr_2 + CO_2 + \tfrac{1}{2}O_2 = BaCO_3 + Br_2$$

(xix) *Barium iodide*, BaI_2, readily obtained by treatment of $Ba(OH)_2$ or $BaCO_3$ with HI or with iodine in the presence of a reducing agent, is a white substance of density 4.92 (anhydrous salt). It is very soluble in water (100 g of water dissolve 170 g of BaI_2 at 0°, 200 g at 20°, 270 g at 100°). It is also very soluble in ethyl alcohol. Deliquescent hydrates with $1-7H_2O$ crystallize from the aqueous solution, the rhombic dihydrate, $BaI_2 \cdot 2H_2O$, being isomorphous with the chloride and bromide.

(xx) *Monohalides of the alkaline earths.* L. Wöhler (1909) thought that he had obtained the monohalides CaF, CaCl and CaI, by heating metallic calcium with the calcium halides, and quenching the melts, as compounds which are unstable at ordinary temperature. The experiments of Guntz and Benoit (1923–24) seemed to confirm this observation. The monohalides of the alkaline earths are described as intensely colored substances, which decompose water with the evolution of hydrogen. It was also believed that their formation had been observed in the electrolysis of melts. However, the heats of solution measured by Benoit, which differ little from those of mixtures of the metal and the normal chloride, led Bichowsky and Rossini (1936) to regard the existence of the monohalides as questionable. More recently, Cubicciotti (1949) has determined the equilibrium diagrams of the systems $CaCl_2 + Ca$, etc., and has shown that although the alkaline earth metals display a considerable solubility in the molten halides, there is no evidence of monohalides either in the solid or the molten state. Schäfer (1952), in a more detailed study of the system $BaCl_2 + Ba$, has proved that there is complete miscibility in the liquid state above 1010°. The extent of miscibility decreases rapidly with falling temperature. However, above 800°, $BaCl_2$ can take up a certain amount of Ba, even in the solid state, whereas Ba can dissolve $BaCl_2$ only in the molten state. At the melting point of $BaCl_2$, the miscibility gap extends from 15 to 93 mole-% Ba.

The work of Schriel (1937) makes it extremely probable that Ba_2O, described by Guntz and Benoit, is only a solution of Ba in BaO.

(b) Nitrates

Like all nitrates, those of the elements of Group IIA are readily soluble in water, and except for strontium and barium nitrates, they are soluble in alcohol also. Beryllium nitrate is hydrolyzed (to the extent of 1.8% in 0.1 N solution at 40°) in aqueous solution. When heated, the nitrates of the Group IIA metals are easily and completely converted to the oxides. They do not show any strong tendency to form double salts with the alkali nitrates, such compounds being known only for magnesium and barium nitrates—e.g., $2KNO_3 \cdot Mg(NO_3)_2$ and $2KNO_3 \cdot Ba(NO_3)_2$. In the quaternary system KNO_3—$NaNO_3$—$Ca(NO_3)_2$—$Mg(NO_3)_2$, in addition to the double salt $2KNO_3 \cdot Mg(NO_3)_2$, the two mixed crystal series $(K,Na)NO_3$ and $(Ca,Mg)(NO_3)_2$ crystallize from the melt (Jänecke, 1942).

The crystal structures of $Ca(NO_3)_2$, $Sr(NO_3)_2$ and $Ba(NO_3)_2$ have been determined by X-rays; if the detailed arrangement of the atoms in the structural group NO_3 is ignored, they show a certain resemblance to the fluorite structure. As in this, the metal ions make up a face centered cubic lattice, and the NO_3- groups lie on the body-diagonals of the 8 small cubes although not, as in fluorite, at their centers. Since $Mg(NO_3)_2$ and $Ca(NO_3)_2$ form a continuous series of mixed crystals, it can be inferred that $Mg(NO_3)_2$ has the same structure.

(i) *Beryllium nitrate*, $Be(NO_3)_2$, is obtainable by double decomposition between $BeSO_4$ and $Ba(NO_3)_2$,

$$BeSO_4 + Ba(NO_3)_2 = Be(NO_3)_2 + BaSO_4,$$

or, technically, by dissolution of beryllium hydroxide in nitric acid. It crystallizes from solution in deliquescent crystals as $Be(NO_3)_2 \cdot 3H_2O$. These melt in their water of crystallization at about 60°; loss of nitric acid commences at about 100°, and the oxide remains if the temperature is raised to 200°. Beryllium nitrate finds limited industrial application as a hardening agent for incandescent gas mantles.

(ii) *Magnesium nitrate* crystallizes from aqueous solution as the 6-hydrate, $Mg(NO_3)_2 \cdot 6H_2O$, forming rhombic rods and needles which deliquesce in moist air and decompose at 89.1°, forming the dihydrate. This melts at 129.5°, and loss of nitric acid sets in at the same time. The pure oxide remains on ignition. A 9-hydrate, $Mg(NO_3)_2 \cdot 9H_2O$, is stable below —17.1°. A (metastable) 4-hydrate (melting at 52°) also exists.

(iii) *Calcium nitrate* crystallizes as the 4-hydrate, $Ca(NO_3)_2 \cdot 4H_2O$, from aqueous solutions as monoclinic prisms, density 1.82, which melt in their water of crystallization a little above 40°. The cooled solutions generally show a very marked sluggishness of crystallization. Lower hydrates probably exist, in addition to the 4-hydrate. By heating above 100°, the anhydrous salt is readily obtained (m.p. 561°). This is very hygroscopic, very soluble in water (121 g of $Ca(NO_3)_2$ in 100 g of water at 18°), and also readily soluble in alcohol.

Calcium nitrate is manufactured by neutralization of calcium carbonate (limestone) or calcium hydroxide with (synthetic) nitric acid. The lime saltpeter so obtained is used in very large amounts as a fertilizer. Its hygroscopic properties, which are undesirable for this purpose, can be almost completely obviated by the addition of quicklime, whereby a basic salt is formed.

When heated above its melting point, calcium nitrate first loses oxygen. Only when more strongly heated is it converted into the oxide by the loss of nitrogen oxides. Solutions of calcium nitrate absorb more ammonia than does pure water, but solid addition compounds of ammonia with calcium nitrate are not known.

(iv) *Strontium nitrate*, $Sr(NO_3)_2$, is most simply prepared by treatment of the carbonate with nitric acid. It may also be obtained by double decomposition between strontium chloride and sodium nitrate. The latter method depends on the fact that in the reciprocal salt pair

$$SrCl_2 + 2NaNO_3 \rightleftharpoons Sr(NO_3)_2 + 2NaCl$$

the salts standing on the right hand side are less soluble than those on the left side. The solubility of strontium nitrate is: 39.5 g of $Sr(NO_3)_2$ in 100 g of water at 0°, 70.8 g at 20°, 101 g at 100°. The monoclinic tetrahydrate, $Sr(NO_3)_2 \cdot 4H_2O$, crystallizes from cold aqueous solutions and has a density 2.25. This can readily be completely dehydrated by heating it to 100°. On evaporation of hot solutions, anhydrous strontium nitrate crystallizes in octohedra and combinations of octohedra and cubes; density 2.93, melting point 645°. At higher temperatures, oxygen is first lost, forming the nitrite, and conversion to the oxide takes place only on strong heating.

Strontium nitrate is sparingly soluble in absolute alcohol (1 part in 8500), and still less soluble (1 part in 60,000) in a mixture of equal volumes of alcohol and ether. Use is made of this for the analytical separation of strontium and calcium. Strontium nitrate is used in pyrotechnics (for Bengal lights, red flares).

(v) *Barium nitrate*, $Ba(NO_3)_2$, is prepared technically by several processes. Most simply, witherite, $BaCO_3$, is dissolved in nitric acid. In place of $BaCO_3$, it is possible to employ BaS, made by the reduction of heavy spar, $BaSO_4$. In this case, the hydrogen sulfide liberated is absorbed in caustic soda. Other processes are based on double decompositions between barium chloride and sodium nitrate (Chile saltpeter) or barium carbonate and calcium nitrate. Both methods involve the behavior of reciprocal salt pairs

$$BaCl_2 + 2NaNO_3 \rightleftharpoons Ba(NO_3)_2 + 2NaCl$$

$$BaCO_3 + Ca(NO_3)_2 \rightleftharpoons Ba(NO_3)_2 + CaCO_3$$

Barium nitrate is much less soluble in water than the other alkaline earth nitrates, and is practically insoluble in alcohol. The solubility in water is 7 g of $Ba(NO_3)_2$ in 100 g of water at 10°, 9.2 g at 20° and 32.2 g at 100°. Below 12°, the dihydrate $Ba(NO_3)_2 \cdot 2H_2O$ is the stable phase in contact with the solution, but at ordinary temperature barium nitrate crystallizes anhydrous, in regular octohedra of density 3.24 and melting point 575°. When more strongly heated, oxygen is first evolved, forming the nitrite. The oxide is obtained only by strong ignition.

Barium nitrate is employed for the preparation of pure barium oxide and peroxide, and also for the production of explosives and in pyrotechnics, for green fire.

A double salt $2KNO_3 \cdot Ba(NO_3)_2$ has been described by Wallbridge (1903). It forms microscopic crystals of tetrahedral habit. No corresponding salt is formed with $CsNO_3$, nor is such obtained from mixtures of KNO_3 and $Sr(NO_3)_2$.

(c) Carbonates

The normal or neutral carbonates, MCO_3, of the alkaline earth group are sparingly soluble in water. The carbonates of calcium, strontium and barium are converted by excess carbonic acid into soluble acid carbonates (hydrogen carbonates), $M(HCO_3)_2$, whereas the neutral carbonates of beryllium and magnesium can exist in contact with water only if excess carbonic acid is present; hydrolysis otherwise occurs, forming *basic carbonates* which are also insoluble. Like other carbonates, those of the alkaline earths are decomposed by acids (including acetic acid) with evolution of carbon dioxide. For solubilities see Table 56.

TABLE 56

SOLUBILITY OF ALKALINE EARTH CARBONATES IN WATER

100 g water at 18° dissolve

9.4 mg $MgCO_3$	1.3 mg $CaCO_3$	1.0 mg $SrCO_3$	1.72 mg $BaCO_3$

(i) *Beryllium carbonate.* The normal salt $BeCO_3$ is formed from aqueous solution only if excess carbonic acid is present. It crystallizes from solution as the tetrahydrate, $BeCO_3 \cdot 4H_2O$, but at 100° it gives up all its water and, at slightly higher temperatures, CO_2 also.

Alkali carbonates precipitate *basic carbonates* (of variable composition) from beryllium salt solutions. These, and also the neutral carbonate, are soluble in excess of alkali carbonate—and especially in concentrated ammonium carbonate solution—forming double salts, which are fairly readily decomposed by water.

(ii) *Magnesium carbonate*, $MgCO_3$, occurs native as *magnesite* (talc spar, bitter spar). [4] The double salt $MgCO_3 \cdot CaCO_3$, *dolomite*, is still more abundant. Both crystallize in the hexagonal system, but occur in nature generally in dense masses in which the eye can no longer detect crystalline character. This is true of magnesite in particular. Dolomite often has a more or less dark color, due to impurities; magnesite is usually white, often yellowish. Its hardness is 3 – 5. Artificially prepared magnesium carbonate is a white powder, density 3.04. The normal carbonate is formed only from aqueous solutions containing a considerable concentration of excess carbonic acid. Below about 16°, it crystallizes with $5H_2O$. The 3-hydrate crystallizes at higher temperatures, and can be converted by dehydration into a monohydrate. The neutral carbonate is gradually decomposed when it is boiled with water, forming basic carbonates. Basic carbonates are also formed when alkali carbonates are added to magnesium salt solutions; these have variable composition, and are in general not definite compounds.

According to Menzel (1930), only *one* basic magnesium carbonate is formed, by precipitation from solutions, as a definite compound with its own crystal structure, namely $4MgCO_3 \cdot Mg(OH)_2 \cdot 5H_2O$. This can lose $1H_2O$ fairly easily by drying, without any change in crystal structure. Products of other composition are mixed crystals, or often mere mixtures, of this compound with the normal carbonate or with magnesium hydroxide.

Basic magnesium carbonate occurs native in a few places in the crystalline state as *hydromagnesite*, with the approximate composition $4MgCO_3 \cdot Mg(OH)_2 \cdot 4H_2O$. The artificially prepared basic carbonate, known as *magnesia alba*, generally has a similar composition. The basic carbonate and the normal carbonate give up their carbon dioxide even on relatively gentle heating—e.g., magnesite can be converted at a gentle red heat into practically pure oxide (cf. Table 57, p. 277). Magnesium carbonate is very sparingly soluble in water, though more soluble than the hydroxide. The solubility (cf. Table 56) depends on the carbonic acid content of the water, and is increased thereby. The $CO_3^=$ ion content of an aqueous solution of carbon dioxide is very small, since carbonic acid is dissociated chiefly according to

$$H_2CO_3 \rightleftharpoons H^+ + HCO_3^- \tag{1}$$

Since the $CO_3^=$ ions which are in equilibrium with solid $MgCO_3$, according to

$$MgCO_3 \rightleftharpoons Mg^{++} + CO_3^= \tag{2}$$

react with H^+ ions, forming HCO_3^-, the equilibrium (2) is disturbed. To restore it, the solid phase must go into solution until the solubility product

$$K_{sp} = [Mg^{++}] \cdot [CO_3^=]$$

has been restored to the value it possessed in pure aqueous solution. To achieve this the decrease in $CO_3^=$ ion concentration must be compensated by an increase in Mg^{++} ion concentration.

Magnesite and dolomite are used on a large scale for the production of refractory bricks, which are obtained by 'burning' them (conversion to oxide). Such bricks are used, for example, to line the converter in the Thomas, or basic Bessemer, process (cf. Vol. II). Magnesite is also used to produce pure carbon dioxide for the manufacture of mineral waters. An especially compact magnesite, occurring in Bosnia, is used under the name of 'Bosnian meerschaum' for the manufacture of carved pipe bowls. Magnesite also constitutes the chief source, next to the magnesium compounds of the 'abraum salts', for the preparation of other magnesium compounds, and especially of Epsom salts.

Artificially prepared *basic magnesium carbonate* is used not only as a starting material for the preparation of other magnesium salts (since it dissolves in acids much more rapidly than does magnesite), but for the production of tooth powders and cleaning powders, and as a filler for paper, rubber and pigments. It is used in medicine under the name 'magnesia alba'.

Magnesium carbonate forms double salts with the alkali carbonates and hydrogen carbonates, of the type $M_2CO_3 \cdot MgCO_3$ —e.g., $K_2CO_3 \cdot MgCO_3 \cdot 4H_2O$, rhombic prisms; $KHCO_3 \cdot MgCO_3 \cdot 4H_2O$, white powder; $Na_2CO_3 \cdot MgCO_3$, small tetragonal crystals.

(iii) *Calcium carbonate*, $CaCO_3$, occurs in nature in the forms of *limestone*, *chalk*, and *marble*. Crystallized in the hexagonal system, it is known as *calcite* or *Iceland spar*. This occasionally occurs native in well formed crystals, usually rhombohedra, and calcite crystals of very considerable size have been found. Calcite is doubly refracting, as can be seen especially easily with the large, clear transparent rhombohedra, such as occur in Iceland. It has the density 2.72. An unstable modification of crystallized calcium carbonate less often met with in nature, is the rhombic *aragonite*, density 2.93. At 970°, calcite changes into another modification, also belonging to the hexagonal system. Aragonite also changes at high temperatures. *Vaterite* is yet another unstable modification.

The crystal structure of calcite, as also that of magnesite, is the same as that of sodium nitrate (Fig. 50, p. 217). Aragonite, like potassium nitrate, forms a rhombic layer lattice. Vaterite has another hexagonal structure. The structure of dolomite can be derived from the crystal lattice of calcite, in that every alternate basal plane occupied by Ca^{++} ions is replaced by one occupied by Mg^{++} ions.

Calcium carbonate is sparingly soluble in water. It is therefore deposited from aqueous solutions as a white precipitate when Ca^{++} ions and $CO_3^=$ ions mix. If precipitation is effected from hot, dilute solutions the precipitate consists at first of very fine aragonite crystals, which are slowly transformed in the cold into calcite. If calcium carbonate is precipitated cold, it at first appears to be amorphous, but also passes gradually into calcite when it stands in contact with the solution. Limestone and marble also consist of more or less fine calcite crystals. Pure calcium carbonate is white or colorless; variegated marble contains impurities, chiefly iron oxide. The yellowish or grey color of limestone is likewise due to impurities. The most important impurity in limestone is *clay*. Limestone rich in clay is known as *marl,—clay marl* or *chalk marl* according to the greater or lesser clay content. Chalk is a soft, earthy form of limestone, made up predominantly of the remains of shells of microscopic, prehistoric molluscs.

The solubility of calcium carbonate in water (Table 56) is appreciably raised by ammonium salts. Calcium carbonate can be completely decomposed by boiling it with ammonium chloride solution;

$$CaCO_3 + 2NH_4Cl = CaCl_2 + 2NH_3 + H_2O + CO_2$$

Such a decomposition is not brought about by alkali salts, nor do these augment the solubility. Calcium carbonate unites with excess carbonic acid, forming the fairly soluble hydrogen carbonate.

$$CaCO_3 + H_2CO_3 = Ca(HCO_3)_2 \qquad (3)$$

It is therefore fairly soluble in water containing carbonic acid. The 'temporary hardness' of water is due to the presence of calcium hydrogen carbonate. If such water is boiled, carbon dioxide is lost, and neutral calcium carbonate is deposited as a result of the reversal of eqn (3). Neutral calcium carbonate also remains on the evaporation of the solution at ordinary temperature. This is the mechanism whereby stalactites are formed.

At low temperatures, calcium carbonate is deposited from solution as the 6-hydrate. This forms rhombic crystals, of density 1.77, which rapidly decompose in air at ordinary temperature, forming the anhydrous salt. Krauss (1930) obtained a monohydrate as an unstable intermediate in the isobaric degradation ($p_{H_2O} = 7$ mm) of the 6-hydrate compound in the tensieudiometer.

When ignited, calcium carbonate loses carbon dioxide:

$$CaCO_3 = CaO + CO_2;$$

although this is considerably more firmly bound than in magnesium carbonate, as may be seen from the following Table 57 which compares the dissociation pressures of magnesium, calcium and barium carbonates. The CO_2-pressure of calcium carbonate reaches 1 atmosphere at 897°*. The thermal decomposition of calcium carbonate is carried out industrially on a very large scale (lime burning) to produce calcium oxide for mortar and other purposes. *Portland cement* is made by burning limestones together with clay (or limestones naturally rich in clay). Naturally occurring chalk is generally purified by levigation to make it suitable for technical purposes—e.g., as a paint (whitewash) or for chalks. Levigated chalk is also used in tooth powders, adhesives and as a polishing powder. Pure, finely divided calcium carbonate, made by precipitation from solution, is used in medicine—e.g., for indigestion. It is also used to deacidify wines.

TABLE 57

DECOMPOSITION PRESSURES OF ALKALINE EARTH CARBONATES

$t =$	400	450	500	540	700	900	1000	1100	1200	1300° C
$MgCO_3$ $p =$	0.1	6.8	100	747	—	—	—	—	—	— mm Hg
$CaCO_3$ $p =$	0	0	0.1	0.3	22.2	793	2942	8739	21800	— mm Hg
$BaCO_3$ $p =$	0	0	0	0	0	0.2	2.7	17.7	92	382 mm Hg

(iv) *Strontium carbonate*, $SrCO_3$, occurs native as *strontianite*, rhombic, isomorphous with aragonite and witherite. For this reason the mineral generally contains calcium carbonate, and frequently barium carbonate also, as impurities. It is used technically chiefly as a source for strontium oxide and hydroxide, for the de-sugaring of molasses. Since the supply of native strontianite is insufficient for this purpose, strontium carbonate is also prepared artificially from celestine, $SrSO_4$, which is fused with calcined soda in rotary kilns:

$$SrSO_4 + Na_2CO_3 = Na_2SO_4 + SrCO_3$$

Pure strontium carbonate is obtained by precipitation of strontium from aqueous solutions by means of ammonium carbonate. Strontium carbonate is very sparingly soluble in water; the solution has a basic reaction, because of partial hydrolysis. The solubility is raised considerably by the presence of carbonic acid in the solution. Strontium carbonate is less readily decomposed by heating than is calcium carbonate, but CO_2 is completely evolved at 1100°, according to Conroy.

Rhombic $SrCO_3$ changes to a hexagonal form at 929°. If the loss of CO_2 is prevented, this form melts in the neighborhood of 1500°.

(v) *Barium carbonate*, $BaCO_3$, occurs in nature as *witherite*, the most important source being certain lead glance deposits in Cumberland, England. It generally forms rhombic crystals, isomorphous with aragonite and strontianite. Above 811°, a hexagonal form is stable, and above 982°, a cubic form. It is thrown down from aqueous solutions, as a white precipitate, by the union of Ba^{++} and $CO_3^=$ ions. It is manufactured industrially from heavy spar, $BaSO_4$, which is reduced to barium sulfide by heating it with carbon to 600—800°: $BaSO_4 + 2C = BaS + 2CO_2$. Carbon dioxide is passed into the aqueous solution, whereupon barium carbonate

---

* The decomposition of calcium carbonate can be catalytically accelerated by other gases. Cf. G. Hüttig, *Z. anorg. Chem.*, 255 (1948) 223.

is precipitated: $BaS + CO_2 + H_2O = BaCO_3 + H_2S$. Another process is to heat finely divided heavy spar with concentrated potassium carbonate solution under pressure, when double decomposition takes place: $BaSO_4 + K_2CO_3 = BaCO_3 + K_2SO_4$. Barium carbonate loses CO_2 at considerably higher temperatures than the carbonates of the other alkaline earth metals, and the CO_2-pressure reaches 1 atmosphere only above 1400°. Since barium carbonate begins to sinter below this temperature, and loses its porosity in consequence, it is difficult to decompose barium carbonate completely by heating alone (cf. Table 57).

Barium carbonate is used in the glass industry for the production of easily fusible, heavy glasses, of high refractive index, and also in the ceramic industry, as well as for the preparation of other barium compounds. The preparation of BaO, which is important as a starting material for the production of BaO_2, is best carried out by heating natural witherite with carbon. For other purposes, artificially prepared barium carbonate is preferred because of its lower cost.

Barium carbonate is rather more soluble in water than are the carbonates of calcium and strontium. The saturated aqueous solution has an alkaline reaction, because of hydrolysis. An aqueous suspension of barium carbonate reacts with salts which have a strongly acid reaction because of hydrolysis,—e.g., the salts of aluminum, trivalent iron and chromium. CO_2 is evolved, and the element in question is precipitated as its hydroxide or oxide. Use is often made of this in analytical chemistry, to separate the trivalent metals from the bivalent metals in the ammonium sulfide group. The solubility of barium carbonate, (like that of the other alkaline earth carbonates,) is somewhat raised by ammonium salts, and considerable quantities dissolve in water containing carbon dioxide (through the formation of hydrogen carbonate).

(d) Sulfates

The sulfates of beryllium and magnesium are readily soluble. Those of the alkaline earths are sparingly soluble, the solubility decreasing not merely from magnesium to calcium sulfate, but progressively to barium sulfate. 100 g of water at 18° dissolve 35.6 g of $MgSO_4$, 0.202 g of $CaSO_4$, 0.014 g of $SrSO_4$ and 0.00022 g of $BaSO_4$.

With the exception of barium sulfate, the sulfates of the Group IIA elements form double salts with the alkali sulfates, of the type $M^I_2SO_4 \cdot M^{II}SO_4$.

(i) *Beryllium sulfate*. The tetrahydrate of beryllium sulfate, $BeSO_4 \cdot 4H_2O$, crystallizes out in colorless octohedra when a solution of beryllium oxide or hydroxide in an excess of dilute sulfuric acid is evaporated. A dihydrate and hexahydrate exist, as well as the tetrahydrate. Beryllium sulfate dissolves readily in water, but is insoluble in absolute alcohol. The aqueous solution is hydrolyzed, but less so than that of the chloride (0.56 % in 0.1-N solution at 40°). Basic products of variable composition separate when solutions which contain no excess of sulfuric acid are evaporated. The anhydrous sulfate, $BeSO_4$, is obtained by heating the hydrates to about 220°, and begins in its turn to lose sulfur trioxide when more strongly heated. Pure beryllium oxide remains after prolonged heating at a white heat.

With the alkali metal sulfates, beryllium sulfate forms double salts—e.g., the beautifully crystalline $K_2SO_4 \cdot BeSO_4 \cdot 2H_2O$.

(ii) *Magnesium sulfate*, $MgSO_4$, when anhydrous, is a white powder of density 2.65. It combines with water, forming hydrates, of which the monohydrate (*kieserite*) and the heptahydrate (*Epsom salt*) are the most important. It forms several hydrates other than these, and compounds are known with 1,2,4,5,6,7 and $12H_2O$. Of these, the hydrates with 2,4 and $5H_2O$ are unstable. The stability ranges of the others are as follows:

—3.9 to 1.8°	1.8 to 48.3°	48.3 to 68°	above 68°
$MgSO_4 \cdot 12H_2O$	$MgSO_4 \cdot 7H_2O$	$MgSO_4 \cdot 6H_2O$	$MgSO_4 \cdot H_2O$
	rhombic	monoclinic	monoclinic

An unstable tetragonal modification of the hexahydrate can be obtained by inoculation of a supersaturated magnesium sulfate solution with crystals of a mixture of zinc and copper sulfates.

The heat of solution of the anhydrous salt is 20.3 kcal., of the monohydrate 13.3 kcal., and of the heptahydrate —3.8 kcal. The *solubility* is:

at	0°	10°	20°	30°	40° C
	26.9	31.5	36.2	40.9	45.6 g of $MgSO_4$ in 100 g of water

Magnesium readily forms double salts with the alkali metal sulfates. The following double salts occur in the potassium salt deposits.

$KCl \cdot MgSO_4 \cdot 3H_2O$	kainite	$3Na_2SO_4 \cdot MgSO_4$	vant'hoffite
$K_2SO_4 \cdot MgSO_4 \cdot 4H_2O$	leonite	$K_2SO_4 \cdot MgSO_4 \cdot 2CaSO_4 \cdot 2H_2O$	polyhalite
$K_2SO_4 \cdot MgSO_4 \cdot 6H_2O$	schönite	$K_2SO_4 \cdot MgSO_4 \cdot 4CaSO_4 \cdot 2H_2O$	krugite
$K_2SO_4 \cdot 2MgSO_4$	langbeinite	$Na_2SO_4 \cdot MgSO_4 \cdot 2\frac{1}{2}H_2O$	löweite
$Na_2SO_4 \cdot MgSO_4 \cdot 4H_2O$	astrakanite		

The double salt $K_2SO_4 \cdot MgSO_4 \cdot 6H_2O$ is also prepared artificially and sold as a fertilizer under the name 'potash magnesia'. It is particularly suitable for those plants that are sensitive to chloride ion.

Kieserite, $MgSO_4 \cdot H_2O$, occurs in large quantities in the German potash salt deposits, especially as a constituent of the mixture known as 'Hartsalz' (hard salt), but also admixed with the crude carnallite and other potassium salts. In spite of the considerable solubility of magnesium sulfate, the monohydrate, kieserite, passes only very slowly into solution, and therefore remains as a sandy residue when the potassium salts are leached out. On standing, this sets completely solid, since the adherent water combines with it, to form higher hydrates. If, therefore, the wet leaching residue from the potassium salts is filled into moulds, it solidifies in the course of a day into blocks as hard as stone. These can be despatched to the magnesium salt works if they are not to be worked up for Epsom salts on the spot.

Pure kieserite forms colorless monoclinic crystals, of density 2.57 and hardness 3.8. Unlike the compact form of the monohydrate, or the blocks made from it, which dissolve only very slowly in water, the salt dehydrated by heating ('calcined kieserite') dissolves fairly rapidly in water. It is used in water treatment.

Epsomite, $MgSO_4 \cdot 7H_2O$, is found in nature in the dried up residues of lakes, and dissolved in many mineral springs (bitter springs). It also occurs in potassium salt deposits, as a transformation product of kieserite. In this case it is known also as *reichardtite*.

Epsom salt crystallizes from aqueous solutions at ordinary temperature in four-sided rhombic prisms (of density 1.68), isomorphous with zinc vitriol and nickel vitriol. It is stable in moist air, but effloresces in dry air. It has a repulsively bitter taste. It was formerly prepared from mineral waters—first by Nehemiah Grew (1695), from the mineral springs of Epsom. Hence it was called *sal anglicum*, or Epsom salt. It is now prepared chiefly from the kieserite residues in the extraction of potassium salts[5]. These residues are dissolved in hot water, and the solution is clarified by allowing it to stand in tanks, well insulated against loss of heat. Epsom salt crystallizes out in glistening needles on cooling. It is also obtained by dissolution of magnesite in sulfuric acid.

Magnesium sulfate is used in industry as a weighting material for cotton and silk, to impregnate muslin to make it non-inflammable, as a mordant in dyeing, as a filler for paper, and in medicine as a purgative.

(iii) *Calcium sulfate* is found widespread in nature, as the dihydrate, *gypsum* (*selenite*), $CaSO_4 \cdot 2H_2O$, and anhydrous as *anhydrite* (*karstenite, muriacite*). It frequently occurs dissolved in drinking waters, and then gives rise to the permanent hardness (i.e., that not removed by boiling). The solubility of calcium sulfate in water is only slight, however, being 202 mg of $CaSO_4$ in 100 g of water at 18°, and changes but little with temperature. The solubility curve passes through a flat maximum between 30° and 40°. The solubility is decreased by addition of sulfates, but quite considerably increased by other salts and by acids, including sulfuric acid. With sulfuric acid, calcium sulfate forms fairly soluble addition compounds, which can be isolated—e.g., $CaSO_4 \cdot H_2SO_4$ and $CaSO_4 \cdot 3H_2SO_4$. Sparingly soluble double salts are formed with the alkali metal sulfates; these also occur naturally— e.g., *glauberite* $Na_2SO_4 \cdot CaSO_4$ and *syngenite*, $K_2SO_4 \cdot CaSO_4 \cdot H_2O$.

Below 66°, calcium sulfate always crystallizes as the dihydrate, in six-sided monoclinic prisms of density 2.32. Crystals of gypsum are very prone to form twins (arrow head or swallow tail twins). Gypsum occurs in very large amounts in nature, sometimes in the form of beautiful, large, well-formed crystals, but generally in the form of a fibrous (satinspar), granular or quite compact rock, made up of small or minute crystals of gypsum. Gypsum rocks are found in all geological formations, but especially in the Permian or Dias, the Trias and the Tertiary, occasionally in vast beds and deposits. It is easily recognizable because of its low hardness ($1\frac{1}{2}$—2) and its excellent cleavage. Like all monoclinic minerals it is doubly refracting. Varieties of gypsum are *selenite* (or 'Marienglas') and *alabaster*. The latter looks very like white marble, but unlike marble does not feel cold to the touch, because gypsum has a very low thermal conductivity. Pure gypsum is colorless, or white, as a crystal aggregate. It may, however, be colored grey, yellowish, brown, reddish or even black by impurities.

When heated to 100°, gypsum gives up $\frac{3}{4}$ of its crystal water and changes into the (metastable) hemihydrate $CaSO_4 \cdot \frac{1}{2}H_2O$. This absorbs water again at ordinary temperature, with a considerable evolution of heat. When mixed to a paste with water, it solidifies fairly rapidly into a solid mass ('plaster of Paris') consisting of fine fibrous gypsum crystals, felted together. On this is based the use of plaster in building, and in sculpture (for castings). The burned gypsum used for industrial purposes ('plaster of Paris') generally contains still less water than the hemihydrate, but it must not be completely dehydrated. If it is so completely burned that all the water is eliminated, it has practically no remaining capacity for rehydration, and is said to be 'dead burned'. The naturally occurring, anhydrous calcium sulfate, *anhydrite*, does not hydrate and set, although when in contact with water for a long time it is likewise ultimately converted into gypsum. A considerable proportion of the naturally occurring gypsum has been formed originally from anhydrite in this way. Conversely, anhydrite has frequently been formed from gypsum. Anhydrite crystallizes from pure aqueous solutions above 66°, but if other salts are present in the solution, anhydrite may be deposited at much lower temperatures. For example, calcium sulfate crystallises as anhydrite at temperatures above 30°, from a solution saturated also with sodium chloride. There is also another modification of anhydrous calcium sulfate, in addition to anhydrite. This is more soluble than anhydrite, and is accordingly unstable.

Native anhydrite is found laid down in beds in rock salt deposits, and between rock salt and potash salt deposits. It is widely distributed elsewhere, however, and is encountered in almost every geological formation, generally mixed with gypsum which has been formed from it. Anhydrite forms rhombic crystals and has a good cleavage, although not as perfect as that of gypsum. It is harder than gypsum $(3—3\frac{1}{2})$ and denser $(d = 2.8—3)$. It is colorless or white when pure, but impurities often confer a blue-grey or other color.

If gypsum or anhydrite are heated above 1000°, they begin to lose sulfur trioxide. The resulting product (a solid solution of CaO in $CaSO_4$) has recoverd the ability to take up water, and when mixed with a little water it hardens faster than a lime-sand mortar, to a very hard, dense and weather-proof mass. This is the basis of the use of strongly burned gypsum (generally heated over 1300°) as a builder's mortar (gypsum mortar, Estrich plaster), which was already known to the ancient Egyptians. Apart from its use as mortar, plaster of Paris is used extensively for making moulds in ceramics, especially for casting porcelain ('slip casting'), for which its porosity makes it very suitable. Finely ground, unburned gypsum is used as an addition to mineral pigments, especially for wall paper printing and in paper-making.

The decomposition pressure of crystalline $CaSO_4$ (anhydrite) is 3 mm at 1200°, 40.5 mm at 1360°. At high temperatures, (over 1000°), dehydrated gypsum is converted into anhydrite. Until this transformation is complete, its dissociation pressure is much higher than that of anhydrite (Zawadski, 1932).

Gypsum has a strong tendency to form supersaturated solutions. The dependence of solubility upon the particle size is also particularly marked for gypsum, and this must be taken into account if use is made of the resistance of a saturated calcium sulfate solution to calibrate conductivity vessels. After making a deduction for the conductivity of the water, the specific conductivity K of a saturated solution of gypsum is 0.001867 at 18° C (Melcher) and 0.002206 at 25° (Hulett). These values are correct so long as the particles of gypsum are not less than 2μ in diameter; above this limit, the conductivity is practically independent of particle size. Finer particles can be removed by repeated leaching, and they disappear spontaneously (the larger particles growing at the expense of the smallest) if the solution is allowed to stand about 3 days.

(iv) *Strontium sulfate*, $SrSO_4$, occurs in nature as *celestine*, which forms the starting material for the technical preparation of most strontium compounds. Although not so widely distributed as heavy spar, celestine is of fairly common occurrence, usually in granular, flaky or dense aggregates, but sometimes in well formed crystals. It is colorless when pure, but is often discolored by impurities. Its hardness is $3—3\frac{1}{2}$, density 3.9—4, solubility 11.4 mg in 100 g of water at 18°. As with calcium sulfate, the solubility in water is raised by the addition of various substances. It forms sparingly soluble double salts with the alkali metal sulfates e.g., with K_2SO_4 and $(NH_4)_2SO_4$. The ordinary rhombic modification of strontium sulfate is transformed above $1152°$ into another, probably monoclinic form. It melts at a white heat, but commences to lose SO_3 before doing so.

(v) *Barium sulfate*, $BaSO_4$, is widely distributed as a mineral (heavy spar or *barytes*). It is often found in well-formed rhombic crystals, which are notable for the wealth of crystal forms displayed. Hardness $3—3\frac{1}{2}$. Density 4.48. It generally forms fibrous, granular, compact or earthy aggregates. It is colorless, but is often colored by the impurities present. Barytes often contains considerable amounts of admixed strontium sulfate, with which it is isomorphous, and in a few places is found containing calcium sulfate (lime barytes of Freiberg, Germany and Derbyshire, England).

Barium sulfate has a rhombic crystal lattice, made up from Ba^{2+} and SO_4^{2-} ions. The sulfates of strontium and lead have the same structure. A series of perchlorates ($M[ClO_4]$, M = K, Rb, Cs, NH_4, Tl^I), fluoroborates ($M[BF_4]$, M = Rb, NH_4) and potassium permanganate, $K[MnO_4]$ also crystallize according to the same structural type ('barytes type').

Barium sulfate is almost insoluble in water, only 0.22 mg of $BaSO_4$ dissolving in 100 g of water at $18°$. It is thrown down as a white precipitate from barium salt solutions by the addition of SO_4^-ions, and from solutions containing sulfuric acid or sulfates, by the addition of Ba^{++}ions. The solubility is markedly raised by strong acids. Barium sulfate is, in fact, quite soluble in concentrated sulfuric acid, up to a content of $10—12\%$. The solubility is here due to the formation of complexes. Because of its insolubility, barium sulfate serves for the gravimetric determination of $SO_4^=$ and of Ba^{++}ions. It must be noted that $BaSO_4$ can carry down significant quantities of other substances which are present in solution, and especially those with ions of high valence—e.g., ferric, chromic, and aluminum sulfates. It is not yet known definitely whether this is due merely to adsorption, or whether mixed crystal formation plays any part.

According to Grimm (1924), $BaSO_4$ can take up very considerable amounts of $KMnO_4$ (up to 60 mol%), by forming mixed crystals (see also KARAOGLANOW, *Z. anorg. Chem.*, 222 (1935) 249). It follows that the possibility of forming mixed crystals depends not so much on 'chemical similarity' as on the fulfilment of the following conditions: (i) analogy of chemical formula (e.g., $Ba[SO_4] — K[MnO_4]$; $Ca[CO_3] — Na[NO_3]$); (ii) identity of crystal type; (iii) similarity of interionic distances in the crystal lattice (the differences should not exceed about 5 % at room temperature).

Barium sulfate is notable for its absolute stability against heat, the atmosphere, and injurious gases present in the atmosphere (especially hydrogen sulfide). It is therefore used as a pigment (permanent white, blanc fixe). It is, however, seldom used alone, since, it has only poor *covering power*. On the other hand, it is used on a large scale as an addition to mineral pigments and organic lakes. It is also used widely as a filler for paper, especially for wallpaper, cartons, colored papers, and especially for photographic papers, and confers an excellent glaze. For all these purposes, barium sulfate must be in a state of extremely fine subdivision. The finest grinding of barytes yields a product unsatisfactory for many purposes, and large quantities of barium sulfate are therefore made artificially. Either witherite or natural barytes can serve as the starting material. The latter is first reduced to the sulfide by heating it with carbon; the sulfide so obtained, or the native carbonate, is dissolved in hydrochloric acid, and barium sulfate is precipitated by addition of sodium or potassium sulfate. The commercial product is a paste with 15—20% water content, since its covering power is lost again if it is fully dried out. Barium sulfate is obtained as a by-product in the preparation of hydrogen peroxide by the action of sulfuric acid on barium peroxide. Because of the high absorption of X-rays by barium sulfate, it is used in medicine to render visible the contours of the stomach and intestinal tract in radiological examinations. Its excessively low solubility explains why it is quite non-poisonous.

9. Analytical

In the course of the usual analytical separation, *calcium, strontium* and *barium* are precipitated together in the form of carbonates, by treatment of the solution with ammonium carbonate in the presence of ammonium chloride, after removal of all the heavy metals. The three elements are best separated from each other by utilization of the differences in solubility of their nitrates and chlorides in ether and alcohol. Only calcium nitrate dissolves readily in an alcohol-ether mixture; barium chloride is the only chloride insoluble in absolute alcohol. Barium may also be separated by taking advantage of the fact that it is precipitated alone, as barium chromate, from acetic acid solutions, by means of potassium dichromate. The solubility of barium chromate is about 1 part in 300,000. Strontium chromate is also sparingly soluble (about 1 part in 800), but its solubility product, $K_{sp} = [Sr^{++}]\cdot[CrO_4^=]$, is so much larger than that of barium chromate that the small concentration of $CrO_4^=$ ions, present in equilibrium with $Cr_2O_7^=$ ions in acetic acid solution, does not suffice to bring about precipitation. Barium and strontium salt solutions give precipitates with calcium sulfate solution, for since the solubility products of $BaSO_4$ and $SrSO_4$ are so much smaller than that of $CaSO_4$, the concentration of $SO_4^=$ ions in calcium sulfate solution is sufficient to precipitate them, even—in the case of Ba^{++} ions—when their concentration is quite small. The alkaline earth metals may be identified by observation of the flame colorations or, better, by use of a spectroscope.

In the qualitative analytical group separation, *magnesium* is separated from the filtrate after precipitation of the alkaline earth carbonates, and after removal of all ammonium salts by gentle ignition. It is commonly precipitated as hydroxide, by means of baryta water. It is most conveniently identified by converting it to magnesium ammonium phosphate, $Mg(NH_4)PO_4\cdot6H_2O$, which crystallizes in characteristic form.

In recent years, organic reagents such as quinalizarin and diphenylcarbazide have been much used for the detection of magnesium. As a result of adsorption, these impart a blue or red coloration, respectively, to magnesium hydroxide precipitated in their presence; this color is very apparent even when precipitation is carried out in very dilute solutions.

Beryllium accompanies aluminum in the usual course of analytical separation, and is not very readily distinguished from it. Although beryllium sulfate forms no alums, it does form other crystalline double salts with alkali metal sulfates. Beryllium differs from aluminum in that it is not precipitated, in the cold, by a sufficiently large excess of concentrated ammonium carbonate. On boiling, beryllium is precipitated as basic carbonate. The solubility of beryllium chloride in a mixture of equal volumes of saturated aqueous and ethereal hydrogen chloride can also be used for a separation from aluminum, the hydrated chloride of which is insoluble in this solvent. Beryllium is determined quantitatively in the same way as aluminum—precipitation as hydroxide, and weighing as oxide.

Magnesium is generally determined by precipitation as ammonium magnesium phosphate; this is converted by ignition into magnesium pyrophosphate, $Mg_2P_2O_7$, and weighed as such. *Calcium* is usually separated in quantitative analysis as the oxalate, CaC_2O_4, since this is the most insoluble calcium salt. Calcium oxalate is converted to the oxide by ignition, and is weighed as such. *Strontium* is often precipitated as oxalate, but better as carbonate, and is weighed as the oxide. It can also be determined gravimetrically as the sulfate. *Barium* is usually determined as the sulfate.

Magnesium can also be determined volumetrically by first precipitating it as magnesium ammonium arsenate, $Mg(NH_4)AsO_4$. This is dissolved in hydrochloric acid, and arsenic acid is determined iodometrically (Daubner, 1935). Magnesium can be separated from alkali metal ions and (in the presence of ammonium salts) simultaneously from alkaline earth ions, by precipitation with 8-hydroxyquinoline ('oxine') (Berg, 1927). The precipitate can be dried and weighed directly, or can be titrated bromometrically. Calcium can be determined volumetrically by precipitation of the oxalate, liberation of oxalic acid from this by means of dilute sulfuric acid, and titration with permanganate. Very small amounts of strontium in the presence of calcium can best be determined spectroscopically, by means of an electric arc struck over a solution of the salt (Ruthardt, 1931).

Methods used for the quantitative separation of the alkaline earth metals from one another follow the processes employed in qualitative analysis. However, in order to achieve a complete separation of the three related elements, such as is required for their quantitative determination, it is necessary to follow special procedures, such as are given in handbooks of rock analysis.

References

1 A. BECK (Editor), *Technology of Magnesium and Its Alloys*, 2nd Ed., London 1941, 512 pp.
2 W. BULIAN and E. FAHRENHORST, *Metallographie des Magnesiums und seine technischen Legierungen*, 2nd Ed., Berlin 1949, 139 pp.
3 V. D. I., *Werkstoff Magnesium*, Berlin 1939, 164 pp.
4 R. BANCO, *Der Magnesit und seine Verarbeitung*, Dresden 1932, 64 pp.
5 F. RUNGE, *Organomagnesiumverbindungen*, Stuttgart 1932, 328 pp.
6 F. J. NORTH, *Limestones, Their Origins, Distributions and Uses*, London 1930, 467 pp.
7 A. B. SEARLE, *Limestone and Its Products, Their Nature, Production and Use*, London 1935, 709 pp.
8 F. RAULS, *Der Gips*, Vienna 1936, 267 pp.
9 Union Minière de Haut Katanga, *Le Radium: Production, Propriétés générales, Applications thérapeutiques, Appareils*, Brussels 1931, 366 pp.
10 R. BRUNNGRABER, *Radium*, Milan 1937, 338 pp.
11 E. V. PANNELL, *Magnesium, Its Production and Use*, 2nd Ed., London 1948, 189 pp.

CHAPTER 9

CONSTITUTION AND PROPERTIES

1. General

The compounds of the elements hitherto considered have all been extremely simple in structure. Metals of the first two main groups of the Periodic System form *salts*, or '*salt-like*' *compounds* (i.e., compounds which correspond in structure to the salts) in the great majority of cases. The typical salts are built up out of positively and negatively charged ions; when these ions are 'inert gas'-like (i.e., have the same electronic configuration as the neutral inert gas atoms), compounds of the types AB and AB_2 formed from them tend to have simple structures, as is shown by the examples already encountered. Except for differences in solubility, the salts (and salt-like compounds) discussed in the earlier chapters display a considerable similarity in properties. Those compounds of the metals of Main Groups I and II which are not salt-like in character (i.e., the *covalent compounds* of these elements) are also simple in structure, with a few exceptions; they mostly consist of diatomic or triatomic molecules, AB or AB_2, in which the atoms are bound by electron-pair bonds. The compounds of this class also display a general similarity in properties, and only in exceptional cases (e.g., beryllium hydride and magnesium hydride) have we hitherto dealt with compounds which could not at once be assigned to the appropriate class, on the basis of their properties and chemical behavior.

In subsequent chapters, it will be necessary to consider compounds which display a much greater diversity in properties. In addition to the classes of compound already discussed, other types will be met with, and it will be found that the physical and chemical properties of any compound are determined not only by its *composition*, but to a marked degree by its *constitution*. It is therefore desirable to discuss the relationship between *the constitution and the properties* of compounds, before dealing with the elements and compounds of the later groups of the Periodic System.

A proper understanding of this relationship is one of the most important objectives of chemistry. [1–4] It may be used to predict the existence and properties of compounds which are as yet unknown, and the conditions under which they may be formed; it enables us to modify the properties of known substances, by suitably changing their chemical constitution, or to produce new materials, with any desired properties, by intelligent and systematic chemical synthesis. Within the domain of organic chemistry, this aim has long been realized to a considerable extent. Inorganic chemistry, at the present time, is in an earlier stage of development, which may lead to the same end.

The classical methods used in organic chemistry for the determination of constitution can be applied only to certain fields of inorganic chemistry, and then usually only to a restricted extent.* Purely empirical knowledge of the connection between constitution and properties, based on such methods, cannot necessarily be transferred from organic to inorganic substances. Even when the most modern theoretical and experimental techniques of physics have been applied to such researches, their application to organic compounds is often much simpler than to inorganic compounds. The older methods of determining structure, based essentially on the study of chemical properties, have generally proved quite useless in the case of inorganic compounds. The reason for this is that organic compounds are almost invariably built up from *molecules*, which contain only a limited number of atoms, and which can be transferred to the gaseous state, or brought into solution, without any significant change in constitution. Most inorganic compounds, on the other hand, are built up in the solid state from *atoms* or *ions* in quite *indefinite number*. When such substances are vaporized, or brought into solution, the atoms or ions are removed from the fields of force which are operative between them in the solid state, and it is no longer possible to infer what sort of structure existed previously. Moreover, organic compounds involve, for the most part, only one type of valence bond. There is no *fundamental* difference between the different types of carbon-carbon bond (single and double or triple bonds, aromatic bonds), nor between these bonds and the bonds which link other atoms to carbon in the typical organic compounds. In inorganic compounds, on the other hand, a distinction must be drawn between several fundamentally distinct types of chemical bond. Between these types there are various transitions, which complicate the matter further.

Laue's discovery of the diffraction of X-rays by crystals (p. 206 *et seq.*) opened up the possibility of studying the structure of solids built up from atoms or ions in infinite array. It became possible to understand the typical differences between the various kinds of chemical bond met with in inorganic compounds as the theory of atomic structure (pp. 82 *et seq.*, 118 *et seq.*) and the quantum mechanical theory of valence (cf. pp. 131 *et seq.*) were developed.

2. Types of Bond [5]

The following types of bond are recognized:
 (i) Electrovalence (ionic binding, heteropolar binding)
 (ii) Covalence (atomic binding, homopolar binding)
 (iii) Metallic bond
 (iv) Van der Waals' forces (molecular forces).
 (v) Hydrogen bridge bonds.

 (i) *Electrovalence* arises from the operation of forces of attraction between oppositely charged ions. If we can regard the ions very approximately as spheres, charged uniformly over their surfaces, the magnitude of the force is defined by Coulomb's law (p. 83). This force may be intensified if the distribution of charge on the ions is altered (deformation, polarization—cf. pp. 48, 215).

At a certain distance between the centers of the ions, this given by the sum of their effective radii, this attraction is just compensated by a repulsion, which increases very rapidly if

* One example of the very successful application of methods borrowed from organic chemistry to structural studies in a wide and important section of inorganic chemistry, is provided by Werner's interpretation of the three-dimensional structure of complex compounds, such as cobaltammines and chromammines, etc., on the basis of the phenomena of isomerism and optical activity (cf. Vol. II, Chap. 5) The type of chemical binding involved in the compounds concerned (the strong complexes or 'penetration' complexes) is essentially covalent, and closely related to that which operates in organic compounds (cf. p. 412).

the interionic distance is yet further diminished, and which equally rapidly drops to a vanishingly small magnitude if the distance is increased. The resultant variation in the potential energy of a pair of oppositely charged ions, as the distance between them alters,

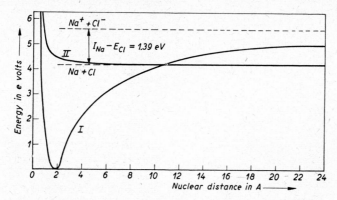

Fig. 64. Relation between potential energy and nuclear distance for a couple of ions ($Na^+ + Cl^-$, curve I) and a couple of atoms (Na + Cl, curve II).

is represented by curve I, Fig. 64. The position of the minimum in this potential energy curve gives the distance between the centers of the ions in the molecule that they form.

(ii) It follows from wave mechanics (cf. pp. 91 *et seq.*, 131 *et seq.*) that uncharged atoms can also exercise attractive forces on one another, and these may be as strong as, or even much stronger than, the forces between oppositely charged ions. A bond resulting from these forces between uncharged atoms (exchange forces) is a *covalent bond*.

Curve II, Fig. 65, shows the potential energy curve for a pair of H atoms which are capable of combining, as calculated by quantum mechanics. For comparison, curve I gives the potential energy curve for an H_2 molecule built up from oppositely charged H^+ and H^- ions, making due allowance for the repulsion which operates at very small internuclear distances. It is evident that the energy minimum of this curve lies far higher than that of curve II, and that only the H_2 molecule built up from neutral atoms is stable; that built up from H^+ and H^- ions would dissociate exothermically into H atoms.

As has already been discussed, the usual condition for the formation of a covalence is the 'pairing' of electron spins. An attraction is set up between two H atoms only if their electrons have opposed spins. If the electron spins are in the same direction, the forces are repulsive. For the same reason, there can be no covalent forces between atoms of the inert gases, but only repulsive forces, since these atoms contain no unpaired electrons. The same is true of ions with the inert gas configuration—i.e., Na^+ and Cl^- ions. These would exert repulsive forces on each other, if it were not that the weak exchange forces are outweighed by the Coulomb attractive forces. The free atoms, Na and Cl, also exert repulsions, as can be seen from curve II, Fig. 64. It is clear from this figure that *electrovalent* binding is the stable form for the NaCl molecule.

Curves I and II of Fig. 64 and Fig. 65 differ from one another in a striking way. The potential energy curves for the *ions* (curve I in each case) approach asymptotically and relatively slowly to a limiting value, representing the potential energy of the system when the ions are an infinite distance apart. These values are shown by the dotted lines, labelled $Na^+ + Cl^-$ and $H^+ + H^-$, respectively. The potential energy curves for the *atoms* attain the corresponding limit (shown by the lines labelled Na + Cl and H + H, respectively) very much more rapidly. Electrostatic attractions thus operate over considerably greater distances than do the exchange forces involved in covalence. The difference in energy between these limiting values is also significant in the two cases shown in Fig. 64 and Fig. 65, being 1.39 e.v. and 12.88 e.v., respectively. This corresponds in each case to the energy difference between the state of free ions, $A^+ + B^-$, and of free atoms, $A + B$, and is given by $I_A — E_B$, where I_A is the ionization energy of the atom A, and E_B is the electron affinity of atom B. It can be shown that if the difference $I_A — E_B$ is small, an ionic bond will be formed, whereas if $I_A — E_B$ is large, a molecule will be formed from the neutral atoms.

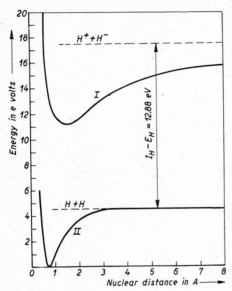

Fig. 65. Relation between potential energy and nuclear distance for $H^+ + H^-$ (curve I) and $H + H$ (curve II).

(iii) In the *metallic bond,* as follows from the theory of the metallic state, (cf. Vol. II, Chap. 1) the structural units exercise a collective or cooperative action, such that the atoms are positively charged, whereas electrons, in a more or less free state, are located between them.

(iv) The name of *Van der Waals' forces* [6] is given to those binding forces which give rise also to the deviation of gases from the ideal gas laws, and to the liquefaction of gases at sufficiently low temperatures. Since these forces act between molecules which are saturated in the ordinary sense, the resulting forces are sometimes referred to as molecular forces.

As has already been mentioned (p. 102), the Van der Waals forces arise partly from polarization effects, and partly from wave mechanical resonance forces. The latter are spoken of as 'dispersion forces', since they are to be ascribed to the same sort of effect as the dispersion of light.* They differ from the resonance forces concerned in the covalent bond by their much greater range of action. In general the Van der Waals forces are much weaker than the ordinary valence forces. The potential energy curve which represents their action thus shows only a shallow minimum. Since they are operative at distances where the exchange forces responsible for covalent bonding have vanished, Van der Waals forces can bring about attraction between atoms which would repel one another by the action of the ordinary exchange forces—e.g., a pair of inert gas atoms, or two H atoms with parallel spins. These cannot combine to form a molecule since the repulsions are too strong at small interatomic distances, but the Van der Waals forces bring about a weak attraction even between such atoms at greater interatomic distances.

In chemical compounds, the several sorts of bond enumerated above are often not realized in their pure forms. There are intermediate bond types, which cannot be strictly assigned to one kind or the other.

For example, if the potential energy curve for an ion-pair molecule A^+B^- lies only a little above that for the atom-molecule AB (cf. Fig. 65), it may be deduced from quantum me-

---

* Dispersion, in optics, originally meant the decomposition of light by a prism. The name is also applied, however, to the phenomenon upon which this decomposition depends, namely the change of refractive index of any medium with the wave length of light.

chanics that the possibility of a transition to the state of the ion-pair molecule leads to a considerable increase in the stability (i.e., to a stronger bond).* Correspondingly, if the ionic bond confers the greater stability, it may be yet further strengthened by the possibility of transition to a covalence. The contribution made by the ionic state to the binding energy has been called by Mulliken (1936) the *ionicity* of the bond in question. A bond is to be regarded as a *pure electrovalence* if its ionicity is practically 100 %, and as a *pure covalence* if its ionicity does not differ markedly from zero. All molecules with a finite dipole moment involve bonds of perceptible ionicity.

Just as there are transitional types between the elctrovalent and covalent bond, so there is also a continuous transition between these forms and the metallic bond.

In principle, the Van der Waals' forces always act. They are not usually significant, however, when other valence forces are operative, since their effects are quite outweighed.

(v) Certain compounds containing hydrogen bound to highly electronegative atoms—i.e., especially compounds in which H is attached to O or N—have a strong tendency to associate, which results not only in a considerable measure of association in the liquid state, but also to the persistence of association when the compounds are dissolved in inert solvents, and often even when they are vaporized. The kind of bond whereby the simple molecules are linked together to form associated molecules ('super-molecules') is known as the *hydrogen bond*. Hydrogen bonds may be formed not only between similar molecules, but also between unlike molecules—e.g., the combination of NH_3 with H_2O is due to hydrogen bonding, as also is the linkage of water to oxalic acid in oxalic acid dihydrate**.

The idea that hydrogen could link two molecules which were already saturated in the ordinary sense, had been proposed by Werner, who suggested that hydrogen could have the coordination number two. P. Pfeiffer (1913) explained the reduced reactivity of phenolic hydrogen in certain compounds, in terms of Werner's coordination theory, on the assumption of formation of an intramolecular auxiliary valence (i.e., intramolecular complex formation) by hydrogen, leading to ring closure in the same way as a cyclic structure arises through the 'auxiliary valence' of nickel in nickel dimethylglyoxime (the type of ring closure now known as *chelation*, from χηλή, a crab's claw). The name 'hydrogen bond' was introduced by Latimer and Rodebush in 1920, who based their ideas on the Lewis-Langmuir valence theory, and advanced the electronic structure

$$R_3N : H : \overset{..}{\underset{..}{O}} : H$$

for tertiary amine hydrates, with a coordinative covalence between the nitrogen and one hydrogen atom. This drew the first clear distinction between this type of binding and the so-called 'onium-salt' link, in which a neutral molecule acquires a positive charge by adding on a hydrogen *ion* (e.g., to form $[NH_4]^+Cl^-$). This interpretation of the hydrogen bond, however, is incompatible with the Pauli principle. It has also been suggested that the bond arises from mesomerism or resonance between two electronic structures (for example, in the HF_2^- anion, the simplest hydrogen bonded structure, resonance might be invoked be-

* This resonance-stabilization can be crudely interpreted on the assumption that the resonating electrons spend more time in the neighborhood of nucleus B than of nucleus A. As a result, superimposed on the exchange forces between the atoms, is an additional force arising from the fact that the atoms periodically bear opposite charges.

** The occurrence of hydrogen bonding in the latter case is shown by the very short $O \leftrightarrow O$ distance between the carboxyl group oxygen and the water molecules in the crystalline dihydrate — 2.52 Å, as compared with the usual $O \leftrightarrow O$ distance of about 3 Å between unlinked oxygen atoms; cf. Fig. 82, p. 452.

tween F—H F$^-$ and F$^-$ H—F), but this concept is also unsatisfactory. It has been pointed out by Briegleb (1943) and others that strong hydrogen bonds are formed only between molecules which have a strongly dipolar character, or contain strongly polar groups. Furthermore, only the non-metal atoms of small radius can participate in hydrogen bonding. Thus H_2O molecules enter into indefinite polymerization through hydrogen bonding, as also do HF molecules, but HCl molecules do not. The mutual attraction between dipoles is only a prerequisite for the establishment of a hydrogen bond; there are other interactions of the electronic systems of the atoms, of which the polarizing action of the hydrogen (as positive end of the dipole) is only a part. These interactions lead to a very pronounced shortening of the interatomic distances between the linked atoms. The heat of formation of a hydrogen bond is usually about 5 to 10 kcal per mol and is thus very much greater than the energy involved in van der Waals attractions.

3. Resonance (Mesomerism)

Just as there are interactions between the covalent and the ionic states, so it is possible for mutual interaction, or *resonance*, to be set up between states of a molecule having the same, or nearly the same, energy, but involving different bond arrangements (i.e., different distributions of electron density) within the molecule. The most familar example of such a system is the benzene ring. This can be represented by five bond patterns

The first two of these (Kekulé forms) are equivalent in energy. The other three (Dewar forms) are also equivalent to each other, but represent a stucture of slightly higher energy content than the Kekulé formula. According to the principles of quantum mechanics, the electronic structure of the molecule does not correspond exactly to any one of these. The electron density distributions represented by the different formulas combine together to form an intermediate configuration (*resonance hybrid*), of lower energy than any of the so-called *canonical (resonance) forms*. It is fundamentally different from tautomerism or other forms of isomerism. Isomeric substances (including tautomeric substances) differ structurally—i.e., in the positions of their atoms. In resonance (mesomerism), the atomic positions are identical in all the bond distributions. The properties of a resonance hybrid are therefore not the average properties of a mixture of isomers: there is no mixture, but only a single molecular form, which cannot be represented by a unique bond pattern.

The enhanced strength of chemical bonds which results from mesomerism is due to the operation of exchange forces (cf. p. 131), i.e., to quantum mechanical resonance. This resonance results in a mesomeric state only if certain conditions are fulfilled.

(i) Different states (or bond arrangements) of a compound can give rise to resonance only if the distinction between them is solely that of *electronic arrangement*; the atomic *nuclei must occupy practically the same positions* in all the canonical formulas.

(ii) The different states must be *equivalent, or nearly equivalent, in energy* when considered by themselves.

(iii) The *number of unpaired electrons must be the same* in the several states. In most cases, this number is zero—i.e., the resonance structures in most (but not all) cases contain only paired electrons.

It should not be concluded, from the choice of the benzene molecule as an example of resonance, that the phenomenon is chiefly of importance in organic chemistry. It is very frequently involved in the constitution of inorganic compounds, although not in compounds of simple salt-like type.

As an example of the occurrence of resonance in inorganic compounds, we may consider the constitution of the $[SO_4]^{2-}$ ion. This can be represented by the following dispositions of the valence electrons (among others)

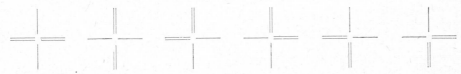

Formula [II] corresponds to the classical formulation of this ion, with two double and two single bonds. The electron distribution represented by formula [II] alone can be present in six resonance forms, differing in the way in which the single and double bonds are distributed among the four O atoms.

In addition to the 6 forms corresponding to formula [II], and the two forms corresponding to formulas [I] and [III], four more forms must be considered, which correspond to a state intermediate between I and II, with one double and three single bonds, and also another four forms with three double and one single bond, corresponding to a state intermediate between II and III. There are thus no fewer than 16 forms in all for the $[SO_4]^{2-}$ ion, between which there may be resonance. In addition, it is possible that there may be resonance between forms with purely covalent bonds (i.e., in which the valence electrons are shared equally between the linked atoms), and forms in which the atoms are to some extent oppositely charged (i.e., in which the bonding electrons are more closely associated with one atom than the other, in each bond). It cannot be affirmed which, and to what extent these possibilities of resonance contribute to a strengthening of the S—O bonds by resonance, since the energies of the canonical forms have not been calculated. There is, however, no doubt that the bonds are strengthened by resonance, since the S—O distance is shortened. This fact alone excludes the possibility that formula [I], based on the octet theory of valence, satisfactorily represents the constitution of the $[SO_4]^{2-}$ ion (cf. Table 58).

TABLE 58

INTERATOMIC DISTANCES IN SOME OXYACID ANIONS

(after Pauling)

Ion	$[SiO_4]^{4-}$	$[PO_4]^{3-}$	$[SO_4]^{2-}$	$[ClO_4]^{-}$
Distance R–O, observed	1.60	1.55	1.51	1.48 Å
Distance R–O, calc. for single bond	1.83	1.76	1.70	1.65 Å
Difference	—0.23	—0.21	—0.19	—0.17 Å

The data of Table 58 show that the considerations applying to the $[SO_4]^{2-}$ ion apply also to the ions of the oxyacids of the other elements of the second row in the Periodic Table. In the absence of knowledge of their actual energy contents, it is impossible to state which of the formally possible resonance forms actually contribute to the establishment of the bonds. In the circumstances, Pauling recommends that these classical formulas for the constitution of these ions should be retained in use, especially as this formula itself adequately expresses the possibility of several resonating forms. This has been exemplified by formula [II] for

$$\left[\begin{array}{cc} O & O \\ & S \\ O & O \end{array}\right]^{2-}$$

above. Only in the case of the $[SiO_4]^{4-}$ ion is the existence of several resonance forms not at once evident from the classical formulation, since the latter contains no double bonds. However, the contribution made by ionic bonding, in addition to covalent bonding, is quite appreciable in the case of the $[SiO_4]^{4-}$ ion (cf. p. 499). The shortening of the interatomic distances, and the stronger bonding which is thereby indicated, can be adequately accounted for by the assumption of resonance between ionic bonds and purely covalent structures.

However, the classical formulation is not in satisfactory agreement with modern determinations of constitution in every case. For example, the constitution of the compound BF_3 is not adequately represented by the classical formula

but by

(three resonance forms), as will be shown in the next chapter.

4. Structure and Constitution

The properties of any substance are determined by the masses and nuclear charges of its constituent atoms, their spatial configuration, and the type of bonds formed between them. The masses of the atoms (except in the case of hydrogen and deuterium compounds) determine relatively few of the properties, except density, and such properties of the gaseous state as depend directly on molecular velocities. The majority of the properties depend rather on the spatial arrangement of the atoms and on the number and energy states of the electrons. These are ultimately determined by the nuclear charges. It is the spatial arrangement of the atoms that is meant by the *structure* of a compound. A distinction is drawn between information of this kind and statements as to the states of the electrons (their distribution between the various atoms, their mode of interaction and the types of bonds formed, the steric arrangement of bonds), which are said to define the *constitution* of the substance.

Any complete description of the constitution includes also an account of the structure. In many cases, however, in inorganic chemistry, it is less important to define the steric arrangement of the atoms in a compound than to know the types of bonds between them. It is then permissible to use simple but incomplete constitutional formulas, which merely express the relation of the various atoms to one another—as, for example, the constitutional formulas of complex compounds such as $K_2[Ni(CN)_4]$. This formula states no more than that four CN groups are linked to a nickel atom, and that the resultant complex is bound by electrovalence to two potassium ions. It cannot be inferred from the constitutional formula what type of bond is involved between nickel and CN groups (whether covalent or electrovalent), how the CN groups are arranged about the nickel (whether planar or tetrahedrally), or how, in the crystalline compound, the K^+ ions are disposed about the

$[Ni(CN)_4]^{2-}$ ions.* In the majority of cases the complete structure of complex compounds is not known. On the other hand, there are very many inorganic compounds for which the structure is known, without having a complete knowledge of their constitution.

The influence of structure shows itself, for example, with polymorphic substances, in the different properties of the several modifications. This is especially evident in the case of diamond and graphite, where the change of structure is associated also with a significant change in the bond type. The very great influence of bond type upon properties is shown by the example of cesium iodide and β'-brass (ZnCu). These two compounds are identical in structure, but there is no resemblance whatever in their properties. It has been possible to discern relatively simple relations between structure and properties in the case of the silicates (p. 503 *et seq.*)

5. Constitution as a Basis for Classification

It is possible to classify compounds, on the basis of their constitution, into the five classes set out in Table 59. If the significant properties of typical members of the several classes are compared, the relationship between the properties and the constitution of the compounds will be evident.

Salt-like compounds are those which have the structures of typical salts, such as sodium chloride—i.e., have an ionic crystal lattice, mostly of the coordination lattice type (cf. pp. 212, 216). Oxides, such as CaO, and other ionic compounds with coordination structures are included in this group. The group does not, however, include other compounds such as Hg_2Cl_2 or $HgCl_2$, which might be considered as salts from their mode of formation, but which possess clearly defined molecular lattices. Only a compound can have a salt-like structure, whereas all the other classes of substances can include both compounds and elements.

The typical salt-like compounds are those built up from elements which form ions with the inert gas configuration. The forces of attraction between them are then almost pure Coulomb forces, and since these are exerted equally in all spatial directions, each ion tends to surround itself with the greatest possible number of oppositely charged ions. This is why, in the crystallization of a solid, ions associate together in indefinite, sensibly infinite numbers, to build up a coordination lattice, rather than linking up in finite groups to form molecules.

The uniform structure of the field of force around each ion can become distorted by polarization effects. This often leads to the formation of *layer lattices* (p. 216 *et seq.*), especially in salt-like compounds involving cations of the elements in the B subgroups of the Periodic System. Among compounds of the Main Group elements, layer lattices are often found for the hydroxides (cf. pp. 217, 261), since the hydroxyl ions have a high dipole moment. Layer lattices are also found in substances of essentially covalent character, and it is therefore not possible to draw inferences as to bond type from this characteristic alone.

The involatility of salt-like compounds can be attributed to the fact that, in their free (ion-pair) molecules, the valence forces are not completely saturated. Nevertheless, the composition is identical in the solid state and in the gaseous state, and the composition of pure salt-like compounds is invariably constant, corresponding to simple whole number ratios of atoms. This is the consequence of the law of elec-

---

* It is, inherently difficult to express by formulas the structure of inorganic compounds which are based upon ionic or atomic crystal lattices. The structural formula must be replaced by a description of the unit cell.

TABLE 59

CLASSIFICATION OF SUBSTANCES ON THE BASIS OF THEIR CONSTITUTION

Class of substance	Type of bonds	Structure	Characteristic properties	Examples
Salt-like compounds	Electrovalences (Heteropolar)	Ionic crystal lattice	Not very volatile. Electrolytic conductors in the molten state and in solution. Dissolve most readily in electrolytically dissociating solvents of high dielectric constant.	NaCl, KBr, CaF_2, CaO
Simple molecular substances	Covalent	Free molecules; molecular crystal lattice in solid state.	Gaseous or readily volatile. Non-conductors. Dissolve readily in indifferent solvents as well as (or rather than) in dissociating solvents.	H_2, O_2, Cl_2, CH_4, NH_3, N_2O
Adamantine substances	Covalent	Atomic crystal lattice	Not volatile, non-conducting, insoluble in all solvents, often extremely hard.	Diamond, SiC, B_4C
Metals and quasi-metallic compounds	Metallic bonds	Metallic crystal lattice	Metallic conductors, high optical reflectivity, great tendency to form mixed crystals.	Cu, ZnCu, $Zn_8\,Cu_5$, $Sn_8\,Cu_{31}$
Van der Waals compounds	Van der Waals forces	Molecular aggregates.	Decompose readily when the temperature is raised.	Inert gas hydrates, associated liquids.
Association products and addition compounds of highly polar hydrogen compounds.	Hydrogen bonds	Molecular aggregates	Exist in the solid and the liquid state, and very frequently in solution. Occasionally persist in gaseous state.	$(HF)_n$, $(H_2O$ $(NH_3 \cdot H_2O)$ $(AlH_3)_x$, $(BH$

troneutrality—in any compound the number of positively charged ions must always be equivalent to the number of negatively charged ions.

The high electrolytic conductivity displayed by salt-like compounds in the molten state and in solution can at once be understood since they are built up from ions.

That substances built up from *covalent molecules* have no electrical conductivity when they are pure is also a direct consequence of their constitution. In some circumstances they may furnish electrically conducting solutions, but only when a change of constitution results from interaction with the solvent and the solvation of the resultant species. Since their ions are solvated, the compounds thereby also undergo a change of composition.

Covalent bonds, unlike electrovalences, can be described as directed forces, which may be symbolized by a line between linked atoms. The range of the forces concerned is short, so that their action is restricted, in general, to immediately adjacent, linked atoms. One consequence of this is that it is possible to form discrete *molecules*, containing a finite number of atoms, whereas the only forces acting between these molecules are the van der Waals forces. Van der Waals forces bind the molecules together in the solid state. Since they are relatively weak forces, they are readily overcome by thermal energy, and substances built up from covalent molecules are consequently volatile.

A simple relation appears to exist between the volatility of substances built up from covalent molecules and their molecular volume in the liquid state. Van't Hoff first suggested that the boiling points of liquids must be proportional to their molecular volumes, but found that the hypothesis was not borne out by observation. However, according to Billig (1931), there is a true proportionality between boiling point and molecular volume in the liquid state when allowance is made for association.

The molecular volume of covalent compounds, and especially of carbon compounds, is an approximately additive function of certain values which are characteristic of the individual constituents (Kopp's law). Deviations from the simple covalent bond (i.e., multiple bonds, resonance systems) must be taken into account by the addition of suitable 'bond increments'. It is also essential to compare the molecular volumes at corresponding states of the liquids—e.g., at their boiling points. It has been found that the parameter introduced by Sugden (1924) [7], termed the *parachor* is more strictly additive for covalent compounds than Kopp's molecular volumes. The parachor is the product of the molecular volume and the fourth root of the surface tension; its essential feature is that it provides a better defined 'corresponding state'. Even in this case, different bond types must be allowed for by appropriate increments, but it is conversely possible, with the exercise of due caution, to make certain deductions about bond type from the magnitude of the parachor.

It is not necessary that covalent bonding should invariably lead to the formation of discrete molecules. It is also possible for the linking of one molecule to another to proceed indefinitely in all directions—i.e., so that an atomic lattice is formed*. Since carbon, in the form of diamond, provides an excellent illustration of this, substances which form atomic lattices of this type are known as *adamantine* substances. The interatomic forces in structures of this kind are very strong, so that substances with this constitution are characterized by extreme involatility (and infusibility), almost absolute insolubility, and often by great hardness.

Covalent compounds, whether with molecular or atomic lattice types, display constancy of composition, like the ionic compounds. In the case of covalent compounds this is because any given bond arrangement always requires the same number of electrons.

The characteristic property of intermetallic and quasimetallic compounds is their metallic conductivity**. The association of this type of conductivity with a particular bond type follows from the theory of the metallic state (Vol. II, Chap. I). Since the ions which are bound together by the 'electron gas' are similarly charged, it is possible for a compound to display continuous variability of composition, within certain limits, when the interatomic forces are of metallic type.

---

* The crystal lattices of metals and quasi-metallic substances are not included among the atomic lattices in this sense.

** This applies in the sense that whenever metallic conductivity is observed, metallic binding forces must be operative. However, it is not necessary that binding forces of the metallic type must be associated with metallic conductivity. This follows from the theory of insulators (cf. Vol. II, p. Chap. 1).

The principal limitations to such variability of composition arise on geometrical grounds.* The interaction between the metallic ions and the surrounding electron gas tends to establish the densest and most symmetrical packing possible. If the ions differ considerably in radius, this can usually be achieved only for given relative numbers of the ions. There is also some evidence of a tendency to attain certain values for the electron density (i.e., certain ratios between the number of electrons and the number of atoms), as is shown by the Hume-Rothery rules. However, *when the binding forces are metallic in type, there is no reason for the law of constant proportions to be generally valid.*

Van der Waals' forces can also, in themselves, not lead to the aggregation of molecules in definite ratios. Once again, however, steric considerations may result in the existence of compounds with a constant, stoichiometrically simple composition—e.g., because only a definite number of molecules can come into immediate contact with some one other molecule. In the same way, if foreign molecules are incorportated in a crystal lattice by the operation of Van der Waals forces, their number may be determined by the number of positions available for their occupation.

Classification of substances in the manner shown in Table 59 is of value in that it brings out the connection between the constitution and a number of striking properties. This is especially true of simple examples, although the relation is not at all obvious for all compounds. By no means all compounds can therefore be unambiguously assigned to one or other of the classes shown in Table 59, on the basis of the properties there given. The fact that there are transitions between the different bond types inevitably means that the boundaries between the various classes of substance are not sharp, so that properties typical of different classes may be exhibited by the same substances**. It is also possible for different types of binding force to act in different spatial directions, in one and the same substance.

Thus the carbon atoms in the basal plane (hexagonal network) of graphite are bound together by essentially covalent forces. In the perpendicular direction, however, the forces between sheets are probably Van der Waals forces, since they are so much weaker in that direction. This is evident from the interatomic distances (cf. Fig. 80, p. 423) and from the topochemical reactions of graphte. At the same time, the bonds have a certain degree of metallic character, since graphite exhibits electronic conductivity.*** It is possible to take account of the anisotropy of bond types by a further subdivision of the classification, but the scheme so obtained is directly applicable only to binary compounds (cf. GRIMM, *Angew. Chem.*, 47 (1934) 53. Substances composed of more than two elements often involve bonds of completely different type between different pairs of atoms. Thus in the alkali metal salts of organic acids, the alkali ions are bound by electrovalent forces while all the atoms of the anions are covalently bound.

One very important class of compound, in which bonds of different type are often present simultaneously, is that made up of the *compounds of higher order*. Complex compounds which are electrolytes generally involve ionic bonds, even in the

---

* On this matter see F. LAVES, *Z. anorg. Chem.*, 250 (1942) 110.
** This is why, in broad terms, some substances regarded as ordinary valence compounds (i.e., either ionic or covalent in structure) are not strictly constant in composition. This is possible if the structure involves other binding forces (metallic bonds or van der Waals forces) in addition to electrovalent or covalent bonds. Thus it may be assumed that in cubic copper(I) sulphide (Vol. II, Chap. 8) the bonds partake in some measure of metallic bond character.
*** In terms of quantum mechanics, an extended, resonating system of double bonds has certain analogies with a metallic system, in that the electrons can be described as moving in a regular periodic field set up by all the carbon atoms of a conjugated system of double bonds. See N. S. BAYLISS, *Quart. Rev. Chem. Soc.*, 6 (1952) 319.

solid state, between those components of the compound which go into solution as ions (without decomposition of the complex). *Within* the complex, however, the bonds between the central atom and the ligands may be of various types:—they may be electrovalences, colences, or Van der Waals forces (See Chap. 11 for further discussion). In certain complexes formed in liquid ammonia (cf. p. 573) they may approach metallic bonding in their character.

It may generally be assumed that neutral molecules are bound by Van der Waals forces, in those addition compounds of salts with electroneutral components which are either very weakly complexed, or in which the neutral molecules are built into the crystal lattice without being definitely coordinated with any central atom. The most familiar examples of such compounds are given by the hydrated salts*. Van der Waals forces cannot, in themselves, lead to combination in stoichiometric ratios. Where the crystal structure does not require the H_2O molecules to be bound in definite positions, it is possible for water of crystallization to be lost continuously as the temperature is raised, or the pressure lowered. The easily decomposed ammoniates belong to the same class of compounds as the hydrated salts.

Double salts which dissociate into their components in solution do not differ in any essential way from the simple salts in their bond type.

6. Determination of the Constitution of Inorganic Compounds

(a) Introduction

The most obvious properties of compounds, considered above—their volatility, solubility and electrical conductivity—can, at most, enable the bond type of a series of compounds to be recognized. They do not permit of any quantitative deductions as to the strength of the binding forces, however, nor do they provide other information whereby the constitution can be more closely defined. Bond strengths can be deduced from thermochemical measurements, or from equilibrium measurements (e.g., of dissociation pressures). A number of special physical measurements can be invoked for the determination of other properties which characterize the constitution of inorganic compounds. The most important of these may now be briefly discussed. [9]

As has already been mentioned, the methods used in classical organic chemistry, based essentially on the investigation of chemical reactions, are in most cases not applicable to the determination of the constitution of inorganic compounds. They can only be used for substances which can be converted to the gaseous state, or brought into solution, without change of molecular constitution.** In addition, it must be assumed that the bonds are unchanged in position and in type during the reactions. It is only relatively seldom that both these assumptions are valid for the reactions of inorganic compounds. There are, however,

---

* It has been shown from crystal structure determinations that, in many cases, the water molecules are bound to each other by *hydrogen bonds* in heavily hydrated salts. They may also be attached to oxygen atoms of oxyacid anions in the same way.

** Even for organic compounds, the application of new physical techniques has proved absolutely indispensable in working out their constitution, in cases where the above conditions is not fulfilled —e.g., in the case of cellulose.

certain groups of atoms (radicals) which retain their identity during chemical changes, and their existence has therefore long been recognized. It is only in exceptional cases, however, that it has been possible to determine the *constitution* of such groups from the evidence of their reactions (with the exception of the strong complexes—cf. footnote, p. 286).

For volatile substances, or substances which dissolve without electrolytic dissociation, a first step in elucidation of their constitution consists in determining the molecular weight in the gaseous or dissolved state. From the foregoing section, it may be generally assumed that for such substances, the molecules existing in the gaseous state or in solution are identical with those making up the solid state. Substances which undergo electrolytic dissociation in solution often contain radicals which do not change in composition when they enter solution. In this case, measurements of ionic weight are important. However, it is only legitimate to infer that the groups identified as ions in the solution are also the groups present in the solid compounds when the typical radicals are involved—e.g., groups of atoms known to persist through chemical changes. In other cases, the groups present in solution are quite different from those in the solid compounds (as, for example, with the thiosulfato-compounds of silver, Vol. II, Chap. 8). Even in these cases, a determination of the ionic weight in solution is of some interest, since it often provides some evidence as to the interactions between the ions in the solution.

Another important requirement is the determination of the *structure*. For solids, this is generally ascertained by the methods of X-ray crystallography. A variety of other methods is important for determining the molecular structure of liquids or gases (see below). Further information as to bond type can be derived from electrical and magnetic measurements, and also from the study of band spectra. The latter, while limited essentially to molecules of simple structure, provides very detailed evidence as to their constitution.

(b) Molecular Weights

The molecular weights of substances which are gaseous, or can be converted to the vapor state without decomposition, are derived from gas- or vapor densities, by application of Avogadro's law. Molecular weights of dissolved substances are generally found from the freezing point depression or boiling point elevation of their solutions.

When measurements have to be made on solutions which boil below ordinary temperature, it is often better to measure directly the difference in vapor pressure between the solution and the pure solvent. This method was used, for example, by Joannis (1892) and Kraus (1908), to determine the molecular weight of alkali metals dissolved in liquid ammonia, and by Stock (1925) for boron hydrides in liquid ammonia solution.

The method of *isothermal distillation* is based on the difference in vapor pressure between solution and solvent. This was first used by Barger (1904); it enables measurements to be carried out at room temperature. In the form developed by Ulmann (1931–33), accurate molecular weight determinations can be carried out even when the molar concentration of the solute is very small. For this last reason, it is a suitable method for determining the molecular weights of high polymers, for which cryoscopic or ebullioscopic measurements fail. [*8*]

With high molecular weight substances, it is often possible to measure the osmotic pressure directly, with the use of a cell with a selectively permeable membrane. In principle, direct determination of osmotic pressure should give far higher accuracy of measurement than the indirect methods. Thus a substance of molecular weight 250, present as a 0.5 weight-% solution in water, gives a freezing point depression of 0.038°, a boiling point elevation of 0.0104°, but an osmotic pressure equivalent to a column of water 497.10 cm high. For a substance with a molecular weight 10,000, the corresponding figures are 0.00095°, 0.00026° and 12.43 cm. However, it is only possible to measure the osmotic pressure, directly and accurately, if a membrane is available which is permeable to the solvent, but completely impermeable to the solute. This condition is easier to realize for high molecular weight substances than for substances of low molecular weight. For the determination of the osmotic pressure of colloidal substances see Vol. II, Chap. 16.

The *ultracentrifuge* has also proved useful for the determination of the molecular weight of high polymers (cf. Vol. II, Chap. 16).

(c) Ionic Weights

Measurements of *diffusion rates* can also be used for determining the molecular weights of substances in solution. In recent years they have been used, especially, to determine *ionic* weights. The most reliable method is the calculation of the molecular weight from the velocity of 'free diffusion' (cf. Vol. II, Chap. 18, eqns. (2) and (2a)).—i.e., diffusion in a system which involves no membrane. G. Jander (1925—1942), in particular, employed this method successfully in investigations of the hydrolysis equilibria of basic salts and of isopolyacids (examples will be met with in Vol. II). Interesting results have also been obtained by Brintzinger (1928 and later) from measurements of *permeation*, i.e., diffusion through a porous membrane, instead of free diffusion. Brintzinger called this the 'dialysis' method, since the apparatus used was based on Graham's dialyser. It was found empirically that, for substances which were not too dissimilar structurally and constitutionally, the dialysis (or permeation) coefficients—i.e., the coefficients of diffusion through the membrane—were proportional to the square root of the molecular or ionic weights, as for free diffusion*.

Even for processes of free diffusion, the theoretical basis of this relationship is very questionable. In practice, a comparison must be made between the diffusion coefficient of a substance and that of same reference material, and it is assumed that the molecular weight (or ionic weight) of the reference substance is known. Effects of solvation can then be only arbitrarily allowed for. Furthermore, in condensed systems, it is certain that the *size* (i.e., effective collision diameter) of a diffusing ion or molecule must be important, as well as its translational velocity. This is why the empirical validity of the relation $D = k \cdot MW^{\frac{1}{2}}$ must be hedged round with serious limitations.

It must be noted that, when ions form addition compounds with the molecules of the solvent, the diffusion rates lead to the ionic weights of the solvated species, whereas the osmotic properties (being dependent only on the *number* of solute atoms or ions) give the molecular weights of the unsolvated species.

In measurements of free diffusion, or of membrane permeation, it is normally necessary to observe the diffusion of ions along a concentration gradient. The behavior of ions in solution depends upon the ionic strength of their environment, and if an electrolyte diffuses into pure water, both anion and cation must necessarily diffuse at the same speed. In order to maintain uniformity of ionic strength, and to eliminate diffusion potentials so that it is possible to observe the diffusion rates of individual ionic species, it is customary to carry out diffusion measurements in the presence of a high concentration of some indifferent electrolyte. Even in a uniform solution, however, the ions are executing random motions, and it

* Jander found that with the very fine-pored cellulose sheet (cellophane and cuprophane) used by Brintzinger, the proportionality did not extend up to high molecular or ionic weights, but did so with the more open-pored 'cella-filter'. The rate of diffusion through cellafilter was inversely proportional to the viscosity of the solution, as for free diffusion. This relationship did not hold for cellophane or cuprophane membranes.

is of some interest to compare the rates of migration of dissolved ions in a *uniform environment*, with no concentration gradients. This is particularly important when (as in studies of the aggregation of isopolyacid anions), there must be a shift of chemical equilibrium as the concentration gradient changes. By use of compounds labelled with radioactive isotopes, it becomes possible to measure the rate of *self-diffusion* of ions or molecules in uniform solutions. A convenient method for carrying out such measurements has been described by SADDING-TON and ANDERSON, *J. Chem. Soc.*, (1949) 381.

A procedure which can often be used to advantage for determination of ionic weights is that of '*salt cryoscopy*'. This is based on the fact that the melting points of salt hydrates (e.g., $CaCl_2 \cdot 6H_2O$), transition points of salt hydrates (e.g., of $Na_2SO_4 \cdot 10H_2O$) and the eutectic temperatures of cryohydric mixtures (e.g., ice + KNO_3) are all lowered by other substances dissolved in the liquid phase. The depression of the phase-reaction temperature follows the Raoult law. Salt cryoscopy has advantages over ordinary cryoscopic measurements for the determination of ionic weights, in that the depression is effected only by those ions of the added substance which are not common to both the added substance and solvent salt. Substances which have been chiefly used for salt cryoscopy up to the present are the following:

$Na_2SO_4 \cdot 10H_2O$ Transn.pt. $32.38°$ $D = 3.21°$ (Löwenherz, 1895, Darmois, 1923)
$CaCl_2 \cdot 6H_2O$ M.p. $30.2°$ $D = 4.13°$ (Morgan, 1907, Darmois, 1930–32)
Ice + KNO_3 Cryoscop.temp. — $2.84°$ $D = 1.76°$ (Cornec and Müller, 1932)
Ice + $NaNO_3$ Cryoscop.temp. —$18.5°$ $D = 1.55°$ (Jahr, 1952)

(Transn.pt. = transition point; Cryoscop.temp. = temperature of eutectic; D = molar depression of phase-reaction temperature, referred to 1000 g of 'solvent').
The following substances have also been found suitable for salt cryoscopy: $LiNO_3 \cdot 3H_2O$, $Na_2S_2O_3 \cdot 5H_2O$, $Na_2CrO_4 \cdot 10H_2O$, $Ca(NO_3)_2 \cdot 4H_2O$, $Zn(NO_3)_2 \cdot 3H_2O$, $Mn(NO_3)_2 \cdot 3H_2O$ (Morgan, 1907, Boutaric, 1911).

(d) Structure Determinations

X-ray methods for determination of the structure of solids have been briefly reviewed in Chap. 7. It is also possible to determine the atomic arrangement (except for the position of hydrogen atoms) in liquids and gases by means of X-ray diffraction. The diffraction pattern consists of very much broadened maxima and minima of intensity, from which it is possible to infer the interatomic distances, as was first shown by Debye (1929). The *diffraction of electrons* is widely used, in place of X-ray diffraction, for elucidation of the structure of gas molecules [10]. In terms of wave mechanics, the two processes are very similar, but the intensity of the diffracted beams is very much greater for electrons than for X-rays.

(e) Determination of Bond Type

In many cases, the criteria set out in Table 59 are sufficient to provide well-founded evidence as to the binding forces in any compound. Thus if a soluble substance dissolves with a low heat of hydration, to form an electrolytically conducting solution, it is safe to infer that it is built up in the solid state from ions. This conclusion would not, however, be justifiable if the substance were sparingly soluble, or if its heat of hydration were large.

The hydrogen ion is noteworthy for its very high heat of hydration. There is no doubt that in most, if not all, anhydrous acids (except in so far as auto-ionization takes place, as in H_2SO_4 or HNO_3, q.v.), hydrogen is bound in a manner which is more like a covalent bond than an electrovalence, even though the dipole moment is considerable (i.e., the ionicity of the bond is not zero).

In certain cases, reliable conclusions about bond character can be drawn from X-ray diffraction evidence. This method, which is generally laborious, can be used when the first mentioned criteria fail. It is possible to determine the distribution of electrons around individual atoms since the scattering of X-rays is brought about by the electrons. It has already been pointed out (p. 213) how this fact can lead to a distinction between an ionic structure and an atomic lattice, in the particularly simple case of a binary compound between elements which would be isoelectronic if they were present as ions. Even when a less simple system is involved, accurate measurements of the intensity of diffracted X-rays can be used to derive the distribution of electrons around the various atoms. The number of electrons associated with any atom can then be compared with that expected on the

assumption that each ion bears a charge equal to its valence. These numbers may not be identical. The effective charge computed from X-ray data (which may not be an integral charge) may be less than that corresponding to the valence, indicating that the bond concerned is not purely ionic.

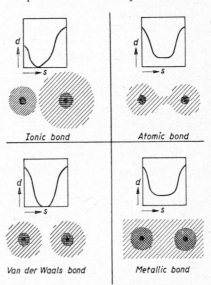

This method has been greatly refined by application of techniques of Fourier analysis* to the observational data, largely as a result of the work of Duane (1915), Havighurst (1915), W. L. Bragg (1933), Robertson, and Patterson. Grimm, Brill, Hermann and Peters (1939) used this procedure to map out the electron density throughout the whole space between atoms in relatively simple crystals. Fig. 66 represents, diagrammatically, the electron density distributions so found, corresponding to the different kinds of bond. They correspond very closely to what would be expected from the views which had been formed up to that time. The curves in Fig. 66 show how the electron density varies along a line between the centers of adjacent atoms, while the sketches give, very roughly, the electron density distribution in any plane. It may be seen that, with ionic bonding, the electron density falls practically to zero between the ions, so that each ion has its own electron cloud, and shares no electrons with its neighbors. The same is true of van der Waals binding, except that the electron density falls to zero over a rather wider free space between the electron clouds (corresponding to the rather larger interatomic distances between atoms which act on each other only by Van der Waals forces), whereas the electron clouds of the ions 'touch'. In a covalent bond, the electron density between atomic nuclei does not fall to zero; the elec-

Fig. 66. Röntgenographic estimation of the distribution of electron densities in different types of bonds (schematically). The curves represent the relation of the electron densities (d) and the distance (s) from the nuclear centrum on the line of communication between the two atomic nuclei.

* 'Fourier analysis' is a procedure of mathematical physics, based on a theorem by Fourier, that any periodic function $F(x)$ can be represented as a sum of trigonometric functions: $F(x) = a_0 + a_1 \sin x + b_1 \cos x + a_2 \sin 2x + b_2 \cos 2x + \ldots\ldots$ A crystal lattice can be considered as periodic in three dimensions, and so can be represented by a triple Fourier series of electron density. For a review of the application of the method to X-ray crystal analysis and electron densities, see A. Sommerfeld, *Naturwissenschaften*, 28 (1940) 769.

tron clouds of adjacent atoms merge into each other, but only along the direction of the shortest interatomic distance. There are regions of clearly localized high electron density, forming 'bridges'. With metallic binding, the valence electrons are distributed uniformly through the space between the metallic ions.

Fourier analysis of crystals, if carried out with the utmost refinement, is a time-consuming and difficult technique. It can, however, give a virtually complete representation of the distribution of electrons, and can reveal fine points of bond character which could not have been predicted. The very exact measurement of bond lengths, based on electron density maps, has provided a basis for determining the bond-order of covalent bonds in inorganic and organic molecules. It has been shown that the ideas discussed above under resonance are widely applicable, and that in many molecules the actual electron density distribution does not coincide with any one bond pattern. It is interesting that this method also enables the position of hydrogen atoms to be located, from their influence on the electron density in their immediate neighborhood, which is not directly possible by other X-ray techniques (see, for example, the determination of the structure of polyethylene, by BUNN, *Trans. Faraday Soc.*, 35 (1939) 482).

(f) Magnetochemistry [*11–14*]

It is often possible to derive rather direct evidence as to bond type, valence and other constitutional questions from measurements of *magnetic susceptibility*. This procedure has been successfully applied during recent years to the constitution of many compounds (particularly compounds of the transition metals) by a number of workers, especially Klemm and Selwood.

Substances are classified as *diamagnetic, paramagnetic* or *ferromagnetic*, according to their behavior in a magnetic field. A measure of their behavior is given by the magnetic susceptibility, which defines their receptiveness for lines of magnetic force. A substance which is less permeable to the lines of force than is empty space has a negative susceptibility, and is said to be *diamagnetic*. Substances which have a positive susceptibility (i.e., in which the lines of force tend to crowd together) are *paramagnetic*. If introduced between the poles of a strong magnet, the former would set itself across, and the latter along, the lines of force (cf. Fig. 67). Substances with an especially high susceptibility are said to be *ferromagnetic*.

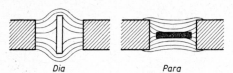

Dia   *Para*

Fig. 67. Diamagnetic and paramagnetic rod in the magnetic field.

The magnetic susceptibility $\varkappa$ is defined by $\varkappa = I/H$, where I is the magnetic intensity (i.e., the magnetic moment per unit volume), and H is the magnetic field strength. The magnetic moment is the turning moment exerted by a magnetic field on a bar magnet or a coil through which a current is flowing. Unit force (1 dyne) is exerted on a unit magnetic pole in a field of unit strength (thereby defining the unit of field strength, the *gauss*). A unit magnetic pole is that which exerts a force of 1 dyne on a similar pole at a distance of 1 cm. The susceptibility may be alternatively defined in terms of unit mass instead of unit volume, the mass susceptibility being given by $\chi = \varkappa/d$ (d = density). The *molar susceptibility* χ_M is obtained by multiplying χ by the molecular weight.

For diamagnetic substances, the susceptibility is independent of the field strength, and usually independent of the temperature also. The susceptibility of paramagnetic substances is also independent of field strength, but is strongly temperature-dependent. For many paramagnetic substances it is inversely proportional to the absolute temperature: $\chi T = C$ (Curie's law), where T is the absolute temperature, and C is a constant. For a ferromagnetic substance, the susceptibility at first increases with field strength, at low field strengths, and then decreases. At sufficiently high field strengths, the product of magnetic susceptibility and field strength (i.e., the magnetic intensity I) becomes almost independent of the field strength, so that the susceptibility is then inversely proportional to field strength. With rise of temperature, the susceptibility of ferromagnetic substances diminishes—gradually at first, and then very sharply at a definite temperature, the Curie temperature. At the Curie temperature, ferromagnetism changes to paramagnetism.

Substances which obey Curie's law are said to display normal paramagnetism. More commonly, Weiss' law is obeyed— $\chi(T - \theta) = C$, where θ is a constant. For ferromagnetic substances, θ is equal to the Curie temperature in degrees absolute, but for many substances θ has a negative value. This case is described as 'antiferromagnetism'.

The temperature dependence of paramagnetism can be interpreted as arising from the thermal agitation of 'elementary magnets' (see below), which the external field strives to align. In order to determine the true moment of the individual atoms, it is therefore necessary to eliminate the dis-orienting effect of thermal agitation. The magnetic moment μ per gram atom is given by $\mu = \sqrt{3RC}$, where R is the gas constant, and C the Curie law constant, per g atom.

Magnetic behavior has been explained in terms of 'elementary magnets', even in the older magnetic theories (going back to Ampère). These, in turn, have been attributed to electrical 'molecular currents', i.e., the circulation of electricity within the atom. With the development of atomic theory, these ideas have been given precise meaning, in that the molecular currents are identified with the orbital motion and spin of the electron [15, 16]. The electrons within any one atom can diminish or augment each other's effects, according as their individual moments are oppositely or similarly oriented. If their resultant effect is zero, the substance is diamagnetic: in the other case it is paramagnetic (or, in certain cases, ferromagnetic).

If the moments of the individual electrons cancel out, so that the resultant moment of the atom is zero, an external magnetic field induces in the atom a moment of sign opposite to that of the field. The case is similar to that of a coil of wire introduced into a magnetic field, whereby the induced current gives rise to a magnetic moment opposed to that of the field. Whereas the induced current in the wire rapidly falls off, because of the ohmic resistance, the moment induced in the atom persists, and is done away with only as a result of the contrary induction effect when the atom is removed from the magnetic field. That part of the magnetic moment of the atom which is independent of the external field, is known as the 'permanent' moment, to distinguish it from the induced moment. In paramagnetic substances the atoms or molecules possess a permanent magnetic moment, whereas the permanent moment of diamagnetic substances is zero. When a paramagnetic substance is introduced into a magnetic field, the permanent individual moments are oriented along the field, to an extent which depends upon the temperature. At the same time, magnetic induction gives rise to an induced magnetic moment which is opposite in sign to the external field, and superposed on the permanent moment of every individual atom. Thus, in the case of a paramagnetic substance, the quantity measured is actually the difference between its paramagnetism and its diamagnetism. The latter is so small compared with the former, however, that it can generally be neglected.

One consequence of the foregoing is that all free atoms or ions with complete electron shells must be diamagnetic, since the resultant orbital angular momentum and spin momentum are both zero. Electrons with the subsidiary quantum number $l = 0$ (s electrons) have, according to wave mechanics, no orbital moment, but only a spin moment. If both s states in any quantum shell are occupied, their resultant spin moment is zero. Hence atoms or ions with this configuration will also be diamagnetic. All other *free* ions must be paramagnetic.

Hence diamagnetism will be exhibited by

[i] all inert gases and all ions with the inert gas configuration,

[ii] all ions in which, in addition to inert gas shells, there are *complete d* or *f* levels (i.e.,

completed sub-shells with the subsidiary quantum number $l = 2$ or 3)—i.e., ions like Ag^+, Cd^{2+}, In^{3+}, Sn^{4+}, etc., and also Hf^{4+}, Ta^{5+}, W^{6+}, Re^{7+}.

[iii] All atoms or ions which contain only a pair of s electrons external to an inert gas core— e.g. Be, Zn, Ga^+, Pb^{2+}, P^{3+}, Se^{4+}.

Paramagnetism is exhibited, for example, by the ions of the rare earths (except La^{3+}, Ce^{4+}, Yb^{2+}, Lu^{3+}, in which all the electrons are in complete shells), and very often by the ions of elements in the center of the long periods of the Periodic Table, i.e., transition elements, when these are in valence states different from the group number, and when the electron orbits are not strongly perturbed by the fields of force of neighboring atoms or ions.

The magnitude of the magnetic moment μ of a free ion can be calculated from the theory of magnetism.* For the orbital moment of an electron with the angular quantum number $l = 1$ (i.e., a p electron), theory gives the value $\mu_l = \sqrt{l(l+1)}\,\dfrac{e \cdot h}{4\pi m}$ (where e and m are the charge and mass respectively of the electron, in e.m.u., and h is Planck's quantum). The quantity $e.h/4\pi m$ is used in magnetochemistry as the unit in which magnetic moments are measured, and is given the name of the *Bohr magneton*. It has the value $9.26 \cdot 10^{-21}$ gauss cm^3 (for 1 atom) or 5580 gauss cm^3 per gram atom. The magnetic moment due to electron spin, for each electron, is $\mu_s = \sqrt{3}\,\dfrac{e \cdot h}{4\pi m}$, i.e., 1.73 Bohr magnetons. The total magnetic moment of an atom or ion can be calculated as the vector sum of the orbital moments and spin moments of all the electrons**. This is exemplified by Table 60, which gives the magnetic moments of the rare earth ions as calculated by Van Vleck, on the assumption that the $4f$ orbits first become occupied by electrons with parallel spin, and that electronic states of opposite spin become filled up only when all the available orbits are singly occupied***. The agreement between calculated and observed values indicates that the assumption is valid.

It is also apparent from Table 60 that ions with the same total number of electrons (La^{3+} and Ce^{4+}, Ce^{3+} and Pr^{4+}) have the same magnetic moment. This corresponds to a general rule, stated by Kossel—the magnetic moment of any ion is the same as that of the ion formed by the preceding element in the Periodic Table, but with one unit of positive charge less, or that of the following element, with a charge one unit greater (Kossel's displacement rule).

TABLE 60

MAGNETIC MOMENTS OF LANTHANIDE IONS
(in Bohr magnetons) at $T = 300°$ K

.	La^{3+} Ce^{4+}	Ce^{3+} Pr^{4+}	Pr^{3+}	Nd^{3+}	Pm^{3+}	Sm^{3+}	Eu^{3+} Sm^{2+}	Gd^{3+} Eu^{2+}
Calc. (Van Vleck)	0	2.56	3.62	3.68	2.83	1.6	3.45	7.9
Observed (Klemm)	0	2.58	3.61	3.66	—	1.5	3.4	7.9

	Tb^{3+}	Dy^{3+}	Ho^{3+}	Er^{3+}	Tm^{3+}	Yb^{3+}	Lu^{3+} Yb^{2+}
Calc. (Van Vleck)	9.7	10.6	10.6	9.6	7.6	4.5	0
Observed (Klemm)	9.7	10.5	10.5	9.55	7.3	4.55	0

* The magnitude of the permanent moment is found experimentally from Curie's law.
** For further information on the mode of deriving the resultant magnetic moment, reference should be made to the monographs on magnetism cited at the end of the chapter [*11–16*].
*** It is only for the case of s orbits that electrons tend to couple spins. A more general rule, propounded by Hund, is that all available states are occupied by electrons with parallel spins, before any orbits become occupied by pairs of electrons with opposed spins. This implies that there will be the maximum possible number of parallel spins in the state of lowest energy.

Compounds containing only diamagnetic ions are diamagnetic both in solution and in the solid state. Compounds containing paramagnetic ions are paramagnetic if the binding forces are purely electrovalent. *Formation of covalences can lead to the complete disappearance of paramagnetism.*

Paramagnetism disappears, in particular, when a covalent bond is formed by the sharing of s electrons, since these necessarily have opposed spin, and therefore have opposed magnetic moments. Covalent molecules therefore usually conform to G. N. Lewis' rule—if the total number of electrons is even, the magnetic moment of the molecule is zero; if the number of electrons is odd, the moment is 1.73 Bohr magnetons, corresponding to one unpaired electronic spin. (The O_2 and NO molecules, among others, form exceptions to this rule.) Crystal lattices made up from atoms or highly deformed ions present a more complicated problem, and the factors that modify the resultant paramagnetism are by no means fully understood. It does, however, seem to be established that the phenomena of ferromagnetism and antiferromagnetism are due to a mutual 'magnetic coupling' of the atoms; the direction of the spin of the unpaired electrons in each atom is determined by interaction with neighboring atoms, and the whole effect is equivalent to the establishment of covalent bonds (although weak bonds, since the interatomic distances are such that the overlap of wave functions is very small) throughout the crystal lattice. *Ferromagnetism* arises if the electrons are so coupled as to have their spin vectors parallel (contrary to the usual case for bonding). Spin coupling throughout the crystal lattice such that electrons on neighboring atoms have opposed spins (antiparallel spins) gives rise to *antiferromagnetism*.

Although not fully explained in all points, magnetic susceptibility measurements often afford unambiguous evidence concerning the structure of inorganic compounds. This may be illustrated by a few examples.

(i) *Valence.* The blue peroxychromates, like the pyridine addition compound of the blue chromium peroxide (Vol. II, Chap. 5), display a weak, temperature-independent paramagnetism; the $[CrO_4]^{2-}$ ion likewise does so*. The red peroxychromates, by contrast, have a normal, temperature-dependent paramagnetism, with a magnetic moment of 1.73 Bohr magnetons. This would correspond to the presence in the compound of one unpaired electron. The analytical evidence would permit the formulation of both the blue and the red peroxychromates as binuclear compounds of $+6$ chromium, or as mononuclear compounds of $+5$ chromium. It is clear from the magnetic evidence that the former structure is correct for the blue peroxychromates, and the latter for the red compounds.

Magnetic measurements also prove that hypophosphoric acid is correctly formulated as $H_4P_2O_6$, since the compound is diamagnetic. If the formula were H_2PO_3, it would have one unpaired electron spin, and would be paramagnetic.

The bivalence of silver, in the complex salts discussed in Vol. II, Chap. 8, is also proved by magnetic susceptibility measurements. If they contained univalent silver, they would be diamagnetic since the Ag^+ ion contains only completed electron shells. Dipositive silver, with one unpaired electron, should have the same paramagnetism as bivalent copper. Measurements on the compounds of supposedly bivalent silver (Sugden, 1931, Klemm, 1931) proved that this was, in fact, the case.

(ii) *Transformations in the solid state.* The compound FeS undergoes transformations at 130° and 315°,

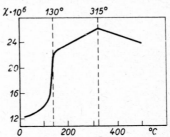

Fig. 68. Relation between temperature and magnetic susceptibility of FeS (after H. Haraldsen).

which do not show up very clearly by thermal effects, but which are very evident in the variation of paramagnetism with temperature (Fig. 68). In other cases also where solid transformations are accompanied by changes in the magnetic properties, magnetic measurements have been used to determine the transition temperatures accurately.

* Temperature-independent paramagnetism (usually very weak) is often observed for compounds involving inert gas-like ions of the transition elements—e.g. K_2CrO_4, $KMnO_4, UF_6$. It is to be explained as the result of polarization.

In the so-called 'active state' (Vol. II, Chap. 18), many substances show a considerably enhanced magnetic susceptibility. In certain cases, this may rise to values usually found only for ferromagnetic substances (cf. Fig. 69).

(iii) *Magnetism of the metals.* In the metals, the magnetism of the ions or atomic cores has superposed on it the magnetism of the 'electron gas' (Vol. II, Chap. 1). The contribution of the electron gas to the magnetism of the whole is practically independent of temperature, since the magnetic field can orient just so many electrons as can be raised to vacant energy states by the resulting gain of energy. This number is proportional to the field strength but approximately independent of the temperature (Pauli, 1927). In simple cases it is possible to calculate the magnetic susceptibility of the metal by algebraic addition of the paramagnetism due to the conduction electrons to the magnetism of the atomic cores. Even if the ions have complete electron shells, the resultant susceptibility may be positive. Thus the alkali and alkaline earth metals are paramagnetic, although the ions of these metals are diamagnetic.

In the long periods of the Periodic Table, containing the transition elements, the paramagnetism of the metals increases steadily with rise in atomic number. Thus in the first long period ($Z = 19$ to $Z = 28$), the atomic susceptibilities found at room temperature are:

$10^6 \cdot \chi_A =$	$_{19}K$	$_{20}Ca$	$_{21}Sc$	$_{22}Ti$	$_{23}V$	$_{24}Cr$	$_{25}Mn$	$_{26}Fe$	$_{27}Co$	$_{28}Ni$
	+20	+22	—	+59	+100	+150	+535	ferromagnetic		

The appearance of ferromagnetism in the metals of the iron group appears as the regular continuation of the rapidly increasing paramagnetism found for the preceding metals. Cr and Mn also exhibit ferromagnetism in certain alloys.*

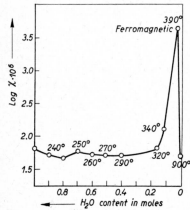

Fig. 69. Magnetism of α-Fe_2O_3, H_2O as dependent on the state of dehydration (after Hüttig).
Ordinate: susceptibilities at room temperature of samples pre-heated at temperatures indicated alongside the curve.
Abscissa: water content of samples in mol H_2O per mol Fe_2O_3.

The elements following the transition metals are diamagnetic. In general, the atomic susceptibility falls progressively with increasing atomic number in each period. Thus χ_A falls from $-5.4 \cdot 10^{-6}$ to $-25 \cdot 10^{-6}$ in passing from $_{29}Cu$ to $_{34}Se$. In the same way, the atomic susceptibility decreases on going down each group—e.g., in the copper group from $-5.4 \cdot 10^{-6}$ for Cu to $-22 \cdot 10^{-6}$ (Ag) and $-27.3 \cdot 10^{-6}$ (Au). There are some irregularities in the course of the susceptibility-atomic number relationship, which are to be explained as arising from the effect of differences in crystal structure. The susceptibility is considerably influenced by the crystal structure, as is evident in the case of tin. Cubic tin is diamagnetic ($\chi_A = -36 \cdot 10^{-6}$), whereas the tetragonal modification is paramagnetic ($\chi_A = +2.4 \cdot 10^{-6}$). Bismuth is noteworthy for the extremely low value of its susceptibility ($\chi_A = -219 \cdot 10^{-6}$ in the direction of the c-axis, $-311 \cdot 10^{-6}$ perpendicular to the c-axis). Palladium has the highest susceptibility of any metal that does not display ferromagnetism ($\chi = +572 \cdot 10^{-6}$ at room temperature). In contrast with this, the homologues of iron are surprisingly weakly paramagnetic ($\chi_A \cdot 10^{-6} = +43.5$ for Ru, $+9.5$ for Os). The susceptibilities of these metals increase with rising temperature, whereas that of palladium diminishes. Pd is dia-

---

* It has already been noted, that whereas paramagnetism is an *atomic* property, ferromagnetism is regarded in the theory of magnetism as a property of the *crystal lattice* of a particular crystalline solid. It results from a 'cooperative' effect, in which the elementary magnets mutually orient themselves throughout a certain volume (or 'domain') of the solid. It is possible to infer the 'true' susceptibility of isolated Fe, Co and Ni atoms from their magnetic properties in dilute solid solutions with other metals.

magnetic in its mixed crystals with Au (Vogt, 1932). It is to be inferred from this that Pd fills up its $4d$ level by taking electrons from the Au. There is also other evidence from alloy phases that the metals of groups VII and VIII have a tendency to acquire electrons from other metals, to fill up their vacant d states. Thus, the metals Mn, Fe, Co and Ni fit into the Hume-Rothery rules (to be considered in Vol. II, Chap. 1) if it is assumed that they have the valence 0, and function as 'metals of the IInd kind'.

The strong paramagnetism of the transition metals of groups V to VIII is probably due chiefly to the fact that their crystal lattices are built up from ions with incompletely filled d levels, and the paramagnetism of these is superimposed on that of the electron gas.

It is generally accepted, following the suggestion of Weiss (1907), that ferromagnetism [17, 18] is really only a special case of paramagnetism, arising from the orientation of the elementary magnets in an exceptionally strong molecular field. This molecular field, according to Heisenberg (1928), arises from quantum mechanical resonance between the freely mobile metallic electrons. Whereas, in general, an antiparallel arrangement corresponds to the state of lowest energy, Heisenberg showed that in special cases, and at not too high a temperature, it is possible for a parallel orientation to be the most stable for a certain group of electrons. This case appears to be that which obtains in ferromagnetic metals and alloys. The Curie point is then the temperature at which the coupling between electrons disappears. Sommerfeld and Bethe (1933) considered that ferromagnetism could only arise when incomplete electron shells of high quantum number were present, and when the interatomic distances in the crystal lattice were large compared with the radii of these incomplete shells. This condition is fulfilled not only in Fe, Co and Ni, but in certain rare earth metals also. It has, in fact, been found that these metals become ferromagnetic at sufficiently low temperatures (Vol. II, Chap. 10). The ferromagnetism of the Heusler alloys (q.v.) is possibly connected with the fact that the distances Mn—Mn, Cr—Cr in these structures are increased by the presence of the atoms of the other alloy components.

(g) **Band Spectra and Molecular Structure**

One of the most important techniques for investigation of the constitution of molecules— and especially of simple molecules—is provided by the study of *band spectra* [19–23]. A polyatomic molecule not only possesses translational energy, but also rotational energy of the molecule, vibrational energy of the atoms within the molecule, and energy of excitation of the electrons. Alterations in the content of any of these forms of energy are associated with the emission or absorption of light (including ultraviolet and infrared electromagnetic waves in that heading). The energies required to excite these several forms of potential energy are very different. Excited electronic states involve excitation energies of the order of several electron volts as for free atoms. Absorption (or emission) of light by transitions between electronic states therefore gives rise to visible or ultraviolet spectra. Excited vibrational states involve the absorption of about 0.1 to 1 electron volt, so that molecules absorb in the near infrared, by reason of excitation to states of higher vibrational energy. Differences in energy between different rotational states involve much smaller quanta of energy. Pure rotational spectra can be observed in the far infrared (wave lengths of the order 0.1 mm or greater). The infrared absorption spectra of molecules are *rotation-vibration* spectra—i.e., the rotational and vibrational energy are inevitably changed simultaneously. Similarly, the visible or ultraviolet spectra are electronic rotation-vibration spectra. A molecule is excited to higher vibrational (and rotational) states in the act of electronic excitation. The effect of this is to replace the sharp lines of atomic spectra by bands, made up of many lines representing the various possible transitions between closely spaced states of different vibrational and rotational energy. If the band spectra can be completely evaluated, the moments of inertia of molecules can be calculated, and thence the interatomic distances. In addition, the vibrational frequencies enable the complete potential energy—interatomic distance curve to be determined, with the restoring force constants, whereas the electronic band spectra furnish the binding energy of the electrons in the normal state and the excited state of the molecule, just as in atomic spectra. The study of band spectra constitutes a special branch of chemical physics, but a considerable amount of attention has been paid during recent years to the interpretation of infrared (vibration-rotation) spectra. The measurement of infrared spectra, and their use in constitutional studies and in chemical analysis, has become a matter of routine in many chemical laboratories.

Until recently, the study of pure rotational spectra was of little importance in chemistry, owing to the great experimental difficulties involved. Rotational absorption bands of gaseous molecules are generally found at frequencies corresponding to wave lengths of 0.5 to 5 cm, at which detection and measurement by the methods of classical optics is very difficult. These frequencies fall, however, into the range ('microwave' region) opened up by modern work on radiowave propagation, and the techniques devised for radar work have been applied to measurements of microwave absorption spectra. These have special advantages in the study of molecular structure. The absorbed quanta are very small, and frequencies can be measured with extreme precision. This makes it possible to investigate fine details of molecular structure—moments of inertia, bond lengths, isotope effects, dipole moments, internal electric fields of molecules and nuclear spins. In liquids and solids, quantized energy levels disappear, but it is possible to study phenomena involving intermolecular forces, such as the energy barriers hindering molecular rotation. For a review, see WHIFFEN, *Quart. Rev. Chem. Soc.*, 4 (1950) 131.

Closely related to the infrared spectra of molecules are *Raman spectra*, which provide a valuable source of information (complementary to, but not replacing infrared spectra) for studies of constitution.

(h) Raman Spectra [24–28]

When light passes through a transparent medium (a gas or liquid), some light is scattered in all directions (Rayleigh scattering). The intensity of the scattered light is inversely proportional to the fourth power of the wave length. Most of it undergoes no change of frequency in the act of scattering but—as was predicted by Smekal in 1923, and confirmed experimentally by C. V. Raman in 1928—a small fraction of it has a different frequency from the incident light. The reason for this is that some of the energy of the light is converted to vibrational energy of the molecule*. The latter is thereby altered by a single quantum, $h\nu_V$ (ν_V = frequency of atomic vibration). The relation (1) then holds.

$$\nu_0 - \nu_R = \frac{\Delta E}{h} = \nu_V, \tag{1}$$

where ν_0 = frequency of incident light, ν_R = frequency of light scattered in Raman effect, ΔE = change in vibrational energy per molecule, h = Planck's quantum, and ν_V = vibrational frequency (characteristic frequency) of the molecule.

Thus, when monochromatic light is used, spectrographic examination of the scattered light (provided the exposure is sufficient) reveals one or more extra lines (Raman lines) in addition to the spectral line with the same frequency as the incident light. The frequency difference between the incident and Raman lines is independent of the frequencey of the incident light, and is determined solely by the vibrational frequency of the irradiated molecules. If there are several vibrational states that can be excited by this means** several Raman lines are observed.

---

* This can be thought of as an inelastic collision of a photon with the molecule.
** Not all vibrational states of molecules can be excited by absorption of light, and not all are active in producing Raman scattering. In general, 'Raman-active' fundamental frequencies give rise to no infrared absorption, and vice versa. The two effects are therefore complementary, and both infrared spectra and Raman spectra must be studied for a complete elucidation of molecular constitution. Reference should be made to the monographs cited at the end of the chapter, for information as to the 'selection rules' governing excitation.

The fundamental frequences of the excited vibrations are at once obtained from the spacing between Raman lines and exciting line, and it is possible to obtain from them direct evidence on molecular structure and bond strength.

Strictly speaking, each fundamental vibration frequency should give rise to two Raman lines, since ΔE in eqn (1) could be either positive or negative (i.e., the molecule could give up energy to the incident quantum of light, instead of receiving energy). The second Raman line would be displaced from the exciting line by the same frequency difference as the first but on the violet side, instead of on the red side. It is, however very much weaker than the first. It should also be possible for energy of incident light to be converted into rotational energy, and it has, indeed, been found that a sufficiently high spectral resolving power shows the presence, not of single lines, but of closely spaced lines, differing in frequency by an amount that corresponds to the different rotational quantum numbers of the molecule.

Raman spectra can therefore be used like (or in conjuction with) infrared spectra, for the complete investigation of molecular structure. Reference is made below, and in subsequent chapters, to specific examples of the application of Raman spectra to chemical problems. Possible applications include studies of bond type and bond strength; determination of the stereochemistry of molecules and ions; the identification of the molecular species present in solutions and in liquids; and quantitative studies of chemical equilibria. [29]

Fundamental vibration frequencies depend upon the nature (i.e., the masses) of the atoms, and on the strength and kind of the bonds linking the atoms together. The vibrations may be either bond-stretching vibrations, in which a pair of atoms vibrates along the line of centers (or bond direction), or bond-bending vibrations, in which the motion of the atoms is at right angles to the bond directions, so that the angles included between the bond directions (the *valence angles*) oscillate about the equilibrium value. The restoring force in the latter case depends upon the directional properties of the covalent bond.

To a first approximation, molecular vibrations resemble those of a simple harmonic oscillator. The *frequency of vibration* is therefore given by

$$2\pi\omega_0{}^2 = \sqrt{f/M} \tag{2}$$

where ω_0 is the fundamental vibration frequency, M the reduced mass of the vibrating system ($= m_1 m_2/(m_1 + m_2)$, where m_1, m_2 are the masses of the separate atoms), and f is the *restoring force constant* which opposes the particular type of deformation involved.

The force constant f of a stretching vibration is a direct measure of the energy which must be expended in order to separate the atoms, and is roughly proportional to the heat of dissociation of the bond—i.e., to the depth of the minimum in the potential energy curve of Fig. 65. For a given pair of linked atoms (e.g., C—C or C—O in a range of organic compounds), the reduced mass is constant, and the infrared or Raman frequency ω is proportional to the square root of the force constant f, and thus affords a measure of the bond strength. In accordance with this, it is found that the fundamental vibration frequencies of molecules fall roughly into a number of distinct groups:

$\nu > 2800$ cm^{-1}	Correspond to H–X vibrations, for which the reduced mass M is small.
$\nu \sim 2400$–1900 cm^{-1}	Vibration frequencies of triple-bonded systems (—C≡C—, —C≡N, also C≡O itself in the free state and in metal carbonyls).

$\nu \sim 1800\text{–}1300 \text{ cm}^{-1}$ Vibration frequencies of double bonded systems ($>C=C<$ $>C=O$ etc.).

$\nu \sim 1200\text{–}800 \text{ cm}^{-1}$ Vibrations of single bonds.

ν small   Bending frequencies.

The force constants f for single, double, and triple bonds are almost in the ratio $1 : 2 : 3$. The bond frequencies are not completely independent of the nature of the rest of the molecule. Nevertheless, they are usually affected only to a rather small extent by adjacent atoms, and it is possible to recognize the presence of certain definite atomic groupings from the occurrence of characteristic frequencies in the Raman spectra.

The *number of independent vibrational modes* observed in the Raman spectrum and the infrared spectrum provides direct evidence as to the spatial configuration of molecules. It can be shown that a molecule consisting of N atoms should have $3N - 6$ independent vibrational modes; several of these may prove to be the same ('degenerate'), for reasons of symmetry. The symmetry also determines which of the vibrations shall be observed in infrared absorption, and which in the Raman spectrum. Observation of the Raman spectrum thus makes it possible to state whether, for example, a molecule or ion $[XY_4]$ is tetrahedral, pyramidal, or planar in configuration. If every fundamental mode of vibration can be correctly identified from the infrared and Raman spectra, it becomes possible to derive a very complete picture of the constitution of the molecule.

An example of the use made of Raman spectra is the proof of Hantzsch's assumption that the 'ester form' of nitric acid, $HO-NO_2$ is present in the anhydrous acid (p. 599). This was proved by the fact that anhydrous nitric acid showed the Raman frequencies of the $-OH$ group and the $-NO_2$ group. In the same way, it has been shown that anhydrous sulfuric acid is practically unionized, and is therefore a hydroxyl compound $(HO)_2SO_2$. Even anhydrous perchloric acid shows the typical Raman frequency of the $-OH$ group, and is thus $HO-ClO_3$. As mentioned above, the number of Raman lines which may be found depends upon the molecular symmetry. Since anhydrous perchloric acid gave 6 lines in addition to the $-OH$ line, it can be deduced that the molecules in anhydrous perchloric acid have the oxygen atoms arranged in the form of a pyramid of high symmetry, with a three-fold axis, the Cl atom being at the center of gravity and the H atom attached to the oxygen at the apex. This configuration should give the number of fundamental (Raman-active) vibrations observed. A tetrahedral arrangement of oxygens, on the other hand, would give 4 different vibrational states. It is, in fact, found that solutions of perchloric acid, which contain the tetrahedral ClO_4^- ion, actually give a Raman spectrum with 4 lines, the frequencies being somewhat different from those of the anhydrous acid.

(i) **Dipole Moment and Polarization** [30–33]

Measurements of dipole moments provide important evidence on molecular structure. A dipole with the charges $+e$ and $-e$, at a distance l apart, would possess a dipole moment μ given by $\mu = e.l$ (cf. p. 48, Fig. 9).

Dipole moments give evidence of bond type and of structure. If a molecule has a dipole moment considerably greater than zero*, all the bonds cannot be described as *purely covalent* (cf. p. 289). On the other hand it can be stated with certainty that asymmetric molecules (and this definition includes all diatomic molecules made up of atoms of different atomic number) can contain no ionic bonds if they do not have a correspondingly high dipole moment. Symmetrical molecules would possess no dipole moment even if the bonds were purely ionic. Hence it follows that molecules with a permanent dipole moment must be asymmetric in structure. For example, a molecule AB_2 with purely ionic bonding would have no dipole moment if the atoms were collinear—as in $B^-\!-\!A^{++}\!-\!B^-$. The vector sum of the two separate moments $B^-\!-\!A^+$ and $A^+\!-\!B^-$ is zero. If, however, the bonds form an angle with each other, $\underset{B^-\diagup\quad\diagdown B^-}{A^{++}}$, the moments do not cancel each other, and the

---

* An asymmetric molecule always has a small dipole moment.

resulting molecule has dipolar character. A molecule made up of two different elements has a certain dipole moment even if the bonds are purely covalent, since the electron density around two nuclei of different nuclear charge is never identical.

Measurements of dipole moments have shown, for example, that the molecules CO_2 and CS_2 are linear, whereas in H_2O, H_2S and SO_2 the atomic nuclei occupy the corners of a triangle. In the same way it has been shown that the H atoms in NH_3, PH_3 and AsH_3 are located at the basal corners of a three-sided pyramid, with the N, P or As atom at the vertex.*

Dipole moments can be measured by determination of the dielectric constant. Allowance must thereby be made for the fact that the *permanent* dipole moment of a polar molecule in an electric field has superposed on it an *induced* moment. The latter is the result of *polarization*, i.e., the displacement of the electric charges in the molecule under the action of the field. An induced dipole moment is also produced in non-polar molecules (i.e., molecules with zero dipole moment), and is proportional to the field strength— $\mu_i = a.\varepsilon$, where μ_i is the induced dipole moment and ε is the field strength. The factor a gives a measure of the *polarizability*.

The phenomenon of electric polarization corresponds exactly to the induction of a magnetic moment by a magnetic field (cf. p. 303), except that the induced electric moment is oriented in the same direction as the field, whereas the induced magnetic moment is opposed to the external field. As in the case of the magnetic moment, the induced moment is independent of the temperature, whereas the permanent electric moment (the dipole moment in the ordinary sense) is temperature dependent** and follows the same laws as the permanent magnetic moment. This different temperature dependence enables the two effects to be separated, from each other.

The induced electric moment, or the polarizability a, which determines its magnitude, can also be determined independently, from an equation derived by Lorentz from the electromagnetic theory of light.

$$a = \frac{3}{4\pi N_A} \cdot \frac{n^2 - 1}{n^2 + 2} V \tag{3}$$

where N_A is Avogadro's number, V the molecular volume, and n the refractive index for light of infinitely long wave length. The term $(n^2 - 1)/(n^2 + 2)$ is termed the *molecular refraction*.

Measurements have shown that the molecular refraction (and therefore also the polarizability, in accordance with eqn. 3) increases very considerably with increasing size of the electron cloud, for inert gas atoms and free ions with inert gas configuration. For any one electron configuration, it decreases strongly as the nuclear charge increases—e.g., in the 'neon-like' series O^{2-}, F^-, Ne, Na^+, Mg^{2+}. The polarizability of ions is quite considerably influenced by compound formation. Hence the molecular refraction of ionic compounds is not derived additively from that of the constituent ions, as is the case with organic compounds.

Thus the Cl^- ion in the free state (or in aqueous solution) has the molecular refraction 9.00, whereas in the compounds NaCl, LiCl, HCl the values are only 8.0, 7.4 and 6.67, respectively. The molecular refractions of the free Br^- and I^- ions are 12.67 and 19.24 as compared with 9.14 and 13.74 in the compounds HBr and HI. The molecular refraction of the free O^{2-} ion is 7. In the compounds or composite ions:

	CaO	MgO	BeO	PO_4^{3-}	SO_4^{2-}	ClO_4^-
the value is	6.1	4.2	3.2	4.05	3.62	3.3

* The positions of the atoms (distances and valence angles) may be fixed by combining the results of dipole moment measurements with moments of inertia obtained spectroscopically. The angle can often be fixed independently, from band spectrum data alone.

** This applies, of course, to the permanent electric moment per unit of volume of an assembly of molecules (gas, liquid or crystal). For each *isolated* molecule, the permanent electric or permanent magnetic moment is, of course, not dependent on temperature.

This example shows that the polarizability of an elementary anion is diminished when it forms a compound with a positively charged atom; the extent of this effect increases as the ionic radius of the cationic element gets smaller, as its ionic charge gets larger, and also as the polarizability of the anion rises. The more strongly the cation polarizes the anions with which it is combined, the less can the anion undergo additional distortion by the action of external fields.

7. The Color of Inorganic Compounds

A substance is colored if it has the property of absorbing visible light (or certain wave lengths in the visible spectrum)*. Absorption of visible (or ultraviolet) light can occur only if electrons are present which are capable of being excited to states of higher energy by the absorption of a quantum of energy defined by $\varepsilon_2 - \varepsilon_1 = h\nu$, where ν corresponds to the vibrational frequency of light in the visible spectrum. Substances which appear colorless generally require a larger increment of energy, as corresponds to absorption in the ultraviolet. Electrons which can be excited to higher states by absorption of visible light are therefore said to be 'loosely bound'. However, the important quantity is not the binding energy of the electrons, but the difference in binding energy between the different electronic states concerned in the act of absorption.

Substances composed of inert gas-like ions are in general colorless**. Color may arise, however, in compounds which contain electropositive ions of very high polarizing power. Thus (if it is legitimate to consider them as polar structures), the compound ions CrO_4^{2-} and MnO_4^- are colored, although they can be formally considered as built up from inert gas-like ions. Since the polarizing power rises as the charge on an ion increases, and also as the ionic radius diminishes, ions such as Cr^{6+}, in which both conditions are fulfilled, have very great polarizing power.

Compounds built up from ions with incomplete electron shells are usually colored. Thus compounds derived from the elements of the Sub-groups, exercising valences lower than the group number, are almost invariably colored. The same is true of the 'sub-compounds' of the Main Group elements, when these contain easily polarizable anions. In general, the color of a series of compounds becomes deeper as the polarizablity of the anion is increased—i.e., the absorption bands shift towards the red.

Examples of this behavior are shown in Table 61. The polarizability of the halogens increases from F^- to I^-, and the depth of color increases correspondingly from the fluorides to the iodides. The same is true of the chalcogenides. In the sulfates and perchlorates, the polarizability of oxygen (which is not present in these compounds strictly as O^{2-} ions) is reduced by its combination with sulfur or chlorine, and the sulfates and perchlorates shown in the table are colorless or only weakly colored.***

*As used here, 'colored' is the opposite of 'colorless', and therefore includes black. In the narrower sense, substances are colored only if they have a selective absorption of visible light. In view of the increasing importance of spectrophotometry in chemical laboratories, it is convenient to include under the heading of 'colored' substances those which have absorption bands in the near ultraviolet region of the spectrum.

** The excitation energies of such ions are of the order of 5 electron volts or greater, corresponding to absorption in the far ultraviolet region below 2000 Å. The absorption spectra usually show complete absorption below such wavelengths, with no band structure.

*** All the factors involved in these phenomena are not yet understood. The act of absorption undoubtedly involves the $3d$ levels of Cu and Ni respectively. These levels are thus apparently affected by polarization, in the same way as by complex formation.

TABLE 61

COLORS OF ANHYDROUS NICKEL AND COPPER COMPOUNDS

	F	Cl	Br	I	O	S	Se	SO$_4$	ClO$_4$
Ni^{II}	Yellowish	Yellow brown	Dark brown	Black	Dark green	Black	Grey blue	yellow	yellow
Cu^{II}	Colorless	Yellow brown	Brown black	—	Black	Dark blue black	Greenish black	colorless	—

The colors shown in Table 61 are those of the *anhydrous* salts. From the fact that the depth of color runs parallel to the polarizability of the anion, and that there is no visible color if the anion is practically unpolarizable, it is reasonable to assume that it is the anion, and not the cation, which actually causes the color of these compounds. It is, in fact, probable that the absorption spectra of these substances are *charge transfer spectra*, and that the act of absorption can be represented

$$X^- \to X + e; \quad M^{2+} + e \to M^+ \quad \text{(electron trapped in } d \text{ level)}$$

The quite different, but characteristic absorption spectra of the solutions and hydrated ions of these substances are not yet so readily interpreted. Polarization of the water molecules by the cations undoubtedly modifies the energy levels of the H$_2$O molecules; the coordination of H$_2$O around these transition metal cations certainly modifies the d levels of the metal ion (see ORGEL, *J. Chem. Soc.*, (1952) 4756). It is therefore possible that the excited electron may be in the H$_2$O molecule, rather than in the metal ion. The same considerations apply to ammoniates.

On this basis it is possible to interpret a number of puzzling phenomena. Thus many of the elements of the Main Groups of the Periodic System form colored, anhydrous compounds of salt-like character, (e.g., PbO, PbS, Bi$_2$O$_3$, Bi$_2$S$_3$), although the relevant cations are completely colorless in aqueous solution. The polarizability of the O^{2-} ion is greatly diminished by compound formation (cf. p. 311); it is therefore plausible that certain cations should polarize O^{2-} ions in oxide structures, to the extent of giving charge transfer spectra, but should not polarize H$_2$O molecules to a sufficient degree. In general, the ions of the Sub-group elements are smaller in radius than those of the Main Groups, and have a greater polarizing power. They are therefore able to bring about the necessary mutual modification of energy levels which is necessary if the hydrated (or ammoniated) ions are to have excitation energies of 1 to 3 electron volts, so as to give charge transfer absorption spectra in the visible region.

References

1 R. KREMANN and M. PESTEMER, *Zusammenhänge zwischen physikalischen Eigenschaften und chemischer Konstitution*, Dresden 1937, 225 pp.

2 E. FERMI, *Moleküle und Kristalle* (translated from Italian by M. Schön and K. Birus), Leipzig 1938, 234 pp.

3 C. H. D. CLARK, *The Electronic Structure and Properties of Matter*, London 1934, 374 pp.

4 R. GLOCKER, *Materialprüfung mit Röntgenstrahlen unter besonderer Berücksichtigung der Röntgenmetallkunde*, 2nd. Ed., Berlin 1936, 386 pp.

5 L. C. PAULING, *The Nature of the Chemical Bond and the Structure of Molecules and Crystals; An Introduction to Modern Structural Chemistry*, 2nd Ed., Ithaca 1940, 450 pp.

6 G. BRIEGLEB, *Zwischenmolekulare Kräfte und Molekülstruktur*, Stuttgart 1937, 308 pp.

7 S. SUGDEN, *The Parachor and Valency*, London 1930, 224 pp.

8 M. ULMANN, *Molekulgrössen-Bestimmungen hochpolymerer Naturstoffe*, Dresden 1936, 200 pp.

9 H. A. STUART, *Molekülstruktur, Bestimmung von Molekülstrukturen mit physikalischen Methoden*, Berlin 1934, 388 pp.

10 H. MARK and R. WIERL, *Die experimentellen und theoretischen Grundlagen der Elektronenbeugung*, Berlin 1931, 126 pp.

11 P. W. SELWOOD, *Magnetochemistry*, New York 1943, 287 pp.

12 W. KLEMM, *Magnetochemie*, Leipzig 1936, 262 pp.

13 E. C. STONER, *Magnetism and Matter*, London 1934, 575 pp.

14 S. S. BHATNAGAR and K. N. MATHUR, *Physical Principles and Applications of Magnetochemistry*, London 1935, 375 pp.

15 F. BLOCH, *Molekulartheorie des Magnetismus*, Leipzig 1934, 110 pp.

16 J. H. VAN VLECK, *The Theory of Electric and Magnetic Susceptibilities*, Oxford 1932, 384 pp.

17 R. BECKER and W. DÖRING, *Ferromagnetismus*, Berlin 1939, 440 pp.

18 W. S. MESSKIN and A. KUSSMANN, *Die ferromagnetischen Legierungen und ihre gewerblichen Verwendungen*, Berlin 1932, 418 pp.

19 H. SPONER, *Molekülspektren und ihre Anwendung auf chemische Probleme*, Vol. I, Berlin 1935, 154 pp.; Vol II, Berlin 1936, 506 pp.

20 G. HERZBERG, *Molekülspektren und Molekülstruktur*, Vol. I, *Zweiatomige Moleküle*, Dresden 1939, 404 pp.

21 W. FINKELNBURG, R. MECKE, O. REINKOBER and E. TELLER, *Molekül- und Kristallgitterspektren*, Leipzig 1934, 408 pp.

22 W. FINKELNBURG, *Kontinuierliche Spektren*, Berlin 1938, 368 pp.

23 C. SCHAEFER and F. MATOSSI, *Das ultrarote Spektrum*, Berlin 1930, 400 pp.

24 G. B. B. M. SUTHERLAND, *Infra-Red and Raman Spectra*, London 1935, 112 pp.

25 G. PLACZEK, *Rayleigh-Streuung und Raman-Effekt*, Leipzig 1934, 170 pp.

26 C. SCHAEFER and F. MATOSSI, *Der Raman-Effekt*, Berlin 1930, 52 pp.

27 P. DAURE, *Introduction a l'Etude de l'Effet Raman; Ses Applications Chimiques*, Paris 1933, 90 pp.

28 K. W. F. KOHLRAUSCH, *Der Smekal-Raman-Effekt*, Berlin 1931, 392 pp.; *Supplement*, Berlin 1938, 288 pp.

29 G. KORTÜM, *Das optische Verhalten gelöster Elektrolyte*, Stuttgart 1936, 106 pp.

30 P. DEBYE, *Polar Molecules*, New York 1929, 172 pp.

31 P. DEBYE and H. SACK, *Theorie der elektrischen Molekulareigenschaften*, Leipzig 1934, 136 pp.

32 R. J. W. LeFÈVRE, *Dipole Moments, Their Measurement and Applications in Chemistry*, London 1938, 110 pp.

33 C. P. SMYTH, *Dielectric Behavior and Structure*, New York 1955, 441 pp.

CHAPTER 10

THIRD MAIN GROUP OF THE PERIODIC SYSTEM: BORON-ALUMINUM GROUP

Atomic numbers	Names of the elements	Symbols	Atomic weights	Densities	Melting points	Boiling points	Specific heats	Valences
5	Boron	B	10.82	2.34	ca. 2300°	—	0.252	III
13	Aluminum	Al	26.98	2.70	660.2°	2270°	0.214	III
31	Gallium	Ga	69.72	5.9	29.78°	ca. 2000°	0.0918	I? II III
49	Indium	In	114.76	7.31	156.4°	ca. 2300°	0.0570	I II III
81	Thallium	Tl	204.39	11.83	302.5°	ca. 1450°	0.0316	I III

1. General

The IIIrd Main Group of the Periodic System comprises first two widely distributed elements, boron and aluminum; and then three rare elements, gallium, indium and thallium. These last are among the elements which are, indeed, fairly widely distributed in the earth's crust, but almost invariably in minimal concentrations, as isomorphous replacements in certain minerals formed by other elements. In this sense they are rare elements—gallium and indium, indeed, exceedingly rare; all were discovered by spectroscopic analysis.

The elements of the third main group have a maximum valence of three in their compounds, in accordance with their group number. With *boron* and *aluminum*, the formation of compounds in which they are present in a lower valence state is very exceptional. Gallium, indium and thallium, however, can readily be reduced to lower valence states. Gallium and indium are less stable in their lower valence states than in the trivalent state, whereas with thallium the univalent state is the favored one. The compounds of trivalent thallium show little resemblance to those of aluminum, the typical element of the third family. The resemblance of indium, and especially of gallium, to aluminum is very much closer. Thus their hydroxides are amphoteric, like aluminum hydroxide, and their salts are hydrolyzed to some extent in solution like the salts of aluminum. Gallium and indium also form alums, $M^IM^{III}(SO_4)_2 \cdot 12H_2O$, although the range of compounds formed is not so extensive as that from aluminum. Trivalent thallium, on the other hand, forms double sulfates of a different type, $M^IM^{III}(SO_4)_2 \cdot 4H_2O$. Aluminum can also form double salts of this type, and under the appropriate conditions their crystallization can be initiated by inoculation with the corresponding thallium(III) double sulfate.

Gallium, indium and thallium differ from boron and aluminum not only in the ease with which they exercise lower valences, but also in the smaller heats of formation of their oxides and the ease with which they are reduced to the metals. The metals are soft, and have low melting points, in sharp contrast with aluminum, and especially with boron.

Boron is definitely non-metallic in character. Its homologues, however, are typical metals. In accordance with its non-metallic nature, boron tends to form covalent compounds, whereas the homologues of boron preferentially form ionic compounds. They are all able to exist in solution in the form of free (that is, hydrated) tripositive ions. Boron cannot so function.

In the IIIrd Main Group, the electropositive character of the elements does not show the same regular increase with atomic number as it does with the elements of Main Groups I and II. It first increases very greatly from boron to aluminum, but decreases again from aluminum, and then falls off still further from gallium to thallium. This irregular trend in electropositivity is connected with the fact that whereas boron and aluminum follow immediately after the elements of the alkaline earth group, gallium indium and thallium do not; interposed between them are the elements in which the d-shells are filled (see Table II, Appendix). From beryllium to boron and from magnesium to aluminum, the nuclear charge increases by only one unit, whereas from calcium to gallium it increases by 11 units. Since the principal quantum number remains the same in this process, the result is a considerable increase in the strength of binding of the electrons in the gallium atom, and a consequent marked weakening in electropositive character. Similar considerations apply to indium and thallium.

Boron is a distinctly *acid-forming* element. Whereas aluminum oxide can act as an acid anhydride towards strong bases, it generally functions as a basic oxide. Gallium(III) and indium(III) oxides are also amphoteric substances of predominantly basic character. The amphoteric nature of thallium(III) oxide shows up less clearly, because of the extreme insolubility of this oxide. Its basic character, however, is not more strongly developed than that of the oxides of aluminum, gallium(III) or indium(III). *Thallium(I) oxide*, on the other hand, is *strongly basic* in nature. The hydroxide derived from it resembles the alkali hydroxides in this respect.

The strongly basic nature of thallium(I) hydroxide fits the general rule that the oxides and hydroxides derived from the lower valence states of any element have a stronger basic (or more weakly acidic) character than those derived from the higher valences of that element.

The decrease in acidic character from boron to aluminum oxide, and the associated increase in basic character, are to be explained in the same way as the increase in basic character from lithium to cesium hydroxides, or from beryllium to barium hydroxides.

In the hydroxides of the Group IIIA elements, the triply charged positive ions attract the oxygen ions strongly, and exert a relatively strong repulsion for the hydrogen ions. This greatly weakens the attraction of the oxygen atoms for the hydrogen atoms, if, the centers of the H^+ ion and the M^{3+} ion are close together, as is the case with ions of small radius. The result is that with the hydroxide of the ion of smallest radius in this group (i.e., the hydroxide of boron), H^+ ions are more easily split off than OH^- ions. This repulsive action of tripositively charged boron upon the hydrogen ion explains the inability of boron to exist in solution as a free positive ion. The so-called elementary ions are invariably present in aqueous solution as *ion hydrates*. Such a hydrate—for example $[M^{III}(OH_2)_3]^{+++}$—would at once break up, in the case of boron, into $B(OH)_3$ and three H^+ ions.

The radius of the Al^{3+} ion is considerably greater than that of the B^{3+} ion. The aluminum ion thus exerts a smaller repulsion on the H^+ ions, and binds the OH^- ions appreciably less firmly. This is the reason why $Al(OH)_3$ displays a weaker acidic and a much stronger basic character than $B(OH)_3$. The same applies to the homologues of aluminum.

The *amphoteric* character of aluminum hydroxide—i.e., its ability to form salts with bases as well as with acids, does not necessarily depend on any ability to dissociate by ionizing off H^+ ions as well as by ionizing off OH^- ions. The formation of salts with bases may be due to the ability of $Al(OH)_3$ *to add on additional OH^- ions*. It can, indeed, be shown that energy would be liberated by the reaction $Al(OH)_3 + OH^- = [Al(OH)_4]^-$ (cf. Chap. 11). Experiment has shown that this, and not the reaction $Al(OH)_3 = [AlO(OH)_2]^- + H^+$, is the process which takes place when aluminum hydroxide is brought into contact with strong bases in aqueous solution. Thus $Al(OH)_3$ functions in alkaline aqueous solutions not as an *acid*, but as an *anhydroacid*, and forms *hydroxo-salts*—e.g., $K[Al(OH)_4]$, potassium hydroxoaluminate. On the other hand, H^+ ions are able to abstract OH^- ions from the ion $[Al(OH)_4]^-$, so that equilibria depending on the hydrogen ion concentration are set up:

$$[Al(OH)_4]^- + H^+ \rightleftharpoons Al(OH)_3 + H_2O \qquad \text{(a)}$$

$$Al(OH)_3 + 3H^+ \rightleftharpoons Al^{+++} + 3H_2O \qquad \text{(b)}$$

These equilibria characterize the amphoteric behavior of aluminum hydroxide.

The equilibrium constant for reaction (a) is

$$\frac{[Al(OH)_4]^-[H^+]}{[Al(OH)_3]} = 2\cdot10^{-11} \qquad \text{or} \qquad \frac{[Al(OH)_4]^-}{[Al(OH)_3][OH^-]} = 2\cdot10^3.$$

The electrolytic mobility of the $[Al(OH)_4]^-$ ion, from the work of Bode (1952), appears to be about the same as that of the hydrated Li^+ ion. It appears that in sodium hydroxoaluminate solutions there are no colloidal particles or macromolecular ionic aggregates.

Of the *hydrogen compounds* of the IIIrd Main Group elements, those of boron and gallium are highly volatile. In this and other respects they show resemblance to the hydrides of the elements lying further to the right in the Periodic Table. The hydrides of aluminum and indium are highly polymerized solids. They are related to the hydrides of beryllium and magnesium, and do not have the salt like character of the alkali and alkaline earth metal hydrides. Thallium is the only element in the whole of the Main Groups of the Periodic System (except for the inert gases) which appears to be unable to form a hydrogen compound that is capable of being isolated*. However, thallium hydride is obtainable in in the form of double compounds. A feature common to all the hydrogen compounds of the elements of the group is that in the free state they exist only in a polymerized form (e.g. $(BH_3)_2$, $[AlH_3]_x$). This is due to the linkage of the monomeric molecules to one another through *hydrogen bridge bonds*.

Main Group III also provides a typical example of the rule that the first member of a Main Group tends to resemble elements of the next Group, and that the second member resembles the Sub-group of the same family. Apart from its valence, boron has little in common with its higher homologues. In its nature as an acid-forming element it stands much closer to the neighboring carbon and silicon. Aluminum is more closely related to the elements of Sub-group III than to boron. Indeed, its relation to these is hardly less marked than its relation to its heavier homologues in the main group, and in many respects it occupies a more definite

---

* Pietsch (1933) found that gallium, indium and thallium would combine with atomic hydrogen to form hydrides. He found that indium and gallium formed solid hydrides by this means, whereas thallium, like the neighboring element, lead, formed an extremely unstable gaseous hydride.

transitional position between boron and the Sub-group elements than between boron and the Main Group elements. For example, the electropositive character increases regularly from boron, through aluminum, to lanthanum, whereas (as has already been mentioned) this is not the case in the sequence boron—aluminum —gallium—indium—thallium. The heats of formation of the chlorides and oxides also increase regularly from boron, through aluminum to lanthanum, whereas they diminish from aluminum to thallium (cf. Fig. 1, p. 13). The relationship between aluminum and its heavier homologues of the Main Group is most clearly displayed in the *identical structure of their hydrogen compounds*. Gallium and indium also share with aluminum the characteristic property of *forming alums*.

On the basis of the Lewis-Kossel theory, the electropositive trivalence of the elements of Main Group III follows from their relation to the inert gases preceding them in the Periodic System. As neutral atoms, boron and aluminum each have three electrons more than the preceding inert gas, and these 3 electrons are lost relatively easily. They are therefore lost to elements such as oxygen, chlorine, etc., which can acquire a negative charge, leaving trivalent positive ions. Gallium, indium, and thallium correspond to the elements boron and aluminum in the structure of their outer shells and accordingly in their chemical behavior also.

We have reliable evidence as to the atomic structure of boron and aluminum, from their spectra. We learn from the term analysis of the spark spectra of boron and aluminum that the first electron bound by the atomic cores B^{3+} and Al^{3+} is bound in exactly the same manner as in the lithium or sodium atom. The binding is, however, 9 times as strong, corresponding to the fact that the core has 3 times the charge. We conclude that, apart from the higher positive charge contained in the nucleus, the atomic cores B^{3+} and Al^{3+} correspond in their electron configuration to the cores Li^+ and Na^+, respectively, and therefore to the inert gas atoms He and Ne. The electron orbits in these atomic cores obviously shrink more and more together as the nuclear charge increases. This follows from eqn. (9) of Chap. 3, and is clearly shown in the decrease of radii in the series $He—Li^+—Be^{2+}—B^{3+}$, and $Ne—Na^+—Mg^{2+}—Al^{3+}$ (cf. Fig. 3, p. 16).

The close analogy between the spectral term systems of Na, Mg^+ and Al^{2+} is shown by Fig. 70a. The energy levels corresponding to the spectral terms are represented in the manner

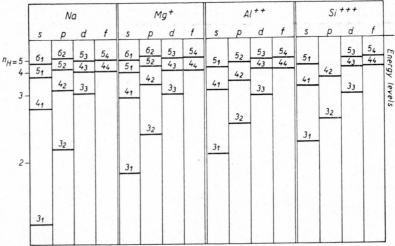

Fig. 70a. The energy levels as given by the arc spectrum of sodium, compared with the energy levels deduced from the spark spectra of magnesium, aluminum and silicon. All spectra relate to atoms or atomic cores with the same number of electrons (11 electrons).

used previously (p. 252). The numbers marked above the various energy levels represent the Bohr-Sommerfeld quantum numbers n_k (as also in Fig. 70b). The scale on the left hand side shows the energy levels for $n = 2, 3$, etc. for the hydrogen atom. To eliminate the influence of nuclear charge, the ordinates of the energy levels of Mg^+ are reduced in scale to $\frac{1}{4}$, and those of Al^{2+} to $\frac{1}{9}$, as compared with those of sodium. The energy levels for the atomic core of silicon, with one circulating electron, are also shown, and are reduced in scale by a factor $\frac{1}{16}$. We shall return to this spectrum in Chap. 12.

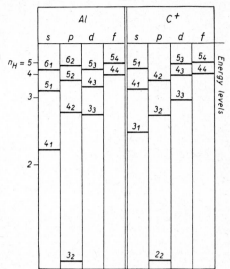

Fig. 70b. Energy levels as given by the arc spectrum of aluminum and the spark spectrum of carbon.

Fig. 70b represents (on the *same* scale as for sodium in Fig. 70a) the energy levels corresponding to the various modes of binding of the third electron bound by Al^{3+}, as indicated by spectral emission. These refer, accordingly, to the *neutral* aluminum atom. Next to them are represented, for purposes of comparison, the energy levels deduced from the spark spectrum of singly ionized carbon, which will be discussed later. It may be seen that these two spectra are considerably different in structure from those represented in Fig. 70a. In the Al^{2+} ion, the electron in its normal state moves in a $3s$ orbit, but in the neutral Al atom, in a $3p$ orbit. The energy level corresponding to the quantum state $3s$ is absent from the spectral terms of the neutral Al atom. This means that the last electron taken up by the aluminum atom cannot drop to any orbit lower than the $3p$ orbit. The available $3s$ orbits are occupied by the first two electrons to be taken up— the second of these is bound just as in the alkaline earth metals, as follows from the term system of Al^+.* The three outer electrons of aluminum are thus not identical in binding, and the same is true for boron. Two are in $3s$ orbits, and one in a $3p$ orbit. The extent to which this determines the formation of *covalent* compounds of boron and aluminum will be discussed later. It is of no importance for the formation of ionic compounds unless the different modes of binding (i.e., either as an s- or a p-electron) has a strong influence on the *binding energy* of the electron. This is not the case with B or Al. Transitions to the ionic state depend less on the mode of binding of the electrons in the atom than on the work which must be expended to remove them.

The atomic spectra of *gallium*, *indium* and *thallium* correspond closely in structure to that of aluminum. They lead to the conclusion that in the free atoms of these elements, in the ground state, the 'outermost' electrons have a configuration which is essentially similar to those of boron and aluminum—i.e., with 2 electrons in s orbits, and 1 electron in a p orbit. From the magnitudes of the spectral terms, it follows that these three electrons, and not others, can relatively easily be removed from the atom. The maximum positive trivalence of the element is thereby explained.

The *total* energy expended in the removal of three electrons from an atom of any element of Main Group III is less than that required to remove yet one further electron, as the data of Table 62 show. The work expended in ionizing off three electrons can be performed at the expense of the lattice forces, but not the work required to abstract 4 electrons. This increases in considerably greater degree than does the lattice energy. From eqn. (11), p. 147, we can calculate that for a compound of 4-valent aluminum, which was isotypic with a compound of 3-valent aluminum and had the same lattice dimensions, the lattice energy would be greater only in the ratio $1.8 : 1$. The energy to be expended in removing 4 electrons from an Al-atom, however, is greater in the ratio $3.25 : 1$ than that needed to remove

* It follows from the Pauli principle that not more than 2 electrons in an atom can occupy each s level.

3 electrons. This explains why aluminum is never quadrivalent in ionic compounds, but always behaves as trivalent.

TABLE 62

IONIZATION ENERGIES OF ELEMENTS OF MAIN GROUP III

	Boron	Aluminum	Gallium	Indium	Thallium
I M → M⁺ + e	190.3	137.4	137.7	132.8	140.2 kcal/g atom
II M⁺ → M²⁺ + e	576.1	431.8	435.9	387.5	468.2 ,,
III M²⁺ → M³ + e	869.7	652.4	705.3	643.7	685.0 ,,
Sum I + II + III	1636.1	1221.6	1278.9	1164	1293 kcal/g atom
M³⁺ → M⁴ + e	5946.5	2751.4	1474	1222	— kcal/g atom

It can also be shown, from the cyclic processes discussed on p. 146 *et seq.*, that, in general, we cannot expect boron and aluminum to form ionic compounds, stable at ordinary temperature, in which they exert a *lower* valence than 3*. Nevertheless, calculations of stability are necessary for each individual case if reliable predictions are to be made. Moreover, it cannot necessarily be concluded that instability of compounds in the solid state implies that they will be unstable in solution, for stability in aqueous solution is determined substantially by the heats of hydration of the ions. The following example illustrates the importance of this factor.

The energy which must be expended to transform 1 g atom of metallic Al into the hydrated ion Al^{+++} can be calculated, by means of eqn. (12), p. 149, from the heat of sublimation (57 kcal per g atom of Al, p. 322), the energy of ionization (Table 62) and the heats of hydration. From the heats of solution of aluminum salts, the heat of hydration is found to be 1065 kcal per g ion of Al^{+++}, when the heat of hydration of the H^+ ion is taken as 250 kcal.** The energy of formation of the hydrated Al^{++} ion is obtained similarly, except for the inknown quantity $Q_{Al}{}^{++}$, the heat of hydration of this ion:

$$[Al] \xrightarrow{+57} (Al) \xrightarrow{+1222} (Al^{3+}) \xrightarrow{-1065} Al^{+++}$$

$$[Al] \xrightarrow{+57} (Al) \xrightarrow{+569} (Al^{2+}) \xrightarrow{-Q_{Al}{}^{++}} Al^{++}$$

Work *done on* the system is here written as *positive*. Using the cyclic process, we obtain for the free energy changes in the reactions:

$$[Al] \rightarrow Al^{+++} + 3e + 214 \text{ kcal} \tag{1}$$

$$[Al] \rightarrow Al^{++} + 2e + (626 - Q_{Al}{}^{++}) \text{ kcal} \tag{2}$$

We have further:

$$\tfrac{1}{2}(H_2) + \tfrac{1}{2}(Cl_2) = H^+ + Cl^- - 31.4 \text{ kcal} \tag{3}$$

The free energy of this reaction is obtained directly from the normal potential of chlorine (−1.36 volt), from eqn. (3), p. 142. Further, from the data of Table 7, p. 39, we obtain

$$\tfrac{1}{2}(H_2) \xrightarrow{+51.4} (H) \xrightarrow{+312.2} (H^+) \xrightarrow{-250} H_3O^+$$

* At high temperatures it is possible to obtain *gaseous* compounds in which B and Al are present with a valence less than 3.

** The hydration energy of the H^+ ion subsequently cancels out of the calculation. The conclusions are therefore not affected by the uncertainties in the value assumed.

i.e., $$\tfrac{1}{2}(H_2) + H_2O = H_3O^+ + e + 113.6 \text{ kcal} \tag{4}$$

Eqn. (1) + 3 × [eqn. (3) — eqn. (4)]:

$$[Al] + \tfrac{3}{2}(Cl_2) = Al^{+++} + 3Cl^- - 221 \text{ kcal} \tag{5}$$

Eqn. (2) + 2 × [eqn. (3) — eqn. (4)]:

$$[Al] + (Cl_2) = Al^{++} + 2Cl^- + (336 - Q_{Al^{++}}) \text{ kcal} \tag{6}$$

Eqn. (5) — eqn. (6):

$$Al^{++} + \tfrac{1}{2}Cl_2 = Al^{+++} + Cl^- + (Q_{Al^{++}} - 557) \text{ kcal} \tag{7}$$

From eqn. (5), with eqn. (3) p. 142, the decomposition potential of $AlCl_3$, referred to a solution l-molar in Al^{+++} and Cl^- ions, is found as $Z_5 = -3.2$ volts, in satisfactory agreement with the value (—3.1 volts) found by addition of the normal potentials*. The free energy change of eqn. (7), converted to e-volts, or the numerically equal decomposition potential in volts (Z_7), for the dissociation of $AlCl_3$ into $AlCl_2$ and $\tfrac{1}{2}Cl_2$, works out as

$$Z_7 = 4.34 \cdot 10^{-2} \cdot Q_{Al^{++}} - 24.2$$

The Al^{2+} ion, not being inert-gas like, has a stronger polarizing action and therefore probably a higher heat of hydration than the Mg^{2+} ion. On the assumption that its heat of hydration is the same as that of the Mg^{2+} ion ($Q = 437$ kcal), we obtain $Z_7 = -5.2$ volts. Since this value is numerically larger than Z_5, the potential required to deposit aluminum as the metal, this assumption about the heat of hydration excludes any possibility of forming Al^{++} ions. If, on the other hand, we assume that the hydration energy of the Al^{2+} ion is the same as that of the Be^{2+} ion ($Q = 570$ kcal), we obtain $Z_7 = +0.5$ volt. This would mean that the reaction of eqn. (7) would proceed spontaneously from right to left; an $AlCl_3$ solution would decompose, forming $AlCl_2$ with the evolution of chlorine. It may be seen from this how the contribution made by the hydration energy is of decisive influence for the stability of the Al^{++} ion. Since its energy of hydration presumably lies somewhere between that of the Mg^{2+} and Be^{2+} ions, the possibility must be considered that Z_7 is, indeed, not positive, but is still numerically smaller than Z_5. In this event, when an increasing potential was applied across an $AlCl_3$ solution the potential Z_7 would be reached before the potential Z_5—i.e., Al^{+++} ions would first be discharged to form Al^{++} ions, and only at a higher potential, given in this case by eqn. (6), would the deposition of aluminum as metal take place.

For the formation of *covalent* compounds, the type of binding of the electrons within the atom plays an important role. The evidence of the spectra shows that the elements of Main Group III have only one unpaired electron in the normal state of the atom. According to the Heitler-London theory, we may therefore expect them to display *univalence* in covalent compounds. The molecules BH and AlH have, indeed, been recognized by band spectroscopic means. Their free energy of formation from the *free atoms* amounts to —79.6 and —71.0 kcal, respectively. Their free energies of formation from crystallized B or Al and H_2 molecules, however, have strongly positive values (+87 and +35 kcal per mol. respectively). Thus these monohydride molecules do not represent stable compounds in the *chemical* sense. However it is also found from the spectra that only an extremely small expenditure of energy is needed to 'excite' the atoms of B and Al—i.e., to charge an electron from an *s*-state to an adjacent *p*-state—namely 0.04 and 0.32 kcal per g atom, respectively. The excited atoms (with the configuration *sp²*) contain 3 unpaired electrons, and according-ly can function as *trivalent*. However, the wave mechanical calculation of the stability of compounds derived from this state presents difficulties. Chemical experience shows that the covalent compounds of the elements of Main Group III have a low stability, as compared with most of the ionic compounds. The heat of formation of B_2H_6 is 44 kcal per mol, accor-

---

* The difference is due to neglect of the variation with temperature of the values used. If their temperature variation is allowed for, there is agreement within a few hundredths of a volt.

ding to Roth (1937)—i.e., only $\frac{1}{8}$ that of B_2O_3. Nevertheless, the ability to form covalent compounds is much greater in boron than in the elements of Main Groups I and II. It is apparently smaller for aluminum, however, and diminishes yet further as one proceeds to thallium.

The decreased ability of the elements to form covalent compounds in going from Al to Tl shows itself in the *organometallic compounds*, in that trialkyl compounds MR_3 are obtained from Al, Ga and In, by the reaction of their chlorides with the Grignard reagents MgRCl, whereas only the dialkyl compounds of thallium, TlR_2Cl, can be prepared in this way (cf. p. 387). In the case of the aryl compounds, even In yields only the diaryl compounds InR_2Cl, whereas the triaryl compounds of Al and Ga can be obtained. The power of forming *addition compounds with ether* also follows the same gradation in this series of metals. Such compounds are known for the various trialkyls of Al and Ga. Only one such compound of indium (trimethyl indium etherate) is known, and no compound of Tl.

These ether addition compounds have the composition $MR_3 \cdot (C_2H_5)_2O$. They are probably formed through the ether oxygen sharing its lone pair of electrons with the incomplete octet of the metal atom:

$$(C_2H_5)_2O + Al(CH_3)_3 = \begin{array}{c} C_2H_5 \diagdown \\ \qquad\quad O{\longrightarrow}Al(CH_3)_3 \\ C_2H_5 \diagup \end{array}$$

The heats of formation of some compounds of B, Al, Ga, In and Tl are set out in Table 63.

TABLE 63

HEATS OF FORMATION OF COMPOUNDS OF ELEMENTS OF MAIN GROUP III

(kcal per g equivalent)

Elements	Oxides	Fluorides	Chlorides	Bromides	Iodides	Sulfides	Nitrides
Boron	58.2	99.0	40.7	40?	—	—	ca. 12
Aluminum	66.41	125.3	55.8	40.3	23.4	—	21.3
Gallium	42.9	—	41.7	30.8	17.0	—	8.3
Indium	37.1	—	42.8	32.4	18.8	—	1.5
Thallium	20?	—	26.7	—	—	—	—

The heats of formation of the compounds of the elements of the third Main Group are, on the whole, considerably lower than in the first two Main Groups, when referred to the equivalent amounts. This is in part the result of the considerably higher ionization energy of the electrons (cf. Table 62). For the compounds of boron and aluminum, this increase is roughly compensated by the increase in the lattice energy. The decrease in the heat of formation per gram equivalent of compound, which is found in the series Li —Be —B and Na—Mg—Al is to some extent the consequence of the substantial rise in the heats of sublimation of the elements along these series. It is true that, for most of the elements cited, the heats of sublimation have not been measured directly. It follows, however, from the considerable rise in boiling points, together with Trouton's rule, that they must increase very greatly from Li to B and from Na to Al. Trouton's rule states that, for substances of high boiling point, the molecular heat of evaporation λ is approximately proportional to the absolute boiling point T_s. The ratio λ/T_s (Trouton's constant) is usually about 21.5, so that the approximate heat of evaporation of any substance, in cal. is $21.5 \cdot T_s$. In the case of aluminum the heat of evaporation is so calculated as $2543 \cdot 21.5 = $ ca. 55,000 cal per g atom. The heat of fusion of aluminum has been measured, and is 92 cal per g, or 2500 cal per g atom. The approximate value of the heat of sublimation is found by adding the latent heats of fusion and evaporation*.

* The heat of fusion is generally given at the melting point, and the heat of evaporation at the boiling point. To calculate the heat of sublimation accurately, both should be corrected to that temperature for which the heat of sublimation should apply—usually $0°K$.

2. Boron (B)

(a) Occurrence

Boron never occurs free in nature, but always combined with oxygen. It is present in this form in *boric acid* H_3BO_3, which is found especially in the waters of hot springs and in volcanic regions, e.g., at Sasso, in Tuscany, from which it is named *sassolin*. Numerous salts of boric acid also occur native, their occurrence also being rather localized. The best known is *borax* or *tincal*, $Na_2B_4O_7 \cdot 10H_2O$. Technical importance also attaches to *boracite*, $2Mg_3B_8O_{15} \cdot MgCl_2$ (which occurs, for example, in the Stassfurt potash deposits), *pandermite*, $Ca_2B_6O_{11} \cdot 3H_2O$, *colemanite*, $Ca_2B_6O_{11} \cdot 5H_2O$ and especially *kernite* $Na_2B_4O_7 \cdot 4H_2O$, which was discovered in enormous deposits in California in 1928, and which has become the most important raw material in the world's production of borax and boric acid.

Among other minerals derived from boric acid there may be mentioned *borocalcite*, $CaB_4O_7 \cdot 4H_2O$, *boronatrocalcite*, $NaCaB_5O_9 \cdot 6H_2O$ *franklandite* $Na_2CaB_6O_{11} \cdot 7H_2O$, *hydroboracite* $MgCaB_6O_{11} \cdot 6H_2O$, *boromagnesite* $2Mg_5B_4O_{11} \cdot 3H_2O$, *ascharite* $3Mg_2B_2O_5 \cdot 2H_2O$, *pinnoite* $MgB_2O_4 \cdot 3H_2O$, *larderellite* $(NH_4)_2B_8O_{13} \cdot 4H_2O$, *lagonite* $2FeB_3O_6 \cdot 3H_2O$, *sulphoborite* $2Mg_2B_2O_5 \cdot 2MgSO_4 \cdot 9H_2O$, *lüneburgite* $Mg_2BO_4 \cdot 2MgHPO_4 \cdot 7H_2O$, and *sinhalite*, $MgAlBO_4$.

The radical of boric acid also occurs together with silicic acid—e.g., in *datolite* $CaBSiO_4(OH)$ and *axinite* $Ca_3Al_2BSi_4O_{15}(OH)$, and also in the *tourmalines*, which are apparently isomorphous mixtures of borates and silicates.

It is generally assumed that the volatility of boric acid in the presence of steam played some part in the formation of many boron minerals. This raises the question whether, under the conditions involved, such a volatilization process could have resulted in any alteration in the relative abundance of the isotopes B^{11} and B^{10}. The atomic weight determinations carried out by H. V. A. Briscoe (1925—27) on borax minerals from different, known sources are of some interest in this connection. He found that boron from Californian colemanite had a perceptibly higher atomic weight (10.840 ± 0.003) than that from Tuscan sassolin (10.823) or boracite from Asia Minor (10.819 ± 0.004).

(b) History

Borax has long been known, being mentioned (as a flux) as early as the manuscripts attributed to Geber. Homberg in 1702 liberated boric acid, by heating borax with sulfuric acid, and boric acid soon entered the pharmacopeia as 'sal sedativum'. Impure elementary boron was first prepared in 1808 by Gay-Lussac and Thenard, by reduction of the oxide with potassium, and soon afterwards, electrolytically, by Davy. Pure crystalline boron was obtained for the first time in 1909 by Weintraub, by melting 'amorphous boron' in a vacuum.

(c) Preparation

Pure crystalline boron is obtained directly by the method of Van Arkel and De Boer, described under zirconium in Vol. II, and also, according to Hackspill (1933), by decomposition of BCl_3 by high frequency sparks between tungsten electrodes in the presence of hydrogen. The so-called amorphous boron is obtained in the form of a brown powder by reduction of B_2O_3 by means of metallic sodium or magnesium. This may be freed from impurities by boiling it first with dilute hydrochloric acid, and then treating it with hydrofluoric acid. It is doubtful, however, whether the element can be obtained in the pure state by this means.

A crystalline product can be obtained from 'amorphous boron' by melting it with aluminum, and was formerly known as 'quadratic boron'. The same substance can be obtained directly by reduction of B_2O_3 with Al. 'Quadratic boron' always contains aluminum, and in some circumstances carbon also. It is, in fact, a compound AlB_{12}, with which is mixed a compound containing carbon, probably a double compound $3AlB_{12}.2B_4C$, which can be separated mechanically (von Náray-Szabó, 1936).

(d) Properties

The so-called amorphous boron is a tasteless, odorless, brown powder of density 1.73. Pure crystalline boron is blackish-grey in color, density 2.34, and has hardness 9 (Richards, Hackspill, Laves). The melting point of boron is very high. It begins to volatilize in the elctric arc. It is a poor conductor of electricity, but the conductivity increases with rise of temperature. It is very resistant chemically, even towards oxidizing agents. It occurs in the form of apparently monoclinic needles, or as hexagonal plates (Laubengayer, *J. 'Am. Chem. Soc.*, 65 (1943) 1924).

The mean specific heat of boron between 0° and 100° is 0.252, giving the value 2.73 for the atomic heat capacity, which deviates greatly from the Dulong-Petit law. The atomic heat capacity decreases still further with decreasing temperature. If the temperature is raised, however, it approximates more and more the value required by the Dulong-Petit law, to which it nearly conforms at 900°, with a value of 5.5 according to the experiments of Magnus and Danz.

Amorphous boron is stable in air at ordinary temperature, but ignites when heated in air at 700° and burns with a reddish flame. If it is burned in oxygen, a higher temperature is reached. The boron vaporises in traces, and colors the flame green. When heated, it also combines directly with chlorine, bromine and sulfur, but not perceptibly with hydrogen. From about 900° upwards, it can also combine with nitrogen, forming the nitride BN, so that this is also formed as a by-product of the combustion of boron in air.

Amorphous boron is oxidized to boric acid or alkali borate by means of aqua regia or concentrated nitric acid, and by fusion with alkali, though not by fused potassium nitrate at 400°. Concentrated sulfuric acid first reacts at 250°. Phosphoric acid is reduced by boron only above 800°, elementary phosphorus being formed. Boron is oxidized by steam at a red heat, liberating hydrogen. It reacts with nitric oxide at red heat also, forming boron trioxide and boron nitride, and is also able to reduce carbon monoxide and silica at very high temperatures. Because of its high affinity for oxygen and other electronegative elements boron can displace metals from their oxides, sulfides and chlorides. The heats of formation of the simplest boron compounds are given in Table 63, p. 322.

(e) Uses

Elementary boron is used to a limited extent as a deoxidizer for casting copper. It is also used in the steel industry, mostly in the form of *ferroboron*—i.e., an iron alloy with 10—20% B. Even an addition of 0.001 to 0.003% of boron considerably raises the hardenability of steel. Because of its extreme hardness, approaching that of the diamond, the so-called 'quadratic boron' has been suggested as an abrasive. Boric acid and its compounds, especially borax, find extensive applications. The borate peroxyhydrates are of considerable importance in the manufacture of bleaching materials.

3. Boron Compounds

Boron forms compounds normally having the expected formulas with the halogens, oxygen, nitrogen, and sulfur. The compounds formed with hydrogen are unusual in composition and structure.

The most important boron compounds industrially are boric acid, H_3BO_3, and its salts, especially borax. The boric acids show a great tendency to form complex compounds or coordination compounds (cf. Chap. 11). The simplest coordination compounds of boron are fluoroboric acid, $H[BF_4]$ and its salts, the fluoborates.

Boron forms coordination compounds with many organic radicals, corresponding in type to the fluoroborates. The dipyrocatechoborates, prepared by Hermans, are among these, and also the corresponding compounds of the type $M^I[BR_4]$

prepared by Böeseken. In these compounds the boron is electrochemically trivalent, and coordinatively quadrivalent, just as in the fluoroborates. The concept of *formal valence* (cf. p. 9) cannot be applied to the coordination compounds (cf. pp. 391, 399 *et seq.*).

Boron also exerts its coordinative quadrivalence in boron triphenyl, $(C_6H_5)_3B$, which can add on 1 atom of sodium.

The normal metallic borides, such as Mg_3B_2, can be regarded as salt-like compounds, with boron as the electronegative constituent. However, the borides very often have compositions which do not correspond to the normal valences, and display peculiar crystal structures—e.g., AlB_{12} and CaB_6. Thus CaB_6 (also SrB_6, BaB_6, LaB_6, CeB_6) form a crystal lattice which might be regarded as derived from that of CsI, by replacement of the Cs atoms by Ca atoms, and substitution for each I atom of an octahedral group of 6 closely adjacent B atoms (v. Stackelberg,

SURVEY OF THE MOST IMPORTANT BORON COMPOUNDS

Boron hydrides	*Organo-boron*	*Halides*	*Oxygen compounds*
B_2H_6	*compounds*	BF_3	B_2O_3
B_4H_{10}, etc.	alkyls BR_3	$H[BF_4]$	BO
borohydrides	alkyl oxides	BCl_3	boric acids
$M^I[BH_4]$	OBR	BBr_3	H_3BO_3
$M^{II}[BH_4]_2$		BI_3	HBO_2
hydroxoborane			borates
salts			e.g., borax
$M^I{}_2[B_2H_4(OH)_2]$			$Na_2B_4O_7 \cdot 10H_2O$

	Metal borides	*Nitride* BN
(a)	salt-like, e.g., Mg_3B_2	*Sulfide* B_2S_3
(b)	not salt-like, e.g., CaB_6	*Borophosphoric oxide* BPO_4
		(boron phosphate)
		Borazole $B_3N_3H_6$

1931). It is, however, probably better to regard these borides as based upon a continuous three-dimensional framework of boron atoms (similar in principle to the linking of carbon atoms in diamond, cf. p. 421), with metal atoms occupying interstices of the boron lattice. The great hardness of the borides, like that of diamond, is then associated with the rigid structure of covalent bonds by which the whole crystal lattice is knit together.

(a) Boron Hydrides (Boranes) [1, 20]

If metallic magnesium is heated with boron trioxide, the product (so-called magnesium boride) liberates a gas, when it is treated with acid; this consists mostly of hydrogen, but is admixed with some other volatile substance, as is evident from the characteristic, highly unpleasant smell. Stock, in 1912, showed that this volatile substance is in reality a mixture of various boron hydrides. With his collaborators, he was able to isolate pure compounds from it for the first time. The most important of the boron hydrides discovered by Stock are shown in Table 64. Their formulas were established by analysis and gas density determinations. The chemical behavior of the boron hydrides has also been worked out largely by Stock and his pupils (especially Wiberg), and more recently by Schlesinger and Burg and their collaborators in the United States. Their results may be applied to the interpretation of the constitution of these compounds, which originally presented great difficulties from the standpoint of valence theory.

TABLE 64

BORON HYDRIDES

Formula	B_2H_6	B_4H_{10}	B_5H_{11}	B_5H_9	B_6H_{10}	$B_{10}H_{14}$
Boiling point	$-92.5°$	$+18°$	—	—	—	ca. $+213°$
Vap. press. at 0°	—	388 mm	57 mm	66 mm	7.2 mm	—
Melting point	$-165.5°$	$-120°$	$-128.6°$	$-45.6°$	$-65.1°$	99.7
Gas density (obs.; hydrogen = 1)	13.9	26.7	32.1	32.0	37.25	61.0
Density in liquid state	0.447 (at $-112°$)	0.59 (at $-70°$)	—	0.61 (at 0°)	0.70 (at 0°)	0.78 (at 100°)

The preparation of the boron hydrides, and work with them, is made extremely difficult by their great sensitiveness towards water, which breaks them down (by way of intermediate products) into boric acid and hydrogen. Difficulty was also formerly experienced in separating them from silicon hydrides, which are almost always present in the crude gas as a consequence of the presence of silicon, if only in traces, as an impurity in the magnesium. Formation of these can be avoided by use of beryllium boride in place of magnesium boride. For work with the boron hydrides, it is necessary to use a special apparatus which completely excludes the entry of air, moisture and the vapors derived from the usual tap greases.

The principal product resulting directly from the decomposition of 'magnesium boride' with acids (8-N phosphoric acid is most suitable) is B_4H_{10}. This is a gas, condensing at 18°, with a peculiarly nauseating smell, and spontaneously inflammable in air in the liquid state. The compound itself is unstable. It gradually decomposes, forming chiefly B_2H_6, together with less volatile boron hydrides, such as B_5H_9.

B_2H_6, the simplest boron hydride yet found, can be obtained pure by heating

B_4H_{10} at 100° for a few hours. It remains gaseous down to quite low temperatures, is stable for months with practically no decomposition, and the gas does not ignite spontaneously in the air. It does not react with hydrogen sulfide, but is even more sensitive towards water than is B_4H_{10}.

Large quantities of *diborane*, B_2H_6, can be prepared by the method of Schlesinger (1931), by the action of the electric discharge on a mixture of H_2 and BCl_3 or, better, BBr_3 (Stock 1934). *Monobromodiborane* is first formed under the influence of the discharge—$2BBr_3$ + $5H_2 = B_2H_5Br + 5HBr$—and this disproportionates according to the equation

$$6B_2H_5Br = 5B_2H_6 + 2BBr_3$$

The best method of preparation of B_2H_6, however, is by the reduction of BCl_3 or BF_3 with *lithium aluminum hydride*, $LiAlH_4$ (see below, p. 362).

$$4BCl_3 + 3LiAlH_4 = 3LiCl + 3AlCl_3 + 2B_2H_6$$

This reaction takes place almost quantitatively in ether solution.

Other boron hydrides are now generally obtained by the pyrolysis of B_2H_6, which furnishes B_4H_{10} and B_5H_{11} as the main products at 100—120°.

B_5H_{11}, B_5H_9, and B_6H_{10} are colorless liquids, considerably more stable towards water then the boron hydrides just considered.

There is also a series of solid compounds, obtained by the decomposition of B_4H_{10}—e.g., $B_{10}H_{14}$, colorless crystals, volatile without decomposition, insoluble in water but soluble in carbon disulfide; also a non-volatile yellow solid boron hydride. The latter can be obtained in the pure state by the action of a silent electric discharge on B_2H_6 (Stock, 1936). It is insoluble in solvents which do not decompose it, (CCl_4, CS_2, etc.), has the composition $[BH]_x$, and in its properties and behavior can be included in the series of non-volatile hydrides of the elements standing close to boron (C, Si and Ge—cf. p. 411).

(b) Properties and Constitution of the Boron Hydrides

The volatile boron hydrides may be divided into two groups, having the general formulas B_nH_{n+4} (e.g., B_2H_6, B_5H_9, B_6H_{10}, $B_{10}H_{14}$) and B_nH_{n+6} (e.g., B_4H_{10}, B_5H_{11}), respectively. The latter are less stable than the compounds of the first group. All are decomposed at a red heat into boron and free hydrogen, inflame very readily when mixed with oxygen (burning to B_2O_3 and H_2O), and are decomposed by water and, more rapidly, by alkalis: e.g., $B_2H_6 + 6H_2O = 2H_3BO_3 + 6H_2$.

Whereas the boron alkyls have the molecular formula BR_3, the molecular weight of the simplest boron hydride corresponds to the formula B_2H_6, and not to the formula BH_3. In accordance with this, its crystal structure is related to that of ethane, C_2H_6 (Mark, 1926). However, the detailed structure of B_2H_6 is not the same as that of ethane, and the dimerization of the BH_3 radical is to be explained in terms of a non-polar type of hydrogen bond, which is described below.

The chemical properties of B_2H_6 are better known than those of the other compounds. The density of B_2H_6 shows no evidence of dissociation into BH_3, the expected hydride of boron. However B_2H_6 reacts with trimethylamine, carbon monoxide, and several other substances to give derivatives of the 'borine' radical—e.g., the addition compounds $BH_3 \cdot N(CH_3)_3$; $BH_3 \cdot CO$; $BH_3 \cdot O(CH_3)_2$; and the substitution product $BH_2 \cdot N(CH_3)_2$.

Diborane also reacts reversibly with boron alkyls, forming a series of alkyl-substituted boranes which are of some interest in relation to the structure of B_2H_6. Successive methylation reactions yield $CH_3BH_2 \cdot BH_3$, $(CH_3)_2BH \cdot BH_3$, $(CH_3)_2BH \cdot BH_2CH_3$ and $(CH_3)_2BH \cdot BH(CH_3)_2$. The symmetrical dimethyl diborane can be made from monomethyl diborane by reaction with dimethyl ether:

$$2CH_3BH_2 \cdot BH_3 + 2(CH_3)_2O = 2(CH_3)_2O \cdot BH_3 + CH_3BH_2 \cdot BH_2CH_3$$

The constitution of all these compounds has been proved from the nature of the *methyl boric acids* formed when they are hydrolyzed

$$CH_3BH_2 \cdot BH_3 + 5H_2O = CH_3B(OH)_2 + B(OH)_3 + 5H_2$$
$$(CH_3)_2BH \cdot BH_3 + 4H_2O = (CH_3)_2BOH + B(OH)_3 + 4H_2$$
$$(CH_3)_2BH \cdot BH_2CH_3 + 3H_2O = (CH_3)_2BOH + CH_3B(OH)_2 + 3H_2$$

It is noteworthy that not more than two of the hydrogen atoms attached to any one atom of boron can be replaced by alkyl groups.

From its position in the Periodic System, it would be expected that boron would form a hydride BH_3, in which all three valence electrons of the boron were utilized for boron co-valent bonds, with no electrons available for any other bonding. The molecule of B_2H_6 contains too few valence electrons for us to represent its structure in terms of the concepts which have been presented hitherto. It was formerly suggested (Wiberg, 1936) that B_2H_6 has unsaturated properties, and should be regarded as in some ways analogous to ethylene, with two of the hydrogen atoms ionically bound (I)

(I)   (II)

However, the properties of diborane are those of a completely non-polar compound, and it is now believed that these compounds involve a kind of non-polar double hydrogen bond (II), two of the hydrogen atoms being so vital to the structure that they cannot be replaced by alkyl groups. The structure (II) can be considered as a resonance hybrid of the structures IIa—IId—i.e., in the quantum mechanical description of the molecule, the wave equation

IIa   IIb   IIc   IId

is compounded of the wave equations representing the separate forms IIa—IId (see Bell and Eméleus, *Quart. Rev. Chem. Soc.*, 2 (1948) 132; Pitzer, *J. Am. Chem. Soc.*, 67, (1945) 1126; Price, *J. Chem. Phys.*, 15, (1947) 614; Wiberg, *Z. anorg, Chem.*, 256, (1948) 285). So called 'electron deficient' bonds of similar type (Rundle, *J. Am. Chem. Soc.*, 69, (1947) 1327) are involved in the metallic borohydrides (see below) and in the hydrides of aluminum (p. 361) and gallium (p. 374) also.

Hydrogen bridges are present in the structure of all the boranes. It is believed that B_4H_{10} has the structure III, and B_5H_{11} may be represented by (IV).

(III)   (IV)

The evidence for structure (III) lies in the fact that B_4H_{10} can be formed from B_2H_5I, by a reaction analogous to the Wurtz synthesis of hydrocarbons (Stock, 1926):

$$B_2H_5I + 2Na + IB_2H_5 = B_2H_5 - B_2H_5 + 2NaI$$

It has been shown that B_5H_9, B_6H_{10} and $B_{10}H_{14}$ possess cyclic structures. In every case, the hydrolysis of these compounds yields one molecule of hydrogen for each B—H and each B—B bond in the structure.

The reactions of the boron hydrides frequently involve the opening of the hydrogen 'bridges', and the formation of addition compounds. As well as forming borine derivatives, diborane reacts with ammonia to form a white, solid ammoniate, $BH_3 \cdot NH_3$, which was formerly considered to be the salt $(NH_4)_2[B_2H_4]$ derived from Wiberg's formula for B_2H_6. However, it has been shown by Burg (1947) that there is no exchange of hydrogen between B_2H_6 and deutero-ammonia, ND_3 (cf. Vol. II). Hence B_2H_6 does not act as an acid. B_4H_{10} forms the addition compound $B_4H_{10} \cdot 4NH_3$. Thermal decomposition of the compounds $BH_3 \cdot NH_3$ and $B_4H_{10} \cdot 4NH_3$ leads to loss of hydrogen and formation of *borazole*, $B_3N_3H_6$ (see below).

B_2H_6 reacts quantitatively with dimethylamine, $(CH_3)_2NH$, at about 200°, forming $BH_2 \cdot N(CH_3)_2$ (white crystals, m.p. 73.5°, insol. in water). This is dimeric at ordinary temperature, but is completely dissociated into $BH_2 \cdot N(CH_3)_2$ molecules at higher temperatures (Wiberg).

The unsaturated character of diborane shows itself in its reaction with, e.g., the alkali metals (in the form of their amalgams). According to Wiberg, salt-like compounds are formed by the reaction $B_2H_6 + 2K = K_2[B_2H_6]$. The identity of these products is perhaps not finally established; it is possible that the substances later identified as sodium and potassium borohydride (see below) may have been formed.

The reaction of diborane with alkali or alkaline earth hydroxides (in place of the metal) leads to the substitution of H atoms by negative OH-groups: hydrogen is evolved, and *dihydroxodiborane salts* (formerly called 'hypoborates') are formed:

$$B_2H_6 + 2M^IOH = M^I_2[B_2H_4(OH)_2] + H_2$$

The boranes also undergo *substitution* reactions with the hydrogen halides, especially in the presence of aluminum halides, which act as catalysts. E.g.,

$$B_2H_6 + HCl \xrightarrow{\text{Heat with } AlCl_3} B_2H_5Cl + H_2$$

Replacement of hydrogen is also effected by the free halogens, but if there is any excess of free halogen present, complete oxidation takes place:

$$B_2H_6 + 6Cl_2 = 2BCl_3 + 6HCl$$

(c) The Borohydrides (Boranates)

It was discovered by Schlesinger (1940) that diborane reacts with metal alkyls, forming a new type of compound, the *metal borohydrides*, or *boranates*. E.g.,

$$Al(CH_3)_3 + 2B_2H_6 = B(CH_3)_3 + Al(BH_4)_3 \tag{1}$$

Metal borohydrides can also be obtained by the reaction of diborane with salt-like metal hydrides, or by the action of metal hydrides and hydrogen (under high pressures) upon boron alkyls.

$$2NaH + B_2H_6 = 2Na(BH_4) \tag{2}$$

$$NaH + B(C_2H_5)_3 + 3H_2 = Na(BH_4) + 3C_2H_6 \tag{3}$$

The borohydrides of Li, Na, K, Be, Mg, Zn, Cd, Cu, Ag, Ga, Ti, Zr, Hf, Th, U, Np and Pu have been prepared. Most of these are salt-like in character, but a few of them are volatile and have the properties of covalent substances.

$Al(BH_4)_3$ and $Be(BH_4)_2$, in particular, are definitely not salt-like. Electron diffraction measurements on the vapor (Bauer, 1946, 1950) show that the molecule is built up from two BH_4 tetrahedra, which are probably linked to the Be atom through hydrogen bridges like those in the parent boranes:

$$\begin{array}{ccccccc} H & \diagdown & H & & H & \diagup & H \\ & B & & Be & & B & \\ H & \diagup & H & & H & \diagdown & H \end{array}$$

(i) *Aluminum borohydride* has an essentially similar structure, with three BH_4 tetrahedra linked to the central Al atom. It has been prepared by the reaction of diborane with aluminum methyl (eqn. (1)) or directly with aluminum hydride (eqn. (2)) (Wiberg, 1942), but is more conveniently prepared by reaction of aluminum halides with lithium borohydride (Schlesinger, 1952):

$$AlX_3 + 3Li[BH_4] = Al(BH_4)_3 + 3LiX \tag{4}$$

Aluminum borohydride is a volatile, colorless liquid ($d_4° = 0.569$, m.p. $-64.5°$, b.p. $+44.5°$, vap. press. 119.5 mm at 0°). It is the most volatile aluminum compound known. It is spontaneously inflammable in air, and has an extraordinarily high heat of combustion (989 kcal per mole). The combustion of aluminum borohydride therefore liberates 3750 kcal per kilogram of combustible mixture (AlB_3H_{12} + $+ 6O_2$), as compared with only 2550 kcal per kilogram of combustible mixture (C_5H_{12} + $8O_2$) for the combustion of the hydrocarbon pentane, which has almost the same molecular weight, the same hydrogen content, and the same specific gravity. It has accordingly been suggested that this liquid aluminum compound could find a use as a very high-grade fuel—e.g., for rocket motors. Aluminum borohydride reacts vigorously with water:

$$Al(BH_4)_3 + 12H_2O = \tfrac{1}{2} Al_2O_3 \cdot 3H_2O + 3H_3BO_3 + 12H_2 + 192 \text{ kcal}$$

(ii) *Magnesium borohydride*, $Mg(BH_4)_2$ was obtained by Wiberg and Bauer (1950) as a white solid (m.p. ca. 180°) insoluble in ether, by the action of diborane on magnesium diethyl in ether solution.

$$4B_2H_6 + 3Mg(C_2H_5)_2 = 3Mg(BH_4)_2 + 2B(C_2H_5)_3$$

It reacts with HCl to form diborane $[Mg(BH_4)_2 + 2HCl = MgCl_2 + 2H_2 + B_2H_6]$, and with methanol to give a solid methoxy derivative, $Mg[B(OCH_3)_4]_2$. The latter decomposes into magnesium methoxide and boric ester when it is more strongly heated:

$$Mg[B(OCH_3)_4]_2 = Mg(OCH_3)_2 + 2B(OCH_3)_3$$

(iii) *Lithium borohydride*, $LiBH_4$, is a white, non-volatile solid, soluble in water and ether (m.p. 279°). It can be prepared by several methods, but most conveniently by the reaction of lithium hydride with boron trifluoride (Winternitz, 1950; Wittig, 1951) or with boric acid ester (Schlesinger, 1950):

$$4LiH + BX_3 = LiBH_4 + 3LiX \quad (X = F \text{ or } -OR).$$

It is a weaker hydrogenating agent than lithium aluminum hydride (q.v.). It is used chiefly for the preparation of other borohydrides (reaction (4), above). It can also be used in a very simple preparation of borazole:

$$3LiBH_4 + 3NH_4Cl = B_3N_3H_6 + 9H_2 + 3LiCl$$

(Schaeffer, Anderson and Schlesinger, 1949—51).

Wittig (1951) has reported that the action of HCN on $LiBH_4$ in ether solution produces lithium cyanoborohydride, $Li[BH_3(CN)]$. This is remarkably stable, and is soluble without decomposition in water and dioxane.

(iv) *Sodium borohydride*, $NaBH_4$, is a white, crystalline, salt-like solid. It can be prepared by heating sodium hydride with boron trioxide:

$$4NaH + 2B_2O_3 = 3NaBO_2 + NaBH_4$$

but is more conveniently obtained pure by the reaction between sodium hydride and methyl borate at 250°:

$$4NaH + B(OCH_3)_3 = 3NaOCH_3 + NaBH_4$$

The first product of reaction is sodium trimethoxyboranate, $Na[BH(OCH_3)_3]$: $NaH + B(OCH_3)_3 = Na[BH(OCH_3)_3]$, a white solid which undergoes disproportionation at 230°, forming the tetramethoxyboranate and borohydride:

$$4Na[BH(OCH_3)_3] = 3Na[B(OCH_3)_4] + NaBH_4$$

For a more extensive discussion of these compounds, their use as reducing agents, and for the preparation of diborane and other metal borohydrides, reference should be made to Schlesinger, *J. Am. Chem. Soc.*, 75 (1953) 186, *et seq.*

The salt-like character of $NaBH_4$ and $LiBH_4$ has been proved by crystal structure determinations. The Na^+ and $[BH_4]^-$ ions in $NaBH_4$ are arranged in a NaCl-type structure ($a = 6.15$Å; Soldate, *J. Am. Chem. Soc.*, 69, (1947) 987). This compound is remarkably stable, and does not decompose at 400° in vacuum. $Li[BH_4]$ crystallizes rhombic; each Li^+ ion is surrounded by $4[BH_4]^-$ ions (Harris and Meibohm, *J. Am. Chem. Soc.*, 69, (1947) 1231). Zachariasen has also shown that the crystal structure of $U[BH_4]_4$ is that of an ionic compound.

(d) Borazole

Thermal decomposition of the diborane-ammonia addition compound, $B_2H_6 \cdot 2NH_3$ at 200° gives rise to the remarkable substance *borazole*, $B_3N_3H_6$. This is also very conveniently formed by the reaction between lithium borohydride and ammonium chloride (see above).

This is a colorless, mobile liquid (m.p. — 58° b.p. 55°), which is remarkably stable; it has an aromatic smell and is a good solvent. It has been proved by electron diffraction methods that borazole has the cyclic structure (V), and is isosteric with benzene, which it resembles in its physical properties. Methyl-substituted boranes and methylamine similarly form the various possible B- and N-substituted methyl borazoles. Of these, the two isomeric B- and N-trimethylborazoles, $B_3N_3H_3(CH_3)_3$, closely resemble mesitylene (1, 3, 5 trimethyl benzene), and hexamethylborazole, $B_3N_3(CH_3)_6$, is the analogue of hexamethyl benzene. The distance B—N in borazole is 1.45 Å, and the bonds have less double bond character (i.e., are bonds of lower order) than the bonds in benzene—30% double bond character in borazole as compared with 50% in benzene (Spurr and Chang, *J. Chem. Phys.*, 19 (1951) 518).

Borazole is more reactive than benzene. It reacts with polar compounds as if it were unsaturated, forming first addition compounds and ultimately substituted borazoles. Borazole has very aptly been called the 'inorganic benzene' (cf. E. Wiberg, *Ber.*, 73 (1940) 209; *Z. anorg. Chem.*, 255 (1947) 141; 256 (1948) 177.

(V)

(e) Borazane, borazene and borazine

In the preparation of borazole by the reaction $3B_2H_6 + 6NH_3 = 2B_3N_3H_6 + 12H_2$, the following substances are formed as intermediate products.

$$H_3B \leftarrow NH_3 \qquad H_2B \Longleftarrow NH_2 \qquad HB \Longleftarrow NH$$
borazane   borazene   borazine

The names express the isosterism of the compounds with the hydrocarbons $H_3C—CH_3$ (ethane), $H_2C=CH_2$ (ethene) and $HC≡CH$ (ethine). Methyl-substituted borazanes, etc., are also known, and it has been shown by Goubeau from Raman spectroscopic studies that the strength of the B—N bond increases progressively in the sequence borazane→borazole→borazene, just as the C—C bond order increases in the sequence ethane→benzene→ethylene.

Borazenes, like the olefins, are unsaturated in character. They can add on HCl, HBr, and H_2O, but not Br_2. Their unsaturation is also shown in their slight tendency to dimerize e.g.,

(f) Boroxoles and Borosulfoles

The name boroxoles is given to a group of compounds of the general formula $B_3O_3R_3$, which contain a ring system analogous to that in the borazoles. The parent member of the group, $B_3O_3H_3$, is unknown, but its derivatives can be prepared in various ways. Thus *trimethyl boroxole*, $B_3O_3(CH_3)_3$ (colorless liquid, b.p. 80°, m.p. —38°) is formed by heating B_2O_3 and $B(CH_3)_3$ under pressure at 300—330° (Goubeau, 1951), or by elimination of water from methyl boric acid, $CH_3B(OH)_2$ (Snyder, 1938):

The stability of the boroxoles diminishes as the electronegative character of the groups bound to the ring increases. Whereas the alkyl boroxoles are fairly stable, the compound $B_3O_3[N(CH_3)_2]_3$ (m.p. 64°, b.p. 224°) is rapidly hydrolyzed in moist air, and $B_3O_3(OCH_3)_3$ (m.p. about 10°) cannot be distilled without decomposition. The halides $B_3O_3F_3$ and $B_3O_3Cl_3$ (see below) are still less stable.

It is probable that the ring in boroxoles has a resonance structure to which the following forms contribute:

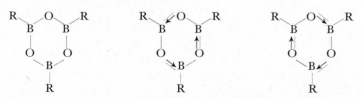

The cyclic structure of trimethylboroxole was proved by Bauer and Beach (1940) by electron diffraction methods. The X-ray work of Tazaki (1940) has shown that metaboric acid, HBO_2 or $H_3B_3O_6$, also has the boroxole structure. According to Goubeau (1951) this is true also of the compound $B_3O_3Cl_3$, which can be regarded as trichloro boroxole. The compound $B_3O_3F_3$ prepared by Baumgarten (1940) is probably structurally similar. The compounds $B_3O_3Cl_3$ and $B_3O_3F_3$ are formed by heating B_2O_3 with BCl_3 or BF_3, but are stable only at high temperature in equilibrium with the reactants:

$$B_2O_3 + BX_3 \rightleftharpoons B_3O_3X_3$$

It is therefore impossible to isolate them by partial hydrolysis of the boron halides.

Structures analogous to those of the boroxoles are found for the compounds $B_3S_3R_3$. These are known as *borosulfoles*.

(g) Boron Halides

(i) *Boron trifluoride and fluoroboric acid.* [21] Boron trifluoride, BF_3, is a colorless gas with a suffocating smell; density 2.37 (air = 1), liter weight 3.07 g, b.p. —101°, m.p. —128°. It is not combustible, but is decomposed by water and therefore fumes in air.

In the molecule of BF_3 all the atoms are coplanar, with the distance $B \leftrightarrow F = 1.30$ Å (Brockway, 1937). The alkyl derivatives $B(CH_3)F_2$, $B(CH_3)_2F$ and $B(CH_3)_3$ are also planar, with $B \leftrightarrow C = 1.56$ Å (Bauer, 1942).

Boron trifluoride can be prepared by heating boron trioxide with fluorspar and sulfuric acid(1). It is best obtained pure, according to Hellriegel (1937), by reaction of potassium fluoroborate with boron trioxide (2).

$$B_2O_3 + 3CaF_2 + 3H_2SO_4 = 2BF_3 + 3CaSO_4 + 3H_2O \qquad (1)$$

$$KBF_4 + 2B_2O_3 = K[B_4O_6F] + BF_3 \qquad (2)$$

The immediate products of decomposition by water are boric acid and hydrofluoric acid (3). The latter, however, at once combines with as yet undecomposed boron trifluoride, forming fluoroboric acid (4), so that the over-all reaction is represented by (5)

$$BF_3 + 3HOH = B(OH)_3 + 3HF \quad \text{(hydrolysis)} \qquad (3)$$

$$BF_3 + HF = H[BF_4] \qquad (4)$$

$$4BF_3 + 3H_2O = H_3BO_3 + 3H[BF_4] \qquad (5)$$

The fluoroboric acid resulting from the union of BF_3 with HF is a strong acid, considerably stronger than hydrofluoric acid. It does not attack glass in the cold, even in aqueous solution. It is decomposed by heating it with water, however, oxo-fluoroboric acids being formed. The well-crystallized salts of fluoroboric acid, the fluoroborates, most of which were already obtained by Berzelius, can be prepared by dissolution of the corresponding metallic oxides, hydroxides or carbonates in aqueous fluoroboric acid, or by treatment of the metallic fluorides with boric acid in hydrofluoric acid. Except for potassium fluoroborate, which is deposited as a gelatinous precipitate when fluoroboric acid is neutralized with potassium hydroxide, they are mostly fairly soluble. After drying, potassium fluoroborate, $K[BF_4]$, is a white powder, of density about 2.5. It can be recrystallized from water, but is gradually decomposed on prolonged standing with water, as the acid reaction acquired by the solution shows. The fluoroborates are decomposed by heating, BF_3 being liberated.

The alkali metal fluoroborates are isotypic with the corresponding perchlorates (Hoard, *J. Am. Chem. Soc.*, 57 (1935) 1985, Klinkenberg, 1937). The same is true of the hydroxofluoroborates—e.g., $Na[B(OH)F_3]$.

Fluoroborates of the alkaline earth metals and the heavy metals are known, as well as of the alkali metals. It is found, however, that the fluoroborates of the bivalent metals are stable only in the form of hydrates, ammoniates, etc. It can be deduced theoretically that this should be so, on the basis of the Born-Haber cycle (De Boer and Van Liempt, 1927). Potassium fluoroborate has been suggested as a welding flux. Fluoroboric acid is poisonous, and inhibits fermentation even in great dilutions.

BF_3 combines not only with HF, but with other molecules also. With water, it forms a compound $BF_3 \cdot 2H_2O$, which can be regarded as a hydroxonium hydroxofluoroborate, $[OH_3][B(OH)F_3]$, corresponding to the hydroxonium perchlorate (cf. p. 809) (Klinkenberg, 1935). With ammonia, BF_3 forms the solid addition compound $BF_3 \cdot NH_3$, with a heat of formation 41.3 kcal per mol at 0°. This compound dissolves in water without being hydrolyzed (solubility, 36.0 g in 100 g of water at 25°), and decomposes when heated above 125°; $4(BF_3 \cdot NH_3) = 3[NH_4][BF_4] + BN$ (Laubengayer, *J. Am. Chem. Soc.*, 70 (1948) 2274).

The liquid addition compounds of ammonia, $BF_3 \cdot 2NH_3$ and $BF_3 \cdot 3NH_3$ are much more readily decomposed. Organic amines, and also nitric oxide, can add to BF_3. The addition compound with PH_3 is stable only below room temperature.

Boron trifluoride acts as a catalyst for the introduction of HF into unsaturated organic compounds; it can bring about the exchange of other halogen atoms for fluorine, as well as addition of HF (Henne and Arnold, 1948).

The distance B—F is 1.30 Å in the BF_3 molecule, and 1.43 Å in the $[BF_4]^-$ ion. This increase in interatomic distance parallels a decrease in the vibrational force constant k_{BF}, as determined from the Raman spectra, from $6.86 \cdot 10^5$ to $5.28 \cdot 10^5$ dyne cm^{-1} (Goubeau, 1952). This considerable decrease can be understood if it may be assumed that $[BF_4]^-$ is isosteric with CF_4, and that in BF_3 there is the possibility of a double bond (distributed by resonance over all three B—F bonds):

$$
\begin{array}{ccc}
\ddot{:}\text{F}\ddot{:} & \ddot{:}\text{F}\ddot{:} & \ddot{:}\text{F}\ddot{:} \\
:\text{F}:\text{C}:\text{F}: & \left[\;:\text{F}:\text{B}:\text{F}\;\right] & :\text{F}:\text{B}::\text{F}: \\
:\text{F}: & :\text{F}: & \\
& & \text{(3 resonance forms)}
\end{array}
$$

This hypothesis is supported by Goubeau's observation that there is *no* difference in the

B—C bond strength in the compounds $B(CH_3)_3$ and $H_3N \cdot B(CH_3)_3$. There can be no double bonding in $B(CH_3)_3$, since the CH_3-radical lacks the requisite second pair of electrons. The dimerization of $CH_3O \cdot BF_2$ and the instability of $(CH_3O)_2BF$ can be explained by the same assumption.

At high temperatures it is possible to detect the existence of the compound BF. In this compound, k_{BF} is $7.73 \cdot 10^5$ dyne cm^{-1}, as determined from band spectroscopy by Chrétien (1950).

(ii) *Boron trichloride*, BCl_3, may be obtained by heating boron in chlorine, or in dry hydrogen chloride, or by passage of chlorine over a heated mixture of boron trioxide and carbon. It is a colorless, mobile, highly refractive liquid, fuming in moist air; density 1.43, boiling point 17.5—18.5°. (Mazzetti), melting point —107° (Stock). The vapor density is found to be 4.065 (air = 1), from which it follows that it is monomolecular in the vapor state (calc. 116.0, observed 117 for he molecular weight).

Boron trichloride is decomposed by water, with elimination of HCl (6)

$$BCl_3 + 3HOH = B(OH)_3 + 3HCl \tag{6}$$

Other compounds containing oxygen also effect the exchange of oxygen for chlorine—e.g., with sulfur trioxide at 120°

$$2BCl_3 + 3SO_3 = B_2O_3 + 3SO_2Cl_2$$

Like boron fluoride, boron trichloride can form addition compounds, especially with chlorides such as PCl_3, $POCl_3$, NOCl. Ammonia reacts by substitution, HCl being split out—e.g.

$$BCl_3 + 3NH_3 = B(NH_2)_3 + 3HCl, \text{ or } BCl_3 + 6NH_3 = B(NH_2)_3 + 3NH_4Cl.$$

Reaction with alcohols leads to the exchange of the Cl atoms for alkoxy groups, RO-, and ormation of hydrogen chloride (7)

$$BCl_3 + 3C_2H_5OH = B(OC_2H_5)_3 + 3HCl \qquad \text{(alcoholysis)} \quad (7)$$

Alcoholysis may take place in stages. Thus it is possible to isolate boron methoxy dichloride (dichloroboric methyl ester), $B(OCH_3)Cl_2$, m.p. —15°, b.p. 58.0°; boron dimethoxychloride (monochloroboric dimethyl ester), $B(OCH_3)_2Cl$, m.p. —87.6°, b.p. 74.7°; and trimethyl borate, $B(OCH_3)_3$, m.p. —29°, b.p. 68.7°. These are all colorless liquids, hydrolyzed by water to H_3BO_3, CH_3OH and HCl. The partially alkoxylated chlorides disproportionate:

$$2B(OR)Cl_2 \rightleftharpoons BCl_3 + B(OR)_2Cl \tag{8}$$

and
$$2B(OR)_2Cl \rightleftharpoons B(OR)Cl_2 + B(OR)_3 \tag{9}$$

if the equilibria are displaced by the addition of substances which form addition compounds with one or other species (Wiberg 1931, 1935). Apart from this, the compounds are remarkably stable, even at their boiling points.

BCl_3 reacts with ethylene oxide, C_2H_4O, to form the esters Cl_2B—OC_2H_4Cl, $ClB(OC_2H_4Cl)_2$ and $B(OC_2H_4Cl)_3$; the first two readily undergo disproportionation (Schmeisser, 1952). Ethylene oxide reacts in a similar manner with $AsCl_3$ (Malinowski) and $SiCl_4$ (Nitzsche).

(iii) *Diboron tetrachloride* (tetrachlorodiborine), B_2Cl_4, was first isolated by Stock (1929) It is most conveniently prepared by passing BCl_3 vapor through an electric glow discharge struck between mercury electrodes. It is a liquid, with a vapor density corresponding to the molecular complexity B_2Cl_4. It decomposes slowly at ordinary temperature and is complete-ly hydrolyzed by dilute caustic soda at 70°, according to the equation $B_2Cl_4 + 6OH^- =$ $= 2BO_2^- + 4Cl^- + H_2 + 2H_2O$. It dissolves in water without the evolution of hydrogen, however. The solution has strongly reducing properties, which are attributed to the pre-sence of the acid $B_2(OH)_4$, formed by hydrolysis.

(iv) *Boron tribromide*, BBr_3, is a colorless, viscous, strongly fuming liquid; density 2.65, b.p. 90.5°, solidifying at —46°. The heat of formation is 43.2 kcal per mol. It is very similar to the trichloride in properties.

(v) *Boron triiodide*, BI_3, is solid at ordinary temperature. It can be very conveniently prepared by the action of iodine upon sodium borohydride, $NaBH_4$. It forms colorless, very hygroscopic leaflets melting at 49°; boiling point 210°, density of liquid 3.3. at 50°. It is soluble in carbon disulfide, carbon tetrachloride and benzene. It burns in oxygen with a brilliant flame. The tendency to form addition compounds, which is so strongly developed in the other boron halides, is exhibited only to a very slight extent by boron triiodide.

Boron triiodide vapor is decomposed when it is passed through an electrodeless glow discharge at 1—3 mm pressure. The principal product is *diboron tetraiodide*, B_2I_4, a pale yellow crystalline solid, which decomposes slowly at ordinary temperature into BI_3 and a black highly polymerized *monoiodide*, $[BI]_x$. These lower iodides are soluble in water, and form solutions with strongly reducing properties, from which silver nitrate gives a mixed precipitate of metallic silver and silver iodide.

(h) Boron Alkyls

Boron alkyls are formed by double decomposition between boron halides and zinc alkyls—e.g.,

$$2BCl_3 + 3Zn(CH_3)_2 = 2B(CH_3)_3 + 3ZnCl_2$$

(i) *Boron trimethyl*, $B(CH_3)_3$, is a colorless gas, *boron triethyl* a liquid boiling at 95°. Vapor density determinations show the compounds to be monomolecular. According to Krause, they are best prepared from boron fluoride, by reaction with organo-magnesium compounds in ether solution. The boron alkyls are very readily oxidized, those derived from the lowest hydrocarbons being spontaneously inflammable in air. They can be kept in sealed glass vessels, but with gradual access of air they are oxidized to alkyl boron oxides, $R·BO$, which dissolve in wa-ter to form alkyl boric acids, $R·B(OH)_2$. The latter crystallize excellently, and are very volatile in steam; they have an aromatic smell and sweetish taste. They are soluble in organic solvents, and reduce alcoholic silver nitrate only on heating.

(ii) *Boron triphenyl*, $B(C_6H_5)_3$, has similar properties. An ethereal solution of this gradually turns an intense ruby red when metallic sodium is added, and sodium triphenyl boron, $NaB(C_6H_5)_3$ ultimately separates out in yellow crystals.

(i) Oxygen Compounds of Boron

(i) *Boron trioxide*, B_2O_3, the product of combustion of boron, is generally prepar-ed by ignition of boric acid. It forms a colorless, brittle, vitreous, hygroscopic mass, of density 1.844, which melts at a red heat, or softens forming a viscous mass which can be drawn out into threads. It is very stable to heat, and is not reduced by car-bon even at a white heat. If, however, substances which can replace oxygen (e.g., chlorine or nitrogen) are present as well as carbon, the oxide is decomposed.

Boron trioxide is a non-conductor of electricity. It has a weakly bitter taste. It dissolves in water with considerable evolution of heat —100 g of boron trioxide added to 125 g of water raises the temperature to the boiling point. The large heat effect arises from the combination of boron trioxide and water to form boric acid:

$$\tfrac{1}{2}[B_2O_3] + \tfrac{3}{2}H_2O + aq. = H_3BO_3 \cdot aq. + 4.0 \text{ kcal}$$

The value given represents the heat of solution in a large volume of water. If the oxide is dissolved in but little water, the heat evolved is much greater, since the heat of dilution of boric acid is strongly negative. The heat of solution of $[H_3BO_3]$ in much water is —5.1 kcal per mol, so that the heat of formation of $[H_3BO_3]$ from $\tfrac{1}{2}[B_2O_3]$ and $\tfrac{3}{2}H_2O$ amounts to +9.1 kcal. Metaboric acid exists in several modifications; the heat of formation of β-$[HBO_2]$ is +4.4 kcal, and its heat of solution is —0.56 kcal (von Stackelberg, Roth, 1937).

Boron trioxide usually solidifies as a glass. It can be obtained crystalline only by the dehydration of α-HBO_2. The dehydration of H_3BO_3 usually produces vitreous B_2O_3, since the β- or γ-modifications or HBO_2 are usually formed as intermediate stages in the process (cf. p. 338) Crystalline boron trioxide ($d = 2.460$, m.p. 450°) is optically uniaxial, and has a rhombohedral structure.

Vitreous B_2O_3 consists of an irregular network of equilateral triangles of O-atoms, with the B-atoms at their centers. The distance $B \leftrightarrow O$ is 1.39 Å.

(ii) *Boron monoxide*, BO. Zintl has shown (*Z. anorg. Chem.*, 245 (1940) 8) that if a mixture of boron and zirconium dioxide is heated, boron volatilizes in the form of *boron monoxide* (cf. Klemm, 1948). The existence of BO molecules in the gaseous state had already been established by Mulliken, from band spectroscopic evidence in 1925. The free energy change of the reaction B + O = BO is about —210 kcal per mol; that of the reaction Al + O = = AlO (cf. p. 355) about — 97 kcal per mol.

(iii) *Boric acid*. The normal boric acid, orthoboric acid H_3BO_3, forms flaky white, transparent, six-sided leaflets of density 1.46. It is generally prepared by the addition of hydrochloric or sulfuric acid to solutions of borates, especially borax. It may be recrystallized from hot water. The aqueous solubility is, at 0° 19.47 g; at 20° 39.92; and at the boiling point of the saturated solution (102°), 291.2 g of H_3BO_3 per liter. The solubility increases very greatly with rise of temperature because of the large negative heat of solution. Boric acid is volatile in steam. It is therefore contained in the steam which issues from the ground, mixed with other gases, in the volcanic district of the Tuscan; the lagoons from which the acid arises are known as 'soffioni' or 'fumaroli'. It is also present in many mineral waters and in traces in berries and fruits, in hops and often in wine.

Orthoboric acid is a very weak acid. The specific conductivity of a 0.94-normal solution is $\varkappa_{18} = 1.1 \cdot 10^{-5}$. The dissociation constants are $K_1 = 7.3 \cdot 10^{-10}$, $K_2 = 1.8 \cdot 10^{-13}$, $K_3 = 2 \cdot 10^{-14}$ (at 18°). Since dissociation is so slight for the first stage, even the primary salts of boric acid are hydrolyzed, and tertiary salts of boric acid cannot be prepared from aqueous solution.

When heated, boric acid (and also fused boron trioxide) very readily dissolves metallic oxides, forming salts. It can displace volatile acids from their salts. With silicic acid, it forms complex silicates (borosilicates), which are decomposed by hydrochloric acid, and it is therefore used in analytical chemistry to open up silicates. The boric acid can subsequently be removed readily, since it forms volatile esters with alcohols—e.g., trimethyl borate, $B(OCH_3)_3$.

Large quantities of boric acid are utilized in Europe for the production of borax. At the present time, the enamel industry is the chief consumer of boric acid. Considerable quantities are used as a preservative in the food industry, although in several countries this is not permitted. Boric acid has a disinfectant action, and is used medically as a dusting powder for wounds, for the impregnation of dressings, and to prevent excessive perspiration. Many pharmaceutical preparations are manufactured in which boric acid is employed. In the candle industry it is used to stiffen wicks. It is also employed in tanning, and in the manufacture of pigments.

When it is heated, boric acid loses water and is changed into *metaboric acid*, HBO_2. This at once reverts to orthoboric acid when dissolved in water. Further dehydration (by heating) converts metaboric acid directly to vitreous B_2O_3, containing some water, without the formation of any *tetraboric acid*, $H_2B_4O_7$, as an intermediate stage. The latter is known only in the form of its salts (e.g., borax), but is present to a very minute extent in equilibrium with orthoboric acid in aqueous solution. According to Thygesen (*Z. anorg. Chem.*, 237 (1937) 101), the concentration of undissociated $H_2B_4O_7$ in a 0.6-molar solution of H_3BO_3 is $3.5 \cdot 10^{-5}M$, that of $B_4O_7^=$-ions is $8 \cdot 10^{-5}M$, while, because of the excessively small degree of dissociation of H_3BO_3, the concentration of monoborate ions is only $4 \cdot 10^{-6}M$. With increasing dilution, the equilibria shift in favor of monoborate ions. Thus, in 0.1-molar solution these are present at a concentration of $8 \cdot 10^{-6}M$, while the concentration of $B_4O_7^=$-ions has fallen to 3×10^{-6}, and that of undissociated $H_2B_4O_7$ to $6 \cdot 10^{-8}M$.

Only the acid H_3BO_3 is volatile in steam, and this is the form which volatilizes even from highly concentrated solutions in equilibrium with solid HBO_2. The very much smaller volatility of HBO_2 and B_2O_3 is explained in that these, unlike H_3BO_3, are coordinatively unsaturated in their monomeric forms (v. Stackelberg, 1937). In the crystal structure of H_3BO_3 it is possible to discern the existence of separate H_3BO_3 molecules, held together by hydrogen bonds.

(iv) *Metaboric acid*, HBO_2, exists in three modifications, as is evident from the solubility diagram determined by Kracek, Morey and Merwin (1938), which is reproduced in Fig. 71. As in the case of $CaCl_2 \cdot 4H_2O$ (Fig. 13, p. 72), the product separating when a solution of suitable concentration is cooled is not as a rule the stable α-modification, but the least stable, modification, γ-HBO_2. This slowly changes into the β-modification. Transformation of the latter into α-HBO_2 takes place only after heating for weeks in an autoclave. This is the explanation of the congruent melting displayed by H_3BO_3, although as Fig. 71 shows, a melt of this composition is unstable with respect to α-HBO_2, which should separate as the stable solid phase. α-HBO_2 (cubic, $d = 2.486$) melts congruently at 236°. The quadruple point, at which the four phases α-HBO_2, crystalline B_2O_3, solution, and water vapor can coexist in equilibrium, occurs at 235° and 1.9 atm. α-HBO_2 attains a water vapor pressure of 1 atm. at 225°. It happens, by chance, that the temperatures at which the two metastable modifications reach a water vapor pressure of 1 atm. are close to their

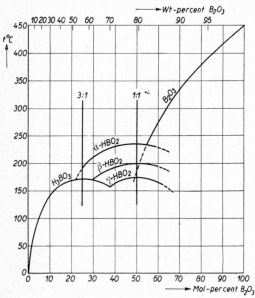

Fig. 71. Solubility diagram of the system $B_2O_3 — H_2O$.

melting points. β-HBO_2 (monoclinic, $d = 2.044$) melts congruently at 200.9°, and γ-HBO_2 (rhombic, $d = 1.78$) also melts congruently, at 176°. It is surprising that although they are unstable, these two forms should have congruent melting points, and it may be assumed that the atomic arrangements present in them persist over small regions in the molten state also. From the chemical standpoint, this means that the different forms of metaboric acid can be regarded as different, structurally isomeric compounds. This might involve similar types of isomerism to that found among the metaphosphoric acids (cf. p. 634). In this connection an observation made by Kracek, Morey and Merwin is of interest—that γ-HBO_2, unlike the other modifications, has an exceptionally good cleavage (perpendicular to the c-axis). Even a quite small pressure is sufficient to cleave a crystal into thin flakes.

Boric acid is also soluble in many organic solvents, especially in hydroxylic liquids. The conductivity of an aqueous solution of boric acid is enhanced by the addition of organic substances, and particularly of polyhydric alcohols such as mannitol, dulcitol, or glycerol. The strength of boric acid is raised by such additions to such an extent that it can be titrated with phenolphthalein as indicator. This is attributed to complex formation with the organic compounds (see next chapter).

Esters of boric acid (e.g., $B(OCH_3)_3$) form addition compounds with ammonia and organic amines. The simplest of these are of the type $R_3N \cdot B(OR)_3$, and can be regarded as derivatives of borazane (p. 332). Addition compounds of greater complexity also exist (Goubeau, 1951). If boric acid esters are heated with triethanolamine, $N(C_2H_4OH)_3$, triethanolamine boric ester, $N(C_2H_4O)_3B$, is obtained, as rhombic crystals, m.p. 231°. This compound is monomeric in its solution in nitrobenzene, but has a strong tendency to polymerize. It is partially hydrolyzed by water. It behaves as an anhydrobase towards acids and ansolvo-acids (e.g., $SnCl_4$, $SbCl_3$, or $HgCl_2$). Only the N-atom is a site at which addition takes place, however, whereas with the analogous aluminum compound, $N(C_2H_4O)_3Al$, the aluminum atom can also add on to other substances (Hein, 1952).

Inorganic substances, and especially acid residues, may also add on to boric acid, forming so-called *heteropolyacids* (see Vol. II).

(v) *Salts of boric acid.* The *borates* are derived for the most part neither from orthoboric acid nor from metaboric acid, but from other forms still poorer in water, which have not been detected as free acids. Examples of such salts are provided by the minerals mentioned in the first section. The more abundant of these are utilized for the manufacture of borax, technically the most important compound of boron.

The alkali borates are all fairly soluble in water, and their solutions have a strongly basic reaction, as a result of hydrolysis. The remaining borates are insoluble in water.

Many borates share with boric acid the property of solidifiying from their melts in glassy form. Lithium-beryllium borate glasses—which have a very low coefficient of absorption for X-rays since all the constituents have very low atomic numbers—are used for the windows of X-ray tubes, and for specimen tubes used in X-ray crystallography (cf. Menzel, 1942) ('Lindemann glass').

Borax, $Na_2B_4O_7 \cdot 10H_2O$, the sodium salt of tetraboric acid, forms large colorless, transparent monoclinic crystals, which effloresce superficially in dry air. Their density is 1.72. The aqueous solution is strongly alkaline, and absorbs 1 mol of CO_2 per mol of $Na_2B_4O_7$ in the cold; the carbon dioxide is driven out again on heating. The solubility of borax in water is:

at	0°	10°	30°	60°	80°	100° C	
	1.23	1.58	3.75	16.7	23.9	34.3	g of $Na_2B_4O_7$ in 100 g of solution.

Borax occurs in Tibet, under the name of *tincal*, and in California. Tincal was formerly of great importance, but today by far the greatest proportion of borax is
extracted from other minerals, especially kernite.

In addition to ordinary borax, there is a 5-hydrate, $Na_2B_4O_7 \cdot 5H_2O$, known as
octahedral or jewellers' borax. This crystallizes from solutions above 60°, whereas
ordinary, prismatic borax crystallizes at room temperature. $Na_2B_4O_7 \cdot 5H_2O$ forms
rhombohedral crystals of density 1.81. It is harder than ordinary borax, and does
not effloresce.

The course followed in the dehydration of borax depends upon whether it has been
previously moderately heated (to about 50°, e.g., in the course of drying) or not. In the
former case, its vapor pressure at 20° is 10 mm, and it passes reversibly into the 5-hydrate.
This can be reversibly dehydrated to the dihydrate, then to the monohydrate and ultimately to the anhydrous salt. If, however, the borax has not previously been heated, its vapor
pressure is only 1.6 mm at 20°, and it is quite stable in air under ordinary conditions. In a
vacuum desiccator over phosphorus pentoxide it changes very slowly and irreversibly into
the 2-hydrate. The 4-hydrate, which occurs as the mineral *kernite* or *rasorite* does not appear
in the course of dehydration of the 10- or 5-hydrate, but can be crystallized from supersaturated solutions (Menzel, 1935—40, especially *Z. anorg. Chem*, 245, (1940) 157).

Anhydrous sodium tetraborate, $Na_2B_4O_7$ (density 2.37), is obtained by heating
borax to 350—400°. The melting of the borax in its own water of crystallization
can be avoided by an initial slow heating. Sodium tetraborate melts at 878° to a
glassy mass, and the melt is a solvent for metallic oxides. Use is made of this in
analytical chemistry (borax beads).

Borax finds extensive uses for the preparation of low-melting glazes for earthenware and procelain, and especially for enamel ware. It is also used in the production of special types of glass, resistant to changes of temperature (e.g., Pyrex lamp
glasses), and for optical glasses. Because of its ability to dissolve metallic oxides, it
is used as a flux in welding and soldering (jeweller's borax). Large quantities are
used in laundering, and it is important as the starting material for the manufacture of the borate peroxyhydrates. Several applications utilize its preservative
action. In the laboratory, it is used as a primary volumetric standard, for the
standardization of acids, and for the preparation of buffer solutions (cf. p. 833), as
well as for borax bead tests.

(vi) *Peroxyborates and borate peroxyhydrates.* The *peroxyborates* are derived from the
borates through the replacement of O atoms by the 'peroxy'-group —O—O—,
which is present in hydrogen peroxide. The peroxyborates were formerly known
as perborates. Not all of the compounds known as 'perborates' however, are true
peroxy-salts; some, at least, are merely *borate peroxyhydrates*—i.e., addition compounds of H_2O_2 with ordinary borates. These are generally obtained by the action
of alkali peroxides in aqueous solution, or caustic alkalis containing hydrogen
peroxide, upon borates.

Since the true peroxyborates are also decomposed hydrolytically in aqueous solution,
with the formation of H_2O_2, they are not easily distinguished from the hydrogen peroxide
addition compounds. Differentiation between peroxy-salts and peroxyhydrates is generally
based upon the Riesenfeld reaction, which depends on the fact that true peroxy-salts, as
distinct from H_2O_2 and its addition compounds, liberate I_2 from neutral KI, even in the
cold*.

* The validity of the Riesenfeld test has been questioned by Liebhafsky (*Z. anorg. Chem.*,
221 (1934) 25.

Although the Riesenfeld reaction can be applied to readily soluble peroxy-salts, such as the peroxysulfates, it is possible that sparingly soluble compounds, or solids which dissolve slowly, may undergo very extensive hydrolysis before a sufficient concentration of peroxy-anions is built up in the solution. If this occurred, such a compound would be wrongly characterized as a peroxyhydrate. For other aids in distinguishing peroxy-salts and peroxy-hydrates see Le Blanc, Z. *Elektrochem.*, 29 (1923) 193; Menzel, Z. *phys. Chem.*, 105 (1923) 402*.

Since OH$^-$ ions formed by hydrolysis may interfere with the reaction, it is advisable to buffer the solution by the addition of phosphate (Liebhafsky, 1934). Even under these conditions, however, the Riesenfeld reaction does not reliably distinguish between peroxybora-tes and borate peroxyhydrates, for the reasons given.

Among the true peroxyborates, compounds known definitely include $NaBO_3$ (prepared by Le Blanc by rubbing NaOOH with H_3BO_3), $KBO_3 \cdot \frac{1}{2}H_2O$ and $(NH_4)BO_3 \cdot \frac{1}{2}H_2O$. The latter can be dehydrated in vacuum, without appreciable decomposition. According to Menzel, the water is constitutional, and the salts may be formulated as

$$\left[\begin{array}{c} HOO \\ \diagdown \diagup OOH \\ B{-}O{-}B \\ O \diagup \diagdown O \end{array} \right] M_2 \qquad (M = NH_4, K)$$

These compounds contain one atom of 'active' oxygen per atom of boron. There is also some evidence for the existence of true peroxyborates of the types $M^I BO_4 \cdot H_2O$ (boron: active oxygen = 1 : 2; M^I = Na, Li), and $KBO_5 \cdot H_2O$ (boron: active oxygen = 1 : 3) (Parting-ton and Fathallah, 1949).

Since the borate peroxyhydrates break up in aqueous solution, re-forming hydrogen peroxide, they behave just like H_2O_2 in that they readily lose oxygen or pass it on to other substances. This is the basis of the great importance of the commercial 'perborates' (which are actually peroxyhydrates) as washing and bleaching agents for wool, silk, straw, ivory, and many other substances. They are also widely used in cosmetics as bleaching agents (for hair), and as disinfectants. Mixed with other substances which catalytically decompose them, or the liberated H_2O_2, so that oxygen is evolved, they are used for the preparation of oxygen baths.

Care must be exercised in handling or working with borate peroxyhydrates containing more than 15% of active oxygen, since they may *decompose explosively* as a result of friction.

One of the best known borate peroxyhydrates is *perborax*, $Na_2B_4O_7 \cdot H_2O_2 \cdot 9H_2O$, a white crystalline salt obtainable by dissolution of a mixture of boric acid and sodium peroxide in water.

$$4H_3BO_3 + Na_2O_2 = Na_2B_4O_7 \cdot H_2O_2 + 5H_2O.$$

When this is recrystallized from water, it is transformed into the most widely used borate peroxyhydrate, $NaBO_2 \cdot H_2O_2 \cdot 3H_2O$

$$Na_2B_4O_7 \cdot H_2O_2 + 4H_2O = NaBO_2 \cdot H_2O_2 + NaH_2BO_3 + 2H_3BO_3.$$

* Many borate peroxy-hydrates do not liberate H_2O_2 when they are dissolved in water; it often remains bound to the borate ion in a complex, and this makes the distinction between borate peroxyhydrates and peroxyborates yet more difficult. There are many uncertainties, accordingly, as to the constitution of the individual 'perborates'. It is convenient to call the compounds in which H_2O_2 is recognizably bound in a complex with the borate ion, *peroxyhydrato-borates*, as distinguished from *borate peroxyhydrates* in the narrower sense, which merely contain 'hydrogen peroxide of crystallization' which is split off when they dissolve in water.

$NaBO_2 \cdot H_2O_2 \cdot 3H_2O$ is generally prepared directly by addition of hydrogen peroxide to sodium borate solutions, or by melting sodium peroxide with boric acid or borates. It is a white crystalline salt, soluble in about 40 parts of water, and with a negative heat of solution. The commercial product contains about 10.4% of active oxygen, which is determined analytically in the same manner as in hydrogen peroxide. The washing and bleaching agent 'Persil' contains 10% of sodium borate peroxyhydrate, together with soap and soda.

The peroxyhydrates of the borates of the bivalent metals are of importance, as well as those of the alkali metal borates. These substances—e.g., the alkaline earth peroxyhydrates—are only sparingly soluble, unlike the alkali metal compounds, and are therefore used as disinfectant powders and in dentifrices.,

(vii) *Hypoboric acid*, $H_4B_2O_4$, has been obtained by Wiberg (1937) by the hydrolysis of its ester, as a solid, white, oxidizable substance. The ester was prepared from chloroboric ester by the Wurtz synthesis—a reaction first applied in boron chemistry by Stock:

$$2B(OR)_2Cl + 2Na = B(OR)_2\text{—}B(OR)_2 + 2NaCl$$

It follows that the constitution of hypoboric acid is:

$$\begin{matrix} HO & & OH \\ & B\text{—}B & \\ HO & & OH \end{matrix}$$

The hypoboric esters are colorless liquids, with an unpleasant smell. The methyl ester melts at $-24°$ and boils at $+93°$.

(viii) *Borous acid*, H_3BO_2, was obtained by Schlesinger (1933) in the form of its methyl ester $H \cdot B(OCH_3)_2$, by the reaction of B_2H_6 with methanol. The ester is a colorless liquid, b.p. 25.9°, which is decomposed by water with the evolution of hydrogen:

$$H \cdot B(OCH_3)_2 + 3HOH = H_2 + B(OH)_3 + 2CH_3OH$$

(j) Other Compounds of Boron

(i) *Boron sulfide*, B_2S_3, results as a vitreous mass when boron is heated in sulfur vapor, or when carbon disulfide vapor is passed over a heated mixture of boron trioxide and coal. It is somewhat soluble in phosphorus trichloride, and can be obtained therefrom in fine needles of m.p. 310° and density 1.55. It forms crystalline addition compounds with ammonia and with boron halides. It is decomposed by water, forming boric acid and hydrogen sulfide.

A compound of B_2S_3 with H_2S, which is soluble in benzene and has the composition $H_2B_2S_4$, can be obtained as white needles by the action of hydrogen sulfide on boron tribromide.

(ii) *Boron nitride*, BN, is formed when boron is heated to a white heat in the presence of nitrogen. It is a white compound, of density 2.34, which feels like talc. It is unaltered by heating in air, in hydrogen or in hydrogen sulfide. It can be melted by heating to 3000° under pressure. It is decomposed by steam at a red heat, forming ammonia, and also by fusing it with alkali hydroxides.

The crystal lattice of boron nitride is related to that of graphite (see p. 423); in each hexagonal sheet of the structure, it may be considered that one half of the carbon atoms are replaced by B, and the other half by N atoms, making $a = 2.51, c = 6.69$ Å.

(iii) *Boron phosphoric oxide* ('boron phosphate'), BPO_4, a double compound of B_2O_3 and P_2O_5, is formed as a white powder by melting phosphoric acid and boric acid together. In aqueous solution, phosphoric acid and boric acid do not mutually precipitate each other, but they do so in concentrated sulfuric or glacial acetic acid. The double compound formed as ordinary temperature, and subsequently dehydrated by moderate heating, is fairly rapidly decomposed by water. That which has been ignited is practically insoluble in water, but is structurally identical (Gruner 1934). BPO_4 forms sharply defined hydrates with 3, 4, 5 and $6H_2O$. The 3-hydrate gives almost the same X-ray diffraction pattern as does H_3BO_3. The crystal structure of BPO_4 (and of $BAsO_4$, which is isomorphous) can be derived from that of cristobalite (see p. 489) by replacement of the Si atoms alternately by B- and P-atoms. (Schulze, 1934). It is thus not a boron or boryl phosphate, but a double oxide.

4. Analytical (Boron)

The detection of boron is generally based upon the green coloration which is imparted by its volatile compounds to an alcohol flame or a non-luminous gas flame. For this purpose, boric acid or its salts are converted to trimethyl borate, $B(OCH_3)_3$, by treatment with methyl alcohol and concentrated sulfuric acid.

$$B(OH)_3 + 3H—O—CH_3 \rightleftharpoons B(O—CH_3)_3 + 3HOH$$

The sulfuric acid serves to combine with the water, whereby the equilibrium is displaced and the reaction is made to proceed from left to right. Minerals are tested for boric acid by mixing them with calcium fluoride and sulfuric acid and introducing them into (or better, close to the lower margin of) the flame, so that volatile boron trifluoride is formed, and gives a green coloration. Boric acid itself also colors the flame green, because of its volatility.

With yellow turmeric paper, free boric acid (and therefore acidified solutions of borates also) gives a red-brown coloration, which appears only after the paper is dried. If the paper is then moistened with ammonia, it is temporarily colored blue-black.

Neutral solutions of borates also give white precipitates, soluble in acetic acid, with silver nitrate and barium nitrate. The precipitate of $Ba(BO_2)_2$ usually appears only upon the addition of a few drops of ammonia. The silver borate, $AgBO_2$, goes brown when the solution is boiled, because of hydrolysis. It is not possible to distinguish the various boric acids by qualitative reactions, since they readily pass one into another.

Boric acid can be determined quantitatively by conversion to methyl borate. This can be allowed to react with $Ca(OH)_2$, and is thereby saponified

$$2B(OCH_3)_3 + Ca(OH)_2 + 2H_2O = Ca(BO_2)_2 + 6CH_3OH$$

If a weighed amount of CaO is used, the increase in weight after ignition of the borate represents directly the amount of B_2O_3 taken up.

Boric acid can also be determined volumetrically. For this purpose, the alkali content is first determined by titration, with *methyl orange* serving as indicator. Hydrochloric acid is then added, in amount exactly equivalent to the alkali content; this is therefore enough exactly to set free all the boric acid. Then, by addition of glycerol—or better, mannitol or fructose (or invert sugar)—, the weak boric acid is converted into a stronger monobasic acid, which can be titrated with sodium hydroxide, with *phenolphthalein* as indicator.

5. Aluminum (Al)

(a) Occurrence

Aluminum ranks as the third most abundant of all the elements, and the most abundant metal. It occurs principally in the form of double silicates, as in the *felspars* and *micas*, and in the products of their weathering, the *clays*. It never occurs as the native metal. The oxide, Al_2O_3, occurs as *corundum* and *emery*. Of the oxide hydrates, bauxite, $AlO(OH)$ is important as being at present the chief raw material for the extraction of aluminum. *Cryolite*, Na_3AlF_6, which occurs especially in Greenland, is important for the same reason.

Of the *double silicates*, the most important are: *potash felspar* or *orthoclase*, $K[AlSi_3O_8]$, a principal constituent of igneous rocks such as granite, gneiss, syenite, porphyry, basalt; *soda felspar* or *albite*, $Na[AlSi_3O_8]$; *lime felspar* or *anorthite*, $Ca[Al_2Si_2O_8]$; *plagioclase* (isomorphous mixtures of lime and soda felspar, to which belong *oligoclase, andesine* and *labradorite*); and also the *micas*, (*biotite, muscovite, zinnwaldite* and *lepidolite*), which are also contained in the igneous rocks mentioned. *Nepheline*, $Na[AlSiO_4]$, and *leucite*, $K[AlSi_2O_6]$, are related to the felspars, and often replace them in rocks. Of the calcium aluminosilicates there may also be mentioned *zoisite*, $Ca_2Al_3[SiO_4]_3[OH]$; *epidote*, $Ca_2(Al, Fe)_3[SiO_4]_3[OH]$; and *vesuvianite*, $Ca_{10}Mg_2Al_4[SiO_4]_5[Si_2O_7]_2[OH]_4$. *Cordierite*, $Mg_2Al_3[AlSi_5O_{18}]$, is a magnesium aluminosilicate. *Cyanite, sillimanite* and *andalusite* are pure aluminum silicates, all with the formula Al_2SiO_5 (for their structure see pp. 499 and 501). The gem stone *topaz* is an aluminum silicate containing fluorine, $Al_2(OH,F)_2[SiO_4]$.

Kaolin, (china clay) a hydrated aluminum silicate of the composition $Al_2O_3 \cdot 2SiO_2 \cdot 2H_2O$, can be formed by the weathering of felspars. The usual weathering product of felspathic rocks, however, is *clay* (p. 507 *et seq.*). A clay containing much calcium and magnesium is known as *clay-marl*, and one contaminated by much iron(III) oxide and sand as *loam*.

(b) History

The name *aluminum* is derived from *alumen*, alum, and this name in its turn, according to Isidorus (7th century) from the use of alum as a mordant in dyeing—'quod lumen coloribus praestat tingendis'. Pliny describes alum and its application, and it is mentioned under the name of στυπτηρία as early as the time of Herodotus (5th century B.C.). Alum in its present sense, as potassium aluminum sulfate, $KAl(SO_4)_2 \cdot 12H_2O$, was, however, not at that time distinguished from other substances which had a similar action—e.g., iron vitriol. Alum appears to have been first prepared in the pure state by the alchemists. The earth (i.e., the metallic oxide) forming the basis of alum was recognized for the first time as a distinct one by Marggraf, in 1754. Davy tried unsuccessfully (1808—1810) to liberate the metal from which alumina was derived by the electrolytic method. The first to succeed in so doing was Oersted (1825), who reduced anhydrous aluminum chloride, which he had discovered, by heating it with potassium amalgam. This process, which succeeds only under certain experimental conditions, was improved by Wöhler in 1827, by the use of pure metallic potassium in place of the amalgam. Wöhler described the properties of the metal accurately for the first time, and he is usually cited as its discoverer, since Oersted did not furnish any unequivocal proof that the substance he prepared was, in fact, the pure metal. The industrial production of aluminum was first worked out by St. Claire Deville, on the basis of Wöhler's process, and he was able to lower the price from £120 to £10 per kilo-gram, so that a large block of the 'silver from clay' could be shown at the Paris World's Fair in 1855. The present price of aluminum is about 19 cents per pound in the United States. In the meanwhile, Bunsen succeeded (1854) in preparing aluminum by the electrolysis of sodium aluminum chloride. This forms the basis of the technical process now used, except that a solution of aluminum oxide in molten cryolite is employed in place of the volatile chloride. Up to 1883, metallic aluminum was produced industrially only in France; then production in the United States, Germany and England followed. Since 1890, considerable quanti-ties of aluminum have been extracted in Switzerland, where cheap water power is available for the electrolysis. The United States is today the chief producer of aluminum (see Smithells, *J. Roy. Soc. Arts*, 98 (1950) 822).

(c) Manufacture [4–6]

The processes generally used today for the industrial production of aluminum are all based on the electrolytic decomposition of aluminum oxide dissolved in molten cryolite. Retort graphite is usually employed for the electrode material. The bath is maintained molten by the heating effect of the current; its temperature should not exceed 1000°. The metallic aluminum separating at the cathode collects in the liquid state on the bottom of the furnace. At the anode, which dips into the bath from above, the oxygen combines with the carbon, forming carbon monoxide, which at once burns to carbon dioxide. The latter is also formed directly to some extent, at the anode.

The decomposition potentials, at 950°, are; —4.5 volts for NaF, —3.5 v. for AlF_3 and —2.18 v. for Al_2O_3. By use of a carbon anode, the decomposition potential of Al_2O_3 should be reduced theoretically to only —1.00 volts, because of the formation of CO, but it rises to more than —2 volts at high current densities, since the formation of CO lags behind the discharge of oxygen. If the applied voltage is not too high, and the alumina content of the melt is not too small, then it is substantially only the aluminum oxide that is decomposed. Formation of CF_4 at the anode takes place simultaneously to quite a small extent. If the electrolysis is carried out properly, sodium is not deposited at all. Since a considerable pro-portion of the applied voltage is absorbed in overcoming the ohmic resistance of the cell, it is considered in practice that 5—6 volts are needed at a current density of 70—90 amps per square decimeter of anode surface. As soon as the cell voltage begins to rise appreciably, fresh aluminum oxide must be added.

It is only very rarely indeed that the aluminum oxide required for the process occurs in nature in a state of sufficient purity. It is extracted at the present time almost entirely from bauxite, which occurs chiefly in France, Hungary, Dalmatia, Istria, Russia, India, Arkan-sas, and British and Dutch Guiana. The cryolite needed as solvent is, in part, prepared synthetically. Other fluxes, such as calcium fluoride, are commonly added, to lower the melting point. By this means, a bath with 20—30% Al_2O_3 content can conveniently be maintained liquid at 800—900°. If the temperature is too high, the difference in density between the liquid metal and the melt becomes less, so that the aluminum no longer sinks to the bottom, but reaches the surface and burns.

The naturally occurring bauxites are contaminated to a greater or lesser extent by iron oxide and silica. To obtain aluminum oxide of the purity required for the electrolytic pro-duction of the metal [2], the bauxites are therefore attacked either by heating them with CaO and Na_2CO_3 (dry process), or by heating them with caustic soda in autoclaves (Bayer pro-cess). In both cases, the aluminum oxide goes into solution as aluminate, and this can then be decomposed either by passing in carbon dioxide, or by 'seeding' with previously prepared aluminum hydroxide. In the former case, the decomposition is

$$2[Al(OH)_4]^- + CO_2 = 2Al(OH)_3 + CO_3^= + H_2O$$

Decomposition by 'seeding' depends upon the fact that the aluminate solutions obtained by autoclave heating are metastable after cooling and dilution. The added aluminum hydroxide accelerates the decomposition of the aluminate, since it provides nuclei for the crystallization of the hydroxide formed by the reaction $[Al(OH)_4]^- = Al(OH)_3 + OH^-$.

A variant of the Bayer process is the tower-leaching process worked out by Ver. Aluminumwerke. In this process, the bauxite is not dissolved in autoclaves, but sodium hydroxide solution is allowed to trickle through tall towers in which the bauxite is stacked.

About 32% of the costs of production of aluminum are due to the Al_2O_3, 4% to the cryolite, 12% to the electrode carbon and 25% to the electrical energy, at typical costs per kWh. The price of bauxite makes up only about 6% of the total costs of aluminum production. Nevertheless, the possibility of production from other raw materials is of importance, especially in countries (e.g., Germany) which are dependent upon imported bauxite. Much attention has therefore been given to the extraction of aluminum from *clays*. The high silica content of clay makes processes involving alkaline attack, such as the Bayer process, rather unsuitable. To prevent silica from being taken into solution, attack on the clay is effected by acids. The process is so controlled that the dissolution of iron oxide is avoided as far as possible, since the subsequent separation of large amounts of iron from aluminum presents considerable difficulty.

In the Buchner process (Nuvalone process), the separation of Al_2O_3 from SiO_2 and Fe_2O_3 (also from TiO_2) is effected by heating the clay in autoclaves with nitric acid, in suitable amount and concentration. The aluminum nitrate passing into solution contains substantially only alkali and alkaline earth nitrates as impurities, and is separated from these by fractional crystallization. The nitric acid is recovered from the thermal decomposition of the hydrated aluminum nitrate, and Al_2O_3 remains in a very pure state.

In the Goldschmidt process, aqueous solutions of the very much cheaper sulfurous acid are used to attack the clay. However, this usually takes considerable amounts of iron into solution also, but as aluminum forms a well crystallized basic salt, a relatively simple separation of aluminum from iron can be achieved by fractional crystallization of the basic sulfite, provided that the amounts of iron are not excessive.

The Haglund process has a completely different basis. This was worked out originally for the preparation of aluminum oxide from bauxite, but is also applicable to many clays. In this process, aluminum is partly converted to the sulfide by heating the oxidic ore with coke and pyrites in the electric furnace. The sulfide forms a very fluid slag with aluminum oxide and thus because of its low density, this separates from the iron-silicon alloy formed simultaneously. After cooling, the slag, consisting of 80% Al_2O_3 and 20% Al_2S_3 is treated with hydrochloric acid, whereby $AlCl_3$ is formed and H_2S is evolved (from which sulfur can be recovered), while crystalline Al_2O_3 remains undissolved.

Electrolytic aluminum is generally purified by remelting it in the reverberatory furnace. Commercial aluminum is usually about 99.5% pure. The impurities consist principally of silicon and iron, occasionally with traces of copper.

The *recovery of aluminum from scrap* is a process of great importance, and the scale of the process may amount to as much as 50% of the production of new metal. It is generally carried out by re-melting, sometimes with the addition of magnesium or zinc; these can subsequently be removed by vacuum distillation. *Chlorine refining* is simple and efficient. By blowing chlorine through the molten light metal, the common impurities (other than copper) are converted to chlorides, which collect on the surface of the melt. In doing so, they carry up the oxides, nitrides, and carbides suspended in the melt. Only about 2 to 3% of the aluminum is thereby lost by conversion to $AlCl_3$. At the end of the treatment, which lasts about 10 minutes, nitrogen is blown through, to carry away the chlorine gas.

Aluminum with a purity of at least 99.99% can be obtained from the metal, from the molten salt electrolysis (usually 99.5% pure), or from aluminum scrap, by 'three layer electrolysis'. This is a molten salt electrolysis, in which an anode consisting of a relatively dense aluminum alloy (Al—Cu) covers the bottom of the cell. This is covered by an electrolyte, the density of which is so controlled by the addition of BaF_2 or $BaCl_2$ that it is lighter than the molten anode alloy, but denser than pure molten aluminum. The latter accordingly floats on top of the electrolyte, and is thereby prevented from becoming admixed with the constituents of the anodic alloy.

Efforts to replace the electrolytic production of aluminum by direct *thermal reduction* of aluminum oxide have not hitherto met with any practical success. If carbon is used as reducing agent, it is apparently not possible to avoid the formation of aluminum carbide, Al_4C_3. It has been shown by Kohlmeyer (*Z. anorg. Chem.*, 260 (1949) 208) that this can be reduced in extent if the Al_2O_3 is fused very rapidly, and restricted in the duration of its contact with the carbon used for reduction. In laboratory experiments it was possible to prepare aluminum with purity up to 93% by this means.

(d) Properties

Aluminum is a silver-white metal of density 2.70, m.p. 660.2°, and boiling point 2270°. It crystallizes face-centered cubic, with $a = 4.0414$ Å. The thermal conductivity of aluminum, $\lambda = 0.5$ at ordinary temperature, is about three times as great as that of wrought iron, and half that of copper. The specific electrical conductivity of drawn aluminum wire is about 60% of that of copper wire. The specific heat (0.23 cal per g at 100°) is high compared with that of other metals, being $2\frac{1}{2}$ times that of copper or zinc, and more than twice that of iron. The latent heat of fusion is also high. Hence, in spite of its low melting point, aluminum is less readily brought to the molten state than copper, but when once melted it remains fluid longer the other metals. It is very ductile; it may be drawn into very fine wires, rolled into thin sheets, or hammered into extremely thin foils (aluminum leaf). The tensile strength of pure aluminum is about $\frac{1}{4}$ of that of copper, but can be increased considerably by the addition of a few percent of copper. The chemical resistance is reduced thereby, however.

Pure aluminum does not corrode in air, since it becomes covered with a thin film of the oxide which protects it from further attack. For the same reason, it is indifferent towards water, and even towards steam at high temperatures. It is also unattacked by hydrogen sulfide, but dissolves in most acids, however, and also in caustic alkalis. Dilute organic acids, such as acetic acid and citric acid, hardly attack it in the cold, but do so at 100°. The addition of sodium chloride accelerates its dissolution. Concentrated acetic acid does not attack it, however, neither do fats or fatty acids. Nitric acid, both dilute and concentrated, is also without action in the cold, but a vigorous reaction eventually sets in on heating.

The solubility of aluminum in acids is due to the fact that, because of its position in the electrochemical series, it discharges hydrogen ions: $Al + 3H^+ = Al^{+++} + \frac{3}{2}H_2$. The reason that it is not attacked by water or dilute, weakly dissociated acids, is that the formation of the extremely insoluble aluminum hydroxide inhibits the reaction, and in solutions of low hydrogen ion concentration the ion product never falls below the solubility product of the hydroxide. Aluminum hydroxide dissolves in caustic alkalis, however, forming hydroxo-aluminates (cf. p. 355). Since the concentration of the Al^{+++} ion in the alkaline solutions is very minute, the potential of the metallic aluminum with respect to such solutions is strongly displaced, making the metal still more base. Caustic alkalis therefore attack metallic aluminum vigorously.

The resistance of aluminum towards attack by acids which readily give up oxygen (such as nitric acid) is observed also in the case of many other metals with a high affinity for oxygen. The phenomenon is known as passivation (see Vol. II).

If the surface of aluminum is rubbed with mercury, the formation of a coherent oxide film, which protects the surface from attack by air and water, is prevented. White threads, consisting of aluminum hydroxide formed by the action of atmospheric moisture, quickly sprout from the surface of an amalgamated aluminum sheet on standing in air. The aluminum hydroxide obtained in this way, and the oxide formed from it by ignition, are extraordinarily voluminous, and highly surface-active. The oxide has been suggested by Wislicenus for the determination of tannins, and is sold commercially under the name of 'fibrous alumina'.

Use is also made of the activation of aluminum by amalgamation with mercury when aluminum is used as a reducing agent in organic chemistry. It is then used in the form of a granular amalgam, obtained by the treament of aluminum turnings with mercuric chloride.

The protection conferred by the oxide skin can be considerably increased by building up an oxide layer on the aluminum electrochemically, of much greater thickness (0.02 mm) than the natural oxide skin (*anodization* or *eloxal* process) [*13, 14*].

The oxide film which usually covers the surface of aluminum is not due to the reaction of the metal with atmospheric oxygen, but rather to reaction with atmospheric water vapor. The hydrogen which is thereby liberated is partly taken up in solid solution by the aluminum. It is present in the metal in atomic form, as is proved by the manner in which the solubility varies with the pressure of hydrogen.

If finely divided aluminum is ignited, it burns with great brilliance to aluminum oxide. 7.47 kcal are thereby liberated per g of aluminum. Because of its high affinity for oxygen, aluminum is used to liberate metals from oxides which are difficult to reduce, and for the attainment of high temperatures—e.g., for the welding of iron. The mixtures of aluminum turnings with the oxides of other metals—e.g., Fe_3O_4—used for this purpose are known as *thermite*. According to von Wartenberg (1936), temperatures up to 2400° C. may be reached in the iron-thermite reaction. The process introduced by Goldschmidt for the preparation of metals, by reduction of their oxides with aluminum, is known as *aluminothermy* [*15*].

Aluminum also unites with chlorine with the evolution of much heat. It catches fire in liquid bromine, combines directly with iodine also, but unites with sulfur only at a red heat. It combines with nitrogen only at very high temperatures, such as are obtained in the partial combustion of aluminum powder in oxygen. Aluminum may react with explosive violence when it is heated with selenium and tellurium. The compounds Al_2Se_3 and Al_2Te_3 are thereby formed.

(e) Uses [*4–9*]

Since it has become possible to produce aluminum cheaply, it has found extensive applications. Industrial apparatus and numerous domestic objects—e.g., saucepans and other cooking utensils—are manufactured from it.

The brownish tarnish frequently seen on aluminum utensils arises from the impurities contained in the metal. For polishing aluminum ware, a paste of rouge and tallow is used, which is obtainable commercially. The inside of cooking vessels is cleaned with wood ash, fine sand or levigated emery, or by boiling with dilute solution of alum and potassium tartrate (1 spoonful per liter of a mixture of 1 part of cream of tartar and 2 parts of alum).

Scrap aluminum sheet is used for the production of aluminum powder, which is employed as a paint [*16*], in lithography, and in explosives and pyrotechnics. The metal has found increasing uses in the electrical industry, in place of the more expensive copper. Although its conductivity is only 60% that of copper, this is more than compensated by its low density, which enables thicker wires to be used. For the same conductivity, the weight of an aluminum wire is only about half that of a copper wire. The use of aluminum, in the form of alloys [*10*], in the aircraft and automobile industries, is also based on its lightness. Such alloys are, for example, *magnalium* (containing 10—30% magnesium) and *duralumin* [*11, 12*], an alloy which can be hardened, with 93—95% aluminum, 2.5—5.5% copper, 0.5—2% magnesium, 0.5—1.2% manganese and 0.2—1% silicon. *Scleron*, an aluminum alloy containing about 12% Zn, 3% Cu, 0.6% Mn, up to 0.5% Fe, up to 0.5% Si and 0.08% Li, has similar properties. *Hydronalium* is an alloy with 3—12% magnesium, resistant to sea water. The metal is used in the iron and steel industry as a deoxidizer, to free molten iron from dissolved oxide. Coarsely granular aluminum powder (aluminum turnings) is used for the preparation of such metals as chromium, manganese and titanium by the Goldschmidt process [*15*], and for the preparation of thermite mixtures for the welding of railway rails, etc. Rolled into thin foils, aluminum is employed in place of tinfoil for wrappings—e.g., for chocolate and confectionery.

Aluminum foils about 0.5 μ thick are used in photography, in vacuum flash bulbs.

The practice has recently been introduced of protecting iron objects against rusting by covering them with a layer of aluminum (aluminizing). Because of the smaller difference in thermal expansion, the aluminum coatings should be more durable than zinc coatings. Aluminum is increasingly employed for the production of *mirrors*, (especially for telescopes and optical instruments). The reflecting layer is applied to the glass by evaporating the aluminum, a process originated by Pohl (1912). It retains its high reflecting power much better than a silvered surface, and has the advantage of an increased reflectivity in the ultraviolet.

Attempts to deposit aluminum electrolytically from aqueous or non-aqueous solutions of its salts have met with no success. According to Menzel (1940), it is possible, however, to electrodeposit the metal from baths containing organo-aluminum compounds of the types AlR_2X or $AlRX_2$ (X = halogen).

6. Compounds of Aluminum

The normal compounds of aluminum are derived without exception from the triply positively charged ion Al^{3+}. They therefore correspond to the type AX_3, X being a univalent radical. The solid aluminum hydride $[AlH_3]_X$ recently prepared by Wiberg has, however, a constitution differing from the normal aluminum compounds (see below).

SURVEY OF THE MOST IMPORTANT ALUMINUM COMPOUNDS

Oxygen compounds	*Halides*	*Other salts*	
Al_2O_3	AlF_3	$Al_2(SO_4)_3$	Sulphide Al_2S_3
	Fluoroaluminates	Alums	
	e.g., $Na_3[AlF_6]$	e.g., $KAl(SO_4)_2 \cdot 12H_2O$	Nitride AlN
$Al(OH)_3$			
Aluminates			
e.g., $Na[Al(OH)_4]$	$AlCl_3$	$Al(NO_3)_3$	Organo-aluminum
	$AlBr_3$	$Al(SCN)_3$	compounds—AlR_3
	AlI_3	$Al(C_2H_3O_2)_3$	AlR_2X
			$AlRX_2$

Of the oxygen compounds, the crystalline *oxide* is notable for its hardness and high heat of formation and the *hydroxide* for its amphoteric character. The *salts* of aluminum mostly crystallize with a high water content. In the foregoing summary, the water of crystallization (the amount of which may depend upon the temperature and pressure) has in general been left out. The salts of aluminum are colorless, and those of strong acids are all easily soluble in water. Among the salts of moderately strong and weak acids, the *phosphate*, *borate* and *silicate* are insoluble. Aqueous solutions of aluminum salts contain colorless Al^{+++}-ions, which are highly hydrated. Brintzinger (1935) found that in dilute solutions each Al^{+++}ion is associated with $18H_2O$. Although their solubility increases with rise of temperature, the anhydrous aluminum salts have for this reason a large positive heat of solution.

The salts of aluminum, even with strong acids, are all hydrolyzed in aqueous solution, in accordance with the weakly basic character of aluminum hydroxide. This does not react at all with very weak acids, such as H_2S. If the corresponding compounds—e.g., Al_2S_3—are prepared by dry methods, and introduced into water, they are at once completely hydrolyzed. With the salts of the stronger acids, however, the hydrolysis comes to a standstill when a small fraction has been split up. Thus the degree of hydrolysis of aluminum chloride is 2.0% and of the sulfate 1.3%, in 0.1–N solution at 25°.

Aluminum has a tendency to form *complex ions* with excess halogen or $SO_4^=$ ions. This shows itself not only in the crystallization of double salts from the corresponding solutions, but also in that the potential of aluminum, at equi-molecular concentrations of the salts, is appreciably higher in solutions of the chloride and sulfate than, e.g., in nitrate solutions.

Aluminum salts are generally prepared from the sulfate as starting material. This is prepared principally from the oxide hydrate, for which in turn bauxite is at present the most important raw material. Apart from the oxide, which constitutes the starting material for the preparation of the metal, the sulfate $Al_2(SO_4)_3 \cdot$ $\cdot 18H_2O$, and alum, $KAl(SO_4)_2 \cdot 12H_2O$, derived from it, are the most widely used aluminum compounds.

(a) Oxygen Compounds of Aluminum

(i) *Aluminum Suboxides*. If Al_2O_3 mixed with boron is heated in a vacuum to 1300°, or Al_2O_3 mixed with aluminum powder heated to 1450°, it is found that it volatilizes relatively rapidly. Zintl (1941) interpreted this as being due to the formation of *aluminum(II) oxide*, AlO, or *aluminum(I) oxide*, Al_2O, at these temperatures:

$$Al_2O_3 + B \rightleftharpoons 2AlO + BO \quad \text{and} \quad Al_2O_3 + Al \rightleftharpoons 3AlO \quad \text{or} \quad Al_2O_3 + 4Al \rightleftharpoons 3Al_2O$$

The reactions proceed from right to left once more on cooling.

According to Grube (1949), Al_2O_3 vaporizes when heated with silicon to 1800° in high vacuum, silicon monoxide and *aluminum(I) oxide* being formed:

$$Al_2O_3 + 2Si \rightleftharpoons 2SiO + Al_2O$$

Grube obtained two distinct sublimates in his experiments, one of them corresponding in composition to the compound SiO, and the other to the compound Al_2O. It was not conclusively established whether the sublimates actually consisted of these compounds, or whether they were mixtures of the substances formed by disproportionation of the lower oxides during cooling ($2SiO = Si + SiO_2$, $3Al_2O = Al_2O_3 + 4Al$).

(ii) *Aluminum oxide*, alumina, Al_2O_3, is formed by the combustion of aluminum and by igniting aluminum hydroxide, or the aluminum salts of volatile acids. It is a white amorphous powder, which is insoluble in water and—if ignited at a high temperature—insoluble in acids also. It is found crystalline in nature as *corundum*. This mineral, which is notable for its great hardness, is generally opaque, and drab in color because of impurities. The coloration of the varieties of corundum which are prized as gem stones—ruby (red) and sapphire (blue)—is due to the traces of impurities present, which can also produce other colors—e.g., green (oriental emerald), violet (oriental amethyst) and yellow (oriental topaz). Since corundum can be obtained artificially by the Goldschmidt process, these gem stones can also be prepared by suitable methods (synthetic gems [3]).

In the form of corundum and emery (a granular variety of corundum, contaminated with iron oxide and quartz), aluminum oxide is used as an abrasive and polishing agent. Corundum is also used for the preparation of highly refractory materials—e.g., the 'Dynamidone bricks' used as furnace linings in cement kilns. Large quantities of aluminum oxide obtained by the dehydration of the hydroxide are used as raw material for the production of the metal.

Aluminum oxide is extremely refractory. It vaporizes slowly at very high temperatures, its vapor pressure being about 10^{-6} atm. at 1950°, and it has been estimated that it would reach 1 atm. at 3500°. It is probable that alumina does not vaporize as Al_2O_3, but dissociates to the suboxide AlO:

$$Al_2O_3 \text{ solid} \rightleftharpoons 2AlO_{vapor} + O_g$$

At the moment of condensation from the vapor phase, alumina is deposited in a vitreous-amorphous form, which immediately becomes crystalline. Under certain conditions, it may thereby form a 'quenched' modification which cannot be obtained by any other method (von Wartenberg, 1952). Alumina vaporizes much more rapidly in the presence of metallic aluminum (Zintl, 1941), and it has been concluded from the composition of the sublimate obtained that, under these conditions, the evaporation involves the reversible reaction

$$Al_2O_3 + 4Al \rightleftharpoons 3Al_2O.$$

Alumina usually crystallizes in the rhombohedral hemihedral class of the hexagonal system (α-alumina), density 3.99. Its crystal structure is represented in Fig. 72. This might be regarded as a *molecular* lattice, in that it is possible to distinguish Al_2O_3 molecules in it. Each aluminum atom is, however, surrounded by 6 oxygen atoms in all, and those belonging to different 'molecules' are, indeed, rather closer than those belonging to the same 'molecule'. This can be attributed to the repulsion exerted on each other by like atoms within any one molecule. As shown by Fig. 72, 1 molecule of Al_2O_3 lies completely inside the unit cell, and the others (of which only 2 are shown complete) project outside the unit cell, which contains 2 molecules of Al_2O_3 in all. The upper aluminum atom of the Al_2O_3 molecule which projects outside the unit cell on the right, is 0.14 Å nearer to the oxygen atom above it, and to its left, than to any of the three oxygen atoms belonging to its own molecule. In the same way, every oxygen atom is surrounded in the corundum structure by 4 aluminum atoms, of which two, not belonging to the same molecule, are closer to it.

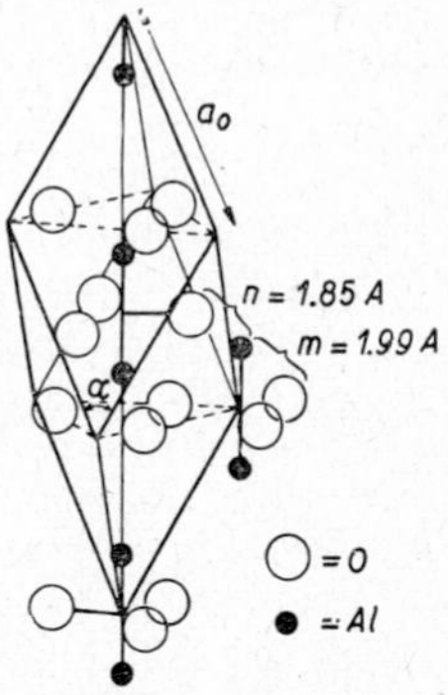

Fig. 72.
Corundum, Al_2O_3
$a_0 = 5.12$ Å;
$\alpha = 55° 17'$;
Al—Al = 1.36 Å

Another so-called hexagonal modification of aluminum oxide (β-alumina), density 3.30, is occasionally formed during the slow cooling of the fused oxide. It is, however, only formed in the presence of certain other metallic oxides, and it has been shown by Beevers (1936) that these added substances are essential constituents. 'β-alumina' is, in fact, not another modification of alumina itself but is a double oxide—e.g., $Na_2O \cdot 11Al_2O_3$ or $NaAl_{11}O_{17}$. Similar 'β-aluminas' are formed with other cations in place of sodium.

A third cubic modification of aluminum oxide (γ-alumina), density 3.4, is obtained by the dehydration of hydrargillite (gibbsite), or of bauxite, or by gentle heating of the amorphous oxide hydrate. According to the mode of preparation, γ-Al_2O_3 may be obtained with very different particle sizes and, in some cases, with considerable lattice distortion. These differences may give rise to differences of up to 10 kcal per mole in the heat of solution (Fricke, 1946). Above 1000°, γ-Al_2O_3 changes irreversibly into α-Al_2O_3. (Heat of transformation 20.6 kcal/mole). There are also other unstable modifications: e.g. δ-Al_2O_3 (Parravano, 1928) and ζ-Al_2O_3 (Barlett, 1932).

γ-Al_2O_3, like γ-Fe_2O_3, has a structure like that of the spinels. The oxygen atoms occupy the positions of the spinel lattice, but $\frac{1}{9}$ of the places belonging to the metal atoms are left unoccupied. The vacant positions are distributed essentially over those sites which, in the structure of spinel ($MgAl_2O_4$) itself, are occupied by Al atoms. Another modification, known as γ'-Al_2O_3, has a very similar structure. It differs from γ-Al_2O_3 in that the vacant cation sites in its crystal lattice are distributed over *all* the metal atom positions—i.e., those occupied both by Mg and by Al in $MgAl_2O_4$. (Verwey, *Z. krist.*, A, 91 (1935) 65, 317) γ'-Al_2O_3 is converted into γ-Al_2O_3 at about 900°

Recent work (Russell and Stumpf, 1950) has shown that the transformation of aluminum oxide hydrate into aluminum oxide may produce not only the modification properly known as γ-Al_2O_3, as a metastable phase, but possibly five unstable varieties which change successively and ultimately into α-Al_2O_3 (see below).

Aluminum oxide prepared under certain specified conditions—e.g., by the autoxidation of amalgamated aluminum—is highly surface-active. It is therefore widely employed as an adsorbent, as a catalyst, or as a carrier for other catalytically active materials. It also exhibits cation-exchange properties towards other inorganic cations. A commercial preparation specially suited for chromatography (cf. Vol. II) is Woelm alumina.

The adsorptive properties of surface-active alumina are enhanced by the fact that it not only picks up foreign ions on its surface, but also builds them superficially into its crystal lattice to a certain extent, in place of aluminum ions which go into solution. Aluminum oxide thus possesses ion-exchanging properties (sometimes known as 'permutoid' behavior), and that part of its behavior which is due to such properties is referred to as 'chemical adsorption'. Fricke (1948—50) has shown that the ion-exchange properties of surface active alumina are not the consequence of a certain alkali metal content, as had been assumed. If alumina that contains alkali (e.g., Merck's 'Brockmann process' aluminum oxide) is used, the alkali ions take part in the exchange. However, a pure γ-Al_2O_3 prepared by Fricke had more than twice as great an adsorptive capacity as Brockmann alumina towards Cu^{++} ions, and the proportion of the total adsorption due to chemical adsorption was about the same in both cases (about 40%).

(iii) *Aluminum oxide hydrate and aluminum hydroxide.* When aqueous ammonia is added to solutions of aluminum salts, a gelatinous precipitate is thrown down. When precipitated in the cold, this consists at first of a hydrogel of amorphous aluminum oxide, which slowly changes, in contact with the solution (more rapidly when it is warmed), into the crystalline hydroxide, *aluminum metahydroxide*, AlO(OH). The latter, generally mixed with amorphous aluminum oxide hydrogel, is also obtained when aluminum salts are boiled with substances which can bind the H$^+$ ions formed by hydrolysis—e.g., with carbonates, acetates, thiosulfates, etc. No precipitate is obtained with alkali hydroxides in excess, since formation of soluble hydroxoaluminates occurs at higher hydroxyl ion concentration.

The materials of variable water content, and generally of heterogeneous character, obtained by precipitation from acid solutions, are known as *oxide hydrates*, as distinct from the hydroxides of well defined composition, *aluminum orthohydroxide*, $Al(OH)_3$, and *aluminum metahydroxide*, AlO(OH).

(iv) *Aluminum orthohydroxide*, $Al(OH)_3$, is obtained as a white precipitate, which is shown by X-ray investigation to have a crystalline texture, when precipitation is effected from aluminate solutions—e.g., by passing in carbon dioxide:

$$2[Al(OH)_4]^- + CO_2 = 2Al(OH)_3 + CO_3^= + H_2O$$

Al(OH)$_3$ occurs crystalline (monoclinic) in nature, as *hydrargillite* or *gibbsite*. The crystallized hydroxide (density 2.42) differs from the amorphous oxide hydrate in dissolving only with difficulty in acids, and in losing no water when it is heated for several hours at 100°. If it is heated for a long time—e.g., for 14 days at 150° in a sealed tube, it is transformed into the metahydroxide, which is also crystalline, and which occurs native in large deposits as *bauxite* (named after the occurrence at Les Baux, in France). Bauxite is formed in tropical climates by the weathering of rocks containing aluminosilicates, whereas outside the tropics the decomposition of the same rocks leads usually to the formation of clay or kaolin. The artificially prepared crystalline metahydroxide corresponding to bauxite is named *böhmite*, to distinguish it from the mineral.

Another crystalline modification of aluminum metahydroxide, AlO(OH), is the rhombic *diaspore*, which is also found as a mineral. When heated to 420°, diaspore loses all its water and passes directly into corundum, whereas böhmite changes into anhydrous aluminum oxide at temperatures as low as 300°, but into the cubic γ-alumina (or the other recently recognized metastable forms), and not directly into corundum (cf. Table 65, which summarizes the transformations in the system Al$_2$O$_3$—H$_2$O as shown from the results of a number of independent workers).

TABLE 65

SUMMARY OF THE TRANSFORMATIONS OF THE VARIOUS CRYSTALLINE
MODIFICATIONS OF THE ALUMINUM OXIDES AND HYDROXIDES

[after Laubengayer and Weisz (1946), Brown, Clark,
and Elliott (1952) and Day and Hill (1952)]

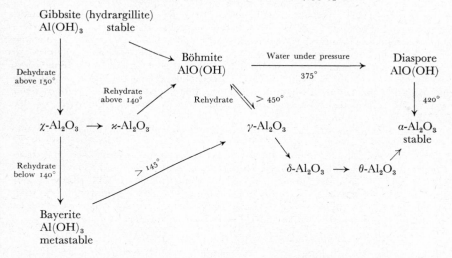

The gelatinous precipitates obtained by the action of aqueous ammonia, etc. on aluminum salt solutions lose varying amounts of water on drying. The conclusion was formerly drawn that a large number of hydrates of aluminum oxide existed, some with complex compositions. It was shown by Van Bemmelen, however, in 1888, that gels of this kind may lose water discontinuously even although they contain no definite hydrates (cf. under silicic acid). This is because the gels are penetrated by numerous, interconnected, very fine pores, in which water is bound by the action of capillary forces. The dehydration curve of such a gel is very similar to that of a chemical compound. The binding of water by capillary forces cannot take place, however, if the formation of menisci is prevented; this is the case if the gel is not simply dried in vacuum, but by extraction of the water by treatment of the gel with a liquid which has a strong affinity for water. In this way, Wilstätter and Kraut (1923-24) by extraction with acetone, and Biltz (1928), using liquid ammonia, were able for the first time to demonstrate the existence of definite hydroxides in the gels precipitated from aqueous solutions of aluminum salts by means of ammonia. X-ray investigations by Haber and Böhm (1925) and by Biltz and Meisel (1928) led to the same results. If aluminum salts are precipitated hot by means of aqueous ammonia, precipitates are obtained which clearly furnish the diffraction patterns typical of bauxite or böhmite. The precipitates obtained from hot solutions have the same slimy consistency as those precipitated at room temperature in which, when they are freshly precipitated, no crystalline structure can be detected. According to Hüttig, the latter should be regarded, when fresh, as hydrogels of amorphous aluminum oxide, which gradually pass into hydrargillite $Al(OH)_3$ by binding part of the water chemically and recrystallizing at the same time. If aluminum hydroxide is allowed to form slowly at ordinary temperature from solutions of *aluminates*, *bayerite* or *hydrargillite* may be formed, according to the rate of precipitation (Table 66). The latter may, in some circumstances, be obtained in crystallites clearly recognizable under the microscope.

TABLE 66

SURVEY OF THE DIFFERENT TYPES OF ALUMINUM HYDROXIDE PRECIPITATES

(a) Precipitated from *acid solutions* by ammonia
   (i) In the cold.   Amorphous aluminum oxide hydrate
   (ii) From hot solution.   *Böhmite*, γ-AlO(OH)   (usually mixed with amorphous aluminum oxide hydrate)

(b) Precipitated from *aluminate solutions* (e.g., by passing in CO_2)
   (i) By slow precipitation: *gibbsite* or *hydrargillite*, α-$Al(OH)_3$
   (ii) By rapid precipitations: *bayerite*, γ-$Al(OH)_3$

Bayerite, γ-$Al(OH)_3$, can also be obtained by the aging of amorphous aluminum oxide hydrates, and of böhmite, under water (which must have no trace of acidity) or under dilute alkali solutions. It can also be conveniently and reliably obtained by treating amalgamated sheet aluminum with conductivity water. Aluminum oxide hydrate, amorphous towards X-rays, is first formed, and this passes over into very well crystallized bayerite in the course of a few days (Fricke and Schmäh, 1945). The amorphous gel first obtained in this preparation, and that obtained by the hydrolysis of aluminum ethylate, is strongly surface-active, and can be used for the chromatographic separation of cations. Its solubility is less than 10^{-6} mol per liter of pure water.

The action of boiling water or steam on amalgamated aluminum sheet furnishes böhmite (Fricke, 1947).

The different forms of aluminum hydroxide display quite specific differences in their adsorptive and ion-exchange properties. Hüttig (1931) and Fricke (1936) showed that the basic character of the surface diminished, and the acidic character became stronger, in the series amorphous aluminum oxide hydrate→böhmite→bayerite→hydrargillite→γ-Al_2O_3. This shows up especially clearly in the isoelectric points of particles of the various modifications suspended in water. Fricke (1949) found the following isoelectric points (at 20°).

	amorphous aluminum oxide hydrate	böhmite	bayerite	γ-Al_2O_3
pH	9.45	9.45—9.40	9.20	8.00

In studies of adsorptive power and catalytic efficiency, it is therefore not sufficient to characterize a preparation simply as 'aluminum oxide' or 'aluminum hydroxide', but it is absolutely essential to determine by X-ray study the modification which is present. In addition to these chemical differences, account must also be taken of the influence of particle size and perfection of order upon the surface activity (cf. Vol. II).

For the primary and secondary particle size of bayerite, böhmite, and γ-Al_2O_3, and the specific surface, porosity and micro-structure of such preparations, see Fricke, *Z. anorg. Chem.*, 265 (1950) 21, 41.

Aluminum hydroxide forms the starting material for the preparation of other aluminum salts, especially the sulfate, and for the production of pure aluminum oxide for the electrolytic extraction of the metal. For this purpose, the hydroxide must be transformed into the oxide by strong ignition, as it otherwise takes up water again from the air on cooling. The amorphous oxide hydrate precipitated from cold solutions of aluminum salts also finds technical applications [19].

Bauxite is used for the production of bauxite bricks, as well as for the extraction of aluminum. These are obtained by firing a mixture of bauxite and clay.

(v) *Aluminates*. It has already been mentioned that aluminates are formed by dissolution of aluminum or aluminum hydroxide in caustic alkalis. In these aluminum functions as an acidic element. Werner suggested that the hydrated aluminates contain the anion $[Al(OH)_4]^-$, $[Al(OH)_5]^=$ and $[Al(OH)_6]^\equiv$, so that they should be regarded as *hydroxoaluminates*—e.g.,

$$Al(OH)_3 \cdot KOH = K[Al(OH)_4]$$

$$Al(OH)_3 \cdot Ba(OH)_2 = Ba[Al(OH)_5]$$

$$2Al(OH)_3 \cdot 3Ca(OH)_2 = Ca_3[Al(OH)_6]_2$$

This assumption has been substantiated for the salts $Ca_3[Al(OH)_6]_2$ and $Sr_3[Al(OH)_6]_2$ by the determination of their crystal structure (Brandenberger, 1933).

In addition to these true hydroxoaluminates, aluminum hydroxide can combine with certain hydroxides, especially $Mg(OH)_2$ and $Ca(OH)_2$, to form what are really double hydroxides, built up from alternating layers of $Al(OH)_3$ and $Mg(OH)_2$ or $Ca(OH)_2$ layers ('Double layer lattices', cf. Vol. II) Thus by addition of sodium hydroxide to mixed solutions of aluminum and magnesium chlorides, precipitates are obtained which, in their fine structure, are essentially similar to the green cobalt(II, III) hydroxide (Vol. II). The OH^- ions in these double layer lattices may be partially replaced by Cl^-, and H_2O molecules can be in part replaced by $Al(OH)_3$. The composition of these compounds can thus vary (Feitknecht, 1942). The naturally occurring *hydrocalumite* $2Ca(OH)_2 \cdot Al(OH)_3 \cdot 2\frac{1}{2}H_2O$ was also found by Megaw (1934) to have a double layer lattice structure. No compounds of this kind are formed with $Sr(OH)_2$ and $Ba(OH)_2$.

So-called 'anhydrous aluminates' can be obtained by fusion processes. These correspond principally to the type $M^{II}Al_2O_4$. Compounds of this type occur native in the crystalline state, and are known as *spinels*—e.g., ordinary spinel $MgAl_2O_4$; zinc spinel or *gahnite* $ZnAl_2O_4$; *hercynite*, $FeAl_2O_4$; iron spinel (*pleonast*) $(Mg,Fe)(Al,Fe)_2O_4$; chrome spinel (*picotite*), $(Fe,Mg)(Al,Cr,Fe)_2O_4$. As the formulas show, aluminum may be partly replaced by trivalent iron and chromium.

The determination of their crystal structure by means of X-rays has shown that the spinels are not really aluminates $M^{II}[AlO_2]_2$ at all. They do not contain the group $[AlO_2]^-$ as a structural unit. Each aluminum atom in spinel is surrounded by 6 oxygen atoms at the same distance (2.02Å), and each oxygen atom is equidistant from 3 aluminum atoms. Thus the oxygen atoms cannot be assigned to definite aluminum atoms. The oxygen atoms in spinel and similar compounds occupy positions corresponding closely to those of cubic closest packing. In this arrangement, there are two sorts of positions in the interstices, available for the metallic atoms: one set in which each is surrounded by six oxygen atoms, and another set in which each is surrounded by four oxygen atoms. In the cubic cell of $MgAl_2O_4$, containing 32 O-atoms, 8 Mg-atoms and 16 Al-atoms, there are 16 'octahedral' and 8 'tetrahedral' positions. The magnesium atoms are located in the latter interstices, so that each has four oxygen atoms close to it, at equal distances of 1.75 Å. In so far as the Mg—O distance are shorter than the Al—O distances, we can consider that [MgO$_4$] groups are present in the structure, and formulate the spinel accordingly as Al$_2$[MgO$_4$]. It has been established by X-ray analysis that the structure of spinel is in essence the same as that of salts such as Ag$_2$[MoO$_4$], K$_2$[Zn(CN)$_4$] and the like; this is expressed by the formula Al$_2$[MgO$_4$]. However, the atomic or ionic grouping [MgO$_4$] does not stand out as a clearly defined structural unit of the crystal, and it must not be regarded as a radical or a complex in the chemical sense. On chemical grounds, it is therefore better to speak of spinel not as a salt but as a *double oxide*.

Alumina also gives rise to other double oxides of more complex nature. Thus with CaO it forms the compounds 3CaO·Al$_2$O$_3$, 12CaO·7Al$_2$O$_3$ (or 9CaO·5Al$_2$O$_3$?), CaO·Al$_2$O$_3$, CaO·2Al$_2$O$_3$ and 3CaO·16Al$_2$O$_3$. With BaO, it forms only the compound BaO·Al$_2$O$_3$, but this, and also 3CaO·16Al$_2$O$_3$ and 3SrO·16Al$_2$O$_3$, are all isotypic with the so-called β-alumina (Westgren, 1937).

The crystalline 'anhydrous aluminates' are practically unattacked by acids, whereas the 'hydrated aluminates' are easily decomposed by acids. If alkali metal aluminate solutions are diluted, crystalline aluminum hydroxide slowly separates out (von Bonsdorff), and if the decomposition takes place over a period of months, it is possible to obtain microscopically visible crystals (Fricke), showing the diffraction pattern of hydrargillite.

(b) Halides of Aluminum

(i) *Aluminum fluoride and fluoroaluminates*. Aluminum fluoride, AlF$_3$, is formed by passing hydrogen fluoride at a red heat over aluminum (1) or alumina (2). It can also be prepared by fusion of cryolite with aluminum sulfate (3), and leaching out with water the sodium sulfate formed.

$$Al + 3HF = AlF_3 + \tfrac{3}{2}H_2 \tag{1}$$

$$Al_2O_3 + 6HF = 2AlF_3 + 3H_2O \tag{2}$$

$$2Na_3AlF_6 + Al_2(SO_4)_3 = 4AlF_3 + 3Na_2SO_4 \tag{3}$$

The anhydrous fluoride so obtained is a white powder, of density 3.10, and is insoluble in water, acids and caustic alkalis. It has some technical importance, being used as an addition to the melt in the electrolytic extraction of aluminum.

Complex salts are formed by the combination of alkali metal fluorides and other fluorides with aluminum fluoride—the *fluoroaluminates*. These are white crystalline powders, insoluble in water. Three main types of fluoroaluminates are known— M^{I}AlF$_4$, M$^I_2$AlF$_5$ and M$^I_3$AlF$_6$. The last corresponds to the industrially important mineral *cryolite*. X-ray analysis has shown that in the fluoroaluminates, and also in AlF$_3$ itself, aluminum always has the coordination number 6. In AlF$_3$ each F is shared between two AlF$_6$ octahedra. In the compound TlAlF$_4$, Brosset (1937) found that 4 of the six F that build up each octahedron are shared with neighboring AlF$_6$ octa-

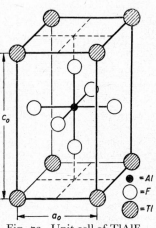

Fig. 73. Unit cell of TlAlF$_4$ (tetragonal, $a_0 = 3.61$ Å, $c_0 = 6.37$ Å).

● = Al
○ = F
◑ = Tl

hedra (see Fig. 73). In Tl_2AlF_5 there are chains of octahedra, formed by the sharing of a diagonally opposite pair of F atoms in each octahedron with its neighbors. Free $[AlF_6]^{3-}$ octahedral groups are present in the fluoroaluminates of the type M_3AlF_6. The compound *chiolite*, with the unusual composition $Na_5Al_3F_{14}$ has a structure corresponding to that of the tetrafluoroaluminates $MAlF_4$. It is derived from this by replacement of one AlF_6 octahedron in every four by a Na^+ ion.

Cryolite, Na_3AlF_6, occurs native almost exclusively, in very large quantities, in western Greenland. In the pure state, it forms snow-white crystals belonging to the monoclinic system, but with a pseudo-cubic habit and an excellent cleavage along cube faces. The density is 2.95, and hardness $2\frac{1}{2}$—3. It is relatively easily fusible (m.p. 1000°). It is resistant towards acids, but is readily decomposed by boiling with caustic alkalis or milk of lime. It was formerly used, accordingly, for the preparation of pure aluminum oxide and soda, and considerable quantities of cryolite are used for this purpose in the United States (Pennsylvania). Cryolite is also now prepared artificially, by dissolving alumina and soda in aqueous hydrofluoric acid, or by Loesekann's process. In the latter, potassium fluoride is first formed by melting calcium fluoride together with potassium sulfate and carbon (4). By double decomposition with sodium sulfate, this is transformed into sodium fluoride, and potassium sulfate is recovered (5). Cryolite is obtained from the sodium fluoride by double decomposition with aluminum sulfate (6).

$$CaF_2 + K_2SO_4 + 4C = CaS + 2KF + 4CO \tag{4}$$

$$2KF + Na_2SO_4 = 2NaF + K_2SO_4 \tag{5}$$

$$12NaF + Al_2(SO_4)_3 = 3Na_2SO_4 + 2Na_3AlF_6 \tag{6}$$

Cryolite is used as an opacifier for milk glass and enamel, as well as in the electrolytic preparation of aluminum.

(ii) *Aluminum hydroxyfluoride*, which may have a composition varying between $AlF(OH)_2$ and $AlF_2(OH)$, is formed from mixed solutions of aluminum fluoride and sulfate by the addition of ammonia. In the cubic crystal lattice of the compound (which contains 16 molecules of the hydroxyfluoride and 6 molecules of H_2O), each Al^{3+} ion is surrounded octahedrally by six F^- or OH^- ions. The crystal lattice is remarkably stable, and persists unchanged even after heating to 500°, although a considerable proportion of the water present in the form of OH^- ions, as well as the crystal water, is then driven off (Cowley and Scott, *J. Am. Chem. Soc.*, 70 (1948) 105).

(iii) *Aluminum monofluoride*. Unlike the other aluminum halides, aluminum fluoride is very involatile. Its volatility is considerably increased however, when it is mixed with aluminum powder. As was shown by Klemm (*Z. anorg. Chem.*, 251 (1943) 233), this is due to the formation of aluminum monofluoride, AlF, which can, however, exist only in the gaseous state; it decomposes into the metal and AlF_3 in the process of condensation. In a similar manner (by heating Al_2O_3 mixed with boron), Zintl (1940) was able to volatilize aluminum as the suboxide AlO or Al_2O (cf. Klemm, *Z. anorg. Chem.*, 255 (1948) 292).

(iv) *Aluminum chloride*, $AlCl_3$, is obtained anhydrous by heating aluminum in a stream of chlorine or hydrogen chloride; by passing chlorine over a heated mixture of aluminum oxide and carbon; or by treating aluminum oxide with a mixture of chlorine and carbon monoxide. It is a colorless crystalline mass (or yellowish, if ferric chloride is present as an impurity), of density 2.44; it is perceptibly volatile at ordinary temperature, and sublimes at 183°. The melting point (192.6°) can be reached only under pressure. Near its sublimation point, aluminum chloride is bimolecular in the vapor state, but is completely dissociated into $AlCl_3$ molecules at about 800°. The dimeric molecule Al_2Cl_6 has been found from electron diffraction measurements (Pauling, Balmer and Elliot, *J. Am. Chem. Soc.*, 60 (1938) 1852) to have the structure shown in Fig. 74. The molecules Al_2Br_6 and Al_2I_6 are similarly constituted. Aluminum chloride is soluble in almost all organic solvents. The molecular weight in ethereal solution has been

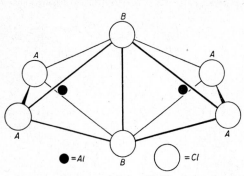

Fig. 74. Structure of the Al_2Cl_6 molecule.

found to correspond to the formula $AlCl_3$. Aluminum chloride fumes strongly in air, being hydrolyzed by atmospheric moisture: $AlCl_3 + 3HOH = Al(OH)_3 + 3HCl$.

It is very soluble in water. The aqueous solution has a strongly acid reaction because of hydrolysis, which takes place to the extent of 2% in 0.1-N solutions, and about 4.5% in 0.001 molar solutions. Soluble basic salts may exist as intermediate products in the hydrolysis reaction. From aqueous solutions (most simply obtained by dissolving the metal or the oxide in hydrochloric acid), the salt crystallizes upon evaporation as the hexahydrate, $AlCl_3·6H_2O$, in colorless deliquescent crystals. If ammonia gas is passed over anhydrous aluminum chloride, the hexammoniate, $AlCl_3·6NH_3$, is formed. This is a white powder which begins to lose ammonia at about 180°. Anhydrous aluminum chloride also forms addition compounds with many organic substances—e.g., with acid chlorides, ethers, esters, etc.; it generally adds on 1 molecule of these substances per $AlCl_3$. These compounds are at once decomposed by water. The same is true of the addition compounds of $AlCl_3$ with other substances, such as H_2S, SO_2, SCl_4, PCl_3, and also of the *complex salts* with alkali chlorides. These have compositions corresponding to those of the complex fluorides, but are less stable.

Monomeric $AlCl_3$ exerts a very strong polarizing action (its dipole moment is about $5.3 · 10^{-18}$ c.g.s. unit). This accounts for its strong tendency to form addition compounds, and also underlies its catalytic action, since it strongly modifies the binding forces within the molecules with which it forms addition compounds.

Anhydrous aluminum chloride is used in preparative organic chemistry for Friedel-Crafts reactions—i.e., for the introduction of alkyl radicals, originally linked with halogen, into the benzene nucleus, whereby hydrogen halide is simultaneously eliminated (e.g., $CH_3Cl + C_6H_6 \rightarrow C_6H_5·CH_3 + HCl$). It is also used as a condensation agent, as a halogen carrier, and as catalyst for various reactions such as dehydrogenations and fissions. In recent times it has been manufactured industrially on a considerable scale, and used, for example, in the dyestuff and essence industries, and for the cracking of petroleum. [*17*]

The decomposition potential of aluminum chloride dissolved in molten sodium chloride is —1.31 volts at 800° according to Neumann, and increases numerically by 1.132 millivolts for every degree decrease in temperature. Corresponding figures for $LaCl_3$ are —1.62 volts and 1.882 millivolts per degree.

Pure molten aluminum chloride is a very poor conductor of the electric current. According to Biltz (1923), solid crystalline aluminum chloride is a better conductor; if this is raised to its melting point, its conductivity instantly vanishes. This is very striking, since, as a rule, molten salts are better conductors than the crystalline materials. Biltz assumes, therefore, that solid aluminum chloride has a different structure from the liquid, in that the former is composed of ions, the latter from molecules. This would account also for the abnormally great increase in volume (almost twofold) which takes place when aluminum chloride is melted.

If aluminum is boiled with a quantity of dilute hydrochloric acid insufficient for its complete dissolution, it goes into solution as a *basic chloride*, $Al_2(OH)_5Cl$. This can be thrown out of solution by the addition of NaCl, or may be converted to the basic sulfate, $Al_4(OH)_{10}SO_4$, by double decomposition with Na_2SO_4 (Tenk and Bauer, 1951). Feitknecht (1950) has prepared another basic chloride, in the form of the double compound $2Ca(OH)_2 \cdot Al(OH)_2Cl \cdot 2H_2O$.

(v) *Aluminum monochloride*, AlCl. Metallic aluminum vaporizes in aluminum chloride vapor at high temperatures, as a consequence of the formation of aluminum monochloride: $AlCl_3 + 2Al \rightleftharpoons 3AlCl$. Aluminum monochloride is capable of existence only in the gaseous state, and reverts quantitatively to metallic aluminum and the trichloride on cooling. It has been proposed to refine aluminum, or to recover the metal from scrap, by means of this reaction. The heat of formation of gaseous AlCl from its elements is $+11.6$ kcal per mol. (Gross, 1946). When aluminum trichloride vapor is brought into equilibrium with the metal at $1400°K$, about 13% of the $AlCl_3$ undergoes reaction to form AlCl, if the total pressure is maintained at 1 atmosphere (Russell, Martin and Cochran, 1951).

(vi) *Aluminum bromide*, $AlBr_3$, colorless, brilliant leaflets, density 3.20, melting point 97.5, boiling point 265°, decomposes with effervescence when brought into contact with a little water, and burns with a flame in oxygen. With much water, it forms a hydrate, $AlBr_3 \cdot 6H_2O$. The complex bromides known so far correspond to the types $M^I[AlBr_4]$ and $M^I[Al_2Br_7]$.

(vii) *Aluminum iodide*, AlI_3, density 3.95, melting point 191°, boiling point 382°, is very similar to the bromide. Complex iodides of the type $M^I[AlI_4]$ are known.

(c) Organoaluminum Compounds [18]

When mercury alkyls are allowed to react with metallic aluminum at 100°, reaction occurs with the formation of aluminum alkyls—e.g.,

$$2Al + 3Hg(CH_3)_2 = 2Al(CH_3)_3 + 3Hg$$

The aluminum alkyls are colorless liquids; aluminum methyl boils at 130°, aluminum ethyl at 194°. They are spontaneously inflammable in air, and are decomposed vigorously by water into aluminum hydroxide and the corresponding hydrocarbon. Freezing point measurements, and vapor density determinations, have shown that the alkyls are associated to a considerable extent to dimeric molecules Al_2R_6, both in solution and in the vapor state in the neighborhood of the boiling point.

The aluminum alkyls can be used for the step by step synthesis of high-molecular weight paraffins and olefins from ethylene, and for the preparation of benzene derivatives from ethylene (Ziegler, *Angew. Chem.*, 64 (1952) 323), and they have therefore become of some technical importance. For their technical preparation, the most important reactions are

$$Mg_3Al_2 + 6C_2H_5Cl = 2Al(C_2H_5)_3 + 3MgCl_2,$$

$$AlCl_3 + 3NaH + 3C_2H_4 = Al(C_2H_5)_3 + 3NaCl,$$

$$2Al + 3C_2H_5Cl + 3NaH + 3C_2H_4 = 2Al(C_2H_5)_3 + 3NaCl$$

The last two reactions do not take place directly, but can be brought about indirectly—e.g., by the process

$$AlCl_3 + 2Al(C_2H_5)_3 \quad = 3(C_2H_5)_2AlCl,$$

$$3(C_2H_5)_2AlCl + 3NaH = 3(C_2H_5)_2AlH + 3NaCl,$$

$$3(C_2H_5)_2AlH + 3C_2H_4 = 3Al(C_2H_5)_3$$

Thus 2 mols of $Al(C_2H_5)_3$ are indirectly converted to 3 mols, so that by repetition of the reaction cycle, indefinite quantities of $AlCl_3$, NaH and C_2H_4 can be converted into aluminum triethyl.

The alkyl groups in aluminum alkyls may be partially replaced by halogen atoms, and it was shown by Brockway and Davidson (1941) that in $Al_2(CH_3)_4Cl_2$ and $Al_2(CH_3)_4Br_2$, as in Al_2Cl_6, the Al-atoms are linked by a 'bridge' of two halogen atoms. The structure of $Al_2(CH_3)_6$ is not yet fully understood. Brockway assumed that the two Al-atoms were directly linked, as are the Si-atoms in $Si_2(CH_3)_6$, but Rundle (*J. Am. Chem. Soc.*, 69 (1947) 1327) considers that the electron diffraction measurements would be equally in agreement with a structure in which the two Al atoms were linked through two methyl groups (cf. Pitzer and Gutowski, *J. Am. Chem. Soc.*, 68 (1946) 2204). A direct Al—Al bond would be difficult to understand on the basis of any valence theory. The heat of dissociation of $Al_2(CH_3)_6$ is about 20 kcal per mol. The interatomic distances are as given in Table 67, where M = Al or Si, A = halogen atom or methyl group at end of molecule, B = halogen atom or methyl group acting as bridging group.

TABLE 67

INTERATOMIC DISTANCES

Compound	$M \leftrightarrow A$	$M \leftrightarrow B$	$A \leftrightarrow B$	$M \leftrightarrow M$	$\angle B$-Al-B	$\angle A$-Al-A
Al_2Cl_6	2.06 Å	2.21 A	3.56 Å	3.41 Å	80°	118°
Al_2Br_6	2.21 ,,	2.33 ,,	3.78 ,,	3.39 ,,	87°	115°
Al_2I_6	2.53 ,,	2.58 ,,	4.24 ,,	3.24 ,,		
$Al_2(CH_3)_4Cl_2$	1.85 ,,	2.31 ,,	3.43 ,,		89°	ca. 130°
$Al_2(CH_3)_4Br_2$	1.90 ,,	2.42 ,,	3.59 ,,		90°	ca. 125°
$Al_2(CH_3)_6$	2.01 ,,	2.2 ,,		ca. 3.2 ,,		
$Si_2(CH_3)_6$	1.90 ,,			2.34 ,,		$\angle A$-M-M 109°

The alkyl groups may also be replaced partially or completely by hydrogen. Thus Wiberg (1939—42) obtained the compounds $Al_2H(CH_3)_5$ and $Al_2H_2(CH_3)_4$ by the action of the electric glow discharge on a mixture of aluminum methyl and hydrogen. These compounds are colorless, not very volatile liquids of high viscosity. They are extraordinarily sensitive towards moisture, and ignite in air with explosive violence, burning with a violet flame. They readily form addition compounds with trimethylamine—e.g., $AlH_2(CH_3) \cdot N(CH_3)_3$. From this, by the action of aluminum chloride, the compound $Al_2H_4(CH_3)Cl$ can be obtained. Solid substances are formed as well as the liquid compounds, and from these the addition compound $AlH_3 \cdot 2N(CH_3)_3$ can be isolated by treatment with trimethylamine; when this is heated it loses trimethylamine, and a residue of non-volatile $[AlH_3]_x$ remains.

(d) Aluminum Hydrides

(i) *Aluminum hydride*, $[AlH_3]_x$ is a white amorphous mass, which is undoubtedly highly polymerized and which begins to decompose above 105° with loss of hydrogen. It reacts with B_2H_6 according to the equation

$$2AlH_3 + 3B_2H_6 = 2Al(BH_4)_3 \text{ (aluminum borohydride)}$$

The compound so formed, which was also obtained by Schlesinger (1940) by the reaction of $Al(CH_3)_3$ with B_2H_6, is a colorless liquid having a vapor pressure of 119.5 mm at 0°. The constitution of aluminum hydride and the borohydride is not known with certainty. The original suggestion of Wiberg, that they should be regarded as containing AlH radicals which were formally equivalent to Si atoms, is not readily interpreted in valence theory, and it is probable that in these, as in the boron hydrides, there is some kind of hydrogen bridge—e.g.,

It is not clear why aluminum should form the high-molecular weight hydride whereas boron and gallium form simple dimeric hydrides. According to Schlesinger (*J. Am. Chem. Soc.*, 69 (1947) 1199), however, AlH_3 is obtained in ethereal solution, when $AlCl_3$ is added to a solution of $Li[AlH_4]$ in ether, and is slowly deposited from solution as polymerization takes place. Deposition of the polymer may be inhibited, however, by the addition of substances with which AlH_3 forms fairly stable addition compounds—e.g., $N(CH_3)_3$. Stable aluminum hydride solutions may be prepared in this way, and may then be used for hydrogenation reactions.

The polymerization and flocculation of aluminum hydride from its ether solutions can be inhibited by addition of aluminum halides, as well as by trimethylamine. In this case, halogen derivatives of aluminum hydride ('halogeno-alanes') are formed—e.g., $2AlH_3 + AlX_3 = 3AlH_2X$; $AlH_3 + 2AlX_3 = 3AlHX_2$. These halogeno-alanes are obtained in the form of ether addition compounds when the ethereal solutions are evaporated, and can in some cases be distilled in vacuum without decomposition (Wiberg, 1951). Ethereal solutions of halogeno-alanes can be used for the selective hydrogenation of organic compounds, in the same way as aluminum hydride or lithium aluminum hydride (see below).

The addition compound $AlH_3 \cdot 2NR_3$ (R = CH_3) is monomeric in benzene solution, whereas $AlH_3 \cdot NR_3$ is dimeric. Both compounds are monomeric in ether solution. Wiberg has concluded that dimerization involves the formation of hydrogen bridge compounds, as with B_2H_6, but that ether inhibits, or at least retards, their formation by being itself coordinated:

This would explain why AlH_3 can exist as the monomer in ether solution:

The $Al \leftarrow OR_2$ bond, being weak by comparison with the $Al \leftarrow NR_3$ bond, is gradually repalced by hydrogen bridge bonds, thereby leading to the deposition of solid aluminum hydride. Whereas in boron hydride the hydrogen bridging leads only to the dimer, in aluminum hydride the higher coordinative valence of aluminum results in the creation of an indefinite network of Al atoms linked through H-bridges, and therefore to a solid product.

(ii) *Complex aluminum hydrides* ('*alanates*'). Aluminum hydride, like boron hydride, can form complex hydrides, for which Wiberg has proposed the name of *alanates*. The most important of these is lithium aluminum hydride (lithium alanate), $LiAlH_4$, because of the ease with which it is prepared and its versatile possibilities as a hydrogenating agent in organic chemistry.

The alanates, like the metallic borohydrides (boranates), fall roughly into the two classes of *salt-like* and *non-salt-like* compounds. Lithium aluminum hydride, $Li[AlH_4]$, is salt-like, whereas the compounds $MgH_2 \cdot 2AlH_3$ and $BeH_2 \cdot 2AlH_3$ (Wiberg and Bauer, 1950—51) are examples of the covalent type. Alanates of Ag, Ga and In are also known.

(iii) *Lithium aluminum hydride*, $LiAlH_4$, was first prepared by Schlesinger by the reaction of $AlCl_3$ with LiH, in ether solution:

$$4LiH + AlCl_3 = Li[AlH_4] + 3LiCl$$

Wiberg (1952) states that it is advantageous to use $AlBr_3$ instead of $AlCl_3$, since LiBr is soluble in ether, and the reaction of the LiH is therefore not blocked by the deposition of LiCl. Lithium aluminum hydride is a white solid, which decomposes into LiH, Al and H_2 when heated in high vacuum to 125°. It is soluble in ether; in 0.08 molar solution it is dimeric in solution, and in 0.8 molar solution it is trimeric. Metallic dust catalytically decomposes the dissolved lithium aluminum hydride, which is therefore metastable in solution.

Lithium aluminum hydride has proved to be a valuable reagent in inorganic chemistry for the preparation of hydrogen compounds which are difficult to obtain in a pure state by other means. Chlorides and alkyls of other elements are converted to the corresponding hydrides. Thus BCl_3 is quantitatively converted to B_2H_6, $SiCl_4$ to SiH_4:—e.g., $4BCl_3 +$ $+ 3LiAlH_4 = 2B_2H_6 + 3LiCl + 3AlCl_3$. Si_2Cl_6 correspondingly gives rise to Si_2H_6, and silicon alkyl chlorides furnish alkyl silanes. With alkyl tin chlorides, alkylated stannanes are formed (e.g., $(CH_3)_2SnH_2$ from $(CH_3)_2SnCl_2$). Germanium and tin hydrides can also be obtained in relatively good yield by the reaction of $GeCl_4$ and $SnCl_4$ with $LiAlH_4$. BeH_2, MgH_2, AlH_3 and the double hydrides derived from them can also most conveniently be prepared by the use of $LiAlH_4$. The corresponding deuterium compound, $LiAlD_4$, can often be used with advantage for the preparation of other deuterium compounds—e.g., $LiAlD_4 + 4Cl_2 = LiCl + AlCl_3 + 4DCl$; $LiAlD_4 + SiCl_4 = LiCl + AlCl_3 + SiD_4$.

In *organic chemistry*, lithium aluminum hydride has recently found very extensive uses as a *hydrogenating agent*. It converts not only aldehydes and ketones, but also the carboxylic acids and their derivatives, directly into alcohols. Alkyl halides yield hydrocarbons, and acid amides and nitriles are converted to primary amines. Aliphatic nitro compounds are converted to amines and aromatic nitro compounds to azo compounds. On the other hand, alcohols and ethers are usually not attacked by lithium aluminum hydride, whereas double or triple bonds are hydrogenated only if they are polar or at least polarizable. This makes valuable *selective hydrogenations* possible. Hydrogenations with lithium aluminum hydride all follow the general scheme of attachment of hydrogen to the carbon atom, and of the electronegative hetero-atom of the double bond to the aluminum. Thus $LiAlH_4 + 4O=CR_2$

$= Li[Al(—O—\overset{\displaystyle R}{\underset{\displaystyle H}{C}}—R)_4]$. Hydrolysis of the intermediate finally gives the desired product.—

$Li[Al(—O—CR_2H)_4] + 4HOH = 4HO—CR_2H + Li[Al(OH)_4]$. The $C=C$ double bond can be hydrogenated by lithium aluminum hydride in certain cases. Thus Ziegler (1952) found that ethylene, $CH_2=CH_2$, added smoothly to $LiAlH_4$ at temperatures a little above 100°, with the formation of $LiAl(C_2H_5)_4$; hydrolysis of the latter produces ethane, C_2H_6. Other olefins with terminal double bonds react in the same way as ethane.

$Mg(AlH_4)_2$ may be obtained by the reaction of $LiAlH_4$ with $MgBr_2$ in ether solution:

2LiAlH$_4$ + MgBr$_2$ = 2LiBr + Mg(AlH$_4$)$_2$. The compound is known only in ether solution, but can be used as a hydrogenating agent in organic chemistry, in the same way as LiAlH$_4$ (Wiberg, *Z. Naturforsch.*, 5b (1950) 397).

(e) Aluminum Sulfate. Alums

(i) *Aluminum sulfate*, Al$_2$(SO$_4$)$_3$ is a white powder, of density 2.71 when anhydrous. It crystallizes in hydrated form from aqueous solution, with 18 molecules of water of crystallization at ordinary temperature. The hydrate Al$_2$(SO$_4$)$_3$·18H$_2$O forms colorless acicular crystals, density 1.62, with a sour astringent taste. The solubility in water is:

at 0°	10°	20°	30°	50°	100° C
31.2	33.5	36.2	40.5	52.2	89.1 g of Al$_2$(SO$_4$)$_3$ in 100 g of H$_2$O

The solution has an acid reaction, because of partial hydrolysis:

$$Al_2(SO_4)_3 + 6HOH \rightleftharpoons 2Al(OH)_3 + 3H_2SO_4$$

The hydrolysis products of aluminum sulfate may also be isolated in the crystallized state. Thus the basic sulfate (AlO)$_2$SO$_4$·9H$_2$O occurs native as *aluminite* (*websterite*). Conversely, the acid salt AlH(SO$_4$)$_2$·1½H$_2$O crystallizes from solutions containing a great excess of acid.

From warm solutions of aluminum sulfate, the 16-hydrate usually crystallizes (rhombic prisms). In addition, hydrates with 27, 10 and 6 H$_2$O exist. Dehydration is complete if the hydrates are heated above 340°. Loss of SO$_3$ commences at about 600° (Krauss, 1927).

Aluminum sulfate finds widespread industrial uses, especially in paper making. It is used in the so-called sizing of the paper, being added to the paper pulp, mixed with sodium resinate. The aluminum resinate formed by double decomposition cements the paper fibers together. For this application of aluminum sulfate it is important that it should be free from iron salts. This applies even more to its use in tanning skins (white tanning), and in dyeing, where it is used as a mordant. Aluminum sulfate is also used as the starting material in the preparation of other aluminum salts, which can conveniently be prepared from it by double decomposition with the corresponding lead salts.

The use of aluminum sulfate as a mordant depends on the fact that aluminum hydroxide, formed hydrolytically in aqueous solution in the utmost fineness of subdivision, is picked up on and retained by the wool fiber. The aluminum hydroxide, in turn can adsorb organic dyestuffs (forming so-called *lakes*). Other salts with a strong tendency for hydrolysis, such as chromic sulfate and stannic chloride, have a similar action. The wool fibers, previously treated with such salts ('mordanted'), can thus be dyed with substances which will not ordinarily take on the fiber, because the adhering metal hydroxide can be dyed. Cotton, unlike wool, is unable to take up aluminum hydroxide directly from the (hot)solution. For cotton, therefore, a precipitate of aluminum hydroxide is produced *within* the fibers, by impregnating then first with a solution of aluminum sulfate, and then subjecting them to the action of alkali (e.g., soda).

Aluminum sulfate is best prepared technically by dissolution of pure (iron-free) aluminum hydroxide in hot concentrated sulfuric acid. Bauxite or clay can be directly dissolved by treatment with sulfuric acid, but the difficulty then arises of freeing the aluminum sulfate from iron in a simple manner.

(ii) *Alums*. Compounds of aluminum sulfate with alkali metal sulfates, with the general formula M^{I}Al(SO$_4$)$_2$·12H$_2$O are known as *alums*. They usually crystallize in regular octahedra, or in certain circumstances in cubes. They belong to the class of *double salts* or *mixed salts*. The ammonium radical, or univalent thallium can also serve as the univalent metal, and in place of aluminum the alums can contain other tervalent metals, such as iron or chromium. The general formula of the alums is thus $M^I M^{III} (SO_4)_2 \cdot 12H_2O$.

Crystal structure of the alums. The atoms of the univalent and tervalent metals together form a lattice of rock-salt type (edge of the unit cell $a = 12.13$ Å in the case of potassium aluminum alum). In each of the 8 small cubes into which the unit cell of the NaCl structure may be divided, a sulfur atom is situated on the body-diagonal, and is surrounded tetrahedrally by 4 oxygen atoms. 6 molecules of water are grouped around each metal atom—both the univalent and the tervalent. Thus the water of crystallization is not coordinated only around the atom of the tervalent metal, in the form of double molecules, $(H_2O)_2$, as Werner assumed.

Ordinary *alum*, after which the group of salts is named, is potassium aluminum sulfate, $KAl(SO_4)_2 \cdot 12H_2O$. It forms colorless octahedra (or cubes), with a sour astringent taste, and density 1.75. It melts at 92.5° in its own water of crystallization, and may be easily and completely dehydrated by further heating ('burned alum'). Unlike aluminum sulfate, which has a positive heat of solution even in the hydrated form, alum dissolves in water with the absorption of heat (—10.12 kcal. per mol of $KAl(SO_4)_2 \cdot 12H_2O$). The solubility in water is quite small at low temperatures, but increases considerably with rise of temperature. Values are:

at 0°   15°   30°   60°   92.5°   100° C
   2.95   5.04   8.40   24.8   119.5   154 g of $KAl(SO_4)_2 \cdot 12H_2O$ in 100 g of H_2O

Alum is insoluble in absolute alcohol.

It is now generally prepared from clay, which is first dehydrated and then brought into solution by treatment with hot concentrated sulfuric acid. The sulfuric acid solution is concentrated by evaporation, and after addition of potassium sulfate (or other potassium salts, such as the chloride, if sufficient sulfuric acid is present), alum crystallizes out on cooling. To avoid inclusions of mother liquor, the solution is stirred, so that a fine-grained powder (alum meal) is obtained.

Alum was formerly extracted from the naturally occurring basic double sulfate *alunite* (alum stone), $K(AlO)_3(SO_4)_2 \cdot 3H_2O$, or else from *alum shale*, a mixture of clay or marl with pyrites, FeS_2. The roasting of alum shale furnished a product from which, after long standing and repeated roasting, alum could be deposited by addition of potassium sulfate.

Alum is used industrially for the same purposes as is the simple aluminum sulfate, but is being increasingly replaced by the latter. Where, however, particular importance is laid upon freedom from iron salts (e.g., for use as a mordant and in tanning), alum is still often preferred. Alum is used in medicine as an astringent and mild caustic, and for many other purposes.

(f) Other Aluminum Compounds

(i) *Aluminum nitrate*, $Al(NO_3)_3$, is prepared by dissolution of aluminum hydroxide in nitric acid, or by double decomposition between aluminum sulfate or alum and lead nitrate. It is used as a mordant for alizarin red in dyeing, and in the gas mantle industry. It is considerably hydrolyzed in aqueous solution. The hydrated salt can be obtained in colorless, hygroscopic crystals by evaporation of strougly acid solutions. The solubility is 63.7 g of $Al(NO_3)_3$ in 100 g of water at 25°.

(ii) *Aluminum thiocyanate*, $Al(SCN)_3$, formed by reaction of a solution of aluminum sulfate with barium or calcium thiocyanate, is used as a mordant in alizarin printing. For this purpose it is sold commercially as the aqueous solution. By evaporating this *in vacuo*, the anhydrous salt can be obtained as a gummy mass. Any trace of iron impurity reveals itself by a red coloration (formed by the ferric thiocyanate reaction).

(iii) *Aluminum acetate*, $Al(C_2H_3O_2)_3$, can be obtained in aqueous solution by double decomposition of aluminum sulfate with barium or lead acetate, or by dissolving aluminum hydroxide in acetic acid. It is much more extensively hydrolyzed than the other aluminum salts discussed. A precipitate of aluminum oxide hydrate, or of basic aluminum acetate, therefore slowly deposits from the aqueous solution, and the basic salt remains when any attempt is made to evaporate the solution, acetic acid being lost. The electrical conductivity of the neutral solution is strikingly small, being actually less than that of the equivalent amount of acetic acid in highly dilute solutions. Aluminum acetate is used as a mordant in cotton dyeing. If fabrics soaked in the solution are hung up in a warm, moist place, the acetate decomposes through evaporation of the acetic acid, so that the hydroxide remains in the fiber. The solutions of aluminum acetate much used in medicine (e.g., as disinfectant and antiseptic) contain basic aluminum acetate.

(iv) *Aluminum sulfide*, Al_2S_3, can be obtained by the direct union of the components at a high temperature. [Mixtures of aluminum powder and sulfur attain a very high temperature when reaction is initiated, and are sometimes used as 'boosters' in thermite reactions]. The crystalline sulfide has a density of 2.37 and a melting point of 1100°. It is completely hydrolyzed by water, and therefore cannot be obtained from aqueous solution.

Al_2S_3 and Al_2Se_3 are isotypic with Ga_2S_3 and Ga_2Se_3 (Geiersberger, 1952).

The double sulfides $CuAlS_2$ and $AgAlS_2$ are more stable than Al_2S_3 itself. They crystallize with the *chalcopyrite* structure. The structure of chalcopyrite, $CuFeS_2$, resembles that of zinc blende (cf. Vol. II): every S atom is surrounded tetrahedrally by 4 metal atoms, and every metal atom by 4S atoms. The chalcopyrite structure differs from the zinc blende structure in that it contains two sorts of metal atoms, and that it is slightly compressed along the *c*-axis, so that the resulting structure has tetragonal symmetry. The compounds $CuAlSe_2$, $AgAlSe_2$, $CuAlTe_2$ and $AgAlTe_2$ have the same structure. These were obtained by Hahn (1951) by heating Al_2Se_3 and Al_2Te_3 with the copper(I) and silver chalcogenides, in sealed silica tubes. Gallium and indium form similar double chalcogenides $M^IM^{III}X_2$ (X = S, Se, Te) with the chalcopyrite structure. With thallium, Hahn obtained only the double sulfide $CuTlS_2$, also with chalcopyrite structure. Some of these double chalcogenides have a non-Daltonide character. The structure of $AgInS_2$ changes at high temperatures from the chalcopyrite type to a structure similar to that of wurtzite (cf. p. 258).

(v) *Aluminum nitride*, AlN, is formed when alumina is heated with coke and nitrogen in an electric furnace:

$$Al_2O_3 + 3C + N_2 = 2AlN + 3CO$$

It is decomposed, forming sodium aluminate and ammonia, when heated with caustic soda in pressure vessels. The attempt has been made to utilize these reactions for the production of ammonia from atmospheric nitrogen (Serpek process), but the process has not achieved any importance. Aluminum nitride decomposes below its melting point when it is heated under ordinary pressure. It melts at 2200° under 4 atmospheres pressure.

The crystal structure of AlN is like that of BeO (wurtzite type), $a = 3.11$Å, $c = 4.98$Å, Al—N $= 1.87$ Å.

(vi) *Aluminum carbide*, Al_4C_3, is formed by the direct union of the constituents at high temperatures. It is soluble in molten aluminum, and separates out on cooling in the form of golden leaflets. Al_4C_3 vaporizes above 2200°, with partial decomposition, the vapor consisting of a mixture of Al_4C_3 and Al. The total vapor pressure reaches 1 atm. at about 2400°; if decomposition of Al_4C_3 could be repressed, its vapor pressure would be 1 atm. at about 2800°. For the behavior of Al_4C_3 towards water, cf. p. 467.

7. Analytical (Aluminum)

In the usual course of the separation of the cations, aluminum is precipitated as the oxide hydrate in the ammonium sulfide group.

The precipitation of aluminum oxide hydrate depends on the fact that the hydrolysis equilibrium of aluminum salts,

$$2AlX_3 + 3H_2O + aq. \rightleftharpoons Al_2O_3 \cdot aq. + 6H^+ + 6X^-,$$

is pushed completely from left to right, because the concentration of H^+ ions is reduced to a minimal value through the formation of H_2S:

$$2H^+ + S^= = H_2S.$$

All substances which sufficiently decrease the hydrogen ion concentration therefore act in a similar way—e.g., sodium thiosulfate, iodide-iodate mixtures (cf. p. 813), barium carbonate as well, naturally, as alkali metal carbonates, ammonium carbonate, and ammonia.

Aluminum salts at first give a precipitate with caustic alkalis, but this redissolves in an excess of precipitant, through the formation of aluminates.

The aluminum salts of volatile acids are converted to aluminum oxide on strong ignition. If the oxide is moistened with a little cobalt nitrate solution, and once more ignited, a fine blue color is obtained (Thenard's blue).

Conversion to cesium alum, $CsAl(SO_4)_2 \cdot 12H_2O$ provides a convenient means of identifying aluminum. The reaction is sufficiently sensitive to permit of the detection of small amounts of the metal.

Traces of aluminum can be detected in solution through the green fluorescence which appears on addition of an alcoholic solution of morin. This highly sensitive reaction depends on the formation of a compound analogous to the color lakes, between aluminum oxide hydrate present in the solution through hydrolysis, and the morin.*

In *quantitative* analysis, aluminum is generally precipitated as oxide hydrate and weighed as the oxide Al_2O_3. Since γ-Al_2O_3 is hygroscopic, it is best converted to the non-hygroscopic α-Al_2O_3 by igniting it above 1000°.

If aluminum oxide hydrate is precipitated by means of ammonia, it is necessary to maintain a specified hydrogen ion concentration (pH between 4 and 7), since the oxide hydrate is perceptibly soluble both in more strongly alkaline and in more strongly acid solutions. According to Fricke (Z. *anorg. Chem.*, 188 (1930) 127), aluminum can be conveniently and accurately determined by precipitation from alkaline solutions by means of CO_2.

Aluminum may be determined *volumetrically* (also in presence of iron) by precipitation of the *arsenate* $AlAsO_4$, and iodometric titration of the arsenic acid (Daubner, *Angew. Chem.*, 48 (1935) 589; 49, (1936) 137.

Small amounts of aluminum may be determined *colorimetrically* by means of the reaction with eriochromcyanine (Alten, *Angew. Chem.*, 48 (1935) 273).

The method of G. Jander (Z. *anorg. Chem.*, 199 (1931) 48) is suitable for the determination of the oxide content of metallic aluminum (an analysis which is important technically). It involves passage of dry hydrogen chloride over the metal at about 200°, whereby $AlCl_3$ and $SiCl_4$ volatilize and leave behind the Al_2O_3 (together with $FeCl_2$).

* Morin is a yellow dyestuff—a tetraoxy-flavonol, $C_{15}H_6O_3(OH)_4$—contained in fustic, (the wood of Morus tinctoria), and used as a yellow dye for wool.

8. Gallium (Ga), Indium (In), Thallium (Tl)

(a) General

Gallium, indium, and thallium differ both in their physical properties and their chemical behavior from their lighter homologues to a much more marked extent than is found with the corresponding elements of the Ist and IInd Main Groups. The reason for this can be understood from their atomic structure. Gallium, indium, and thallium are structurally analogous to boron and aluminum in the sense that their outermost electron shells are similar in configuration (with two s-electrons and one p-electron). They differ, however, in that the s^2p shell in gallium, indium and thallium immediately follows a d^{10} shell (cf. Table II, Appendix), instead of the s^2p^6 shell ('inert gas' configuration) found in boron and aluminum. This difference shows itself in the properties both of the free elements and of their compounds.

Gallium, indium and thallium, differ from aluminum, e.g., in their low melting points, and in their extreme softness. Whereas the hardness and the melting point fall sharply from boron, through aluminum to gallium, they increase again from gallium to thallium. In Main Groups I and II, by contrast, both hardness and melting point diminish continuously from the lightest element to the heaviest. Gallium, indium, and thallium differ very markedly indeed from boron and aluminum in the ease with which they are reduced to lower valence states. This property is most strongly developed in the case of thallium, which is *univalent* in its most important compounds. The weaker electropositive character of gallium, indium, and thallium, as compared with aluminum, has already been noted in the introduction to this chapter.

Of these three elements, the one most nearly related to aluminum is gallium, its immediate neighbor in the Periodic System. This shows itself, for example, in that gallium (but not indium or thallium) almost invariably accompanies aluminum as a minor consitutent of its minerals, although it is only present in traces on account of its rarity. Gallium also resembles aluminum in that its sesquisulfide is hydrolyzed by water, so that it cannot be precipitated from aqueous solution, whereas the sulfides of indium and thallium can be precipitated from solution by hydrogen sulfide, although not by ammonium sulfide. Unlike the corresponding oxides of indium and thallium, gallium(III) oxide can crystallize with the corundum structure. In this respect also, gallium is more closely related to aluminum than are the other two metals.

The properties of the oxides, sulfides and other chalcogenides of gallium, indium and thallium are now well known as a result of modern work. These metals form chalcogenides of the general compositions M_2X and M_2X_3, and in some cases those of the type MX.

Chalcogenides of the composition MX have been reported for all three metals. In so far as their magnetic properties have been investigated (Klemm and von Vogel, 1934), these compounds have proved to be diamagnetic. This implies that the metals are not present in them in the electropositive bivalent state, since this would involve the presence of an unpaired s-electron which would confer paramagnetic properties. The diamagnetism might be explained by the pairing of these electrons—i.e., by the formation of a covalence between the metallic ions, such as is found in the Hg_2^{2+} ion. However, it is more likely that these chalcogenides can be regarded as double compounds of the type $M^IM^{III}X_2$. This has been confirmed by structure determinations in the cases of the compounds TlS and TlSe; the hypothesis also appears to be confirmed by evidence based on consideration of atomic volumes for the chalcogenides of Ga and In (Hahn and Klingler, Z. anorg. Chem., 260 (1949) 110).

Little is as yet known of the structures of the chalcogenides M_2X, derived from the metals in the univalent state. Only the crystal structure of Tl_2S is known; this forms a distorted layer lattice of the brucite type. All three metals form *oxides, sulfides, selenides* and *tellurides* of the composition M_2X, with the one exception that Ga_2Te does not exist.

Of the chalcogenides of the type M_2X_3, only the *oxides* are known for all three metals. The corresponding *sulfides, selenides* and *tellurides* are formed only by Ga and In. They are known in the case of Tl only in the form of certain double compounds, such as $K[TlS_2]$, $Tl[TlS_2]$ and $Tl[TlSe_2]$.

Ga_2O_3 exists in several modifications (cf. p. 351). The usual form (α-Ga_2O_3) is rhombohedral, and isotypic with corundum. In_2O_3 and Tl_2O_3 crystallize with the cubic Sc_2O_3 ('rare earth type C') structure. This can be derived from the crystal lattice of CaF_2 (cf. Fig. 62, p. 265) by taking out one quarter of the anions, in an ordered manner, and making certain small displacements of the remaining ions. The change-over from corundum type to Sc_2O_3 type is a result of the increase in radius of the cations.

Among the *sulfides* M_2S_3, Ga_2S_3 is dimorphous like ZnS, the two forms being derived from the wurtzite and the zinc blende types, respectively. In_2S_3 is also dimorphous, forming structures related to that of spinel. All these substances have structures based upon a cubical closest packing of S atoms, with the metallic atoms contained in the interstices of the lattice. The Ga atoms distribute themselves over the *tetrahedral* holes. These are too small for the atoms of In, and in In_2S_3 the metal atoms are present mostly in *octahedral* holes, although some tetrahedral positions are also occupied.

Of the *selenides* and *tellurides*, Ga_2Se_3, Ga_2Te_3, and In_2Te_3 also have crystal structures derived from that of zinc blende. In_2Se_3 has a unique structure, since even with Se atoms, the tetrahedral positions in an assembly with cubical closest packing are too small to contain the large In atoms. In the corresponding packing of Te atoms, the space is sufficient. Hence In_2Te_3 is isotypic with Ga_2Se_3 and Ge_2Te_3, but In_2Se_3 is not.

The sulfides Ga_2S_3 and In_2S_3 are related in structure and properties to the sulfides of zinc and cadmium. They are therefore associated with these elements in nature, as trace constituents of naturally occurring sulfide minerals. There are thus certain relationships between the homologues of aluminum in Group III and the elements of Sub-group II, which immediately precede them in the sequence of atomic numbers. This relationship is exemplified by other properties. Thus thallium resembles mercury in forming a very large number of intermetallic compounds with the alkali and alkaline earth metals.

Gallium, indium and thallium resemble the alkali and alkaline earth metals, in that when they are introduced into the flame of the Bunsen burner, in the form of sufficiently volatile compounds, they produce flame colorations as a result of the excitation of characteristic spectral lines.

Gallium and its homologues are therefore usually detected, in the course of analysis, by observation of the spectra excited by introducing the oxides, moistened with hydrochloric acid, into the Bunsen flame, or (better) by striking a small arc between a platinum or iridium point and a solution of the chloride. The spectra obtained by these methods differ considerably from those which are characteristic of the free atoms, as emitted by an arc or spark struck between electrodes of the respective metals. These latter are much richer in lines than the chloride spectra usually observed by the analyst, which contain only a few lines, listed in Table 68.

TABLE 68

CHLORIDE SPECTRA OF THE HEAVY ELEMENTS OF MAIN GROUP III

Gallium	$\lambda = 417.1$ mμ and 403.1 mμ (both violet)
Indium	$\lambda = 451.1$ mμ (indigo blue), 410.1 mμ (violet)
Thallium	$\lambda = 535.1$ mμ (green)

Only a weak indication of the line at 417.1 mμ is obtained when gallium salts are introduced into the Bunsen flame, but both the lines given in Table 68 appear very clearly if a spark or arc is struck between a platinum wire and the solution. The green thallium line almost coincides with the barium band at 534.7 mμ, but may be clearly distinguished by its sharpness and intensity.

(b) Crystal Structure

Gallium has a unique rhombic structure, in which the atoms are associated in pairs (Ga↔Ga = 2.437 A). *Indium* is isotypic with γ-manganese (tetragonal face-centered lattice; each indium is surrounded by four neighbors at 3.24 Å, and eight others at 3.37 Å). *Thallium* exists in two modifications; the form stable at ordinary temperature (a-Tl) has a close packed hexagonal structure (Mg type), $a = 3.45$ Å, $c = 5.52$ Å. The form stable at high temperatures, which can be obtained in the metastable state at room temperature by quenching (β-Tl) has the face-centered cubic structure, $a = 4.84$ Å.

9. Gallium (Ga)

(a) Occurrence

Gallium is found in nature associated with zinc in many blendes, although invariably in extremely small quantities (0.002% or less). In traces, it also almost invariably accompanies aluminum, and can be detected spectrographically in commercial aluminum from all sources. The mineral with the highest known gallium content is the *germanite* from Tsumeb, in South West Africa, which contains 0.6—0.7% gallium (Pugh and Sebba, 1937).

(b) History

Lecoq de Boisbaudran discovered gallium spectrographically, in 1875, in a zinc blende from Pierrefitte in France. He named the element after his native land (Gallia).

Mendeléeff, in 1871, had already predicted the existence of an as yet unknown homologue of aluminum ('eka-aluminum'), to account for a corresponding gap in the Periodic Table. Gallium proved to be identical with eka-aluminum.

(c) Preparation

The discoverer and his pupil, Jungfleisch, worked out various methods for the isolation of gallium. It is usually most convenient, on the laboratory scale, to precipitate gallium as its ferrocyanide, which is converted by ignition to a mixture of Ga_2O_3 and Fe_2O_3. The mixture of oxides is fused with potassium hydrogen sulfate, and iron is removed by precipitation from a hydrochloric acid solution with a large excess of potassium hydroxide solution, in which gallium oxide is soluble. Gallium can then be electrolytically deposited from the alkaline solution. Keil (1926) recommended fusing the ferrocyanide with caustic potash in a silver crucible. The melt was dissolved in water (whereby iron remained as the hydroxide) and gallium was precipitated with ammonia, as hydroxide, after acidification with hydrochloric acid. The hydroxide can be ignited to the oxide, which furnishes the metal by reduction with hydrogen. Richards (1923) described a method for working up the lead which remains as a residue when zinc (from ores containing gallium) is refined by distillation. Gallium can be isolated simply and quantitatively from germanite by a method described by Berg and Keil (1932), which depends upon the solubility of gallium(III) chloride in ether (cf. p. 371).

Gallium has been produced industrially for a number of years at Leopoldshall, from the residues obtained in smelting the Mansfeld copper shales. According to Feit (1933), the residues containing gallium, which also contain aluminum and the heavy metals, chiefly in the form of sulfates, phosphates and molybdates, are first treated with sodium hydroxide. After filtration of the undissolved heavy metal hydroxides, the solution is neutralized. The precipitate thrown down at this stage consists essentially of gallium, aluminum and tin, in the form of their phosphates and sulfates. Some enrichment of gallium is first effected by repeated dissolution in sulfuric acid and fractional precipitation by the addition of water. Such molybdenum and tin as remain are then removed from the sulfuric acid solution by precipitation with hydrogen sulfide. The solution is then treated with solid sodium hydroxide, whereby phosphoric acid is separated in the form of trisodium phosphate, which is almost insoluble in caustic soda. The solution is finally electrolyzed. With careful control of the operations, the process furnishes practically pure gallium.

(d) Properties

Gallium [22] is a rather soft, ductile metal, with a white luster tinged with blue-grey. It melts at 29.78° (heat of fusion 19.16 cal. per g). Molten gallium does not

TABLE 69
PROPERTIES OF GALLIUM, COMPARED WITH THOSE PREDICTED
FOR EKA-ALUMINUM

Eka-aluminum	Gallium
Atomic weight about 68	Atomic weight 69.72
Density about 6.0	Density 5.9
Atomic volume about 11.5	Atomic volume 11.8
Melting point of metal low	Melting point of metal 29.78°
Stable in air	Stable in air
Able to form alums	Gallium can form alums
The chloride should be more volatile than zinc chloride.	Gallium(III) chloride boils at 201°; $ZnCl_2$ boils at 730°
The metal is easily obtained by reduction	Metallic gallium can be prepared by heating the oxide in hydrogen, or from aqueous solution by electrolysis.
The sulfide is not precipitated by hydrogen sulfide	In the absence of other metals, gallium is not precipitated by hydrogen sulfide. It is almost quantitatively carried down, however, by a number of other metallic sulfides when they are precipitated from alkaline or acetic acid solution.

at once solidify when it is cooled, but may remain for months in a supercooled state unless seeded with particles of the solid metal.

It can be employed as a thermometric liquid in silica-glass thermometers, for use at high temperatures.*

Some properties of gallium are compared in Table 69 with those predicted by Mendeléeff for eka-aluminum, on the basis of the Periodic System.

Gallium is stable in air at ordinary temperature, but is vigorously attacked by chlorine and bromine in the cold. It combines with iodine when heated. For the crystal structure, see p. 369.

Gallium is readily deposited electrolytically from aqueous solution, although it is difficult to make the deposition quantitative. The normal potential of gallium (for $Ga = Ga^{3+} + 3e$), relative to the normal hydrogen electrode, is $+0.52$ volts.

10. Compounds of Gallium

Gallium is ordinarily positive and trivalent in its compounds. The salts are colorless, and are even more strongly hydrolyzed in solution than the salts of aluminum, which they closely resemble in properties.

Gallium, like aluminum is precipitated as the white oxide hydrate from solutions of its salts by any substances which disturb the hydrolysis equilibrium by reducing the hydrogen ion concentration. Tartaric acid hinders the precipitation, by forming complexes with gallium, as it does with aluminum.

For *bivalent* gallium, the halides $GaCl_2$ and $GaBr_2$, and the chalcogenides GaS, GaSe, and GaTe are known, but gallium(II) oxide has not yet been prepared. This is remarkable, since the other gallium(II) chalcogenides are relatively stable. Gallium(II) compounds of salt-like type are not very stable. They readily undergo oxidation, or disproportionation to give gallium(III) compounds and gallium metal.

Gallium(I) compounds are still less stable, and in only a few cases has their existence been definitely proved. The sulfide Ga_2S (Brukl and Ortner, 1930) and the selenide Ga_2Se (Klemm and von Vogel, 1934) exist, but may be metastable at ordinary temperature.

Gallium(III) and gallium(I) compounds are diamagnetic, as also, rather remarkably, are the well defined gallium(II) compounds. The diamagnetism of the latter compounds must imply that they are not built up from Ga^{2+} ions, but either involve Ga—Ga bonds or contain Ga^+ and Ga^{+++} ions in random distribution.

The existence of a volatile hydride of gallium is notable.

(a) Gallium(III) Compounds

(i) *Gallium(III) chloride*, $GaCl_3$, forms long white crystals ($d = 2.47$). It may be obtained by heating gallium in a stream of chlorine or of hydrogen chloride. It melts at 77.9° and boils at 201.3° (heat of evaporation at the boiling point, 11.4 kcal per mol, heat of fusion 5.2 kcal per mol, heat of sublimation 17.0 kcal per mol at the m.p.). It can be purified by redistillation in an atmosphere of nitrogen or chlorine. Vapor density determinations have shown that in the neighborhood of the boiling point, the vapor consists of double molecules, Ga_2Cl_6 (cf. Al_2Cl_6). At 600° the vapor density corresponds to that required by the simple formula $GaCl_3$, and at higher temperatures the incipient dissociation of the $GaCl_3$ molecules becomes evident. Fused gallium chloride is practically a non-conductor of electricity. The heat of formation of $GaCl_3$ is 125 kcal per mol. Gallium(III) chloride combines very exothermically with water. It fumes in moist air, since it forms hydrogen chloride by hydrolysis. Its aqueous solution is acidic in reaction, and readily deposits the hydroxide.

Gallium(III) chloride is very soluble in ether, without any decomposition. It can be extracted by means of ether from its solutions in aqueous (e.g. $6N$) hydrochloric acid, and this property can be used to effect a separation of gallium from other elements (Swift, 1924).

Ulich (1941) found that $GaCl_3$, like $AlCl_3$, acts as a catalyst in certain synthetic organic reactions. It is, indeed, a more efficient catalyst than $AlCl_3$ for some hydrocarbon syntheses.

---

* For the use of gallium in dental fillings, see Vol. II.

(ii) The *bromide*, $GaBr_3$, ($d = 3.69$, m.p. $121.5°$, b.p. $279°$, heat of formation 92.4 kcal per mol) and the *iodide*, GaI_3 ($d = 4.15$, m.p. $212°$, b.p. $346°$, heat of formation 51 kcal per mol) resemble the chloride in properties. Electron diffraction measurements show that both the chloride and bromide exist as dimeric molecules in the vapor state, with bond distances being $Ga—Cl = 2.22$ Å and $Ga—Br = 2.38$ Å. GaI_3 is only slightly associated ($Ga—I = 2.49$ Å). The anhydrous *fluoride*, GaF_3 ($d = 4.47$, sublimes $950°$) is only very slightly soluble in water or dilute hydrochloric acid. The hydrate, $GaF_3 \cdot 3H_2O$, obtained by dissolution of gallium(III) oxide hydrate in aqueous hydrofluoric acid, is quite soluble in dilute hydrochloric acid, and the gallium must therefore be complexed. Addition of NH_4F to the solution yields octahedral crystals of $(NH_4)_3[GaF_6]$. Pugh (1937) has prepared a series of other fluorogallates. Most of these are of the type $M^I_2[GaF_5(H_2O)]$.

(iii) *Gallium(III) oxide*, Ga_2O_3, is best obtained by heating the nitrate or sulfate. It is a white powder, which like aluminum oxide, loses its solubility in acids and caustic alkalis when it is strongly ignited. It is reduced to the metal when it is heated in hydrogen; Ga_2O may be formed as an intermediate stage.

Gallium(III) oxide is polymorphic, like alumina. The modification stable below about $600°$ (α-Ga_2O_3) has the corundum structure; the form stable at high temperatures (β-Ga_2O_3) is either rhombic or monoclinic. Ignition of gallium nitrate at low temperatures yields a modification (δ-Ga_2O_3) with the same structure as In_2O_3, Tl_2O_3 and the rare earth oxides (Mn_2O_3 or 'rare earth type C' structure, cubic). The relation between these modifications, as given by Roy, Hill and Osborn (*J. Am. Chem. Soc.*, **74** (1952) 719) is as follows:

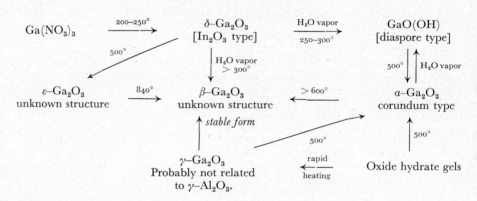

Gallium(III) oxide also forms modifications related to the so-called β-alumina, in which complex oxide structures are stabilized by the presence of a small proportion of alkali or alkaline earth oxide. (Foster and Stumpf, *J. Am. Chem. Soc.*, **73** (1951) 1590). Ga_2O_3 also forms double oxides of spinel structure with MgO and ZnO.

(iv) *Gallium oxide hydrate and gallium hydroxide.* Substances which lower the hydrogen ion concentration throw down a white gelatinous precipitate from Ga(III)salt solutions. This precipitate is amorphous to X-rays and has a variable water content (gallium oxide hydrate, $Ga_2O_3 \cdot xH_2O$). It dissolves both in acids and in strong bases, and differs from aluminum oxide hydrate in being markedly soluble in strong ammonia solutions. It is therefore more acidic than aluminum oxide hydrate. As the precipitate ages, its solubility in caustic alkalis diminishes. The solubility also depends on the quantity of solid phase present—an indication that gallium oxide goes into solution to some extent in colloidal dispersion, as well as in molecular dispersion (Ostwald's rule, Vol. II, Chap. 16).

A microcrystalline gallium(III) hydroxide of definite composition can be obtained by very slow precipitation from both alkaline and acidic solutions. This is *gallium metahydroxide*, GaO(OH), which was shown by Boehm (1938) to have the α-AlO(OH) or diaspore structure. The same compound is formed by hydration of α- and δ-Ga_2O_3 with steam under certain conditions (see above).

Alkaline solutions of gallium hydroxide contain salts known as *gallates*. Gallium hydroxide appears to have stronger acidic and weaker basic properties than aluminum hydroxide. A preparative separation of gallium from aluminum can be based on this gradation of properties (p. 370).

(v) *Gallium(III) sulfide*, Ga_2S_3, was first isolated by Brukl (1930), by direct union of the elements*. Reaction proceeds to completion only at high temperatures (1200°). Gallium(III) sulfide is yellow, with a m.p. of about 1225° and $d = 3.48$. It is decomposed by water, with evolution of H_2S.

Gallium(III) sulfide is reduced by hydrogen at 800° to gallium(II) sulfide, GaS**. This is a yellow solid, with a m.p. of about 965° and $d = 3.75$. It is stable towards water, but decomposes when it is heated in a high vacuum: $4GaS = Ga_2S_3 + Ga_2S$. The gallium(I) sulfide so formed is a grey black sublimate ($d = 4.22$), which decomposes into $Ga_2S_3 + Ga$ when it is heated again.

(vi) *Gallium sulfate*, $Ga_2(SO_4)_3$, crystallizes as an 18-hydrate from its solutions (cf. $Al_2(SO_4)_3 \cdot 18H_2O$), as white lamellae or star-shaped clusters. It is very soluble in water. It can be dehydrated by heating, and decomposes above 520° with loss of SO_3. It appears to be perceptibly volatile at high temperatures, and therefore loses weight when it is fumed down with sulfuric acid. Marchal (1926) found the dissociation pressure of SO_3 to be 391 mm at 700°, but the total vapor pressure to be about 1 atm.

With ammonium sulfate, gallium sulfate forms the double salt $(NH_4)Ga(SO_4)_2 \cdot 12H_2O$, *ammonium gallium alum*. The double selenate $CsGa(SeO_4)_2 \cdot 12H_2O$, also isomorphous with the alums, has been prepared by Dennis (1918). Among the *basic double sulfates*, the compound $(NH_4)(GaO)_3(SO_4)_2 \cdot 4H_2O$ is isotypic with the mineral alunite, $K(AlO)_3(SO_4)_2 \cdot 3H_2O$.

(vii) *Gallium nitrate* crystallizes from nitric acid solutions as the octahydrate, $Ga(NO_3)_3 \cdot 8H_2O$, in colorless, highly refractive deliquescent prisms (m.p. ca 65°). The hydrate is converted to the anhydrous salt, $Ga(NO_3)_3$, when it is warmed to about 40° in dry air.

(viii) *Gallium nitride*, GaN, is obtained as a dark grey powder ($d = 6.10$) by heating the metal in ammonia at 900—1000°. It is unattacked by most acids, but is slowly converted to Ga_2O_3 when it is heated in air (W. C. Johnson, 1932) GaN and also InN (p. 377) have the hexagonal wurtzite structure ($a = 3.18$, $c = 5.17$Å for GaN).

(b) Gallium(II) and Gallium(I) Compounds

(i) *Gallium(II) chloride*, $GaCl_2$, is formed by the incomplete combustion of gallium in chlorine, or by heating gallium(III) chloride with metallic gallium. It forms colorless transparent crystals (m.p. 170.5°, b.p. about 535°). The melt supercools very readily. Unlike gallium(III) chloride, the molten dichloride has a good electrical conductivity. At 1000°, the vapor density of gallium(II) chloride corresponds to the molecular complexity $GaCl_2$. (Nilson and Pettersson, 1888). At lower temperatures, polymeric molecules are also present (Laubengayer and Schirmer, 1940), and at higher temperatures some dissociation occurs.

Gallium(II) chloride dissolves in benzene. It reacts vigorously with water, hydrogen being evolved. Evolution of hydrogen is also observed on dilution of a solution of gallium in a small quantity of concentrated hydrochloric acid. It is probable that such solutions contain gallium(II) chloride.

Gallium(II) bromide resembles the chloride.

(ii) The existence of *gallium(I) oxide* is not certain. Dupré (1878) claimed to have prepared it in the impure state by heating Ga_2O_3 to redness in hydrogen. Brukl (1931) considered

---

* Lecoq de Boisbaudran, in 1881, observed the formation of a white precipitate when H_2S was passed through a weakly acid solution of a gallium salt, and supposed that gallium sulfide had been formed. According to Brukl, however, the precipitate must probably have been gallium oxide hydrate, formed by hydrolysis. Gallium(III) sulphide is not stable in the presence of water, and could not be precipitated under the conditions used.

** This compound is better obtained synthetically, by combination of the calculated amounts of the elements (Klemm, 1934).

that it was formed by heating an intimate mixture of gallium and gallium(III) oxide. It is described as a dark brown powder (d = 4.77), which sublimes in high vacuum above 500° and decomposes into Ga_2O_3 + Ga above 700°. However, X-ray studies and measurements of the heat of oxidation of Ga metal and of the supposed Ga_2O did not unambiguously confirm the existence of the lower oxide (Klemm and Schnick, 1936).

(c) Gallium Alkyls and Gallium Hydride

(i) *Gallium triethyl*, $Ga(C_2H_5)_3$, was obtained by Dennis (1932) by the action of metallic gallium on mercury diethyl: $3Hg(C_2H_5)_2 + 2Ga = 2Ga(C_2H_5)_3 + 3Hg$. It is a colorless, unpleasant smelling liquid (d = 1.058, m.p. —82.3°, b.p. 142.8°). It inflames in air, and is decomposed by water with explosive violence. At low temperatures it reacts with water giving successively the bases $(C_2H_5)_2GaOH$ and $C_2H_5Ga(OH)_2$:—e.g., $(C_2H_5)_3Ga + H_2O = C_2H_6 + (C_2H_5)_2GaOH$. Addition compounds with ammonia and ether, $(C_2H_5)_3Ga·NH_3$ and $(C_2H_5)_3Ga·O(C_2H_5)_2$, are more stable than gallium triethyl itself.

(ii) *Gallium trimethyl*, $Ga(CH_3)_3$, (m.p. —15.8°) can be obtained in the same way as the triethyl.

(iii) *Gallium hydride*. Gallium trimethyl reacts with hydrogen, under the influence of the electrical glow discharge, to form *tetramethyl digallane*, $Ga_2H_2(CH_3)_4$, a colorless, viscous liquid. When this is heated above 130°, it decomposes into $Ga(CH_3)_3$, Ga and H_2. Acting on the hypothesis that this reaction involved a disproportionation into $Ga(CH_3)_3$ and Ga_2H_6, Wiberg and Johansen (1941) found that the addition of trimethylamine, with which gallium trimethyl forms the very stable compound $(CH_3)_3Ga·N(CH_3)_3$, brought about reaction at ordinary temperature. Gallium hydride could then be isolated without decomposition. It is a colorless, mobile liquid (m.p. —21.4°, b.p. 139°), which decomposes fairly rapidly above 130° into Ga and H_2. The molecular weight corresponds to the formula Ga_2H_6 (*digallane*), corresponding in complexity and structure (hydrogen bridged structure) to diborane.

Reaction of gallium trimethyl with B_2H_6 at —45° produces *dimethylgallium borohydride*, $(CH_3)_2Ga(BH_4)$ (m.p. 1.5°, b.p. 92°).

Wiberg and Schmidt (1952) found that the action of an excess of lithium hydride on $GaCl_3$ in ether solution formed lithium gallium hydride, or *lithium gallanate*, $LiGaH_4$:— $GaCl_3 + 4LiH = 3LiCl + Li[GaH_4]$. Lithium gallium hydride reacts with $GaCl_3$ in ether to give an ether solution of gallium hydride: $GaCl_3 + 3Li[GaH_4] = 3LiCl + 4GaH_3$. As in the corresponding reaction of $Li[AlH_4]$, the product is initially monomeric in ether, but is slowly converted to a white solid polymeric hydride, $[GaH_3]_n$, which decomposes slowly above 140°.

The first product of decomposition seems to be a lower hydride $[GaH]_x$ (?), since formation of metallic gallium is not observed below 380°. The solid gallium hydride is fairly stable towards water, but is vigorously decomposed by dilute acids.

The reaction of lithium gallium hydride with $AgClO_4$ in ether yielded the very unstable *silver gallanate*, $AgGaH_4$, which decomposes above —75°, with deposition of silver. *Thallium gallanate*, $Tl(GaH_4)_3$, is still less stable; it is formed as an insoluble white solid by reaction of $TlCl_3$ with $LiGaH_4$ in ether at —115°, and decomposes into thallium, hydrogen and gallium hydride at —90°.

11. Indium (In)

(a) Occurrence and History

Indium occurs in minute traces, in the form of its sulfide, as an impurity in many zinc blendes—especially in the dark-colored zinc blendes, containing iron, which are considered to have been formed at relatively high temperatures. It is also frequently present in stannite and other complex sulfides of tin, and traces occur associated with tin in many cassiterites.

Indium was discovered in 1863 by Reich and Richter in the residues from Freiberg zinc blende, which they were examining spectroscopically for the presence of thallium. The new element was revealed by, and named after, its characteristic line in the indigo-blue of the spectrum.

Indium was originally believed to be a *bivalent* element. Mendeléeff, however, assigned it to its proper place in the Periodic Table from a consideration of its properties, and maintained that it must be trivalent. This conclusion was promptly confirmed by a determination of its specific heat (which leads directly to the atomic weight, from the Dulong-Petit law, and then by comparison with the equivalent weight, to the valence), by its atomic volume, and by the observation that indium forms alums. There can be no doubt as to the correctness of the place assigned to indium in the Periodic System, since its atomic number (49) has been determined from the wave-length of its characteristic X-ray spectrum (cf. Moseley's law).

(b) Preparation

The most convenient raw materials for the extraction of indium are certain intermediate products from the smelting of zinc and lead ores that contain traces of indium*. The procedure adopted for working these up depends upon the nature of the material. Thus, if a zinc of relatively high indium content is available, it is possible to treat this with a quantity of hydrochloric acid insufficient for complete dissolution. The indium then remains undissolved in the residual slime containing the less base impurities. Most of the heavy metals which are likely to be present can be precipitated by means of hydrogen sulfide from an acid solution of the slime. Indium is then precipitated from solution as the oxide hydrate, by means of ammonia. It is usually admixed with iron, which must be previously oxidized to an iron(III) salt; the method chosen to separate indium from iron depends upon the quantity of the latter. No difficulty is encountered in preparing metallic indium, either by reduction of the oxide with hydrogen, or by electrodeposition.

(c) Properties

Indium is a silver-white, lustrous metal. It is very soft—probably the softest of the elements—and can easily be cut with a knife. Its melting point is very low (156.4°), but its boiling point is high (about 2300°) ($d = 7.31$, spec. heat 0.057). For crystal structure see p. 369.

Indium does not tarnish in dry air. When heated, it becomes covered with an oxide film, but begins to oxidize rapidly only above its melting point. It burns vigorously in chlorine, and combines directly with the other halogens and with sulfur. The standard potential of indium (for $In = In^{3+} + 3e$), relative to the normal hydrogen electrode, is +0.34 volts.

12. Compounds of Indium

Indium is trivalent in most of its compounds, but can also function as bivalent and univalent, especially in its compounds with the halogens and chalcogens. A characteristic property of compounds of the lower valence states is that they disproportionate into indium (III) compounds and free metal in aqueous solution.

* In the fractional distillation of crude zinc by the New Jersey process described in Vol. II, Chap. 9, certain residues are obtained in which a relatively high degree of enrichment of indium is achieved. Quite large quantities of indium are now extracted industrially from this source (Ensslin, 1940—41).

Most indium(III) salts are colorless. Salts derived from the common acids (except the oxalate, phosphate and sulfide) are water soluble. They are hydrolyzed to a considerable extent in solution. Oxy-salts salts containing indium in the anion (*indates*) are formed with a large excess of hydroxyl ions. Indium can also form acido-compounds. However, indium forms no complexes with ammonia in aqueous solution, although a pyridine addition compound of indium chloride, $InCl_3.3C_5H_5N$, can be obtained from alcoholic solution (fine white needles, m.p. 253°).

The most important compounds of indium are surveyed in Table 70.

TABLE 70

SUMMARY OF THE COMPOUNDS OF INDIUM

m.p.		m.p.		m.p.	
InF_3	1170° colorless	InF_2	colorless		
$InCl_3$	586° colorless, yellow molten	$InCl_2$ ca. 235°	colorless, yellow-brown molten	$InCl$ 225°	lemon yellow. Deep red above 120°. Dark red molten.
$InBr_3$	436° colorless, light brown molten	$InBr_2$ —	horny, almost colorless, dark yellow molten	$InBr$ —	carmine. Very deep red molten
InI_3	210° yellowish, dark brown molten	InI_2 —	—	InI —	—
In_2O_3	— yellow			In_2O —	black
In_2S_3	1050° yellow and red	InS	692° wine red	In_2S 653°	brown black

Salts e.g., indium nitrate, $In(NO_3)_3 \cdot 4\frac{1}{2}H_2O$; indium sulfate, $In_2(SO_4)_3 \cdot 5H_2O$
Double salts (acidosalts) e.g., $K_3InCl_6 \cdot 1\frac{1}{2}H_2O$ (potassium hexachloroindate (III))
$(NH_4)In(SO_4)_2 \cdot 12H_2O$ (indium ammonium alum).

In general, the indium(I) and indium(II) compounds are more stable than the corresponding compounds of gallium. Indium(II) compounds are diamagnetic, like the indium(I) and indium(III) compounds, and it is likely that they are built up from In^{3+} and In^+ ions, and do not involve truly bivalent indium.

Heats of formation $InCl$ 44.6, $InCl_2$ 86.8, $InCl_3$ 128.5, $InBr_3$ 97.2, InI_3 56.5, In_2O_3 222.5 kcal per mole.

Solubilities in 100 g of H_2O at 22°: InF_3 8.50 g, $InCl_3$ 183 g, $InBr_3$ 536 g, InI_3 1090 g. in 100 g of absolute ethanol: $InCl_3$ 114 g, $InBr_3$ 285 g (Ensslin, 1942).

(a) Indium(III) Compounds

(i) *Indium(III) chloride and chloroindates. Indium trichloride*, $InCl_3$, can be prepared by burning indium metal in chlorine, or by heating a mixture of indium oxide and charcoal in chlorine. It forms white, nacreous crystal leaflets, which sublime readily ($d = 3.46$ at 25°; m.p. 586°). At about 800° the vapor density corresponds to the formula $InCl_3$. Indium chloride dissolves with evolution of much heat. Its solution, which has an unpleasant inky taste, can be prepared directly by dissolving indium in hydrochloric acid. Indium(III) chloride is strongly hydrolyzed in solution (hydrolysis constant about $1.2 \cdot 10^{-5}$). A tetrahydrate, $InCl_3 \cdot 4H_2O$, crystallizes from solution. Double (complex) salts (chloroindates) are formed with the alkali chlorides and with chlorides of organic bases—e.g., $(NH_4)_2InCl_5$·

·H_2O, $K_3InCl_6 \cdot \frac{3}{2}H_2O$, $[CH_3NH_3]_4InCl_7$, $[C_9H_7NH]_4InCl_7$ (quinolinium heptachloroin-date).

(ii) *Indium(III) bromide* ($d = 4.75$) and indium(III) iodide ($d = 4.68$) are very similar to the chloride. There is a red modification of InI_3, as well as the yellow form listed in Table 70.

(iii) *Indium(III) fluoride* is prepared anhydrous by heating In_2O_3 or $(NH_4)_3InF_6$ in hydrogen fluoride (Klemm, 1936) ($d = 4.39$, m.p. 1170°). Unlike gallium(III) fluoride, it is not volatile. It is extremely sparingly soluble in water, whereas the trihydrate $InF_3 \cdot 3H_2O$, which separates from a solution of indium(III) oxide hydrate in hydrofluoric acid, is fairly soluble in water (cf. p. 376). With ammonium fluoride, InF_3 forms the fairly soluble double salt $(NH_4)_3InF_6$. When this is heated it decomposes to form indium nitride, InN, which has the wurtzite structure like GaN ($a = 3.533$Å, $c = 5.693$ Å, Juza, 1938). Sodium fluoroin-date, Na_3InF_6, (solubility 9.1 g in 100 g of H_2O) has also been prepared.

(iv) *Indium(III) oxide*, In_2O_3, is obtained as a yellow powder ($d = 7.28$) by ignition of the sulfate, hydroxide or nitrate. It darkens in color reversibly on heating. It is soluble in acids but insoluble in alkali hydroxides and in ammonia. It becomes crystalline on prolonged strong ignition, and can readily be reduced to the metal by heating in hydrogen or in dry ammonia gas to 200—300°.

Indium(III) oxide has the cubic Mn_2O_3 or 'rare earth type C' structure (cf. p. 387).

In_2O_3 forms double oxides with MgO (spinel type) and with CaO and CdO. The latter also correspond to the spinel formula, and are apparently related structurally to the spinels (Passerini, 1930).

(v) *Indium oxide hydrate*, $In_2O_3 \cdot xH_2O$, is precipitated from indium salt solutions by the addition of alkali hydroxides, ammonia or other substances (e.g. potassium nitrite) which displace the hydrolysis equilibrium by reaction with H_3O^+ ions. It forms a white gelatinous precipitate. The oxide hydrate is insoluble in ammonia solution, but is readily peptized colloidally by ammonia. Peptization can also be brought about by pure water, and this may cause trouble in washing the precipitated oxide hydrate (e.g., in analytical work). The oxide hydrate dissolves easily in acids, and also in excess caustic alkalis, and thus behaves as if it were amphoteric. However, a portion of the dissolved oxide hydrate is reprecipitated from the alkaline solutions after some time, possibly as a result of the formation of crystalline indium hydroxide with a smaller solubility product.

(vi) *Indium nitrate* crystallizes from solutions of the oxide hydrate in a large excess of nitric acid. It forms long colorless prisms or needles, with the composition $In(NO_3)_3 \cdot 4\frac{1}{2}H_2O$. A double salt crystallizes from solutions containing ammonium nitrate.

(vii) *Indium sulfate*. The compound $HIn(SO_4)_2 \cdot 3\frac{1}{2}H_2O$ crystallizes when a solution of indium, indium oxide, or the oxide hydrate in sulfuric acid is concentrated by evaporation. The neutral salt $In_2(SO_4)_3$ (with 6 or $12H_2O$, depending on the temperature) crystallizes from dilute sulfuric acid solutions. Anhydrous indium sulfate, $In_2(SO_4)_3$ is obtained by heating either the neutral hydrate or the acid sulfate to 200°.

Precipitation from concentrated sulfate solutions as the compound $HIn(SO_4)_2 \cdot 3\frac{1}{2}H_2O$ affords a convenient method of freeing indium from impurities, especially from iron and tin (Ensslin). Indium sulfate forms double salts with the alkali metal sulfates. *Ammonium indium sulfate* crystallizes from aqueous solutions above 36° as the tetrahydrate $(NH_4)In(SO_4)_2 \cdot 4H_2O$. Below 36° it forms crystals of the alum type, $(NH_4)In(SO_4)_2 \cdot 12H_2O$. Rubidium and cesium also form alums with indium, but the potassium and sodium double sulfates crystallize only as tetrahydrates.

Indium sulfate also forms double salts with organic-substituted ammonium bases—e·g., $(CH_3NH_3)In(SO_4)_2 \cdot 2H_2O$, $(C_2H_5NH_3)In(SO_4)_2 \cdot 3\frac{1}{2}H_2O$, $(C_6H_5CH_2NH_3)In(SO_4)_2 \cdot 3H_2O$.

(viii) *Indium(III) sulfide*, In_2S_3. Yellow indium sulfide, which is readily soluble in nitric acid, is precipitated when hydrogen sulfide is passed into weakly acid solutions of indium salts. Digestion of this material with dilute acids transforms it into a brick red modification which is less soluble in nitric acid. A cinnabar-red sulfide is obtained when indium and sulfur are melted together. The In_2S_3 precipitated from solution can be dehydrated by heating it in a vacuum, but remains hygroscopic unless it is transformed into the high temperature modification. According to Hahn and Klingler (1949) the two modifications of indium sulfide (a-In_2S_3 and β-In_2S_3) differ in just the same way as do γ'-Al_2O_3 and γ-Al_2O_3. a-In_2S_3

changes monotropically into β-In_2S_3 above 300°. β-In_2S_3 is isotypic with γ-Al_2O_3 (a = 10.72 Å, $d_{pycnom.}$ = 4.63), and is hard and brittle.

When neutral solutions of indium salts are treated with ammonium sulfide or alkali sulfide solution, white precipitates are formed, which may in some circumstances redissolve in excess of precipitant. These are considered to be *thiosalts* of indium. Schneider found that a white precipitate, formed slowly from an aqueous extract of a melt of indium oxide with soda and sulfur, had the composition $NaInS_2 \cdot H_2O$. He also obtained $KInS_2$ (hyacinth-pink crystals). Hahn has shown that indium forms the double sulfides $CuInS_2$, $AgInS_2$, with chalcopyrite structure.

The compounds In_2Se_3 and In_2Te_3 are formed when indium is melted with selenium and tellurium. The selenide has a complex crystal structure, and is soft like graphite (compare the sulfide and telluride). The telluride, which is hard and brittle like the sulfide, has the same crystal structure as Ga_2S_3, Ga_2Se_3 and Ga_2Te_3. This is derived from the structure of zinc blende, by leaving one third of the metal-atom sites vacant, the empty sites being randomly distributed.

(b) Indium(II) and Indium(I) Compounds

(i) *Indium(II) chloride.* When indium is heated in hydrogen chloride to about 200°, $InCl_2$ is formed as an amber liquid which solidifies to colorless crystals (d = 3.64 at 25°, m.p. 235°, b.p. 570°). Above 1300°, the vapor density corresponds to the formula $InCl_2$. The compound is decomposed by water into metallic indium and indium(III) chloride. This disproportionation takes place in two stages: (a) $2InCl_2 = InCl + InCl_3$; (b) $3InCl = 2In + InCl_3$. The indium(I) salt is formed only in very small amounts as an intermediate in the decomposition of the chloride, but when indium(II) bromide (d = 4.22) is decomposed by cold water the reaction comes to a halt at the first stage, and only slowly proceeds further.

Solid $InCl_2$ is isomorphous with $SnCl_2$. It is diamagnetic, whereas it would be expected that a compound containing In^{2+} ions should be paramagnetic. Aiken, Haley and Terrey (1936) therefore suggest that the compound should be formulated as $In^I[In^{III}Cl_4]$ (and stannous chloride similarly as $Sn^{II}[Sn^{IV}Cl_4]$), with In^+ and In^{3+} occupying the cation positions in the crystal lattice. The crystal structure of $InCl_2$ has, however, not been elucidated.

(ii) *Indium(I) chloride*, InCl, was first prepared by Nilsson and Pettersson (1888). It can be prepared by passing $InCl_3$ vapor over heated indium. It forms a blood-red liquid, which solidifies to a red mass (d = 4.18, m.p. 225°). At 1100° the vapor density is a little higher than corresponds to the formula InCl; it is said that it *increases* with further rise of temperature.

Indium(I) chloride is immediately decomposed by water into indium(III) chloride and the free metal.

(iii) *Indium(II) fluoride*, InF_2, was obtained by Klemm (1936) by gently heating InF_3 in hydrogen (reduction to metal occurs if it is more strongly heated). It is colorless, but gradually turns grey in moist air, since water causes decomposition into $InF_3 + 2In$.

(iv) *Indium(II) oxide*, InO. When In_2O_3 is heated in a high vacuum or in reducing gases, it may be turned white superficially. According to Thiel (1928) this is due to the formation of the monoxide, InO. The existence of this compound has not been proved however; Klemm and von Vogel observed only the X-ray diffraction patterns of In_2O_3 and In_2O from the supposed InO.

(v) *Indium(I) oxide*, In_2O, which is volatile at high temperatures, is formed (together, perhaps, with non-volatile InO) when In_2O_3 is heated below 400° in hydrogen or is decomposed thermally in a high vacuum. It is deposited in the cooler parts of the apparatus as a yellow transparent layer when thin, or as a black, brittle deposit (d = 6.31) when thick. It is not attacked by cold water, but dissolves readily in hydrochloric acid, with evolution of hydrogen. It becomes incandescent when heated in air, and is oxidized to In_2O_3 (Thiel, 1928).

(vi) *Indium(II) sulfide*, InS, is best prepared by heating the components in appropriate proportions. Its existence was proved by thermal analysis (Thiel, 1928). It is a wine-red substance ($d = 5.18$), and volatilizes when heated to $850°$ in a high vacuum, being decomposed to In_2S and free sulfur.

(vii) *Indium(I) sulfide*, In_2S, is formed as a black powder when indium(III) sulfide is heated in hydrogen. It may be sublimed at high temperatures in hydrogen, and is deposited in yellow transparent crystals which appear black by reflected light. It is not readily obtained pure by this method, since it is reoxidized to indium(III) sulfide when it is heated in an atmosphere of hydrogen sulfide. The reaction $In_2S_3 + 2H_2 \rightleftharpoons In_2S + 2H_2S$ is thus distinctly reversible. Indium(I) sulfide is best prepared by heating the metal with the requisite quantity of sulfur.

Indium(I) sulfide ($d = 5.90$, m.p. $653°$) sublimes at temperatures above its melting point. In properties and appearance it resembles indium(I) oxide. It is decomposed by dilute acids, with deposition of red In_2S_3.

Indium(I) selenide, In_2Se, and telluride, In_2Te, have also been prepared.

(c) Indium Alkyl Compounds and Indium Hydride

(i) *Indium trimethyl*, $In(CH_3)_3$, was prepared by Dennis (1934) by the same method as was used for gallium triethyl (p. 374). It forms colorless, highly refractive crystals ($d = 1.568$, m.p. $88.4°$, b.p. $135.8°$), readily soluble in organic solvents. Indium trimethyl is unstable in air. It reacts with dry oxygen:

$$2In(CH_3)_3 + \tfrac{1}{2}O_2 = [(CH_3)_2In]_2O + C_2H_6$$

With water, two methyl groups are eliminated as methane [$(CH_3)_3In + 2HOH = CH_3In(OH)_2 + 2CH_4$], and dilute acids remove all three methyl groups. giving an indium (III) salt. In these reactions indium trimethyl differs considerably from the analogous gallium compound. With ether, indium trimethyl forms the addition compound $(CH_3)_3In \cdot O(C_2H_5)_2$ (colorless liquid, $d = 1.241$ at $20°$, m.p. $-15°$, b.p. $139°$).

(ii) *Other indium alkyls*. Indium triethyl, $In(C_2H_5)_3$, ($d = 1.260$ at $20°$, m.p. $-32°$, b.p. $144°$) and *indium tripropyl*, $In(C_3H_7)_3$, ($d = 1.187$ at $20°$, m.p. $-51°$, b.p. $178°$) do not form ether addition compounds which are stable enough to be distilled without decomposition. The only known *aryl* compounds of indium are those with two aryl groups—e.g., *diphenyl indium bromide*, $(C_6H_5)_2InBr$, colorless rhombic prisms.

(iii) *Indium trimethoxide*, $In(OCH_3)_3$, is obtained by double decomposition between $InCl_3$ and sodium methoxide in methyl alcohol solution. It forms colorless crystals, readily soluble in methyl alcohol and in benzene (Runge, *Z. anorg. Chem.*, 267 (1951) 39.

(iv) *Indium hydride*, $[InH_3]_x$, was prepared by Wiberg (1951), by the reaction of indium (III) chloride with lithium hydride in ether solution at ordinary temperature:

$$InCl_3 + 3LiH = InH_3 + 3LiCl$$

It is a white, highly polymerized substance, which is insoluble in ether and is not volatile. Above $80°$ it decomposes into its elements.

By the reaction of $InCl_3$ with $LiAlH_4$, in ether solution at $-70°$, Wiberg obtained indium aluminum hydride (indium alanate):

$$InCl_3 + 3LiAlH_4 = 3LiCl + In(AlH_4)_3$$

This is a white solid, insoluble in ether. Above $-40°$, this complex hydride decomposes into aluminum hydride, indium and hydrogen. It is probable that monomeric InH_3 is formed as a primary reaction product, but decomposes into its elements before it can be stabilized by polymerization at this low temperature. The reaction between $InCl_3$ and $LiAlH_4$ proceeds by way of intermediate compounds in which not all the Cl atoms are replaced by —AlH_4 groups—e.g., dichloroindium aluminum hydride, $InCl_2[AlH_4]$, which is stable up to 100°.

13. Thallium (Tl)

(a) Occurrence

Thallium is one of the elements which is almost invariably found in very low concentration, although its occurrence is widespread. Thus it is commonly associated with zinc, copper and iron in blendes and pyrites. In addition, it can frequently be detected in potash salts and micas. These two distinct fields of occurrence correspond with the two different aspects of its chemical character, whereby on the one hand it belongs to the typical heavy metals, but is closely related also, in many ways, to the alkali metals.

Actual thallium minerals are extraordinarily rare. A thallium arsenic sulfide, $TlAsS_2$ (*lorandite*), has been reported from Macedonia as an overgrowth on realgar. An isomorphous mixture of the selenides of thallium, copper, and silver, relatively rich in thallium, is the *crookesite* of Skrikerum, in Sweden. *Berzelianite*, found at Skrikerum and at Leibach (Ober-Harz), is similar in composition, but poorer in thallium.

(b) History

Thallium was discovered in 1861 by Crookes, in the lead chamber sludge of a sulfuric acid factory in the Harz. It was discovered spectroscopically, and named after the characteristic green line of its spectrum, and its green flame coloration (θάλλος, a green twig).

(c) Preparation

The starting material usually available for the extraction of thallium is the flue dust arising from the roasting of blendes or pyrites containing traces of thallium. These are leached with water, and the thallium is either thrown out of solution by means of metallic zinc, or is precipitated as the chloride by addition of hydrochloric acid. It is purified by reconversion to the sulfate and precipitation (repeated, if necessary) as the chloride. The metal is finally electrodeposited from its sulfuric acid solution.

Commercial thallium commonly contains lead as an impurity. This can be removed by dissolving it in nitric acid, and precipitating the lead by means of hydrogen sulfide from the weakly acid solution.

(d) Properties

A freshly cut surface of thallium is white and lustrous, but at once acquires a grey tarnish in air. It is softer than lead, and its tensile strength is even lower (m.p. 302.5°, heat of fusion 1.03 kcal per g atom, b.p. about 1450°, $d = 11.83$).

Ordinary hexagonal thallium (α–Tl) is converted at 232.2° into a cubic modification (β–Tl); the heat of transformation = 82 cal per g atom. β–thallium can be obtained in the supercooled state at ordinary temperature. It differs very little in density ($d = 11.86$) from α–thallium, since both modifications have close-packed structures.

Thallium can be distilled in hydrogen. It is oxidized superficially in air, forming, thallium(I) and thallium(III) oxides. Ozone and hydrogen peroxide oxidize it to thallium(III) oxide. The metal is attacked only to a very slight extent by air-free water at ordinary temperatures, but it dissolves slowly in alcohol, liberating hydrogen and forming thallium ethoxide. If there is an ample supply of oxygen, no hydrogen is evolved, but reaction follows the equation

$$2Tl + 2C_2H_5OH + \tfrac{1}{2}O_2 = 2C_2H_5OTl + H_2O$$

The thallium ethoxide thus formed is a dense, yellowish oil ($d = 3.52$ at $20°$), which solidifies at $-3°$. It dissolves readily in alcohol and ether, but is decomposed by water. It is also decomposed at temperatures above $130°$. The properties of thallium ethoxide and the other alkoxides $Tl \cdot OR$ (R = alkyl radical) are quite different from those of the salt-like alkali metal alkoxides. They appear to be covalent compounds, and are polymerized (up to $[Tl \cdot OR]_4$) in organic solvents (see Sidgwick and Sutton, *J. Chem. Soc.*, (1930) 1461).

The best solvent for thallium is dilute nitric acid. The metal dissolves less readily in sulfuric acid, and still less readily in hydrochloric acid, because the products formed with these acids are more or less insoluble. The relatively sluggish reaction of the metal with hydrofluoric acid is noteworthy, in view of the solubility of the fluoride and the position of thallium in the electromotive series.

The standard potential of thallium, in contact with Tl^+ ions, is $+ 0.336$ volt, relative to the normal hydrogen electrode. Thallium thus falls between cadmium and cobalt in the electrochemical series. It displaces lead, copper, mercury, silver and gold from solutions of their salts. Thallium(III) salt solutions are unstable in the presence of metallic thallium; the reaction $Tl^{+++} + 2Tl = 3Tl^+$ proceeds almost to completion.

Thallium reacts with the halogens at ordinary temperature. It also reacts with sulfur, selenium, and tellurium when heated, and alloys with arsenic and antimony. Phosphorus appears to react only superficially.

Thallium does not combine with boron or silicon, and it does not form alloys or mixed crystals when melted with these elements.

The direct reaction of thallium with sulfur, selenium and tellurium yields the following compounds, according to Hahn and Klingler (1949): Tl_2S, Tl_4S_3 (or $Tl^I_3Tl^{III}S_3$?), TlS (or $Tl^ITl^{III}S_2$), TlS_2; Tl_2Se, $TlSe$ (or $Tl^ITl^{III}Se_2$); Tl_2Te, $TlTe$. Tl_2S forms a distorted brucite type layer lattice structure. $TlSe$ appears from the structure analysis to be a double compound of univalent and trivalent thallium. The same is true of TlS, and probably of Tl_4S_3. $TlTe$ has a more complicated structure. Its composition can vary over a wide range of homogeneity, and it appears to be related rather to the intermetallic compounds.

Thallium does not react with molecular hydrogen, nitrogen or carbon dioxide. It is also insoluble in liquid ammonia.

Thallium and its compounds have found only limited applications. It is occasionally employed in the laboratory as a source of monochromatic green light. For the therapeutic uses of thallium see p. 382.

14. Compounds of Thallium

Thallium exhibits the $+1$ and $+3$ valence states, the compounds of univalent thallium being the more stable. Thallium(III) compounds are easily reduced to thallium(I) compounds, and are therefore fairly strong oxidizing agents.

Compounds of univalent thallium display many similarities to those of the alkalis. Thus the hydroxide is very soluble, and is a strong base. The carbonate, which is also soluble, resembles soda and potash. Like the alkali metals, univalent thallium forms many well crystallized salts. These are mostly colorless, and volatile when heated. Many of the salts crystallize anhydrous, like the salts of the heavier alkali metals. Thallium(I) salts derived from weak acids have a basic reaction in aqueous solution, as a result of hydrolysis.

Many of the salts of univalent thallium fit into the sequence which gives progressively less soluble salts in passing from potassium to cesium (e.g., the nitrates, sulfates and hexachloroplatinates). Like the alkali metals, univalent thallium forms well crystallized salts with complex anions, and also forms polysulfides (e.g., Tl_2S_5) and polyiodides (e.g., TlI_3).

In other respects, however, univalent thallium is closely related to silver—e.g., in the color and the insolubility of its halides, chromate and sulfide, and in the color of the oxide. The thallium(I) ion shares with the silver ion the peculiarity of forming yellow precipitates with the ions of many weak acids—e.g., with nitrous and orthoarsenic acids—but differs from the silver ion in that it does not form ammine complexes in aqueous solution. In this respect, thallium is more like lead, its neighbor in the Periodic Table. It resembles lead in other ways also, especially in the metallic state.

The electrolytic mobility of the Tl^+ ion at 18° is 66.0—i.e., only a little less than that of the potassium ion (p. 157).

Many compounds of trivalent thallium have a strong tendency for complex formation (e.g., the chloride). Those that do not do so are extensively hydrolyzed in aqueous solution—e.g., the nitrate and sulfate, which gradually deposit brown thallium(III) oxide hydrate when the solid salts stand in moist air.

Whereas thallium(I) salts impart a *basic* reaction to their solutions if they are hydrolyzed, the hydrolysis of thallium(III) salts confers an *acidic* reaction to the solutions.

Complex salts of trivalent thallium are mostly of the types $M^I[TlX_4]$ and $M_3[TlX_6]$.

Thallium readily gives rise to compounds containing the element in an apparent intermediate valence—i.e. compounds containing both thallium(I) and thallium(III). These are often notable for their insolubility and intensity of color, as compared with the simpler compounds.

The potential of the Tl^+ — Tl^{+++} electrode is — 1.25 volts. The oxidation potential is naturally raised by the presence of substances which form complexes with the Tl^{+++} ion. Hence oxidation of thallium(I) compounds to thallium(III) compounds is most readily effected in the presence of complexing agents.

Thallium compounds are *rather highly toxic* [23]. They not only cause damage to the nervous system and digestive organs, but cause the hair to fall out. They are therefore used as depilatories in diseases of the scalp. They occasionally find other uses in medicine.

Thallium(III) compounds are readily reduced to thallium(I) compounds in solution, (e.g., by sulfurous acid). Sufficiently strong oxidants, such as chlorine or bromine water, conversely oxidize thallium(I) compounds quantitatively to thallium(III) compounds.

(a) Thallium(I) Compounds

TABLE 71

THALLIUM(I) HALIDES

	TlF	TlCl	TlBr	TlI
Color	white	white	yellowish	α yellow β red
Density	8.36	7.00	7.5	7.29*
Melting point	327°	430°	456°	440°
Boiling point	655°	806°	815°	824°
Solubility in water (millimoles/liter, 18°)	very sol.	12.7	1.48	0.17*
Heat of formation (kilocal per mol)	77.1	48.5	41.3	30.2*

* These values are for the modification stable at ordinary temperature.

(i) *Thallium(I) chloride*, TlCl, is thrown down as a curdy white precipitate when Cl⁻ ions are added to a solution of a thallium(I) salt. It can be obtained in the form of small cubes by recrystallization. Dry chlorine converts it, on warming, into the compound Tl_2Cl_3 or $Tl_3[TlCl_6]$, and chlorine water oxidizes it to thallium (III) chloride, $TlCl_3$.

(ii) *Thallium(I) bromide*, which is yellowish in color, closely resembles the chloride.

(iii) *Thallium(I) iodide* exists in two modifications, yellow and red respectively. The latter is stable above 168°, but remains unchange for long periods if it is cooled down to ordinary temperature.

TlCl, TlBr and the red form of TlI (β-TlI) have the cesium iodide type structure (p. 211, Fig. 48). Dimensions of the cubic cells are: TlCl $a = 3.834$Å; TlBr $a = 3.97$Å; β-TlI $a = 4.18$ Å. The yellow modification of TlI (α-TlI, stable at room temperature) forms a rhombic layer lattice, composed of TlI molecules.

(iv) *Thallium(I) fluoride*, TlF, differs from the other halides in being very soluble in water (1 part in 1.25 parts of water). Its aqueous solution is basic in reaction. An acid fluoride, $TlH_2F_3 \cdot \frac{1}{2}H_2O$, crystallizes in octahedra from solutions containing excess hydrofluoric acid.

TlF crystallizes in rhombic leaflets. It has a deformed NaCl structure ($a = 5.180$Å, $b = 5.495$Å, $c = 6.080$ Å). The deformation of the crystal lattice (which results in a preferred cleavage of the crystals perpendicular to the c axis) is attributable to the polarizing effect of the F⁻ ions on the highly polarizable Tl⁺ ions (Ketelaar, 1935). For thallium(I) triiodide, see p. 386.

(v) *Thallium(I) cyanide*, TlCN, is best obtained in the pure state by double decomposition of barium cyanide with thallium(I) sulfate. It forms a white crystalline powder or glistening leaflets, which are rather soluble in water (15.2 g in 100 g of water at 14°). The solution is strongly basic in reaction. The cyanide reacts with iodine according to the equation $TlCN + I_2 = TlI + ICN$.

(vi) *Thallium(I) thiocyanate*, TlSCN, is precipitated in small glistening tetragonal crystals when SCN⁻ ions are added to solutions of thallium(I) salts. Solubility, 0.32 g of TlSCN in 100 g of water at 20°.

(vii) *Thallium(I) hydroxide and thallium(I) oxide.* When a solution of thallium(I) sulfate is treated with the calculated amount of barium hydroxide, a solution with a strongly basic reaction is obtained, which exactly parallels the caustic alkalis in its behavior. This contains thallium(I) hydroxide, TlOH, which is very soluble in water and in alcohol. It is obtained as a yellow crystalline mass when the solution is evaporated. Thallium(I) oxide, Tl_2O, is formed as a black hygroscopic powder by heating the hydroxide to 100°. It melts at about 300°, and strongly attacks glass. It is readily reduced to the metal by means of carbon monoxide. The heat of formation of the oxide is 42.1 kcal per mol; the heat of formation of TlOH from Tl_2O is only 1.6 kcal per mol of TlOH.

(viii) *Thallium(I) sulfide*, Tl_2S, is formed as a black precipitate by the action of ammonium sulfide or hydrogen sulfide on weakly acid solutions of thallium(I) salts. It is best obtained pure, in a finely divided state, by the reaction of H_2S with thallium(I) ethoxide, $TlOC_2H_5$, in alcohol solution. The latter compound is readily prepared by the action of alcohol on thallium turnings, in the presence of oxygen. The precipitated sulfide becomes crystalline when it is heated to 150—200° in ammonium sulfide solution. The crystalline material is most readily prepared by direct reaction of Tl with S, in the atomic ratio 2 : 1. Thallium(I) sulfide is precipitated (mixed with sulfur) by the action of hydrogen sulfide on thallium(III) salt solutions: $2Tl^{+++} + 3H_2S = Tl_2S + 2S + 6H^+$.

Tl_2S is very slowly oxidized by dry oxygen at ordinary temperature to form thallium(I) oxide and thiosulfate: $2Tl_2S + 2O_2 = Tl_2O + Tl_2S_2O_3$. Tl_2O_3 is ultimately formed from the Tl_2O by further absorption of oxygen. $Tl_2S_2O_3$ is stable at room temperature, but decomposes above 130°. If oxygen is allowed to react with Tl_2S at higher temperatures (250°), or if the equimolecular mixture of Tl_2O and $Tl_2S_2O_3$ is heated, a yellow to greenish substance is obtained, with a face-centered cubic crystal structure; this has a composition corresponding to the formula Tl_2SO_2, but its nature has not yet been explained. It is decomposed by water: $3Tl_2SO_2 = Tl_2S + 2Tl_2SO_3$. The further action of dry oxygen at 250° ultimately converts it to Tl_2SO_4 (Reuter, 1949—52; von Hippel, 1946). It is possible by the controlled, partial autoxidation of Tl_2S to obtain substances with efficient photoelectric properties, which are especially sensitive in the long wave region (Case, 1917; von Hippel, 1946).

(ix) *Thallium(I) sulfate*, Tl_2SO_4, formed by dissolving thallium in dilute sulfuric acid, crystallizes from solution in large, colorless rhombic prisms ($d = 6.7$—6.8, m.p. 632°), isomorphous with potassium, rubidium and cesium sulfates. It is moderately soluble in water (4.87 g in 100 g of water at 20°). The salt begins to volatilize above a dull red heat. Acid sulfates crystallize from solutions containing a large excess of sulfuric acid.

With aluminum sulfate, thallium(I) sulfate forms a double salt of the alum type—$TlAl(SO_4)_2 \cdot 12H_2O$, *thallium alum.*

(x) *Thallium(I) nitrate*, $TlNO_3$, crystallizes at ordinary temperature in rhombic prisms from a solution of the metal, or its hydroxide or carbonate, in nitric acid. A trigonal modification is stable above 61°, and a cubic form above 143.5°; it melts at 206°, and decomposes above 300°. The solubility of the salt in water increases to an unusual degree with rise of temperature: 100 g of water dissolve 9.55 g of $TlNO_3$ at 20°, and 594 g of $TlNO_3$ at 105° (the boiling point of the saturated solution).

(xi) *Thallium(I) chlorate*, $TlClO_3$, obtained by dissolution of the metal in chloric acid, or by double decomposition between barium chlorate and thallium(I) sulfate, crystallizes in long colorless needles ($d = 5.05$). It forms mixed crystals with potassium chlorate (miscibility gap between 36.3 and 97.93 mol-per cent $KClO_3$). Its solubility increases steeply with temperature (from 2.00 g in 100 g of water at 0° to 57.3 g in 100 g of water at 100°). The hot solutions gradually decompose.

Solubility of the bromate: 0.35 g of $TlBrO_3$ in 100 g of water at 20°.
Solubility of the iodate: 0.058 g of $TlIO_3$ in 100 g of water at 20°.

(xii) *Thallium(I) perchlorate*, $TlClO_4$, prepared like the chlorate, forms rhombic crystals, isomorphous with potassium perchlorate. Like the latter, it changes reversibly into a cubic form when heated (transition points: for $KClO_4$, 299.5°; for $TlClO_4$, 266°). The solubility increases from 10 g of $TlClO_4$ in 100 g of water at 15° to 166.6 g in 100 g of water at 100°.

(xiii) *Thallium(I) acetate*, $TlC_2H_3O_2$, forms silky, deliquescent needles ($d = 3.9$, m.p. 110°); prepared by dissolution of the carbonate in acetic acid. Very soluble in hot alcohol.

(xiv) *Thallium(I) carbonate*, Tl_2CO_3, is precipitated when alcohol is added to a solution of thallium(I) hydroxide which has been saturated with CO_2. It is fairly soluble in water (5.2 g in 100 g of H_2O at 18°, 22.4 g in 100 g of H_2O at 100.8°). Its solution, like that of an alkali metal carbonate, is strongly alkaline, as a result of hydrolysis. The salt crystallizes in long monoclinic needles ($d = 7.16$; m.p. 272—273°). It decomposes, with loss of CO_2, when strongly heated.

(xv) *Thallium(I) oxalate*, $Tl_2C_2O_4$, crystallizes in small nacreous prisms from a solution of thallium(I) carbonate in boiling aqueous oxalic acid. It is insoluble in alcohol; 100 g of water dissolve 1.44 g of $Tl_2C_2O_4$ at 15°. The solubility is greater in the presence of excess oxalic acid, as a result of the formation of the hydrogen oxalate, $TlHC_2O_4$ (solubility 5.35 g in 100 g of H_2O at 15°).

(xvi) *Thallium(I) phosphate*, Tl_3PO_4, forms colorless, sparingly soluble crystals (solubility 0.497 g in 100 g of H_2O at 15°). The acid phosphate TlH_2PO_4 is quite soluble. It forms pearly monoclinic leaflets, melting at about 190°.

(b) Thallium(III) Compounds

(i) *Thallium(III) chloride and chlorothallates*. Treatment of thallium(I) chloride with chlorine yields a solution which, if evaporated in a current of chlorine at 60°, deposits the hydrated thallium(III) chloride $TlCl_3 \cdot 4H_2O$. The hydrate forms large, colorless six-sided plates, which are deliquescent in moist air. It loses first $3H_2O$, and ultimately all of its water of crystallization in a vacuum over sulfuric acid. The anhydrous chloride, $TlCl_3$, melts at 25°. Solutions of thallium(III) chloride are strongly acidic in reaction, as a result of hydrolysis, and deposit brown thallium(III) oxide hydrate when they are greatly diluted.

Thallium(III) chloride readily forms complexes with excess chloride ions. Thus *chlorothallates* crystallize from solutions containing thallium(III) chloride and alkali chlorides. These are mostly of the types $M^I_3[TlCl_6]$ and $M^I_2[TlCl_5(H_2O)]$. *Chlorothallic acid* (hydrogen tetrachlorothallate(III)), $HTlCl_4 \cdot 3H_2O$, crystallizes in hair-fine needles from solutions of thallium(III) chloride in excess hydrochloric acid. The 'intermediate' chlorides formed by incomplete oxidation of thallium(I) chloride, or by the reduction of thallium(III) chloride, are also chloro-salts of this type, and should be regarded as thallium(I) chlorothallate(III) salts. The compound of this type which is most readily obtained pure is Tl_2Cl_3, or $Tl^I_3[Tl^{III}Cl_6]$; shimmering yellow leaflets ($d = 5.9$), very sparingly soluble in water (0.0033 mols per liter at 25°). When thallium(I) chloride or thallium metal is cautiously heated in chlorine, a pale yellow mass is obtained with the composition $TlCl_2$ or $Tl^I[Tl^{III}Cl_4]$. This decomposes when more strongly heated:

$$3Tl^I[Tl^{III}Cl_4] = Tl^I_3[Tl^{III}Cl_6] + 2TlCl_3$$

Thallium(III) chloride also readily forms addition compounds with neutral ligands. Thus the anhydrous salt combines with ammonia to form the triammine $TlCl_3 \cdot 3NH_3$, which can also be prepared from alcoholic (but not from aqueous) solution. The compounds $TlCl_3 \cdot C_2H_5OH$ and $TlCl_3 \cdot O(C_2H_5)_2$ separate out in large crystals when solutions of thallium(III) chloride in alcohol, or ether are evaporated.

(ii) *Thallium(III) bromide and bromothallates*. Thallium(III) bromide is far less stable than thallium(III) chloride, but its compounds with thallium(I) bromide are more readily obtained than are the corresponding chlorine compounds. Thallium(III) bromide loses bromine spontaneously, and passes over into the double compound $TlBr_2$ or $Tl[TlBr_4]$ (long yellow needles). In contact with water, this breaks up into $Tl_3[TlBr_6]$ (red hexagonal leaflets) and $TlBr_3$. Most bromothallates are of the type $M^I[TlBr_4]$ (often $2H_2O$).

Rather unstable mixed salts, $TlBr_2Cl \cdot H_2O$ and $TlBrCl_2 \cdot 4H_2O$, are obtained by the action of bromine on thallium(I) chloride, or of chlorine on thallium(I) bromide. Mixed chlorobromothallates have also been described.

Among the halogenothallates, the compounds $Cs_3Tl_2Cl_9$, $Rb_3TlBr_6 \cdot H_2O$ and $K_3TlCl_6 \cdot 2H_2O$ are of interest from the standpoint of their crystal structures. Binuclear complex anions, $[Tl_2Cl_9]^{3-}$, clearly exist as structural groups in the crystal lattice of $Cs_3Tl_2Cl_9$, which crystallizes in the hexagonal system. They contain 6-coordinated thallium atoms:

$$\begin{bmatrix} Cl & Cl & Cl \\ Cl-Tl & Cl-Tl & Cl \\ Cl & Cl & Cl \end{bmatrix}^{3-}$$

(Hoard, 1935). The unit cell of the tetragonal $Rb_3TlBr_6 \cdot \frac{9}{7}H_2O$ contains 14 formula-weights of the compound—i.e., 16 molecules of water. Of the $42Rb^+$ cations in each unit cell, only 2 occupy positions surrounded by H_2O molecules, and these are so placed that each of the Rb^+ cations is surrounded by $8H_2O$ in roughly cubic arrangement. Each Tl^{3+} is surrounded by 6Br atoms, forming $[TlBr_6]^{3-}$ structural groups (similar to the octahedral $[MX_6]$ anionic groups in the $[PtCl_6]^{2-}$ or $[AlF_6]^{3-}$ complexes; cf. Fig. 74, p. 358). Thus the crystal lattice is built up from $[TlBr_6]^{3-}$ complex anions, $[Rb(H_2O)_8]^+$ complex cations, and anhydrous Rb^+ ions which occupy 'holes' in the crystal lattice. The difference between this structure and that of the more highly hydrated compound $K_3TlCl_6 \cdot 2H_2O$ is not merely that a large proportion of the alkali ions is surrounded by water molecules. The unit cell of the more highly hydrated compound (which also contains 14 formula-weights of compound, or $28H_2O$ in all) contains an extra $12H_2O$, which are linked to the two K^+ ions which occupy positions corresponding to those of the $[Rb(H_2O)_8]^+$ ions in the bromo-complex just considered. An octahedron is built up from these H_2O molecules, around each K^+ ion. This has the effect of pushing apart the two end faces of the group of H_2O molecules which, in the rubidium complex, formed the cubic $[Rb(H_2O)_8]^+$ group. The resultant structure is such that, in $K_3TlCl_6 \cdot 2H_2O$, the groups are stacked along the c-axis in the following manner: $[TlCl_6]^{3-}$ octahedron, $(H_2O)_4$ square group, $[K(H_2O)_6]^+$ octahedron, $(H_2O)_4$ square group, $[TlCl_6]^{3-}$ octahedron... etc. The structure is in essence similar to that of $Rb_3TlBr_6 \cdot \frac{9}{7}H_2O$. These halogeno-salts illustrate the fact that a complicated gross formula does not necessarily correspond with a complicated structure, but may arise from the operation of simple structural principles.

(iii) *Thallium(III) triiodide.* When potassium iodide is added to thallium(III) salt solutions, a black precipitate with the composition TlI_3 is formed. By recrystallization, the compound can be obtained in rhombic prisms which are *isomorphous with the alkali triiodides*. This would imply that *solid* thallium triiodide must be formulated as a polyiodide of *univalent* thallium, $Tl^I[I_3]$. In accordance with this, it loses two atoms of iodide with extreme ease—e.g., by treatment with organic solvents. Conversely, the triiodide may be formed by addition of iodine to thallium(I) iodide.

Abegg states that thallium(I) iodide is the stable solid phase in contact with solutions containing not more than $0.76 \cdot 10^{-5}$ mols per liter of free iodine. Higher concentrations of iodine convert it first into a black substance of the composition Tl_3I_4 (i.e., $TlI_3 \cdot 5TlI$) which crystallizes in the rhombic system. The triiodide is formed from this only if the iodine concentration exceeds $3.3 \cdot 10^{-4}$ mols per liter. In solution, there is an equilibrium between the two isomeric forms of the triiodide:

$$Tl^I[I \cdot I_2] \rightleftharpoons Tl^{III}[I]_3$$

Thallium triiodide forms dark red complex salts with alkali metal iodides. These are mostly of the type $M^I[TlI_4]$, in which thallium is definitely present in the trivalent state, as proved by Abegg's equilibrium measurements.

(iv) *Thallium(III) fluoride*, TlF_3, can be obtained by heating Tl_2O_3 in fluorine (Klemm, 1936). It is a white powder ($d = 8.36$), which decomposes when heated in air, although it may be melted in an atmosphere of fluorine (m.p. about 550°). It is completely decomposed by water at ordinary temperature, being hydrolyzed to thallium(III) oxide hydrate.

(v) *Thallium(III) oxide hydrate* is precipitated on addition of ammonia to thallium(III) salt solutions as a voluminous red-brown substance, readily soluble in acids. It has the approximate composition $Tl_2O_3 \cdot H_2O$ after drying at ordinary temperature, although water is lost continuously during isobaric dehydration, without any sign of a discontinuity (Hüttig, 1930). Material filtered immediately after precipitation loses all its water when it is heated to 300° under 10 mm water vapor pressure. Precipitates which have 'aged' by allowing them to stand for some time in contact with the precipitation solution are more readily dehydrated, and a practically anhydrous oxide is obtained if the oxide hydrate is boiled with the precipitation solution. The hydrated precipitate has the same X-ray diffraction pattern as the oxide. There is thus no thallium(III) hydroxide of well defined composition or distinct structure. The firm retention of water by the freshly prepared oxide hydrate does, however, make it probable that H_2O molecules are initially either incorporated in a random fashion in the crystal structure of the precipitated oxide, or are 'chemisorbed' on the enormous surface of the newly formed precipitate.

(vi) *Thallium(III) oxide*, Tl_2O_3, forms a dark brown to black powder in the finely divided state—e.g., as obtained by digestion of the oxide hydrate under the precipitant. It can be obtained in black leaflets ($d = 10.19$) by the thermal decomposition of the nitrate. It is cubic, with the same crystal structure as In_2O_3, Mn_2O_3 and the so-called 'type C' rare earth oxides (see Vol. II, Chap. 10); Tl_2O_3 has $a = 10.57$ Å, In_2O_3 has $a = 10.12$ Å). When heated, Tl_2O_3 decomposes into Tl_2O and oxygen. Evolution of oxygen from the finely divided oxide begins at temperatures as low as 100°, whereas the compact black oxide does not display a measurable dissociation pressure of oxygen below a red heat.

(vii) *Thallium(III) sulfate or disulfatothallic acid. Sulfatothallates.* Solutions of thallium(III) oxide in dilute sulfuric acid usually deposit crystals of the 'acid sulfate', $HTl(SO_4)_2 \cdot 4H_2O$. It is probable that this should be regarded as *disulfatothallic acid*, rather than as an acid thallium(III) salt. Complex salts (*disulfatothallates*) crystallize from mixed solutions of thallium(III) sulfate and alkali metal sulfates. These are of the type $M^I[Tl(SO_4)_2]$; some crystallize with 4 H_2O. *Trisulfatothallates*, $M^I_3[Tl(SO_4)_3]$, are also known. Analogous complex sulfates are obtained from solutions containing both thallium(I) sulfate and thallium(III) sulfate.

Thallium(III) sulfate is used as a rat poison.

(viii) *Thallium(III) nitrate* crystallizes in brilliant, colorless deliquescent crystals, with the composition $Tl(NO_3)_3 \cdot 3H_2O$, when a solution of thallium(III) oxide in concentrated nitric acid is evaporated. If the solutions also contain alkali metal nitrates or thallium(I) nitrate, complex nitrates are obtained. These are of the type $M^I_2[Tl(NO_3)_5]$—e.g., $Tl^I_2[Tl^{III}(NO_3)_5]$, colorless prisms, m.p. 150°.

(ix) *Thallium(III) acetate*, $Tl(C_2H_3O_2)_3$, crystallizes out in lustrous colorless leaflets when a solution of thallium(III) oxide in boiling glacial acetic acid is cooled. Although this compound is decomposed even by traces of moisture, the complex salt (or acido-compound) $NH_4[Tl(C_2H_3O_2)_4]$ (ammonium tetraacetatothallate) is stable in moist air.

(x) *Thallium(III) oxalate, dioxalatothallic acid, and oxalatothallates.* Precipitates of variable composition are obtained when oxalic acid is added to acidified thallium(III) salt solutions. Pure neutral thallium(III) oxalate has not been isolated, but *dioxalatothallic acid* (hydrogen dioxalatothallate), $H[Tl^{III}(C_2O_4)_2]$, has been obtained by various methods; it crystallizes with $3H_2O$. The corresponding acido-salts, $M^I[Tl(C_2O_4)_2]$, dioxalatothallates are known, and also some trioxalatothallates, $M^I_3[Tl(C_2O_4)_3]$.

The compound $K_3[Tl(C_2O_4)_2(NO_2)_2] \cdot H_2O$ (potassium dioxalatodinitrothallate), which is unstable towards water, can be isolated in yellow crystals from a solution of dioxalatothallic acid in concentrated potassium nitrite.

(c) Thallium Alkyl Compounds

Thallium trialkyl compounds are not very stable, and are difficult to isolate. Thallium dialkyl compounds of the type TlR_2X (R = alkyl radical, X = acid radical) are much more readily prepared. These relatively stable substances resemble the thallium(I) compounds rather than the thallium(III) compounds. They can be regarded as being derived from univalent radical $[TlR_2]^+$, with definitely electropositive character, which can exist as free ions in aqueous solution. The compounds formed by them with various acids are very stable, and display some analogies (e.g., in solubility) to the salts of univalent thallium. Thus they are not decomposed by boiling with water or with caustic alkalis.

These compounds are best prepared from thallium(III) chloride, by reaction with the corresponding alkyl magnesium chloride, in ether solution (R. J. Meyer):

$$2CH_3MgCl + TlCl_3 = (CH_3)_2TlCl + 2MgCl_2.$$

Other salts can be prepared from the chlorides by double decomposition. The acid sulfides, [TlR$_2$]SH, which are insoluble in water, are formed as white precipitates on the addition of ammonium sulfide. The bases from which the dialkyl thallium salts are derived can be prepared by reaction between the iodides and a hot aqueous suspension of silver oxide. These hydroxides, [TlR$_2$]OH, are comparable in properties to the alkali hydroxides.

(i) Trialkyl thallium compounds were first prepared by Groll (1930). *Thallium triethyl*, Tl(C$_2$H$_5$)$_3$, was obtained by the reaction of diethyl thallium chloride with lithium ethyl in petroleum ether solution: [Tl(C$_2$H$_5$)$_2$]Cl + LiC$_2$H$_5$ = LiCl + Tl(C$_2$H$_5$)$_3$. It is a yellow, mobile liquid ($d = 1.97$), which can be distilled without decomposition only under low pressures. It is not attacked by dry oxygen, but is immediately decomposed by water, forming diethylthallium hydroxide: Tl(C$_2$H$_5$)$_3$ + H$_2$O = [Tl(C$_2$H$_5$)$_2$]OH + C$_2$H$_6$.

(ii) *Thallium trimethyl* is formed from thallium(I) halides by reaction with lithium methyl and methyl iodide (Gilman and Jones, 1939): 2CH$_3$Li + CH$_3$I + TlX = (CH$_3$)$_3$Tl + + LiI + LiX. It forms a colorless solid (m.p. 38.5°) which explodes when heated above 90°. Like indium trimethyl, it is monomeric in benzene solution.

(d) Thallium hydrides

Thallium is the only element in this group which forms no isolable hydride. The reaction of TlCl$_3$ with LiAlH$_4$ yields only thallium metal and hydrogen, even at −115°; TlH$_3$ may be formed transiently in this reaction. Complex hydrides with aluminum and gallium are rather more stable, and have been isolated by Wiberg and Schmidt (1951): 3TlCl$_3$ + + 8AlH$_3$ = 3TlCl(AlH$_4$)$_2$ + 2AlCl$_3$ (at −115°); TlCl$_3$ + 3LiGaH$_4$ = 3LiCl + Tl(GaH$_4$)$_3$ (at −115°). These complex hydrides are very unstable white solids, which decompose at −90°.

15. Analytical (Gallium, Indium, Thallium)

Gallium, indium and thallium are generally detected by spectroscopic methods (cf. p. 369). They may be determined gravimetrically by precipitation with ammonia, and are then weighed in the form of their oxides. Thallium can be precipitated with ammonia only if it is present in the trivalent state, and must first be oxidized if present as a thallium(I) compound. The solubility of thallium(I) hydroxide, however, enables thallium to be separated readily and completely from gallium, indium, aluminum and the other trivalent elements.

Very small amounts of gallium and indium can be detected by color reactions with quinalizarin, which are given by their oxide hydrates formed hydrolytically (Pietsch, 1933—34). Limits of detection: 0.02γ of Ga^{+++} or 0.05γ of In^{+++} per ml. The reaction can be carried out in the presence of salts of aluminum or zinc (the oxide hydrates of which also give colored lakes with quinalizarin), provided that the Al^{+++} or Zn^{++} ions are held in solution by suitable complex-forming substances (pyridine or ethylamine); the sensitivity of the reaction is diminished under these conditions, however.

Very small amounts of indium can be determined polarographically, as well as by spectrographic methods (Takagi, 1928; Ensslin, 1941). If the polarographic method is applied to the determination of indium in crude zinc, it is necessary, according to Rienäcker (1952) to separate the indium from cadmium. This can be effected by precipitation of indium as the oxide hydrate by means of ammonia, in the presence of ammonium salts. The oxide hydrate is dissolved in 10% tartaric acid for the preparation of the polarogram.

References

1 A. Stock, *Hydrides of Boron and Silicon*, Ithaca 1933, 250 pp.
2 W. Fulda and H. Ginsberg, *Tonerde und Aluminium, Ergebnisse und Erfahrungen aus der Betriebspraxis 1920–1950*. Vol. I, *Die Tonerde*, Berlin 1951, 226 pp; Vol. II, *Das Aluminium*, Berlin 1953, 356 pp.
3 H. Michel, *Die künstlichen Edelsteine*, 2nd Ed., Leipzig 1926, 477 pp.
4 A. von Zeerleder, *Technology of Light Metals*, (translated by A. J. Field), Amsterdam 1949, 366 pp.
5 Aluminium-Zentrale, *Aluminium-Taschenbuch*, 10th Ed., Düsseldorf 1951, 628 pp.
6 N. F. Budgen, *Aluminium and Its Alloys*, 2nd Ed., New York 1947, 369 pp.
7 C. Panseri, *La Fonderia d'Alluminio*, Milan 1934, 582 pp.
8 N. F. Budgen, *Heat-Treatment and Annealing of Aluminium and Its Alloys*, London 1932, 341 pp.
9 E. Herrmann and E. Zurbrügg, *Die Bearbeitung des Aluminiums*, 3rd. Ed., Leipzig 1943, 210 pp.
10 V. Fuss, *Metallographie des Aluminiums und seiner Legierungen*, Berlin 1934, 219 pp.
11 P. L. Teed, *Duralumin and Its Heat Treatment*, London 1936, 116 pp.
12 A. Becco, *Il Duralluminio nelle costruzioni aeronautiche*, Turin 1936, 143 pp.
13 A. Jenny, *Die elektrolytische Oxydation des Aluminiums und seiner Legierungen*, Dresden 1938, 224 pp.
14 L. Lux, *Die elektrolytische Schutzoxydation von Aluminium nach dem Eloxal-Verfahren*, Berlin 1941, 44 pp.
15 K. Goldschmidt, *Aluminothermie*, Leipzig 1925, 174 pp.
16 J. D. Edwards, *Aluminum Paint and Powder*, 2nd Ed., New York 1936, 216 pp.
17 G. Kränzlein, *Aluminiumchlorid in der organischen Chemie*, Berlin 1932, 143 pp.
18 J. Schmidt, *Organo-Metallverbindungen*, Stuttgart 1934, 376 pp.
19 F. Krczil (= Kainer), *Aktive Tonerde, ihre Herstellung und Anwendung*, Stuttgart 1938, 274 pp.
20 D. T. Hurd, *Introduction to the Chemistry of the Hydrides*, New York 1952, 231 pp.
21 H. S. Booth and D. R. Martin, *Boron Trifluoride and Its Derivatives*, New York 1949, 315 pp.
22 E. Einecke, *Das Gallium*, Leipzig 1937, 155 pp.
23 J. J. G. Prick, W. G. Sillevis Smitt and L. Muller, *Thallium Poisoning*, Amsterdam 1955, 155 pp.

CHAPTER 11

COORDINATION THEORY

1. General

(a) Radical

It has long been known that certain groupings of chemical elements—e.g., NO_3, SO_4, CO_3, NH_4—persist throughout the most varied chemical changes, so that they can be transferred without alteration from one compound to another, and from this to yet other compounds, in the course of substitution reactions or double decompositions. The name *radicals* has been introduced for such groupings. Instead of the expression radical, they are often called simply *groups*.

In most double decomposition reactions, radicals behave just like chemical elements. The following examples, in which the radicals are enclosed in square brackets to make them obvious, may serve to illustrate the analogy.

$$2HCl + ZnO \qquad = ZnCl_2 + H_2O$$
$$2H[NO_3] + ZnO \qquad = Zn[NO_3]_2 + H_2O \tag{1}$$

$$2NaCl + H_2[SO_4] \qquad = 2HCl + Na_2[SO_4]$$
$$2Na[NO_3] + H_2[SO_4] = 2H[NO_3] + Na_2[SO_4] \tag{2}$$

$$Na_2[SO_4] + BaCl_2 \qquad = Ba[SO_4] + 2NaCl$$
$$[NH_4]_2[SO_4] + BaCl_2 = Ba[SO_4] + 2[NH_4]Cl \tag{3}$$

Most inorganic radicals exist in aqueous solution in the form of free ions. Many organic radicals also exist as free ions, especially the radicals of the acids, such as acetic acid $H[C_2H_3O_2]$, and those of organic-substituted ammonium compounds.

It should be noticed, however, that the name 'acid radical' is more commonly used in organic chemistry to denote the residues obtained by splitting off the —OH group from a carboxylic acid. These 'acyl' groups retain their identity through many reactions (e.g., formation of acid chlorides, etc.), whereas the ions, formed by dissociation of H^+ions (or more strictly, by transfer of a proton to the solvent, giving H_3O^+ ions), do not. It is therefore permissible to regard either the group $C_2H_3O_2$— or the group C_2H_3O— as the radical of acetic acid, depending on the reactions considered. It is generally more useful to apply the name to the C_2H_3O— group; in organic chemistry it is therefore customary to refer to the group $C_2H_3O_2$, where it occurs charged (as in solutions and in metallic salts), simply as the acetate ion. Similar considerations apply with respect to other organic acids.

It was long disputed whether radicals could also occur free, and uncombined with other substances. The fact that most of the inorganic radicals can exist as free ions was then interpreted on the idea that ions are so closely analogous to chemical compounds that they can be regarded as 'compounds of atoms or radicals with positive or negative charges'. Such a concept is really possible only in a formal sense, and misses the essential point. It is precisely because they are electrically charged that the radicals occurring as free ions tend to associate with other substances, also present in the ionized state, to form compounds.

Nevertheless, it must be noticed that electrically uncharged radicals seldom occur free. Chemical reactions which might be expected to give rise to free, uncharged radicals almost invariably furnish products which can be regarded as originating from the union of radicals (usually two radicals). The elements behave in a similar way. Hydrogen, chlorine, etc. are formed only transitorily as free atoms by appropriate reactions, and immediately join together to form diatomic molecules. The same is essentially true of the metals at ordinary temperatures, except that these do not combine to form molecules built up of two, or a few, atoms, but associate to form crystal lattices built up of very large numbers of atoms.

(b) Structural Group

Although the concept of radicals is derived from chemical reactions, the idea of the *structural group*, introduced in Chapter 7, is a crystallographic one. In general, however, it may be said that atoms or ions which make up a coherent whole in processes of chemical change tend to do so also within the crystal, so that as a rule radicals exist in the crystal lattice as structural groups.

It does not necessarily follow, however, that a structural group which stands out as a unit in the crystal lattice always persists as such when the crystal is dissolved, or its crystal lattice broken down. In particular, it cannot be assumed that this will be so when even in the crystal the mutual relation between the components of the structural group is not very close. In such cases, dissolution of the crystal lattice may be accompanied by a complete disruption of the structural groups into their individual components, so that the structural group passes neither into the aqueous solution, as an ion, nor into another compound, unless the latter happens to have a crystal structure completely analogous to the first.

(c) Complex [2, 3]

Atomic groupings of fairly complicated composition are found to exist in many substances, and persist through many chemical reactions as radicals do; like radicals, they display characteristic reactions sometimes differing quite considerably from those of their components. These complicated entities are termed *complex groups* or *complexes*. The stability of complex groups (i.e., their immutability during chemical reactions) is often not as great as that of the typical radicals*.

Complexes which are comparable in stability to the typical radicals are known as *strong complexes*; those which readily break up into their components as *weak complexes*. In so far as complexes are built up from ions, the strong complexes, do not give in aqueous solution, the reactions characteristic of the ions concerned, since they do not dissociate into these ions, to any perceptible extent. Weak complexes give the reactions characteristic of their components; the weaker the complex and the more sensitive the reactions used, the more evident is their dissociation.

The complete or partial disappearance of the reactions typical of the components is so characteristic of complex formation that whenever typical reactions fail as a result of the formation of some particular chemical compound, complex formation is said to occur. It is not necessary for this purpose that the group characterizing such a complex should, invariably, persist also through other chemical changes, as a radical does. Thus the fact mentioned in the preceding chapter, that aluminum is not precipitated from solutions which contain certain organic substances such as tartaric acid, is attributed to the formation of complex compounds of the aluminum salts with these substances. It is assumed that the aluminum

---

* Many complexes should be regarded as radicals in the wider sense, according to the above definition.

ion associates with the organic compound to form ions of more complicated structure. It does not follow that the complex sulfates, chlorides, nitrates, etc., which it may be possible to isolate from such solutions should have the same composition as the species present in the solution. When metallic ions, (e.g., aluminum ions) fail to give reactions which are otherwise characteristic of them, the reactions are said to be *masked* by complex formation.

(d) Compounds of Higher Order

Simple binary compounds (e.g., ClH, OH_2, NH_3, CH_4) are known as *compounds of the first order*, as also are the compounds derived from them by the process of direct substitution. Thus the following compounds can be regarded as formally derived from those mentioned above, by atom for atom substitution:

$$\text{I—H} \qquad S\!\!\begin{array}{c} H \\ H \end{array} \qquad \overset{H}{\underset{H}{P}}\!\!-H \qquad \overset{H}{\underset{H}{Si}}\!\!\overset{H}{}$$

$$\text{I—Cl} \qquad S\!\!\begin{array}{c} Cl \\ Cl \end{array} \qquad \overset{Cl}{\underset{Cl}{P}}\!\!-Cl \qquad \overset{Cl}{\underset{Cl}{Si}}\!\!\overset{Cl}{}$$

These are all compounds of the first order, and this class includes almost all purely organic compounds.

Compounds which are formed by the union of molecules of compounds of the first order with one another, and which cannot be derived from compounds of the first order by substitution, are known as *compounds of higher order*.

Examples of compounds of higher order are, in the first place, the *hydrates, ammoniates, acid addition compounds*, and analogous compounds formed by the addition of organic molecules, and also *double salts*, such as *carnallite* $KCl \cdot MgCl_2 \cdot 6H_2O$. Even apart from the water which it contains, the latter is to be regarded as a compound of higher order, since it can (formally) be regarded as arising from the union of the molecules KCl and $MgCl_2$. Finally, *complex salts*, such as $KBF_4 (= KF + BF_3)$ are also compounds of higher order.

The water, ammonia, or whatever other compound may be added on, may be relatively firmly bound, so that it is not readily split off (especially upon dissolution in water); the added molecule, or *ligand*, may in that case modify the reactions of the element with which it is combined, or may affect the color or other properties characteristic of the original compound. In such a case the ligand is said to be *complex-bound*. If the complex-bound molecule is one that would dissociate in aqueous solution, only the one constituent dissociates off, and the other remains bound in a complex.

Thus fluoroboric acid is to be considered a compound of *higher order* because it is formed by the association of BF_3 and HF. It is to be regarded as a *complex* compound because the HF molecule, or at least one part of it, the F^- ion, is very firmly bound to the BF_3. In aqueous solution the compound dissociates in the following way:

$$HF \cdot BF_3 = H^+ + [F \cdot BF_3]^-$$

—i.e., only one constituent of the complex-bound molecule, the H^+ ion, is split off by electrolytic dissociation.

Fluoroboric acid is an example of a *complex acid*. Complex salts are formed from such acids by replacement of the dissociable hydrogen by metals. Complex compounds which dissociate to form hydroxyl ions when they are dissolved in water are to be regarded as *complex bases*.

Since complex *salts* are frequently more stable than the parent acids, we may define them independently. We regard a double salt (defined as above) as a *complex salt* if, when it is dissolved in water, it does not break up into the ions of the salts of which it is composed, or does so only to a small extent. Salts which dissociate electrolytically to give ions that contain firmly bound neutral ligands are also to be regarded as complex salts. Water may also be bound as a neutral ligand. It is usual, however, to class the hydrated ions among the complex compounds only when the water is bound especially firmly, and in a clear stoichiometric ratio.

Double salts which break up completely, or to a considerable extent into the ions of their component salts, when they are dissolved in water, are regarded as double salts in the stricter sense, as distinct from complex salts. It is not possible to draw an absolutely sharp distinction between these two classes, except when crystal structure determinations show that the crystal lattice is built up only from the ions of the component salts.

(e) Principal and Auxiliary Valence

The formation of compounds of higher order was formerly attributed to special forces, differing from the normal valences. These were called *auxiliary* valences, and were represented by dotted lines in structural formulas, to distinguish them from the ordinary valences, or *principal* valences, represented by full lines. The formation of the compound $HF \cdot BF_3$ would then be attributed to the fact that the boron exerts one auxiliary valence towards the hydrogen fluoride:

$$B{\Large\langle}{\overset{\displaystyle F}{\underset{\displaystyle F}{F}}} + H\!-\!F = H\!-\!F\cdots B{\Large\langle}{\overset{\displaystyle F}{\underset{\displaystyle F}{F}}} \tag{4}$$

It is better expressed, however, in the form: boron can exert more than its usual three valences towards fluoride ions:

$$BF_3 + F^- = [BF_4]^- \tag{5}$$

This mode of expression, and the representation of the process by eqn. (5), contains no special assumption about the mode of binding of the fourth fluoride ion. We shall see later on that it is not necessary to assume a special sort of force, different from that involved in the exercise of the ordinary valence.

Even in cases where we have, as yet, no insight into the ultimate reason why complex compounds are formed, it has been possible to show that, in the completed molecule at least, there is no fundamental difference between principal and auxiliary valences.

(f) Coordinative Binding

The term *coordinative binding* is applied to the particular way in which valence forces are exercised in the formation of compounds of higher order. It is supposed that the atoms capable of complex formation tend to arrange around themselves (latin: coordinare) a definite number of atoms, molecules or radicals, exceeding the number which would be bound by the normal exercise of valence forces. The atom of a complex around which the coordinatively bound atoms, molecules or radicals are arranged, is called the *central atom* of the complex. (See below for complexes with more than one central atom; for 'two-shelled complexes', in which a complex ion takes the place of a central atom, see Vol. II.)

The number specifying how many atoms, molecules or groups are coordinatively bound by the atom of any element is known as the *coordination number* or *coordinative valence* of the element in the compound concerned. The greatest number of atoms, etc., of any kind which may be coordinated around any central atom is the *maximum coordination number* of the central atom towards atoms, etc. of this particular kind. Compounds in which the central atom functions with its maximum coordination number are said to be coordinatively saturated.

In fluoroboric acid, boron is the central atom of the complex, and has the coordination number 4 (or is coordinatively quadrivalent) since it has combined coordinatively with 4 F^- ions. It follows that coordinative combination must be operative since the number of F^- ions exceeds the valence which boron is otherwise able to exercise towards electronegative elements; furthermore, the compound $HF \cdot BF_3$ is a compound of higher order.

The coordination number most frequently encountered is 6, but it seems that the preference for this number is restricted to certain classes of compounds. After the coordination number 6, coordination numbers of 4 and 3 are most common.

The principles governing the compounds of higher order constitute the *coordination theory*, and compounds of higher order are also called *coordination compounds*. The concept of coordination compounds is wider than that of the complex compounds alone. Coordination theory embraces not only the complex compounds, but also the double salts and loose addition compounds, which break up into their components in aqueous solution, but which, in the solid state, are in many cases similar in structure to the complex compounds. We have already seen in Chapter 7 that the coordination number is a useful concept for the understanding of crystal structure, as well as in the field of coordination compounds proper.

(g) Historical

The coordination theory was founded by Alfred Werner [1], and had its origin in the metal-ammonia compounds. It was not possible to interpret the composition of these in terms of the older valence theory—i.e., by attempts to refer them to compounds of the first order. Werner showed that these, and many other compounds could be interpreted without any special arbitrary hypotheses on the single assumption that atoms may exercise additional valences after satisfying the valence deduced from their ordinary compounds. In a large number of cases, this assumption follows directly as a conclusion from the observed facts as, in the example already cited, when boron trifluoride adds on a further fluoride ion. In the field of the metal-ammonia compounds and their derivatives, Werner was able to draw certain conclusions as to the *configuration* of the compounds, and so arrived at a theory of inorganic isomerism and a stereochemistry of inorganic compounds. This received its most striking confirmation through the discovery of optical isomerism in compounds, e.g., of chromium, cobalt and platinum, as predicted from the theory (see Vol. II further upon this point). The basic concepts of the 'Werner theory' have become of the utmost importance throughout the whole field of chemistry. Among those who have contributed to the further development of Werner's theory, into the current doctrine of coordination, with its broad scope, special mention should be made of Weinland, Pfeiffer and Bjerrum on the experimental side, and of Sidgwick and Pauling on the theoretical side.

The idea that the extraordinarily numerous compounds of higher order are really 'coordination compounds' brings a unifying viewpoint into a vast field which, without some ordering principle, would be too complex to disentangle. Until recently, apart from explaining the examples of isomerism, the coordination theory was restricted substantially to systematizing an otherwise intractable field of study—a task important in itself, and brilliantly fulfilled. More recently, however, it has become possible to get some idea of the basic courses to which the formation of compounds of higher order can, in the last analysis, be assigned. Introduction of the idea of coordinative combination is, in itself, no more than a description of the *fact* that compounds of higher order are formed; it does not provide any explanation of *why* this should be so.

(h) Classification of Coordination Compounds

As far as their structure is known, the coordination compounds may be subdivided into (i) compounds with a *complex cation*; (ii) compounds with a *complex anion*; (iii) *non-electrolytes* (compounds with an uncharged complex, having an electrochemical valence of zero); (iv) compounds built up from complex cations and complex anions; examples of the last will be found in Vol. II.

Werner originally classified complex compounds into substitution or penetration compounds and addition compounds (German 'Einlagerungsverbindungen' and 'Anlagerungsverbindungen'). The simplest illustration for the formation of a substitution compound is the formation of a hydrate when a salt is dissolved in water:

$$CaCl_2 + 6H_2O = [Ca(H_2O)_6]Cl_2$$

The water is interposed *between* the Ca^{++} ion and the Cl^- ions, whereby the latter are forced away from the Ca^{++} ion.

We have already encountered one of the simplest examples of the formation of an addition compound, in the formation of fluoroboric acid from boron trifluoride and hydrogen fluoride (see above). No expulsion of any fluoride ion takes place here when hydrogen fluoride and boron trifluoride come together; the hydrogen fluoride is not introduced in some way in between, but is simply added on.

Not all complex compounds can be assigned unambiguously to one or other of these classes, however. Thus tetramethyl ammonium iodide, $[N(CH_3)_4]I$ can equally well be regarded as a penetration compound of $N(CH_3)_3$ in CH_3I, as an addition compound of CH_3I with $N(CH_3)_3$. With the development of modern valency theory, the distinction between these two classes of complex compounds has lost its significance.

2. Detection and Structure of Complexes

(a) Detection of Complexes

Strong complexes can be identified in the same manner as ordinary radicals—by their ability to pass unaltered from one substance to another. Thus, in fluoroboric acid, the hydrogen ion may be replaced by potassium or other metal ions The group $[BF_4]$—which can therefore also be called the radical of fluoroboric acid—thereby remains unchanged.

Strong complexes may also be recognized by the electrical conductivity of solutions of their salts. The molar conductivity of a dilute salt solution depends substantially only on the *number of ions* into which the salt dissociates when it is dissolved in water; the nature of the constituents from which the salt is formed has relatively little effect upon the molar conductivity of a normal salt at high dilutions. By measuring the molecular conductivity of highly dilute aqueous solutions (it is usual to choose the molecular concentration $^1/_v = {}^1/_{1024}$—see below) it is possible to deduce the number of ions into which the salt has dissociated.

In Table 72 are compared the molecular conductivities μ at 25°, and at the dilution $v = 1024$ (i.e., in solutions containing 1 g mol of the salt in 1024 liters of solution). It may be seen that the molecular conductivity at the given concentration falls into a certain range of values, according to the type to which the salt belongs. For salts dissociating into 2 univalent ions, it is close to 130, for salts which dissociate into 2 univalent ions and 1 bivalent ion, it is around 270, for salts forming 3 univalent ions and 1 trivalent ion, about 400, and for salts forming 4 univalent ions and 1 quadrivalent ion, in the neighborhood of 550. Thus it is possible to deduce the type of salt from the molecular conductivity, and, with salts of unknown consitution, to decide how many ions they form by dissociation. The composition of the individual ions must be deduced from other data—e.g., from the chemical reactions.

TABLE 72

MOLAR CONDUCTIVITIES OF SALTS, AS DEPENDENT ON THE NUMBER OF IONS

(Molar conductivities at 25° C and dilution $v = 1024$)

Salt	Formula type	Type	Number of ions	Molar conductivity
NaCl		uni-univalent electrolyte		124
KClO$_3$	AB		2	136
AgNO$_3$				132
BaCl$_2$		bi-univalent electrolyte		270
Mg(NO$_3$)$_2$	AB$_2$		3	251
K$_2$SO$_4$				298
AlCl$_3$	AB$_3$	tri-univalent electrolyte	4	384
LaCl$_3$				397
Na$_4$[Fe(CN)$_6$]	AB$_4$	quadri-univalent electrolyte	5	514
K$_4$[Fe(CN)$_6$]				583

Salts of the same valence type agree still more closely in the manner that the molecular conductivity varies with dilution. It is therefore, usual to carry out a series of measurements at increasing dilutions, doubling the dilution at each stage. Starting from a 1-molar solution ($v = 1$), one thus obtains a series with the dilutions $v = 2, 4, 8, 16, 32, 64, 128, 256, 512, 1024$, etc. For strong complex salts, the dilution curve is completely analogous to that of a simple salt. In the case of weaker complexes, it is often possible to infer the mode of dissociation from the form of the dilution curve.

A further possibility for recognizing both the presence of a complex and (in conjunction with other methods) its nature, is provided by the observation that chemical reactions typical of the substances present in solution do not take place.

Thus, in the solutions of fluoroboric acid or its salts, there is no reaction with water, to form boric acid, a reaction typical of one of the constituents, boron trifluoride.* Similarly, the solution does not give the reactions characteristic of fluoride ions—e.g., precipitation of calcium fluoride upon addition of calcium ions. Thus the solution contains neither boron trifluoride nor fluoride ions in appreciable amounts. In conjunction with measurements of the electrical conductivity, it may be concluded that a complex ion [BF$_4$]$^-$ is present in the solution.

* Since the fluoroborate complex is not a very strong complex, a partial decomposition with water does generally occur. It is completely inhibited, however, if the dissociation of the complex according to the equation [BF$_4$]$^- \rightleftharpoons$ BF$_3$ + F$^-$ is repressed by the addition of an excess of fluoride ions.

It is inferred similarly that complex ions are formed in aluminum salt solutions containing certain organic substances, such as tartaric acid, from the fact that such solutions fail to give many reactions typical of aluminum, or the aluminum is but incompletely precipitated. In this instance it has not yet been found possible to determine the actual composition of the complexes formed.

Those properties of solutions which depend upon the osmotic pressure, such as the lowering of the freezing point and elevation of the boiling point, also afford evidence of the existence and composition of complexes in solution, inasmuch as the number of molecules present in the solution is affected thereby. Measurements of the freezing point depression and boiling point elevation have often been used to confirm, by an independent method, the conclusions drawn from conductivity measurements. Their field of application is, however, far wider than that of conductivity measurements, since they also make it possible to detect the formation of complexes between ions and neutral ligands, which cannot be detected from conductivity measurements.

Other methods which furnish information as to the *number of molecules* or the *concentration of some molecular species* in a solution may be used to detect complexes— e.g., potentiometric measurements, and equilibrium studies.

Evidence for the presence of complex ions can often be derived from measurements of electrolytic transference (direction of ionic migration). If the transference numbers show that the electric current has been transported to the *anode* by some element which, by itself, would necessarily exist in the solution in the form of *positive ions* (as is true of all the metallic elements), it follows that the element in question must be present in solution in the form of negatively charged (anionic) complexes.

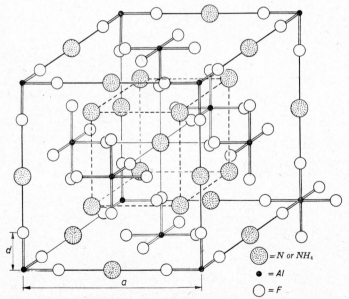

= N or NH$_4$

= Al

= F

Fig. 75. Unit cell of ammonium hexafluoro-aluminate, (NH$_4$)$_3$[AlF$_6$].
$a = 8.40$ Å;   Al—F $= d = 1.66$ Å

It may be recalled that conclusions as to the direction in which the components of an electrolyte are transported by the current can be drawn only from the changes in concentration in the neighborhood of the anode and cathode, and not from the site at which electrodeposition takes place. Lead is deposited at the anode from concentrated nitric acid solutions because it is there oxidized to the insoluble dioxide, and not because it is transported there as a complex ion. Quite weak (highly dissociated) complexes can often be detected by means of transference measurements.

During recent years, these methods have been supplemented by X-ray determinations of the crystal structures of crystalline coordination compounds. As stated in Chap. 7, one result of structure analysis has been to show that the chemical radicals generally function in crystals as distinct structural groups—e.g., the radicals $[NO_3]^-$ and $[SO_4]^{2-}$ (cf. Figs. 50 and 51, pp. 217, 218).

Fig. 75, which represents the unit cell of ammonium hexafluoro-aluminate, $(NH_4)_3[AlF_6]$, shows how a complex radical in the stricter sense is met with also as a structural group in the crystal lattice. It may be seen that the F-atoms or ions in this are so arranged that 6 of them surround every Al atom octahedrally. Only those parts of each octahedron which belong to a single unit cell are shown in the figure; only one octahedron (at the bottom right hand side) is completed by the addition of three fluorine atoms lying outside the unit cell. If the N-atoms (with the H atoms belonging to them) were taken away from the middle of the cell edges and the cube center, one would have the unit cell of ammonium hexafluorosilicate, $(NH_4)_2[SiF_6]$, which will be discussed in the next chapter. Many compounds of analogous composition—e.g., potassium hexachloroplatinate, $K_2[PtCl_6]$—have been found to possess the same structure.

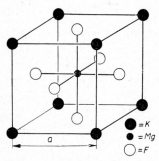

Fig. 76 represents the crystal structure of potassium magnesium fluoride, $KF \cdot MgF_2$. This compound provides an example of a double salt which is almost completely dissociated into its components (or their ions) in aqueous solution, although, with certain limitations, it is possible to speak of complex formation in the crystalline state. All the fluoride ions can definitely be assigned more closely to the magnesium ions than to the potassium ions. Each magnesium ion is surrounded by 6 fluoride ions at

Fig. 76. Crystal structure of potassium magnesium fluoride, $KF \cdot MgF_2$ (perovskite type). $a = 4.00$ Å

a distance $a/2 = 2.00$ Å, whereas the shortest distance between potassium ions and fluoride ions is $\sqrt{2}\, a/2 = 2.83$ Å—i.e., as great as the distance of the fluoride ions from each other. The coordination number of magnesium in the crystalline compound is 6. It must be noticed, however, that each of the 6 fluoride ions which surround a magnesium ion is shared also with a second magnesium ion situated in an adjoining unit cell, so that stoichiometrically there are only 3 fluoride ions to each magnesium ion. The lattice type represented in Fig. 76 is fairly frequently met with among compounds of the general formula MRX_3, and especially in a whole series of oxygen salts MRO_3—e.g., in calcium titanate $CaTiO_3$ (perovskite). It is usually called the 'perovskite type' after the latter compound.

Since numerous instances are known in which sparingly soluble compounds become soluble through complex formation, the converse inference is frequently assumed to be a valid generalization,—that complex formation must have occurred whenever the solubility is increased by the addition of certain salts to the solvent. This conclusion is not necessarily justifiable for *small* increases in solubility, however, since modern theories of strong electrolytes lead to the conclusion that it is possible for the solubility of sparingly soluble salts to be raised by the addition of electrolytes with a common ion. On the classical theory, on the other hand, such an increase could occur, in the absence of complex formation, only through the addition of electrolytes furnishing different ions.

The formation of complexes is frequently revealed by changes in *color*. Thus anhydrous copper sulfate is white, whereas the usual hydrate, is blue, as is the aqueous solution. It is therefore assumed that in the blue salt the copper ion contains complex-bound water.

The water contained in crystallized compounds is generally termed *water of crystallization*. It is often very difficult to decide to what extent this is 'complex bound'. In the last analysis, the same forces are responsible for the structure of the crystal as for the formation of compounds of higher order, and to a restricted sense it is permissible to speak of the 'coordinative binding' of water of crystallization. More strictly, however, this term, or the equivalent expression 'complex-bound' water, should only be applied to the binding of water in crystals when the water molecules can be assigned unambiguously to a particular cation, around which they are grouped, so that together they make up a more or less self-contained structural unit. Evidence for this can be provided by X-ray structure analysis in many instances. In general, it might be expected that the water molecules in such structural groups would be more firmly bound than when incorporated in other ways in the crystal lattice. Water that is readily eliminated is therefore ordinarily regarded as water of crystallization in the narrower sense, and that which is removed with more difficulty is considered to be complex bound. It is impossible, however, to draw a definite distinction on solely the grounds of greater or lesser ease of dehydration. The same applies to those ammoniates which are known only in the solid state.

(b) Structure of Complexes [4, 5]

Before Werner's work, attempts were made to assign structural formulas to complex compounds, in terms of the older valence theory, i.e., considering them as compounds of the first order. This led to chain-like formulas, such as were proposed by Blomstrand, in particular for the metal ammines.

For the case of fluoroboric acid, Blomstrand's interpretation would lead to the structure:

$$\begin{array}{c} F \\ \diagdown \\ B—F—F—H. \\ \diagup \\ F \end{array}$$ Underlying this formula is the quite arbitrary assumption that two of the

fluorine atoms in the compound act as bivalent, which is in contradiction with the conclusion drawn from all the simple compounds of fluorine, that that element is invariably univalent. Similar considerations apply in many other instances.

Before discussing the structure of such compounds on the basis of Werner's ideas, it may be well to summarize briefly what statements can be made from a direct interpretation of the experimental evidence in terms of the ionic theory. We have established that boron trifluoride can add on a fluoride ion. The neutral boron trifluoride molecule is thereby converted into a singly negatively charged ion, $[BF_4]^-$, which we call a complex ion—$BF_3 + F^- = [BF_4]^-$.

The complex ion so formed can associate itself with various cations—H^+, K^+, $[NH_4]^+$, Ba^{++}, etc.—to form compounds; the hydrogen compound, fluoroboric acid, is a stronger acid than hydrofluoric acid.

Since the cation with which the complex ion is combined can be varied at will within certain limits, and the free complex ion can be detected in solutions, we can consider it without reference to any particular cation.

What then is the structure of the complex anion $[BF_4]^-$? Since we can reject the assumption that the fluoride ions are linked together in chains, as being incompatible with the univalence of fluorine, there remains only the assumption that *all four fluorines are bound to the boron*. The question next arises whether they are equivalent, or differ in binding. In no case has evidence been found for differences in binding in complexes containing only one kind of atom or group bound to the central atom. We therefore conclude that the binding of fluorine atoms is equivalent.

In Werner's time, valence forces were regarded as directed, individual forces. If strictly applied, this conception does not permit of additional valences (Werner's auxiliary valences) being exerted in the formation of coordination compounds. Once this view is rejected, by setting up 'central' formulas in place of chain formulas, no valid basis remains for assuming that the valence forces originally exercised in the simple compound remain in some way differentiated in the complex. Evidence for such a view might have come to light as the experimental material was extended. Werner proved the contrary, however, from the examples of the metal ammines and their derivatives. From a study of the possible types of isomerism he showed that the coordinatively bound groups are arranged in spatially equivalent positions around the central atom, and so are all bound with equal strength. At the present time, the hypothesis of valence as a directed individual force is recognized as applicable only to a certain field, so that the assumption of equivalent binding is the most obvious one, to be made in every case unless experimental evidence points directly to the contrary.

The components bound to the complex—i.e., the hydrogen ion in the case of fluoroboric acid, potassium ion in potassium fluoroborate—are dissociated off as ions when the compounds are dissolved in water. Werner therefore spoke of them as ionically or ionogenically bound, as distinct from the non-ionogenically bound atoms, ions or groups directly surrounding the central atom. Werner assumed that the ionogenically bound groups were not bound to any particular atom of the complex, but to the complex as a whole. This also follows as an obvious inference from the experimental evidence. In solution, an electrically charged complex can act at a distance only as a unit, and it is to be expected that it will behave in substantially the same way after the ions have come together to form a crystal. Such X-ray structure analyses as have hitherto been made of strongly complex coordination compounds fully substantiate this assumption. A strong complex ion occupies a lattice site in a crystal in just the same way as does a NO_3 or a SO_4 radical (cf. Fig. 75).

Werner assumed that atoms which function as the central atoms of complexes tend to surround themselves with a definite number of other atoms or groups. Just as the ordinary, or formal, valence determines the formation of the simple compounds, so, in Werner's view, the coordination number or coordinative valence determines the formation of coordination compounds, and it makes no difference whether the coordinative valences are satisfied by simple atoms or by groups. A difference exists only in so far as there are certain groups that can occupy two coordination positions, which a simple atom can never do.

Groups capable of filling two coordination positions (groups with the coordination valence 2, bi-dentate groups or chelate groups—greek $\chi\eta\lambda\dot\eta$, a crab's claw) are, especially, hydrazine, ethylenediamine and the oxalate ion. [6] Their formulas show how it is that they can occupy two coordination positions.

$$H_2N—NH_2 \qquad \begin{array}{c} H_2C—CH_2 \\ | \quad | \\ H_2N \quad NH_2 \end{array} \qquad \left[\begin{array}{c} OC—CO \\ | \quad | \\ O \quad O \end{array}\right]^=$$

They possess two spatially separate positions from which the coordinative valences can act. A group which occupies 2 coordination positions is called a two-fold group. It may be noticed that an atom with the *coordination number n* is said to be *coordinative y n-valent*; a group with a *coordination valence* of *n* is said to be an *n-fold group*.

The fact that the coordination number 6 is especially favored by many elements, whereby the non-ionogenically bound groups are arranged in a completely symmetrical manner, at the corners of a regular octahedron, led Werner to the conclusion that, in the formation of coordination compounds, particular importance attaches to the tendency to attain the most symmetrical steric arrangement. The development of the quantum mechanical theory of valence has provided an explanation why certain coordination numbers (4 and 6) and certain symmetrical configurations (tetrahedral, square planar and octahedral) should be favored in strong complexes, and why the tendency to form strong complexes should be most marked among the elements in the center of the long periods of the Periodic Table (see Vol. II).

3. Theoretical

(a) The Origin of Coordinative Valence

The formation of coordination compounds can be explained equally well either on the assumption that they arise from the forces of attraction (Coulomb forces) between oppositely charged atoms, or on the assumption that the forces involved are the same wave mechanical resonance forces which give rise to covalent bonds. It has, in fact, emerged that *both interpretations* are valid. For many coordination compounds, the assumption that they are formed by the attraction between oppositely charged ions, or between ions and polar molecules, provides a satisfactory explanation of their existence, and the properties and chemical behavior of the compounds are in accordance with what would be predicted on the basis of this assumption. There are many coordination compounds, however, which it is impossible to explain on this hypothesis. Their composition, properties and chemical behavior all show that the chemical bonds in them are of quite a different kind, and are in fact covalent bonds.

Many coordination compounds could equally well be interpreted from either standpoint. This is the case with the *fluoroborates*, and both theories of the formation of coordination compounds will therefore be developed with reference to this complex ion.

(b) Application of Kossel's Theory to the Fluoroborate Ion

Kossel's theory treats atoms of different electrochemical character essentially as spherical particles, charged according to their valence, and acting on one another in accordance with Coulomb's law. The existence of complexes such as the fluoroborate ion is then found to be not only comprehensible, but is a *necessary consequence* of the properties of the system. If it is permissible to regard boron as the electropositive constituent of the compound BF_3, it may be shown that the same force responsible for the formation of boron trifluoride from electropositive boron and electronegative fluorine must also enable the boron to bind a further fluoride ion, and to form the complex ion $[BF_4]^-$.

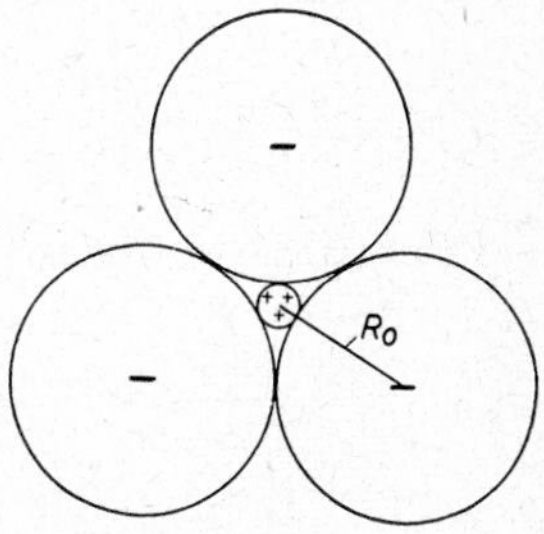

Fig. 77.
Model of boron trifluoride.

A structure consisting of a triply positively charged atom, surrounded by three singly negatively charged atoms (Fig. 77), behaves essentially as an electrostatically neutral group at large distances. On negatively charged atoms in its immediate neighborhood, however, it exerts an attractive force, the magnitude of which can be calculated from Coulomb's law (cf. p. 83). Let us assume that Fig. 77 represents a molecule of boron trifluoride. The radius of B^{3+} is so smal that the three F^- ions will be in contact with each other. Neglecting the compressibility and mutual polarization of the ions, and taking the dielectric constant as that of empty space ($D = 1$), we may calculate the energy of formation E_0 for a boron trifluoride molecule from B^{3+} and $3F^-$ on the basis of this model as

$$E_o = A_1 - A_2 = \frac{3 \cdot 3e^2}{R_0} - \frac{3e^2}{R_0 \sqrt 3} = \frac{7 \cdot 27}{R_0} e^2$$

where

$A_1 = $ energy gained, from the mutual attraction of oppositely charged ions
$A_2 = $ energy expended in overcoming repulsion of similarly charged ions
$R_0 = $ distance between centers of the boron and the fluorine atoms.

Taking for the radius of the fluoride ion the value it has in crystalline compounds (1.33 Å), it follows from the geometry of Fig. 77 that $R_0 = 1.54$ Å.

We then obtain for the energy of formation of the BF_3 molecule

$$E_0 = 4.71 \cdot 10^8 e^2 \text{ ergs.}$$

We can calculate the energy of formation of the complex ion $[BF_4]^-$ in the same way, on the assumption that this is built up from B^{3+} and four F^- arranged tetrahedrally about it. We may denote the distance between the centers of B^{3+} and F^- as R_1; this is rather greater than R_0, namely 1.63 Å.

We then obtain for the energy of formation E_1 of the complex ion $[BF_4]^-$, from B^{3+} and $4F^-$

$$E_1 = \frac{4 \cdot 3e^2}{R_1} - \frac{3\sqrt 6 e^2}{2R_1} = \frac{8.33}{R_1} e^2 = 5.11 \cdot 10^8 e^2 \text{ ergs}$$

E_1 is larger than E_0. There is thus a gain in energy associated with the formation of $[BF_4]^-$ from BF_3 and F^-. Hence the reason for the formation of the complex ion—i.e., for the exercise of the 'auxiliary' valence—is at once to be understood. If, in the same way, the energies of formation of $[BF_5]^{2-}$ and of $[BF_6]^{3-}$ from B^{3+} and 5 or 6 F^-, respectively, are calculated, the values obtained are $4.53 \cdot 10^8 e^2$ ergs and $4.27 \cdot 10^8 e^2$ ergs respectively. Since the energy of formation of these complexes is smaller than that of the formation of the $[BF_4]^-$ ion, there is no reason for these complexes to be formed. They have, in fact, not been observed.

The formation of *hydroxo-ions*—e.g., of $[Al(OH)_4]^-$, $[Al(OH)_5]^{2-}$ and $[Al(OH)_6]^{3-}$ from $Al(OH)_3$ can be explained in a similar way. By taking account of polarization, the binding of neutral molecules, such as H_2O, NH_3, etc. can also be interpreted as the consequence of purely electrostatic attraction. This treatment is only valid in certain cases, however, namely when only loose addition of the neutral molecules is involved.

(c) Application of the Theory of the Covalent Bond to the Fluoroborate Ion

The first attempt to apply the theory of the covalent bond to coordination comn pounds was made by Sidgwick, in 1923. The essential feature was the assumptioo that the two paired electrons which constitute the covalent bond between tw, atoms might both be contributed by the same atom ('coordinate covalence' 'dative covalence', 'semipolar bond'—see p. 138), instead of one from each atom.

If this is assumed, then the formation of the ion $[BF_4]^-$ from BF_3 and F^- is to be expected even if the B—F bonds are pure covalences. The formation of the fourth bond would be most simply explained if BF_3 had the constitution corresponding to formula[I]. In this case, the boron in BF_3 would have an incomplete outer electron shell, in which an additional electron pair could be accommodated. However, the constitution of BF_3 is perhaps better represented by formula [II] (see p. 334). On this basis, the coordination of the F^- ion to the BF_3 molecule is a consequence of the unsaturated character of the latter, associated with the double bond:

$$
:\overset{..}{\underset{..}{F}}:\overset{..}{\underset{..}{B}}:\overset{..}{\underset{..}{F}}: \ + \ :\overset{..}{\underset{..}{F}}:^- \ = \ \left[\begin{array}{c} :\overset{..}{F}: \\ :\overset{..}{\underset{..}{F}}:\overset{..}{\underset{..}{B}}:\overset{..}{\underset{..}{F}}: \\ :\overset{..}{\underset{..}{F}}: \end{array}\right]^- \qquad :\overset{..}{F}:\overset{..}{\underset{..}{B}}:\overset{..}{\underset{..}{F}}: \ + \ :\overset{..}{F}: ^- \ = \ \left[\begin{array}{c} :\overset{..}{F}: \\ :\overset{..}{\underset{..}{F}}:\overset{..}{\underset{..}{B}}:\overset{..}{\underset{..}{F}}: \\ :\overset{..}{\underset{..}{F}}: \end{array}\right]^-
$$

$$
[I] \qquad\qquad\qquad\qquad\qquad\qquad [II]
$$

According to Pauling, the B—F bond has about 63% ionic character. In the ions $[SO_4]^{2-}$ and $[ClO_4]^-$, of analogous composition, the bonds are predominantly covalent, with only about 22% and 7% ionic character, respectively. The predominantly covalent consitution of these ions has also been proved by physical investigations. The theory of the covalent bond enables the formation of the $[SO_4]^{2-}$ and $[ClO_4]^-$ ions to be interpreted in just the same way as that of the $[BF_4]^-$ ion. Applications of Kossel's theory also leads, purely formally, to the same result; nevertheless, it is not really justifiable to use the purely electrostatic model in such cases.

(d) Relations between Ordinary and Coordinative Valences

In terms of the foregoing discussion, the difference between ordinary and coordinative valence is contained in the statement that in the formation of (ionic) compounds of the first order from their elements, two processes occur; (i) transfer of electrons between the atoms, which thereby become charged, (ii) binding of the oppositely charged atoms to each other. In the formation of coordination compounds, since the first process has already taken place, only the second occurs, and as many atoms or groups associate together as can do so with a gain in energy.

The same happens in the formation of crystals, except that here the energy of formation is given not by the association of a very few atoms, to form a complex, but of all the ions, to form a crystal lattice.

4. Penetration Complexes

(a) 'Penetration' Complexes and 'Normal' Complexes

When covalent bonds are involved, it is often found that extremely strongly complexed compounds are formed. Not only radicals, but also neutral groups such as NH_3, may be very firmly bound to the central atom by covalent bonds. By contrast, coordination compounds in which the binding forces are merely ion-ion attractions, or ion-dipole forces (without any strong contribution from covalent bonding forces) generally have the character of weakly complexed compounds. This is a characteristic difference between complex compounds of the two classes. Those in which complex formation is due to the operation of covalent bonds are often termed *penetration complexes*, as proposed by Biltz, since the establishment of a covalence involves the mutual interpenetration of the electron shells of the central atom and the ligands. The existence of covalent bonds between the central atom and the ligands has in many cases been proved directly by *magnetic measurements* (see below).

Biltz applied the term 'normal' complex to coordination compounds in which the ligands were bound by electrostatic or Van der Waals forces. It is probable that systems in which compounds of non-electrolyte character are formed are generally penetration complexes (e.g., $[CrCl_3(NH_3)_3]$).

A characteristic difference between 'penetration' complexes and normal complexes is that the coordination number of the central atom is more or less constant in the former type, whereas in 'normal' complexes the coordination number may vary; the composition of the complex ions formed under given conditions may vary with the proportions in which the components are present.

Like most of the other criteria, this last distinction between normal and penetration complexes is not sharp. The coordination number may also be a constant in 'normal' complexes, as is evident from the calculation given above, for the energy of formation of the $[BF_4]^-$ ion as calculated on the assumption of purely ionic bonding. Furthermore, variability of coordination number is not readily established. It is not always possible to draw conclusions as to the constitution of the compounds formed in a system from the gross composition of the solid compounds which may be isolated, especially if full structural knowledge is lacking. Where a metallic salt forms several hydrates or ammoniates, it does not follow that all the H_2O or NH_3 molecules are coordinated to the metal atom; this is well known in the case of H_2O-rich hydrates, and is probably true also of ammoniates with high NH_3 content. Furthermore, it cannot be assumed that the coordination number of the metal atom changes in passing from the higher to the lower hydrates or ammoniates in the series. In solution, it sometimes is possible (see below) to detect the existence of coordination compounds with varying numbers of ligands, even in systems (e.g., the complex iron(II) phenanthroline compounds, cf. Vol. II) in which the solid compounds isolated from the system are typical penetration complexes. This does not necessarily imply, however, that the coordination number is variable. In aqueous solution, water molecules are present in large excess, and every metallic ion is necessarily coordinated with solvent molecules, held by ion-dipole forces. Other ligands can form coordination complexes only by competition with and displacement of H_2O molecules, and the species that exist in solution are present in proportions that depend upon the position of equilibrium in the displacement reactions. The difference in stability between normal and penetration complexes therefore reflects the much greater free energy change accompanying formation of covalent bonds than that arising from creation of ion-dipole bonds.

Valuable information on the thermodynamic properties, constitution, and mode of formation of complex cations or complex anions in solution is therefore afforded by equilibrium studies. Consider the equilibria established in a solution, containing a metallic ion M^{n+} and a ligand A (either a neutral molecule or an acid radical); the resulting complex $[MA_N]$ may be a complex cation or a complex anion, depending on the charges borne by the cation M^{n+} and the ligand A, respectively. There will then be a set of successive equilibria:

$$M + A \rightleftharpoons MA \qquad\qquad k_1 = \frac{[MA]}{[M][A]}$$

$$MA + A \rightleftharpoons MA_2 \qquad\qquad k_2 = \frac{[MA_2]}{[MA][A]}$$

$$MA_{N-1} + A \rightleftharpoons MA_N \qquad\qquad k_N = \frac{[MA_N]}{[MA_{N-1}][A]}$$

The over-all equilibrium constant, $K = k_1 k_2 k_3 \cdots k_N = \dfrac{[MA_N]}{[M][A]^N}$

measures the total free energy of the complex forming reaction, $\Delta G^\circ = -RT \ln K$.

If the ligand A is a weak Brönsted base (e.g., if it is the anion of a strong acid, or is NH_3, an organic amine, etc.), it is involved in a second equilibrium with hydrogen ion: $A + H_3O^+ \rightleftharpoons AH^+ + H_2O$. and it is possible to measure the amount of free (i.e., uncomplexed) ligand in the system [A]. The set of successive equilibria, although mathematically complex, can be analyzed by the methods devised by Bjerrum ('*Metal ammine formation in aqueous solution*, Haase, Copenhagen, 1941; see Burkin, *Quart. Rev. Chem. Soc.*, 5, (1951) 1, Irving and Williams, *J. Chem. Soc.*, (1953) 3192). Such analyses of the process of complex formation in solutions have shown that, in general, it is reversible, taking place by successive steps. The equilibrium constants in aqueous solution for different metallic central ions, and for different ligands, cover a wide range of magnitudes, however. Those compounds (largely compounds of the transition metals) which have been classified by the magnetic criterion and otherwise as penetration complexes are largely those for which the over-all equilibrium constants correspond to virtual irreversibility of formation; the solution equilibria and solution reactions of complexes thus afford a measure of the strength of the interaction between the central ion and the ligands.

(b) Characterization of Penetration Complexes by Magnetic Measurements

Simple salts of bivalent iron, such as $FeCl_2$ and $FeSO_4$, have magnetic moments of about 5 Bohr magnetons, as also have the compounds $[Fe(H_2O)_4]Cl_2$, $[Fe(NH_3)_2]Cl_2$, $[Fe(NH_3)_6]Cl_2$ and other weakly complexed salts of iron. On the other hand, the compounds $K_4[Fe(CN)_6]$ and $K_2[Fe(CN)_5(CO)]$ have magnetic moments which are practically zero. It may be concluded that in the compounds cited first, iron is present in the same form as in its simple salts, and that the ligands are bound only by weak ion-dipole (or Van der Waals) forces. Some other kind of binding force is evidently operative in the two latter compounds, involving changes in the population of electron orbits. It can be shown theoretically that diamagnetism would be expected if each CN^- group formed a covalent bond with the iron. Magnetic measurements can be taken as supporting the view put forward by Biltz that the complex compounds fall into two classes—the normal complexes, in which the ligands are held by ionic attraction or ion-dipole forces, and penetration complexes, in which the ligands form covalent bonds with the central atom. The $[FeF_6]^{3-}$ complex is a normal complex, with the magnetic susceptibility of the Fe^{3+} ion, If the $[Fe(CN)_6]^{3-}$ is a penetration complex, its magnetic moment should be 1.73 Bohr magnetons; the value found is about 2 Bohr magnetons. Similarly, the metal carbonyls $Cr(CO)_6$, $Fe(CO)_5$, $Ni(CO)_4$ are all diamagnetic, as would be expected theoretically if they are covalent compounds.

5. Formation of Complexes of the Transition Elements

Many of the transition elements (Vol. II) display a very striking capacity for forming strong complexes. Their observed behavior can be interpreted on the basis of quantum mechanics, as Pauling has shown. The ions of the transition metals are characterized by the presence of incompletely filled *d* electron levels. Some or all of the electrons in the incompletely filled *d* levels are unpaired*. Thus the Cr^{3+} ion has *three* unpaired electrons, the Mn^{2+} ion has *five* unpaired electrons, the Fe^{2+} has two paired and *four* unpaired electrons. Under the influence of ligands which interact sufficiently strongly with these ions, these electrons may pair up, whereby a larger or smaller number of the *d*-orbits originally singly occupied, become vacant. The energy of these *d*-levels is very little lower than that of the *s*- and *p*-levels of the next higher electron shell; it is therefore possible for a set of new levels to be set up by *hybridization* from the *d*-, *s*- and *p*-levels (e.g., the 3*d*, 4*s* and 4*p*-levels in the first transition series), for occupation by electrons of the ligands. The energy which must be expended in 'pairing up' the originally unpaired electrons** is made up from the resonance energy associated with the hybridization. The electrons which effect the covalent bond are all supplied by the ligands; the central atom only provides energy levels (hybrid levels) in which they can be accommodated. Thus the Co^{3+} ion has two paired and four unpaired 3*d* electrons. When these four electrons pair up, two 3*d* levels are freed. With the 4*s* level and the three 4*p* levels, these combine to form a hybrid set of levels in which $2 \times 6 = 12$ elec-

---

* In the first transition series, at least, the ions tend to conform to Hund's principle of *maximum multiplicity*—i.e., as many electron orbits as possible are singly occupied before any pairing of electrons takes place.

** It is clear that in this case energy must be *expended* in order to couple the electron spins, since in the normal state of the ions the electrons are unpaired.

trons can be accommodated. They may be utilized, for example, to form covalent bonds with 6 NH_3 molecules, each of which donates its 'lone pair' of electrons:

$$Co^{3+} + 6NH_3 \rightarrow \begin{bmatrix} H_3N & & NH_3 \\ H_3N & \rightarrow Co \leftarrow & NH_3 \\ H_3N & & NH_3 \end{bmatrix}^{3+} = \begin{bmatrix} H_3N & & NH_3 \\ H_3N & Co & NH_3 \\ H_3N & & NH_3 \end{bmatrix}^{3+}$$

The first way of writing the formula brings out the way in which the compound originates, the electrons being donated in pairs by the nitrogen atoms of the NH_3 (the free NH_3 molecule has a 'lone pair' of electrons). The second way of writing it emphasizes the fact that the covalent bonds present in the complex ion are in no way different in kind from covalences in any other compound.

It might appear that the mechanism of bond formation just considered would lead to an improbable distribution of positive and negative charges within the complex. This is, however, not necessarily the case, as Pauling has shown. The covalent bonds between Co and N, and between N and H, are formed between atoms of considerably different electronegativity. It follows (Chap. 5) that they will partake of some ionic character. The result of this will be so to redistribute the polarities within the complex that the cobalt and nitrogen atoms are left with little residual charge, and the positive charge of the cation as a whole is distributed very largely among the H atoms, which constitute the positive ends of the N—H dipoles.

Pauling was able to show that, as in the case of carbon, the hybridization of d, s, and p levels gives rise to bonds which are *strongly directed in space*. Hybridization of two d-levels with an s- and three p-levels produces six $[d^2sp^3]$ bonds, *directed towards the vertices of a regular octahedron*. This explains the octahedral configuration of the typical strong complexes of coordination number 6, which Werner was able to deduce from the phenomena of isomerism that they display (see Vol. II, Chap. 5). Hybridization of three d-levels with an s-level, or (as in carbon) of an s-level with three p-levels produces *four* bonds ($[d^3s]$ or $[sp^3]$ bonds) directed towards *the vertices of a regular tetrahedron*. Hybridization of one d-level with an s-level and two p-levels produces *four coplanar bonds*, directed towards the corners of a square. This square planar configuration is, in fact, that which is observed in certain complex compounds—e.g., some of those of nickel(II), those of palladium(II), platinum(II) and gold(III). This is of significance in demonstrating the importance of directed covalences in complex formation, since the electrostatic force model necessarily leads to the tetrahedral configuration as the equilibrium configuration of a system with four ligands.

The phenomena of isomerism which are associated with the structure of complex compounds will be discussed in Vol. II, in conjunction with the compounds of chromium, cobalt, platinum, etc., since most of the studies of these phenomena have been concerned with the compounds of those elements.

6. Nomenclature

A survey of the enormous number of coordination compounds known is made much simpler by introducing a suitable *nomenclature* for them, indicating simply and clearly their composition and structure. Werner worked out a uniform scheme for this purpose. Assignment of a rational name to a coordination compound presupposes that the essentials of its structure are known. If this is not the case, it is necessary to retain the terminology used in naming double salts and other kinds of addition compounds.

To express the fact that any element or group is coordinatively bound, it is given the ending —o, and is made to precede the name of the element to which it is bound. Ammonia is an exception to this rule, and recieves the designation *ammine* (note a*mm*ine, to be distinguished from the a*m*ino group —NH_2) in the names of coordination compounds. Coordinatively bound water is denoted by *aquo*-, oxygen by *oxo*-, sulfur by *thio*- and the singly bound hydroxyl group —OH by *hydroxo*-.

Others are *chloro-*, *fluoro-* (or *fluo-*), *sulfato-*, *carbonato-*, *nitrato-*, *oxalato-* and other *acido-* groups. The ethylene diamine group, NH_2—CH_2—CH_2—NH_2, frequently introduced into complexes, is indicated by the abbreviation *en*.

In the name and formula of a complex, the negatively charged, coordinatively bound atoms or groups are written first, and then the neutral groups; of the latter, H_2O (aquo) is written first, then the substituted amines, and finally the NH_3 (ammine) groups.

The number of coordinatively bound groups is shown by greek numeral prefixes

1	2	3	4	5	6	7	8	9	10	12
mono	di	tri	tetra	penta	hexa	hepta	octa	ennea	deka	dodeka

When several different groups are present, each receives its appropriate numerical prefix. When these numerical prefixes are not essential, they are omitted; *mono-* is generally omitted.

Examples:

$[Ca(H_2O)_6]Cl_2$	hexaquocalcium chloride
$[Ca(NH_3)_6]$	hexamminecalcium
$[CrCl(H_2O)_2(NH_3)_3]SO_4$	chlorodiaquotriamminechromium(III) sulfate
$[Cr(OH)(H_2O) \, en_2](NO_3)_2$	hydroxoaquodiethylenediaminochromium(III) nitrate

A negative complex always receives the termination *-ate* in the names of salts. E.g., $K[BF_4]$ is potassium tetrafluoroborate or potassium fluoroborate. The corresponding acid, in Werner's nomenclature, is tetrafluoroboric acid, or fluoroboric acid*. This name brings out the full analogy to boric acid, and not to hydrofluoric acid.

The electrochemical valence (electrovalence) of the central atom in complex or coordination compounds is shown, in accordance with internationally agreed rules, in the same way as in simple compounds—i.e., by roman figures. In cationic complexes and in complex acids, the figure denoting the valence is written immediately after the name of the element. For the anionic complexes of salts, it is written after the suffix -ate at the end of the name.

Examples		
	$[Cr(NH_3)_6]Cl_3$	hexamminechromium(III) chloride
	$H_2[PtCl_6]$	hexachloroplatinum(IV) acid
	$K_2[PtCl_6]$	potassium hexachloroplatinate(IV)
	$K_4[Fe(CN)_6]$	potassium hexacyanoferrate(II)

These agreed rules of nomenclature are, however, not in universal usage. Werner proposed that the valence of the central atom should be shown by characterisitic endings e.g.:

for the valence	1	2	3	4	5	6	7	8
the ending	a	o	i	e	an	on	in	en

In naming complex groups in salts, the ending denoting the valence was interposed between the name of the central atom and the suffix -ate—e.g., hexamminechrom*i*chloride, potas-

---

* Abbreviations of this kind, in which numerical prefixes are omitted are permissible for well known compounds, and where no confusion can arise. In the same way, with elements which can function in only one valence state it is permissible to omit an indication of their valence.

sium hexachloroplateate. These endings have not been found convenient, and have in general not been used in English and American chemical literature*. They have accordingly gone out of use.

By giving the valence of the central atom, and the number and kind of the coordinatively bound groups, the composition of a complex is unambiguously shown. The charge on the complex ion is given by the algebraic sum of the charge on the central atom and the charges on the attached groups. The charge on the complex ion in turn at once determines the number of atoms or groups bound ionically—i.e., outside the complex.

If the valence of the central atom is not known with certainty—as for example, in compounds containing NO in the complex—it is not stated, and the number of atoms or groups bound outside the complex is then stated explicitly. In the case of uncharged complexes (non-electrolytes) it is not necessary to state the valence of the central atom.

There are also coordination compounds which contain *more than one* central atom. These are called *polynuclear*, as distinguished from the ordinary *mononuclear* compounds. If the central atoms are linked together through hydroxyl groups—which are then attached at both sides, instead of by one side only, as ordinarily—this is denoted by the syllable *-ol*. Compounds containing such doubly linked hydroxyl groups are then sometimes known as ol-compounds. Examples of such compounds will frequently be encountered later in this book. Other groups which can join two central atoms to each other are indicated by prefixing μ to their name—e.g., the decammine-μ-amido dicobalt(III) salts:

$$[(NH_3)_5\ Co\text{—}NH_2\text{—}Co\ (NH_3)_5]X_5.$$

* Their use in the German literature cut across the usual means of indicating the higher and lower valence states of the metals, so that some would interpret 'cuprocompounds' as being compounds of univalent copper, and some as those of bivalent copper. Similar confusion arose in several other cases.

References

1 A. WERNER, *New Ideas in Inorganic Chemistry* (translated by E. P. Hedley), London 1911, 268 pp.
2 R. WEINLAND, *Einführung in die Chemie der Komplexverbindungen*, 2nd Ed., Stuttgart 1924, 537 pp.
3 R. SCHWARZ, *Chemie der anorganischen Komplexverbindungen*, Berlin 1920, 71 pp.
4 W. HÜCKEL, *Structural Chemistry of Inorganic Compounds* (translated by L. H. Long), Vol. I, Amsterdam 1950, 438 pp.; Vol. II, Amsterdam 1951, 656 pp.
5 F. HEIN, *Chemische Koordinationslehre*, Zurich 1950, 683 pp.
6 A. E. MARTELL and M. CALVIN, *Chemistry of the Metal Chelate Compounds*, New York 1952, 613 pp.

CHAPTER 12

FOURTH MAIN GROUP OF THE PERIODIC SYSTEM: CARBON-SILICON GROUP

Atomic Numbers	Names of the Elements	Symbols	Atomic Weights	Densities	Melting points	Boiling points	Heats of Sublimation, kcal per g atom	Valence States
6	Carbon	C	12.011	3.51*	—	3850°	125	IV, III, II
14	Silicon	Si	28.09	2.33	1413°	—	85	IV, II
32	Germanium	Ge	72.60	5.35	958.5°	—	—	IV, II
50	Tin	Sn	118.70	7.28	231.8°	2362°	78	IV, II
82	Lead	Pb	207.21	11.34	327.4°	1750°	47.5	IV, II

* Diamond

1. General

Group IVA of the Periodic System contains the elements carbon, silicon, germanium, tin and lead. In the sequence of atomic numbers, carbon and silicon follow immediately after boron and aluminum, the first two elements of Group IIIA. They are, like these, separated from the previous inert gases—helium and neon respectively—by only a few places. Germanium, tin and lead follow the elements gallium, indium and thallium. These are widely separated from their foregoing inert gases by the 10 elements of the Sub-groups of the Periodic Table. (In the case of thallium, in fact, the 14 lanthanide elements are also interposed.) Just as the elements of Group IIIB show in their compounds a close similarity to the compounds of aluminum (Group IIIA), so the elements of Group IVB—titanium, zirconium and hafnium—show similarities in their compounds to the corresponding compounds of silicon, resemblances which are occasionally closer than those borne by the compounds of germanium, tin and lead. However, only these last three elements share with carbon and silicon the ability to act as quadrivalent towards both electropositive and electronegative elements, and to form volatile compounds with hydrogen. This property is especially characteristic of carbon the most important member of Group IVB. It is not so strongly marked in the case of silicon, with which the tendency to form oxygen compounds is dominant. The similarity between silicon and the elements assigned to the Sub-group, as regards certain classes of compounds, accords with the rule noticed in the preceding families, that the second member provides the transition between the elements of the Main group and those of the Sub-group.

The rule that the first element provides a transition to the next Main group is also followed in Group IV. The simplest oxygen compounds of carbon are volatile, like those of nitrogen, but unlike those of the homologues of carbon. In the carbonates, carbon has the maximum coordination number of 3, like that of nitrogen in the nitrates, whereas silicon functions in the silicates with the coordination number 4. The hydrocarbons resemble the hydrogen compounds of nitrogen in that they are not decomposed by water, as are the hydrides of silicon and its homologues.

Towards oxygen, halogens and other electronegative elements, the elements of Group IVA exert a maximum valence of 4, corresponding to the group number. In addition, they can function as *bivalent*. As well as CO_2 and CS_2, carbon forms CO and CS, though the latter is very unstable. The compounds SiO and SiS also have a low stability. Germanium has a greater tendency to be bivalent, forming a dichloride as well as GeO and GeS. The tendency is still stronger in tin, for which the stability of the bivalent and quadrivalent states is about equal, whereas with lead the bivalent state predominates over the quadrivalent.

The decrease in acidic character (or increase in basic character), from the lighter elements to the heavier, which is apparent in all the main Groups, is strengthened in Group IVA by an increase in the stability of the lower valence state, which runs in the same direction. It is quite general that the basic character is manifested more strongly in the compounds of lower valence, and the acidic character in those of higher valence. Where an element forms several oxides, the higher oxides invariably tend more towards acidic and less towards basic properties than do the lower.

Carbon and silicon are elements of distinctly non-metallic character,—*acid forming* elements. Germanium is also an acid forming element, although this property is exceedingly weakly developed in the bivalent state. Elementary germanium is generally considered to be a metal. Tin and lead are typically metallic in their physical properties. In its compounds, quadrivalent tin is predominantly acid forming; in the bivalent state it is amphoteric. With lead, the acidic character of the oxide is only weakly developed in the quadrivalent state; in the bivalent state lead is predominantly base-forming, though even here the capacity for acid formation is not quite lost.

The increase in metallic character from carbon to lead shows up very strikingly in the behavior of the elements towards metals, as well as towards non-metals, e.g., in their behavior towards the alkali metals. *Tin* and *lead* form typical *intermetallic compounds* with the alkali metals (cf. Table 87, p. 576). Some of these, however, can evince a measure of salt-like character under certain conditions, as is shown in the formation of 'polyanionic salts', such as $[Na(NH_3)_9]_4[Pb_9]$ (cf. Vol. II). *Carbon*, however, forms compounds of definitely salt-like character with the alkali metals—the *acetylides* (cf. p. 466). In addition to these, carbon in the form of graphite can react to form the peculiar compounds M^IC_8 and M^IC_{16} (for structure see Vol. II). *Silicon* and *germanium* occupy an intermediate position. They combine directly with the alkali metals to form intermetallic compounds of particularly simple composition, MX. However, Klemm and Hohmann (1948) found that K, Rb and Cs also form compounds of the compositions MSi_8 and MGe_4 which form a sort of transition from the compounds MC_8 and MC_{16} (also formed only by K, Rb and Cs) to the compounds KSn_4 and KPb_4, listed in Table 87.

The elements carbon, silicon, germanium, tin and lead are invariably quadrivalent towards hydrogen. The simplest hydrides (with the general formula RH_4) are volatile. Their stability decreases in an extremely marked manner from carbon to silicon, and diminishes even further for the heavier elements. Little is yet known of the nature of the non-volatile hydrides of this group, with a low content of hydrogen—e.g., cuprene $[CH]_X$, polysilene $[SiH_2]_X$, and the polygermenes $[GeH]_X$ and $[GeH_2]_X$.

The following summary shows the analogies between the elements of Group IVA.

Hydrogen Compounds	Oxides		Sulfides		Chlorides		Negative Ions		Positive Ions	
CH_4	CO	CO_2	CS	CS_2	—	CCl_4	$CO_3^=$	$CS_3^=$	—	
SiH_4	(SiO)	SiO_2	SiS	SiS_2	—	$SiCl_4$	$SiO_3^=$	$SiS_3^=$	—	
GeH_4	GeO	GeO_2	GeS	GeS_2	$GeCl_2$	$GeCl_4$	$GeO_3^=$	$GeS_3^=$	?	
SnH_4	SnO	SnO_2	SnS	SnS_2	$SnCl_2$	$SnCl_4$	$\begin{cases} SnO_3^= & SnS_3^= \\ Sn(OH)_6^= \end{cases}$		Sn^{++}	Sn^{++++}
PbH_4	PbO	PbO_2	PbS	—	$PbCl_2$	$PbCl_4$	$\begin{cases} PbO_3^= & — \\ Pb(OH)_6^= \end{cases}$		Pb^{++}	Pb^{++++}

Table 73 summarizes the heats of formation of the most important simple compounds

TABLE 73

HEATS OF FORMATION OF COMPOUNDS OF THE ELEMENTS OF GROUP IVA

(in kcal per g equivalent of carbon, silicon, etc.)

	Tetra-hydride	Mon-oxide	Di-oxide	Di-chloride	Tetra-chloride	Mono-sulfide	Di-sulfide
Carbon (diamond)	4.56	13.42	23.65	—	6.5	— 6.5	—3.9
Silicon	2.2	—	52.1	—	38.5	—	+8.5
Germanium	—	—	32.0	—	31.6	—	—
Tin	—	33.6	34.5	40.5	32.0	—	+9.3
Lead	—	26.20	16.1	42.8	—	+11.6	—

From the standpoint of Kossel's theory, the behavior of these elements is a consequence of their position in the Periodic System. Thus carbon and silicon are four places removed both from the preceding and from the following inert gas. They therefore have the possibility of attaining the 'inert-gas like' state either by losing four electrons or by acquiring four electrons. The former, according to Kossel, they can achieve by reacting with more electronegative elements—e.g., with oxygen or halogens—; the latter results from reaction with strongly electropositive metals or hydrogen. This interpretation assumes, however, that in their compounds with oxygen, halogens etc., carbon and silicon can be regarded—in a certain sense at least—as the electropositive constituents, whereas they are the electronegative components in their hydrides, and in their compounds with the alkali and alkaline earth metals.

Most carbon compounds, however, (and especially the hydrocarbons and their derivatives) possess to a very marked extent the characteristics of covalent compounds, so that the Kossel theory can be applied to them only with very considerable reservations. However, in view of the fact that even in the so-called covalent compounds one component can generally be regarded* as bearing more of the positive charge, and the other more of the negative, the origin of this polarity can be interpreted in terms of the Kossel theory. Thus

* This is true even for molecules which have no permanent electric moment, such as CH_4, CCl_4, CO_2.

the formation of methane can be explained on the assumption that the carbon atom acquires a fourfold negative charge, because of its tendency to assume the inert gas configuration, in that it takes the electrons from four hydrogen atoms, and then binds the resulting protons electrostatically. Because they are so small, the protons penetrate the outer electron shell into the inner part of the atoms. However, in view of the properties of the compound, this view must be modified at least to the extent that the electrons are *not completely* removed from the hydrogen atoms, and that in consequence, the constituent atoms are not held together only by their opposite charges, but other forces (wave mechanical resonance forces) are also involved. The effect of these latter is that more energy is released in the formation of a covalent compound (or one which is, at least, not purely electrovalent) than by the formation of the purely electrovalent compound which, at first sight, appears possible on Kossel's hypothesis. Similar remarks apply to the formation of silane, SiH_4, and to the hydrides of the other elements of the group.

The formation of carbon tetrachloride can be formally attributed, by similar considerations, to the fact that the C-atom relatively easily parts with those electrons which it possesses over and above the electron number of the preceding inert gas (helium). It therefore gives them up to 4 chlorine atoms, which each require 1 electron in order to attain the argon configuration. The oppositely charged atoms which thereby result can then unite by purely ionic bonds. However, since carbon tetrachloride also has the character of a covalent rather than of an ionic compound, it is probable in this case also that the 4 electrons are not completely removed from the carbon atom, and that pure ionic bonds are not involved.* The same holds for other compounds of carbon with elements of electronegative character.

Apart from these considerations, Kossel's theory does explain why carbon and its homologues are quadrivalent towards both positive and negative elements. The same conclusion regarding the valence is reached, however, on the basis of the Lewis theory, as the following formulas show:

$$\cdot \overset{\displaystyle \cdot}{\underset{\displaystyle \cdot}{C}} \cdot + 4H = H : \overset{\displaystyle H}{\underset{\displaystyle H}{C}} : H \; ; \qquad \cdot \overset{\displaystyle \cdot}{\underset{\displaystyle \cdot}{C}} \cdot + 4 : \overset{\displaystyle \cdot \cdot}{\underset{\displaystyle \cdot \cdot}{Cl}} : = : \overset{\displaystyle : \overset{\cdot \cdot}{Cl} :}{\underset{\displaystyle : \underset{\cdot \cdot}{Cl} :}{Cl}} : \overset{\cdot \cdot}{\underset{\cdot \cdot}{C}} : \overset{\cdot \cdot}{\underset{\cdot \cdot}{Cl}} :$$

This interpretation possesses the advantage that it at once makes it clear that carbon (and to some extent its homologues also) can be combined *at the same time* with elements of positive and negative character (e.g., in $CHCl_3$). By a generalization of the results obtained by Heitler and London for atoms with fewer outer electrons, it is widely assumed today that the pairing of electrons which is regarded as the origin of the bond is to be explained by the so-called 'spin coupling' (p. 133). Since there is already one pair of electrons with coupled spins present in the free neutral carbon atom (see below), it is assumed by Heitler and London that the compounds of quadrivalent carbon are derived from an *excited* state of the atom, in which one electron is raised from the orbit with the subsidiary quantum number $l = 0$ (2s orbit) to an orbit with $l = 1$ (2p orbit) and that all four outer electrons then have parallel spins.

It has not yet been possible to carry out exact calculations of the stability of carbon compounds on the basis of this theory, but the energy required to excite the carbon atom from the ground state into this excited state is about 65 kcal per g atom. The heat of sublimation of carbon (at 0° K), to give C-atoms in the ground state, is still a matter of dispute; the probable values are either about 125 kcal or about 170 kcal per g atom, whereas for the formation of *excited* C-atoms, the heat of sublimation is either about 190 or 235 kcal per g atom. This difference has to be taken in account if the sublimation energy of carbon is compared with the bond energies in carbon compounds, in which carbon is quadrivalent, and therefore present in the excited state (LONG and NORRISH, *Proc. Roy. Soc.*, A 187 (1946) 337).

It is not certain whether carbon dioxide should be regarded as a strictly covalent compound or not. Judged by its chemical behavior, at least, it stands close to the ionic com-

* According to Pauling's conceptions (p. 135), 11% of the total energy of formation can be attributed to the ionic component of the O—Cl bond. The residual charges on the atoms are similarly responsible for 28% of the single C—O bond, 41% of the C—F bond, 2% of the C—Br bond, 0% of the C—I bond and 7% of the C—H bond.

pounds, although as has already been pointed out, these are not sharply distinguished from covalent compounds.

The compounds of silicon with the electronegative elements are already related more closely to the definitely ionic compounds than to the definitely covalent. Silicon tetrachloride, $SiCl_4$, which has the volatility of covalent compounds, differs characteristically from carbon tetrachloride in its strong tendency for hydrolysis. It is more like the ionic chlorides of the elements preceding it in the Periodic Table—NaCl, $MgCl_2$, $AlCl_3$— which dissociate into ions in aqueous solution, and undergo hydrolysis in increasing measure as the charge on the cation increases. Silicon dioxide, SiO_2, has predominantly the character of an ionic compound, both in its chemical behavior and in its physical properties.

Germanium, tin and lead differ from carbon and silicon in that they are not closely preceded in the series by an inert gas. Reference to Fig. 28, p. 127 shows that the electron numbers possessed by Ge and Sn, in their quadruply charged state, are repeated also in a whole series of neighboring elements. The same is true of Pb, which is not shown in Fig. 28. It may be inferred that the electronic system present in these quadrivalent ions, like that of the inert gases, is characterized by a great stability as compared with other configurations.

Thus the fact that the particular number of 28 electrons occurs successively in the ions of univalent copper, zinc, gallium etc., up to sexavalent selenium, suggests that this number corresponds to a configuration of particular stability in the positively charged atoms, even if not in the free atoms. The same is true of the number of 46 electrons, present in each of the ions Ag^+, Cd^{2+}, In^{3+}, Sn^{4+}, Sb^{5+} and then in the positively charged atoms as far as iodine; likewise of the number of 78 electrons, present in Au^+, Hg^{2+}, Tl^{3+}, Pb^{4+} and Bi^{5+}. The position is actually simpler than in the case of carbon and silicon in so far as with tin and lead, the fourfold positively charged atoms can be identified in the form of ions.

The electron numbers of 30, 48 and 80, possessed by the elements Ge, Sn and Pb in their positive bivalent state, are also repeated in a series of adjacent elements—e.g., in Ga^+, Ge^{2+}, As^{3+}, Se^{4+}, Br^{5+}, which occur in compounds, at least, although not all exist as free electrolytic ions.

The formation of *acid ions* by the dioxides and disulfides of the Group IVA elements can be explained on the Kossel theory in that energy is released by the addition of a negative ion, such as O^{2-} or S^{2-}, to a dioxide or disulfide with a quadruply positively charged central atom—corresponding exactly to the addition of F^- to BF_3 (cf. Chap. 11). In some circumstances yet more energy may be released by addition of a further O^{2-} ion (formation of SiO_4^{4-}); this depends on the size of the central atom. The larger this is, the more ions (other things being equal) may be added on with a gain in energy.

For this reason, tin dioxide and lead dioxide are able to add on a maximum of *four* O^{2-} ions (which may carry hydrogen ions with them also, for reasons to be discussed below), over and above the number of O^{2-} ions contained in the neutral molecules. This can be expressed by the statement that *the maximum coordination number towards oxygen rises* from 3 to 6 in the series from carbon to tin. Werner showed that the oxyacids and their salts (as also the thioacids and thiosalts) belong to the class of coordination compounds.

The negative ions with the compositions mentioned can unite with hydrogen ions to form acids, and with other positive ions to form salts. If we calculate the work which must be expended to separate either a hydrogen ion or a hydroxyl ion from an acid $H_4R^{IV}O_4$— which can also be represented as a hydroxide $R^{IV}(OH)_4$—as was done in Chap. 6 for compounds of the formula R^IOH, it is found that the work expended to remove a hydrogen ion is very much less than that to remove a hydroxyl ion. This is because of the strong repulsion exerted by the quadruply positively charged central atom on the positively charged hydrogen atom. As the radius of the central atom increases its repulsive effect diminishes, in accordance with the Coulomb law, so that with increasing radius of the central atom, the hydrogen ion is removed with greater difficulty, the hydroxyl ion with greater ease. There is thus explained, first, the acid character of the hydroxides in this group, and, second, the decrease in acidic character and simultaneous appearance of basic properties, as we proceed from the lighter elements of the group (with the smaller atomic

and ionic radii) to the heavier elements (those with the larger ionic radii). Similar considerations apply to the hydroxides $H_2R^{IV}O_3$, or $OR^{IV}(OH)_2$ formed by the group.

Stannic acid, has a composition corresponding to the formula $Sn(OH)_4(OH_2)_2$, and is able
to ionize off H^+ ions with relative ease only as long as it contains oxygen atoms which are
bound to more than one hydrogen atom. In such H^+ atoms, the repulsion of the other H^+
atom, bound to the same oxygen, is added to the repulsion of the central ion. The hydrogen
ions are apparently so firmly bound in the ion $[Sn(OH)_6]^{2-}$ that further dissociation can
furnish only OH^- ions. When the compound has thereby passed into the form $Sn(OH)_4$, its
further dissociation conforms to that already described. Conditions are similar in the case
of lead.

The more strongly developed basic character of the hydroxides derived from the doubly
charged atoms is also a consequence of these views.

The ions Sn^{2+}, Pb^{2+} should not, however, be regarded as equivalent to simple alkaline
earth ions of about the same size, since unlike the latter, they exert a strong *polarizing action*.
This is shown, for example, by the deep color of their oxides, sulfides, etc. (cf. Vol. II). The
polarization increases the strength of the bonds. The O^{2-} ions, and therefore the OH^-
groups, in $Sn(OH)_2$ and $Pb(OH)_2$, are therefore more firmly bound than in the corresponding alkaline earth hydroxides, and consequently $Sn(OH)_2$ and $Pb(OH)_2$ are only
weak bases. The energy of separation of the H^+ ion from them is not much greater than
that necessary to remove the OH^- ion. For this reason these hydroxides, and especially that
of tin, can function as acids towards strong bases.

The atomic and ionic radii of the elements of Group IVA, derived from their crystalline
compounds, are collected in the following table (Tab. 74).

TABLE 74

ATOMIC AND IONIC RADII OF ELEMENTS OF GROUP IVA, Å

Charge	C	Si	Ge	Sn	Pb
—4	—	1.98	—	2.15	2.15
0	0.77	1.18	1.22	1.40	1.74
+2	—	—	—	—	1.32
+4	≈0.1	0.39	0.44	0.74	0.84

Spectroscopic data lead to conclusions as to the structures of the atoms, which are important for the chemical behavior of the elements. From the spark spectrum of triply ionized
silicon, Si^{3+}, the energy levels of which have already been illustrated in Fig. 70a, p. 318, it
follows that, except for the higher nuclear charge and the consequent contraction of the
electron orbits, the atomic core of silicon is the same as that of the preceding elements, and
therefore like that of neon, the preceding inert gas. There is a similar uniformity in the
arrangement of energy levels in the series Li^+, Be^+, B^{2+}, C^{3+}, from which it may be inferred
that, except for the nucleus, the atomic core of carbon is the same as that of helium. We
concluded from the level scheme of Fig. 71, compared to that of Fig. 70, that in aluminum
the third electron taken up by the Al^{3+} core is bound in a manner different from that of the
first two. The right half of Fig. 71 shows that in singly ionized carbon C^+, the radiating
electron in its ground state is present in a $2p$ level. According to the concepts of the Bohr
theory it thus describes a circular orbit, and not an ellipse ($2s$ orbit) like the first two
electrons taken up after completion of the helium shell. The spectral evidence is thus that
the fourth outer electron is to be found in a $2p$ level. Thus, of the four outer electrons of
carbon, two are in a $2p$ level and two in a $2s$ level, the two latter having antiparallel spins.

The spectral terms indicate that, in silicon, the outer orbits are analogous to those of
carbon—i.e., two electrons in $3s$ orbits and two in $3p$ orbits. Corresponding assignments
apply also to the remaining homologues of carbon which,—just as the relationships encountered in Group IIIA,—are the analogues of the two lightest elements of the group, not
only chemically, but also spectroscopically.

The energy to be expended for the separation of a fifht electron is very considerably greater than that required to remove four electrons (cf. Table 23, p. 118). This is especially true of carbon and silicon, but even for the higher homologues of the group it is so large that it is not to be expected that they could enter into stable chemical compounds with a higher charge than $+4$ units.

2. Carbon (C)

(a) Occurrence

(i) *Carbonates and Carbon dioxide*. Carbon occurs in the mineral kingdom chiefly in the form of *carbonates*, compounds derived from carbonic acid H_2CO_3. Not only the alkali and alkaline earth metals, the carbonates of which have been met in the preceding chapters, but also many of the heavy metals frequently occur as carbonates—e.g., iron as *siderite*, $FeCO_3$, manganese as *rhodochrosite*, $MnCO_3$, zinc as *smithsonite*, $ZnCO_3$, etc.

Carbon is an essential material for the whole vegetable and animal kingdom. There is no living organism, vegetable or animal, in the structure of which carbon compounds do not play a vital part. For this reason, the compounds of carbon, and especially the more complex substances, which were formerly believed to be capable of formation only through some special force, operating in living organisms, are therefore also called *organic compounds*. The branch of chemistry which is particularly concerned with their study, *organic chemistry*, has long been a most important branch of the science.

Carbon, in the form of *carbon dioxide*, CO_2, is present in the atmosphere in low concentrations (0.03 volume per cent on the average). Green plants abstract it from the air; with the aid of their green pigment, chlorophyll, they are able, from carbon dioxide and water, to build up carbohydrates $(C_6H_{10}O_5)_x$, such as starch and cellulose, oxygen being eliminated. The requisite energy is furnished by sunlight. The carbohydrates and their transformation products are then degraded again to carbon dioxide and water in vegetable and animal organisms (the latter obtaining the from the plants) by a process—*respiration*—which liberates energy. Thus carbon dioxide enters into a continuous cycle between the atmosphere and the organic world. Although the concentration of carbon dioxide in the air is very small, its total amount in the atmosphere considerably exceeds that present in the entire vegetable and animal world, because of the great volume of the atmosphere.

The CO_2 of the atmosphere contains a definite proportion of radioactive carbon, ^{14}C. This ^{14}C is produced continually in the atmosphere by the nuclear reacion $^{14}_7N + ^1_0n = {}=^{14}_6C + ^1_1H$. The neutrons (symbolized by n) entering into this reaction are formed by the effects of cosmic rays, which produce approximately 1 neutron per second per cm³ of the atmosphere. Since almost all of these neutrons react with nitrogen to form ^{14}C, the half-life of which is known (5300 years), it is possible to calculate the quantity of ^{14}C continuously present on the earth. This amounts to about 20 tons. CO_2 is at once formed in the atmosphere from the radioactive carbon, which is thereby introduced into the cycle of biological processes. The total quantity of carbon participating in this cycle of changes—i.e., the carbon content of the biosphere and the oceans—has been estimated as about $8 \cdot 10^{13}$ tons, of which $6 \cdot 10^{11}$ tons is present in the atmosphere in the form of CO_2. From these figures it may be calculated that about 1 part in 10^{13} of the carbon in the biological cycle is radioactive. The proportion found experimentally is in fair agreement with this estimate (ANDERSON and LIBBY, *Phys. Rev.*, 72 (1947) 931). The experimental determination of this extremely minute quantity is facilitated by converting the carbon to methane, and enriching this in

$^{14}CH_4$ by thermal diffusion (cf. Vol. II). In any substance which is withdrawn from the biological cycle, the radioactivity decays with the half-life of ^{14}C. Hence, by determining the ^{14}C content of any organic substance, it is possible to determine its 'age'—i.e., the time which has elapsed since it was removed from the cycle of biological processes [*60, 61*]. It is, however, extremely difficult to carry out accurate measurements of the ^{14}C content of materials that are more than 40,000 years old, since their radioactivity has decayed too much. Thus it is hardly possible to detect any remaining ^{14}C in coals. For substances of organic origin, dating from the later prehistoric, or ancient historical times, on the other hand, it is possible to fix their age fairly exactly by determinations of their ^{14}C content (W. F. Libby, 1947).

The ^{14}C-content of the carbon of all organic substances which have not long been withdrawn from the biological life cycle makes it possible to distinguish them from materials obtained synthetically from coal or coke. Fresh biological carbon displays an activity of 15.6 β-particles per gram per minute, as an average value (Anderson and Libby, 1951), whereas fossil carbon is practically inactive. It has been shown by Faltings (1952) that by measurement of the ^{14}C radioactivity, it is possible not only to distinguish e.g., natural acetic acid from the synthetic product, but also to determine the proportions in a mixture of natural and synthetic material with fair accuracy.

We find the products of decay of primeval vegetation in nature as *coal*, of which carbon is the principal constituent. *Petroleum, ozokerite* (natural wax) and *asphalt* are also carbon compounds, the origin of which is to be sought in primeval living organisms, from which they have been formed by decay. Also to be mentioned are *amber*, a fossil resin with the approximate composition $C_{40}H_{64}O_4$, and *honey stone* or *mellilite*, $Al_2C_{12}O_{12}·18H_2O$, the aluminum salt of benzene hexacarbonic acid, $C_6(COOH)_6$, named mellitic acid after the mineral. The salts of oxalic acid also occasionally occur as minerals—e.g., *whewellite*, $CaC_2O_4·H_2O$ and *oxalite*, $FeC_2O_4·1\frac{1}{2}H_2O$.

Unlike the elements of Groups IA, IIA and IIIA, discussed in Chaps. 6, 8 and 10, carbon also occurs in nature in the elementary state, as *diamond* and *graphite*.

(ii) Coals. The term 'coal' is applied to certain black or brown, carbon-rich decomposition products of organic substances. They may be formed artificially by heating in the absence of air 'charcoals', or obtained naturally, from the plants of former geological ages, by similar processes of decomposition, proceeding at much lower temperatures for long periods of time. According to the material from which they have been prepared, products of the first kind are known as *wood charcoal, bone charcoal, blood charcoal*, etc. To the natural coals or mineral coals [*1*] belong the various types of *bituminous coal* and *brown coal* [*2*], or *lignite*, to which may be added *peat* [*3*], as a still very incompletely decomposed product.

The process of '*carbonization*' consists of a decomposition whereby the other elements originally combined with carbon such as hydrogen, oxygen and nitrogen, are in large measure eliminated, so that carbon is more and more enriched in the residue. Most organic substances undergo such a carbonization process when they are heated strongly enough, as can be seen from the resulting black coloration, characteristic of finely divided carbon or soot. In the course of geological time, such carbonization processes have taken place with primeval forests, and our present coal deposits have originated in this way. In many cases, especially with the younger coals, the structure of the coal clearly reveals its vegetable origin.

The geologically oldest coals, from which the elements originally combined with carbon have been most extensively removed, are known as *anthracite*. This consists of over 90% carbon, and contains only a few per cent of oxygen and hydrogen, and only traces of nitrogen. Anthracite is deep black to iron-grey in color. It burns only in a strong draft of air, and with a short flame, since it evolves but little gas. Of rather younger date is *coal* in the narrower sense, which is also black, but which, unlike anthracite, often has a bituminous or lustrous appearance and burns with a luminous smoky flame. This is because of the greater content of gaseous constituents which escape when it burns. Coals contain about 4–6% H, 5–18% O, 0.5–1.5% N, and only 75–90% C. Different varieties of coal are recognized, according to their texture and properties—bituminous coal, non-bituminous coal, gas coal, open burning coal, open burning gas coal, caking coal, close burning or non-caking coal, etc.

Their vegetable origin can be distinguished even more clearly in the brown coals, or *lignites*, originating in the Tertiary epoch, than in ordinary coal. Lignite is brown in color, or often black, but is always much softer than coal and anthracite. While these have a hardness of 2–2½, lignite has a hardness of only 1.1–1.4. It contains only 65–75% C, with 5–6% H, 20–30% O, and 1–2% N. The youngest member of the series is *peat*, which can clearly be seen to consist of plant remains felted together. The formation of peat can be observed at the present time. It contains about 55–65% C, 5.5–7% H, 30–40% O, and 1–2% N, so that its composition is not far removed from that of wood (with about 50% C, 6% H, 44% O and 0.1–0.5% N) and other vegetable fibres.

The *calorific value* of the various kinds of coal is of great importance for their industral uses. The combustion of 1 kg liberates for

anthracite	bituminous coal	lignite	peat	wood
8000–9000	7000–8000	6000–7000	5000–6000	4000 kcal.

These quantities of heat, called the *upper calorific value*, refer to combustion yielding liquid water as a product. The *lower calorific values* are obtained from these by subtracting the heat of condensation of the water formed in the combustion.

(*iii*) *Petroleum.* Petroleum has also probably resulted from the decomposition of organic substances, and, principally those of animal origin in particular. It consists primarily of a mixture of hydrocarbons, the nature of which can vary widely, according to the source of the petroleum. The *Caucasian* petroleum, already known in ancient times, and obtained principally from the neighborhood of Baku, on the Caspian Sea, consists predominantly of saturated cyclic hydrocarbons, the so-called *naphthenes*, whereas American petroleum consists almost exclusively of hydrocarbons of the paraffin series. In addition to the places mentioned, petroleum is found in enormous quantities in South America, especially in Venezuela, in Mexico, and also in Iran and Rumania. Iraq and the shores of the Persian Gulf are also important oil regions, as are Burma, the East Indies (Borneo and Sumatra) and, more recently, Alberta in Canada. Galicia was formerly an important oil-producing region. Small amounts have been raised in many other parts of Europe, including the United Kingdom.

In 1948, the world production of petroleum amounted to over 3,400 million barrels, Of this production, upwards of 2,000 million barrels were produced in the United States; 490 million barrels in Venezuela; 415 million barrels from Persia, Arabia and the Middle East; 218 million barrels in the U.S.S.R., and about 40 million barrels in Europe (including 300,000 barrels from British wells).

Petroleum is extracted by boring wells into the fissures or cavities containing it. It is frequently expelled from the wells with great violence, by the gases occurring with it, but more often it must be raised to the surface by pumping. Crude petroleum, called naphtha in Russia, is a yellow-brown to black, often fairly viscous liquid, with a density 0.8–0.9. It is used to some extent as a fuel without any refining. Generally, however, it is separated by fractional distillation into several portions, which find different applications. There is thereby first obtained (i) crude benzine, boiling up to 150°, density about 0.7, (ii) lighting paraffin, boiling point 150–300°, density 0.75–0.87, (iii) gas oil, or middle oil, boiling point 300–350°. From the constituents boiling above 350°, *lubricating oils* with a density about 0.9 can be obtained by vacuum distillation with superheated steam. There remains a mixture of black, resinous materials, *asphalt* [4], a product that occurs also in Nature, where it is formed as the residue of the evaporation of mineral oil. The crude benzine, called naphtha in the United States, is generally treated with sulfuric acid and with caustic soda, and separated yet further. There are thus obtained *petroleum ether* or *gasoline* (boiling point 30–70°), density generally between 0.64 and 0.66), *ligroin* or *extraction benzine* (boiling point 70–110°) and *heavy benzine* (boiling point 100–140°, density about 0.75). The boiling points and densities given furnish only an approximate indication of the nature of the products concerned. These vary considerably, as between the different commercial varieties, and since they are not products of uniform boiling point, it is not possible to give even the limits with accuracy. Benzine or petrol is used principally as fuel for internal combustion engines, and as an extraction solvent and washing solvent; middle oil for the production of oil gas, the carburation of water gas and as fuel for heavy oil engines. The uses of the other petroleum distillates are shown by their names. Benzine, gasoline and petroleum have about the same calorific value (about 10,000 kcal per kg). A further product of the petroleum industry, *paraffin wax*, is obtained from the further refining of lubricating oils. It separates as a coarsely crystalline mass when they are cooled to —5 to —10°. Paraffin wax, a waxy, white translucent material, is a mixture of hydrocarbons, like all the other constituents of petroleum. Hard and soft paraffin waxes are recognized. Hard paraffin, also called *ceresine*, melts at about 52–56°. It is used chiefly for the manufacture of candles. The lower melting soft paraffins are used in match manufacture, the dressing of fabrics, the preparation of waxed paper, as an insulator in the electrical industry and in pharmacy for the preparation of salves. Liquid paraffin (paraffin oil), which is liquid at the ordinary temperature and forms a clear, colorless, non-fluorescent, oily liquid, boiling above 360°, density 0.88, is used for similar purposes.

(b) Properties and Uses of Carbon

Pure carbon exists in two modifications—*diamond*, which crystallizes in the regular system, and *graphite*, crystallizing in the hexagonal system. Whereas diamond is colorless and graphite is grey, carbon obtained by the thermal decomposition of its compounds is generally a deep black (e.g., lamp black). It was formerly considered that black carbon was a distinct *amorphous* modification, but according to modern investigations, its fine structure is essentially the same as that of graphite (cf. p. 423).

In Table 75 are compared some characteristic properties of the various forms of carbon. Lamp black and 'Glanzkohlenstoff' have here been selected as examples of the black carbon, formerly considered as amorphous. Unlike most of the other materials included under the term 'amorphous carbon', these are practically free from impurities.

TABLE 75

PROPERTIES OF SOME FORMS OF CARBON

	Diamond	Graphite	Black carbon	
			Carbon black	'Glanzkohlen-stoff'
Color and transparency	Colorless, clear and transparent	Grey, opaque	Black, opaque	
Specific gravity	3.514	2.22	1.85	1.86–2.07
Heat of combustion, cal per g C	7873	7856	8129	8148–8051

The properties of black carbon depend to a considerable extent on its mode of preparation, and are also altered by subsequent treatment. When it is heated to high temperatures, black carbon becomes more like graphite without undergoing any change in outward appearance; its density increases and its heat of combustion simultaneously diminishes.

The view that black carbon is identical in structure with graphite was first expressed by Debye and Scherrer (1917), who found that it gave the same X-ray diffraction lines as graphite, though very diffuse. It was at first held by many that this assumption could not be reconciled with the heats of combustion of the various types of black carbon, which greatly exceed that of graphite, or with the characteristic high surface activity of charcoal, which is not possessed by graphite to any marked extent. However, Hofmann was able to show (1931) that these properties of black carbon are due merely to the extreme smallness, and therefore the exceptionally great surface development of its crystallites. Its higher energy content, as compared with graphite, and its high surface activity (adsorptive power and catalytic efficiency) can both be explained by the unsatisfied valences of the carbon atoms lying in the surfaces. The surface forces can only become effective, however, if the preparation of the material is carried out under such conditions that the surfaces are accessible. Hence for the activity, although not for the heat of combustion, the *orientation* of the crystals is important as well as their small size. The activity is manifest only if the crystallites are not cemented together to form dense aggregates (as for example in the case of retort graphite), but form a loose powder or a rigid skeleton penetrated by very numerous capillary channels. The properties of black carbon pass continuously into those of ordinary graphite as the crystallite size increases (lamp black → Glanzkohlenstoff → retort carbon → Acheson graphite). The maximum hardness is obtained with retort carbon, since hardness increases with crystal size if the crystallites are randomly oriented, but decreases with increasing orientation of the crystallites.

In the gaseous state, carbon is present almost entirely in the monatomic form, although the dissociation energy of the C_2 molecule (which has been identified spectroscopically) is 3.6 ev. (= 83 kcal per mol), i.e., greater than that of F_2. However, at temperatures such that the C_2 molecule would be only slightly dissociated, the vapor pressure of carbon is excessively small, and at temperatures where the vapor pressure approaches atmospheric, the thermal energy leads to almost complete dissociation of the C_2 molecules. For example, the calculations of Zeise (1940) for equilibrium in the reaction $C_2 \rightleftharpoons 2C$ show that at 2000° K the degree of dissociation of the C_2 molecule would be only 1.7%, and at 3000° 56%, if carbon vapor could exist at these temperatures under normal pressure. The actual vapor pressure of carbon at 2000° K is about 10^{-8}, and at 3000° K $0.9 \cdot 10^{-3}$ atm, and the resulting degree of dissociation is therefore 100.0% and 99.9% accordingly.

In whatever form carbon is present, it is tasteless and odorless, extremely infusible and non-volatile, and insoluble in all the usual solvents. It dissolves, however, in many molten metals—e.g., in iron, cobalt, nickel and the platinum metals—and separates from them again in the form of graphite on cooling.

At ordinary temperature, and expecially in the form of diamond or graphite, carbon is extremely inert chemically. There are some varieties of black carbon which ignite in oxygen when heated relatively gently. Fluorine reacts with charcoal even at ordinary temperature. At high temperatures, carbon unites with many elements—e.g., with hydrogen, sulfur, silicon, boron and numerous metals. The compounds with metals, and with other elements more electropositive than carbon, are known as *carbides*. Carbon does not combine directly with nitrogen, though it does so in the presence of hydrogen, forming hydrogen cyanide.

(*i*) *Diamond*. In the form of diamond, carbon forms extremely hard, brillant and highly refractive crystals, of density 3.51, which are colorless and perfectly transparent when pure. The natural crystal facets are often those of the regular octahedron, but other forms of the regular system also occur, including the tetrahedron—an indication that the diamond belongs to the tetrahedral hemihedral class of the regular system. The crystals of diamond often have curved faces, so that the domed crystals can actually be said to be characteristic of diamond.

For its use as a jewel, other faces are cut artificially on the diamond, and these are so chosen that the gem receives the greatest possible degree of 'fire', and the liveliest play of color, with the minimum possible loss of material. This is most completely achieved by the so-called *brilliant cut* (Fig. 78). In this, the facets are so situated that the greatest possible

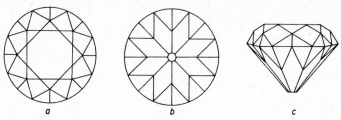

Fig. 78. Brilliant cut (*a*) from above, (*b*) from beneath, (*c*) from the side.

fraction of the light falling on it from above undergoes total internal reflection. The refractive index of diamond for red light (Fraunhofer's B line) is $n = 2.407$, and that for violet light (the H line) is $n = 2.465$. The spread of the colors (the dispersion) is thus considerable, and is the reason for the well known play of color, the 'fire', of diamonds or brilliants.

Diamonds generally occur in so-called 'placers', i.e., in water-concentrated, fragmented rocks which contain other gem stones and precious metals also, for example in the East Indies, Borneo, New South Wales, British Guiana, California, Brazil, Mexico and especially in the Congo region, on the Gold Coast, and in South and South-West Africa. In South Africa they are also found in primary ore deposits, namely in eruptive olivine rocks (kimberlites). Their extraction was formerly carried out by elutriating the rock in casks, and sorting the residue by hand. At the present day, in the chief producing areas of the Cape, the rock, after being allowed to weather, is levigated on centrifuges. The residue, containing the components of higher specific gravity, then passes over vibrating sieves, coated with fat. These retain the diamonds, which are wetted by fat, although not by water. In the richest reef of the Cape, the Kimberley reef, there is, on the average, 1 g of diamonds for every 2 tons of rock, but rocks are now worked up which are twenty times poorer in diamonds. The weights of diamonds are generally expressed in *carats*. 1 carat = 0.205 g, though this unit varies somewhat in different countries. The world production of diamonds in 1935 amounted to 6.2 million carats, of which 51.2% came from the Belgian Congo, 21.8% from the Gold Coast, and 13.0% from South Africa. The largest diamond which has yet been found is the *Cullinan* diamond, from South Africa, with a weight of 3024 carats. Other well known large diamonds are the Regent (193 carats), the Orloff (193 carats) and the Great Mogul (186 carats).

The diamonds found in the South African diamond field are seldom as perfectly clear as the Brazilian and Indian diamonds, and are therefore often not so suitable for jewels. Cloudy, almost lead-grey diamonds are known as *boart*; most of the diamonds which are extracted are of this class. There are also perfectly transparent colored diamonds—yellow, brown, violet, pale blue and light green. Deep black diamonds, *carbonados*, are found in Brazil; these often contain 2–4% of impurities. According to Roth (1925), the heat of combustion of carbonado is significantly higher than that of ordinary diamond. Perfectly transparent diamonds ('of the first water') are used as jewels, but only a small proportion of diamonds (about 5%) possess the qualities requisite for this. The rest are used for industrial purposes [5]: for boring and grinding especially hard materials (e.g. 'diamond drilling' of rocks in mining exploration), for cutting glass, for the bearings of precision instruments, and as dies for wire drawing. They are suitable for these purposes because of their great hardness, in which respect the diamond is superior to all other substances. It can therefore be ground only by means of its own powder. On the other hand, it is quite brittle, and can readily be reduced to powder in a steel mortar.

Carbon in the form of diamond is extremely resistant chemically. It is attacked neither by acids nor by alkalis, unless these readily give up oxygen. However, it burns to carbon dioxide when it is heated to above 800° in oxygen. It is transformed into graphite by heating it in the absence of air.

It is probable that diamond has never been prepared artificially. In 1897 Moissan dissolved charcoal in molten iron, which he then chilled in the form of droplets, so that the carbon separated from solution in the interior under very high pressure. He obtained minute hard crystals, which he considered to be diamonds, and since then numerous experiments of this kind have been carried out. In more recent work magnesium silicate has been used as the solvent, since the diamonds in South Africa occur chiefly in olivine rocks. Minute crystals have occasionally been so obtained, displaying properties like those of the diamond, but it appears that in no case has unambiguous proof been advanced (e.g., by X-ray examination) that these were diamonds (See D. P. MELLOR, *Research*, 2 (1949) 314).

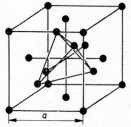

Fig. 79.
Lattice of diamond.
$a = 3.560$ Å.

The diamond was among the first substances for which the structure was determined by X-ray diffraction. As was shown in 1913 by W. H. and W. L. Bragg, and repeatedly confirmed by later investigations, it forms a crystal lattice of the kind represented by Fig. 79. This is a face-centered cubic lattice, in which four additional atoms are inserted in tetrahedral positions. These also belong to a face-centered cubic lattice, which is displaced by $\frac{1}{4}$ of the body diagonal with respect to the first. Every carbon atom is equidistant from four others, which surround it tetrahedrally. The distance is $\sqrt{3}a/4$. The space-filling is only 34.01%.

(*ii*) *Graphite* [6] occurs native in many places in considerable quantities, especially in the island of Ceylon, also in Madagascar, in the United States, in New Zealand, in Siberia, Korea, and in the Bohemian mountains. Less extensive deposits occur in many other places—e.g., in Cumberland, and at various places in Germany, such as Passau, where graphite was mined in the Middle Ages. Graphite forms grey, opaque masses, sometimes flaky, sometimes earthy. Well formed crystals are rare. They form six sided tablets, with triangular striations and excellent cleavage parallel to the basal plane. Graphite is soft (hardness 1 on Mohs' scale), is flexible, with a greasy feel, and has a weak metallic luster. It gives a lead-grey streak on paper, and so is used for the manufacture of 'lead pencils'. The density of natural graphite is 2.1–2.3. Unlike the diamond, graphite is a good conductor of heat and electricity. It is more easily attacked chemically than is diamond, and burns in the flame of the Bunsen burner, though with some difficulty. Natural graphite is often very impure, so that is leaves a residue of 20% or more of ash.

When it is heated with potassium chlorate and sulfuric acid, graphite furnishes the so-called *graphitic acid*, a product containing oxygen which deflagrates when it is rapidly heated. When they are moistened with fuming nitric acid and then heated to a red heat, many specimens of graphite swell up, while others remain unaltered. Graphite swells when it is treated with concentrated sulfuric acid in the presence of oxidizing agents, and at the same time takes on a blue color. Graphite anodes show a similar behavior under certain circumstances when used for the electrolysis of concentrated solutions of H_2SO_4, $HClO_4$, HNO_3 or HF (Thiele, 1934). The so-called 'amorphous' carbon shows this phenomenon to a much smaller degree. The processes involve what are termed 'topochemical' reactions (cf. Vol. II).

In these reactions, which have been extensively studied by Hofmann and Rüdorff (1928 and subsequently), the sheets of carbon atoms of the graphite layer lattice behave as if they were semi-metallic in character. They enter into reaction as 'giant cations', without any severance of the carbon-carbon bonds in the graphite structure. The compounds are thus *graphite salts*. Acid anions enter *between* the sheets (thereby bringing about a growth of the structure along the *c* axis). The number of such anions is determined in part by geometrical factors; a number of the compounds conform to the type $[C_{24}]^+X^-$ — e.g.,

$$[C_{24}]^+HSO_4^-.\ 2H_2SO_4.$$

The heat of combustion of graphite is 0.20 kcal per g atom lower than that of diamond, from which it follows that graphite is the more stable modification of carbon. Graphite is therefore always formed whenever carbon separates under such conditions that crystallization can occur. For example, the carbon dissolved by molten iron separates out in part as graphite when the melt is cooled. This is responsible for the grey color of cast iron.

According to Roth, certain natural graphites (namely those which have been subjected to high pressures in the course of their geological history) have a heat of combustion which is 9 cal per g C lower than that of ordinary graphite. He therefore calls these α-graphite and ordinary graphite β-graphite. This view is, however, not accepted by all thermochemical investigators.

In 1896, Acheson succeeded in preparing graphite artificially on a large scale. He observed that it was formed in the manufacture of carborundum, SiC, by heating coke with quartz in an electric furnace. The process is used on a very large scale today by the Acheson Graphite Company, at Niagara. It is carried out by heating siliceous coke for 12–24 hours in an electric furnace, by a current of several thousand amperes. It was formerly assumed that SiC was formed as an intermediate product, and that carbon was deposited, crystallized in graphitic form, by the decomposition of this. According to Arndt (1931), however, the process consists essentially in the growth of graphite crystals which already exist in the coke. The process is therefore comparable to the recrystallization which occurs in metals when they are heated.

Graphite is used for making crucibles for the casting of metals. As a highly refractory material, with good thermal conductivity and resistance towards rapid changes of temperature it is particularly suited for this purpose. The crucibles are either turned out of blocks of Acheson graphite, or are moulded from laminar graphite, mixed with a little clay and silica sand, and then fired. Graphite is used for the manufacture of pencils, the hardness of which can be varied by the addition of clay. It is also used as a heat-resistant black for stoves, and (mixed with linseed oil) as rust-protective paint for iron. Finely divided graphite is used as a lubricant. Colloidal graphite prepared artificially for this purpose is known as 'Dag' (Deflocculated Acheson Graphite). Thus the 'oildag' used as a cylinder lubricant in internal combustion engines is a lubricating oil to which colloidal graphite has been added. 'Aquadag' and 'kollag' (the latter prepared from natural graphite) are similar. The good electrical conductivity of graphite, combined with its other properties, leads to many applications in the electrical industry—e.g., in galvano plastics, for coating matrices with a conducting surface, for the brushes of dynamos, and for microphone carbons. Many types of arc lamp carbon—e.g., for searchlights—and electrodes, such as those for the electrolysis of alkali chlorides, are either prepared from Acheson graphite, or are graphitized by the Acheson process [7].

The fine structure of graphite was explained by the X-ray investigations of Debye and Scherrer (1917) and of Hassel and Mark (1924). Fig. 80a represents the unit cell of the graphite crystal lattice, and in Fig. 80b, to illustrate the structure better, additional lattice points are shown, as well as those of the unit cell. It may be seen that a *layer lattice* is formed.

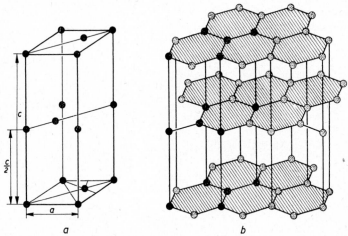

a   *b*

Fig. 80. Lattice of graphite. $a = 2.46$ Å, $c = 6.69$ Å

Each C-atom is surrounded by 3 others at a distance of 1.45 Å, which lie in the same plane as itself. 6 other C atoms in the same plane are situated at rather greater, but equal, distances (2.46 Å) from the carbon atom in question. Well defined, parallel planes, densely occupied by atoms, are clearly to be distinguished in the crystal lattice, fairly widely spaced from each other—namely at 3.345 Å. They are displaced relatively to one another, in such a way that a C atom of an adjacent plane lies above the center of each hexagon, the planes following each other at the doubled distance (6.69 Å) being occupied by atoms in an identical manner.

It is characteristic of the graphite lattice, and in contrast to that of the diamond, that the carbon atoms are arranged in planar *hexagons*. Whereas carbon exercises four valences in the diamond structure, directed towards the corners of a regular tetrahedron just as in the aliphatic compounds, the valence forces of the carbon in the graphite structure are restricted, essentially, to three directions. These lie (perhaps only approximately) all in the one plane. On the basis of their chemical behavior, the so-called *aromatic* compounds, of which benzene is the simplest representative, are considered to have an exactly similar structure. Some indication of the structure of the two principal classes of organic compounds is thus already hinted at in the crystal structures of the two modifications of carbon, as diamond and graphite, respectively.

Graphite reacts with fluorine at about 420°, F-atoms being built in between the sheets of carbon atoms, to form *carbon monofluoride*, a grey solid which contains a maximum of 1 F to each C atom. The sheets of carbon atoms thereby become puckered, and assume an 'aliphatic' character (Ruff, 1934; Rüdorff, *Z. anorg. Chem.*, 253 (1947) 281). Gaseous fluorides are formed at higher temperatures (cf. p. 438).

(*iii*) *Black carbon.* If carbon is deposited at not too high temperatures, it forms deep black masses which, as has been mentioned, were formerly regarded as amorphous carbon. The black carbon, generally known as *charcoal*, has its properties determined largely by the material from which it was prepared and by the conditions of its formation. The different kinds of charcoal generally contain considerable amounts of impurities, which often modify their properties, and it is often not possible to decide to what extent these are only impurities, or how far they are chemically combined. In the formation of natural coal, it is possible for compounds of carbon with hydrogen, oxygen and nitrogen, present in the vegetation, to be transformed into others with a higher carbon content, without elementary carbon necessarily being formed. It is not possible to say when, and to what extent, the formation of free carbon has taken place in the ageing of coals. A similar situation obtains when charcoals are produced artificially—e.g., by heating wood or other organic substances in the absence of air. It is usual to include all these products under the heading of *amorphous carbon*.

The kinds of amorphous carbon which are most important technically are *coke, wood charcoal, bone charcoal, animal charcoal* and *lamp black*.

(iv) *Coke* [*13, 14*]. The name coke is applied to the solid residue which remains when coal is strongly heated ('coked', or 'carbonized') in the absence of air. A distinction is drawn between *gas coke* and *coke-oven coke*. The former is produced in the gas works, by heating the varieties of coal which are suitable for the manufacture of coal gas (gas coals), and is generally porous, with a high ash content. As a rule, therefore, gas cokes can be used only as fuel. Coke-oven coke is produced from appropriate types of coal (coking coals), which evolve less gas and sinter strongly together, so that a relatively dense and strong coke is formed, such as is used in blast furnaces. The coal to be carbonized, in a ground up and moistened condition, is introduced into vertical retorts in which it is strongly heated; it sinters solidly together, giving up about 20–30% of the weight of the dry coal in the form of gases (coke oven gas). The coke is thus obtained as grey-black lumps, fissured like basalt, and consisting of about 90% C, 1% H, 3% O, 0.5–1% N and about 5% of 'ash'—i.e., of incombustible constituents. It burns without any deposition of soot, with a short bluish flame, and furnishes about 7000–8000 kcal per kilogram of heat.

The low temperature carbonization of brown coals (*lignites*), yields so-called lignite coke. This consists of dull black, porous granules, several millimeters in size, with an ash content of 15–25%, and a calorific value of 6000–7000 kcal per kilogram. This 'coalite' glows without flame or smoke when it is ignited. It can be employed for the filtration of water and for the production of black pigments, as well as for a fuel.

(v) *Wood charcoal*. Wood charcoal is obtained by heating wood in the absence of air. This was formerly carried out in the charcoal burners' mounds, but at the present day this method is employed only in remote districts. Carbonization is now generally carried out in large iron retorts, which enable the valuable distillates to be recovered (cf. Table 76, p. 425). Wood charcoal is a porous black material, in which the cellular structure of the wood can still be discerned. It is easily ignited, and continues to glow when once lit. Its ash content is low—generally about 1%. It is also free from sulfur, unlike coke. Wood charcoal is therefore used for many metallurgical purposes for which the purity of the charcoal is of particular importance—e.g., in the refining of copper. It is also used for the removal of fusel oil from alcohol, as a deodorant in the treatment of wounds, and for other medicinal purposes. A form of wood charcoal which is particularly easily ignited, porous and soft is produced by carbonizing the wood of the buckthorn, poplar or alder, generally at 300–400°, and is used in the manufacture of black powder.

(vi) *Bone charcoal*, or *bone black* is made by heating bones, previously freed from fat, in retorts. Ammonia solution and tar (Dippel's bone oil) are obtained as by-products. Since about two-thirds of the bone-substance consists of incombustible materials, bone charcoal has a very high ash content (90% or more), but possesses excellent decolorizing properties. It is therefore widely employed for the removal of coloring matters and other impurities from solutions, and especially for the purification of sugar syrups. It is also used in large quantities as a black pigment (bone black), especially for shoe polishes, as a leather black,

TABLE 76

SUMMARY OF PRODUCTS OF THE DESTRUCTIVE DISTILLATION OF COAL AND WOOD

Starting material	Coal	Wood
Residue	Coke	Charcoal
Gas evolved	Illuminating gas After removal of NH_3 and H_2S, contains chiefly H_2, CH_4, CO, C_2H_4, C_2H_2, C_2H_6 Calorific value about 5500 kcal/cu.m	Wood gas Weakly luminous flame Contains chiefly CO_2, CO, H_2 and CH_4 Calorific value about 3000 kcal/cu.m.
Distillate	Ammonia liquor and coal tar (The latter contains benzene, phenol, naphthalene, anthracene, pyridine, etc.)	Wood alcohol (methyl alcohol) Pyroligneous acid (acetic acid) Wood tar.

hoof blacking and for stencil colors. For fine artists colors *ivory black* is used, being obtained by calcining ivory chips and waste.

(*vii*) *Blood charcoal* and *animal charcoal* are obtained by heating blood and other animal offal with potash, and leaching the product of calcination. They find medicinal applications, in particular—e.g., for the elimination of certain poisons, or of excessive accumulations of gas in the bowel.

The sort of uses which have been mentioned for wood-, bone-, blood- and animal-charcoal depend primarily on the high adsorptive capacity displayed by charcoal towards many dissolved substances and gases. A carbon with a high adsorptive power is known as an '*active charcoal*' [8–11]. The activity of the charcoal depends to a considerable extent not only on the starting material but on the way in which it was prepared. The activity may often be increased subsequently by special treatment—e.g., by heating it with certain inorganic salts. The adsorptive capacity is very different towards different substances. In general, gases which are difficult to liquefy, are less readily adsorbed, but there is no complete parallelism between the adsorbability and the boiling point or critical temperature of the liquefied gases. According to Remy (1932), the lower the vapor pressure of a gas in the liquid state, the larger is the quantity of that gas taken up by a highly active charcoal. The amount of gas adsorbed increases greatly as the temperature is lowered (see Vol. II).

(*viii*) *Lamp black* or *soot* [12] is formed by the thermal decomposition of many gaseous or vaporized hydrocarbons. It is manufactured by burning tar, tar oil, naphthalene, paraffin oil or natural gases rich in hydrocarbons in an insufficient supply of air, and chilling the flame by water-cooled metal plates or other means. It has recently been enprepared from acetylene also. Under the name of 'carbon black', it is used in very large quantities as a black pigment. It is used primarily for the manufacture of printer's ink, and also for drawing inks and artists' pigments, for coloring patent leather and rubber products—e.g., rubber shoes. Certain grades of carbon black are also beneficial in modifying the mechanical properties of vulcanized rubber, and are used in very large amounts in the manufacture of pneumatic tires.

Carbon black, wood charcoal and other varieties of carbon obtained by heating organic substances at relatively low temperatures differ from graphite in their fine structure, not only in the small size of their crystallites (the mean diameter in the direction perpendicular to the *c*-axis is generally between 20 and 60 Å), but also in that the individual sheets of hexagonal rings of C-atoms are not stacked regularly one upon the other, in the way shown by Fig. 79. Although they have approximately the same spacing as in the true graphite lattice along the *c*-axis direction, the sheets are irregularly displaced with respect to one another in the two directions normal to this axis [Arnfelt, 1932; Berl, 1932–33; Warren, 1934; Hofmann, especially *Z. Elektrochem.*, 42 (1936) 504]. Structures such as those of these varieties of carbon, which occupy an intermediate position between the crystalline and the amorphous, have been called *mesomorphic* structures.

(*ix*) *Glanzkohlenstoff.* If the flame of burning illuminating gas or methane is allowed to impinge on a surface—e.g., of glazed porcelain—heated to 650°, a highly lustrous form of carbon, 'glanzkohlenstoff', is deposited, as was found by K. A. Hofmann. The evidence of X-ray analysis shows it to consist of extremely minute crystals of graphite, which can be detected only by this means, U. Hofmann has recently been able to obtain it in a finely divided form, as a porous powder. For any temperature of preparation, the crystallite size, is the same (30–40 Å) in the finely divided and the compact glanzkohlenstoff, but the finely divided form has a very much larger adsorptive capacity. This is because, in the compact form, only a relatively small proportion of the surface of the crystallites is accessible to adsorbents.

(*x*) *Retort carbon* is formed in essentially the same manner as glanzkohlenstoff but at considerably higher temperatures. This form of carbon is obtained as a by-product in the process of carbonization, since the gases expelled from the coal decompose to some extent as they sweep over the very hot walls of the retorts, and deposit carbon there as a very hard, dense mass. Retort graphite differs from ordinary, or true, graphite in its lower density (about 2), its great hardness, and the absence of any crystalline texture detectable by ordinary means. It possesses the good electrical conductivity and chemical inertness of ordinary graphite. It is therefore used for the production of carbon rods for arc lamps, and for electrodes in galvanic cells [7].

X-ray analysis shows that retort carbon has exactly the same crystal structure as true graphite. Its crystallites are of very much smaller size, however (mean diameter about 60 Å), so that they lie far below the limit of microscopic detection. Since, as in glanzkohlenstoff also, the crystallites are oriented at random, these types of carbon lack the cleavage which is so characteristic of graphite. The hardness of retort carbon is also to be explained through the random arrangement of its crystallites.

Most of the forms of 'amorphous' carbon give X-ray diffraction patterns related to that of graphite, but with very much broadened lines, corresponding to the presence of a graphitic structure extending over very small regions (i.e., very small crystallite size), Gibson, Holohan and Riley (*J. Chem. Soc.*, (1946) 456) have described the preparation of carbon by the decomposition of C_6I_6 at low temperatures, or by the action of sodium on C_6Cl_6, which are apparently truly amorphous in character; from the smoothness with which the carbon was formed, at temperatures as low as 150°, it appears that the carbon rings of the benzene structure must be retained in the product. Instead of being linked to build up an extended coplanar structure, however, Riley considers that in these carbons the rings are completely cross-linked to form a three dimensional structure, with no long-range order.

For a discussion of the structure of amorphous carbon and the nature of the process of recrystallization to form graphite, see RILEY, *Quart. Rev. Chem. Soc.*, 1 (1947) 60.

3. Compounds of Carbon

The multiplicity of compounds formed by carbon is enormous, even although almost all of them are derived from a *single valence state*, namely from quadrivalent carbon. Trivalent carbon exists only in some compounds of complex structure, which fall within the domain of organic chemistry. Compounds of bivalent carbon are also known only in very few instances. Apart from carbon monoxide, in which the carbon is only formally bivalent, they have very low stability.

The great variety of compounds formed by carbon arises, in the first place, from the ability of carbon to unite both with electropositive and with electronegative elements, and to do so simultaneously. Because of this, the hydrogen atoms in such a compound as methane, CH_4, can be replaced partly or completely by almost any other element, and especially by those of electronegative character. Associated with this is the second peculiarity of carbon, which is responsible for the great number of carbon compounds—namely the ability of carbon atoms *to unite with other carbon atoms*. In this way there may be formed so-called *carbon chains*, which may in some instances have many members:

$$H_3C—CH_3 \qquad H_3C—\overset{\displaystyle H}{\underset{\displaystyle H}{C}}—CH_3 \qquad H_3C—\overset{\displaystyle H}{\underset{\displaystyle H}{C}}—\overset{\displaystyle H}{\underset{\displaystyle H}{C}}—CH_3 \qquad H_3C—\overset{\displaystyle H}{\underset{\displaystyle H}{C}}—\overset{\displaystyle H}{\underset{\displaystyle H}{C}}—\overset{\displaystyle H}{\underset{\displaystyle H}{C}}—CH_3$$

ethane   propane   butane   pentane   etc.

'Branched' chains may also be formed—e.g.,

$$CH_3—CH(CH_3)—CH(C_2H_5)—CH_2—CH_3 \qquad \text{methylethylpentane}$$

It is possible for the third, fourth, fifth, sixth, etc. carbon atom to be joined again to the first; in this way 'rings' are built up, and further chains or other rings may in turn be joined to these rings. Furthermore, reactions such as the elimination of halogen, or of halogen acid, which would result in a pair of carbon atoms becoming joined together, if they took place between carbon atoms of different molecules or between non-adjacent carbon atoms in the same molecule, can also take place between carbon atoms which are already linked. A *double* or *triple* bond is thereby formed between the carbon atoms. When it is remembered that in such chains or rings, just as in the simple hydrocarbon CH_4, the hydrogen atoms may be replaced by numerous other 'ligands' (from the Latin *ligare*, to bind—i.e., any atom or radical which may be capable of combination), and moreover, that multivalent ligands, such as N, O, S can act as the members of chains or rings (for example, the oxygen atom in ether, $H_3C—CH_2—O—CH_2—CH_3$), we may glimpse the enormous variety of compounds which owe their origin to the two unique properties of carbon mentioned above. The discussion of these compounds constitutes the domain of *organic chemistry*.

SURVEY OF THE SIMPLEST COMPOUNDS OF CARBON

Hydrocarbons	Oxygen compounds	Sulfides
Examples: CH_4 Methane	(a) Oxides	CS_2 Carbon disulfide
C_2H_6 Ethane	CO_2 Carbon dioxide	COS Carbon oxysulfide
C_2H_4 Ethylene	CO Carbon monoxide	CS Carbon monosulfide
C_2H_2 Acetylene	C_3O_2 Carbon suboxide	C_3S_2 Carbon subsulfide
Halides	(b) Acids	Carbides
CF_4 Carbon tetrafluoride	H_2CO_3 Carbonic acid. Its	Examples: CaC_2 }Acetylides
CCl_4 Carbon tetrachloride	salts $M^I{}_2[CO_3]$: carbonates	Ag_2C_2 }
$COCl_2$ Phosgene	$H_2C_2O_6$ Peroxycarbonic acid.	Al_4C_3 Decomposed by
CBr_4 Carbon tetrabromide	Its salts $M^I{}_2[C_2O_6]$: peroxy-	dilute acid
CI_4 Carbon tetraiodide	carbonates	SiC }
	$H_2C_2O_4$ Oxalic acid. Its salts	B_6C } Stable towards acid
	$M^I{}_2[C_2O_4]$: oxalates	
	$HCHO_2$ Formic acid. Its salts	
	$M^I[CHO_2]$: formates	
$(CN)_2$ Cyanogen; HCN	$HC_2H_3O_2$ Acetic acid. Its	
Hydrocyanic acid. Its	salts $M^I[C_2H_3O_2]$: acetates	
salts $M^I[CN]$: cyanides		
$(SCN)_2$ Thiocyanogen;		
HSCN Thiocyanic acid.		
Its salts $M^I[SCN]$: the		
thiocyanates		

(a) Hydrocarbons

The simplest 'saturated' hydrocarbons are *methane* CH_4 and *ethane* C_2H_6; the simplest 'unsaturated' hydrocarbons are *ethylene* C_2H_4 and *acetylene* C_2H_2. These latter are called unsaturated because they have the property of readily adding on other substances, such as the halogens, whereby the double or triple bonds are transformed into single bonds.

Table 77 provides a survey of the most important constants of these compounds. The liter weight shown in the Table gives the weight of 1 liter of the gas in question at 0° and 1 atmosphere pressure. If this is divided by the weight of 1 liter of oxygen (1.4289 g) or of air (1.293 g) the density of the hydrocarbon, relative to oxygen = 1 or air = 1 is obtained. Table 77 also shows the calorific values of the gases. These are referred, in each case, to 1 cu.m (1000 liters) of gas at 0° and 760 mm pressure—i.e., they represent the quantity of heat, in kcal, obtained by burning 1 cu.m of gas.

TABLE 77

PHYSICAL CONSTANTS OF THE SIMPLEST HYDROCARBONS

	Methane	Ethane	Ethylene	Acetylene
Formula	CH_4	$H_3C—CH_3$	$H_2C=CH_2$	$HC\equiv CH$
Boiling point	—164°	— 88.3°	—103.9°	—83.6°
Freezing point	—184°	—172.1°	—169.4°	—81.8°
Critical temperature	—82.5°	+ 32.1°	+ 9.5°	+35.9°
Critical pressure, in atmos.	45.7	48.8	50.7	61.6
Liter weight in g	0.7168	1.3564	1.2604	1.1708
Max. calorific value, kcal per cu.m	9512	16750	15510	14010
Volumes of gas dissolved by 100 vols. of water at 0°	5.56	9.87	22.6	173
Same at 20° (Gas volumes reduced to 0°)	3.31	4.72	12.2	103
Volumes of gas dissolved by 100 vols. alcohol, at 0°	52.3	ca. 150	360	ca. 600

The saturated hydrocarbons (methane, ethane, propane, etc.) are known collectively as *alkanes*, unsaturated hydrocarbons with one double bond as *alkenes*, and those with one triple bond as *alkynes*. In accordance with this general and systematic nomenclature, the name *ethene* is also applied to ethylene, and acetylene is called *ethyne*.

Unlike the simplest hydrocarbons, those with a larger number of carbon atoms (the 'higher' hydrocarbons) are liquid or even solid. As has been mentioned, lighting paraffin, gasoline, lubricating oils, etc. are composed of mixtures of liquid hydrocarbons. Benzene, C_6H_6, is a homogeneous higher hydrocarbon which is also liquid. The liquid hydrocarbons have become of great importance, especially as fuels for internal combustion engines. For this reason, the problem of manufacturing them by the so called '*liquefaction of coal*' has assumed great importance.

(i) *Liquefaction or Hydrogenation of Coal*. The liquefaction of coal [15, 16] is the term popularly applied to the conversion of coal into liquid hydrocarbons, or into hydrocarbon derivatives containing oxygen, by combining it with hydrogen (hydrogenation). The hydrogenation can be carried out

[a] *directly*: by the action of hydrogen on coal at high pressures (200 atm.), at about 450°, in the absence of catalysts (Bergius process), or in the presence of catalysts (I. G. Farbenindustrie process);

[b] *indirectly*: by the catalytic reaction between carbon monoxide and hydrogen, forming either *methanol* (I. G. process) or hydrocarbons (Fischer process): also

by the catalytic reaction between carbon monoxide and steam (Rheinpreussen A. G. process).

The conversion of coal into liquid hydrocarbons was first achieved by Berthelot, by the action of hydrogen iodide on coal under pressure. It was later found by Bergius (1911) that hydrogen at high pressures was taken up directly by coal at temperatures above 400°. The process can be visualized as occurring through the thermal breakdown of the hydrocarbon compounds of which the coal consists which are relatively poor in hydrogen, and the hydrogenation of the products. In the large scale application of the Bergius process, the difficulty arose that considerable quantities of tarry and asphaltic by-products were formed, through incomplete hydrogenation and inability to control the reaction correctly. It was found by the I. G. Farbenindustrie, in 1925, that the hydrogenation, which otherwise takes place sluggishly, is accelerated by the use of suitable catalysts, and can be so controlled that the reactions proceed in the required sense (e.g., the formation of readily volatile hydrocarbons).

The procedure in the large scale production of gasoline is as follows. The finely ground coal, made into a paste with oil, and with small quantities of a contact substance (which is lost with the ash), is first heated with highly compressed hydrogen ('sump phase'). The primary oil so obtained is then passed, as vapor, ('gas phase') over a contact mass (for example, molybdenum or tungsten compounds mixed with other substances) packed in the reaction vessel. The conditions of operation—such as the pressure, temperature, type and arrangement of the catalyst—can be varied over a wide range, and so worked that lubricating oil, fuel oil, diesel oil, illuminating oil, etc. are produced, instead of gasoline exclusively. Both hydrogenation processes take place in large pressure vessels at about 400–450° and about 250 atm. pressure. Brown coal, bituminous coals and other carbonaceous materials such as tars and oils can all be used as raw materials. Thus, in the United States, high pressure catalytic hydrogenation is used for the production of high grade lubricating oils from heavy petroleum fractions. In England, gasoline is produced by the hydrogenation of coal and coal tars. In Germany, brown coal, brown coal tar and, on a smaller scale, petroleum and coal tar oils have been hydrogenated to gasoline for a number of years. Gasoline produced synthetically was first marketed in 1927, and in Germany the proportion of synthetic material consumed rose during the period 1932–36 from about 10% to 27%, at the same time as the total consumption of gasoline increased considerably.

In the *methanol* process, also worked out by the I.G., (Pier and Mittasch, 1923), carbon monoxide is hydrogenated to methanol (methyl alcohol, CH_3OH) at about 400° and 200 atm. pressure, by means of catalysts containing zinc oxide. Methanol may be used either as a fuel (mixed with gasoline), as a solvent, or as the starting point for chemical syntheses. According to the nature of the catalyst and the conditions of operation, it is also possible to obtain higher alcohols by this process.

In the hydrogenation of carbon monoxide by the process of Fischer, the catalysts used consist of the elements of the iron group, in conjunction with the alkalis and possibly other oxides also. By working under pressure, mixtures are obtained which consist chiefly of oxygenated hydrocarbon derivatives ('*synthol*' process); at atmospheric pressure, hydrocarbons are obtained ('Fischer-Tropsch' process). The carbon monoxide must be carefully purified, especially from sulfur compounds, since the catalyst is very readily poisoned. Smaller amounts of gas oil and paraffin wax are obtained, as well as gasoline. These can either be used as such, or may be converted to gasoline by cracking. High-melting paraffin waxes are a by-product of this process, and their oxidation furnishes fatty acids which can be used for the synthetic production of soap.

Fatty acids can also be made directly by high pressure syntheses, starting with olefins, and bringing them into combination with CO and H_2O by the use of nickel carbonyl as catalyst.

$$RCH=CH_2 + CO + H_2O = RCH_2-CH_2-CO-OH$$

Fatty acid esters can be synthesized in a similar manner.

$$R^1CH=CH_2 + CO + R^2OH = R^1CH_2-CH_2-CO-OR^2$$

Aldehydes are obtained from olefins, CO and H_2, by high pressure synthesis in the presence of a catalyst of ThO_2 containing some Co.

$$RCH=CH_2 + CO + H_2 = RCH_2-CH_2-CHO$$

Table 78 lists a few compounds which can be obtained through the reaction of carbon monoxide with hydrogen, This table (which gives only a selection of the processes which are now available) shows how the choice of catalyst and other experimental conditions can influence the carbon monoxide-hydrogen reaction in the most varied way.

It was shown by Kölbel (1951) that water can be used in place of hydrogen for the hydrogenation of CO. By the use of Fe, Co and Ni (with certain additions), containing nitrogen, as catalyst, he obtained a good yield of light petroleum, higher paraffins and other hydrocarbons, from the reaction of CO with H_2O at 210 to 260° and normal pressure. Operation at high pressures (e.g., 100 atm.) led to the formation of alcohols (especially ethyl alcohol), as well as hydrocarbons.

Carbon dioxide can also be hydrogenated by hydrogen to gaseous, liquid and solid hydrocarbons. It was first shown by Sabatier and Senderens, in 1902, that CH_4 is formed when either CO or CO_2 is passed with hydrogen over finely divided, heated nickel.

(*ii*) *Methane*, CH_4, is a colorless, odorless gas, which burns with a weakly luminous flame. It is frequently found in coal mines where, mixed with air, it forms the explosive *fire damp*. It is formed in the decay of vegetable matter under anaerobic conditions—e.g., in marshes, for which reason it is also known as *marsh gas*. It was discovered as such by Volta in 1778. In many localities, and especially near oil wells, methane escapes in large volumes from fissures in the ground—e.g., at Baku, in the Caucasus, and in Pennsylvania, Ohio and Indiana. It is also formed in the destructive distillation of wood, coal, lignite, etc. (cf. Table 76).

The quantity of methane continuously supplied to the atmosphere by decay processes and from natural gases is not known with certainty. It would be of interest to know the quantity involved, since this would make it possible to calculate exactly the relative extents to which the air masses undergo mixing by convection and turbulence, and unmixing through diffusion. At present it is only known that the effects of convection far outweigh those of diffusion in the neighborhood of the earth's surface, and it may be surmised that the effects begin to change in importance at about 70 km altitude (cf. Harteck, *Angew. Chem.*, 63, (1951) 1; Paneth, *J. Chem. Soc.*, (1952) 3651).

It is prepared in the laboratory either [i] by heating a mixture of sodium acetate and soda lime (a mixture of CaO and NaOH):

$$NaC_2H_3O_2 + NaOH = Na_2CO_3 + CH_4;$$

or [ii] by the action of water on aluminum carbide, Al_4C_3:

$$Al_4C_3 + 12H_2O = 4Al(OH)_3 + 3CH_4$$

Methane is lighter than air (density = 0.558 relative to air = 1). It is but slightly soluble in water (cf. Table 77), more soluble in alcohol. With air or oxygen, it forms mixtures which explode when ignited, since it combines with oxygen when heated above 670°, forming carbon dioxide and water —

$$CH_4 + 2O_2 = CO_2 + 2H_2O_{liq} + 212.3 \text{ kcal (at const. volume)}$$

From this equation, according to Avogadro's law, 1 volume of methane sould require 2 volumes of oxygen, or 10 volumes of air, for its complete combustion. A mixture of

TABLE 78

REACTION OF CARBON MONOXIDE WITH HYDROGEN UNDER VARIOUS CONDITIONS

Catalyst	Temperature and press.	Schematic equation for reaction	Chief product of reaction
Ni or Co	200–300° normal press.	$CO + 3H_2 = CH_4 + H_2O$	Methane
Fe	300–400° normal press.	$2CO + 2H_2 = CH_4 + CO_2$	Methane
Co activated with MgO or ThO$_2$	170–200° normal press.	$nCO + 2nH_2 = C_nH_{2n} + nH_2O$ $nCO + (2n + 1)H_2 = C_nH_{2n+2} + nH_2O$	Gasoline
Fe activated with Cu + KOH + CaO	225° 10 atm.	e.g., $8CO + 10H_2 = C_6H_{12} + 2CO_2 + 4H_2O$ $9CO + 10H_2 = C_6H_{14} + 3CO_2 + 3H_2O$	Gasoline
Fe activated with Cu, Al$_2$O$_3$ and KOH	230° 20 atm.	$3nCO + 3nH_2 = 2C_nH_{2n} + nCO_2 + nH_2O$	Olefins
Ru activated with K$_2$CO$_3$	180–200° 300 atm.	$nCO + (2n + 1)H_2 = C_nH_{2n+2} + nH_2O$	Paraffin wax
ThO$_2$ activated with Al$_2$O$_3$ and K$_2$CO$_3$	450° 300 atm.	e.g., $10CO + 9H_2 = C_6H_{14} + 4CO_2 + 2H_2O$	Branched chain hydrocarbons
ZnO activated with Fe and KOH	450° 200 atm. low flow rate	$nCO + 2nH_2 = C_nH_{2n+1}OH + (n - 1)H_2O$	'isobutyl oil' (= methanol + isobutanol and other aliphatic alcohols)
ZnO activated with Cu and Al$_2$O$_3$	200–400° 200–300 atm.	$CO + 2H_2 = CH_3OH$	Methanol
Fe activated with Al$_2$O$_3$ and K$_2$CO$_3$	200° 20 atm.	$17CO + 20H_2 = C_6H_{14} + C_6H_{13}OH + 5CO_2 + 6H_2O$	'synthol' (mixture of hydrocarbons and compounds containing oxygen)

methane and air in the ratio 1 : 10 therefore explodes with great violence. The mixtures are, however, still explosive when the methane content is very much less than this. This is the reason for the mine explosions which occur from time to time, in spite of numerous safety precautions.

It follows further from the above equation that 1 volume of CO_2 is formed by the complete combustion of 1 volume of methane. The decrease in the volume of the mixture is twice the volume of methane, if the gases are measured at ordinary temperature and the volume of the water is neglected. When a mixture of hydrogen with air or oxygen is burned, the decrease in volume is $1\frac{1}{2}$ time the hydrogen content. If, therefore, we burn or explode a mixture of hydrogen and methane, to which an excess of air has been added (igniting it with an electric spark or an incandescent wire), and determine the CO_2 content after the explosion (by absorption in KOH), this gives the CH_4 content directly. If we deduct twice this amount from the total contraction produced by the explosion, we obtain the decrease in volume which must be attributed to the hydrogen present in the original mixture. The hydrogen content is then $\frac{2}{3}$ of this amount.

In the presence of catalysts—e.g., Ni activated by Al_2O_3 — CH_4 may be partially oxidized according to the equations

$$CH_4 + \tfrac{1}{2}O_2 \;\rightleftharpoons\; CO \;+\; 2H_2 +\; 6.5 \text{ kcal}$$
$$CH_4 + H_2O \;\rightleftharpoons\; CO \;+\; 3H_2 -\; 48.9 \text{ kcal}$$
$$CH_4 + 2H_2O \;\rightleftharpoons\; CO_2 + 4H_2 -\; 38.5 \text{ kcal}$$

These reactions are of some importance in the technical production of hydrogen (p. 25) An equilibrium is also set up between CH_4 and CO_2 in the presence of Ni as catalyst:

$$p_{CO}^2 \cdot p_{H_2}^2 / p_{CH_4} \cdot p_{CO_2} = K_p. \quad \text{Log } K_p = -12.662/T + 13.875 \quad \text{(Fischer, 1931)}$$

Since the H-nuclei in CH_4 penetrate the outer electron shell of the C atom, the gas has physical properties resembling those of an inert gas. Its melting point and boiling point are close to those of argon. It forms a continuous series of mixed crystals with krypton (von Stackelberg, 1936). Solid CH_4 has a transition point at $20.4°$ K, attributed to some intra-molecular transformation (Clusius, 1930).

After hydrogen, methane is the principal constituent of the *illuminating gas* obtained by the destructive distillation of coal; pure coal gas contains on the average 30–33 volume-% of methane, and about 50% of hydrogen. Coal gas also contains about 4% of so-called 'heavy hydrocarbons' (chiefly ethylene, but with some acetylene and benzene also), and about 9% of CO, 2% of CO_2, 1% N_2 and traces of oxygen. Such coal gas has a maximum calorific value of about 5500 kcal/cu.m.; water gas (cf. p. 445) is generally mixed with coal gas, however, whereby the calorific value is reduced to about 4200 kcal/cu.m. ('town's gas').

5–6 cu.m of air are required for the combustion of 1 cu.m of coal gas proper, but only 2.4 cu.m of air for 1 cu.m of water gas. For town's gas, the air requirement may be taken as about 4 cu.m.

Coke oven gas contains about 55% H_2, 25% CH_4, 2% heavy hydrocarbons, 4–6% CO, 2% CO_2 and 10–12% N_2. Its composition differs from that of coal gas chiefly in that, in coke ovens, the coal is more completely carbonized than in the gas works. The higher the temperature of carbonization the greater is the proportion of hydrogen as compared with methane and the heavy hydrocarbons. Since the hydrocarbon content has considerable influence on the calorific value, the calorific power of coke oven gas (4000–4500 kcal maximum value) is necessarily always less than that of coal gas proper.

(*iii*) *Ethane*, C_2H_6, is also a colorless, odorless gas. It was first obtained by Kolbe (1848), by the electrolysis of a concentrated solution of potassium acetate.

In this process it is evolved at the anode, the acetate ions (or acetic acid radicals) coming together with the elimination of CO_2 when they are discharged

$$
\begin{array}{cc}
CH_3\text{—}CO.O^- & CH_3 \\
= & | \quad + 2CO_2 + 2e \\
CH_3\text{—}CO.O^- & CH_3 \\
\text{acetate ions} & \text{ethane}
\end{array}
$$

It was originally believed by Kolbe to be the radical of methane, *methyl*, CH_3. It follows, however, from the density of the gas that it has the doubled molecule weight — C_2H_6. The methyl radical is very unstable, and free methyl radicals at once combine in pairs, to form the compound $H_3C\text{—}CH_3$.

Ethane can also be obtained by the action of metallic sodium on methyl iodide in ethereal solution (at 100°), or by heating methyl iodide with metallic zinc in a sealed tube:

$$
\begin{array}{cccc}
CH_3\text{—}\;I\;\;Na & CH_3 & \qquad CH_3\text{—}\;I & CH_3 \\
+ & = | \quad + 2NaI; & \qquad \qquad + Zn = | \quad + ZnI_2 \\
CH_3\text{—}\;I\;\;Na & CH_3 & \qquad CH_3\text{—}\;I & CH_3
\end{array}
$$

If the methyl radical were stable, its formation would be expected in these processes also.

Paneth, in 1929, succeeded in obtaining free *methyl* radicals for the first time, by the thermal decomposition of $Pb(CH_3)_4$. It was then found that they are so unstable that in an atmosphere of 2 mm pressure of hydrogen, for example, the concentration of free methyl has fallen to one half its value after 0.006 sec, and has practically vanished after 0.1 sec. The *ethyl*, *propyl* and *phenyl* radicals which are likewise extraordinarily unstable, have also been obtained in the free state (Paneth 1931, Purcell 1935, Polanyi 1934) [62]. Hein had rather earlier (1924) deduced that free ethyl radicals have a momentary existence as intermediates in the electrolysis of sodium ethyl NaC_2H_5, dissolved in zinc ethyl, $Zn(C_2H_5)_2$. Such a hypothesis would explain the formation of equivalent amounts of ethylene and ethane at the anode:

$$2C_2H_5^- = 2e + 2C_2H_5 \rightarrow C_2H_4 + C_2H_6 \text{ (disproportionation)}$$

The *n*-propyl radical has a still stronger tendency for disproportionation and also decomposes directly at higher temperatures, forming methyl radicals:

$$C_3H_7 \rightarrow C_2H_4 + CH_3$$

Ethane is prepared technically by the hydrogenation of ethylene. It is used in refrigeration engineering. Ethane is sometimes found in nature in the gases issuing from oil wells. Like methane, ethane is a *saturated* hydrocarbon. It is not attacked by bromine water. It is but slightly soluble in water, but appreciably soluble in alcohol (Table 77). It burns with a faintly luminous flame, like methane, but differs from methane in being much more readily condensable. Its critical point is above room temperature, whereas methane can be liquefied by pressure only if it is also strongly cooled.

C_2H_6 forms a 'gas hydrate' (cf. p. 219), of the composition $C_2H_6 \cdot 5H_2O$ (found $5.8 \pm 0.5\ H_2O$, von Stackelberg, 1949). This has a dissociation pressure of 5.2 atm. at 0°. CH_4 forms a similar hydrate (decomposition pressure 26 atm. at 0°).

(iv) *Ethylene*, C_2H_4, is present to the extent of about 4% in coal gas. It is generally prepared in the laboratory by heating a mixture of ethyl alcohol with excess concentrated sulfuric acid, to about 160°; the addition of some coarse grained sand is advantageous. The reaction can be regarded as a dehydration, brought about by the sulfuric acid:

$$C_2H_6O - H_2O = C_2H_4$$
ethyl alcohol   ethylene

The detailed mechanism of the reaction involves the initial reaction of the sulfuric acid with alcohol, forming an acid ester, the so-called ethylsulfuric acid. Since this is not very stable towards heat, it decomposes into ethylene and sulfuric acid:

$$C_2H_5 - O - H + HO - SO_3H = C_2H_5-O-SO_3H + H_2O$$
ethylsulfuric acid

$$C_2H_5-O\,SO_3H = C_2H_4 + H_2SO_4$$

Ethylene is prepared industrially by Ipatieff's process—the passage of alcohol vapor over amorphous aluminum oxide at 360°. It is a colorless gas, with a peculiarly sweetish smell. When ignited, it burns with a much more luminous flame than do methane and ethane. Unlike these gases, it is absorbed by fuming sulfuric acid and by bromine water; ethyl sulfuric acid and ethylene dibromide, $C_2H_4Br_2$, are thereby formed. The former decomposes again into ethylene and sulfuric acid when it is heated. Ethylene may be regenerated from the ethylene dibromide by the action of zinc dust on the alcoholic solution:

$$C_2H_4Br_2 + Zn = ZnBr_2 + C_2H_4$$

Ethylene is moderately soluble in water, but readily soluble in alcohol (Table 77). In the presence of a catalysts (platinum black or nickel), it can also add on hydrogen, forming ethane:

$$C_2H_4 + H_2 = C_2H_6$$

Addition of hydrogen in this manner is known as *hydrogenation*. The ability of ethylene to undergo addition reactions is attributed to the presence of a *double bond* between the two carbon atoms, this being changed to a single bond through the addition of the other atoms concerned:

Ethylene is an endothermic compound, but its heat of decomposition is only small:

$$C_2H_4 = 2C_{soot} + 2H_2 + 10\ kcal$$

The name ethylene is also customarily used for the corresponding bivalent radical CH_2-CH_2. The radical isomeric with this, $>HC-CH_3$, is known as the *ethylidene* radical.

With ethylene and hydrocyanic acid as starting materials, the *acrylic resins* can be obtained. These have found extensive application as synthetic plastics—e.g., as the middle lamina in safety glass, for electrical insulating material, pharmaceutical plaster, artificial leather, lacquers, adhesives, etc.

(*v*) *Acetylene* [*17–19*], C_2H_2 is a colorless, poisonous gas which usually has a peculiar unpleasant smell, due to the impurities contained in it. When ignited (ignition temperature 428°) it burns with a strongly luminous flame. The flame is a brilliant white when burners are used which give a broad, thin flame, in which a copious supply of air makes it possible to burn completely the particles of carbon which are present in the flame in considerable amounts as a result of thermal decomposition. Under other conditions much soot is deposited.

Acetylene is a strongly endothermic compound. It can be made to detonate by means of mercury fulminate, and decomposition can also be initiated by the electric spark, proceeding according to the equation;

$$C_2H_2 = 2C_{soot} + H_2 + 54 \text{ kcal}$$

Compressed acetylene, expecially when liquefied, often explodes for very little cause. It will, however, withstand a pressure of a few atmospheres when mixed with other gases or when dissolved in acetone. One part of acetone dissolves 25 volumes of acetylene at 15° and 1 atmosphere pressure, or 300 volumes under 12 atmospheres pressure. The solubility in water is also quite appreciable (about 1 : 1 at room temperature), and is still greater in alcohol (Table 77).

Use is made of the solubility of acetylene in acetone for the commercial transport of the gas. It is sold as 'dissolved acetylene' in steel cylinders, which contain the gas under a pressure of 15 atm, dissolved in acetone which is absorbed by kieselguhr.

Acetylene is hydrogenated by hydrogen, in the presence of platinum black, forming ethane. It requires 4 atoms of hydrogen per molecule of acetylene for this—i.e., 2 atoms more than for ethylene. There is a *triple* bond between the carbon atoms in acetylene, and this is converted to a single bond by the addition of 4 H atoms:

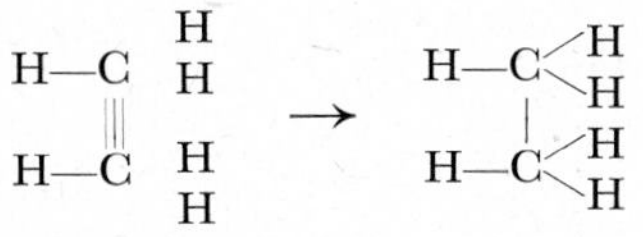

The halogens are added on even more readily than is hydrogen, the maximum amount again being 4 atoms. By suitable methods, however, it is possible to bring about a partial halogenation, in which only 2 halogen atoms are added on. In the former case, halogen derivatives of ethane are formed—e.g., $CHCl_2$—$CHCl_2$, tetrachlorethane; in the latter case, the products are halogen derivatives of ethylene, in which one double bond is still present—e.e., $CHCl=CHCl$, dichlorethylene. Acetylene is oxidized by potassium permanganate in alkaline solution to formic acid, carbonic acid and oxalic acid; in acid solution it yields principally formic acid and carbonic acid.

The heat of combustion of acetylene is 311.5 kcal per mol at constant volume. Acetylene-air mixtures are *highly explosive*. Even 3.5 vol.-% of acetylene suffices to produce a dangerously explosive mixture.

Acetylene has the property of reacting with the salts of many heavy metals, forming insoluble compounds. Thus, from alcoholic $AgNO_3$ solution it precipitates white Ag_2C_2, and from ammoniacal $CuCl$ solution, brown-red Cu_2C_2. These compounds are called *acetylides*, and can be regarded as salts of acetylene. The latter has no perceptible acidic character in other respects. The heavy metal acetylides mentioned are very explosive when they are dry, and detonate even on gentle friction. The acetylides of the strongly electropositive metals are not explosive. These, however, can be prepared only in the dry way, since they are at once decomposed by water. The heavy metal acetylides can be isolated from solution since they are very insoluble. Mercury(II) acetylide can be obtained in a similar manner.

Acetylides of the alkali metals can be obtained by the action of acetylene on the free metals. The primary, or acid, acetylides are first formed, and these are converted to the normal or neutral acetylides by loss of acetylene when more strongly heated, e.g.,

$$K + C_2H_2 = C_2HK + \tfrac{1}{2}H_2; \qquad 2C_2HK = C_2K_2 + C_2H_2$$

The acetylides ZnC_2, CdC_2, BeC_2 and $Al_2(C_2)_3$ have also been obtained by passing acetylene over the heated metals.

Calcium acetylide, usually known as *calcium carbide*, which is prepared by heating calcium oxide and carbon in the electric furnace [20], is of particular importance for the manufacture of acetylene. Acetylene is generally prepared by decomposing calcium carbide by water, a process in every respect analogous to the hydrolysis of the salts of other weak acids

$$CaC_2 + 2HOH = C_2H_2 + Ca(OH)_2 + 32 \text{ kcal}$$

The acetylides of the other alkaline earths react with water in the same manner as calcium carbide. The alkali acetylides, however, are decomposed by water with explosive violence, so that the acetylene liberated is at once dissociated, with the deposition of carbon. The normal hydrolysis takes place, however, when water vapor is admitted slowly:

$$K_2C_2 + 2HOH = 2KOH + C_2H_2$$

For the preparation of small quantities of acetylene, water is allowed to drop on to calcium carbide, and the rate of evolution of gas is controlled by the rate of admission of water. The apparatus used in larger installations enables lumps of calcium carbide to be fed gradually from a hopper into a large quantity of water. This prevents any dangerous rise of temperature of the acetylene, and avoids the nuisance of the continued generation of gas after the apparatus has been shut down. 'Dry gasification' may also be employed. In this, the decomposition is effected by a fine spray of water, which absorbs the heat of reaction by its evaporation. The $Ca(OH)_2$ is then obtained as a dry powder, which can be pressed into pellets, fired, and returned to the carbide furnace. Crude acetylene contains hydrogen sulfide and phophine as impurities. In some circumstances the last mentioned can make the gas spontaneously inflammable. The impurities are removed by washing with calcium chloride solution, which retains the ammonia, and then by treating the gas with a mixture of calcium oxide and bleaching powder. The latter oxidizes hydrogen sulfide and phosphine to the corresponding oxy-acids, which are at once bound by the lime.

Acetylene has recently been prepared industrially to an increasing extent by the *thermal decomposition of saturated hydrocarbons*, such as CH_4. The hydrocarbon to be decomposed is heated, for example, to over 5000° in the electric arc, and is then chilled, as in the nitrogen fixation process. It has been demonstrated spectroscopically that free $C\equiv C$ radicals are formed at the high temperatures, and these combine with H atoms, forming C_2H_2.

Acetylene is used today on a large scale for the synthesis of organic compounds (see below). It is also used for autogenous welding and as an illuminant. Precautions must be taken, in view of the extraordinarily explosive nature of acetylene–air mixtures, already mentioned. The most explosive is a mixture of 1 volume of acetylene with $2\tfrac{1}{2}$ volumes of oxygen or $12\tfrac{1}{2}$ volumes of air, as follows from the equation

$$2C_2H_2 + 5O_2 = 4CO_2 + 2H_2O$$

If acetylene is passed through gently heated 15% sulfuric acid, containing mercuric sulfate, acetaldehyde is formed according to the equation

$$C_2H_2 + H_2O = C_2H_4O$$

This, and the acetic acid formed from it by oxidation, are the starting materials for many important products of the organic chemical industry, which are produced on a large scale. In addition to dyestuffs and pharmaceuticals, lacquers, films, fibers, plastics and synthetic resins are becoming of increasing importance among these. Catalytic hydrogenation of acetaldehyde yields ethyl alcohol. The quantity of ethyl alcohol now manufactured from calcium carbide ('carbide spirit') is very considerable. Combination of C_2H_2 with acids, alcohols, phenols or amines leads to *vinyl compounds*, i.e., compounds of the radical $CH_2=CH—$, which are also starting materials for the production of important derivatives—e.g., various types of high-grade synthetic materials, which now find very widespread uses for waterproofing fabrics, for protection against chemical warfare agents, as insulating material for electrical apparatus, coverings for cables and wires, as substitutes for metal and wood, as lacquers and non-inflammable paints, and as plastic masses for the manufacture of buttons, boxes, cups, etc. Vinylacetylene, $CH_2=CH—C\equiv CH$ can be obtained from acetylene under the catalytic influence of Cu^I compounds, as was found by Nieuwland; from this *butadiene* $CH_2=CH—CH=CH_2$ is obtained by hydrogenation, and can be polymerized to form *Buna-rubber*. Alternatively, by addition of HCl (with CuCl as catalyst), *chloroprene*, $CH_2=CH—CCl=CH_2$ is obtained; polymerization of this yields the *Duprene rubber* manufactured especially in the United States and Russia. These synthetic rubbers attained very great importance during the 1939–45 war. Apart from this, however, products made from them may be superior ot those fabricated from natural rubber in many properties, such as resistance to heat, oil and abrasion. The Duprene rubber, in particular, is notable for its resistance to oil.

(*vi*) *Butadiene*, which is required for the synthesis of Buna-type synthetic rubbers, was originally (from 1929 onwards) prepared by the 'four stage' process.

$$2C_2H_2 \xrightarrow{+2H_2O} 2CH_3CHO \rightarrow \underset{\text{aldol}}{CH_3.CH(OH).CH_2.CHO} \xrightarrow{+H_2}$$

$$\underset{\text{butylene glycol}}{CH_3.CH(OH).CH_2.CH_2(OH)} \xrightarrow{-2H_2O} \underset{\text{butadiene}}{CH_2=CH—CH=CH_2}$$

In this, and in the 'vinyl synthesis' subsequently devised by Nieuwland (see above), 2 molecules of the relatively expensive acetylene are consumed for each molecule of butadiene produced. Only 1 molecule of acetylene is required for the '*alkinol synthesis*' worked out by Reppe (1937). This synthesis involves the addition of acetylene to aldehydes, with copper acetylide, Cu_2C_2, as catalyst; depending on the experimental conditions this can lead to monohydric or dihydric alcohols with a $C\equiv C$ bond (alkinols). Thus the addition of formaldehyde (obtained synthetically from water gas) to acetylene (at 100°, 5 atm.) first produces butyne diol, $HO.CH_2—C\equiv C—CH_2OH$. This is easily hydrogenated in aqueous solution, with Ni as catalyst, to butane diol, $HOCH_2—CH_2—CH_2—CH_2OH$. Butadiene, $CH_2=CH—CH=CH_2$, is formed from this by elimination of water (with phosphoric acid and sodium phosphate as catalyst).

By means of the alkinol synthesis, a large number of other organic compounds, can be prepared technically from acetylene and formaldehyde or other aldehydes and ketones. Such products find wide applications—for the production of plastics, solvents, waxes, scents, and pharmaceutical materials, etc. Analogous syntheses of compounds from acetylene and amines are also important industrially.

By polymerizing acetylene, with nickel compounds as catalysts, it is possible to produce cyclic hydrocarbons, and especially *cyclooctatetraene*, C_8H_8, a highly reactive substance from which a wide range of other organic compounds can be obtained. Bicyclic hydrocarbons can also be prepared directly by this route from acetylene—e.g., the deep blue hydrocarbon *azulene*, $C_{10}H_8$, consisting of a five-membered and a seven-membered ring system.

CO and H_2O can also be added directly to acetylene (with metal carbonyl compounds serving as catalysts). In the simplest case, *acrylic acid* is formed ($C_2H_2 + CO + H_2O =$ $= H_2C=CH—COOH$) or, if H_2O is replaced by alcohols, amines, etc., the corresponding derivatives of acrylic acid are produced. Under rather different conditions, the product is hydroquinone, $C_6H_4(OH)_2$, whereas C_2H_2 and CO alone (without water) form cyclic ketones such as hydrindone, C_9H_8O.

(*vii*) *Calcium acetylide* (See also p. 466). *Calcium carbide*, CaC_2 (more correctly *calcium acetylide*, being a derivative of acetylene), generally known as *carbide*, forms a colorless crystalline mass, of density 2.2, when pure. It is insoluble in all the usual solvents, but is decomposed by water, forming acetylene. At high temperatures it reacts with metallic oxides, either liberating the metals, or else undergoing double decomposition and forming the carbide of the metal and calcium oxide. When heated in nitrogen it is converted to calcium *cyanamide*, $CaCN_2$ (cf. p. 462).

Calcium carbide is produced industrially on a very large scale [20], generally by heating burned lime with coke or wood charcoal in the electric furnace. The electrodes consist of large blocks of carbon, through which currents of up to 50,000 amperes may be passed in large furnaces.

In the large scale manufacture, about 1 kg of quick lime, 0.7 kg of coke and 2.8–3.5 kWh of electricity are consumed for each kg of 85% carbide.

Calcium carbide can be prepared on a small scale, for demonstration purposes, by taking a hessian crucible of moderate size, and grinding away the bottom on an emery wheel until a hole is produced through which a 13–15 mm diameter arc carbon can just conveniently be introduced (Fig. 81). The flat end of the carbon is allowed to project about 3 mm into the crucible, and is fixed in this position with cement. It is then connected through a resistance and an ammeter to the negative pole of a direct current supply. For the anode, a pointed carbon rod, 10–15 cm long and 7 mm diameter, is used, fixed to an insulated stand which has a rack and pinion for the convenient raising and lowering of the carbon. The reaction mixture is prepared by intimately mixing 15 g of finely powdered quicklime with 20 g of wood charcoal. When just a little of this mixture has been added to the crucible, the circuit is closed and the current regulated to about 12–14 amps; alternating current may also be used in place of direct current. Fresh portions of the reaction mixture are repeatedly added during the experiment, which lasts about 10–15 minutes. The crucible is kept covered by means of a perforated asbestos board. The distance between the carbon rods, initially 1 cm, is progressively increased during the experiment to 5–6 cm. The amount of carbide prepared by means of this simple experimental arrangement suffices to prepare several liters of acetylene gas.

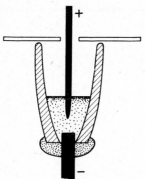

Fig. 81. Small scale preparation of calcium carbide.

Apart from the production of acetylene, calcium carbide is used chiefly for the manufacture of calcium cyanamide [21]. It is also used in metallurgy as a reducing agent.

According to Franck (1937), pure CaC_2 has a crystal structure different from that of the technical carbide, which is contaminated with impurities. The latter has a face-centered tetragonal structure, as have pure SrC_2 and BaC_2 (see Fig. 84, p. 466). Calcium carbide containing $CaCN_2$ has still another crystal structure. Conversion to $CaCN_2$ takes place most readily with the pure CaC_2, and least readily with the technical material.

(b) Halides

(*i*) *Carbon tetrafluoride*, CF_4, is a colorless gas (b.p. —128°, m.p. —183.6°, liter-weight = 3.94 g) which may be obtained by the direct union of its components, or by the action of AgF on CCl_4 at 300°. It is very slightly soluble in water, is extremely stable towards heat, and is chemically inert. It is therefore very suitable as a manometric liquid at low temperatures.

A number of fluorides richer in carbon have recently been prepared:

	C_2F_6	C_3F_8	C_4F_{10}	C_5F_{10}	C_6F_{12}	C_7F_{14}
b.p.	$-78.2°$	$-38°$	$-4.7°$	$+23°$	$+51°$	$+80°$
m.p.	$-100.6°$	$-183°$	about $-12°$		$+58°$	

(*ii*) *Carbon tetrachloride*, CCl_4, is a colorless, highly refractive, non-inflammable liquid, with a not unpleasant sweet smell, and a narcotic action. Its density is 1.593; it boils at 76.7° (vapor pressure$=33.4$ mm at 0°), and melts at $-22.87°$. It has a transition point at $-47.66°$ which has been recommended by Johnston (1934) as a thermometric fixed point. Its solubility in water is very small (0.08 g in 100 g water at 20°). The dielectric constant is low (2.3 at 18°) and the refractive index is high ($n = 1.463$ for yellow light).

Carbon tetrachloride is chemically inert at ordinary temperature. It reacts neither with bases nor with acids, including concentrated sulfuric acid. It does, however, corrode many metals appreciably—e.g., iron or aluminum. It is also gradually decomposed by water in the precence of these metals, according to the equation $CCl_4 + 2H_2O = CO_2 + 4HCl$, whereas it is otherwise not detectably hydrolyzed by water at ordinary temperature.

Carbon tetrachloride is notable for its good solvent properties towards organic substances. It is completely miscible with alcohol and other organic liquids. It is much used both in the laboratory and in industry, as a solvent for fats, oils, resins, etc. It is also used as a fire extinguishing material. It is used in medicine as an anaesthetic and as an anthelmintic.

It is prepared by chlorinating carbon disulfide, CS_2 in the presence of halogen carriers, such as manganese(II) chloride:

$$CS_2 + 3Cl_2 = CCl_4 + S_2Cl_2$$

The sulfur monochloride, S_2Cl_2, so formed can also be made to react with CS_2 by moderate heating (to about 60°) in the presence of suitable catalysts, such as FeS, $-2S_2Cl_2 + CS_2 = CCl_4 + 6$ S. Under suitable conditions both reactions proceed concurrently. The sulfur which is rejected then goes back into the manufacturing cycle for the preparation of CS_2. The carbon tetrachloride is purified by washing it with caustic potash, followed if necessary by fractional distillation.

(*iii*) *Carbon tetrabromide*, CBr_4, is formed from CCl_4 and $AlBr_3$ at 100°. It separates from solution in monoclinic crystals, which change at 47° into the cubic system; m.p. 93.7°, b.p. 189.5°. It is but slightly soluble in water. When heated with water to 200°, in a sealed tube, hydrolysis takes place forming CO_2 and 4HBr.

(*iv*) *Carbon tetraiodide*, CI_4, can be obtained by the double decomposition of CCl_4 with AlI_3 or BiI_3. It forms dark red cubic crystals (m.p. 171°, density 4.32). It can be sublimed in a vacuum at 90–100°, but elimination of iodine takes place if it is more strongly heated. This also occurs in sunlight, C_2I_4 being formed.

For mixed halides— $CClF_3$, CCl_2F_2, etc., see RUFF, *Z. anorg. Chem.*, 201 (1931) 245; 210 (1933) 173; HAUPTSCHEIN, *J. Am. Chem. Soc.*, 74 (1952) 1347. CCl_2F_2, which is manufactured by the action of HF on CCl_4 in the presence of $SbCl_3$, is used as a refrigerant under the name of 'Freon'.

Other carbon-fluorine compounds ('fluorocarbon' derivatives) are becoming increasingly important technically, e.g., trifluoracetic acid, $CF_3.COOH$ (colorless liquid, b.p. 72.5°), which is prepared technically by electrochemical fluorination in liquid HF. The 'perfluoro' compounds (that is, compounds in which all hydrogen atoms are replaced by fluorine, except those bound as OH groups) are often notable for their thermal stability and their great chemical inertness. By polymerization of the compounds C_2F_4 (b.p. —76°) and C_2ClF_3 (b.p. —28°), derived from ethylene, it is possible, according to the conditions, to obtain liquid to waxy products which can be used as hydraulic fluids, lubricants or plasticizers, or else solid plastics (e.g., Kel F) which have proved particularly useful for the construction of chemical apparatus. Organic dyes containing CF_3-groups in place of CH_3-groups have been found to be very fast towards light. As examples of perfluorinated carbon compounds with relatively long chains, mention may be made of perfluoroglutaric and perfluoroadipic acids,

$$HO.CO.CF_2.CF_2.CF_2.COOH \quad and \quad HO.CO.CF_2.CF_2.CF_2.CF_2.COOH,$$

which were prepared by Hendricks (1951). The perfluorinated amines, $CF_3.CF_2.NF_2$ (perfluoroethylamine) and $CF_2=NF$ (perfluoroethylene-imine) may also be mentioned (Bigelow, 1951). Trifluoracetic anhydride, $(CF_3CO)_2O$ (colorless liquid, b.p. 39°) has proved to be a useful reagent in organic chemistry.

(c) Oxygen Compounds

(i) *Oxides.* Carbon forms two simple oxides. *Carbon dioxide*, CO_2, is obtained as the product of complete combustion; if combustion is incomplete, *carbon monoxide*, CO, is formed.

The heats of formation of the two oxides (at 20° and constant pressure) are

$$C_{graphite} + \tfrac{1}{2}O_2 = CO + 26.64 \text{ kcal}$$
$$CO + \tfrac{1}{2}O_2 = CO_2 + 67.75 \text{ kcal}$$
$$\overline{C_{graphite} + O_2 = CO_2 + 94.39 \text{ kcal}}$$

The heat liberated by the formation of carbon monoxide from carbon and oxygen is thus very much less than that of the further oxidation of the monoxide to the dioxide. This is because a very considerable expenditure of energy is involved in disrupting the linking o the carbon atoms to one another on the graphite. Both the very substantial energy of sublimation of graphite (about 170 kcal) and the energy of dissociation of the oxygen molecules must be substracted from the energy set free when C-atoms and O-atoms come together to form CO molecules.

An oxide of carbon with still less oxygen was obtained by Diels by elimination of the elements of water from malonic acid by means of phosphorus pentoxide:

O H   OH
‖
C—C—C $-2H_2O = O=C=C=C=O$
| ‖
HO H   O

malonic acid   carbon suboxide

The constitution of this compound follows from its mode of formation and its reactions. Electron diffraction measurements and observations of the Raman spectrum (Chap. 9) have confirmed the linear structure. Carbon suboxide is a colorless gas with a suffocating smell (b.p. 7°, m.p. —107°), which polymerizes very readily when it is impure, forming a red amorphous substance,. It is quite stable when pure. Under certain conditions it was found by Klemenc (1934) to decompose according to $C_3O_2 \rightleftharpoons CO_2 + C_2$; a carmine-red coloration then appears in the gas phase, and it is uncertain whether this should be attributed to carbon, forming as an intermediate stage in the gaseous state (*dicarbon*, C_2), or whether it is due to the formation of an aerosol of carbon. Solid carbon with a purple color (excessively finely divided graphite) is deposited on the wall.

1. *Carbon dioxide* [22], CO_2, is a colorless, incombustible gas, with a weakly acid smell and taste. It is considerably heavier than air (liter-weight = 1.9768 g at 0° and 760 mm); its density, relative to air = 1, is 1.529. Carbon dioxide therefore accumulates near the floor in spaces where it is being evolved—(brewhouses, well shafts, etc.); such places must therefore be entered cautiously, since carbon dioxide cannot support respirations. It is easily liquefied, requiring a pressure of 56.5 atm. at 20°, and forms a colorless mobile, liquid, of density 0.766. The critical temperature is 31.3°, the critical pressure 72.9 atm., and the critical density 0.464. Carbon dioxide sublimes at —78.48° under ordinary pressure, and the solid melts at —56.7° under a pressure of 5 atm.

Crystalline carbon dioxide forms a *molecular lattice*, in which the C atoms occupy the points of a face-centered cubic lattice, with the edge length $a = 5.63$ Å. The two oxygen atoms of each molecule are collinear with the carbon atom, and closely adjacent to it (C—O = 1.05 Å), lying along the direction of the cube diagonals. The free molecule also has a linear symmetrical structure: O=C=O, with the interatomic distances C—O 1.13 Å.

The heat of fusion of carbon dioxide is 2.24 kcal per mol., the heat of evaporation 6.0 kcal (at —56°). Liquid carbon dioxide is a non-conductor of electricity, and has only slight solvent properties.

Considerable quantities of carbon dioxide gas can be absorbed by water. 100 parts of water dissolve

at	0°	10°	20°	25°	60° C
	171	119	88	75.7	36 volumes of CO_2 at 1 atm.

Carbon dioxide is fairly inert chemically. As the anhydride of *carbonic acid*, it combines vigorously only with strong bases (cf. p. 448), forming carbonates. It also reacts at high temperatures with the strongly electro-positive metals, such as K, Mg, Zn, giving up its oxygen to them partially or completely. When passed over red hot charcoal, it forms carbon monoxide—the 'Boudouard equilibrium':

$$CO_2 + C \rightleftharpoons 2CO$$

Form the measurements of Boudouard and of Meyer, the composition of the gas in equilibrium, under 1 atm. pressure, and in contact with solid carbon, is

at $t =$	450°	600°	650°	700°	750°	800°	900°	1000° C
CO_2	98	77	61.5	42.3	24.7	6.0	2.8	0.7 %
CO	2	23	38.5	57.7	75.3	94.0	97.2	99.3 %

At lower temperatures the equilibrium can be established only in the presence of a catalyst such as iron carbide, Fe_3C.

Carbon dioxide is not poisonous in low concentrations. It is always present in the blood, being formed in the organism by oxidation processes, and eliminated by respiration. The carbon dioxide contained in the blood has a stimulating effect on the respiratory centers. However, if the blood is continuously over-charged with carbon dioxide, there is a deleterious action. Hence if large quantities of carbon dioxide are inhaled, the consequence is not merely suffocation due to lack of air, but disturbances are set up in the organism. These may continue to operate, and may lead to death, even after the acute danger has been removed.

As has already been mentioned, carbon dioxide is present in sparse concentration (3 : 10,000) in the air. It issues from the earth in many regions, especially in the neighborhood of volcanoes. Many mineral springs furnish water which is considerably supersaturated with carbon dioxide at ordinary pressure, so that it escapes for the most part when the water issues at the surfaces. Carbon dioxide is contained to a small extent in all natural waters. Combined with metallic oxides, enormous quantities are present in nature as carbonates, and especially as calcium and magnesium carbonate (cf. Chap. 8).

In the laboratory, carbon dioxide is generally prepared by the decomposition of marble with hydrochloride acid in a Kipp's apparatus, or frequently by heating sodium hydrogen carbonate or magnesite.

Large quantities of carbon dioxide are available industrially as a by-product of lime burning, but the gases so obtained contain much nitrogen. To extract pure carbon dioxide, the gas is passed through towers down which sodium- or potassium carbonate solution trickles over coke. This absorbs the carbon dioxide, forming the hydrogen carbonate, and liberates it again on boiling. A still better absorbent is an aqueous solution of ethanolamine $HO.CH_2.CH_2.NH_2$; when this is warmed it likewise gives up the CO_2 which it has absorbed at ordinary temperature ('Gerbotol' process). Carbon dioxide can be obtained in the same way from the gases formed by burning coke. In many places the natural gas wells (e.g., at Herste in Westphalia, and Hönningen on the Rhine), or the mineral springs (in the Eifel and Taunus) are utilized for the production of carbon dioxide. Large quantities are also liberated in the course of alcoholic fermentation. This is also utilized in some countries.

Large quantities of carbon dioxide are used in the production of mineral waters and in the retail of beer; it is sold liquefied in steel bottless for the latter purpose. In medicine, mixtures of carbon dioxide and oxygen are used when oxygen is administered (see effect on respiratory centers, above); it is also employed for the preparation of carbonic acid baths. It is used in fire extinguishers, both to supply pressure for spraying the extinguishing liquid, and directly for stifling fires. In the chemical industry, petroleum industry, etc., cylinders of liquid carbon dioxide are frequently employed in automatic installations, to produce a blanket of the gas over any equipment in which fire breaks out. It is also very suitable for use as a pressure gas for the handling of inflammable liquids.

Solid carbon dioxide has found increasing use as a refrigerant, under the name of 'dry ice'. [23] Weight for weight it extracts twice as much heat as ordinary ice, and has the advantage, among others, that it leaves no residue on evaporation. The use of carbon dioxide as a fertilizer has been proposed, since it is the nutriment which the plants extract from the air. It has been demonstrated that enrichment of the air with carbon dioxide considerably promotes the growth of plants. Fertilizing with carbon dioxide might, in some circumstances, be practicable for closed glasshouses.

2. *Carbon monoxide*, CO, results from the combustion of carbon in an insufficient supply of air. It is a colorless, odorless, *poisonous* gas, which will not support combustion, but is itself combustible. The liter-weight of carbon monoxide is 1.25001 g (at 0° and 760 mm), and the density, relative to air = 1, is 0.967; it is thus rather lighter than air. Its solubility in water is small (3.3 volumes of CO

in 100 parts of water at 0°, 2.3 vols. at 20°), but it is more soluble in alcohol (20.4 volumes in 100 parts of alcohol between 0° and 25°). It cannot be liquefied at ordinary temperature; its critical temperature is —140.2°, critical pressure 34.6 atm., and critical density 0.311. Under atmospheric pressure, carbon monoxide condenses at —191.5° and solidifies at —204.0°.

Carbon monoxide (like N_2) crystallizes in a body-centered cubic molecular lattice, $a = 5.63$ Å. The interatomic distance C—O is 1.06 Å in the solid, 1.26 Å in the gas. The dipole moment is $1.2 \cdot 10^{-19}$ e.s.u. The cubic form changes into a hexagonal modification at —211.6°. The heat of transformation is 0.151, heat of fusion 0.200, and heat of evaporation 1.44 kcal per mol.

The toxicity of carbon monoxide arises from the fact that it combines with *hemoglobin*, the pigment of the blood, which thereby loses its ability to combine with oxygen. The affinity of hemoglobin for carbon monoxide is considerably greater than that for oxygen; the blood takes up equal amounts of oxygen and carbon monoxide from air containing 0.1% CO—i.e. CO and O_2 in the ratio 1 : 200. Thus even quite small concentrations of carbon monoxide cause suffocation, in that they suffice to diminish considerably the ability of the blood to take up oxygen. On the other hand, if air free from carbon monoxide, or better, oxygen, is inhaled, the carbon monoxide is gradually given up again, and if this happens at an early enough stage, carbon monoxide poisoning generally leaves no permanent ill effects. The poisonous nature of coal gas, and of the gases formed when combustion takes place in insufficient air, is due to the carbon monoxide contained therein. Successful experiments have been carried out in recent years to make coal gas non-poisonous, by removing the carbon monoxide (catalytic reaction with steam, p. 444) [25]. In breathing apparatus, such as that used by fire brigades, a catalyst known as *hopcalite*, consisting of a mixture of the oxides of manganese and copper, is employed to remove CO from the air inhaled, by oxidising it to CO_2. The method was worked out in the United States during the 1914–18 war. It also makes it possible to *determine very small quantities* of CO, either by absorbing the CO_2 formed in $Ba(OH)_2$ solution and titrating it, or by measuring the heat of oxidation (usually by automatically recording the rise of temperature). CO mixed with oxygen can also be oxidized continuously by passing it into an ammoniacal solution of copper carbonate; $(NH_4)_2CO_3$ is thereby formed, but the reaction is slow (Leschewski, 1935–38).

Carbon monoxide is not strictly an acid anhydride. It begins to react with alkali hydroxides only at elevated temperatures. It then combines with them to give *formates*, and reacts similarly with the alkali methoxides forming acetates—salts which, unlike carbon monoxide itself, are derivatives of formally *quadrivalent* carbon

$$CO + NaOH = Na[CO_2H]; \qquad CO + NaOCH_3 = Na[CO_2.CH_3]$$
$$\text{sodium formate} \qquad\qquad \text{sodium methoxide} \quad \text{sodium acetate}$$

If carbon monoxide, mixed with steam, is passed over calcium oxide at 400°, it forms calcium carbonate, hydrogen being set free

$$CO + H_2O + CaO = CaCO_3 + H_2$$

It is also absorbed by strongly heated carbide, forming carbonate

$$CaC_2 + 3CO = CaCO_3 + 4C$$

Carbon monoxide ignites in air at about 700°, and burns to carbon dioxide

$$2CO + O_2 \rightleftharpoons 2CO_2$$

However, the reaction leads to an equilibrium, which is displaced towards the left at high temperatures.

The constant K_c of the Mass Action equation

$$K_c = \frac{[CO]^2 \cdot [O_2]}{[CO_2]^2}$$

from the measurements of Nernst and von Wartenberg, and of Bjerrum, has the following values

at $t =$	$1205°$	$2367°$	$2606°$	$2843°$ C
K_c	$10^{-12.85}$	$10^{-4.51}$	$10^{-3.00}$	$10^{-1.96}$
$100a$	0.032	21.0	51.7	76.1 for $p = 1$ atm.

The last row gives the degree of dissociation of the carbon dioxide, in volume-per cent, deduced from the measurements. From this, the dissociation constant can be calculated, according to formula (13), p. 53, derived for an analogous equilibrium. It may be seen that at atmospheric pressure, carbon dioxide is already half dissociated at temperatures above 2600°. The degree of dissociation is decreased if the pressure is raised; this follows directly from the reaction equation, in accordance with Le Chatelier's principle. The dependence of the degree of dissociation upon pressure and temperature is represented quantitatively by eqn. (13), p. 53. Above 5000°, CO_2 under a pressure of $\frac{1}{10}$ atmosphere is almost completely decomposed into CO and O_2.

A corresponding equilibrium holds also for the reaction with steam

$$CO + H_2O \rightleftharpoons CO_2 + H_2. \qquad K = \frac{[CO] \cdot [H_2O]}{[CO_2] \cdot [H_2]} \quad \text{(water gas equilibrium)}$$

$K =$	0.60	0.90	1.0	1.7	2.6	3.45
at	$700°$	$800°$	$830°$	$1000°$	$1200°$	$1300°$ C

Since this reaction takes place without any change in volume, the constant K remains the same if the gas pressures are substituted in place of the concentrations. At 830°, equal quantities of the four reactants are stable in presence of one another. At lower temperatures, the equilibrium is displaced towards the right. Thus carbon monoxide is the strongrer reducing agent under these conditions, but at higher temperatures hydrogen is the stronger.

Carbon monoxide unites with sulfur when heated, forming carbon oxysulfide, COS, and with chlorine, when illuminated, or in the presence of catalysts (spongy platinum or animal charcoal), forming carbonyl chloride (phosgene), $COCl_2$. When mixed with hydrogen and passed over finely divided nickel at 250°, it forms methane —

$$CO + 3H_2 = CH_4 + H_2O$$

Carbon monoxide unites with many metals, and especially with Fe, Co and Ni, when heated with them under pressure, forming *metal carbonyls* e.g.,

$$Ni + 4CO = Ni(CO)_4 \quad \text{(nickel carbonyl)}$$

Carbon monoxide reacts with metallic potassium at temperatures as low as 80°, forming the so called 'potassium carbonyl', really a benzene derivative of the formula

This forms white crystals, and is easily transformed into 'triquinone', C_6O_6, isomeric with carbon monoxide.

Carbon monoxide reduces many metallic oxides to the metals at high temperatures. Palladous chloride, $PdCl_2$, in aqueous solution, reacts with carbon monoxide even at room temperature:

$$CO + PdCl_2 + H_2O = CO_2 + Pd + 2HCl$$

The occurrence of reduction may be detected even in minute amounts by the dark coloration caused by the deposition of metal, so that the reaction provides a sensitive means of detecting CO. CO is absorbed by CuCl, in ammoniacal solution or dissolved in concentrated hydrochloride acid, forming addition compounds.

The addition compound formed in hydrochloric acid solution, and readily dissociated again, has the composition $CuCl \cdot CO \cdot 2H_2O$. It was isolated by Manchot. The absorption in ammoniacal cuprous chloride, which is more complete than by the hydrochloric acid solutions, is used to remove CO from gas mixtures—e.g., in gas analysis.

The CO-compounds of the metals of the Sub-groups are discussed further in Vol. II.

Carbon monoxide is usually prepared in the laboratory by adding formic acid drop by drop to concentrated sulfuric acid, heated to 100°. $HCO_2H = = CO + H_2O$. Alternatively, concentrated sulfuric acid may be allowed to react with potassium cyanide or powdered potassium hexacyanoferrate(II):

$$2KCN + 2H_2SO_4 + 2H_2O = K_2SO_4 + (NH_4)_2SO_4 + 2CO$$

$$K_4[Fe(CN)_6] + 6H_2SO_4 + 6H_2O = 2K_2SO_4 + FeSO_4 + 3(NH_4)_2SO_4 + 6CO$$

Carbon monoxide is obtained as a by-product of various industrial processes— e.g., the preparation of hydrogen by the partial oxidation of methane (cf. p. 25) and in the manufacture of calcium carbide (p. 438). It is produced technically on a very large scale in the form of *water gas* and *producer gas* [24].

3. *Producer gas* is a mixture, obtained by the incomplete combustion of coal, coke or charcoal, in large shaft furnaces or 'producers'. It consists, on the average, of about 25% CO, 70% N_2, 4% CO_2 and small amounts of H_2, CH_4 and O_2. Its calorific value amounts to 800–1100 kcal per cu.m according to whether it was obtained from coke or from coal. Producer gas formed from the latter has a higher calorific value, because of its higher content of hydrogen and methane.

The stack gases of blast furnaces are similar in composition to producer gas. Their calorific power is about 750 kcal per cu.m.

4. *Water gas* [26] is prepared by blowing steam over coke or anthracite at a bright red or white heat. The reaction occurring is

$$C + H_2O + 28.1 \text{ kcal} = CO + H_2 \qquad (1)$$

The heat of reaction quoted is for the reaction of black carbon with steam.

The CO formed reacts with steam to some extent, forming CO_2. At high temperatures, however, the reaction

$$CO + H_2O \rightleftharpoons CO_2 + H_2 \qquad (2)$$

is pushed fairly far over to the left, as may be seen from the numerical value for the equilibrium constant given on p. 444. Moreover, the concentration of steam in the mixture is vanishingly small because of the occurrence of reaction (1). 'Ideal' water gas therefore contains only traces of CO_2, and consists of equal parts by volume of hydroge and carbon monoxide. It has a calorific value of about 2800 kcal per cu.m. In parctice, water gas generally consists of about 40 volumes of CO, 50 volumes of H_2, 5 volumes of CO_2, 4–5 volumes of N_2 and a little CH_4. Its formation involves a considerable absorption of heat, so that when the process ('cold blowing') has proceeded for a few minutes, it is necessary for the coal to be 'hot-blown' once more. This is done with air, and according to the rate at which the air is blown through, and the height of the fuel bed, either producer gas is obtained, and is collected separately, or else combustion is complete, forming carbon dioxide, which is rejected. The latter is usually preferable, since less coal is then consumed during hot blowing because of the greater liberation of heat during complete combustion. For the use of the Winkler generator for producing water gas, see p. 705.

Producer gas is widely used in industry for the operation of gas-fired installations. It is not suitable for use in place of illuminating gas, because of its low calorific power. Water gas is employed in industry where a very hot flame is required. Although it has only about half the calorific power of coal gas, its flame is very much hotter, and will, for example, melt a platinum wire, which an ordinary coal gas flame will not do. The higher temperature of the water gas flame is due to the fact that the products of combustion of water gas occupy less space than those of coal gas. The heat is therefore concentrated within a smaller volume. The calorific value of water gas can be raised considerably by carburation. Carburated water gas is widely used in the United States for illuminating purposes. Some water gas is now generally mixed with coal gas in Europe also.

5. *Mixed Gas.* The two processes of 'hot blowing' and 'cold blowing' needed for the production of water gas can be combined, by blowing steam and air over the coke simultaneously. The mixed gas so obtained is also known as 'Dowson gas', after the inventor of the process, or as 'power gas', because it is used especially for driving gas engines. It consists, as a rule, of about 30% CO, 15% H_2, 5% CO_2 and 50% N_2, and has a calorific value of about 1300 kcal per cu.m.

Pure carbon monoxide can be obtained from producer gas by bringing it into contact with a hydrochloric acid solution of copper(I) chloride, under pressure. This absorbs CO from the gaseous mixture, and liberates it again when the pressure is reduced. Pure carbon monoxide is also prepared industrially by passing carbon dioxide over heated carbon.

Apart from heating purposes, in the form of illuminating gas, producer gas, water gas and mixed gas, carbon monoxide is employed for the reduction of ores, the refining of nickel (Vol. II) and for the preparation of phosgene and of anhydrous metallic chlorides (e.g., $AlCl_3$). At the present time, however, it is even more important as the raw material for a series of *large scale syntheses*. When CO, mixed with H_2, is passed over suitable catalysts, a variety of hydrogenation products may be obtained, according to the experimental conditions. Whereas at ordinary pressure, with nickel as catalyst, hydrogenation leads to methane; the use of other appropriate catalysts at ordinary pressure yields liquid hydrocarbons (benzine synthesis), or, at elevated pressures, either mixtures of higher alcohols, aldehydes, ketones, etc. suitable for use as motor fuels (Fischer's 'Synthol'), or methyl alcohol (methanol) (cf. p. 429).

(*ii*) *Acids and their Salts.* The simplest oxy-acids of carbon are: *carbonic acid,* H_2CO_3; *oxalic acid,* $H_2C_2O_4$; *formic acid* $HCHO_2$; and *acetic acid,* $HC_2H_3O_2$.

Carbonic and oxalic acids are *dibasic*—i.e., they are able to neutralize 2 equivalents of base; they contain two hydrogen atoms which are capable of replacement by metals, and can split off two H^+ions in solution. Formic and acetic acids are *monobasic*—they can furnish only *one* H^+ion. This fact is expressed in the formulas above, by writing one hydrogen atom separately from the others.

The formulas of the corresponding salts are

M^IHCO_3 hydrogen carbonates (acid carbonates)	$M^IHC_2O_4$ hydrogen oxalates (acid oxalates)
$M^I_2CO_3$ (normal) carbonates	$M^I_2C_2O_4$ (normal) oxalates
M^ICHO_2 formates	$M^IC_2H_3O_2$ acetates

The following ions are present in aqueous solutions of the normal salts and, at sufficiently high dilutions, of the acid salts also,

$[CO_3]^=$ carbonate ion $[C_2O_4]^=$ oxalate ion

$[CHO_2]^-$ formate ion $[C_2H_3O_2]^-$ acetate ion

In CO_2, carbon is saturated with respect to its formal valence, but is coordinatively unsaturated. CO_2 is therefore capable of adding on another O^{2-} ion:

$$O=C=O + O^{2-} = \begin{bmatrix} O \\ \ C \ O \\ O \end{bmatrix}^=$$

If one positive charge is abstracted from CO_2 by treating it with a strongly electropositive metal (e.g., sodium), or if it receives a negative charge, the carbon in the resulting radical is no longer saturated with respect to its formal valence. The radical $—CO_2^-$, which displays an unsatisfied valence on the carbon atom, can either associate with a similar radical to form an oxalate ion:

$$\begin{matrix} O=C=O \\ O=C=O \end{matrix} \ + \ 2e \ = \ \begin{matrix} O=C—O^- \\ | \\ O=C—O^- \end{matrix} \ = [C_2O_4]^=,$$

or it can combine with a neutral H atom to form the formate ion:

$$O=C=O + e + H \ = \ \begin{matrix} O=C—O^- \\ | \\ H \end{matrix} \ = [CHO_2]^-$$

If one supposes the H atom of the formate ion to be replaced by a methyl group, the acetate ion is obtained

$$\begin{matrix} O=C—O^- \\ | \\ CH_3 \end{matrix} \ = [CH_3—CO_2]^-$$

Oxalic acid, formic acid and acetic acid can thus be derived from carbon dioxide by reduction (abstraction of positive charges or donation of negative charges*). Its conversion into oxalic acid represents reduction by one stage, that into formic acid reduction by 2 stages, for in the former case one electron is added per CO_2, but in the second case 2 electrons per CO_2, as may be seen by writing the above equation in the form

$$CO_2 + 2e + H^+ = [CHO_2]^-$$

The converse process—oxidation of these acids to CO_2—is more easily achieved than the reduction of the latter.

1. *Carbonic acid and the Carbonates.* An aqueous solution of carbonic acid behaves as the solution of a weak acid. It reddens litmus, though only weakly. It may be concluded from this that the carbon dioxide is present in the solution, to some extent, in the form of carbonic acid, H_2CO_3, which in turn dissociates into its ions:

$$CO_2 + H_2O \rightleftharpoons H_2CO_3; \qquad H_2CO_3 \rightleftharpoons 2H^+ + CO_3^= \qquad \text{(1a) and (1b)}$$

Carbonic acid can react with either one or two equivalents of a strong base, to form *primary* or *acid* carbonates (hydrogen carbonates) and *secondary* or *neutral* (*normal*) carbonates:

$$H_2CO_3 + M^IOH = M^IHCO_3 + H_2O; \qquad H_2CO_3 + 2M^IOH = M^I_2CO_3 + 2H_2O$$
$$\text{primary carbonate} \qquad\qquad\qquad\qquad \text{secondary carbonate}$$

* See Chap. 16

As a dibasic acid, carbonic acid dissociates in two stages

$$H_2CO_3 \rightleftharpoons H^+ + HCO_3^-; \qquad HCO_3^- \rightleftharpoons H^+ + CO_3^= \qquad \text{(2a) and (2b)}$$

The dissociation constants for the two stages have the following values, according to Harned 1943 and Kauko 1936.

$$K_1 = \frac{[H^+] \cdot [HCO_3^-]}{[CO_2 + H_2CO_3]} \qquad ; \qquad K_2 = \frac{[H^+] \cdot [CO_3^=]}{[HCO_3^-]}$$

at	0°	10°	15°	18°	20°	25°	30°	40°	50°
$K_1 =$	2.65	3.43	3.80	4.01	4.15	4.45	4.71	5.06	5.16 · 10⁻⁷
K_2				5.2		5.7 · 10⁻¹¹			

K_1 relates to the equilibrium of the ions with the *total amount* of carbon dioxide present in the solution, both that which is dissolved as such and that present in the form of undissociated carbonic acid, H_2CO_3. This total concentration is represented by the expression $[CO_2 + H_2CO_3]$. According to Harned and Davis [*J. Am. Chem. Soc.*, 65 (1943) 2030] the molar solubility of CO_2 in water, L, can be represented, as a function of temperature by

$$\log L = \frac{2385.73}{T} - 14.0184 + 0.0152642\, T$$

The ratio of the amount of CO_2 present in the form of H_2CO_3 to that dissolved as CO_2 has been determined by Thiel and Strohecker, by utilizing the fact that whereas the undissociated acid H_2CO_3, like its free ions, reacts instantaneously with alkalis, the CO_2 present unchanged in the solution requires some time for reaction. Thus, if an aqueous solution of carbonic acid is reacted very rapidly with alkali, only just as much base is consumed in the first instant as is required to neutralize the carbonic acid (H_2CO_3 and its ions) actually present in the solution. If more alkali is taken, a perceptible time is needed for it to be neutralized. By determination of the amount of alkali which reacts instantaneously, the content of dissociated and undissociated carbonic acid in the solution is obtained. This amounts to less than 1% of the total CO_2 content, and the 'true dissociation constant' of carbonic acid, K_a:

$$K_a = \frac{[H^+] \cdot [HCO_3^-]}{[H_2CO_3]},$$

is of the order of magnitude $5 \cdot 10^{-4}$. It is thus larger than the dissociation constant of formic acid, and much greater than that of acetic acid (cf. p. 454). In practice, however, the 'apparent dissociation constant', K_1 is always used. Equilibrium between CO_2, H_2O and H_2CO_3 is, indeed, established so rapidly that for all reactions which last more than a fraction of a second the constant K_1 is the one which governs the attainment of equilibrium. To all intents and purposes, therefore, an aqueous solution of carbon dioxide behaves like the solution of a very weak acid. The apparent degree of dissociation of carbonic acid in 0.1–N aqueous solution is 0.12%.

The salts of carbonic acid, the *carbonates* are hydrolyzed in solution, since the following equilibria are set up:

$$M^I_2CO_3 + H_2O \rightleftharpoons M^IOH + M^IHCO_3; \qquad M^IHCO_3 + H_2O \rightleftharpoons M^IOH + H_2CO_3$$

In consequence, the carbonates have a basic or alkaline reaction, and this is true not only of the secondary or 'neutral' carbonates, but also of the primary or 'acid' carbonates. The primary carbonates (hydrogen carbonates) display an 'acid' reaction in the cold (0° or a little above) only towards those indicators, such as phenophthalein, for which the color change basic $\rightarrow$ acidic takes place while the solution is still feebly basic (see Chap. 18 for further discussion).

According to Auerbach, the hydrolysis of (secondary) sodium carbonate amounts to 3.5% in 0.1–N solution, 12.4% in 0.01–N solution, at 18°. Thus in a 0.1–N sodium carbonate solution at 18°, the hydroxyl ion concentration is $3.5 \cdot 10^{-3}$ mol per liter. In a solution of sodium hydrogen carbonate at the same temperature it is $1.5 \cdot 10^{-6}$ mol per liter.

Primary carbonates (hydrogen carbonates) are known for the alkalis, alkaline earths, and a few other bivalent metals. They are all freely soluble in water, with the exception of sodium hydrogen carbonate. The Solvay process of soda manufacture is based on the smaller solubility of this salt. When their solutions are boiled, the hydrogen carbonates lose CO_2 and pass into the normal carbonates.

Secondary or *normal carbonates* are formed chiefly by the univalent and bivalent metals. With the exception of the alkali carbonates, the normal carbonates are sparingly soluble in water. In addition to the carbonates of the alkalis, ammonium carbonate is readily soluble, and the carbonate of univalent thallium is also fairly soluble.

All carbonates are decomposed by non-volatile acids. Very weak acids (such as boric acid, silicic acid and their anhydrides) bring about decomposition only at a red heat.

If the H^+–ion concentration of the aqueous solutions is increased, the reactions (1b) and (1a), p. 447, proceed from right to left. If, in consequence, the solution becomes supersaturated with CO_2, this escapes from the solution, and the reaction proceeds quantitatively from right to left in consequence. Even if the solution does not become supersaturated with CO_2 at ordinary temperature, this supersaturation frequently takes place on heating.

The alkali carbonates may be melted without decomposition. The other carbonates decompose when they are heated, CO_2 being lost:

$$M^{II}CO_3 = M^{II}O + CO_2$$

This decomposition is favored either by removal of the CO_2 which is thus formed (lowering the pressure), or by removal of the oxide $M^{II}O$ from the mixture. The latter can be achieved by addition of thermally stable acids or their anhydrides, which form salts with the basic oxide. This is why carbonates are decomposed at a red heat by the anhydrides of exceedingly weak, but thermally stable, acids such as boric acid or silicic acid.

2. *Phosgene.* As has already been mentioned, carbon monoxide can combine with chlorine to form carbon oxychloride or *carbonyl chloride*, $COCl_2$. This compound is generally called *phosgene* (from φῶς, light, and γεννᾶν, to produce) because its formation was first observed in a mixture of carbon monoxide and chloride exposed to sunlight (by Davy, 1812).

Combination takes place in the presence of catalysts, as well as in light. In industry, active charcoal is used at 125–150°. In the dark and in the absence of a catalyst, CO and Cl_2 react with each other only at temperatures above 500°, at which the equilibrium $CO + Cl_2 \rightleftharpoons COCl_2 + 26.2$ kcal is displaced considerably towards the left.

Phosgene is a colorless, extremely poisonous gas, with a suffocating smell. It can easily be liquefied by cooling or pressure; boiling point 7.56°; melting point —118.8°; density 1.41 at 4°; critical temperature 182°; critical pressure 56 atm.; critical density 0.52. It is very soluble in benzene and toluene.

Phosgene can be regarded as the chloride of carbonic acid. It accordingly reacts with water, to form carbonic acid and hydrochloric acid

$$COCl_2 + 2HOH = H_2CO_3 + 2HCl$$

This reaction takes place only slowly at ordinary temperature, however, since phosgene is not very soluble in water. But it reacts very readily with organic hydroxyl compounds, and also with amides, forming carbonic acid esters and urea derivatives, respectively. For this reason it is frequently used in the dyestuffs industry and for synthetic work in scientific laboratories. Considerable quantities were used as a poison gas in the Great War of 1914–18.

3. *Carbonyl fluoride*, COF_2, obtained by passing CO over AgF_2 (cf. Vol. II), is a colorless gas with a pungent smell (b.p. —83.1°, m.p. —114.0°). It is vigorously decomposed by water, to form CO_2 and 2HF (Ruff, 1934) The COF_2 molecule is planar, $O \leftrightarrow C = 1.225$ Å, $C \leftrightarrow F = 1.313$ Å, angle F—C—O $= 110°$ (I. L. Karle and J. Karle, 1950).

4. *Peroxycarbonic acid* and *Peroxycarbonates*. In the electrolysis of concentrated solutions of alkali carbonates, the formation of *peroxycarbonate* ions takes place at the anode:

$$2CO_3^- - 2e = C_2O_6^=$$

The *peroxycarbonates*, $M^I_2[C_2O_6]$, can be considered as derivatives of hydrogen peroxide, as follows from their mode of decomposition on treatment with acid

$$\begin{bmatrix} & O—O & \\ OC & & CO \\ & O \quad O & \end{bmatrix}^= + 2H^+ = \begin{matrix} O—O \\ | \quad | \\ H \quad H \end{matrix} + 2CO_2$$

They are decomposed into hydrogen peroxide and hydrogen carbonate merely by dissolving them in water:

$$C_2O_6^{2=} + 2H_2O = H_2O_2 + 2HCO_3^-$$

Aqueous solutions of the peroxycarbonates therefore react, in general, as mixtures of hydrogen peroxide and hydrogen carbonates. It has not yet been possible to isolate the free acid, peroxycarbonic acid. The peroxycarbonates can also be obtained by various chemical reactions, as well as by electrolysis. They may be distinguished from the carbonate-peroxyhydrates by the criteria discussed on p. 340 *et seq.*

5. *Oxalic acid and the Oxalates*. Oxalic acid, $H_2C_2O_4$, is produced, in the form of its sodium salt, by passing carbon dioxide over metallic sodium heated to about 350°.

The yield in this reaction, however, is usually small. According to Haupt (1913), it is possible to obtain almost complete reaction by finely dispersing liquid sodium by means of heated carbon dioxide gas. If sodium *amalgam* is used in place of the metal, as much as 80% of the sodium may be converted to sodium oxalate at room temperature (Henglein, 1952); the rest reacts substantially to form carbonate. Henglein considers that the mechanism of the reaction involves the initial addition of Na to the CO_2, so as to form the radical $NaO—C\diagup^O_{\diagdown O}$. Two such radicals then either unite to form $Na_2C_2O_4$, or else the radical reacts further with Na:

$$NaO—C\diagup^O_{\diagdown} + Na = Na_2O + CO$$

whereupon carbonate is formed by the reaction $Na_2O + CO_2 = Na_2CO_3$

Sodium oxalate is generally prepared industrially by heating synthetic sodium formate, $NaCHO_2$, to 360°. Hydrogen is thereby eliminated, and sodium oxalate $Na_2C_2O_4$ is formed

$$
\begin{array}{c}
H\text{---}\text{---}CO_2Na \\
|\phantom{\text{---}CO_2Na} = H_2 + \\
H\text{---}\text{---}CO_2Na
\end{array}
\qquad
\begin{array}{c}
CO_2Na \\
| \\
CO_2Na
\end{array}
$$

Sodium oxalate is not directly converted into free oxalic acid in technical practice, but in order to recover the sodium formate it is first treated with milk of lime, and producer gas is then passed in under 15 atm. pressure. Sodium formate and calcium oxalate are thereby obtained:

$$Na_2C_2O_4 + Ca(OH)_2 + 2CO = 2NaCHO_2 + CaC_2O_4$$

The calcium oxalate, being sparingly soluble, is precipitated, and is converted to oxalic acid by treatment with sulfuric acid.

Sodium oxalate is also prepared by heating saw-dust with very concentrated caustic soda to about 285°. It is then converted to oxalic acid by way of the calcium salt.

Oxalic acid crystallizes from aqueous solutions with 2 molecules of water, $H_2C_2O_4 \cdot 2H_2O$, in monoclinic prisms, melting at 101.5°, with a density of 1.653. The hydrate can be dehydrated by heating it to 100° in a current of dry air. The anhydrous acid crystallizes in rhombic bipyramids, melting point 189.5°. When rapidly heated, or when heated with concentrated sulfuric acid, it decomposes into carbon dioxide, carbon monoxide and water:

$$
\begin{array}{c}
O=C\text{---}O\text{---}H \\
| = CO_2 + CO + H_2O \\
O=C\text{---}O\text{---}H
\end{array}
$$

Oxalic acid is moderately soluble in cold water, and very soluble in hot. It can therefore be recrystallized very well. 100 g of water

at	0°	10°	20°	50°	90° C
dissolve	3.6	5.3	10.2	32.1	120.0 g of $H_2C_2O_4$

It is still more soluble in alcohol. It is less soluble in ether than in water.

The X-ray analysis of the crystallized dihydrate of oxalic acid (Brill, Hermann and Peters, 1942; DUNITZ and ROBERTSON, J. Chem. Soc., (1947) 142) has shown the presence of well-defined hydrogen bond bridges (cf. Chap. 9, p. 289). The arrangement of atoms is shown in Fig. 82 in which the hydrogen bonds are shown dotted. These have various lengths, that with the length 2.52 Å being the shortest hydrogen bond yet observed between oxygen atoms. If the hydrogen is replaced by deuterium, the crystal lattice of oxalic acid dihydrate undergoes a striking expansion, as a result of changes in the length of these hydrogen bonds [ROBERTSON, Proc. Roy. Soc., A170 (1939) 221, 241].

Oxalic acid is a moderately strong dibasic acid. Since it is less volatile than is hydrochloric acid, it can expel hydrochloric acid when heated with chlorides—e.g., common salt.

As a dibasic acid, oxalic acid dissociates in two stages:

$$H_2C_2O_4 \rightleftharpoons H^+ + HC_2O_4^-; \qquad HC_2O_4^- \rightleftharpoons H^+ + C_2O_4^=$$

The dissociation constants for these two stages, at 25°, are

$$K_1 = \frac{[H^+] \cdot [HC_2O_4^-]}{[H_2C_2O_4]} = 5.9 \cdot 10^{-2}; \qquad K_2 = \frac{[H^+] \cdot [C_2O_4^=]}{[HC_2O_4^-]} = 6.4 \cdot 10^{-5}$$

The conductivity quotient Λ_c/Λ_0 for oxalic acid in normal solution is 0.15, in 0.1–N solution 0.31. The (apparent) degree of dissociation is accordingly 15% and 31% respectively. The

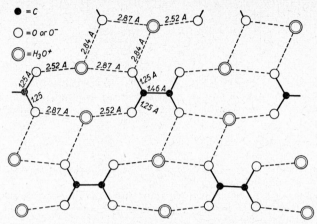

Fig. 82. Structure of oxalic acid dihydrate (projection of atom sheets perpendicularly to the b-axis)

values for the two dissociation constants are closer together than is usual for other dibasic acids (e.g., H_2CO_3 or H_2S). This is associated with the fact that the two dissociable hydrogen atoms are bound to two spatially separated groups (Treadwell, 1928).

Oxalic acid finds technical applications as a mordant in dyeing, as a bleaching agent for stearine and straw, as a condensation agent in preparative organic chemistry, and for various other purposes. In the gas mantle industry it is used as a precipitant for the rare earths. It is also used in analytical chemistry as a precipitant, especially for the detection and determination of calcium. Oxalic acid or its sodium salt are widely used in quantitative analysis for the standardization of permanganate solutions.

This application depends on the reaction, occurring quantitatively in sulfuric acid solution, between the permanganate ions, MnO_4^-, and the oxalate ions, which are thereby oxidized to carbon dioxide, while the permanganate ions are reduced to manganese(II) salts.

$$2MnO_4^- + 5C_2O_4^= + 16H^+ = 2Mn^{++} + 10CO_2 + 8H_2O$$

Oxalic acid is poisonous in fairly high concentrations. It occurs in many plants (usually in the form of its potassium salt), especially in the wood sorrel (oxalis acetosella) and in sour dock (rumex acetosa).

As a dibasic acid, oxalic acid forms two series of salts—primary or acid oxalates, and secondary or neutral oxalates. With the exception of the alkali oxalates, the oxalates are insoluble in water, but generally dissolve readily in strong acids. Oxalic acid has a strong tendency to form double or complex satls.

Sodium oxalate, $Na_2C_2O_4$, crystallizing anhydrous, in colorless, non-hygroscopic crystals, is especially suitable for the standardization of permanganate solutions. It has the advantage, over free oxalic acid, that it does not take up ammonia from the air of the laboratory.

In addition to the *neutral potassium oxalate*, $K_2C_2O_4 \cdot H_2O$ (monoclinic prisms) and the *acid oxalate*, $KHC_2O_4 \cdot H_2O$, there is also a 'super-acid' salt, $KHC_2O_4 \cdot H_2C_2O_4 \cdot 2H_2O$, a beautifully crystalline addition compound of 1 molecule of oxalic acid with 1 molecule of acid oxalate. It is used, among other purposes, for the removal of ink and rust stains, and is known in commerce as 'salt of sorrel', although sorrel actually contains not this 'tetroxalate', but the ordinary acid potassium oxalate.

The alkaline earth oxalates, of which calcium oxalate, CaC_2O_4, is of special analytical importance because of its insolubility (cf. p. 284), decompose irreversibly when heated, according to the equation

$$M^{II}C_2O_4 = M^{II}CO_3 + CO \qquad (Wöhler, 1932)$$

They crystallize as hydrates from aqueous solution. CaC_2O_4 forms a monohydrate (monoclinic), a $2\frac{1}{2}$ hydrate (triclinic) and a 3-hydrate (tetragonal). Their stability increases in the sequence triclinic → tetragonal → monoclinic (Kohlschütter, 1930).

According to Brintzinger (1935), the $C_2O_4^=$ ion in aqueous solutions is firmly bound to $4H_2O$.

6. *Formic acid and the Formates.* Formic acid, $HCHO_2$, so named because it occurs in, and was first discovered in the ant (*formica*), is fairly frequently found in organic nature—e.g., it occurs also in the hairs of the stinging nettle. It is manufactured at the present time on a large scale, since it finds extensive industrial applications. In particular, it is used in the leather industry for de-liming leather, and for acidifying the dye baths in dyeing, since it is less harmful to the fiber than sulfuric acid, which is otherwise used. Formic acid is also used as a disinfectant—e.g., for beer and wine barrels—and as an antiseptic, e.g., for the conservation of fruit juices. It also finds medicinal applications.

The preparation of formic acid is based on the ability of carbon monoxide to react with alkali hydroxides at elevated temperatures, producing the alkali formates. For example, producer gas is allowed to react at 120–130° and 6–8 atm. pressure with well stirred, finely ground caustic soda. Formic acid is liberated by the action of sulfuric acid on the sodium formate so obtained.

Newer processes use a mixture of $Ca(OH)_2$ and Na_2SO_4, instead of NaOH, or $Ca(OH)_2$ alone, at higher pressures (60 atm.). Under a pressure of 100 atm. it is possible to react CO with $NH_3 \cdot H_2O$ also. The $[NH_4][CHO_2]$ so produced gives the fertilizer $[NH_4]_2SO_4$, as a valuable by-product of its reaction with H_2SO_4.

Pure formic acid is a colorless, pungent smelling liquid, which blisters the skin. It solidifies on moderate cooling, melts at 8.43°, boils at 100.6° and has the density 1.220. It is completely miscible with water. An aqueous solution containing about 75% formic acid boils without change of composition at atmospheric pressure.

The dimensions of the formic acid molecule have been determined by means of electron diffraction measurements (Schomaker, 1947), and the interatomic distances have been found to be: C=O 1.21 Å; C—O 1.37 Å; O ↔ O 2.28 Å; angle $C\underset{O}{\overset{O}{<}}$ 124°.

Formic acid is a strong reducing agent. Its vapor is combustible. It is decomposed into hydrogen and carbon dioxide at ordinary temperature by finely divided rhodium:

$$HCHO_2 = H_2 + CO_2$$

It precipitates the noble metals from their salts, and reacts with mercuric chloride according to the equation

$$HCHO_2 + 2HgCl_2 = CO_2 + 2HCl + Hg_2Cl_2$$

According to Scala, the reaction is quite quantitative in the presence of an excesse of mercuric chloride.

Formic acid is a fairly strong monobasic acid; the dissociation constant

$$K = \frac{[H^+] \cdot [CHO_2^-]}{[HCHO_2]} = 1.8 \cdot 10^{-4} \text{ at } 18°.$$

The degree of dissociation is 1.4% in a 1-normal solution at 18°, 4.4% in 0.1-N solution.

The salts of formic acid, the *formates*, are all soluble in water. Some of them— e.g., aluminum formate $Al(CHO_2)_3$, obtained by double decomposition between aluminum sulfate and sodium formate—have technical uses.

7. *Acetic acid and the Acetates.* Acetic acid, $HC_2H_3O_2$, is a colorless liquid with a pungently acid smell, and a sour taste even in extreme dilutions. Completely anhydrous acetic acid (density 1.049) solidifies very readily (m.p. 16.7°) and boils at 118.1°. Acetic acid is also perceptibly volatile at ordinary temperature. It is completely miscible with water, alcohol, ether, chloroform and glycerol.

Acetic is a weak acid. As a weak electrolyte, it conforms very well to the Ostwald dilution law (cf. p. 66).

The degree of dissociation at 18° is 0.45% in a 1-N solution, and 1.4% in 0.1-N solution. The value of the dissociation constant is $1.750 \cdot 10^{-5}$ at 18°, $1.754 \cdot 10^{-5}$ at 25°.

Acetic acid has many uses. Its 80% aqueous solution is used, as 'essence of vinegar' for the preparation of vinegar. Pure acetic acid is used in medicine as a caustic, e.g., for the removal of warts and corns. Dilute acetic acid is much used as a reagent. Acetic acid is used in industry as a solvent for celluloid and collodion, for the preparation of lacquers and polishes. It is also used in the manufacture of white lead, artificial silk (acetate silk), synthetic pharmaceuticals such as antifebrin, aspirin and antipyrine, of scents, which are often esters of acetic acid, and also of organic dyestuffs.

Acetic acid is obtained to some extent, especially for edible purposes, by the fermentation of alcohol by acetic bacteria (*Bacterium aceti*). For industrial purposes it is prepared largely from the 'pyroligneous acid' obtained by the dry distillation of wood. This is a mixture, chiefly of methyl alcohol and acetic acid, and is treated with milk of lime. On evaporation, the methyl alcohol distils and is recovered as a by-product, and ultimately (impure) calcium acetate is obtained. Acetic acid is liberated from this by the action of sulfuric acid. Considerable quantities of acetic acid are now obtained from synthetically prepared acetaldehyde (cf. p. 437), which is oxidized by atmospheric oxygen under the catalytic action of manganese salts

$$CH_3{-}CHO + \tfrac{1}{2}O_2 = CH_3{-}CO(OH)$$

Acetic anhydride, $(CH_3.CO)_2O$, which is also widely used in preparative organic chemistry, is prepared technically by the reaction of sodium acetate with sulfuryl chloride, or by passing phosgene into acetic acid in the presence of $MgCl_2$, whereby HCl is obtained as a by-product.

Most of the salts of acetic acid, the *acetates*, are readily soluble in water. Only silver and mercurous acetates have a poor solubility in cold water. Acetic acid is readily displaced from its salts by non-volatile acids. Because they are readily soluble and crystallize well, the acetates find wide applications, especially sodium acetate. This crystallizes as a hydrate, $NaC_2H_3O_2 \cdot 3H_2O$, in clear, colorless monoclinic prisms. It is very soluble in water, and quite soluble in alcohol; it melts in its water of crystallization at $75°$, and loses all its water at about $120°$; the anhydrous salt melts at $319°$. It is used as a precipitant in analytical chemistry, and is much used to 'buffer', or lower the acidity of, solutions of acetic acid. The concentration of acetate ions is greatly augmented when sodium acetate is added to a solution containing free acetic acid, since the salt is almost completely dissociated in solution. The H^+-ion concentration must then be diminished proportionately.

The acetates, including those of the alkali and alkaline earth metals, have a marked tendency to form double salts and complex compounds. Thus, sodium acetate can combine with 1 or 2 molecules of acetic acid, forming so-called *acid acetates*. These are actually addition compounds of acetic acid with the neutral acetate.

Since acetic acid is a weak acid, aqueous solutions of the acetates have a basic reaction, as a result of hydrolysis:

$$M^IC_2H_3O_2 + H_2O \rightleftharpoons M^+ + OH^- + HC_2H_3O_2$$

In 0.1-N solution, *one* molecule in every $15,000$ of $NaC_2H_3O_2$ is hydrolyzed at $18°$.

Acetic acid is capable of forming salts with *acids*, as well as with bases, but since compounds of this kind are decomposed by water, they can be formed only by reactions in which no water is set free. Thus *acetyl fluoroborate*, $[CH_3—CO]^+[BF_4]^-$ can be obtained by the action of acetyl fluoride, $CH_3 \cdot COF$, with BF_3. It has been shown by conductivity measurements (in SO_2 as solvent) that this is a compound with salt-like properties (SEEL, *Z. anorg. Chem.*, 250 (1943) 331). Acetyl chloride has also long been known to form an addition compound with aluminum chloride, which can also be formulated as a salt $[CH_3—CO]^+[AlCl_4]^-$. This hypothesis receives support from experiments using artificially radioactive chlorine as tracer (cf. Vol. II, Chap. 12), which showed that the chlorine originally present in the acetyl chloride undergoes exchange with that present originally in the $AlCl_3$.

The acetate ion has a tendency to form complex ions (acetato-ions) with the heavy metals, and these are predominantly *cationic* complexes—e.g., the type $[M^{II}_2(C_2H_3O_2)_2]^{++}$ formed with bivalent metals (Weinland).

(d) Sulfides, Selenides, Tellurides

The normal sulfide of carbon is *carbon disulfide*, CS_2. The lower sulfides of carbon are very unstable.

According to Dewar, (1910–11), *carbon monosulfide* is formed at $-185°$, as a white deposit, by the action of a silent electric discharge on carbon disulfide, and is transformed explosively at $-180°$ into a brown polymeric substance, $[CS]_x$. Klemenc (1930), however, regards the white deposit as a mixture of ordinary CS_2 molecules and molecules activated by the electric discharge, the latter decomposing into $[CS]_x$ and S under the catalytic influence of the wall of the reaction vessel, when the temperature is raised. The occurrence of CS in the gaseous phase appears to be unlikely on the basis of Klemenc's experiments; the CS molecule must, in any case, be extremely unstable.

Carbon subsulfide, C_3S_2, is not quite so unstable. Its reactions show it to be structurally analogous to carbon suboxide, and it is accordingly formulated as $S=C=C=C=S$.

Carbon oxysulfide (carbonyl sulfide), COS, occupies an intermediate position between carbon dioxide and carbon disulfide.

(*i*) *Carbon disulfide* [27], CS_2, is prepared technically by passing sulfur vapor over white hot wood charcoal

$$C + 2S \rightleftharpoons CS_2.$$

It forms a colorless liquid, which possesses a pleasant aromatic smell when pure, but generally has a repulsive smell, because of the presence of impurities. Its density is 1.262 (at 20°), boiling point 46.2° (latent heat of vaporization 6.35 kcal per mol.); it freezes only when strongly cooled (—111.6°). Its vapor pressure is 127.1 mm at 0°, 298 mm at 20°. The vapor density corresponds to that required by the molecular the formula CS_2. The vapor is extremely inflammable, the ignition temperature being only 236°, so that it can be ignited by a hot-running bearing. It burns ($CS_2 + 3O_2 = CO_2 + 2SO_2 + 258$ kcal) with a light blue flame which is very rich in actinic rays, but has so low a temperature that it fails to char paper.

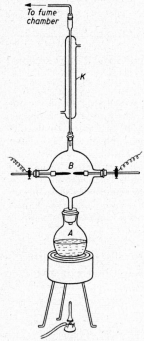

To fume chamber

Carbon disulfide is an endothermic compound, and liberates 15.4 kcal per mol of CS_2 vapor by its decomposition. Decomposition can take place explosively if it is initiated by a detonator such as mercury fulminate, but the explosion does not propagate itself far through the vapor. It decomposes slowly when exposed to light or to the silent electric discharge; for this reason carbon disulfide turns yellow when exposed to light. Since the equilibrium $C + 2S \rightleftharpoons CS_2$ is shifted to the right by rise of temperature, carbon disulfide is only partially decomposed when it is heated to a high temperature. Substances poorer in sulfur are thereby formed—e.g., carbon subsulfide, C_3S_2, a red liquid of density 1.27, with a very repulsive smell and strongly irritant action on the eyes. C_3S_2 freezes at —0.5°, and readily polymerizes to a black material. It is prepared by passing an electric arc between carbon or graphite electrodes through liquid carbon disulfide or carbon disulfide vapor. For a laboratory experiment the simple apparatus shown in Fig. 83 may be used. Carbon disulfide is boiled in the wide-necked flask A, standing in a water bath, and the vapors are condensed by the reflux condenser K. Carbon rods are inserted into the glass bulb B through two of its four tubulations; the arc is struck between these. The carbon rods are fixed to copper rods, which are mounted in corks fitting the tubulures so that, although gas tight, they can be moved longitudinally. The products of reaction are filtered, and then treated with copper turnings to remove dissolved sulfur. If the carbon disulfide is evaporated in a rigorously dried stream of air, a few cubic centimeters of a deep red liquid are left behind, and this can be identified, from its reactions, as carbon subsulfide.

Fig. 83. Preparation of carbon subsulfide C_3S_2.

Carbon disulfide is *poisonous* (respiratory and cutaneous poison). It is decomposed by water (above 150°), according to the equation $CS_2 + 2H_2O = = CO_2 + 2H_2S$. It is more rapidly decomposed by heating it with baryta water.

Carbon disulfide is decomposed by strong oxidants, such as permanganate, sulfur being deposited. It reacts with sulfur trioxide, forming carbon oxysulfide, COS:

$$CS_2 + 3SO_3 = COS + 4SO_2$$

Carbon disulfide can burn in an atmosphere of nitrogen dioxide. It reacts with chlorine monoxide, forming phosgene:

$$CS_2 + 3Cl_2O = COCl_2 + 2SOCl_2$$

Altogether, carbon disulfide very readily exchanges sulfur either completely or partially for oxygen at higher temperatures—e.g., when heated with metallic oxides. Sulfur may be replaced step by step by treatment with chlorine, and may be replaced by hydrogen through the action of powerful reducing agents.

CS_2 reacts with hydrogen at elevated temperatures, forming H_2S. In contact with MoS_2 or Pt as catalyst, the reaction is

$$CS_2 + 4H_2 \rightleftharpoons CH_4 + 2H_2S \; ;$$

$K_p = (p_{CH_4} \cdot p^2_{H_2S}) / (p_{CS_2} \cdot p^4_{H_2}) = 2.2$ at 600°, $3.0 \cdot 10^{-5}$ at 900°   (Terres, 1934) This reaction can be used to remove CS_2 from town's gas.

Just as CO_2 adds on an oxygen ion to form the ion $[CO_3]^=$, so CS_2 can add on a sulfide ion, to form the ion $[CS_3]^=$. The salts derived from this ion, $M^I_2[CS_3]$, are called *thiocarbonates*, and can readily be formed by the reaction $CS_2 + S^= = [CS_3]^=$. Thus *potassium thiocarbonate*, K_2CS_3, which is used for protection against the vine louse, is prepared by shaking a concentrated solution of potassium sulfide with carbon disulfide. Free thiocarbonic acid, an oily liquid, soluble in water, can be obtained from its salts by the action of hydrochloric or sulfuric acid.

There are also compounds which occupy a position intermediate between ordinary thiocarbonic acid, H_2CS_3, (*trithiocarbonic acid*) and carbonic acid, H_2CO_3, in that the oxygen atoms of carbonic acid are only partially replaced by sulfur. *Dithiocarbonic acid*, H_2COS_2, is of this type; its monoethyl ester, $HC(OC_2H_5)S_2$ is known as *xanthic acid*. If, in sodium xanthate, the radical of ethyl alcohol is replaced by the radical of cellulose, and one H-atom of the latter is also replaced by Na, the so-called *cellulose xanthate*, $Na[CS_2(OC_6H_8O_4Na)]$ is obtained. The formation of this substance makes it possible to dissolve cellulose, and it is therefore of great importance in the manufacture of *artificial silk* ('Viscose' process).

Carbon disulfide is notable for its high refractive index and fairly high dispersion. The refractive index for yellow light (Na_D) at 20° is 1.6276; this is strongly temperature-dependent—about 20 times as much so as the refractive index of water.

Carbon disulfide is an excellent solvent for fats, oils, waxes, resins, rubber, and also for sulfur, phosphorus, iodine and many other substances. It is completely miscible with alcohol, ether, and chloroform. It is only slightly soluble in water, although at low temperatures (below —3°) it forms a hydrate $2CS_2 \cdot H_2O$. 100 g of water dissolve 0.26 g of CS_2 at 0°, 0.10 g at 20°.

Carbon disulfide is widely used in industry as a solvent and extraction medium. It is also used as an insecticide and for other purposes. The largest quantities are employed in the production of *viscose silk*, an artificial silk made by impregnating cellulose (from wood) with sodium hydroxide, and then treating it with carbon disulfide. The water-soluble cellulose xanthate is so formed, and the concentrated, very viscous solution of this is extruded through fine nozzles ('spinnerets') into salt solutions containing sulfuric acid. The acid decomposes the compound, and throws down cellulose once again, which is thus obtained in the form of fine fibers with the lustre of silk.

(ii) *Carbon diselenide*, CSe_2, can be obtained by the reaction of CCl_4 with H_2Se at 500° (Grimm, 1936). It is a dark yellow liquid (density 2.68, b.p. 124°, m.p. —45.5°), not readily inflamed, and insoluble in water. It is still more endothermic than CS_2 (heat of formation —34 kcal per mol), and has a strong tendency to polymerize. (Cf. also Ives, Pittman and Wardlaw, *J. Chem. Soc.*, (1947) 1080). It has not yet been possible to prepare CTe_2, which would be expected to have a still lower stability. CSSe (yellow, density 1.98, m.p. —85°, b.p. +84°) and CSTe (red, m.p. —54°) are also rather unstable (Stock, 1914).

(iii) *Carbon oxysulfide*, carbonyl sulfide, COS, is a colorless, odorless, inflammable gas, which liquefies at —50.2° and solidifies at —138°. It is formed by passing a mixture of sulfur vapor and carbon monoxide through a red hot tube:

$$S + CO = COS$$

According to Stock, it is best obtained pure by the action of acids on ammonium thiocarbaminate. [Thiocarbaminic acid, $(HO)CS(NH_2)$ is derived from monothiocarbonic acid, H_2CO_2S, by replacement of one OH-group by the amino group, —NH_2]:

$$NH_4\left[{}^S_O{>}C{-}NH_2\right] + 2HCl = 2NH_4Cl + COS$$

Carbon oxysulfide dissolves readily in carbon disulfide, toluene and alcohol, but is less soluble in water (0.55 volumes of COS in 1 volume of water at 20°). It is practically insoluble in concentrated sodium chloride solution. It is gradually decomposed by water, forming carbon dioxide and hydrogen sulfide, and smells of the latter if it is moist:

$$COS + H_2O = CO_2 + H_2S$$

It can be freed from its decomposition products by washing it with concentrated sodium hydroxide solution, by which it is but little absorbed. It is stable in the absence of moisture. When heated it decomposes, the reactions being

$$2COS \rightleftharpoons CO_2 + CS_2 \qquad \text{and} \qquad 2COS \rightleftharpoons 2CO + S_2$$

Ammonium thiocarbaminate, which can readily be obtained from COS and NH_3, is converted into urea when it is gently heated, hydrogen sulfide being lost.

$$NH_4[OCS(NH_2)] = CO(NH_2)_2 + H_2S$$

This reaction, which takes place very smoothly, was proposed by Klemenc (1930) for the technical preparation of urea.

(e) Cyanogen Compounds

Compounds which contain the univalent radical —CN are formed in the pyrolysis of organic substances containing nitrogen—such as wool, leather or horn—in the absence of air, and especially if alkali is present. The compound of this class which has been longest known is *prussian blue* (see below), discovered by Dippel and Diesbach in 1704. The radical —CN has accordingly been named the *cyanogen* radical (from κυάνεος, dark blue). The simplest acid derived from the cyanogen radical, HCN, is called *hydrocyanic acid*, or *prussic acid*. Its salts are the *cyanides*.

Many other compounds are derived from the cyanogen radical, as well as prussic acid and its salts. These compounds will not be discussed in this book, since they belong to the field of organic chemistry.

The cyanogen radical is not stable in the free state—i.e., when it is neither joined to another atom nor electrically charged. If the charge is removed from the CN^-ion, two cyanogen radicals link up, by means of the valence becoming free on the carbon atom, to form *dicyanogen*, $(CN)_2$. Dicyanogen is usually called *cyanogen*.

(*i*) *Cyanogen*, $(CN)_2$, is a colorless poisonous gas, with an unpleasant smell, reminiscent of bitter almonds. It burns with bluish-tinged flame, of peach-blossom color. It is readily liquefied, by a pressure of 3.3 atm. at $15°$, and at ordinary pressure if it is cooled to —20.7°. It solidifies at —34.4°. The vapor density corresponds to the molecular weight required for the formula $(CN)_2$.

Cyanogen is fairly soluble in water, alcohol and ether. Its solutions in these solvents soon decompose, however, forming a variety of products which include oxalic acid among others:

$$NC—CN + 4H_2O = HO—CO—CO—OH + 2NH_3$$

Under different conditions, cyanogen polymerizes to a friable brown black mass, known as *paracyanogen*, which is insoluble in water and alcohol, and which is reconverted into ordinary cyanogen if it is heated to 860° in the absence of air.

Cyanogen can be prepared by heating mercuric cyanide, $Hg(CN)_2$, which is preferably mixed with mercuric chloride, since paracyanogen is formed as a by-product if cyanogen is strongly heated. The decomposition of mercuric cyanide is an endothermic reaction; the reaction with mercuric chloride is weakly exothermic. It is therefore necessary to heat more strongly if mercuric chloride is not present.

$$Hg(CN)_2 + 11 \text{ kcal} = Hg + (CN)_2;$$

$$Hg(CN)_2 + HgCl_2 = Hg_2Cl_2 + (CN)_2 + 0.4 \text{ kcal}$$

Cyanogen is most conveniently prepared by the reaction between copper sulfate and potassium cyanide in fairly concentrated solution:

$$Cu^{++} + 2CN^- = Cu(CN)_2 \rightarrow CuCN + \tfrac{1}{2}(CN)_2$$

To make the reaction complete, the mixture is finally warmed gently.

The copper(I) cyanide thrown down as a precipitate can be filtered off, and worked up for cyanogen by treating it with ferric chloride solution:

$$Cl^- + CuCN + Fe^{+++} = CuCl + Fe^{++} + \tfrac{1}{2}(CN)_2$$

(*ii*) *Hydrogen cyanide and the Cyanides.* Hydrogen cyanide (prussic acid) is a colorless, mobile liquid of density 0.697, boiling at 26.5°. It has a paralyzing taste, resembling that of bitter almonds when it is very dilute. It solidifies at —15° to a mass of fibrous crystals.

Hydrogen cyanide is notable for its high *dipole moment*, which measures $2.88 \cdot 10^{-18}$ c.g.s. units for the gaseous molecule.

Hydrocyanic acid is *extraordinarily poisonous* (antidote – hydrogen peroxide). As little as 50 mg is sufficient to kill a man, and death may ensue in a few seconds. Because it is so toxic, hydrocyanic acid is used as an insecticide for pest destruction. It is also widely used in the synthesis of organic compounds. Being volatile, it is readily liberated by the action of other acids upon its salts, the cyanides. To prepare it in the anhydrous state, a mixture of equal parts of concentrated sulfuric acid and water is allowed to drop on to lumps of potassium cyanide.

Small amounts of prussic acid, or of compounds which readily hydrolyze with the formation of prussic acid, are frequently met with in the vegetable kingdom. The most familiar is its occurrence in bitter almonds, in the form of *amygdalin* (from ἀμυγδάλη, almond), a compound which is hydrolyzed by the action of emulsin, an enzyme also present in bitter almonds, into fructose, benzaldehyde (oil of bitter almonds) and hydrocyanic acid. Hydrocyanic acid is always present in crude coal gas (usually 0.1 to 0.3% by volume), and is removed therefrom by passage through 'purifiers' (containing hydrated ferric oxide—'bog iron ore'—or spent pyrites, cf. Vol. II). Complex iron cyanides are thereby formed, and are utilized to some extent for the production of alkali cyanides. Direct union of the elements H_2, N_2 and C, to form HCN, takes place only at temperatures above 1800°. NH_3 is more readily converted to HCN—e.g., by reaction with CO at 500–700° on the surface of a catalyst such as Al_2O_3 or $CeO_2(NH_3 + CO = HCN + H_2O)$, or by partially oxidizing a mixture of CH_4 and NH_3 with atmospheric oxygen, by passing the mixture over a heated platinum gauze:

$$CH_4 + NH_3 + \tfrac{3}{2}O_2 = 3H_2O + HCN + 115 \text{ kcal}$$

The preparation of HCN from carbon monoxide and ammonia is also of considerable technical importance. Formamide, $HCONH_2$, is thereby obtained, in methyl alcoholic solution under pressure, and this decomposes into HCN + H_2O when it is heated to 400° in the presence of catalysts.

Hydrocyanic acid is completely miscible with water, alcohol and ether. In its aqueous solution it is dissociated to some extent into ions

$$HCN \rightleftharpoons H^+ + CN^-$$

although only to a very small degree. Prussic acid is an *extremely weak acid*, much weaker even than carbonic acid.

At 18°, HCN is 0.002% dissociated in 1-N solution, 0.007% dissociated in 0.1-N solution; dissociation constant $K = 4.8 \cdot 10^{-10}$ at 18°. The cyanides accordingly are strongly hydrolyzed in solution. Thus potassium cyanide, KCN, is 0.38% hydrolyzed in 1-N solution, 1.20% in 0.1-N solution, at 18°.

The salts of hydrocyanic acid, the cyanides, have the general formula M'CN. In many of their reactions they resemble the chlorides—e.g. in their behavior towards silver nitrate, which throws a curdy white precipitate of silver cyanide from cyanide solutions

$$Ag^+ + CN^- = AgCN$$

Although hydrocyanic acid is a weak acid, the precipitate is not very soluble in dilute strong acids.

Silver cyanide is practically undecomposed by strong acids, because of its extreme insolubility. The solubility product of silver cyanide at 17.5° (according to Morgan) is

$$K_{spAgCN} = [Ag^+][CN^-] = 10.07 \cdot 10^{-14}$$

In the solution obtained by treating silver cyanide with an acid, the silver ion concentration is given by

$$[Ag^+] = [HCN] + [CN^-] \qquad (1)$$

By substituting

$$[HCN] = \frac{[H^+]\,[CN^-]}{K} = \frac{[H^+]\,[CN^-]}{4.8 \cdot 10^{-10}}$$

and

$$[CN^-] = \frac{10.07 \cdot 10^{-14}}{[Ag^+]}$$

in (1), we obtain

$$[Ag^+]^2 = 2.09 \cdot 10^{-4}\,[H^+] + 10.07 \cdot 10^{-14};$$

$$[Ag^+] = 1.45 \cdot 10^{-2}\,\sqrt{[H^+] + 4.8 \cdot 10^{-10}}$$

This means that less than 0.2 g of silver cyanide can dissolve in 100 cc of 2-N sulfuric acid ($[H^+] \sim 1$). On the other hand, if silver cyanide is treated with hydrochloric acid it is converted into silver chloride, since the solubility product of this is low enough ($K_{sp\,AgCl} = 2.0 \cdot 10^{-10}$) to be reached with the concentration of silver ions that can be attained in the acid solution.

The cyanides of the alkalis and alkaline earths are readily soluble in water; the cyanides of the heavy metals are mostly insoluble (exception: mercuric cyanide, $Hg(CN)_2$). The solutions of alkali and alkaline earth cyanides have a strongly basic reaction, as a result of hydrolysis, and smell of hydrocyanic acid.

The cyanide ion has a great tendency for complex formation. Almost all the heavy metals that form cyanides at all form complex ions with an excess of cyanide ion. The best known examples of these are the *cyanoferrate(II)* (*ferrocyanide*) and *cyanoferrate(III)* (*ferricyanide*) ions, $[Fe(CN)_6]^{4-}$ and $[Fe(CN)_6]^{3-}$. The corresponding potassium salts are potassium ferrocyanide, $K_4[Fe(CN)_6]$ ('yellow prussate of potash') and potassium ferricyanide, $K_3[Fe(CN)_6]$ ('red prussate of potash'). These, and also the alkali salts of the other complex cyanides, are soluble in water. Heavy metal cyanides which are insoluble in water dissolve in excess of alkali cyanide, as a result of complex formation with excess cyanide ions. This applies also to silver cyanide, so that no precipitate is obtained when a little silver nitrate is added to a solution containing a large quantity of cyanide ions.

When a solution of potassium ferrocyanide is mixed with a ferric salt solution, a deep blue precipitate is obtained—*prussian blue*. The composition of prussian blue is discussed under *iron* (Vol. II).

1. *Sodium cyanide*, NaCN, is prepared technically chiefly by the Castner-Kellner process, from *sodium amide*, $NaNH_2$. This is obtained by passing ammonia gas over molten sodium:

$$NH_3 + Na = NaNH_2 + \tfrac{1}{2}H_2$$

When heated with charcoal at 600°, sodamide is converted first into sodium cyanamide, Na_2N_2C:

$$2NaNH_2 + C = Na_2N_2C + 2H_2$$

On stronger heating with carbon, this reacts to form sodium cyanide:

$$Na_2N_2C + C = 2NaCN$$

Sodium cyanamide is derived from cyanamide, NC—NH$_2$, and this in turn from hydrocyanic acid, by substitution of the amido group —NH$_2$ for the H atom.

Newer processes for the preparation of sodium cyanide start from calcium cyanamide. A good yield is also obtainable from the reaction of NH$_3$ and CO with Na$_2$CO$_3$ at 600°:

$$2NH_3 + 3CO + Na_2CO_3 \rightleftharpoons 2NaCN + 2H_2O + H_2 + 2CO_2 - 29 \text{ kcal} \quad \text{(Henglein, 1930)}$$

Sodium cyanide forms colorless crystals, very soluble in water and alcohol. It is used chiefly in the extraction of gold by the cyanide leaching process (cf. Vol. II). It is also used in photography and in electroplating and galvanoplastics, for the preparation of plating baths, especially for the noble metals*.

2. *Potassium cyanide*, KCN, crystallizes from solution in colorless octahedra. It is hygroscopic, and very soluble in water, but much less soluble in alcohol. It is prepared for the most part from the iron cyanide compounds present in 'spent oxide' from gas purification plant. It is also recovered from the gases containing trimethylamine which are obtained by destructive distillation of the molasses residues, which constitute a waste product of the beet sugar industry:

$$N(CH_3)_3 \xrightarrow{\text{at 1000°}} HCN + 2CH_4; \quad HCN + KOH = KCN + H_2O$$

The uses of potassium cyanide are in general the same as for sodium cyanide, which is replacing the potassium salt to an increasing extent.

The CN⁻-content of technical material is often expressed in 'per cent potassium cyanide'. 100 g of anhydrous sodium cyanide are equivalent to 132.65 g of potassium cyanide. The '100 per cent potassium cyanide' of commerce is, accordingly, not necessarily pure potassium cyanide, but merely has a CN⁻ content corresponding to that of pure KCN; the lowering of CN⁻-content brought about by the presence of impurities such as potassium carbonate may be compensated by the presence of sodium cyanide. In the same way, 'sodium cyanide, 100%', is not pure sodium cyanide, but a product with the same CN⁻-content as pure KCN.

According to Bozorth (1922), potassium cyanide has a crystal structure corresponding to that of KCl or NaCl, except that the centroids of the CN radicals occupy the place of the Cl atoms. The long axes of the CN groups lie along the body diagonals of the eight small cubes into which the unit cell may be subdivided. Edge length of unit cell: $a = 6.55$ Å; distance C $\leftrightarrow$ N = 1.15 Å; shortest distance between K and C (or K and N) = 3.0 Å.

3. *Calcium cyanide* Ca(CN)$_2$, m.p. 640°, is used as a pesticide.

4. *Calcium cyanamide*, CaCN$_2$ [21] is prepared by the reaction of finely ground calcium carbide with nitrogen at a high temperature:

$$CaC_2 + N_2 \rightleftharpoons CaCN_2 + C + 72 \text{ kcal}$$

The reaction was discovered in 1898 by Rothe, and developed into an industrial process by Frank and Caro. It leads to a (bivariant) equilibrium (Cochet), which is displaced to the left by a rise of temperature. The heating (to about 1100°) need be carried out at only one point, since the reaction, being strongly exothermic, then propagates itself further. Admixture with dry calcium chloride accelerates the reaction. The last traces of carbide are removed by treatment with a fine spray of water. The maximum permitted carbide content of calcium cyanamide is 0.1%.

* It has been proposed (Göckel, 1934) that the harmless thiourea should be used in place of the poisonous cyanide for plating baths.

Calcium cyanamide is obtained as a grey powder. It is derived from cyanamide, $NC \cdot NH_2$, by replacement of the hydrogen atoms by calcium $NC—NCa$. It is decomposed by superheated steam, forming ammonia:

$$CaCN_2 + 3H_2O = CaCO_3 + 2NH_3$$

The same decomposition is gradually brought about by water at ordinary temperature (by way of intermediate products), through the action of bacteria in the soil. Calcium cyanamide has therefore become of great importance as a nitrogen fertilizer (and was also for a time, important in the production of ammonia—cf. p. 607).

In the use of calcium cyanamide as a fertilizer, its property of dusting, and the caustic properties resulting from the presence of free lime, are undesirable. It has, however, been possible to overcome these by incorporating small amounts of oil or fat. The production and use of calcium cyanamide has been developed especially in Germany (German production in 1936, 678,000 tons, almost half the world production).

Pure calcium cyanamide is colorless. It forms rhombohedral crystals, which sublime at about 1300° without melting. It may be obtained by the reaction

$$CaO + 2HCN \rightleftharpoons CaCN_2 + CO + H_2$$

and HCN can be prepared, conversely, by the action of CO and H_2 on $CaCN_2$ (Franck, 1931).

Calcium cyanamide is decomposed by cold water primarily to form $Ca(HCN_2)_2$:

$$2CaCN_2 + 2H_2O = Ca(HCN_2)_2 + Ca(OH)_2 + 161.2 \text{ kcal}$$

Free cyanamide is formed by the action of acids. This is therefore the product of the action of carbon dioxide and water on calcium cyanamide at ordinary temperature. When cyanamide solution is warmed to 70° urea is obtained:

$$CaCN_2 + H_2CO_3 = CaCO_3 + CN—NH_2$$

$$CN—NH_2 + H_2O = CO(NH_2)_2$$

This reaction was for some time of technical importance for the preparation of urea.

5. *Magnesium cyanamide*, $MgCN_2$, can be prepared by the reaction of MgO with HCN at a high temperature (Franck, 1931; Hartmann, 1951). It is a white, finely crystalline powder, $d = 2.170$. It is stable in dry air, but gradually decomposes in moist air, with the formation of NH_3. At 1370° it breaks up into C, Mg and N_2.

(f) Thiocyanogen Compounds

The thiocyanogen compounds contain the univalent thiocyanogen radical —SCN. Like the cyanogen radical, this can exist as a negative univalent ion, SCN^-. From it are derived *thiocyanic acid*, HSCN, and its salts, the *thiocyanates*, M'SCN. Among the thiocyanates, the complexes with ferric iron are notable for their intense blood-red color.

Thiocyanogen compounds occur in small amounts in the human organism, especially in the saliva.

Unlike the cyanogen compounds, they are not poisonous in small amounts. They are considered to have a germicidal action and to promote digestion.

The thiocyanogen radical is incapable of free existence in the uncharged state. As with cyanogen, two such radicals associate to form dithiocyanogen (more commonly *thiocyanogen*), $(SCN)_2$.

(*i*) *Thiocyanogen*, $(SCN)_2$, was obtained in the free state by Söderbäck, in 1919, by the action of bromine, dissolved in carbon disulfide, upon silver thiocyanate:

$$2AgSCN + Br_2 = 2AgBr + (SCN)_2$$

It is stable only at low temperatures, and forms light yellow crystals melting at —2° to —3° The melt spontaneously decomposes after a little time, emitting a yellow smoke and forming a brick-red, amorphous solid. It is rather more stable in solution. Thiocyanogen is extremely soluble in alcohol and ether, and is also soluble in carbon disulfide and carbon tetra-chloride. It is immediately decomposed by water. Solutions in organic solvents decompose gradually at low temperatures, or immediately at room temperature, forming a yellow amorphous substance which was first obtained by Liebig in attempts to prepare free thio-cyanogen by oxidizing cyanides with chlorine in aqueous solution.

Free thiocyanogen closely resembles iodine in its chemical behavior. Thus it combines with metals, and has about the same oxidizing power as iodine. Thiocyanogen can liberate iodine from iodides, and is itself set free by an excess of iodine. There is thus a definitely reversible equilibrium,

$$I_2 + 2SCN^- \rightleftharpoons 2I^- + (SCN)_2$$

It has not been possible to determine the vapor density of thiocyanogen, on account of its instability. However, Lecher (1919) found a molecular weight corresponding to the formula $(SCN)_2$ from measurements of the freezing point depression in bromoform, $CHBr_3$.

Thiocyanogen, dissolved in ether, reacts with hydrogen sulfide forming sulfur thio-cyanate, $S(SCN)_2$, a compound analogous in constitution to sulfur dichloride, SCl_2:

$$2(SCN)_2 + H_2S = S(SCN)_2 + 2HSCN$$

Sulfur thiocyanate was observed as long ago as 1828, by Lassaigne, as a product of the action of sulfur dichloride on mercury thiocyanate:

$$SCl_2 + Hg(SCN)_2 = HgCl_2 + S(SCN)_2$$

It has also been possible to prepare the compound $S_2(SCN)_2$, corresponding to S_2Cl_2, (by the action of $Hg(SCN)_2$ on S_2Cl_2 dissolved in carbon tetrachloride or carbon disulfide:

$$Hg(SCN)_2 + S_2Cl_2 = HgCl_2 + S_2(SCN)_2)$$

Nitrosyl thiocyanate, the analogue of nitrosyl chloride, also exists. It is a very unstable, in-tensely red-brown compound, which readily breaks down into nitric oxide and thio-cyanogen, and therefore shows the reactions of the latter. Thus it reacts with white copper(I) thiocyanate, forming the black copper(II) thiocyanate:

$$NO—SCN + CuSCN = NO + Cu(SCN)_2$$

Nitrosyl thiocyanate is formed by treating a concentrated solution of an alkali thiocyanate, acidified with sulfuric acid, with alkali nitrite, and can also be obtained by the direct union of nitric oxide with free thiocyanogen.

(*ii*) *Thiocyanic acid and the Thiocyanates.* Thiocyanic acid is a colorless, oily, very volatile and pungent liquid, which easily solidifies (m.p. 5°). It decomposes with extreme ease in the pure state, and can be kept only at low temperatures (freezing mixture) or in dilute solutions (less than 5%). Hydrocyanic acid is formed in the decomposition, together with a solid yellow product, the so-called isoperthiocyanic acid, $H_2C_2N_2S_3$.

Thiocyanic acid is completely miscible with water; the aqueous solution is readily prepared by the action of acids on the thiocyanates or by treatment of a solution of ammonium thiocyanate with a cation-exchange resin (previously treated with HCl, and therefore in its 'acid' form) such as Zeocarb or Dowex.

The anhydrous compound is obtained when mercury or lead thiocyanate is gently warmed in a stream of hydrogen sulfide:

$$Pb(SCN)_2 + H_2S = PbS + 2HSCN$$

Thiocyanic acid is a *strong acid*. It is completely, or almost completely dissociated in aqueous solution.

The salts of thiocyanic acid, the thiocyanates, are readily formed from the cyanides, by combination with sulfur. They resemble the chlorides to a considerable extent in their reactions, giving with silver nitrate a precipitate of silver thiocyanate, insoluble in water and in dilute acids. A characteristic and very sensitive reaction for thiocyanates is the red coloration already mentioned, as a result of the formation of ferric thiocyanate complete, by the action of Fe^{+++} with SCN^- ions. The thiocyanate ions themselves are colorless as are their salts with colorless cations. Most thiocyanates are soluble in water; silver, mercury, copper, and gold thiocyanates are insoluble. Lead thiocyanate is sparingly soluble and is decomposed by boiling water.

Thiocyanates are decomposed by moderately concentrated ($1 : 1$) sulfuric acid, COS being evolved:

$$M^ISCN + 2H_2SO_4 + H_2O = COS + (NH_4)HSO_4 + M^IHSO_4$$

Many thiocyanates, and also the SCN^--ion in solution, combine additively with SO_2. This property can be utilized for the removal of SO_2 (and H_2S) from gases, and for the preparation of pure SO_2 (Hansen, 1933).

Thiocyanates, have some industrial uses, especially in dyeing. Ammonium thiocyanate is the principal technical product formed by the action of ammonia in aqueous solution on CS_2, at about $110°$, under pressure:

$$2NH_3 + CS_2 = NH_4SCN + H_2S$$

The evolution of H_2S can be prevented by adding slaked lime:

$$H_2S + Ca(OH)_2 = CaS + 2H_2O$$

Ammonium thiocyanate is a colorless salt, crystallizing in plates or prisms, of density 1.31 and m.p. 159°. It dissolves very readily in water, with the absorption of a considerable quantity of heat. 100 g of water dissolve 122 g of NH_4SCN at 0°, and 162 g at 20°. It is also freely soluble in alcohol. It is used in the laboratory as a reagent for iron(III) salts, and for the determination of silver by Volhard's method.

1. *Potassium thiocyanate*, KSCN, crystallizes in colorless prisms, of density 1.9. It melts at 161°. The molten salt turns blue at 430°, and becomes colorless once more on cooling. It is extremely soluble in water, with the absorption of much heat—100 g of water dissolve 177 g of KSCN at 0°, 217 g at 20°, and 239 g at 25°. Potassium thiocyanate is formed by melting potassium cyanide with sulfur, or by melting potassium ferrocyanide, potash and sulfur together. It has the same uses as ammonium thiocyanate.

2. *Sodium thiocyanate*, NaSCN, which crystallizes in unhydrated, but very deliquescent, colorless, rhombic plates, is not so often used.

The constitution of the thiocyanate group has been discussed by Goubeau [*Ber.*, 73 (1940) 127].

(g) Carbides

All the binary compounds of carbon might be included under the term carbides in its widest sense. The name is, however, generally restricted to the compounds of carbon with the metals and with boron and silicon (see also p. 438).

These compounds can be grouped into two classes: those which can be decomposed by the action of cold, or warm water, and those which are not attacked by water. Those carbides which are unaffected by water are, as a rule, resistant also to the action of acids, including even the concentrated strongly oxidizing acids. They are, however, easily oxidized, with the formation of carbonate, when they are melted with alkali hydroxide in the presence of air.

(*i*) *Carbides Decomposed by Water or Dilute Acids.* These include the following classes.

1. *Carbides forming acetylene* (acetylides). These can be regarded as the salts of acetylene (cf. p. 435), and they correspond in composition to the general formulas $M^I_2C_2$, $M^{II}C_2$, and $M_2^{III}(C_2)_3$. They are hydrolyzed by water or dilute acids, forming acetylene. Their salt-like character was demonstrated by the electrolysis of solid Na_2C_2 (von Antropoff, 1932). Acid acetylides, M^IHC_2, are known in the case of the alkali metals. In the crystal lattice of the acetylides the radicals —C≡C— constitute clearly marked structural units. Most of the acetylides crystallize with the barium carbide type of structure (Fig. 84) or structures related to it.

Acetylides are formed by all the metals of the first two groups of the Periodic Table, and also by Al, Zn and Cd. According to von Stackelberg (1930), acetylides are normally formed when the radius of the metal atom exceeds 0.8 to 1.0 Å. The metals of the rare earths, thorium and uranium also form acetylides, in which they appear to be bivalent (YC_2, LaC_2, CeC_2, etc., ThC_2, UC_2). Since the hydrolysis of these latter compounds also liberates hydrogen as a result of the change in valence of the metal ($M^{2+} + H^+ = M^{3+} + H$), the acetylene (or its radical) is partially hydrogenated to CH_4 and C_2H_4 (J. Schmidt, 1934). For this reason, the action of dilute acids on the acetylides of the rare earth metals and thorium does not furnish pure acetylene, but considerable quantities of methane (20—30%), with some hydrogen and ethylene as well. The explosive acetylides of Cu^I, Ag^I and Hg^{II} have been mentioned previously (p. 435). Hg^I acetylide, $Hg_2C_2 \cdot H_2O$, is not explosive but decomposes at 100°.

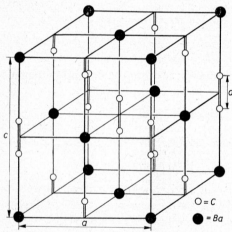

Fig. 84. Unit cell of barium carbide
$a = 6.22$ Å, $c = 7.06$ Å, $d = 1.4$ Å

O = C
● = Ba

2. *Carbides forming allylene.* Only Mg_2C_3, prepared by Novák, is known as a representative of this group. This furnishes *pure allylene* CH≡C—CH₃ when hydrolyzed:

$$Mg_2C_3 + 4HOH = 2Mg(OH)_2 + C_3H_4$$

Mg_2C_3 is formed from MgC_2 above 500°, by elimination of carbon. It gradually decomposes into its elements above 600°.

3. *Carbides forming methane*. The carbides Be_2C and Al_4C_3 are decomposed by warm water or dilute acids, forming *pure methane*:

$$Be_2C + 4HOH = 2Be(OH)_2 + CH_4; \qquad Al_4C_3 + 12HOH = 4Al(OH)_3 + 3CH_4$$

Isolated carbon atoms are present in the crystal lattices of these compounds; Be_2C is anti-isotypic with CaF_2.

The carbide Mn_3C probably has a related character, since its hydrolysis furnishes only hydrogen and methane:

$$Mn_3C + 6HOH = 3Mn(OH)_2 + CH_4 + H_2$$

4. *Carbides which form mixtures of hydrocarbons, but no acetylene*. This class includes, especially, the carbides of the iron group, and also U_2C_3, which is remarkable in that it gives principally *liquid* hydrocarbons as hydrolysis products. These carbides appear to be more closely related in structure to the intermetallic compounds, and are discussed further in Vol. II.

The heats of formation of some carbides of the classes (i)1 to (i)4 are given in the following table.

Li_2C_2	Na_2C_2	MgC_2	Mg_2C_3	CaC_2	BaC_2	Al_4C_3	
$+13.9$	-4.1	-21	-19	$+14.5$	$+5.0$	$+40$	kcal per mol
Ag_2C_2	ThC_2	UC_2	Mn_3C	Fe_3C	Co_3C	Ni_3C	
-83.6	$+46$	$+29$	$+3.6$	-2.5 or -5.8	-9.1	-9.2	

(ii) Carbides Not Decomposed by Dilute Acids. The acid-resistant carbides have excellent powers of crystallization (as also have most of the non-acid-resistant carbides). They are, in some cases, extremely hard and brittle, and include *boron carbide* and *silicon carbide*, compounds which approach the diamond in hardness.

Only a few of the carbides of this class have simple formulas which reflect the valences of their components—e.g., SiC, TiC, ZrC. Many of them, such as B_4C, Cr_4C, Cr_3C_2 do not correspond at all in composition to the normal valences of their elements. They are related rather to the intermetallic compounds, which will be considered in the next chapter. TiC and ZrC also have the character of typical intermetallic compounds, according to Klemm (1931) and von Stackelberg (1934), whereas SiC belongs to the *adamantine* compounds—i.e., covalent compounds with an atomic lattice. TiC, ZrC, VC, NbC, TaC have crystal structures of sodium chloride type. The structure of SiC is considered on the next page.

The carbides are generally prepared by heating the corresponding oxides with carbon, which abstracts and replaces the oxygen. At very high temperatures the carbides decompose into their constituents; in some cases this occurs on relatively gentle heating—e.g., with the alkali carbides. These must, accordingly, be prepared in a different manner, as was indicated earlier.

Lithium carbide, Li_2C_2, occupies an intermediate place, since its dissociation temperature is fairly low (about 600°), but is above the temperature of formation.

1. *Silicon carbide, carborundum*, SiC is prepared by heating a mixture of silica sand and coke to about 2000° in the electric furnace.

The reaction sets in at about 1600°, but leads to the formation of crystallized SiC only at temperatures above 1900°. Decomposition takes place above 2200°, silicon evaporating away and leaving a residue of graphite. Carborundum is manufactured principally at the works at Niagara Falls, founded by Acheson, the inventor of the process. Heating currents up to 10,000 amperes are employed.

The heat of formation of SiC from its elements (reduced to room temperature) is 28 kcal per mol.

When quite pure, silicon carbide forms colorless crystals (density 3.20) nearly as hard as the diamond. It is extremely resistant chemically, and even at high temperatures it is only superficially attacked by chlorine and not at all by oxygen or sulfur. It is not attacked by strong acids, even by a mixtures of fuming nitric and hydrofluoric acids, which will dissolve crystalline silicon. It is, however, fairly readily decomposed by melted caustic alkalis in the presence of air:

$$SiC + 4KOH + 2O_2 = K_2SiO_3 + K_2CO_3 + 2H_2O$$

It is also completely oxidized by red hot dichromate.

Silicon carbide closely resembles the diamond and silicon in its crystal structure although it crystallizes in the hexagonal system, and these elements in the cubic system. The crystal lattice of silicon carbide is obtained if, in a somewhat expanded diamond lattice, one half of the carbon atoms are exchanged for silicon atoms. The distance $C \leftrightarrow Si$ is 1.90 Å, and is thus almost the arithmetic mean of the distances $C \leftrightarrow C$ and $Si \leftrightarrow Si$ in diamond and crystalline silicon, $\frac{1}{2}(1.54 + 2.34) = 1.94$ Å. There is also a cubic modification of carborundum (formerly believed to be amorphous) with the zinc blende type of structure, with $a = 4.37$ Å, $C \leftrightarrow Si$ also 1.90 Å. Other modifications have also been described. All have in principle the same structure, differing only in the sequence in which planes containing Si and C atoms respectively are repeated throughout the crystal lattice.

Technical carborundum has a dark color, because of the impurities it contains. Its great hardness makes it very valuable abrasive. It is also used for the manufacture of refractory bricks, etc. *Silundum* and *silite* are ceramic bodies, consisting substantially of carborundum, though manufactured by other processes. They are used as insulating packing in electric furnaces, and as resistance elements for furnaces.

2. *Boron carbide*, B_4C, is obtained by heating boron or boron trioxide with carbon to about 2500°. It forms black, lustrous crystals, of density 2.52. It is so hard that it can be used for grinding diamonds, though according to Leon (1937) it can accomplish only about one quarter as much cutting as can diamond powder when used for this purpose. For grinding hard metals, however, it approaches the performance of diamond powder, and is better than carborundum.

It is completely resistant to attack by potassium chlorate and nitric acid, and is only slowly attacked by chlorine and oxygen at temperatures below 1000°. It was formerly believed to have the formula B_6C, but investigations by Laves and by Ridgway (1934) point to the composition B_4C. This has now been confirmed by the determination of its crystal structure (Clark and Hoard, 1943).

4. Analytical (Carbon)

Elementary carbon in its various forms is easily recognized by its physical properties. Carbon and its compounds are detected chemically by conversion to *carbon dioxide*. Carbonates evolve carbon dioxide, with effervescence, on the addition of dilute mineral acids. Other carbon compounds can be converted to carbon dioxide by heating them with suitable oxidizing agents—copper oxide is ordinarily used. Carbon dioxide is detected by means of baryta water, in which it produces a white precipitate of barium carbonate.

There are some other volatile compounds which give white precipitates with baryta water, namely SO_2 and HCN. These can, however, readily be distinguished from CO_2—e.g., by their smell.

Carbon dioxide in carbonates is determined quantitatively either *indirectly*, by determining the loss in weight on ignition or on decomposition with acids, or *directly*, by acid decomposition and absorption of the liberated CO_2 in concentrated alkali hydroxide solution or soda lime. The carbon content of other carbon compounds is found by *combustion*, usually with copper oxide as oxidizing agent or platinum as an oxygen carrier. The CO_2 so formed is absorbed in soda lime or caustic potash and weighed. ('Ultimate analysis').

Very small quantities of carbon—e.g., in samples of steel—can be determined simply and accurately, by passing the CO_2 formed by their combustion into a very dilute solution of $Ba(OH)_2$. The $BaCO_3$ which is precipitated can then be titrated potentiometrically (Oelsen and others, *Angew. Chem.*, 63 (1951) 557).

A sensitive reagent for *carbon monoxide* is palladium(II) chloride. Carbon monoxide is generally determined, in gas analysis, by absorbing it in ammoniacal copper(I) chloride. *Methane* is determined by combustion to CO_2 and absorption of the latter. *Unsaturated hydrocarbons* are absorbed by fuming sulfuric acid or bromine water. *Acetylene* may be detected by the formation of a precipitate when it is passed through alcoholic silver nitrate or ammoniacal copper(I) chloride solution.

The reaction with iodine pentoxide is suitable for the determination of small amounts of CO (under 1%):

$$5CO + I_2O_5 = I_2 + 5CO_2$$

Either the I_2, or the CO_2, which is collected in $Ba(OH)_2$, may be titrated [Pieters, *Z. anal. Chem.*, 85 (1931) 50]. The hopcalite method is still more convenient.

Oxalic acid is recognized, among other reactions, in that it decolorizes warm, acidified permanganate solution, with evolution of CO_2. The reaction with permanganate is also suitable for its quantative determination. The reduction of salts of the noble metals is a characteristic reaction of *formic acid*. The reaction with $HgCl_2$, mentioned on p. 453, is suitable for the determination of formates. *Acetic acid* is identified by its smell, and by the still more characteristic smell of its ester with ethyl alcohol. The cacodyl reaction (p. 661) is more sensitive.

The property of many organic compounds, especially those containing hydroxyl groups, of hindering the precipitation of many heavy metal salts or rendering it incomplete, is also important in analysis. The organic compounds can be removed by fuming down with concentrated sulfuric acid.

5. Silicon (Si)

(a) Occurrence

Next to oxygen, silicon is the most abundant of the elements. It never occurs free in Nature, but always in the form of its compounds, almost invariably in compounds derived from the *dioxide*, SiO_2. Free silicon dioxide is itself very frequently found. In the form of *quartz* it is a constant constituent of the mixtures making up *granite* and other widely distributed rocks. As these undergo weathering, quartz in the form of grains and lumps passes into the rivers and seas, and is washed back from these to the land as beach sand.

The two other constituents of granite, *felspar* and *mica*, are *silicates*—i.e., compounds also derived from silicon dioxide. It is chiefly the most electropositive of the metals which are widely distributed in nature in the form of silicates. A number of such silicates have been shown in the tables on pp. 163 and 242.

Among the heavy metals, iron in particular often occurs in silicates and especially in certain isomorphous mixtures of silicates, which occur very commonly—e.g., in the *hornblendes* and *augites*, and also the *micas*—in which iron replaces magnesium to a greater or lesser extent.

Altogether, silicon makes up about $\frac{1}{4}$ and oxygen about $\frac{1}{2}$ of the weight of the earth's crust, inclusive of the oceans and the atmosphere. The quantity of silicon, in the part of the earth's crust that is accessible to us, is thus *equal to the combined total weight of all the other* elements, with the exception of oxygen. Silicon is of subordinate importance for the organic world. However, almost all plants contain small amounts of SiO_2, and the straw of grasses, especially the cereal grasses, is in fact relatively rich therein. It acts in these as a structural material. In the animal organism, SiO_2 occurs especially in the connective tissues. In the human body, at least, its quantity decreases with advancing age.

(b) History

Flint (*silex*), quartz sand, rock crystal and other minerals rich in silica were already used in antiquity for the manufacture of glass. In the 17th century of the present era it was concluded that the suitability of these minerals for the manufacture of glass was due to the presence of a particular substance in them, later called siliceous earth. Silicon was first obtained in the elementary state by Berzelius, in 1822, by reducing silicon fluoride, SiF_4, with metallic potassium. Berzelius recognized that silicon was turned into 'siliceous earth' by combustion, and was thus the element underlying it. The fluorine compounds of silicon—fluorosilicic acid, H_2SiF_6, and the silicon fluoride, SiF_4, from which it is derived—had already been discovered half a century earlier, by Scheele, during his researches on hydrofluoric acid.

(c) Preparation and Uses

The preparation of silicon depends on the reduction of silicon dioxide, or of the silicon halides, by means of heated magnesium, aluminum or carbon. Alkali metals are less suitable, even although the first preparation was carried out by means of potassium. The products obtained may differ considerably in appearance and properties, according to the experimental conditions; these materials were formerly taken to be different modifications of silicon. In fact, they are identical in crystal structure, but differ in respect of their particle size, surface development, and impurity content (especially SiO_2). Relatively pure silicon can be obtained, in beautiful crystal leaflets, by the aluminothermic reduction of K_2SiF_6:

$$3K_2[SiF_6] + 4Al = 3Si + 2K[AlF_4] + 2K_2[AlF_5]$$

This process dates back to Wöhler. It is convenient to carry it out by heating a mixture of 125 g of aluminum turnings with 40 g of $K_2[SiF_6]$ at a bright red heat for half an hour, and then allowing it to cool slowly. The regulus, consisting of a silicon-aluminum alloy, is separated from the slag, roughly broken up and treated first with dilute and then with concentrated hydrochloric acid, whereupon silicon remains undissolved. It is essential that an excess of aluminum should be employed, since this acts as a solvent for the silicon, which crystallize out on cooling. So-called 'amorphous' silicon can be recrystallized by melting it with aluminum.

SiO_2 may also be reduced by aluminum under suitable conditions. According to a patent of Kühne (1903), 400 g of aluminum turnings are mixed with 500 g of powdered sulfur and 360 g of quartz sand, covered with a mixture of aluminum and sulfur or some other suitable igniting mixture, and fired by means of a red hot iron rod.

Silicon is obtained in a very reactive form by the process of Gattermann-Winkler, the reduction of silicon dioxide by magnesium.

$$SiO_2 + 2Mg = Si + 2MgO$$

To prevent too violent a reaction, the mixture, in the proportions of the theoretical equation, is diluted with one-fourth part of magnesium oxide. An excess of magnesium is to be avoided, as it results in the formation of silicide. It is important that the materials used for the reaction should be pure and absolutely dry. The fireclay crucible containing the mixture is introduced into a furnace already heated to redness, whereupon reaction ensues in a few minutes, the mass becoming brightly incandescent. In this way, a brown powder is obtained, which was formerly considered to be amorphous silicon, but which differs from crystallized silicon only in that it invariably contains considerable amounts of SiO_2, but is structurally the same.

Genuinely amorphous silicon can be prepared by the thermal decomposition of silicon monochloride (p. 484), as has been shown by Schwarz.

Silicon is prepared industrially by reducing quartz with carbon in the electric furnace, in the presence of iron. Silicon alloys with the latter, forming *ferrosilicon*, and the formation of carbide (cf. p. 467) is thereby suppressed. However, this process furnishes iron-silicon alloys, and not pure silicon. The product known in industry as pure silicon or metallic silicon contains at least 2%, and usually 3–5% of iron. It is more usual to produce ferrosilicon containing 25% of iron or more. Ferrosilicon with less than 20% silicon can be smelted in the ordinary blast furnace (cf. Vol. II).

For the electrothermal preparation, furnaces are used which resemble those employed for carbide and carborundum. CaO is often added to the reaction mixture. In this case CaC_2 is formed as an intermediate product and, being an excellent reducing agent, facilitates the production of silicon.

Ferrosilicon is used chiefly in the production of iron and steel. It decreases the solvent power of iron for carbon, and so leads to an increased deposition of the carbon in the form of graphite. This is frequently desirable in the production of cast iron. In the Bessemer process, ferrosilicon is used as a deoxidizer, to reduce the iron oxide formed by the burning. Iron alloys rich in silicon are finding increasing use, because of their outstanding resistance towards attack by acids.

Technically pure silicon is used as deoxidizer for copper alloys (cf. Vol. II). It has been used for the generation of hydrogen in the field (cf. p. 24). It has also been proposed as a reducing agent to replace aluminum.

Among the compounds of silicon, certain *silicates*, in particular, are of great industrial importance. *Organic compounds of silicon* are also finding technical applications on an increasing scale. Thus the *esters of silicic acid* and the *polysilicic acids* are used as binders in the ceramic industry, for the preparation of lacquers which adhere to glass, for the production of particularly finely divided silica gel, and for other purposes. Certain *alkoxy-aminosilanes*—e.g.,

$$\begin{array}{l} (CH_3)_3C\text{—}O \\ (CH_3)_3C\text{—}O \end{array}\!\!\!\! \diagdown\!\!\!\diagup Si(NH_2)_2$$

are used for the impregnation of paper, textiles, ceramic products and asphalt carpets in road construction, since they possess water-repellant properties. The *silicones* (p. 478), in particular, have acquired considerable technical importance. The silicones are notable for their thermal stability and high chemical resistance. They are also excellent electrical insulators. They are used to coat glass surfaces so that they are not wetted by water. Some of the silicones are highly elastic (silicone rubbers), and retain their elastic properties even at relatively low temperatures (to —65°C). Other silicones are used for the production of synthetic resins (silicone resins) which are stable up to high temperatures and are very resistant chemically.

(d) Properties

Pure crystalline silicon forms deep grey, very lustrous, opaque octahedra of the regular system, often distorted into plates. It is fairly hard (7 on Mohs' scale), and can scratch glass. It is brittle, however, and can easily be powdered. Its density is 2.36, and melting point 1413°. It is therefore easily fused in the electric arc. It is an electrical conductor like graphite, its conductivity increasing with rise of temperature. The specific heat at the ordinary temperature is 0.171. The atomic heat calculated from this value is 4.80, — significantly lower than the value required by the Dulong and Petit law. It approximates more and more closely to this as the temperature is raised.

Coarsely crystalline silicon is not particularly reactive chemically. It is practically insoluble in all acids, including hydrofluoric acid. It burns only with difficulty, but at very high temperatures (white heat) it combines not only with oxygen but with nitrogen and, in the electric arc, with hydrogen. Most metals alloy with silicon on heating; although lithium is the only alkali metal to do so. In many cases, definite compounds are formed—e.g., with magnesium, the silicide Mg_2Si. In other cases, it is only dissolved by the molten metal—e.g., aluminum—but separates again on cooling. Silicon can therefore be recrystallized by melting it with aluminum, and also with silver.

On slow cooling, the element crystallizes in well formed octahedra, from melts rich in silicon, but crystallizes in plates resembling graphite in appearance on more rapid cooling. Silicon crystallizes in very finely crystalline form from melts having a composition corresponding to the eutectic; with the alloys mentioned, the eutectic point (p. 564) corresponds to a low silicon content. The 'HF-soluble' silicon observed by Moissan is formed in this way. The silicon made by Gattermann's method is still more reactive. It is a brown or grey-brown loose powder, easily combustible, very readily dissolved by hydrofluoric acid and attacked by halogens. All these modifications have the same crystal structure, as shown by X-ray investigations (Gerlach, Debye and Frauenfelder, Küstner and Remy), namely a diamond lattice (Fig. 79, p. 421), with $a = 5.42$ Å, and a shortest interatomic distance of 2.35 Å. Even though the differences in reactivity between the various forms in which silicon may be obtained are not due to different crystalline modifications, but only to secondary factors (particle size, surface area, impurities), they are none the less of considerable practical significance. They may determine the suitability of preparations of silicon for any particular purpose.

All forms of silicon react very readily with caustic alkalis, even when these are very dilute. Water acquires the property of dissolving silicon, at least on boiling, merely by contact with a glass which loses alkali. The reaction is due to the tendency of silicon to form silicate ions:

$$Si + 4OH^- = SiO_4^{4-} + 2H_2$$

Only quite a small concentration of alkali is required for the reaction to proceed, since the OH$^-$ ions used up are at once regenerated in weakly alkaline solution, because of the extreme weakness of silicic acid:

$$SiO_4^{4-} + 4H_2O = H_4SiO_4 + 4OH^-.$$

6. Compounds of Silicon

The most stable and most important compounds of silicon are *silicon dioxide* and the numerous *silicates* derived from it. The affinity of silicon for oxygen is particularly great, and determines also the behavior of the other silicon compounds. Most of them decompose if the silicon is offered the chance of combining with oxygen—e.g., by reaction of the compound with water. In view of the great affinity of silicon for oxygen, it is noteworthy that bonds between silicon and *sulfur*, the homologue of oxygen, should have such low stability.

The compounds of silicon with *hydrogen* appear from their properties to be very much less stable than the hydrocarbons. Silicon compounds rich in hydrogen inflame in air, and are rapidly decomposed by water containing alkali. This is, however, the consequence not so much of the smaller affinity of silicon for hydrogen, as compared with carbon, but of the particularly great affinity of silicon for oxygen (cf. Table 73, p. 411).

The *silicon alkyl compounds*—i.e., the compounds of the type SiR_4, where R is an organic radical—are considerably more stable than the silicon hydrides. A considerable number of these is known. They resemble the corresponding carbon compounds rather than the silicon hydrides, and are not spontaneously inflammable in air. If the four valences of the silicon are linked to four different alkyl groups, the compounds can display *optical activity*. The *carbide* of silicon (cf. p. 467) which is not notable for a particularly large heat of formation, is strikingly inert chemically.

The silicon halides are relatively stable, but they are hydrolyzed by water. The action of alcohols gives rise to alkoxy-compounds, and Grignard reagents produce the alkyl compounds SiR_4, mentioned above.

The Si-N bond is relatively weak. It is usually ruptured by water if the N is linked also to other elements. The *binary* silicon-nitrogen compounds, however, are relatively stable,— e.g., Si_3N_4, formed by passing N_2 over finely crystalline Si at 1400°, according to Weiss (1910) and Funk (1924). (However, the composition of this compound is not quite certain, since it has never been possible to prepare it pure. Other silicon nitrides, poorer in nitrogen, also appear to exist.) The calcium silicides combine fairly readily with nitrogen, forming the compounds $CaSiN_2$ and $Ca(SiN)_2$, analogous to calcium cyanamide and calcium cyanide (Wöhler, 1924).

With phosphorus, silicon forms the very stable compound SiP; the compounds SiAs and $SiAs_2$ are formed with arsenic. Silicon does not combine with Sb or Bi.

The ability of silicon atoms to join up, forming chains without the interposition of atoms of some other element, is not nearly so strongly developed as in the case of carbon. The higher molecular silicon hydrides, such as Si_4H_{10}, decompose easily. In the majority of cases the Si-Si bond is severed by the action of water, even in the cold.

The linkage Si-O-Si, present, for example, in the siloxanes, is extremely stable however. Silicon-oxygen compounds have a strong tendency to condense, with the formation of oxygen bridges. This is especially evident in the silicates. (cf. pp. 498 *et seq.*).

(a) Silicides

The *silicides* are the binary compounds of silicon with the more electropositive elements, and especially the metals. The silicides are closely related to the carbides, but resemble the intermetallic compounds more closely than do the latter. Most of them have a metallic appearance. They are superior to the carbides in crystallizing power.

Most of the silicides are stable towards water and dilute acids, but the *alkaline earth silicides* and *lithium silicide*, Li_6Si_2 (silicides of the other alkali metals are not known) are decomposed. The silicides are related to the intermetallic compounds in composition: like most of the carbides, the majority of silicides do not have compositions which reflect the normal valence properties of their constituents.

The silicides are prepared either by melting the components together, or by reducing silicon dioxide with a sufficient excess of the metal concerned.

Magnesium silicide, Mg_2Si, (m.p. 1070°) crystallizes with the fluorspar structure $a = 6.37$ Å; Mg_2Ge ($a = 6.38$ Å), Mg_2Sn ($a = 6.77$ Å), and Mg_2Pb ($a = 6.84$ Å) have the same structure. MgSi is also stable at high temperatures, as Wöhler showed by quenching. With calcium, silicon forms the compounds Ca_2Si (m.p. 920°), CaSi (m.p. 1220°) and $CaSi_2$ (m.p. 1020°). Calcium-silicon alloys (generally with 30–35% Ca by weight) are used as deoxidizers in the steel industry, especially for stainless steels. Mono- and disilicides are also formed by Sr and Ba; Ba also forms $BaSi_3$ and probably still higher silicides. Heats of formation in kcal per mol are: CaSi 85.3; $CaSi_2$ 161.5; SrSi 112.8; $SrSi_2$ 147.4; BaSi 181.5; $BaSi_3$ 399.2 (Wöhler, 1932).

Many of the silicides—e.g., magnesium and manganese silicides—furnish silicon hydrides, in addition to free hydrogen, when they are decomposed by acids.

(b) Silicon Hydrides (Silanes)

The formation of silicon hydrides was first observed by Wöhler, in 1857, when aluminum, containing silicon, was dissolved in hydrochloric acid. As was found by Wöhler, magnesium silicide, Mg_2Si, is more suitable for the preparation; this can be made, for example, by firing a mixture of magnesium with finely divided quartz sand:

$$4Mg + SiO_2 = 2MgO + Mg_2Si$$

The decomposition of magnesium silicide by acids has been shown by Schwarz [*Z. anorg. Chem.*, 215 (1933) 288] to follow a rather complex course. As with magnesium boride, it furnishes much hydrogen, together with a mixture of various silicon hydrides which is spontaneously inflammable in air, and which was more closely investigated by Stock in 1916 and the subsequent years. Following Stock's proposal, the silicon hydrides are known as the *silanes*. Stock was able to prepare the following silanes in the pure state: — *monosilane* SiH_4, *disilane* Si_2H_6, *trisilane* Si_3H_8, *tetrasilane* Si_4H_{10} (cf. Table 79). In addition, *pentasilane* Si_5H_{12} and *hexasilane* Si_6H_{14} were probably present in his mixture. The relative amounts of the various constituents diminishes rapidly as their molecular weight increases. Stock separated the various compounds from each other by fractional distillation in an apparatus which permitted of working with the complete exclusion of air. This is necessary, since the silanes, like the boranes, are extremely sensitive to air. They inflame in air, and explode with violence.

TABLE 79

SILICON HYDRIDES (SILANES)

Formula	SiH_4	Si_2H_6	Si_3H_8	Si_4H_{10}
Boiling point	$-112°$	$-15°$	$+53°$	*ca.* $85°$
Melting point	$-184.7°$	$-132.5°$	$-117°$	$-93.5°$
Gas density, obs.	16.02	31.7	46.5	61.0
(Hydrogen = 1)	(at 19°)	(at 21°)	(at 24°)	(at 25°)
Density in liquid state	0.68	0.686	0.725	0.79
	(at $-185°$)	(at $-25°$)	(at 0°)	(at 0°)

1. *Monosilane* is a colorless gas, quite stable in the absence of air. When dilute, it has a weak musty smell. It is generally spontaneously inflammable in air; if its higher homologues are present it invariably inflames. Caustic alkalis bring about decomposition, the reaction being

$$SiH_4 + 4OH^- = SiO_4^{4-} + 4H_2$$

A similar decomposition occurs, although only slowly, with water alone; alkali-free water in a quartz vessel is without action. SiH_4 decomposes into Si and H_2 when it is heated above 400°. Decomposition also takes place under the influence of a silent electrical discharge, a golden lustrous deposit being formed, which consists of solid unsaturated silicon hydrides (Schwarz, 1935).

2. *Disilane*, Si_2H_6, is very similar to monosilane in behavior. Unlike B_2H_6, it begins to decompose only at about 300°. It reacts slowly with water:

$$Si_2H_6 + 4H_2O = 2SiO_2 + 7H_2$$

It reacts with carbon tetrachloride and chloroform, often with inflammation, in contrast with monosilane which is not markedly attacked by these substances.

Large quantities of mono- and disilane can best be prepared, according to Johnson (1935) by the decomposition of magnesium silicide with ammonium bromide dissolved in liquid ammonia, in which its reactions are analogous to those of an aqueous acid.

3. *Trisilane*, Si_3H_8, and *tetrasilane*, Si_4H_{10}, are considerably more unstable than the simpler silanes, and this is true to an increasing degree of the still higher members.

4. *Halogenosilanes*. By the action of bromine, or by gently heating with hydrogen bromide in the presence of aluminum bromide, it is possible to exchange the hydrogen atoms of the silanes step by step for bromine atoms. Thus from SiH_4 there result the compounds SiH_3Br, SiH_2Br_2, $SiHBr_3$ and $SiBr_4$, the first a gas and the others all volatile liquids. The H atoms of SiH_4 can be replaced by Cl, F or I in a similar manner. Mixed types, such as $SiHF_2Cl$ and $SiHFCl_2$ are also known. Some properties of the halogenosilanes are given in the appended table.

TABLE 79a

HALOGENOSILANES

	Density (at t °C)		Melting point	Boiling point
SiHF$_3$	—		−131.2°	− 97.5°
SiH$_3$Cl	1.145	(−113°)	−118°	− 30°
SiH$_2$Cl$_2$	1.42	(−120°)	−122°	+ 8°
SiHCl$_3$	1.34		−128.2°	+ 31.5°
SiH$_3$Br	1.533	(0°)	− 94°	+ 2°
SiH$_2$Br$_2$	2.17	(0°)	− 77°	+ 64.0°
SiHBr$_3$	2.7		− 73.5°	+111.8°
SiH$_3$I	2.035	(14.8°)	− 57.0°	+ 45.4°
SiH$_2$I$_2$	2.746	(15.1°)	− 1°	149.5°
SiHI$_3$	3.286	(23°)	+ 8°	*ca.* 220°

Some other properties of SiH$_3$I and SiH$_2$I$_2$ are:

	Heat of evaporation	Surface tension at 15°	Parachor
SiH$_3$I	7.13 kcal mol	30.50 dyne/cm	182.1
SiH$_2$I$_2$	8.05	44.1	267

(Emeleus, *J. Chem. Soc.* (1941) 353).

The '*silyl*' compounds SiH$_3$X (X = Cl, Br, I) are to some extent analogous to the methyl halides in their reactions. Thus SiH$_3$Cl reacts with zinc methyl (1) water (2) and ammonia (3):

$$SiH_3Cl + Zn(CH_3)_2 = CH_3ZnCl + SiH_3.CH_3 \qquad (1)$$

$$2SiH_3Cl + H_2O \quad = (SiH_3)_2O + 2HCl \qquad (2)$$

$$3SiH_3Cl + 4NH_3 \quad = (SiH_3)_3N + 3NH_4Cl \qquad (3)$$

(SiH$_3$)$_3$N, the silyl analogue of trimethylamine, has no basic properties, and is decomposed by hydrogen chloride, SiH$_3$Cl being re-formed. Silyl alkyl ethers, silyl alkyl amines and other derivatives have also been prepared, and there is some evidence that silicon analogues of the metal alkyls can also exist.

(*ii*) *Polysilenes* [SiH$_2$]$_x$ were obtained by Schwarz (1935), by decomposing CaSi with glacial acetic acid or with absolute alcoholic hydrochloric acid, as light brown solids, spontaneously inflammable in air. Water decomposed the solids—without any formation of silanes:

$$SiH_2 + 2H_2O = SiO_2 + 3H_2$$

The latter were formed, however, when polysilene was decomposed thermally in a high vacuum.

(*iii*) *Siloxanes.* The halogen-substituted silanes are rapidly hydrolyzed by water. Some interesting hydrogen-oxygen compounds of silicon, in which atoms of silicon are linked together through oxygen bridges (—Si—O—Si—) may be obtained in this way from the partially substituted silanes. Stock called these compounds *siloxanes.* Thus *disiloxane,* (SiH$_3$)$_2$O is immediately formed when bromomonosilane and water are brought together. The reaction proceeds practically quantitatively, according to

$$H_3Si—\boxed{Br \quad H} \qquad H_3Si\diagdown$$
$$\qquad\qquad O \qquad = \qquad\qquad O + 2HBr$$
$$H_3Si—\boxed{Br \quad H} \qquad H_3Si\diagup$$

Disiloxane is a colorless, odorless gas, which is not spontaneously inflammable in air, but burns with the evolution of a white smoke of SiO_2 (b.p. —15.2° m.p. —144°, density 0.881 at —80°). It is indefinitely stable at ordinary temperature, in the absence of air, and decomposes rapidly only at a red heat. It ignites explosively when mixed with oxygen, burning to SiO_2 and H_2O. It is rapidly hydrolyzed by alkalis:

$$(SiH_3)_2O + 8OH^- = 2SiO_4{}^{4-} + H_2O + 6H_2$$

With water, the reaction proceeds in stages, however, *prosiloxane*, H_2SiO, being formed as an intermediate product; this is a gas which readily polymerizes to a white, amorphous solid.

The liquid tetrasiloxane, $(Si_2H_5)_2O$, is obtained in a similar manner from Si_2H_5Br. Organic derivatives of many substances of this type are known, containing alkyl or aryl groups in place of H atoms.

(*iv*) *Siloxene*. Another interesting silicon-oxygen-hydrogen compound is *siloxene*, $[Si_6O_3H_6]x$, obtained by treating $CaSi_2$ with hydrochloric acid:

$$3CaSi_2 + 6HCl + 3H_2O = Si_6O_3H_6 + 3CaCl_2 + 3H_2$$

This is a solid, insoluble in all solvents, sensitive towards oxygen, water and light, and possessed of extremely high surface activity. It has been investigated especially by Kautsky. The following reactions of siloxene are typical of its behavior:

$$Si_6O_3H_6 \quad + I_2 \quad = Si_6O_3H_5I \qquad + HI \tag{1}$$

$$Si_6O_3H_6 \quad + HBr \quad = Si_6O_3H_5Br \qquad + H_2 \tag{2}$$

$$Si_6O_3H_6 \quad + 3Br_2 \quad = Si_6O_3H_3Br_3 \qquad + 3HBr \tag{3}$$

$$Si_6O_3H_3Br_3 + 6NH_3 = Si_6O_3H_3(NH_2)_3 + 3NH_4Br \tag{4}$$

All these compounds are solids, which are completely insoluble in all solvents, and are more or less intensely colored. Thus trioxysiloxene, obtained by hydrolyzing tribromosiloxene, is violet red; hexaoxysiloxene, $Si_6O_3(OH)_6$, is deep black. All the compounds are strongly activated by light, and many show marked luminescence phenomena. On the basis of the reactions, Kautsky proposed for siloxene the following formula:

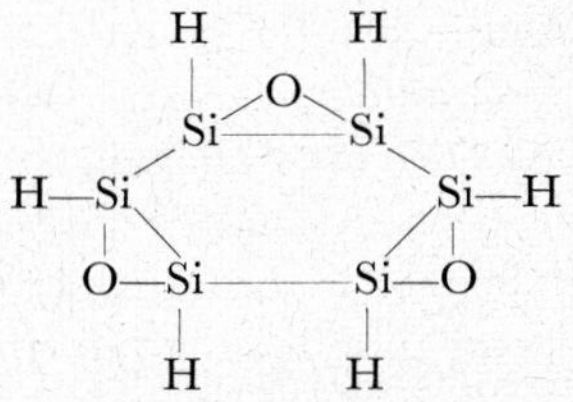

$CaSi_2$ is known from X-ray analysis to have a layer structure, with sheets of linked silicon atoms. It is likely, although not proven, that these sheets of silicon atoms remain intact in the hydrolysis of $CaSi_2$ to siloxene, so that the latter is a polymer of the formula proposed by Kautsky, i.e., $[Si_6O_3H_6]x$. Exhaustive chlorination of siloxene converts it to hexachlorosiloxane, $Cl_3Si—O—SiCl_3$ and alcohols yield esters of disilicic acid, $(RO)_3Si—O—Si(OR)_3$.

(c) Silicon Alkyl Compounds

Considerably more stable than the silicon hydrides are the *silicon alkyl compounds*, i.e., compounds containing organic (alkyl) radicals linked to silicon. Their general type is SiR_4, a few compounds of the type Si_2R_6 are also known. The organic radicals may be derived from the hydrocarbons of either the aliphatic or the aromatic series, or radicals of both types may be linked to silicon simultaneously.

Unlike the silicon hydrides, the silicon alkyls are not spontaneously inflammable in air, but are relatively stable. They are not decomposed by water, in which they are insoluble. The pure aliphatic silicon compounds are liquids, readily soluble in organic solvents. The purely aromatic compounds are solid at the ordinary temperature, and are usually well crystallized.

The original mode of preparation was from the zinc alkyls, ZnR_2, by reaction with silicon halides—e.g., $SiCl_4$. It is better to employ the alkyl magnesium halides. It is also possible to apply a reaction analogous to the Wurtz and Fittig syntheses of organic chemistry, namely the action of sodium on mixtures of silicon and alkyl halides:

$$SiCl_4 + 4RCl + 8Na = SiR_4 + 8NaCl$$

This is particularly successful with the compounds of the aromatic series.

Silicon alkyls can also be prepared directly from SiH_4 and hydrocarbons. Fritz (1951) showed that the compounds $C_2H_5.SiH_3$, $(C_2H_5)_2SiH_2$ and $(C_2H_5)_3SiH$ were produced when SiH_4 was heated to 400° with ethylene. Compounds with modified alkyl groups are formed at the same time—e.g., $\frac{CH_3}{C_2H_5}{>}SiH_2$ and $\frac{CH_3}{C_3H_7}{>}SiH_2$. On prolonged heating, non-volatile oily products are formed, together with a yellow solid, $(SiCH_3)_x$. SiH_4 also reacts with saturated hydrocarbons and with vinyl chloride in the same manner as with ethylene. The reaction of SiH_4 with ethylene oxide and with acetone yielded alkoxy compounds—e.g., $C_2H_5O{-}SiH_3$ and $C_3H_7O{-}SiH_3$. SiH_4 and its derivatives can undergo reaction with organic compounds even at low temperatures, if suitable catalysts are employed (Nitzsche, 1951).

(d) Silicones

The decomposition of trialkyl halogen silanes by means of water leads to the formation of alkyl substituted siloxanes—e.g.,

$$2(CH_3)_3SiCl + H_2O = (CH_3)_3Si{-}O{-}Si(CH_3)_3 + 2HCl$$

If the same reaction is applied to compounds of the type SiR_2X_2 (R = alkyl radical, X = halogen), *silicols* are obtained as intermediates (compounds in which the halogen atoms are replaced by —OH groups), and these condense, by loss of water, to form substances containing a greater or smaller number of —SiR_2- groups, linked together through O atoms. Following the pioneering work of Kipping (1901 and later), these have been designated *silicones* [63, 64]. Their chemistry and technology were developed to a considerable extent in the United States during the war of 1939–45, and they have attained very considerable practical importance. Chemically and thermally, the silicones are very stable substances. The compounds of lower molecular complexity, in which the SiR_2O groups are joined together either in chains or (more often) in rings, are water-clear, mobile liquids. The high molecular compounds, which contain long chains (branched or unbranched, according to the mode of preparation) are viscous, either oily or pasty, depending on the molecular weight. They are excellent lubricants, since their viscosity varies but little with temperature. With still greater chain length, rubbery products are obtained. Under suitable conditions, instead of a linear condensation of SiR_2O groups, three dimensional networks can be built up, similar in principle to those present in the silicates with network structures (cf. p. 502). Such three dimensional networks are present in the *silicone resins*, which are especially suitable for electric insulating material.

In the preparation of the silicones, alkyl or aryl halogen silanes are first prepared. This may be accomplished either from $SiCl_4$ or from ethyl orthosilicate, $Si(OC_2H_5)_4$, or—of great technical importance—by the action of alkyl halides on elementary silicon, usually present as a silicon-copper alloy. The alkyl halogen silanes are hydrolyzed to the corresponding silicols, and these are condensed to silicones by suitable treatment—e.g. by heating, or by the action of acids. The length of the chains produced in the condensation can be controlled by adding *tri*alkylhalogen silanes to the *di*alkylhalogen silanes, since the univalent SiR_3-groups necessarily form the terminal groups of silicone chains. E.g.,

$$SiR_2Cl_2 + 2H_2O = SiR_2(OH)_2 + 2HCl; \quad SiR_3Cl + H_2O = SiR_3OH + HCl$$

$$
\begin{array}{cccc}
R & R & R & R \\
HO-Si-OH + HO-Si-OH & = & HO-Si-O-Si-OH + H_2O \\
R & R & R & R
\end{array}
$$

$$
\begin{array}{cccccc}
R & R & R & R & R & R \\
HO-Si-O-Si-OH + 2SiR_3OH & = & R-Si-O-Si-O-Si-O-Si-R + 2H_2O \\
R & R & R & R & R & R
\end{array}
$$

(e) Silazanes

Compounds in which silicon atoms are linked together by imide groups, —NH—, are known as *silazanes*; cyclic compounds of this type are called *cyclosilazanes*. These substances may be exemplified by the compounds prepared by Larson and Smith (1949) from NH_3 and $Si(CH_3)_2Cl_2$:

$$
\begin{array}{c}
\quad\quad NH \\
(CH_3)_2Si \quad\quad Si(CH_3)_2 \\
| \quad\quad\quad | \\
HN \quad\quad NH \\
\quad Si \\
(CH_3)_2
\end{array}
\quad\text{and}\quad
\begin{array}{c}
\quad\quad (CH_3)_2 \\
\quad NH-Si \\
(CH_3)_2Si \quad\quad NH \\
| \quad\quad\quad | \\
HN \quad\quad Si(CH_3)_2 \\
\quad Si-NH \\
(CH_3)_2
\end{array}
$$

hexamethyl cyclotrisilazane   octamethyl cyclotetrasilazane

These are exactly analogous to the simplest cyclic silicones. More recently, Schwarz has prepared similar compounds, which contain RO— radicals in place of the CH_3-groups (R = isopropyl or phenyl). He showed that the diamides $(RO)_2Si(NH_2)_2$ generally have a stronger tendency to condense to cyclosilazanes, with elimination of ammonia than the monamides $(RO)_3SiNH_2$. The latter split out ammonia, to form disilazanes, $(RO)_3Si—NH—Si(OR)_3$. Schwarz has also prepared compounds of the type $(RO)_3Si—NR'_2$, which he calls *silazines*—e.g.

$$
(C_6H_5O)_3Si-N\underset{CH_2-CH_2-CH_2}{\overset{CH_2-CH_2-CH_2}{\diagup}}
\quad\text{and}\quad
\left(\begin{array}{c} H_3C \diagdown \quad H \\ \quad C \\ H_3C \diagup \diagdown O \end{array}\right)_3 Si-N\underset{CO-CH_2-CH_2}{\overset{CO-CH_2-CH_2}{\diagup}}
$$

Triphenoxy-cyclohexyl   triisopropoxy-cyclotetramethylene
silazine   diketosilazine

The starting points for the preparation of these compounds were hexamethylene diamine and the acid chloride of adipic acid, respectively. Under the influence of the silicon atom, ring-closure occurred, instead of the formation of linear high polymeric compounds, as would be anticipated on the basis of experience in pure organic chemistry.

(f) Silicon Halides

The simple silicon halides, corresponding to the general formula SiX_4, can be obtained by the direct union of the constituent elements. The fluoride is also formed by the action of HF on SiO_2. The compounds are themselves quite stable, but are immediately decomposed by water, even at ordinary temperature, in accordance with the great affinity of silicon for oxygen. The stability increases progressively in going from the iodide to the fluoride. The fluoride is able to add on additional fluoride ions, forming complex fluorosilicate ions, and in particular the hexafluorosilicate ion:

$$SiF_4 + 2F^- = [SiF_6]^=$$

This ion is stable in aqueous solution. It is thus more strongly complexed than the fluoroborate ion.

The other silicon halides are not capable of adding on further halogen ions.

The silicon halides can also be regarded as substituted silanes. Corresponding to this view, silicon forms halogen compounds in which several silicon atoms are linked together, as well as the simple halides SiX_4. These halides form some interesting hydrolysis products. The same is true of the silanes in which substitution of halogen is incomplete. These have already been mentioned under the silicon hydrides and in the following section only the trihalogen silanes, $SiHX_3$ will be briefly discussed; these are more easily prepared by other means than those used for the other halosilanes.

Table 80 provides a summary of some characteristic properties of the simplest silicon halides.

TABLE 80

HALIDES OF SILICON

Compound	SiF_4	$SiCl_4$	$SiBr_4$	SiI_4
State of aggregation at room temperature	gaseous	liquid	liquid	solid
Melting point	—	$-70°$	$+5.2°$	$+120.5°$
Boiling point	sublimes at $-95°$	$+57.5°$	$+152.8°$	$287.5°$
Heat of formation (kcal per mol)	360	154.0	88.5	6.7
Products of decomposition by cold water	H_2SiO_3 and $H_2[SiF_6]$	H_2SiO_3 and HCl	H_2SiO_3 and HBr	H_2SiO_3 and HI

Compound	Si_2F_6	Si_2Cl_6	Si_2Br_6	Si_2I_6	Si_3Cl_8
State of aggregation at room temperature	gaseous	liquid	liquid	solid	liquid
Melting point	$-18.7°$ (780 mm)	$+2.5°$	—	$+250°$	$-67°$
Boiling point	$-19.1°$ (subl.)	$+147°$	ca. $240°$	—	ca. $215°$
Products of decomposition by cold water	$H_2Si_2O_4$, HF. H_2SiO_3, $H_2[SiF_6]$ and H_2	$H_2Si_2O_4$ and HCl	$H_2Si_2O_4$ and HBr	$H_2Si_2O_4$ and HI	$H_4Si_3O_6$ and HCl

Within recent years, Schwarz (1937, 1947) has succeeded in characterizing silicon chlorides with very long chains—e.g., $Si_{10}Cl_{22}$ and $Si_{25}Cl_{52}$. The latter is obtained by a hot-and-cold-tube method from the pyrolysis of $SiCl_4$ vapor in an atmosphere of nitrogen at $1250°$; it is a resinous, plastic substance. Under rather different conditions (hydrogen atmosphere), Schwarz obtained $Si_{10}Cl_{20}H_2$, and from this he prepared silicon chlorides of cyclic structure. The first product, obtained by elimination of HCl when $Si_{10}Cl_{20}H_2$ is gently heated, is the bicyclic compound $Si_{10}Cl_{18}$, a transparent yellowish substance with the consistency of vaseline. On stronger heating, the lower silicon chlorides (e.g., $SiCl_4$, Si_2Cl_6, Si_3Cl_8, Si_4Cl_{10}) are split out, with the formation of first resinous and ultimately brittle, vitreous substances, involving many-membered cyclic structures. All these are quite soluble in benzene. By progressive condensation, the compound *silicon monochloride*, $[SiCl]_x$, is finally formed. This has an infinite (but not necessarily perfectly regular) two-dimensional network structure. It is amorphous to X-rays, and is converted into dark grey *amorphous* elementary silicon (again with the splitting, out of the lower silicon chlorides) when it is heated to 800° (Schwarz, 1951–52). $Si_{10}Cl_{22}$ is a colorless, viscous oil which may be distilled in a high vacuum at 215°, and which burns in air. It reacts vigorously with water, forming $Si_{10}(OH)_{22}$, a white powder, which is slowly oxidized in air, the Si—Si bonds being gradually replaced by Si—O—Si bridges.

(*i*) *Silicon fluoride*, SiF_4, is obtained by warming a mixture of powdered fluorspar and quartz sand with concentrated sulfuric acid. The latter first liberates HF from the fluorspar, and this then reacts with the quartz to form silicon tetrafluoride:

$$CaF_2 + H_2SO_4 = CaSO_4 + 2HF; \qquad SiO_2 + 4HF = SiF_4 + 2H_2O$$

Pure SiF_4 can most readily be obtained by strongly heating barium fluorosilicate, which is thereby decomposed into barium fluoride and silicon fluoride:

$$Ba[SiF_6] = BaF_2 + SiF_4$$

Silicon fluoride is a colorless, pungent gas, which forms dense fumes in moist air. The liter weight is 4.693 g, and thus the molecular weight corresponds with the formula SiF_4. When strongly cooled under ordinary pressure, silicon tetrafluoride passes directly from the gaseous to the solid state (sublimation temperature —95°, heat of sublimation 6.18 kcal per mol.) It can be liquefied under pressure (critical temp. —1.5°, crit. pressure 50 atm.).

The compound has a high heat of formation (360 kcal per mol.,) and is therefore very stable. It is not decomposed by the electric spark. Sodium and potassium will burn in the gas, however, forming their fluorides. Dry alkali carbonates or borates are not attacked in the cold, nor is glass, provided that no trace of moisture is present. Water hydrolyzes silicon fluoride, however, forming silica gel:

$$SiF_4 + 2H_2O = SiO_2 + 4HF$$

The hydrofluoric acid thereby formed unites with as yet unhydrolyzed silicon fluoride, to form *fluorosilicic acid*, H_2SiF_6. Silicon fluoride gas must be collected over mercury, since this decomposition takes place immediately on contact with water.

Silicon tetrafluoride forms loose addition compounds with many substances—e.g., $SiF_4 \cdot 2NH_3$ formed with ammonia. SiF_4 reacts more or less quantitatively with $AlCl_3$ at 180–190°, to form $SiClF_3$, $SiCl_2F_2$, $SiCl_3F$ and $SiCl_4$. By raising the temperature it is possible to carry out the reaction so as to produce only $SiCl_4$. Analogous reactions take place with $AlBr_3$ and AlI_3. According to Schmeisser (1952), SiF_4 differs from the other silicon halides in that it does not react immediately to form silicic esters, although it does so with aluminum alkoxides. It also differs from the other silicon halides in other reactions. Thus ethylene oxide, C_2H_4O, is converted by SiF_4 into dioxane, $C_4H_8O_2$, whereas $SiCl_4$ reacts with ethylene oxide to form chlorinated silicic esters (cf. p. 485).

(*ii*) *Fluorosilicic acid and the Fluorosilicates.* Fluorosilicic acid (hexafluorosilicic acid, formerly called silicofluoric acid), H_2SiF_6, is formed by the combination of hydrogen fluoride with silicon fluoride:

$$SiF_4 + 2HF = H_2SiF_6$$

Its aqueous solution is usually prepared by passing silicon fluoride into water ($3SiF_4 + 2H_2O = SiO_2 + 2H_2[SiF_6]$), and filtering off the silica gel which is thrown down. The anhydrous compound may be prepared by the action of concentrated sulfuric acid on its salts—e.g., on barium fluorosilicate, $Ba[SiF_6]$. It then comes off in the gaseous state, but is largely dissociated into its components, $SiF_4 + (HF)_2$, even at ordinary temperature (more than 50%). The aqueous solution, however, contains no detectable amount of free hydrofluoric acid, and therefore does not etch glass, provided, of course, that the ratio F : Si is not greater than 6. A solution containing 13.3% of the acid can be distilled unchanged. Hydrates separate from the concentrated solutions on cooling—e.g., $H_2SiF_6 \cdot 2H_2O$, hard, colorless crystals which melt at 19°.

Fluorosilicic acid has excellent disinfecting properties, even in quite dilute aqueous solutions. It therefore finds widespread application, in the brewing of beer, for the sterilization of copper and brass vessels.

Fluorosilicic acid is a *strong acid*, much stronger than hydrofluoric acid, and about as strong as sulfuric acid. Its apparent degree of dissociation at 25° (as measured by the conductivity coefficients) is 53% in 1-N solution, 76% in 0.1-N solution. It neutralizes metallic oxides or hydroxides, forming *fluorosilicates*. $2M^IOH + H_2SiF_6 = M^I_2SiF_6 + 2H_2O$.

The fluorosilicates can be obtained by decomposing the carbonates with fluorosilicic acid, or by mixing solutions of the components. In the latter case, fluorosilicates of different types are frequently obtained, corresponding to no known free acids.

Thus, in addition to $(NH_4)_2SiF_6$, the compound $(NH_4)_3SiF_7$ has been obtained, but it has been shown by X-ray analysis (Hoard, 1942) that this crystalline salt is built up from NH_4^+, $[SiF_6]^{2-}$ and F^- ions as structural units. It is thus a double salt of $(NH_4)_2[SiF_6]$ and NH_4F, and not a fluorosilicate of different type. The crystal structure of $(NH_4)_2[SiF_6]$, discussed on p. 398, is found also in the hexafluorosilicates of K, Rb, Cs and Tl, as well as in $Cs_2[GeF_6]$, in the hexachlorostannates of NH_4, K, Rb, Cs and Tl, and in the hexachloroplumbates of NH_4, Rb and Cs.

Organic substituted ammonium fluorosilicates such as $[CH_3 \cdot NH_3]_2[SiF_6]$ (colorless crystals, sol. in water, m.p. 232°) and $[C_6H_5 \cdot NH_3]_2[SiF_6]$ (blue grey, water sol. crystals, m.p. 230°) find some application as pesticides.

The alkali metal fluorosilicates, other than those of lithium and ammonium, are relatively sparingly soluble in water. Barium fluorosilicate is still more insoluble (0.0268 g in 100 g of water at 17°). The other fluorosilicates are mostly soluble. Many of the fluorosilicates crystallize in characteristic habit, and this is occasionally utilized in analysis (inorganic microanalysis).

1. *Potassium fluorosilicate*, $K_2[SiF_6]$, forms as a finely divided opalescent precipitate when a solution of fluorosilicic acid is neutralized with potassium hydroxide. The solubility is 0.120 g in 100 g of water at 17.5°, but increases considerably with rise of temperature. The salt is practically insoluble in alcohol.

2. *Sodium fluorosilicate*, $Na_2[SiF_6]$ is rather more soluble than the potassium salt (0.65 g in 100 g of water at 17.5°). It is frequently used in the enamel industry as an opacifier, in place of cryolite.

In many cases, sodium fluorosilicate can advantageously be used in chemical reactions in place of SiF_4. It is produced in considerable quantities, as a by product of the superphosphate industry, since the phosphate rock (phosphorite) on which the industry is based generally contains apatite.

3. *Magnesium fluorosilicate*, $Mg[SiF_6] \cdot 6H_2O$, and *zinc fluorosilicate*, $Zn[SiF_6] \cdot 6H_2O$, both quite soluble in water, find a use in the waterproofing of cement. Their operation depends on the fact that they react with the calcium oxide present in the cement, forming calcium fluoride and silicic acid, which are deposited in very finely divided form, and block up the pores of the cement.

(*iii*) *Disilicon hexafluoride*, Si_2F_6 (liter weight 7.778 g) was prepared by Schumb (1931), by warming Si_2Cl_6 with ZnF_2. Mixed halides can be prepared by means of its reaction with Cl_2 and Br_2—e.g., SiF_3Cl, SiF_2Cl_2, SiF_3Br, etc. Hydrolysis of Si_2F_6 takes place with evolution of hydrogen, since the primary reaction products, formed by the process

$$Si_2F_6 + 4H_2O = H_2Si_2O_4 + 6HF$$

react further to some extent, according to

$$H_2Si_2O_4 + 6HF = H_2SiF_6 + H_2SiO_3 + H_2O + H_2$$

(*iv*) *Silicon tetrachloride*, $SiCl_4$, can be obtained by heating silicon, or a mixtur, of silicon dioxide and carbon, in a stream of chlorine. It forms a colorless, mobile pungent smelling liquid which fumes strongly in air; density 1.49. It is hydrolyzed by water:

$$SiCl_4 + 2H_2O = SiO_2 + 4HCl$$

The formation of fumes in the air is due to this reaction; the mist obtained is still denser if ammonia is allowed to evaporate at the same time (formation of ammonium chloride fog). Use is made of this in the generation of smoke clouds—e.g., by warships.

Silicon tetrachloride also reacts with metallic oxides, exchanging oxygen for chlorine—e.g.,

$$3SiCl_4 + 2Al_2O_3 = 3SiO_2 + 4AlCl_3$$

Under some circumstances the reaction with aluminum oxide proceeds so vigorously that the silicon tetrachloride is heated to its boiling point. With most metallic oxides, however, reaction only takes place upon prolonged heating in a sealed tube. With P_2O_5 it forms $POCl_3$, and SO_3 forms $S_2O_5Cl_2$. Silicon tetrachloride reacts with alcohols, forming silicic acid esters. In other respects it is quite stable. It reacts with the alkali metals only at a red heat (forming 'amorphous' silicon), it is without any reaction upon most sulfides, and with hydrogen sulfide it forms $SiCl_3(SH)$. Exchange of chlorine for other halogen atoms occurs when it is heated with HBr and HI. Mixed halogen compounds of silicon, such as $SiCl_3Br$, $SiCl_2Br_2$, $SiClBr_3$, $SiCl_3I$, etc. can be obtained in this way. The compounds $SiFCl_3$, SiF_2Cl_2, SiF_3Cl and SiF_4 can be formed by treating $SiCl_4$ with SbF_3 (in the presence, of $SbCl_5$ as catalyst) (Booth, 1935). It unites with ammonia, to form $SiCl_4 \cdot 6NH_3$, and also, at low temperatures, with PH_3. When $SiCl_4$ vapor, mixed with air, is passed through a strongly heated tube, oxychlorides—e.g., Si_2OCl_6, $Si_4O_4Cl_8$ and others—are formed. These are colorless, volatile compounds, of low stability.

Schumb (1950) obtained oxychlorides of quite high molecular weight (e.g., $Si_6O_5Cl_{14}$, $Si_7O_6Cl_{16}$) by adding water to an ethereal solution of $SiCl_4$ at —78°.

Trost (1952) found that $SiCl_4$ formed addition compounds with organic amines, in the ratio 1 : 2 and 1 : 4. Compounds of the latter type are colorless, and more stable than the 1 : 2 addition compounds. $SiCl_4$ can also form addition compounds with alcohol, ketones and other organic compounds containing oxygen.

The vapor density of silicon tetrachloride (at 239°) is 9.7 (air = 1); this value leads to a molecular weight corresponding to the formula $SiCl_4$.

(v) *Silicon monochloride*, $[SiCl]_x$, was obtained by Schwarz (1939), by the thermal decomposition of $Si_{10}Cl_{22}$, as yellow solid, (leaflets, stable towards dry oxygen at ordinary temperature, but igniting in air at 98°). It is insoluble in inert solvents, but dissolves in caustic potash with vigorous evolution of hydrogen:

$$2SiCl + 10OH^- = 2Cl^- + 2SiO_4^{4-} + 2H_2O + 3H_2$$

Hydrolysis takes place in acid solutions (at low temperatures), forming a *silicon hydroxy-compound*:

$$4SiCl + 6HOH = Si_4(OH)_6 + 4HCl + H_2$$

(vi) *Silicon monoiodide*, $[SiI]_x$, discovered by Schwarz in 1942, undergoes hydrolysis in rather different manner, forming an exceedingly easily oxidizable compound $[Si_2O_3H_4]_x$ (given the name 'hyposilicic acid' by Schwarz), which probably contains —Si—Si— chains of indefinite length. When it is heated in vacuum, this ivory colored compound is degraded in stages, by way of lemon yellow $[Si_2O_3H_3]_x$ and red-brown $[Si_2O_3H_2]_x$ to a mixture of SiO_2 and Si.

(vii) *Silicon tetrabromide*, $SiBr_4$, obtained by methods similar to those for the chloride, forms a colorless liquid, fuming strongly in air (density 2.789, b.p. 152.8°, m.p. 5.2°). Unlike the chloride, it reacts very violently when warmed with metallic potassium. It reacts with ammonia, combining with either 4 or 6 molecules, according to the conditions. Addition compounds with other substances are also known.

(viii) *Silicon tetraiodide*, SiI_4, formed by passing iodine vapor, in a stream of carbon dioxide, over red hot silicon, forms colorless, regular octahedra, isomorphous with CI_4 and extremely soluble in carbon disulfide. The value 19.12 (air = 1) has been found for the vapor density, corresponding to the formula SiI_4. The heated vapor of silicon tetraiodide inflames in air, and burns with a red flame. In its reaction with alcohol it differs from the other halides: it forms ethyl iodide and hydrogen iodide, together, with silicic acid.

(ix) *Silicon hexa-iodide, -bromide, -chloride*. If the tetraiodide is heated to 290–300° with finely divided metallic silver, *silicon hexaiodide*, Si_2I_6, is obtained:

$$2SiI_4 + 2Ag = 2AgI + I_3Si—SiI_3$$

The compound can conveniently be recrystallized from hot carbon disulfide in which, unlike the tetraiodide, it is only moderately soluble in the cold. It forms colorless, doubly refracting, six-sided prisms, which sublime in a vacuum at 250° with partial decomposition. If bromine is allowed to react with the carbon disulfide solution, it displaces iodine and forms the *hexabromide*:

$$Si_2I_6 + 3Br_2 = Si_2Br_6 + 3I_2$$

The *hexachloride*, Si_2Cl_6, is obtained in the same way, by treating the hexaiodide with chlorine. It is a colorless liquid of density 1.58 at 0°. This compound is also obtained when $SiCl_4$ vapor is passed over molten silicon, and the reaction products are suddenly chilled ('hot and cold tube' method). The silicon hexahalides are actually unstable at ordinary temperature; the only reason that they do not decompose is that the velocity of the reaction is too small. Decomposition occurs on moderate heating. Thus Si_2Cl_6 decomposes above 350° and below 1000°, into $SiCl_4$ and Si. The silicon hexahalides are extraordinarily sensitive towards water. Decomposition by cold water forms the so-called *silico-oxalic acid* $[H_2Si_2O_4]_x$, an extremely easily decomposed white mass, which has a strong reducing action, and dissolves in alkalis with evolution of hydrogen:

$$H_2Si_2O_4 + 8OH^- = 2SiO_4^{4-} + 4H_2O + H_2$$

The so-called *silico-mesoxalic acid*, $[H_4Si_3O_6]x$, which is also exceedingly unstable, is obtained in a similar manner by the hydrolysis of Si_3Cl_8.

(x) *Trichlorosilane (Silicochloroform)*. If dry hydrogen chloride gas is passed at temperatures below a red heat (best at 380°) over silicon, the compound trichlorosilane, $SiHCl_3$, is obtained. It is also known as *silicochloroform*, being the analogue of chloroform in composition. It can also be obtained by chlorination of monosilane. Silicochloroform is a colorles, very mobile liquid, a non-conductor of electricity, fuming strongly in moist air (b.p. 31.5°, m.p. —128.2°). It is soluble without decomposition in carbon disulfide, carbon tetrachloride, chloroform and benzene. It reacts with absolute alcohol, producing $SiH(OC_2H_5)_3$. It reacts with water in the cold to form *dioxo-disiloxane* ('*silico-formic anhydride*'), $[Si_2O_3H_2]x$, a voluminous white substance which only inflames in air when it is fairly strongly heated, but acts otherwise as a strong reducing agent. Thus it decolorizes permanganate, deposits silver from silver nitrate and sulfur from sulfurous acid. Dioxo-disiloxane, and hence trichlorosilane also, are decomposed by caustic alkalis, evolving hydrogen:

$$Si_2O_3H_2 + 8OH^- = 2SiO_4{}^{4-} + 3H_2O + 2H_2$$

$$SiHCl_3 + 7OH^- = SiO_4{}^{4-} + 3Cl^- + 3H_2O + H_2.$$

The compounds $SiHBr_3$ and $SiHI_3$ are largely similar to $SiHCl_3$, but are rather more readily decomposed. $SiHF_3$ (b.p. —97.5°, m.p. —131°), first prepared by Ruff by the action of SbF_3 on $SiHCl_3$, undergoes gradual decomposition even at very low temperatures. The compounds $SiHCl_2F$ and $SiHClF_2$, prepared by Booth (1934) are more stable.

(g) Silicic Acid Esters

Silicic acid esters, $Si(OR)_4$ (R = organic radical), are formed by the reaction of the silicon halides with alcohols or phenols. Many of these have become important industrially.

Among the phenolic esters of silicic acid, special mention may be made of the derivatives of the dioxybenzenes, investigated by Schwarz (1951). These are resinous substances, which differ in structure according to the orientation of the OH groups about the benzene ring. The pyrocatechol ester is cyclic; the resorcinol ester has a chain structure. If $SiCl_4$ and hydroquinone are brought together in appropriate molar proportions (e.g., 1 : 2 or 1 : 2.5), the hydroquinone ester has a ribbon structure, since the chains are then bound together

in pairs by —O—⟨benzene ring⟩—O— bridges.

(h) Silicon Thiocyanate

$Si(SCN)_4$, (white needles, melting at 143° and boiling at 314°), is obtained by double decomposition between $SiCl_4$ and lead thiocyanate (preferably in benzene solution).

$$SiCl_4 + 2Pb(SCN)_2 = Si(SCN)_4 + 2PbCl_2$$

The compound is stable in dry air, but is hydrolyzed by water, acids and alkalis, with the deposition of silica gel or formation of silicates.

(i) Sulfides of Silicon

Silicon disulfide, SiS_2, is obtained by melting 'amorphous' silicon with a threefold excess of sulfur. When purified by sublimation under reduced pressure it forms colorless, silky, feathery crystals. It is stable in dry air at ordinary temperature. It is hydrolyzed by water:

$$SiS_2 + 2H_2O = SiO_2 + 2H_2S$$

The crystal structure of silicon disulfide was determined by Zintl (1935). He obtained the compound by Tiede's method, i.e., by heating Al_2S_3 with SiO_2 to 1100°. It was found that the SiS_2 crystal is built up from chains:

of indefinite length, in which each Si-atom is surrounded tetrahedrally by 4 S-atoms, and each S-atom joined to 2 Si-atoms. Within each chain the interatomic distance $Si \leftrightarrow S$ is 2.14 Å, the distance $Si \leftrightarrow Si$ 2.77 Å. Adjacent S-atoms of each tetrahedron are 3.62 Å apart. The distance of a Si-atom from the nearest S-atom of a neighboring chain, however, is 4.18 Å. Such chains, complete in themselves, are characteristic of substances with a *fiber structure*, the chains running along the fiber direction. The SiS_4 chains of silicon disulfide differ from the SiO_4 chains of the fibrous silicates in that their constituent tetrahedra share *edges*, whereas SiO_4 tetrahedra are generally able to share only *corners*. (cf. p. 500). Unlike the SiO_4 chains, the charges on the atoms are neutralized within the SiS_4 chains, so that each can be regarded as a molecule of infinite length. The SiS_2 structure thus occupies a place intermediate between a molecular lattice in the ordinary sense, such as is present in solid CO_2, and the coordination structures formed by SiO_2 and the silicates.

According to Hempel and von Haasy, SiS_2 can combine with alkali sulfides, forming *thiosilicates*, $M^1{}_2SiS_3$, which have been but little studied. Colson has found that at very high temperatures a monosulfide SiS is formed. Unlike the disulfide, this evolves hydrogen when it is dissolved in alkalis.

(j) The Oxygen Compounds of Silicon

The normal oxide of silicon is *silicon dioxide*, SiO_2. This, combines with basic oxides, to form *silicates*, with a wide variety of compositions. Of the *silicic acids*, formed by the chemical combination of SiO_2 with H_2O, only a few well defined examples are known, and these can be obtained only under certain experimental conditions (cf. pp. 494 *et seq.*). Silicon dioxide displays a marked tendency to assume the colloidal state, and to form *gels* with water. These are referred to as *silica gels*, or *silicon dioxide hydrates*, irrespective of whether or not part of the water is bound chemically to the silicon dioxide; it is, indeed, often impossible to determine this with certainty.

It was shown by Bonhoeffer (1927), by observations of the absorption spectrum, that gaseous *silicon monoxide*, SiO, is formed when silicon dioxide is strongly heated with charcoal. It is also formed when SiO_2 is heated with Si. In consequence of this, volatilization takes place fairly rapidly when a mixture of these two substances is heated to 1300–1400° *in vacuo* (Biltz, 1938). The monoxide disproportionates again into SiO_2 + Si when it is condensed from the gaseous phase. Zintl [*Z. anorg. Chem.*, 245, (1940) 1] has suggested that the formation of volatile SiO could be applied to the elimination of silica from silicates, and the preparation of metals: e.g.,

$$Al_2Si_2O_7 + 2Si = 4SiO + Al_2O_3$$

$$\lfloor Nb_2O_5 + 5Si = 5SiO + 2Nb$$

According to Schäfer and Hörnle (1950), the vapor pressure of SiO over mixtures of silicon and SiO_2 is represented by

$$\log p = \frac{-77000}{4.57\,T} + \frac{37.1}{4.57} + 2.88T$$

The heat of reaction between silicon and SiO_2 is thus —77 kcal per mol., and the heat of formation of SiO from its elements is +26 kcal per mol. at 1125°, and 23.3 kcal per mol. at 25°.

(*i*) *Silicon dioxide* is found in nature in three different crystalline forms—as *quartz*, as *tridymite*, and as *cristobalite*. It is also found hydrated (in the form of silica gel) as *opal*, and in earthy form as *kieselguhr* or *diatomaceous earth*.

1. *Quartz* is the crystalline form stable below 870°. In the range of temperature between 870° and 575° it belongs to the trapezohedral-hemihedral class of the hexagonal system; below 575° to the trapezohedral tetratohedral class of the trigonal system. Quartz is frequently found in very well formed crystals, which often attain a considerable size. Quartz is doubly refracting, and occurs in two forms, one of which rotates the plane of polarization of polarized light anti-clockwise, the other rotating it clockwise. They are accordingly distinguished as *left-handed* quartz and *right-handed* quartz, respectively. These two forms may often be distinguished also by the crystal faces present on them—e.g., in the examples illustrated in Figs. 84 and 85, by the object and mirror image relation between the shaded faces *s*.

The faces marked *s* on Figs. 85a and 85b are those of *trigonal bipyramids*. These enable left- and right-handed quartz to be distinguished only if, as shown in the diagram, they are striated. When present, *trapezohedral* faces enable the two forms to be distinguished most readily. In the absence of

Fig. 85a.
Left-handed
quartz.

Fig. 85b.
Right-handed
quartz.

suitable crystal faces, right-handed and left-handed quartz can be distinguished not only by their optical rotation, but also by means of the *etch figures* produced by treatment with hydrofluoric acid.

Varieties of quartz are, *rock crystal*, water clear; *smoky quartz* (also known as smoky topaz), dark brown, *morion*, black; *citrin*, yellow; *rose quartz*, pink; *chryso-prase*, leek green; *amethyst*, usually violet. The latter, in particular, in the form of clear transparent pieces, is used as a gem stone. Quartz is among the most widely distributed minerals. It is encountered extremely frequently as a component of the mixtures constituting the eruptive rocks (granite, porphyry, liparite) and the crystalline schists (gneiss, mica schist). Sandstone and quartzite, among the sedimentary rocks, also consist entirely of quartz grains, as do many sands (quartz sand).

If quartz is heated above 575°, it undergoes a change into another modification, as was first shown by Le Chatelier in 1889 from measurements of thermal expansion; this form belongs to the trapezohedral-hemihedral class of the hexagonal system. It reverts to the original form again upon cooling, so that the transformation is enantiotropic. The form stable at ordinary temperature is known as α-quartz, or *low quartz*, the other as β-quartz or *high quartz**. The axial ratio only changes by a very small amount as a result of this transformation, so that it is only slightly noticeable in external properties. The crystal form remains practically unaltered, and even the arrangement of the individual atoms changes only slightly, as X-ray examination has shown.

Figures 86 and 87 represent the unit cells of left-handed and right-handed β-quartz. Each

* The designation by the letters α and β is not uniformly applied. The same applies for tridymite and cristobalite.

Si-atom is surrounded by 4 O-atoms at a distance of 1.55 Å, in almost regular tetrahedral arrangement. The atoms are arranged in a spiral about the axis shown. In left-handed quartz the direction of rotation is the reverse of that in right-handed quartz. (The 4 O-

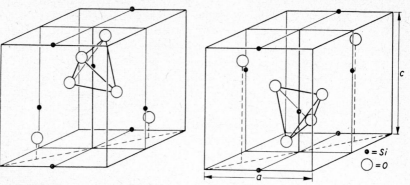

Fig. 86. Unit cell of left-handed β-quartz.   Fig. 87. Unit cell of right-handed β-quartz.

$$a = 5.01 \text{Å}, \, c = 5.47 \text{Å} \text{ (at 600°)}$$

atoms which make up the tetrahedron lying on the axis lie vertically above the points on the basal plane represented by the coordinates $\frac{1}{5}a, \frac{2}{5}a; \frac{4}{5}a, \frac{3}{5}a; \frac{2}{5}a, \frac{1}{5}a; \frac{3}{5}a, \frac{4}{5}a$. The back left-hand corner of the basal plane is here taken as the origin of the coordinate system, and the coordinate direction pointing to the right is cited first. In left-handed quartz the first pair of coordinates refers to the projection of the lower pair of O-atoms of the tetrahedron considered, and in right-handed quartz to the upper pair of O-atoms).

2. *Tridymite* is the form of crystalline silicon dioxide stable between 870° and 1470°, though it occurs also as a metastable form at ordinary temperature. This form of silicon dioxide also occurs frequently in nature, both in meteorites and in rocks, though generally in small quantities. It is of interest in that its presence in rocks may given information about the history of their formation.

In order to transform quartz into tridymite, it is in general not sufficient merely to heat it above 870°. Even at very high temperatures the velocity of transformation is far too small. If, however, a suitable solvent is used, (a so-called mineralizing agent, such as fused sodium tungstate) the behavior is different*, and tridymite can then be obtained in the form of small crystal platelets, having forms characteristic of the hexagonal system. On cooling, the tridymite reverts to quartz, as being more stable below 870°, but again only in the presence of a suitable solvent. Even then the transformation into quartz proceeds only very slowly, and requires, for example, about 3 days at 825° in the presence of fused sodium tungstate. If hexagonal tridymite is cooled down, either by itself or too rapidly for conversion to quartz to occur, it changes below 117°, without any externally recognizable change of form, into a rhombic modification. This form, known as α-tridymite, is optically biaxial, as distinct from the optically uniaxial modification stable at higher temperatures. It is stable below 117° with respect to β-tridymite, but unstable or metastable with respect to quartz; α- and β-tridymite are thus enantiotropic modifications. The rhombic (α-)tridymite crystals found in nature have been formed from hexagonal (β-)tridymite, as their structure clearly reveals. They are reconverted to β-tridymite by heating them above 117°.

Hexagonal β-tridymite seems to have yet another transition point at about 163°. The two hexagonal forms may be distinguished as $β_1$- and $β_2$-tridymite respectively; $β_2$-tridymite, stable above 163°, differs but little from $β_1$-tridymite, however.

* Although mineralizing agents are not present, the transformation of quartz into cristobalite and tridymite has been shown to take place in Dinas bricks exposed for a long time to very high temperatures, as the lining of a Siemens-Martin furnace.

The crystal structure of β-tridymite (Fig. 88) is much simpler than that of quartz. The tridymite lattice is obtained if all the lattice points in a wurtzite lattice (Fig. 60, p. 258) with $a = 5.03$, $c = 8.22$ Å, are first filled with Si-atoms, and then an O-atom is fitted in between every pair of Si atoms. A structure is thus obtained in which every Si-atom is surrounded tetrahedrally by 4 O-atoms, while each O-atom is equidistant from 2 Si-atoms (Si—O = 1.54 Å). Each Si-atom is thus also surrounded tetrahedrally by 4 other Si-atoms at a distance of 3.08 Å, sharing an oxygen atom with each. A similar structure is possessed by (ordinary) ice, as has already been mentioned. The crystal structure of α-tridymite is more complicated, and has not yet been fully worked out.

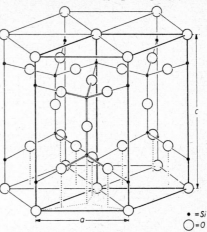

Fig. 88. Crystal lattice of β-tridymite. $a = 5.03$ Å, $c = 8.22$ Å. The unit cell is emphasized by heavier lines.

3. *Cristobalite* (named after the mountain of San Cristobal in Mexico) is the form of silicon dioxide stable above 1470°, and up to the melting point. In the absence of mineralizers to accelerate the transformation, cristobalite is metastable below the transition point. It is frequently found in nature in small crystals, enclosed in lava. In general, it resembles tridymite in its occurrence.

β-cristobalite belongs to the regular system. Its crystal lattice differs from that of tridymite in the same way as that of zinc blende does from wurtzite. The silicon atoms make up a diamond structure, with $a = 7.12$ Å; each Si-atom is surrounded by 4 oxygen atoms, in a regular tetrahedron, at a distance of 1.54 Å. The arrangement of oxygen atoms about the silicon is thus substantially the same in quartz, tridymite and cristobalite.

Strictly speaking, two forms must be distinguished in the case of cristobalite also. That stable at ordinary temperature is referred to as the α-form. The crystal structure described above is that of β-cristobalite. According to Nieuwenkamp (1935), α-cristobalite has a similar structure, except that the SiO$_4$ tetrahedra in it are somewhat twisted relatively, so that the Si—O—Si bond directions are not collinear, as in β-cristobalite, but bent. The α–β transition in cristobalite always occurs at a lower temperature during cooling than during heating, and the transition temperature range depends upon the previous treatment of any sample. It lies roughly between 200° and 270°.

Below 1470°, cristobalite is not only less stable than quartz, but also less stable than tridymite. When it is heated to very high temperatures for long periods, even in the absence of mineralizing agents, quartz begins to change gradually, and is thereby converted into cristobalite, even when the transformation occurs below 1470°—i.e., in a range in which tridymite, not cristobalite, is the stable species. This experimental observation at first sight appears paradoxical, but is in accord with *Ostwald's rule of successive transformations*. This states that a substance which exists in several modifications is formed in stages, the less stable modifications being formed first, and these then undergoing transformation into the stable form. As the less stable modification, cristobalite is formed first, and can persist since it is converted into tridymite at a measurable speed only if a solvent or mineralizer is present.

Silicon dioxide in the form of cristobalite melts at 1705°. Quartz melts about 150° lower. The cooled melt remains vitreous and amorphous. The vitreous silicon dioxide is also metastable at ordinary temperature. On strong and protracted heating, crystallization ('devitrification') slowly takes place. From the foregoing discussion it may be understood why, in the absence of mineralizers this leads to cristobalite.

The ranges over which these forms of silicon dioxide are actually stable are thus delimited by the following transition temperatures.

$$\alpha\text{-quartz} \xrightleftharpoons{575°} \beta\text{-quartz} \xrightleftharpoons{870°} \beta\text{-tridymite} \xrightleftharpoons{1470°} \beta\text{-cristobalite} \xrightleftharpoons{1705°} \text{melt.}$$

In addition, there are the following transition points in regions where the modifications in question are actually metastable:

$$\alpha\text{-tridymite} \xrightleftharpoons{117°} \beta\text{-tridymite;} \qquad \alpha\text{-cristobalite} \xrightleftharpoons[\text{ca. } 230°]{\text{ca. } 260°} \beta\text{-cristobalite.}$$

Figure 89 sets out the ranges of existance of the different modifications of SiO_2, as established especially by the work of Fenner and Wietzel. In this diagram, the abscissas represent temperatures, and the ordinates the vapor pressures exerted by the various forms, in

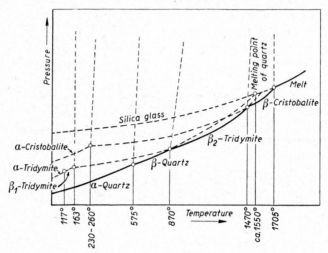

Fig. 89. Phase diagram of silicon dioxide.

accordance with their stability. The actual magnitude of these vapor pressures is, indeed, not known but, as was disccussed previously (p. 51), the less stable of the two modifications of any substance must have the higher vapor pressure. The least stable form at ordinary temperature, and up to about 1550°, is the vitreous solidified melt, silica glass. As has been stated, this devitrefies upon prolonged heating, especially in the presence of impurities, generally forming cristobalite. It can be seen from the diagram how this exemplifies Ostwald's rule since, after vitreous silica, cristobalite is the most unstable form at the usual temperatures of ignition.

4. *Vitreous* amorphous solidified silicon dioxide *(silica glass)* is, as X-ray studies have shown, not built up from SiO_2 molecules, but from Si atoms linked together by O-atoms (or ions—cf. p.499) in such a way as to make up a continuous network. This consists of tetrahedra of O-atoms surrounding the Si-atoms, and sharing corners, just as in the crystalline modifications. In silica glass, however, the SiO_4 tetrahedra are arranged completely at random, and not in a regular manner as in the crystalline forms (Warren, 1933). Silica glass has a remarkably small coefficient of thermal expansion, amounting to $5.85 \cdot 10^{-7}$ at 100°, $6.2 \cdot 10^{-7}$ at 500°, and $5.45 \cdot 10^{-7}$ at 1000°.

It has been found that a binary compound can exist in the vitreous state only if the following conditions are fulfilled:

[i]   The coordination number of electronegative atoms about the electropositive atoms must be 4 or 3.

[ii] The coordination tetrahedra or triangles, formed by the electronegative atoms around the cations, must share only corners with one another.

[iii] Every atomic grouping (i.e., every coordination polyhedron) must share at least three corners with neighboring groups.

[iv] Each electronegative atom must be bound to only two electropositive atoms.

It is only possible for these conditions to be fulfilled in compounds of the compositions A_2B_3, AB_2 and A_2B_5, but not in compounds of the types AB, AB_3, A_2B_7, AB_4, etc. The best known examples of binary compounds which form glasses are B_2O_3, As_2O_3, Sb_2O_3, BeF_2, SiO_2, GeO_2, P_2O_5 and As_2O_5. These fulfil the conditions set out above, but it cannot be inferred that every compound which fulfils these conditions can be obtained in the vitreous state. Smekal considers that this is possible only for substances of *mixed bond type*, in which there are both directed and undirected valence forces operative: this is the state when the bond type is intermediate between ionic and covalent bonding. This is probably the case with all the binary compounds listed above. It is also possible for the atoms to exercise binding forces of different types towards different neighbors—e.g., covalent bonds and metallic forces. This is the case in certain elements which can exist in the vitreous state—e.g., As, Sb, and Se. In *organic* substances which form glasses there is also a mixture of bond types operative—intramolecular covalent bonds, and intermolecular Van der Waals forces (or, in polyhydric alcohols, such as the glassy state of glycerol, hydrogen bonds).

5. *Kieselguhr.* Amorphous silica can also be obtained in the form of a white powder by dehydrating silica gels deposited from solution (see below); the last portions of water are, however, difficult to remove completely. Silicon dioxide occurs in nature in this form, as *kieselguhr* [*28*]. This consists of the remains of the silica skeleton of former 'infusoria' (diatoms), and is therefore known also as *infusorial earth*, or *diatomaceous earth*. It is notable for its excellent capacity for absorbing liquids, and is therefore used, e.g., as a packing material for acid carboys. Dynamite is prepared by saturating kieselguhr with nitroglycerine, of which it can soak up three times its own weight. Kieselguhr is also used for lagging steam pipes, because of its thermal insulating properties. It is also employed for sound-insulating flooring, and for many other purposes.

6. *Opal.* Compact *xerogels* (cf. p. 494) of silicon dioxide are found in nature in the form of *opal*; beautiful specimens of this are used as gems.

Silicon dioxide and its gels, as precipitated in the ordinary manner, furnish no X-ray diffractions indicative of a crystalline structure. The same is true of opals formed by the deposition of SiO_2 from water at ordinary temperature. Opals formed, however, from hot magmatic waters do give the X-ray diffractions characteristic of quartz or of cristobalite, according to their source.

7. *Chalcedony* is aged opal. It therefore has a lower water content than the latter (and is often anhydrous), and displays a crystalline structure recognizable by ordinary means. Thus under microscopic examination, especially by polarized light, it shows a fibrous structure. Varieties of chalcedony are *agate*, *onyx*, *carneol*, *heliotrope* and *jasper*, used as gem stones, as well as *flint*, which was used in the Stone Age for tools and weapons, and later as a means of making fire. It still finds uses— e.g., in ball mills and in ceramics. The colors of these minerals are due to traces of impurities contained in them. Thus flint is colored black through impregnation with carbon.

Amorphous silicon dioxide, in both the precipitated powder and the vitreous forms, is completely converted into quartz if it is heated for several days, in an autoclave at 400°–500° in sodium carbonate solution. It is, indeed, possible, (by this means) to grow synthetic quartz crystals of a size large enough to be used for the regulation of radiofrequency equipment. Tridymite and cristobalite are also converted to quartz under these conditions.

8. *Properties and Uses of Silicon dioxide.* Silicon dioxide is extremely resistant chemically. Hydrofluoric acid is the only acid to dissolve it, forming silicon tetrafluoride or fluorosilicic acid. It is practically insoluble in water. As the anhydride of silicic acid it is, however, readily converted into silicates by fusion with alkalis.

$$SiO_2 + 2NaOH = Na_2SiO_3 + H_2O$$

It reacts in a similar manner on fusion with alkali carbonates, from which it displaces carbon dioxide:

$$SiO_2 + Na_2CO_3 = Na_2SiO_3 + CO_2$$

Amorphous silica can be readily brought into solution by boiling it with solutions of alkali hydroxide or carbonate, soluble alkali silicates being thereby formed.

Silicon dioxide also goes into solution in very small amount if it is treated at elevated temperatures and high pressures with methanol containing some sodium methoxide:

$$SiO_2 + 4CH_3OH = Si(OCH_3)_4 + 2H_2O$$

From the extremely slow rate at which this reaction takes place, it can be inferred that a very high energy of activation is required to break the Si—O—Si bonds in solid SiO_2. The Si—O—Si bonds in polysilicic esters can be ruptured very much more easily. Thus $(RO)_3Si$—O—$Si(OR)_3$ (R = *iso*propyl) forms $Si(OR)_4$ in good yield when it is treated with *iso*propyl alcohol containing sodium isopropoxide (Kautsky and Daubach, 1950–53). In this connection it is worth noting that Toptschijewa and Ballod (1951) found that methyl orthosilicic ester was formed when methanol was adsorbed on silica gel.

In the form of quartz sand, silicon dioxide has extensive uses. It is used in building, for the preparation of lime mortar, and is also mixed with cement. The purest obtainable quartz sand is needed for the glass and porcelain industries. Chemical apparatus is fabricated from quartz, sintered at high temperatures, and thereby transformed into vitreous silica; the small coefficient of thermal expansion of silica glass enables such apparatus to withstand abrupt changes of temperature, and it can also be heated to high temperatures without melting. This is true in still greater degree of apparatus made from transparent vitreous silica—i.e., from completely melted quartz [*38*]. In working with such apparatus, the sensitiveness of silica glass towards alkali must be borne in mind. One of the main difficulties in the manufacture of silica glass originally lay in avoiding crystallization at the very high temperatures, close to the melting point of quartz, at which it is necessary to carry out the blowing and working of the glass. Devitrification may be initiated by traces of alkali; even minute amounts will suffice, such as are transferred to the material by touching it with sweaty or greasy hands before melting and blowing. A gradual recrystallization of silica glass also takes place in the course of its use. The higher the temperature to which the silica glass is exposed, the more rapidly does this devitrification occur.

Clear rock crystal is used in many optical instruments, and as a gem stone. Many of the varieties of quartz are also valued as gems, as has been mentioned.

(ii) *Silicic acid and the Silicates*. *Silicates* are obtained by melting silicon dioxide with basic oxides. The salts of acids which are volatilized on heating can be used in place of basic oxides, silica expelling the more volatile acids by reason of its own great thermal stability. Silica unites in very varied proportions with the oxides, to form well defined compounds. The simples silicates are the *meta-* and *orthosili-cates*, which result from the union of their components in accordance with the reactions:

$$SiO_2 + M^I_2O = M^I_2SiO_3 \quad \text{and} \quad SiO_2 + 2M^I_2O = M^I_4SiO_4$$
$$\text{metasilicates} \qquad\qquad\qquad \text{orthosilicates}$$

(constitution of the silicates—cf. pp. 498 *et seq.*).

Silicates which contain 2, 3, etc. SiO_2 for each M^I_2O are known as *di- tri-* etc. *silicates*; in general, silicates containing more than one SiO_2 radical for each M^I_2O are grouped together as *polysilicates*.

Whereas the silicates are formed according to very varied ratios, *free silicic acid* is known only in very few well-defined forms. These include, *metasilicic acid*, H_2SiO_3, and *disilicic acid* $H_2Si_2O_5$.

Silicic acid is an *extremely weak acid*; it is electrolytically dissociated only to a very slight extent. As a rule, i.e. when it is present in the condensed state, its solubility is also extremely small. It is therefore displaced from the soluble silicates by even the weakest acids.

Of the salts, only the *alkali silicates* are soluble in water. Many of the silicates which are practically insoluble in water are soluble with decomposition (deposition of silica gel) in strong acids. Many silicates, however, are not decomposed by strong acids, but only by hydrofluoric acid, which attack them because of the strong tendency of silicon to form silicon fluoride. All silicates undergo double decomposition with fused alkali carbonate, alkali silicate being formed and, in some cases, carbon dioxide being liberated. E.g.,

$$MgSiO_3 + Na_2CO_3 = MgCO_3 + Na_2SiO_3$$

$$KAlSi_3O_8 + 3Na_2CO_3 = 3Na_2SiO_3 + KAlO_2 + 3CO_2$$

Aqueous solutions of the alkali silicates are strongly alkaline in reaction, the silicates being extensively hydrolyzed in them.

According to Schwarz (1927), hydrolysis of sodium silicate by the reaction:

$$2Na_2SiO_3 + H_2O = Na_2Si_2O_5 + 2NaOH$$

occurs to the extent of 14% in 1-N solution, 28% in 0.1-N and 32% in 0.01-N solution; that of sodium disilicate in 1-N and 0.1-N solution being 2.4% and 6%, respectively. Ions other than SiO_3^{2-} (or SiO_4^{4-}) and $Si_2O_5^{2-}$ have not as yet been identified in the solutions.

If silicate solutions are acidified, silicic acid is at once liberated; it is usually not immediately precipitated, however, but remains at first in solution. Flocculation commences only after a longer time. This is due in part to the fact that silicic acid can exist in a monomolecular, water-soluble form; this loses water and condenses rapidly or slowly, according to the experimental conditions, to higher molecular, and ultimately practically insoluble, polymeric aggregates. However, even after the silicic acid has been converted completely into the insoluble form, precipitation need not necessarily occur, since it may remain dispersed as a *colloid*. Silicic acid has *a great tendency to form colloidal dispersions* (*silicic acid sols*). It is stable in the colloidal state both in acid and in neutral or weakly basic solutions. Addition of electrolytes does not usually bring about coagulation, but provided the solution is not too dilute, the whole solution sets to a jelly. Slimy precipitates are deposited from dilute solutions. Baryta water or concentrated aluminum sulfate solution in particular bring about immediate precipitation. Since, however, most other electrolytes slowly cause precipitation or gelation, they must be removed in order to obtain colloidal dispersions which are stable for a long period. This is done by *dialysis*, i.e., by placing the solution in a vessel ('dialyzer') separated from the pure solvent by a membrane of parchment paper, pigs' bladder, collodion or similar material. Dissolved substances, in a state of molecular dispersion ('crystalloids') diffuse through such a membrane, while colloidally dispersed substances are retained, or at least penetrate the membrane much more slowly (cf. Vol. II, Chap. 16).

Silica sols which are not too dilute solidify to jellies on standing; others do so on evaporation, or on addition of electrolytes. Silica gels rich in water are transparent, soft and fairly elastic.

Freshly prepared silica gels may contain 330 mols. of H_2O per mol. of SiO_2. Upon prolonged standing (in air saturated with water vapor), they begin to shrink gradually, water being expressed. This phenomenon occurs also with other jellies, and is known as *syneresis*. A considerable amount of the water may be expelled by rubbing and pressing. Larger quantities of water are given up by exposure to dry air, and silica jellies and silica gels in general* ultimately dry out in a desiccator to completely solid masses, which may be rubbed to powder, but which still contain considerable amounts of water—e.g., $6H_2O$ per SiO_2. A very considerable shrinkage of the gel accompanies the drying. At a certain stage, however, the contraction suddenly ceases; at the same time the gel, hitherto transparent, becomes opaque, and chalk-white. Van Bemmelen, who was the first to investigate these processes, called the point at which opacity set in the *end point*. The end point may, for example, occur at a water content of $2H_2O$ per SiO_2. Since from that point onwards the water vapor pressure of the gel often remains essentially constant until practically all the water has been given up (when a second 'end point' occurs, the gel suddenly becoming transparent once again), it was natural to assume that the composition of the gel at the end points represented definite chemical compounds. It is characteristic of a compound which may be decomposed with the evolution of a gas that, at any temperature, it exerts a definite vapor pressure, which remains constant so long as any portion of the compound remains undecomposed (cf. p. 74). Van Bemmelen found, however, that the composition corresponding to the end point was variable within fairly wide limits. According to him the phenomena can be explained at least as well, or in some respects better on the view that the water is bound mechanically, and not chemically, and that the discontinuity in the dehydration curve is due to the fact that the water is very firmly bound in the smallest cavities of the gel by capillary forces ('imbibed water'). In fact, Zsigmondy showed that after the water filling the capillary spaces had been given up, any other liquid, such as benzene, may be taken up by the gel, to an extent which corresponds in *volume* to the water removed (and not to any molecular proportion), and that on repeated drying the phenomena characteristic of the end point occurred with other liquids also. Since it can not be supposed that liquids such as

---

* The expression 'gel' is used to describe both the voluminous precipitates obtained by the flocculation of colloids and the jellies formed by the uniform solidification of sols.

benzene are chemically bound by SiO_2, and since moreover, there is no indication of any stoichiometric proportions, but only of imbibition by the capillary spaces, there is, on the whole, no cogent ground for assuming that the water in such a gel is bound chemically.

Under certain definite experimental conditions, it is however, possible to obtain preparations with a water content which obviously conforms to simple stoichiometric proportions. This is most easily achieved by the action of fairly concentrated acids (e.g., 80% sulfuric acid) on *crystalline silicates*. Schwarz showed that silica hydrates of definite stoichiometric composition could be obtained in this way, after drying by a suitable procedure (Willstätter's acetone method—cf. under aluminum hydroxide). The products are the acids H_2SiO_3 (metasilicic acid) and $H_2Si_2O_5$ (disilicic acid). The former is obtained from a metasilicate starting material, the latter from a disilicate. Melts, containing alkali and silica in other proportions, yield mixtures of these two acids, or in some cases, preparations containing free silicon dioxide. It cannot be deduced, merely from the composition alone, whether a preparation with a composition such as e.g., $2SiO_2 \cdot H_2O$ represents a definite acid $H_2Si_2O_5$ or a mixture of H_2SiO_3 with SiO_2. Biltz, however, was able to characterize the two acids as definite compounds on the basis of their differing behavior towards ammonia. Both form addition compounds when treated with liquid ammonia. Whereas, however, metasilicic acid gives up ammonia in stages on warming and evacuation, successively forming several ammonia addition compounds of simple stoichiometric composition, the ammonia combined by disilicic acid is given off continuously. X-ray examination shows disilicic acid to be crystalline, whereas metasilicic acid is amorphous. Metasilicic acid is also much less stable than disilicic acid, and is obtained only at temperatures of about 0°. It decomposes at ordinary temperature:

$$2H_2SiO_3 = H_2Si_2O_5 + H_2O.$$

At 150° disilicic acid also decomposes, into SiO_2 and H_2O.

In the decomposition of silicates by concentrated acids, silicic acids are obtained in the form of powders, relatively poor in water. These have, indeed, the nature of gels, although much less clearly marked than in the case of the silica gels prepared from aqueous solutions. It is not at all uncommon to find that a chemical reaction leads to products with differing properties, depending on whether it is carried out in solution or with crystallized solids. Reactions of this kind, in which the properties of the solid products of reaction are determined essentially by the fact that reaction takes place on a solid substance—i.e., is spatially localized—are known, following the suggestion of Kohlschütter, as *topochemical reactions* (Greek τόπος, a place).

By using the method of acetone drying, Schwarz was able to show that metasilicic acid, H_2SiO_3, is present in the gels produced at low temperatures (0°) by the hydrolysis of $SiCl_4$ or SiF_4, and $H_2Si_2O_5$ in those obtained by the decomposition of silicate solutions.

Another method of detecting definite silicic acids was introduced by Thiessen (1929). He started from the consideration that, with gels of sufficiently wide pore diameter, the interfering effects due to imbibed water would be eliminated, and that according to the laws of crystal growth, it should be possible to achieve the preparation of gels of the largest possible pore size by producing the compound building up their skeleton at the slowest possible rate. He succeeded, in fact, by the extremely slow hydrolysis of *orthosilicic acid ethyl ester*, $Si(OC_2H_5)_4$, in obtaining preparations which lost water in a distinctly discontinuous manner during isothermal dehydration (at 13°). The first break in the pressure-concentration curve came at the composition corresponding to *orthosilicic acid*, H_4SiO_4. When more water was abstracted, further discontinuities occurred at the compositions $2SiO_2 \cdot 3H_2O$, $SiO_2 \cdot H_2O$ and $2SiO_2 \cdot H_2O$. The existence of these compounds can, however, not be regarded as proved with certainty. Weiser, in a repetition of Thiessen's experiments, could find only one break in the pressure-concentration curve, corresponding to metasilicic acid, H_2SiO_3.

Mylius and Groschuff (1906) showed that when alkali silicate solutions are decomposed by hydrochloric acid, silicic acid is liberated initially in the crystalloidal state. The freshly formed silicic acid rapidly diffuses through a parchment membrane. It also differs from aged material, which is no longer crystalloidal, but is colloidally dispersed, in that it does not bring about coagulation of albumen. The aging is to be attributed to growth of particle size which takes place, according to Willstätter, through continual condensation of the primary H_4SiO_4, with elimination of water. Willstätter found the rate of aggregation of the particles to be markedly dependent upon the hydrogen ion concentration; it took place most slowly at pH = 3.2. By passing $SiCl_4$ vapor into water, and removing the hydrolytically formed hydrochloric acid by reaction with silver oxide in order to maintain the optimum H^+-ion concentration, it was possible to obtain solutions in which at least 75–80% of the H_4SiO_4 (or SiO_2) was present in the monomolecular form, as evidenced by freezing point determinations.

As was shown by Van Nieuwenburg (1930), silicic acid is distinctly volatile in steam above the critical temperature. This observation is important for the explanation of certain geological phenomena.

The separation of solid silicon dioxide hydrate, from solutions of crystalloidal (molecular disperse) silicic acid, as a result of particle growth, can be prevented in another way. At a certain stage, when condensation has not proceeded too far, the hydroxyl groups which bring about the condensation are esterified by a method described by Kirk (1946). Polysilicic esters are obtained by this means; they have acquired some industrial importance. The solid esters are obtained when the solvent is evaporated off, and can be brought into solution again. From measurements of the viscosity of their solutions (Iler and Pinkney, *Ind. Eng. Chem.*, 39 (1947) 1379), it has been concluded that their molecules or micelles are almost spherical.

The glycol ester of orthosilicic acid is interesting, in that it has an acidic reaction in alcohol solution, and forms addition compounds with metal alkoxides. In these, it would appear that silicon has a coordination number of five.

$$
\begin{array}{c}
CH_2-O \diagdown \quad \diagup O-CH_2 \\
| \qquad Si \qquad | \qquad + \; M^I OR \quad = \\
CH_2-O \diagup \quad \diagdown O-CH_2 \\
\end{array}
\quad
\begin{array}{c}
\left[\begin{array}{c}
CH_2-O \diagdown \quad \diagup O-CH_2 \\
| \qquad Si \qquad | \\
CH_2-O \nearrow\!\!\uparrow\; O-CH_2 \\
\end{array}\right] M^I \\
O \\
| \\
R
\end{array}
$$

Because of their highly developed internal surfaces, silica gels possess excellent adsorptive powers, similar to charcoal. This is particularly evident for the dried-out gels: Freundlich has proposed the general term *xerogels* (ξηρός, dry) for such dry residues of jellies. Silica xerogels have accordingly found increasing application in recent times for the adsorption of vapors, the refining of liquids (especially for the removal of sulfur-containing impurities from crude petroleum), and for catalytic purposes. In this connection it is important that by incorporating special substances it is possible to impart to the gels a specific affinity for certain materials, and further, that the gel structure is not altered, or only to an unimportant extent, by moderate heating (to dull redness). It is often possible, therefore, simply by heating, to remove the substances taken up by the gel, and to regenerate the gel once more. The artificially produced silica xerogels have the advantage over infusorial earth—which is also to be included under the heading of the xerogels—that their structure can be influenced by controlling the way in which they are prepared, and can be adapted to the purpose for which they are to be used. In surgery and dermatology, powdered silica xerogels, in some cases impregnated with liquids such as ichthyol, peru balsam, etc., find use as deodorizing, disinfectant dusting powders, with drying properties. The electrolyte in galvanic cells and accumulators is often gelatinized by means of silicic acid, to prevent splashing.

Silica xerogel, impregnated with cobalt chloride, is used in the laboratory as an efficient drying agent for gases ('blue gel'). Deterioration in its drying power is shown up by a color change of the cobalt chloride, from blue to pink. The gel can be regenerated by warming it.

Silicic acid finds very extensive uses in the form of the *silicates*. The porcelain industry is based upon the hydrated aluminum silicate occurring in nature as kaolin, or china clay. The glass industry depends on the property of many silicate melts, of solidifying in vitreous form. The setting of cement, and of hydraulic mortar prepared from it, depends chiefly upon the formation of silicates. The *permutits* used in water softening are silicates, as also are the widely used ultramarine pigments. Certain clay-like aluminosilicate xerogels occuring in Nature, with selective adsorptive properties, are used, under the name of fullers' earth, [*30*], for decolorizing mineral, vegetable and animal oils and fats. Viscous aqueous solutions prepared form alkali silicate melts also find extensive industrial uses, under the name of water-glass. [*29*]

1. *Water-glass* is made by melting quartz sand with potash or soda, or with sodium sulfate and wood charcoal (3–5 mols. of SiO_2 are usually taken per mol. of alkali oxide). It forms lumps which are colored grey or greenish by traces of iron salts, which it contains as impurities, but is generally sold in the form of aqueous solutions, since its dissolution is attended with some difficulty. This is carried out technically by treating it with steam, or else by first treating the powdered glass with a little water. The finely ground gel then readily dissolves. Water-glass solutions are used for impregnating paper fabrics, for sizing paper, for weighting silk, for stiffening bandages for fractures of the bones, for the preparation of cements and adhesives, for affixing labels to glass, for fire-proofing wood and textiles and for the preservation of eggs (4–10% solution). Powdered water-glass gel is used as a filler in soap. Solutions of water-glass are not permanently stable, but gradually undergo decomposition and deposit silicic acid. The solution often solidifies to a jelly during this process.

2. *Sodium silicate* can be obtained by fusion of powdered quartz with sodium carbonate (1), or by the reaction of freshly precipitated hydrated silica with the calculated quantity of caustic soda (2).

$$SiO_2 + Na_2CO_3 = Na_2SiO_3 + CO_2 \qquad (1)$$

$$SiO_2 + 2NaOH = Na_2SiO_3 + H_2O \qquad (2)$$

It solidifies in crystalline form from the pure melt (m.p. 1088°). (Mixed melts of sodium silicate—e.g., with calcium silicate, which also solidifies crystalline when it is pure—solidify to glasses.) It crystallizes as a hydrate from aqueous solution. Hydrated sodium silicate comes into commerce under the name of *alkasil*. It is used in slip-casting of porcelain and for the same purposes as waterglass.

In addition to the ordinary silicate (the metasilicate Na_2SiO_3), the compounds Na_4SiO_4, $Na_6Si_2O_7$ and $Na_2Si_2O_5$ can be obtained from the melts (D'Ans, 1930). Hydrated sodium silicates crystallize from the solutions [see Lange and von Stackelberg, *Z. anorg. Chem.*, 256 (1948) 273]. They are extremely soluble, as also, in still higher degree, are the potassium silicates, which are therefore difficult to crystallize from solutions. K_2SiO_3, $K_2Si_2O_5$, and $K_2Si_4O_9$ can be isolated from the melt. The lithium silicates have excellent crystallizing properties.

(iii) *The Nature of Hydrated Alkali and Alkaline Earth Silicates*. Thilo (1951) regards the hydrated sodium silicates as *acid* silicates. This view is based chiefly on studies of their chemical properties and measurements of particle size. It would appear from these that solutions of these compounds contain mono-nuclear silicate ions, namely $[HSiO_4]^{\equiv}$,

$[H_2SiO_4]^=$ and $[H_3SiO_4]^-$. Thilo also finds that Ca^{++} ions are capable of forming acid silicates, such as $Ca[H_2SiO_4]$ and $NaCa[HSiO_4]$ (see *Angew. Chem.*, 63, (1951) 201). These findings are remarkable, since previously no evidence for the existence of acid silicates had been forthcoming from X-ray structure determinations. Recently, however, these have been detected by this method. Therefore the question arises whether many other silicates containing water, which have hitherto been formulated as hydrates, should not be regarded as acid salts. Examples are the minerals afwilite ($3CaO \cdot 2SiO_2 \cdot H_2O$), riversideite ($2CaO \cdot 2SiO_2 \cdot H_2O$), crestenoreite ($2CaO \cdot 2SiO_2 \cdot 3H_2O$), okenite ($CaO \cdot 2SiO_2 \cdot 2H_2O$), gyrolite ($2CaO \cdot 3SiO_2 \cdot 2H_2O$), foshagite ($5CaO \cdot 3SiO_2 \cdot 3H_2O$), xonotlite ($4CaO \cdot 4SiO_2 \cdot H_2O$) and tobermorite ($4CaO \cdot 5SiO_2 \cdot 4H_2O$).

For the interpretation of the structure of the alkali and alkaline earth silicates, the following types of reaction furnish significant information:

[1] Hydrolytic splitting, following the pattern of salt hydrolysis, e.g.,

$$Na_2Ca[SiO_4] + HOH = NaCa[HSiO_4] + NaOH, \text{ or}$$

$$[SiO_4]^{4-} + HOH = [SiO_3(OH)]^{3-} + OH^-$$

[2] Hydrolytic degradation of the anions of polyacids into simpler species, e.g.,

$$[Si_2O_7]^{6-} + HOH = 2[SiO_3(OH)]^{3-}, \text{ and}$$

$$[Si_4O_{11}(OH)_2]^{8-} + 3HOH = 4[SiO_3(OH)]^{3-} + 4H^+$$

[3] The converse process to [2]—the condensation of acid anions, with elimination of water

e.g., $$2[SiO_3(OH)]^{3-} = [Si_2O_7]^{6-} + H_2O$$

[4] The rearrangement of infinite structures based on SiO_4 tetrahedra—e.g., SiO_2 (3-dimensional network) → $[Si_3O_9]$ rings.

Each one of these types of reaction corresponds to a certain range of temperature, within which that reaction goes to completion. Thilo found that the lowest temperature at which the calcium silicates would undergo these changes bore a practically constant ratio to the temperature (in degrees K) at which the corresponding transformations take place in the *phosphates of sodium* (sodium and calcium differ very little in ionic radius). This is strongly suggestive of a close relationship between the two systems, in both structure and reaction mechanisms.

(iv) *Silicate Peroxyhydrates.* Peroxysilicic acids, or their salts, appear to be non-existent. A few silicates, however, form well defined addition compounds with H_2O_2. Of these, the *sodium metasilicate di-peroxyhydrate* $Na_2SiO_3 \cdot 2H_2O_2$ was first prepared by Krauss (1932–35). He obtained the hydrate $Na_2SiO_3 \cdot 2H_2O_2 \cdot H_2O$, a white, finely crystalline powder, by the vacuum evaporation of a solution of sodium metasilicate, treated with H_2O_2. This salt could be dehydrated to the anhydrous compound without any considerable loss of active oxygen. The alkali silicate peroxyhydrates are readily soluble in water, and are stable in the absence of moisture. The alkaline earth silicate peroxyhydrates, and silicon dioxide peroxyhydrate, $SiO_2 \cdot H_2O_2$, on the other hand, are sparingly soluble and rather unstable.

7. Crystal Structure of the Silicates

Only within recent years, has the fine structure of the silicates been explained, principally through the application of X-ray crystallography [32]. X-ray crystal structure determinations have shown that crystalline silicates form coordination lattices, which can be regarded, to a first approximation, as being built up from ions*.

All the crystalline silicates hitherto investigated are based on a common structural principle, the formation of tetrahedra of oxygen atoms around each Si^{4+} ion with the distance $Si \leftrightarrow O = 1.22\text{Å}$. If, for stoichiometric reasons, there are fewer than 4 oxygen ions for each Si^{4+} ion, this principle is still maintained, by the possibility of sharing O^{2-} ions between neighboring tetrahedra. The positive ions involved in the structure of the crystal are situated between the negative complexes, formed either by individual SiO_4^{4-} tetrahedra, or by tetrahedra linked together through sharing of O^{2-} ions, in such a manner that they lie as far apart as possible from each other, and—as is general in coordination structures—have as many negative ions as possible grouped closely around them (in this case the O^{2-} ions of the SiO_4-tetrahedra). According to their fine structure, the silicates are subdivided into those with island structures (isolated anions), with chain or band structures, with sheet structures (layer lattices) and with 3-dimensional network structures.

(a) Island Structures

This term is applied when 'island' groups, made up of a finite number of Si^{4+} and O^{2-} ions, are present—e.g., $[SiO_4]^{4-}$, $[Si_2O_7]^{6-}$, $[Si_3O_9]^{6-}$. The SiO_4 tetrahedra can exist as independent and discrete structural groups, $[SiO_4]^{4-}$, only in those silicates in which there are at least 4 O atoms for every Si atom—i.e., in the orthosilicates such as phenacite, $Be_2[SiO_4]$, forsterite, $Mg_2[SiO_4]$, olivine, $(Mg,Fe)_2[SiO_4]$, zircon, $Zr[SiO_4]$. The unit cell of olivine is represented in Fig. 90, as an example of the structure of such orthosilicates with discrete

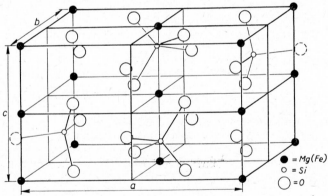

Fig. 90. Unit cell of olivine, $(Mg,Fe)_2[SiO_4]$. $a = 10.21$ Å, $b = 4.755$ Å, $c = 5.985$ Å

- = Mg(Fe)
- = Si
- = O

SiO_4 tetrahedra (cf. also the zircon structure, discussed in Vol. II). Corresponding structures have also been found for other compounds of the type $M_2[RO_4]$. Thus, in addition to other orthosilicates, $Li_2[BeF_4]$, $Li_2[MoO_4]$ and $Li_2[WO_4]$ are all isotypic with phenacite; not only $Mg_2[SiO_4]$, $Fe_2[SiO_4]$, $MgCa[SiO_4]$, etc., but also $LiFe^{II}[PO_4]$ and $Al_2[BeO_4]$ are isotypic with olivine; $Y[PO_4]$ and $Ca[CrO_4]$, etc. are isotypic with zircon.

In addition to the $[SiO_4]^{4-}$ radicals, other negative ions may participate in the structure of such lattices—e.g., F^-, $[OH]^-$, O^{2-}. Examples are cyanite (disthene), $Al_2O[SiO_4]$ and titanite, $CaTiO[SiO_4]$, also topaz, $Al_2(OH,F)_2[SiO_4]$ and humite, $3Mg_2[SiO_4]$, $Mg(OH,F)_2$.

* The presence of ions in the silicate crystal lattices follows both from their electrical conductivity and ionic migration, and from the additivity of refractive indices and specific heats. There are grounds, however, for the view that in addition to the electrovalent forces, forces attributable to covalent interactions also play a part in the particularly close linkage of the O-atoms and Si-atoms.

The structures of humite and related minerals are derived from that of olivine by the introduction of sheets of $Mg(F,OH)_2$ (or of Mg^{2+} and OH^- ions which may be replaced in any proportion by F^- ions) in an ordered manner between layers having the structure proper to olivine. The structure of topaz is derived in a similar way from the crystal lattice of cyanite, by introduction of sheets of $Al(OH,F)_3$.

If the number of O^{2-} ions does not suffice for the formation of discrete $[SiO_4]^{4-}$ groups, the result is that tetrahedra link up by sharing adjacent O_2^- ions. Experience* shows that neighboring SiO_4 tetrahedra generally share only *one* O^{2-} ion—in other words, the tetrahedra are in contact always through their vertices, and never along edges or even faces. Linkage of tetrahedra in pairs leads to the formation of $[Si_2O_7]^{6-}$ groups (examples are $Na_2Ca_2[Si_2O_7]$, $Ca_4(OH)_2[Si_2O_7]$, and the mineral *thortveitite*, $Sc_2[Si_2O_7]$). Three or more tetrahedra frequently link up to form rings, as represented diagrammatically by the following formulas, in which the tetrahedron edges which are directed towards one another are shown by dotted lines:

The *cyclic* linking up of tetrahedra in this fashion leads to groups with the general formula $[Si_nO_{3n}]^{2n-}$. Rings of this kind have been identified, for example, in the crystal structures of *wollastonite*, $Ca_3[Si_3O_9]$, *benitoite*, $BaTi[Si_3O_9]$ and *beryl*, $Al_2Be_3[Si_6O_{18}]$. In this case also, substitutional structures can exist.

If a finite number of $[SiO_4]$ tetrahedra join up without the formation of rings, the resultant structural groups have the general formula $[Si_nO_{3n+1}]^{2(n+1)-}$. Thus, five tetrahedra may associate in such a way that one of them shares each of its four oxygens with four others. This form of grouping has been identified in *zunyite*,

$$Al[Al_{12}O_4(OH,F)_{18}][Si_5O_{16}]Cl.$$

The structure of this compound, with its complex composition, can be derived from the relatively simple crystal lattice of zinc blende (cf. Vol. II), if the Zn^{2+} and S^{2-} ions are replaced respectively by the structural groups, made recognizable by enclosing them in square brackets in the formula as written above. The ions Al^{3+} and Cl^-, which enter into the composition of the compound as well as these structural groups, and which serve to balance the excess charges, are arranged in a regular manner in interstices of the structure.

There are also some silicates known in which silicate groups of different composition enter simultaneously into the structure—e.g., *vesuvianite*,

$$Ca_{10}Mg_2Al_4[SiO_4]_5[Si_2O_7]_2(OH)_4.$$

(b) Chain and Ribbon Structures

The SiO_4 tetrahedra can also be linked together in the way represented in Fig. 91 *a-d*, so that 'chains' or 'bands' are formed. Since the chain is of indefinite length, in chain structures each Si^{4+} ion is associated stoichiometrically with 3 (i.e., $2 + {}^2/_2$)O^{2-}. Silicates with chain structures are therefore *metasilicates* as far as their composition is concerned**. Since

---

* Pauling has also given some theoretical justification—cf. *J. Am. Chem. Soc.*, 51, (1929) 1010).

** Certain of the silicates with island structures also have the stoichiometric composition of metasilicates.

however, it gives a false impression of the coordination number of silicon in the anion to formulate the metasilicates as $M^{II}[SiO_3]$, it is customary to denote the silicates with chain structures by the general formula $M^{II}_2[Si_2O_6]$. Examples are *enstatite*, $Mg_2[Si_2O_6]$, *diopside*, $CaMg[Si_2O_6]$, *spodumene* $LiAl[Si_2O_6]$, and also *pseudowollastonite*, $Ca_2[Si_2O_6]$, which is formed from wollastonite (see above) by a monotropic transformation at high temperatures. *Pectolite*, which is isomorphous with wollastonite, and which is considered by Thilo to be an acid silicate:

$$NaCa_2[Si_3O_8(OH)],$$

undergoes a corresponding transformation when it is heated.

In building up a 'ribbon' out of two chains, as in Fig. 91c, one additional O^{2-} ion is eliminated from every four tetrahedra. Thus a ribbon $[Si_4O_{11}]^{6-}$—also of indefinite length—is formed from two $[Si_2O_6]^{4-}$ chains. It is conceivable that two $[Si_4O_{11}]^{6-}$ ribbons might join up to form a fourfold ribbon, $[Si_8O_{21}]^{10-}$, but such has not hitherto been found. Examples of ribbon structures are *tremolite*

$$Ca_2Mg_5[Si_8O_{22}](OH)_2,$$

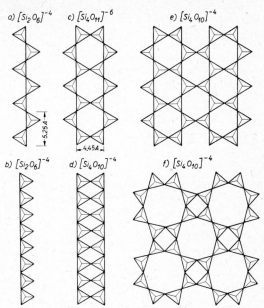

a) $[Si_2O_6]^{-4}$ c) $[Si_4O_{11}]^{-6}$ e) $[Si_4O_{10}]^{-4}$

5.25 Å   4.45 Å

b) $[Si_2O_6]^{-4}$ d) $[Si_4O_{10}]^{-4}$ f) $[Si_4O_{10}]^{-4}$

Fig. 91 Chains (*a* and *b*), ribbons (*c* and *d*), hexagonal sheet (*e*) and tetragonal sheet (*f*) of SiO_4-tetrahedra.

riebeckite $Na_2Fe^{II}_3Fe^{III}_2[Si_8O_{22}](OH)_2$, *glaucophane* $Na_2(Mg,Fe^{II})_3(Al,Fe^{III})_2[Si_8O_{22}](OH)_2$; *hornblende* has the same structure*. An example of a substitutional ribbon structure is *chrysotile*, fibrous serpentine, $Mg_3[Si_4O_{11}] \cdot 3Mg(OH)_2 \cdot H_2O$.

It is also possible, however, for two chains to link up so that one O^{2-} ion is left out for each pair or tetrahedra, so that a $[Si_4O_{10}]^{4-}$ ribbon is produced from two $[Si_2O_6]^{4-}$ ribbons (cf. Fig. 91d). This structure is present in *sillimanite*, $Al_2[Al_2Si_2O_{10}]$, except that instead of consisting exclusively of SiO_4 tetrahedra, the chains consist of alternate SiO_4 and AlO_4 tetrahedra, linked in such a way that in the double chain a SiO_4 tetrahedron and an AlO_4 tetrahedron come opposite one another each time. The structure of *andalusite* is intermediate between this structure and that of cyanite (p. 499).

(c) Sheet Structures (Layer Lattices)

The continuous broadening of the $[Si_4O_{11}]^{6-}$ ribbons would produce a planar network, or '*sheet*' (Fig. 91e). In such a sheet, each Si^{4+} ion shares *three* O^{2-} ions with its neighbors, giving a stoichiometric ratio of $1 + {}^3/_2 = 2{}^1/_2 O^{2-}$ ions per Si^{4+} ion. This structure would therefore be represented by the formula $[Si_2O_5]^{2-}$ or some multiple thereof. The *micas*, and similar silicates with a sheet-like cleavage are made up of such planar networks, more or less loosely piled together. Examples are *talc*, $Mg_3[Si_4O_{10}](OH)_2$, *antigorite* ('leafy serpentine'), $Mg_3[Si_4O_{10}](OH)_2 \cdot 3Mg(OH)_2$ (a substitutional sheet structure), *kaolinite*

$$Al_2[Si_4O_{10}](OH)_2 \cdot 2Al(OH)_3$$

* In all these examples the structures involve $[Si_4O_{11}]^{6-}$ ribbons. If it is necessary to double the formula of the band to preserve whole-number atomic indices, it is, of course, pointless to denote this by an index 2 outside the bracket, which does not indicate the formula of a closed group but of an anion of infinite length. The individual atomic indices are therefore doubled *inside* the bracket.

Between the two structural types of serpentine (*chrysotile* and *antigorite*) represented by the formulas $Mg_3[Si_4O_{11}] \cdot 3Mg(OH)_2 \cdot H_2O$ and $Mg_3[Si_4O_{10}](OH)_2 \cdot 3Mg(OH)_2$, there is a continuous transition series. This is because the Si_4O_{11} ribbons can broaden more and more by condensation with one another, until they eventually attain the structure of the Si_4O_{10} sheets. The $Mg(O,OH,H_2O)$ ribbons simultaneously grow until they become brucite layers. In the case considered, the existence of the sheet structure is not associated with a flaky cleavage, as it commonly is in other silicates. As Noll has shown, the difference in 'mesh' between the Si_4O_{10} network and the brucite sheet, which is condensed with it to form a double layer, can lead to such a strong curvature of the double layer structure that it curls up to form fibers with a capillary structure. Noll has shown by electron microscopy that these are often present in fibrous serpentine (chrysotile). Fibrous serpentine may thus have the fine structure corresponding to either of the above formulas, whereas flaky serpentine (antigorite) always has the fine structure corresponding to the formula:

$$Mg_3[Si_4O_{10}](OH)_2 \cdot 3Mg(OH)_2$$

Isomorphous with chrysotile is *garnierite*, which is found native as a (generally more or less impure) nickel-magnesium silicate. Noll (*Naturwissenschaften*, 39, (1952) 233) obtained it artificially, by hydrothermal synthesis, in the pure state, corresponding to the formula $Ni_3[Si_4O_{10}](OH)_2 \cdot 3Ni(OH)_2$; chrysotile had been obtained similarly. Electron-microscopic investigation showed the presence of capillary fibers in garnierite, as in chrysotile, as might be expected from the similarity in mesh between the $Mg(OH)_2$ and $Ni(OH)_2$ networks.

The minerals *montmorillonite* $Al_2[Si_4O_{10}](OH)_2 \cdot nH_2O$, *hectorite* $Mg_3[Si_4O_{10}](OH)_2 \cdot nH_2O$ and *nontronite*, $(Fe^{II}_3, Fe^{III}_2)[Si_4O_{10}](OH)_2 \cdot nH_2O$, which are isomorphous with one another, are built up from metal silicate sheets (with the OH^- ions incorporated in a regular manner), between which variable number of water molecules can be inserted. This is the basis of the cation-exchanging properties of these minerals, and of their unusually great swelling power. By mixing magnesium chloride with sodium silicate solutions, Hofmann obtained slimy precipitates which, when prepared under suitable conditions, exhibited the X-ray diffraction pattern characteristic of the silicate sheets of hectorite. However, the diffraction pattern showed that the silicate sheets were rather irregularly stacked, and only took up a parallel orientation when the precipitates were digested with caustic alkali solutions at about 200° under pressure. Stronger heating with potash (but not with caustic soda) brought about a conversion into mica. Precipitates with the antigorite structure could be obtained by altering the proportions in which the reagents were mixed. These did not undergo any structural alterations as a result of hydrothermal treatment (i.e., when strongly heated with water or alkali under pressure).

As has already been shown by one example, it is possible in certain structures for the Si^{4+} ions within the tetrahedra to be partially replaced by Al^{3+} ions without disturbing the structure. The negative charge on the fabric of tetrahedra is, of course, increased by one unit for each atom of Al, as a result of this substitution. Thus $[AlSi_3O_{10}]^{5-}$ and $[Al_2SiO_{10}]^{6-}$ are derived from $[Si_4O_{10}]^{4-}$. By this means one passes from the silicates of the talc group, such as *pyrophyllite* $Al_2[Si_4O_{10}](OH)_2$, on the one hand to the micas, such as *muscovite* $KAl_2[AlSi_3O_{10}](OH)_2$, and on the other hand to the *brittle micas*, such as *margarite*, $CaAl_2[Al_2Si_2O_{10}](OH)_2$. The inclusion of the K^+ and Ca^{2+} ions respectively corresponds to the increase in the charge of the negative group.

The flat network of tetrahedra shown in Fig. 91e can also be thought of as the result of the indefinite linking together of hexagonal rings of tetrahedra. The analogous linking of rings of eight tetrahedra would build up a flat tetragonal network, consisting of alternate rings of four and eight tetrahedra (cf. Fig. 91f). Such a structure is present, for example, in *apophyllite*, $Ca_4[Si_8O_{20}][K(H_2O)_8](OH,F)$.

(d) Three Dimensional Networks

If every SiO_4 tetrahedron shares *all four* of its O^{2-} ions with neighboring tetrahedra, a three dimensional skeleton or network is built up. In this case, $^4/_2 = 2O^{2-}$ ions belong to each Si^{4+} ion. We thus arrive at silicon dioxide, SiO_2, the crystal structure of which has already been discussed. If, in some of the tetrahedra, Si^{4+} is replaced by Al^{3+}, or other small ions of lower valence, such as B^{3+} or Be^{2+}, positive ions must be incorporated simultaneously in the structure to compensate the resulting charge. Thus *nepheline* $Na[AlSiO_4]$ (hexagonal) possesses the same skeleton of tetrahedra as tridymite; *carnegieite* $Na[AlSiO_4]$ (cubic) has that of cristobalite.

Other networks of tetrahedra, differing from those present in the various modifications of SiO_2, can also be built up with a portion of the Si^{4+} replaced by Al^{3+}. Such are present, for example, in the *felspars*, such as $K[AlSi_3O_8]$, in the silicates of the *natrolite* group, such as natrolite, $[Na(H_2O)]_2[Al_2Si_3O_{10}]$ and *skolecite*, $[Ca(H_2O)_3][Al_2Si_3O_{10}]$, in *analcime*, $[Na(H_2O)]_4[Al_4Si_8O_{24}]$, and in *scapolite* $Na_4[Al_3Si_9O_{24}]Cl$. In all these compounds, the aluminosilicate groups—which can be only arbitrarily delimited, since the anionic group is of indeterminate size—would be changed into some multiple of SiO_2 if the Al were replaced by Si.

Examples of substitutional structures with three dimensional networks are the minerals of the sodalite group, such as *sodalite*, $Na_8[Al_6Si_6O_{24}]Cl_2$ and *nosean*, $Na_8[Al_6Si_6O_{24}][SO_4]$. Replacement of the $[SO_4]^{2-}$ groups of the latter by S_2^{2-} groups, gives *ultramarine* (p. 514).

Between the silicates with three dimensional networks and those with island structures it is possible to perceive transitions. For example, in the compound Na_2CaSiO_4, which crystallizes cubic, there is the same three dimensional network as in cristobalite, if one ignores the considerable distortion and stretching of the CaO_4 tetrahedra. However, if the O^{2-} ions are assigned wholly to the Si^{4+} ions, on the grounds that the $Ca\leftrightarrow O$ distance (average 2.00Å) is much larger than the $Si\leftrightarrow O$ distance (1.58 Å), and not much smaller than the smallest $Na\leftrightarrow O$ distance (2.37 Å), the compound is to be considered as a double silicate with discrete anions. Its fine structure could legitimately be represented by the formula $Na_2[CaSiO_4]$, as well as by $Na_2Ca[SiO_4]$.

e) Relation between Structure and Properties of the Silicates [31]

As has been shown, the unravelling of the structures of the silicates has revealed principles for classifying them, on the basis of which the analogies in chemical composition between structurally related compounds stand out clearly. It has, moreover, become possible for the first time to assign chemical formulas denoting the idealized compositions of many silicates occurring in Nature, which—because of wide variations in composition brought about by isomorphous replacement— could not formerly be formulated. In addition, important relations between the crystal structures of silicates and their physical and chemical properties have become apparent.

Even though the investigation of these relations is still in an early stage, it has already been found that many properties, such as the hardness, strength, cleavage, thermal stability, and also the optical properties of the silicates, are clearly related to their fine structure. This is most evident with respect to the cleavage. It has already been mentioned that silicates with sheet structures usually have a particularly good cleavage, parallel to the sheets, so that substances with this structure, such as mica, can often readily be cleaved into fine leaves. In a similar manner, silicates with chain or ribbon structures are found to have a very good cleavage parallel to the direction of the chains. The rod-like or fibrous habit often exhibited by these silicates (e.g., by *asbestos*) is connected with this fact. The silicates with island structures usually have a much less well developed cleavage; in so far as such can be distinguished at all, it is usually parallel to those planes most thickly occupied by O^{2-} ions. Silicates with uniform three dimensional networks, such as are present in quartz, tridymite and cristobalite, usually lack any clear cleavage directions, as would be expected from their structures. When Si^{4+} ions are partially replaced by Al^{3+} ions, lattice networks of less homogeneous structure frequently result, which then often have a distinct, or even a perfect, cleavage, corresponding to the different density of packing in different directions. This is observed, for example, in the felspars.

Considerable difficulty is often found in artificially preparing the same silicates and double silicates as occur in nature. This is particularly the case with experiments designed to prepare them, not from melts, but by chemical reactions in the presence of water (*hydrothermal syntheses*), by analogy with the conditions under which they were produced naturally. *Potash felspar*, which is extremely widely distributed in Nature, is one such compound, which could not at first be prepared successfully by hydrothermal synthesis. Recently, however, Barrer and Hinds, by heating synthetic leucite, $K[AlSi_2O_6]$, to $200°$ in an autoclave in an aqueous solution of K_2CO_3 and Na_2CO_3, have been able to obtain a good yield of minute crystals, of refractive index 1.521, which were identified by X-ray examination as potash felspar.

8. Silicate Products of Special Technical Importance

(a) Glasses and Glazes

(*i*) *Glass*. A *glass* in the widest sense is a melt which solidifies in the amorphous state—i.e., without crystallizing.

The glasses were formerly regarded simply as 'supercooled liquids', but according to modern investigations there are some significant differences between a glass and a super-cooled liquid [*33*]. The temperature at which the properties of a glass pass over into those of a super-cooled liquid is known as the 'transition point'.

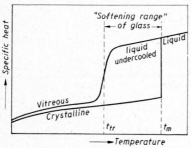

Fig. 92. Relation between temperature and specific heat of liquid, vitreous and crystalline material.

t_{tr} = transition point
t_m = melting point

Within a few degrees, in the neighborhood of the transition point, marked changes take place in the specific heat, the coefficient of expansion, the temperature coefficients of the dielectric constant and the refractive index, and also, in some cases, the temperature coefficient of the electrical conductivity (cf. Fig. 92). Below the transition point, typical glasses are absolutely rigid, so that cracks are started by scratching; above that temperature they are plastic or viscous. On further increase in temperature above the transition point, the properties mentioned above change slowly and steadily into those of the liquid which is obtained by melting the crystallized material. Below the transition temperature, these properties generally differ but little from those possessed by the same substance in the crystalline state. In recent years the use of X-ray methods, supplementing the study of changes in physical properties, has given valuable information as to the nature of the vitreous state. (See especially the work of B. E. Warren).

The glasses can be regarded as aperiodic network structures, such that any domains of large extent exhibit no symmetry. Symmetrical arrangements of atoms are to be found only in the immediate proximity of certain atomic groupings; the resulting 'short range order' corresponds to the same coordination numbers as are found in crystalline compounds. Atomic groupings exhibiting short range order are the SiO_4 and PO_4 tetrahedra in silicate and phosphate glasses, the BO_3 triangles in vitreous B_2O_3. These are linked together by sharing corners, as in crystalline network structures, but in an irregular manner so that a disordered network is produced instead of a crystal lattice.

There is no *fundamental* structural difference between the vitreous and the liquid states of the same substance. In the liquid state, especially close to the crystallization temperature, atomic arrangements showing short term order are to be found over small domains, as has been proved by X-ray diffraction. On purely structural grounds, therefore, it is not possible to draw a sharp distinction between the liquid and the glassy state. Thermodynamic considerations, however, do lead to such a distinction.

From the thermodynamic point of view, glasses differ from supercooled liquids in that, unlike the latter, they are not in inner thermodynamic equilibrium. That is to say, when glasses are cooled to lower temperatures they retain the disorder (i.e., the entropy) present at the transition point, whereas in a supercooled liquid an increasing degree of order in the atomic arrangement sets in as the temperature is lowered. If supercooling could be taken down to the absolute zero, the entropy of a supercooled liquid would approach the value zero. It must not be overlooked that just at the transition point (or more exactly, in the transition range) there is a fairly extensive ordering of the atomic arrangement in a glass. The glassy state is more closely related, structurally, to the crystalline state than to the liquid state. This has been directly demonstrated, for example, by Richter's X-ray determination of the structure of vitreous As and Se (1951–52).

The property of solidifying from the melt in the vitreous-amorphous state is possessed in a special degree by certain *double silicates*. Glass in the narrower sense therefore usually consists of a mixture of silicates. It generally contains sodium or potassium silicate, along with calcium silicate.

The so called 'soda glass' or 'normal glass' has the approximate composition $Na_2O \cdot CaO \cdot 6SiO_2$. Ordinary glass, such as window glass and bottle glass, generally approaches this composition. The melting point and the chemical resistance are raised by replacing Na by K (potash glass, hard glass, Bohemian or Jena glass). Such glass is used, for example, for combustion tubes for organic ultimate analysis. If Ca is replaced at the same time by Pb, glasses are obtained which are characterized by a high optical refractive power and high specific gravity; these are used for cut glass ware and ornamental objects (lead glass, lead crystal). *Flint glass* is a lead glass, which finds application especially in optical instruments for lenses and prisms. For this purpose it is usually combined with *crown glass*, which has a smaller optical dispersion; this has a high content of P_2O_5 (up to 70%). Glasses rich in lead have a relatively low melting point, and a relatively low resistance towards chemical reagents. A glass particularly rich in lead, and approaching the diamond in its refractive power, is that known as paste or 'Strass', after the Viennese jeweller, Strasser. It is used for imitation gems, but soon loses its luster because it is so soft.

Colored glasses are obtained by adding small amounts of those oxides which form colored silicates: e.g., blue with cobalt oxide, green with chromium or copper(II) oxide, and red with copper(I) oxide. Iron(II) silicate in high concentration gives a black color, but small amounts produce a dirty green (the color of beer bottles). Fe_3O_4, iron(II, III) oxide, produces particularly intense coloration; that due to pure trivalent iron is much weaker (yellow green to brownish yellow). The color produced by minimal amounts of iron can be removed by the addition of pyrolusite to the glass melt, this forming manganese(III) silicate which is violet, and complementary in color to the yellow green of iron(III) silicate. This is the origin of the old name of pyrolusite: 'glass maker's soap'. In combination with much iron oxide however, pyrolusite produces the brown color of certain bottle glasses.

Light-sensitive glasses have also been produced in recent years ('photographic glass'). The glass melts for this purpose may either have a composition similar to that used for copper ruby glass (cf. Vol. II), or may contain gold and some sensitizing agent such as cerium oxide. When a glass of this kind is exposed to ultraviolet light through an ordinary negative, a latent image is produced. This can be developed by heating to 600–700°.

Many glasses contain aluminum oxide also (generally only a few per cent), and boron oxide is also frequently added. Certain glasses of this type have a particularly low coefficient of thermal expansion. Such glasses include the Jena Normal Thermometer glass, Schott's 'Geräteglas', Pyrex and similar glasses. The fabrication of thick glassware, which can be heated without imposing strains due to unequal thermal expansion, has made it possible to use glasses of this type for the construction of large-scale chemical equipment, as well as for domestic glass ware which can be used for cooking and baking in ordinary stoves.

Protective glasses recently developed for use in atomic energy work contain cadmium oxide and fluoride, as well as borosilicates (to confer a high absorption for slow neutrons), and others contain tungsten phosphate, or as much as 50% by weight of lead oxide, for the absorption of γ-rays.

Glass is manufactured [34] by melting silicon dioxide (quartz or flint) with calcium carbonate (in the form of limestone, marble or calcite) and calcined soda, or with sodium sulfate and coke. Potash is used in place of soda for potash glasses, and lead oxide in place of calcium carbonate for lead glass, etc. Melting is carried out in tanks or in large refractory crucibles, the glass pots. The furnaces are usually fired with producer gas, by means of Siemens regenerative heating, in which the hot flue gases are used to heat up chambers filled with refractory bricks; the gases to be burned are subsequently passed through these, so that they arrive in the furnace pre-heated to a high temperature.

Electric furnace melting has been introduced into glass manufacture during recent years. *Optical glass* (i.e., glass designed for the manufacture of optical appliances) is now melted to an increasing extent in *platinum clad* pots and crucibles in modern glassworks.

Certain borosilicate glasses have the property of gradually precipitating alkali borate, at temperatures above their transition point. As a rule, this remains suspended in the glass mass, in the form of very fine droplets. It is possible, however, to carry out the process in such a way that it forms a coherent phase, which can subsequently be removed by leaching the cold glass with dilute acid. When this is done, a spongy mass remains behind, with a very high silica content (up to 96%). If this mass, in turn, is sintered at about 1200°, a homogeneous glass is produced, which resembles pure quartz glass in its properties, and softens only above 1500° (Corning 'Vycor' glass, of Hood and Nordberg). In the Vycor process, the vessels may be shaped from the original alkali borosilicate glass at the usual working temperatures. The vessels are subsequently heat treated, leached and sintered; a considerable amount of shrinkage occurs, but they retain their shapes.

Use has also been made of the Vycor process in the production of highly porous, catalytically active materials (e.g., for the cracking of petroleum). In this case, no sintering is carried out after the leaching, and the composition of the original melt is modified to give the desired properties (e.g., by adding CuO or NiO).

The properties of glass [35–37] can be modified to a considerable extent by changes in the nature and proportions of the constituents. In general, the 'acid' glasses, rich in silica, are the least fusible and the most resistant towards attack by chemical agents. Increase in the basic constituents increases the fusibility and at the same time diminishes the chemical resistance. Glasses rich in alkali are markedly attacked by boiling water. Water to which phenolphthalein has been added turns red when it is boiled in an ordinary beaker which has not previously been used. When some of the alkali has been leached out of the surface by frequent boiling, the resistance is greater. An increase in resistance is more rapidly achieved by steaming out. Glass is much more strongly attacked by caustic alkalis than by water and acids.

There is a current trend in glass manufacture (especially in the United States) to diminish the alkali content of the cheaper sorts of glass (bottle glass and window glass). This is because sodium carbonate is the most expensive constituent of the glass mixes used for the production of the commoner glasses, and this can be replaced only to a limited extent by sodium sulfate (e.g., for brown glass). The practice is to add cheap barium and fluoride minerals and fluxes, and to lower the SiO_2 content together with the alkali content.

When hot and relatively strongly alkaline solutions are employed for the cleaning and disinfection of glassware (e.g., milk bottles), it is important that their aggressive action should be inhibited by the addition of aluminates or traces of beryllia, which have an anti-catalytic action.

In addition to its transparency, a property of special importance possessed by glass is that of softening long before it melts, when heated to a red heat. On this depends the possibility of working hot glass (glass-blowing). A less favorable property of glass is its sensitiveness towards rapid changes of temperature, which often leads to the cracking of glass apparatus. Glass heated to a high temperature must be very slowly cooled, as internal strains are otherwise produced, and these can easily lead to cracking after it has cooled down. In contrast to this, sudden chilling can be the means of imparting great toughness and resistance towards changes of temperature to certain borosilicate glasses (Durax, Resista).

A very large optical glass industry has grown up in the United States since the 1914–18 war, and as a result of intensive research the range of optical glasses has been considerable extended. The work of Morey (1937) marked a new development, with the preparation of the *rare metal-borate* glasses. These contain the borates, tantalates and tungstates of zirconium, thorium and lanthanum. They are characterized by particularly high refractive indices. Glasses with a high fluoride content (*fluoride glasses*) have low refractive indices and particularly low chromatic dispersion.

At elevated temperatures, and especially above the transformation point, glass has a considerable electrical conductivity. Under conditions such that metal ions (or H^+) can migrate into the glass, the conductivity is purely electrolytic, and depends, in fact, on the mobility of sodium ions. In potash glasses, the potassium ions take over the transport of current only if no sodium ions are present as well. The silicate ions play no part in the electrolytic conduction. (Schwarz, 1932–33, Manegold 1932–35).

(ii) *Glazes and Enamels*. The term *glazes* is applied to thin coatings of glass on objects of sufficiently refractory character. Glazes which have been rendered opaque by adding certain 'opacifying agents' are called *enamels*. [52] Enamelling is used especially for iron ware. It is carried out as follows. The thoroughly clean article is covered with glazing composition by dipping it in a suspension made by finely grinding a suitable glass (containing felspar and boric acid) with water; after drying, this is melted on by strong heating. The object can be covered with the glazing composition by spraying the suspension instead of by dipping, or by air blowing with the dry powder. The coating enamel is melted on to the ground glaze in a similar manner. Melting is carried out in two (or sometimes more) layers, because enamels generally contain SnO_2 as an opacifier. At the high temperature of fusion, this would oxidize the carbon contained in the iron to CO, if it were to come into contact with the iron, and bubbles would be formed in the enamel in consequence. For the glazing of porcelain objects cf. p. 510.

(b) Clay, Kaolin and Ceramic Products

(i) *Clay and Kaolin* [39]. The term *clay* is applied to a group of substances, derived from the weathering of felspathic rocks, which are earthy and soft when dry, sticking to the tongue, but form more or less plastic masses when they are moist. The constituents making up the clays may vary in nature, according to the kind of rock from which they were formed by weathering. Purely analytically (without specifying how they are combined), the principal constituents are found to be Al_2O_3, SiO_2, and H_2O. It was formerly assumed that *kaolin* constituted the true clay substance, i.e., that the various clays were all kaolins, contaminated to a greater or less extent by impurities. According to modern investigations, however, clay and kaolin are substances of quite different character, even though they happen to have the same composition. The valuable *ceramic clays* do, nevertheless, contain considerable amounts of kaolin mixed with them, which has been elutriated from its primary deposits and deposited again later, together with other colloidally dispersed minerals, which also occur on their own as clay forming materials.

The chief constituent of kaolin is *kaolinite*, a compound which is shown by its X-ray pattern to be crystalline, having the composition $Al_2O_3 \cdot 2SiO_2 \cdot 2H_2O$. This compound is either absent from the clays, or present only as an accidental impurity. Pure kaolin is white in color, and has a relatively low plasticity. Since it is used for the production of porcelain (or china) it is also known as *china clay*.

The *clays*, which are often considerably superior to kaolin in plasticity, are used for the manufacture of stoneware, pottery, faience, and majolica. They are generally yellowish, grey or blueish in color, but may also be pure white. Clays rich in iron oxide turn brown or red when they are ignited ('fired'). Ordinary earthenware pots and terracotta are made from these. Clay containing much sand as impurity, as well as iron oxide, is called *loam*. It is used primarily for the production of bricks and tiles. Clays contaminated with much calcium and magnesium carbonate are termed *marls*. It is not possible to use these for ceramic purposes. They are, however, used to some extent in the cement industry.

Formation of clay takes place through the weathering of silicate rocks, which is accompanied by a far reaching *mechanical* subdivision of the rock (reduction to colloidal subdivision). Along with this, a *chemical* process takes place, namely the hydrolysis of a larger or smaller proportion of the silicates—especially the felspars—to form amorphous aluminasilica gels. The latter, which are known as *allophanes*, may be mere mixtures of aluminum oxide and silicon dioxide hydrates, or amorphous, hydrated aluminum silicates of the type of the so-called 'pro-kaolin'. This is probably a definite compound of the formula $Al_2O_3 \cdot 2SiO_2$. It contains variable amounts of water which are not chemically combined with it, as they are in kaolinite, but are present as gel water. Constituents of rocks which have undergone only mechanical subdivision, and which are therefore still crystalline, are also generally present in clays in a state of colloidal subdivision. (Correns, 1936).

The special properties of clays are brought about by certain constituents which have layer lattice structures, with sheets built up from hexagonal networks of silicon-oxygen tetrahedra. Like the permutits (cf. p. 513), they display a more or less extensive capacity for cation-exchange (cf. Vol. II, Chap. 16). The most important of these are *kaolinite* and substances related to it (e.g., halloysite, $Al_2O_3 \cdot 2SiO_2 \cdot 4H_2O$), *montmorillonite* and certain mica-like minerals (Hofmann, *Z. Krist.*, (A) 98, (1937) 31; *Die Chemie*, 55, (1942) 283). The relationship between the crystal structures of these clay-forming minerals is noteworthy. The amorphous constituents of clays (allophanes), usually admixed with the (chiefly colloidal) crystalline minerals, do not seem to be of great significance for the properties of the clays.

The formation of *kaolin* from felspar, as was shown by Schwarz (1933 and later), occurs under the special conditions of elevated temperatures and pressures, and is considerably promoted by the presence of strong acids, though not of carbonic acid. Kaolinization is a purely chemical reaction, represented by the over-all equation

$$2K[AlSi_3O_8] + 7H_2O = Al_2[Si_2O_5](OH)_4 + 4H_2SiO_3 + 2KOH$$
$$\text{felspar} \qquad\qquad \text{kaolinite}$$

Kaolin can be formed not only directly out of felspar, but also subsequently out of the pro-kaolin which is the first product of ordinary weathering. Only in laboratory experiments, however, is the formation of kaolin from felspar restricted to elevated temperatures and high pressures (Noll, 1935 and later). It can also take place at ordinary temperatures during geological ages. Strong acids favor the formation of kaolin because they accelerate the hydrolysis of felspar, but they are not essential. Starting with alkali-free hydrolysis products of felspar, the formation of kaolin may be observed even in the absence of acids. Thus Noll, by heating with water under pressure, was able to prepare kaolin synthetically by starting, for example, with a mixture of amorphous SiO_2 and böhmite or bayerite. Heating in the presence of sodium hydroxide yielded montmorillonite. It would thus appear that, in Nature, the formation of kaolinite takes place when the alkalis and alkaline earths have been completely leached out of the original rock, montmorillonite being formed otherwise. It follows that the natural formation of kaolin must be favored especially by the intensive action of water, and by the acid reaction of the water involved. As has been stated, kaolin is transformed into montmorillonite, $AlSi_2O_5(OH)$, by heating it under pressure in a weakly alkaline medium (in contact with alkali carbonate solutions), whereas in strongly alkaline media *zeolites* are formed.*

When heated, kaolinite first loses its water (under a pressure of 10 mm at 430°), the manner in which the water is given up indicating that it is chemically bound. On stronger heating, the dehydrated kaolinite (metakaolinite) first decomposes into Al_2O_3 and SiO_2; at still higher temperatures these form *mullite*, $3Al_2O_3 \cdot 2SiO_2$, together with tridymite.

X-ray analysis has shown that kaolinite is built up from $[Si_2O_5]^{2-}$ sheets, between which are arranged pairs of $[Al(OH)_2]^+$ sheets. The minerals *dickite* and *nacrite*, occurring in many kaolins, have the same composition as kaolinite; they give different X-ray diffraction patterns from kaolinite, but appear to be very similar in structure (Kerr, 1930; Gruner, 1932).

Pure clay is used in medicine as a powder, under the name of 'bolus alba' (βῶλος, lumps of earth).

Articles fabricated from naturally occurring or artifically produced plastic mixtures of clay or kaolin with other substances are known as 'ceramic products' (from κεραμικός, earths or clays). A mass is said to be 'plastic' when it is viscous, but can be moulded under gentle pressure into any desired shape, and retains that shape when the pressure is removed. The most important ceramic products, with their principal distinguishing characteristics, are summarized in Table 81. [40–42]

TABLE 81

THE MOST IMPORTANT CERAMIC PRODUCTS

Porcelain	Body dense, transparent, white
Stoneware	Body dense, not transparent, white or colored (grey, yellow, brown)
Whiteware	Body porous, not transparent, white or nearly white. Not so strong as the foregoing.
Pottery	Body porous, not transparent, colored
Bricks	Porous, rather coarse grained, generally reddish in color
Refractories	Porous or dense, not melting below 1600° C

(*ii*) *Porcelain* [*42*], known in China since very early times, was first manufactured in Europe at Meissen** (since 1710).

It is obtained by strongly heating ('firing') plastic masses made by kneading together kaolin with powdered felspar and quartz, with the addition of a little water. If the temperature is not raised too high in the firing, the shape of the article is preserved, but its volume diminishes considerably: the porcelain shrinks

---

* Montmorillonite is the principal constituent of *bentonite*, which occurs in great deposits in the United States. This mineral finds technical applications, not only for cosmetic purposes and as a filler for soap, but for plasticizing non-plastic substances and, especially, for stabilizing suspensions, since it is characterized by especially great swelling power and dispersive properties. It is therefore of very great economic importance in the 'drilling muds', used in drilling for petroleum. These serve to carry out, in suspension, the debris from the drilling operations. The origin of the properties of montmorillonite have been explained, to a great extent, by the chemical colloid and X-ray investigations of Hofmann (*Z. Elektrochem.*, 41, (1935) 469). On this basis he was able to prepare, from European minerals, materials with properties similar to American bentonites.

** The physicist Ehrenfried von Tschirnhaus must be regarded as the inventor of European porcelain, which he succeeded in making for the first time after years of systematic experiments. Because of the unsettled nature of the times, the porcelain factory which he had already planned in 1703 was first erected, after his death, by the alchemist Böttger. Böttger, a man of questionable character, whom von Tschirnhaus first brought into his ceramic experiments in 1707 as an assistant, professed to be the discoverer of porcelain after the death of his master in 1708. This claim did not go uncontested even at the time.

during firing. At the same time the mass (the 'body') becomes dense (impervious to water) and coherent.

About 50% kaolin, 25% felspar and 25% quartz are generally used for hard porcelain. On firing, the kaolin first gives up its water of constitution, decomposing into Al_2O_3 and SiO_2; these are dissolved by the felspar, which softens vitreously. On further increase of temperature, the felspar also dissolves the coarsely crystalline quartz in increasing amount. In proportion as it becomes enriched in SiO_2, mullite is deposited, since the solvent power of felspar for mullite decreases with rising SiO_2 content. The finished porcelain therefore consists of a vitreous ground mass, interspersed with intimately felted needles of mullite, undissolved grains of quartz and minute air bubbles. For tableware it is usual to perform two or three firings. After the first, or 'rough firing', which is carried out at about 900°, a transparent *glaze* is applied: the still porous 'body' obtained by the first firing is immersed in a glazing dip, consisting of a suspension of kaolin, clay, felspar and marble in water. This, on heating, forms a high-melting glass. After drying, the glaze is fired on ('glost-firing') at about 1450°. A third firing frequently follows, carried out in a muffle at a red heat, after pigments—i.e., finely powdered colored glasses, ground up in turpentine oil—have been applied to the glazed body. Under-glaze or hard-fired colors are more durable, applied to the still unglazed body, but this is possible only for a few pigments. 'Biscuit' is the name applied to hard fired, unglazed porcelain. Electrical, chemical, sanitary and other wares are produced in Europe by processes similar to that for tableware, but in the United States the body and the glaze are developed for such objects in the same firing process. A distinction is usually drawn between the two- or three-fire *china process* and the one-fire *porcelain process*.

Instead of moulding the porcelain mass by means of its plastic properties, it can be liquefied by the addition of small amounts of alkali. The resulting 'slip' can be poured into plaster moulds, in which solidification takes place rapidly, as a consequence of the absorption of water by the burned plaster. In place of a dipping process, glaze can be applied to the objects by spraying.

As well as being used for househould ware and ornaments, porcelain finds widespread application for chemical equipment and for electrical insulators, since porcelain has excellent insulating properties.

'Soft porcelain', used chiefly as a material for the manufacture of ornamental objects, differs from ordinary or hard porcelain in its smaller kaolin content, and a correspondingly greater proportion of 'fluxes', such as felspar, or chalk. In accordance with its greater fusibility, it is fired at a lower temperature (usually 1200–1300°). Because of these properties, it is more readily covered with underglaze painting in many colors.

(*iii*) *Stoneware*, like porcelain, is dense, resonant and hard, so that it is not scratched by steel. It is likewise very resistant to chemical attack. It is prepared form clay, has a lower firing temperature than hard porcelain (1200–1300°), but is not translucent and generally not white but grey, yellow or brown. It is often covered only with a thin 'salt glaze', produced by throwing common salt into the furnace; this evaporates forming a vitreous sodium double silicate on the surface of the earthenware. Unglazed earthenware vessels are also much used in the chemical industry.

Fine earthenware serves as the material for vases and other ornamental products, and in architecture for reliefs and facades. The antique German ware (drinking mugs and the like) in grey, ornamented in blue, was made of earthenware. Examples of articles made from coarse brown earthenware are water pipes and drain pipes. Many articles made in brown earthenware are to be found used in the chemical industry—tourils, cooling coils, pipes, troughs and other vessels.

(*iv*) *Whiteware*, like porcelain, is white or nearly white, but is softer (it is scratched by steel), more brittle and porous. For most uses, therefore, it must be glazed. The glaze is prepared from a mixture of clay, quartz, alkali and red lead, and pigmenting oxides are often added also. Whiteware is fired twice: first, without the glaze, at 1200–1300°, and then, rather less strongly, with the glaze (glost firing). White sinks, baths and similar objects are made out of whiteware. Coarser varieties of whiteware (e.g., toilet sets) are often given a pale cream color by means of titanic acid. Examples of unglazed whiteware are porous cells, clay pipes, etc.

(v) *Faience* possesses an unattractive, dirty grey porous body. It is therefore covered with a white glaze, made opaque by the addition of stannic oxide. It was formerly widely used for cheap utensils, but has been almost completely displaced from household uses by the chinaware discovered by Josiah Wedgwood in England. Finer sorts of faience are suitable for artistic ceramics. *Majolica* ware, usually covered with colored glazes, is related to faience.

(vi) *Ordinary pottery*—e.g., flower vases, earthenware cooking vessels—also has a porous body. The glaze usually contains lead, and is generally colored by the addition of metallic oxides—green colors with copper, yellow with iron oxide, or brown if manganese dioxide is also added.

(vii) *Tiles and bricks* are fired from loam ('brick earth'), and are usually red in color through the presence of iron oxide. Since they are fired at relatively low temperatures, they are highly porous. Strongly fired, dense and very strong bricks are known as hard brick.

(viii) *Refractory Materials* [*43*]. Materials are termed 'refractories' if they can withstand high furnace temperatures (at least 1600°) without melting. One of the most commonly used refractories is *fire clay* ('schamotte'), which consists of a mixture of as refractory a clay as possible (the actual fire clay), fired to the point of sintering, with fresh, plastic clay (clay binder). Certain deposits of clay lend themselves particularly to the production of fire clays. Fireclay, usually containing about 42–45% Al_2O_3 and 50–54% SiO_2, is used especially for lining hearths, blast furnaces and hot-blast heaters (Couper stoves). For coke ovens, ceramic kilns and for furnaces in the steel industry (e.g., Siemens-Martin furnaces) it is usual to employ *dinas bricks*, which were first manufactured in England. They are produced by firing a coarse-grained quartz sand, mixed up with a little of a suspension of lime, or with clay. Clay-dinas bricks contain 15–17% Al_2O_3 and 80–83% SiO_2. They begin to soften at 1350°, but melt only above 1650°. They are surpassed in refractory properties by the lime-dinas or *silica* bricks (with 1.5–4% CaO, 0.3–2% Al_2O_3 and 94–96% SiO_2), which begin to melt only at 1700–1750°. These are used especially for open-hearth furnaces. The so-called *sillimanite* bricks are more refractory still. These are prepared from sillimanite, cyanite or andalusite (minerals of the same composition, Al_2SiO_5, but different crystal structure—cf. p. 501), by firing at a high temperature, whereby *mullite* $3Al_2O_3 \cdot 2SiO_2$, is formed; this has already been mentioned as a constituent of hard porcelain.

Among refractory materials which do not contain silica, or only in minor amounts, there may be mentioned bauxite bricks, dynamidone bricks, magnesite and dolomite bricks. Magnesia, zirconia and—in the absence of air—especially graphite are highly refractory.

(c) Cement

If limestones containing clay, or intimate mixtures of limestone and clay, are fired at a high temperature, and are then finely ground up, the products have the property of setting to very strong masses when mixed into a paste with water. Unlike lime mortar, they do this without requiring the access of air containing carbon dioxide, and will thus set under water. They are called *cements*, and find application [*44–46*] for the production of the so-called *hydraulic* mortars, for use under water. Cement is also widely used, however, above ground, and its greatest importance lies in its use in the preparation of *concrete* [*45–51*].

Cement mortar is prepared by mixing cement with 1–2 parts of sand for use under water or, for use, in air, with 3 parts of sand; or with 6 parts of sand and $\frac{1}{2}$ part of quick lime, adding water to make the whole of a pasty consistency. *Concrete* is obtained by mixing cement mortar with gravel or stone chippings. Among its other properties, the ability of concrete to adhere tightly to iron is important. Concrete is therefore widely used as a structural material, in conjunction with inserted steel rods, which increase its strength yet further (reinforced concrete). Iron is not merely unattacked by concrete, but is actually protected from rusting by it.

The best cement is *Portland cement*. Other sorts which may be mentioned are *Puzzolana cement, Roman cement, slag cement* and *hydraulic lime*.

(*i*) *Portland cement* is a greenish grey, very finely powdered material, consisting of a mixture of strongly basic calcium silicates, aluminates and ferrites. It is prepared by burning a finely divided mixture of limestone and clay, which is heated up to about 1400° so that sintering takes place with the formation of a dense clinker which is subsequently ground up to a fine dust.

The average composition of Portland cement is roughly represented by the following figures:

CaO	63%	Alkali	0.5%
MgO	1.5%	SO_3	1.5%
SiO_2	21%	Insol. in conc. HCl	0.5%
Al_2O_3	7%	$CO_2 + H_2O$ (ignition loss)	2%
Fe_2O_3	3%		

Experience shows that the best ratio of the total quantities of 'basic' to 'acidic' constituents in the finished Portland cement lies around 2 : 1. This ratio is termed the '*hydraulic modulus*' of the cement. In the above example this is 2.0 : 1; it should be at least 1.7 : 1. The principal basic constituent is calcium oxide, but Portland cement usually contains small quantities of magnesium oxide and alkali as well. The principal acidic constituent is silica, and after this aluminum oxide; a few per cent of iron oxide is usually present as well. The setting of cement depends essentially on the fact that the basic calcium silicate that it contains is gradually decomposed by the added water, to form monocalcium silicate, $CaSiO_3$ (or an acid silicate, $Ca[SiO_2(OH)_2]$, see p. 498), and calcium hydroxide. The aluminates and ferrites are decomposed simultaneously. In course of time, the calcium hydroxide, deposited as a primary product, is converted to calcium carbonate. It appears that the formation of a compound with the composition $2CaO \cdot SiO_2 \cdot H_2O$ also plays an important role in the setting of cement. This is found in nature as the rare mineral *hillebrandite*. According to Thilo, it can be regarded as both a basic and an acidic salt

$$Ca_2(OH)[SiO_3(OH)].$$

This formula is based on a variety of chemical evidence—for example, the fact that it gives up its constitutional water in two stages when it is heated (one half at 350°, the rest at 600°). The products of decomposition and transformation, formed in the setting of cement, separate out in the form of crystals of microscopic size, intimately felted together, and probably also separate to some extent in an amorphous state, forming a gel. A texture thus develops which becomes increasingly strong with the growth of the crystallites. The setting of lime mortar (air mortar) differs from that of hydraulic mortar primarily in that the essential role is played in the latter by the formation of calcium *silicates*, and in the former by the formation of calcium *carbonate*.

(*ii*) *Slag Cement*. If blast furnace slag is quenched with cold water, a sandy product is obtained which, when finely ground, is also 'hydraulic'—i.e., it has the property of setting with water. To improve the quality, and to bring the composition up to that of Portland cement, ground limestone is added to the ground slag, which initially contains 45–50% CaO. The mixture is burned to a clinker, like Portland cement, and is very finely ground. To make it cheaper, a further 30% of ground slag is mixed in. The product is sold under the name of slag cement or 'iron Portland cement'. Only slags originating in the smelting of *grey* cast iron are suitable for its production.

(iii) *Roman cement* was originally prepared by burning naturally occurring argillaceous limestones. The products obtained by firing artifical mixtures are now also known as Roman cement, which differs from Portland cement in its lower CaO content (down to about 50%), and in that it is burned below the sintering limit. Roman cement is yellowish in color; it is inferior in strength to Portland cement.

(iv) *Puzzolana cement. Tufa*,—i.e., volcanic lava quenched by water—often displays hydraulic properties in Nature. The ancient Romans prepared a cement by grinding up such tufa (pulvis Puteolanus, Puzzolana earth, from its occurrence at Puteoli). The Greek Santorin earth, and the *trass* obtained by grinding up the tuff stones of the Eifel, are also natural cements of this type.

(v) *Hydraulic lime.* By the careful burning of limestones with only 10–20% clay content, products are obtained which can be slaked in lumps, because of their high CaO content; they thereby break up into a powder so that they do not need to be ground up. Such 'hydraulic limes' are cheap to manufacture, but are usually not used alone, but admixed with cements rich in silicic acid, such as puzzollana or trass.

(d) Permutits and Zeolites

By fusing quartz with kaolin and sodium carbonate (in the proportions by weight 1 : 2 : 4, for example), a glassy product is obtained. When this is granulated and treated with water, it changes into a yellowish white flaky mass, which has the property of exchanging the sodium that it contains for other metals, when introduced into their solutions. It is therefore known as a *permutit* (*permutare*, to exchange), or base-exchanger. The preparation can also be carried out by wet methods, by adding sodium silicate to a warm solution of sodium aluminate. Such a base-exchanging silicate takes up calcium or magnesium ions from the solution if it is placed in water containing lime or magnesia, an equivalent quantity of sodium ions from the silicate going into solution. Extensive use is made of this property for softening boiler feed water, and water for laundering and dyeing.

Not only the calcium content, but also the presence of iron and manganese, even in traces, is objectionable in laundering and dyeing. To remove these elements completely, a cation exchanger containing manganese is employed. The manganese is first oxidized by means of potassium permanganate, and is then able to transfer its oxygen to the iron(II) and manganese(II) salts dissolved in the water. The hydrated oxides of iron(III) and manganese(IV) are thereby thrown out of solution.

If, after taking up calcium ions, etc., a base exchange material is allowed to react with a solution of sodium chloride, the calcium (or other) cations are once more replaced by sodium. This exchange can be repeated indefinitely.

The *zeolites* are a class of crystalline compounds occurring in Nature, and possessing properties similar to those of the permutits. Thus *gmelinite* (sodium chabazite), which is one of the zeolite group, can have its sodium exchanged for magnesium by immersing it in magnesium chloride solution. It is also characteristic of zeolites that they can lose water continuously, without suffering any destruction of the crystal structure.

The permutits also give up water continuously on heating. Like the zeolites, these substances do not display a step-wise increase of decomposition temperature as the water content diminishes, as is characteristic of most crystalline hydrates (cf. p. 74). Instead they yield curves of the type represented in Fig. 89, with inflection points instead of steps. This is connected with the fact that, in these compounds at least some of the water molecules do not occupy fixed positions in the crystal lattice. The same is true of the alkali cations. Such constituents of a compound, which do not occupy definite positions have been said to 'wander' through the crystal lattice (Hüttig, 1924). The particular readiness with which the alkali cations in the zeolites and permutits undergo exchange depends not only on their lack of uniquely defined positions, (or, in the case of the amorphous permutits, on their undefined location within the aluminosilicate skeleton), but also on the peculiar structure of the alumino-silicate skeletons themselves. The X-ray diffraction data show that the SiO_4 and AlO_4 tetrahedra in the zeolites are linked in rings, which are so stacked one over the other as to form channels, running right through the crystal structure. The water molecules and exchangeable cations are located within these channels, and since they do not occupy fixed positions are able to diffuse out easily. Part of the water in permutits, and in most zeolites, is more firmly bound than the rest. Thus from *heulandite*, $CaAl_2Si_7O_{18} \cdot 6H_2O$, only part of the water can be expelled by heating without destroying the crystal lattice (Rinne 1924, Wyart 1933), and only a part can be removed by extraction with liquid ammonia (Gruner, 1933). The firmer binding of a portion of the water expresses itself in the thermal degradation curve of heulandite, in that there are inflection points (at rather less than a content of $3\frac{1}{2}H_2O$ and in the neighborhood of $2\frac{1}{2}H_2O$) (Fig. 93). The behavior of other zeolites and of permutits is similar. The water is not freely mobile in all the zeolites, however (Glemser, 1944). There are zeolites, such as natrolite and analcime, in which it occupies definite lattice positions. It may, however, be equally readily removed from the crystal lattice, re-introduced or replaced by other substances, without involving major changes in the crystal structure, as a consequence of the special structure of the aluminosilicate skeleton of the zeolites.

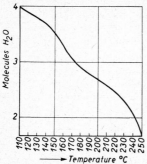

Fig. 93. Thermal degradation curve of heulandite (isobaric, $p = 7$ mm), as an example of the behavior of a zeolite when water is expelled by heating.

Singer observed in 1910 that blue to blue-grey substances could be obtained by treating base exchanging silicates with sulfide solutions, in the presence of atmospheric oxygen. These products were apparently related to the ultramarines. According to Gruner (1932), the first process, in the absence of oxygen, is an anion-exchange, replacing OH^- ions by SH^- ions. The resulting 'sulfide permutits' are colorless, but on exposure to air are changed nto colored 'polysulfide permutits', through the production of polysulfide groups, S_2^{2-}, by he oxidation of pairs of SH^- ions:

$$2SH^- + \tfrac{1}{2}O_2 = S_2^{2-} + H_2O$$

If the reactions are carried out at high temperatures, true ultramarines are obtained.

(e) Ultramarine

Ultramarine is a mineral pigment, which occurs in nature as *lasurite*, (*lapis lazuli*). It is prepared artically by fusing together clay, anhydrous sodium carbonate and sulfur; or clay, sodium sulfate and charcoal. Lapis lazuli is still valued as a gem stone, because of its fine blue color. The color of artificially prepared ultramarine is variable, according to the manner in which it has been prepared. Ultramarine blue possesses the greatest importance, and is one of the most important mineral pigments. Ultramarine violet and ultramarine red also find applications; ultramarine green less often today. Ultramarine blue is used as a pigment in oil and distemper paints, as a color in book and wallpaper printing, and in paper making. Because of its power of masking yellowish tints, making them appear white (by the action of the complementary color), it is used extensively for 'blueing' washing, paper, sugar, starch and many other materials. Good varieties are completely fast towards light and soap. Ultramarine is insoluble in all solvents. It is attacked, however, even by weak aqueous acids, being decolorized with the evolution of hydrogen sulfide and the deposition of silicic acid. It is therefore not suitable as a blue for preserving sugar. It is stable towards dilute alkalis at ordinary temperature.

If kaolin, sodium carbonate and sulfur are heated together in the absence of air, a colorless ultramarine is obtained, but green ultramarine is formed if air has access. The latter was formerly converted into blue ultramarine by repeated ignition. A method of preparing blue ultramarine by a single firing is now known. If finely powdered quartz or kieselguhr is added to the reaction mixture, an ultramarine with a reddish hue is obtained, which is stable towards alum. Prolonged, mild heating of silica-rich ultramarine blue with ammonium chloride yields a violet, which is converted into red ultramarine by heating it in chlorine.

The artificial production of ultramarine seems to have been first suggested by Goethe (in his 'Italian Journey'). It was first accomplished in 1828 by Gmelin, by a wet process, and the manufacture of the material was taken up industrially by the Meissen porcelain works in the following year.

Analytical data indicate that natural ultramarine, or lapis lazuli, is a double silicate of aluminum and sodium, containing sodium polysulfide. The composition corresponds roughly to the formula $Na_8Al_6Si_6O_{24}S_2$, but may be subject to considerable variations (see below). The *structure* of ultramarine has been explained by X-ray investigations, chiefly by Jaeger (1927), and Podschus, U. Hofmann and Leschewski (1936). They show that an aluminosilicate skeleton is present in ultramarine, consisting of a three dimensional network of SiO_4 and AlO_4 tetrahedra forming equivalent, hollow cavities throughout the structure (a honeycomb), bounded by polyhedron surfaces made up of Si and Al atoms (linked together by oxygen), joined in rings. Each honeycomb cell, which has the roughly spherical form shown in Fig. 94, includes $^{12}/_2 = 6Si$ atoms, and the same number of Al atoms. Around each of

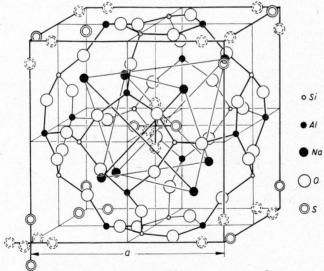

o Si

● Al

● Na

◯ O

◎ S

Fig. 94. Unit cell of ultramarine. $a = 9.06$ Å

these, in tetrahedral arrangement, are 4 O atoms, of which one, in each case, lies outside the unit cell represented—i.e., belongs to an adjacent cell. Within each honeycomb cell are 8 Na atoms, in the form of two interpenetrating tetrahedra, and in the center of each is an S_2 group, so placed that the S-atoms are distributed 'statistically' (i.e., ranging over all the possible positions, quite at random) on the corners of an octahedron which encloses the center of the cell. One of these octahedra is shown in the middle of the unit cell represented in Fig. 94. Similar octahedra, enclosing the point where eight adjacent cells meet, are situated at the corners of each unit cell. (In Fig. 94 the possible positions for sulfur atoms in one of these are shown.) The S_2 groups may consist either of neutral S_2 molecules or of disulfide radicals S_2^{2-}. It is also possible for a single S^{2-} ion to occupy the midpoint of some of the honeycomb cells. In this way the sulfur content of the ultramarines may vary within certain limits; it is usually between 2 and 4S per unit cell*. The constitution of ultramarine may accordingly be represented by the formula $Na_8[Al_6Si_6O_{24}]S_{2-4}$. The ultramarines often have a smaller (but never a higher) alkali content than corresponds with this formula. They then contain correspondingly less negative sulfide ions**. As was shown by Gruner (1935), the electroneutral sulfur can be driven out by heating ultramarine in a vacuum, without destroying the crystal structure. The S^{2-} ions can also be leached out of the lattice by melting with KCN, but an equivalent amount of Na^+ ions is removed at the same time. In this way, from an ultramarine of the composition $Na_8Al_6Si_6O_{24}S_2$, Gruner obtained a colorless compound $Na_6Al_6Si_6O_{24}$ in which, as X-ray examination showed, the aluminosilicate skeleton was still intact. It was transformed only at about 1050° into nepheline, which has the same composition. It follows from the X-ray work, and also from chemical investigations (Leschewski, 1932 and later), that the color of ultramarine is not attributable to colloidal sulfur, as was formerly frequently assumed. It is probably connected with the fact that in ultramarine the sulfur is present in the same form as in the polysulfides (cf. p. 738). Substantially the same structure as that of ultramarine is found in a series of other minerals—e.g., *nosean* (grey or colorless), $Na_8[Al_6Si_6O_{24}][SO_4]$, *hauyne* (blue), $Na_6Ca_2[Al_6Si_6O_{24}][SO_4]_2$ (Machatski, 1934), *sodalite* (colorless, or occasionally blue), $Na_8[Al_6Si_6O_{24}]Cl_2$, and *helvine*, (honey yellow) $(Mn,Fe)_8[Be_6Si_6O_{24}]S_2$ (Pauling, 1930). It is possible to impart a blue color to nosean by heating in sulfur vapor.

It is also possible to prepare ultramarines in which Na^+ ions are replaced by other cations—e.g., K^+ or Ag^+. The Na^+ ions of ultramarine can also be be exchanged, either wholly or incompletely, (e.g., by heating them with solutions of salts of Tl(I), Ca, Sr, Ba, Zn, etc.). The substituted ultramarines obtained in this way have various colors, according to the nature of the metal introduced in place of Na^+ ions. This implies that the positive ions situated within the aluminosilicate skeleton are also of influence upon the color.

9. Analytical (Silicon)

Silicon is almost invariably detected, and determined gravimetrically, in the form of SiO_2. A characteristic reaction for silicon is its conversion by treatment with hydrofluoric acid into a volatile compound (SiF_4) which is decomposed by water, depositing a white product.

The test is most sensitive if the SiF_4 vapors are evolved in a small crucible (of platinum or lead), which is provided with a perforated lid covered with a moist, black filter paper. White hydrated silica is then formed, according to the equation $SiF_4 + 2H_2O = SiO_2 + 4HF$, in such a way that a white spot is formed on the black paper, of the same size as the hole in the crucible lid (Biltz).

Quartz, in the presence of silicates, can be determined by heating to 220° with 85 per cent phosphoric acid. The silicates go into solution under this treatment, whereas the quartz remains unattacked. (Talvitic, *Anal. Chem.*, 23, (1951) 623).

10. Germanium (Ge)

(a) Occurrence

Germanium occurs in nature in the form of exceedingly rare minerals, such as *argyrodite*, $4Ag_2S \cdot GeS_2$ (found at Freiberg, Saxony and in Bolivia), and *germanite*,

---

* The amount of 'sulfide sulfur'—of S^{2-} ions— is almost always smaller than that of the electroneutral or 'polysulfide' sulfur. The former is found from the quantity of hydrogen sulfide evolved during dissolution in acids.

** The S^{2-} ions can be replaced not only by S_2 or S_2^{2-} groups, but also to some extent by H_2O molecules.

a copper iron thiogermanate. Germanite is known to occur only at one place—at Tsumeb, in South West Africa. It is found there in fairly large amounts, as inclusions in tetrahedrite, from which it clearly stands out by reason of its pink color. It was discovered in 1920 by Schneiderhöhn. As a minor impurity, germanium is also present in some other rare minerals, such as samarskite, tantalite and gadolinite. Small traces of germanium are found also in some North American zinc blendes. Germanium becomes enriched in certain smelting residues which, as was found by Buchanan in 1916, can contain up to 0.25% or more of GeO_2. Germanium is also often considerably enriched in the ashes of coals of low ash content. Ash from certain seams of coal in the Northumberland (England) coalfield, in particular, is rich enough to provide a source for the extraction of the element. (Morgan, 1937).

(b) History

Germanium was discovered in 1885 by Clemens Winkler. He noticed that in the analysis of a new silver ore, *argyrodite*, from Freiberg, the sum of the constituents found was invariably 6–7% too low, and recognized that this was due to the presence of a hitherto unknown element which he named *germanium*, after its occurrence in Germany. Subsequent investigation proved that it was the *ekasilicon* predicted in 1871 by Mendeléeff, on the basis of the Periodic Table.

The far reaching agreement between the properties predicted by Mendeléeff for ekasilicon, and those found for germanium may be seen from the comparison set out in Table 82.

TABLE 82

PROPERTIES OF GERMANIUM,
COMPARED WITH MENDELÉEFF'S PREDICTIONS FOR 'EKA-SILICON'

	Ekasilicon (Es)	Germanium
Atomic weight	72	72.60
Specific gravity	5.5	5.35
Formula of oxide	EsO_2	GeO_2
Density of the oxide	4.7	4.70
Formula of the chloride	$EsCl_4$	$GeCl_4$
Boiling point of the chloride	$< 100°$	83°

After the researches of Winkler, the chemistry of germanium remained practically untouched for more than 30 years, since the necessary material was lacking on account of the great rarity of the element. More extensive investigations became possible only after the discovery of germanite in South Africa, and of the presence of germanium in the smelting residues from North American zinc ores. These researches have been carried out in particular by Dennis (since 1921) and Schwarz (since 1929).

(c) Preparation

Germanite can be attacked by treating the finely powdered mineral with a nitric acid-sulfuric acid mixture. In this process, most of the germanium separates out as the dioxide. It is separated from other contaminating elements by dissolving it in 20 per cent aqueous hydrochloric acid, and distilling out the volatile tetrachloride, which is received in water. The pure germanium dioxide formed by hydrolysis is dehydrated; it is reduced to the metal by fusion with potassium cyanide and wood charcoal, or by heating in a current of hydrogen. According to Johnson, it is better in working up large quantities of ore to employ dry methods of attack. This is carried out by heating the finely powdered germanite to 800° in a current of oxygen-free nitrogen, whereby admixed arsenic sulfide and sulfur are removed. Ammonia gas is then passed over the residue at 825°, whereupon GeS sublimes out. This can readily be converted to the dioxide by treating it with nitric acid. The metal is then obtained by reduction with hydrogen or carbon.

(d) Properties

Germanium is a very brittle, grey-white, lustrous metal. It crystallizes in the cubic system. Its hardness is about 6.5, density 5.35 (at 20°) and melting point 958°. Compact germanium is unaltered in air. It combines with oxygen above a red heat. It neither unites directly with hydrogen nor occludes hydrogen to a detectable extent. At high temperatures it alloys readily, however, with platinum, gold, silver, copper and other metals. A eutectic Ge-Au alloy, with 24 atom-% Ge, has a melting point of 359°, remarkably low for an alloy of gold. Germanium is insoluble in hydrochloric acid, and also in dilute sulfuric acid. It dissolves, however, in hot concentrated sulfuric acid, with the evolution of SO_2. It is converted into the hydrated dioxide by moderately concentrated nitric acid, as is tin. It is not attacked by dilute caustic potash, but very readily by an alkaline solution of hydrogen peroxide.

It is also readily brought into solution by anodic oxidation, and under these conditions is converted directly to the quadrivalent state (Jirsa, 1952). Germanates are thereby formed in alkaline solutions, and germanium(IV) salts in acid solutions.

As accords with its position, intermediate between Si and Sn, the properties of germanium and its compounds are related in many respects to those of silicon, whereas in other ways it resembles tin. Like silicon, it crystallizes with the diamond structure (with the cube edge $a = 5.63$ Å). Under certain conditions (high vacuum evaporation), germanium can also be obtained in the *amorphous* state. As in crystalline germanium, every Ge atom in amorphous germanium is surrounded tetrahedrally by four others (distance Ge↔Ge = 2.40 Å) (Fürst, Glocker and Richter, 1949), whereas in liquid germanium every atom has 8 neighbors (at a mean distance of 2.70 Å) (Hendus, 1947). Germanium forms several volatile hydrogen compounds, and also highly unsaturated solid hydrides. The latter are more readily obtained than the corresponding silicon compounds. Like silicon, germanium is also capable of serving as the central atom in heteropolyacids (cf. Vol. II). GeO_2, like SiO_2, can solidify from the melt in vitreous form. Whereas it usually crystallizes with the same structure as quartz (although this form is metastable below 1033°), it is also capable of crystallizing with the rutile structure, like SnO_2. The germanates and fluorogermanates are completely isomorphous with the corresponding compounds of silicon. Germanium, like tin, can function as a positive ion in aqueous solutions, and forms typical salts. It can also be precipitated from aqueous solutions as the sulfide (which is able to form thiogermanates). Its compounds are also easily reduced.

11. Compounds of Germanium

Halides, oxides and sulfides of both bivalent and quadrivalent germanium are known. Germanium is also quadrivalent towards hydrogen and alkyl radicals.

The compounds of bivalent germanium are quite unstable. They have the tendency to be oxidized to germanium(IV) compounds.

(a) Germanium(IV) Compounds

(i) *Germanium hydrides.* If zinc is allowed to react with sulfuric acid which contains germanium, the escaping hydrogen contains some *germanium hydride*, GeH_4, as was found by Voegelen in 1902. If the gas mixture is passed through a strongly heated glass tube, the germanium is deposited on the wall as a bright metallic mirror, which appears red by transmitted light.

Voegelen deduced the formula of the compound from the composition of the silver germanide formed by passing the gases through silver nitrate solution (1), and also from the proportion in which hydrogen sulfide and germanium sulfide were formed by passage over finely divided sulfur, when exposed to light (2).

$$GeH_4 + 4AgNO_3 = Ag_4Ge + 4HNO_3 \qquad (1)$$

$$GeH_4 + 4S = GeS_2 + 2H_2S \qquad (2)$$

The formula was confirmed in 1922 by Paneth, who separated the germanium hydride from the associated hydrogen, by condensing it by means of liquid air, and decomposed it by heating. $GeH_4 = Ge + 2H_2$. The formula GeH_4 then follows from the ratio of the hydrogen formed to the germanium deposited.

Germanium hydride is obtained in relatively good yield by the action of sulfuric acid on the alloy of composition Mg_2Ge, obtained when germanium and magnesium are melted together. In addition to the ordinary gaseous germanium hydride (*monogermane*), m.p. —165°, b.p. —90°, formed as the main product, Dennis (1924) was able to identify two germanium hydrides which are liquid at ordinary temperature—*digermane*, Ge_2H_6 (m.p. —109°, b.p. +29°), and *trigermane*, Ge_3H_8 (m.p. —105.6°, b.p. 110.5°).

Monogermane is now most conveniently prepared by the reduction of $GeCl_4$ with $LiAlH_4$: in ether medium:

$$GeCl_4 + LiAlH_4 = GeH_4 + LiCl + AlCl_3$$

In addition to these volatile compounds, germanium forms highly unsaturated solid hydrides (*polygermenes*) $[GeH]_x$ and $[GeH_2]_x$. The former was obtained by Dennis, in the hydrolysis of NaGe. It forms a dark brown powder, which deflagrates upon admission of air when it is dry, is converted by oxidizing agents into germanium(IV) compounds, but does not react with dry hydrogen chloride. Schwarz obtained the hydride $[GeH_2]_x$ by decomposing CaGe both with dilute hydrochloric acid and with caustic soda. It is yellow in color, stable in the absence of air when dry, but at once ignites explosively in the air. With concentrated hydrochloric acid, it forms the series of volatile germanium hydrides—GeH_4, Ge_2H_6 and Ge_3H_8—together with $GeCl_2$ and H_2. It is possible, from this reaction, to draw some conclusions as to the mechanism by which silane is formed in the decomposition of Mg_2Si by concentrated hydrochloric acid.

(ii) *Germanium alkyls.* Germanium forms compounds of the general formula GeR_4 with organic radicals. Thus *germanium tetraethyl* is obtained by the action of zinc ethyl. $Zn(C_2H_5)_2$, on germanium tetrachloride:

$$2Zn(C_2H_5)_2 + GeCl_4 = 2ZnCl_2 + Ge(C_2H_5)_4$$

This is a colorless liquid, immiscible with water, with a weak garlic-like smell; b.p. 163°, density 0.991. It is possible to obtain the asymmetric compound $(C_2H_5)(C_3H_7)Ge(C_6H_5)Br$ in optically active forms, just as is the case with asymmetric carbon compounds (Schwarz, 1931).

(iii) *Germanium(IV) halides.* The following table indicates the most important physical properties of the germanium(IV) halides.

	Color	Melting point	Boiling point
GeF_4	colorless	sublimes at $-36.6°$	
$GeCl_4$	colorless	$-49.5°$	$+83.1°$
$GeBr_4$	colorless	$+26°$	$+186°$
GeI_4	orange	$144°$	above $300°$

Mixed halides are also known—e.g., GeF_3Cl (b.p. $-20.3°$), GeF_2Cl_2 (b.p. $-2.8°$), and $GeFCl_3$ (b.p. $+37.5°$).

1. *Germanium tetrafluoride and Fluorogermanates.* The action of concentrated hydrofluoric acid on germanium dioxide yields a clear solution, from which colorless, very hygroscopic crystals of hydrated germanium tetrafluoride, $GeF_4·3H_2O$, separate. If the attempt is made to dehydrate the salt by heating, hydrolytic decomposition takes place. Part of the germanium simultaneously volatilizes as the anhydrous fluoride.

The anhydrous fluoride is gaseous at ordinary temperature (liter weight 6.650 g). When strongly cooled, it condenses to a white flocculent mass which sublimes without melting when it is warmed.

Potassium fluorogermanate, $K_2[GeF_6]$, separates from a solution of germanium tetrafluoride to which potassium fluoride is added, in white hexagonal prisms or plates. It is not hygroscopic and is sparingly soluble in cold water, insoluble in alcohol. It is decomposed when it it heated to a red heat.

$K_2[GeF_6]$ and $(NH_4)_2[GeF_6]$ appear to be isomorphous with the corresponding fluorosilicates. The isomorphism of $Cs_2[GeF_6]$ with the alkali fluorosilicates has been proved by X-ray methods. Hydroxylammonium fluorogermanate, $[NH_3(OH)]_2[GeF_6]$ crystallizes with $2H_2O$, as does $[NH_3(OH)]_2[SiF_6]$, and hydrazinium fluorogermanate, $[N_2H_6][GeF_6]$, like $[N_2H_6][SiF_6]$ crystallizes anhydrous (Dennis, 1933).

2. *Germanium tetrachloride,* made by burning germanium in a stream of chlorine, or better prepared pure by warming germanium dioxide with fuming hydrochloric acid (preferably in a pressure flask—Bauer, 1933), is a mobile colorless liquid, of density 1.88, b.p. 83°, solidifying at $-49.5°$; distance $Ge \leftrightarrow Cl = 2.10Å$. It is slowly hydrolyzed by water, with the formation of hydrated, finely divided germanium dioxide.

In its concentrated hydrochloride acid solutions, germanium tetrachloride is present in the form of chlorogermanic acid, $H_2[GeCl_6]$, as is shown by the migration of the germanium towards the anode upon electrolysis. Laubengayer (1940) found that cesium chlorogermanate, $Cs_2[GeCl_6]$ (light yellow octahedral crystals of density 3.45) is isotypic with $(NH_4)_2[PtCl]_6$.

If germanium tetrachloride is heated with germanium, it undergoes partial reduction to the dichloride:

$$GeCl_4 + Ge \rightleftharpoons 2GeCl_2$$

By heating $GeCl_4$ in a hot and cold tube, Schwarz (1952) prepared the germanium compound, $[GeCl]_x$, corresponding to silicon monochloride. The unstable Ge_2Cl_6 was formed at the same time, and also the dichloride $GeCl_2$.

3. *Germanium oxychloride,* $GeOCl_2$, is formed by the reaction of $GeHCl_3$ (cf. p. 523) with Ag_2O, as a colorless liquid solidifying at $-56°$. It decomposes into Cl_2 and GeO when it is heated. The GeO is thereby obtained in a lemon yellow form, which passes into the ordinary black monoxide at 650° (Schwarz, 1932). Another oxychloride Ge_2OCl_6, which decomposes with formation of GeO_2 when it is warmed, was obtained by Schwarz (1931) by the action of O_2 on $GeCl_4$ at 950°. It is a colorless liquid, solidifying at $-60°$.

4. *Germanium tetrabromide*, GeBr$_4$, (colorless regular octahedra, $d = 3.13$, m.p. 26°, b.p. 185.9°), and *germanium tetraiodide*, GeI$_4$ (orange crystals, $d = 4.32$, m.p. 144°; for structure see p. 531) can be prepared by methods similar to those for the chloride. They are much more vigorously decomposed by water than is the chloride, and therefore fume strongly in the air. The iodide begins to decompose into GeI$_2$ and I$_2$ a little above its melting point.

(iv) Germanium dioxide, Germanic acid and Germanates. *Germanium dioxide*, GeO$_2$, is formed by strongly heating germanium or germanium sulfide in a current of oxygen, or by oxidizing these substances with concentrated nitric acid. It is a white, sandy powder which melts completely at 1115° after gradually softening, and solidifies from the melt as a glass; this devitrifies more readily than silica glass. Germanium dioxide is perceptibly volatile above 1250°. It is only moderately insoluble in water (solubility about 0.4 g in 100 g of water at 20°, about 1 g in 100 g at 100°).

As was shown by Schwarz (1931), the solubility of germanium dioxide is not independent of the quantity of solid phase present. This is because germanium dioxide is present in solution not only in molecular dispersion, but simultaneously in a colloidal state, and it accords with Ostwald's rule (Vol. II). The formation of a sol also explains the fact that although the solubility in the cold is considerably smaller than when hot, no turbidity results when a solution of germanium dioxide, saturated hot, is cooled down. The solutions display a perceptibly acid reaction, and have an appreciable electrical conductivity. The primary dissociation constant of the *germanic acid* present in them is about 10^{-9} at 20°, according to Pugh (1929) and Dennis and Gulezian (1932). Schwarz was able to show from dialysis measurements that, at a pH of 8.4–8.8, the ions present were Ge$_5$O$_{11}^=$, and not GeO$_3^=$ as in strongly alkaline solutions.

Germanium dioxide separates out in microscopic crystals when an aqueous solution is evaporated. It is dissolved with difficulty by acids, but readily by caustic alkalis. The acidic character of germanium dioxide thus considerably outweighs its basic character. The compounds of germanium dioxide with strongly basic metallic oxides are called *germanates*; they can be obtained both from aqueous solution and by melting the components together, and display some interesting relationships with the silicates (Schwarz, *Angew. Chem.*, 48, (1935) 219). The existence of ortho-, meta-, di- and tetragermanates has been detected from melting point diagrams—e.g., Li$_4$GeO$_4$ (m.p. 1298°), Li$_2$GeO$_3$ (m.p. 1239°), Na$_2$GeO$_3$ (m.p. 1083°), Na$_2$Ge$_2$O$_5$ (m.p. 799°), Na$_2$Ge$_4$O$_9$ (m.p. 1052°). Metagermanates—mostly as hydrates—usually crystallize from aqueous solutions.

Germanium dioxide crystallizes not only in the water-soluble, hexagonal modification, having the low-quartz structure ($a = 4.97$ Å, $c = 5.65$ Å), but also in a tetragonal modification, insoluble in water, with the structure of rutile ($a = 4.39$ Å, $c = 2.86$ Å). This changes extremely slowly into the hexagonal modification above 1033°. The converse transformation (hexagonal→tetragonal) may also be strongly inhibited; it can be accelerated by adding ammonium fluoride (Schwarz, 1943). The *vitreous* form of germanium dioxide has a solubility still greater than that of the hexagonal form.

If germanium dioxide is prepared by hydrolysis—e.g., of the tetrachloride—it separates out in hydrated form, as a *gel*. However, since GeO$_2$ has a considerably greater tendency to crystallize than has SiO$_2$, the particles of oxide gradually undergo aggregation on standing under water. If it has aged sufficiently, the oxide loses its water simply on standing in the air. The water is given off without any intermediate steps, from which it follows that GeO$_2$ does not form hydrates (Schwarz, 1931). Thus germanic acid, like carbonic acid, exists only in aqueous solution.

(v) Germanate Peroxyhydrates. As in the case of silicon, *peroxydrates* are obtained by the action of H$_2$O$_2$ on alkali germanate solutions. Schwarz (1930) prepared the germanate peroxyhydrates K$_2$Ge$_2$O$_5$ · 2H$_2$O$_2$ · 2H$_2$O, Na$_2$Ge$_2$O$_5$ · 2H$_2$O$_2$ · 2H$_2$O, and Na$_2$GeO$_3$ · · 2H$_2$O$_2$ · 2H$_2$O in the crystalline state, and showed (1935) that these were hydrogen peroxide addition compounds and not peroxysalts.

(*vi*) *Germanium disulfide and Thiogermanates.* If hydrogen sulfide is passed into an aqueous solution of germanium dioxide, no immediate precipitation is observed. Only when a sufficient quantity of a strong acid has been added is germanium disulfide, GeS_2, formed as a white precipitate.

The reason for this is that only in the presence of a large amount of hydrogen ions is the equilibrium $GeO_2 + 4H^+ \rightleftharpoons Ge^{++++} + 2H_2O$ displaced far enough to the right for the solubility product of GeS_2, $K_{sp} = [Ge^{++++}] \times [S^=]^2$, to be attained. The sulfide exhibits a relatively low solublility in water (0.455 g in 100 g of water); its aqueous solution gradually decomposes, hydrogen sulfide being evolved by hydrolysis:

$$GeS_2 + 2H_2O \rightleftharpoons GeO_2 + 2H_2S.$$

The disulfide dissolves readily in ammonium sulfide, forming thiogermanate ions:

$$GeS_2 + S^= \rightleftharpoons [GeS_3]^=$$

On the addition of acid (conversion of $S^=$ to H_2S), the reaction proceeds from right to left again, but goes to completion only at high H^+-ion concentrations.

By pouring alkaline thiogermanate solutions into acetone, Schwarz (1930) was able to isolate the compounds $Na_6Ge_2S_7 \cdot 9H_2O$ and $K_6Ge_2S_7 \cdot 9H_2O$ (very hygroscopic, long needle shaped crystals). In the solutions, however, the ions present are not $[Ge_2S_7]^{6-}$, but $[GeS_3]^=$, as was shown by Brintzinger (1934), who determined their ionic weights by the dialysis method. This is but one example of a principle following from observations on other substances also—that radicals often exist as structural units in crystalline compounds, which are not present in significant amounts in the solutions with which they are in equilibrium.

The thiogermanates are related to the naturally occurring double sulfides of germanium —the minerals *germanite* (cf. p. 516), rose colored radiating prisms, *argyrodite*, Ag_8GeS_6, regular holohedral crystals with a metallic luster, *canfieldite* $Ag_8(Ge,Sn)S_6$ and *ultrabasite* (a lead-germanium double sulfide). Germanite has a crystal lattice similar to that of zinc blende.

(*vii*) *Nitrogen Compounds of Germanium.* $GeCl_4$ reacts with liquid ammonia, forming *germanium imide*:

$$GeCl_4 + 6NH_3 = Ge(NH)_2 + 4NH_4Cl$$

When this is heated to about 150° in an atmosphere of nitrogen, it loses ammonia and is transformed into *germanam*, Ge_2N_3H. This decomposes above 300°, forming the *nitride* Ge_3N_4, which in turn dissociates into Ge and N_2 at about 1000°. These compounds are less stable than the corresponding compounds of silicon. Germanium forms no compound corresponding to silicon amide $Si(NH_2)_4$ (which itself decomposes at 0°) (Schwarz, 1930). On the other hand, in addition to those mentioned, it also forms nitrogen compounds in which germanium functions as *bivalent*—GeNH and Ge_3N_2. The former is hydrolyzed by water, forming germanium(II) oxide hydrate:

$$GeNH + H_2O = GeO + NH_3;$$

it is converted into germanium(II) nitride when heated:

$$3GeNH = Ge_3N_2 + NH_3$$

(Johnson, 1934).

(*viii*) *Germanium(IV) sulfate*, $Ge(SO_4)_2$, obtained as a white powder of density 3.92 by Schwarz (1931), by the action of SO_3 on $GeCl_4$, is hydrolyzed by water. It reacts vigorously and exothermically with caustic soda forming Na_2GeO_3 and Na_2SO_4. Thermal decomposition sets in at 200°.

(b) Germanium(II) Compounds

(*i*) *Germanium(II) chloride*, $GeCl_2$, is formed by passing $GeCl_4$ vapor over heated metallic germanium. It is a solid, which is very sensitive towards moisture and atmospheric oxygen. It reacts with water according to:

$$GeCl_2 + H_2O = GeO + 2HCl$$

(see below on formation of germanium(II) oxide hydrate); and with oxygen:

$$2GeCl_2 + O_2 = GeO_2 + GeCl_4$$

$GeCl_2$ is nearly colorless when cold, but turns orange when warmed. Decomposition occurs on stronger heating:

$$2GeCl_2 \rightleftharpoons Ge + GeCl_4.$$

(ii) *Trichlorogermane*, $GeHCl_3$, corresponds in composition to chloroform, and is therefore also known as *germanium chloroform*. It resembles chloroform in many of its physical properties, but is quite different chemically. The compound was first obtained by Winkler, by passing hydrogen chloride over gently heated powdered germanium:

$$Ge + 3HCl = GeHCl_3 + H_2.$$

A better process (Dennis, 1926) is to pass $GeCl_4$ over heated germanium, and to treat the product, consisting chiefly of $GeCl_2$, with HCl:

$$GeCl_2 + HCl = GeHCl_3.$$

Trichlorogermane forms a colorless not readily condensable liquid, which turns milky on exposure to air, through the formation of oxychloride; b.p. $75.2°$, m.p. $-71°$. Germanium must be present in this compound in the *bivalent* state, for it is decomposed by much water, depositing yellow germanium(II) oxide hydrate. Iodine has an oxidizing action:

$$GeHCl_3 + I_2 = GeICl_3 + HI.$$

(iii) *Germanium(II) bromide* (colorless needles or leaflets, m.p. $122°$) can be obtained by the action of HBr on metallic germanium at $400°$; $GeHBr_3$ is formed simultaneously, and may be reduced with zinc. The dibromide is soluble in alcohol and acetone, but insoluble in benzene. It is hydrolyzed by water, with deposition of $Ge(OH)_2$ (Brewer and Dennis, 1927).

$GeBr_2$ combines with HBr, to form tribromogermane, $GeHBr_3$, which decomposes again at high temperatures:

$$GeBr_2 + HBr \rightleftharpoons GeHBr_3.$$

Tribromogermane is a colorless liquid, solidifying at $-24.5°$.

The reaction of HBr on GeH_4 in the presence of $AlBr_3$ yields GeH_3Br ($d_0^{30} = 2.34$; m.p. $-32°$; b.p. $+52°$) and GeH_2Br_2 ($d_0^{\circ} = 2.80$; m.p. $-15°$; b.p. $+89°$), whereas the reaction with HI under the same conditions forms GeI_2 without any intermediate products (Dennis, 1929).

(iv) *Germanium(II) iodide*, GeI_2 (yellow hexagonal plates; crystal structure: brucite or CdI_2 type, $a = 4.13$ Å, $c = 6.79$ Å; sublimes at $240°$ in vacuum) is most simply prepared by the action of concentrated hydriodic acid on $Ge(OH)_2$. It is obtained only in very poor yield by passing GeI_4 vapor over heated Ge, since it disproportionates when it is heated, according to the equation $2GeI_2 = Ge + GeI_4$ (Brewer and Dennis, 1927).

(v) *Germanium(II) oxide*. The decomposition of germanium(II) chloride with water or sodium hydroxide yields a yellow precipitate, which turns rust red, being partially oxidized, when boiled with a little caustic soda. The yellow precipitate is probably a hydrogel of the yellow modification of germanium(II) oxide. The hydroxide $Ge(OH)_2$ appears to exist only in solution. The precipitate—which is known as germanium(II) oxide hydrate, since the mode of binding of the water is unknown—is redissolved by a larger excess of sodium hydroxide. According to Hantzsch (1902), the conductivity of the solution is perceptibly less than that of pure sodium hydroxide of the same concentration, showing that salt formation takes place with alkalis. Saturation of the sodium hydroxide solution with an exactly equivalent amount of hydrochloric acid does not precipitate germanium(II) hydroxide. The solution, containing only sodium chloride in addition to this compound, has a perceptibly acid reaction. From the magnitude of its electrical conductivity, it follows that germanium(II) hydroxide is definitely an acid, although weaker than acetic acid. If the precipitated oxide hydrate is heated to $650°$ in nitrogen, black crystalline germanium(II) oxide, GeO, is obtained (Dennis, 1930). Yellow GeO has been mentioned above (p. 520).

(vi) *Germanium(II) sulfide* can be obtained, beautifully crystallized, by heating germanium(IV) sulfide, GeS_2, not too strongly, in a stream of hydrogen. It forms leaflets, which are grey black, with metallic luster, by reflected light, and bright red by transmitted light. It is thrown down as a red brown precipitate by the action of hydrogen sulfide on the hydrochlorid acid solution of germanium(II) chloride. It is then soluble in caustic potash and —unlike the sulfide prepared by dry methods— also dissolves in hot concentrated hydrochloric acid and in ammonium polysulfide. In the latter case, through the oxidizing action of the dissolved sulfur, thiogermanates with 4-valent germanium are formed:

$$Ge^{II}S + S^= + S = [Ge^{IV}S_3]^=.$$

White GeS_2 is precipitated by acidifying the thiogermanate solutions, but red-brown GeS is thrown down unchanged from the solution in caustic potash. The latter has a strong tendency to form colloidal dispersions. Germanium(II) sulfide forms rhombic crystals; it has a unique structure, which may be described as a highly deformed rock salt crystal lattice.

(vii) *Germanium selenides*. GeSe (brown-black, tetragonal; density 5.30, m.p. 667°) and $GeSe_2$ (yellow, rhombic; density 4.56, m.p. 707°) have been prepared by Ivanov-Emin (1940). They can be obtained, like the sulfides, by precipitation from hydrochloric acid solutions, or by direct union of the components at 500°.

12. Analytical (Germanium)

A characteristic test for germanium is the white color of the disulfide, which can be precipitated from strongly acid solutions, and which is soluble in ammonium sulfide. In determining germanium gravimetrically as the dioxide, it should be noticed that when this is precipitated from sulfuric acid solutions it carries down considerable quantities of the acid. This can be removed by fuming down with concentrated nitric acid, subsequently igniting, and extracting with ammonia. By adherence to specified conditions, it is possible to deposit germanium, together with tin, quantitatively by electrolysis, (Schwarz, 1936).

13. Tin (Sn)

(a) Occurrence

The most important—almost the only—ore of tin is *tinstone* (*cassiterite*), SnO_2. In its primary ore deposits, the mineral is found interspersed in other rocks, especially in granite ('reef tin'), and in secondary deposits ('stream tin') in small grains, mixed up with large amounts of clay or sand; the tin content of the ores actually coming into consideration for the extraction of the metal is, in consequence, often only a few per cent, although the pure dioxide contains 78.62% of tin.

Stannite, $Cu_2S \cdot FeS \cdot SnS_2$, is found rather rarely, accompanying cassiterite. It is said that small amounts of tin are occasionally found native, together with gold.

The principal deposits of tin ores (i.e., of rocks bearing tin oxide) occur on the high plateau of Bolivia, in the Malayan peninsula (Malacca, Kuantan) and in Indonesia (Banka and Billiton). Very important deposits have also been found in Australia (New South Wales and Tasmania), and in Mexico, Siam and China. The continent of Europe is almost entirely dependent upon the importation of tin ores for its tin smelting, since the tin ores of Cornwall (England) and of other regions (e.g., the Saxon Erzgebirge) can now be worked economically only when the price of the metal rises to a very high figure. The recovery of tin from tin plate residues is of considerable importance.

The world production of tin (in the form of tin ores) rose from 99,600 tons in 1932 to 182,000 tons in 1936, and was 172,000 tons in 1950. In the latter year, Indonesia produced 32,000 tons, Malaya 57,500 tons, and Bolivia 31,000 tons.

(b) History

Tin is among the metals that have been longest known. It was already in use—in the form of bronze at least—in the earliest periods of human culture (Bronze Age).

In the Old Testament, in the book of Numbers, tin is mentioned as an object of value under the name of 'bedil'. It is referred to as 'trapu' in the ancient Indian writings (Veda, Mahabharata). The Greeks (Homer) called it κασσίτερος (it is not quite certain, admittedly, whether this word may not also have signified lead, since at that time tin and lead were not always clearly distinguished from one another.) Caesar, who reported the occurrence of tin in Britain, spoke of it as *plumbum album*, as distinct from ordinary lead (*plumbum nigrum*), as also did Pliny later. Up to some time in the 12th century, England (with the tin deposits of Cornwall) was the only European source of tin. The mines of the Erzgebirge (in Saxony, and later in Bohemia) then attained great importance for the extraction of tin, up to the time of the Thirty Years War. This almost completely destroyed the mining in that region, and only isolated smelters were operated again afterwards. In recent years, renewed attention has been paid to these and to the Cornish workings.

(c) Preparation

Tin is prepared by reducing cassiterite with coal in the blast furnace or reverberatory furnace:

$$SnO_2 + 2C = Sn + 2CO$$

Before it is introduced into the furnace, the tin ore is 'dressed'—i.e., freed from the most important impurities by mechanical means or, to some extent by chemical treatment (removal of sulfur and arsenic by roasting).

The slags arising from the reduction of tinstone are generally rich in tin, which can be extracted either by 'reduction smelting'—i.e., by melting them in a reverberatory furnace with limestone and coal (1)—or by 'precipitation smelting', i.e., fusion with scrap iron and coal (2).

$$SnSiO_3 + CaO + C = Sn + CaSiO_3 + CO \tag{1}$$

$$\left. \begin{array}{l} SnSiO_3 + Fe = Sn + FeSiO_3 \\ SnO_2 + 2C = Sn + 2CO \end{array} \right\} \tag{2}$$

As first obtained, the crude tin is still highly contaminated with iron, and to a lesser extent, with other metals also. It is first freed from iron by remelting it, on the inclined hearth of a reverberatory furnace ('liquation'), and then from other metals by an oxidizing fusion ('poling'). The former process depends on the fact that when tin containing iron is heated just above its melting point the iron remains behind in the form of a relatively infusible alloy (liquation dross). In 'poling', a billet of fresh green wood is plunged into the molten tin. The gases escaping from the wood as it chars, stir the tin vigorously, bringing it sufficiently well into contact with the oxygen of the air for the impurities to be oxidized and to collect (together with the oxidized portion of the tin) as a scum ('tin dross') on the surface of the tin.

For the recovery of tin from tin plate scrap, use is made either of *electrolysis*, whereby the scrap, in iron wire baskets, is made the anode in a solution of caustic soda, or of *detinning with chlorine*, the latter being based on the fact that tin, unlike iron, is readily attacked by dry chlorine.

(d) Properties

Tin is a silver-white, lustrous metal which is very soft but has considerable ductility, so that it can be rolled out into very thin sheets ('tin foil'). Its density is

7.28 and melting point 231.8°. It vaporizes to a considerable extent at temperatures as low as 1200°, although its boiling point is 2362°. Tin usually solidifies from the melt in tetragonal crystallites; the crystalline texture is very evident if the surface is etched with hydrochloric acid ('watered' tin). When a rod of tin is bent, a creaking noise arises from the friction of the crystallites one upon the other (the 'cry of tin'). Above 161°, tin changes into a *rhombic* modification. In this form it is very brittle, so that it can be powdered (best at about 200°), and breaks into pieces if dropped from some little height. The production of the so-called 'grain tin' depends on this brittleness. A third modification, the powder *grey tin* (density at 18° = 5.75) is stable below 13.2°. The transformation into this form, and the reverse process, occur as a rule with infinite slowness. Objects made of tin are completely destroyed by the transition of the tin into the grey modification, since at the places affected by the transformation they crumble into powder ('tin pest').

Under prolonged intense cold it may happen that the transformation of tin into the grey form starts spontaneously, occurring first at isolated spots. Grey pustules, consisting of grey tin in powder form, then grow on the affected objects. The particles of powder then serve as crystallization nuclei for other spots, so that the destructive change spreads like an infectious disease once it has started at one point. The name 'tin pest' given to the phenomenon is thus very apt. The tendency for transformation to occur is naturally greater, the further below 13.2° the tin is cooled. On the other hand, the transformation also follows the rule, that the rates of reactions diminish with falling temperature. There is thus a certain temperature at which the velocity of transformation has a maximum value. This is about —48°. The transformation is accelerated by an alcoholic solution of 'pink salt', and inhibited (according to Farup) especially by bismuth and antimony salts. The thermal and mechanical treatment to which the metal has previously been subjected, and the number of transformations which it has already undergone, also have a considerable influence on the rate of transformation (Cohen, 1935).

The *crystal structure* of grey tin (α-Sn) is like that of diamond, but with $a = 6.46$ Å. Ordinary tetragonal tin (β-Sn) has a structure not found in other elements, which could be described as a diamond lattice compressed along the direction of the c axis. The structure of rhombic tin (γ-Sn) is not yet known.

The transformation α-Sn → β-Sn is accompanied by both a considerable increase in density and a very marked increase in metallic character. These two effects are interrelated, since a particularly dense packing of the atoms (high coordination number) is a characteristic property of the metallic state (cf. Vol. II, Chap. 1). Non-metals, by contrast, have only low coordination numbers (4 or less), because of their directed covalent binding forces, and therefore have relatively open structures. The contraction which sets in when the directed binding forces are weakened by increasing thermal agitation is observed not only for the case of tin, but also for other elements occupying positions in the Periodic System just on the boundary between metals and non-metals—i.e., Ga, Si, Ge, Bi and Te. In the case of these elements, however, the contraction does not occur in the solid state, but only on melting—in the case of Te, indeed, rather above the melting point.

Tin is stable towards both air and water at ordinary temperature. It is oxidized, however, at higher temperatures. The vapor of tin can readily be burned completely to the dioxide. Tin combines with the free halogens, forming the tetrahalides. It reacts with chlorine and bromine, even at ordinary temperature, and with iodine when gently warmed. It does not react to any marked extent with fluorine at ordinary temperature, but does so very violently at 100° catching fire. It also unites vigorously with sulfur, selenium and tellurium on heating. It does not combine directly with nitrogen, but does so with phosphorus when heated.

Tin is only slowly attacked by dilute acids, in accordance with the small difference between its normal potential and that of hydrogen (cf. p. 529). It dissolves best in concentrated hydrochloric acid:

$$Sn + 2HCl = SnCl_2 + H_2.$$

(This equation does not completely express the process, inasmuch as the tendency of tin to form chloro complexes—such as $[SnCl_3]^-$—also plays a role). Tin reacts vigorously with concentrated nitric acid, being thereby converted into *b-stannic acid*, a white powder insoluble in water. Tin reacts much more slowly with concentrated sulfuric acid, in which it dissolves with evolution of SO_2.

When boiled with caustic alkalis, tin goes into solution with the formation of *hydroxostannate* ions:

$$Sn + 4H_2O + 2OH^- = [Sn(OH)_6]^= + 2H_2.$$

It also dissolves when it is made the anode in concentrated caustic soda:

$$Sn - 4e + 6OH^- = [Sn(OH)_6]^=.$$

However, if the current density in this process rises above a certain value, the tin suddenly becomes 'passive'—i.e., it behaves as an unattackable electrode (cf. Vol. II).

One property of tin, which leads to important applications, is the ease with which it alloys with other metals. As a practically insoluble substance, metallic tin is not poisonous. However, soluble tin compounds may be present in old and highly acidic conserves, and can cause digestive disturbances; these are, however, not serious as a rule.

(e) Uses

Before the invention of porcelain, tin was an important material for utensils, plates, cans, and jugs. Today, almost half the production of tin goes into the manufacture of tin plate.

Tin plate is made by cleaning iron sheets with dilute sulfuric acid ('pickling'), and dipping them in molten tin. Because it is so resistant towards acids, a coherent coating of tin protects the iron from corrosion. If, however, the coating is perforated, an accelerated attack takes place at the point in question, since a galvanic cell is set up in which the iron is the anode. Where tin plate is required for the canning of food, only lead-free tin should be used for tinning, since the lead is not only harmful in itself, but its presences increases the corrosibility of the tin by dilute acids (such as, for example, the organic acids present in fruits). The inside of tins intended for the canning of fruit is often covered with a thin coating of lacquer ('varnish') which gives a golden yellow color to the bright silvery surface.

Tin is a constituent of many important alloys, such as the *bronzes*, *britannia metal*, and *soft solder*. The latter consists of 40–70% tin and 60–30% lead. *Bearing metals* are also mostly tin alloys. The tinware usual today, including organ pipes, invariably contains lead. It can generally be recognized as such by reason of its darker color as compared with pure tin.

Britannia metal comprises alloys of much tin with a little antimony (and usually some copper), which are used for the production of utensils—e.g., spoons and forks—by *stamping*. *Bearing metals* are alloys used for making the bearings of moving parts of machinery—shafts, etc. A distinction is drawn between *white bearing metals* and *bronze bearing metals*. The former consist of 90–50% tin, 7–20% antimony, and usually a few per cent of copper. In the latter, copper preponderates, being present to the extent of 75–90%, as compared with an average of 10% tin, and smaller amounts of zinc or lead. Since World War I, lead-base bearing metals have been used to an increasing extent in place of tin-base bearing metals (cf. p. 542). The *bronzes*, which are also copper-tin alloys, will be discussed under copper.

Tin foil, formerly used for the packaging of many foodstuffs, consists of pure tin. Thin aluminum foil is now generally used in its place.

Compounds of tin which find application are, especially, tin dioxide, stannic chloride, tin salt and mosaic gold. Some tin salts of organic acids are used in dyeing—e.g., tin(II) acetate, $Sn(C_2H_3O_2)_2$, and thiocyanate, $Sn(SCN)_2$, as reducing agents in print dyeing and tin oxalate, SnC_2O_4, as a mordant.

14. Compounds of Tin

In its compounds tin can act as

(1) negatively quadrivalent, or electroneutral and formally quadrivalent
(2) positively quadrivalent
(3) positively bivalent.

Negatively quadrivalent tin may perhaps be present in *tin hydride* SnH_4, though this is probably a purely covalent compound. The *tin alkyls* are also probably covalent in structure.

As in the case of carbon, compounds of tin are known which dissociate in dilute solution into free radicals SnR_3, with trivalent tin. There are also some indications that tin can function as trivalent in inorganic compounds also, at least transiently and to a small extent (Ball, 1935).

Tin can act as positively quadrivalent, in the elementary cation Sn^{++++}. In this state, however, the predominant tendency is to form *anions*, such as $[SnCl_6]^=$, $[Sn(OH)_6]^=$, $[SnO_3]^=$ and $[SnS_3]^=$. These are present in the hexachloro-, hexahydroxo-, trioxo- and trithiostannates. The two last are usually known simply as stannates and thiostannates.

In the positively bivalent state, tin exists principally in salts, as the cation Sn^{++}. Complex anions derived from bivalent tin are formed especially by the halogens [halogenostannites, or halogenostannates(II)]. The addition of hydroxyl ions to $Sn(OH)_2$, forming *hydroxostannites*, has also been observed.

The ions Sn^{++} and Sn^{++++} are colorless in solution, as also are the salts derived from them (unless colored by other components). The complex ions derived from tin are also mostly colorless. Some binary compounds of tin are colored; thus the monoxide is blue-black, the monosulfide brown, and the disulfide yellow.

The compounds of bivalent tin [tin(II) compounds] were formerly known as *stannous* compounds, those of quadrivalent tin [tin(IV) compounds] as *stannic* compounds.

The oxidation potential Sn^{++}/Sn^{++++}, relative to the normal hydrogen electrode, is —0.154 volt at 25° (Huey, 1934); that is, a platinum foil, immersed in a solution containing equal amounts of tin(II) and tin(IV) ions, is at a potential 0.154 volts lower than a normal hydrogen electrode. On completion of the circuit, a current would therefore flow through the wire from the electrode dipping in the Sn^{++}/Sn^{++++} solution to the normal hydrogen electrode. Hydrogen would go into solution at the latter, forming hydrogen ions, while an equivalent quantity of Sn^{++++} ions would disappear at the other electrode by conversion into Sn^{++} ions.

The potential of tin itself, in contact with a solution 1-molar with respect to Sn(II) ions (i.e., the *standard potential* of tin) is $+0.158$ volt (Prytz, 1934). In this case, the current flows through the wire to the tin, which goes into solution as bivalent ions while hydrogen is evolved at the other electrode. The solubility of tin in acids depends in part on the ability of tin to discharge H^+ ions. However, because its normal potential differs but little from that of hydrogen, tin has only quite a small tendency to dissolve. As a rule therefore it is markedly attacked only by those acids, such as HCl, which can augment this tendency by forming complexes.

a) Tin(IV) Compounds

(*i*) *Tin hydride* SnH_4. The existence of tin hydride was proved in 1919, by Paneth. He prepared it originally by decomposing a tin-magnesium alloy, containing the compound Mg_2Sn, by means of 4-normal hydrochloric acid, but later, with better yield, by the cathodic reduction of tin salt solutions at electrodes with a high overvoltage (best at lead electrodes).

In either case, hydrogen is the principal gas evolved. This is mixed, however, with tin hydride in small amount, as may be simply shown by passing the gas, filtered through a plug of cotton wool, through a heated glass tube. A metallic mirror is deposited in this, and can be identified as a mirror of tin by its reactions. Thus, after converting the metal into the dichloride by treating it with hydrogen chloride gas, it gives the reactions typical of tin dichloride, with gold chloride solution, (formation of purple of Cassius—p. 538), and with mercuric chloride (precipitation of calomel). Unlike the mirrors obtained by the decomposition of arsenic, antimony and bismuth hydrides, it is also insoluble in cold concentrated nitric acid but dissolves readily in cold concentrated hydrochloric acid. By condensing it in liquid air, Paneth was later able to collect rather larger quantities of pure tin hydride, and to determine its melting point ($-150°$), boiling point ($-52°$) and composition (by a method similar to that used for germanium hydride). It can now be prepared much more conveniently by the reduction of $SnCl_4$ with lithium aluminium hydride, $LiAlH_4$ (p. 362).

Tin hydride is a relatively stable compound; this is rather surprising in view of the difficulty with which it is formed. It can be stored for days at ordinary temperature in clean glass vessels. The decomposition is accelerated to an extraordinary degree, however, by traces of metallic tin. It decomposes immediately at 145–150°. Tin hydride can be passed through moderately strong caustic alkalis, dilute acids and most salt solutions without decomposition occurring, but is decomposed by silver nitrate and mercury(II) chloride solutions, and also by solid alkalis, soda lime and by concentrated sulfuric acid. Tin hydride is *very toxic*.

(ii) *Tin tetrafluoride and Fluorostannates*. Tin tetrafluoride, SnF_4, can be prepared, according to Ruff, by introducing tin tetrachloride into anhydrous hydrogen fluoride, and heating gently. For the reaction $SnCl_4 + 4HF \rightleftharpoons SnF_4 + 4HCl$ to proceed to the right, the hydrogen fluoride must be present in excess. A double compound $SnF_4 \cdot SnCl_4$ is formed as an intermediate. This decomposes between 130° and 220°, leaving pure SnF_4.

Tin tetrafluoride forms a colorless mass of radiating crystals; density 4.78, b.p. 705°. It combines with water with a hissing noise, because of the large evolution of heat. The solution is more readily obtained by dissolving freshly precipitated tin dioxide hydrate in aqueous hydrofluoric acid; a hydrated, gummy mass is left when the solution is evaporated. Crystalline double fluorides, or *fluorostannates* can readily be obtained from the solution, however. These were investigated especially by Marignac, and correspond in almost every case to the type $M^I_2[SnF_6]$. With NH_4F, however, a salt of the composition $(NH_4)_4SnF_8$ is also formed. With KF, Marignac obtained the compound K_3HSnF_8, in monoclinic needles, as well as $K_2SnF_6 \cdot H_2O$, from solutions of high hydrofluoric acid content.

(iii) *Tin tetrachloride and Chlorostannates.* Tin tetrachloride [stannic chloride, tin(IV) chloride], $SnCl_4$, is manufactured chiefly by treating tin plate scrap with chlorine. It forms a colorless liquid which fumes in air (*spiritus fumans Libavii*), density 2.229 (at 20°), freezing point —36°, boiling point 114°. It is miscible with carbon disulfide in all proportions, and can dissolve phosphorus, sulfur, iodine, arsenic triiodide, antimony triiodide and tin tetraiodide.

Tin tetrachloride dissolves in water with considerable evolution of heat. According to the conditions, several hydrates may crystallize from the solution. The pentahydrate, $SnCl_4 \cdot 5H_2O$, (stable between 19° and 56°) forms white, opaque, deliquescent crystals, melting at about 60°; it is very soluble in water.

The aqueous solution of tin tetrachloride is hydrolyzed to a considerable degree, substantially according to:

$$SnCl_4 + 2H_2O = SnO_2 + 4HCl.$$

The resulting tin dioxide remains in solution in colloidal form. The hydrochloric acid formed simultaneously combines to some extent with undecomposed tin tetrachloride, to form *hexachlorostannic acid*, $H_2[SnCl_6]$. This acid can be isolated, by passing hydrogen chloride gas into concentrated aqueous solutions of tin tetrachloride. It crystallizes in leaflets, of composition $H_2SnCl_6 \cdot 6H_2O$, melting at 19.2°. The salts derived from this acid are usually known as *chlorostannates*, and correspond to the general formula $M^I_2[SnCl_6]$, where M^I is an alkali metal or the equivalent amount of a bi- or trivalent metal. (For crystal structure of the alkali chlorostannates, see p. 482). *Ammonium chlorostannate*, $(NH_4)_2[SnCl_6]$, is important industrially; it forms colorless octahedra, of density 2.387, readily soluble in water. It is also known as 'pink salt', being used as a mordant in dyeing. Its aqueous solution, unlike that of tin tetrachloride, is neutral in reaction and, if sufficiently concentrated, is not decomposed even by boiling.

Many other substances are also able to form addition compounds with tin tetrachloride— e.g., NH_3, PH_3, PCl_5, $POCl_3$, SCl_4, $NOCl$, organic sulphides R_2S and ethers R_2O.

Niebergall (1951) has shown that $SnCl_4$ reacts rapidly with C_2H_5I, in the presence of $AlCl_3$ as catalyst, to form SnI_4 ($SnCl_4 + 4C_2H_5I = SnI_4 + 4C_2H_5Cl$). It is noteworthy that the corresponding reaction takes place only very slowly with $GeCl_4$, and not at all with $SiCl_4$.

Tin tetrachloride is used as a mordant dyeing. It is also used for weighting silk, as a catalyst for chlorinations and as a condensing agent in organic chemistry. It is also the principal constituent of the 'rose salt', prepared by dissolving tin in aqua regia, and also used in dyeing.

(iv) *Tin tetrabromide and Bromostannates.* Tin tetrabromide, $SnBr_4$, prepared by direct union of the elements, forms a snow-white crystalline mass of density 3.35 (at 35°), m.p. 33° and b.p. 203°. It crystallizes from its aqueous solution at ordinary temperature as the tetrahydrate, $SnBr_4 \cdot 4H_2O$. *Hexabromostannic acid*, $H_2[SnBr_6] \cdot 8H_2O$, forms fine, colorless, deliquescent prisms; its salts (hexabromostannates, $M^I_2[SnBr_6]$) are known.

(v) *Tin tetraiodide*, SnI_4, obtained either by direct combination of tin with iodine (preferably dissolved in carbon disulfide) or by precipitation from a concentrated solution of $SnCl_4$ by means of KI, forms yellow to yellow brown, highly refractive octahedral crystals, not isomorphous with those of tin tetrabromide; density 4.46, m.p. 146°, b.p. 346°. It dissolves readily in methylene iodide, CH_2I_2, carbon disulfide, alcohol, ether, benzene and other organic solvents, and also in arsenic tribromide. The saturated solution of tin tetra-iodide in arsenic tribromide is notable for its high density (3.731 at 15°). Freezing point measurements on such solutions give a value for the freezing point depression twice as great as that corresponding to the simple molecular weight. This cannot be attributed to electro-lytic dissociation, since the solutions are practically non-conductors, and it is therefore assumed that the dissociation follows the scheme:

$$SnI_4 \rightleftharpoons SnI_2 + I_2$$

Tin tetraiodide does not form clear aqueous solutions, because of hydrolysis. *Iodo-stannates* have not as yet been isolated from aqueous solutions, although Rosenheim was able to obtain such salts from alcoholic solutions. Tin tetraiodide possesses a *molecular* crystal lattice. Each Sn is surrounded by 4I in a regular tetrahedron. GeI_4 has the same structure. Inter-atomic distances:

$$Ge \leftrightarrow I = 2.57 \text{ Å}; \quad Sn \leftrightarrow I = 2.65 \text{ Å}.$$

(*vi*) *Tin alkyls.* Dialkyl tin iodides can be obtained by heating tin with alkyl iodides. Thus tin reacts with ethyl iodide, forming diethyl tin iodide.

$$Sn + 2C_2H_5I = \begin{array}{c} I \\ I \end{array}\!\!>\!Sn\!<\!\!\begin{array}{c} C_2H_5 \\ C_2H_5 \end{array}$$

This can be treated with zinc ethyl, $Zn(C_2H_5)_2$, to form tin tetraethyl, $Sn(C_2H_5)_4$. It is more convenient to start from the tin halides, and to replace the halogen atoms in these either partly or completely, by the Grignard method. Numerous tin alkyls and aryls can be obtained in this way. They have many similarities to the corresponding compounds of silicon.

The tin alkyls display a much smaller capacity for forming addition compounds than do the tin halides, and this capacity diminishes progressively as halogen atoms are successively replaced by alkyl groups.

If an aqueous-alcoholic solution of diethyltin iodide is treated with ammonia, *diethyl tin oxide*, $(C_2H_5)_2SnO$, is thrown down as a white precipitate, which dissolves in acids, forming salts. Diethyltin oxide thus functions towards acids as if it were a base. The same applies to *triethyltin hydroxide*, $(C_2H_5)_3SnOH$.

(*vii*) *Tin dioxide* is found in nature, crystallized in the tetragonal system, as *cassiterite*. It can also exist in rhombic and hexagonal forms, and is thus trimorphous. The action of heat on its hydrates, or on tin oxalate yields 'amorphous' tin dioxide—i.e., it is obtained in the form of a powder with no detectable crystalline structure. Pure tin dioxide is white. It sublimes above 1800° without melting. It is insoluble in water, and hardly attacked by either acids or caustic alkalis. It can readily be brought into solution by fusion with alkali hydroxide or with a mixture of soda and sulfur. In the former case it is converted to stannate, in the latter to thiostannate:

$$SnO_2 + 2NaOH = Na_2SnO_3 + H_2O \tag{1}$$

$$SnO_2 + 2Na_2CO_3 + 4S = Na_2SnS_3 + Na_2SO_4 + 2CO_2 \tag{2}$$

It is reduced to metallic tin by heating it with carbon, or in a stream of hydrogen. When it is heated in chlorine, tin tetrachloride is formed.

Tin dioxide has widespread technical uses, in the production of all types of white enamels and glazes. It is also used as a polishing agent. It is prepared on a large scale by burning metallic tin, heated to a high temperature, in a current of air. The naturally occurring tin dioxide (usually colored brown by the impurities present) is the raw material for the extraction of metallic tin.

Tetragonal tin dioxide has the same crystal structure as magnesium fluoride—the 'rutile' structure (cf. Fig. 63, p. 265), $a = 4.72$ Å, $c = 3.17$ Å, $d = 2.06$ Å.

(*viii*) *Stannic acid and Stannates.* A rather soluble salt of the composition $Na_2O \cdot SnO_2 \cdot 3H_2O$ crystallizes in hexagonal plates from concentrated solutions of the melt of SnO_2 and NaOH. This is used, under the name of 'preparing salt', for the pre-treatment of textiles to receive mordant dyes. Bellucci and Parravano (1905) concluded from the firmness with which the water is bound in this compound that it has the constitution $Na_2[Sn(OH)_6]$. The salt should thus systematically be called sodium hexahydroxostannate. Other salts of this type exist, as well as the sodium salt—e.g., $K_2[Sn(OH)_6]$, $Ca[Sn(OH)_6]$, $Sr[Sn(OH)_6]$, $Pb[Sn(OH)_6]$. The three last mentioned, which are insoluble in water, are formed by the reaction of the potassium salt with soluble salts of calcium, strontium and lead. These reactions, and also an X-ray determination of the structure (Wyckoff, 1928), confirm the view that a radical $[Sn(OH)_6]^{2-}$ underlies the stannates obtained from aqueous solutions.

The *acid* corresponding to this radical, $H_2[Sn(OH)_6]$, cannot be obtained in the free state. If alkali (hydroxo)stannate solutions are treated with limited quantities of dilute, strong acids, or if carbon dioxide is passed in, voluminous white precipitates are obtained; these are soluble both in an excess of strong acid, and in alkalis. The precipitates have the character of gels, and contain variable amounts of water. Analogous precipitates are obtained by treating solutions of tin tetrachloride with ammonia or ammonium carbonate. As in the case of silicic acid, it is still not certain whether the freshly formed precipitates, which are not detectably crystalline, contain chemically bound water, or whether they are merely tin dioxide gels. When through aging (by prolonged standing in contact with the solution, or by heating), they develop a crystalline structure detectable by X-rays, they give the diffraction pattern characterisitic of tin dioxide. At the same time, they become progressively more insoluble in acid. The precipitates are ordinarily called 'stannic acids', the acid-soluble and acid-insoluble varieties being distinguished as a- and b-stannic acid respectively, or as 'ordinary stannic acid' and 'metastannic acid'. The latter is obtained directly by treating metallic tin with concentrated nitric acid.

The b-stannic acid apparently goes into solution when it is treated with concentrated hydrochloric acid, and subsequently diluted. This does not involve a true dissolution, however, but the formation of a *colloidal* solution, as a consequence of *peptization* by the hydrochloric acid. The b-stannic acid can be reprecipitated by concentrated hydrochloric acid, and also by other electrolytes. On the other hand, tin tetrachloride can be distilled from a solution of a-stannic acid in concentrated hydrochloric acid.

The a- and b-stannic acids differ in the following reactions, as well as in their behavior towards acids. a-Stannic acid is readily soluble both in potassium carbonate and in caustic potash of any concentration; b-stannic acid is insoluble in potassium carbonate solution, and does not dissolve in concentrated potassium hydroxide. Only when this is diluted does it go into (colloidal) solution. It is not possible to prepare crystalline salts from such solutions. Amorphous substances are obtained by evaporating alkaline solutions of b-stannic acid;

these were formerly considered to be salts of 'metastannic acid', and called 'metastannates', but they are merely adsorption products of alkali on hydrated tin dioxide. Potassium or sodium sulfate in the cold precipitates only b-stannic acid, and not a-stannic acid, from its hydrochloric acid or colloidal solutions. The gel of a-stannic acid has its water more firmly bound, and has a higher adsorptive power (e.g., for phosphoric acid and for organic dye-stuffs) than the b-stannic acid.

The different properties of the two modifications of stannic acid were already recognized by Berzelius (1817), who introduced the concept of *isomerism* into chemistry in this connection.

According to Mecklenburg (1909), the significant differences in the behavior of the a- and b-stannic acids can be explained by the assumption that the two substances are *different colloidal varieties* of insoluble tin dioxide or its hydrates, distinguished from one another by their particle size. On this view ordinary stannic acid (a-stannic acid), consists of relatively small, and b-stannic acid of relatively large colloidal particles (which are, however, still below the lower limit of visibility in the ultramicroscope). This assumption is reasonable, in view of the continuous change of one form into the other as the gel ages. Mecklenburg was able to show, further, that the stannic acid produced by hydrolysis of tin(IV) sulfate at 0° approached a-stannic acid in its properties, but that it took on in increasing degree the typical properties of b-stannic acid as it was prepared at successively higher temperatures. This confirms the view that the b-stannic acid consists of coarser particles since, quite generally, larger particles are produced by precipitation from hot solutions than in the cold.

It remains an open question, however, whether the difference between the a- and b-stannic acids is conditioned by the difference in particle size *alone*. By suitable methods of preparation and drying (acetone drying—see p. 354), Willstätter (1924) was able to obtain preparations which contained SnO_2 and H_2O in simple stoichiometric ratios, and which showed the same composition over wide ranges of temperature. The most active modification, precipitated by ammonia from $SnCl_4$ solutions in the presence of ammonium chloride, had the approximate composition $SnO_2 \cdot 3H_2O$ when dried by acetone in the temperature range —35° to —10°. When treated with acetone at higher temperatures, it split off the remaining water, apparently continuously. However, when preparations, which had stood for a long time in contact with water, and were thus already somewhat 'aged'—their solubility in acids, and especially in nitric and sulfuric acids had already distinctly diminished—were dried with acetone, their water content became constant between 0° and +15°, at a composition corresponding roughly to $2SnO_2 \cdot 3H_2O$, and showed a second range between +30° and 56° (the boiling point of acetone), corresponding to the formula $4SnO_2 \cdot 5H_2O$. Willstätter inferred from this that during aging a *condensation* took place—i.e., simpler molecules joined up to make higher molecules, with the elimination of water. According to Zsigmondy it is probable that both condensation, in Willstätter's sense, and particle growth by polymerization, as envisaged by Mecklenburg, play their role simultaneously in the change of a- into b-stannic acid.

Willstätter also established that when a-stannic acid, first dried by acetone at ordinary temperature, was subsequently dried in a high vacuum over phosphorus pentoxide, it lost just enough water to bring its composition to the formula H_2SnO_3, and that it retained this composition in a current of dry air up to 65°. It was later concluded quite independently by Thiessen and by Simon (in 1931), that a definite compound of the composition H_2SnO_3 existed, from the (admittedly not very clearly marked) steps obtained in the isothermal or isobaric degradation of tin dioxide gels. According to Simon the great ease with which this decomposes into SnO_2 and H_2O is due, as in the case of silicic acid, to the small heat of binding of the water in it, (about 30 kcal per mol.). According to Thiessen, other hydrates of tin dioxide, in addition to $SnO_2 \cdot H_2O$, exist as definite compounds. No hydrates of tin dioxide have as yet been detected by X-ray methods, however, and their existence has therefore been disputed by Weiser (1932).

Among the *anhydrous salts of stannic acid*, the compounds Mg_2SnO_4, Zn_2SnO_4 and Co_2SnO_4 have had their structure investigated. These all form *spinel* crystal lattices (cf. p. 355), and thus (in so far as it is proper to consider them as salts at all) should be designated *orthostannates*. Analogous preparations which had the composition of metastannates, $M^{II}SnO_3$, were shown by X-rays to be mixtures of the orthostannates with SnO_2 (Natta, 1929, Taylor, 1930).

Gels of stannic acid or tin dioxide are used in dye printing and turkey red dyeing, and also for the preparation of tin oxalate, which also finds application in dyeing.

(*ix*) *Tin disulfide and Thiostannates.* Tin disulfide (stannic sulfide), SnS_2, is thrown down by hydrogen sulfide from not too strongly acid solutions of tin(IV) salts, or by the acidification of thiostannate solutions, in the form of a yellow precipitate. It is usually prepared technically by dry methods—e.g., by heating a mixture of tin foil (or, better, broken up tin amalgam) with flowers of sulfur and ammonium chloride. By this method it is obtained in the form of golden yellow scales (density 4.51), belonging to the hexagonal system. It is known commercially as *mosaic gold* (*aurum mosaicum*, gold for mosaic work), and is used for bronzing, under the name of 'tin bronze'.

Tin disulfide crystallizes with the brucite structure (Fig. 61, p. 261), $a = 3.62$ Å, $c = 5.85$ Å, $Sn \leftrightarrow S = 2.55$ Å.

Whereas stannic sulfide prepared by dry methods is insoluble in hydrochloric acid and nitric acid, that precipitated from aqueous solution dissolves even in moderately concentrated, warm hydrochloric acid:

$$SnS_2 + 4HCl \rightleftharpoons SnCl_4 + 2H_2S.$$

It dissolves readily in alkali sulfide solutions, forming readily soluble alkali *thiostannates*—e.g., ammonium thiostannate, $(NH_4)_2[SnS_3]$. The formation of the thiostannates arises from the ability of SnS_2 to add on additional $S^=$ ions; by combination with *one* $S^=$ ion, the *metathiostannate* ion $[SnS_3]^=$ is formed, by union with *two* $S^=$ ions the orthothiostannate ion $[SnS_4]^{4-}$.

$$SnS_2 + S^= = [SnS_3]^= \tag{1a}$$

$$SnS_2 + 2S^= = [SnS_4]^{4-} \tag{1b}$$

Thiostannates are also formed by the dissolution of brown tin monosulfide in alkali polysulfides:

$$SnS + S + S^= = [SnS_3]^=$$

Tin disulfide also dissolves in alkali hydroxide solutions, forming in this case both hydroxostannate ions and thiostannate ions:

$$3SnS_2 + 6OH^- = 2[SnS_3]^= + [Sn(OH)_6]^= \tag{2}$$

The processes represented by the eqns (1a and b) and (2) are reversible. Tin disulfide is precipitated again if solutions of thiostannates are acidified. This will be apparent if the law of Mass Action is applied to the two equations concerned—directly in the case of eqn (2), and for eqn (1) also when it its remembered that the S^{2-} ion concentration in acid solutions is very minute, because of the small degree of dissociation of hydrogen sulfide.

The alkali thiostannates crystallize in highly hydrated form. Jelley (1933) found that part of the water was more firmly retained than the rest on heating. He concluded from this that this portion of the water was complex-bound within the anion of the salts, and accordingly formulated the anion of the hydrated metathiostannates as $[SnS_3(H_2O)_3]^=$ or $[Sn(SH)_3(OH)_3]^=$, and the anion of the hydrated orthothiostannates as $[SnS_4(H_2O)_2]^{4-}$ or $[SnS_2(SH)_2(OH)_2]^{4-}$. Brintzinger (1934) arrived at the same formula for the orthothiostannate ion in solution, from determinations of the ionic weights by the dialysis method. He also found ionic weights corresponding to the diaquo-ion in the case of the orthothioarsenate and orthothioantimonate ions.

(x) *Tin(IV) sulfate* can be obtained from the solution of freshly precipitated tin dioxide hydrate in dilute sulfuric acid, in the form of colorless needles of the composition:

$$Sn(SO_4)_2 \cdot 2H_2O$$

(b) Tin(II) Compounds

(i) *Tin(II) chloride* (stannous chloride), $SnCl_2$, can be prepared in the anhydrous state by heating tin in a current of hydrogen chloride, as a white, lustrous mass, of density 3.95, m.p. 247° and b.p. 605°. Close to the boiling point, the vapor density shows partial association of the molecules to take place; only above 1100° is the vapor entirely monomeric. According to Castoro, however, the solution in urethane has a freezing point depression corresponding to the simple unimolecular formula.

Stannous chloride is readily soluble in water and also in alcohol, ether, acetone and ethyl acetate. It crystallizes from the aqueous solution as $SnCl_2 \cdot 2H_2O$, in clear monoclinic prisms or bipyramids which are not deliquescent, melt at 40.5°, and have a density of 2.710. This is the 'tin salt' of commerce. It is manufactured by dissolving tin turnings in hydrochloric acid. The salt dissolves in a small amount of water to form a clear solution; the solution becomes turbid when it is diluted however, through the precipitation of a basic salt:

$$SnCl_2 + H_2O = Sn(OH)Cl + HCl$$

The hydrated crystalline tin(II) chloride can be dehydrated by heating it to redness in a current of hydrogen chloride.

Tin dichloride is a *fair reducing agent*. It precipitates gold and silver from their solutions in the form of the metals. It can also precipitate mercury as the metal or, if present in insufficient amount for this, can reduce salts of bivalent mercury to those of the univalent metal. It also reduces iron(III) salts to iron(II) salts, arsenates to arsenites, chromates to chromium(III) salts, permanganates to manganese(II) salts, nitrocompounds to amines, diazonium salts to hydrazine salts. It is slowly oxidized in aqueous solution by atmospheric oxygen:

$$3SnCl_2 + \tfrac{1}{2}O_2 + H_2O = SnCl_4 + 2Sn(OH)Cl$$

This oxidation is prevented by the addition of metallic tin to the solution.

Stannous chloride has many uses. As well as being a reagent in qualitative and quantitative analysis, it is employed industrially as a reducing agent for organic nitro-, azo- and diazonium compounds, and also especially as a mordant in dye printing. It is also used in the preparation of lake colors.

Dry tin dichloride adds on 1 molecule of NH_3 when it is warmed in a stream of this gas. By heating it in hydrochloric acid, Engel obtained the complex acid $HSnCl_3 \cdot 3H_2O$ as a liquid, freezing at —27°. Crystalline *chlorostannites* (or chlorostannates(II)), $M^I[SnCl_3]$ and $M^{II}[SnCl_4]$ have been obtained from solutions of stannous chloride to which alkali chlorides have been added.

The other tin(II) halides are substantially similar to the chloride in their behavior. Tin(II) fluoride forms white monoclinic prisms, soluble in water. The bromide is a pale yellowish mass, density 4.92, m.p. 232°, b.p. 619°. Tin(II) iodide forms orange octahedra, density 5,28, m.p. 320°, b.p. 720°. These halides form complexes corresponding to those derived from the dichloride, though in the case of the diiodide only the complexes of type $M_I[SnI_3]$ are at present known.

(ii) Tin(II) oxide and Tin(II) hydroxide. If the solution of a tin(II) salt is treated with alkali carbonate, or with a little alkali hydroxide, *tin(II) hydroxide*, $Sn(OH)_2$, is thrown down as a white precipitate, very sparingly soluble in water

$$Sn^{++} + 2OH^- = Sn(OH)_2.$$

After careful drying it has the composition corresponding to this formula. In contact with tin(II) salt solutions, or with alkali, it readily changes into the dark, anhydrous *tin(II) oxide* (stannous oxide, tin monoxide), SnO, especially at elevated temperatures. The hydrated white precipitate is soluble both in acids and in caustic alkalis. In the former case it forms the corresponding tin(II) salts—e.g., stannous chloride, by dissolution in hydrochloric acid

$$Sn(OH)_2 + 2HCl = SnCl_2 + 2H_2O$$

since tin(II) hydroxide functions as a base towards acids. Towards strong bases, however, it behaves as an anhydro-acid.

$$Sn(OH)_2 + OH^- \rightleftharpoons [Sn(OH)_3]^-$$

According to Goldschmidt, the equilibrium constant, referred to the process

$$Sn(OH)_2 + H_2O \rightleftharpoons Sn(OH)_3^- + H^+,$$

has the value

$$K = \frac{[Sn(OH)_3^-] \cdot [H^+]}{[Sn(OH)_2]} = 4 \cdot 10^{-10}$$

The salts derived from tin(II) hydroxide, in its function as an anhydroacid, are called *hydroxostannites* (or hydroxostannates(II)). Their solutions decompose readily, the anhydrous oxide SnO separating from them as a dark powder. Only recently, therefore, have attempts to isolate the crystalline salts been successful (Scholder, 1933).

Scholder obtained $Na[Sn(OH)_3]$, $Ba[Sn(OH)_3]_2$, $Ba[Sn(OH)_3]_2 \cdot 2H_2O$ and $Sr[Sn(OH)_3]_2 \cdot 2H_2O$, as colorless, crystalline compounds. The alkaline earth hydroxo-stannites, when heated to 100–110°, first change into yellow oxo-tetrahydroxo-distannites, $M^{II}[(HO)_2Sn—O—Sn(OH)_2]$. On stronger heating, these furnish anhydrous stannites, $M^{II}[SnO_2]$. The oxotetrahydroxodistannites can also be obtained directly from warm solutions. The stannites $Ca[SnO_2]$, $Sr[SnO_2]$ and $Ba[SnO_2]$ may also be obtained by the action of SnO, in the gaseous state, upon the strongly heated alkaline earth oxides (Tamaru, 1931).

Anhydrous tin monoxide may be prepared by the method of Sandall. Tin dichloride is rubbed together with a small excess of sodium carbonate, and heated on a sand bath, with constant stirring, until the mixture has turned black. The blue-black finely divided tin monoxide (a-SnO) has the same crystal structure as red lead oxide (p. 545). It changes at about 550° into a second modification (β-SnO) when heated in a vacuum. When heated to redness in air it burns with incandescence to tin dioxide. Tin(II) hydroxide gives a diffraction pattern different from that of SnO. It is thereby proved to be a definite compound, and not merely a gel of SnO.

When tin(II) oxide is heated in air, oxidation begins below 300°; it becomes incandescent, and burns to tin dioxide. The stability of tin(II) oxide under non-oxidizing conditions is rather uncertain, however. Some have stated that it changes into a second modification (β-SnO) when heated to 550° in vacuum, as mentioned above, but other investigators have reported that it decomposes:

$$2SnO = Sn + SnO_2$$

(disproportionation). Ditte (1882), who first observed this decomposition, found that the product of decomposition contained tin(II) oxide as well as tin and tin(IV) oxide, in the ratio Sn : SnO_2 : SnO = about 1 : 1 : 2, and he concluded that decomposition yielded an intermediate oxide Sn_3O_4 : $4SnO = Sn + Sn_3O_4$. Spandau and Kohlmeyer (1947) have recently confirmed Ditte's observations. In the two component system Sn-O, however, a mixture of Sn + SnO_2 + SnO(solid) + SnO(vap.) would be *invariant*, and so could be in equilibrium only at a single temperature. Since Spandau and Kohlmeyer obtained the same products from the decomposition of tin(II) oxide at 400°, 600° and 1000°, they concluded that Ditte correctly interpreted the thermal decomposition as leading to Sn_3O_4.

SnO is stable in the gaseous phase, and is much more volatile than either SnO_2 or metallic tin. A mixture of Sn and SnO_2 vaporizes relatively rapidly at 1300–1500°, through the formation of SnO. Metallic tin and SnO_2 also give a homogeneous melt at 1300°; this evolves the vapor of SnO, but decomposes into its components when it solidifies at about 1050°.

Because of their tendency to pass into the corresponding tin(IV) compounds, tin(II) hydroxide and stannite solutions have strong reducing properties. This is the basis of their technical uses—e.g., in vat dyeing and dye printing.

(*iii*) *Tin(II) sulfide*, stannous sulfide, SnS, results from the direct union of tin and sulfur, and is thrown down as a dark brown precipitate from tin(II) salt solutions by the action of hydrogen sulfide. The precipitate is insoluble in dilute strong acids. It is also insoluble in colorless ammonium sulfide, but dissolves readily in ammonium polysulfide.

Since *bivalent* tin is less prone to form acids than is *quadrivalent* tin, it gives rise to no thiosalts. The formation of thiosalts, and consequent dissolution of the sulfide, therefore takes place only when neutral sulfur is also present, in addition to $S^=$ ions, to oxidize bivalent tin to the quadrivalent state. This is the case in ammonium polysulfide. If the thiosalt solution so obtained is made acid, brown tin(II) sulfide is not precipitated from it, but rather yellow tin(IV) sulfide, as a result of this oxidation.

Tin(II) sulfide can be sublimed unchanged in hydrogen. When crystalline, it forms leaflets with a metallic luster, of density 5.27, m.p. 882° and b.p. about 1230°.

SnS crystallizes rhombic. Its crystal lattice can be regarded as a highly deformed rock salt structure. Each Sn is surrounded by 6 S atoms, and each S by 6 Sn. However, the octahedra so formed around each atom are highly deformed, three of the corners being much nearer to the central atom than the other three. A layer-like structure results, which corresponds to the platy habit of the crystals (Hofmann, 1935). The double compound PbS · SnS (*teallite*) has the same structure.

(*iv*) *Tin(II) sulfate* crystallizes from a solution of tin in a mixture of 1 volume of concentrated sulfuric acid, 2 volumes of concentrated nitric acid and 3 volumes of water. It forms white needles, readily soluble in water.

(*v*) *Tin(II) nitrate*. A mixture of tin(II) nitrate and tin(IV) nitrate is obtained by the action of dilute nitric acid on tin. According to Weber, pure tin(II) nitrate is obtained, in the form of white, very deliquescent leaflets, of composition $Sn(NO_3)_2 \cdot 20H_2O$, when a solution of tin monoxide in nitric acid (of density 1.2) is cooled to —20°. The compound is extremely readily decomposed. The basic nitrate, $Sn_2O(NO_3)_2$, is rather more stable; this crystallizes anhydrous, and can be heated to almost 100° without undergoing decomposition. It detonates with a violent report, however, on stronger heating,—especially if heated rapidly—and also through shock or friction.

15. Analytical (Tin)

Since it can be precipitated by hydrogen sulfide from acid solutions, and its sulfides are soluble in ammonium polysulfide, tin belongs analytically with arsenic and antimony to the group of 'acid' sulfides—i.e., those capable of forming thiosalts. The state of oxidation of the tin can often be inferred simply from the color of the precipitated sulfide (SnS brown, SnS_2 yellow). The presence of tin(II) salts is also shown by their reducing properties. These also make it possible to determine tin(II) salts readily by volumetric methods—e.g., with potassium permanganate or iodine solution.

A very sensitive reaction for tin salts is the deep purple red coloration ('purple of Cassius') appearing when a few drops of gold chloride are added to a weakly acidic tin(II) salt solution.

The reaction depends on the fact that the bivalent tin reduces gold chloride to the metal, while at the same time tin dioxide hydrate is formed, according to the equation:

$$3Sn^{++} + 2Au^{+++} + 6H_2O = 2Au + 3SnO_2 + 12H^+.$$

The stannic oxide remains in colloidal dispersion, and adsorbs the gold, which is in the same fine state of subdivision and in this state forms a purple colored sol. The purple of Cassius is thus the result of the mutual adsorption of two colloids. If it is prepared in less highly dilute solutions, it is deposited as a deep violet red precipitate, which can readily be converted to a colloidal dispersion again ('peptized') by treatment with ammonia, and boiling. Purple of Cassius is employed as a splendid red color in porcelain and glass painting.

Tin(II) salts are often tested for by means of mercuric chloride solution, in which, depending on the proportions of the reagents, either a white precipitate of calomel (mercury(I) chloride) or a black precipitate of mercury is produced. If tin is present in the quadrivalent state, a suitable test for its identification is provided by converting it into rubidium chlorostannate, $Rb_2[SnCl_6]$, which crystallizes in characteristic form. Carried out as a *microreaction*—i.e., with a small droplet of solution on a microscope slide, and observing it by means of a microscope, this reaction permits the detection of 0.2 γ* of tin.

The behavior of tin compounds before the blowpipe is also characteristic. The metal appears with ease in the reducing flame; at once burns again, at least superficially, if exposed to air while hot, forming the white dioxide.

In *gravimetric analysis*, tin is generally precipitated as the sulfide, and weighed as the dioxide. If it is present in the form of alloys, it is usually separated as 'b-stannic acid' by treating the alloys with concentrated nitric acid. The stannic acid can then be converted to the dioxide by ignition, and weighed as such. The separation of tin can also be carried out electrolytically, either from acid oxalate solution, or from ammoniacal thiostannate solution. In both cases, the tin is deposited as metal on the cathode, and can be weighed as such.

16. Lead (Plumbum, Pb)

(a) Occurrence

The most widely distributed and most important ore of lead is *lead glance, galena*, PbS. This occurs in many places throughout the world. It is found at a

---

* 1 γ = 0.001 mg.

number of places in Britain, and in the Harz, Erzgebirge, in Carinthia and other localities in Europe. Outside Europe, the large deposits in Broken Hill (New South Wales), in the United States and in Mexico are of great economic importance.

Among decomposition products of galena, *anglesite* (lead vitriol) $PbSO_4$, and *cerussite* (white lead ore) $(PbCO_3)$ are frequently met with; also, less often, *pyromorphite* (green lead ore), $PbCl_2 \cdot 3Pb_3(PO_4)_2$, and *mimetesite*, $PbCl_2 \cdot 3Pb_3(AsO_4)_2$. Among other lead minerals which have no significance as ores, there may be mentioned *crocoite* (kallochrome, red lead ore), $PbCrO_4$, *wulfenite*, (molybdenum lead spar, yellow lead ore), $PbMoO_4$, and *stolzite*, $PbWO_4$.

(b) History

Lead is among the metals which have been known since earliest times. It may be shown that the ancient Egyptians knew it, and very probably the Israelites did so also. The Greeks called it μόλιβος or μόλυβδος. It has already been mentioned that in ancient times it was not clearly distinguished from tin. At the time of the Punic Wars, numerous lead mines already existed in Spain; these had been started there by Greek and Phoenician colonists, and had been taken over by the Romans, who later used lead chiefly for making water pipes. The method of fastening of iron and bronze clamps into stone blocks, by casting lead in the holes, was mentioned as early as Herodotus. Lead preparations such as litharge, PbO, minium, Pb_3O_4, and lead white (basic lead carbonate) were already known to the Greeks and Romans.

(c) Preparation

The metallurgical extraction of lead from galena is carried out principally by methods, distinguished as the *roast-reaction* process, the *roast-reduction* process, and the *precipitation* process.

(i) *Roast-Reaction Process.* In this method of extraction, the procedure is to *roast the galena incompletely* in a reverberatory furnace at a relatively low temperature (500–600°)—i.e., to heat with access of air, but in such a way that only a portion of the lead sulfide is oxidized to oxide and sulfate, according to eqn. (1), while the rest remains unchanged.

$$\left. \begin{array}{l} PbS + \tfrac{3}{2}O_2 = PbO + SO_2 \\ PbS + 2O_2 = PbSO_4 \end{array} \right\} \text{ Roasting process} \qquad (1)$$

The subsequent 'reaction process' consists of heating further, with as complete exclusion of air as is possible, whereupon the as yet unchanged lead sulfide reacts, according to eqn. (2), with the oxide and sulfate, forming metallic lead.

$$\left. \begin{array}{l} PbS + 2PbO = 3Pb + SO_2 \\ PbS + PbSO_4 = 2Pb + 2SO_2 \end{array} \right\} \text{ Reaction process} \qquad (2)$$

The process is also carried out in Scotch hearth furnaces. In this case, processes (1) and (2) take place simultaneously.

(ii) *Roast-Reduction Process.* The first operation is roasting, in this process also, but carried out so as to oxidize as much lead sulfide as possible to lead oxide. It is now usual to carry this out by the wind- or blast-roasting process invented by Huntington and Heberlein, which consists of blowing or sucking air through a mixture of galena with limestone and quartz. The oxidation of galena to lead oxide is considerably promoted by the presence of the limestone. Lead sulfate, formed at the same time, which would otherwise be reconverted to lead sulfide in the subsequent reduction stage, is converted by the quartz into silicate (eqn. (3c)). The mixture of lead oxide and lead silicate, obtained by the blast roasting, is reduced in the second part of the process by means of coke (or the carbon monoxide from its combustion) in a shaft furnace or blast furnace. Carbon monoxide does not, indeed, react directly with lead silicate, but lead oxide is first liberated from the silicate by the action of the admixed calcium oxide. The oxide then reacts with carbon monoxide according to eqn. (4a). The most important chemical reactions of the roast reduction process are thus represented, in all, by the following equations:

$$PbS + \tfrac{3}{2}O_2 = PbO + SO_2 \qquad (3a)$$
$$PbS + 2O_2 = PbSO_4 \qquad (3b) \Big\} \text{ Roast process}$$
$$PbSO_4 + SiO_2 = PbSiO_3 + SO_2 + \tfrac{1}{2}O_2 \qquad (3c)$$

$$PbO + CO = Pb + CO_2 \qquad (4a)$$
$$PbSiO_3 + CaO + CO = Pb + CaSiO_3 + CO_2 \qquad (4b) \Big\} \text{ Reduction process}$$

The roast reduction process is applicable to all lead ores, and is suitable, in particular, for those which cannot be subjected to the roast-reaction process because of their impurity content.

(iii) *Precipitation Process*. This process, which finds only limited application, depends on the fact that lead may be set free directly from the sulfide ('precipitated') by the action of metallic iron:

$$PbS + Fe = Pb + FeS \qquad (5)$$

Metallic iron as such is not used for the process, however, but the galena is heated in a blast furnace together with a mixture of iron oxide and coke. The primary reaction is then the reduction of iron, which reacts in turn with the lead sulfide. The precipitation process has the disadvantage that unless it is operated at a very high temperature, the process represented by eqn. (5) does not go to completion, but that much lead is lost by evaporation if it is very strongly heated.

Lead can be extracted from ores poor in lead by leaching with sodium chloride solution, after a sulfatizing roast. The lead can then be deposited from the solution electrolytically (Tainton process).

(iv) *Refining of Lead*. The crude lead or 'base bullion', as at first smelted by the usual processes, is still contaminated with other elements, and especially copper, antimony, arsenic and sulfur. It also generally contains a not inconsiderable amount of silver. Since the desilverization of lead is of great importance for the extraction of silver, the methods used for this purpose will be discussed under silver. The removal of the remaining impurities is carried out by remelting. In so far as air is allowed to enter, arsenic and antimony are oxidized, with the formation of lead arsenate and antimonate, which form a scum on the surface of the melt. Copper forms a relatively infusible alloy, containing little lead ('matte' or 'speiss'), which also separates out and at the same time takes up all the sulfur from the lead. The remelting is often carried out by heating the lead on an inclined hearth, so that it slowly runs down. This process is known as 'liquation', and the relatively infusible constituents remain behind as a liquation dross.

A process was worked out by Williams, in 1932, for carrying out the refining of lead continuously, by an adaptation of the Parkes process. 400–800 tons are refined daily by this process at Port Pirie (South Australia), which furnishes lead of 99.993% purity.

Lead containing bismuth is most conveniently refined electrolytically. It is deposited from a solution in fluorosilicic acid, containing gelatin (Betts process). The addition of gelatine serves to bring about the deposition of lead in a compact form. The process is in use especially in North America, but is worked also in England (Newcastle) and in Germany (Hamburg).

World production of lead in 1950 amounted to 1.66 million tons; the countries with the biggest production were the United States (390,000 tons), Mexico (238,000 tons), Australia (222,000 tons) and Canada (154,000) tons.

(d) Properties

Lead is a bluish white metal, with a bright luster on a freshly cut surface, but quickly tarnishing in air to a dull matte grey blue. It is the softest of the ordinary heavy metals, being considerably softer than tin. It can be cut with a knife, and even scratched with a finger nail. Because of its low hardness and high ductility, lead can easily be rolled into sheet, but cannot be drawn into fine wires because its tensile strength is too low. Its density is 11.34, m.p. 327.4° and b.p. 1750° (Fischer, 1934).

From the measurements of von Wartenberg, the vapor of lead at 1870° is monatomic. Lead crystallizes cubic, differing from its lighter homologues in forming a face centered cubic lattice with $a = 4.91$ Å. Its specific heat is 0.0299 at 18°, and its atomic heat is accordingly 6.2, in agreement with the Dulong and Petit rule. The thermal conductivity of lead is relatively low, being but 8.5% that of silver. The specific, electrical conductivity at 18° is $\varkappa = 4.8 \cdot 10^{-4}$, or 7.8% that of silver.

Lead readily alloys with other metals. It forms amalgams with mercury, which are liquid if their lead content is small. Lead is also monatomic in the amalgams.

Lead stands immediately above hydrogen in the electrochemical potential series. Its standard potential, relative to the normal hydrogen electrode, is +0.130 volts. Although this implies that it is somewhat baser than hydrogen, lead does not, in general, dissolve in dilute acids. This is in part because hydrogen is evolved on pure lead only at a considerable *overvoltage* (cf. p. 31). Furthermore in many cases the lead is protected from dissolution by the formation of an insoluble coating on the surface, which protects the surface from further attack—e.g. by lead sulfate in contact with sulfuric acid, or by lead fluoride in hydrofluoric acid. Its insolubility in moderately concentrated sulfuric acid is important for the use of lead in accumulators and in the sulfuric acid industry, where the dilute acid obtained in the lead chamber process is evaporated in lead pans to 60° Baumé (78% H_2SO_4 by weight). The acid so prepared is, admittedly, not completely free from lead. Lead is also practically unattacked by hydrochloric acid. Nitric acid, however, because of its strong oxidizing power, dissolves it readily. According to Garre (1933), the resistance of lead towards sulfuric acid can be increased by alloying it with small amounts of $AgCd_4$.

In the presence of atmospheric oxygen, lead is attacked by all acids, including very weak acids and even water:

$$Pb + \tfrac{1}{2}O_2 = PbO; \qquad PbO + 2H^+ = Pb^{++} + H_2O.$$

The relatively great susceptibility of lead to attack by acetic acid in the presence of air is striking. In this case, the tendency to dissolve is probably augmented through the formation of complexes.

Although compact lead, at ordinary temperature, is attacked only superficially by atmospheric oxygen, finely divided lead is pyrophoric. When it is melted, lead first becomes covered with a grey oxide layer, the so-called lead dross. If heating is continued, this changes first into yellow litharge, PbO and—with an ample supply of air and not too high a temperature—finally into minium, or red lead, Pb_3O_4. Lead also combines directly when it is heated with sulfur, selenium and tellurium and with the halogens.

The compounds of lead are extremely poisonous. Even traces of lead, if continuously ingested, can lead to severe illness and death, since the lead accumulates in the organism. Alloys rich in lead are therefore not employed for household utensils. Workers in the lead industries are exposed to a particular hazard, and special regulations therefore apply to them.

Symptoms of lead poisoning ('plumbism') are slate grey edges to the gums ('lead gums'), pallor of the face and lips, constipation and loss of appetite. In serious cases, severe body pains, ('lead colic'), lameness or pains in the limbs occur, and finally cramps, loss of consciousness and other signs of disease of the brain. However, provided the disease has not gone too far, recovery is possible by suitable treatment.

When dissolved in acids, lead passes into the bivalent state, with the formation of Pb^{++} ions. It can be converted to the quadrivalent state by strong oxidizing agents, and especially by electrolytic methods. The normal potential corresponding to the increase of charge $Pb^{++} \rightarrow Pb^{++++}$ is -1.8 volts, relative to the normal hydrogen electrode (cf. under tin). The tendency to revert from the quadrivalent to the bivalent state is thus much greater in the case of lead than with tin. Lead(II) salts, unlike the tin(II) salts, therefore do not act as reducing agents.

(e) Uses [*57, 58*]

Metallic lead has many applications—e.g., as piping, for the coating of cables, in the form of sheet for roofing, and for chemically resistant coverings, e.g., for the lead chambers of sulfuric acid plants. It is also used in the manufacture of crucibles, dishes and evaporating pans, and for the plates of accumulators. Because of its high density, it is used for the cores of bullets, and for making shot.

A little arsenic (about 0.3%) is alloyed with the lead used for making gunshot. The addition of arsenic makes the lead more fluid in the molten state, leads to better pouring properties and more spherical shot, and confers greater hardness after solidification. Among other alloys of lead may be mentioned *typemetal*, usually with 70–90% lead together with antimony and usually some tin also; and *solder* (soft solder). Low-melting alloys of lead and tin are used for this purpose. The lowest melting point (181°) is found for an alloy with 64% tin and 36% lead. However, solders richer in lead are frequently used; although for soldering containers intended for the preservation of foodstuffs—e.g., preserve cans—a maximum content of 10% of lead is specified for the solder. Bearing metals containing lead have already been mentioned under tin. The true lead base bearing metals [*53*] contain lead as the major constituent. To harden this, either antimony is added (often with some tin, copper, arsenic, etc., as well), or hardening is achieved by alloying with alkali and alkaline earth metals. The 'railway metal' used for the axle bearings on the German State Railways belongs to the second type, and consists of lead containing about 0.7% calcium, 0.6% sodium and 0.04% lithium. At temperatures below 65°, this is superior to the tin-base bearing metals. The lead-antimony bearing metals generally contain 60–80% lead, together with antimony, or antimony and tin in about equal amounts. Lead-antimony alloys are known as *hard lead* (antimonial lead), as distinct from *soft lead*, the ordinary pure lead.

The *lead pigments* must be considered as the most important compounds of lead technically and are amongst the oldest mineral colors known to us. The most important are *white lead* (the basic carbonate), *minium* (Pb_3O_4), and *chrome yellow* ($PbCrO_4$). In using these colors, the painter has to bear in mind the high toxicity of lead. Once the oil paint has dried upon its base, no further hazard is to be feared. Of the other compounds of lead, litharge (lead(II) oxide), lead dioxide, lead chloride, lead nitrate and lead acetate (sugar of lead) have technical uses.

The use of lead glass (crystal glass) for decorative and other purposes has already been mentioned (p. 505).

With the organic acids present in resin and in linseed oil lead forms compounds which have the property of accelerating to a considerable degree the resinification, and therefore the hardening, of the oils used in paints. The preparations which bring about such a rapid drying of the paint are known as 'siccatives' or 'driers' and their solutions in linseed oil as 'varnishes'. Manganese oxide and zinc oxide are chiefly used, in addition to lead oxide, for the production of driers and varnishes.

17. Compounds of Lead

In its compounds, lead functions as predominantly positive-bivalent. However, like its lighter homologues, it can also be quadrivalent, although it is much less stable in this state than they are.

Like its lighter homologues, lead in the quadrivalent state forms a volatile hydride, *lead hydride*, PbH_4, and compounds with alkyl radicals having corresponding compositions—the *lead alkyls*, PbR_4. These compounds may be regarded as covalent in constitution.

In so far as it is permissible to speak of polarity in such compounds, lead functions in the lead alkyls as an electropositive atom. This is shown not only by their mode of formation (p. 556) but also by the ability of the oxides belonging to this class, $PbOR_2$, to form well marked salts with strong acids.

In aqueous solutions, quadrivalent lead exists as a free positive ion only in minimal concentrations. It tends, however, to combine with additional negative ions, forming complex anions.

Thus the chloroplumbate ion, $[PbCl_6]^=$, is derived from lead tetrachloride by combination with Cl^- ions; the ions $[PbO_3]^=$ and $[PbO_4]^{4-}$ which are present as structural groups in the *meta-* and *orthoplumbates*, are derived from PbO_2; and the hexahydroxoplumbate ion, $[Pb(OH)_6]^=$, is formed by the addition of two OH^- ions to the hydrated oxide $Pb(OH)_4$.

Free Pb^{++++} ions cannot be present in appreciable concentrations because they react so easily with water (hydrolysis of lead(IV) salts):

$$Pb^{++++} + 2H_2O = PbO_2 + 4H^+$$

Numerous salts are derived from positive bivalent lead; their solutions contain the colorless cation Pb^{++}. The *sulfate* and the *halides* of bivalent lead are sparingly soluble, as also are the lead salts of other acids which form sparingly soluble salts with other metals.

The lead(II) salts are notable for their tendency to form well defined and often well crystallized 'basic salts' by the addition of lead oxide or lead hydroxide. In spite of this, the degree of hydrolysis of the lead salts of strong acids in aqueous solution is only extremely small, and in some cases hardly detectable. 'White lead', $Pb_3(OH)_2(CO_3)_2$, may be cited as a basic salt constituting a product of the greatest technical importance.

In the view of Werner, the 'basic salts' of bivalent lead belong to the class of '*ol-compounds*'. Compounds of the type $PbX_2 \cdot Pb(OH)_2$, such as the basic chloride $PbCl_2 \cdot Pb(OH)_2$ referred to on p. 550, are thus to be formulated as

$$\begin{array}{c} X \\ \diagdown \\ Pb \\ \diagup \\ X \end{array} \begin{array}{c} OH \\ \diagdown \\ \diagdown \\ OH \end{array} Pb \quad \text{or} \quad \left[Pb \begin{array}{c} OH \\ \diagdown \\ \diagdown \\ OH \end{array} Pb \right] X_2$$

On this view it would be expected that basic salts of this type should dissociate in aqueous solution, e.g.:

$$[Pb(OH)_2Pb]X_2 = [Pb(OH)_2Pb]^{++} + 2X^-.$$

Reiff (1936) proved by conductivity measurements, for the case of the basic perchlorate $[Pb(OH)_2Pb](ClO_4)_2$, that this is indeed the case. The view that diol-dimetal ions, $[M^{II}(OH)_2M^{II}]^{++}$ are formed in solution has also been advanced by Hayek (1934), on the basis of observations of the solubility of the oxides and hydroxides of bivalent metals in salt solutions. The crystal structure of salts of this type is considered in Vol. II.

The binary compounds of lead—e.g., the oxides and sulfide—are frequently colored. The halides, however, except for the iodide, are colorless.

The compounds of bivalent lead (lead(II) compounds) are unsystematically known as plumb*ous* compounds, and those of quadrivalent lead as plumb*ic* compounds.

In the following section the oxides of lead will be considered first, and then the compounds which rank next in importance—the lead(II) salts. The discussion of the relatively unstable lead(IV) salts (which are less important than the lead(II) salts for this reason), and of other lead(IV) compounds, is deferred to the end of the chapter.

a) Oxides

Lead forms two simple oxides and at least one mixed oxide:

simple oxides		mixed oxide	
PbO	Lead oxide, litharge	Pb_3O_4	minium (lead(II,IV) oxide)
PbO_2	lead dioxide		

It was formerly assumed that a 'suboxide', Pb_2O, existed as well. More recent investigations, using X-ray crystallographic methods (Van Arkel, Le Blanc, Fricke, Darbyshire) have shown, however, that the products considered to be 'lead suboxide' were only mixtures of Pb and PbO. The existence of a mixed oxide of the composition Pb_2O_3 is also assumed by many, but has not been established; it is not formed by the thermal degradation of PbO_2. There is, however, at least one other oxide intermediate between PbO and PbO_2, although it has not been fully characterized (Clark, 1942, A. Byström, 1946, Katz, 1949)

The mixed oxide Pb_3O_4 results from the union of the simple oxides. This is possible in that lead(II) oxide is essentially basic, and lead(IV) oxide predominantly acidic in character; these oxides may thus combine to form a salt. There is no difference in principle between the combination of the two lead oxides, and that of any other basic oxide with any acidic oxide (acid anhydride) to form a salt:

$$2CaO + SiO_2 = Ca_2[SiO_4]; \qquad 2PbO + PbO_2 = Pb_2[PbO_4]$$

The salts derived from lead dioxide, acting as an acidic oxide, are called *plumbates*. Lead monoxide is also capable, in slight degree, of acting as an acidic oxide. The compounds in which it so functions are known as *plumbites*.

(*i*) *Lead oxide* (lead monoxide), PbO, is manufactured, under the name of 'litharge', by oxidizing molten lead by a blast of air. It is generally obtained as a product of the 'cupellation' of lead from silver. Lead oxide melts at 884°. If, in course of its preparation, it is heated above this temperature, it crystallizes in compact rhombic flakes when it solidifies. By more careful heating, and also by thermal decomposition of lead carbonate or nitrate, it is obtained in the form of a loose, yellow powder, *massicot*. It was formerly used as a pigment. At the present time it is used chiefly for the preparation of minium.

Lead oxide is obtained in the wet way by boiling lead hydroxide with caustic soda. By this means, depending on the concentration, either *yellow* or *red* lead oxide is obtained; both forms have the same composition. The red form is the stable modification at ordinary temperature, and is less soluble than the yellow form accordingly. The yellow oxide is transformed into the red by prolonged boiling with water.

The solubility of yellow lead oxide in water is 1.2 mg in 100 g of water at 20°; that of the red form is about half as great. Red lead oxide (X-ray density = 9.36) belongs to the tetragonal system, the yellow form (X-ray density = 9.66) to the rhombic system. The transition temperature is 488°. According to Dickinson (1924) and Pauling (1941), the red oxide forms the layer lattice represented in Fig. 95. (Tin oxide has the same structure with $a_0 = 3.77$ Å, $c_0 = 4.82$ Å, $d = 1.13$ Å). The yellow form, according to Halla, has a rhombic lattice built up from double molecules

$$\text{Pb} \underset{O}{\overset{O}{<}} \text{Pb}.$$

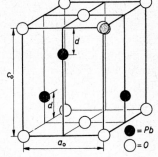

Fig. 95. Unit cell of red lead oxide. $a_0 = 3.95$ Å, $c_0 = 4.99$ Å, $d = 1.19$ Å

The density of ordinary litharge is about 9.4. It is perceptibly volatile even below its melting point. When heated cautiously with access of air, lead(II) oxide can be oxidized to red lead, Pb_3O_4. Only the loose, powdery modification of lead oxide, *massicot*, is suitable for this purpose.

When reducing agents, such as hydrogen, carbon, carbon monoxide, potassium cyanide, are heated with lead oxide, they reduce it readily to the metal.

Lead oxide dissolves easily in acids, forming salts. It is only slightly soluble in sodium hydroxide, unless this is very concentrated. Nevertheless, the solubility of lead oxide in water is markedly increased by sodium hydroxide, as a result of salt formation, *hydroxoplumbites* being formed, as has been proved by Scholder (1934). Lead monoxide therefore functions towards strong bases, as the anhydride of an anhydro-acid, $Pb(OH)_2$. The solution of lead oxide turns red litmus paper blue, in accordance with the dominantly basic nature of the oxide.

If lead(II) salts are precipitated from solution by the addition of alkali, *lead(II) hydroxide* is obtained as a white precipitate:

$$Pb^{++} + 2OH^- = Pb(OH)_2$$

In general, the precipitate has a gel-like character and therefore, like tin(II) hydroxide, contains variable amounts of water. However, Hüttig (1931) has shown, by isobaric dehydration and by X-ray methods, that it contains lead hydroxide $Pb(OH)_2$ as a definite compound. If it is dehydrated at 100°, red lead oxide is formed, but at lower temperatures the *yellow* form is obtained, in accordance with Ostwald's rule.

The dissociation constant for the first stage dissociation of lead oxide as a base—i.e., $Pb(OH)_2 \rightleftharpoons Pb(OH)^+ + OH^-$—has the value

$$K_{B_1} = \frac{[Pb(OH)^+] \cdot [OH^-]}{[Pb(OH)_2]} = 4 \cdot 10^{-5} \quad \text{(Pleissner and Auerbach)}.$$

For the reaction of lead hydroxide as an anhydro-acid, $Pb(OH)_2 + HOH \rightleftharpoons Pb(OH)_3^- + H^+$, the equilibrium constant is

$$K_{A_1} = \frac{[Pb(OH)_3^-] \cdot [H^+]}{[Pb(OH)_2]} = \text{about } 1.10^{-12} \quad \text{(Berl and Austerweil)}.$$

These values imply that, as a *base*, lead oxide has about the same strength as ammonia. As an *anhydro-acid*, it is even weaker than phenol, which has a dissociation constant about a hundred times as great as the equilibrium constant for the acid reaction of $Pb(OH)_2$.

In addition to being used for the preparation of other lead compounds, lead oxide is used for the production of lead glasses (crystal, flint and strass) and of glazes, and as a flux in glass and porcelain painting. It is also used for the manufacture of varnishes and plasters.

(ii) *Lead dioxide* (lead(IV) oxide), PbO_2, is obtained by oxidizing lead(II) salts electrolytically or by means of chlorine, bromine, or hypochlorite. It is manufactured by the oxidation of lead acetate with bleaching powder (1), or by dissolving out lead(II) oxide from red lead by means of dilute nitric acid (2)

$$Pb(C_2H_3O_2)_2 + Ca(OCl)Cl + H_2O = PbO_2 + 2HC_2H_3O_2 + CaCl_2 \qquad (1)$$

$$Pb_3O_4 + 4HNO_3 = PbO_2 + 2Pb(NO_3)_2 + 2H_2O \qquad (2)$$

The electrolytic preparation is also carried out industrially—e.g., by the electrolysis of a sodium chloride solution, in which lead oxide is suspended. The latter is oxidized to the dioxide by the hypochlorite produced electrolytically.

Lead dioxide is a brown powder consisting of fine flakes, with a density 8.9–9.2 (X-ray density 9.42). It is practically insoluble in water, but quite appreciably soluble in acids. It thus has a weakly basic character. Its acidic character is more clearly developed. When heated with strongly basic oxides, it combines with them, forming *plumbates*.

Three series of plumbates are distinguished: *orthoplumbates*, (tetroxoplumbates(IV)), $M^I_4[PbO_4]$, *metaplumbates* (trioxoplumbates(IV)), $M^I_2[PbO_3]$, and *hydroxoplumbates*, $M^I_2[Pb(OH)_6]$.

Calcium orthoplumbate, $Ca_2[PbO_4]$, is obtained by heating PbO_2 with CaO, or a mixture of PbO with CaO in the presence of air. The compound is decomposed even by very weak acids—e.g., by carbonic acid:

$$Ca_2(PbO_4) + 2CO_2 = 2CaCO_3 + PbO_2$$

Calcium plumbate is stable towards water, in which it is only very slightly soluble. Its suspension in water, when treated with sodium peroxide, yields *calcium metaplumbate*, $Ca[PbO_3]$. When lead dioxide is fused with alkalis, however, plumbates of the type $M^I_2[Pb(OH)_6]$ are obtained.

On the basis of their composition alone, the hydroxoplumbates at first appear to be merely hydrated metaplumbates, $M^I_2PbO_3 \cdot 3H_2O$. Their constitution was first deduced, by Bellucci and Parravano, from the isomorphism of the potassium salt, with potassium hexahydroxoplatinate, $K_2[Pt(OH)_6]$, and potassium hexahydroxostannate, $K_2[Sn(OH)_6]$. Further support for this view of their structure was provided by the observation that potassium hexahydroxoplumbate cannot be dehydrated without complete decomposition (loss of oxygen, and formation of a mixture of KOH and PbO). It was, indeed, later found by Grube that the sodium salt can be dehydrated without undergoing decomposition beyond the metaplumbate state. This was confirmed by Simon (1928), who showed that at 110° the sodium salt lost water and was converted to sodium metaplumbate, Na_2PbO_3. The constitutional formula proposed by Bellucci and Parravano was not thereby disproved, even in the case of the sodium salt, since a change in constitution may well have been brought about on heating.

Lead dioxide is a strong oxidant. It evolves oxygen when warmed with concentrated sulfuric acid, and chlorine with concentrated hydrochloric acid. It gives off oxygen even when gently heated. If it is ground up with easily combustible substances, such as sulfur or red phosphorus, inflammation occurs. This is the basis of its use in the manufacture of matches, for which calcium plumbate (mixed with potassium chlorate) is also used. In Dennstedt's method of combustion analysis, lead dioxide is used for the absorption of NO_2, SO_2, Cl_2, HCl, Br_2 and HBr, all of which are quantitatively held back by gently warmed PbO_2.

No hydrogen peroxide is formed in the decomposition of PbO_2 by acids. Lead dioxide is therefore not derived from hydrogen peroxide, as are the 'peroxides' discussed earlier—e.g., barium peroxide. The name 'lead peroxide', still occasionally used for the compound, is therefore misleading. The constitution of the compound may be represented by the formula* $O=Pb=O$, with quadrivalent lead.

Except in dimensions ($a = 4.93$ Å, $c = 3.37$ Å), the crystal structure of lead dioxide is the same as that of tin dioxide. Lead dioxide in the dry state usually still contains small amounts of water, which cannot be removed completely by heating in air or in a vacuum without some simultaneous loss of oxygen. By drying it in a current of oxygen at about 250°, however, Krustinsons (1934–37) was able to obtain absolutely anhydrous lead dioxide. The partial pressure of oxygen reached 1 atmosphere at 344°. This was for relatively coarse-grained lead dioxide. In a state of very small particle size—as obtained for example, by electrolytic deposition—it exerts considerably higher partial pressures of oxygen.

In Scotland and in Idaho, lead dioxide occurs native, in very fine crystals of grey-black color, which have a brown streak (*plattnerite, heavy lead ore*).

(*iii*) *Red lead*, minium, Pb_3O_4, is formed as a brilliant red powder when finely divided lead oxide is heated in air to about 500°. Mixed with linseed oil, it is used very extensively as an oil paint for the protection of iron against rusting, and also as a cement for luting joints in steel plates and tubes. It is usually prepared technically from massicot (powdery lead oxide), which is heated, with

---

* It is possible to regard the PbO_2 lattice, like all structures of the rutile type, at least in a formal sense as a molecular lattice (cf. Fig. 63, p. 265). If it is considered to be a coordination structure, then there are $^6/_3$ O atoms, not linked to each other, for each Pb atom, again leading to the view that lead is quadrivalent in the electrochemical sense.

continuous stirring, in a current of air. A minium of still brighter orange red color. (Paris red, saturn cinnabar) is obtained by starting with the oxide obtained from the decomposition of lead carbonate or nitrate.

Red lead is practically insoluble in water. It dissolves, however, in molten potassium nitrate, and crystallizes out in small, doubly refracting prisms. The density of ordinary, loose red lead is between 8.8 and 9.2. It turns dark when it is heated, but the original color is regained upon cooling.

When red lead is heated in a vacuum, it begins to give off oxygen at about 400°. The oxygen pressure reaches $\frac{1}{5}$ of an atmosphere—i.e., the partial pressure of oxygen in the air—at around 550°. Decomposition takes place in air above this temperature, accordingly. If the dissociation is prevented, by raising the oxygen pressure, red lead melts at 830°.

Red lead is decomposed by dilute nitric acid, forming lead dioxide and lead(II) nitrate. The compound may probably be formally considered as lead(II) ortho-plumbate, $Pb_2[PbO_4]$; it is not merely a mixture of PbO and PbO_2, as follows from the difference in decomposition pressure between red lead and PbO_2. This is shown also by the crystal structure of Pb_3O_4, which has been determined by Gross (1943).

According to Le Blanc (1932), X-ray evidence shows that a black modification of Pb_3O_4 exists, as well as the red modification. It can be obtained both by the union of PbO with oxygen, and by the degradation of PbO_2, if the reactions are performed below 389°. At 390° the black minium changes (monotropically) into the red form.

(iv) *The Lead Accumulator.* Accumulators or *storage batteries* [54–56] are used to store up electrical energy. During 'charging', electrical energy is transformed into chemical energy by means of a reversible chemical process. During 'discharge' the same process, but pro-ceeding in the reverse direction is used to regenerate the electrical energy thus stored up.

The most important type of accumulator is the *lead accumulator*. In principle, it consists of two lead plates, perforated in a lattice pattern, which dip into sulfuric acid of density 1.15–1.20. One of the plates is packed with lead dioxide, the other with spongy lead. In-stead of two plates, several are frequently employed, arranged alternately, not too far apart, with like plates joined together. If the plates of the two different kinds are joined by a wire, a current flows from the lead dioxide plate (recognizable by its brown color) to the grey lead plate. The electric current is produced through the tendency of the positive quadri-valent lead in the lead dioxide to become discharged, giving up positive electricity (in other words, by accepting electrons):

$$PbO_2 + 4H^+ + 2e = Pb^{++} + 2H_2O$$

at the positive plate.

At the negative plate, simultaneously, metallic lead goes into solution as the positive bivalent ion:

$$Pb = Pb^{++} + 2e.$$

The Pb^{++} ions formed in the two cases combine with $SO_4^=$ ions of the sulfuric acid to form lead sulfate, which is insoluble, and is therefore deposited in place of the PbO_2 and Pb which have disappeared. The over-all processes at the two electrodes can therefore be represented by the equations:

$$PbO_2 + 4H^+ + SO_4^= + 2e = PbSO_4 + 2H_2O$$

$$Pb + SO_4^= \qquad\quad = PbSO_4 + 2e$$

Combining the two equations, we obtain for the total chemical reaction in the accumulator during the discharge process:

$$PbO_2 + Pb + 2H_2SO_4 = 2PbSO_4 + 2H_2O.$$

The potential difference between the two plates, maintained by the processes cited, amounts to about 2 volts. If an external potential larger than this is applied in the reverse direction—i.e. so that the positive pole of the external source of current is joined to the positive plate—these processes are reversed at the two electrodes. At the plate joined to the external positive pole, Pb^{++} ions are charged up again to Pb^{++++} ions, which at once react with water to form PbO_2:

$$Pb^{++++} + 2H_2O = PbO_2 + 4H^+$$

At the other plate, lead is redeposited, by the discharge of Pb^{++} ions.

During discharge, the acid concentration falls, since water is formed; during charging, it increases again, since water is consumed. The state of charge of an accumulator can thus be controlled by determining the specific gravity of the acid. If the charging process is continued after all the lead sulfate deposited on the plates has been used up, so that no more Pb^{++} ions are available, hydrogen is liberated at the lead electrode, and oxygen is evolved at the lead dioxide electrode—the accumulator 'gases'. Since this requires a higher potential than for the discharge or increase of charge of the Pb^{++} ions, (at their normal concentrations), the terminal voltage rises considerably towards the end of the charging process. During discharge, the voltage rapidly falls to 2 volts, and then remains for a long time practically constant —the weaker the discharge current drawn from it, the more constant is the voltage. If the discharge current is larger, the surroundings of the positive plate become depleted in acid, since what is consumed cannot be replaced fast enough by diffusion. The fall of voltage upon discharge is due to this, since the discharge tendency of lead dioxide is greater in a strongly acid than in a weakly acid solution.

For the reasons given, and also because electrical energy must be expended to overcome the ohmic resistance of the acid (during both charge and discharge), the charging voltage is always higher than the discharge voltage. Losses of electrical energy are therefore unavoidably associated with storage in accumulators. In practice, losses of 20–25% are not unusual.

(b) Lead(II) Salts

Metallic lead and litharge are the usual starting materials for the preparation of lead(II) salts. Lead is converted by nitric acid into soluble lead nitrate, or by acetic acid, in the presence of air and carbon dioxide, into white lead (basic lead carbonate). Insoluble salts are obtained from the soluble salts by double decomposition.

(i) *Lead chloride*, $PbCl_2$, forms white rhombic crystals with a silky luster; density 5.9, m.p. 498°, b.p. 954°. It is moderately sparingly soluble in water (cf. Table 83). It is therefore precipitated from not too dilute solutions of lead(II) salts when Cl^- ions are added.

Lead chloride is much more insoluble in alcohol than in water. It dissolves more readily in glycerine than in water (2.04 g in 100 g).

Lead chloride is prepared technically either by reacting lead oxide or basic lead carbonate in hydrochloric acid (1a and 1b), or by dissolving granulated lead in dilute nitric acid, and precipitating with hydrochloric acid (2); the nitric acid regenerated in the second part of the process can be used over again for the dissolution of the lead.

$$PbO + 2HCl = PbCl_2 + H_2O \tag{1a}$$
$$Pb_3(OH)_2(CO_3)_2 + 6HCl = 3PbCl_2 + 2CO_2 + 4H_2O \tag{1b}$$

$$\left. \begin{array}{l} 3Pb + 8HNO_3 = 3Pb(NO_3)_2 + 2NO + 4H_2O \\ Pb(NO_3)_2 + 2HCl = PbCl_2 + 2HNO_3 \end{array} \right\} \tag{2}$$

Lead chloride can be distilled without decomposition in a current of carbon dioxide. When heated in the air, it suffers hydrolysis through the action of moisture present in the air. It is reduced to the metal when it is heated in hydrogen, or with carbon in the presence of water vapor.

Fused lead chloride solidifies to a horny mass (horn lead). In the molten state, lead chloride has a considerable electrical conductivity, as it also has in aqueous solution at ordinary temperature. From the conductivity, the saturated solution has an apparent content of 6% of undissociated salt at 25°; about half of the remaining lead chloride has undergone primary dissociation ($PbCl_2 = PbCl^+ + Cl^-$), and half secondary dissociation ($PbCl_2 = Pb^{++} + 2Cl^-$). The solution reacts acid towards litmus. The degree of hydrolysis is only small, however (about 0.6% in 0.01–N solution at 100°, according to Ley)

TABLE 83

CRYSTAL STRUCTURE AND SOLUBILITY OF LEAD(II) HALIDES

	PbF_2	$PbCl_2$	$PbBr_2$	PbI_2
Crystal structure	(a) Lead chloride type (b) Fluorite type	Lead chloride type	Lead chloride type	Brucite type
Solubility, millimoles per liter at 25°	5.5	38.8	26.5	1.6
Same, at 100°	—	119	124	9

Lead chloride forms a rhombic crystal lattice, in which each Pb is surrounded by 9 Cl atoms. Two of the Cl atoms are closest to the Pb (distance 2.67 Å), a third Cl is at a distance of 2.88 Å, and the remaining 6 Cl are at distances between 3.05 and 3.29 Å.

$PbCl_2$ is capable of combining with additional Cl^- ions, forming complex anions. The solubility of lead chloride is therefore increased by the addition of large quantities of chlorides or of HCl; it is diminished by the addition of small concentrations of Cl^- ions. The lead(II) halogeno salts (halogenoplumbates(II), or halogenoplumbites) known in the solid state correspond for the most part to the types $M^I_2PbCl_4$ or M^IPbCl_3; representatives of other types—e.g., $M^IPb_2Cl_5$—are also known.

Lead chloride also readily forms basic chlorides, which can be considered to be compounds of $PbCl_2$ with PbO and with $Pb(OH)_2$. Many of these occur native—e.g., $PbCl_2 \cdot 2PbO$ (*mendipite*). By precipitation of a hot concentrated solution of lead chloride with lime water, the basic salt $PbCl_2 \cdot Pb(OH)_2$ is obtained; this is an article of commerce, as an artists' pigment (Pattinson's lead white). For its constitution, see p. 544.

Lead chloride also forms complex compounds with pyridine and thiourea.

(*ii*) *Lead bromide*, $PbBr_2$, obtained by methods similar to those for lead chloride, crystallizes from hot water in white, silky, rhombic needles, of density 6.6. It melts at 373°, to a red liquid, which solidifies again on cooling to a horny mass; b.p. 916°. Lead bromide slowly blackens when exposed to light, as a result of deposition of metal.

Lead bromide forms double salts and other addition compound resembling those derived from the chloride. It crystallizes as the hydrate $PbBr_2 \cdot 3H_2O$ from solutions containing much hydrobromic acid. It may also add on HBr.

(*iii*) *Lead iodide*, PbI_2, is formed as a yellow precipitate when I^- ions are added to lead salt solutions; this changes into beautiful golden sparkling leaflets when recrystallized from hot water; density 6.2, m.p. 412°, b.p. about 900°. Lead iodide turns first brick red and then brown red when heated; its original color is restored on cooling.

The solubility of lead iodide in water is considerably less than that of the chloride (cf. Table 83). Although the compound is intensely yellow in color, its aqueous solution is quite colorless. This is to be explained in that, in solution, it is present almost completely in the dissociated state—i.e., in the form of the ions Pb^{++} and I^-, both of which are colorless. The addition compound $HI \cdot PbI_2 \cdot 5H_2O$ crystallizes from solutions containing hydriodic acid. Numerous well crystallized double salts of lead iodide are also known; most of them are hydrated.

Lead iodide forms a layer lattice structure, of brucite type (cf. Fig. 61, p .261; $a = 4.59$ Å, $c = 6.86$ Å).

(iv) *Lead fluoride*, PbF_2, is the least soluble of the lead(II) halides, with the exception of the iodide. It is obtained as a white powder of density 8.24 (X-ray density = 8.37) by the action of hydrofluoric acid on lead carbonate or hydroxide; m.p. 818°, b.p. 1292°; solubili-ty—cf. Table 83; heat of formation 159 kcal. per mol. Lead fluoride is dimorphous. The rhombic α-PbF_2, stable at ordinary temperature, changes above 220° into cubic β-PbF_2. α-PbF_2 has the same structure as $PbCl_2$ (X-ray density 8.37); β-PbF_2 has the fluorspar struc-ture ($a = 5.93$ Å, X-ray density = 7.68 at 25°). The mutual transformation of the two forms usually takes place very slowly.

(v) *Lead cyanide*, $Pb(CN)_2$, results as a white precipitate from the addition of CN^- ions to lead salt solutions. It is insoluble in water, but soluble in acids, with decomposition.

(vi) *Lead thiocyanate*, $Pb(SCN)_2$, is formed in yellow crystals, very sparingly soluble in water, by the reaction between lead acetate and potassium thiocyanate.

Both lead cyanide and lead thiocyanate tend to form double compounds with the lead halides.

(vii) *Lead acetate*, $Pb(C_2H_3O_2)_2$, is prepared by dissolving lead oxide in acetic acid. It is very soluble in water (50 g in 100 g of water at 25°, 200 g at 100°). It crystallizes as the hydrate, $Pb(C_2H_3O_2)_2 \cdot 3H_2O$ when the solution is evaporated, forming clear monoclinic crystals of density 2.50, melting at 75°. The anhydrous salt melts around 280°, and has a density of 3.25. Because of its intensely sweet taste (which is followed by an unpleasant, metallic after-taste), it is also known as *sugar of lead*. It is *intensely poisonous*.

Lead acetate is much used industrially, as the starting point for the preparation of other compounds. Dilute lead acetate solutions are used in medicine for poultices and washes.

Paper impregnated with lead acetate smoulders like tinder when it is inflamed. Aqueous solutions of lead acetate can dissolve considerable amounts of lead oxide. Solutions prepared in this way are known as vinegar of lead, and contain *basic acetates*—e.g., $Pb(OH)(C_2H_3O_2)$ and $Pb_3(OH)_4(C_2H_3O_2)_2$. Other lead compounds, especially the halides, also readily form addition compounds with lead acetate. It also combines with other acetates, forming complex salts—e.g., with sodium acetate, to form $NaPb(C_2H_3O_2)_3 \cdot 3H_2O$ (sodium triacetatoplumbite).

(viii) *Lead formate*, $Pb(CHO_2)_2$, brilliant rhombic prisms, very soluble in water (16 g in 100 g at 16°), is very similar to lead acetate in its behavior. It differs from the latter in being sparingly soluble in alcohol.

(ix) *Lead oxalate*, PbC_2O_4, forms as a white precipitate when $C_2O_4^=$ ions are added to a lead salt solution. Its solubility is about 2 mg per liter of water at 25°.

(x) *Lead nitrate*, $Pb(NO_3)_2$, is prepared by dissolving litharge, white lead, or granulated lead in hot dilute nitric acid, and is crystallized from the solution containing a slight excess of nitric acid, to prevent contamination with basic salt. It is obtained as a by-product in the preparation of lead dioxide from red lead. It is used in the match industry, and also as a starting material in the preparation of other lead compounds.

Lead nitrate separates from solution in large, water-clear crystals, belonging to the regular system; density 4.53. Its solubility in water is

at	0°	20°	40°	60°	80°	100°
	38.8	56.5	75	95	115	139 g in 100 g of water.

When heated, lead nitrate decomposes:

$$Pb(NO_3)_2 = PbO + 2NO_2 + \tfrac{1}{2}O_2$$

Lead nitrate also readily forms addition compounds, especially with lead oxide (basic nitrates). No hydrolysis is directly detectable in the aqueous solutions of lead nitrate.

Lead nitrate solutions have a strongly corrosive action in metallic lead. According to Thiel (1920), this is because the lead nitrate oxidizes the lead, being itself converted to the yellow nitrite. In doing so, it attacks the more impure regions of the lead preferentially —i.e. the 'intercrystalline' substance, as it was called by Tammann (p. 562),—which is present between the individual crystallites and in which all the impurities of the metal are accumulated, since it consists of the 'eutectic', the last portions of the melt to solidify (cf. Chap. 13).

(*xi*) *Lead carbonate*, $PbCO_3$, occurs native as *cerussite* (white lead ore), iso-morphous with aragonite, strontianite, and witherite. It is prepared artificially as a crystalline precipitate, density 6.53, by passing carbon dioxide into a cold dilute solution of lead acetate, or by adding lead acetate or nitrate to ammonium carbonate solution. Its solubility in water is extremely small; it depends on the carbon dioxide content, but is, in practice, about 0.3 mg in 100 g of water.

Lead carbonate is readily decomposed into lead oxide and carbon dioxide when it is heated. The CO_2 pressure reaches 1 atmosphere just above 300°.

If lead salt solutions are precipitated hot with alkali carbonate *basic lead carbonate* is obtained. This is also formed from the neutral carbonate by boiling it with water while passing in a current of carbon dioxide. In the pure state it has the composition $Pb(OH)_2 \cdot 2PbCO_3$. It is very widely used as pigment, under the name of white lead.

(*xii*) *White Lead*. The important pigment, *white lead* [59], basic lead carbonate, is manufactured by several processes, which operate either 'in the dry way', as in the Dutch and the German processes, or 'in the wet way', as in the French and the electrochemical processes.

[1] The *Dutch process* is the oldest, but is still operated on a large scale. In this process, thin sheets of lead, rolled into spirals, are exposed to the vapors of acetic acid, which is contained in pots embedded in dung or other fermenting matter. The fermentation provides the warmth needed for a slow evaporation of the acetic acid, and at the same time produces the carbon dioxide required to convert the basic lead acetate first formed into the basic carbonate. The essential reactions in the formation of white lead can be represented by the following equations:

$$Pb + 2HC_2H_3O_2 + \tfrac{1}{2}O_2 = Pb(C_2H_3O_2)_2 + H_2O$$

$$Pb(C_2H_3O_2)_2 + H_2O = Pb(OH)(C_2H_3O_2) + HC_2H_3O_2$$

$$6Pb(OH)(C_2H_3O_2) + 2CO_2 = Pb_3(OH)_2(CO_3)_2 + 3Pb(C_2H_3O_2)_2 + 2H_2O$$

The various reactions take place concurrently, until the evolution of carbon dioxide ceases, or until all the lead is used up.

[2] In the *German process*, the lead sheets are hung up in chambers, into which are passed carbon dioxide, evolved by burning coke, together with the vapors of acetic acid. Substantially the same reactions occur as in the Dutch method of preparation.

[3] The *French process* operates in the 'wet way', and depends on the formation of basic lead carbonate when carbon dioxide is passed into solutions of basic lead acetate. The latter is produced by dissolving lead oxide in boiling lead acetate solution. The *English process*, and the '*Hebrew*' process commonly used in Russia work on the same principle, except that these operate with thick preparations of a doughy consistency, containing only a little water, instead of with solutions.

[4] In the *electrochemical process*, lead sheets are suspended, as anodes, in suitable electrolyte solutions—e.g., a solution of sodium chloride, with sodium carbonate added—and carbon dioxide is led in during the passage of the current. The lead passing into solution is thereby at once precipitated as the basic carbonate:

$$Pb = Pb^{++} + 2e$$

$$3Pb^{++} + 2OH^- + 2CO_3^= = Pb_3(OH)_2(CO_3)_2.$$

White lead forms a heavy, compact amorphous powder, which combines very intimately with oil. It has the greatest covering power of all the white pigments. A disadvantage is its property of gradually darkening in air containing traces of hydrogen sulfide (due to the formation of lead sulfide). Barium sulfate is generally mixed with the white lead of commerce, and a content of up to 20% barium sulfate does not adversely affect the covering power of white lead, but increases its property of 'brushing out'. Larger amounts of barium sulfate are often added, however, for the sake of cheapness.

(*xiii*) *Lead chromate*, $PbCrO_4$, also forms an important pigment (chrome yellow).

It is manufactured by mixing a solution of lead acetate with a solution of potassium chromate, acidified with sulfuric acid, at ordinary temperature. Under these conditions, lead chromate is precipitated as a brilliant yellow powder.

$$2Pb^{++} + Cr_2O_7^= + H_2O = 2PbCrO_4 + 2H^+$$

If a neutral chromate solution is used for precipitation, darker shades are obtained.

Lead chromate occurs native in yellowish-red, monoclinic needles or prisms, of density 5.9–6.0, as *crocoite*. It is extremely insoluble in water (about 0.01 mg. in 100 g of water), but readily soluble in nitric acid and fairly soluble in caustic alkalis also. When heated, it at first melts without decomposition, but can give up a portion of its oxygen to other substances which are not very easily oxidized. It is therefore used in organic ultimate analysis, for the combustion of substances containing sulfur.

By triturating moist, finely ground lead chromate with lead oxide or by precipitating solutions of basic lead salts with neutral chromates, *basic chromates* are obtained, some of which are used as pigments (chrome red). A basic lead chromate of the composition $2PbO \cdot PbCrO_4$ occurs in nature as *phoenicite (melanochroite)*, in the form of dark red crystal leaflets; density 5.75.

(*xiv*) *Lead sulfate*, $PbSO_4$, also known as *lead vitriol*, is precipitated as a white powder when dilute sulfuric acid is added to solutions of lead salts. It is sparingly

soluble in water (about 40 mg per l), but is considerably more soluble in concentrated strong acids (hydrochloric, nitric or sulfuric acids). The greater part is
re-precipitated on dilution.

Solutions of sodium and ammonium acetate can also take up considerable
amounts of lead sulfate. Its solubility in ammoniacal ammonium tartrate solution
is utilized in analysis, to distinguish and separate lead sulfate from other insoluble,
white powders, such as barium sulfate, or silicon dioxide, which are insoluble in
ammonium tartrate.

The augmented solubility in the reagents mentioned is due in part to complex formation,
and in part to the formation of feebly dissociated compounds with the added ions—e.g., to
the formation of feebly dissociated lead acetate, in the presence of acetate ions.

Lead sulfate also dissolves in concentrated caustic alkalis. In this case, *alkail
hydroxoplumbites* are formed.

Lead sulfate is found native, often in well formed, large, rhombic crystals,
density 6.1–6.4, as *anglesite*. In the pure state, these are glass-clear (and are therefore also called 'lead glass'), but they often contain impurities which confer a
color.

Lead sulfate exists in a monoclinic modification, as well as in the rhombic form (for
structure see p. 282). It also occurs native in this form in a few isolated places (e.g., in
Sardinia).

The melting point of lead sulfate is between 1150° and 1200°; it cannot be determined
accurately since the compound begins to decompose at 1000°.

Lead sulfate can add on lead oxide, to form basic salts. It also forms double salts—e.g.,
$(NH_4)_2Pb(SO_4)_2$, small transparent crystals.

(*xv*) *Lead sulfide* is thrown down as a black precipitate by passing hydrogen
sulfide into a solution of a lead salt. It may also be obtained by direct combination
of the constituents (by heating lead in sulfur vapor). It is found in large quantities
in nature as *galena*, lead glance, often in large well developed crystals of the cubic
system (generally in cubes and octahedra), with a lead-grey color, strongly
metallic luster, and excellent cleavage parallel to the cube faces, density 7.58.
The crystal structure is like that of rock salt, with $a = 5.93$ Å, Pb↔S $= 2.97$ Å.

When it is heated in air, lead sulfide is oxidized to lead sulfate and lead oxide.
This property is utilized in the 'roasting' of lead ore in the extraction of lead.
Lead sulfide is gradually reduced to the metal by heating it in hydrogen, but
carbon monoxide reacts with it only poorly. Chlorine, at elevated temperatures,
forms $PbCl_2$ and SCl_2. A mixture of lead sulfide and lead nitrate begins to burn at
about 50°. When it is melted with soda, with free access of air, the metal is
liberated from the sulfide, since oxidation first occurs, forming the sulfate, which
then reacts with a further portion of the sulfide to form SO_2 and Pb.

The solubility of precipitated lead sulfide in water is about 0.8 mg per liter.
It is also practically insoluble in dilute hydrochloric and sulfuric acids, but
dissolves readily in dilute nitric acid, which oxidizes the sulfur. Concentrated
hydrochloric acid decomposes it, with liberation of hydrogen sulfide, as also does
citric acid, probably because of a tendency for complex formation.

Lead sulfide melts at about 1110°. Molten lead sulfide is miscible in all proportions with iron(II) sulfide, copper(I) sulfide and silver sulfide. The eutectics (cf.
p. 564) are at 863°, 30% FeS; 540°, 51% Cu_2S; and 630°, 77% Ag_2S respectively.

(c) Lead(IV) Compounds

The lead(IV) compounds may be divided into two classes:

[1] Those compounds which may be regarded as *the lead(IV) salts of acids*. The most important compounds of this class are the lead(IV) halides, lead(IV) sulfate, and lead(IV) acetate. In addition to these, which are described below, a few other lead(IV) salts are also known—e.g., the (primary) phosphate, the chromate, and the lead(IV) salts of various organic acids.

[2] *Lead(IV) compounds without any salt character*. To this group belong lead hydride PbH_4 and the lead alkyl compounds of the type PbR_4 (R = alkyl). On the other hand, the lead alkyl compounds of the type PbR_3X, or PbR_2X_2 (X = an acidic group) are distinctly saline in character, and belong to the former class.

The base (or anhydride) from which the lead(IV) compounds are derived is lead dioxide, PbO_2. Lead(IV) salts are formed, accordingly, when lead dioxide is dissolved in acids. The electrolysis of solutions of lead(II) salts in concentrated sulfuric acid is often a more suitable method for their preparation.

Lead(IV) salts are hydrolyzed by water, lead dioxide being deposited. They have a strong tendency to form *acido-compounds*, by adding on the salts of stronger bases, e.g.,

$$PbCl_4 + 2KCl = K_2[PbCl_6]$$

potassium hexachloroplumbate, isomorphous with K_2PtCl_6.

The acido-plumbates are considerably more stable than the free lead(IV) salts, and exist in some cases even where the free salts are unstable.

(i) *Lead(IV) halides and Halogenoplumbates*. 1. *Lead tetrachloride*. Lead dioxide dissolves in concentrated hydrochloric acid, forming lead tetrachloride. This is very unstable, however, and readily breaks up with liberation of chlorine.

$$PbO_2 + 4HCl = PbCl_4 + 2H_2O; \qquad PbCl_4 \rightleftharpoons PbCl_2 + Cl_2$$

The process represented by the second equation is reversible—i.e., lead dichloride can take up chlorine in aqueous solution, forming lead tetrachloride. Attempts to isolate lead tetrachloride from solution were for a long time unsuccessful, owing to its instability. Its preparation in the pure state was achieved by Friedrich (1893), who first separated the complex salt $(NH_4)_2[PbCl_6]$ from the solution obtained by the action of chlorine on a suspension of lead(II) chloride in hydrochloric acid. This complex was then decomposed, by adding it to well cooled, concentrated sulfuric acid:

$$(NH_4)_2[PbCl_6] + H_2SO_4 = (NH_4)_2SO_4 + 2HCl + PbCl_4.$$

Lead tetrachloride, being insoluble in concentrated sulfuric acid, separated as a heavy, highly refractive yellow liquid.

Lead tetrachloride, $PbCl_4$, density 3.18 at 0°, is liquid at ordinary temperature and solidifies at about —15° to a yellow crystalline mass. It fumes strongly in air, since it is hydrolyzed by water. Thus the first of the two reactions given above is also distinctly reversible.

2. *Halogenoplumbates*. The *chloroplumbates*, $M^I_2[PbCl_6]$, derived from lead tetrachloride by combination with other chlorides, are distinctly more stable than $PbCl_4$ itself.

In addition to the ammonium salt mentioned, other well crystallized salts of this type have been isolated, some of which may be heated to 200° without undergoing decomposition. They are all hydrolyzed by pure water at ordinary temperature, however, forming PbO_2. The alkali hexachloroplumbates are isomorphous with the alkali hexachlorostannates and platinates (cf. p. 482). By the reactions of chloroplumbates with potassium bromide and iodide, *bromoplumbates*, $M^I{}_2[PbBr_6]$ and *iodoplumbates*, $M^I{}_2[PbI_6]$ have been obtained. The lead(IV) halides giving rise to these compounds are themselves not stable.

3. *Lead tetrafluoride*, PbF_4, was prepared by von Wartenberg (1940), by passing F_2 over PbF_2 above 250°. It forms colorless, tetragonal needles; density 6.7, heat of formation (from the elements) 222 kcal per mol. Complex salts of the type $M^I{}_3HPbF_8$ are derived from lead tetrafluoride; they are isomorphous with K_3HSnF_8, mentioned previously.

(ii) *Lead(IV) azide*, $Pb(N_3)_4$, can be prepared in solution by the action of HN_3 on Pb_3O_4, according to Möller (1949). The red solution rapidly fades, with the evolution of nitrogen and deposition of lead(II) azide, $Pb(N_3)_2$. Reaction of PbO_2 with HN_3 results only in decomposition, without formation of lead(IV) azide. Möller obtained a pungent smelling oil by the reaction of $(NH_4)_2[PbCl_6]$ with NaN_3. This deposited dark red needles when it was evaporated, but these exploded so readily that analysis was not possible; they were presumably an ammonium lead(IV) complex azide.

(iii) *Lead(IV) sulfate*, $Pb(SO_4)_2$, is best obtained, according to Elbs, by the electrolysis of about 80% sulfuric acid between lead electrodes, using a porous clay cell as diaphragm.

A current density of about 2–6 amp. per sq. decimeter is used, and the temperature is kept below 30°, but is allowed to rise to 50° during the last hour of electrolysis. When the solution is cooled, lead(IV) sulfate separates out as a yellowish crystalline powder. It is decomposed by water, the colorless compound $Pb(OH)_2SO_4$ being thereby formed as an intermediate. On further dilution, this latter decomposes, and deposits PbO_2. Lead(IV) sulfate is decomposed by cold concentrated hydrochloric acid, forming a yellow solution from which the alkali chloroplumbates are slowly deposited when alkali chlorides are added.

(iv) *Lead(IV) acetate*, lead tetraacetate, $Pb(C_2H_3O_2)_4$, crystallizes in white needles (melting point 175°) from a solution of lead(IV) sulfate in moderately warm glacial acetic acid.

It can be prepared more simply by treating red lead with warm glacial acetic acid. Half of the lead(II) acetate formed simultaneously can be transformed into a further quantity of lead(IV) acetate, by passing in chlorine, while the rest is precipitated as lead(II) chloride:

$$Pb_3O_4 + 8HC_2H_3O_2 = Pb(C_2H_3O_2)_4 + 2Pb(C_2H_3O_2)_2 + 4H_2O$$

$$2Pb(C_2H_3O_2)_2 + Cl_2 = Pb(C_2H_3O_2)_4 + PbCl_2.$$

The salts of other organic acids—propionic, butyric, stearic acids, etc.—with quadrivalent lead can be prepared in an analogous manner. They melt without decomposition, but they are not stable towards water.

(v) *Lead alkyls*. The lead alkyl compounds, which are likewise derivatives of quadrivalent lead, are of interest because of their mode of formation. They are the products of reaction of zinc alkyls, or of organomagnesium compounds with lead(II) chloride. It might be expected that the reactions which occur would be:

$$PbCl_2 + ZnR_2 = ZnCl_2 + PbR_2 \quad \text{or} \quad PbCl_2 + 2Mg{<}^R_{Br} = MgBr_2 + MgCl_2 + PbR_2$$

However, in place of lead(II) alkyls, PbR_2, the lead(IV) alkyls are obtained. One lead(II) atom reduces a second to the metal, itself becoming quadrivalent (disproportionation):

$$2Pb^{II}R_2 = Pb + Pb^{IV}R_4.$$

As examples of lead alkyls may be mentioned *lead tetramethyl*, $Pb(CH_3)_4$, colorless liquid, density 2.03, b.p. 110° (stable when protected from light, explosive in some circumstances when ignited); *lead tetraethyl*, $Pb(C_2H_5)_4$, colorless liquid, density 1.652 (at 20°), which boils without decomposition only under reduced pressure (vapor pressure 0.056 mm at 0°, 0.260 mm at 20°, 19.0 mm at 90°, 290 mm at 152°); it is insoluble and unaffected in water. *Lead tetraphenyl*, $Pb(C_2H_5)_4$ forms white needles, melting at 224°.

Lead tetraethyl is employed as an anti-knock in motor fuels (added to the extent of 0.08%). Gasoline to which lead tetraethyl is added is colored red in commerce, to enable health hazards, which might otherwise result, to be avoided.

The lead alkyls react with hydrogen chloride or nitric acid, forming substitution products such as PbR_3X and PbR_2X_2 ($X = Cl$ or NO_3), which are completely salt-like in character. The bases (or basic oxides) from which they are derived are known in some cases.

(vi) Lead hydride, PbH_4, is considerably more difficult to prepare than is tin hydride. It was obtained in 1920 by Paneth under special experimental conditions, which enabled hydrogen to be electrolytically liberated at a lead cathode at a high current density, and at the same time produced some sputtering of the electrode. As in the case of tin hydride, the compound was detected by the formation of a metallic mirror when it was passed through a heated tube. A thick cotton wool filter was interposed, to hold back lead, transported in fine suspension as a result of the sputtering.

By analogy with tin hydride, it was to be expected that lead hydride would be formed by decomposing the compound Mg_2Pb with acid. Although it could not be thus detected with certainty, evidence was obtained for the formation of the hydrogen compound of *the lead isotope, thorium B*, when magnesium, alloyed superficially with thorium B, was treated with hydrochloric acid. In this case an invisible metal deposit was formed in the glass tube, but because of the very great sensitivity of radioactive methods it was possible to identify it as consisting of thorium B, after the activity due to thorium C had been allowed to decay away.

Schultze (1929) later observed the formation of volatile lead hydride by the action of atomic hydrogen on lead. Unlike the lead hydride described by Paneth, however, the lead hydride so obtained could not be condensed to the liquid state, but decomposed completely when the attempt was made to condense it.

18. Analytical (Lead)

Lead compounds, when ignited with soda on a charcoal block, give a soft ductile, metallic bead and a yellow incrustation attributable to the oxide. In the systematic separation of the cations, lead occurs in the *basic sulfide* group of the precipitate formed with H_2S (i.e., the group insoluble in ammonium sulfide). It is separated from mercuric sulfide which accompanies it by means of its solubility in warm, moderately dilute nitric acid (1 part of nitric acid, specific gravity 1.4, $+ 2$ parts water), and from the other accompanying sulfides through the sparing solubility of lead sulfate. Characteristic compounds of lead are lead(II) chloride and lead(II) iodide.

The double salt $Cs_2PbCu(NO_2)_6$, which is readily formed from the components, is very suitable for the microanalytical detection of lead. The formation of this salt permits the detection of as little as 0.03 γ of lead. If lead is deposited as PbO_2, it can then most simply be identified by means of Arnold's reagent [tetramethyl diaminodiphenyl methane, $(CH_3)_2N \cdot C_6H_4 \cdot CH_2 \cdot C_6H_4 \cdot N(CH_3)_2$], which gives a blue-violet color with the dioxide.

Lead is usually determined quantitatively as the *sulfate*, $PbSO_4$. It may alternatively be deposited anodically as the dioxide, PbO_2, by the electrolysis of a solution of lead nitrate, weakly acidified with nitric acid. It is then weighed as such*, or it may be weighed as the oxide after gentle ignition.

Very small quantities of lead in aqueous solution can be determined *colorimetrically*. The solution to be tested is treated with aqueous hydrogen sulfide, or with a few drops of a colorless solution of sodium sulfide. The yellow to yellow-brown color of the colloidal dispersion of lead sulfide obtained in this way is dependent on the concentration. It is compared with that of a solution of known lead content, prepared in the same way. Minimal quantities of lead (down to 20 γ) can be reliably and simply determined by the method of Schmidt; PbO_2 is deposited electrolytically, dissolved in glacial acetic acid containing Arnold's reagent, and the resulting color determined colorimetically. G. Jander (1937) states that amounts of lead down to as little as 1 γ can be determined with adequate accuracy by conductometric titration.

* In doing so, it must be remembered that lead dioxide can be completely dehydrated only with difficulty. It is best to dry it at not too high a temperature, and to make allowance for the water content of the dioxide by multiplying its weight by an empirical factor (0.9978, if the dioxide is dried at 200°).

References

1 P. Erasmus, *Über die Bildung und den chemischen Bau der Kohlen*, Stuttgart 1938, 121 pp.

2 A. Fürth, *Braunkohle und ihre chemische Verwertung*, Dresden 1926, 135 pp.

3 J. Steinert, *Der Torf und seine Verwertung*, Berlin 1925 (Sammlung Göschen), 148 pp.

4 E. J. Fischer, *Die natürlichen und künstlichen Asphalte und Peche*, Dresden 1928, 114 pp.

5 P. Grodzinski, *Diamant-Werkzeuge, Anwendung des Diamanten in Industrie und Gewerbe*, Berlin 1936, 214 pp.

6 O. Kausch, *Der Graphit*, Halle 1930, 247 pp.

7 K. Arndt, *Die künstlichen Kohlen für elektrische Öfen, Elektrolyse und Elektrotechnik*, Berlin 1932, 336 pp.

8 G. Bailleul, W. Herbert and E. Reisemann, *Aktive Kohle und ihre Verwendung in der chemischen Industrie*, 2nd Ed., Stuttgart 1937, 114 pp.

9 A. Ramat, *Les Charbons Actifs*, Paris 1943, 186 pp.

10 O. Wohryzek, *Die aktivierten Entfärbungskohlen*, Stuttgart 1937, 99 pp.

11 C. L. Mantell, *Industrial Carbon*, 2nd Ed., New York 1946, 472 pp.

12 H. Hadert, *Der Russ, seine Herstellung, Eigenschaften und Verwendung*, Berlin 1933, 83 pp.

13 H. Kurz and F. Schuster, *Koks, ein Problem der Brennstoffveredelung*, Leipzig 1938, 382 pp.

14 A. Spilker, *Kokerei und Teerprodukte der Steinkohle*, 5th Ed., edited by O. Dittmer and O. Kruber, Halle 1933, 198 pp.

15 F. Fischer, *The Conversion of Coal into Oils* (transl. by H. Borns, edited by R. Lessing), New York 1925, 284 pp.

16 E. Galle, *Hydrierung der Kohlen, Teere und Mineralöle*, Dresden 1932, 111 pp.

17 J. A. Nieuwland, *Chemistry of Acetylene*, New York 1945, 219 pp.

18 J. H. Vogel, *Das Acetylen, seine Eigenschaften, seine Herstellung und Verwendung*, 2nd Ed., Leipzig 1923, 424 pp.

19 L. Bloch-Sée, *La Groupe des Industries de l'Acétylène et de la Soudure Autogène*, Paris 1936, 176 pp.

20 R. Taussig, *Die Industrie des Calciumcarbides*, Halle 1930, 519 pp.

21 H. H. Franck, W. Makkus and F. Janke, *Der Kalkstickstoff in Wissenschaft, Technik und Wirtschaft*, Stuttgart 1931, 213 pp.

22 E. L. Quinn and Ch. L. Jones, *Carbon Dioxide*, New York 1936, 294 pp.

23 I. M. Kuprianoff, *Die feste Kohlensäure (Trockeneis), Herstellung und Verwendung*, Stuttgart 1939, 102 pp.

24 J. Schmidt, *Das Kohlenoxyd, seine Bedeutung und Verwendung in der technischen Chemie*, Leipzig 1935, 235 pp.

25 F. Schuster, *Stadtgas-Entgiftung*, Leipzig 1935, 167 pp.

26 P. Dolch, *Wassergas, Chemie und Technik der Wassergasverfahren*, Leipzig 1936, 268 pp.

27 O. Kausch, *Der Schwefelkohlenstoff*, Berlin 1929, 265 pp.

28 F. Krczil (= Kainer), *Kieselguhr, ihre Gewinnung, Veredlung und Anwendung*, Stuttgart 1936 (Sammlung Ahrens, N. F. Heft 32), 197 pp.

29 H. Mayer, *Das Wasserglas, seine Eigenschaften, seine Herstellung und Verwendung*, 2nd Ed., Braunschweig 1939 (Sammlung Vieweg No. 79), 125 pp.

30 O. Eckart and A. Wirzmüller, *Die Bleicherde, ihre Gewinnung und Verwendung*, 2nd Ed., Braunschweig 1929, 63 pp.

31 W. Eitel, *Physikalische Chemie der Silikate*, 2nd Ed., Leipzig 1941, 826 pp.

32 W. L. Bragg, *The Structure of Silicates*, Leipzig 1930, 69 pp.

33 G. Tammann, *Der Glaszustand*, Leipzig 1933, 123 pp.

34 R. Dralle und G. Keppeler, *Die Glasfabrikation*, 2nd Ed., 2 Vols., Munich 1926/1931, 1487 pp.

35 W. Eitel, M. Pirani and K. Scheel, *Glastechnische Tabellen*, Berlin 1932, 714 pp.

36 H. Thiene, *Glas*, 2 Vols., Jena 1931/1939, 363 + 756 pp.

37 H. G. Bodenbender, *Sicherheitsglas (Verbundglas, Panzerglas, Hartglas, Kunstdrahtglas)*, Berlin-Steglitz 1933, 320 pp.

38 B. Alexander-Katz, *Quarzglas und Quarzgut*, Braunschweig 1919 (Sammlung Vieweg No. 46), 52 pp.

39 A. B. Searle, *The Chemistry and Physics of Clays and Other Ceramic Materials*, 2nd Ed., London 1933, 738 pp.

40 H. Salmang, *Die physikalischen und chemischen Grundlagen der Keramik*, Berlin 1933, 229 pp.

41 A. Berge, *Chemische Technologie der Tonwaren*, Halle 1925, 47 pp.

42 R. Rieke, *Das Porzellan*, 2nd Ed., Leipzig 1928, 163 pp.

43 C. Koeppel, *Feuerfeste Baustoffe silikatischer und silikathaltiger Massen*, Leipzig 1938, 296 pp.

44 L. Mazzocchi, *Calci e Cementi*, 6th Ed., Milan 1932, 270 pp.

45 F. M. Lea and C. H. Desch, *Chemistry of Cement and Concrete*, New York 1935, 429 pp.

46 F. Wecke and W. A. Kaminsky, *Zement*, 3rd Ed., Dresden 1950, 230 pp.

47 O. Graf, *Der Aufbau des Mörtels und des Betons*, 3rd Ed., Berlin 1930, 150 pp.

48 R. Grün, *Der Beton*, 2nd Ed., Berlin 1937, 498 pp.

49 E. Probst, *Handbuch der Betonsteinindustrie*, 4th Ed., Halle 1936, 840 pp.

50 G. Kupelian, *Béton Armé*, Paris 1936, 156 pp.

51 L. Santarella, *Il Cemento Armato*, Vol. I, 14th Ed., Milan 1951, 505 pp.; Vol. II, 13th Ed., Milan 1953, 739 pp.; Vol. III, 8th Ed., Milan 1951, 394 pp.

52 L. Stuckert, *Lehr- und Handbuch für die Emailindustrie*, 2nd Ed., Berlin 1941, 300 pp.

53 H. N. Bassett, *Bearing Metals and Alloys*, London 1937, 428 pp.

54 R. Albrecht, *Elektrische Akkumulatoren und ihre Anwendung*, Leipzig 1937, 159 pp.

55 K. Drucker and A. Finkelstein, *Galvanische Elemente und Akkumulatoren*, Leipzig 1932, 425 pp.

56 G. W. VINAL, *Storage Batteries*, 2nd Ed., New York 1930, 427 pp.

57 A. BURKHARDT, *Blei und seine Legierungen*, 2nd Ed., Berlin 1940, 103 pp.

58 W. HOFMANN, *Blei und Blei-Legierungen, Metallkunde und Technologie*, Berlin 1941, 293 pp.

59 E. ZIMMER, *Bleiweiss und andere Bleifarben*, Dresden 1926, 132 pp.

60 W. F. LIBBY, *Radiocarbon Dating*, Chicago 1952, 124 pp.

61 M. CALVIN, C. HEIDELBERGER, J. C. REID, B. M. TOLBERT and P. E. YANKWICH, *Isotopic Carbon, Techniques in its Measurement and Chemical Manipulation*, New York 1949, 376 pp.

62 W. A. WATERS, *Chemistry of Free Radicals*, Oxford 1946, 295 pp.

63 E. G. ROCHOW, *Introduction to the Chemistry of the Silicones*, 2nd Ed., New York 1954.

64 R. SODER, *Die Silicone*, Zurich 1948, 60 pp.

CHAPTER 13

ALLOYS

1. Introduction

Alloys are mixtures of metals which are homogeneous in the molten state, the term usually being applied to the product which solidifies from a homogeneous melt. Molten alloys are, essentially, solutions of metals in one another, but it is possible for molten compounds to be present also.

Solidified alloys may vary widely in constitution. They may be quasi-homogeneous (see below) or quite inhomogeneous, and may consists of mixed crystals or of compounds of the metals, or of both together. The metals forming the alloy may have unmixed during solidification, so that the solidified alloy is a more or less coarse grained mixture of the two components; they may have unmixed only partially, or not at all, or may have united, completely or to some extent, during cooling, to form compounds which are incapable of existence at high temperature; these compounds, in their turn, may also form mixed crystals. The multiplicity of behavior is so great that the study of the nature of alloys has developed into a special branch of science, *metallography*. [*1–5*]. Three chief methods are used in metallography for the investigation of the structure of metals and alloys. The first of these is '*thermal analysis*', which will be more fully discussed below; this is supplemented by the second technique—the *microscopic examination of surfaces*, which are ground and polished, and then etched with suitable media. To these, the third method, that of *X-ray structure analysis*, has been added in recent years.

Among other techniques useful in metallography may be named the determination of *electrical conductivity*, of *thermoelectric properties*, of *magnetic properties*, of *solution potential*; also measurements of *hardness* and of *thermal expansion* (dilatometry). For technical purposes, these are supplemented by studies of how the properties of practical importance depend upon the composition and treatment of the alloys—e.g., tests under continuous loading (fatigue), etc. [*6*]. To detect the formation of *intermetallic compounds*, and to determine their composition (see below), chemical analysis may often be invoked, in addition to the methods just mentioned. Compounds are often more resistant towards chemical reagents (e.g., acids) than are their components. In applying the method of 'residue analysis', however, it is essential to be certain (e.g., by microscopic investigation) that the crystals are free from inclusions, as only in this case will the analytical results correctly represent their composition.

2. The Texture of Metals and Their Alloys

(a) The Process of Crystallization

The solidification of a metal or of an alloy is always a *process of crystallization*. Metals or their alloys never solidify to form *amorphous* or *isotropic** products, as do the glasses.

The *crystalline structure* of many metals is clearly visible to the unaided eye, especially after surfaces have been etched with acids, etc. In other cases it may be

detected only by microscopic examination or, in exceptional cases, only from the X-ray diffraction pattern. In a metal which solidifies completely, as distinct from the process of crystallization from a solution, the surfaces of the individual crystals are often unable to develop freely** and their forms are not identical with those which would result from free crystallization. The growth of every particle is constrained by contact with its neigbors. The particles of such a conglomerate, which are thus shaped by contact with other particles, instead of being bounded by free surfaces, are called crystallites. Although their facets are not those of the natural crystal growth, the crystallites are still fairly regular in structure. This is because the rate

Fig. 96. Etched metal surface, showing the boundaries between the crystallites.

of crystal growth varies in different spatial directions. On a polished surface, suitably etched by some reagent, the boundaries between the crystallites generally appear as fine lines (Fig. 96). The impurities contained in the metal are accumulated, as 'inter-crystalline material', in the boundary surfaces, revealed by these lines. The size of the crystallites depends on the *power of spontaneous crystallization*, and the *rate of crystal growth*. These, in turn, are dependent on the rate of cooling and the degree of supercooling of the melt.

The number of crystallization centers formed in unit volume during unit time gives a measure of the spontaneous crystallizing power (rate of nucleation). The *crystallization center*, or *nucleus*, is the term applied to the point from which the growth of a single crystal originates. The larger the number of crystallization centers produced in unit time, and the smaller the velocity with which each crystal grows out in all directions from its center, the more fine grained is the texture of the solidified melt. The fine grained structure of most metals results, accordingly, largely from their great power of spontaneous crystallization.

(b) Single Crystals

If the crystallization of a metallic melt is allowed to take place in long capillary tubes, it is possible, with a very small degree of supercooling, to arrange for

* A substance is said to be *isotropic* when it has uniform properties along all directions. Crystals are *anisotropic*, since they do not have identical properties in all directions. An *amorphous* substance (greek ἄ-μορφος, without form) is one which cannot occur in regular forms. The concepts 'amorphous' and 'isotropic', as applied to the solid state, overlap in so far as anisotropy in the solid state leads to the development of regular forms when conditions permit of these being freely formed.

** If the metal does not solidify uniformly, the forms of the first constituents to crystallize out may be preserved. In this case they display the faces characteristic of free growth. They may, however, also subsequently undergo deformation.

crystallization to begin at a single point only. By this means one *single crystal* is obtained, filling the whole tube to a length of 20 cm or more. Its boundary surfaces, like those of the ordinary crystallites, are quite fortuitous. But whereas the lattice planes of the individual crystallites are quite randomly oriented, all the lattice planes of the single crystal have the same orientation. As its name implies, the single crystal represents one continuous crystal, even although it may often be bounded by accidental surfaces—i.e., by faces unrelated to its crystal structure.

Another method of preparing single crystal wires—several meters in length— will be described under tungsten. It depends upon transiently heating a short length of wire, starting from one end, almost to the melting point. As the heated zone moves along the wire, the crystallite texture of the wire is transformed into a single crystal.

Another method whereby single crystals may be produced was given by Czochralski (1917); it consists of withdrawing a crystal grain from the melt at the same speed as crystallization proceeds.

Single crystals of considerable size—up to a weight of several kilograms—can be obtained by a method first described by Tammann. In this, a melt is allowed to solidify slowly, starting from a single point. This is achieved by containing the melt in a glass or porcelain tube, which is drawn out to a point at the lower end, and which is slowly lowered out of an electric furnace. [Straumanis, *Z. phys. Chem.*, A, 147 (1930) 163, has described a simple experimental arrangement].

The alkali metals potassium, rubidium, and cesium at ordinary temperature are already so close to their melting points that they always form single crystals. Unlike crystallite conglomerates, they therefore give no X-ray diffraction patterns by the Debye-Scherrer method at ordinary temperature.

Single crystals are distinguished from *granular* crystallite agglomerates of the same material by their much greater breaking strength. Since the impurities included between the crystallites are absent, it is possible, moreover, to obtain single crystals in a particularly high state of purity.

Single crystals are relatively soft, however. The tensile strength of single crystal wires is usually considerable inferior to that of *drawn* wires. This is because the single crystals possess glide planes parallel to the crystallographic cleavage planes (cf. Fig. 97). Single crystal wires are therefore extraordinarily ductile. They do not undergo a uniform diminution in diameter when they are stretched, however, but flatten. The broken pieces obtained by rupture under tension are not granular or fibrous, as with ordinary wires, but are smooth and highly lustrous. If several fragments are produced by tensile rupture, all their end surfaces are parallel. Under torsion, single crystal wires do not display the phenomenon known as elastic hyste-

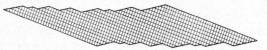

Fig. 97. Schematic representation of part of a stretched single crystal wire with glide planes.

resis, (i.e., failure to return immediately to the equilibrium position when the torque is removed). An important property of single crystal wires is that they retain their strength unaltered even at high temperatures (below the melting point), whereas the strength of drawn wires is lost on heating.

(c) Solidified Alloys [7–10]

We shall call a substance such as a chemically simple metal, which has solidified from the melt into an agglomerate of crystallites which are all of the same kind, 'quasi-homogeneous'. Solidified alloys are rarely quasi-homogeneous. They generally consist of agglomerates of crystallites of differing composition. The simplest case is that of the solidification of an alloy consisting of two metals which are completely miscible in the liquid state, but completely immiscible in the solid state, and incapable of forming a chemical compound with each other. What happens in this case may be illustrated by reference to the tin-lead alloys, which were among the first alloy systems to be scientifically investigated.

(d) Tin-lead Alloys. Eutectic

If a fused lead-tin alloy is allowed to cool gradually, tin crystallizes first from a melt rich in tin, and lead first if the melt contains lead in excess. As a consequence, the melt is progressively enriched in the other component. At the same time, the temperature of solidification falls. The melting point (or freezing point) of tin (232°) must be depressed by lead dissolved in the melt, and the freezing point of lead (327°) is lowered by tin—in each case to an increasing extent as the amount of the substance dissolved in the melt is raised. The way in which the melting or freezing point depends on the composition of the melt is represented by Fig. 98. The composition (in atom–per cent) is represented as abscissa, and the corresponding freezing temperatures as ordinate. It may be seen that the two freezing point or melting point curves must intersect at some point, which corresponds to a content of 26.1 atom–per cent lead and 73.9 atom–per cent tin, and a melting point of 183.3°. This point is known as the *eutectic point*. The alloy having a composition corresponding to this point has the lowest melting point or freezing point of any possible lead-tin alloy. It is known as the *eutectic alloy* or, briefly, as the *eutectic* (εὐ τεκτικός, easily workable), and its melting point as the *eutectic temperature*. A eutectic alloy solidifies as a whole, without any change of composition to a mixture of the crystallites of its components—often very fine grained in texture. From any melt richer in tin, tin crystallizes out, and from any melt richer in lead, lead crystallizes until, in either case, the composition has been changed to that of the eutectic. On a polished and etched section it is often clearly possible to see, by microscopic examination, that the (usually fairly large) crystallites of the one component which were first formed are embedded in the fine grained or foliated eutectic deposited later.

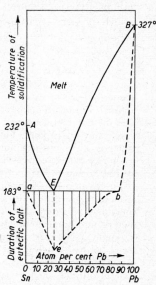

Fig. 98. Melting point diagram of tin-lead alloys (after Degens, 1909).

If the crystallization of a pure metal takes place at a relatively high temperature, the boundary surfaces are often strongly rounded, and the grain forms often almost spherical. This is because, at high temperatures, the influence of surface tension (tending to produce spherical forms) outweighs the directive forces of crystallization, which produce polyhedral forms.

3. Cooling Curves and Phase Diagrams

(a) General

Information about the processes which occur during cooling, and therefore also about the products of solidification, is furnished by observations of the course of the cooling process, as well as by the microscopic examination of etched sections. Diagrams, such as Fig. 99, in which the temperature of the melt, as it is allowed to cool, is plotted against the time are known as *cooling curves*. Analogous curves are obtained for the rise of temperature during uniform heating and are known as *heating curves*.

The curve *a* shown in Fig. 99 is characteristic of a melt consisting of a single substance, cooling without transformation—a so-called *unary system*. According to Newton's law of cooling, the rate of cooling of any body is proportional to the difference between its temperature and that of its surroundings, so that the temperature of the body falls regularly, although at a continually decreasing rate. It will be observed, however, that in Fig. 99 the cooling was interrupted for a certain length of time (H_1-H_2); the curve displays a so-called *arrest* (i.e., the temperature t_H corresponding to the line H_1-H_2). The interruption in cooling is due to the *latent heat of crystallization* liberated during the transition to the solid state. A similar arrest point would be found at the same temperature in the heating curve (provided that superheating was avoided in this case, and supercooling in the former case), because of the heat absorbed during melting (latent heat of fusion).

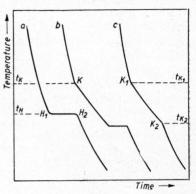

Fig. 99. Cooling curves: (*a*) with arrest; (*b*) with discontinuity and arrest; (*c*) with two discontinuities.

Other transformations which do not involve any change in composition (transition of one modification into another) are evidenced by arrest points in the curves, in the same way as solidification and liquefaction, provided that the heat effects associated with these transitions are not too small. Transformations of this kind may take place both above and below the temperature of solidification.

A different picture is provided by the cooling curve if the melt changes in composition during cooling, as is the case with the tin-lead alloys other than the eutectic alloy. In this case, the temperature of solidification falls during crystallization, which can continue, therefore, only as the melt cools. However, the rate of cooling will be decreased, through the liberation of the latent heat of crystallization. The cooling curve of such a melt does not display an arrest point, but only a discontinuity (at *K* in curve *b*, Fig. 99). Only when the melt has reached the eutectic composition does an arrest appear, since crystallization of the remaining melt takes place at constant temperature. From the length of the horizontal portion of the curve—the duration of the *eutectic halt*—it is possible to infer the amount of the eutectic contained in a solidified alloy. The cooling curve of an alloy which is originally of the eutectic composition has the same form as that of a pure simple metal.

The lengths of the eutectic halts for melts of various compositions are plotted (downwards) as ordinates in Fig. 98. The point on the composition axis corresponding to the peak of the curve *aeb*, joining the ends of these ordinates (shown dotted), gives the proportions of the components in the eutectic, and provides a check upon the composition deduced from the temperature of intersection of the two freezing point curves *AE* and *BE*.

At the lead-rich end of the tin-lead alloy systems, the eutectic halt curve has already met the corresponding abscissa axis (*a–b*) at about 88 atom per cent lead; alloys containing less than 12 atom per cent tin deposit no eutectic. The reason for this is that lead is able to 'dissolve' up to 12 atom per cent of tin (recent investigations indicate as much as 29.7 atom per cent) in the solid state (mixed crystal formation). The dotted curve *bB* is connected with this fact; it is further discussed below.

Melting point diagrams analogous to Fig. 98 are furnished by other binary alloys of metals which form neither compounds nor any extensive range of mixed crystals with each other.

Some tin and lead alloys of this type are listed in Table 84, which records both the eutectic temperature and the eutectic composition. In a few instances the eutectic contains the second component in such small amount that it has not been possible to observe any effect on the melting point of the pure metal. As examples of alloys with a very small degree of miscibility there may also be mentioned the alloys of aluminum with silicon (eutectic temp. 577°, 11.7 wt-per cent Si) and with beryllium (eutectic temp. 644°, 1.4 wt. per cent Be). Other examples will be met with later.

TABLE 84

TIN AND LEAD ALLOYS WITH NO MISCIBILITY (OR ONLY SLIGHT MISCIBILITY) IN THE SOLID STATE

Alloy	Sn–Al	Sn–Si	Sn–Bi	Sn–Zn	Sn–Cd	Sn–Hg	Sn–Ga	Sn–Tl
Eutectic temp.	229°	232°	137°	204°	177°	—39°	20°	170°
Wt.-per cent Sn	99.5	100	42	89	67	0	8	56.5

Alloy	Pb–Ge	Pb–Sn	Pb–As	Pb–Sb	Pb–Bi	Pb–Cd	Pb–Hg	Pb–Ag
Eutectic temp.	327°	183.3°	288°	247°	125°	248°	—42°	304°
Wt.-per cent Pb	100	38.1	97	87	43.5	82.5	0.35	97.5

b) Magnesium-Lead Alloys. Formation of a Compound

The melting point diagram of the magnesium-lead alloys (according to Grube, 1905) is reproduced in Fig. 100, as an example of the case where the two components of the alloy form a *chemical compound* with each other. It can be seen that in this case, the melting point curve is made up of *two curves* of the type represented in Fig. 98. There are, correspondingly, two eutectics, into both of which the compound enters as one constituent—in one case together with crystals of pure magnesium, in the other case together with crystals of pure lead.

The *composition* of the compound—Mg_2Pb—follows from the position of the maximum (at *Mx*). A melt with the composition corresponding to this (80.94 parts by weight of lead and 19.06 parts by weight of magnesium) shows only an arrest, and not a change of slope in its cooling curve (curve of type *a*, Fig. 99)—i.e., the substance crystallizes completely at a single temperature. The composition of the compound can also be deduced from the curve of eutectic times (shown dotted in Fig. 100, the duration of the eutectic halt being once more plotted downwards in Fig. 100). Since, at this point, no eutectic appears, the compound must correspond to the composition for which the two curves of eutectic times intersect the composition axis. The melting point diagram of the magnesium-tin alloys is exactly similar. In this case also a compound, Mg_2Sn, is formed, and there are accordingly two eutectics.

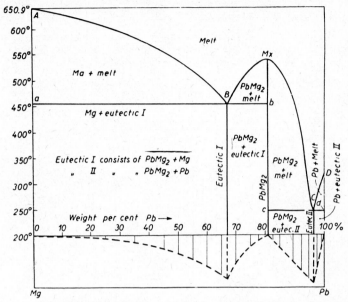

Fig. 100. Phase diagram of the magnesium-lead alloys.

In the diagram reproduced in Fig. 100, the curves *A-B-Mx-C-D*, representing the variation of freezing point with temperature; the straight lines *ab* and *cd* joining the temperatures of the halt points, and the ordinates through the eutectic points and the maximum, together divide the diagram up into certain fields which indicate the state of the system in the corresponding temperature-concentration range. Above the melting point curves, the whole system is in the molten state. Between the melting point curves and the lines *ab* and *cd*, which join the eutectic halts, are the fields in which crystallization takes place to an increasing extent as the temperature falls. In these ranges, liquid and crystals are present together. Below the temperatures indicated by the halt points, the whole has solidified. Such a diagram, which represents the various states of the system, is known as a *phase diagram*. Other examples of phase diagrams are given in Figs. 101, 102 and 104.

If several compounds exist, a corresponding number of maxima appear. Thus Fig. 101 shows this for the case where two compounds occur. Fig. 102 gives an example where the formation of four compounds is revealed by the melting point diagram.

It may happen that the maximum corresponding to a compound is concealed, this

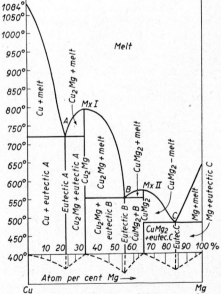

Fig. 101. Phase diagram of the magnesium-copper alloys (after Sahmen).

being so if the compound decomposes on melting. In such a case, the relevant curves intersect before the maximum is reached. The composition of the compound may then be deduced from the curves of eutectic halts. Compounds which decompose when they are melted are said to melt *incongruently*.

(c) Mixed Crystal Formation

In the last example cited, (the sodium-lead alloys) the curves for the times of eutectic halts do not in every case intersect the axis of abscissas at the same point. In so far as the deviations lie outside the experimental error, this is a sign of the *formation of mixed crystals* (cf. Fig. 102).

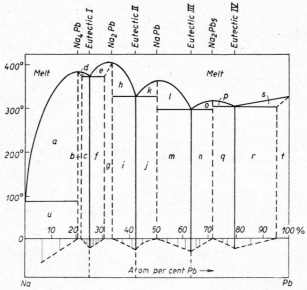

Fig. 102. Phase diagram of sodium-lead alloys (after Mathewson): a = melt + Na_4Pb; b = mixed crystals (Na_4Pb/Na_2Pb); c = mixed crystals (Na_4Pb/Na_2Pb) + eutectic I; d = melt + mixed crystals (Na_4Pb/Na_2Pb); e = melt + mixed crystals (Na_2Pb/Na_4Pb); f = mixed crystals (Na_2Pb/Na_4Pb) + eutectic I; g = mixed crystals (Na_2Pb/Na_4Pb); h = melt + Na_2Pb; i = Na_2Pb + eutectic II; j = NaPb + eutectic II; k = melt + NaPb; l = melt + NaPb; m = NaPb + eutectic III; n = Na_2Pb_5 + eutectic III; o = melt + Na_2Pb_5; p = melt + Na_2Pb_5; q = Na_2Pb_5 + eutectic IV; r = mixed crystals (Pb/Na_2Pb_5) + eutectic IV; s = melt + mixed crystals (Pb/Na_2Pb_5); t = mixed crystals (Pb/Na_2Pb_5); u = Na_4Pb + Na.

As a result of further research the phase diagram has been found to be rather more complicated than is indicated here. Not Na_4Pb but $Na_{15}Pb_4$ mixed crystals separate. In melts with 28–31 atom-% Pb not Na_2Pb but Na_5Pb_2 mixed crystals are deposited. Na_2Pb is only formed by a gradual transformation (in phases with 33–49 atom-% Pb) in the solid state at 182°. Further, not the compound Na_2Pb_5 is formed but mixed crystals containing 68–72% Pb, consisting of $NaPb_3$, in the lattice of which a variable quantity of Pb atoms is replaced by Na atoms. The compound NaPb also forms mixed crystals (within very narrow limits).

Only the simplest possible case of mixed crystal formation will be discussed here, namely that of complete miscibility in both the liquid and the solid state. It might be thought that in this case the cooling curve would be analogous to that found for a simple substance, or for a mixture having the composition of any compound that separates, but in general this is not so. It is true at the most of certain systems and for a quite definite concentration ratio of the components—namely, if the crystals which separate out have the same composition as the melt. As a rule, the melt and the crystals present in equilibrium with it differ in composition. During solidification, therefore, the composition of the melt alters, and with it the temperature of solidification. In the ideal case, one obtains a cooling curve of the type represented by curve c of Fig. 99. If one joins up the points at which solidification begins

(K_1) and ends (K_2), for melts of different composition, a double curve is obtained, of the kind shown in Fig. 103. The upper, or *liquidus* curve, gives the temperature at which solidification commences for the compositions given by the abscissas. The lower, or *solidus* curve, which joins the points marking the completion of solidification, represents the composition of the crystal species which is in equilibrium with the liquid at temperatures corresponding to each point on the liquidus curve. Thus, mixed crystals having the composition *b* are in equilibrium with a melt of composition *a*. The composition of the crystals should accordingly alter continuously as the temperature falls. Since, however, with ordinary rates of cooling, the crystals do not generally change appreciably in composition once they have been formed, the liquidus curves actually found always differ more or less considerably from the ideal form. A theoretical treatment of all the possible cases that can arise in the formation of mixed crystals from melts was given by Roozeboom in 1899.

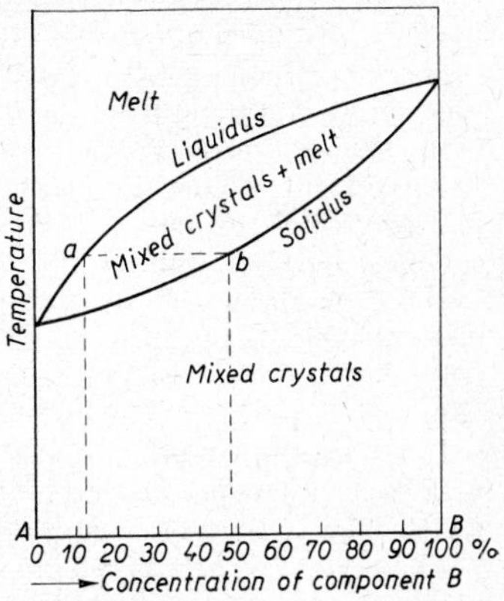

(d) Incomplete Miscibility

There are numerous instances in which miscibility is incomplete, in the molten or in the solid state. The range of concentrations over which there is no complete miscibility is termed a *miscibility gap*.

Fig. 103. Ideal melting point diagram with unlimited miscibility in the solid and liquid states. Melts, and the crystals in equilibrium with them, have a varying composition within the entire range.

The case of incomplete miscibility in the solid state may be illustrated by reference to the freezing point diagram drawn up by Lewkonja for the *thallium-lead alloys* (Fig. 104). Up to a content of 5% of lead, the cooling curves of thallium-lead alloys show 2 discontinuities, like curve *c*, Fig. 99. In this part of the melting point diagram, accordingly, we find the double curves (liquidus and solidus curves) typical of mixed crystal formation. In the interval from 5 to 24% of lead the cooling curves have not only a break, but a halt point, like curve *b* Fig. 99. This corresponds to the formation of a eutectic. In this range of compositions, therefore, the alloys solidify to form inhomogeneous crystallite conglomerates, as is confirmed by microscopic examination of etched sections. The range of mixed crystal formation thus extends only up to a lead content of 5%. The eutectic halt times are plotted (downwards) on the horizontal line a_1b_1. The curve of eutectic times intersects the abscissa at the point b_1, corresponding to a lead content of 24%; this accordingly markes the end of the range over which eutectic alloys are formed. At higher concentrations, cooling curves of the type of Fig. 99c are observed—i.e., the formation of pure mixed crystals again. The miscibility gap in the lead-thallium alloys thus extends from 5 to 24% of lead. Over the subsequent range, the cooling curves again display only two breaks, and a liquidus curve and a solidus curve are accordingly found in the diagram from that point onwards. In the example under consideration, however, they display a peculiarity: they touch one another at one point (B), which also constitutes a maximum. If the liquidus and solidus curves touch one another, it follows from what has been said that for this particular ratio of the components the crystals separating have the same composition as the melt. As was shown by Roozeboom, the two

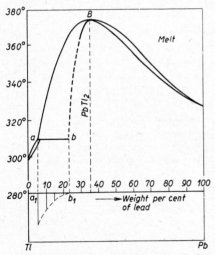

Fig. 104. Phase diagram of the thallium-lead alloys.

curves can touch only in a maximum or a minimum, or in an inflection point of the curve, except at the beginning and the end points of the diagram, where the liquidus and solidus curves must necessarily run together, since these points correspond to melts of the pure components. (An inflection point is the point at which a convex segment of a curve passes into a concave segment). In this case, it is not possible without further evidence to infer the existence of a compound from the occurrence of a maximum, even though at the composition of the maximum the cooling curve shows a halt point just as for a pure substance (Fig. 99a). In the present instance, it may be concluded that a chemical compound does exist, since the maximum falls exactly at the proportion of the two components corresponding to a simple chemical formula, namely $PbTl_2$.

We have already met with an example of very limited mixed crystal formation, in the melting point diagram of the tin-lead alloys (Fig. 98). The dotted curve bB there represents the solidus curve. According to the diagram as shown, the miscibility gap extends from 0–88 atoms per cent Pb*.

If two substances are also mutually insoluble in the *liquid* state, and form no compounds with each other, *two* arrest points are observed. These correspond to the temperatures at which the two sorts of crystals are deposited from the melt.

The examples cited all refer to alloys between two metals (binary systems). For ternary systems, quaternary systems, etc. similar principles apply, except that the relationships are less simple.

4. Influence of Mechanical Working on the Properties of Metals and Alloys

The properties of a metal do not follow immediately from its chemical nature, nor are those of an alloy given by its chemical composition alone. Rather are they dependent, to a high degree, upon its physical texture—i.e., by the size and arrangement of the crystallites from which the metal is built up. The texture, however, is influenced to a considerable extent by *mechanical working* performed in the cold, or at moderately elevated temperatures ('cold working')—i.e., by hammering, rolling, pressing and—in the highest degree—by wire drawing. In the case of copper, for example, it has been possible to increase the tensile strength up to 14 times its original value by mechanical working. [*11*]

The changes undergone by the metal as a result of cold working are the consequence of changes of texture. Glide lines, recognizable under the microscope, appear on polished surface, arising from the displacement of individual parts of the crystallites relative to one another. The glide lines can be seen only when displacement has occurred along the direction of observation, so that grooves appear on the smooth surface turned towards the observer (Fig. 105). The glide lines must not be confused with the cracks resulting from excessive deformation. The planes along which slip takes place are known as glide planes. The greater the ability of any substance to form glide planes, the greater is its *ductility*; the less

---

* More recent investigations have shown that, at the eutectic temperature, lead can take up to 29.7 atom-% tin, and tin up to 1.5 atom-% lead in solid solution. That a much more restricted solubility was found previously is due to the fact that the miscibility gap is considerably wider, even a few degrees above or below the eutectic temperature, and that at this temperature equilibrium is only slowly established. At 150° the solvent power of tin for lead has already dropped to zero, and that of lead for tin has fallen to 18.9 atom-%. It may also be observed, in connection with the diagrams shown in Figs. 100 and 101, that at high temperatures magnesium forms mixed crystals to a limited extent with lead and copper. Lead is taken up by magnesium to the extent of 26 wt. per cent, at the melting point of the eutectic between the compound Mg_2Pb and the magnesium-rich mixed crystals (468°), and up to 2.8 wt. per cent of magnesium is taken up by copper at 722°.

readily it forms glide planes, the more *brittle* it is. Whereas most metals, and also, in general, their mixed crystals, readily form glide planes, the *compounds* of the metals with one another are invariably brittle (except at temperatures close to their melting points).

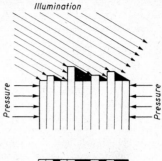

Tammann and others have shown that in mechanical working it is not the properties of the individual crystallites that are altered. Their hardness remains the same, and their crystal structure is unchanged. Only their *shape* alters. It is assumed that this is associated with a diminution of the internal stresses present in the metal. In general, fine grained textures, such as rapidly cooled alloys or eutectics, are apt to have a greater strength than the coarser grained. The texture is made finer by cold working. If the grain size of the texture is allowed to grow again, by prolonged heating (annealing), the original properties are restored. This phenomenon is known as *recrystallization*. By wire drawing, the

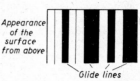

Fig. 105. Schematic representation of the origin of glide lines.

crystallites which were previously randomly oriented are, in addition, aligned more or less parallel to one another. This confers to wires a more or less fibrous structure (cf. the schematic Fig. 106). A similar structure is also often obtained by rolling. A fibrous structure is, however, particularly beneficial to the tensile

Fig. 106. Fibrous structure of a drawn wire.

Fig. 107. Granular structure of a wire after recrystallization by heating.

properties of a material. If a wire is heated, it becomes mechanically weaker, since the granular structure possessed by the metal before it was drawn into wire is thereby restored (cf. the scheme of Fig. 107). For this reason, single crystal wires, which have no deformed crystallite texture and so cannot recrystallize, retain their mechanical properties unaltered even at high temperatures.

5. Intermetallic Compounds and Mixed Crystals

(a) General

Where metals are involved—as also in many other cases where mixed crystals are formed—some difficulty arises in determining and clearly defining the concept of a *chemical compound*. The criterion that chemical compounds are less readily resolved into their components than are physical mixtures often fails here—and is, indeed, inadequate in many other cases also. The definition of the chemical compound based on the laws of constant and multiple proportions is also too narrow, as the more recent developments of chemistry have shown.

On the basis of the laws of constant and multiple proportions, the following definition has been advanced: the product of the union of two or more elements is to be considered as a chemical compound if the atomic ratio in which the components have combined is simple, and expressible in whole numbers, and does not vary continuously with changes in the external conditions. This rule, indeed, still remains valid; however, in addition to the substances which would be considered as chemical compounds on this basis, there are also substances which, by other criteria, must be considered as compounds even *although their composition is variable within certain limits*. The attempt was formerly made to interpret these as mixed crystals of compounds with other elements or compounds. This view is, however, not generally applicable (see Vol II). The concept that chemical compounds must *invariably* display constancy of composition must therefore be given up. Since they do not conform to the principles which led Dalton to the formulation of the atomic theory, compounds of variable composition are known as *non-Daltonide*, or *Berthollide*, compounds as distinguished from the *Daltonide* compounds, of constant composition, which obey these laws. Berthollide compounds are not at all uncommonly formed, even outside the alloys. Thus the compounds with constituents 'wandering' in the crystal lattice (cf. p. 514) are to be included under this head. The Berthollide compounds formed in intermetallic systems play a specially important role, however, and particularly those in systems involving certain metals of the Sub-groups of the Periodic System. As far as is at present known, it is very exceptional for compounds made up exclusively of metals of the Main-Groups to display variability of composition. For this reason, the nature of Berthollide compounds, and their relation to the mixed crystals will be treated later, in conjunction with the discussion of the elements of the Sub-groups.

Apart from that borderline region—not very important in the case of alloys composed only of metals of the Main Groups—where the concepts of mixed crystals and of crystallized chemical compound merge into one another*, compounds and mixed crystals generally differ from one another in typical ways, even in the case of metals. The properties of mixed crystals generally correspond to those of their components, whereas the properties of compounds may differ considerably from those of their constituents. It has already been mentioned that the crystals of compounds are almost invariably brittle; the conductivities of intermetallic compounds are also usually poorer than those of their components. In general, the more the constituents of the compound differ in electroaffinity, the greater is the difference in properties likely to be between a compound and its components; simultaneously, the intermetallic compounds approximate increasingly closely in composition to the ordinary chemical compounds. With metals differing widely in electrochemical properties the composition of compounds frequently corresponds, in numerical ratio, to the compounds of the metals in their typical salts; in other words, metals of very different electrochemical character often exert the same (formal) valences in their compounds with one another as in their salts. However, the valences exercised in the salts have no special significance in by far the largest number of intermetallic compounds. It is quite commonly found that two metals may form several compounds with one another without any of them conforming in composition to the valences of the elements in their salts. In cases where, among a large series of compounds formed between some pair of metals, the composition of one or two compounds does correspond to that required by the normal valences, it is questionable whether any special significance attaches to this fact. Even where two elements form only *one* compound, with composition conforming exactly to that derived from the salt valences, it is not necessarily permissible to relate this composition to the operation of normal valence forces, as exerted in salts and other compounds regarded as ionic in

* In such borderline cases, it is often possible to utilize calorimetric determinations, density determinations and, in some cases, tensimetric measurements, to decide whether a chemical compound is involved, or merely mixed crystals. X-ray methods have become particularly important for this field of work (cf. Vol. II).

nature. In accordance with what has been said, such an inference is most justifiable in the case of metals which differ widely in electrochemical character. Magnesium, in particular, apparently tends to form compounds according to the normal valence rules, with metals which are far removed from it in electrochemical nature.

It is common to observe in intermetallic compounds compositions which differ considerably from simple integral atomic ratios. Thus ratios such as 5 : 6, 3 : 8, 4 : 15, 12 : 13, 12 : 17 and 8 : 25 occur.

Measurements of the *heats of formation* of intermetallic compounds are of importance in interpreting the nature of intermetallic compounds. The baser the components of such compounds, the greater, as a rule, are their heats of formation* (Biltz)—i.e., the stronger are the binding forces.

If a pair of metals forms several compounds, the successive addition of further atoms of one kind brings about the liberation of progressively smaller amounts of heat in successive stages of combination. Thus, in the series of compounds Ca_4Zn, Ca_2Zn_3, $CaZn_4$ and $CaZn_{10}$, containing 1, 6, 16 and 40 atoms of zinc respectively for every 4 atoms of calcium, the heat of formation per gram atom of zinc amounts to 32, 13, 7.4 and 4.8 kcal, respectively; the greater the amount of zinc already combined, the smaller is the energy liberated by further combination with zinc. Similar relations apply in other cases. The same regularity is thus observed as with the addition compounds—e.g., of ammonia with salts. Thus there is a noteworthy connection between the intermetallic compounds and the complex compounds or coordination compounds, which suggests that intermetallic compounds can be considered from the standpoint of the coordination law (Biltz, 1924).

Kraus (1907) has also made the interesting observation that many intermetallic compounds, dissolved in liquid ammonia, dissociate into electrolytic ions, the process corresponding exactly to the dissolution of a salt in water. Zintl (1932–33) was able to detect the existence in these solutions of salt-like compounds, of the type of polysulfides—i.e., compounds composed of atomic cations (which may be solvated) and complex anions, the latter formed by the addition of neutral atoms of one component to negatively charged atoms of the same kind. It is significant that such electronegative ions are formed only by the metals of Main Groups IV to VII of the Periodic System—i.e., by metals standing 4 to 1 places before the inert gases. Proof is thereby afforded that these, and these only, can become negatively charged, by taking up as many units of negative charge as are needed to confer on them the electron configuration of the inert gas. It is also noteworthy in this connection that the incorporation of strongly electropositive metal atoms, such as Na, in the crystal lattice of lead produces a *shrinkage* of the crystal lattice, and not an expansion, although the Na atom is larger than the Pb atom. This observation is also of significance in explaining the hardening of lead by alloying with small amounts of alkali and alkaline earth metals (cf. p. 542).

Many non-metals—especially the semi-metal arsenic, but also phosphorus, silicon, and sulfur (also selenium and tellurium), and occasionally boron and carbon also—form compounds with the metals, which correspond exactly in character with the compounds of the metals with one another. The compounds of these elements with one another are also similar to the intermetallic compounds in many respects.

(b) Behavior of the Metals of Main Groups I-IV Towards Each Other

Tables 85 to 88 show how far the metals of the first four Main Groups are capable of dissolving one another, in the molten and in the solid state, and what binary compounds they form. For the sake of comparison, the non-metals of these groups are also included in part.

* Since they are solids, the free energies of formation (affinities) can be taken as roughly equal to the heats of formation.

TABLE 85

MISCIBILITY AND COMPOUND FORMATION BETWEEN ELEMENTS OF IST AND IIND MAIN GROUPS

	Li	Na	K	Rb	Cs	Mg	Ca	Sr	Ba
Li		liq > 0 / sol 0 / X	liq 0 / sol 0 / X	liq 0 / sol 0 / X	liq 0 / sol 0 / X	liq ∞ / sol < ∞ / $LiMg_2$	— / Li_2Ca* 225° / $LiCa_2$* 225°	—	—
Na			liq ∞ / sol 0 / KNa_2	liq ∞ / sol 0 / X?	liq ∞ / sol 0 / $CsNa_2$	liq < ∞ / sol 0 / X	liq ∞ / sol 0 / X	—	—
K				liq ∞ / sol ∞ / X	liq ∞ / sol ∞ / X	liq < ∞ / sol 0 / X	—	—	—
Rb					liq ∞ / sol ∞ / X				
Mg							liq ∞ / sol ∞ / $CaMg_2$	liq ∞ / sol ~ 0 / SrMg? / $SrMg_2$ 680° / $\overline{SrMg_4}$* 598° / $SrMg_9$ 603°	liq ∞ / sol ~ 0 / $BaMg_2$ 607° / $\overline{BaMg_4}$* 598° / $BaMg_9$ 707°
Ca								sol ∞ / X	sol < ∞ / X
Sr									sol < ∞ / X

Italicized notes refer to miscibility. Compounds shown in ordinary type. Compounds marked * melt incongruently. Non-Daltonide compounds are denoted by a stroke over the formula. The symbol [] denotes that the relevant compound is formed by a reaction taking place in the solid state.

Miscibility   ∞: complete miscibility
       < ∞: restricted miscibility
       > 0: very small miscibility
       0: immiscible.

     liq: refers to liquid state
     sol: refers to solid state.

Compound formation. X denotes no compounds formed.
           — phase diagram unknown or incompletely known.

Numbers written below the formulas show the melting points or decomposition temperatures of the compounds.

TABLE 86

MISCIBILITY AND COMPOUND FORMATION BETWEEN ELEMENTS OF THE
THIRD MAIN GROUP AND THE METALS OF MAIN GROUPS I TO III

	B	Al	Ga	In	Tl
Li	— — —	*liq* ∞ *sol* > 0 Li₂Al LiAl	— — LiGa 625°	— *sol* < ∞ LiIn	— *sol* > 0 Li₄Tl* Li₃Tl 381° 447° Li₅Tl₂ Li₂Tl [] 448° 381° LiTl 508°
Na	— — NaB₂	*liq* < ∞ *sol* 0 X	— — compd.	— — NaIn	— *sol* > 0 Na₆Tl* Na₂Tl* 77·4° 154° NaTl 305°
K	— — —	*liq* < ∞ *sol* 0 X	— — —	— — —	— *sol* 0 K₂Tl* KTl 242° 335°
Mg	— — Mg₃B₂ Mg₅B₂	*liq* ∞ *sol* > 0 Mg₁₇Al₁₂ Mg₂Al₃ MgAl?	— *sol* ~ 0 Mg₅Ga₂ Mg₂Ga 456° 441° MgGa MgGa₂ 373° 285°	— *sol* > 0 Mg₅In₂* Mg₂In MgIn $\overline{\text{MgIn}_3}$	— *sol* < ∞ Mg₅Tl₂ Mg₂Tl* 413° 392° MgTl 358°
Ca	— — CaB₆	*liq* ∞ *sol* ~ 0 CaAl₂ CaAl₃	— — —	— —	— *sol* < ∞ CaTl Ca₃Tl₄* 970° 557° CaTl₃ 524°
Sr	— SrB₆	*sol* ~ 0 SrAl SrAl₄	—	—	— SrTl
Ba	BaB₆	BaAl₄	—	—	BaTl
B		— AlB₂ AlB₁₂	—	—	*liq* 0 : *sol* 0 X
Al			*sol* > 0 X	—	*liq* ~ 0 : *sol* 0 X
Ga				*sol* > 0 X	—
In					*liq* < ∞

For meaning of symbols refer to Table 85.

TABLE 87

MISCIBILITY AND COMPOUND FORMATION BETWEEN METALS OF MAIN GROUP IV
AND THE METALS OF MAIN GROUPS I TO III

	Li	Na	K	Mg	Ca	Sr	Ba	Al	Ga	In	Tl
				sol ~ 0							
Ge	—	NaGe	KGe	Mg₂Ge	CaGe	—	—	sol > 0	—	—	—
			KGe₄	1115°				X			
	sol 0	sol 0	sol 0	sol > 0	sol 0	sol 0?	sol 0?	sol 0	sol 0	<u>sol ∞</u>	sol ∞
	Li₄Sn	Na₁₅Sn₄	K₂Sn	Mg₂Sn	Ca₂Sn			X	X	In₃Sn?	X
Sn	Li₇Sn₂	Na₂Sn	KSn		CaSn	SrSn				<u>InSn₁₅?</u>	
	Li₅Sn₂	Na₄Sn₃	KSn₂		CaSn₃	SrSn₃	BaSn₃				
	Li₂Sn										
	LiSn	NaSn	KSn₄			SrSn₅	BaSn₅				
	LiSn₂	NaSn₂									
	sol > 0	sol > 0	sol 0	sol > 0	sol ~ 0	—	sol ~ 0	liq < ∞	liq < ∞		
								sol 0	sol 0	<u>sol < ∞</u>	sol <
	Li₄Pb	Na₁₅Pb₄						X	X	In₁₅Pb*	
Pb	Li₇Pb₂	Na₅Pb₂								172°	
	Li₃Pb	Na₂Pb	K₂Pb?	Mg₂Pb	Ca₂Pb		<u>Ba₂Pb</u>			In₁₃Pb₂*	PbTl₂l
	Li₅Pb₂									159°	380°
	LiPb	NaPb	KPb?		CaPb		BaPb*				
		NaPb₃	KPb₂		CaPb₃	SrPb₃	BaPb₃				
			KPb₄								

Meaning of symbols as in Table 85. Unless otherwise specified, miscibility in the liquid state is comple

TABLE 88

MISCIBILITY AND COMPOUND FORMATION BETWEEN
THE ELEMENTS OF MAIN GROUP IV

	Si	Ge	Sn	Pb
	—	—	—	—
C	Si₂C	X?		
	SiC			
		liq ∞	liq ∞	liq 0
Si		sol ∞	sol 0	sol 0
		X	X	X
			liq ∞	liq ∞
Ge			sol 0	sol 0
			X	X
				liq ∞
Sn				sol > 0
				X

Meaning of symbols as in Table 85.

Among those combinations of the metals of Main Groups I–IV which are not given in the tables, the only phase diagrams at present known are those of the two systems Be–Al (miscibility very small in the solid state; no compounds) and Ba–Al (no perceptible miscibility in the solid state; compound $BaAl_4$), and of the alloys formed between Ca, Sr and Ba.

One peculiarity of the metals of the first four Main Groups, which is immediately obvious from a glance at the Tables, is that in general they have a very small capacity for forming mixed crystals with each other. Among the systems investigated, except for K–Rb, K–Cs and Rb–Cs, in which there is unrestricted miscibility, and the systems Ca–Sr, Ca–Ba and Sr–Ba, there is no instance of appreciable miscibility at ordinary temperature. Indeed, the metals are in many cases only partially miscible in the *molten* state. It is also striking that, as far as is at present known, the *alkali metals* (other than Li) form no compounds with metals of the IInd Main Group, although by contrast they form a particularly large number of compounds with Sn and Pb (IVth Main Group). Furthermore, the Tables substantiate the rule that, in combining with one another, the metals exert valences quite different from those shown in salt-like compounds. The Tables also furnish examples of the rather complex atomic ratios often observed in intermetallic compounds.

When intermetallic compounds deviate especially widely from simple numerical ratios in their composition, X-ray structure determinations have often provided the only means of assigning the correct formula. Thus the compound $Mg_{17}Al_{12}$ (Laves, 1934) was formerly written with the composition Mg_3Al_2 or Mg_4Al_3, and the compounds $Na_{15}Sn_4$ and $Na_{15}Pb_4$ (Zintl, 1936) as Na_4Sn and Na_4Pb. It is possible that X-ray structure determinations may also result in more complicated formulas being assigned to other compounds (e.g., Li_4Sn and Li_4Pb) than are at present assumed. It has already been mentioned that *compounds of variable composition* rarely occur in the binary systems formed between the elements of the first four Main Groups. This is probably to be associated with the small capacity of these elements for forming mixed crystals with one another. Among the elements with a strong tendency towards mixed crystal formation—these are chiefly the elements of the Sub-groups of the Periodic System—intermetallic compounds of variable composition occur very frequently. Among certain types of compounds of these elements, it is the ratio of the *total number of valence electrons* to the *total number of atoms* which determines their composition (and their structure)—Hume-Rothery's rule, cf. Vol. II. For a very large number of intermetallic compounds, however, including those involving only the elements of the Main Groups, the valence electrons have no influence upon the composition. The composition, is then determined purely by *geometrical factors* (cf. p. 296).

As an example of the structure of a typical intermetallic compound, Fig. 108 represents the crystal lattice of $CaSn_3$. If all the lattice points of this were occupied by the *same kind of atoms*, it would be turned into the face-centered cubic lattice (Fig. 46) which is very frequently found for pure metals. The same structure as in $CaSn_3$

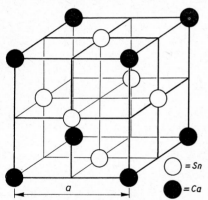

○ = Sn
● = Ca

Fig. 108. Crystal lattice of $CaSn_3$.
$a = 4.732$ Å;   Ca↔Sn $= 3.35$ Å

is possessed by $CaPb_3$ ($a = 4.89$ Å), $NaPb_3$ ($a = 4.87$ Å), $SrPb_3$ (but somewhat distorted tetragonally, with $a = 4.96$ Å, $c = 5.03$ Å), $LaSn_3$ ($a = 4.77$ Å) and $LaPb_3$ ($a = 4.89$ Å) as well as by a series of compounds of analogous composition with metals of the lanthanides and Sub-groups (Zintl, 1931–33; Rossi, 1933). In contrast with these, it is noteworthy that the compounds Mg_2Ge, Mg_2Sn, and Mg_2Pb, which correspond in composition to the salt-like compounds of these elements, have a structure typical of ionic compounds—the fluorspar structure (Mg_2Ge, $a = 6.38$ Å; Mg_2Sn, $a = 6.77$ Å; Mg_2Pb, $a = 6.84$ Å).

We have discussed in this chapter only the *binary* intermetallic compounds. *Ternary* intermetallic compounds have also been observed, but they do not appear to be very numerous.

References

1 G. Tammann, *Lehrbuch der Metallkunde*, 4th Ed., Leipzig 1932, 536 pp.

2 W. Hume-Rothery, *The Metallic State*, Oxford 1931, 371 pp.

3 R. S. Williams and V. O. Homerberg, *Principles of Metallography*, 5th Ed., New York 1948, 319 pp.

4 A. F. Collins, *The Metals, Their Alloys, Amalgams and Compounds*, New York 1932, 310 pp.

5 R. Debar, *Einführung in die Leichtmetallkunde*, 2nd Ed., Leipzig 1941, 187 pp.

6 L. Guillet, *Les Méthodes d'Étude des Alliages métalliques*, 2nd Ed., Paris 1933, 859 pp.

7 W. Hume-Rothery, *The Structure of Metals and Alloys*, 2nd Ed., London 1939, 120 pp.

8 M. Hansen, *Der Aufbau der Zweistofflegierungen*, Berlin 1936, 1100 pp.

9 E. Jänecke, *Kurzgefasstes Handbuch aller Legierungen*, Leipzig 1937, 493 pp.

10 G. H. Gulliver, *Metallic Alloys, Their Structure and Constitution*, 5th Ed., London 1933, 439 pp.

11 A. Kochendörfer, *Plastische Eigenschaften von Kristallen und metallischen Werkstoffen*, Berlin 1941, 312 pp.

[See also literature references following Vol. II, Chapter 1.]

CHAPTER 14

FIFTH MAIN GROUP OF THE PERIODIC SYSTEM: NITROGEN-PHOSPHORUS GROUP

Atomic numbers	Elements	Symbols	Atomic weights	Densities	Melting points	Boiling points	Latent heats of fusion (kcal/g atom)	Valences
7	Nitrogen	N	14.008	0.96[1]	—210.0°	—195.8°	0.085	I, II, III IV, V
15	Phosphorus	P	30.975	1.82[2]	44.1°[2]	280°[2]	0.155	I, III, IV V
33	Arsenic	As	74.91	5.72[3]	817°[4]	633°	5.1	III, V
51	Antimony	Sb	121.76	6.69	630.5°	1640°	4.8	III, V
83	Bismuth	Bi	209.00	9.80	271.0°	1560°	2.64	III, V

[1]) Solid nitrogen, at the m.pt.   [2]) White (cubic) modification.   [3]) Grey metallic (rhombohedral) modification.   [4]) Under 36 atm. pressure.

1. General [1]

The elements of Group VA—nitrogen, phosphorus, arsenic, antimony and bismuth—exhibit a maximum oxidation state of five towards oxygen, but otherwise their principal state is three; towards hydrogen they are exclusively trivalent. Most of these elements can exert valences of both five and three towards other electronegative elements, especially fluorine, chlorine, bromine, and sulfur. As the atomic weight increases, the tendency to exhibit an oxidation state of three towards oxygen and the halogens becomes increasingly favored as compared with the five state. At the same time, the stability of the hydrides diminishes.

The situation can be most readily summarized by regarding the elements of Group VA as positively charged in their compounds with oxygen, halogens, etc., but as negatively charged in their hydrides. The rule then holds that the elements of Group VA have a maximum positive valence of five, but frequently function as positively trivalent; the tendency to acquire only three positive changes increases with increasing atomic number. In addition, there is the possibility of acting as negatively trivalent, and this tendency decreases strongly from nitrogen to bismuth.

Nitrogen, phosphorus, arsenic and antimony can exist in several modifications in the solid state. The ordinary modification of arsenic, antimony and bismuth is metallic; its structure is represented in Fig. 109. It can be thought of as made up of two inter-penetrating rhombohedra, and also as a deformed rock salt crystal lattice. Black phosphorus has a unique crystal structure. Its crystal lattice is built up from double layers, and each layer consists of zig-zag chains lying alongside each other

$$\ldots\!\diagdown\!\underset{P}{\overset{P}{\diagup}}\!\diagdown\!\underset{P}{\diagup}\!\overset{P}{\diagdown}\!\underset{P}{\diagup}\!\overset{P}{\diagdown}\!\underset{P}{\diagup}\!\overset{P}{\diagdown}\!\diagup\ldots \qquad \text{(Angle P–P–P } 99°, \text{ P}\leftrightarrow\text{P} = 2.17 \text{ Å).}$$

The P atoms of the chains which make up the lower layer form links ($\text{P}\leftrightarrow\text{P} = 2.20$ Å) with the atoms of the chains lying above them, to the right and to the left alternately. In this way, the bonds running cross-wise between the chains bind the whole into a double layer with a network structure. Every P atom has three neighbors in all, two belonging to the same chain, and one to another chain in the layer above or below.

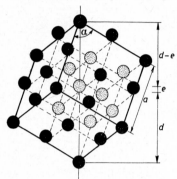

Fig. 109. Antimony type lattice.

P, As and Sb exist as *tetratomic* molecules in the vapor state at moderately high temperatures, and Bi as a diatomic molecule. The Sb_4 molecule is known only in the gaseous state, whereas As_4 molecules may be condensed out as solid yellow arsenic. This is metastable, however, and readily undergoes transformation into other modifications (cf. p. 652). White phosphorus, which is made up of P_4 molecules, is also metastable at ordinary temperature, but is stable at high temperatures (above the melting point of red phosphorus—see p. 627).

	As	Sb	Bi
a	5.60	6.20	6.56 Å
a	84°36′	86°58′	87°34′
$d{-}e$	2.51	2.87	3.10 Å
$d{+}e$	3.15	3.37	3.47 Å

A striking characteristic of the elements of Group VA is their low melting and boiling points as compared with the elements on their left in the Periodic System. This is true especially of the lighter elements. Whereas carbon is excessively difficult to melt and to volatilize, nitrogen is a gas, difficult to condense. Even between phosphorus and silicon, the differences in melting points and volatilities are enormous.

The elements arsenic, antimony and bismuth can exist in aqueous solution, in the trivalent state, as positive ions: As^{+++}, Sb^{+++}, and Bi^{+++}. However, the salts derived from these ions have a strong tendency to be hydrolyzed. In the case of arsenic, a definite equilibrium exists—$As^{+++} + 3H_2O \rightleftharpoons [AsO_3]^{3-} + 6H^+$—such that only in strongly acid solutions is this element present to any appreciable extent as As^{+++} ions. The equilibrium is displaced further to the left with antimony than with arsenic, and still further in the case of bismuth.

This accords with the rule that the basic character of the hydroxides increases from top to bottom in every Main Group of the Periodic Table. The fact that it is impossible to detect any free positive ions at all with nitrogen and phosphorus can be explained on the assumption that the equilibria here are displaced even further to the right.

The hydroxides formed by these elements in the +5 state all have the character of *acids*:

$NO_2(OH)$	$PO(OH)_3$	$AsO(OH)_3$	$H[Sb(OH)_6]$	$Bi_2O_5 \cdot H_2O$?
nitric acid	phosphoric acid	arsenic acid	antimonic acid	bismuth pentoxide hydrate, bismuthic acid?

According to the rule just cited, the *acidic* character of the hydroxides in any Main Group decreases with increasing atomic number of the group element. This holds here also: nitric acid is very strong, phosphoric acid only a moderately strong acid, and the other acids all weak or very weak acids. It is uncertain whether bismuth oxide hydrate is a compound of definite composition at all, and whether it is capable of forming salts of definite composition (cf. p. 677). As compounds with the compositions cited, arsenic acid and antimonic acid exist only in solution, or in the form of their salts (cf. pp. 657 and 667).

TABLE 89

THE NORMAL OXIDES OF MAIN GROUP 5A, AND THEIR HEATS OF FORMATION (KCAL PER MOL)

N_2O	NO	N_2O_3	NO_2	N_2O_5
dinitrogen oxide gaseous —19.6 kcal liquid —15.6 kcal	nitrogen monoxide or nitric oxide gaseous —21.6 kcal	nitrogen trioxide gaseous —20.0 kcal liquid —10.6 kcal	nitrogen dioxide —8.0 kcal N_2O_4 dinitrogen tetroxide gaseous —2.4 kcal liquid +6.8 kcal	nitrogen pentoxide gaseous — 1.2 kcal liquid + 3.6 kcal solid +11.9 kcal
—	—	P_2O_3 Phosphorus trioxide	—	P_2O_5 phosphorus pentoxide +370 kcal
—	—	As_2O_3 Arsenic trioxide +156 kcal	—	As_2O_5 arsenic pentoxide +219 kcal
—	—	Sb_2O_3 antimony trioxide +163 kcal	—	Sb_2O_5 antimony pentoxide +230 kcal
—	—	Bi_2O_3 bismuth trioxide +138 kcal	—	Bi_2O_5 bismuth pentoxide

TABLE 90

OXY ACIDS OF THE 5TH MAIN GROUP

[Heats of formation given in the table are per mol of solid compound]

$H_2N_2O_2$ hyponitrous acid	H_2NO_2 nitroxylic acid (hydronitrous acid)	HNO_2 nitrous acid	—	HNO_3 nitric acid $+42.2$ kcal
H_3PO_2 hypophosphorous acid $+140$ kcal	—	H_3PO_3 phosphorous acid $+228$ kcal	$H_4P_2O_6$ hypophosphoric acid	H_3PO_4 phosphoric acid $+304$ kcal
—	—	$HAsO_2$ arsenious acid	$-+$	H_3AsO_4 arsenic acid
—	—	H_3SbO_3 or $Sb(OH)_3$ antimonious acid	—	$H[Sb(OH)_6]$ antimonic acid
—	—	$Bi(OH)_3$ bismuth hydroxide $+138$ kcal	—	$Bi_2O_5.H_2O$ or $HBiO_3$? bismuthic acid?

Some of the elements of Main Group V can function in valence states other than the two already mentioned. This is true especially of nitrogen and phosphorus. Nitrogen can behave towards oxygen as *formally* uni, bi- tri-, quadri-, and quinquevalent. Although the oxides of phosphorus are derived only from the normal valence states, oxyacids are also formed from two other states (I and IV).

Tables 89 and 90 give a brief summary of the oxides and oxyacids of the Vth Main Group. These give only the normal oxides and acids, or those which define special valence states. The *tetroxides*, formed by the union of trioxides and pentoxides, such as Sb_2O_4 or $[Sb_2O][Sb_2O_7]$ (cf. p. 668), are not given in the tables, nor are the peroxyacids. The molecular complexity in the vapor state has been taken into account only in the case of compounds which are gaseous at ordinary temperature, or observable in solution. The heats of formation are given below the names of the individual compounds; these are for 1 g mol of the compound, and relate to the solid state unless otherwise specified. Some of the acids shown in Table 90 are known only in solution or in the form of salts.

The following summary gives a survey of the *chlorides* and *sulfides* of Main Group V

	Chlorides					Sulfides				
N_3Cl	—	NCl_3	—	—	N_4S_4	—	—	—	N_2S_5
—	P_2Cl_4	PCl_3	PCl_5	P_4S_3	—	P_4S_5	—	P_4S_7	P_2S_5
—	—	$AsCl_3$	—	As_4S_3	As_4S_4	—	As_2S_3	—	As_2S_5
—	—	$SbCl_3$	$SbCl_5$	—	—	—	Sb_2S_3	—	Sb_2S_5
—	Bi_2Cl_4	$BiCl_3$	—	—	Bi_4S_4	—	Bi_2S_3	—	—

In the sulfur compounds, the very considerable increase in stability from nitrogen to bismuth is noteworthy. The sulfur compounds of nitrogen decompose explosively when heated; those of phosphorus may be distilled without decomposition in the absence of air, but inflame in the air even when only gently heated. The sulfides of arsenic, antimony and bismuth are considerably more stable, and are frequently found native accordingly. The principal natural mode of occurrence of the two last mentioned elements in fact, is in the form of the sulfides. The sulfur compounds of nitrogen furnish *ammonia* on hydrolysis, together with oxyacids of sulfur. On the other hand, the hydrolysis of the phosphorus sulfides produces *hydrogen sulfide*, along with oxyacids of phosphorus. This indicates that nitrogen is the more negative constituent in the former, and sulfur in the latter. As their mode of formation shows the same is true of the sulfides of arsenic, antimony and bismuth, which are not decomposed by water and by dilute acids because of their extreme insolubility.

Table 91 summarizes the properties of the simplest *hydrogen compounds* of Group VA. Their stability decreases very greatly from NH_3 to BiH_3.

TABLE 91

PROPERTIES OF THE SIMPLEST HYDROGEN COMPOUNDS OF GROUP 5A

Formula and Name	NH_3 Ammonia	PH_3 Phosphine	AsH_3 Arsine	SbH_3 Stibine	BiH_3 Bismuth hydride
Boiling point	$-33.3°$	$-87.4°$	$-58.5°$	$-17.0°$	$+22°$
Melting point	$-77.7°$	$-132.5°$	$-111.2°$	$-88.5°$	—
Heat of fusion (kcal per mol)	5.65	3.83	4.34	5.08	—
Trouton's constant	23.6	20.6	20.3	19.8	—
Critical temperature	$132.5°$	$52.0°$	—	—	—
Density (of liquid at b.p.)	0.681	0.765	1.621	2.204	—
Surface tension at b.p. (dynes per cm)	34.25	20.59	21.98	24.19	—
Dipole moment of gas molecule (in c.g.s. units)	$1.44 \cdot 10^{-18}$	$0.55 \cdot 10^{-18}$	$0.15 \cdot 10^{-18}$	—	—

It is notable that the boiling point, melting point, heat of fusion, and surface tension diminish from NH_3 to PH_3, and thence rise again to BiH_3. This is due to the fact that ammonia, unlike its analogues, is associated to a considerable extent in the liquid state. This association is indicated, among other evidence, by the abnormally high value of the Trouton constant (cf. p. 322) of ammonia.

The NH_3 molecule behaves as if it were coordinatively unsaturated. It is therefore able to add on another hydrogen ion, forming the positive radical $[NH_4]^+$ (ammonium radical), which also exists in solution as a free ion. PH_3 is also able to from a phosphonium radical $[PH_4]^+$, but to a much smaller degree. With the higher homologues, the ability of the hydrides to form such radicals is lost, although it persists as a property of the *alkyl compounds*.

If all the hydrogen atoms in the hydrides listed in Table 91 are replaced by alkyl radicals, (R), the following compound types are obtained:

	NR_3	PR_3	AsR_3	SbR_3	BiR_3
(trialkyl-)	-amines	-phosphines	-arsines	-stibines	-bismuthines

Except for the bismuth alkyls these are also coordinatively unsaturated, and add on alkyl halides, once more with the formation of univalent positive radicals. For example:

$$N(CH_3)_3 + CH_3I = N(CH_3)_3 \cdot CH_3I, \text{ or } [N(CH_3)_4]^+I^-$$
trimethylamine   tetramethyl ammonium iodide.

The *nitrides, phosphides, arsenides, antimonides,* and *bismuthides,* with the general formula M_3X, derived from these hydrogen compounds by exchanging the hydrogen for a strongly electropositive metal, have a *salt-like structure*. Li_3N possesses a definitely ionic lattice (Zintl, 1935), and an ionic lattice may also be inferred in the case of Li_3Bi from the atomic positions. Li_3Bi crystallizes cubic, and according to Zintl (1935) is anti-isotypic with BiF_3, the structure of which is closely related to that of CaF_2. The structures of all the magnesium compounds of the Group VB elements are known. They are the same as those found for many oxides of trivalent metals, and especially of the rare earths—i.e., for compounds of undoubted ionic nature—except that the positions of anions and cations are interchanged with one another (*anti-isotypy*). The following table summarizes the structures found for the compounds of bivalent metals with the elements of Group VB.

Sc_2O_3 type cf. Vol. II			La_2O_3 type Fig. 15, Vol. II		Zn_3P_2 type related to Sc_2O_3 type		Other types Not yet fully determined	
Be_3N_2	Be_3P_2	—	—	—	Zn_3P_2	Zn_3As_2	—	Sr_3N_2
Mg_3N_2	Mg_3P_2	Mg_3As_2	Mg_3Sb_2	Mg_3Bi_2	Cd_3P_2	Cd_3As_2	—	Ba_3N_2
α-Ca_3N_2	—	—	—	—	—	—	β-Ca_3N_2	—

The change of structure in passing from Mg_3As_2 to Mg_3Sb_2 is to be explained, like the transition from Y_2O_3 to La_2O_3, by the manner in which the lattice type, or the coordination number, depends on the ratio of the ionic radii.

The metallic elements of the Vth Main Group (including arsenic) form no compounds with one another. However, As with Sb, and Sb with Bi form a complete series of mixed crystals. As and Bi, however, are not capable of forming mixed crystals with one another. In the molten state these elements are completely miscible; they are also as a rule completely miscible with the metals of the preceding Main Groups (the only known exception is provided by the system Al/Bi). They form no mixed crystals in the solid state with the metals of the first two Main Groups, but form compounds in every case (cf. Table 92). The compounds richest in the alkali and alkaline earth metals all correspond in composition to the same type—namely to that of the hydrides, i.e., RH_3. They are to be regarded accordingly as 'true valence compounds', and are apparently related to the salt-like compounds. This is evidenced, among other things, in that the melting point maxima in the

TABLE 92

MISCIBILITY AND COMPOUND FORMATION OF ELEMENTS OF THE 5TH MAIN GROUP WITH THE METALS OF 1ST TO 4TH MAIN GROUPS

	Li	Na	K	Mg	Ca	Al	Ga	In	Tl	Ge	Sn	Pb	As	Sb
As	— Li_3As	Na_3As	K_3As K_3As_2? KAs_2	Mg_3As_2	Ca_3As_2	$AlAs$	$GaAs$	—	*liq* $<\infty$ *sol* o X	*sol* $<\infty$ $GeAs$ 737° $GeAs_2$ 732°	*sol* $<\infty$ Sn_3As_2 596° $SnAs$ 605°	*sol* $>$ o X		
Sb	— Li_3Sb	*sol* o Na_3Sb 856° $NaSb$* 465°	*sol* o K_3Sb 812° KSb 605°	*sol* o Mg_3Sb_2 1228°	Ca_3Sb_2? eutec. 583°	*sol* o $AlSb$ 1050°	*sol* o $GaSb$	$InSb$	*sol* $<\infty$ Tl_7Sb_2 226°	*sol* $>$ o X	*sol* $<\infty$ $SnSb$* 425°	*sol* $>$ o X	*sol* ∞ X	
Bi	*sol* o Li_3Bi 1145° $LiBi$* 800°	*sol* o Na_3Bi 775° $NaBi$ 446°	*sol* o K_3Bi 671° K_3Bi_2 420° KBi_2 553°	*sol* ~ o Mg_3Bi_2 800°	*sol* o Ca_3Bi_2 928° $CaBi_3$ 507°	*liq* $<\infty$ *sol* ~ o X	*liq* $<\infty$ *sol* o X	*sol* $<\infty$ alloys —	*sol* $<\infty$ Tl_8Bi 301.5° Tl_3Bi [ ] 90° Tl_3Bi_5 213°	*sol* ~ o X	*sol* $>$ o X	*sol* $<$ o Pb_2Bi* ?	*sol* o X	*sol* ∞ X

All symbols have the same meaning as in Table 85 (q.v.).

phase diagrams of the corresponding systems corresponding to these compounds, are particularly steep. In addition to these, compounds are frequently formed, bearing no recognizable relation to the valences usually exercised. In every case, these compounds have considerably lower melting points than the 'normal' compounds mentioned first, and in some cases they melt incongruently. The metals of the IIIrd and IVth Main Groups are less inclined to form compounds with As, Sb and Bi than are the alkali and alkaline earth metals, but have a greater capacity for forming mixed crystals. From Table 92 it can clearly be seen how there is a transition here, towards the behavior displayed by the metals of Group VA itself towards each other. In the system Sn/Sb, a non-Daltonide compound occurs. It forms a rhombohedrally distorted rock salt lattice (Hägg, 1935). From this it follows that the ideal composition is analogous to that of NaCl, i.e. SnSb. However, a proportion of the lattice points properly occupied by Sn atoms may be occupied instead by Sb atoms, or some of the Sb lattice points by Sn atoms. The composition may vary, in consequence, within certain limits, namely from 45 to 55 atom per cent Sb. To indicate that a compound is variable in composition, a stroke will be written above the formula corresponding to its ideal composition, as was suggested by Biltz—e.g., $\overline{\text{SnSb}}$. A non-Daltonide compound is said to occur also in the system Pb/Bi (with 67–75 atom per cent Pb).

TABLE 93

MISCIBILITY AND COMPOUND FORMATION OF THE ELEMENTS OF
THE 5TH MAIN GROUP WITH SILICON AND GERMANIUM

	N	P	As	Sb	Bi
Si	Si_3N_4	SiP	*liq* ∞; *sol* 0 SiAs 1083° $SiAs_2$ 944°	*liq* ∞; *sol* 0 X	*liq* ~ 0; *sol* 0 X
Ge	Ge_3N_4	GeP	*liq* ∞; *sol* > 0 GeAs 737° $GeAs_2$ 732°	*liq* ∞; *sol* > 0 X eutec. 590°	*liq* ∞; *sol* ~ 0 X eutec. 271°

The chief characteristics of the elements of Group VA can be understood satisfactorily in terms of the Kossel theory. The maximum positive valence of the elements of atomic number 7, 15, 33, 51 and 83 follows from the fact that they possess 5 electrons in excess of the very stable configurations, with 2, 10, 28, 46 and 78 electrons respectively (cf. Fig. 28, p. 127). It is true that only in the cases of nitrogen ($Z = 7$) and phosphorus ($Z = 15$) are the stable configurations which 'tend' to be formed those of inert gases (helium and neon respectively). However, in the other cases, where inert gas configurations are not involved, it may be inferred that these configurations are also of special stability, from the fact that the preceding elements (of the IInd to IVth families) accord with Group VA in giving up, as a maximum, just so many electrons as will lower their electron number to 28, 46 or 78. The trivalence exhibited towards electropositive elements such as hydrogen can also be explained by the tendency to form electronic configurations of especial stability, namely those present in the *succeeding* inert gas. Whether the nitrogen, phosphorus, or other atom thereby takes on a negative charge, as assumed by Kossel, so that the atom with its triple negative charge binds an equivalent amount of other electropositive elements, or whether there is the sharing of electrons to form octets, as supposed by Lewis and others, is not in every case clear. In terms of Kossel's views, the exercise of a fourth, coordinative valence —e.g., by nitrogen in ammonia—is at once explicable, since it may be calculated by the application of Coulomb's law, as was done on p. 402, that a substantial amount of energy is liberated by the reaction $NH_3 + [OH_3]^+ = [NH_4]^+ + H_2O$.

On the basis of the theory of Lewis and Sidgwick, the formation of the ion $[NH_4]^+$ can be explained by the presence in the NH_3 molecule of an unshared pair of electrons. This enables the nitrogen to exercise a dative covalence and to bind a proton, in the same way as a BF_3 molecule, by exerting a dative covalence, binds a fluoride ion, as discussed in Chap. 11 (cf. also p. 612).

2. Nitrogen (N)

(a) Occurrence

The atmosphere consists of elementary nitrogen, to the extent of about $\frac{4}{5}$ by volume. In the air are also found traces of ammonia, NH_3, arising from the decay of nitrogenous organic substances, and also nitric acid after thunderstorms. Inorganic compounds of nitrogen scarcely occur in nature in large amounts, apart from sodium nitrate, $NaNO_3$, which—presumably resulting from the decomposition products of vegetable and animal substances—has been accumulated in large deposits in many places, especially on the coast of Chile. Nitrogen is an indispensable constituent for living organisms, since it is among the substances from which albumen and other proteins, essential to life, are built up.

(b) History

It was discovered in the second half of the eighteenth century that the air contained a constituent which was incapable of supporting combustion or respiration. This was most clearly stated by Scheele, in his 'Treatise on the air and on fire', which appeared in 1777. Lavoisier gave the name of *azote* (Greek ἀζωτικός, no life), the asphyxiating gas (cf. the German Stickstoff), to the gas which Scheele called 'foul air'. When it was discovered that nitric acid was derived from 'azote', Chaptal proposed the name of *nitrogen*, the 'saltpeter former' (from *nitrum*, salpeter, and γεννᾶν to form).

The most important compounds of nitrogen, nitric acid and ammonia, were already known, in the form of their salts, to the Arabic alchemists. The preparation of free nitric acid was described as early as in the writings ascribed to Geber, whereas the preparation of free ammonia, in the gaseous state, was first achieved by Priestley (1774).

The *utilization of atmospheric nitrogen* for the production of ammonia and nitric acid on a large scale was first carried out successfully at the beginning of the present century— the production of nitro-lime by the Rothe-Frank Caro method since about 1904, the preparation of nitric acid by the Birkeland and Eyde process since 1905, the production of nitric acid by catalytic oxidation of ammonia (Ostwald) since 1906 and the synthesis of ammonia by the Haber-Bosch process since 1909.

(c) Preparation

Pure nitrogen may be prepared in the laboratory by warming a concentrated solution of ammonium nitrite, or of a mixture of ammonium chloride and sodium nitrite, to about 70°. Decomposition then occurs:

$$NH_4NO_2 = N_2 + 2H_2O.$$

To remove traces of admixed oxides of nitrogen, the gas is washed with a mixture of dichromate and sulfuric acid.

Particularly pure nitrogen is obtained by the thermal decomposition of ammonia (by passage over nickel powder at 1000°), and separation of the N_2 from H_2 by freezing it out. (Harteck, 1930).

Generally, however, nitrogen is prepared in the laboratory by abstracting the oxygen from air, by passing it over red hot copper. So-called 'atmospheric nitrogen', containing argon, is obtained by this method, since the air contains 1% of argon. For most purposes, however, the argon content is not disadvantageous.

Nitrogen is prepared industrially in an analogous manner, except that carbon (coke) is used principally as the means of abstracting oxygen. The carbon dioxide produced from this by complete combustion can readily be separated from the nitrogen (by washing with water under pressure, or by condensation), and is recovered as a by-product.

Nitrogen is now produced industrially, on a very large scale, by the liquefaction and fractional distillation of air. Oxygen and argon, and in some circumstances neon and helium as well, are thereby obtained as valuable by-products.

Cylinder nitrogen (which generally contains between 0.1% and 1% oxygen) can be freed completely from oxygen either by passing it through a 'copper tower' as described by Fricke (p. 607), or by using $Na_2S_2O_4$ as absorbent. The form of apparatus described by Kautsky (1926) is most convenient for this purpose, the gas being brought into very intimate contact with the solution by introducing it through an unglazed porcelain 'filter candle' or a fine glass frit.

(d) Properties

Nitrogen [2] is a colorless gas, with no taste or smell, and lighter than air. The liter-weight of pure nitrogen at 0° and 760 mm pressure is 1.2505 g, that of 'atmospheric nitrogen', containing 1.185 per cent of argon by volume, is 1.2567 g, that of air being 1.2928 g. Nitrogen is difficult to condense (critical temp. —147.1°, crit. press. 33.5 atm., crit. density 0.3110). The boiling point of liquid nitrogen is —195.8°, and the melting point of the solid —210.5°. Nitrogen is less soluble in water than is oxygen, 1 liter of water at 0° dissolving 23.6 cc. of 'atmospheric nitrogen' or 23.2 cc. of pure nitrogen.

The specific heats of pure nitrogen at constant volume and constant pressure, respectively, are $c_v = 0.178$, $c_p = 0.249$, both values relating to ordinary temperature; $c_p/c_v = 1.40$. The specific heat of liquid nitrogen, in the neighborhood of the boiling point, is $c_p = 0.46$. The latent heat of evaporation is 47.74 cal/g.

Below —237.7° nitrogen changes into another modification, as was first deduced by Keesom and Kamerlingh Onnes in the course of experiments on its specific heat. The modification stable at low temperatures luminesces brightly when bombarded with cathode rays, the spectrum of its glow being substantially that of the *aurora borealis*. Above the transition temperature, nitrogen is not readily made to luminesce, and then furnishes a different spectrum (Vegard, 1924).

In the gaseous state, nitrogen consists of diatomic molecules, N_2. Even at 3000° these are not dissociated to any measurable extent—an indication that the energy of dissociation must be particularly large. The most probable value for this has been found spectroscopically to be 7.34 ev. (169.3 kcal per mol) (Herzberg and Sponer, 1934). Nitrogen may be dissociated into atoms to an appreciable extent at low pressures by the action of the electrical glow discharge. This was first observed by Strutt (1903). Atomic nitrogen is highly reactive chemically; thus it combines even at ordinary temperature with mercury (forming Hg_3N, according to Tiede, 1935), and also with sulfur and phosphorus. The recombination of the atoms is associated with a yellow afterglow, which persists for a short time after the cessation of the electric discharge.

Ordinary nitrogen is a very inert gas at room temperature. Its reactivity is considerably augmented, however, by raising the temperature; at high temperatures it combines with numerous substances, forming nitrides.

Under the influence of the electric spark, nitrogen combines with hydrogen to form traces of ammonia, NH_3. The equilibrium

$$N_2 + 3H_2 \rightleftharpoons 2NH_3 \qquad (1)$$

is, however, displaced completely to the left at very high temperatures. For this reason, as soon as a minimal concentration of ammonia is exceeded, it is decomposed again by the electric spark. Larger amounts of ammonia can be obtained, however, if steps are taken to bring about the union of its constituents at the lowest possible temperature. This may be achieved by the use of a suitable catalyst. Furthermore, the equilibrium represented by eqn. (1) can be displaced from left to right by raising the pressure.

An appropriate experimental artifice is also necessary in order to make nitrogen combine directly with atmospheric oxygen to any considerable extent. In this case, the product of combination, nitric oxide, NO, is indeed stable at very high temperatures, but with decreasing temperature the equilibrium

$$N_2 + O_2 \rightleftharpoons 2NO \qquad (2)$$

shifts increasingly to the left. This may be deduced directly from the negative heat of formation of nitric oxide, shown in Table 89. That nitric oxide is nevertheless capable of existence at ordinary temperature is due to the fact that its rate of decomposition is practically zero at room temperature: it is metastable. When heated, however, it decomposes, even before reaching that temperature range in which the equilibrium (2) is displaced appreciably towards the right. According to Nernst, about 1.2 volume per cent of nitric oxide is in equilibrium at 2000° with nitrogen and oxygen, present in the proportions of ordinary air, and 5.3 volume per cent at 3000°. Very high temperatures must therefore be employed in order to obtain sufficient concentrations of nitric oxide, and the reaction mixture must be 'quenched' in order to prevent decomposition during cooling—i.e., it must be cooled as suddenly as possible from temperatures around 3000° to something below 1000°. Suitably high temperatures may be reached by use of the electric arc. The sudden quenching can be achieved, for example, by drawing the arc out, by means of a powerful electromagnet, into a disc through which the gas mixture (air) is blown. The 'combustion of air' carried out in this way enjoyed for a time a considerable importance for the production of nitric acid (Birkeland and Eyde process).

It had already been observed by Priestley that a mixture of nitrogen and oxygen, confined over water, experienced a diminution in volume when sparks were passed through it continuously. Nitric oxide is first produced under these conditions also, and from this there is formed nitrogen dioxide and finally nitric acid, the formation of which was already known to Cavendish (1784). Since the resulting nitric oxide is continually removed from the reaction mixture, it is ultimately possible to convert all the nitrogen into nitric acid by long continued sparking in the presence of an excess of oxygen. This tedious process is, of course, of no interest for the preparation of nitric acid, but it served Lord Rayleigh for the first preparation of argon. (cf. p. 103).

(e) Uses

Nitrogen is often used in the laboratory as an inert protective gas, to prevent the oxidation of easily oxidizable substances through exposure to air. It is also used for similar purposes industrially—e.g., for the handling of gasoline and other inflammable liquids, and for gas-filled mercury thermometers. Uses of this kind are, however, now quite insignificant compared with its use the production of nitrolime, ammonia, and nitric acid for the fertilizer industry [3, 4].

The conversion of atmospheric nitrogen into compounds usable as fertilizers also occurs on a very considerable scale by natural means, through the agency of micro-organisms. Many species of these—e.g., *azotobacter*—are present in soils containing humus, and it has been estimated that by the action of these bacteria 19.5 kg of nitrogen per acre of soil may be assimilated annually. Also of importance for the storage of nitrogen in agricultural soils are the bacteria occurring in nodules on the roots of leguminous plants, such as clover. The cultivation of legumes is widely used to replenish the nitrogen of soils. The legumes may either be completely ploughed under, or their roots only may be left in the soil as fertilizer. In light soils, and with favorable climatic conditions, it is possible for more than 80 kg, of nitrogen per acre per year to be stored up by this means.

Fixation of nitrogen by microorganisms [12] ('biological' fixation of nitrogen) is, like the assimilation of carbon dioxide, a process of fundamental importance for the whole of terrestrial life. Research and trials (especially the work of Virtanen, 1931 onwards) have shown that if the biological fixation of nitrogen were systematically exploited, it would be possible to bring about a considerable increase in the fertility of the soil and the agricultural production of the whole world.

Rain water also transports nitrogen to the soil (in the form of nitric and nitrous acids), since a certain amount of chemical combination of nitrogen takes place in the atmosphere under the influence of electrical discharges and photochemical reactions. Nitrogenous fertilization of the soil from this source is very much less than that brought about by micro-organisms; nevertheless, when the whole of the earth's surface is considered it is vastly greater than the nitrogen supplied in artificial fertilizers. According to Lipman and Cony-beare, the quantity of nitrogen supplied annually to that area of the United States which is under agriculture was about 16.5 million tons in 1935. About 60% of this was due to the action of microorganisms, 21.5% to rain water, 15.5% to natural fertilizers, and only 3% to artificial fertilizers—even though this region accounted for about 20% of the entire world consumption of artificial fertilizers[2]. In the European region, with more intensive agricultural production, a considerably higher proportion of the nitrogen supplied to the soil comes as artificial fertilizer, especially in the production of cereal crops. When the total vegetation of the earth is considered, however, the proportion of the nitrogen supplied by the chemical industry is only a vanishingly small fraction.

Ammonia, nitric acid and other nitrogen compounds also find very wide-spread uses for many other purposes besides the production of fertilizers. More will be said on this in discussing the compounds concerned; at this stage, however, attention may be drawn to the large number of *organic nitrogen compounds* which are important in technology and in ordinary life, including, for example, a large proportion of the most important dyestuffs and drugs.

3. Compounds of Nitrogen [10]

(a) Oxides

A summary of the oxides of nitrogen is provided by Table 89, p. 581 (cf. also p. 595).

NO and NO_2 also occur as radicals in compounds. Inorganic compounds containing NO as a *univalent radical* are known as *nitrosyl* compounds*, e.g., NOCl, nitrosyl chloride; NaNO, sodium nitrosyl. On the other hand, compounds in which NO is assumed to be present as

* The suffix *-yl* is occasionally omitted, as in *nitros-amide* and *nitr-amide* (pp. 603, 604).

a neutral constituent (bound by Van der Waals' or similar forces), are termed *nitroso-compounds*—e.g., $[Fe(NO)]SO_4$, nitroso-ferrous sulfate. The organic compounds of the NO radical are also known as *nitroso*-compounds. The inorganic compounds of the univalent radical NO_2 with electronegative substances are called *nitryl* compounds—e.g., NO_2Cl, nitryl chloride. For terminology of the NO_2 radical in salts, organic compounds and complexes, see p. 603.

(*i*) *Dinitrogen oxide* (nitrous oxide), N_2O, discovered by Priestley in 1776, is prepared by heating solid ammonium nitrate:

$$NH_4NO_3 = N_2O + 2H_2O.$$

Decomposition sets in at about 170°, and proceeds exothermically. Care must be taken, by removing the source of heat at the right moment, that the decomposition does not become too violent (cf. p. 613).

Dinitrogen oxide is a colorless gas, with a feeble, pleasant smell, and a sweetish taste. Inhaled in small amounts it induces intoxication and convulsive desire to laugh ('laughing gas'). In larger amounts it acts as a narcotic. Over prolonged periods, however, it should be inhaled only in admixture with oxygen, since it is not itself capable of supporting respiration. Dinitrogen oxide is easily liquefied (b.p. −89.5°, m.p. −102.4°, crit. temp. +36.5°, crit. press. 71.7 atm.). It is sold commercially in steel cylinders, since it is used for anaesthesia.

The liter-weight of N_2O is 1.9804 g at 0° and 760 mm. The density of the liquid is 1.2257 at the boiling point, and 0.45 at the critical point. Its solubility in water is considerable, 1 volume of water absorbing 1.3052 volumes at 0°, 0.5962 volumes at 25°. Its solubility in alcohol is considerably greater still.

Substances with a high affinity for oxygen—e.g., carbon, or many metals—can burn in dinitrogen oxide with as much vigor as in oxygen. The combustion is, however, harder to initiate, since even although the N_2O molecule is metastable, it begins to decompose only at fairly high temperatures. A mixture of equal volumes of dinitrogen oxide and hydrogen explodes with violence when brought into contact with a flame:

$$N_2O + H_2 = N_2 + H_2O_{gas} + 77.5 \text{ kcal.}$$

A mixture of dinitrogen oxide with ammonia also explodes with great violence when ignited:

$$3N_2O + 2NH_3 = 4N_2 + 3H_2O_{gas} + 210 \text{ kcal.}$$

Solid dinitrogen oxide has the same crystal structure as solid carbon dioxide. The edge of the unit cell ($a = 5.72$ Å) is also very similar to that of the CO_2 crystal lattice. It has been shown by X-ray structure analysis that in solid dinitrogen oxide the atoms in the N_2O molecule are *collinear*. Recent physical investigations (e.g., electron diffraction experiments) also make it probable that the configuration is linear in the gaseous state. Since N_2O has the same number of electrons as CO_2, and displays a close similarity to the latter in its physical properties, it was already assumed by Langmuir (1919) that the two compounds had the same electronic configuration (isomerism). The formula proposed by Langmuir:

$$\overset{..}{\underset{..}{N}} :: \overset{..}{N} :: \overset{..}{\underset{..}{O}} \quad \text{corresponding to} \quad \overset{..}{\underset{..}{O}} :: C :: \overset{..}{\underset{..}{O}}$$

is substantiated by modern observations (e.g., the Raman spectrum) although this electronic configuration may be mesomeric with the configuration $: N \overset{..}{\underset{..}{:}} N : O :$. The interatomic distances in the molecule are $N \leftrightarrow N = 1.12$ Å, $N \leftrightarrow O$ 1.19 Å =. (Schomaker, 1942).

The dipole moment of the N_2O molecule is $0.14 \cdot 10^{-18}$ e.s.u., and its moment of inertia $66 \cdot 10^{-40}$ g cm².

(*ii*) *Nitric oxide* (nitrogen monoxide), NO, the product of combination of nitrogen and oxygen at very high temperatures, is a colorless gas, b.p. —151.8° m.p. —163°. As soon as it comes into contact with the air, however, it forms brown fumes as a result of oxidation to the dioxide, NO_2. It is but slightly soluble in water (1 vol. of water dissolves 0.07 vol. of nitric oxide at 0°), and is practically insoluble in saturated brine. It is generally formed when nitric acid or nitrous acid is treated with reducing agents. The method for determining the concentration of nitric acid by means of the Lunge nitrometer is based on the formation of nitric acid when mercury is shaken with nitric acid and concentrated sulfuric acid:

$$6Hg + 2HNO_3 + 3H_2SO_4 = 2NO + 3Hg_2SO_4 + 4H_2O.$$

It is generally prepared in the laboratory by the action of moderately concentrated nitric acid on copper:

$$-\ 8HNO_3 + 3Cu = 3Cu(NO_3)_2 + 4H_2O + 2NO$$

Nitric oxide is a fairly reactive substance. It reacts with chlorine and bromine, forming nitrosyl halides (cf. p. 621). It forms nitrosyl sulfuric acid with concentrated sulfuric acid, if oxygen also has access

$$2H_2SO_4 + 2NO + \tfrac{1}{2}O_2 = 2HSO_4(NO) + H_2O.$$

It is oxidized to nitric acid by chromic acid, acidified permanganate solution, or hypochlorous acid. It is reduced to nitrous oxide by sulfur dioxide in the presence of water, to ammonia by chromium(II) salts in neutral solution, or to hydroxylamine in acid solution. The last mentioned results also from the reduction by a hydrochloric acid solution of tin(II) chloride. Substances with a high affinity for oxygen, such as charcoal, phosphorus, or magnesium can burn vigorously in nitric oxide; burning sulfur is extinguished, however. A mixture of equal volumes of nitric oxide and hydrogen explodes when it is ignited. Nitric oxide can enter as a neutral constituent into many metallic salts. Thus it reacts with ferrous sulfate, $FeSO_4$, in aqueous solution, forming the deep brown colored nitroso-ferrous sulfate, $[Fe(NO)]SO_4$. It can be completely expelled from this solution again by warming. It forms similar loose compounds with copper sulfate dissolved in concentrated sulfuric acid, and with copper chloride dissolved in alcohol. Most of these compounds of relatively simple composition are stable only in solution. Considerably more stable are certain strongly complex nitroso compounds, of rather more complicated formula, of which sodium nitroprusside, $Na_2[Fe(CN)_5NO]$, may be taken as an example at this point.

With HCl, NO forms an addition compound of deep red color, NO · HCl. According to Rodebush and Yntema (1923) this is salt-like in character, and so should be regarded as nitrosonium chloride, [NOH]Cl. The formation of this compound (first observed by Briner in 1909) has several times in recent years, in experiments with vacuum apparatus, given rise to wrong deductions—e.g., to the supposed discovery of 'inert gas halides', and of a 'red modification of hydrogen chloride'. It forms from the components only at very low temperatures, and in the presence of traces, at least, of higher nitrogen oxides. It is generally stable only in the solid state, and can be melted only under high pressures.

NO contains 11 orbital electrons in all, and like all gases with an odd number of electrons, it is *paramagnetic*. Its constitution may be represented roughly as $: N \ {\overset{..}{..}} \ O \ :$. Watson (1934) found the *dipole moment* to be $0.16 \cdot 10^{-18}$ e.s.u. The *moment of inertia* (Hulthén, 1927) is $11.7 \cdot 10^{-40}$ g cm^2.

NO is polymerized in the liquid state, as is evidenced by the large and highly tempera-ture-dependent specific heat of the liquid, its high entropy of vaporization, and its magnetic properties. Smith, Keller and Johnston (1951) inferred from the infrared and Raman spectra that the liquid must be completely dimerized, probably with the formation of bent ONNO molecules.

(iii) Nitrogen trioxide (strictly, dinitrogen trioxide), N_2O_3, the anhydride of nitrous acid, is prepared by dropping nitric acid of sp. gr. 1.3 on to coarsely powdered, vitreous arsenic trioxide, or by warming nitric acid with starch powder. It forms a deep blue liquid which solidifies to a pale blue mass when strongly cooled (m.p. —102°). Even below 0°, nitrogen trioxide decomposes to a consider-able extent, according to

$$\sim N_2O_3 \rightleftharpoons NO + NO_2 \quad \text{or} \quad 2N_2O_3 \rightleftharpoons 2NO + N_2O_4 \tag{1}$$

According to E. Abel (1929) 10% of undissociated N_2O_3 is present at 25° and atmospheric pressure in a mixture which contains NO and NO_2 in equal proportions. From the equili-brium constants, the heat of formation from NO and NO_2 is calculated to be 9.6 kcal per mol of N_2O_3 at room temperature. The free energy of formation, however, is only —0.44 kcal. per mol at 25° (Verhoek, 1931).

Since reaction (1) is readily reversible, a gaseous mixture of equal volumes of NO and NO_2 behaves in most chemical reactions just like the compound N_2O_3. Thus it reacts smoothly with alkali, forming nitrite

$$N_2O_3 + 2NaOH = 2NaNO_2 + H_2O.$$

Nitrous acid is formed as primary product by reaction with water, but rapidly decomposes, forming nitric acid

$$N_2O_3 + H_2O = 2HNO_2; \qquad 3HNO_2 = HNO_3 + 2NO + H_2O \tag{2}$$

(iv) Nitrogen dioxide, NO_2, and *Dinitrogen tetroxide*, N_2O_4. As has already been mentioned, nitric oxide readily combines with oxygen, forming nitrogen dioxide

$$NO + \tfrac{1}{2}O_2 = NO_2 + 13.6 \text{ kcal.}$$

This compound is produced on a large scale from the nitrous gases obtained by the catalytic oxidation of ammonia, and can readily be isolated therefrom in the solid state by cooling to a low temperature. It can also be absorbed in concen-trated sulfuric acid (formation of nitrosyl sulfuric acid, cf. p. 592) and liberated again by warming. Nitrogen dioxide is most conveniently prepared in small amounts by heating well dried lead nitrate, preferably mixed with quartz sand to obtain a regular evolution of gas.

Nitrogen dioxide is a brown-red, *very poisonous* gas, with a characteristic smell. It can be condensed to a red brown liquid (b.p. 22.4°), which turns paler on cooling, ultimately becoming colorless and solidifying at —10.2° to colorless crystals. Conversely, the color deepens when the gas is heated. As vapor density determinations have shown, the change in color is due to the fact that the nitrogen dioxide polymerizes on cooling to the colorless dinitrogen tetroxide. A strongly temperature dependent equilibrium is set up

$$2NO_2 \rightleftharpoons N_2O_4 + 14.7 \text{ kcal (at } 25°)$$

such that when the gas mixture is under a pressure of 1 atmosphere,

at 27°	50°	100°	135°
20%	40%	89%	98.7%

of the tetroxide is split up into the dioxide.

The *free energy* of formation of N_2O_4 from 2 NO_2, $\triangle F°$, is —1.15 kcal at 25°, —0.25 kcal at 45° (Verhoek, 1931). It thus decreases rapidly with temperature, whereas the *heat of formation* changes but little (by about 1%) over the same temperature range.

The dielectric constant of liquid nitrogen dioxide is 2.42 and the conductivity of the liquid is $1.3 \cdot 10^{-12}$ ohm^{-1} cm^{-1} at 17° (Addison, 1951). This small conductivity is believed to be due to self-ionization ('auto-ionization') of the N_2O_4, according to the equation

$$N_2O_4 \rightleftharpoons NO^+ + NO_3^-$$

since liquid nitrogen dioxide has certain properties characteristic of ionizing solvents. Thus it reacts with diethylammonium chloride in a manner strictly analogous to hydrolysis reactions in aqueous solution:

$$[(C_2H_5)_2NH_2]Cl + N_2O_4 = [C_2H_5)_2NH_2](NO_3) + NOCl$$

$$\text{cf. } RCl + [H_3O]^+[OH]^- = [H_3O]Cl + ROH$$

It also attacks metallic zinc, forming zinc nitrate:

$$Zn + 2N_2O_4 = Zn(NO_3)_2 + 2NO$$

$$\text{or} \qquad Zn + 2NO^+ = Zn^{2+} + 2NO.$$

When the nitrogen dioxide is evaporated, the zinc nitrate is isolated as a double compound $Zn(NO_3)_2 \cdot 2N_2O_4$, which is believed to be the nitrosyl salt of a complex nitrate, $[NO^+]_2[Zn(NO_3)_4]^{2-}$ (Compare the amphoteric nature of zinc hydroxide in aqueous systems,

$$Zn(OH)_2 + 2OH^- \rightleftharpoons [Zn(OH)_4]^=).$$

NO_2 is paramagnetic, N_2O_4 diamagnetic. This makes it probable that the union of $2NO_2$ to form N_2O_4 is to be attributed to the tendency to 'pair up' the unpaired electron of the NO_2. That NO_2 does contain one unpaired electron follows from the odd number of electrons in the molecule. The color of NO_2 also appears to be due to the presence of the unpaired electron. According to Broadley and Robertson (1949), the crystal structure of solid N_2O_4 shows that the molecule has the structure

and not the alternative   O = N—O—N⟨O⟩O .

The distances N↔O are each 1.17 Å, and N↔N = 1.64 Å, O—N—O = 126°. In the molecule of gaseous NO_2, N↔O = 1.20 Å, and O—N—O = 120°.

Nitrogen dioxide is a powerful oxidant. Potassium, phosphorus, charcoal, and sulfur will burn in it, and a mixture of nitrogen dioxide and carbon disulfide can be made to detonate violently. Hydrogen reacts with nitrogen dioxide in the presence of catalysts, such as platinum or finely divided nickel, forming ammonia and water.

Nitrogen dioxide reacts with water, first forming nitric acid and nitrogen trioxide; the latter then reacts according to eqn (2) above. The total reaction thus follows the equation

$$3NO_2 + H_2O = 2HNO_3 + NO \qquad (3)$$

If air is present, the NO so formed is at once oxidized again to NO_2, so that in this case the NO_2 is ultimately and completely converted into nitric acid. Caustic alkalis, however, absorb nitrogen dioxide forming a mixture of nitrate and nitrite (disproportionation)

$$2NO_2 + 2NaOH = NaNO_3 + NaNO_2 + H_2O \qquad (4)$$

(v) *Nitrogen pentoxide*, N_2O_5, the anhydride of nitric acid, can be obtained by treating nitric acid with phosphorus pentoxide:

$$2HNO_3 + P_2O_5 = 2HPO_3 + N_2O_5$$

In the pure state it forms hard, colorless crystals (rhombic prisms) which deliquesce in air; density 1.63, m.p. 30°, b.p. 45–50°. It is very unstable, and may explode without any apparent external stimulus. It combines avidly with water to form nitric acid: $N_2O_5 + H_2O = HNO_3 + 10.4$ kcal; $HNO_3 + aq = HNO_3 \cdot aq + 7.5$ kcal (at 20°).

It has been shown both from optical evidence (Ingold, Millen and Poole, 1950) and by X-ray crystallography (Grison, Ericks and de Vries, 1950), that solid nitrogen pentoxide has the ionic structure $[NO_2^+][NO_3^-]$. Nevertheless, its volatility proves that the ionic crystal lattice very readily passes over into the covalent molecular constitution of the vaporized oxide. It is usual for ionic solids to have very high heats of sublimation. The volatility of nitrogen pentoxide is not necessarily incompatible with an ionic structure in the solid state if, in this case, the energy gained by the formation of an N—O covalency just about compensates for the energy expended in separating the oppositely charged ions of the crystal lattice. Very similar considerations apply to PCl_5 and PBr_5 (q.v.), which as solids have the ionic constitutions $[PCl_4^+][PCl_6^-]$ and $[PBr_4^+]$ $[Br^-]$ respectively.

Schmeisser (1952) has found that N_2O_5 combines with BF_3, to form the addition compound $N_2O_5 \cdot BF_3$, which is stable at ordinary temperature.

It has been deduced from a variety of observations that nitrogen can form some oxide still richer in oxygen than N_2O_5. This is usually considered to have the formula NO_3. The compound is formed as an intermediate product in the reaction of N_2O_5 (and also of NO_2) with ozone, as Schumacher (1928 and later) showed by spectroscopic and reaction-kinetic measurements. It rapidly decomposes, however, according to $2NO_3 = 2NO_2 + O_2$. It can therefore be obtained in a certain stationary concentration in a reacting gas mixture, but cannot be isolated. On the basis of its reactions, Schwarz (1935) considers it to be a

peroxide, nitrosyl peroxide, $\quad O = N\underset{\displaystyle O}{\overset{\displaystyle O}{<}}\cdot\cdot$.

(b) Oxyacids of Nitrogen

The most stable and most important of the oxy-acids of nitrogen is *nitric acid*, HNO_3. Its anhydride is nitrogen pentoxide, and its salts, M^1NO_3 are known as *nitrates*. *Nitrous acid*, HNO_2, is very unstable, and is only known in very dilute aqueous solution. Its salts, the *nitrites*, $M^1[NO_2]$, are more stable. The anhydride from which nitrous acid is derived is nitrogen trioxide, N_2O_3—capable of existing at ordinary temperature only in equilibrium with NO and NO_2.

Nitrogen may be regarded as *electropositively quinquevalent* in nitric acid, and *electropositively trivalent* in nitrous acid. By a further stage of reduction we obtain *hyponitrous acid*, $H_2N_2O_2$, in which, in so far as the nitrogen may be regarded as charged, it is electropositively univalent Formally and structurally, however, nitrogen is trivalent in hyponitrous acid, which, from its mode of formation and properties, is to be formulated as $H—O—N=N—O—H$. Hyponitrous acid may be isolated in the free state, but is very unstable (explosive); its salts are the *hyponitrites*. A place intermediate between nitrous acid and hyponitrous acid, so far as the oxidation state of the nitrogen is concerned, is taken by *nitroxylic acid (hydronitrous acid)*

$$H_4N_2O_4 \quad \text{or} \quad \begin{matrix} HO \\ HO \end{matrix}\!\!>\!N—N\!<\!\!\begin{matrix} OH \\ OH \end{matrix},$$

which is known only in the form of its salts (cf. p. 603). A peroxy acid should be derived from nitric acid by exchanging the —OH group of nitric acid, $O_2N—OH$, for the —OOH group of hydrogen peroxide—i.e., peroxynitric acid, $O_2N—O—OH$, or HNO_4. According to Schwarz (Z. *anorg. Chem.* 256, (1948) 3), this is obtained by the action of 100% H_2O_2 on N_2O_5 at —80°, but is very unstable and decomposes explosively even at —30°. In solutions formed from equimolar amounts of H_2O_2 and HNO_3, however, it is stable for some time, and is safe. The solution is hydrolyzed by further dilution, hydrolysis being complete in the 20% solution. A pungent smelling unstable compound is also formed by the action of F_2 on nitrite solutions (Fichter), and was formerly considered to be peroxynitric acid. In this, and in some other processes giving rise to a peroxy acid, it is possible that the product is *pernitrous acid*, $ON—O—OH$; the probable formation of this as an intermediate in certain reactions has been shown particularly by some work of Gleu (p. 603).

(*i*) *Nitric acid*, HNO_3, when pure, is a colorless liquid, of density 1.522, melting at —41.65° (Forsythe and Giauque, 1942) and boiling at 84°; it thereby undergoes slow decomposition. In light, it decomposes even at ordinary temperature. The nitrogen dioxide, formed according to

$$2HNO_3 = 2NO_2 + H_2O + \tfrac{1}{2}O_2$$

remains dissolved, and colors the acid yellow or, at higher concentrations, red. Nitric acid is miscible with water in all proportions. The specific gravity and boiling points of the solutions are:

Spec. Grav. (d_4^{20})	1.150	1.200	1.300	1.400	1.410	1.420	1.480	1.500
Weight-% HNO_3	25.5	32.9	48.4	67.0	69.2	71.6	89.0	98.2
Boiling point	106.8°	108.8°	115.5°	121.7°	121.8°	121.2°	100.6°	86°

The maximum boiling point is found for a concentration of 69.2%. A solution of this concentration is obtained by distillation until the boiling point is constant, whether one starts with a more dilute or a more concentrated solution. The solutions known as 'Concentrated Nitric Acid' are therefore usually of about this concentration. Solutions of higher concentration fume in air, since they give off nitrogen pentoxide, which forms a fog with atmospheric moisture.

Mixtures such as 69% nitric acid, which undergo no change of composition when boiled or distilled, are known as *azeotropic mixtures* (Greek ζέειν, boil and τροπή, change). The concentration at which 'constant boiling' occurs depends upon the pressure, as also therefore does the composition of the vapor given off from the constant boiling mixture, whereas with chemical compounds which boil without decomposition, the composition of the vapor is independent of the pressure. Hydrochloric acid, aqeous alcohol, and nitric acid, provide the best known examples of azeotropic mixtures.
With water, nitric acid forms two hydrates:

$$HNO_3 \cdot H_2O \text{ (m.p. } —37.85°) \quad \text{and} \quad HNO_3 \cdot 3H_2O \text{ (m.p. } —18.5°).$$

Their existence is shown by maxima in the melting point diagram of the system HNO_3–H_2O (just as in the system SO_3 – H_2O, cf. Fig. 115, p. 710). The existence of the hydrates is not revealed by the boiling point or vapor pressure curves of aqueous nitric acid solutions, since they cannot be vaporized without decomposition. The compound $HNO_3 \cdot H_2O$ may perhaps be 'orthonitric acid' (or trihydronitric acid) H_3NO_4 or $ON(OH)_3$. In any case, it appears not to be a hydronium nitrate, $[H_3O][NO_3]$, corresponding to the case of $HClO_4 \cdot H_2O$ (p. 796). Whereas the latter, $[H_3O][ClO_4]$, has the same X-ray diffraction pattern as $[NH_4][ClO_4]$, that of $HNO_3 \cdot H_2O$ is completely different from that of $[NH_4][NO_3]$ (Zintl, 1935). By melting $NaNO_3$ with Na_2O, Zintl was able to prepare a salt, Na_3NO_4, trisodium nitrate, formally derived from trihydronitric acid. An organic derivative of this acid, trihydroxy diacetyl nitric acid, $(CH_3 \cdot CO_2)_2N(OH)_3$ (b.p. 45° at 15 mm) was described by Pictet as long ago as 1902. This compound has a strong nitrating action, and, according to Bacharach (1931), it is formed as an intermediate product in the nitration of organic compounds by nitrates dissolved in glacial acetic acid or acetic anhydride. Nitric acid dissolved in glacial acetic acid has considerably stronger nitrating (and oxidizing) properties than aqueous nitric acid of the same concentration, as was proved by Briner (1935), by measuring the velocities of reaction.

Nitric acid is a strong oxidizing agent, especially when concentrated. It converts sulfur to sulfuric acid, phosphorus first to phosphorous acid and then to phosphoric acid. It is particularly aggressive towards metals—only gold, platinum and some of the platinum metals are resistant to its action. The reactions proceed substantially with the formation of nitric oxide—e.g.,

$$8HNO_3 + 3Cu = 3Cu(NO_3)_2 + 4H_2O + 2NO$$

The nitric acid acts upon the copper first as an oxidizing agent. A further portion then at once reacts further with the copper oxide primarily formed, giving the salt.

$$2HNO_3 + 3Cu = 3CuO + 2NO + H_2O$$

$$3CuO + 6HNO_3 = 3Cu(NO_3)_2 + 3H_2O$$

Iron is dissolved by very dilute nitric acid (e.g., $0.2N$) in the cold (0°) principally with the formation of Fe^{II} nitrate. Higher concentrations of nitric acid, and warmer solutions, produce Fe^{III} nitrate.

It is a peculiar fact that many metals which are attacked with great vigor by dilute nitric acid are stable towards the *very concentrated* acid—e.g., iron, chromium, aluminum and calcium. This is due to the phenomenon of *passivation*.

Organic compounds are either oxidized or nitrated by concentrated nitric acid. In the latter case, the residue —NO_2, the *nitro* group, enters the compound in place of hydrogen. Thus nitrobenzene, $C_6H_5 \cdot NO_2$ is formed from benzene, C_6H_6. The yellow colors produced by nitric acid with many organic substances, such as the skin, are due to such nitrations; most nitro compounds are yellow. A mixture of concentrated nitric acid and concentrated sulfuric acid has an especially powerful nitrating action, and is known as *nitration acid*. Because of its passivating action, nitric acid can be handled in iron vessels.

The widespread technical applications of nitric acid are based chiefly on the two properties of *oxidizing* power and *nitrating* action. It is used as an oxidant, e.g., in preparing phosphoric acid from phosphorus, oxalic acid from carbohydrates, and sulfuric acid by the lead chamber process. The dyestuffs industry, in particular, utilizes its nitrating properties. Most of the organic dyestuffs containing nitrogen involve nitric acid for their preparation. Nitric acid is also used in the manufacture of nitroglycerine from glycerol, of nitrocellulose (gun cotton and collodion) from cellulose, and of picric acid, and indeed almost all the nitrogen-containing explosives. It is employed also in the manufacture of nitrates and as a chemical solvent for most metals. Under the name of *parting acid* it is used for the separation of gold and silver.

Gold and platinum do not dissolve in nitric acid, but do so in a mixture of concentrated nitric acid and concentrated hydrochloric acid (*aqua regia*). The two acids are usually mixed in the proportion of 1 : 3 by volume. Aqua regia evolves nitrosyl chloride, NOCl, and free chlorine

$$HNO_3 + 3HCl = NOCl + Cl_2 + 2H_2O$$

and the action of these converts the metals into their chlorides.

Up to the beginning of the present century, nitric acid was prepared almost exclusively by the same process as was used by Glauber, in the middle of the 17th century—namely by heating alkali nitrates with moderately concentrated sulfuric acid, whereby nitric acid distils off:

$$2M^INO_3 + H_2SO_4 = 2HNO_3 + M^I_2SO_4,$$

Glauber used potassium nitrate. This was later replaced by sodium nitrate, which is cheaper and better to work up. In the technical preparation, the reaction was taken to the half-way stage:

$$M^INO_3 + H_2SO_4 = HNO_3 + M^IHSO_4.$$

This takes place with only gentle heating (to about 150°), whereas complete decomposition takes place only at temperatures at which the nitric acid sustains appreciable decomposition.

Nitric acid is now produced on a *large scale* almost exclusively by the *catalytic combustion of ammonia* (Ostwald process). In this, ammonia mixed with air is passed over heated platinum or platinum-rhodium. Ammonia is burned, according to the equation

$$4NH_3 + 5O_2 = 4NO + 6H_2O + 215 \text{ kcal}, \tag{1}$$

forming nitric oxide and water. As the gases cool, the NO is oxidized further, to NO_2, by the excess of atmospheric oxygen present. The nitrous gases are at once brought into contact with water in scrubber towers, air being passed in simultaneously. Nitric acid is thereby formed, as described on p. 595.

It had already been observed by Kuhlmann (1839) that a mixture of ammonia and air produced oxides of nitrogen when it was passed over heated platinum. The technical application of the reaction, however, hinged upon an observation by Ostwald—that it is essential that the mixture should remain in contact with the catalyst for only a very short time (about $1/1000$ second) in order to secure a good yield. The nitric oxide decomposes otherwise, since it is an unstable compound, and the gases issuing from the contact plant then contain only elementary nitrogen. The reaction of eqn.(1) represents only the over-all process, and according to Bodenstein [*Z. Elektrochem.*, 47, (1941) 501]. it proceeds by way of the following intermediate steps,

(a) $$NH_3 + O \text{ (on catalyst)} = NH_3O$$
(b) $$NH_3O + O_2 = HNO_2 + H_2O$$
(c) $$HNO_2 + O_2 = HNO_4$$
(d) $$HNO_4 \rightarrow NO + O_2 + OH$$
(e) $$2OH = H_2O + O \text{ (taken up by catalyst)}$$

Reaction (a) may be followed, however, by other reactions leading to N_2O or NO—e.g., $NH_3O + O = H_2O + HNO$, $2HNO = H_2O + N_2O$, and $HNO + NH_3O = 2H_2O + N_2$ To achieve the short contact time required for a high yield of NO, the gases are blown at a high speed through a wire gauze of fine mesh. Once the reaction has been initiated, this is maintained at a red heat by the heat of reaction. The 40–50% solution of HNO_3 obtained by direct absorption of the nitrous gases can be employed for most purposes (e.g., fertilizer manufacture) without further treatment. Concentrated acid, up to 69%, can be obtained by distillation. If a more concentrated acid is required, dehydrating agents must be employed (conc. sulfuric acid or phosphorus pentoxide). Nitric acid of high concentration (up to 100%) can be obtained directly, by the action of a pure NO_2-O_2 mixture on water or dilute nitric acid.

The first *synthetic* process, that of burning the atmospheric nitrogen, was introduced at the beginning of this century, its oldest form of application being the process invented by Birkeland and Eyde in 1903, which has already been briefly described. The atmospheric combustion process consumes very large amounts of electrical energy [5]. It was thereby restricted at the outset to those countries, such as Norway and Switzerland, with especially cheap water power for the generation of electricity. Even there, it has subsequently been displaced by the process of oxidation of ammonia.

Nitric acid ranks among the strongest of the acids. Its apparent degree of dissociation (measured by conductivity coefficients) is 82% in 1-N solution, or 93% in 0.1-N solution (at 18°).

On the basis of Schaefer's, optical, measurements Hantzsch advanced the hypothesis that nitric acid exists in two isomeric forms, the esters being derived from one, and the salts from the other:

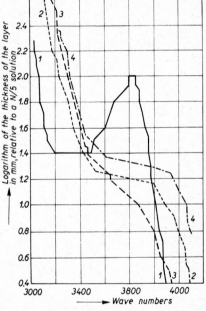

Fig. 110. Absorption spectra of nitric acid and its salts and esters.
Curve 1 HNO_3 and KNO_3 in dilute aqueous solutions
  ,,   2 $NO_2 \cdot OH$ in ether solution
  ,,   3 $NO_2 \cdot OC_2H_5$ (ethyl ester of nitric acid)
  ,,   4 HNO_3 nearly anhydrous.

ethyl nitrate   potassium nitrate

The difference lies in that, in the ester, the ethyl group is linked to the nitrogen through the agency of one particular oxygen atom, whereas in the typical salts, the negative charge of the whole $[NO_3]^-$ radical acts on the positive ion of the metal. In dilute aqueous solution, where the radical of nitric acid is present as the free $[NO_3]^-$ ion, its charge acts upon the hydrogen or hydronium ions, $[H_3O]^+$, again as a whole. Almost anhydrous nitric acid, however, gives an absorption spectrum which, as shown in Fig. 110, is much more like that of the esters than that given by nitric acid in dilute aqueous solution. The latter is identical with that of the salts. A solution of the acid in ether gives almost the same absorption curve as the absolute acid. Since, according to Hantzsch it is a principle (the basis of which cannot be discussed here) that appreciable changes in the absorption spectrum are always due to changes in chemical constitution, it follows that nitric acid, in the anhydrous state and in ether solution, has a structure different from that possessed in aqueous solution, and corresponding, in fact, to that of its esters—i.e., O_2N—OH. This form was called by Hantzsch '*pseudonitric acid*'.

Nevertheless, the curves 2 and 4 in Fig. 110 differ quite materially from the curve 3, given by the ester. This, according to Hantzsch, is because anhydrous nitric acid is not strictly homogeneous, but consists of a mixture of the 'pseudoacid' and a salt of the 'true acid', the cation of this salt being formed by the addition of a proton to the undissociated acid.

Subject to certain modifications, this hypothesis is consistent with present knowledge of nitric acid and its solutions.

Hantzsch was able to show by preparative methods that the postulated cations do exist, since he was able to prepare what he believed to be two crystalline perchlorates, $[ON(OH)_2][ClO_4]$ and $[N(OH)_3][ClO_4]_2$, derived from the *dihydroxonitronium* and the *trihydroxonitronium* ions, respectively:

$$O_2N(OH) + H^+ = [ON(OH)_2]^+$$

$$O_2N(OH) + 2H^+ = [N(OH)_3]^{++}$$

However, a careful repetition of Hantzsch's work (Goddard, Hughes and Ingold, 1950) has shown that the sole product of the action of absolutely anhydrous nitric acid or nitrogen pentoxide upon perchloric acid is *nitronium perchlorate*, $[NO_2][ClO_4]$:

$$HNO_3 + 2HClO_4 = [NO_2][ClO_4] + [H_3O][ClO_4]$$

$$[NO_2][NO_3] + HClO_4 = [NO_2][ClO_4] + HNO_3.$$

The same nitronium perchlorate has been obtained by the action of ozone upon a mixture of nitrogen dioxide and chlorine dioxide (Gordon and Spinks, 1940). It forms colorless, monoclinic crystals, which may be recrystallized from anhydrous nitric acid. Other nitronium salts, $[NO_2][HS_2O_7]$, $[NO_2][FSO_3]$ (cf. Chap. 15) have also been isolated by the action of nitrogen pentoxide on sulfur trioxide and fluorosulfonic acid. In all these cases, the presence of the $[NO_2]^+$ ion (together with the corresponding anion) is proved by its very simple and characteristic Raman spectrum (Chédin, 1935). Concentrated sulfuric acid solutions of nitric acid similarly contain the rather soluble nitronium hydrogen sulfate, $[NO_2][HSO_4]$. This is of importance in that Hantzsch showed by optical methods that these sulfuric acid solutions contain compounds similar to the supposed hydroxonitronium salts which he isolated.

The crystal structure of nitronium perchlorate has been determined by Cox, Jeffery and Truter, confirming that it is correctly formulated as $[NO_2][ClO_4]$. The $O=N=O^+$ ion is linear, the N—O distance being rather shorter than in the neutral molecule of NO_2.

The ideas of Hantzsch and Ingold also afford an explanation for the high electrical conductivity of practically anhydrous nitric acid, which would otherwise be difficult to understand. From accurate measurements of the effect of small additions of water and of N_2O_5 upon the freezing point of nitric acid, Ingold, Hughes and Gillespie (1950) concluded that absolute nitric acid undergoes auto-ionization by two processes:

$$HNO_3 + HNO_3 \rightleftharpoons H_2NO_3^+ + NO_3^- \tag{1}$$

$$H_2NO_3^+ + HNO_3 \rightleftharpoons NO_2^+ + H_3O^+ + NO_3^- \tag{2}$$

Reaction (1), which forms the dihydroxonitronium ion postulated by Hantzsch, is responsible for perhaps 15% of the total ionization. The formation of nitronium ion by reaction (2) is thus more important, as is confirmed by the large cryoscopic effect of nitric acid upon the freezing point of absolute H_2SO_4. This approximates to the effect produced by an electrolyte furnishing four ions by the dissociation reaction

$$HNO_3 + 2H_2SO_4 \rightleftharpoons NO_2^+ + H_3O^+ + 2HSO_4^- \tag{3}$$

Nitrogen pentoxide in sulfuric acid furnishes solutions of identical properties and identical constitution, the freezing point depression corresponding, in this case, to the formation of six ions per molecule of N_2O_5

$$N_2O_5 + 3H_2SO_4 \rightleftharpoons 2NO_2^+ + H_3O^+ + 3HSO_4^- \tag{4}$$

The conclusion that the anhydrous acid contains the OH group, is not restricted to nitric acid. Studies of Raman spectra have shown that a considerable number of other acids (H_3AsO_4, H_2SO_4, H_2SO_5, H_2SeO_3, H_6TeO_6, $HClO_4$, HIO_3) in the undissociated state contain hydrogen bound in the form of hydroxyl groups. It is a reasonable inference that *all oxy-acids, in the undissociated state, are hydroxyl compounds*. In those exceptional cases (e.g., phosphoric acid) where it has not been possible to prove the presence of hydroxyl groups by Raman spectroscopy, the result has been explained as being due to the effect of hydrogen bonds (cf. p. 632).

(*ii*) *Nitrates.* The salts of nitric acid, $M^I[NO_3]$, can be obtained by dissolving the metals in nitric acid, or by neutralizing nitric acid with the corresponding hydroxides or carbonates. On a large scale, nitrates are generally prepared by the latter reaction.

If nitrous gases from the catalytic combustion of ammonia are allowed to act directly upon bases in aqueous solution, nitrites and nitrates are formed, in accordance with eqn. (4), p. 595. The former can subsequently be oxidized completely up to nitrates only by rather unsatisfactory methods. However, according to Briner (1929), nitrite-free calcium nitrate can be obtained by the reaction of *dry* nitrogen dioxide with calcium carbonate

$$3NO_2 + CaCO_3 = Ca(NO_3)_2 + NO + CO_2 + 27\ kcal$$

Briner states that, in the presence of water, NO_2 also reacts with $Ca_3(PO_4)_2$ without formation of nitrite,

$$3NO_2 + Ca_3(PO_4)_2 + H_2O = Ca(NO_3)_2 + 2CaHPO_4 + NO$$

The nitrates derived from *simple inorganic bases* are *all freely soluble in water*. The nitrates of some complex bases are moderately sparingly soluble. An organic (anhydro-)base of the composition $C_{20}H_{16}N_4$, known as *nitron*, forms a practically insoluble nitrate, which is often employed to precipitate the NO_3^- ion for the gravimetric determination of nitrates. Nitron nitrate has the composition $C_{20}H_{16}N_4 \cdot HNO_3$.

All nitrates are decomposed when they are heated, with the evolution of oxygen. The alkali nitrates thereby lose only one third of their oxygen, and are converted to nitrites—eqn. (1)—a reaction which can be greatly accelerated by adding substances with a weak reducing action, such as metallic lead.

$$KNO_3 = KNO_2 + \tfrac{1}{2}O_2 \qquad\qquad (1)$$

In addition to reaction (1), decomposition proceeds simultaneously by reaction (1a)

$$2KNO_3 = K_2O + N_2 + \tfrac{5}{2}O_2 \qquad\qquad (1a)$$

Both of these reactions are endothermic. If the nitrates are heated in the presence of metal with a high heat of combustion, however, it is possible for reactions such as

$$3KNO_3 + 5Al = 3KAlO_2 + Al_2O_3 + \tfrac{3}{2}N_2$$

to be initiated; these are strongly exothermic, and may follow an explosive course. To avoid serious accidents when sodium-potassium nitrate mixtures are used for the tempering of metals and alloys by heat treatment (Vol. II, Chap. 1), it is therefore necessary to take care that certain maximum temperatures are not exceeded, above which combustion of the metal sets in.

Since the corresponding nitrites are unstable when heated, the nitrates of the heavy metals undergo a more complete decomposition than those of the alkalis, losing nitrogen dioxide and oxygen (eqn (2)).

$$Hg(NO_3)_2 = HgO + 2NO_2 + \tfrac{1}{2}O_2 \qquad (2)$$

The nitrates are excellent oxidizing agents at high temperatures, because of their property of losing oxygen when heated. Their aqueous solutions on the other hand, have in general no oxidizing properties. Nitrates are reduced in solution only by strong reducing agents—e.g., by 'nascent' hydrogen. In this case, reduction may proceed as far as the formation of ammonia, which can be obtained in quantitative yield, for example, by the dissolution of aluminum, zinc, or Devarda's alloy in alkaline nitrate solutions.

By heating sodium nitrate with Na_2O, Zintl (1937) was able to prepare *sodium ortho*s *nitrate*, Na_3NO_4. This formed a white powder, which was deliquescent in air, and wat completely hydrolyzed by water into $NaNO_3$ and $2NaOH$. It was more stable towards hea- than $NaNO_3$, and also had a higher melting point.

(*iii*) *Nitrous acid and the Nitrites.* As has already been stated, the alkali nitrites are formed by absorbing a mixture of nitric oxide and nitrogen dioxide in alkali hydroxide solution, and also by the mild reduction of nitrates—e.g., sodium nitrite, $NaNO_2$, is obtained by melting sodium nitrate with lead

$$NaNO_3 + Pb = NaNO_2 + PbO$$

Nitrous acid, HNO_2, is liberated from the nitrites by means of acids; it is stable, however, only in cold, very dilute aqueous solutions. Decomposition occurs on warming (or even in the cold, if sand, fragments of glass or other sharp-edged objects are present)

$$3HNO_2 = HNO_3 + 2NO + H_2O$$

As befits this strong tendency to decompose, nitrous acid is a very reactive sub- stance. It is oxidized to nitric acid by strong oxidants, such as potassium per- manganate. It can be reduced, by alkali amalgams or electrolytically, to compounds poorer in oxygen, such as nitric oxide, hydroxylamine, hyponitrous acid or ammonia. Its property of *diazotizing* organic amines is very important, and sodium nitrite, manufactured as given above, therefore finds extensive in- dustrial application in the manufacture of azo dyestuffs.

Sodium nitrite, $NaNO_2$, is generally supplied for laboratory purposes in the form of pale yellow sticks. It is hygroscopic, and dissolves very readily in water (100 g of water dissolve 85.5 g of $NaNO_2$ at 25°), but only to a very slight extent in alcohol. *Potassium nitrite*, KNO_2, is deliquescent, and even more soluble in water than sodium nitrite (100 g of water dissolve 314 g of KNO_2 at 25°). It is precipi- tated by the addition of alcohol. Most of the other nitrites are also freely soluble. Silver nitrite, $AgNO_2$, often used as a preparative reagent, is sparingly soluble in cold water (1 part in 300).

The alkali nitrites can be fused without decomposition. Other nitrites decom- pose when heated—e.g., barium nitrite above 220°, silver nitrite at 140°, mercury nitrite at 75°.

Nitrous acid is a moderately weak acid, its dissociation constant in very dilute solutions having been found as $K = 4.5 \cdot 10^{-4}$. In 0.1-normal solution, its degree of dissociation would accordingly be about 6.5%. Its salts are not very stable in aqueous solution because of hydrolysis.

Nitrous acid exists, in its compounds at least, in two isomeric forms

$$HO—N=O \qquad \text{and} \qquad H—N{\displaystyle {\overset{\textstyle O}{\underset{\textstyle O}{<}}}}.$$

Two series of organic derivatives are derived from it accordingly—the nitrous esters, $RO—N=O$ (I), and the nitro-compounds (II) $R—N{\overset{O}{\underset{O}{<}}}$. The one series, (I), is easily saponified, reforming nitrous acid and alcohol. The action of 'nascent' hydrogen on these compounds also regenerates the alcohols and forms ammonia. The other compounds, (II), cannot be saponified, and their reduction yields organic amines $R—NH_2$ —i.e., the nitrogen is not split off from the alkyl group. We must conclude from this that in the nitro compounds the nitrogen is linked directly to the alkyl group, as is expressed by formula (II).

This isomerism between the two forms of the negative radical $—NO_2$ also plays a role in many complex compounds. Those which contain the nitrous acid residue in the form $—ONO$ are termed *nitrito*-compounds, and those containing the residue $—NO_2$, *nitro*-compounds. The nitro compounds are considerably more stable than the nitrito complex salts.

The *amide* of nitrous acid, *nitrosamide*, $ON—NH_2$ was obtained by Schwarz (1934), by the action of ammonia on N_2O_3 at very low temperatures, in the form of an intensely red solution in liquid ammonia. It decomposes with extreme ease, according to

$$2ON.NH_2 = [NH_4] [NO_2] + N_2$$

The organic derivatives of this compound, long known under the name of nitrosamines, $ON—NHR$, are much more stable.

Alkali nitrites dissolved in liquid ammonia combine with alkali metals, to form *nitro-xylates* (hydronitrites), the salts of *nitroxylic acid*, $N_2(OH)_4$, which is not known in the free state—e.g.,

$$2NaNO_2 + 2Na = Na_4[N_2O_4].$$

The magnetic properties of the compounds show that the nitroxylate (hydronitrite) anion is correctly formulated as $[O_2N—NO_2]^{4-}$, and not as $[NO_2]^{2-}$. The alkali nitroxylates are deep yellow in color. They are stable at ordinary temperature in the absence of oxygen, but are highly reactive. They are vigorously decomposed by water, disproportionation occurring. They are insoluble in liquid ammonia (Maxted, 1917; Zintl, 1928).

(iv) *Peroxynitrous acid.* As was first observed by Raschig [6], a mixture of hydrogen peroxide and a nitrite immediately liberates bromine from bromides when it is acidified with dilute sulfuric acid. Neither hydrogen peroxide, nitrite, nor nitrate alone will give this reaction, which is due, according to Baeyer (1901), Schmidlin (1910), Trifonow (1922) and Gleu (1929–35), to the intermediate formation of *peroxynitrous acid*, $O=N—O—OH$. According to Gleu this compound, an isomer of nitric acid, is also formed, by the action of ozone on an aqueous solution of alkali azide. Its aqueous solution is orange red. It is highly unstable, and it appears impossible to prepare it in the pure state even in solution; it occurs, rather, as an intermediate in the reactions cited and others. Although its alkali salts are more stable it has not yet been possible to isolate even these in the pure state.

(v) *Hyponitrous acid and the Hyponitrites.* Free hyponitrous acid, $H_2N_2O_2$, forms white crystalline leaflets, which are extremely explosive when they are dry. It is very soluble in water, and also in alcohol, but less soluble in ether. Conductivity measurements have shown that it is dissociated electrolytically only to a small extent in aqueous solution. Freezing point measurements have yielded a molecular weight corresponding to the formula $H_2N_2O_2$.

With phenolphthalein as indicator, only one hydrogen of hyponitrites is neutralized by alkali, (cf. carbonic acid). In addition to the *acid hyponitrites*, $M^I H[N_2O_2]$, which are formed in this way, hyponitrous acid also forms *neutral hyponitrites*, $M^I_2[N_2O_2]$, which are more stable, when isolated, than the very unstable acid salts. They are extensively hydrolyzed by water, however. In aqueous solution, the free acid decomposes, chiefly according to the reaction

$$H_2N_2O_2 = H_2O + N_2O.$$

This process is not reversible, and nitrous oxide, N_2O, cannot therefore be considered to be the anhydride of hyponitrous acid.

Hyponitrous acid is formed when the salts of nitric or nitrous acids, or nitric oxide, are reduced in aqueous solution by sodium amalgam. It may also be obtained (but in poor yield) by the careful oxidation of hydroxylamine (1), or by condensing hydroxylamine with nitrous acid (2).

$$\begin{array}{c} Ag_2O \quad H_2 \, NOH \\ + \\ Ag_2O \quad H_2 \, NOH \end{array} = \begin{array}{c} N-OH \\ \| \\ N-OH \end{array} + 2H_2O + 4Ag \qquad (1)$$

$$HONH_2 + ONOH = HO-N=N-OH + H_2O \qquad (2)$$

The two modes of formation last mentioned lead to the formula $HO-N=N-OH$ for hyponitrous acid. An isomer is *nitramide*, discovered by Thiele in 1895. As its name implies, this compound is usually regarded as the amide of nitric acid, O_2N-NH_2, but its constitution has not been finally established. Because of its acid character (Brönsted gives its dissociation constant $K = 2.6 \cdot 10^{-7}$), Hantzsch regarded it as *imidonitric acid*, $HN=NO(OH)$, whereas Pedersen (1934), from a study of its catalityc decomposition by alkalis in aqueous solution ($H_2N_2O_2 \rightarrow N_2O + H_2O$), suggested that the two forms exist in equilibrium, in which, however, the amido-form considerably predominates.

(c) Hydrogen Compounds of Nitrogen, and Their Derivatives

The following hydrogen compounds are derived from nitrogen.

(*i*) HN_3, *hydrazoic acid*. It is very unstable (explosive). Its salts are the *azides*.

(*ii*) NH_3, *ammonia*. This can add on a hydrogen ion, to form the positive ion $[NH_4]^+$. It accordingly forms salts, $[NH_4]X$, with acids—the ammonium salts. Aqueous solutions of ammonia have a basic reaction, by reason of the equilibrium set up in them:

$$NH_3 + HOH \rightleftharpoons [NH_4]^+ + OH^-$$

Ammonia itself is not a base, however (since it contains no hydroxyl group), but an *anhydro-base* (cf. pp. 61 and 62).

(*iii*) N_2H_4, *hydrazine*. This can also combine with acids, forming salts, the *hydrazinium* salts (cf. p. 616).

(*i*) *Hydrazoic acid and the Azides.*

1. *Hydrazoic acid (azoimide)*, HN_3, is a colorless, sharp smelling liquid, of density 1.126, boiling at 35.7° and freezing at —80°. It is a highly endothermic compound, and its vapor, which is poisonous and which irritates the mucous membrane, explodes with great violence when brought into contact with a hot object:

$$2HN_{3 \text{ gas}} = 3N_2 + H_2 + 141.8 \text{ kcal. (at const. volume)}.$$

Hydrazoic acid is stable in aqueous solution. It is a rather weaker acid than acetic acid, with the dissociation constant $K = 1.2 \cdot 10^{-5}$, and is dissociated to the extent of about 1% in 0.1-N solution at 20°. Being volatile, it is readily expelled from its salts, the *azides*, by less volatile acids.

2. The *azides*, M^IN_3, resemble the halides with respect to their solubilities. Most are soluble, the sparingly soluble salts including the azides of silver, bivalent lead and univalent mercury. The usual starting material for the preparation of other azides, and of the free acid, is *sodium azide*, NaN_3, which is obtained by passing nitrous oxide into fused sodamide:

$$NaNH_2 + ON_2 = NaN_3 + H_2O$$

The reaction proceeds smoothly from left to right, since the water is at once removed from the reaction mixture by reaction with unchanged sodium amide.

$$HOH + NaNH_2 = NaOH + NH_3$$

Sodium azide can be melted without decomposition, and deflagrates only when it is more strongly heated. The azides of the other alkalis and of the alkaline earths behave similarly. The azides of heavy metals, however—e.g., lead azide and silver azide—detonate violently when heated, and especially through shock. An explosion initiated by them is readily transferred to other explosives, and for this reason lead azide finds application as an intial detonator for ammunition and blasting explosives. Ammonium azide, $[NH_4][N_3]$, is interesting for its composition.

Whereas hydrazoic acid itself is decomposed by iodine (in the presence of some thiosulfate), with evolution of nitrogen, according to

$$2HN_3 + I_2 = 2HI + 3N_2,$$

silver azide reacts with iodine to form *iodazide*

$$AgN_3 + I_2 = IN_3 + AgI.$$

3. *Iodazide*, IN_3, is an extremely explosive solid, which is probably colorless when pure, though it is usually yellowish white. *Bromazide*, BrN_3, is liquid at ordinary temperature, and *chlorazide*, ClN_3, which can be obtained by the reaction of sodium azide with sodium hypochlorite:

$$NaN_3 + NaOCl + H_2O = ClN_3 + 2NaOH$$

is a colorless gas which explodes violently on contact with a flame.

4. *Fluorazide*, FN_3, was prepared by Browne and Haller (1942) by the action of fluorine on HN_3, in an atmosphere of nitrogen. It is a greenish yellow gas (b.p. —82°, m.p. —154°), which gradually decomposes, forming N_2F_2:

$$2N_3F = 2N_2 + N_2F_2.$$

The liquid is explosive.

The heat of formation of HN_3 liquid is —63.0 kcal per mol (at constant pressure), the heat of evaporation is 7.29, and the heat of solution is 9.73 kcal per mol (Günther, 1935–36). When azide solutions are electrolyzed, the N_3^- ion is discharged at the anode, and nitrogen is evolved. The discharge potential of the N_3^- ion (on smooth platinum) is 1.18 volts at 18° (Riesenfeld, 1935).

5. *Constitution.* Physical investigations (Raman spectrum, band spectrum, moment of inertia) make it probable that the N-atoms in hydrazoic acid are linear. The structure is essentially the same as that of N_2O, except that an NH group replaces the O atom e.g., $O=N \rightleftharpoons N$ and $H—N=N \rightleftharpoons N$, respectively. Interatomic distances are $N \leftrightarrow N = 1.14$ Å, $N \leftrightarrow NH = 1.25$ Å (Schomaker, 1942). The linear structure of the N_3 group in crystalline azides has been proved by X-ray structure determinations.

If HN$_3$ is passed, at a low pressure (about 0.1 mm) through a quartz tube heated to about 1000°, it decomposes according to the equation HN$_3$ = N$_2$ + NH. The *imine radical*, NH, which is so formed, has a mean half life of about 1.3 · 10^{-3} seconds. It can be obtained in the form of a blue (polymerized) solid, [NH]$_n$, by freezing it out on a surface cooled with liquid air (Rice, 1951).

The structure of the solid compound [NH]$_n$ is not known. It is paramagnetic, but a non-conductor of electricity. When it is allowed to warm up to —125° C, it undergoes transformation into colorless ammonium azide, NH$_4$N$_3$.

The existence of the NH radical in the gaseous state had been detected previously by band spectroscopic methods. It is formed not only in the thermal decomposition of HN$_3$, but in the thermal decomposition of ammonia at about 2000°, by the action of 'active nitrogen' on HN$_3$, and probably in the photochemical decomposition of HN$_3$.

(ii) Ammonia.

1. *Properties.* Ammonia, NH$_3$, is a colorless gas, with a pungent smell and biting taste. It is considerably lighter than air (density 0.5963, relative to air = 1; liter weight = 0.7713 g). It may be condensed by a pressure of 8.46 atm. at 20° to a mobile, colorless, highly refractive liquid; b.p. —33.4°; crit. temp. 132.4°; crit. press. 112 atm.; crit. density 0.236; m.p. —77.7°; heat of fusion 81 cal per g; heat of evaporation 327 cal per g at the b.p., and 302 cal per g at 0°.

The large latent heat of vaporization of ammonia is important for the use of ammonia in refrigeration. It arises, essentially, from the fact that although monomolecular in the vapor state, ammonia polymerizes through hydrogen bonding, when it is liquefied, as does water. The depolymerization occurring on vaporization requires the expenditure of a considerable amount of heat.

Liquid ammonia displays many similarities to water in its physical behavior. It is a good solvent for many substances, including numerous salts, which dissociate electrolytically in liquid ammonia, just as in water. Double decomposition reactions between substances dissolved in liquid ammonia frequently follow a different course from that in aqueous solution, since many substances are soluble in liquid ammonia which are insoluble in water (see further Vol. II).

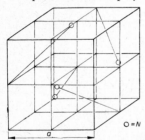

The dielectric constant of liquid ammonia is 21–23 at —34°; the viscosity η = 0.0026. The auto-conductivity of absolutely pure liquid ammonia is extraordinarily small; Carvallo was able to reduce the specific conductivity to about 4 · 10^{-10} (at —15°).

Fig. 111. Crystal lattice of ammonia a = 5.15 Å. (The position of the N-atoms only has been represented.)

Ammonia crystallizes in the cubic system, and its structure was worked out by Mark, by X-ray methods. The position of the N-atoms is represented in Fig. 111. As is invariably the case, the hydrogen atoms give rise to no detectable X-ray interferences, and cannot be located.

From spectroscopic measurements (absorption spectrum, Raman spectrum) it may be deduced that the H-nuclei in the NH$_3$ molecule are disposed in the form of an equilateral triangle, with sides 1.6 Å in length. The N-nucleus lies 0.38 Å vertically above the mid point of this triangle. The NH$_3$ molecule has a fairly large dipole moment (μ = 1.5 · 10^{-18} e.s.u.). As follows from the even number of electrons in the molecule, ammonia is diamagnetic.

Ammonia is avidly absorbed by cold water; 1 cc of water at 0° dissolves 1176 cc (= 0.907 g) of ammonia, and 702 cc at 20°. The ammonia is completely expelled from solution on boiling. Aqueous solutions of ammonia behave like solutions of a weak base, coloring red litmus paper blue and requiring a definite quantity of acid for their neutralization. Evaporation of the solutions after neutralization with acids yields salts—the ammonium salts (see below).

The heat of solution of ammonia in much water is 8.4 kcal per mol at 20°. Concentrated ammonia solutions freeze only at very low temperatures. Ammonia is also freely soluble in alcohol, acetone, chloroform, benzene and other organic solvents, though not so soluble as in water.

2. *Formation and Preparation.* Ammonia is formed in the decomposition of most nitrogenous organic materials. It is therefore found in nature in the form of its salts (ammonium salts), although generally only in small amounts. In certain places—e.g., in the neighborhood of volcanoes—such compounds are found in quantity. These occurrences are of no practical significance for the production of ammonia. The *ammonia liquor* (gas liquor) obtained as a by-product from coke ovens, from the production of coal gas, and from the gasification of brown coal, however, is an important source for the extraction of ammonia. The quantities available from these sources are, however, quite insufficient to meet the demands for ammonia. It is therefore of very great importance that it has been found possible to produce ammonia directly from atmospheric nitrogen, by the catalytic combination of nitrogen and hydrogen (the Haber-Bosch process).

3. *Ammonia from Gas Liquor.* Gas liquor, obtained chiefly from the destructive distillation of coal in gas works and coke ovens, is a colorless liquid which smells of ammonia, tar and hydrogen sulfide; it contains ammonia (usually between 1 and 3% by weight) in the form of various salts—principally as the carbonate, but in part also as the sulfide, sulfate, thiosulfate, chloride, thiocyanate, etc. The ammonia is first driven off as completely as possible from this solution by boiling. The ammonia which is present in the form of salts that are hydrolyzed by boiling is thereby expelled, and the remainder is liberated by adding milk of lime.

The ammonia vapors, which pass over, admixed with steam and other volatile materials, are freed from the principal impurities, in particular CO_2 and H_2S, by treatment with milk of lime. The water is condensed by cooling, and the gas so dried is freed from empyreumatic substances by passing it over wood charcoal, or by washing it with paraffin oil. The ammonia is then either liquefied by compression or is absorbed in distilled water, to obtain the solution known as 'spirit of ammonia'.

Ammonia is also obtained by the destructive distillation of the molasses residue—which is a waste product of the sugar industry—in this case chiefly in the form of ammonium carbonate and ammonium sulfate. Ammonia and ammonium salts were formerly prepared from the most varied nitrogenous organic waste materials, and especially from decomposed urine. Sal ammoniac (ammonium chloride) was prepared in this way as long ago as by the Arabic alchemists.

For some time, the calcium cyanamide process of Rothe, Frank and Caro was also of considerable importance for the production of ammonia. This ammonia synthesis was based on the decomposition of heated calcium cyanamide, by steam, or by water under pressure. The fixation of atmospheric nitrogen by strongly heated calcium carbide, and the liberation of ammonia from this, have already been discussed on p. 462 *et seq.* At the present time, calcium cyanamide is, indeed, manufactured on a large scale by the Rothe-Frank-Caro process, but it is almost exclusively used directly as a fertilizer, and hardly to any extent for the production of ammonia.

4. *Ammonia Synthesis.* The synthesis of ammonia by the Haber-Bosch process hinges upon the possibilities (a) of bringing nitrogen and hydrogen into reaction with one another, by using a catalyst, at temperatures substantially below those at which the equilibrium (1) lies almost completely to the left;

$$\tfrac{1}{2}N_2 + \tfrac{3}{2}H_2 \rightleftharpoons NH_3 + 11.05 \text{ kcal*}; \qquad (1)$$

* This is the value for the heat of formation at 20°. At 500°, the heat of formation is 12.7, at 700° 13.2, and at 0° 10.95 kcal per mol. The free energy of formation, $\Delta F = -3.91$ kcal per mol of NH_3 at 0°.

and (b), of exploiting the possibility of displacing the equilibrium, by pressure, in such a direction as to increase the concentration of ammonia.

The qualitative effect of temperature and pressure on the equilibrium (1) follows directly from the Le Chatelier principle. This indicates that the equilibrium is displaced to the left by raising the temperature, and to the right by increasing the pressure. Quantitatively, the manner in which equilibrium in reaction (1) depends upon the partial pressures of the constituents, and on the total pressure given by their sum, is represented by the Mass Action Law expression (2).

$$\frac{p_{NH_3}}{p_{N_2}^{1/2} \cdot p_{H_2}^{3/2}} = K_p \tag{2}$$

p_{NH_3}, p_{N_2}, p_{H_2} = partial pressures of the gases NH_3, N_2, H_2. K_p = equilibrium constant.
The temperature-dependence of K_p is given by Van't Hoff's Reaction Isochore (cf. p. 37). Values for the equilibrium constant K_p, calculated in this way by Haber for various temperatures, are shown in Table 94. The Table also gives concentrations of ammonia at various total pressures, calculated by means of eqn (2) for a mixture initially consisting of 3 parts of hydrogen and 1 part of nitrogen by volume.

TABLE 94

PERCENTAGE OF AMMONIA IN EQUILIBRIUM IN A MIXTURE OF
3 PARTS HYDROGEN + 1 PART NITROGEN, BY VOLUME

t	K_p	Volume per cent NH_3 at equilibrium				
		at 1 atm	at 30 atm	at 100 atm	at 200 atm	at 1000 atm
200° C	0.660	15.3	67.6	80.6	85.8	98.3
300°	0.070	2.18	31.8	52.1	62.8	92.6
400°	0.0138	0.44	10.7	25.1	36.3	79.8
500°	0.0040	0.129	3.62	10.4	17.6	57.5
600°	0.00151	0.049	1.43	4.47	8.25	31.4
700°	0.00069	0.0223	0.66	2.14	4.11	12.9
800°	0.00036	0.0117	0.35	1.15	2.24	—
900°	0.000212	0.0069	0.21	0.68	1.34	—
1000°	0.000136	0.0044	0.13	0.44	0.87	—

It is apparent from the table that the attainable yield of ammonia depends not only on the pressure, but also on the temperature of reaction—i.e., on the efficiency of the catalyst. Ruthenium, osmium, and uranium are notable for particularly high catalytic efficiency, being already effective at temperatures in the neighborhood of 400°, but they are too costly for industrial use. *Iron*, utilized by Haber in some of his experiments, has proved to be an industrially practicable contact substance. It is ordinarily active only above 650°, but it is possible to enhance its catalytic efficiency by means of suitable additions. Iron oxide or iron cyanide, with certain activating additions, have also proved usable. A pressure of 200 atm. and a temperature of about 500° are commonly employed. The exceptional demands placed upon the equipment by the use of such high pressures, in conjunction with high temperatures, were overcome technically by Bosch. Since heat is liberated by the union of nitrogen and hydrogen, the temperature can be maintained in contact furnaces of suitable construction without any additional heating. The hydrogen for ammonia synthesis is generally prepared by the water gas contact process described on p. 26. The cost of generating hydrogen is of critical importance for the cost of manufacture of synthetic ammonia.
The measurements of the equilibrium in the formation and decomposition of ammonia, which are fundamental for the synthetic ammonia process, were carried out in 1905–1910, chiefly by Haber. The first works for the production of synthetic ammonia (Oppau) came

into operation in the summer of 1913, and was designed for the production of about 4000 tons of NH_3 per annum. The shortage of raw materials during the first World War led to an unanticipated expansion within a very short time. By 1916, the production of synthetic ammonia rose to 40,000 tons, and in 1917 more than 60,000 tons. After the War, processes, such as the Claude process in France, were developed in other countries, on the same basic principles. The Claude process operates at a pressure of 1000 atm., and the resulting increase in the concentration of ammonia in the gases leaving the contact mass simplifies the operation. Against this advantage must be balanced the greater cost of the equipment, and other disadvantages. In France and Italy, the processes of Fauser and Casale are also employed, working at 250 and 800 atm. respectively. Other processes, such as the Mont Cenis process, employ lower pressures (90–100 atm.), and attain a favorable yield by operating at the lowest possible temperature (400°).

By far the greatest part of the entire world production of ammonia is now synthetic. Of the world production of combined nitrogen (including the extraction of Chile saltpeter) in 1935, 11.3% came from calcium cyanamide, 63.2% from synthetic ammonia and nitrogen compounds prepared from it, 15.5% from by-product ammonium sulfate (chiefly from gas works and coking ovens), 1% from other nitrogen compounds obtained as by-products (potassium ferricyanide, etc.), and 9% from Chile saltpeter. In 1935 there were 159 plants for the synthetic production of nitrogen compounds, with an annual capacity of 3.25 million tons of combined nitrogen. The actual output of combined nitrogen (including calcium cyanamide) by synthetic processes amounted in that year to 1.5 million tons, of which 35% was produced in Germany, 17% in Japan, 8% each in France and the United States.

In the laboratory, ammonia is generally prepared by decomposing ammonium salts (e.g., ammonium chloride) with strong bases, such as calcium hydroxide

$$2NH_4Cl + Ca(OH)_2 = CaCl_2 + 2NH_3 + 2H_2O \tag{3}$$

Large quantities of ammonia are best taken from the steel cylinders of the liquefied gas, which are obtainable commercially. Small amounts of pure, dry ammonia gas are very conveniently obtained by heating the addition compound of ammonia with silver chloride: $AgCl \cdot 3NH_3 = AgCl + 3NH_3$.

5. *Uses.* The chief field of application of ammonia is in the fertilizer industry [*3, 4*]. The nitrogen in nitrogenous fertilizers is generally present in the form of ammonium salts, or of nitrates. These latter, however, as has already been mentioned, are produced industrially almost exclusively from ammonia.

Urea, $CO(NH_2)_2$, which has been increasingly used as a fertilizer in recent years, is prepared from synthetic ammonia, as well as from calcium cyanamide by the method given on p. 463.

Ammonia combines with carbon dioxide, forming ammonium carbamate $[NH_4]$ $[CO_2.NH_2]$ (see p. 614). This is converted into urea when heated above 115° under pressure.

$$2NH_3 + CO_2 = [NH_4][CO_2.NH_2]; \qquad [NH_4][CO_2.NH_2] = CO(NH_2)_2 + H_2O$$

Urea is now prepared technically by the direct action of CO_2 on aqueous ammonia at 130–140° and 100 atm. pressure. Important plastics are obtained by the condensation of urea with formaldehyde (Pollak, 1921)—e.g., *kaurite* (a water-soluble adhesive), *plastopal* (shellac substitute), and *pollopase* (a widely used plastic).

Ammonia also finds extensive applications in other ways in chemical industry. Aqueous ammonia is a precipitant widely used in analytical chemistry. The dilute solution has household uses as a cleansing agent under the name of spirit of ammonia. It is also used in medicine. Because of its great latent heat of evaporation, anhydrous ammonia is very important in the refrigeration industry (cold stores, ice works).

In scientific work, liquid ammonia has become increasingly important as a solvent and medium for chemical reactions (cf. Vol. II), and as a dehydrating agent (ammonia extraction process of Biltz, cf. pp. 354 and 495).

6. *Chemical Properties.* Ammonia is stable toward decomposition at ordinary temperature. It is decomposed into its elements by the electric spark, and partial decomposition sets in at relatively low temperatures in the presence of catalytically active materials. Ammonia is not ordinarily combustible in air, although certain ammonia-air mixtures can be inflamed. Ammonia also burns when it is introduced into a gas flame burning in air. Ammonia burns in pure oxygen with a yellowish flame, mostly to nitrogen and water: $2NH_3 + \frac{3}{2}O_2 = N_2 + 3H_2O$.
At high pressures, mixtures of ammonia and pure oxygen are explosive. The combustion of ammonia-air mixtures can be brought about at relatively low temperatures ($300-500°$) in the presence of catalysts, and under suitable conditions, oxides of nitrogen are the sole products. As has been seen (p. 598), this is used technically for the production of nitric acid (Ostwald process).

It is probable that *nitrosyl hydride*, HNO, is formed as an intermediate in the catalytic combustion of ammonia. Harteck (1930–33) was able to obtain this substance, which is unstable in the gaseous state, as a bright yellow solid deposit at the temperature of liquid air, by the action of atomic O on NH_3, or by the addition of atomic H to NO, *Sodium nitrosyl*, NaNO, can be obtained as a colorless, cryptocrystalline compound, by the action of NO on sodium dissolved in liquid ammonia (Joannis, 1894). According to Gehlen (1939), it is also formed by the action of NO, diluted with 4 parts of nitrogen, on metallic sodium at $170-180°$. Zintl (1933) showed by X-ray methods that this compound is distinct from sodium hyponitrite, which is identical in composition. When it is dissolved in water, NaNO vigorously evolves N_2O (sodium hydroxide and sodium hyponitrite being formed simultaneously). KNO and the alkaline earth nitrosyls can also be prepared in liquid ammonia.

Ammonia can be oxidized in aqueous solution by strong oxidants, such as hydrogen peroxide, chromic acid, and potassium permanganate. Chlorine and bromine react energetically—e.g., $3Cl_2 + 2NH_3 = N_2 + 6HCl$.
Ammonia is also generally decomposed when it is passed over heated metals or metal oxides. In many instances, the nitrogen of the ammonia thereby combines with the metal, forming the nitride. The alkali and alkaline earth metals substitute one H-atom, forming *amides*—e.g.,

$$NH_3 + Na = \tfrac{1}{2}H_2 + NaNH_2 \text{ (sodium amide)}.$$

Ammonia has a very great tendency to add on, as a whole molecule, to other substances, forming *ammoniates*—e.g.,

$$AgCl + 3NH_3 = AgCl \cdot 3NH_3 \text{ (triammine silver chloride)}.$$

Ammonia shares this great tendency with water. It is to be explained in part by the high dipole moment of the two molecules; the molecules PH_3 and H_2S, which have considerably smaller dipole moments (cf. Tables 91 and 99, pp. 583 and 687), form addition compounds only to a limited extent, and are in general unable to combine with substances that have typical ionic crystal structures (cf. Vol. II).

Above all, ammonia combines avidly with acids, forming compounds which, from their whole behavior, must be considered as *true salts*. These contain the positive $[NH_4]^+$ ion.

The formation of the ammonium ion, which may most simply be represented by the equation

$$NH_3 + H^+ \rightleftharpoons [NH_4]^+ \tag{1}$$

also takes place in aqueous solutions of ammonia. Because of the extremely slight dissociation of water, however, the equilibrium is in this case displaced strongly to the left. There are, accordingly, only a few ammonium ions present in an aqueous solution of ammonia, and correspondingly few hydroxyl ions, since these latter—apart from the vanishingly small concentration present in pure water—are formed by the reaction

$$NH_3 + H_2O \rightleftharpoons [NH_4]^+ + OH^- \tag{2}$$

For this reason, ammonia solutions give only a weakly basic reaction or, alternatively, behave like solutions of a weak base.

We may suppose the process (2) to be resolved into two partial reactions

$$NH_3 + H_2O \rightleftharpoons NH_3 \cdot H_2O \rightleftharpoons [NH_4]^+ + OH^- \tag{2a}$$

That some compound is formed when ammonia is dissolved in water may be deduced from the very high solubility of ammonia, and its strongly positive heat of solution. The compound, *ammonia hydrate*, $NH_3 \cdot H_2O$, may also be isolated in crystalline form at low temperatures (m.p. —79.0°); the hemihydrate, $NH_3 \cdot \frac{1}{2}H_2O$ (m.p. —78.8°) crystallizes from solutions of especially high NH_3 concentration. On the other hand, *undissociated ammonium hydroxide*, $[NH_4]OH$ is not present in the solutions, in equilibrium with the ions $[NH_4]^+$ and OH^-; Briegleb (1942) has shown that the compound $[NH_4]OH$ is not stable in the undissociated state. It differs from the ammonia hydrate in its constitution. The latter is an *addition compound*, containing the H_2O combined as such with the NH_3 (presumably through the operation of a 'hydrogen bond'). Ammonium hydroxide however, would be a *coordination compound*, with a structure corresponding to the ammonium salts $[NH_4]X$—i.e., built up from the positive $[NH_4]^+$ and the negative OH^- ions. Ammonia hydrate, which dissociates to a slight extent according to eqn. (2a), is a *weak base*. On the other hand, since ammonium hydroxide exists only in the completely dissociated state, it must be regarded as a *strong base*.

In sufficiently dilute solutions, for which the concentration of water is practically unaffected by the reaction (2), application of the Mass Action law gives

$$\frac{[NH_4^+] \cdot [OH^-]}{[NH_3]} = K \tag{3}$$

At 18°, $K = 1.75 \cdot 10^{-5}$. It follows that K may also be spoken of as the electrolytic dissociation constant of ammonia hydrate, but it is logically incorrect to refer to it as the 'dissociation constant of ammonium hydroxide', as was formerly frequently the practice.

A 1-molar solution of ammonia, at 18°, contains 0.4% of the ammonia in the form of dissociated ammonium hydroxide, and a 0.1-molar solution contains 1.3%.

In terms of the Kossel theory, the process represented by eqn. (1) can be regarded as the consequence of the interaction of the threefold-negatively charged nitrogen atom on the positive ions in its immediate neighborhood. According to Kossel, the nitrogen atom acquires 3 negative charges from hydrogen, in the formation of the ammonia molecule. In addition to binding the three positively charged hydrogen nuclei formed in this process, making the neutral ammonia molecule, it may be shown, by application of Coulomb's law, that the addition of a *fourth* hydrogen nucleus to the nitrogen atom would be exothermic; the energy would suffice to remove the proton from the hydrated ion $[H_3O]^+$:

$$NH_3 + [H_3O]^+ = [NH_4]^+ + H_2O$$

As has been stated, the Kossel approximation is in certain respects artificial. The binding of the fourth H-nucleus by the NH_3 molecule can also be interpreted in terms of the Lewis-

Langmuir theory, and a more precise treatment of the bonding is provided by wave mechanics. According to this, it must be supposed in this and in other similar cases that quantum mechanical resonance forces are superimposed on the electrostatic attractions, as ordinarily understood. These exchange forces increase the strength of binding of the H-atoms to the N-atom, both in ammonia and in the $[NH_4]^+$ ion. Electronically, the structure of the latter is closely related to that of the CH_4 molecule, and the process of its formation can be represented thus:

$$\begin{array}{ccc} H & & \left[\begin{array}{c} H \\ \overset{\cdot\cdot}{} \end{array}\right]^+ \\ H:\overset{\cdot\cdot}{\underset{\cdot\cdot}{N}}: \quad + \quad H^+ \quad = \quad H:\overset{\cdot\cdot}{N}:H \\ H & & H \end{array}$$

The tendency of the NH_3-molecule to form an $[NH_4]^+$-ion by addition of a proton is due to the formation of a *dative covalence* by the unshared pair of electrons present in the NH_3-molecule. The formation of the ions $[RNH_3]^+$, $[R_2NH_2]^+$ $[R_3NH]^+$ and $[R_4N]^+$ (R = alkyl or aryl radical) from substituted organic amines is explained in a corresponding manner.

7. *Ammonium Amalgam. The Free Ammonium Radical.* When a concentrated ammonium salt solution is decomposed with sodium amalgam (Davy, 1808), or is electrolyzed using a mercury cathode (Seebeck, Berzelius, 1808), a pasty mass is obtained which has strongly reducing properties when first prepared. Thus it reduces solutions of copper salts at 0°. It rapidly decomposes, however, ammonia and hydrogen being formed in the ratio of 1 : 2 by volume. From the mode of formation of the material, it has been regarded by many as an *amalgam* of the quasi-metallic ammonium radical, $[NH_4]$.

$$NH_4^+ + Na-Hg_x \rightarrow Na^+ + NH_4-Hg_x$$

$$NH_4^+ + e \xrightarrow{\text{at Hg surface}} NH_4-Hg_x$$

$$2NH_4 = H_2 + 2NH_3$$

The pasty froth is easily compressible, and its volume roughly obeys Boyle's law. For this reason, some have considered that it is merely an emulsion of hydrogen and ammonia gases in mercury. However, Rich and Travers (1906) found that 'ammonium amalgam' had a lower freezing point than mercury, the freezing point being depressed as if by a metallic solute. Moreover, the amalgam may be preserved for long periods at —78° without loss of its power of reducing potassium iodate, although it progressively decomposes at —30°. Recent investigations (Ubbelohde, 1951) have shown that the behavior of the ammonium ion, when freshly discharged at a mercury surface, closely parallels that of the alkali metal ions, and there is little doubt that a true ammonium amalgam is initially formed, although it rapidly decomposes at ordinary temperature. A free molecule NH_4, with one electron in excess of the neon configuration would undoubtedly be unstable. A metal, however, can be regarded as built up from cations embedded in a 'gas' of quasi-free electrons (cf. Vol II). In amalgamating with mercury, therefore, the ammonium radical can be considered as giving up one electron to the electron levels of the metallic solvent, so that it virtually enters the mercury as the stable NH_4^+ ion.

(*iii*) *Ammonium Salts.* Ammonium salts show a great resemblance to the salts of the alkali metals in their solubilities and other properties. They differ, however, in being volatile when heated, and in that ammonia is liberated from them by the action of strong bases (alkali or alkaline earth hydroxides)—cf. p. 609, eqn. (3).

Ammonium salts are *prepared* [7] chiefly by passing ammonia gas into solutions of the corresponding acids, or by mixing the acids with solutions of ammonia or ammonium carbonate.

1. *Ammonium chloride*, sal ammoniac, NH_4Cl, is prepared by neutralizing ammonia or ammonium carbonate solution with hydrochloric acid:

$$NH_3 + HCl = [NH_4]Cl, \quad \text{or} \quad [NH_4]_2CO_3 + 2HCl = 2[NH_4]Cl + H_2O + CO_2.$$

The gas liquor obtained from coal gas manufacture and from coke ovens is the starting material for the technical production. It is first concentrated by a single distillation. The concentrated gas liquor, containing 12–25% of NH_3, is mixed with concentrated hydrochloric acid, and the sal ammoniac solution so obtained is evaporated down. The salt is purified by recrystallization or sublimation.

Ammonium chloride is a colorless salt, with a bitter-salty taste, very soluble in water, less soluble in alcohol, and easily sublimed, being perceptibly volatile at 100°. The vapor of ammonium chloride is partially dissociated into NH_3 and HCl. When deposited from the vapor state, it forms feathery crystalline masses. It crystallizes in regular octahedra from aqueous solutions (heat of solution = —3.84 kcal per mol at 18°). The density is 1.52.

It has long been known that ammonium chloride crystallizes in beautiful large crystals from solutions which are contained in wooden vats, although it otherwise tends to form small, feathery branched crystals. The greater growth of the crystals is due to the action of pectic substances given up by the wood. This fact is made use of, by adding pectin or other substances with a similar action to the solutions, in order to get as large crystals of sal ammoniac as possible.

At ordinary temperature, the crystal sturcture of ammonium chloride corresponds to that of cesium iodide, (Fig. 48, p. 211), except that the position of the cesium atom is occupied by the atom of nitrogen (or the ammonium radical). The side of the cubic unit cell is $a = 3.859$ Å. Above 184.5°, ammonium chloride crystallizes with the sodium chloride structure. At about —30°, it undergoes a further transformation which is due not to a change in crystal structure, but to the fact that at this temperature the free rotation of the $[NH_4]$-radical about its axis ceases. (cf. p. 218).

The solubility of ammonium chloride in water is,

at	0°	20°	60°	100°	115.6°
	29.7	37.2	55.2	77.3	87.3 g of NH_4Cl in 100 g of water

Solutions of ammonium chloride have essentially a neutral reaction at ordinary temperature, but become acid on boiling, because ammonia, which is formed in traces by hydrolysis, is lost by vaporization.

Sal ammoniac is used as a flux in soldering since it reacts with metallic oxides to form volatile chlorides, and so cleans up the surface of the metals. It is also used for the electrolyte in Leclanché and dry cells. It finds numerous uses in industry and in the laboratory.

2. *Ammonium nitrate*, $[NH_4]NO_3$, prepared from ammonia and nitric acid, is a colorless salt (density —1.73), which deliquesces in moist air. It usually crystallizes in the rhombic system, melts at 169.5°, and decomposes into nitrous oxide and water at rather higher temperatures.* It is used for the preparation of nitrous oxide, but mainly for the production of safety explosives. Mixtures of ammonium nitrate with calcium carbonate are used as a fertilizer.

Ammonium nitrate dissolves in water with considerable absorption of heat (6.2 kcal per mol). The solubility is, at 0° 118, at 25° 214, and at 100° 871 g of NH_4NO_3 in 100 g of water.

---

* N_2 is also formed in larger or smaller amounts, as well as N_2O, depending on the conditions. Since its formation gives rise to free nitric acid:

$$5NH_4NO_3 = 4N_2 + 9H_2O + 2HNO_3,$$

and since the latter accelerates the decomposition catalytically, it is possible for explosive decomposition to set in when large quantities of ammonium nitrate are melted. According to Tramm (1934), this may be inhibited by neutralizing the acid, e.g., with NH_3.

Ammonium nitrate can exist in six different modifications, five of these being stable under ordinary pressure. The transformation temperatures are

$$\text{Hexagonal} \; \rightleftharpoons \; \alpha\text{-rhombic} \; \rightleftharpoons \; \beta\text{-rhombic} \; \rightleftharpoons \; \text{tetragonal} \; \rightleftharpoons \; \text{cubic}$$
$$-18° \qquad\qquad 32.3° \qquad\qquad 84.2° \qquad\qquad 125.2°$$

3. *Ammonium carbonate*, $[NH_4]_2CO_3$, is prepared by bringing ammonia and carbon dioxide together in aqueous solution (1), or by heating a mixture of ammonium sulfate and precipitated chalk (2):

$$CO_2 + 2NH_3 + H_2O = [NH_4]_2CO_3 \qquad\qquad (1)$$

$$[NH_4]_2SO_4 + CaCO_3 = [NH_4]_2CO_3 + CaSO_4 \qquad\qquad (2)$$

The salt which sublimes away in process (2) contains both acid ammonium carbonate (ammonium hydrogen carbonate), $[NH_4]HCO_3$, and ammonium carbamate, $[NH_4][CO_2(NH_2)]$, as well as the normal ammonium carbonate.

Dry ammonia gas combines with dry carbon dioxide to form ammonium carbamate— i.e., the ammonium salt of *carbamic acid*, $CO(NH_2)(OH)$. This is derived from carbonic acid, by the exchange of an OH group for the NH_2 group (amino group). The carbamate is transformed into the neutral carbonate by water, but the process is reversible.

$$[NH_4] \begin{bmatrix} O \\ C(NH_2) \\ O \end{bmatrix} + H_2O \rightleftharpoons [NH_4]_2 \begin{bmatrix} O \\ C=O \\ O \end{bmatrix}$$

Ammonium hydrogen carbonate can be converted into the neutral salt by addition of ammonia, or by warming the solution (whereby CO_2 is driven off)

$$[NH_4]HCO_3 + NH_3 = [NH_4]_2CO_3; \qquad 2[NH_4]HCO_3 = [NH_4]_2CO_3 + CO_2 + H_2O$$

When exposed to the air, the neutral salt loses ammonia, and passes into the hydrogen carbonate. At about 60° it rapidly decomposes, into ammonia, carbon dioxide and water.

The 'hartshorn salt' obtainable commercially consists chiefly of the double compound of ammonium hydrogen carbonate and ammonium carbamate, $[NH_4]HCO_3 \cdot [NH_4][CO_2(NH_2)]$.

4. *Ammonium sulfate*, $(NH_4)_2SO_4$, is now prepared not only by neutralizing sulfuric acid with ammonia gas, but by the important gypsum-ammonium sulfate process, based on the double decomposition of ammonium carbonate with calcium sulfate. This is operated by passing ammonia and carbon dioxide into an aqueous suspension of finely ground gypsum:

$$CaSO_4 + 2NH_3 + CO_2 + H_2O = CaCO_3 + [NH_4]_2SO_4.$$

Ammonium sulfate forms colorless, rhombic crystals of density 1.77. It is very soluble in water (at 0° 71.0, at 20° 76.3, and at 100° 97.5 g of salt in 100 g of water). Its most important application is as a fertilizer.

Ammonium sulfate can be recovered from coal gas and coke oven gas by *Feld's polythionate process*, based on the fact that NH_3 and H_2S react with polythionates in aqueous solution, to form ammonium thiosulfate—e.g.,

$$2(NH_4)_2S_4O_6 + 2(NH_4)_2S_3O_6 + 6NH_3 + 6H_2S = 7(NH_4)_2S_2O_3 + 6S + 3H_2O$$

The polythionates are regenerated by burning the sulfur, and treating the sulfur dioxide with thiosulfate solution:

$$2(NH_4)_2S_2O_3 + 3SO_2 = (NH_4)_2S_4O_6 + (NH_4)_2S_3O_6.$$

That portion of the solution, enriched in thiosulfate, which is not required for the above reaction, is heated to 120–130° in a pressure vessel, whereby ammonium sulfate is formed and sulfur is deposited:

$$2(NH_4)_2S_2O_3 + (NH_4)_2S_4O_6 = 3(NH_4)_2SO_4 + 5S$$

Sulfur is also recovered by this process, and can be either used as such, or may be converted into sulfuric acid.

Ammonium sulfate decomposes at 357° when heated at atmospheric pressure, losing ammonia gas, and forming a melt which consists of a mixture of unchanged ammonium sulfate with the acid ammonium sulfate (ammonium hydrogen sulfate), $(NH_4)HSO_4$. Pure ammonium hydrogen sulfate melts at 251° and boils at 490° without decomposition.

5. *Ammonium hydroperoxide*, $[NH_4]OOH$, and *ammonium peroxide*, $[NH_4]_2O_2$, which may be regarded as the primary and secondary ammonium salts of hydrogen peroxide, may be isolated by passing ammonia into a well cooled, concentrated ether solution of hydrogen peroxide (D'Ans, 1913).

(iv) Derivatives of Ammonia. If one H-atom were abstracted from NH_3, the univalent group $-NH_2$ (the amido- or amino-group)* would remain. Two amido groups can unite to form a *diamide*, H_2N-NH_2 *(hydrazine)*. If an $-OH$ group is linked to the $-NH_2$ radical, *hydroxylamine*, NH_2-OH, is obtained. Replacement of 1 H-atom of ammonia by the equivalent amount of a metal gives *metal amides*, such as *sodamide*, $NaNH_2$. If it is replaced by chlorine, the very unstable *chloramine*, NH_2Cl, is formed. If the group $-NH_2$ is introduced into acids, in place of the $-OH$ group, then depending on whether all, or only a portion of the $-OH$ groups are so replaced, either *acid amides*—e.g., acetamide, sulfamide—or *amido-acids* (e.g., amidosulfonic acid) are obtained.

$$CH_3-C\begin{smallmatrix}O\\\\NH_2\end{smallmatrix} \qquad O\!\!=\!\!\underset{O}{\overset{}{S}}\!\!\begin{smallmatrix}NH_2\\\\NH_2\end{smallmatrix} \qquad O\!\!=\!\!\underset{O}{\overset{}{S}}\!\!\begin{smallmatrix}NH_2\\\\OH\end{smallmatrix}$$

acetamide	sulfamide	amidosulfonic acid, sulfamic acid
(amide of acetic acid)	(diamide of sulfuric acid)	(monoamide of sulfuric acid)

The bivalent group $HN<$ from ammonia is known as the *imido-* or *imino*-group, and the compounds derived from it are the *imides* or *imines*. By abstracting all three H-atoms from ammonia there remains the trivalent radical $N\equiv$, known in certain compounds as the *nitrilo*-group (e.g., nitrilosulfonic acid, $N(SO_3H)_3$). Binary compounds of nitrogen with electropositive elements are called *nitrides* (e.g., AlN, aluminum nitride).

Chloramine, NH_2Cl, can be stabilized by the substution of organic radicals for hydrogen. Thus the sodium salt of *p*-toluenesulphon-chloramide, $CH_3 \cdot C_6H_4, SO_2 \cdot NHCl$, is so stable that it can be recrystallized without decomposition from hot water. This compound (which

---

* It is usual to apply the name ami*do*-compounds, or ami*des* to the compounds formed by substituting the group $-NH_2$ for the $-OH$ groups of *acids*, and to the compounds of the $-NH_2$ group with metals. Most other compounds of the group $-NH_2$ are usually called ami*no*-compounds, or ami*nes*. A similar distinction is drawn between imi*do* compounds, or imi*des*, and imi*no*-compounds.

comes into commerce under the name of 'Chloramine T') is an excellent disinfectant and bleaching agent. It can also be used for oxidimetric titrations [Noll, 1924, cf. Poethke and Wolf, *Z.anorg. Chem.*, 268 (1952) 244.] Its oxidizing action arises from the fact that in aqueous solution it is very slightly hydrolyzed to form hypochlorite:

$$[CH_3 \cdot C_6H_4 \cdot SO_2 \cdot NCl]^- + H_2O \rightleftharpoons CH_3 \cdot C_6H_4 \cdot SO_2 \cdot NH_2 + [ClO]^-.$$

Like ammonia, hydrazine and hydroxylamine have the ability to add on H^+ ions, and to combine with acids forming salts. As in the case of ammonia, these arise from the coordinative quadrivalence of nitrogen. Hydrazine, which contains two coordinatively quadrivalent nitrogen atoms, is accordingly capable of forming two series of *hydrazinium* salts:

$$H_2N—NH_2 + HCl = [H_2N—NH_3]Cl \qquad H_2N—NH_2 + 2HCl = [H_3N—NH_3]Cl_2$$
$$\text{hydrazinium chloride} \qquad\qquad \text{hydrazinium dichloride}$$

Only one series of salts is derived, however, from hydroxylamine—the *hydroxylammonium* salts, e.g., $H_2NOH + HCl = [H_3NOH]Cl$, hydroxylammonium chloride.

1. *Hydrazine*, [*11*] N_2H_4, was first prepared in 1889, by Curtius, from organic substances. It can also be obtained, however, from inorganic materials, e.g., by the reduction of hyponitrous acid with ammonium sulfide or, better, by a method due to Raschig, the oxidation of ammonia with sodium hypochlorite in the presence of gelatin. The probable function of the gelatin is to combine with traces of heavy metal ions (especially copper) which may be present, since these accelerate the decomposition of NH_2Cl, the primary intermediate in the process, to give N_2. According to Pfeiffer (1947) the gelatin may be replaced by other substances which are especially effective in binding copper ions.

The primary reaction of sodium hypochlorite with ammonia is to form chloramine, NH_2Cl:

$$NH_3 + NaOCl = NaOH + NH_2Cl.$$

This very unstable compound at once undergoes further reaction, usually evolving nitrogen:

$$2NH_2Cl + NaOCl + 2NaOH = N_2 + 3NaCl + 3H_2O.$$

However, if this reaction is suppressed by gelatin, the chloramine reacts with an additional molecule of ammonia, forming hydrazine.

$$NH_2Cl + HNH_2 + NaOH = H_2N—NH_2 + NaCl + H_2O$$

The hydrazine may be precipitated from the solution as hydrazinium sulfate $[N_2H_6]SO_4$, and can be liberated from this salt by concentrated potassium hydroxide solution. It is thereby obtained as *hydrazine hydrate*, $N_2H_4 \cdot H_2O$. The anhydrous compound may be obtained from this by distillation over caustic soda.

Hydrazine, is a colorless liquid, fuming strongly in air; density 1.011 at 15°, m.p. 1.4°, b.p. 113.5°. It is miscible in all proportions with water and alcohol. *Hydrazine hydrate*, $N_2H_4 \cdot H_2O$, is also liquid at ordinary temperature, but has a lower melting point and higher boiling point than the anhydrous compound (m.p. —40°, b.p. 118.5° at 739.5 mm pressure, density 1.030).

Hydrazine can combine with 1 or 2 equivalents of acid to form the hydrazinium salts; e.g., with hydrochloric acid it forms the two chlorides $[N_2H_5]Cl$ and $[N_2H_6]Cl_2$. The sulfate, $[N_2H_6]SO_4$, is sparingly soluble in cold water, and readily soluble in hot. It is therefore readily recrystallized. It forms thick colorless plates or long prisms. It is insoluble in alcohol.

In the form of its salts, hydrazine is used in the preparation of azides, and as a reducing agent in analytical chemistry. A characteristic property of hydrazine, and of alkaline solutions of hydrazine salt is the deposition of the noble metals from solutions of their salts.

According to Giguère and Schomaker (1943), the dimensions of the N_2H_4 molecule are $N \leftrightarrow N = 1.47$ Å, $N \leftrightarrow H = 1.04$ Å, with the angle H—N—N about 108°.

2. *Hydroxylamine*, NH_2OH, can be prepared by the electrolytic reduction of nitric acid at a lead cathode:

$$H—O—NO_2 + 6H^+ + 6e = H—O—NH_2 + 2H_2O$$

In addition to this method an older process, due to Raschig, is used for the technical preparation of hydroxylamine. In this, sodium hydroxylaminedisulfonate is first prepared, by the action of sodium nitrite on a strong acid solution of sodium hydrogen sulfite at 0° (1). The sparingly soluble potassium salt is then precipitated, by the action of potassium chloride (2), and this is finally hydrolyzed by boiling it with water (3), whereby hydroxylamine sulfate is formed.

$$NaNO_2 + NaHSO_3 + SO_2 = HO—N(SO_3Na)_2 \tag{1}$$

$$HO—N(SO_3Na)_2 + 2KCl = HO—N(SO_3K)_2 + 2NaCl \tag{2}$$

$$2HO—N(SO_3K)_2 + 4H_2O = [HO—NH_3]_2SO_4 + 2K_2SO_4 + H_2SO_4 \tag{3}$$

As a rule, only the salts are prepared, since the preparation of the free anhydro-base is difficult, and not without hazard under some conditions. Free hydroxyl-amine has a tendency to decompose explosively when it is heated. Quite a small quantity, heated in a test tube over a free flame, detonates with a report like a gun shot.

Solutions of hydroxylamine have a strongly basic reaction. Hydroxylamine gradually decomposes in alkaline solution, usually with the evolution of nitrogen according to $3NH_2OH = 3H_2O + NH_3 + N_2$. In some circumstances (e.g., in the presence of platinum black), hyponitrite may be formed.

$$4NH_2OH + 2NaOH = 4H_2O + 2NH_3 + Na_2N_2O_2$$

In either case, the primary product is nitroxyl, NOH:

$$2NH_2OH = NH_3 + H_2O + NOH$$

This immediately undergoes further reaction—either with hydroxylamine:

$$NOH + NH_2OH = N_2 + 2H_2O$$

or alternatively its anion dimerizes to that of hyponitrous acid.

$$NOH + OH^- = NO^- + H_2O \quad 2[\ddot{N} :: \ddot{O}]^- = [: \ddot{O} : \ddot{N} :: \ddot{N} : \ddot{O} :]^=.$$

The NO^- ions can be quantitatively trapped by $[Ni(CN)_4]^=$ ions, with the formation of the intensely red-violet tricyanonitrosyl nickelate ion:

$$[Ni(CN)_4]^= + NO^- = [Ni(CN)_3NO]^= + CN^-.$$

$[Ni(CN)_4]^=$ ions in alkaline solution are therefore a very sensitive reagent for the presence of nitroxyl. No red-violet coloration is produced when nitramide or hyponitrite decompose in alkaline solution in the presence of $[Ni(CN)_4]^=$ ions. It follows that these reactions do not involve formation of nitroxyl as intermediates (Nast, 1948). Another reaction for detection of nitroxyl was discovered by Angeli and Rimini. It depends on the observation that NOH reacts with aldehydes to form hydroxamic acids, $R.CO.NH(OH)$, which give a deep red coloration on the addition of $FeCl_3$ solution.

The hydroxylammonium salts, $[NH_3(OH)]X$, are mostly soluble in water. Like the anhydrobase, they have a strong reducing action. It is to be noted that they are *very poisonous*.

Compounds derived from hydroxylamine by substitution of the two H-atoms linked to nitrogen, i.e., containing the radical $>N$—OH, are called oximes. These compounds—e.g., aldoximes, $R \cdot CH=NOH$, and ketoximes, $R_1R_2C=NOH$—are of importance in organic chemistry.

3. *Hydroxylamine Sulfonic Acids.* If one or more of the H-atoms in hydroxylamine are replaced by the sulfonic acid radical —$SO_2(OH)$, the resulting compounds are known as hydroxylamine sulfonic acids ($R = $ —$SO_2(OH)$).

$$\begin{array}{ccccc} R & R & H & R & R \\ {>}N{-}OH & {>}N{-}OH & {>}N{-}OR & {>}N{-}OR & {>}N{-}OR \\ H & R & H & H & R \end{array}$$

All five of these compound types are known, either as such or in the form of their alkali salts. They differ widely from one another in their stability towards alkalis, and also in the mechanism of the reactions involved in their hydrolysis (Nast, 1952).

4. *Metal Amides.* The metal amides have the general formula M^INH_2. Amides of the alkali and alkaline earth metals are obtained by the direct action of ammonia on the metals. Thus *sodamide*, $NaNH_2$, is prepared by passing dry ammonia gas over molten sodium:

$$NH_3 + Na = NH_2Na + \tfrac{1}{2}H_2$$

Sodium amide is formed even in the cold when a solution of metallic sodium in liquid ammonia stands for a prolonged period. This reaction is catalyzed by various substances, e.g., Fe, Fe^{++}, etc. The amides of the heavy metals can be prepared from solutions of their salts in liquid ammonia, by double decomposition. Unlike the rather stable alkali and alkaline earth amides, most of them are explosive.

Sodamide forms a colorless crystalline mass, m.p. $210°$. when pure. It can be sublimed in a vacuum, but undergoes decomposition at about $500°$, or even above $300°$ in a vacuum. The melt acts as an energetic reducing agent and simultaneously as a condensation reagent since sodamide reacts vigorously with water:

$$NaNH_2 + H_2O = NaOH + NH_3$$

Sodamide accordingly finds applications in organic synthetic work—especially for the technical synthesis of indigo—and also for the preparation of sodium cyanide and sodium azide.

5. *Imides.* The alkaline earth amides are converted into imides, $M^{II}NH$, when they are heated. These crystallize with face-centered cubic structures. On heating in a high vacuum, the yellow alkaline earth imides are transformed into reddish yellow or reddish brown *pernitrides*, $M^{II}_3N_4$. The decomposition of these compounds by dilute acids furnishes one mo of N_2 and two moles of NH_3 per mole of pernitride e.g.:

$$Sr_3N_4 + 8HCl = 3SrCl_2 + 2NH_4Cl + N_2$$

(Hartman, Fröhlich and Ebert, 1934).

6. *Nitrides*. This term is applied to the compounds of nitrogen with the more electropositive elements—i.e., principally with the metals. Many metals unite directly with nitrogen when heated. The nitrides are frequently better prepared by heating the metals, or their oxides or chlorides, in a current of ammonia. The nitrides of the alkaline earth metals can also be prepared by the decomposition of their amides.

Those nitrides which, from their composition, can be regarded as substitution products of ammonia—e.g., Li_3N, Mg_3N_2, Ca_3N_2, Sr_3N_2, Ba_3N_2, Zn_3N_2, AlN—are mostly decomposed by water, with formation of ammonia (causic potash solution is better with Zn_3N_2 and AlN). There are also nitrides which cannot be interpreted as simple substitution products of ammonia—e.g., Mn_5N_2, W_2N_3, ZrN. As a rule, these are considerably more resistant towards water. Chromium nitride, CrN, is also not decomposed by water, even at 220°.

In general, the nitrides are notable for their high thermal stability. Like the oxides, most of them are relatively infusible.

Double nitrides include compounds such as LiMgN, LiZnN, Li_3AlN_2, and Li_3GaN_2 (Juza, 1948). They crystallize with coordination lattice structures, and are salt-like in character. The same is true of the corresponding compounds formed by the homologues of nitrogen. Laves, Nowotny and Juza have prepared the following compounds of these types, and have investigated their structure.

LiMgP	LiMgSb	LiZnP	Li_3AlP_2
LiMgAs	LiMgBi	LiZnAs	Li_3AlAs_2

(d) Halogen Compounds of Nitrogen

Two classes of halogen compounds of nitrogen are known. The compounds of the one class have the general formula XN_3, and those of the other NX_3 (X = halogen). The former compounds are derivatives of hydrazoic acid, and have already been described in that connection. The latter can be regarded as *halogen substitution products of ammonia*, and their mode of formation agrees with this. Thus nitrogen trichloride, NCl_3, can be obtained by the action of chlorine on ammonium chloride in concentrated aqueous solution. The *fluoride* of nitrogen, a colorless gas condensing only at —129°, was discovered by Ruff in 1928. It is an exothermic and remarkably stable compound. In contrast, *nitrogen chloride*, NCl_3, and *nitrogen iodide*, NI_3, (which exists as ammoniates) are excessively explosive substances. It has not been possible to isolate the corresponding bromine compound at all, but only its ammoniate, $NBr_3 \cdot 6NH_3$, which is likewise extremely unstable. The *oxyhalides* of nitrogen are in general more stable than the trihalides. The most important properties of the compounds are listed in Table 95.

(i) Nitrogen trifluoride, NF_3, was prepared by Ruff, by electrolysing molten, anhydrous $[NH_4]HF_2$. It is a colorless gas, which may be condensed when strongly cooled to a clear, light mobile liquid. Nitrogen trifluoride is practically insoluble in water and caustic potash. It does not attack dry glass or mercury, but reacts slowly with water vapor when ignited by means of the electric spark:

$$2NF_3 + 3H_2O = 6HF + N_2O_3.$$

It reacts quite violently with hydrogen when ignited:

$$2NF_3 + 3H_2 = 6HF + N_2$$

The NF_3 prepared by the electrolysis of $[NH_4HF_2]$ is mixed not only with considerable quantities of nitrogen, but also with small amounts of the highly explosive compounds NH_2F, NHF_2, and N_2F_4. These may be removed by passing the gas over MnO_2. N_2F_4 (b.p. $-125°$) is derived from N_2H_4, in the same way that NF_3 is derived from NH_3.

The molecule of NF_3 forms a trigonal pyramid, with N at the apex; the distance $N \leftrightarrow F = 1.37$ Å, and the F—N—F angle is $103°$ (Schomaker, 1947).

In addition to the compounds just considered, dinitrogen difluoride, N_2F_2 also exists. It is a colorless gas, and is highly explosive. Electron diffraction measurements (Bauer, 1947) have shown that the molecule exists in two forms:

$$F\diagdown N=N\diagup F \quad (\textit{cis}\text{-form}) \qquad \text{and} \qquad F\diagdown N=N\diagdown F \quad (\textit{trans}\text{-form})$$

The $N \leftrightarrow F$ distance is 1.44 Å, the $N \leftrightarrow N$ distance 1.25 Å, and the N=N—F angle $115°$ in both forms.

Fluorine nitrate, NO_3F, which was discovered by Cady in 1934, and more fully investigated by Ruff in 1935, is notable for its constitution. It may be considered to be derived from oxygen difluoride, F—O—F, by exchange of one F-atom by the *nitryl* radical —NO_2. It is a musty smelling, poisonous gas, which explodes when heated. In contrast to NF_3, NOF and NO_2F, NO_3F thus appears to be an endothermic compound.

(*ii*) *Nitrogen trichloride*, NCl_3, forms a dark yellow, volatile oil of density 1.65, which smells pungent and strongly attacks the eyes and mucous membranes. It explodes with the greatest violence when warmed to $93°$, or if it is brought into contact with substances such as rubber or turpentine, on which it exerts a chlorinating action. Its solutions in carbon disulfide, benzene, ether, chloroform, or carbon tetrachloride are stable if protected from light. It reacts with ammonia, according to the equation

$$NCl_3 + 4NH_3 = N_2 + 3NH_4Cl.$$

It is slowly hydrolyzed by water

$$NCl_3 + 3HOH = NH_3 + 3ClOH$$

It can be prepared either by the action of hypochlorite on aqueous ammonia or ammonium salts, which is a reversal of this hydrolysis, or—a better method for

TABLE 95

TRIHALIDES AND OXYHALIDES OF NITROGEN

	NF_3	NCl_3	NI_3	NOF	NO_2F
Color	colorless	yellow	black	colorless	colorless
Melting point	$-208.5°$	liquid at	solid at	$-132.5°$	$-166°$
Boiling point	$-129.0°$	ordinary temp.	ordinary temp.	$-59.9°$	$-72.4°$
Heat of formation			strongly	probably	probably
(kcal/mole)	$+26$	-54.7	negative	positive	positive
Liter wt. (g, at $0°$ C)	3.17	—	—	2.231	2.971
Density at b.p.	1.885	1.65	—	1.326	1.796

	NO_3F	NOCl	NO_2Cl	NOBr
Color	colorless	red-yellow	colorless	black-brown
Boiling point	$-175°$	$-5.8°$	$-145°$	$-55.5°$
Melting point	$-45.9°$	$-61°$	$-15.0°$	$+19°$
Heat of formation (kcal/mole)	negative	-7.2	—	-16.9
Liter wt. (g, at $0°$ C)	—	2.992	2.81 (at $100°$)	—
Density at b.p.	1.507	1.592	1.41	—

demonstration purposes—by the electrolysis of a saturated ammonium chloride solution, on to which is poured a little turpentine. As droplets of the dangerous substance rise to the surface, they at once detonate, and it is not possible for appreciable amounts to collect.

(iii) Nitrogen iodide. The action of iodine on a concentrated solution of ammonium iodide forms only ammonium polyiodides, and not nitrogen iodide (cf. NCl_3). However, if iodine is allowed to react with aqueous or alcoholic solutions of ammonia, brown-black products are obtained which, when dried, are much more explosive even than nitrogen trichloride, and detonate even when gently touched. They are customarily called nitrogen iodide, but the composition of these products is variable. The *ammonia addition compounds of nitrogen iodide* are more readily obtained pure. The best known of these is the mono-ammoniate, $NI_3 \cdot NH_3$. This crystallizes in copper colored needles (density 3.5) which explode at the slightest touch when they are dry. The reactions of this compound show that it is to be regarded as an ammoniate of nitrogen triiodide, NI_3—e.g., with zinc ethyl:

$$NI_3 \cdot NH_3 + 3Zn(C_2H_5)_2 = NH_3 + N(C_2H_5)_3 + 3Zn(C_2H_5)I$$
$$\text{zinc ethyl} \qquad \text{triethylamine} \quad \text{ethyl zinc iodide}$$

(iv) Nitrosyl chloride, $NOCl$, the formation of which in various reactions has already been mentioned, is not explosive although it is an endothermic compound (cf. NCl_3). This is because it generally decomposes into chlorine and nitric oxide, and not into its elements, and its formation from these constituents is an exothermic reaction:

$$NO + \tfrac{1}{2}Cl_2 = NOCl + 14.4 \text{ kcal.}$$

Nitrosyl chloride is a reddish yellow, easily condensable gas. It forms blood red crystals below —61°. It is hydrolyzed by water to nitrous acid and hydrogen chloride:

$$NOCl + HOH = NO(OH) + HCl$$

The reaction is reversible. In the presence of dehydrating agents, nitrous anhydride reacts with hydrogen chloride to form nitrosyl chloride

$$N_2O_3 + 2HCl = 2NOCl + H_2O$$

This reaction can be used to prepare pure nitrosyl chloride, by adding P_2O_5 to liquid N_2O_3, and passing in HCl gas.

The hydrogen of other acids may be replaced by the positive radical $[NO]^+$ by reaction with N_2O_3, in a similar manner. In this way are formed *nitrosyl perchlorate*, discovered by K. A. Hofmann, 1909, $N_2O_3 + 2HClO_4 = 2[NO]ClO_4 + H_2O$, heat of formation 41.8 kcal per mole (Cruse 1950), and *nitrosyl hydrogen sulfate*,

$$N_2O_3 + 2H_2SO_4 = 2[NO]HSO_4 + H_2O.$$

The latter is usually called nitrosyl sulfuric acid, or 'chamber crystals', but has a salt-like structure, like nitrosyl perchlorate, as was first suggested by Hantzsch (1930). This is true also of nitrosyl fluoroborate, $[NO][BF_4]$, which is isotypic with ammonium fluoroborate, $[NH_4][BF_4]$, and of nitrosyl chlorostannate, $[NO]_2[SnCl_6]$ (isotypic with $[NH_4]_2[SnCl_6]$). All these compounds are immediately decomposed by water in the same manner as NOCl.

In the gaseous state, the molecules of $O{=}N{-}Cl$ are bent, with the angle 116°, and interatomic distances $O \leftrightarrow N$ 1.04 Å, $N \leftrightarrow Cl$ 1.95 Å (Ketelaar (1943)). The dipole moment is $1.83 \cdot 10^{-18}$ e.s.u. The moment of ONBr is $1.87 \cdot 10^{-18}$ e.s.u. These high dipole moments show that the compounds are not purely covalent. Ketelaar assumed that there was resonance between the two forms $O{=}N{-}Cl$ and $[ON]^+Cl^-$. This would lead also to the observed lengthening of the N—Cl distance, and shortening of the N—O distance as compared with that expected for covalent bonds.

(v) Nitrosyl and Nitryl Compounds. The *nitrosyl ion*, NO^+, and the *nitryl ion*, NO_2^+, are isosteric with the molecules CO and CO_2 respectively.

Schumacher (1929) found that the action of ozone on $NOCl$ led to the formation of the rather unstable *nitryl chloride*, NO_2Cl. It has since been found that NO_2Cl can be prepared by the action of chlorsulfonic acid on concentrated nitric acid (sp. gr. 1.5). The compound decomposes when warmed

$$NO_2Cl \rightleftharpoons NO_2 + \tfrac{1}{2}Cl_2,$$

the reaction being reversible. Hasenbach reported in 1871 that NO_2Cl was formed by the action of Cl_2 on NO_2 under pressure, but the observation was subsequently doubted. According to Brintzinger (1948) NO_2Cl can be used to introduce the NO_2-group and Cl simultaneously into organic compounds.

Nitryl perchlorate, $[NO_2][ClO_4]$, was prepared in 1940 by Gordon, by the action of ozone and nitrogen oxides on chlorine dioxide. It forms colorless needles, soluble in $POCl_3$, but insoluble in CCl_4. It reacts vigorously with organic compounds containing hydrogen, and rapidly decomposes when warmed to 120°, but does not explode.

In liquid sulfur dioxide as solvent, it is possible to carry out double decomposition reactions with nitrosyl- and nitryl compounds, exactly parallel to the double decompositions between salts in aqueous solution. Thus

$$[NO][SbCl_6] + [NR_4][SO_3.OR] = [NR_4][SbCl_6] + [NO][SO_3.OR];$$

$$[NO_2][SbCl_6] + [NR_4][ClO_4] = [NR_4][SbCl_6] + [NO_2][ClO_4] \text{ (where R = CH}_3).$$

Nitryl fluoroborate, $[NO_2][BF_4]$ and dinitrosyl sulfate, $[NO]_2[SO_4]$, may be obtained similarly; the latter at once decomposes to form dinitrosyl pyrosulfate, $2[NO]_2[SO_4] = [NO]_2[S_2O_7] + N_2O_3$. If NO is passed into liquid oxygen, yellow *dinitrosyl peroxide*, $(NO)_2O_2$, is formed, together with N_2O_3. Tetramethylammonium azide, $[NR_4][N_3]$, reacts with nitrosyl compounds in non-aqueous solutions (including compounds of non- electrolyte character, or of low solubility, such as $NOCl$, $NOBr$, $NOSCN$, $NO.OCOCH_3$, $Cu(NO)Cl_2$):

$$[NR_4][N_3] + NO{-}X = [NR_4]X + N_2O + N_2.$$

The same reaction is given by nitrogen tetroxide which thereby behaves as nitrosyl nitrate, $NO{-}NO_3$. NO_2Cl, however, behaves not as nitrosyl hypochlorite, $NO{-}OCl$, but as a compound of the nitryl radical, NO_2^+. The same behavior is shown in other reactions— e.g., with antimony pentachloride, $NO_2Cl + SbCl_5 = [NO_2][SbCl_6]$ (Seel, 1951–52).

It has been found (Seel, 1953) that all nitrosyl compounds add on NO under pressure, to form salts of a dinitrosyl cation, $[N_2O_2]^+$ —e.g., $[N_2O_2][SbCl_6]$, $[N_2O_2][AlCl_4]$. These are all intensely blue in color. The so-called 'violet acid' or 'blue acid' of the lead chamber sulfuric acid process, which was formerly believed to have the composition $(NO)H_2SO_4$ is actually the hydrogen sulfate of this complex dinitrosyl ion, $[N_2O_2]HSO_4$.

Nitrosyl fluoride, NOF, may be prepared by heating $NO[BF_4]$ (Balz and Mailänder, 1934). *Nitryl fluoride*, NO_2F, is obtainable similarly from nitryl fluoroborate, $NO_2[BF_4]$. The latter can be prepared from KNO_3, B_2O_3 and BrF_3 (Emeleus, 1950) or by passing HF into a solution of $N_2O_5 \cdot BF_3$ in nitromethane (Schmeisser, 1952; cf. p. 595):

$$N_2O_5 \cdot BF_3 + HF = HNO_3 + NO_2[BF_4].$$

The reaction

$$2N_2O_5 + SiF_4 + 2HF = 2HNO_3 + [NO_2]_2[SiF_6]$$

similarly furnishes nitryl fluorosilicate. Lehmann (1951) prepared *dinitryl trisulfate*, $[NO_2]_2[S_3O_{10}]$ by a similar reaction. He obtained the compound $N_2O_5 \cdot 4SO_3$ by the action of SO_3 on $Sr(NO_3)_2$. This compound lost SO_3 when it was heated to 80° an a vacuum:

$$N_2O_5 \cdot 4SO_3 = [NO_2]_2[S_3O_{10}] + SO_3.$$

Above 120°, dinitryl trisulfate loses oxygen, and is converted into nitryl nitrosyl trisulfate, $[NO_2][NO][S_3O_{10}]$.

e) Compounds of Nitrogen and Sulfur

Nitrogen combines with sulfur to form two compounds, S_4N_4 and S_4N_2 (formerly considered to be N_2S_5), and also the univalent radical $[S_4N_3]^+$.

(*i*) *Tetrasulfur tetranitride* ('Sulfur nitride'), S_4N_4, is formed by the reaction of sulfur with liquid ammonia:

$$10S + 16NH_3 \rightleftharpoons 6(NH_4)_2S + S_4N_4.$$

If the ammonium sulfide which is formed be removed by some suitable double decomposition, the reaction proceeds to the right: e.g.,

$$(NH_4)_2S + 2AgI = Ag_2S + 2NH_4I.$$

Sulfur nitride is moderately soluble in organic solvents—e.g., 15 g of S_4N_4 dissolve in 1 kg of carbon disulfide at the boiling point. The compound crystallizes from the solution on cooling in golden yellow crystals, m.p. 178°. Cryoscopic measurements have led to the formula S_4N_4. Sulfur nitride is a strongly endothermic compound, which can decompose explosively through shock or when heated:

$$S_4N_4 = 4S + 2N_2 + 128 \text{ kcal.}$$

It is insoluble in water, but is attacked when its solution is boiled for a long time, whereby it is hydrolyzed to ammonia and oxyacids of sulfur. This indicates that nitrogen is the more electronegative constituent—i.e., the compound is to be regarded as a nitride of sulfur.

A peculiar reaction sets in if sulfur monochloride, S_2Cl_2, dissolved in chloroform, is treated with sulfur nitride. It then forms a compound of the composition S_4N_3Cl, in which the chlorine atom can be replaced by bromine, iodine, and acid radicals such as —SCN, —NO$_3$. A univalent radical, *thiotriazyl*, $[S_4N_3]^+$, is thus present in these substances.

(*ii*) *Tetrasulfur dinitride*, S_4N_2, forms deep red crystals, m.p. 23°. The supercooled liquid has a density of 1.71 at 20°. The compound is formed by the action of carbon disulfide on S_4N_4 at 100° under pressure. It is also formed as a by-product in the decomposition of S_4N_4. It explodes when heated, and undergoes slow decomposition at ordinary temperature. S_4N_2 is soluble in many organic solvents (e.g., in benzene, ether, CS_2, CCl_4), and is quite stable for long periods in such solutions at low temperatures. S_4N_2 is moderately soluble in alcohol, but insoluble in water. It is gradually decomposed by water, with the deposition of sulfur and formation of ammonia.

An impure S_4N_2 was formerly considered to be *nitrogen pentasulfide*, S_5N_2. Its true nature, as tetrasulfur dinitride, was first established by Meuwsen (1947–51). It can be formed not only by the reactions given above, but also by the following process:

$$Hg_5(NS)_8 + 4S_2Cl_2 = 4S_4N_2 + 3HgCl_2 + Hg_2Cl_2.$$

S_4N_2 forms a notably stable, solid, ocher-colored addition compound with BF_3, with the formula $S_4N_2 \cdot 2BF_3$. Goehring (1952) has shown from hydrolytic experiments that S_4N_2 contains one S atom in the $+4$ valence state, one in the $+2$ valence state, and two electroneutral S atoms, whereas the N atoms are negatively charged. The compound is considered to have a cyclic structure, and its absorption spectrum and diamagnetism are compatible with this conclusion.

4. Analytical (Nitrogen)

Compounds of nitrogen worthy of note from the analytical standpoint are, firstly, ammonia and nitric acid, and next in importance, nitrous acid.

Ammonia may readily be expelled from its salts (the ammonium salts) by caustic alkalis or quick lime, and can easily be recognized by its characteristic smell and alkaline reaction on moist litmus paper. Very small traces of ammonia— e.g., in the investigation of drinking water—are tested for and determined colorimetrically by means of Nessler's reagent. Ammonia is otherwise usually determined by alkalimetry. The usual methods for the determination of ammonia are not applicable when it is bound in a complex since alkalis often expel the ammonia incompletely, if at all. In such cases it is often possible to effect a determination by dissolving the compound in a known quantity of acid, and determining the excess of acid conductimetrically (Jander, Z. anal. Chem., 131 (1950) 89).

A brown ring forms at the junction of the two layers, when concentrated sulfuric acid is added to form a layer beneath a solution of nitric acid or a nitrate, to which ferrous sulfate has been added. The reaction depends on the reduction of nitric acid by the ferrous sulfate, forming nitric oxide, which adds on to the excess of ferrous sulfate giving nitroso-iron(II) sulfate. Nitrites give the brown coloration at once on addition of acetic acid. If nitrites are present, they must therefore be destroyed completely before testing for nitrates by boiling with ammonium chloride, or by addition of urea.

Nitric acid and nitrates may be determined quantitatively either by precipitation with nitron, or alkalimetrically after reduction to ammonia by means of Devarda's alloy. It can also be determined by gas analysis, by reducing the nitric acid to nitric oxide with mercury or with iron(II) chloride.

Nitrous acid is also frequently determined in the last mentioned way, but can also be titrated with potassium permanganate. When very small quantities are involved (e.g., in water analysis), nitrous acid is determined colorimetrically, through its diazotising action on organic amines, whereby intensely coloured dyestuffs are produced.

5. Phosphorus (P)

(a) Occurrence

Phosphorus occurs in nature almost exclusively in the form of salts of phosphoric acid, and chiefly as *phosphorite*, $3Ca_3(PO_4)_2 \cdot Ca(OH)_2$* and as *apatite*, $3Ca_3(PO_4)_2 \cdot Ca(F, Cl)_2$. There are isolated occurrences of the phosphates of iron—e.g., *vivianite*, $Fe_3(PO_4)_2 \cdot 8H_2O$ —, of aluminum—e.g., *wavellite*, $3Al_2O_3 \cdot 2P_2O_5 \cdot 12H_2O$ —and also of the rare earth metals (monazite). Derivatives of phosphoric acid are essential constituents of the animal and vegetable organism. In these, phosphoric acid is partly in organic combination (e.g., in the

* Later investigations have shown that under the conditions leading to the formation of phosphorite in nature, the product is not $Ca_3(PO_4)_2$, as was formerly assumed, but always the double compound, 'hydroxyl apatite', $3Ca_3(PO_4)_2 \cdot Ca(OH)_2$. It is practically certain, therefore, that this compound is also present in phosphorite, in part admixed isomorphously with carbonate apatite and ordinary apatite. See on this point Rathje, *Ber.*, 74 (1941) 349.

form of lecithin in the substance of the brain, and in egg yolk), and partly in the form of hydroxyl apatite (see above) and *carbonate apatite*, $3Ca_3(PO_4)_2 \cdot CaCO_3 \cdot H_2O$, in the bones as a structural material.

Contrary to the view formerly held, the dentine of the teeth does not consist principally of *fluoroapatite*, $3Ca_3(PO_4)_2 \cdot CaF_2$, recent investigations (e.g., Klement and Trömel, 1933) having shown that its inorganic substance is substantially the same in composition as the other bones. The greater hardness of dental enamel is connected with the relatively larger size of the crystals of hydroxyl- and carbonato-apatites in dentine (diameter 10^{-4} cm, as compared with 10^{-5}—10^{-6} cm in the other bones).

Human and animal excreta are also rich in phosphoric acid (about 40% of the ash content). The excreta of animals of former ages, occurring in petrified form (so called coprolites) consist of almost pure calcium phosphate (or hydroxy apatite?). A great part of the present phosphate deposits, such as the mighty deposits of North Africa, can be attributed to the deposition of vast quantities of animal excreta and accumulation of animal remains in earlier geological times. There are places where the formation of such great deposits of dung can be observed today, in the formation of *guano*. Large deposits of guano, originating chiefly from sea birds, occur especially on the islands of the Pacific Ocean, off the coast of Peru. Guano contains more or less important quantities of nitrogen (in the form of ammonia, urea, etc.), as well as calcium phosphate. It occurs only in regions of low rainfall, and in proportion as the nitrogenous constituents are leached out by rain, guano passes over into phosphorite.

(b) History

Phosphorus was discovered in 1669 by the alchemist Hennig Brand, when, in the course of experiments devoted to the search for the Philosophers' Stone, he strongly heated evaporated urine, in the absence of air. Phosphorus was named because of its property of glowing in the dark (φωσφόρος, light bearer). It was first recognized as a chemical element by Lavoisier, who elaborated his theory of combustion with particular reference to phosphorus.

(c) Preparation

The preparation of phosphorus depends essentially on the reduction of the pentoxide, P_2O_5, by carbon at a high temperature:

$$P_2O_5 + 5C = 5CO + 2P.$$

The phosphorus which distils off is received under water. To remove mechanically admixed impurities, the liquefied material is pressed through porous stone plates or soft leather, or is redistilled from iron retorts. It is usually sold in the form of sticks, of the thickness of a finger, which are stored under water.

The raw material for the manufacture of phosphorus is tertiary calcium phosphate, $Ca_3(PO_4)_2$ (or hydroxylapatite), in the form either of phosphorite or of bone ash. This is heated in an electric furnace with silica sand and coke. The SiO_2 liberates P_2O_5, which is then reduced by the carbon.

$$\left. \begin{array}{l} Ca_3(PO_4)_2 + 3SiO_2 = 3CaSiO_3 + P_2O_5 \\ P_2O_5 + 5C = 5CO + 2P \end{array} \right\} Ca_3(PO_4)_2 + 3SiO_2 + 5C = 3CaSiO_3 + 5CO + 2P$$

The principle of this method was stated by Wöhler, as long ago as 1806, but because of the high temperature which is required (above $1300°$) it became technically practicable only through the application of the electric furnace. The annual production of phosphorus by this method amounts to 20,000 tons in Germany alone. The greater part of this is immediately burned to P_2O_5, which is further worked up into phosphate fertilizers.

(d) Properties

Phosphorus exists in several modifications. Ordinary *white* or colorless phosphorus forms a transparent mass, as soft as wax, with a characteristic smell; density 1.82, m.p. 44.1°, b.p. 280°. It dissolves in water only in traces, but is markedly volatile in steam. It is also only slightly soluble in alcohol, but freely soluble in ether, benzene, turpentine and fatty oils, and especially in carbon disulfide, sulfur monochloride, and phosphorus trichloride. It crystallizes from carbon disulfide in regular crystals, usually rhombic dodecahedra.

The molecular weight of white phosphorus in solution corresponds with the molecular complexity P_4, and the same molecules exist in the vapor. Up to about 800° the vapor consists almost entirely of P_4 molecules, but at higher temperatures dissociation into P_2 molecules occurs. It has been shown by electron diffraction that the P_4 molecule is tetrahedral. Each phosphorus atom is bonded to each of the three others, so that the distance $P \leftrightarrow P = 2.21$ Å, and the angle $P—P—P = 60°$.

White phosphorus is extremely reactive chemically. In the finely divided state (as obtained, for example, by evaporating its carbon disulfide solution on filter paper) it inflames spontaneously in the air. Even in compact pieces, white phosphorus inflames a little above 50°, and can therefore be ignited through friction. It should, accordingly, be cut up only under water. It burns with a yellowish white flame to the pentoxide:

$$2P + {}^5/_2O_2 = P_2O_5 + 370 \text{ kcal.}$$

Phosphorus glows in the dark when exposed to air. This is because the vapor, evolved from it in traces, is oxidized by atmospheric oxygen, with the emission of light. Many substances, such as alcohol, ether, turpentine, hydrogen sulfide, sulfur dioxide, chlorine, or ammonia weaken or suppress the phosphorescence. The glow is also inhibited by oxygen at 1 atmosphere pressure, but reappears when the pressure is reduced. The principal product of the slow oxidation of phosphorus in moist air is phosphorous acid, together with hypophosphorous acid.

White phosphorus also combines avidly with the halogens. It ignites spontaneously in chlorine. It also reacts vigorously with sulfur and many metals. Strong oxidants, such as nitric acid, oxidize it to phosphoric acid. It dissolves in warm caustic potash, evolving phosphine and forming potassium hypophosphite.

$$4P + 3KOH + 3H_2O = PH_3 + 3KH_2PO_2.$$

Phosphorus deposits the easily reducible metals, such as gold, silver, copper, and lead from their salts, in some cases combining with them at the same time—e.g., with copper, to form copper phosphide.

White phosphorus is *very poisonous*, a quantity of 0.1 g being sufficient to kill a man if it is introduced into the stomach.

In cases of acute poisoning, immediate removal of the poison by vomiting or by pumping out the stomach is indicated. A very dilute solution of copper sulfate acts as an antidote, the phosphorus being rendered harmless by conversion to copper phosphide; it also induces vomiting. The prolonged inhalation of phosphorus vapor also leads to serious damage, especially to the gums and jaw bones (phosphorus necrosis, 'phossy jaw'). For this reason, the use of white phosphorus has now been forbidden in almost all civilized countries in the manufacture of matches, as was formerly customary.

When white phosphorus is heated to about 260° in a closed vessel, it is transformed into *red phosphorus*, which is considerably less reactive chemically, and is completely non-toxic. This transformation also takes place at the ordinary temperature under the influence of light. Red phosphorus, a powder of dark red color, insoluble in carbon disulfide, does not glow, and ignites only above 400°. Its density is about 2.2—i.e., considerably higher than that of white phosphorus. It reacts with the halogens and with sulfur at higher temperatures than does white phosphorus, and does not displace metals from their solutions. It can readily be oxidized to phosphoric acid, however, by means of nitric acid, and also ignites easily when rubbed up with potassium chlorate.

More reactive than ordinary red phosphorus is Schenck's *scarlet phosphorus*, obtained by boiling ordinary phosphorus with phosphorus tribromide. Like white phosphorus, it dissolves in caustic alkalis, and deposits copper from copper sulfate solution. Its density is 1.88. It may well be only a finely divided form of red phosphorus, and like the latter, it is not toxic.

The ordinary red phosphorus of commerce is not a homogeneous material, even apart from accidental impurities such as the oxidation products which form on prolonged standing in air. According to Roth, De Witt and Smith [*J. Am. Chem. Soc.*, 69 (1947) 2881], red phosphorus exists both in the amorphous state and in 3 (or even 4) crystalline modifications. Three of these (triclinic, hexagonal and tetragonal) can be obtained directly from phosphorus vapor. The principal allotropic form in red phosphorus is that known in the pure state as *violet phosphorus*, or Hittorf's phosphorus, since it was first prepared by Hittorf by crystallization from molten lead. Stock showed that it can also be obtained from a phosphorus-bismuth melt. The solidified alloy is best dissolved electrolytically, the crystals of phosphorus then remain undissolved. The density of violet phosphorus is 2.35, rather higher than that of commercial red phosphorus. It belongs to the monoclinic system. Violet phosphorus is transformed into white phosphorus when heated above its melting point, and especially if it is evaporated. White phosphorus thus appears to be the more stable form at high temperatures, although it is metastable at ordinary temperature. Because of its instability when heated, violet phosphorus has no well defined melting point or vapor pressure. Liquefaction usually occurs at about 600°, and Schenck's phosphorus and ordinary red phosphorus melt at around the same temperature.

When red phosphorus is vaporized at ordinary pressures, it condenses out as white phosphorus (i.e., as P_4 molecules). However, it was shown by Melville (1936) that in a high vacuum red phosphorus may be volatilized and condensed again in the form of red phosphorus, without undergoing transformation. He also found that the vapor of *white* phosphorus, consisting of P_4 molecules, gave a condensate of *red* phosphorus when passed, as a 'molecular beam'* over an incandescent tungsten filament, or through a silica jet heated to 800°. The evidence strongly indicates that the vapor from red phosphorus consists of P_2 molecules, which may also be formed by the dissociation of P_4 molecules at high temperatures. Except at very high temperatures, however, equilibrium in the reaction $2P_2 \rightleftharpoons P_4$ lies far over to the right. If, therefore, P_2 molecules from the vaporization of red phosphorus undergo molecular collisions in the gas phase, they combine to form P_4 molecules which, on condensation, form white phosphorus. For this reason, as was found by Smits (1914–1916), it is not possible to measure a stable vapor pressure over red phosphorus, although the latent heat of vaporization (approx. 25.6 kcal per g atom) is certainly higher than that of white phosphorus (13 kcal per g atom). If, however, P_2 molecules impinge on a cooled

---

* In a sufficiently good vacuum, when the mean free path of a molecule is long as compared with the dimensions of the apparatus used, a jet of vapor can be defined by suitable collimating apertures in such a way that the trajectories of the molecules in the beam lie sensibly parallel. Relatively few collisions then take place in the gas space, and the resulting 'molecular beam' can be condensed on a target, without undergoing much spreading.

surface before they can recombine to form P_4, they may condense in such a way as to yield solid red phosphorus.

The high melting point and involatility of red, violet, and black phosphorus (see below), and their insolubility in solvents, suggest that these are substances of high molecular weight. Indeed, it is only by building up solid structures of an indefinitely polymerized nature that strain-free structures can be built up, compatible with the valence angles of the P atom. Saturated, discreet molecules, such as the tetrahedral P_4 molecule, involve a considerable distortion. The dissociation of P_4 into strain-free, but unsaturated, P_2 molecules at high temperatures may be associated with this fact. The condensation of P_2 molecules on a cold surface then represents a polymerization process. The relation between the various species may be set out as follows.

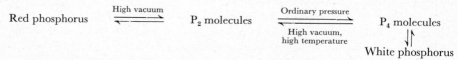

It is probable therefore, that the particles of solid red phosphorus consist of regions of ordered, crystalline high polymer (violet phosphorus, existing possibly in several modifications), with amorphous polymerized material. The latter is thermodynamically unstable with respect to the crystalline polymer. The more fully any sample of red phosphorus is annealed, the greater is the extent to which ordering takes place, and the vapor pressure diminishes accordingly. Above 450°, the amorphous network is sufficiently loosened to enable crystallization to occur.

The concept of red (or violet) phosphorus as a high polymer also enables certain features of the white phosphorus → red phosphorus transformation to be understood. The reaction can take place either by thermal 'cracking' of the discrete molecules, whereby free radical fragments are formed, or catalytically. Catalysts which can initiate the reaction are substances which are able to attack the discrete molecules forming either anionic radicals (e.g., the alkali metals) or unsaturated molecules (e.g., the halogens). By combining with free phosphorus valences, the catalytic substances block a certain proportion of potentially reactive centers in the polymer, and thereby determine the degree to which polymerization can proceed.*

The effect of halogens as chain terminating atoms is shown by the retention of unreactive bromine in Schenck's scarlet phosphorus, and of iodine in red phosphorus made with iodine as catalyst (Brodie, 1953). The bromine content usually found in Schenck's scarlet phosphorus is compatible with the view that between 1 in 10 and 1 in 30 of the phosphorus bonds is satisfied by union with Br, instead of with another P atom. Other 'mixed polymerizates' are also possible—e.g., with oxygen ('phosphorus suboxides') or oxygen and water, to give terminal —OH groups.

A third well defined modification of phosphorus is *black phosphorus*, prepared by Bridgman by heating white phosphorus to 200° under a pressure of 12000 atm. Its density is 2.70 and hardness 2. It is iron grey in color, with a metallic luster. It is also a metallic conductor of electricity, with good thermal conductivity. Its crystal structure has been given on p. 580. According to Gunther (1943), black phosphorus is metastable at ordinary pressure.

There is also evidence for a further modification of white phosphorus (also metastable), since it becomes double-refracting when strongly cooled. Vorländer gives the transition point as about —68°.

Phosphorus is soluble only to a very limited extent in molten bismuth and antimony, its homologues in the Periodic Table. It is completely miscible with arsenic in the molten state, however, and forms solid solutions containing 0 to 43 and 87 to 100% atom per cent As, as well as a *compound* with a range of homogeneity from 53–74 atom per cent As. The latter crystallizes in leaflets of graphitic appearance, and is a good electrical conductor (Klemm and von Falkowski, 1948).

---

* This matter will be more explicitly discussed in Chap. XIV, in connection with the allotropy of sulfur. The case of phosphorus is rather more complicated in that polymerization involves the formation of three covalent bonds per atom of phosphorus, instead of two bonds whereby long chain polymers of sulfur and selenium are formed.

(e) Uses [8, 9]

Because of its toxicity, white phosphorus is used only to a small extent—e.g., in rat poisons and for pharmaceutical preparations. Red phosphorus is used in large quantities in the match industry, which also utilizes Schenck's scarlet phosphorus. Red phosphorus is also used for the preparation of other phosphorus compounds— e.g., the chlorides—and as a halogen carrier, as in the preparation of hydrobromic acid (p. 792).

Of the compounds of phosphorus, the most important are the phosphates, especially as fertilizers* [8]. The calcium phosphate occurring naturally in the form of phosphate rock is not suited for this purpose, since it is insoluble. It is therefore generally converted into the water-soluble primary calcium phosphate, $Ca(H_2PO_4)_2$, by treating it with sulfuric acid. The mixture of primary calcium phosphate and gypsum so obtained is known in commerce as *superphosphate*. If phosphoric acid is used for the attack in place of sulfuric acid,** the product contains no gypsum, and *double superphosphate* is obtained. Another important phosphoric acid fertilizer is the *Thomas meal* obtained by finely grinding basic slag, which is a by-product of steel making. This contains phosphoric acid in a form which can be assimilated by plants, although insoluble in water, principally as a double salt of $Ca_3(PO_4)_2$ and Ca_2SiO_4 (and the corresponding magnesium double salt). The available phosphoric acid content of basic slag can be estimated from the amount which can be extracted by means of citric acid ('citrate soluble phosphoric acid'). Among fertilizers containing nitrogen compounds as well as phosphoric acid ('mixed fertilizers'), mention may be made of *bone meal* and *fish* and *flesh*-meal, as well as of *guano*. These are obtained from the waste products of the fish curing factories and slaughter houses.

6. Compounds of Phosphorus

(a) Oxygen Compounds

(i) *Survey.* The ordinary product of combustion of phosphorus is *phosphorus pentoxide*, P_2O_5. *Phosphorus trioxide*, P_2O_3, is formed if combustion takes place in an insufficient supply of air.

Phosphorus tetroxide, P_2O_4, can be regarded as a compound of the trioxide with the pentoxide (cf. p. 636). It is doubtful whether the 'phosphorus suboxides' frequently mentioned in the literature are really definite compounds.

Phosphorus pentoxide, P_2O_5 or P_4O_{10}, is the anhydride of phosphoric acid, H_3PO_4. Phosphorus trioxide, P_2O_3 or P_4O_6, is the anhydride of phosphorous acid, H_3PO_3. In addition to these, phosphorus forms *hypophosphoric acid*, $H_4P_2O_6$, and *hypophosphorous acid*, H_3PO_2. Anhydrides of these two acids are not known.

* The acid phosphates of zinc and manganese are extensively used in anti-rusting preparations. Cf. Vol. II.
** Attempts to use carbonic acid in place of sulfuric acid have been unsuccessful. Only very small amounts of calcium phosphate dissolve in water saturated with CO_2, even under a considerable pressure. With sulfuric acid, the conversion into primary calcium phosphate is favored by the insolubility of gypsum, whereas with carbonic acid the reaction comes to an equilibrium, which is reached when only a very small proportion of the phosphate has gone into solution.

Various *condensed phosphoric acids* are derived from ordinary *orthophosphoric acid*, H_3PO_4, by the partial elimination of water ('condensation'). The most important of the condensed phosphoric acids are *pyrophosphoric acid* (diphosphoric acid), $H_4P_2O_7$, and the *metaphosphoric acids*. The metaphosphoric acids have the analytical composition HPO_3, but both the acids and their salts are polymerized: a *trimetaphosphoric acid*, $(HPO_3)_3$, and a *tetrametaphosphoric acid*, $(HPO_3)_4$, are known.

H_3PO_4 and $H_4P_2O_7$ are the initial members of the series of *polyphosphoric acids*. These have the general formula $H_{n+2}P_nO_{3n+1}$, e.g., $H_5P_3O_{10}$ (triphisphoric acid) and $H_6P_4O_{13}$ (tetraphosphoric acid). As is considered below, they have chain structures, and exist in the form of salts with very great chain lengths (high values of n). *Isometaphosphoric acid* is an isomer of tetrametaphosphoric acid; for its structure, see below.

Partial elimination of water from *orthophosphorous acid*, H_3PO_3, gives rise to *pyrophosphorous acid*, $H_4P_2O_5$, and *metaphosphorous acid*, HPO_2. It is likely that the latter, like metaphosphoric acid, has a molecular weight corresponding to a multiple of the simple formula, but there is no direct evidence on this point.

Phosphorus can be regarded as $+5$ valent in phosphoric acid, $+4$ valent in hypophosphoric acid, $+3$ valent in phosphorous acid, and $+1$ valent in hypophosphorous acid. These electro-chemical valences are assigned on the assumption of ion binding (although with considerable reservations), since the phosphorus and hydrogen together must bear twice as many positive charges as there are oxygen atoms in the compound.

Orthophosphoric acid, orthophosphorous acid and hypophosphorous acid all have the same number of hydrogen atoms. In orthophosphoric acid alone, however, can all three be replaced by metals; only *two* atoms in phosphorous acid, and *one* in hypophosphorous acid, are replaceable.

The relations between the oxyacids of phosphorus and their salts are indicated below.

Phosphates	$M^IH_2PO_4$ primary or dihydrogen orthophosphates	$M^I_2HPO_4$ secondary or hydrogen orthophosphates	$M^I_3PO_4$ tertiary or neutral orthophosphates
	$M^I_2H_2P_2O_7$ acid or dihydrogen pyrophosphates (diphosphates)	$M^I_4P_2O_7$ neutral pyrophosphates (diphosphates)	
	$[M^IPO_3]_n$ metaphosphates; i.e.	$M^I_3(PO_3)_3$ trimetaphosphates	$M^I_4(PO_3)_4$ tetrametaphosphates
Hypophosphates	$M^I_2H_2P_2O_6$ acid hypophosphates	$M^I_4P_2O_6$ neutral hypophosphates	
Phosphites	$M^IH_2PO_3$ primary or dihydrogen phosphites	$M^I_2HPO_3$ secondary or monohydrogen phosphites	No tertiary salts exist

Example of pyrophosphites: $Na_2H_2P_2O_5$
Metaphosphites M^IPO_2

Hypophosphites $M^IH_2PO_2$ (secondary and tertiary salts do not exist).

(ii) Constitution of the Acids of Phosphorus. If we regard the oxyacids of phosphorus in the first instance as *hydroxyl compounds*, and assume that hydrogen atoms which

are not replaceable by metals must be bound in a different manner from the replaceable hydrogen, we arrive at the following structural formulas

$$O=P{\overset{\displaystyle OH}{\underset{\displaystyle OH}{-OH}}}$$

$$O=P{\overset{\displaystyle OH}{\vert}}\quad O=P{\overset{\displaystyle OH}{\underset{\displaystyle OH}{}}}$$

$$O=P{\overset{\displaystyle H}{\underset{\displaystyle OH}{-OH}}}$$

$$O=P{\overset{\displaystyle H}{\underset{\displaystyle H}{-OH}}}$$

phosphoric acid   hypophosphoric acid   phosphorous acid   hypophosphorous acid

According to these formulas, phosphorus has a formal or structural valence of *five* in all its oxygen compounds, irrespective of the decrease in apparent electrovalence from $+5$ to $+1$ in the series of acids.

The assumption that, in phosphorous and hypophosphorous acids, the H-atoms which are not replaceable by metals are bound directly to phosphorus, was confirmed by Simon (1937), from observations of the Raman spectra. He found in these a line characteristic of the P—H bond, which was indeed the strongest line of the whole spectrum in the case of the acid H_3PO_2. As was shown by Schwarzenbach [*Helv. Chim. Acta, 19* (1936) 1043], the dissociation constants of the acids also reflect the manner in which the hydrogen is bound (cf. Table 96).

Salzer, who first prepared hypophosphoric acid in 1877, represented it with the doubled formula as a condensation product of phosphoric and phosphorous acids. Numerous investigators (Cornec, Parravano, Rosenheim, Treadwell, Nylén and Stelling, F. Bell and others) have established that hypophosphoric acid is correctly represented by the doubled formula $H_4P_2O_6$, although the acid cannot be obtained by the condensation of phosphoric acid with phosphorous acid. From this and other observations, it has been concluded that the structure of hypophosphoric acid is probably not

$$\underset{HO}{\overset{HO}{}}{\Big\rangle}O{=}P{-}O{-}P{\overset{OH}{\underset{OH}{}}}\ (I),\quad\text{but}\quad\underset{HO}{\overset{HO}{}}{\Big\rangle}O{=}P{-}P{=}O{\overset{OH}{\underset{OH}{}}}\ (II)$$

This has been confirmed by study of the Raman spectrum (Baudler, 1952). The *methyl ester* of hypophosphoric acid prepared by Baudler displays quite different properties from the compounds with the composition $P_2O_2(OR)_4$, which were prepared earlier by Arbusow (1931) and Nylén (1933) by different methods. These latter may perhaps have been esters of the compound (I), which is not known in the free state.

The esters of hypophosphoric acid also have molecular weights corresponding to the formula $P_2O_2(OR)_4$ (Arbusow, 1931, Nylén 1933). Some ebullioscopic measurements by Rosenheim (1906–10) seem to suggest, that in appropriate solvents, some of the compounds believed to be esters of hypophosphoric acid may partially dissociate into unsaturated molecules $PO(OR)_2$. Recent work (Remy and Falius, 1955) has shown that the method of Rosenheim yields not pure substances but mixtures of esters of various acids of phosphorus.

TABLE 96

DISSOCIATION CONSTANTS OF THE OXYACIDS OF PHOSPHORUS (at 18°)

	H_3PO_4	$H_4P_2O_7$	$H_4P_2O_6$	H_3PO_3	H_3PO_2
K_1	$7.5 \cdot 10^{-3}$	$> 10^{-1}$	$6.4 \cdot 10^{-3}$	$1.6 \cdot 10^{-2}$	$1.0 \cdot 10^{-2}$
K_2	$2.6 \cdot 10^{-8}$	$3.2 \cdot 10^{-2}$	$1.5 \cdot 10^{-3}$	$7.0 \cdot 10^{-7}$	—
K_3	$1.8 \cdot 10^{-12}$	$1.7 \cdot 10^{-6}$	$5.4 \cdot 10^{-8}$	—	—
K_4	—	$6.0 \cdot 10^{-9}$	$9.4 \cdot 10^{-11}$	—	—

It is perhaps not quite settled whether the oxyacids of phosphorus in the undissociated state are properly represented as hydroxyl compounds, in accordance with the above formulas. Raman spectra (Chap. 9, p. 308) have shown the presence of hydroxyl groups in HNO_3, and a considerable number of other acids (H_3AsO_4, H_2SO_4, H_2SO_5, H_2SeO_3, H_6TeO_6, $HClO_4$, HIO_3). It would therefore be a reasonable assumption that all oxyacids in the undissociated state are hydroxyl compounds. However, investigations of the Raman spectra of the phosphorus acids (Simon 1937–39, 1943; Venkateswaran 1936–38; Mathieu, 1942) have shown that the lines which in other cases characterize the hydroxyl group are either absent, or occur ordinarily weakly. On the other hand, there was a significant decrease in the vibrational frequencies of the O-atoms when D-atoms were substituted for H-atoms. This implies that the H- or D-atoms participate in the vibrations of the O-atoms, and must be bound directly to them. The absence, or marked weakening of the OH-bands is probably to be ascribed to the effect of 'hydrogen bonds' (cf. Chap. 9). In a similar manner the OH frequencies have not hitherto been observed in the Raman spectra of formic and acetic acids, and it is now generally assumed that the tendency of these compounds to form double molecules is due to the formation of hydrogen bonds.

On complete dissociation (e.g., in strongly alkaline solution), the following ions are formed by the (ortho) oxyacids of phosphorus:

$$[PO_4]^{3-}, \quad [P_2O_6]^{4-}, \quad [PO_3H]^{2-} \quad \text{and} \quad [PO_2H_2]^-.$$

Viewed from the standpoint of the coordination theory, it is apparent that in the ions of all these acids, and also in the phosphonium ion $[PH_4]^+$, phosphorus has the same coordination number; 4. The fact that phosphorous acid is only dibasic, and hypophosphorous acid monobasic can then be attributed to the coordinative quadrivalence of phosphorus. However, the coordination theory provides no explanation why hypophosphoric acid attains coordinative saturation by the union of two PO_3 groups and not, like the others, by transfer of one hydrogen atom to the phosphorus. In this respect, the Lewis-Langmuir theory takes us further. This leads to the following electronic formulas for the ions:

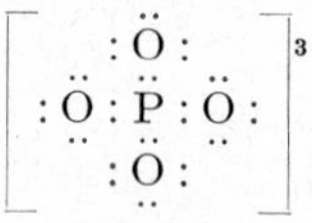

phosphate    hypophosphate 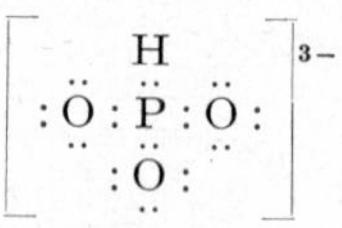   phosphite

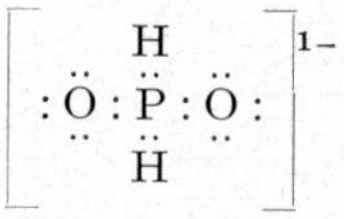

hypophosphite 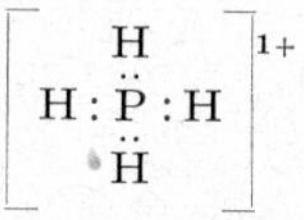   phosphonium

In each case, the valence electrons around the P and O atoms make up complete octets, the bonds being formed by electron pairs. A model satisfying this criterion can be set up neither for a $[PO_3]^{2-}$ nor a $[HPO_3]^-$ ion, and the non-existence of such ions is thus explicable.

As has been proved by structure determinations, the ions shown above exist as discrete structural units in the solid salts, as well as in solutions. We can thus assign the following coordination formulas to the salts.

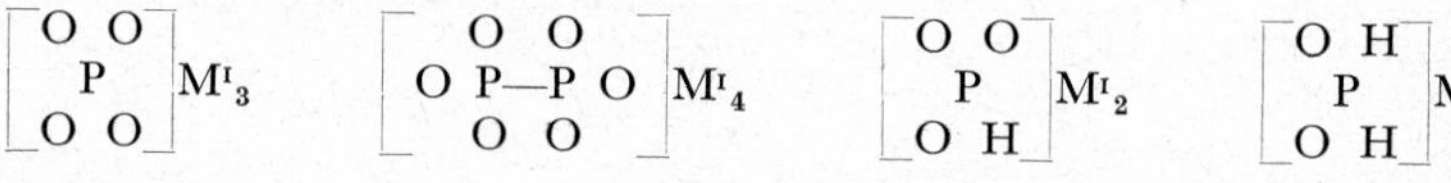

(iii) *Constitution of Condensed Phosphoric Acids and Phosphates.* Phosphorus also has the coordination number 4 in the condensed phosphoric acids and phosphates.

The structure of these compounds, as shown by the known experimental facts, is based upon the following principles:

[α] The compounds are built up entirely from [PO_4] coordination tetrahedra.

[β] Like the [SiO_4] tetrahedra in the silicates (and usually, indeed, all condensed RO_4 tetrahedra as far as is known), the PO_4 tetrahedra can be joined to one another only through a *single* O-atom. Tetrahedra can therefore be linked only through their corners, and not by their edges or faces.

[γ] If several OH-groups are joined to one atom of phosphorus, the first is strongly dissociated, the second weakly, and the third very weakly. Aqueous solutions of the corresponding alkali salts therefore show by their reaction (neutral, basic or strongly basic respectively) whether, in the anion of the salts, there are P-atoms which are linked only to one OH-group each, or whether there are also P-atoms carrying several OH-groups.

[δ] In every oxygen compound of electropositively 5-valent phosphorus there is at least one O-atom which is joined to the P-atom by a double bond, and is therefore unable to exercise additional principal valences.

The *polyphosphates*—i.e., the compounds of general formula $M^I_{n+2}[P_nO_{3n+1}]$, are built up from *chains* of PO_4 tetrahedra:

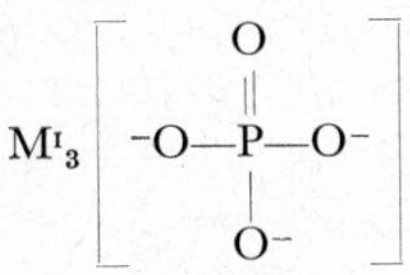

$M^I_3[PO_4]$
monophosphates
(orthophosphates)

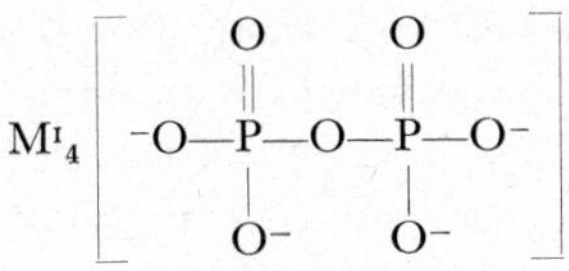

$M^I_4[P_2O_7]$
diphosphates
(pyrophosphates)

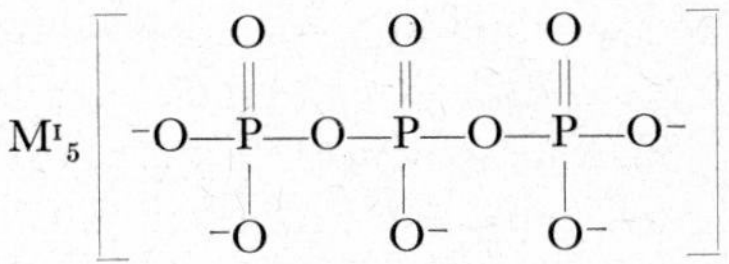

$M^I_5[P_3O_{10}]$
triphosphates

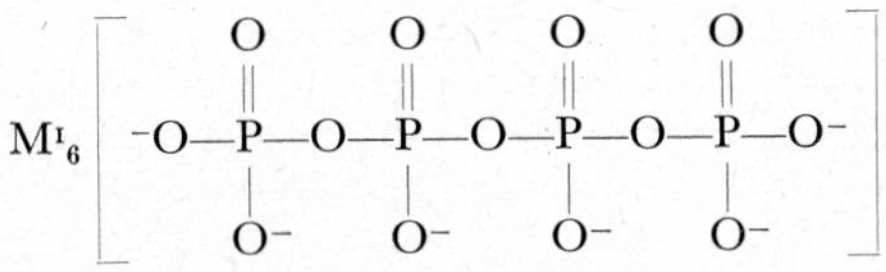

$M^I_6[P_4O_{13}]$ tetraphosphates

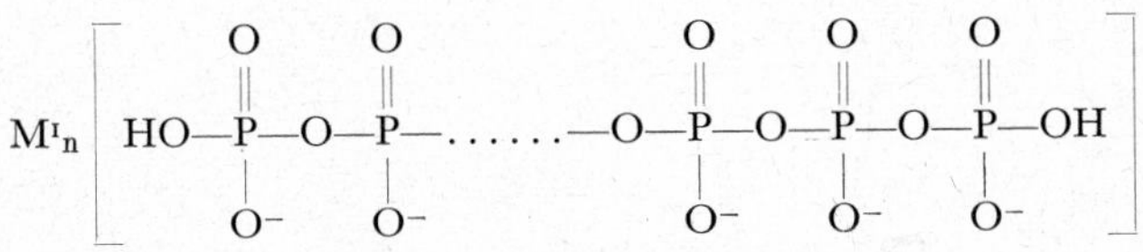

$M^I_n[P_nO_{3n-1}(OH)_2]$ n = 16 to 90
high-polymeric phosphates

Our understanding of the structure of these compounds is due largely to Thilo (1941 and later; see especially *Angew. Chem.*, 63 (1951) 508; 64 (1952) 510; *Z. anorg. Chem.*, 272 (1953) 182; also Topley, *Quart. Rev. Chem. Soc.*, 3 (1949) 345). Thilo showed, for example, that in the hydrolysis of the triphosphate the products were one ion of monophosphate and one of diphosphate, whereas the tetraphosphate yielded two diphosphate ions on hydrolysis.

Examples of *high-polymeric phosphates* are Graham's, Maddrell's, and Kurrol salts (see p. 641). Thilo observed that when NaH_2AsO_4 and NaH_2PO_4 were melted together, and subsequently annealed, it was possible to obtain salts with long-chain anions, in which the AsO_4 tetrahedra were incorporated in regular sequence among the PO_4 tetrahedra, so that hydrolysis yielded only mono-, di-, tri- or tetraphosphate ions, depending upon the ratio P:As. All polyphosphoric acids contain 2 weakly dissociating H atoms, and as many highly dissociating H atoms as there are P atoms. The chain structure of Kurrol salt has been confirmed by the X-ray work of Plieth and Wurster ($Z.$ *anorg. Chem.*, 267 (1952) 49).

Within the errors of analytical work, the high-polymeric phosphates often have the same composition as the metaphosphates—i.e., the apparent empirical formula M^IPO_3. However, the true metaphosphates do not have chain structures, but cyclic structures. Tri- and tetrametaphosphates are known:

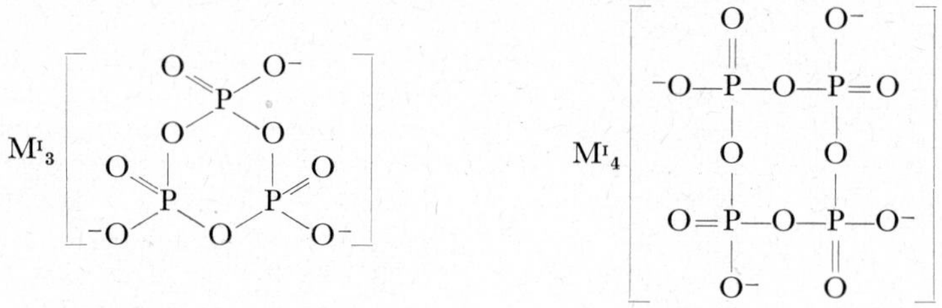

$M^I_3[P_3O_9]$ trimetaphosphates $M^I_4[P_4O_{12}]$ tetrametaphosphates

Hydrolysis of these compounds forms triphosphate and tetraphosphate, respectively, through opening of the ring. Since, in the acids corresponding to the tri- and tetrametaphosphates, there is only *one* OH-group linked to each P-atom, the acids are strong acids, and the metaphosphates have a neutral reaction in aqueous solution. The structures deduced by Thilo, essentially from their chemical reactions, are in harmony with the results of crystal structure determinations on metaphosphates.

Isometaphosphoric acid, $H_4P_4O_{12}$, has the same ring system as trimetaphosphoric acid, but the same molecular weight as tetrametaphosphoric acid. Its esters are formed by the cautious hydrolysis (more accurately, alcoholysis) of phosphorus pentoxide. In its normal (hexagonal-rhombohedral) form, structure determinations have shown the pentoxide to be built up from P_4O_{10} molecules, consisting of four PO_4 tetrahedra, linked together by sharing O-atoms. Hydrolysis can lead to the formation of either tetrametaphosphoric acid or isometaphosphoric acid, depending upon which of the six P—O—P links are ruptured in the process:

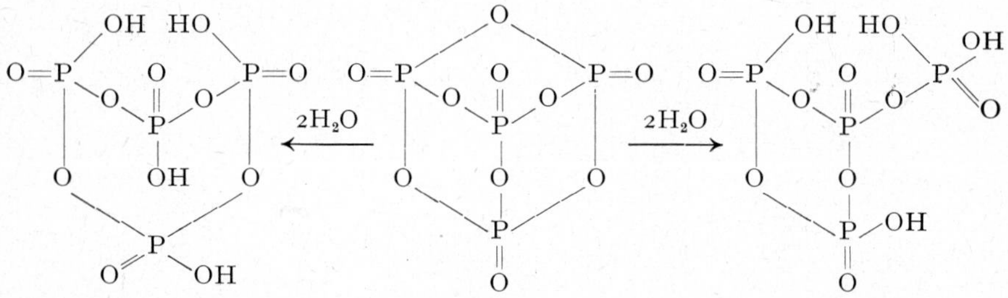

H$_4$P$_4$O$_{12}$ P$_4$O$_{10}$ H$_4$P$_4$O$_{12}$
Tetrametaphosphoric acid   Phosphorus pentoxide   Isometaphosphoric acid

Neither isometaphosphoric acid nor its salts have been isolated in the pure state. Its tetraethyl ester, $P_4O_8(OC_2H_5)_4$ is known, however. This is formed (together with the ester of tetrametaphosphoric acid) by the action of ether, $(C_2H_5)_2O$, on P_4O_{10}. Free isometaphosphoric acid is not stable. In aqueous solution it is very readily hydrolyzed further, breaking off the side chain and yielding trimetaphosphoric acid and monophosphoric acid:

$$H_4P_4O_{12} + H_2O = H_3P_3O_9 + H_3PO_4.$$

Its constitution follows from the fact that hydrolysis of the ester forms 1 molecule of phosphoric acid diethyl ester and 1 molecule of phosphoric acid, with 2 molecules of phosphoric acid monoethylester. (Thilo, 1951).

(*iv*) *Peroxyacids of Phosphorus.* There are also two peroxyacids derived from phosphoric acid: *peroxymonophosphoric acid*, H_3PO_5, and *peroxydiphosphoric acid*, $H_4P_2O_8$, derived from orthophosphoric and pyrophosphoric acids, respectively, by exchange of one oxygen atom for the peroxy group —O—O—. The phosphorus is therefore formally quinquevalent in these acids also.

$$
\begin{array}{cc}
\begin{array}{c}
\text{OH} \\
| \\
\text{O} \Rightarrow \text{P—O—OH} \\
| \\
\text{OH}
\end{array}
&
\begin{array}{c}
\text{OH} \qquad\quad \text{OH} \\
| \qquad\qquad | \\
\text{O=P—O—O—P=O} \\
| \qquad\qquad | \\
\text{OH} \qquad\quad \text{OH}
\end{array}
\\[2em]
\text{peroxymonophosphoric acid} & \text{peroxydiphosphoric acid}
\end{array}
$$

The salts of peroxydiphosphoric acid—e.g., the potassium salt $K_4[P_2O_8]$, which is quite stable in the solid state, can most conveniently be prepared electrolytically.

$$2PO_4^{3-} - 2e = P_2O_8^{4-}.$$

Peroxymonophosphoric acid can be obtained by the reactions

$$H_4P_2O_8 + H_2O = H_3PO_5 + H_3PO_4 \qquad \text{or} \qquad P_2O_5 + 2H_2O_2 + H_2O = 2H_3PO_5.$$

The peroxyphosphates are differentiated from the *phosphate peroxyhydrates*, which have been prepared in great numbers (e.g., $Na_2HPO_4 \cdot H_2O_2$, sodium hydrogen phosphate peroxyhydrate) by a few reactions which are characteristic for them—e.g., the oxidation of Mn^{++} to MnO_4^-.

According to Bockemüller (1934), a *peroxyphosphorous acid* is formed transiently, and in small concentrations, during the autoxidation of sodium hypophosphite in aqueous solution. It differs from the peroxyphosphoric acids in that, unlike them, it gives a yellow coloration with titanous sulfate solution. It is distinguished from hydrogen peroxide, however, in that it does not give the perchromic acid reaction.

(*v*) *Phosphorus pentoxide,* P_2O_5 or P_4O_{10}, the anhydride of phosphoric acid, is prepared technically by burning white phosphorus in iron drums, in an ample supply of air. The commercial product, a white snow-like material, is generally contaminated by lower oxides, from which it can be freed by sublimation in oxygen. It sublimes at 358.9°, and becomes vitreous (amorphous) when heated under pressure to higher temperatures. The amorphous modification melts at 565.6°. There are also three crystalline modifications. The ordinary form is hexagonal-rhombohedral (m.p. 422°), and transforms slowly into a rhombic variety (m.p. *ca.* 560°) when heated. This is unstable, however, and gradually changes into a modification which is stable at high temperatures and is probably tetragonal (m.p. 580°) (cf. Hill, Faust and Hendricks, *J. Am. Chem. Soc.*, 65 (1943) 794). In the vapor state, phosphorus pentoxide has the molecular formula P_4O_{10}. The hexagonal-rhombohedral form of phosphorus pentoxide is built up from P_4O_{10} molecules (see above). Pure phosphorus pentoxide is odorless. After exposure to light, it exhibits a green luminescence, which is stronger at low temperatures. It hisses when brought in contact with water (heat of solution 35 kcal per mol). It at once absorbs moisture from the air, and rapidly deliquesces to a syrup forming so-called 'metaphosphoric acid' (more precisely, a mixture of tetrametaphosphoric acid with trimetaphosphoric acid). Because of its great affinity for water, phosphorus pentoxide is used as a drying agent for gases and liquids, and as a dehydrating agent in chemical reactions, especially in organic chemistry.

(*vi*) *Phosphorus trioxide*, P_2O_3 or P_4O_6, the anhydride of phosphorous acid, is formed when phosphorus burns in a limited supply of air, and can be freed from pentoxide formed at the same time by reason of its greater volatility. It forms a white, waxy, crystalline mass, (density 2.13, m.p. 22.5°, b.p. 173°); the liquid is fairly volatile, its vapor pressure being 37.5 mm at 70°. The molecular weight, in solution and in the vapor state, corresponds to the molecular complexity P_4O_6. Above 210°, it decomposes, forming phosphorus tetroxide by the reaction $2P_4O_6 = 2P + 3P_2O_4$ It is converted to the pentoxide by heating it in a stream of air, and combines slowly with oxygen even at ordinary temperature. At low pressures this oxidation is accompanied by a glow. Phosphorus trioxide vapor produces ionization of the surrounding air. Like phosphorus, the trioxide is *very poisonous*.

Phosphorus trioxide reacts slowly with cold water, forming phosphorous acid. With hot water, vigorous decomposition occurs, phosphine and phosphoric acid being formed. It also reacts vigorously with chlorine and bromine, and, above 150°, with sulfur also. Reaction with hydrogen chloride yields phosphorus trichloride and phosphorous acid.

$$P_4O_6 + 6HCl = 2H_3PO_3 + 2PCl_3$$

(*vii*) *Phosphorus tetroxide*, formed by the thermal decomposition of phosphorus trioxide, forms colorless, lustrous crystals; it dissolves in water with the evolution of considerable heat, to form phosphorous acid and phosphoric acid in equimolar amounts. The reaction can best be considered as the hydrolysis of a compound containing tri- and pentavalent phosphorus (cf. Sb_2O_4, p. 668). In any case, phosphorus tetroxide cannot be regarded as the anhydride of hypophosphoric acid, which cannot be detected in the aqueous solution.

(*viii*) *Phosphoric acid (Orthophosphoric acid)*. As already mentioned, phosphorus pentoxide combines with water to form, first, metaphosphoric acid, HPO_3. In solution, and especially on boiling, or in the presence of hydrogen ions, this is gradually converted into ordinary (ortho-) phosphoric acid, H_3PO_4.

Orthophosphoric acid is frequently prepared by oxidizing phosphorus with nitric acid, as well as from phosphorus pentoxide and water. For technical purposes, it is chiefly prepared by decomposing phosphorite with moderately concentrated sulfuric acid.

Acid prepared by this latter method is very impure, but is good enough for many purposes—e.g., fertilizer manufacture. Pure phosphoric acid is always prepared from phosphorus; either white or red phosphorus may be used. The nitric acid concentration should not exceed 32%, as the reaction otherwise becomes too violent. At the end of the reaction, the excess of nitric acid is expelled by evaporation. If the phosphorus contains any arsenic, the somewhat diluted solution is treated with hydrogen sulfide to precipitate the arsenic. It is then evaporated down—finally in a platinum dish, since porcelain is too strongly attacked by hot concentrated phosphoric acid. When the syrup obtained at 150° solidifies spontaneously on cooling, it is inoculated with a crystal of the acid. During the evaporation of its solutions, orthophosphoric acid begins to condense, forming pyrophosphoric acid, as soon as the concentration exceeds that corresponding to the composition $[H_3O][PO_2(OH)_2]$. This condensation takes place even when evaporation is carried out at ordinary temperature in a high vacuum. It is thus not possible to prepare pure orthophosphoric acid by evaporation alone, although products with the correct composition are obtained. The pure acid may be obtained by filtering off the crystals before the remainder of the liquid solidifies (Simon, 1939). The anhydrous acid is extremely corrosive, and attacks even gold and platinum at 100°.

Orthophosphoric acid crystallizes in hard, colorless rhombic prisms, which deliquesce in air (density 1.88, m.p. 42.3°), forms no hydrates, but is miscible with water in all proportions. It comes into commerce as the syrupy concentrated solution (83–98%). The pure acid has pharmaceutical uses, and is permitted (in Britain and the United States) in foodstuffs (e.g., lemonade). It is added to hydrogen peroxide solutions as a stabilizer (cf. p. 55). Technical phosphoric acid is used not only for fertilizers, but for other purposes, as in dyeing, in enamel manufacture, the preparation of porcelain cement and dental fillings, etc.

Orthophosphoric acid is a moderately strong tribasic acid. Its apparent degree of dissociation (determined from the conductivity) is 6% in normal, and 12% in 0.1 normal solution at 18°.

Values for the individual dissociation constants for the three stages are given in Table 96, p. 631. In the titration of phosphoric acid with sodium hydroxide, the color change is obtained when exactly one equivalent of NaOH has been consumed if methyl orange is used as indicator, and when two equivalents of NaOH have been added, if phenolphthalein is used. Since it dissociates in stages (cf. Chap. 18), phosphoric acid may thus behave as either a monobasic or a dibasic acid, according to the hydrogen ion sensitivity of the indicator, even although it contains 3 replaceable hydrogen atoms altogether.

Towards perchloric acid, phosphoric acid behaves as an anhydrobase. It combines, to form a compound which, as was shown by Simon (1952), can be regarded as a tetrahydroxophosphonium perchlorate, $[P(OH)_4][ClO_4]$.

(ix) Phosphates (Orthophosphates). As a tribasic acid, phosphoric acid forms three series of phosphates. The primary phosphates $M^IH_2PO_4$ are all soluble in water. Of the secondary, $M^I_2HPO_4$, and tertiary, $M^I_3PO_4$, salts only the alkali salts are soluble.

Ignition converts the primary phosphates to metaphosphates, by loss of water, and the secondary salts to pyrophosphates. Tertiary phosphates are unchanged on ignition, unless they contain a positive radical derived from a volatile substance, e.g., $[NH_4]^+$. Thus, the neutral magnesium ammonium phosphate, $[NH_4]MgPO_4$, is converted into magnesium pyrophosphate, $Mg_2P_2O_7$, by ignition, and the secondary salt $[NH_4]NaHPO_4$ ('microcosmic salt') to sodium metaphosphate. Most phosphates are soluble in strong acids, although the phosphates of bismuth, tin, titanium, zirconium, hafnium, and thorium are insoluble.

The alkali phosphates are usually prepared by adding the correct quantity of phosphoric acid to alkali hydroxide or carbonate solutions. Insoluble phosphates are usually obtained from the soluble salts by double decomposition. Thus yellow silver phosphate is precipitated when silver nitrate is added to the solution of any orthophosphate—e.g.,

$$2Na_2HPO_4 + 3AgNO_3 = Ag_3PO_4 + 3NaNO_3 + NaH_2PO_4.$$

If a neutral or weakly basic phosphate solution is treated with a magnesium salt, a gelatinous precipitate of magnesium phosphate is produced.* If ammonium chloride and ammonia are added simultaneously, the crystalline ammonium magnesium phosphate is obtained:

$$Na_2HPO_4 + MgCl_2 + NH_3 + 6H_2O = (NH_4)MgPO_4 \cdot 6H_2O + 2NaCl$$

* On prolonged digestion under the precipitation solution, this becomes crystalline and then has the composition $Mg_3(PO_4)_2 \cdot 22H_2O$.

1. *Ammonium phosphate.* Ordinary ammonium phosphate is the secondary salt, $(NH_4)_2HPO_4$. It forms monoclinic crystals, density 1.62, which effloresce in the air, losing ammonia. It is very soluble in water and the solution is widely used as an analytical reagent. It is also used in pharmacy. It is used technically as a dressing for textiles, to make them non-inflammable, and also as a fertilizer.

2. *Sodium phosphate.* The (secondary) sodium phosphate generally used, $Na_2HPO_4 \cdot 12H_2O$, forms colorless prisms or plates with a cool, saline taste. It loses water and effloresces in air. The powdered salt, exposed in a thin layer to the air, in quantitatively converted in the course of a week or two to the dihydrate, $Na_2HPO_4 \cdot 2H_2O$. On heating, the 12-hydrate melts at 40°, becomes anhydrous at 100°, and changes to the pyrophosphate, $Na_4P_2O_7$ (m.p. 980°) when more strongly heated. It has many uses, similar to those of ammonium phosphate.

Tertiary sodium phosphate (trisodium phosphate) also crystallizes from solution at ordinary temperature as the dodecahydrate, but Menzel (1937) has shown that the crystals vary in composition between the limits $Na_3PO_4 \cdot \frac{1}{10}NaOH$ and $Na_3PO_4 \cdot \frac{1}{4}NaOH$, according to the NaOH content of the solution. The NaOH is built into the crystal lattice of the salt. Above 70°, and in contact with the solution, the NaOH is lost and the salt is transformed into the hexahydrate, $Na_3PO_4 \cdot 6H_2O$, of constant composition, or above 100°, into the hemihydrate, $Na_3PO_4 \cdot \frac{1}{2}H_2O$. Pure trisodium phosphate can thus be manufactured directly, by the evaporation of solutions which contain the components in the proportions corresponding to the formula. It is used as a very efficient water-softening agent, for the feed water of high pressure boilers.

3. *Microcosmic Salt.* When sodium phosphate solution is treated with ammonium chloride, *sodium ammonium hydrogen phosphate*, 'microcosmic salt', $Na(NH_4)HPO_4 \cdot 4H_2O$ is formed in colorless monoclinic crystals. It is used chiefly in analysis for 'microcosmic beads'.

4. *Potassium phosphate.* Primary potassium phosphate, KH_2PO_4, is manufactured on a large scale, as it is used as a fertilizer.

It is prepared by melting a mixture of phosphoric acid and potassium chloride together at a red heat. Potassium metaphosphate, $[KPO_3]_n$, is so formed, hydrogen chloride and water being eliminated. This salt is usually almost insoluble in water and acids, but if it is chilled by pouring it out on cold slabs, it becomes water soluble. The primary phosphate crystallizes from the solution in tetragonal prisms, which are fairly readily soluble in water, and insoluble in alcohol.

5. *Calcium phosphate.* In the form of *apatite*, its double compound with CaF_2 and $CaCl_2$, and of phosphorite (cf. p. 624), the normal or tertiary calcium phosphate constitutes the chief mode of occurrence of phosphorus in nature. It is present in all soils in small amounts, and is indispensible for the growth of plants and for animal life.

Tertiary calcium phosphate is extremely insoluble in water, but is slowly decomposed, by forming hydroxylapatite, $3Ca_3(PO_4)_2 \cdot Ca(OH)_2$ or $Ca_5(PO_4)_3(OH)^-$, which is likewise highly insoluble.

If (secondary) sodium phosphate is added to the neutral solution of a calcium salt, secondary calcium phosphate (calcium hydrogen phosphate) is thrown down as a white precipitate. On long standing under the solution this becomes distinctly crystalline, forming glistening colorless scales with the composition $CaHPO_4 \cdot 2H_2O$. Like the tertiary salt, the secondary phosphate is insoluble in water but readily soluble in acids.

Secondary calcium phosphate is gradually decomposed by water, with the formation of hydroxylapatite and primary calcium phosphate:

$$7CaHPO_4 + H_2O = Ca_5(PO_4)_3OH + 2Ca(H_2PO_4)_2$$

As a result of this hydrolysis, the $CaHPO_4$ precipitated from neutral solutions of calcium salts always contains a certain amount of hydroxylapatite, which remains in the form of slimy curds when the secondary calcium phosphate crystallizes. Formation of hydroxylapatite does not occur, however, if precipitation is effected by means of Na_2HPO_4 from a solution of $Ca(H_2PO_4)_2$:

$$Na_2HPO_4 + Ca(H_2PO_4)_2 = CaHPO_4 + 2NaH_2PO_4$$

(Gericke, *Angew. Chem.*, 65, (1953) 59). If ammonium chloride and ammonia are added before precipitation with sodium or ammonium phosphate, the product obtained is *hydroxylapatite*, $Ca_5(PO_4)_3(OH)$. Tertiary calcium phosphate (tricalcium phosphate) cannot be prepared in the pure state by precipitation from solution. The precipitates formerly regarded as such are variable mixtures of hydroxylapatite and secondary calcium phosphate, depending on the conditions of precipitation (Daneel 1930, Schleede 1932, Trömel 1932). A mixture with the appropriate composition can, however, be converted by ignition into true tricalcium phosphate. In addition to tricalcium phosphate, $Ca_3(PO_4)_2$, the compounds calcium pyrophosphate, $Ca_2P_2O_7$, and tetracalcium diphosphate, $Ca_4P_2O_9$, can also be prepared by fusion of $Ca(PO_3)_2$ with CaO. The last mentioned is converted into a mixture of hydroxylapatite and calcium oxide when this heated in the presence of atmospheric moisture—e.g., when ignited in an open electric furnace. Hydroxylapatite is an extraordinarily stable compound. When pure, it loses the elements of water only above $1500°$, and decomposes into tricalcium phosphate and tetracalcium diphosphate.

Primary calcium phosphate, $Ca(H_2PO_4)_2$, monocalcium phosphate, is readily soluble in water. It is formed by treating tricalcium phosphate or hydroxylapatite with acids, e.g., with sulfuric acid, in amounts corresponding to the reaction

$$Ca_3(PO_4)_2 + 2H_2SO_4 + 4H_2O = Ca(H_2PO_4)_2 + 2CaSO_4 \cdot 2H_2O \qquad \text{or}$$

$$2Ca_5(PO_4)_3(OH) + 7H_2SO_4 + 12H_2O = 3Ca(H_2PO_4)_2 + 7CaSO_4 \cdot 2H_2O$$

Reference has already been made to the importance of this reaction for the 'solubilization' of natural phosphates and bone ash, to increase the availability of phosphate ('superphosphate' fertilizers).

In place of treatment by sulfuric acid or phosphoric acid (p. 629), it is possible to open up phosphorite or hydroxylapatite for fertilizers by sintering them with suitable additions (chiefly calcined soda, chalk and silicate rocks), and grinding them finely. The product, known as 'Rhenania phosphate', contains *rhenanite* as principal constituent, and also $NaCaPO_4$. The former contains phosphoric acid in a form similar to that of basic slag, but involves Na_3PO_4 as well as $Ca_3(PO_4)_2$ and Ca_2SiO_4 in its constitution. Its X-ray diffraction pattern indicates that its structure is related to that of apatite. Like apatite and hydroxylapatite, it is insoluble in water, but it differs from these in that it is attacked by organic acids such as are secreted by the root hairs of plants. The phosphoric acid contained in it is therefore available as plant food. A number of other compounds, differing in composition to a greater or less extent from rhenanite, have a similar crystal structure—e.g., the compound $Ca_{16}(PO_4)_4(SiO_4)_5$ and the high temperature modification of $KCaPO_4$ (unlike $NaCaPO_4$, which is isotypic with K_2SO_4). The

SiO_2 in sodium calcium rhenanite can be partially or completely replaced by CO_2, the 'carbonate rhenanite' having a mean composition $Na_6Ca_4(PO_4)_4CO_3$ (Franck and Bredig, Z. anorg. Chem., 230 (1936) 1). The mixed salt $NaCaPO_4$ is the only compound appearing in the Na_2O–CaO–P_2O_5 system, whereas in the K_2O–CaO–P_2O_5 system the compounds $KCa_4(PO_4)_3$ (with the apatite structure) and $K_2CaP_2O_7$ have been discovered in addition to the mixed salt $KCaPO_4$. This last can incorporate K_2CO_3 and $CaCO_3$ in its structure, and then does not undergo a polymorphic change which otherwise appears at 705° (Franck and Bredig, 1938).

The fertilizer industry represents by far the greatest field of application of the calcium phosphates. Other applications are of apatite, in the production of enamels and opal glasses, and of secondary calcium phosphate as an addition to cattle foods and for pharmaceutical products. A mixture of primary and secondary calcium phosphates is extensively used in place of cream of tartar (potassium hydrogen tartrate) for the production of baking powders.

(x) *Pyrophosphoric acid and the Pyrophosphates.* Pure orthophosphoric acid begins to lose water a little above its melting point. On prolonged heating at 200–300°, it undergoes complete conversion to *pyrophosphoric acid*, $H_4P_2O_7$. This is a colorless, glassy mass, readily soluble in water. It gives a white precipitate with silver nitrate, whereas orthophosphoric acid gives a yellow precipitate. It is only slowly reconverted into orthophosphoric acid by cold water, but rapidly by hot, especially in the presence of nitric acid.

Pyrophosphoric acid is a tetrabasic acid, but forms only *two* series of salts—namely the *acid*, $M^I_2H_2P_2O_7$, and *neutral*, $M^I_4P_2O_7$, pyrophosphates, respectively. The acid pyrophosphates are soluble in water, mostly with a weakly acid reaction. Of the neutral pyrophosphates, only the alkali salts are soluble. Their solutions react weakly alkaline, through hydrolysis. Pyrophosphoric acid is a distinctly stronger acid than orthophosphoric acid (cf. Table 96, p. 631). It is a general rule that the strength of acids increases as they undergo condensation.

(xi) *Metaphosphoric acid and the Metaphosphates.* If phosphoric acid is further heated after the expulsion of one molecule of water, *metaphosphoric acid*, $(HPO_3)_n$ (n = 3 or 4) is gradually formed. This exists as a hard, vitreous mass (acidum phosphoricum glaciale) which dissolves in water with a crackling noise. The solution, like that of pyrophosphoric acid, gives a white precipitate with silver nitrate, but is distinguished by its property of coagulating albumen. The solution is gradually converted to ordinary phosphoric acid—on prolonged standing, or more rapidly on boiling, especially in presence of strong acids. Metaphosphoric acid is perceptibly volatile at red heat.

Its salts, the *metaphosphates*, with the analytical composition M^IPO_3, are obtained by igniting primary phosphates. The formation of sodium metaphosphate by ignition of 'microcosmic salt' is important:

$$(NH_4)NaHPO_4 = NaPO_3 + NH_3 + H_2O.$$

If the metaphosphate is then heated with a metal oxide, the oxide is dissolved, with the formation of a mixed phosphate—e.g.,

$$NaPO_3 + CuO = NaCuPO_4.$$

Since the phosphates of the heavy metals are often characteristically colored, the reaction finds an application in analysis ('microcosmic beads').

As has already been stated, metaphosphoric acid and its salts are not present, either in solution or in the melt, in the form of simple molecules HPO_3 or M^IPO_3, but are always polymerized. The true metaphosphates have the cyclic structures shown on p. 634 (tri- and tetrametaphosphoric acids; tri- and tetrametaphosphates). Solutions of metaphosphoric acid gradually undergo changes due essentially to hydration to form orthophosphoric acid. If hydrolysis is carried out very carefully, however, it may be limited to a hydrolytic opening of the rings, to form polyphosphoric acids—e.g.,

$$(HPO_3)_3 + H_2O = H_5P_3O_{10}$$

trimetaphosphoric   triphosphoric acid
acid

The tri- and tetrametaphosphates are formed from the dihydrogen monophosphates, when water is expelled from the latter by ignition. It appears that the size and the charge on the cation are the factors that determine whether the product is the trimetaphosphate or the tetrametaphosphate. Aqueous solutions of tri- and tetrametaphosphates are neutral in reaction, since the cyclic metaphosphoric acids are strong acids.

$(NaPO_3)_n$ can be obtained from the melt in 3 crystalline forms. $Na_4P_2O_7$, sodium pyrophosphate, probably exists in no less than 5 different crystalline modifications, although only one can be obtained at ordinary temperature. By melting $Na_4P_2O_7$ with $NaPO_3$, the compound $Na_5P_3O_{10}$ is obtained, which exists at ordinary temperature in 2 crystalline forms (Partridge, Hicks and Smith, *J. Am. Chem. Soc.*, 63 (1941) 454). Thilo and Raetz (1949) have shown that $Na_5P_3O_{10}$ can also be formed by hydrolysis of the cyclic trimetaphosphate anion.

(*xii*) *Polyphosphoric acid and the Polyphosphates.* The polyphosphoric acids are derived from orthophosphoric acid (monophosphoric acid) by elimination of water, as are the metaphosphoric acids; the condensation does not involve ring closure, however, but chain formation. The simplest member of the series is pyrophosphoric acid (diphosphoric acid, $H_4P_2O_7$). *Triphosphoric acid*, $H_5P_3O_{10}$, is known in the form of its salts—e.g.,

$$NaZn_2P_3O_{10} \cdot 9\tfrac{1}{2}H_2O \quad\text{and}\quad Na_5P_3O_{10} \cdot 8H_2O.$$

Formation of this latter compound was observed by Fleitmann (1848), but its existence remained doubtful until the more recent work of Huber (1936), Andress and Wüst (1938), and Hicks, Partridge and Smith (1941). It can be obtained by melting sodium pyrophosphate with Graham's salt (Schwarz, 1895), or more conveniently by the action of NaOH on $Na_3P_3O_9$ (sodium trimetaphosphate) (Thilo, 1949). The alkali triphosphates are readily soluble in water; triphosphates of other metals are sparingly soluble.

Sodium tetraphosphate, $Na_6P_4O_{13}$, can be obtained by treating sodium tetrametaphosphate, $Na_4P_4O_{12}$, with sodium hydroxide at 40°. Other tetraphosphates can be obtained from the sodium salt by double decomposition, but the salts are rather unstable.

The polyphosphates include certain salts with very high-molecular weight anions, which can be obtained by following controlled preparative methods. These were formerly thought to be metaphosphates,—namely Graham's salt, Maddrell's salt, and Kurrol salt.

Graham's salt is formed by quenching the melt which is obtained by heating sodium dihydrogen phosphate to 600° or above. It forms a clear, transparent, hygroscopic, vitreous mass. As G. Jander, (1942) and other workers have shown, by measurement of ionic weights, the anion of Graham's salt contains from about 30 to 90 PO_3 groups which, according to Thilo, are strung together in long, unbranched chains of PO_4 tetrahedra. It is quite soluble in water, although it dissolves only slowly at ordinary temperature. Graham's salt is used under the name of 'Calgon', as a water softening agent, especially to prevent the deposition of calcium salts when hard water is used in laundering or for boiler feed water. This use depends on the strong ion-adsorbing power of the high-polymeric compound. It was formerly accepted that it was limited by the formation of a soluble complex compound with calcium, to which was ascribed the analytical composition $Na_2[Ca_2P_6O_{18}]$. From the supposed composition of the calcium complex, Graham's salt was long regarded as sodium 'hexametaphosphate'; the trivial name is still used in industry.

Maddrell's salt and *Kurrol* salt are similar in structure to Graham's salt. Maddrell's salt is obtained by heating sodium dihydrogen orthophosphate or dihydrogen pyrophosphate to 250–500°. It exists in a low temperature form, with unbranched chains of 16–32 PO_4 tetrahedra, and a high temperature form with a chain length of 36–72 PO_4 tetrahedra (transition temperature 330°). X-ray structure analysis has shown that Maddrell's salt has a close analogy to wollastonite. According to a structure determination by Barnik (1936), wollastonite is built up either from $[Si_3O_9]^{6-}$ rings, or from $[SiO_3]_x$ spirals, with 3 SiO_3 groups per turn of the helix. Only the spiral arrangement can be considered for Maddrell's salt, since Thilo has shown that this has a chain structure and not a ring structure. Nevertheless, Maddrell's salt is converted very largely into trimetaphosphate when it is boiled with water—i.e., into a compound with a cyclic anion. Thilo was able to show that arsenatophosphates, in which three or more PO_4 tetrahedra were linked directly to one another, showed the same hydrolytic behavior, although it was certain that they had the chain structure. The assumption that long chains of PO_4 tetrahedra are coiled up into spirals is in harmony with the X-ray evidence, and makes it understandable that triphosphate rings should be formed in the hydrolysis.

Kurrol salt, which is obtained from a $NaPO_3$ melt under certain specified conditions of heating and cooling, also exists in two forms. It first crystallizes from the melt in leaflets ($d^{20} = 2.85$). When ground up, it breaks up into fine woolly fibers ($d^{20} = 2.56$). The same transformation takes place by the action of water, or on exposure to moist air. The two forms differ not only in density, but also in their X-ray diffraction pattern and in their chemical behavior. When the plates are heated to 400°, they form Maddrell's salt ($d^{20} = 2.67$). The fibers are converted into sodium trimetaphosphate ($d^{20} = 2.52$) by heating at 400°. Kurrol salt has ion-exchanging porperties, like the permutits (cf. p. 513).

(*xiii*) *Hypophosphoric acid and the Hypophosphates.* Hypophosphoric acid, $H_4P_2O_6$, is formed (together with phosphorous and phosphoric acids) by the slow oxidation of phosphorus in moist air. It is prepared by placing sticks of white phosphorus in glass tubes open at both ends, and drawn out to a point at the bottom. Several of these are placed in a glass funnel, and the solution which gradually drops from them—so called 'phosphatic acids'—is collected in a vessel placed underneath. When sodium acetate is added the sparingly soluble *acid sodium hypophosphate* $Na_2H_2P_2O_6 \cdot 6H_2O$, crystallizes out in monoclinic plates. The compound is also obtained in good yield (with sodium phosphate) by treating red phosphorus with a mixture of concentrated sodium hydroxide solution and hydrogen peroxide. The lead and barium salts are still less soluble. The free acid may be prepared from these by treatment with hydrogen sulfide or sulfuric acid. It crystallizes as the hydrate, $H_4P_2O_6 \cdot 2H_2O$ (m.p. 70°) in large, hygroscopic plates; the water is given up on drying in a vacuum or over phosphorus pentoxide. Hypophosphoric acid is a weaker reducing agent than phosphorous acid. It is not oxidized by (nitrogen oxide free) nitric acid, which merely catalyses the hydrolysis:

$$H_4P_2O_6 + H_2O = H_3PO_3 + H_3PO_4$$

(Blaser 1933). Hypophosphoric acid is also relatively stable towards reducing agents.

(*xiv*) *Phosphorous acid and the Phosphites. Phosphorous acid,* H_3PO_3, is most conveniently made by the hydrolysis of phosphorus trichloride*:

$$PCl_3 + 3HOH = 3HCl + H_3PO_3$$

It forms colorless crystals (density 1.65, m.p. about 74°), very soluble in water. It is reduced to phosphine by 'nascent' hydrogen, and can also be readily oxidized to phosphoric acid. It therefore precipitates the noble metals from their salts and reduces Hg(II) chloride to Hg(I) chloride. However, Blaser (1931) found that it was not oxidized by nitric acid if the latter were free from the oxides of nitrogen. Heated in the dry state, it decomposes into phosphoric acid and phosphine:

$$4H_3PO_3 = 3H_3PO_4 + PH_3 \qquad \text{(disproportionation)}$$

* A certain amount of disproportionation takes place at the same time.

Phosphorous acid forms only two series of salts, the *primary*, $M^1H_2PO_3$, and *secondary*, $M^1_2HPO_3$, phosphites. It is therefore assumed that one hydrogen atom differs from the others in being linked directly to the phosphorus. The alkali phosphites and calcium phosphite are readily soluble in water. The other phosphites are sparingly soluble.

(xv) *Pyrophosphorous acid*, $H_4P_2O_5$, (dibasic) and *Metaphosphorous acid*, HPO_2, cannot be obtained from the ordinary acid, which decomposes when heated. They can be obtained by other means, but possess no particular importance.

(xvi) *Hypophosphorous acid and the Hypophosphites*. Hypophosphorous acid, H_3PO_2, is obtained, in the form of its salts, together with phosphine, by dissolving phosphorus in potassium hydroxide (cf. p. 626), sodium hydroxide, or barium hydroxide solution. The free acid can be obtained by the action of sulfuric acid on the barium salt, $Ba(H_2PO_2)_2 \cdot H_2O$, which is readily purified by recrystallization. It can also be prepared by passing a solution of sodium hypophosphite through a column of cation-exchanging resin, in its H^+-ion enriched form. The free acid crystallizes in colorless leaflets (density 1.49) from cold concentrated solutions. It melts at 26.5°, or considerably lower if traces of water are present. It is reduced by 'nascent' hydrogen to phosphine. On the other hand, it is itself a powerful reducing agent, and so precipitates gold and silver from solutions of their salts. It is thereby converted first into phosphorous acid. Its reaction with copper salt solutions is characteristic: it precipitates *copper hydride* (cf. Vol. II), whereas phosphorous acid deposits *copper*, but only on boiling.

Hypophosphorous acid forms only one series of salts, the hypophosphites $M^1H_2PO_2$. All are soluble in water, and the alkali salts dissolve also in alcohol.

(b) Phosphorus Halides

(i) *General*. Phosphorus forms three chlorides. The lowest, PCl_2 (or P_2Cl_4?) is not readily obtained, and is of only theoretical interest. The ordinary product of the combustion of phosphorus in chlorine is the *trichloride*, PCl_3. By reaction with an excess of chlorine, this takes up a further molecule of Cl_2, forming *phosphorus pentachloride*, PCl_5.

Phosphorus trichloride is hydrolyzed by water to hydrogen chloride and phosphorous acid (cf. p. 642). Phosphorus can thus be regarded as +3 valent in both phosphorus trichloride and phosphorous acid.

In *phosphorus pentachloride*, phosphorus is pentavalent both structurally and in the electrochemical sense. This has been questioned from time to time, the compound being regarded as an addition compound of the trichloride with chlorine, $Cl_2 \cdot PCl_3$. In the vapor state above 300°, it is completely dissociated into the trichloride and free chlorine, and even although the equilibrium $PCl_3 + Cl_2 \rightleftharpoons PCl_5 + 31$ kcal is displaced considerably towards the right as the temperature is reduced, the products of dissociation are present in perceptible amount even at ordinary temperature. However, the structure as determined by means of electron diffraction (Rouault, 1938) shows definitely that phosphorus pentachloride is a normal valence compound, with pentavalent phosphorus. In *gaseous* phosphorus pentachloride, the five Cl-atoms are arranged around the P-atom in the form of a trigonal bipyramid. Interatomic distances are $P \leftrightarrow Cl$ (basal triangle) 2.10 Å, $P \leftrightarrow Cl$ (apex) 2.25 Å, $Cl \leftrightarrow Cl$ 3.08 Å. The same structure was found by Braune (1937) and Brockway (1938) for PF_5 (with $P \leftrightarrow F = 1.57$ Å). The structure of phosphorus pentachloride is the same in the liquid state also, but in the *solid state*, Raman spectroscopic and X-ray studies agree in finding a structure built up from tetrahedral $[PCl_4]^+$ and octahedral $[PCl_6]^-$ groups. (Moureu, 1937; Powell, 1940). The structural change undergone by PCl_5 in passing from the liquid to the solid state explains a number of peculiar phenomena associated with its fusion and solidification. (MOUREU, *Compt. rend.*, 203 (1936) 257, 205 (1937) 545). Thus it has no sharp melting point or freezing point. Moreover, there is a clearly marked inflection in the heating- or cooling-curve around 115°, implying a maximum in the specific heat at this temperature. Its electrical conductivity is also considerably greater in the solid state than in the melt, and its dielectric constant decreases on melting, without, however, falling to 0. It would appear from this that the two forms PCl_5 and $[PCl_4][PCl_6]$ coexist in the liquid phosphorus pentachloride, This would afford an explanation why, in its reactions with unsaturated organic compounds, PCl_5 often gives up two Cl-atoms, being converted to PCl_3, whereas in other cases it adds on as a —PCl_4 group (Bergmann, 1930). Powell (1941) and van Driel (1943) found a related structure for PBr_5, as did Francois (1941) for $SbCl_5$. In the solid state these compounds are built up from $[PBr_4]^+$ or $[SbCl_4]^+$ tetrahedra and isolated Br^- or Cl^- ions. Phosphorus pentabromide is known in two forms, lemon-yellow and red. The structure mentioned is that of the red (rhombic) modification.

Whereas the group $[PCl_6]^-$ apparently occurs only as a structural group in PCl_5, a whole series of compounds—the *hexafluorophosphates*—is known in which the complex ion $[PF_6]^-$ is present. Free *hexafluorophosphoric acid*, HPF_6, was first prepared by Lange (1928) by the action of hydrofluoric acid on phosphorus pentoxide. Its salts may be formed by neutralizing the acid, or by the reaction of PCl_5 with metallic fluorides. They are more stable than the hexafluoroarsenates and hexafluoroantimonates, which were prepared by Marignac as early as 1868.

The sodium salt of hexafluorophosphoric acid crystallizes from solution as the rhombic monohydrate $(d = 2.39)$. The anhydrous compound $Na[PF_6]$ $(d = 2.51)$ is cubic. The cations and the centres of the $[PF_6]^-$ groups occupy the lattice positions of a structure of NaCl type. $K[PF_6]$ and $Tl[PF_6]$, which are isotypic with one another, differ structurally from $Na[PF_6]$ in the way the $[PF_6]^-$ octahedra are oriented with respect to the body diagonal of the unit cell (Bode, 1951–52). In the reaction between HF and P_2O_5, the *oxofluorophosphoric acids*, $H[PO_2F_2]$ and $H_2[PO_3F]$ are formed, in addition to $H[PF_6]$. Dioxofluorophosphoric acid rapidly undergoes hydrolysis, but trioxomonofluorophosphoric acid is very stable, and the reactions of the $[PO_3F]^=$ ion closely resemble those of the $[SO_4]^=$ ion.

Phosphorus is capable of combining with more than 5 halogen atoms even in neutral molecules. Thus, there are compounds with the composition PBr_7, PCl_2Br_7, PCl_3Br_4, PCl_3Br_8, etc. These undoubtedly belong to the class of polyhalides (cf. p. 797).

Table 97 provides a summary of the simple phosphorus halides.

TABLE 97

PHOSPHORUS HALIDES

—	PF_3 Colorless gas liter wt. 3.926 b.p. —95°, m.p. —160°	PF_5 Colorless gas b.p. —75°, m.p. —83°
PCl_2 $(P_2Cl_4?)$ Colorless liquid m.p. —28°, b.p. 180°	PCl_3 Colorless liquid m.p. —91°, b.p. 75.9°	PCl_5 Colorless tetragonal crystals
PBr_2?	PBr_3 Colorless liquid m.p. —41.5°, b.p. 176°	PBr_5 Orange rhombic needles m.p. ?, b.p. 106°
P_2I_4 Bright orange triclinic prisms m.p. 110°	PI_3 Red hexagonal prisms or plates m.p. 61°	PI_5 Deep red prisms

(*ii*) *Phosphorus dichloride*, PCl_2(or P_2Cl_4), was prepared by Besson, by the action of a silent electric discharge on a mixture of PCl_3 and hydrogen. It is a colorless, oily liquid, smelling strongly of phosphorus (m.p. —28°, b.p. about 180°). It undergoes slow decomposition even at ordinary temperature. This is accelerated by light and heat.

Vapor density determinations indicate a molecular weight corresponding to the formula P_2I_4 for the corresponding iodide. It is probable that the formula of the chloride should be doubled also.

(*iii*) *Phosphorus trichloride*, PCl_3, is prepared by burning white phosphorus in dry chlorine. The phosphorus inflames spontaneously, and burns with a yellowish flame, and the trichloride distils away. It is freed from admixed pentachloride by treating it with white phosphorus. Phosphorus trichloride is a colorless liquid (density 1.57) which fumes strongly in air. It readily adds on to other substances, and thus behaves as if strongly unsaturated. It is converted into $POCl_3$ by substances which readily give up oxygen, or slowly by molecular oxygen itself. Sulfur converts it to $PSCl_3$. It combines with chlorine, forming the pentachloride.

(*iv*) *Phosphorus pentachloride*, PCl_5, is obtained by the action of an excess of chlorine on phosphorus trichloride. It is convenient to allow the trichloride to drop slowly into a wide-mouthed vessel, into which chlorine is led simultaneously, so as always to be in excess. The chlorine inlet tube must be of sufficiently wide bore to avoid becoming choked.

In the pure state, phosphorus pentachloride forms a white crystalline mass; it is ordinarily somewhat greenish because of partial dissociation into the trichloride and free chlorine. It sublimes above 100° without melting, the sublimation pressure reaching 1 atm. at 159°. Heated in a sealed tube, it melts between 159° and 160.5°. With a small quantity of water, it is hydrolyzed first to phosphorus oxychloride, $POCl_3$, and this in turn is converted into phosphoric acid.

$$PCl_5 + H_2O = POCl_3 + 2HCl \qquad\qquad (1)$$

$$POCl_3 + 3HOH = PO(OH)_3 + 3HCl \qquad\qquad (2)$$

Other compounds containing oxygen in the form of hydroxyl groups react in an analogous manner, forming $POCl_3$ and the chloride of the corresponding acid. e.g.

$$CH_3.CO.OH + PCl_5 = POCl_3 + HCl + CH_3.CO.Cl$$
$$\text{acetic acid} \qquad\qquad\qquad\qquad \text{acetyl chloride}$$

$$SO_2(OH)_2 + PCl_5 = POCl_3 + HCl + SO_2(OH)Cl$$
$$\text{sulfuric acid} \qquad\qquad\qquad\qquad \text{chlorsulfonic acid}$$

Many acids form the acid anhydride instead of the acid chloride—e.g., with boric acid,

$$2B(OH)_3 + 3PCl_5 = 3POCl_3 + 6HCl + B_2O_3$$

(*v*) *Phosphorus oxychloride*, $POCl_3$, a colorless, highly refractive liquid, which fumes in moist air (density 1.72, b.p. 108.7°, m.p. 1.3°), is best prepared pure by the action of phosphorus pentachloride on oxalic acid:

$$PCl_5 + H_2C_2O_4 = POCl_3 + 2HCl + CO_2 + CO$$

The pentachloride also reacts with phosphorus pentoxide to form the oxychloride:

$$3PCl_5 + P_2O_5 = 5POCl_3$$

Phosphorus oxychloride reacts with many organic compounds containing hydroxyl groups, in the same way as it does with water (eqn. (2) above):

$$POCl_3 + 3R—OH = PO(OH)_3 + 3R—Cl$$

It thus has a chlorinating action similar to phosphorus pentachloride, but is less energetic—a property which is frequently desirable.

The OH-groups of phosphoric acid may also be replaced by other halogens. Such compounds as POF_3 (phosphorus oxyfluoride, a colorless gas with a suffocating odor, m.p. —68°, b.p. —40°), $POBr_3$ (phosphorus oxybromide, colorless crystals, density 2.822, m.p. 55°, b.p. 193°), $POCl_2Br$ (m.p. 11°, b.p. 137.6°) are thereby formed. Oxyhalides derived from meta- and pyrophosphoric acids are also known. It has not been established whether the oxyacids of the lower oxidation states of phosphorus give rise to similar derivatives.

(c) Sulfides of Phosphorus

The sulfides of phosphorus are obtained by melting the components together in an atmosphere of carbon dioxide. Red phosphorus is used for this purpose, as the reaction of white phosphorus is too violent. Alternatively, sulfur and phosphorus are allowed to react in solution in carbon disulfide. According to the proportions of the components, the compounds P_4S_3, P_4S_7 or P_2S_5 are formed. If a solution of P_4S_3 and S in carbon disulfide is exposed to light, in the presence of a trace of iodine as catalyst, P_4S_5 may also be obtained. All these compounds may be recrystallized from suitable solvents, or by melting and re-freezing. When pure, they form well-formed yellow crystals.

The pentasulfide P_2S_5 is dimerized to P_4S_{10} in solution (CS_2), although the vapor density corresponds to the formula P_2S_5. It exists in two solid modifications, corresponding perhaps to the two formulas. The sulfides P_4S_3 and P_4S_7 have the same molecular complexity in solution and in the vapor state. P_4S_5 dissociates at high temperatures into P_4S_3 and P_4S_7. The other phosphorus sulfides can be distilled unchanged in the absence of air. The boiling points at atmospheric pressure and heats of formation (Treadwell, 1935) are

	P_4S_3	P_4S_7	P_2S_5	
Boiling point	408°	523°	514°	
Heat of formation	29.4	65.7		kcal per mol

The phosphorus sulfides are stable at ordinary temperature in dry air. P_4S_3 is unaffected by moisture. P_4S_7 and P_2S_5 are slowly decomposed by water, however, and therefore smell of hydrogen sulfide when exposed to moist air. In air, or when heated with substances which readily furnish oxygen, the phosphorus sulfides burn with a flame. P_4S_3 ignites in the air at temperatures as low as 100°.

Tetraphosphorus trisulfide, P_4S_3, also known as 'phosphorus sesquisulfide' was for a time used in the manufacture of 'strike anywhere' matches. The high boiling points of the phosphorus sulfides also makes them suitable for use in high temperature heating baths.

(d) Phosphorus Nitrogen Compounds

Of the binary compounds of nitrogen and phosphorus, only P_3N_5 is known in the pure state. Numerous compounds are known however, containing some other negative atoms or radicals bound to the phosphorus, in addition to nitrogen. Among these, the *phosphonitrilic halides*, and the *metaphosphimic acids* derived from them, are of particular interest, because of their well marked capacity for polymerization. The metaphosphimic acids are derived from the phosphonitrilic halides $(PNX_2)_n$ by replacement of the halogen X by hydroxyl. Degradation of these leads to *imidophosphoric acids* i.e. compounds derived from the phosphorus oxyacids by replacement of $=O$ by $=NH$. *Amidophosphoric acids* are derived from the oxyacids of phosphorus by replacing —OH by —NH_2. *Phospham*, $[PN_2H]_x$, may be cited as a phosphorus-nitrogen compound of unknown constitution.

(*i*) *Phosphorus nitride* (triphosphorus pentanitride), P_3N_5, can be obtained in the pure state (Stock, 1907) by treatment of P_2S_5 with ammonia, and prolonged heating of the product in ammonia at 850°. It is a colorless powder, without smell or taste, insoluble in the usual solvents and stable in air; density 2.51. When heated in a vacuum, it decomposes into its elements above 760°, and gives P_4 and NH_3 when heated in hydrogen. It burns when heated in chlorine or oxygen, and is hydrolyzed by water at 180° under pressure, forming H_3PO_4 and NH_3.

(ii) *Phosphonitrilic halides.* Phosphonitrilic dichloride $(PNCl_2)_n$ was prepared as long ago as 1832, by Liebig and Wöhler, as a crystalline sublimate formed by heating the product of reaction of NH_3 with PCl_5. It is better prepared by heating PCl_5 with NH_4Cl in an autoclave at 120° (blowing off the HCl gas liberated, as soon as the pressure exceeds 25 atm.) or, even more conveniently (Schenck, 1924), by reaction of PCl_5 dissolved in tetrachlorethane with NH_4Cl, at 135°:

$$nPCl_5 + nNH_4Cl = (PNCl_2)_n + 4nHCl.$$

As was first recognized by Stokes (1895), the product consists of a mixtures of isomers $(PNCl_2)_3$, $(PNCl_2)_4$, $(PNCl_2)_5$, etc., which can be separated by fractional distillation, and differ also not only in their melting and boiling points but in their crystal form and different solubilities in ether, carbon disulfide, chloroform, benzene, etc. The solid compounds are stable towards acids and caustic alkalis. In ether solution, however, they are hydrolyzed by water, forming metaphosphimic acids (see below).

From their properties, Stokes regarded these substance as cyclic compounds, with the following constitution

tri(phosphonitrilic dichloride)   tetra(phosphonitrilic dichloride)

The foregoing are the two most stable members of the series of cyclic polymers. Their structure was confirmed by Jaeger (1932) by X-ray means, although the question remains open whether the Cl-atoms are all bound to phosphorus, or whether they are evenly distributed between the N and P atoms.

When heated for many hours at 300°, the trimeric phosphonitrilic dichloride is converted into a highly polymerized, insoluble and nonvolatile substance of identical composition, with mechanical properties which markedly resemble those of a lightly vulcanized rubber. Meyer (1936) established by X-ray methods that it was built up from very long zig-zag chains, which take up a parallel orientation when the substance is stretched.

A more careful study of its behavior on extension has shown that the same forces are operative as in the case of rubber.

The *phosphonitrilic fluorides* and *bromides* have similar properties. The fluorine compounds, however, can not only be converted by polymerization into very high-molecular weight rubberlike substances, but may also be depolymerized under other conditions, to give substances of low molecular weight, of types not hitherto obtainable from the compounds containing chlorine or bromine only. Thus tetraphosphonitrilic dichlorohexafluoride $P_4N_4Cl_2F_6$ (colorless liquid, density 1.874, b.p. 105.8°) is converted into the corresponding phosphonitrilic-rubber when heated under high pressure, but is depolymerized when it is heated at atmospheric pressure, apparently according to $P_4N_4Cl_2F_6 \rightarrow 2P_2N_2ClF_3$ (Schmitz-Dumont, 1938).

(iii) *Metaphosphimic acids.* Compounds of the general formula $(PN(OH)_2)_n$ are obtained by the hydrolysis of the cyclic phosphonitrilic halides. Thus *trimetaphosphimic acid*, $(PN(OH)_2)_3$, is obtained from $(PNCl_2)_3$, and *tetrametaphosphimic acid*, $(PN(OH)_2)_4$ from $(PNCl_2)_4$. These are quite stable compounds, forming well characterized salts. Treatment with caustic alkalis, however, opens the ring, and converts them into compounds which differ from the metaphosphimic acids in that they contain an additional molecule of constitutive water. *Amides* of metaphosphimic acids are also known—e.g., $P_3N_3(OH)_4(NH_2)_2$, formed by shaking an ethereal solution of $P_3N_3Cl_4(NH_2)_2$ with sodium hydroxide. The last mentioned compound is obtained by the action of NH_3 on $(PNCl_2)_3$. The constitution of the metaphosphimic acids and their derivatives follows from their mode of formation.

When the salts of the metasphosphimic acids are boiled with 30% acetic acid, degradation occurs, with formation of *imidophosphoric acids*. The simplest of these, *imidometaphosphoric acid*, $O=P(NH) \cdot OH$, can also be obtained directly by the action of NH_3 on P_2O_5. It is a colorless mass, readily soluble in water and alcohol, and gradually hydrolyzed to H_3PO_4 and NH_3 when boiled.

(iv) *Phosphoric acid amides* can be obtained by hydrolysis of their esters, which may be prepared by treating the chlorides of phosphoric esters with ammonia. Thus *amidophosphoric acid*, $O=P(OH)_2 \cdot NH_2$, can be obtained by the reactions:

$$OP(OC_6H_5)_2Cl + 2NH_3 = OP(OC_6H_5)_2NH_2 + NH_4Cl;$$

$$OP(OC_6H_5)_2NH_2 + KOH + HOH = K[O_2P(OH) \cdot NH_2] + 2C_6H_5OH;$$

$$K[O_2P(OH) \cdot NH_2] + H^+ = K^+ + OP(OH)_2 \cdot NH_2$$

(carried out by means of a cation-exchange resin). *Diamidophosphoric acid*, $OP(OH)(NH_2)_2$, can be obtained similarly. Phosphoric acid triamide, $OP(NH_2)_3$, has never been obtained pure, if it exists at all. Amidophosphoric acid is a dibasic acid, crystallizing in colorless prisms which are soluble in water, and sparingly soluble in alcohol. It is gradually hydrolyzed in solution (more rapidly on boiling), forming H_3PO_4 and NH_3, or ammonium phosphate.

(v) *Phospham* $[PN_2H]_x$ is a loose white powder, insoluble in water, obtained when the products of reaction of NH_3 on PCl_5 are heated in the absence of air. Phospham is slowly oxidized to phosphorus pentoxide when heated in the air. It deflagrates with molten caustic potash, forming ammonia and potassium phosphate. The constitution of phospham, and the nature of a series of other phosphorus-nitrogen compounds resembling it in composition and mode of formation, are still unknown.

(e) Phosphorus Hydrides

Phosphorus forms two volatile hydrogen compounds: *phosphine*, PH_3, which is gaseous at ordinary temperature, and *diphosphine*, P_2H_4, which is liquid. It is very doubtful whether the so called solid phosphorus hydride, of supposed composition P_2H or $P_{12}H_6$, is a properly defined compound.

Like ammonia, PH_3 has the capacity to form salts (*phosphonium salts*) with acids—especially halogen acids. These are far less stable than the ammonium salts, however. Thus *phosphonium iodide*, formed by the reaction $PH_3 + HI = [PH_4]I$ (colorless tetragonal crystals, subliming at 80°), is decomposed by water. *Phosphonium bromide*, $[PH_4]Br$ (subl. temp. 30°) and *phosphonium chloride*, $[PH_4]Cl$ (subl. temp. —28°) are completely dissociated into their components in the vapor state, in spite of the low temperatures of sublimation.

P_2H_4 does not combine with acids. The supposed solid hydride P_2H is stated to combine with substances of basic character, such as piperidine, forming salts.

(i) *Phosphine*, PH_3, is obtained in the pure state by the action of water, or better, of potassium hydroxide, on phosphonium iodide:

$$[PH_4]I + H_2O = PH_3 + [H_3O]^+ + I^-; \quad \text{or} \quad [PH_4]I + OH^- = PH_3 + H_2O + I^-.$$

It may also be prepared by decomposing calcium phosphide, Ca_3P_2, or magnesium phosphide, Mg_3P_2, with hydrochloric acid. In this case, however, it contains traces of diphosphine as impurity, as also does the phosphine formed by the action of potassium hydroxide on white phosphorus (p. 626).

Phosphine is a colorless, poisonous gas, with an unpleasant garlic-like smell. The liquid boils at $-87.4°$ and solidifies at $-133°$. It is somewhat soluble in water (0.26 vol. in 1 vol. of water). Pure phosphine ignites in air at about $150°$, and burns to phosphoric acid:

$$PH_3 + 2O_2 = H_3PO_4$$

If the gas contains P_2H_4, it is spontaneously inflammable, but this property disappears when the diphosphine is condensed out by freezing.

Solid phosphine has a face-centered cubic molecular lattice, and displays two transition points, at $-184°$ and $-243°$, respectively.

Phosphine reacts quantitatively with solutions of $HgCl_2$, according to the equation

$$PH_3 + 3HgCl_2 = P(HgCl)_3 + 3HCl.$$

One method for determining phosphine depends on the acidimetric titration of the HCl so formed (Wilmet, 1927). According to Beyer (1943), very small amounts are better determined by iodometric titration of the $P(HgCl)_3$, which is oxidized by iodine in bicarbonate solutions to H_3PO_4.

(ii) *Diphosphine*, P_2H_4, formerly called 'liquid phosphorus hydride', is generally formed as a by-product in the preparation of phosphine. It boils at $51.7°$, melts at $-99°$, is spontaneously inflammable, and tends to decompose into phosphine and the so-called 'solid phosphorus hydride', especially when exposed to light or in contact with porous substances.

(iii) *'Solid phosphorus hydride'*, P_2H or $P_{12}H_6$, a bright yellow powder which is stable in air, is formed by the decomposition of the spontaneously inflammable hydride (i.e., mixtures of PH_3 and P_2H_4)—e.g., by the action of concentrated hydrochloric acid. It is insoluble in water and almost all other media, but dissolves with a yellow color in liquid phosphorus hydride and in molten phosphorus. Schenck found that solutions in the latter solvent, showed a freezing point depression corresponding to the formula $P_{12}H_6$. Royen (1936–38) has questioned, however, whether the 'solid hydride' is a true compound. He regards it rather as an adsorption complex of PH_3 on amorphous phosphorus, and considers the compounds hitherto supposed to be salts of P_2H to be 'polyanionic salts' of PH_3—i.e., as compounds of the type $[NH_4][PH_2 \cdot P_x]$. As he showed, these may be formed directly from NH_3, PH_3 and phosphorus.

(f) Phosphides

The compounds of phosphorus with the strongly electropositive elements are known as phosphides. They fall into two groups, the first of which comprise the compounds derived from phosphine by exchanging the hydrogen for an equivalent amount of a metal. The phosphides of the alkali metals, the alkaline earths and aluminum belong to this group. They are decomposed by water, or more readily by dilute acids, forming phosphine—e.g.,

$$Ca_3P_2 + 6H_2O = 3Ca(OH)_2 + 2PH_3$$

The second group comprises phosphides which resemble the intermetallic compounds in their composition and appearance, and which are for the most part unattacked, or but little attacked, by water and dilute acids. The phosphides of most heavy metals are of this type.

The phosphides are prepared either by treating the metal (or in some instances the metallic oxides or chlorides) with elementary phosphorus, or by reduction of phosphates with carbon or with the metal concerned. Thus magnesium phosphide is obtained when any phosphate, or other phosphorus compounds also, is heated with metallic magnesium.

Most phosphides are high-melting substances, and with the exception of the phosphides of the noble metals they are very stable towards heat.

7. Analytical (Phosphorus)

The presence of phosphorus in a substance in any form may be established by heating the finely powdered material with a few pieces of magnesium ribbon. The magnesium thereby combines with phosphorus, usually with incandescence, forming magnesium phosphide, Mg_3P_2. When this is moistened or breathed upon, it evolves phosphine, which may be recognized by its characteristic smell.

The detection of elementary white phosphorus is of considerable importance in *forensic chemistry*. It is usually effected by the characteristic luminescence of the phosphorus vapors which pass over in the course of steam distillation (Mitscherlich's test). The glow (which is visible in the dark) occurs principally at the point where steam is condensed out in the condenser, since it is at this point that the phosphorus vapors first come in contact with air.

Phosphoric acid (orthophosphoric acid) is usually detected by its reaction with ammonium molybdate (a yellow precipitate from concentrated nitric acid solution). It may also be precipitated in this form for quantitative determinations. Alternatively, magnesium ammonium phosphate $Mg(NH_4)PO_4 \cdot 6H_2O$ may be precipitated from ammoniacal solutions, and this is ignited to magnesium pyrophosphate, $Mg_2P_2O_7$, before weighing.

The phosphate content of phosphorites, fertilizers and other commercial materials may rapidly and conveniently be determined by conductometric titration in perchloric acid, with bismuthyl perchlorate as titrant (G. Jander, *Angew. Chem.*, 49 (1936) 106).

Orthophosphoric acid may be distinguished from pyro- and metaphosphoric acid by the color of the precipitate formed with silver nitrate. This, and the coagulation of albumen characteristic of metaphosphoric acid, have already been mentioned, as also have the differences in reducing power between the acids derived from the various oxidation states of phosphorus.

8. Arsenic (As)

(a) Occurrence

Arsenic occasionally occurs native, as 'Scherbenkobalt', but is found chiefly in the form of its compounds. The most important of these are its compounds with the metals (the *arsenides*), which usually occur in isomorphous mixture with sulfides. The most commonly found is *arsenopyrite* (mispickel), an isomorphous mixture of FeS_2 and $FeAs_2$, usually with the approximate composition FeAsS. Pure iron arsenide, $FeAs_2$, occurs as *arsenical pyrites* or *löllingite*.

Other arsenites worth noting are *cloanthite*, white nickel ore, $NiAs_2$, *niccolite*, red nickel ore, or nickel arsenide, NiAs, *smaltite*, $CoAs_2$, *cobaltite* or *cobalt glance*, CoAsS, and *gersdorffite*, NiAsS.

Another type is that of the arsenical 'Fahlerze' or tetrahydrites—e.g. *tennantite*, $4Cu_2S \cdot As_2S_3$, and *proustite* or light red silver ore, $3Ag_2S \cdot As_2S_3$. These may be considered to be *thioarsenites* (cf. p. 659). *Enargite*, $3Cu_2S \cdot As_2S_5$ or Cu_3AsS_4, is a *thioarsenate*.

Naturally occurring sulfides of arsenic are *realgar*, As_4S_4, and *orpiment* As_2S_3. The oxide As_2O_3 (flowers of arsenic, arsenolite) is found as a product of weathering of arsenical ores.

In general, arsenic is so widely distributed in nature that metals prepared from sulfide ores are almost invariably arsenical, and it is often a matter of considerable difficulty to free them completely from arsenic.

(b) History

The naturally occurring arsenic sulfides, realgar and orpiment, were already known to and mentioned by Aristotle and his pupil Theophrastus, under the names of σανδαράχη (sandarach) and ἀρσενικὸν. The name *auripigmentum*, referring to the golden yellow color of the mineral, is first found in Pliny. Dioscorides (1st century A.D.) already described the roasting of 'arsenic' (meaning the sulfide). The alchemists concerned themselves more closely with the sulfides of arsenic, and generally refer to the product of their roasting as 'white arsenic'. Many considered arsenic to be a fundamental constituent of all metals, similar to sulfur. The preparation of elementary arsenic was first described by Albertus Magnus (13th century). The ability of arsenic to whiten copper, thus apparently turning it into silver, contributed considerably to the strengthening of the belief in the transmutation of the metals which was characteristic of the alchemical period. The alchemists were already aware of the poisonous nature of 'white arsenic' (i.e., arsenic trioxide). The use of arsenic compounds in medicine dates back to the Iatrochemist Paracelsus.

(c) Preparation

Arsenic is prepared chiefly by heating arsenopyrite with exclusion of air, in retorts of refractory clay. The arsenic thereby sublimes off, and condenses in receivers:

$$FeAsS = FeS + As.$$

It is also practicable to employ arsenical pyrites. When heated in the absence of air this decomposes:

$$FeAs_2 = FeAs + As.$$

The iron arsenide which remains is roasted:

$$2FeAs + 3O_2 = Fe_2O_3 + As_2O_3.$$

The arsenic trioxide which sublimes off can be reduced to metallic arsenic by means of carbon. It is not usual to do so, however, since arsenious oxide as such finds far more applications.

(d) Properties

Like phosphorus, arsenic exists in several allotropic modifications. The ordinary form is *metallic* or *gray arsenic*. It forms a steel grey crystalline mass, with a metallic luster, is brittle and rather soft (hardness 3–4 on Mohs' scale), and conducts electricity (its specific conductivity being 4.19% of that of silver at 0°). Its density is 5.72, and magnetic susceptibility $\chi_A = -5.5 \cdot 10^{-6}$. Arsenic sublimes without melting at 633°. It may be melted under 36 atm. pressure, at about 817°. Up to about 800°, the vapor density corresponds to the molecular complexity As_4, and above 1700° to the formula As_2.

Arsenic vapor is colorless. Upon sudden cooling, it deposits *yellow arsenic*, consisting of transparent regular crystals, plastic and soft as wax, with a density of 1.97. Grey or metallic arsenic crystallizes in the rhombohedral system. Yellow arsenic is soluble in carbon disulfide, volatile in steam, and has strongly reducing properties—i.e., is very reminiscent of white phosphorus, but is much more unstable than the latter. It rapidly reverts to metallic arsenic when gently warmed, or under the action of light.

A third form of arsenic, the so-called *black arsenic*, is often formed as an intermediate in the transition of yellow arsenic into the metallic modification. Arsenic separates in this form, as a deposit or mirror on the wall, when arsenic hydride is passed through a heated glass tube. 'Black arsenic' (which may also be gray in color) is vitreous-amorphous, and its density varies from 4.7 to 5.1, according to the mode of preparation. Its magnetic susceptibility is $\chi_A = -23 \cdot 10^{-6}$. Above 270°, it undergoes monotropic transformation into the metallic modification.

A fourth form, *brown* arsenic, (density 3.7–4.1) is obtained by the reduction of solutions of arsenic trioxide in hydrochloric acid—e.g., by means of stannous chloride or hypophosphorous acid. It is not certain whether this represents a special modification, or merely a finer state of subdivision.

Arsenic burns with a bluish flame when it is heated in air, forming a white smoke of arsenic trioxide, As_2O_3. This is accompanied by a characteristic garlic-like smell. Arsenic is oxidized to arsenic acid by concentrated nitric acid or aqua regia, and to arsenious acid by means of dilute nitric acid or concentrated sulfuric acid, and also by boiling caustic alkalis. It combines with, and inflames in, chlorine, and also unites directly with many other elements when heated. It alloys very readily with the heavy metals, even with those with which it forms no compounds (i.e., it dissolves readily in the molten metal). The solidified alloys are mostly brittle, even when the arsenic content is very small. Thus an arsenic content of 1 part in 1000 is sufficient to make gold brittle.

The normal potential of arsenic, relative to the normal hydrogen electrode, is about —0.3 volt. Arsenic is thus less electropositive than hydrogen, and also than antimony or bismuth.

(e) Applications

Metallic arsenic is used largely for the production of gun shot (see under lead for the composition). The compounds of arsenic, especially the trioxide and the sulfides, are more important. The uses of these will be discussed under the compounds.

Care must be taken, because of the great *toxicity* of soluble arsenic compounds. They often prove fatal only after prolonged and very violent pains. On the other hand, the suitability of arsenic for pest destruction, and also, in a certain sense, its use in pharmacy, depend on its poisonous character.

9. Compounds of Arsenic

(a) Oxides and Hydroxides

Arsenic forms two oxides, *arsenic trioxide*, As_2O_3, and *arsenic pentoxide*, As_2O_5; both have the character of acid anhydrides. *Arsenious acid* is derived from arsenic trioxide; it is known only in solution, but numerous salts—the *arsenites*—are known. *Arsenic acid* is derived from arsenic pentoxide. Its salts are the *arsenates*.

The so-called *arsenic tetroxide*, As_2O_4, may be regarded as a mixed anhydride of arsenious and arsenic acids, or as a double compound of the trioxide and pentoxide.

The *heats of formation* (from the elements) are

$$As_2O_3 \quad 153.8 \qquad As_2O_4 \quad 174.3 \qquad As_2O_5 \quad 219.4 \text{ kcal per mol}$$

(*i*) *Arsenic trioxide* (arsenious oxide), As_2O_3 or As_4O_6, (commonly called 'arsenic'), is manufactured on a large scale by roasting arsenopyrite or other native arsenides:

$$2FeAsS + 5O_2 = Fe_2O_3 + 2SO_2 + As_2O_3.$$

It is obtained as a by-product in many smelting plants where arsenical ores are worked. The white smoke, consisting of arsenious oxide, formed in roasting such ores was already known in the Middle Ages as 'furnace fume'.

In order to deposit the arsenic trioxide, the roast gases are passed through long ducts of masonry. The poison meal which is deposited there is purified by sublimation. A loose white powder or a glassy product (vitreous arsenic) is thereby obtained, according to the temperature of condensation.

Arsenic trioxide usually comes into commerce as a colorless, vitreous mass, with a conchoidal fracture (density 3.70). This gradually becomes opaque and porcelain-like, in that it breaks up into an agglomerate of octahedral crystals (density 3.87).

The latter are obtained directly by crystallization from aqueous solution. The act of crystallization from solution is accompanied by a distinct luminescence. The regular octahedral form is the modification of arsenic trioxide stable at ordinary temperature (m.p. 310°, b.p. 465°). A second, monoclinic, crystalline modification also exists, which is stable above 221°. Both forms occur in nature—the cubic as *arsenolite*, and the monoclinic as *claudetite*.

Arsenic trioxide vaporizes readily when it is heated. The vapor condenses in the form of arsenic glass if the walls are at a temperature above 310°, and as a mass of octahedral crystals if the temperature is below this.

The vapor density at 800° corresponds to the formula As_4O_6, and attains the value corresponding to the formula As_2O_3 only above 1800°. Molecular weight determinations in nitrobenzene solution, by the ebullioscopic method, have also led to the doubled formula As_4O_6.

Arsenolite is also built up from As_4O_6 molecules, so arranged that their centroids occupy the lattice points of a diamond-type structure ($a = 11.06$ Å). The modification *claudetite*, which is stable at higher temperatures, forms a layer lattice. This contradicts the general rule, that the modification of any substance which is stable at high temperatures possesses a higher crystal symmetry than the low temperature modification. Arsenic trioxide also deviates from Ostwald's 'law of successive stages', and from the rule that a homogeneous substance must have a sharp melting point. According to Stranski, these anomalies are due to the fact that the sheets of the claudetite layer lattice are built up from atoms or ions in infinite array, and that a relatively high activation energy is involved in forming them by a reorganization of the As_4O_6 molecules (cf. Vol. II, Chap. 17), and in the converse process. In both processes, covalent bonds must be broken, whereas this is not required for the fusion or vaporization of arsenolite; in this, only the weak van der Waals forces between As_4O_6 molecules need be overcome. Traces of water vapor accelerate the transformation arsenolite → claudetite, since the intermediate formation of a compound with water provides the requisite activation energy. Stranski considers that the analogous phenomena which have been observed for Sb_4O_6, P_2O_5 or P_4O_{10}, As_4S_4, P, As and $(CN)_2$ are to be explained in the same way. The activation energy involved in the evaporation process is clearly without influence on the *equilibrium* between solid and gaseous phases. It therefore does not affect the *vapor pressure*. It does, however, exert a strong influence on the *rate* of evaporation. For most substances, the activation energy of the evaporation process is small as compared with the latent heat of evaporation. In some cases, however, the activation energy may be several times as great as the heat of evaporation. The cases just considered are of this class, the one energy quantity being relevant to one modification, and the other to a second modification. In such cases, the rates of evaporation of the two modifications may differ by many orders of magnitude. This can lead to the paradoxical observation that the *rate of condensation* of a vapor is speeded up by heating it. Stranski calls this phenomenon 'forced condensation'. It is readily demonstrated with arsenolite. A sealed, evacuated glass vessel containing a little arsenolite is provided with a spiral of platinum wire a few centimeters above the solid. This is immersed in a thermostat at about 200°, and the platinum spiral is heated to 700° or above; the arsenious oxide thereupon disappears from the bottom of the vessel, and deposits itself on the appreciably hotter zone of the glass wall, just opposite the spiral. The explanation of the phenomenon is as follows. In the neighborhood of the platinum spiral, the As_4O_6 molecules produced by vaporization of arsenolite are 'activated'; that is, some of them are dissociated into As_2O_3 molecules and are thereby enabled to condense to form claudetite. Because of the very high activation energy required for the vaporization of claudetite, this can evaporate again only extremely slowly, whereas the arsenolite vaporizes rapidly because the activation energy is smaller than the heat of evaporation.

Arsenic trioxide is easily reducible to metallic arsenic. Thus, if it is heated with charcoal or potassium cyanide in a small tube (ignition tube), arsenic formed by reduction is deposited in the colder part of the tube as a black shining ring (arsenic mirror)—Berzelius' test for arsenic. Metallic arsenic is thrown down as a black-brown precipitate by the action of stannous chloride on solutions of arsenic trioxide in an excess of hydrochloric acid. The Bettendorff arsenic reaction is based on this reaction. 'Nascent' hydrogen reduces arsenic trioxide in acid solution to arsine. Oxidizing agents convert it to arsenic acid, the oxidation occurring most readily in presence of alkali.

A solution of arsenic trioxide in the presence of sodium hydrogen carbonate consumes exactly as much iodine as is required to convert the arsenic from the trivalent to the penta-valent state.* This fact is utilized for the quantitative determination of the arsenic content of arsenic trioxide solutions.

Arsenic trioxide is moderately soluble in water, the vitreous form being much more soluble than the crystalline. The solubility of the latter is: at 0° 1.2; at 25° 2.1; and at 75° 6.0 g of As_2O_3 in 100 g of water. The solution has a sweetish taste, with an unpleasant metallic after-taste. Reference has already been made to the great toxicity of arsenic trioxide.

Less than 0.1 g of arsenic trioxide can be fatal, if it is taken into the stomach and not quickly removed by vomiting, or made innocuous by conversion to an insoluble compound (e.g., with magnesium oxide), or by adsorption, e.g., on freshly precipitated ferric hydroxide.

* The solution must not be treated with alkali *hydroxide* or *carbonate*, since iodine reacts with these to form hypoiodite (cf. p. 811). The concentration of hydroxyl ions formed by hydrolysis of the hydrogen carbonate suffices to push the equilibrium

$$AsO_3{}^{3-} + I_2 + 2OH^- \rightleftharpoons AsO_4{}^{3-} + 2I^- + H_2O$$

completely over to the right.

The organism has a noteworthy ability to accustom itself to some extent to this poison. Habitual arsenic eaters can in some cases tolerate several times the usual fatal dose without harm to their health.

Arsenic trioxide is widely used as a pesticide. It is also used as a preservative for birds' feathers, animal skins, etc. It is used in glass manufacture as a clarifying agent. It is also used in the manufacture of pigments, especially Schweinfurt green, and for medicinal purposes.

(*ii*) *Arsenious acid and the Arsenites.* Arsenious acid is formed by the union of arsenic trioxide with water. It is not, however, known in the free state, although it is demonstrably present in the aqueous solution, in equilibrium with its dissociation products. The formula to be assigned to arsenious acid has not yet been settled. The possibility that it should be formulated as the *meta-acid* $HAsO_2$ or $AsO(OH)$, or as a *hexahydroxoacid* $H_3[As(OH)_6]$, must also be considered in addition to the customary formulation as the ortho-acid, H_3AsO_3 or $As(OH)_3$ (which will be used for the present here). Arsenious acid can dissociate not only as an acid [eqn. (1)], but also as a base [eqn. (2)]:

$$As(OH)_3 \rightleftharpoons AsO_3^{3-} + 3H^+ \tag{1}$$

and
$$As(OH)_3 \rightleftharpoons As^{3+} + 3OH^- \tag{2}$$

The ratio of As^{3+} ions to AsO_3^{3-} ions is very strongly dependent upon the hydrogen ion (or hydroxyl ion) concentration of the solution, as follows from the application of the Mass Action law to the equilibria in the solution. Combining equations (1) and (2).

$$\frac{[As^{3+}]}{[AsO_3^{3-}][H^+]^6} = \text{const.},$$

since the product $[H^+][OH^-]$ is constant. Thus As^{3+} is present in appreciable amounts only in *strongly acid* solution.

Arsenious acid is a very weak acid, about as strong as boric acid. Its dissociation as a base is far weaker still. The first stage dissociation constants as an acid (K_a) and as a base (K_b) at 25° are

$$K_a = \frac{[H_2AsO_3^-][H^+]}{[H_3AsO_3]} = 6 \cdot 10^{-10} \qquad K_b = \frac{[As(OH)_2^+][OH^-]}{[As(OH)_3]} = \text{about } 10^{-14}$$

Many observations would appear to indicate that arsenious acid is present in solution in the meta-form, i.e., as $HAsO_2$. In this event, the constants given refer to the equilibria

$$AsO_2^- + H^+ \rightleftharpoons HAsO_2 \qquad \text{and} \qquad AsO^+ + OH^- \rightleftharpoons HAsO_2$$

In contrast with this, Brintzinger (1935) found by the dialysis method an ionic weight corresponding to the formula $As(OH)_6^{3-}$ or $H_2As(OH)_6^-$. If, on this basis, arsenious acid is formulated as $H_3As(OH)_6$ or $As(OH)_3(OH_2)_3$, the above dissociation constants can be referred to the equilibria

$$[As(OH)_4(OH_2)_2]^- + H^+ \rightleftharpoons As(OH)_3(OH_2)_3$$

and
$$[As(OH)_2(OH_2)_4]^+ + OH^- \rightleftharpoons As(OH)_3(OH_2)_3 + H_2O.$$

The salts of arsenious acid are the *arsenites*. The alkali arsenites are very soluble in water. The alkaline earth salts are less soluble, and the compounds of arsenious acid with the heavy metals are practically insoluble in water. Arsenious acid may also be retained by *adsorption* in considerable amounts on gelatinous metallic oxide hydrates, such as ferric oxide hydrate. The arsenites are decomposed by acids, since in acid solution the equilibria (1) and (2) above are so far displaced that the concentration of the arsenite ions is no longer great enough to reach the solubility product even of the least soluble salts.

Most of the salts of arsenious acid correspond to the type M^IAsO_2, and are thus derived not from the ortho-acid H_3AsO_3 or $H_3[As(OH)_6]$, but from the meta-acid $HAsO_2$*. This is one piece of evidence supporting the meta-acid formula. It is possible, however, that the predominant formation of primary salts arises from the very weakly acid nature of arsenic hydroxide, which inevitably leads to the hydrolysis of the secondary or tertiary salts in solution, forming the primary salts or their ions—e.g.,

$$K_3AsO_3 + 2HOH = 3K^+ + 2OH^- + H_2AsO_3^-$$

or, alternatively

$$K_3AsO_3 + 5HOH = 3K^+ + 2OH^- + H_2As(OH)_6^-.$$

The yellow sparingly soluble silver arsenite, Ag_3AsO_3 may be cited as an example of an orthoarsenite. The fine green copper arsenite, $Cu(AsO_2)_2$, was formerly used as a pigment (Scheele's green).

(iii) *Arsenic pentoxide, Arsenic acid and the Arsenates.* Unlike phosphorus pentoxide, *arsenic pentoxide* [arsenic(V) oxide], As_2O_5, cannot be obtained directly by oxidation of the trioxide with atmospheric oxygen, but only by the dehydration of its hydrate, $As_2O_5 \cdot 4H_2O$, or of arsenic acid. It is deliquescent in moist air, and recombines with water, forming arsenic acid.

Arsenic acid, H_3AsO_4, is obtained by oxidizing arsenic trioxide or arsenic by means of concentrated nitric acid or other strong oxidants. It separates from the solution, after evaporation, in colorless, deliquescent and usually very small crystals, of the composition $H_3AsO_4 \cdot \frac{1}{2}H_2O$.

Arsenic acid acts as a fairly strong oxidizing agent, though usually only in acid solutions. It is used as an oxidant in the organic dyestuff industry. It is also the starting material for the preparation of organic arsenic compounds which are important in medicine (e.g., salvarsan and neosalvarsan).

Arsenic acid is very soluble. At 20°, 100 g of water dissolve 630 g of arsenic acid, and at higher temperatures still more. It is a tribasic acid of moderate strength, rather weaker than phosphoric acid. Its dissociation constants are $K_1 = 5 \cdot 10^{-3}$, $K_2 = 4 \cdot 10^{-5}$, $K_3 = 6 \cdot 10^{-10}$.

The salts of arsenic acid, the *arsenates*, generally correspond to the phosphates in composition and solubilities. Many are isomorphous with the phosphates—e.g., *magnesium ammonium arsenate* $Mg(NH_4)AsO_4 \cdot 6H_2O$, obtained under conditions similar to those used for magnesium ammonium phosphate, and analogously converted by ignition into *magnesium pyroarsenate*, $Mg_2As_2O_7$. It is very suitable for the gravimetric determination of arsenic acid. The chocolate brown Ag_3AsO_4 is precipitated by silver nitrate (distinguished from the yellow phosphate). Secondary

* Arsenites of more complex composition are known, as well as the ortho- and meta-arsenites. Thus the sodium salts $NaAs_3O_5$, $NaAsO_2$, $Na_4As_2O_5 \cdot 7H_2O$ and $Na_4As_2O_5$ can be crystallized from solutions of varying composition (Nelson, 1941).

sodium arsenate, Na_2HAsO_4, crystallizes from solution as hydrates with the following ranges of stability:

$$\text{Ice} \underset{\longleftarrow}{\overset{-1.14°}{\longrightarrow}} \text{12-hydrate} \underset{\longleftarrow}{\overset{+20.5°}{\longrightarrow}} \text{7-hydrate} \underset{\longleftarrow}{\overset{56.3°}{\longrightarrow}} \text{5-hydrate} \underset{\longleftarrow}{\overset{67.4°}{\longrightarrow}} \text{1-hydrate}$$

$$\underset{\longleftarrow}{\overset{99.5°}{\longrightarrow}} \text{anhydrous salt}$$

Tertiary sodium arsenate (crystallizing as the 12-hydrate) displays a variability of composition similar to that of tertiary sodium phosphate (Menzel, 1937, see p. 638). The secondary sodium salt is used in medicine, and as a pesticide in vineyards in many regions. Other arsenates—e.g., calcium arsenate, $Ca_3(AsO_4)_2$,—also find application as insecticides in agriculture and forestry. Arsenates are prepared by oxidizing arsenites by means of atmospheric oxygen or by chemical reagents, such as hypochlorites, or by electrolytic methods.

Oxidation by atmospheric oxygen can be brought about by heating in air in the dry state, as well as in solution. Thus calcium arsenite becomes incandescent when heated in air, forming arsenate. As was shown by Reissaus (1931), the reaction depends on the decomposition, by a disproportionation process, of the arsenites when heated, e.g.,

$$5Ca_3[AsO_3]_2 = 3Ca_3[AsO_4]_2 + 4As + 6CaO.$$

This reaction may be used to prepare the arsenates, and also arsenic, in the pure state, by heating in a vacuum with the addition of the appropriate quantity of As_2O_3 to combine with the CaO:

$$3Ca_3[AsO_3]_2 + 2As_2O_3 = 3Ca_3[AsO_4]_2 + 4As$$

If it is heated in air, the arsenic is oxidized once more to As_2O_3. This combines with the liberated oxide, forming arsenite which in turn undergoes disproportionation, until all the arsenite has been transformed into arsenate, according to the total equation

$$Ca_3[AsO_3]_2 + O_2 = Ca_3[AsO_4]_2.$$

The interpretation of this cycle of reactions is also of some importance in the metallurgy of arsenical ores. Depending upon the composition of the system, the calcium arsenates deposited from aqueous solution are $CaH_4[AsO_4]_2$, $CaHAsO_4$, $Ca_5H_2[AsO_4]_4$ and $Ca_3[AsO_4]_2$ — the last three in hydrated form—and also a compound corresponding to hydroxylapatite, $Ca_5[AsO_4]_3(OH)$ (Pearce 1936; Guérin, 1941).

The formula H_3AsO_4, assigned above to arsenic acid, is based upon the formulas of the arsenates, which are completely analogous to the phosphates. The crystal structures of KH_2AsO_4 and of Ag_3AsO_4 (Helmholtz, 1942) show the salts to contain a tetrahedral AsO_4^{3-} group. On the other hand, the compound which crystallizes from aqueous solutions of arsenic pentoxide does not have the composition H_3AsO_4, but can be written as a hemihydrate $H_3AsO_4 \cdot \frac{1}{2}H_2O$, or as a tetrahydrate of arsenic pentoxide, $As_2O_5 \cdot 4H_2O$. On dehydration it first yields a compound $H_5As_3O_{10}$ or $3As_2O_5 \cdot 5H_2O$. This begins to decompose into arsenic pentoxide and water at 120°, but retains the last portions of the water so firmly that it must be heated to nearly 500° to dehydrate it completely. No compounds corresponding to the pyro- and metaphosphoric acids are formed in the dehydration. On the other hand, a higher hydrate, $As_2O_5 \cdot 7H_2O$, can be obtained from the solution at about —30° (Simon, 1927). It is not yet known whether this should be taken as a dihydrate of ordinary arsenic acid, $H_3AsO_4 \cdot 2H_2O$, or as a unibasic hydroxo-acid, $H[As(OH)_6]$.

Although no crystalline compound of the formula H_3AsO_4 is known, the existence of this species of arsenic acid in aqueous solutions is indicated by the Raman spectra of the solutions, as compared with those of their acid salts (Simon and Fehér, 1937). The Raman spectrum of almost anhydrous arsenic acid contains a strong band which is to be assigned to the O—H group; thus it is to be inferred that, like nitric acid, arsenic acid exists in two different forms—an 'ester' form, present in concentrated solutions, and a 'salt' form, present in dilute solutions. The same conclusion had previously been reached by G. Jander (1934), who showed that the ultraviolet absorption spectrum of arsenic acid (and also of silicic and telluric acids) was different in the undissociated state and in the form of the dissociated acid or its salts. According to Brintzinger, the arsenate and hydrogen arsenate ions are present in 1-normal solution in the following forms. At pH > 13, $[AsO_4(OH_2)_{12}]^{3-}$ at pH 7–11 $[HAsO_4(OH_2)_6]^{2-}$, and at pH < 6 $[H_2AsO_4(OH_2)_2]^-$.

(b) Halides of Arsenic

Table 98 gives a summary of the halides of arsenic. The lower stability of pentavalent arsenic, as compared with $+5$ phosphorus, shows itself here in the non-existence of the pentachloride, pentabromide and pentaiodide.

All the arsenic halides may be prepared by the direct union of the elements, although—especially in the case of the fluorides—it is frequently more convenient to use other methods of preparation. According to Montignie (1941), arsenic and iodine have only a limited mutual miscibility in mixtures with less than the atomic ratio As: I $= 1 : 2$. There is a eutectic (eutectic temp. 73°) at 88 atom per cent I. Hassel (1941) has shown by electron diffraction that in the molecules of AsI_3 vapor the atoms are arranged at the corners of a trigonal pyramid with As $\leftrightarrow$ I $= 2.54$ Å, angle I—As—I $= 98°\ 30'$.

The halides of arsenic have the property of forming addition compounds with other substances—e.g., ammonia. Arsenic pentafluoride forms double compounds of the types $AsF_5 \cdot M^IF$ and $AsF_5 \cdot 2M^IF$ with metallic fluorides (e.g., with KF). Compounds formed by the other halides are mostly of the type $2AsX_3 \cdot 3M^IX$.

TABLE 98

HALOGEN COMPOUNDS OF ARSENIC

—	AsF_3 Colorless liquid, density 3.01 m.p. —8.5°. b.p. 63°	AsF_5 Colorless gas m.p. —79.8°, b.p. —52.8°
—	$AsCl_3$ Colorless liquid, density 2.17 m.p. —16.0°, b.p. 130.2°	—
—	$AsBr_3$ Colorless columnar crystals m.p. 31.2°, b.p. 221°	—
AsI_2 Dark red prisms m.p. 124°	AsI_3 Red hexagonal plates m.p. 146°, b.p. ca. 400°	—

Arsenic trichloride, $AsCl_3$, a colorless liquid fuming in air, is formed when arsenic burns in chlorine. It is most simply prepared by passing dry hydrogen chloride over arsenic trioxide at 180–200°. It is only incompletely formed by dissolving arsenic trioxide in aqueous hydrochloric acid, since the equilibrium of eqn. (3) is set up in solution

$$As(OH)_3 + 3HCl \rightleftharpoons AsCl_3 + 3H_2O \tag{3}$$

With a sufficient excess of concentrated hydrochloric acid, however, all the arsenic may be quantitatively distilled out of the solution (together with hydrochloric acid vapor). A method for the analytical separation of arsenic from antimony and tin is based upon this procedure; it may be carried out even more successfully using the corresponding *bromides*.

In aqueous solution, arsenic trichloride is dissociated to some extent into the ions As^{+++} and $3Cl^-$. At the same time, it is extensively hydrolyzed [eqn. (3)—compare also eqn. (1), p. 655].

(c) Sulfides and Thiosalts of Arsenic

The sulfides As_4S_3, As_4S_4, As_2S_3, and As_2S_5 may be obtained by melting the components together. Arsenic trisulfide is most readily prepared, however, by passing hydrogen sulfide into a hydrochloric acid solution of arsenic trioxide:

$$2As^{3+} + 3H_2S = As_2S_3 + 6H^+$$

Arsenic pentasulfide can be prepared in a similar manner from a solution of pentavalent arsenic.

$$2As^{5+} + 5H_2S = As_2S_5 + 10H^+.$$

However, the reaction follows this course only under certain well defined experimental conditions. As a rule, hydrogen sulfide acts to some extent as a reducing agent on $+5$ arsenic, so that the precipitate contains a mixture of arsenic trisulfide and sulfur, as well as the pentasulfide:

$$
\begin{aligned}
2As^{5+} + 2H_2S &= 2As^{3+} + 2S + 4H^+ \\
2As^{3+} + 3H_2S &= As_2S_3 + 6H^+ \\
\hline
2As^{5+} + 5H_2S &= As_2S_3 + 2S + 10H^+
\end{aligned}
$$

As_4S_4 occurs native as *realgar*, also known as red arsenic blende, ruby sulfur and sandarach, and As_2S_3 as *orpiment* (yellow arsenic blende). Both are also prepared artificially.

Realgar (mostly prepared artificially) is used in painting and in pyrotechnics. It forms a red, vitreous mass, and an orange powder when ground up. It can be distilled without decomposition, and its vapor density corresponds to the formula As_4S_4. It is decomposed by potassium hydroxide into arsenic trisulfide (which passes into solution, forming thioarsenite and arsenite) and free arsenic:

$$3As_4S_4 = 4As_2S_3 + 4As$$

Arsenic trisulfide forms a fine lemon yellow compound, which melts at $310°$ to a red liquid, and boils at $707°$ without decomposition. It has a notable capacity for forming colloidal dispersions.

Arsenic trisulfide is used as an artists' pigment, under the name of *king's yellow* when pure, and the impure material made by melting arsenic with sulfur as *orpiment* (corrupted from auri pigmentum). The latter always contains unchanged arsenic, and (unlike the pure sulfide) is toxic.

Arsenic trisulfide is insoluble in water and acids,—even in concentrated hydrochloric acid. It dissolves more readily in alkaline reagents, and especially in alkali sulfide solutions (including ammonium sulfide).

It dissolves in these because arsenic can form *thiosalts*:

$$As_2S_3 + 3S^{2-} = 2[AsS_3]^{3-} \quad \text{(thioarsenite ion)} \tag{1}$$

This process corresponds exactly to the formation of the arsenite ion when arsenic trioxide is dissolved in alkalis, except that the equation is then written with $6OH^-$ ions, since O^{2-} ions are virtually absent:

$$As_2O_3 + 6OH^- = 2[AsO_3]^{3-} + 3H_2O \quad \text{(arsenite ion)} \tag{2}$$

Thio-acid ions and oxyacid ions are formed simultaneously by the action of alkalis, including ammonia solution:

$$As_2S_3 + 6OH^- = [AsS_3]^{3-} + [AsO_3]^{3-} + 3H_2O \tag{3}$$

Arsenic trisulfide reacts in a similar manner with alkali carbonates, including ammonium carbonate:

$$As_2S_3 + 3CO_3^{2-} = [AsS_3]^{3-} + [AsO_3]^{3-} + 3CO_2 \tag{4}$$

If arsenic trisulfide is allowed to react with ammonium polysulfide, the sulfur dissolved in the reagent brings about oxidation to *thioarsenate* (just as with tin(II) sulfide, p. 534):

$$As_2S_3 + 3S^{2-} + 2S = 2[AsS_4]^{3-} \quad \text{(thioarsenate ion)} \tag{5}$$

Thioarsenates are likewise obtained when arsenic pentasulfide is dissolved in ammonium sulfide or in alkalis—mixed, in the latter case, with arsenates or with thiooxoarsenates. It is probable that the thioarsenate ion is present in solution as a diaquo-ion, $[AsS_4(OH_2)_2]^{3-}$ (p. 534).

When solutions of thioarsenites and thioarsenates are acidified, arsenic trisulfide and arsenic pentasulfide, respectively, are re-precipitated, since the concentration of S^{2-} ions becomes vanishingly small, because of combination with the added H^+ ions to form almost undissociated hydrogen sulfide. The equilibria are thereby displaced, and the above reactions proceed in the converse direction.

The acids corresponding to the thioarsenites and thioarsenates are not stable in the free state, but their *esters* have been prepared (Klement, 1935)—e.g., the phenyl ester $As(S.C_6H_5)_3$ (colorless needles, m.p. 95°) and p-tolyl ester $As(S.C_6H_4.CH_3)_3$ (light yellow needles, m.p. 76°) of thioarsenious acid, the p-tolyl ester of thioarsenic acid,

$$S{=}As(S.C_6H_4.CH_3)_3 \quad \text{(yellow needles, m.p. 74°)}.$$

(d) Arsenic Hydride and the Arsenides

(*i*) *Arsenic hydride* (arsine), AsH_3, is formed by the action of 'nascent' hydrogen on soluble compounds of arsenic—e.g.,

$$As_2O_3 + 6Zn + 6H_2SO_4 = 2AsH_3 + 6ZnSO_4 + 3H_2O.$$

For the preparation of larger quantities, metallic arsenides (e.g., zinc arsenide) are decomposed with dilute sulfuric acid:

$$Zn_3As_3 + 3H_2SO_4 = 2AsH_3 + 3ZnSO_4.$$

Arsenic hydride is a colorless gas with an unpleasant garlic-like smell. It is *extremely poisonous*. The inhalation even of minute amounts has often proved fatal. The density of the gas corresponds to the formula AsH_3. It liquefies at $-55°$ and solidifies at $-114°$. When ignited, it burns with a bluish flame to arsenic trioxide and water. If the flame is chilled by holding a cold porcelain dish in it, only the hydrogen burns, and the arsenic is deposited as a black or brown spot (Marsh's test). The gas decomposes when it is passed through a heated glass tube:

$$AsH_3 = As + \tfrac{3}{2}H_2 + 44.2 \text{ kcal.}$$

Arsenic is deposited just beyond the heated zone, in the form of a lustrous black metallic mirror. If the mirror of deposited metal is moistened with a drop of sodium hypochlorite solution, or with a mixture of hydrogen peroxide and caustic soda, it dissolves, forming sodium arsenate.

Arsine has strongly reducing properties. It precipitates metallic silver from silver nitrate solution:

$$AsH_3 + 6AgNO_3 + 3H_2O = 6Ag + As(OH)_3 + 6HNO_3,$$

although it reacts with solid silver nitrate to form the yellow double compound $AsAg_3 \cdot 3AgNO_3$. Use is made of this reaction in the Gutzeit test for detection of arsenic.

The Gutzeit test is carried out by covering the substance to be tested with dilute sulfuric acid, in a small test tube. A piece of arsenic-free zinc is dropped in, and after inserting a plug of cotton wool to hold back liquid droplets, the test tube is covered with a piece of filter paper, on which is laid a crystal of silver nitrate. The presence of arsenic is shown by a yellow coloration of the silver nitrate or a blackening if moisture is present. It should be noted that the hydrides of phosphorus and antimony give similar reactions.

(ii) *Organic arsines*—i.e., alkyl derivatives of arsenic hydride—can be obtained by the double decomposition of arsenic trichloride with zinc alkyls—e.g., trimethyl arsine,

$$2AsCl_3 + 3Zn(CH_3)_2 = 2As(CH_3)_3 + 3ZnCl_2.$$

They are poisonous liquids, with repulsive odors and of highly unsaturated character. They are oxidized by oxygen, sulfur or halogens to compounds of the types $(CH_3)_3AsO$, $(CH_3)_3AsS$, $(CH_3)_3AsCl_2$, etc. They combine with alkyl iodides to form tetraalkyl arsonium iodides, $[R_4As]I$. These are converted by moist silver oxide into tetraalkyl arsonium hydroxides,—crystalline, very hygroscopic substances of strongly basic character. With acids, these give rise to typical salts, $[R_4As]X$, of which the tetraalkyl arsonium iodides are examples.

(iii) *Cacodyl compounds.* When arsenic trioxide is heated with potassium acetate, a liquid with a paralyzingly revolting smell distils over, the 'fuming arsenical liquid' of Cadet. The principal constituent of this is *cacodyl oxide*, $[(CH_3)_2As]_2O$. *Cacodyl chloride*, $(CH_3)_2AsCl$, is obtained from this by the action of hydrochloric acid, and may in turn be converted to cacodyl cyanide, $(CH_3)_2AsCN$, cacodyl sulfide, $[(CH_3)_2As]_2S$, etc. All these highly reactive compounds are derivatives of the *cacodyl radical*, $(CH_3)_2As$—, so called from the characteristic and highly unpleasant smell of its derivatives (κακώδης, stinking). Free cacodyl, $(CH_3)_2As$—$As(CH_3)_2$, a liquid boiling at 170°, also with an unbearable smell, can be regarded as the alkyl derivative of the *diarsine*, As_2H_4, discovered by Nast (*Ber.*, 81 (1948) 271). This compound, the analogue of hydrazine H_2N—NH_2 and diphosphine H_2P—PH_2, is even more unstable than the latter, decomposing at —100° into AsH_3 and a red solid arsenic hydride $[AsH]_x$. The formation of the latter when AsH_3 is oxidized by tin(IV) salts had already been observed by Moser and Brukl in 1924.

(iv) *Arsenides.* The compounds of arsenic with the metals are known as *arsenides*. They are generally prepared by melting their components together, combination often being accompanied by the evolution of a considerable quantity of heat. Many arsenides—e.g., Cu_3As_2— may also be precipitated by arsine from solutions of metallic salts. Arsenic compounds of the alkalis and alkaline earth metals, and a few others such as Zn_3As_2, are decomposed either by water or by acids, forming arsine, and may thus be regarded as derivatives of the latter. Other arsenides, from their composition and properties, are to be assigned to the class of intermetallic compounds. They are usually very resistant towards acids. The naturally occurring arsenides are of the latter type.

10. Analytical (Arsenic)

As the transition member between the non-metals and the metals of the fifth Main Group, arsenic combines in itself the characteristic of both classes. This shows itself in its analytical behavior also. Thus it is found both in the tests for acid radicals and in the course of separation of cations. With silver nitrate, it gives a yellow precipitate as arsenite, and a chocolate brown precipitate as arsenate. It is precipitated by hydrogen sulfide as the lemon yellow sulfide. If present in the 5-va ent state, it is not precipitated completely unless the solutions are very stronlgly acidified. Arsenic sulfide dissolves readily in ammonium sulfide solution and also—unlike antimony and tin sulfides—in ammonium carbonate solution; it does not dissolve in concentrated hydrochloric acid.

The Bettendorff test (p. 654) may be used for the identification of arsenic.

For quantitative determination, arsenic is generally precipitated as the sulfide, As_2S_3, if present in the trivalent state, and this may be weighed as such. If the arsenic is 5-valent, it is best precipitated from ammoniacal solution with magnesia mixture* as magnesium ammonium arsenate, $Mg(NH_4)AsO_4 \cdot 6H_2O$, which is converted by ignition to magnesium pyroarsenate $Mg_2As_2O_7$ and weighed in that form. Trivalent arsenic may also be determined by titration with iodine in alkali hydrogen carbonate solution (cf. p. 654).

In forensic medicine, the Marsh test is used for the detection of arsenic, since, when carried out in a suitable manner, it enables extremely minute quantities of arsenic to be identified and determined with certainty.** The Gutzeit test is generally used for the rapid testing of commercial acids for arsenic. Copper has the property of depositing arsenic from hydrochloric acid solutions of the trioxide, in the form of grey copper arsenide, Cu_5As_2, and can also be used for the detection of arsenic (Reinsch's test).

11. Antimony (Stibium) (Sb)

(a) Occurrence

Antimony exists in nature chiefly in the form of the trisulfide, Sb_2S_3 as *stibnite* (*antimonite*). *Valentinite*, (*white antimony*), Sb_2O_3, occurs as a product of decomposition of stibnite. Antimony is also occasionally found native, often in isomorphous mixture with arsenic (*allemontite*). It is also frequently found in the ores of lead, copper, and silver, in the same way as arsenic.

Examples of the naturally occurring compounds of antimony with metals are *breithauptite*, NiSb (isomorphous with wurtzite, ZnS), *ullmannite*, NiSbS (isomorphous with pyrite, FeS_2) and *discrasite*, Ag_2Sb (isomorphous with chalcosite, Cu_2S). The double sulfides or thiosalts of antimony are more numerous, including especially the hexagonal *red silver ore*

* *Magnesia mixture* is a solution containing magnesium chloride, ammonium chloride and ammonia.

** In the form described by Lockemann, the Marsh test permits the unambiguous detection of as little as 0.1 microgram of As. Such minimal quantities of arsenic may even be determined quantitatively with adequate accuracy (about 1%) if the mirror of deposited arsenic is dissolved and titrated conductometrically as arsenite, with extermely dilute iodine solution as titrant (G. Jander, *Angew. Chem.*, 48 (1935) 267).

(*pyrargyrite*, or antimony silver blende), Ag_3SbS_3, and the monoclinic *pyrostilpnite* or *proustite*, of the same composition; also the antimonial tetrahedrites—e.g., *tetrahedrite* itself $4Cu_2S \cdot Sb_2S_3$. Other examples of such minerals are *bournonite*, $3(Pb,Cu_2)S \cdot Sb_2S_3$, *boulangerite*, $5PbS \cdot 2Sb_2S_3$, *jamesonite* $2PbS \cdot Sb_2S_3$, *polybasite*, $9(Ag,Cu)_2S \cdot Sb_2S_3$, *stephanite*, $5Ag_2S \cdot Sb_2S_3$, *silver antimony glance*, *miargyrite* or *hypargyrite*, $Ag_2S \cdot Sb_2S_3$, *copper antimony glance* (chalcostibite), $Cu_2S \cdot Sb_2S_3$, *zinckenite*, $PbS \cdot Sb_2S_3$.

(b) History

Stibnite was already known in ancient times; it was used to darken the eyebrows and lashes, and was known to the Greeks as στίμμι and to the Romans as *stibium*. It later received the name of *antimonium*, probably derived from the Arabic, and this was ultimately applied to the metal obtained from the ore. In his 'Triumphal Chariot of Antimony' the Benedictine monk Basil Valentine, living in the 15th century, fully described* the preparation of metallic antimony, the uses then made of its alloys—e.g., the use of lead antimony alloys for casting type for printing—and a considerable number of antimony compounds. He used the name 'spiess glass' for native antimony sulfide, from which its present name of 'spiess glance' has been derived. In the Iatrochemical period, antimony preparations were among the favorite medicines of the pharmacopoea, e.g., the '*eternal*' pills of metallic antimony. It was also common to allow wine to stand for some time in antimony goblets, and then to take it as an emetic. Antimony compounds now find only restricted applications in medicine, although certain synthetic organic antimony compounds have attained considerable importance as specific remedies for certain tropical diseases.

(c) Preparation

The production of antimony is carried out either by smelting the sulfide with iron (precipitation process):

$$Sb_2S_3 + 3Fe = 2Sb + 3FeS$$

or by roasting the sulfide, and reducing the resulting tetroxide with charcoal (roast-reduction process):

$$Sb_2S_3 + 5O_2 = Sb_2O_4 + 3SO_2; \qquad Sb_2O_4 + 4C = 2Sb + 4CO$$

Only ores of a fairly high degree of purity can be used directly for the precipitation process. However, fairly pure antimony sulfide can be obtained from ores which contain a considerable amount of gangue, by the process of *liquation*. The ore is so heated on a sloping hearth that the comparatively fusible antimony sulfide flows away. The liquated antimony sulfide so obtained is traditionally known as 'antimonium crudum'.

In the roasting stage of the roast-reduction process, antimony tetroxide is obtained only if provision is made for an ample supply of air. It is possible, however, to carry out the roasting in a limited supply of air. In this case, instead of the non-volatile Sb_2O_4, the trioxide Sb_2O_3 is formed. This is volatile at high temperatures, and can be collected in condensation chambers ('volatilizing roast'). In the reduction of the oxide with carbon (wood charcoal or coke), a flux such as sodium carbonate or sulfate is added to cover the melt and to prevent the volatilization of the molten antimony. The flux also exerts a purifying action, in that it slags off impurities such as arsenic, iron, and copper from the antimony.

Antimony of fairly high purity (99.9%) can be obtained by electrolytic refining. Processes have also been worked out for the direct electrolytic extraction of antimony of high purity from sulfide ores.

(d) Properties

Antimony is a silver-white, lustrous, brittle metal, of moderate hardness (3 on Mohs' scale), and therefore esaily powdered. Its electrical conductivity is 3.76% of that of silver at 0°. The density of antimony is 6.69, m.p. 630.5° and b.p. 1640°.

* Some consider that the ascription of this work to 'Basil Valentine' is spurious, and that it was compiled about 1600 by Thoelde.

Like arsenic, antimony is tetratomic in the vapor state (Sb_4 molecules). In the solid state, antimony also resembles arsenic in existing in several modifications. The ordinary grey or metallic antimony belongs crystallographically to the rhombohedral system (structure— see p. 580). Stock obtained *yellow antimony*, corresponding to yellow arsenic and white phosphorus by passing oxygen into liquid antimony hydride at —90°. Yellow antimony is even more unstable than yellow arsenic. It turns black within a short time above —80°, even in the dark; the darkening takes place much more rapidly, or at a lower temperature, in sunlight. The *black antimony* so formed represents a third modification, according to Stock. It is also formed by the action of oxygen or air on liquid antimony hydride at —80°, but is best obtained in the pure state by rapidly cooling antimony vapor. The density of black antimony is 5.3, and is thus markedly lower than that of the gray form. It is also more reactive chemically. Thus black antimony is oxidized, and may even catch fire, in the air at ordinary temperature. It reverts to the gray form when heated in the absence of air.

A fourth form of antimony is the so-called '*explosive*' *antimony*, discovered in 1855 by Gore. Antimony separates in this form when it is deposited cathodically at a sufficiently high current density in the electrolysis of a solution of antimony chloride, bromide or iodide. If the antimony obtained in this way is scratched with a hard object, or rapidly heated, it changes into ordinary antimony with the evolution of sparks and much heat (about 20 cal per g). Böhm first showed by X-rays that explosive antimony was amorphous. More recent work with electron diffraction (Prins, 1933) and monochromatic X-rays (Glocker, 1942) indicates that it is actually vitreous in structure. On the average, every Sb atom has 4 neighbors at a distance of 2.87 Å, and 2 at 3.51 Å, as compared with an environment of 3 neighbors at 2.87 Å and 3 at 3.37 Å in the crystal lattice of ordinary antimony. The evolution of heat represents the heat of crystallization, suddenly liberated when the explosive antimony changes into the ordinary form.

Antimony is stable in air at ordinary temperature, but burns in air when heated above its melting point. The main product of combustion is the trioxide Sb_2O_3, volatile at high temperatures. Antimony is also oxidized at a red heat to the trioxide by water vapor. Finely powdered antimony catches fire in chlorine, and burns to the pentachloride, $SbCl_5$. It also reacts vigorously with the other halogens. It unites with sulfur when melted with it, as it also does with arsenic, phosphorus, and many metals. Antimony powder deflagrates when it is heated with alkali nitrates or chlorates, forming the alkali salts of antimonic acid.

Antimony dissolves in nitric acid, forming the trioxide or the pentoxide, according to the acid concentration. It reacts with hot concentrated sulfuric acid, forming antimony(III) sulfate and sulfur dioxide. It is not dissolved by hydrochloric acid or dilute sulfuric acid.

The insolubility of antimony in non-oxidizing acids is the result of the position of antimony in the electrochemical series. Its normal potential, relative to the normal hydrogen electrode, is about —0.1 to —0.2 volts. It is thus more electropositive than arsenic, but less electropositive than hydrogen.

If an antimony electrode is immersed in any aqueous solution, it assumes a potential which depends upon the hydrogen ion concentration of the solution. This is the basis for the use of the antimony electrode in potentiometric pH determinations, as introduced by Uhl (1923) and Kolthoff (1925). The processes which determine the potential of such an electrode are not yet fully understood. The potentials are not always really reproducible, nor are they usually constant over a period of time. The sources of error arising from these features limit the applicability of the antimony electrode, which otherwise offers the advantages of requiring neither the passage of hydrogen through the solution examined, nor the addition of any other material, such as quinhydrone.

(e) Uses

Metallic antimony is used chiefly for the preparation of alloys. It has the property of hardening soft metals, such as tin or lead, to a marked degree. Lead-antimony alloys are known as *hard lead* or antimonial lead. Type for printing is cast from a lead-antimony alloy with 15–25% of antimony, and generally some tin also (*type metal*). Tin-antimony alloys (pewter) or ternary alloys (e.g., with 90% tin, 8% antimony, 2% copper) such as Britannia metal are used for the manufacture of table utensils; other alloys are used as bearing metals.

Compounds of antimony long played an important role in pharmacy, and many of them, such as antimony pentasulfide and tartar emetic still have medicinal applications. Antimony is now becoming increasingly important for the preparation of complex therapeutic agents. The main consumer of antimony pentasulfide is the rubber industry. Other compounds of antimony which find technical applications are the trisulfide, the trichloride, antimony pentachloride, and lead antimonate (Naples yellow).

12. Compounds of Antimony

Antimony functions in its compounds almost exclusively as a trivalent or pentavalent element. In acid solutions the compounds of pentavalent antimony have a strong tendency to change to those of the $+3$ state. They are thus oxidants—e.g., liberate iodine from iodide solutions.

Formally quadrivalent antimony is present in some complex salts corresponding to the hexachlorostannates $M^I_2[SnCl_6]$ and hexachloroplatinates $M^I_2[PtCl_6]$, and derived from the hypothetical antimony tetrachloride. These compounds are characterized by their intense color, whereas the compounds of 3- and 5-valent antimony with strongly electronegative elements (other than the sulfides) are colorless or weakly colored. It is now believed that these compounds, which may formally be regarded as containing $+4$ antimony, have equivalent quantities of the element in the $+3$ and $+5$ states.

(a) Oxides and Hydroxides

Antimony trioxide, Sb_2O_3 or Sb_4O_6, is derived from trivalent antimony, and *antimony pentoxide*, Sb_2O_5, from the element in the pentavalent state. Antimony tetroxide, Sb_2O_4 is a compound of these two oxides.

The hydroxide corresponding to the $+3$ state, $Sb(OH)_3$, is definitely amphoteric. It dissolves in acids forming *antimony salts*, SbX_3. Basic salts, and especially the *antimonyl salts*, $[SbO]X$, are more often formed than the normal salts SbX_3. Antimony trioxide dissolves in alkalis, forming *antimonites*, $M^I[SbO_2]$.

The antimonyl salts and the antimonites may be derived from the partially dehydrated form of the hydroxide, $SbO(OH)$, which, according to the way in which it dissociates into ions, can be regarded either as antimony metahydroxide or as antimonious acid.

$$SbO(OH) \rightleftharpoons [SbO]^+ + OH^-$$
$$\text{antimony metahydroxide} \quad \text{antimonyl ion}$$

$$Sb(OH)_3 - H_2O = SbO(OH)$$

$$SbOOH \rightleftharpoons [SbO_2]^- + H^+$$
$$\text{antimonious acid} \quad \text{antimonite ion}$$

It is questionable whether any hydroxides of antimony(III) really exist as definite compounds, except in solution. Attempts to prepare them give gels of variable water content, and these on aging pass into crystalline Sb_2O_3, even in contact with water. If antimony(III) hydroxides exist in the solid state at all, they must be extremely unstable, like the silicic acids.

The hydroxides derived from antimony pentoxide have a definitely acidic character. They are known as *antimonic acid*, and their salts as *antimonates*.

The antimonates containing water of crystallization are derived from the *hexahydroxo-antimonic acid* $H[Sb(OH)_6]$ (cf. p. 667). The acid itself cannot be isolated, for the $[Sb(OH)_6]^-$ ions condense even in aqueous solution to triantimonate ions, $[Sb_3O_{10}]^{5-}$, as soon as the hydrogen ion concentration is increased as was shown by G. Jander. With further increase in hydrogen ion concentration, tetraantimonate ions, $[Sb_4O_{13}]^{6-}$, are formed.

Unlike the corresponding arsenic compounds, the oxides and oxide hydrates of antimony are insoluble in water.

(*i*) *Antimony trioxide*, antimony(III) oxide, Sb_2O_3 or Sb_4O_6, is formed when antimony is heated above its melting point in air, in an inclined crucible. It sublimes to the cooler parts, and is deposited in the form of cubic crustals ('flowers of antimony'). It is better prepared by the hydrolysis of *powder of Algaroth*, $Sb_4O_5Cl_2$, by boiling with sodium carbonate solution:

$$Sb_4O_5Cl_2 + Na_2CO_3 = Sb_4O_6 + 2NaCl + CO_2.$$

Antimony trioxide so prepared is a white powder (density 5.2–5.3, m.p. 656°), insoluble in water. It is readily volatilized. It turns yellow when heated, but its white color is restored on cooling. The density of its vapor corresponds to the molecular complexity Sb_4O_6.

In the crystalline state, antimony trioxide, like arsenic trioxide, exists in two forms, both of which occur native. It is found in regular octahedra (density 5.25) as *senarmontite*, isomorphous with arsenolite. It also crystallizes rhombic, as *valentinite* (density 5.7). The cubic modification is stable below 570°, and the rhombic form above 570°. The heat of transformation is 3.24 kcal per mol.

According to the X-ray crystallographic evidence, senarmontite is built up from Sb_4O_6 molecules, occupying the points of a diamond lattice (Fig. 79, p. 421), with a cube edge of 11.14 Å. The centers of adjacent molecules are 4.83 Å apart. Each Sb_4O_6 molecule consists of an octahedron of oxygen atoms, with an interpenetrating tetrahedron of Sb-atoms, the Sb—O distance being 2.22 Å. Arsenolite has the same structure ($a = 11.06$ Å, $s = 4.78$ Å, As—O = 2.01 Å). Valentinite has infinite chains, of the composition Sb_2O_3, without the presence of physically distinct molecules.

Antimony trioxide is insoluble in dilute sulfuric and nitric acid. It dissolves in hydrochloric acid, however, and also in the solutions of many organic acids, e.g., tartaric acid. It also dissolves in alkalis, forming antimonites. When ignited in the air, it combines with oxygen forming the mixed antimony tetroxide, Sb_2O_4. It may readily be reduced to the metal, however—e.g., by heating in hydrogen, or with charcoal or potassium cyanide.

(*ii*) *Antimony trioxide hydrate and the Antimonites.* If a solution of tartar emetic (cf. p. 671) is decomposed at low temperatures (0°) by addition of dilute sulfuric or hydrochloric acid, a voluminous white precipitate is obtained, having a gel-like character, and containing variable amounts of water. By careful drying, this may be brought to a water content roughly corresponding to the composition of antimony(III) hydroxide, $Sb(OH)_3$, but it is doubtful whether this is really present as a well defined compound. Precipitates obtained by decomposing tartar emetic at higher temperatures have a lower water content, after air drying, than corresponds to the formula of the hydroxide. Hydrolysis of antimony trichloride, above

40°, yields the anhydrous crystalline trioxide, Sb_2O_3, directly. The strongly surface-active preparations, precipitated in gel form ('antimony trioxide hydrate'), slowly change into the crystalline oxide, even under water. Antimony trioxide dissolves in acids, forming antimony salts (p. 671). Treatment with alkalis, however, yields *antimonites*, i.e., salts of (meta-)antimonious acid:

$$Sb(OH)_3 + NaOH = NaSbO_2 + 2H_2O.$$

The sodium salt $NaSbO_2 \cdot 3H_2O$, which crystallizes in octahedra, is sparingly soluble in water, in contrast to the extremely soluble potassium antimonite. Antimonite solutions have reducing properties, and thus precipitate metallic silver from ammoniacal silver nitrate.

(*iii*) *Antimony pentoxide*, antimony(V) oxide, Sb_2O_5, is obtained in the hydrated state by oxidizing antimony with concentrated nitric acid. After drying, the reaction product forms a yellowish powder (density about 3.8). It is very sparingly soluble in water, but reddens moist blue litmus paper. It begins to lose oxygen when it is heated, even before all the water has been expelled.

According to Westgren (1937) the compound $Sb^{III}Sb^V_2O_6(OH)$ (isotypic with $BiTa_2O_6F$) is first formed on heating to about 800°, and this passes into Sb_2O_4, antimony tetroxide, only on very prolonged heating. Simon (1927) states that the range of existence of the latter compound (at 10 mm oxygen pressure) extends from about 780–920°.

(*iv*) *Antimonic acid and the Antimonates.* The ordinary, simple antimonic acid (mono-antimonic acid) is known only in solution, but the composition and properties of the salts derived from it show that it is a hexahydroxo acid, $H[Sb(OH)_6]$.

The name 'antimonic acid' was formerly applied to precipitated substances of continuously variable water content, such as are obtained by the hydrolysis of antimony pentachloride (using water saturated with chlorine, to avoid reduction). Since the nature of these substances is unknown, they are best described by the more general name of *antimony pentoxide hydrates*. They are white powders, which dissolve both in hydrochloric acid and in alkali. They are practically insoluble in water, although when freshly precipitated they pass into colloidal dispersion to a considerable extent on treatment with water.

These precipitated 'antimonic acids' are gels. They therefore have a variable water content, according to the conditions of preparation, and do not, as a rule, furnish products of simple stoichiometric composition when they are dehydrated. The definite compound $3Sb_2O_5 \cdot 5H_2O$ or $H_5Sb_3O_{10}$ is formed only at elevated temperatures, as shown by Simon (1927), but more slowly and with much more difficulty than is the corresponding arsenic compound. The antimony pentoxide hydrates are reminiscent of the oxide hydrates of the neighboring element, tin, in that they can exist in two forms corresponding to the a- and b-stannic acids. As in the case of the stannic acids, every intermediate form can be obtained.

Ordinary (hexahydroxo-)antimonic acid is a monobasic acid, the dissociation constant K_a being $4 \cdot 10^{-5}$ at room temperature. The primary dissociation constant of triantimonic acid is $5.3 \cdot 10^{-4}$, thus being greater than that of the mono-acid, as is usual.

Of the salts of antimonic acid, *potassium antimonate*, $K[Sb(OH)_6]$, is used as a reagent for sodium. It forms a granular white powder, rather sparingly soluble in cold water, but rather better soluble in warm. Since the sodium salt, $Na[Sb(OH)_6]$, is considerably less soluble still, potassium antimonate is a precipitant for sodium salts.

Potassium antimonate is prepared by fusing antimony pentoxide with an excess of potassium hydroxide. The melt is dissolved in a little water and allowed to crystallize. The salt, which separates out in crusts, is purified by heating it for a short time with several changes of water.

Potassium hydroxoantimonate was formerly and almost invariably considered to be a hydrated 'acid potassium pyroantimonate', $K_2H_2Sb_2O_7 \cdot 6H_2O$. However, as long ago as 1921, Tomula concluded from conductivity measurements that it must be the neutral salt of a monobasic acid, and not the acid salt of a tetrabasic acid. This view was completely confirmed in 1926 by G. Jander and Brüll, not only by measurements of freezing point depressions and rates of diffusion, but also, in the case of the sodium salt, by determining the influence of electrolytes with a common ion upon the solubility. Hammet (1929) was apparently the first to point out that the sodium salt precipitated from solution had a composition corresponding exactly to the formula $Na[Sb(OH)_6]$, and Pauling showed in 1933 that compatibility with the accepted ionic radius ratios was secured only if these salts were formulated as hydroxosalts, $M^I[Sb(OH)_6]$. The X-ray structure determinations of Beintema (1935) and Westgren (1938) confirmed this formula.

In addition to the salts of mono-antimonic acid, salts of polyantimonic acids—e.g., triantimonates and tetraantimonates—are known.

The *anhydrous* antimonates obtained by dry methods—e.g., by fusing the oxides together —would mostly appear, from their composition, to be salts of acids analogous to the phosphoric acids—i.e., orthoantimonic acid, H_3SbO_3, pyroantimonic acid, $H_4Sb_2O_7$, and metaantimonic acid, $HSbO_3$. As has already been mentioned, there is no evidence for the existence of these acids, and where X-ray structure determinations have been carried out on the salts it has been shown that these are not to be considered as structurally derived from the acids named. The situation is very similar to that of the meta- and polysilicates. Thus the diantimonates, such as $Ca_2Sb_2O_7$ or $Pb_2Sb_2O_7$, frequently formed by bivalent metals, are not based upon any self-contained $[Sb_2O_7]$ radicals, but the group $[Sb_2O_7]^{4-}$ forms a continuous network extending throughout the whole crystal, just as does the group $[AlSi_3O_8]^-$ in the aluminosilicates of felspar type. The structure is, however, somewhat more complex than in the silicates mentioned, since antimony, in accord with its larger ionic radius, has a higher coordination number than silicon, and must be surrounded not by four, but by eight O^{2-} ions. In sodium metantimonate, however, antimony exercises a coordination number of six towards oxygen, as in the hexahydroxoantimonates. Westgren and Schrewelius (1938) showed that $NaSbO_3$ crystallized with the ilmenite structure (Vol. II), although it also has a cubic modification. $AgSbO_3$ is also cubic.

Lead antimonate, made by melting lead nitrate with tartar emetic and sodium chloride, is used, especially in ceramics, as a pigment under the name of Naples yellow.

The antimonates are decomposed by acids, either triantimonic acid or antimony pentoxide gels being deposited, according to the conditions.

(*v*) *Antimony tetroxide*. Both the trioxide of antimony and the (hydrated) pentoxide are converted into *antimony tetroxide* Sb_2O_4 when they are heated in air to 800–900°. The tetroxide is obtained in the pure state from antimony pentoxide hydrate, or from the compound $Sb_3O_6(OH)$ first formed from it, only after prolonged ignition.

Antimony tetroxide forms a white, practically insoluble powder (density 6.59), which turns yellow when heated. When very strongly heated, it loses oxygen and is converted into the trioxide. It may readily be reduced to the metal, by heating it with carbon or potassium cyanide.

On the basis of X-ray structure determinations (Dihlström, 1938), antimony tetroxide is to be regarded as a double oxide of Sb_2O_3 and Sb_2O_5, or as an antimony(III) antimonate, $Sb^{III}Sb^VO_4$. It is isotypic with $SbTaO_4$. In neither compound, however, do SbO_4 or TaO_4 groups exist as structural units, but there is a continuous network extending throughout the structure, made up of alternate $Sb^{III}O_6$ and Sb^VO_6 (or TaO_6) octahedra. Around the Sb^{5+} or Ta^{5+} ions, the O^{2-} ions are arranged in almost regular octahedra, whereas around the Sb^{3+} ions the octahedra are considerably distorted.

By fusing antimony tetroxide with alkali hydroxides or carbonates, compounds are obtained which were formerly given the composition $M^I_2Sb_2O_5$. Corresponding compounds are also found in nature—e.g., romeite, $CaSb_2O_5$, and ammiolite, $HgSb_2O_5$. According to Natta (1933), their structure is similar to that of the diantimonates $M_2^{II}Sb_2O_7$, whereas Zedlitz (1931) considers them to be isotypic with pyrochlore (cf. Vol. II). In the latter case, romeite, for example, should be given the formula $NaCaSb_2O_6(OH)$.

(b) Halides of Antimony

The halides of antimony correspond to the general formulas SbX_3 and SbX_5 (the hypothetical $SbCl_4$ has no independent existence). They may be obtained by the direct union of the components, although they are generally prepared by other methods. They are extensively hydrolyzed by water. Their double compounds with metallic halides, which may best be regarded as complex salts, are more stable in this respect. Such complex *halogenoantimonites* and *halogenoantimonates*, are formed by all the antimony halides, the pentahalides having the greatest tendency to do so.

(i) *Antimony trifluoride*, SbF_3, separates out in colorless, deliquescent needles, when a solution of antimony trioxide in hydrofluoric acid is evaporated. It dissolves in water with an acid reaction, although the extent of hydrolysis is less than with the other halides. It readily combines with alkali fluorides, forming double salts (*fluoroantimonites*), mostly of the types M^ISbF_4 and $M^I_2SbF_5$. It also forms double salts with alkali chlorides and sulfates. The salt formed with ammonium sulfate, $SbF_3 \cdot (NH_4)_2SO_4$, is used as a mordant in dyeing, under the name of 'antimony salt'.

(ii) *Antimony pentafluoride*, SbF_5, can be prepared in the pure state (according to Ruff) by the prolonged action of anhydrous hydrogen fluoride on antimony pentachloride. It is a colorless, oily, very viscous liquid, ($d = 2.99$, m.p. $7°$, b.p. $150°$); it is strongly associated in the vapor state (in contrast to AsF_5, which vaporizes as the monomer). Antimony pentafluoride reacts with water with a hissing sound. With the alkali fluorides it forms double salts (*hexafluoroantimonates*), mostly of the type $M^I[SbF_6]$. It also forms addition compounds readily with other substances, such as SbF_3, P_2O_5 and I_2. SbF_5 dissolves in liquid SO_2 with the evolution of heat; it combines with SO_2 to form $SbF_5 \cdot SO_2$ (colorless crystals, m.p. $57°$). With NO_2 it forms the colorless compound $SbF_5 \cdot NO_2$, which decomposes into SbF_5 and NO_2 again at $150°$. It also adds on to S, Se and Te, to give $(SbF_5)_2S$ (white), $(SbF_5)_2Se$ (yellow) and $(SbF_5)_2Te$ (buff) (Aynsley, Peacock and Robinson, 1951).

(iii) *Antimony trichloride*, $SbCl_3$, forms a soft, colorless mass (density $= 3.06$, m.p. $73.4°$, b.p. $223°$), fuming in air ('butter of antimony'). It is best obtained by dissolving finely ground stibnite in hot concentrated hydrochloric acid:

$$Sb_2S_3 + 6HCl = 2SbCl_3 + 3H_2S.$$

It is used as a caustic in medicine, and also as an etchant for metals—e.g., for the coloring of gun barrels.

Antimony trichloride is a fairly reactive substance. Thus it reacts with concentrated sulfuric acid, forming antimony sulfate and hydrogen chloride:

$$2SbCl_3 + 3H_2SO_4 = Sb_2(SO_4)_3 + 6HCl.$$

It has a strong tendency to form addition compounds, and forms complex salts (*chloroantimonites*) with alkali chlorides. Most of these are of the type $M^I_3[SbCl_6]$, but some are more complex.

(iv) *Oxychlorides of Antimony*. Antimony trichloride dissolves to form a clear solution in a small quantity of water. On dilution, *basic chlorides* are precipitated as a consequence of hydrolysis—e.g., SbOCl, *antimonyl chloride*, and $Sb_4O_5Cl_2$, *powder of algaroth*. The latter is named after the 16th century Italian physician Algarotus, who introduced its use into medicine. It has practically no medical uses today.

(v) *Antimony pentachloride*, $SbCl_5$, can be obtained by treating the trichloride with chlorine. It forms a colorless liquid when pure, but is usually yellowish, with a density 2.35. On cooling, it solidfies to crystals which melt at 2.8°. It boils at 140° under atmospheric pressure, but begins to lose chlorine at this temperature. It reacts with hydrogen sulfide, forming hydrogen chloride and antimony thio-chloride, $SbSCl_3$. It gives up chlorine to many organic substances, being converted to the trichloride, and is therefore frequently used in organic chemistry as a chlorinating agent.

Numerous double compounds of antimony pentachloride are known, especially with organic substances containing oxygen, nitrogen, or chlorine. It also combines with numerous metallic chlorides, forming mostly compounds of the type $M^I[SbCl_6]$ [*hexachloroantimonates*, or strictly systematically, hexachloroantimonates(V)].

(vi) *Antimony Tetrachloride and Chloroantimonates(IV)*. If an equimolecular quantity of antimony pentachloride is added to a hydrochloric acid solution of antimony trichloride, the resulting solution is dark brown in color, in contrast to the colorless or nearly colorless starting materials. The same colored solution is formed when antimony trichloride is converted into the pentachloride by treatment with chlorine. It has been frequently assumed that *antimony tetrachloride*, $SbCl_4$, must be present in this solution, in equilibrium with the other chlorides:

$$SbCl_3 + SbCl_5 \rightleftharpoons 2SbCl_4.$$

It is not possible to isolate the tetrachloride, but by adding the chloride of rubidium or cesium, the complex salts $Rb_2[SbCl_6]$ and $Cs_2[SbCl_6]$ can be prepared. These crystallize in deep violet octahedra, and are isomorphous with the alkali hexachloroplumbates, $M_2[PbCl_6]$, hexachlorostannates, $M_2[SnCl_6]$, and hexachloroplatinates, $M_2[PtCl_6]$. They are thus formally derived from a chloride $SbCl_4$, and it is an obvious assumption that the antimony, like lead, tin or platinum in the isomorphous salts is electropositively quadrivalent. However, quadrivalent antimony would have one unpaired electron and would give rise to paramagnetism, whereas the compounds $M^I_2[SbX_6]$ are diamagnetic. It would appear that the solid salts may contain equivalent amounts of the $[Sb^{III}Cl_6]^{3-}$ and the $[Sb^VCl_6]^-$ anions, although the deep color of the compounds implies a strong interaction between the anionic groups. Such deep colors are not infrequently found in compounds containing the same element in two different valence states.

(vii) Antimony tribromide, $SbBr_3$, forms colorless rhombic crystals (density 4.15, m.p. 96.6°, b.p. about 280°). It is formed, with incandescence, when powdered antimony is added to liquid bromine. It is better prepared by the reaction of antimony with a solution of bromine in carbon disulfide. In its properties, antimony tribromide closely resembles the trichloride. Like the latter, it forms double salts with alkali bromides, but these are mostly of rather complex composition.

(viii) The *pentabromide* of antimony is known only in the form of double compounds. The *bromoantimonates* mostly conform to the type $M^I[SbBr_6]$, and are for the most part hydrated. The free acid $HSbBr_6 \cdot 3H_2O$ (*hexabromoantimonic acid*) is also known.

A compound formally derived from antimony tetrabromide, $Rb_2[SbBr_6]$, was prepared by Weinland (see above, $Rb_2[SbCl_6]$).

(ix) *Antimony triiodide*, SbI_3, is best prepared by introducing antimony powder into a solution of iodine in carbon disulfide. It crystallizes from the solution in ruby-red hexagonal plates, isomorphous with arsenic triiodide and bismuth triiodide. (density 4.85, m.p. 171°, b.p. about 401°). Antimony triiodide also exists in a second, yellow, rhombic modification (density 4.77). It forms complex salts (*iodoantimonites*) with the alkali iodides.

SbI_3 can take a certain amount of excess iodine into solid solution. This can readily be removed, however, by treatment with suitable solvents—e.g., with ether. Modern work shows that no pentaiodide of antimony exists.

(c) Antimony Salts

The antimony halides discussed in the preceding section can be regarded as the antimony salts of the halogen acids. They differ markedly, however, from the typical metallic salts in their volatility and in the small extent to which they are dissociated in aqueous solution. Hydrolysis of these substances is more important than dissociation.

Salts of oxyacids are known only for *trivalent* antimony, the basic character of antimony(V) oxide being so weakly developed that it is incapable of forming salts with nitric acid, sulfuric acid, etc. Even the antimony(III) salts have an extremely strong tendency to hydrolyze. In this process, however, *basic antimony salts* may be formed as an intermediate step, and these contain the radical $[SbO]^+$ in a number of cases (*antimonyl* salts). The basic salts of antimony are mostly insoluble in water, but gradually undergo yet further hydrolytic decomposition in contact with water. The salts of antimony with certain organic acids are notable for their relatively great stability—e.g., those of the type of *tartar emetic*, potassium antimono-tartrate, in which the antimony is bound in a complex.

Most antimony salts can add on alkali salts to form double compounds, or more or less strongly complexed salts. These are all more stable than the simple salts, and the more strongly they are complexed, the higher their stability. As examples, the *sulfatoantimonites*, $M^I[Sb(SO_4)_2]$, and the *oxalatoantimonites*, $M^I[Sb(C_2O_4)_2]$ and $M^I_3[Sb(C_2O_4)_3]$, may be cited.

(*i*) *Antimony sulfate*, $Sb_2(SO_4)_3$, may be obtained by dissolving antimony, antimony trioxide, trisulfide, or powder of algaroth in hot, concentrated sulfuric acid. It crystallizes in colorless, silky, deliquescent needles, density 3.62. With a small quantity of water it forms a hydrate, but is hydrolyzed by a larger quantity. If it is treated with moderately dilute sulfuric acid, it forms *antimonyl sulfate*, $(SbO)_2SO_4$, by partial hydrolysis. The double sulfates (sulfatoantimonites, $M^I[Sb(SO_4)_2]$) formed by combination with alkali sulfates are less unstable than the neutral sulfate.

(*ii*) *Antimony nitrate*. If antimony trioxide is added to cold fuming nitric acid and then diluted with water, a basic antimony nitrate separates out in white small crystals, with a pearly luster. The neutral nitrate cannot be prepared from aqueous solution.

(*iii*) *Potassium antimono-tartrate (tartar emetic)*. If a mixture of antimony trioxide and potassium hydrogen tartrate is boiled until a clear solution is obtained, and allowed to cool, *potassium antimono-tartrate*, $K[C_4H_2O_6Sb(OH_2)] \cdot \frac{1}{2}H_2O$, generally known as tartar emetic, crystallizes out:

$$KHC_4H_4O_6 + \tfrac{1}{2}Sb_2O_3 = K[C_4H_2O_6Sb(OH_2)] \cdot \tfrac{1}{2}H_2O.$$

This compound, which we first find described by Adrian von Mynsicht, a disciple of Paracelsus who lived in the first half of the 17th century, is used in medicine even today. It forms colorless crystals, freely soluble in water, but insoluble in alcohol.

It is obvious that tartar emetic is far less readily hydrolyzed than the inorganic salts of antimony. This alone shows that tartart emetic is *not*—as was formerly generally assumed— an *antimonyl tartrate*—i.e., a tartaric salt containing the $[SbO]^+$ radical as cation. It must be assumed that the antimony is firmly bound to the tartaric acid radical. Indeed, Schiff in 1863 assumed that the antimony replaced not only the H-atom of one carboxyl group, but also the H-atoms of both the alcoholic hydroxyl groups. This assumption has more recently received experimental support from Reihlen (1931), who explains the acidic re-action of tartar emetic in solution on the hypothesis that a H_2O molecule is complex-bound to the antimony, and can ionize to furnish a H^+ ion, so that in solution the following equi-librium is set up

$$
\begin{bmatrix} O{=}C{-}O \\ | \\ H{-}C{-}O{-}Sb\cdots OH_2 \\ | \quad / \\ H{-}C{-}O \\ | \\ O{=}C{-}O^- \end{bmatrix}
\rightleftharpoons
\begin{bmatrix} O{=}C{-}O^- \\ | \\ H{-}C{-}O{-}Sb{-}OH \\ | \quad / \\ H{-}C{-}O \\ | \\ O{=}C{-}O^- \end{bmatrix} + H^+
$$

On this basis, tartar emetic is the potasssium salt of an aquo-antimonotartaric acid*, $H[C_4H_2O_6Sb(OH_2)]$. A series of other salts of aquoantimonotartaric acid is also known. On heating, these first lose the water combined outside the complex ion (differing in amount according to the cation). The complex-bound water is given up on stronger heating; the sodium salt loses this *in vacuo*, even at room temperature, and the free acid does so in process of crystallization.

(d) Sulfides and Thiosalts of Antimony

Antimony forms the two sulfides Sb_2S_3 and Sb_2S_5. These are obtained by melting the components together, or by precipitation by means of hydrogen sulfide, from acidified solutions containing trivalent or pentavalent antimony, respectively. Like the corresponding compounds of arsenic, both sulfides can combine with additional S^{2-} ions to form thio-salts, and dissolve in ammonium sulfide accordingly:

$$Sb_2S_3 + 3S^{2-} \rightleftharpoons 2[SbS_3]^{3-}; \qquad Sb_2S_5 + 3S^{2-} \rightleftharpoons 2[SbS_4]^{3-}$$

thioantimonite ion   thioantimonate ion

The sulfides are reprecipitated by the addition of acid:

$$2[SbS_3]^{3-} + 6H^+ = Sb_2S_3 + 3H_2S; \qquad 2[SbS_4]^{3-} + 6H^+ = Sb_2S_5 + 3H_2S$$

Unlike the corresponding compounds of arsenic, the antimony sulfides are soluble in concentrated hydrochloric acid, but insoluble in ammonium carbonate solution. They dissolve in caustic alkalis in the same way as the arsenic sulfides.

It is probable that the thioantimonates are present in solution in the form of diaquo ions, $[SbS_4(OH_2)_2]^{3-}$ or as hydroxothiohydroxo ions, $[SbS_2(SH)_2(OH)_2]^{-3}$ (cf. p.535). Formation of hydroxothiohydroxo ions also probably occurs when antimony sulfides are treated with alkali hydroxides.

As in the case of arsenic, the free thio-acids are not stable. Their esters are stable how-ever, though much more readily decomposed than the corresponding arsenic compounds—e.g., they are at once hydrolysed by water.

(i) *Antimony trisulfide*, Sb_2S_3, is found in nature as stibnite (antimonite, antimony glance), and is the most important ore of antimony. Stibnite forms grey needles with a strongly metallic luster (density 4.6, m.p. about 550°, heat of formation 38.3 kcal per mol). When it is heated in a sufficient supply of air, stibnite is roasted

---

* The compound $Sb(HC_4H_4O_6)_3$, sometimes wrongly named 'antimono-tritartaric acid' is a true tartrate of antimony, and should accordingly be called (triply acid) antimony tritartrate.

to antimony tetroxide. Finely divided stibnite dissolves in concentrated hydrochloric acid, forming antimony trichloride. Stibnite and the 'antimonium crudum' (p. 663) prepared from it are used in pyrotechnics and in the manufacture of matches and ruby glass, as well as being the raw material for the production of antimony and antimony salts.

Antimony trisulfide precipitated from moderately warm solution has a fine orange red color and has a density of 4.15 after drying. The orange red antimony trisulfide is amorphous. When heated in the absence of air, it changes into the crystalline grey-black modification. The heat of transformation is 7.5 kcal per mol (Fricke).

The so-called *antimony-cinnabar*, used in pharmacy under the name *kermes*, and formerly widely employed as a red pigment, was formerly considered to be a double compound of antimony trioxide with antimony sulfide. It is found in nature as cherry red, monoclinic needles of red spiess glance (antimony blende, pyrostibite). It can be prepared by treating antimony trichloride with sodium thiosulfate. According to more recent work by Dönges and Fricke, it actually consists of Sb_2S_3 mixed with varying amounts of Sb_2O_3 and sulfur, and owes its red color, as compared with the orange of the precipitated sulfide, to its larger particle size.

(*ii*) *Antimony pentasulfide,* Sb_2S_5, also known as 'golden sulfuret of antimony' is prepared technically by decomposing Schlippe's salt (sodium thioantimonate) by means of dilute sulfuric or hydrochloric acid:

$$2Na_3SbS_4 + 3H_2SO_4 = Sb_2S_5 + 3Na_2SO_4 + 3H_2S.$$

It forms an orange powder, insoluble in water. It was introduced into the pharmacopea long ago, and still has uses especially in veterinary medicine. It is much used in the vulcanization of rubber, and imparts a characteristic red color to rubber goods.

Sodium thioantimonate, Na_3SbS_4 ($+9H_2O$), commonly called Schlippe's salt, is made by adding sulfur and powdered stibnite to boiling caustic soda, or to a mixture of milk of lime and soda. The caustic soda dissolves the sulfur to some extent, forming sodium sulfide, Na_2S, and this then reacts further:

$$3Na_2S + 2S + Sb_2S_3 = 2Na_3SbS_4$$

Sodium thioantimonate crystallizes from the solution on cooling, as the hydrate $Na_3SbS_4 \cdot 9H_2O$, in the form of bright yellow tetrahedra which readily effloresce in air (density 1.86).

(e) Antimony Hydride

(*i*) *Antimony hydride* (*Stibine*), SbH_3 is formed, like arsine, by the action of 'nascent' hydrogen upon soluble compounds of antimony.

For the preparation of large quantities, the best method according to Stock is to heat 1 part by weight of antimony with 2 parts by weight of magnesium, and to decompose the resulting powdered alloy by adding it to cold dilute sulfuric acid. The antimony hydride is freed from admixed hydrogen by condensing it by means of liquid air.

Antimony hydride is a colorless, evil-smelling gas, and hardly less toxic than arsine. It is somewhat soluble in water ($\frac{1}{5}$ volume in 1 volume of water), more soluble in alcohol (15 volumes per volume), and still more soluble in carbon disulfide (250 volumes in 1 volume at 0°). 1 liter of the gas weighs 5.685 g (at 0° and 760 mm). The liquefied gas boils at —18° and solidifies at —91.5°.

A mixture of antimony hydride and hydrogen burns with a green flame. A black stain of metallic antimony is formed upon a cold porcelain dish held in the flame, and this may be distinguished from the arsenic stain formed in a similar manner, in that it does not disappear when treated with a drop of sodium hypochlorite solution, or with a mixture, of sodium hydroxide and hydrogen peroxide. The same applies to the antimony mirror formed by passing hydrogen containing stibine through a heated glass tube.

Antimony hydride reacts with silver nitrate in a manner similar to arsenic hydride. The brown-yellow double compound $Ag_3Sb \cdot 3AgNO_3$ is formed with dry silver nitrate, and this is decomposed by access of moisture, with the deposition of black silver antimonide Ag_3Sb.

If decomposition is initiated at one point, pure antimony hydride decomposes with explosive violence, 34 kcal per mol being liberated in this decomposition.

(ii) *Organic stibines.* Just as the organic phosphines and arsines are derived from the hydrides of phosphorus and arsenic, so antimony hydride gives rise to the *organic stibines* SbR_3 (R = alkyl). Thus *trimethyl stibine*, $Sb(CH_3)_3$, is a spontaneously inflammable liquid, obtained by double decomposition between antimony trichloride and zinc methyl:

$$2SbCl_3 + 3Zn(CH_3)_2 = 3ZnCl_2 + 2Sb(CH_3)_3.$$

It reacts with hydrochloric acid, liberating hydrogen, just like a strongly electropositive metal:

$$Sb(CH_3)_3 + 2HCl = Sb(CH_3)_3Cl_2 + H_2.$$

With methyl iodide it forms *tetramethylstibonium iodide*, which behaves as a typical salt, being completely ionized in solution into $[Sb(CH_3)_4]^+$ and I^- ions:

$$Sb(CH_3)_3 + CH_3I = [Sb(CH_3)_4]I.$$

The reaction of moist silver oxide with this compound yields *tetramethylstibonium hydroxide*, $[Sb(CH_3)_4]OH$, which functions as a strong base. The other alkyl stibines behave similarly.

(f) Metallic Compounds of Antimony

The metallic compounds of antimony, the *antimonides*, conform generally in properties to the typical intermetallic compounds. Their composition is often determined by other considerations than the normal valences, this being especially true of the compounds of antimony with the elements of the Sub-groups. In its compounds with the elements of the first three Main Groups, however, antimony preferentially exercises the trivalence which it also exerts towards hydrogen (cf. Table 92, p. 585). Antimony combines with tin, forming the 'non-Daltonide' compound SnSb. It does not combine with lead, and forms mixed crystals to only a very limited extent, although lead and antimony are completely miscible in the molten state. With nickel, antimony forms four compounds: Ni_2Sb_3, NiSb, Ni_5Sb_2 and Ni_4Sb. The compound NiSb is found native as *breithauptite*. Antimony forms two non-Daltonide compounds with silver, which melt incongruently. One of these, Ag_3Sb (20–25 atom-% Sb) occurs in Nature as *discrasite*.

13. Analytical (Antimony)

In the systematic separation of the cations, antimony is precipitated by hydrogen sulfide as the orange sulfide (Sb_2S_3 or Sb_2S_5) which is insoluble in dilute acids. Like the sulfides of arsenic and tin, this dissolves in ammonium sulfide, forming thiosalts. It is distinguished from arsenic sulfide by its color, its insolubility in ammonium carbonate solution, and its solubility in concentrated hydrochloric acid. Antimony is precipitated from the hydrochloric acid solution by means of zinc, in the form of a black powder.

The oxides of antimony yield antimony hydride when treated with zinc and sulfuric acid. This gives metallic mirrors under the same conditions as does arsine, but antimony mirrors may be distinguished from arsenic mirrors by the reactions given above (p. 674).

Antimony compounds yield a brittle metallic bead and a with sublimate before the blow pipe. White precipitates of basic antimony compounds are generally formed when acid solutions of antimony salts are diluted.

For quantitative analysis, antimony is generally precipitated as the sulfide. This may be weighed as such, after drying in a stream of CO_2, or may be converted into antimony tetroxide.

14. Bismuth (Bi)

(a) Occurrence

Bismuth occurs in Nature both native and in the form of compounds. The most important of these are the sulfide, Bi_2S_3, *bismuth glance*, and the oxide, Bi_2O_3, *bismuth ocher*. Bismuth ores are not very widely distributed. In Europe, they are found chiefly, in the Erzgebirge of Saxony and Bohemia, but the principal ore deposits are in South America (Bolivia) and Australia (Tasmania).

Like arsenic and antimony, bismuth is also frequently encountered in nature in the form of double sulfides—e.g., lead bismuth glance, (*galenobismuthite*), $PbS \cdot Bi_2S_3$, *silver bismuth glance* or *argentobismuthite*, $Ag_2S \cdot Bi_2S_3$ (plenargyrite is a mineral of the same composition) and copper bismuth glance or *emplectite*, $Cu_2S \cdot Bi_2S_3$, *wittichenite*, $3Cu_2S \cdot Bi_2S_3$, *lillianite*, $3PbS \cdot Bi_2S_3$ and others. *Selenium bismuth glance*, Bi_3Se_3, and *tetradymite*, Bi_2Te_2S, may also be mentioned. The Australian *maldonite*, Au_2Bi, is a gold ore containing bismuth. The more strongly developed basic character of bismuth, as compared with arsenic and antimony, finds its expression in that bismuth also occurs in nature as the (basic) carbonate, *bismuth spar*.

(b) History

Bismuth was first mentioned in the 15th century by Basil Valentine, as a metal resembling tin. Its real nature was subsequently often in doubt. The two phlogistonists Pott and Bergman were the first to characterize bismuth more precisely, and to establish that it was a distinct element with metallic character. Bismuth oxide was already used as a pigment, and basic bismuth nitrate as a paint ('Spanish white') in the 16th century.

(c) Preparation

In former days, native bismuth was melted out ('liquated') from the ores containing it. It is now produced from oxidic ores by reduction with carbon in crucibles or reverberatory furnaces (reduction smelting). Sulfide ores are either first roasted, to convert them to the oxides, or may be smelted directly to bismuth by means of iron (precipitation smelting).

$$\text{Reduction process} \qquad Bi_2O_3 + 3C = 2Bi + 3CO$$
$$\text{Precipitation process} \qquad Bi_2S_3 + 3Fe = 2Bi + 3FeS$$

The bismuth obtained in this manner is generally contaminated by various impurities, especially arsenic, antimony, lead, iron, copper, and sulfur. It frequently contains silver and gold also. The latter may be extracted from the molten bismuth by means of tin. Copper is converted into copper sulfide, which separates out, by fusion with sodium sulfide after the other impurities have been removed by melting under oxidizing conditions. When high purity is required—as for example, bismuth for pharmaceutical preparations—refining by wet methods is generally necessary also—e.g. by dissolution in nitric acid and crystalliz-ation as the nitrate. Electrolytic refining is also employed on a large scale for the prepara-tion of very pure bismuth.

(d) Properties

Bismuth is a reddish-white, lustrous, brittle, coarsely crystalline metal (density 9.80, m.p. 271.0°, b.p. 1560°). It can be very easily broken up and pulverized. Its hardness is 2.5 (Mohs' scale).

The specific electrical conductivity of bismuth at 18° is 1.37%, and the thermal con-ductivity 1.93% of that of silver. The specific heat at 18° = 0.029 and atomic heat 6.1 cal. In the vapor state, bismuth exists as diatomic molecules, Bi_2.

Bismuth is stable in air at ordinary temperature. It burns with a bluish flame at red heat, forming the yellow oxide Bi_2O_3. Powdered bismuth catches fire in chlorine. It also combines when heated with bromine and iodine, as also with sulfur, selenium, and tellurium. It does not combine directly, however, with nitro-gen and phosphorus. Bismuth is not attacked at ordinary temperature by air-free water. It is slowly oxidized at red heat by water vapor. Bismuth does not dissolve in non-oxidizing acids: even cold concentrated sulfuric acid is without action. It dissolves in hot concentrated sulfuric acid, with the liberation of sulfur dioxide. The best solvent for bismuth is nitric acid.

Bismuth does not dissolve in non-oxidizing acids because of its position in the electro-motive series. The normal potential of bismuth, relative to the normal hydrogen electrode, is about —0.2 volt. Bismuth is thus more noble than hydrogen.

Bismuth readily forms alloys with metal, and many of these are notable for their fusibility. Examples of the formation of *compounds* with other metals are not very numerous, and involve chiefly the metals of strongly electropositive character (cf. Table 92, p. 585).

The best known of the low-melting alloys of bismuth is Wood's metal, consisting of 7–8 parts by weight of bismuth, 4 parts of lead, 2 parts of tin, and 1–2 parts of cadmium. It melts at about 70°. The lowest-melting (eutectic) alloy between the components men-tioned has a melting point of 60°, and consists of 15 parts of bismuth, 8 parts of lead, 4 parts of tin, and 3 parts of cadmium (Lipowitz alloy). Rose's metal, consisting of 2 parts bismuth and 1 part each of lead and tin, is also a eutectic, and melts at 94°.

(e) Uses

Metallic bismuth is used in making low-melting alloys, and also frequently as an addition to Britannia metal and bearing metals. The oxide and the basic nitrate are the most widely used compounds of bismuth. The oxide is used, for example, together with lead oxide, in glass manufacture, to produce optical glasses of high refractive index, and for colored glazes. The basic nitrate is employed in porcelain painting to fire on gilt decoration. It is used, especially, in cosmetic and medicinal preparations. Frequent use is made of bismuth compounds in pharmacy. Thus the basic bismuth salt of gallic acid ('bismutum subgallicum'), $Bi(OH)_2.OCO.$ $C_6H_2(OH)_3$, forms a favored dusting powder for wounds (Dermatol). Bismuth compounds are also used in the treatment of syphilis.

15. Compounds of Bismuth

The compounds of bismuth are derived almost entirely from *tri*valent bismuth, corresponding to the general formula BiX_3. However, bismuth can also act as *penta*valent, and exceptionally (see p. 681), as *di*valent.

In accordance with the definitely basic nature of bismuth oxide, Bi_2O_3, the compounds of bismuth with the more strongly electronegative atoms or radicals are mostly distinctly salt-like in character. Like its lower homologues, bismuth forms a gaseous hydride, which is very unstable.

Compounds of bismuth with the more electropositive metals are of quite minor importance (see above).

(a) Oxides and Hydroxides

(*i*) *Bismuth trioxide*. Ordinary bismuth oxide (bismuth trioxide), Bi_2O_3, is obtained by burning the metal, or by heating the carbonate or nitrate, as a powder (density 8.76, m.p. 820°, heat of formation 137.8 kcal/mol) which is sulfur yellow when cold and red-brown when hot. Platinum is vigorously attacked by molten bismuth oxide, as it is by lead oxide. At red heat, the oxide can easily be reduced to the metal. It dissolves in acids, forming the corresponding bismuth salts. It is insoluble in dilute alkali hydroxides. According to Schumb and Rittner (1943), bismuth trioxide exists in three crystalline modifications.

The insolubility of bismuth oxide in dilute alkalis, as compared with the amphoteric oxides of arsenic and antimony, marks it out as being definitely a basic oxide. However, it forms compounds with other metallic oxides when the components are melted together. Thus, with lead oxide it forms the compounds $2PbO \cdot Bi_2O_3$, $2PbO \cdot 3Bi_2O_3$ and $PbO \cdot 4Bi_2O_3$.

The evidence of modern investigations is that no lower oxide of bismuth exists as a well defined compound. According to Neusser, the products formerly regarded as being bismuth monoxide are, in fact, mixtures of Bi_2O_3 and Bi.

(*ii*) *Bismuth hydroxide*, $Bi(OH)_3$, is precipitated by hydroxyl ions from bismuth salt solutions, as a white flocculent substance. It is difficult to obtain in a state of purity, as it is colloidal in nature, and avidly adsorbs acid anions which are then tenaciously retained. It can also be readily obtained in colloidal dispersion. The precipitate obtained in the usual manner has a composition approximating to $BiO(OH)$ after it has been dried at 100°. Bismuth hydroxide readily dissolves in acids, forming salts, but does not dissolve in dilute alkalis.

The orthohydroxide, $Bi(OH)_3$ has been shown to be a definite chemical entity, in that it gives a characteristic X-ray diffraction pattern. The existence of two other hydrates: $Bi_2O_3 \cdot 2H_2O$ and $Bi_2O_3 \cdot H_2O$ (or $BiO(OH)$) has been shown from the isobaric degradation of the orthohydroxide $Bi(OH)_3$ (which may alternatively be bismuth oxide trihydrate, $Bi_2O_3 \cdot 3H_2O$) (Hüttig, 1931).

(iii) Bismuth pentoxide, Bismuthic acid and the Bismuthates. When bismuth hydroxide, suspended in concentrated alkali hydroxide solutions, is treated with strong oxidants such as chlorine, potassium permanganate, potassium peroxysulfate, potassium ferricyanide, etc., *alkali bismuthates* are obtained—i.e., alkali salts of *bismuthic acid*, $HBiO_3$—although generally in an impure state. Practically pure bismuthates were first successfully obtained in this way by Scholder (1941), who prepared the yellow *sodium metabismuthate*, $NaBiO_3$, brown *sodium orthobismuthate*, Na_3BiO_4 and violet to red *potassium bismuthate*, $KBiO_3$, and from them, by double decomposition with appropriate salts, the bismuthates of the alkaline earths and heavy metals. It was found by Zintl (1940) that sodium orthobismuthate was more readily prepared pure by a dry method, namely by heating Bi_2O_3 with Na_2O_2 in the molecular proportions 1 : 3 ($Bi_2O_3 + 3Na_2O_2 = 2Na_3BiO_4 + \frac{1}{2}O_2$). Under these conditions, the conversion of the bismuth into the pentavalent state takes place with such ease that the formation of bismuthate occurs simply by heating Bi_2O_3 with Na_2O in dry air. *Bismuth pentoxide*, Bi_2O_5 (in the hydrated state), is obtained by treating the alkali bismuthates with nitric acid. The presence of Bi_2O_3 as an impurity does not interfere with the reaction. Loss of oxygen very easily occurs in this process, and the products generally obtained must be regarded as mixtures of the pentoxide with *bismuth dioxide*, BiO_2 (Worsley and Robertson, 1920, Scholder, 1941).

(b) Salts of Bismuth

The typical salts of bismuth are all derived from the trivalent state. They are mostly colorless, and those derived from strong acids are soluble in water. However, dissolution to form a clear solution takes place only if there is an appreciable excess of acid. Basic salts are otherwise formed by hydrolysis, and these are obtained as insoluble precipitates. A number of the basic salts are derived from the bismuthyl radical $[BiO]^+$.

The bismuthyl salts are almost all insoluble. One salt which is soluble in water, and which forms a clear solution even without the addition of excess acid, is *bismuthyl perchlorate* $[BiO][ClO_4]$, easily obtained by dissolving bismuth oxide in moderately dilute perchloric acid (Fichter, 1923). It generally crystallizes as the monohydrate, in small rhombohedra. The rather unstable trihydrate crystallizes in needles from more dilute solutions, but the neutral salt $Bi[ClO_4]_3 \cdot 5H_2O$ crystallizes from concentrated perchloric acid. This is also unstable.

Double salts or acido-salts crystallize from mixed solutions of bismuth salts and alkali salts. These show that bismuth salts have some tendency to add on additional acid anions, to form complex acido-anions—e.g.,

$$BiCl_3 + Cl^- = [BiCl_4]^-$$

The concentration of free bismuth ions, Bi^{+++}, in bismuth salt solutions is therefore determined by a set of simultaneous equilibria such as

$$Bi^{+++} + H_2O \rightleftharpoons [BiO]^+ + 2H^+ \qquad \text{and} \qquad Bi^{+++} + 4X^- \rightleftharpoons [BiX_4]^-$$

and it is often difficult to specify with any accuracy.

(i) Bismuth halides

1. *Bismuth trifluoride and Fluorobismuthates.* Bismuth trifluoride, BiF_3, forms a grey-white crystalline powder (density 5.32), practically insoluble in water. It may be obtained by fuming down bismuth oxide with hydrofluoric acid, or by the addition of potassium fluoride to a bismuth nitrate solution, containing the minimum excess of acid. It is appreciably soluble in concentrated potassium fluoride solution, through the formation of a complex salt. Ammonium tetrafluorobismuthate(III), $[NH_4][BiF_4]$ was obtained by von Helmolt, in transparent lustrous crystals.

It was found by von Wartenberg (1940) that *bismuth pentafluoride*, BiF_5, was formed when BiF_3 was heated to about 500° in a stream of fluorine. It forms colorless rhombic needles, m.p. 725–730°. An oxyfluoride containing pentavalent bismuth had already been prepared in 1908 by Ruff, in the form af the double salt $KF \cdot BiOF_3$.

2. *Bismuth chloride, Bismuth oxychloride and Chlorobismuthates*. Bismuth chloride, $BiCl_3$, was first obtained by Boyle, by heating bismuth with mercuric chloride. It is best prepared by dissolving bismuth oxide in hydrochloric acid, or metallic bismuth in aqua regia. The solution is evaporated, and the residue is distilled in the absence of air. Bismuth chloride forms a snow-white crystalline mass, which is deliquescent in air (density 4.6, m.p. 232°, b.p. 447°).

Bismuth chloride is hydrolyzed by water, forming *bismuth oxychloride*, BiOCl (bismuthyl chloride). The latter forms a white crystalline powder (density 7.72), which turns yellow to brown when warmed, but becomes almost colorless once more on cooling. It also darkens gradually under the influence of light. The same is true of bismuth chloride. In this case, the brown discoloration gradually fades in the dark, whereas the discoloration of oxychloride does not.

With alkali chlorides and ammonium chloride bismuth chloride forms complex salts (*chlorobismuthates*), mostly of the types $M^I[BiCl_4]$ and $M^I_2[BiCl_5]$. It also forms addition compounds with other substances—e.g., with ammonia.

When bismuth trichloride is heated for prolonged periods with metallic bismuth in a sealed tube, *bismuth dichloride*, $BiCl_2$, is obtained, as a very hygroscopic, brown-black, crystalline mass. This is also formed as an intermediate product by the slow action of chlorine on bismuth at ordinary temperature. It is decomposed by water, with the deposition of bismuth oxychloride BiOCl.

3. *Structure of the Bismuth oxyhalides*. BiOF forms a (rather complicated) coordination lattice. The compounds BiOCl, BiOBr, and BiOI all form tetragonal layer lattices. They are built up from extended sheets, with the composition $[BiO]_n$, and one un-neutralized positive charge per metal atom; between these are inserted double layers of halogen atoms (Fig. 112, *a* and *b*). The lanthanum oxyhalides have the same structure. Bismuth forms a series of mixed oxyhalides with other metals, some of which are of complex composition. Most of these crystallize in tetragonal leaflets, like the simple oxyhalides, and they are built up on the same structural principle. Many of them such as $BaBiO_2Cl$ and $LiBi_3O_4Cl_2$ are built up of metal-oxygen sheets, between which is a single layer of halogen atoms (Fig. 112*c*). In other compounds, there is a three-decker layer of halogen atoms between each pair of metal-oxygen sheets, with a layer of metallic cations at the mid-plane of this (Fig. 112*d*). This cationic layer is usually incompletely filled, and this leads to the formation of compounds which do not have a simple stoichiometric composition—e.g., $Ca_{1.25}Bi_{1.5}O_2Cl_3$. The three sequences (*b*, *c* and *d* in Fig. 112) in which the layers may be built up may also be combined with one another, and this often leads to the formation of compounds which may vary in composition between certain limits (non-Daltonide compounds). This is noteworthy, as showing that variability of composition may arise not only in intermetallic compounds, but also in typical ionic or salt-like compounds, with structures fixed by particular geometrical considerations. It is not rational to speak of these as 'mixed crystals', since the compounds from which they must hypothetically be derived are often actually non-existent. Thus there is an oxychloride of the composition $Cd_{2-3x}Bi_{1+2x}O_2Cl_3$ (where x = 0.20 to 0.30), whereas compounds with the limiting compositions $Cd_2BiO_2Cl_3$ (x = 0) and $CdBi_4O_4Cl_6$ (x = 0.5) cannot be obtained (Sillen, *Naturwiss.*, 30 (1942) 318, *Z. anorg. Chem.*, 250 (1942) 173).

The compounds BiSCl, BiSBr, BiSI (as also BiSeBr, BiSeI, SbSBr, SbSI, SbSeBr and SbSeI), which crystallize in rhombic needles, have ribbon structures (Dönges, 1950). The ribbons—e.g., run along the direction of the *c* axis. The crystals consequently have a good cleavage parallel to this axis. The compounds BiTeBr and BiTeI crystallize in layer lattices. These are not tetragonal layer structures, like those of the oxyhalides, but are hexagonal, and are probably related to the structure of brucite; the Te and Br (or I) atoms are distributed in a regular manner among the O-atom positions of brucite.

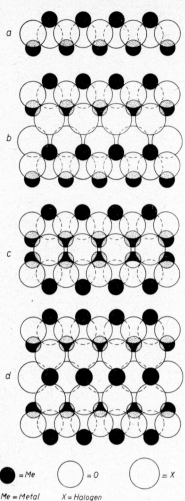

4. *Bismuth bromide*, $BiBr_3$, and *bismuth iodide*, BiI_3, closely resemble bismuth chloride. The bromide forms yellow crystals (density 5.60) which melt at 218° to a deep red liquid. Bismuth iodide forms glistening black to brown crystalline leaflets (density 5.8, m.p. 408°). It is much less soluble in water than the chloride, and so may be precipitated from solution by iodide ions. Bismuth bromide is readily converted by water into the white oxybromide, whereas the iodide, because of its low solubility, is hydrolyzed only on boiling. Bismuth oxyiodide, a heavy brick red powder, finds medical applications. Both bismuth bromide and bismuth iodide form complex salts and other addition compounds.

The tetraiodobismuthic acid, $H[BiI_4]$, (crystallizing with $4H_2O$) is of some analytical importance, since it gives insoluble precipitates with most of the alkaloids (highly toxic basic organic nitrogen compounds, such as strychnine, quinine, cocaine). Its potassium salt is used accordingly as a group reagent for the alkaloids. The 8-hydroxy quinolinium salt $[C_9H_6(OH)NH][BiI_4]$ is excellently suited for the quantitative determination of bismuth, being very insoluble (Berg, 1927).

(*ii*) *Bismuth thiocyanate and Thiocyanatobismuthates*. The yellow bismuth thiocyanate $Bi(SCN)_3$ crystallizes from solutions of bismuth hydroxide in thiocyanic acid. This salt adds on the alkali thiocyanates to form thiocyanato-bismuthates, $M^I_3[Bi(SCN)_6]$.

(*iii*) *Bismuth nitrate* is obtained by dissolving powdered bismuth in nitric acid. Large columnar triclinic crystals, with the composition $Bi(NO_3)_3 \cdot 5H_2O$, crystallize when the solution is evaporated. When these are heated, they lose both water and nitric acid, so that *bismuthyl nitrate*, $BiO(NO_3)$, ultimately remains. This is also formed, with other basic nitrates, according to the conditions, when the nitrate is dissolved in water. Basic bismuth nitrate, prepared according to a specified procedure, the bismuth subnitrate of the pharmacopoea, finds considerable applications in medicine. It has a disinfectant and deodorizing action and is also astringent. It is used as a dusting powder for wounds and for infections of the stomach and bowels, especially for cholera and dysentery. It is also used under the name of pearl white or Spanish white as a non-toxic paint. Its use in porcelain decoration has been mentioned.

● = *Me*   ○ = *O*   ○ = *X*

Me = *Metal*   *X* = *Halogen*

Fig. 112. Layer structure of bismuth oxyhalides.
The compounds crystallize tetragonally. The a-axis is horizontal to the plane of the illustration, the *c*-axis vertical. Examples of metal-oxygen layers without interposed halogen layers are PbO and SnO (cf. p. 545).

The neutral and basic bismuth nitrates constitute the usual starting materials for the preparation of other bismuth compounds.

(*iv*) *Bismuth sulfate*, $Bi_2(SO_4)_3$, crystallises in fine white hygroscopic needles, from a solution of bismuth or its oxide in concentrated sulfuric acid. With a small quantity of water it forms a hydrate, but it is hydrolyzed to a basic salt by larger amounts. It combines the alkali sulfates, forming sulfatobismuthates—e.g., $K[Bi(SO_4)_2]$ and $K_3[Bi(SO_4)_3]$.

(*v*) *Bismuth carbonate*. If solutions of bismuth salts are treated with alkali or ammonium carbonate, *basic bismuth carbonate*, $(BiO)_2CO_3$ (hydrated), is precipitated. It is found native as bismuth spar.

(*vi*) *Bismuth sulfide*. The ordinary sulfide, Bi_2S_3, is thrown down as a dark brown precipitate, soluble only in concentrated acid, when hydrogen sulfide is passed through the solution of a bismuth salt. $2Bi^{+++} + 3H_2S = Bi_2S_3 + 6H^+$. In contrast to the sulfides of arsenic and antimony, it is insoluble in alkali sulfides.

Bismuth sulfide is obtained as a lead-grey fibrous crystalline mass (density 6.5) by melting the constituents together. The brown amorphous bismuth sulfide precipitated from solution gradually passes into the grey crystalline form (more rapidly at elevated temperatures). The bismuth sulfide occurring naturally as bismuth glance (bismuthite) is of this form, existing as long prismatic and vertically striated steel grey to tin-white crystals, which are generally joined into columnar aggregates, and closely resemble stibnite in appearance. It is isomorphous with the latter and with native arsenic trisulfide. Although bismuth glance is somewhat rarely met with in large deposits, it is of considerable importance as an ore of bismuth.

Naturally occurring double compounds of bismuth sulfide with other sulfides have already been listed on p. 675. Double sulfides of similar composition may also be prepared artifically.

It has been stated that bismuth forms a *monosulfide*, BiS, in addition to the ordinary trisulfide. This is said to be produced as a slate grey powder by heating the product formerly regarded as bismuth monoxide, in a current of hydrogen sulfide. Although apparently stable in air at ordinary temperature, it is said to decompose into bismuth and bismuth trisulfide when heated to redness in an atmosphere of dry carbon dioxide.

The compounds Bi_2Se_3 and Bi_2Te_3 (and the isotypic compounds As_2Se_3 and Sb_2Te_3) have rhombohedral layer lattice structures. Sheets, each consisting of only one sort of atom, are stacked perpendicular to the *c*-axis, single and double sheets of Se- or Te-atoms alternate between sheets of Bi-atoms. If the single sheets of Te atoms are replaced by a layer of S atoms, the mineral *tetradymite*, Bi_2Te_2S is obtained. This structure therefore has the sequence of layers —Bi—Te—Te—Bi—S—Bi—, etc.

Bismuth combines with Se to form the compound BiSe (hexagonal) as well as the selenide Bi_2Se_3. Bi_2Te_3 is the only compound in the system Bi—Te (Parravano, Körber, 1930).

(c) Bismuth Hydride

(*i*) *Bismuth hydride* is formed in very small yield (together with much hydrogen) in the decomposition of a powdered magnesium-bismuth alloy with hydrochloric acid. Its stability is very low, and it decomposes slowly even at ordinary temperature, or instantaneously on heating. It can be detected by the formation of a bismuth mirror, in the same way as arsenic and antimony hydrides. The bismuth mirror can be distinguished from an arsenic mirror by its insolubility in sodium hypochlorite, and from an antimony mirror by its insolubility in ammonium polysulfide.

Bismuth hydride can most readily be detected by Donau's luminescence reaction. A small amount of calcium oxide, introduced into a hydrogen flame on a loop of platinum wire, absorbs bismuth from the flame if the burning gas contains bismuth hydride. If the wire is removed from the flame for a moment, so that the calcium oxide cools down, the lime emits an intense cornflower blue luminescence when it is reintroduced into the flame. The reaction, which may be repeated as often as desired, permits the detection of as little as 10^{-10} g of bismuth. Antimony also gives rise to an intense luminescence, but of a sky blue color, whereas arsenic produces only a weak greenish glow.

Earlier workers had attempted to prepare bismuth hydride, but with no success. It was first successfully obtained in 1918 by Paneth, who used one of the radioactive isotopes of bismuth, thorium-C, in his first experiments in place of ordinary bismuth. The use of this 'tracer' had the advantage that its intense radioactivity enabled incomparably smaller quantities of thorium-C hydride to be detected than would have been possible with ordinary bismuth hydride. In this way it was not only relatively easy to establish the existence of the compound sought for, but it was possible by the same method to ascertain its most important properties, and to investigate the most suitable experimental procedure for preparing the hydride, using ordinary bismuth.

(ii) *Bismuth alkyls.* The *bismuth alkyls*, BiR_3, can be regarded as derivatives of bismuth hydride, although they greatly exceed the hydride in stability. They may be obtained by the action of bismuth bromide with zinc alkyls:

$$2BiBr_3 + 3Zn(CH_3)_2 = 2Bi(CH_3)_3 + 3ZnBr_2$$

or by treating bismuth chloride with akyl magnesium bromides. It is noteworthy that they are able to add on chlorine or bromine forming derivatives of pentavalent bismuth—e.g.,

$$(C_6H_5)_3Bi + Cl_2 = (C_6H_5)_3BiCl_2.$$

16. Analytical (Bismuth)

Bismuth is precipitated from solution by hydrogen sulfide, as the brown sulfide, Bi_2S_3, insoluble in ammonium sulfide. This dissolves in 20% nitric acid, forming the nitrate. Ammonia precipitates bismuth as the white hydroxide, $Bi(OH)_3$, from solutions of its salts.

When acid solutions of bismuth salts are diluted with water, white precipitates are generally thrown down, consisting of basic bismuth salts. This is quite a characteristic reaction of bismuth salts, although it must be noted that antimony salts behave similarly.

For quantitative determinations, bismuth is usually precipitated and weighed as the sulfide. It may also be precipitated as the hydroxide or the basic carbonate, and weighed in the form of the metal after reduction with potassium cyanide.

According to Mahr (1932–35), small amounts of bismuth can advantageously be determined by precipitation as $Bi[Cr(SCN)_6]$. Reference has been made (p. 680) to the determination of bismuth as 8-oxyquinolinium tetraiodobismuthate.

In blowpipe reactions, bismuth compounds give a brittle metallic bead and a light yellow incrustation.

Cesium pentaiodobismuthate, $2Cs_2[BiI_5] \cdot 5H_2O$, which is rather sparingly soluble and crystallizes in characteristic habit, is very suitable for the microanalytical identification of bismuth, the limit of detection by this reaction being about 0.1 γ. For the detection of small traces of bismuth by the Donau reaction, see above.

References

1 D. M. YOST and H. RUSSELL, *Systematic Inorganic Chemistry of the Fifth and Sixth Group Nonmetallic Elements*, New York 1946, 423 pp.

2 W. MOLDENHAUER, *Die Reaktionen des freien Stickstoffs*, Berlin 1920, 178 pp.

3 H. GROSSMANN and W. WEICKSEL, *Die Stickstoffindustrie der Welt*, Berlin 1930, 199 pp.

4 B. WAESER, *Die Luftstickstoff-Industrie, mit Berücksichtigung der chilenischen Industrie und des Kokereistickstoffs*, 2nd Ed., Leipzig 1932, 509 pp.

5 H. PAULING, *Elektrische Luftverbrennung*, Halle 1929, 188 pp.

6 F. RASCHIG, *Schwefel- und Stickstoffstudien*, Leipzig 1924, 310 pp.

7 K. DREWS, *Die technischen Ammoniumsalze*, Stuttgart 1938, 200 pp.

8 O. KAUSCH, *Phosphor, Phosphorsäure und Phosphate; ihre Herstellung und Verwendung*, Berlin 1929, 325 pp.

9 M. LABBÉ and M. FABRYKANT, *Le Phosphore; Techniques chimiques, Physiologie, Pathologie, Thérapeutique*, Paris 1933, 396 pp.

10 E. C. FRANKLIN, *The Nitrogen System of Compounds*, New York 1935, 339 pp.

11 L. F. AUDRIETH and B. A. OGG, *The Chemistry of Hydrazine*, New York 1951, 244 pp.

12 P. W. WILSON, *Biochemistry of Symbiotic Nitrogen Fixation*, Madison, Wis. 1940, 302 pp.

CHAPTER 15

SIXTH MAIN GROUP OF THE PERIODIC SYSTEM: OXYGEN-SULFUR GROUP

Atomic numbers	Elements	Symbols	Atomic weights	Densities	Melting points	Boiling points	Heats of fusion	Heats of evaporation	Valences
							kcal/g-atom		
8	Oxygen	O	16.000	1.27^1	$-218.9°$	$-182.96°$	0.053	0.814	II
16	Sulfur	S	32.066	2.06^2	$119.0^{°3}$	$444.60°$	0.35	2.52	II, IV, VI
34	Selenium	Se	78.96	4.82^4	$220.2^{°4}$	688°	1.6	4.35	II, IV, VI
52	Tellurium	Te	127.61	6.25^5	452.0°	1390°	4.27	11.0	II, IV, VI
84	Polonium	Po	210	9.32	—	—	—	—	II, IV, VI

[1] Solid oxygen at the m.p. [2] Rhombic sulfur [3] Monoclinic sulfur. For the melting point of rhombic sulfur see p. 700 [4] Hexagonal-rhombohedral (grey, metallic) selenium [5] Hexagonal-rhombohedral tellurium

1. General [3]

Group VIA of the Periodic System comprises the elements oxygen, sulfur, selenium, tellurium and polonium. The first four of these, which are non-metallic in character, are often collectively known as the 'chalcogens', i.e., the ore-forming elements. All the elements of Group VIA can combine with hydrogen, and are distinctly electronegative in their compounds with the more electropositive elements. The non-metallic character is most strongly developed in oxygen and sulfur. Selenium and tellurium occupy a transitional position between the non-metals and the metals. Thus elementary selenium not only exists in one form resembling the non-metals, but also in a typically metallic modification. In the case of elementary tellurium, the metallic form is the usual one. In their chemical properties, however, these two elements are altogether non-metallic, some resemblance to the metals showing itself only in so far as selenium and tellurium are able to form salts with strong acids, in which they are the electropositive components. This is true of tellurium in particular, although even in this case the salts are not very stable. The last (heaviest) element of the Group—the radioactive and short-lived polonium—has a more marked metallic character. It can exist in solution as an elementary positive ion, and its behavior resembles not only that of tellurium, but also that of its neighbor, bismuth.

Although oxygen is invariably *diatomic* in the gaseous state, even at very low temperatures, its homologues form diatomic molecules only at rather high temperatures. The energies of thermal dissociation of these molecules decrease markedly with increase in atomic weight.

	O_2	S_2	Se_2	Te_2	
Dissociation energy	116.4	75.6	62	53	kcal per mo

The capacity to function as *electronegatively bivalent* in their compounds is common to all the elements of Group VIA. On the Kossel-Lewis theory this is explained by the fact that each stands two places before an inert gas, so that the tendency to assume an 'inert gas configuration' implies that they all readily add on two electrons.

This must not be understood to mean that the acquisition of a two-fold negative charge by these elements is a process liberating energy. In fact, this is not the case. In the case of oxygen, for example, the first stage, whereby *one* electron is added to the free O atom, is *exothermic* (about $+56$ kcal per g atom, according to Briegleb, 1942), but energy must be *expended* to add on a second electron, and the over-all process $O + 2e \rightarrow O^{2-}$ is *endothermic*. The electron affinity of oxygen (to form O^{2-}) is *negative*, the free energy of formation of O^{2-} from O (and *a fortiori* from O_2) is *positive*; it can be calculated by means of the thermo-chemical cycle discussed on p. 146. Briegleb gives -150 kcal per g atom for the electron affinity of O (going to O^{2-}). (A corresponding value for $OH + e \rightarrow OH^-$ is $+45$ kcal per g ion.) The electron affinity of $S_{atomic} \rightarrow S^{2-}$ is similarly found as -84 kcal per g atom; the first stage electron affinity of sulfur ($S_{atomic} \rightarrow S^-$) is found from spectroscopic data as about $+65$ kcal per g atom. The work expended in forming bivalent negative ions is afforded by the *lattice forces* in the case of crystalline compounds, whereas the *hydration energy* of ions furnishes the energy when reactions take place in solution.

The solution potentials provide a measure of the free energy of formation of the electrolytic ions. The normal potentials corresponding to the formation of bivalent negative electrolytic ions at 18° (relative to the normal hydrogen electrode) are,

for	O_2	S	Se	Te	
	$+0.41$	-0.51	-0.77	-0.91	volts.

The value given for oxygen applies to a solution 1-molar with respect to OH^- ions, while for the other values the corresponding solutions are 1-molar in $S^=$, $Se^=$ and $Te^=$ ions. The potentials show that the readiness with which conversion to the negative ions takes place decreases strongly from oxygen to tellurium. If an oxygen electrode is combined with the normal hydrogen electrode, the *positive current* flows from the oxygen electrode to the hydrogen electrode in the outer circuit (i.e., electrons flow from the latter to the former in the outer circuit). Oxygen goes into solution at the former, taking up a double negative charge (the $O^=$ ions so formed at once reacting with H_2O, to form $2OH^-$), whereas hydrogen goes into solution at the latter, forming hydrogen ions. By contrast, when a platinum foil dipping into a sulfide solution (a 'sulfur electrode') is combined with a hydrogen electrode, the positive current flows in the outer circuit from the latter to the former: $S^=$ ions are discharged. The ions $Se^=$ and $Te^=$ behave similarly, except that their tendency to discharge is even greater.

The maximum valence state corresponding to the Group number (six) is encountered with sulfur, selenium and tellurium, and also with polonium, but never for oxygen. *Oxygen is never more than bivalent* in principal valence compounds.* As has been proved by spectroscopic evidence, there are six electrons in the 'outer shell' of oxygen (i.e., in orbits with principal quantum number $n = 2$), but as follows from the high ionization potential of oxygen (cf. Table 23, p. 119) even the

---

* For the valence of oxygen in oxonium compounds, see p. 696.

electrons in the outer shell of this atom are very tightly bound.* For this reason, there is no element which can abstract electrons from oxygen. In all its ionic compounds, oxygen is therefore the negative component. As such it is normally bivalent, but never more than bivalent**.

According to the Kossel theory, the homologues of oxygen, however, can give up electrons, and can therefore be regarded as the electropositive component, in compounds with more electronegative elements.*** If fewer than 6 electrons are given up, the homologues of oxygen exercise a lower positive valence. In doing so, they generally give up the electrons in pairs, as also do the elements of the preceding Main Group. Thus, valences of six, four and—although less frequently— two are displayed in compounds regarded by Kossel as ionic.

Sulfur preferentially exerts a valence of six towards strongly electronegative elements, such as oxygen and fluorine. Selenium and tellurium also tend to be hexavalent towards fluorine, but quadrivalent towards oxygen.

Oxygen is also invariably bivalent in covalent compounds except in certain free radicals (mentioned on p. 695), which exist only in solution and are formed by the dissociation of certain organic peroxides.

That oxygen never exerts a valence higher than 2 in covalent compounds is explained, on the quantum mechanical valence theory of Heitler and London, in the following way. In the oxygen atom there are six outer electrons in an energy level with principal quantum number $n = 2$. According to the Pauli principle (cf. p. 120), this energy level can accommodate a maximum of eight electrons. Each covalent bond involves a pair of electrons, to which each of the linked atoms contributes one, so that in forming two covalences the number of outer electrons of the oxygen atom is made up to the maximum, eight. In order to exert more than two covalences, at least one electron would have to be excited from a 2-quantum to a 3-quantum orbit, requiring a considerable expenditure of energy—about 210 kcal per g atom—as shown by the spectral terms of oxygen. In the homologues of oxygen, more than two covalences can be formed without the necessity of exciting any electrons to levels of higher principal quantum number, since more than eight electrons can be accommodated in the outermost electron levels of their atoms, in accordance with the higher principal quantum number of the outer levels. The homologues of sulfur are bivalent in the elementary state, in both their crystalline and amorphous forms (cf. pp. 703 and 744). In these, each atom has two others as nearest neighbors, so as to build up rings or chains. The distance between atoms belonging to different rings or chains is considerably larger, so that it is permissible to assume that secondary valence forces (Van der Waals forces) operate between them. It may be deduced from the ratio between interatomic distances within and between the chains, that the strength of the secondary valence forces increases

---

* This is the consequence of the high effective nuclear charge, together with the low quantum number (2).

** In hydrogen peroxide and its derivatives, oxygen is electrochemically univalent, although formally bivalent. See p. 695 for the univalent O_2^- ion, present in the alkali and the alkaline earth superoxides.

*** It is very doubtful whether it is permissible to regard any of the compounds of *hexa*valent sulfur, etc. as ionic. Indeed in all the binary compounds containing sulfur, selenium and tellurium in positive oxidation states, these elements are undoubtedly linked by bonds which are predominantly covalent.

from Se to Te. This can be considered as a sign of a transition towards metallic binding.

Towards strongly electropositive elements, all the elements of Group VIA function as negatively bivalent. Their normal compounds with the metals and with hydrogen, accordingly correspond to the general formula M^I_2R or H_2R. The hydrides (cf. Table 99), like all the simple hydrides of the Main Groups on the right hand side of the Periodic System, are volatile, and dissociate electrolytically to a varying degree, liberating hydrogen ions.

TABLE 99
PROPERTIES OF HYDRIDES OF THE CHALCOGENS

Compound	H_2O	H_2S	H_2Se	H_2Te
Melting point	$0.00°$	$-85.60°$	$-60.4°$	$-51°$
Boiling point	$100.00°$	$-60.75°$	$-41.5°$	$-1.8°$
Critical temperature	$366°$	$100.4°$	$137°$	—
Heat of fusion, in kcal per mol (at the m.p.)	1.430	0.5676	—	—
Latent heat of evaporation in kcal mol (at the b.p.)	9.715	4.463	4.75	ca. 5.7
Trouton's constant	25.9	20.9	20.4	ca. 21
Density at the b.p.	0.958	0.993	2.004	2.650
Surface tension at the b.p. (dynes per cm)	58.9	28.7	28.9	30.0
Heat of formation (at 20° and const. pressure), kcal per mol	$+68.35$	$+4.80$	-18.5	-34.2
Free energy of formation at 25°, kcal per mol	-56.72	-8.42	—	—
Electrolytic dissociation constants at 18°: K_1	$*1.3 \cdot 10^{-16}$	$**0.873 \cdot 10^{-7}$	$1.9 \cdot 10^{-4}$	$2.3 \cdot 10^{-3}$
K_2	—	$***0.79 \cdot 10^{-13}$ at 20°	—	—
Dipole moment of the gas molecule (in c.g.s. units)	$1.85 \cdot 10^{-18}$	$1.10 \cdot 10^{-18}$	—	—

* The *dissociation constant of water* usually means the product $[H_3O^+] \cdot [OH^-]$, which has the value $0.74 \cdot 10^{-14}$ at 18° (cf. p. 52). – ** Kubli, 1946. – *** Konopik, 1949.

The volatility of the hydrides increases strikingly on going from H_2O to H_2S, and then declines again with the heavier homologues. The relatively low volatility of water arises from its strongly associated nature in the liquid state (p. 48), which is shown also by the abnormal value of the Trouton constant and the high surface tension. The critical temperatures and melting points follow a course similar to that of the boiling points. Heats of formation diminish markedly from H_2O to H_2Te. The electrolytic dissociation increases strongly in the same direction.

This increase in acidic character from water to hydrogen telluride can be broadly regarded as a consequence of the increasing radius of the negative ion from oxygen to tellurium. The greater this is, the smaller is the amount of work to be performed in order to remove a positively charged hydrogen atom, in accordance with Coulomb's law. Under otherwise comparable conditions, the equilibrium must be proportionately displaced in favor of dissociation. The increase in acidic character of the volatile hydrides of the other groups of the Periodic System with increase of atomic number is explained in the same way.

The compounds of the chalcogens (O, S, Se, Te) with electropositive elements are collectively called chalcogenides. Chalcogenides of strongly electropositive elements form typical ionic crystal structures, and frequently have the same structure as halides of corresponding composition. Thus chalcogenides of the general formula $M^{II}X^{II}$ often have the rock salt structure, like the halides $M^{I}X^{I}$. The alkali hydrogen chalcogenides also form typical ionic lattices, most having a rhombohedrally distorted NaCl structure at ordinary temperature, and the true NaCl structure at high temperatures. CsSH and CsSeH form CsI-type structures.

There are several analogies between the alkali hydrogen chalcogenides $M^{I}HR$ and the alkali halides, $M^{I}X$ —in the trends of their heats of formation (Fig. 113), heats of solution and heats of hydration, and in the proton affinities of the ions HR^{-} and X^{-} (Klemm, 1941).

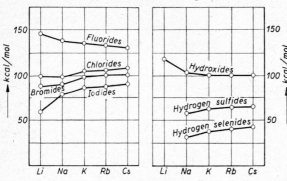

Fig. 113. Heats of formation of the alkali-halides and -chalcogenides.

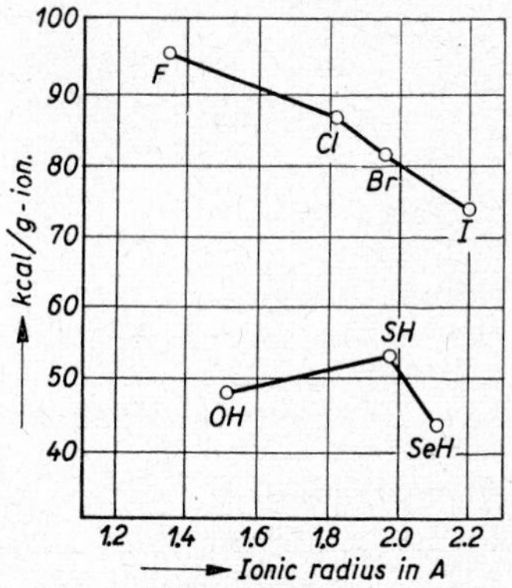

Fig. 114. Electron affinities of the halide- and hydrochalcogenide-radicals.

On the other hand, the proton which remains attached to the residue R^{-} has an influence on the calculated *negative electron affinities* of the free radicals RH. These change in a different manner from those of the halogen atoms, as Fig. 114 shows. The analogies between $M^{I}HR$ and $M^{I}X$, and between HR^{-} and X^{-}, well exemplify the so-called *hydride displacement law* (Grimm, 1925), which states that when an atom standing up to four places before an inert gas combines with m atoms of hydrogen ($m = 1, 2, 3, 4$), its chemical properties alter in such a way that the resultant group of atoms behaves like a 'pseudo-atom' of the element m places to the right in the Periodic Table. This implies that the radicals CH_3, NH_2, and OH are analogues of the F atom or (if negatively charged) of the F^{-} ion. The radicals CH_2^{2-} and NH^{2-} are analogues of the O^{2-} ion, and the neutral molecules CH_4, NH_3, OH_2 and FH are comparable in certain respects with the inert gas Ne.

2. Oxygen (O)

(a) Occurrence

Oxygen is the constituent of the atmosphere which supports respiration and is vital for life. The oxygen content of dry air is 20.946 ± 0.002 per cent by volume, or about 23.0 per cent by weight, and is constant except for very minor local changes due to the consumption of oxygen by combustion or respiration.* Although oxygen is continually consumed by respiration and combustion processes, it is also continuously replaced by the mechanism of assimilation of CO_2 by green

* Older analyses of air, prior to the work of Benedict (1912), seemed to reveal small local variations in oxygen content from place to place. These variations have been traced to consumption of oxygen by oxidation processes between sampling and analysis, and the oxygen content of the atmosphere is now considered to be a geophysical constant (Krogh— cf. Glueckauf *Compendium of Meteorology*, Am. Meteorological Society; Paneth, *Quart. J. Roy. Meteor. Soc.*, 63 (1937) 433).

plants in sunlight. The action of gravity must produce some change in the composition of the atmosphere with increasing height, in the sense that the proportion of the heavier components (oxygen, nitrogen, argon) must diminish relative to the lighter components (helium, hydrogen). This effect shows up, however, only at very great altitudes. Close to the earth's surface it is nullified by the continual mixing of the atmospheric gases by convection and turbulence. It has been found that the effects of turbulence reach to much greater altitudes than was formerly assumed. Paneth, from the investigation of air samples collected from the stratosphere with the aid of rockets, has found the composition of the air up to an altitude of 60 km to be still the same as at the earth's surface (*J. Atm. Terr. Phys.*, 1 (1950) 49; *Nature*, 168 (1951) 358). Only at still greater heights is it possible to detect a change in the relative proportions of the atmospheric gases, due to gravitational separation.

Water contains 88.81% by weight of oxygen, the oceans about 85.8%, and the accessible crust of the earth about 47.3% in the form of oxides and oxy-salts. The total oxygen content of the earth's crust, oceans, and atmosphere together is estimated at about 50% by weight—i.e., oxygen enters into the substance of the earth's crust, including its atmospheric mantle, to as great an extent as all the other elements put together.

As it occurs in Nature, oxygen is a mixture of three isotopes with mass numbers 16, 17 and 18 (see Vol. II, Chap. 12). Urey has found that the ^{18}O content of the ocean water at the present time is about 7% lower than that of the igneous rocks (granites, basalts), and on the average about 7% higher than that of the sedimentary rocks (cf. Vol. II, Chap. 15).

According to Dole (1936), the atomic weight of atmospheric oxygen is higher by 0.00011 atomic weight unit than that of oxygen from sea water. Although this difference (due to differences in the $^{16}O : {}^{17}O : {}^{18}O$ isotope ratio) appears to be very slight, it is important, since the atomic weight of oxygen—or more precisely, the atomic weight of the mixture of oxygen isotopes as it occurs in nature—forms the basis for the atomic weights of all the other elements.

(b) History

Before the discovery of oxygen, the air was held to be a simple substance. It was first explicitly stated by Scheele, in 1777, in his 'Treatise on Air and Fire' that the air must be made up of two components, of which only one was capable of supporting respiration and combustion. The experiments which led to this new perception had already been carried out by Scheele a few years previously. He first obtained pure oxygen by strongly heating saltpeter, and later also prepared it by treating manganese dioxide with concentrated sulfuric acid. Quite independently of Scheele, Priestley in 1774 succeeded in preparing oxygen by heating mercuric oxide and red lead. Lavoisier based his theory of combustion upon the discovery of oxygen. Since he considered oxygen to be an essential constituent of acids, he called it the 'principe acidifiant', or oxygène (from ὀξύς, sour, and γεννάω, I produce), or in latin form, oxygenium. When it was later discovered that there are acids containing no oxygen, Lavoisier's name was nevertheless retained.

(c) Preparation

Oxygen is prepared technically [1] for the most part by the *fractional distillation of liquid air* [2]—the Linde process—or the *fractional liquefaction of air*—the Claude process. Considerable amounts of oxygen are also prepared by the *electrolysis of water* (see under hydrogen). Oxygen obtained by the liquefaction of air usually contains a little argon (3% on the average), the boiling point of which lies close to that of oxygen.

Oxygen comes into commerce compressed, in steel cylinders, and is generally taken from these when wanted in the laboratory. If it is required to prepare it, use may be made either of electrolysis (e.g., of potassium hydroxide solution, using nickel electrodes), or of the thermal decomposition of oxygen compounds—e.g., by heating potassium chlorate (to which manganese dioxide or other catalyst is added, to accelerate the decomposition) or potassium permanganate.

Oxygen may be obtained in exactly calculable amount by dropping potassium permanganate solution into hydrogen peroxide, acidified with sulfuric acid (cf. p. 56). A method of preparation which is convenient for many purposes is by the reaction of bleaching powder with hydrogen peroxide solution, which can be carried out in a Kipp's apparatus.

Traces of oxygen present in gases may be detected by a method due to Kautsky and Hirsch (*Z. anorg. Chem.*, 222 (1935) 126). This is based on the observation that traces of oxygen inhibit the phosphorescent after-glow of trypaflavin (an acridine dyestuff) adsorbed on silica gel. H_2O and NH_3, which can also inhibit the phosphorescence, must be removed before submitting a sample to the test.

For the removal of traces of oxygen from inert gases used in the laboratory for protective atmospheres (hydrogen, nitrogen, argon), the 'copper tower' of Fricke and Meyer is useful (cf. Fricke, *Z. Elektrochem.*, 53 (1949) 76). This contains active copper deposited on kieselguhr. When heated to 160°, this will reduce the oxygen pressure below $3 \cdot 10^{-4}$ mm. After passing through the tower, a gas at atmospheric pressure therefore contains, at the most, only $4 \cdot 10^{-5}$ % oxygen.

(d) Properties and Uses

The most important physical constants of oxygen are collected together in Table 100. Ordinary elementary oxygen consists of diatomic molecules, O_2. (For ozone, O_3, see p. 692.)

Liquid oxygen is pale blue in color. It is fairly strongly paramagnetic. It does not conduct electricity. Solid oxygen is also light blue. It is hexagonal at temperatures a little below the freezing point, but changes its crystal structure when more strongly cooled (to —230°). It undergoes another transformation at —255°.

Except for NO, O_2 is the only diatomic gas molecule which has been proved to be paramagnetic. Whereas, however, the NO molecule contains an odd number of electrons, the O_2 molecule contains an even number. It must be concluded from the paramagnetism that not all the electrons are coupled to form pairs with opposed spins, and it can in fact be deduced from spectroscopic evidence that the O_2 molecule contains two 'unpaired' electrons (cf. p. 132 on this point).

The most characteristic property of oxygen is its ability to unite with most other elements, with the evolution of heat and light. In order to initiate this 'combustion', it is in general necessary to heat to some particular temperature, the *ignition temperature*, since oxygen is a rather inert element at ordinary temperature. In the presence of moisture, however, combination with oxygen frequently occurs slowly even at ordinary temperature ('slow combustion'). The most important example of this is found in the respiration of organisms, but many other slow combustion processes proceed at ordinary temperature in Nature (cf. p. 687 *et seq.*).

Rusting and the tarnishing of metals, the decay of wood, and sundry other processes of rotting are among these slow combustion processes. Since these processes are slow, the heat evolved in them is generally dissipated in the surroundings as fast as it is produced. In some circumstances, however, it may accumulate and lead eventually to ignition—i.e., to a self-propagating combustion process, accompanied by fire. This is the origin of the 'spontaneous combustion' of moist hay, straw, coal and other combustibles. Such substances should never be stored in large amounts if they are moist, but should at least be well ventiltated in order to carry off the heat, and their temperature should be watched.

TABLE 100

PHYSICAL CONSTANTS OF OXYGEN

Liter weight at 0°, 760 mm pressure, and 45° latitude	=	1.42895 g
Density (relative to air = 1)	=	1.1053
Density of liquid oxygen at the boiling point	=	1.118
Density of solid oxygen at —252°	=	1.426
Boiling point	=	—183.00°
Melting point	=	—218.9°
Critical temperature	=	—118.8°
Critical pressure	=	49.7 atm
Critical density	=	0.430
Specific heats of gaseous oxygen (at 20°): c_p	=	0.2184 cal
c_v	=	0.183 cal
c_p/c_v	=	1.395
Molecular heats: C_p	=	6.989 cal
C_v	=	5.84 cal
Specific heat of liquid oxygen at —200°, c_p	=	0.395 cal
Specific heat of solid oxygen at —222°, c_p	=	0.336 cal
at —256°, c_p	=	0.078 cal
Heat of vaporization of liquid oxygen	=	51 cal/g at —183°
Heat of fusion of solid oxygen	=	3.3 cal/g at —219°
Thermal conductivity of oxygen at 0° (cf. Table 7, p. 39)	=	0.0000570 cal cm^{-1}sec^{-1}degree^{-1}
Solubility of oxygen in water: at 0°	=	0.0491
at 20°	=	0.0311
at 100°	=	0.0170 volumes (reduced to 0°) in 1 volume of water
Magnetic susceptibility	=	104.4 · 10^{-6} (at 20°)
Parachor	=	54.0
O_2 *molecule.* Dipole moment	=	0
Moment of inertia	=	19.23 · 10^{-40} g cm^2
Internuclear distance O—O	=	1.20 · 10^{-8} cm
Energy of dissociation	=	5.082 e-Volts (117.2 kcal/g mol)
Ozone, O_3. Boiling point	=	—111.5°
Melting point	=	—251.4°
Critical temperature	=	—5°
Critical pressure	=	92.3 atm
Critical density	=	0.54
Density at boiling point	=	1.46
Molecular heat of vaporization	=	$3500 + 3.499\,T — 0.05817\,T^2$ cal
Magnetic susceptibility	=	0.15 · 10^{-6} (at —183°)
Solubility in water: at 0°	=	0.494
at 18°	=	0.454 volumes (reduced to 0°) in 1 volume of water

Flame temperatures above 2000° can be attained by burning oxygen with hydrogen in a suitably constructed blowpipe, and still higher temperatures may be reached with the oxy-acetylene torch. Such blowpipe flames are employed for the welding and cutting of metals, and for melting platinum, quartz and other refractory substances. Liquid oxygen, or liquid air much enriched in oxygen, is often used in the production of explosives, obtained by saturating voluminous charcoal or other combustible materials such as petroleum, paraffin or naphthalene, with liquid oxygen or liquid air. Liquid oxygen, and especially liquid air, are much used in the laboratory for cooling baths, both for low temperature work and also, for example, to freeze out readily condensable impurities such as H_2O and CO_2 from non-condensable gases. For other uses see [2].

In recent times the use of oxygen in *breathing apparatus* has attained increasing importance —e.g., for the use of airmen, divers, fire brigades, mine rescue teams, and other rescue workers. These devices are either furnished with steel cylinders for compressed oxygen, or else oxygen is evolved by the decomposition of alkali peroxides (cf. p. 177). Appliances in which oxygen is carried in the liquefied state are also in use. Oxygen breathing appliances also contain absorbents for the exhaled carbon dioxide. Special forms of oxygen apparatus are used for the resuscitation of asphyxiated or drowned persons, or for other cases of unconsciousness or apparent death. In the resting state, an adult consumes about 20 liters of oxygen per hour, or several times that amount if performing bodily exertion. Inhalation of pure oxygen is harmless, provided that it is not at a pressure appreciably above 1 atmosphere.

3. Special Forms of Oxygen

(a) Ozone

The formation of ozone can always be observed when air has been subjected to the action of an electrical discharge—e.g., in a room in which an electrostatic machine or sparking coil has been operated. Ozone is recognizable by its peculiar smell, from which it gets its name (Gk. ὄζειν, to smell). As early as 1785, Van Marum noticed the occurrence of this smell when working with the electrostatic machine. Schönbein, who gave it its name, proved (1840) that ozone was formed during the electrolysis of dilute sulfuric acid. It is also formed when moist white phosphorus is slowly oxidized in the air, when substances rich in oxygen, such as permanganate and dichromate, are decomposed by sulfuric acid, and by the action of fluorine on water. It is also produced by the action of ultraviolet light on molecular oxygen, and is therefore almost always detectable near quartz mercury vapor lamps.

Ozone was originally believed to be a higher oxide of hydrogen, but it was shown that it consisted solely of oxygen. It may be obtained from pure dry oxygen, and is reconverted into oxygen once more upon decomposition. In this process, 3 parts by volume of ordinary oxygen are formed from 2 parts by volume of ozone. From this, with Avogadro's law, the formula follows as O_3, as is confirmed by the vapor density.

The question whether ozone is present in the atmosphere close to the earth's surface was, until relatively recently, open to dispute. Older investigations, carried out by chemical methods, had indicated an ozone content of a few millionths per cent. Since the detection and determination of ozone rested upon its oxidizing properties, it was not easy to exclude the possibility that the effects were due to other substances occasionally present in the atmosphere, such as NO_2, and the validity of the results was questioned. More recently, Buisson and Fabry (1930–31) and Götz and Ladenburg (1931) quite independently proved the presence of ozone in the air by measurements of the absorption of light in the ultraviolet. The former found a content of $2.2 \cdot 10^{-6}\%$ in Provence, at 300 meters altitude, and the latter found $2.9 \cdot 10^{-6}\%$ at 2000 meters, in the Alps at Arosa. These agreed in order of magnitude with the older chemical figures. Paneth and Edgar (1938–41) showed that ozone could be condensed and concentrated from the atmosphere by adsorption on silica gel, and proved that the danger of confusion with NO_2 was avoidable. A method for the continuous electrochemical measurement of the oxidizing power of atmospheric ozone was described by Glueckauf, Heal, Martin and Paneth (*J. Chem. Soc.* (1944) 1), and this has enabled our knowledge of atmospheric ozone to be considerably extended. It has been found that the concentration is variable, depending on meteorological conditions and on the time of year. It is greatest in summer (up to about 0.07 parts per million in May), and lowest in winter (0.02 parts per million during daylight in November). These variations are essentially due to variations in atmospheric turbulence and air movements, which transport ozone to the earth's surface from the site at which it is formed.

The ozone content of the atmosphere originates from the photochemical dissociation of O_2 molecules into atoms by ultraviolet light; these atoms form ozone by three-body collisions with other O_2 molecules. $O_2 + O = O_3 + 24$ kcal. Some of the O_3 molecules are decomposed again photochemically, and some react by collision with O atoms to reform O_2 molecules. The interplay of ozone formation and decomposition sets up a stationary concentration of ozone (which depends on the temperature and pressure), and this has its maximum at about 25 km altitude. The number of ozone molecules in the whole atmosphere is about the same as the number of O_2 and N_2 molecules in a layer of air a few millimeters thick under ordinary conditions. Near the earth's surface, ozone is rapidly destroyed under the influence of atmospheric dust, etc. Hence the ozone content of the air at ground level is extremely small. The 'ozone-rich air of pine woods and health resorts' is a complete fable.

Ozone is generally prepared by the action of the silent electric discharge upon oxygen in a Siemens ozonizer. This consists, in essence, of two concentric glass tubes, between which is passed a current of well dried oxygen (or air). The glass tubes have conducting coatings charged with high tension alternating current, in such a way that a 'dark' discharge passes through the oxygen. The oxygen is transformed into ozone to a small extent. A higher concentration of ozone in the oxygen can be obtained by passing the gas through several ozonizers in series. Ozone-oxygen mixtures with upwards of 10% ozone can readily be obtained in this way.

Pure ozone (see Table 100, for physical constants) can be obtained from ozone-oxygen mixtures by liquefaction. It is perceptibly blue even in the gaseous state, and is a very deep blue, almost black, when liquefied. Riesenfeld found that ozone was not miscible with liquid oxygen in all proportions at low temperatures.

A considerable quantity of heat is liberated by the decomposition of ozone:

$$O_3 = \tfrac{3}{2}O_2 + 34.2 \text{ kcal (at const. vol.)}$$

Pure ozone is therefore dangerously explosive.

Dilute ozone decomposes only very slowly at ordinary temperature. The decomposition is strongly accelerated not only by heating, but also by various catalysts such as manganese dioxide, lead dioxide and soda lime. Irradiation with ultraviolet light also accelerates the decomposition.

Work with dilute ozone such as is obtained from an ozonizer is quite free from hazard. It should be borne in mind that ozone, in high concentrations, has a corrosive action on the respiratory organs. Rubber tubing is made unsuitable by ozone in a few seconds, so that joints in tubing should be sealed with mercury. In general, ozone reacts energetically with organic compounds at ordinary temperature. It bleaches dyestuffs and destroys micro-organisms. Many substances, such as ether, alcohol, coal gas, cotton wool soaked in turpentine, inflame in oxygen rich in ozone. The careful treatment of unsaturated organic compounds with ozone frequently yields *ozonides*, in which ozone is added on at the double bond. As a rule, ozonides are highly unstable substances, and are usually decomposed by water in a simple manner which can be readily interpreted. The reaction is often useful, both in the determination of structure and for preparative purposes.

In general, ozone is capable of reacting with many substances with which oxygen cannot react at ordinary temperature. It converts black lead sulfide, PbS, into white lead sulfate, $PbSO_4$, and white lead(II) hydroxide, $Pb(OH)_2$, into brown lead dioxide, PbO_2. Its property of blackening metallic silver, with the formation of silver(II) oxide, is very characteristic. Its reaction with potassium iodide in neutral aqueous solution is generally used for quantitative determination:

$$O_3 + 2KI + H_2O = I_2 + 2KOH + O_2$$

After acidification, the iodine liberated is titrated with thiosulfate.

In addition to its use in preparative chemistry, ozone is used for the purification and sterilization of air—e.g., in slaughterhouses, breweries, cold stores, etc., and for the sterilization of drinking water. However, the ozone process of water purification is everywhere being displaced by the cheaper method of treatment with chlorine.

The ozone molecule is *triangular* in shape. This is shown by the infrared absorption spectrum, and has been confirmed by electron diffraction experiments (Pauling, Shand and Spurr, 1943). The three O-atoms are equidistant (distance $O \leftrightarrow O = 1.26$ Å).

(b) Atomic Oxygen

Like hydrogen, oxygen may be dissociated into free atoms by the action of the glow discharge at low pressures (less than 1 mm Hg), although the yield is smaller. According to Harteck (1930 and later), atomic oxygen is relatively unreactive. Thus only 0.2% of collisions with H_2 molecules lead to the formation of water. It also reacts only to a very small extent with CH_4 (less than 1%), HCN (5%) and CO (5%), but reacts with 100% efficiency with HBr, H_2S, CS_2, CH_3Cl, CH_2Cl_2, $CHCl_3$ and many other organic compounds. When atomic oxygen is allowed to react with organic compounds at very low temperatures (the boiling point of liquid oxygen), the products—as in the low temperature reactions of atomic hydrogen—are frequently species which are not stable at ordinary temperature. Atomic oxygen reacts with molecular oxygen according to the equation $O_2 + O = O_3$. The ozone so formed is reconverted to ordinary oxygen by subsequent collisions $O_3 + O = 2O_2$. For this reason, the atomic oxygen produced at low pressures is not admixed to any great extent with ozone.

(c) Activated Oxygen

The reactivity of oxygen can also be enhanced by activation of the O_2 molecules. As was shown by Kautsky (1931–33), this can occur for example, if the O_2 molecules are brought in contact with dyestuffs, which are excited to fluorescence by irradiation with light. The dyestuff molecules are raised to excited states by the absorption of light quanta. If, before they have re-emitted the absorbed energy in the form of fluorescent radiation, they undergo a collision from an O_2 molecule, it is possible for their excitation energy to be transferred to the O_2 molecule. This is thereby brought into a short-lived state in which it can effect oxidations that can not be brought about by normal oxygen.

4. Compounds of Oxygen

(a) Oxides and Peroxides

The compounds of oxygen with other elements are called *oxides*, and the process itself is known as *oxidation*.

The term oxidation is often used not only for the actual process of combination with oxygen, but in a more general sense, to include processes which are essentially similar in their nature—e.g., for the union of metals with halogens, sulfur and other elements of electronegative character (see the following chapter).

A distinction is made between *normal oxides* and *peroxides*. The former are formally derived from water, as the basic type, by replacing its hydrogen atoms by equivalent amounts of other atoms or radicals. The peroxides are similarly related to hydrogen peroxide, and contain the grouping —O—O— characteristic of this. Each of the atoms of the peroxide group has only one free valence, the other being used in the mutual linkage of the oxygen atoms. Thus oxides are compounds of the general formula R'_2O, and peroxides compounds of the type R'_2O_2.

Examples of normal oxides:

examples of peroxides:

$$\begin{array}{l}
\overset{H}{\underset{H}{\diagup}}O; \quad \overset{Cl}{\underset{Cl}{\diagup}}O; \quad N=O; \quad \overset{O}{\underset{B}{B\diagdown}}O; \quad S\overset{O}{\diagdown O} \qquad \begin{array}{l} H-O \\ | \\ H-O \end{array}; \quad \begin{array}{l} R'-O \\ | \\ R'-O \end{array}
\end{array}$$

In so far as certain of the foregoing oxides are composed of individual molecules, they correspond to water not only formally, but also structurally. Such oxides are usually gaseous or readily volatile, and are mostly formed by *non-metals*. The majority of the oxides of the metals are non-volatile, and salt-like in structure (cf. Chap. 9). Certain oxides such as B_2O_3 and SiO_2 occupy an intermediate position, being non-volatile, but tending to solidify as glasses.

Oxygen is *bivalent* in both the normal oxides and the peroxides. Only in rare and exceptional cases are compounds formed in which oxygen is *univalent*.

If the methyl ester of phenanthrene hydroquinone is oxidized by potassium ferricyanide or lead dioxide, a peroxide is obtained:

$$CH_3O \qquad O\text{————}\quad O \qquad OCH_3$$

This substance is itself colorless, but its solutions are colored. From this, and the evidence of molecular weight determinations, it is to be inferred (Goldschmidt, 1922) that dissociation takes place to some extent upon dissolution, at the point marked ⦙. In the highly unsaturated product of dissociation, oxygen must be considered as formally univalent, and the case is essentially similar to that of the compounds of trivalent carbon (triphenyl methyl). Although the existence of organic compounds of univalent oxygen which exist *only* in the monomolecular form has occasionally been assumed, it has never been substantiated.

The occurrence of the univalent group O_2^- is noteworthy—an ion with an *odd* number of valence electrons formed from two atoms with an *even* number of valence electrons in all. It occurs in the alkali and the alkaline earth hyperoxides (cf. pp. 176, 263).

The principal methods for the *preparation of the oxides* are the two following.

(1) *Direct combination of the elements with oxygen.* Oxygen unites directly with most elements, although not with the noble metals (except in trace amounts at their surfaces) or with the halogens. Most of the alkali metals form higher oxides instead of the normal oxides when they combine directly with oxygen. For the preparation of the alkaline earth peroxides, see Chap. 8.

(2) *By heating the hydroxides, or other easily decomposed oxy-compounds.* Many hydroxides decompose even at ordinary temperature into the oxide and water—e.g., silver hydroxide and mercuric hydroxide. Carbonic acid, which is stable only in aqueous solution, decomposes similarly. Among other easily decomposed oxy-compounds, especially suitable for the preparation of oxides, are the carbonates and nitrates.

Most oxides can combine with water to form *hydroxides*. Oxides which thereby form acids are called *acid anhydrides*. Most of the oxides of the metals behave as the *anhydrides of bases*, but many are *amphoteric*—i.e., the hydroxides derived from them can dissociate either as bases or as acids. However, there are metallic oxides—invariably derived from the highest valence state of the metal in question—which are undoubtedly acid anhydrides—e.g., CrO_3, chromium trioxide, Mn_2O_7, manganese heptoxide. There are also indifferent oxides, such as nitrous oxide, which are the anhydrides of neither acids nor bases.

Most oxides have great thermal stability. Only the oxides of the noble metals are decomposed by gentle heating. Of the peroxides, only those of the alkali metals (other than lithium) are stable towards heat. Most of the oxides can be reduced to the parent element by heating them with carbon or hydrogen.

The preparation of the hydroxides has already been considered in Chap. 2. The hydroxides of most of the heavy metals, as well as those of magnesium and beryllium, are insoluble in water. Those of the alkaline earth metals are much more soluble, and the hydroxides of the alkali metals are very soluble. Hydroxides of the non-metals (the oxy-acids) are also for the most part very soluble in water.

(b) Oxonium Compounds

Many oxides (and peroxides) can combine with other substances to form addition compounds. Water, in particular, has this property, which is responsible for the formation of the numerous hydrates. Other oxygen compounds can also form addition compounds. This is especially true of the hydroxides (and hydrogen peroxide), among inorganic compounds, and the alcohols, ethers, aldehydes and ketones among organic compounds. These oxygen compounds can add on both to metallic salts and to acids—e.g.,

$$(CH_3)_2O + HCl = (CH_3)_2O \cdot HCl$$

The double compounds so formed are known as *oxonium compounds*. They are usually salt-like in character. Their general formula is $(R_2O)_n.MX$, where R_2O is some oxide, and MX a compound of the positive radical M (usually a metal or hydrogen) and a negative radical X (acid radical or hydroxyl). More than 1 molecule of R_2O may be combined with 1 molecule of MX.

According to Werner, the formation of oxonium compounds takes place because oxygen is potentially coordinatively trivalent—i.e., can bind 3 ligands directly, through coordinate bonds. The third ligand may be a neutral molecule, but it is also possible for addition to the oxygen atom to be coupled with a dissociation of the added molecule, so that only its electropositive constituent is added on; the electronegative radical is then bound only ionically ('in the second sphere'). The former case may be represented by formula I, and the other by formula II.

$$\begin{matrix} R \\ \diagup \\ R \end{matrix} O \cdots MX \qquad\qquad \left[\begin{matrix} R \\ \diagup \\ R \end{matrix} O \cdots M \right]^{+} X^{-}$$

$$\text{I} \qquad\qquad\qquad\qquad \text{II}$$

Both cases involve coordinatively trivalent oxygen. The electrochemical valence of oxygen in the oxonium compounds is the same as in the oxides from which they are derived—i.e., —2. Except for the electrovalence of the central atom, and the coordination number, the oxonium compounds are essentially analogous to the ammonium compounds. Although the third ligand in oxonium compounds is ordinarily hydrogen or a metal, the tertiary oxonium salts, $[R_3O]X$, corresponding to the quaternary ammonium salts $[R_4N]X$, have been obtained by Meerwein (1937, 1939).

Some have considered that oxygen is *quadrivalent* in the oxonium compounds. This idea is based chiefly on the view held by Baeyer and Villiger, that both constituents of the compound MX must be linked directly to the oxygen of the oxide R_2O. However, Werner showed that the oxonium compounds can be regarded as coordination compounds, and there is no reason to invoke a quadrivalence of oxygen, which is otherwise unknown, and not in accordance with valence theory. The coordination theory also covers the compounds —of which many examples are known—in which more than one molecule of R_2O is combined with one acid hydrogen (formerly considered to be 'anomalous' oxonium salts). In the Werner theory these would be formulated with hydrogen exerting more than one 'auxiliary valence'—i.e., with a coordination number greater than 2.*

(c) The Hydronium Ion

The *hydronium compounds*, $H_2O \cdots HX$ or $[H_3O]X$, constitute an especially important group of oxonium compounds, based upon the positive ion $[H_3O]^+$. This *hydronium ion* is commonly spoken of as the hydrated hydrogen ion, or simply the *hydrogen ion*. (cf. Chap. 2). It is the essential and invariable component of aqueous solutions of acids**, and is often present in their solid hydrates also (cf. perchloric acid, for example $[H_3O][ClO_4]$ is structurally the analogue of $[NH_4][ClO_4]$).

The formation of the hydronium ion can be formally attributed to the coordinative trivalence of oxygen, as a typical oxonium salt formation. In the case of this simple complex ion, however, it is possible to discern why oxygen should exert just 3 coordinative valences. If we calculate the energy liberated when 1, 2, 3, etc. protons are added on to a negatively charged oxygen atom, O^{2-}, it emerges that the addition of $3H^+$ per O^{2-} is considerably more exothermic than the addition of $2H^+$ per O^{2-}.

A very approximate calculation, ignoring polarization and the probable penetration of the H^+ nucleus into the electron cloud of the oxygen ion (i.e. ignoring quantum mechanical exchange forces), gives the following results (r_0 = radius of oxygen ion in Å).

$$O^{2-} + H^+ = OH^- + 2\frac{e^2}{r_0} \times 10^8 \text{ ergs} \tag{1}$$

$$OH^- + H^+ = OH_2 + 1.5\frac{e^2}{r_0} \times 10^8 \text{ ergs} \tag{2}$$

$$OH_2 + H^+ = [OH_3]^+ + 0.8\frac{e^2}{r_0} \times 10^8 \text{ ergs} \tag{3}$$

$$[OH_3]^+ + H = [OH_4]^{2+} + 0.06\frac{e^2}{r_0} \times 10^8 \text{ ergs} \tag{4}$$

* It should be noted, however, that this formulation is quite speculative, and has not been verified by any experimental evidence. The small effective radius of hydrogen makes it improbable that it can have a coordination number greater than 2.

** The ion $[H_3O]^+$ is probably further hydrated in aqueous solution (see p. 75).

The energy liberated by the addition of a proton to a H_2O molecule is thus of the same order as—although rather smaller than—that liberated by the addition of a proton to a hydroxyl ion. Process (3) can thus compete with (2), as is evidenced by the fact that water is dissociated to some extent. In view of the hydration of the hydrogen ion in aqueous solution, the dissociation equilibrium of water is properly represented by the equation

$$2H_2O \rightleftharpoons [H_3O]^+ + OH^- \tag{5}$$

Although equilibrium in (5) is displaced far towards the left, the dissociation attains a measurable degree.* If substances are present which part with hydrogen ions more readily than does water—i.e., acids—the extent of the dissociation must naturally be much greater. The *real reason for the dissociation of acids* in aqueous solution is thus *the tendency of water to form hydronium ions.*

On the other hand, the calculation given above gives no reason to expect that $[OH_4]^{2+}$ ions should be formed in measurable amount, and no evidence for the existence of such ions has, in fact, been found. In complex oxonium ions, hydronium ions and in coordination compounds formed through oxygen, the maximum coordination number of oxygen is therefore three. In solid structures, where oxygen atoms are joined through hydrogen bonds,—as in the structure of ice (Fig. 10)—oxygen is tetrahedrally coordinated. It is generally considered that the H atom occupies an assymetric position in a hydrogen bond —O—H....O; the environment of the oxygen atom, and the charge distribution, in the tridymite-like structure of ice are quite different from those which would be set up in a free $[OH_4]^{2+}$ ion. The observed tetrahedral coordination around oxygen in ice and similar structures is not therefore in disagreement with the conclusion reached above.

(d) Addition Compounds of Molecular Oxygen

Some oxygen compounds can form addition compounds with molecular oxygen. Traube (1916) considered the hyperoxides of the alkaline earth metals (p. 263) to be of this type, and also the so called 'ozonates' of the alkali metals. The last named are formed by the action of ozone on the alkali hydroxides; they have an intense orange color, and form no hydrogen peroxide when they are dissolved in water or acids, but liberate only gaseous oxygen. On this evidence Traube considered that their constitution should be represented as $(M^IOH)_2 \cdots O_2$. However, it has been shown that these compounds are ozonides of the formula M^IO_3.

5. Sulfur (S)

(a) Occurrence

Sulfur is frequently found native in considerable amounts, often in the neighborhood of volcanoes. In Europe, the most important deposits are in Sicily, but vastly greater deposits are found in the United States (Louisiana and Texas) and in Japan. Hydrogen sulfide, H_2S, often escapes from the ground as a gas in volcanic districts, or comes to the surface dissolved, in sulfur springs. Volcanic gases frequently contain sulfur dioxide also. The sulfides of the metals are very widely distributed indeed, the most abundant being *pyrite*, FeS_2, *chalcopyrite* $CuFeS_2$, *galena*, PbS, and *zinc blende* ZnS. Sulfur is even more frequently met with in the form of sulfates—e.g., calcium sulfate (gypsum and anhydrite), magnesium sulfate (epsom salt and kieserite), barium sulfate (barytes or heavy spar) strontium sulfate (celestine) and sodium sulfate (glauberite).

Sulfur makes up about 0.03 weight per cent of the accessible solid crust of the earth. The oceans contain about 0.09% sulfur, in the form of sulfates. Meteorites

---

* The energy to be expended on the dissociation process is available from the fluctuations of thermal energy.

often contain a certain amount of ferrous sulfide. One theory of the constitution of the earth assumes that, at great depths, below the silicate crust, there is a zone consisting largely of metallic sulfides (cf. Vol II). Sulfur is among the elements which are indispensable to living organisms, since it is a constituent of the albuminous proteins, which contain 0.8–2.4% of sulfur chemically combined. The plants obtain their sulfur requirements from the sulfates in the soil. The foul smells arising in the decay of animal carcasses originate largely from certain sulfur compounds (hydrogen sulfide and mercaptans) which are formed in the decomposition of albumen.

Coal contains an average of 1–1.5% of sulfur, partly bound in organic form, and partly as pyrite.

(b) History

Since it occurs in nature in the free state, sulfur was already known in very ancient times. The use of burning sulfur for disinfection was mentioned in Homer, and Dioscorides described its use in medicine. The alchemists were deeply interested in sulfur, and many of them were acquainted with sulfuric acid. Basil Valentine described its preparation (by heating iron vitriol) in the 15th century. Sulfuric acid was first prepared on a manufacturing scale in England, from the middle of the 18th century onwards (cf. p. 709).

(c) Extraction

Native sulfur is extracted chiefly by melting it out of the rocks in which it is present in the elementary state. [4]

In Sicily, which was the principal producing region until about 1914, the method of extraction formerly used (and still used to some extent) was to pile up the sulfur-bearing rock into large mounds ('calcaroni'), which were penetrated by vertical shafts and covered with marl or spent rock. They were then ignited from the top. The sulfur burned slowly away, while the unburned sulfur melted and flowed away into suitably placed moulds. The liquation of such a mound may take from 1 to 3 months, according to the size. The yield of crude sulfur is about 50–70% of the amount present in the ore, a very considerable amount being lost as sulfur dioxide. In the absence of local supplies of other fuels, however, the operation of the Sicilian process is dependent upon burning a portion of the sulfur. Since sulfur dioxide has a deleterious action on the growth of vegetation, the operation of the sulfur burners is limited by law to the period from August to December.

The crude sulfur liquated in the mounds is refined by distillation from cast iron retorts. The vapors are led into large masonry chambers, where they condense as 'flowers of sulfur', provided the temperature of the chamber is maintained at less than 112°. If it rises above this, molten sulfur collects on the floor of the chamber. This is run off into wooden moulds and sold under the name of 'roll sulfur'.

In Louisiana and Texas, the sulfur deposits are deep below the surface of the earth. The sulfur is extracted by the *Frasch* process, in which superheated water is pumped down a bore hole into the deposit, thereby melting the sulfur underground. The molten sulfur is raised to the surface by means of compressed air, passed down the central one of the set of concentric pipes. The sulfur obtained in this way is very pure, (99.6% S) and does not need to be refined by distillation.

The production of sulfur in the United States began about the beginning of this century, but has since far outstripped that of all other countries. In 1917, the United States produced 77% of the world's sulfur, and Italy 14%. In 1936, production in the United States was 2.05 million tons, in Italy 0.35 million tons, and 0.18 million tons in Japan.

In Japan, the sulfur-bearing rock is liquated in retorts heated with superheated steam, a process which has also been introduced in Sicily. That the wasteful liquation by means of burning sulfur can compete economically with this process is due to the lack of coal in Sicily.

Considerable quantities of sulfur were formerly produced in England from the calcium sulfide which was a waste product of the Leblanc soda process. This was decomposed by treating it with carbon dioxide and water (*Chance* process):

$$CaS + CO_2 + H_2O = CaCO_3 + H_2S.$$

The hydrogen sulfide so formed was mixed with air and passed over a catalyst (bauxite or bog iron ore) in the Claus furnace, and so burned to sulfur:

$$H_2S + \tfrac{1}{2}O_2 = H_2O + S$$

The Chance-Claus process is still of importance for the production of sulfur from gypsum or barytes. These may be converted to calcium or barium sulfide, by heating them with carbon, and the sulfides can then be worked up for sulfur as indicated.

The *recovery of sulfur from coal gas*, coke oven gas, water gas, producer gas, and to some extent from smoke stack gases, has become of considerable importance. As has been mentioned, the crude gas of the gas works and coke ovens contains sulfur, principally in the form of H_2S, which is removed from the gas by the iron oxide purifiers (cf. p. 460). Sulfur can be obtained from the 'spent oxide' by extraction with suitable solvents (carbon disulfide or aqueous ammonium sulfide). When sulfur is to be recovered, however, it is probably better to absorb hydrogen sulfide from the crude gas by a wet process—e.g. the polythionate process discussed on p. 614.

(d) Properties

Sulfur exists in several modifications. The modification stable at ordinary temperature is *rhombic sulfur* (*α*-sulfur). It has the familiar yellow color, density 2.06, hardness 2.5, specific heat 0.172 (at 16°). Its electrical and thermal conductivities are extremely small. It takes up a strongly negative electrical charge when rubbed. It is insoluble in water, but dissolves to some extent in benzene, alcohol, ether and other organic solvents. It is readily soluble in carbon disulfide, of which 100 parts can dissolve 24.0 parts of sulfur at 0°, 46.1 parts at 22°, and 181.3 parts at 55°. Sulfur crystallizes unchanged from the solution.

Piperidine dissolves sulfur to form a red solution (irrespective of which allotropic modification of sulfur is present). This property can be applied to the detection of free elementary sulfur.

Above 110°* rhombic sulfur melts to a yellow, mobile liquid. If this is allowed to partially solidify, and the remaining liquid is poured off, the interior is found to be lined with needle-shaped, almost colorless crystals of the *monoclinic* form of sulfur (*β*-sulfur). This has density 1.96 and melting point 119.0°. It is unstable at ordinary temperature. The needles become opaque after a little time, and are

---

* Various values are found for the melting temperature of rhombic sulfur, depending on the experimental conditions. If it is heated slowly, conversion to monoclinic sulfur generally takes place before fusion. If it is heated rapidly, the constitution of the melt is generally ill defined, since a partial transformation of S_λ into S_μ slowly takes place in the melt (cf. p. 701). Pure rhombic sulfur would be in equilibrium at 112.8° with a liquid consisting solely of S_λ, and this temperature is spoken of as the *ideal* melting point and freezing point of rhombic sulfur. From a melt in which S_λ and S_μ are in equilibrium, rhombic sulfur crystallizes at 110.2°. If it is heated so slowly that equilibrium can be established in the liquid, rhombic sulfur melts at this temperature, provided that it does not previously undergo conversion to the monoclinic form (*natural* melting point and freezing point of rhombic sulfur).

transformed into a mass of minute rhombic crystals, although the aggregate retains the external form of the monoclinic needle; the product is said to be a *pseudomorph* of the original.

The transformation takes place more rapidly if a small crystal of rhombic sulfur is already present, but takes place only below 95.6°. Above 95.6°, conversely, rhombic sulfur changes into the monoclinic form, if a little of the latter is present. Thus 95.6° is the *transition tempera-ture* between the two enantiotropic modifications. The transformation of monoclinic into rhombic sulfur is exothermic, the heat of transformation ΔH being 2.4 cal per g at 0°. (For the free energy of transformation see p. 144.)

Another modification, also monoclinic but with different axial ratios from ordinary β-sulfur, is obtained in the form of yellowish white leaflets, with a mother-of-pearl luster, if a hot, almost saturated solution of sulfur in benzene, turpentine, or alcohol is rapidly cooled ('nacreous' sulfur, Muthmann 1890, Neumann 1934). Sulfur also crystallizes in this form when it is deposited from an ammonium polysulfide solution on standing in the air. The modification is probably unstable at all temperatures, and its formation under the conditions mentioned is an illustration of Ostwald's law of stages.

Molten sulfur undergoes peculiar changes of state when it is more strongly heated. Above 160°, it becomes brown, and increasingly viscous as the temperature is raised. At 200°, it is dark brown, and as viscous as resin. The viscosity begins to diminish again above 250°; at 400° it is quite mobile again, and it boils at 444.60°.

If strongly heated sulfur is poured in a thin stream into cold water, a brown-yellow, plastic, rubbery mass is obtained (*plastic sulfur*) and this product is only partially soluble in carbon disulfide. A loose powder is left, constituting a further modification of solid sulfur. The soluble portion of plastic sulfur is also to be regarded as a distinct modification. The form insoluble in carbon disulfide is also present in flowers of sulfur.

Sulfur suddenly chilled into the solid state, from the liquid or the vapor, is invariably only partially soluble in carbon disulfide, the amount of the insoluble fraction depending upon the temperature from which it was quenched. In the amorphous sulfur formed by rapid supercooling, there are thus 2 modifications present, which must also already have been present in liquid sulfur, since 'plastic sulfur' is merely the supercooled liquid. The amorphous form soluble in carbon disulfide is designated S_λ, and the insoluble species S_μ.* These two are in equilibrium in the melt, and by chilling the molten sulfur it is possible to determine the position of equilibrium at various temperatures. Thus it has been found that the amount of S_μ in the melt at equilibrium** is 4.3% at 150°, 9.1% at 160°, 22.7% at 175°, and 38% at 200°. (Hammick, Cousins and Langford, 1928). Equilibrium is reached only very slowly if no catalysts are present. The halogens, or traces of hydrogen sulfide, catalyze the attainment of equilibrium.

The peculiar phenomena observed on heating liquid sulfur can be explained through the presence of these two modifications, S_λ and S_μ, in the liquid. The gradual transformation of the yellow color into brown occurs because S_λ is yellow, and S_μ is blackish red in color. The changes of viscosity occur because S_μ is much more viscous than S_λ, but its viscosity decreases markedly with rising temperature. Thus the increase in the concentration of S_μ at first raises the viscosity, until the effect of temperature outweighs the increasing concentration of S_μ.

At ordinary temperature S_μ is reconverted into S_λ—though extraordinarily slowly, if catalysts are absent—, and the latter eventually passes into rhombic sulfur, the stable form.

* According to Aten (1912) and Beckmann (1918) there are indeed 3 species of molten sulfur to be distinguished—the form S_π in addition to S_λ and S_μ. The distinction between S_π and S_μ will be ignored here, to avoid complicating the relations too much.

** Freshly melted pure sulfur solidifies in the monoclinic form at 119.0°. If the melt is held a little above the freezing point for some time, the freezing point gradually falls to 114.5°. This is because the quantity of S_μ corresponding to equilibrium gradually builds up in the liquid, thereby lowering the freezing point of S_λ.

Observations on the vapor density of sulfur, and its variation with temperature, show that several molecular species are present simultaneously— S_8, S_6, S_4 and S_2 (Preuner and Schupp, 1909). As the temperature is raised, the relative proportions of these species alter. Thus at 450° and 500 mm pressure, the more recent work of Braune (1951) indicates that 53.9% of the total pressure is due to S_8 molecules, 37.0% to S_6, 4.9% to S_4 and 4.2% to S_2 molecules. At 750° and the same total pressure, the corresponding figures are 0.1%, 0.8%, 7.2% and 91.9% respectively. At about 2000°, the S_2 molecules finally dissociate into single S atoms. These dissociation phenomena can probably be correlated with the color changes observed when sulfur vapor is heated. Close to the boiling point the vapor is orange-yellow. As the temperature is raised the color first becomes more reddish and then lightens once more, so that at 650° sulfur vapor is straw-yellow.

By the sudden chilling of sulfur vapor, from a temperature at which it was dissociated into S_2 molecules down to —195° (liquid nitrogen temperature), Rice obtained a purple modification of sulfur. This was presumed to be built up from S_2 molecules. It was unstable, and rapidly underwent transformation into yellow sulfur when it was warmed to ordinary temperature (energy of activation for the transformation, 3.1 kcal per g atom).

The heat of reaction of the process $S_{rhombic} \rightarrow \frac{1}{2}S_2$, at room temperature and constant pressure, is 14.45 kcal per g atom. This is made up of the following partial heats:

$$S_{rhomb} \xrightarrow{0.075} S_{monocl} \xrightarrow{0.33} S_{liq} \xrightarrow{2.51} S_{8vap} \xrightarrow{0.90} S_6 \xrightarrow{3.52} S_4 \xrightarrow{7.10} S_2$$

The value obtained for the energy of dissociation $S_2 \rightarrow 2S_{atomic}$ is either 102.6 or 75.6 kcal per mole of S_2 (i.e., 51.3 or 37.8 kcal per g atom S), depending on the way the absorption spectrum is interpreted. The smaller value is in harmony with some calculations by Goldfinger (1936), which were based on earlier vapor pressure measurements by von Wartenberg. The higher value agrees better with an extrapolation of the observed vibrational energy levels of the S_2 molecule (Birge-Sponer extrapolation).

Measurements of the molecular weight of sulfur dissolved in several solvents are in accord with the molecular complexity S_8. X-ray structure investigations have shown that crystalline sulfur is also made up of S_8 molecules.

The work of Warren (1935) has shown that the rhombic modification is built up of isolated, puckered 8-membered rings. Burwell (1937) finds the same to be true of monoclinic sulfur, although in this modification a proportion of the S_8 molecules is probably rotating or oscillating about an axis perpendicular to the ring. *Plastic sulfur* consists of zig-zag chains of sulfur atoms, of indefinite length, (Meyer, 1934), which are generally arranged completely at random. If the plastic sulfur is strongly stretched, it behaves in the same way as does india rubber, also a linear high-polymer: the zig-zag chains are oriented in parallel alignment, the structure becomes fibrous, and at the same time the material becomes rigid.

A tendency to form chains is observed not only for elementary sulfur, but also in many of the compounds of sulfur (e.g., the polysulfur chlorides, hydrogen polysulfides). It is understandable on energetic grounds that this should be so, since the bond energy of the S—S single bond is about 63.8 kcal per mol, and is thus comparable with that of the C—C single bond (58.6 kcal per mol).

The hypothesis of the high-polymer nature of plastic sulfur—and therefore of S_μ also— enables us to interpret the thermal transformations with some confidence. The conversion of the mobile yellow sulfur (S_λ) into the dark viscous form is probably due to the fact that the S_8 rings, which probably preponderate greatly in molten sulfur just above the melting point, are broken up into open chains to an increasing extent as the temperature is raised. There is thus a continuous rupture and reformation of bonds. As the sulfur chain is flexible, the re-formation of a sulfur-sulfur bond may take place either intra-molecularly, reconstituting an S_8 molecule, or between two molecules, to give a longer chain. The end atoms of a sulfur chain would be unsaturated, and capable of combination with univalent atoms or groups, such as are provided by iodine (chain termination —S—I) or hydrogen sulfide (chain termination —S—SH); Krebs (1953) has argued that the high polymer stabilizes itself by forming very large cyclic molecules. The concentration of foreign substances which provide chain endings may determine the average length of the polymer chains; and the equilibria between the chains and the terminal groups provide a chemical mechanism for the observed manner in which equilibrium in molten sulfur is catalyzed by impurities.

We are thus justified in identifying the S_λ of the older workers with S_8 ring molecules, and the S_μ with the high polymer chains. Liquid sulfur at high temperatures can be regarded as a solution of the polymer in the S_8-molecular sulfur. It is possible for molecules of the polymer to grow in length or to dissociate by such a reaction as

$$S_x \rightleftharpoons S_{x-8} + S_8 - 4 \text{ kcal.},$$

(where S_x is a polymeric molecule with x atoms of sulfur, S_{x-8} a polymer with $(x-8)$ atoms of sulfur, and S_8 a stable ring molecule). Hence the concentration and mean complexity of the polymer must vary with temperature.

The properties of a solution of this type can be related to the fraction by weight of polymer present, and the average chain length of the polymer, in the same way as has been done for other (organic) high polymer systems [Flory, *Chem. Revs.*, 39 (1946) 127). By so doing, Gee (1952) has shown that the equilibrium concentration of polymer is quite low below a certain temperature, above which it must increase rather abruptly to a maximum of about 75 weight per cent at 380°. The concentration of S_μ in chilled liquid sulfur, as measured by Hammick, Cousins and Langford (1928), and by Smith and Holmes (1905), accord rather well with the high polymer theory, and indicate that the equilibrium concentration of S_μ increases rapidly just above 159°.

The viscosity of sulfur increases enormously over the same temperature range. Bacon and Fanelli (1943) found that the viscosity of sulfur above 160° increased progressively as the sulfur was rigorously purified. Their purest material, which had $\eta = 0.066$ poises at 154°, increased 2000-fold in viscosity over the narrow temperature interval 159° to 166°, and had a maximum viscosity of about 1000 poises at 197°. The viscosity of solutions of long-chain polymers can be correlated with the chain length, and it would appear that the maximum average chain length in the purest sulfur was about 100,000 atoms. Addition of 0.02 weight per cent of iodine lowered the viscosity to an extent which would correspond with a chain length of about 3000 atoms.

It will be found that the ideas developed to account for the thermal transformations of sulfur are also in accord with the allotropy of its homologues, selenium and tellurium.

(e) Suspended and Colloidal Sulfur

If calcium polysulfide solution (obtained by prolonged boiling of sulfur with milk of lime) is decomposed by the addition of dilute acid, a white liquid is obtained, the so-called 'milk of sulfur'. It consists of a suspension of finely divided, amorphous sulfur, in water or aqueous calcium chloride solution, soluble in carbon disulfide. Similar suspensions are obtained when other polysulfide solutions are acidified.

Sulfur is obtained in *colloidal dispersion* when hydrogen sulfide is led slowly into a concentrated cold solution of sulfur dioxide, or when sodium thiosulfate is decomposed by dilute sulfuric acid, according to a procedure described by Raffo (1908). Sulfur can be precipitated from the quite clear and transparent colloidal dispersion by adding electrolytes, but it is generally precipitated only incompletely since, unlike colloidal metals; it is relatively insensitive towards electrolytes. Small quantities of electrolytes, and especially of acids, even have a stabilizing action. After coagulation by addition of electrolytes, the sulfur can be brought into colloidal dispersion again by means of water.

According to Freundlich, hydrosols of sulfur, prepared in the manner described, contain S_μ as the disperse phase. They owe their stability to their content of pentathionic acid, $H_2S_5O_6$, the anions of which are adsorbed by the particles of S_μ, conferring a negative charge. After long standing, rhombic sulfur crystallizes from the sols, and cannot be brought into solution again.

Hydrosols of sulfur, prepared by pouring an alcoholic solution of the element into water, are quite different in character. They are much less stable, turn cloudy after a time, and are irreversibly flocculated by electrolytes, and especially by acids. According to Freundlich they probably contain S_λ as the disperse phase.

The stability of both kinds of sulfur sols can be considerably augmented by the addition of protective colloids, and preparations protected in this manner are manufactured. They are used in the form of solutions or pastes to combat the blight of the grape vine, and also in medicine, for eczema, fungus, etc.

Sulfur is quite a reactive element at even slightly elevated temperatures. It combines directly with almost all the elements, and combination with the metals is often strongly exothermic. Copper and silver are markedly attacked by sulfur even at ordinary temperature, and it combines with mercury even at the temperatures of liquid air.

(f) Uses [4]

Sulfur is used in the manufacture of black powder, fireworks, and matches. Considerable amounts are used for dusting grape vines. The use of burning sulfur as a fumigant is well known; its action depends upon the sulfur dioxide formed by the combustion. Sulfur dioxide is also often used as a bleaching agent—e.g., for wool, silk and gelatine. Large quantities of sulfur are used for vulcanizing rubber, and also for the preparation of carbon disulfide, ultramarine, cinnabar, and organic sulfur dyestuffs. Sulfur ointments and milk of sulfur are used in medicine for the treatment of skin diseases, and flowers of sulfur are used internally—e.g., for chronic digestive disturbances. Mention has already been made of the use of colloidal sulfur preparations for similar purposes. Sulfur is also used for making impressions and moulds, and also as a cement.

Sulfuric acid is by far the most important compound of sulfur, and a considerable proportion of the world's sulfuric acid output is prepared directly from Louisiana sulfur. The uses of sulfuric acid and of other important compounds of sulfur are described under the compounds concerned.

6. Compounds of Sulfur

(a) Oxides and Oxyacids

Sulfur forms four normal oxides: *sulfur trioxide* SO_3, *sulfur dioxide* SO_2, *disulfur trioxide* S_2O_3, and *sulfur monoxide*, SO; also two peroxides, S_2O_7 and SO_4. Sulfur trioxide is the anhydride of *sulfuric acid*, H_2SO_4, and the dioxide that of *sulfurous acid*, H_2SO_3. The salts derived from these acids are the *sulfates* $M^1{}_2[SO_4]$ and *sulfites*, $M^1{}_2[SO_3]$, respectively.

In addition to these, the following oxyacids and oxyacid salts of sulfur are known, to which no known oxides correspond as anhydrides.

Acids		Salts	
$H_2S_2O_3$	thiosulfuric acid (unstable)	$M^I{}_2[S_2O_3]$	thiosulfates
$H_2S_2O_2$	thiosulfurous acid*	—	—
$H_2S_xO_6$	polythionic acids**	$M^I{}_2[S_xO_6]$	polythionates
$H_2S_2O_6$	dithionic acid	$M^I{}_2[S_2O_6]$	dithionates
—		$M^I{}_2[S_2O_4]$	dithionites or hyposulfites
—		$M^I{}_2[SO_2]$	sulfoxylates

Disulfur trioxide, S_2O_3, could, from its formula, be regarded as the anhydride of dithionous acid (which exists only in the form of its salts). However, it does not show any close relationship to the dithionites. Sulfur monoxide similarly does not appear to function as the anhydride of sulfoxylic acid (p. 727).

Of the *peroxides* of sulfur, *sulfur tetroxide* SO_4, does not have the character of an acid anhydride, but two peroxy-acids are derived from *(di)sulfur heptoxide*, S_2O_7 — *peroxymonosulfuric acid* (or *Caro's acid*), H_2SO_5 ,and *peroxydisulfuric acid*, $H_2S_2O_8$. These peroxyacids can be regarded as hydrogen peroxide derivatives of sulfuric acid, and are accordingly discussed in relation to the latter.

The most important of the compounds listed are sulfur trioxide, the dioxide, and the acids derived from them. Sulfur dioxide is the usual product of combustion of sulfur in air.

(*i*) *Sulfur trioxide*, SO_3, is prepared technically by passing the gaseous mixture, obtained by roasting pyrite or other sulfide ores (bornite, zinc blende, galena, etc.), over a heated catalyst—formerly platinum, now generally vanadium oxide. In addition to a considerable proportion of atmospheric nitrogen, the roast gases contain sulfur dioxide and oxygen, which combine in the presence of a suitable catalyst according to

$$SO_2 + \tfrac{1}{2}O_2 \rightleftharpoons SO_{3\text{gas}} + 23.2 \text{ kcal} \qquad (1)$$

Pyrite and other sulfide ores are generally roasted on long hearths, or in furnaces with mechanical rabbles, or rotary kilns. Modern processes include the *fluidized bed, flash roasting* and *Trail* processes.

In the *fluidized bed* ('fluo solid', Fleuss process) process, the ore is fed in the form of fairly coarse grains into a shaft furnace, in which it is held supported on a counter-current blast of the air which brings about combustion and which is blown in from below. The intimate contact between the air and the particles of ore ensures rapid and complete roasting. The fluidized bed process is also used for chloridizing roasting and other chlorinations (manufacture of bleaching powder), for lime burning, etc., as well as in catalytic processes and for the Fischer-Tropsch hydrocarbon synthesis. In the latter two processes, the granular catalyst is maintained in a state of suspension by the reaction mixture. The fluidized bed process was originally worked out (by Winkler, of the Badische Anilin- und Sodafabrik, 1922) for the generation of water gas.

Flash-roasting, which is employed largely in sulfuric acid and sulfite pulp cellulose plants, makes partial use of the counter-current principle. In the Nichols-Freeman furnace, used for the purpose, the very finely ground ore is blown in from above, in a pre-heated blast of air for the combustion. It meets a blast of secondary air, fed in from below, which takes the roasting process to completion in the middle zone of the furnace, where the SO_2 concentration is relatively low.

The roasting process worked out by the Mining and Smelting Company at Trail, (British Columbia) (suspension roasting) is particularly suitable for treating dust-fine zinc blende, such as is obtained from flotation enrichment of ores (cf. Vol. II). The zinc blende is blown into the lower part of the roasting furnace on a jet of compressed air. The dust particles are carried upwards by the air stream; as they pass through the hot zone they are roasted, and a part is carried out of the top of the furnace with the roast gases. Most of the solid, however,

---

* Thiosulfurous acid exists only as an intermediate in reactions cf. p. 720.

** In the polythionic acids and their salts, *x* may be 3, 4, 5, 6 or possibly greater.

sinks down after being suspended for a time which does not suffice to complete the roasting process, which is finished on a hearth below.

Very large quantities of sulfur trioxide are similarly made from the gaseous products obtained by burning the very pure Frasch sulfur.

The reaction represented by eqn. (1) is reversible. Since heat is liberated in the formation of SO_3, the equilibrium is displaced from right to left by raising the temperature—i.e., the decomposition of SO_3 is favored. At too low a temperature, no reaction takes place at all. In the presence of platinum, reaction occurs with adequate speed at temperatures as low as about 400°; iron oxide is effective only at 600°. Starting from a nitrogen-free reaction mixture of the stoichiometric composition, the yield of sulfur trioxide would be 98.1% at 400°, and 76.3% at 600°. Thus a considerably greater yield can be achieved by using platinum than with iron oxide, and for this reason platinum was formerly used almost invariably on the industrial scale for this 'Contact Process'. Vanadium compounds (vanadium pentoxide or vanadyl sulfate) are at the present time used almost exclusively. When these are promoted by appropriate additions, and are supported on suitable carriers (silica gel, zeolites), they almost equal platinum in catalytic efficiency. They are, however, much less costly, and have the advantage that they are not so readily poisoned by arsenic compounds as is platinum.

Applying the mass action law to eqn. (1), we have

$$\frac{p_{SO_2} \cdot p_{O_2}^{\frac{1}{2}}}{p_{SO_3}} = K_p \quad \text{or} \quad \frac{p_{SO_3}}{p_{SO_2}} = \frac{\sqrt{p_{O_2}}}{K_p}$$

The ratio of SO_3 to SO_2 is thus proportional to the square root of the partial pressure of oxygen, and can therefore be improved by admixing air with the roast gases. Starting with roast gases containing 84.85% N_2, 10.10% SO_2, and 5.05% O_2, the yield of SO_3 would be 96.2% at 400° or 59.1% at 600°. If the roast gases are diluted with 4 volumes of air before passage over the catalyst, so that they contain 80% N_2, 2% SO_2, and 18% O_2, the yield of SO_3 is raised to 99.5% at 400° and to 80.5% at 600°.

Sulfur trioxide obtained by means of the contact process is not as a rule condensed as such, but is used for the manufacture of sulfuric acid. When sulfur trioxide itself is desired, it is prepared by heating fuming sulfuric acid. Small quantities of sulfur trioxide are also prepared by this means in the laboratory. It can also be obtained by heating sodium pyrosulfate (or other pyrosulfates):

$$Na_2S_2O_7 = Na_2SO_4 + SO_3$$

Iron(III) sulfate also loses SO_3 when it is heated. The oldest method of preparing sulfuric acid is based upon this reaction.

Sulfur trioxide condenses on cooling into a colorless, transparent mass like ice, which fumes in air, melts at 16.85° and boils at 44.8° (γ-SO_3). Its density is 1.995 at 13°, and 1.97 at 20° (molten). If it is kept for a long time below 25°, it changes into another modification (β-SO_3), consisting of white, silky, felted needles. This 'asbestos-like' form has a higher melting point than the 'ice-like' form. The lower melting, ice-like form can be obtained from it by distillation. In addition, there is yet a third modification (α-SO_3), with a still higher melting point, which can be obtained in the pure state only under special conditions (Smits, 1931). Commercial sulfur trioxide consists of a mixture of α-SO_3 with β-SO_3 (with the latter usually in preponderating amount). It usually begins to melt at about 40°, but the melting point (and also the vapor pressure) is dependent upon the proportions in which the two modifications are present. The proportions generally alter during melting and evaporation, since the β-form passes more readily into the liquid state, and evaporates more rapidly, than the α-form.

If the rate of interconversion is sufficiently slowed down (exclusion of all moisture), a mixture of α- and β-SO_3 behaves like a mixture of two different substances, showing some mutual solubility in the solid state. Thus it does not possess a sharp melting point, but melts over a temperature *range* (which can be quite wide). Similarly, the vapor pressure is observed to diminish as evaporation proceeds. The two modifications can therefore be separated from one another by fractional sublimation. Similar phenomena are also observed with other substances which exist in allotropic modifications—e.g., arsenic trioxide, aluminum chloride, and phosphorus. In order to apply the *phase rule* to systems of this sort, the two modifications must be regarded as different components. Smits [7], who has developed the theory of these phenomena and established it experimentally, applies the name 'pseudo-binary' systems to systems of this kind, which consist of a single substance in the chemical sense, but which behave nevertheless as binary systems. The modifications which make up a uniform mixed phase are termed 'pseudo-components'. For a molecular-kinetic theory of the phenomena, see the discussion of Stranski's views, p. 654.

In the vapor state, the SO_3 molecule is planar in configuration. The oxygen atoms form an equilateral triangle with the sulfur at the center (interatomic distances $S \leftrightarrow O$ 1.43 Å, $O \leftrightarrow O$ 2.48 Å). (Palmer, 1938). The ice-like γ-SO_3 is built up from puckered ring shaped molecules, S_3O_9 (Westrik, 1941).

Sulfur trioxide absorbs water with extraordinary avidity, combining with it to form sulfuric acid:

$$SO_3 + H_2O = H_2SO_4 + 21.28 \text{ kcal (at } 20°) \tag{2}$$

If a drop of water is allowed to fall on to sulfur trioxide, an almost explosive reaction takes place. Because of its great affinity for water, sulfur trioxide can abstract constitutively bound water from many substances. It therefore chars organic materials such as cellulose.

(*ii*) *Sulfuric acid*, H_2SO_4, is the product of combination of sulfur trioxide with water [eqn. (2)]. Sulfuric acid of high concentration (98%) is today manufactured industrially [5, 6] almost exclusively by the *Contact Process*, and moderately concentrated sulfuric acid (78%) still predominantly by the older *lead chamber process*. In 1940 the production of sulfuric acid in the United States was equivalent to about 7,000,000 tons of the absolute (100%) acid.

In the manufacture of sulfuric acid by the contact process, sulfur trioxide is produced by passing roast gases over a catalyst, as already described, and is absorbed in concentrated sulfuric acid. More dilute acid is added, in proportion as it absorbs sulfur trioxide, so that the concentration is kept constant.

Sulfur trioxide produced by the contact process cannot simply be absorbed in water, since a thick fog is thereby formed. The particles of sulfur trioxide borne in the gas stream avidly condense water, forming very fine droplets, and these fine fog droplets remain suspended in the gas. Because of their low mobility, they hardly come in contact with the absorbent. It is therefore almost impossible to absorb the sulfur trioxide completely when it has once been transformed into a fog by the access of water. The gases from the catalyst furnace, charged with sulfur trioxide, are therefore passed into about 98% sulfuric acid. This has only a minimal water vapor pressure, so that the formation of fog is not possible and absorption is complete.

The contact process was invented in 1831 by Peregrine Philipps, in England, but attained practical importance only 65 years later. This development was chiefly due to the work of Winkler and Knietsch, who cleared up the questions which are fundamental to the proper operation of the process (dependence of the SO_2-SO_3 equilibrium upon concentration and temperature, the nature of catalyst poisoning, and measures to avoid it, behavior of SO_3 during absorption). The technological conditions for the process were developed upon this basis. It proved that it was of special importance that the roast gases should be freed from arsenic trioxide, suspended as a colloidal dust (cf. Vol. II), which acts as a strong catalytic poison. The supply of heat must also be properly regulated, in order to maintain the optimum reaction temperature within the catalyst space.

The *lead chamber process*, by which the 78% sulfuric acid used in the fertilizer industry is still mostly manufactured, depends upon the union of sulfur dioxide with oxygen and water under the agency of the *oxides of nitrogen*. The reactions involved can be schematically expressed by the following set of equations; the actual course of the action is discussed below.

$$NO \quad + \tfrac{1}{2}O_2 = NO_2 \tag{1}$$

$$SO_2 \quad + H_2O = H_2SO_3 \tag{2}$$

$$H_2SO_3 + NO_2 = H_2SO_4 + NO \tag{3}$$

Thus nitric oxide, NO, acts as an oxygen carrier. It is oxidized by atmospheric oxygen to nitrogen dioxide, reconverted to nitric oxide by giving up its oxygen to the sulfurous acid, can then take up oxygen once more, and so on. A small amount of nitric oxide could act as carrier for an indefinite quantity of oxygen, if it were not for the occurrence of side reactions, to a very small extent, which gradually convert it to products (nitrous oxide and even nitrogen) which cannot take up oxygen. Apart from these side reactions, the lead chamber process thus also involves a catalytic oxidation of sulfur dioxide. The name 'lead chamber process' comes from the fact that the reactions indicated are usually carried out in large chambers of sheet lead, three being arranged in series as a rule.

Before the roast gases, freed from suspended dust and at a temperature of about 300°, are passed into the first of the chambers, they are first led through a tower (the Glover's tower) in which 'nitrated acid'—i.e., moderately concentrated sulfuric acid, charged with the oxides of nitrogen—is allowed to trickle down over a packing of flints or porous material, counter-current to the gas. By this means the roast gases are cooled and charged with nitrogen oxides, while the acid is at the same time somewhat concentrated. The gases pass into the chambers, into which is also blown steam or a spray of finely dispersed water, and as much nitric acid as is required to make up the losses of nitrogen oxides during operation.

Moderately concentrated sulfuric acid ('chamber acid', about 60%) is formed in the chambers, where it is deposited and collects on the floor. It is removed from time to time. The gases passing out of the chambers, consisting for the most part of nitrogen from the air introduced into the roasting ovens, carry large volumes of nitrous gases with them. To recover the oxides of nitrogen, the effluent gases are passed through the Gay-Lussac towers (usually two in series) before they are led to the stack. In these they are washed, counter-current, with about 80% sulfuric acid. After this has absorbed the oxides of nitrogen, it is mixed with 'chamber acid' and returned to the Glover's tower as 'nitrated acid'. There the nitrogen oxides are liberated for a fresh cycle of reactions.

It is generally assumed that the formation of sulfuric acid in the lead chambers takes place chiefly by way of *nitrosyl sulfuric acid*, $HO.SO_2.ONO$, which is then split up by water, forming sulfuric acid. If the amount of water is insufficient, nitrosyl sulfuric acid can actually be deposited in crystalline form ('lead chamber crystals'). The nitrosyl sulfuric acid is itself formed, however, by way of other intermediate products—e.g.,

$$SO_2 + H_2O + NO_2 = HONO.SO_2.OH \text{ (sulfonitronic acid)}$$

$$2HONO.SO_2.OH + NO_2 = 2HO.SO_2.ONO + NO + H_2O$$
$$\text{nitrosyl sulfuric acid}$$

$$2HO.SO_2.ONO + H_2O = 2H_2SO_4 + NO_2 + NO$$

This view of the mechanism of the reaction has been developed principally by Lunge (1885 onwards), and Berl (1906–7 and 1931–40). Raschig (1887 onwards) also assumed that sulfonitronic acid (which he called nitrosisulfonic acid, and often the 'blue acid' or 'violet acid') was an intermediate, although he assumed another course for its formation and decomposition.* According to Müller (1934) and Abel (1928 onwards), however, neither sulfonitronic acid nor nitrosyl sulfuric acid, but rather *nitrous acid*, HNO_2, is the intermediate essentially responsible for accelerating the oxidation of SO_2. It is, in fact, not yet certain which of the alternative reaction schemes is of the greatest importance in practice. It does appear, however, especially from Bodenstein's kinetic studies (1918, 1935) that the primary step is the oxidation of NO to NO_2 by atmospheric oxygen—probably by the process $2NO \rightleftharpoons N_2O_2$, $N_2O_2 + O_2 = 2NO_2$. The NO_2 then transfers its oxygen to SO_2 in the presence of water, or to H_2SO_3 by way of various intermediates, and possibly by a variety of concurrent reaction sequences, re-forming NO.

The lead chamber process came into use in the middle of the 18th century in England, when Roebuck (1746) built a works at Birmingham, in which the fragile glass bells formerly employed (by Ward, in 1740, at Richmond) were replaced by lead chambers 6 ft. wide. The most important stages in its development were the recovery of the nitrous gases, introduced by Gay-Lussac in 1827, and the method of driving the nitrous gases out of the nitrated acid by means of the hot roast gases, devised by Glover in 1859. By these means, a continuously operating process is made possible. It is now customary, in large installations, to replace the lead chambers by towers of acid-resistant material, similar to the Glover's tower, but of larger dimensions. Between these towers are interposed smaller cells of sheet lead, in which nitrated acid is dispersed as a spray by means of rapidly rotating rollers ('Tower' process).

Sulfuric acid, as produced directly by the lead chamber process is only moderately concentrated. Chamber acid usually has a content of 60–70% H_2SO_4 (density 1.50–1.62), Glover tower acid about 78% H_2SO_4 (sp. gr. 1.71). The latter is usable for many purposes as it is—e.g., for the manufacture of superphosphate. The former must be concentrated by evaporation unless it is transferred to the Glover tower, and is evaporated to a concentration of 78% in lead pans. To concentrate it beyond this point, the evaporation must be continued in platinum vessels, or in porcelain or silica pans. This is now done only to a limited extent, since acid of high concentration is more advantageously prepared by the contact process.

Sulfuric acid generally comes into commerce as the 98% acid (sp. gr. 1.841). The contact process furnishes not only this acid, directly, but also *fuming sulfuric acid* (oleum, Nordhausen sulfuric acid)—i.e., sulfuric acid containing an excess of dissolved sulfur trioxide. The contact process was originally employed solely for the production of the latter.

Sulfuric acid cannot be obtained completely anhydrous by evaporation, the maximum boiling point in the system (338°) corresponding to a concentration of 98.3% H_2SO_4. If it is brought to the stoichiometric composition corresponding to the formula H_2SO_4, by dissolving the calculated quantity of SO_3 in concentrated

---

* According to Seel (1953), the 'blue acid' is actually the (metastable) hydrogen sulfate of the $[N_2O_2]^+$ cation, and can be formed directly from $[NO]HSO_4$ and NO. It may well be formed as a by-product in the lead chamber process, either from $[NO]HSO_4$, or by reduction of some prior intermediate; it now seems quite certain, however, that the Berl-Lunge mechanism of the process will have to be revised again.

sulfuric acid, it solidifies and melts sharply at 10.36°. When heated, however, SO_3 is split off until the composition of the liquid corresponds to 98.3% H_2SO_4.

It is common practice to determine the concentration of sulfuric acid by measuring the specific gravity by means of a hydrometer. The hydrometers used for this purpose were formerly graduated in 'degrees Baumé'. 50° Be correspond to the sp. gr. 1.530 (62.53% H_2SO_4), 60° Be = sp. gr. 1.710 (78.04% H_2SO_4), 65° Be = sp. gr. 1.820 (90.05% H_2SO_4), 66° Be = sp. gr. 1.841 (96–98% H_2SO_4). When the H_2SO_4 content exceeds 90%, however, its determination by means of the specific gravity becomes inaccurate.

The specific gravity of sulfuric acid passes through a flat maximum at about 98% H_2SO_4, as the following figures show.

H_2SO_4 concn.	94.60	95.60	96.38	98.20	98.72	99.12	99.31 %
sp.gr. at 15°	1.838	1.840	1.841	1.841	1.840	1.839	1.838

The melting point curve has a sharp peak at the composition corresponding to the formula H_2SO_4 (Fig. 115). 98% sulfuric acid melts at 3.0°, 100% acid at 10.36°, and acid containing a 5% excess of SO_3 at 3.5°. The electrolytic conductivity falls off progressively as the concentration exceeds 95%, and increases once more as the concentration rises above 100% (cf. Fig. 116). This implies that absolute sulfuric acid can be regarded as an ionizing solvent, both water and sulfur trioxide behaving as electrolytes in sulfuric acid solution. [Hantzsch, 1907; Gillespie, Hughes and Ingold, *J. Chem. Soc.*, (1950) 2473, 2493, 2504]. A wide variety of solutes— e.g., dimethyl sulfate, phosphorus oxychloride, nitromethane—either undergoes electrolytic dissociation, or reacts with sulfuric acid to form ions in solution. Sulfuryl chloride and chlorosulfonic acid, however, are not dissociated in absolute sulfuric acid, and from the freezing point depression produced by these solutes the cryoscopic constant may be calculated. In this way, Ingold, Hughes and Gillespie have recently studied the dissociation of other solutes, and have shown that water behaves as a Brönsted base (see. Vol. II, Chap. 18), and is not quite completely ionized as a binary electrolyte:

$$H_2O + H_2SO_4 \rightleftharpoons H_3O^+ + HSO_4^-$$

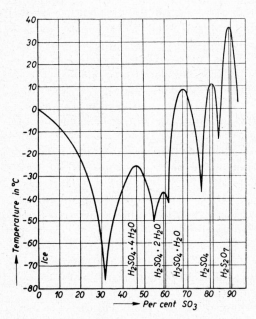

Fig. 115. Melting point diagram of the system SO_3/H_2O.

Ionization is incomplete, because a portion of the water is solvated (with sulfuric acid) without being ionized. They calculated that the dissociation constant of this equilibrium

$$K_b = \frac{[H_3O^+][HSO_4^-]}{[H_2O]}$$

was roughly 1.

Absolute sulfuric acid is not a non-conductor, but owes its conductivity in part to autoionization:

$$2H_2SO_4 \rightleftharpoons H_3SO_4^+ + HSO_4^-$$

$$K_{ai} = [H_3SO_4^+][HSO_4^-] = 1.7 \cdot 10^{-4}$$

and in part to equilibria involving the polysulfuric acids, such as $H_2S_2O_7$:

$$2H_2SO_4 \rightleftharpoons H_2S_2O_7 + H_2O; \qquad H_2O + H_2SO_4 \rightleftharpoons H_3O^+ + HSO_4^-.$$

$$H_2S_2O_7 + H_2SO_4 \rightleftharpoons H_3SO_4^+ + HS_2O_7^-.$$

$$K_{aii} = \frac{[H_3SO_4^+][HS_2O_7^-]}{[H_2S_2O_7]} = \text{about } 0.03.$$

When sulfur trioxide is dissolved in sulfuric acid containing a small excess of water, it thus displaces these equilibria. The first effect is to decrease the concentration of H_3O^+ ions,

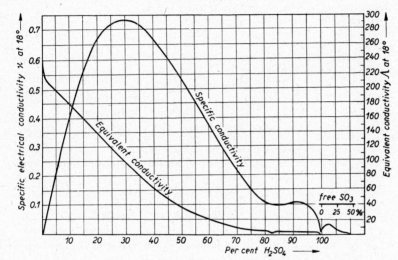

Fig. 116. Variation of the specific electric conductivity and the equivalent conductivity of sulfuric acid with the concentration.

thereby lowering the conductivity, until stoichiometric H_2SO_4 is formed. The further addition of sulfur trioxide forms increasing amounts of disulfuric (pyrosulfuric) acid, which is a slightly stronger acid than H_2SO_4 itself, and which therefore ionizes in the manner shown. There is evidence that solutions more concentrated than 0.1 molar with respect to excess SO_3 also contain more highly condensed polysulfuric acids—e.g., $H_2S_3O_{10}$, $H_2S_4O_{13}$—as well as disulfuric acid.

The existence of several hydrates in solutions of sulfuric acid may be inferred from the maxima in the melting point curve of H_2SO_4-H_2O mixtures (Fig. 115); these are $H_2SO_4.H_2O$ (m.p. 8.5°), $H_2SO_4.2H_2O$ (m.p. —38°) and $H_2SO_4.4H_2O$ (m.p. —27°). In addition to these, Biltz (1934) found that sulfuric acid forms two hydrates which melt incongruently: $H_2SO_4.6H_2O$ (m.p. —54°) and $H_2SO_4.8H_2O$ (m.p. —62°). The melting point diagram also reveals the existence of a compound $H_2SO_4.SO_3$, or $H_2S_2O_7$, pyrosulfuric acid. This forms a transparent, crystalline mass, melting at 36°. Pyrosulfuric acid is completely miscible both with sulfuric acid and sulfur trioxide, most of the mixtures being liquid at ordinary temperature. These constitute the fuming sulfuric acid (oleum) of commerce. Pyrosulfuric acid can be crystallized from mixtures containing more than 18.35% and less than 61.87% of SO_3 in excess of the amount present as H_2SO_4. The vapor pressure curve of fuming sulfuric acid indicates that in addition to $H_2S_2O_7$, some compound exists which is richer in SO_3—probably of the composition $H_2S_4O_{13}$ (Remy, 1942).

Solutions of nitric acid and of the oxides of nitrogen in concentrated sulfuric acid are also important, because of their role both in the lead chamber process of sulfuric acid manufacture, and in the nitration reactions of organic chemistry. Hantzsch (1925) showed that such solutions have a high electrical conductivity, and that the nitric acid produces a large depression of the freezing point. Accurate cryoscopic measurements (Ingold, Hughes and

Gillespie, 1950) have proved that each molecule of nitric acid gives rise to four ions, in accordance with the equation

$$HNO_3 + 2H_2SO_4 = NO_2^+ + H_3O^+ + 2HSO_4^- \tag{1}$$

Solutions of N_2O_5 in sulfuric acid are identical in properties, but *six* ions are formed for each N_2O_5 molecule dissolved:

$$N_2O_5 + 3H_2SO_4 = 2NO_2^+ + H_3O^+ + 3HSO_4^- \tag{2}$$

Hantzsch showed that the ultraviolet absorption spectra of the solutions do not correspond to the presence of either HNO_3 molecules or NO_3^- ions, and it has been since proved by electrolytic transport measurements that the nitric acid gives rise to the positive ion, since it migrates to the cathode during electrolysis. The nature of this cation has now been definitely identified with the nitronium ion NO_2^+, and not the ion $H_2NO_3^+$ or $H_3NO_3^{++}$ postulated by Hantzsch. If the ionization were represented by (3) or (4), giving Hantzsch's ions, not more than *three* ions per molecule of nitric acid should be revealed by cryoscopic measurements:

$$HNO_3 + H_2SO_4 = H_2NO_3^+ + HSO_4^- \tag{3}$$

$$HNO_3 + 2H_2SO_4 = H_3NO_3^{++} + 2HSO_4^- \tag{4}$$

(cf. Chap. 14, p. 600).

Nitrogen dioxide and nitrogen trioxide also form colorless conducting solutions in concentrated sulfuric acid, since they give rise to the nitronium and the nitrosyl cations, respectively, in solution:

$$N_2O_4 + 3H_2SO_4 = NO_2^+ + NO^+ + H_3O^+ + 3HSO_4^- \tag{5}$$

$$N_2O_3 + 3H_2SO_4 = 2NO^+ + H_3O^+ + 3HSO_4^- \tag{6}$$

Pure concentrated sulfuric acid is an oily, viscous, colorless liquid. It is notable for its great affinity for water. Much heat is liberated when it is mixed with water (20.42 kcal per mol of H_2SO_4, on mixing with much water at 20°). Dilution of sulfuric acid must therefore always be carried out cautiously, by pouring the sulfuric acid in a thin stream into water, and never the converse. Sulfuric acid avidly absorbs water vapor from the air, and is therefore used as a drying agent. It is also frequently used as a dehydrating agent in chemical reactions. It has a charring action upon many organic compounds, in that it abstracts the elements of water from them. It also has an oxidizing action, especially when heated, since it is converted into sulfurous acid by splitting off oxygen. Not only the metals, but other elements such as carbon and sulfur are oxidized by concentrated sulfuric acid, sulfur dioxide being liberated. Only those metals which lie above hydrogen in the electromotive series will dissolve in dilute sulfuric acid.

Fuming sulfuric acid has even stronger dehydrating and oxidizing properties than ordinary sulfuric acid. In contact with the skin, it produces severe burns, which heal only with difficulty, unless it is immediately wiped off with a *dry* rag (the burning action would be made yet more severe if the area of contact were moistened).

The behavior of sulfuric acid towards *iron* and *lead* is of great technical importance. Whereas iron dissolves with great ease in the dilute or moderately concentrated acid, it is very resistant towards acid of high concentration, since it becomes passive (cf. p. 347). Sulfuric acid more concentrated than 93% may be stored and heated in cast iron vessels. Cast iron with a high silicon content is particularly resistant, and apparatus for the evaporation of sulfuric acid is therefore now fabricated of such high-silicon cast iron. Wrought iron or mild steel also resists the action of concentrated sulfuric acid at ordinary temperature, so that the acid can be shipped in steel drums and tanks. Mild steel is also unattacked by oleum containing at least 27% of free SO_3.

Lead is only superficially attacked by dilute or by moderately concentrated sulfuric acid, forming lead sulfate which impedes further attack. Lead sulfate is perceptibly soluble, however, in sulfuric acid of high concentration, and lead therefore has little resistance towards such acid, especially when heated.

Sulfuric acid is a dibasic strong acid. Its apparent degree of dissociation (as given by the conductivity quotients Λ_c/Λ_0) is 51% in 1-normal acid, 59% in 0.1-normal acid.

The dissociation of the acid clearly occurs in two stages. The first hydrogen ion is almost completely ionized in solutions of moderate concentration. The ionization of the second hydrogen obeys the mass action law:

$$\frac{[SO_4^-][H^+]}{[HSO_4^-]} = K = 1.29 \cdot 10^{-2} \text{ at } 18° \quad \text{(Hamer, 1934)} \quad (3)$$

If the dissociation of the first stage in 1-normal sulfuric acid is taken as 100%, the dissociation of the second stage amounts to 2.5%.* The average value 51.2% so found agrees satisfactorily with the value deduced from the conductivity quotient (51.4%). For 2-normal sulfuric acid, the average dissociation of the first and second stages dissociation is 50.6%. For this acid $[H^+]$ is therefore 1.012, and pH = —0.005.

Whereas the *specific conductivity* of sulfuric acid shows a well marked maximum at about 30% H_2SO_4 (7.5-normal) — $\varkappa_{18} = 0.739$ (cf. Fig. 116), the equivalent conductivity increases regularly with progressive dilution in this region. If dilution is extended below 10% by weight (2 g equiv. per liter), the rate of increase of conductivity begins to fall off somewhat. Below 0.2 g equiv. per liter, however, the conductivity begins to increase strongly once again, since the *second stage dissociation* begins to increase greatly. If the cube root of the equivalent concentration is plotted (as abscissa) against the equivalent conductivity, the resulting curve shows a distinct inflection between 0.5 and 0.2 g equiv. of H_2SO_4 per liter, corresponding to the onset of the second stage dissociation.

Sulfuric acid is widely used in the laboratory and in various branches of industry. The chief consumer is the fertilizer industry, in which it is used to produce ammonium sulfate (in gas works and coke ovens) and superphosphate. It is also used for the purification of vegetable oils and fats, and of petroleum, for the preparation of other acids, and also for preparing sulfates, esters and ethers. Fuming sulfuric acid (oleum) is used in the organic chemical industry, in addition to concentrated sulfuric acid, especially for *sulfonation*—i.e., the introduction of the group —SO_3H in place of H in organic compounds. Moderately concentrated sulfuric acid (72%–75%) is used in preparing parchment paper. Sulfuric acid of sp. gr. 1.15–1.25 (nearly of the maximum conductivity) is used as accumulator acid.

* In calculating the dissociation of the second stage, repression by the first stage dissociation must be taken into account. If a_1 and a_2 are the degrees of dissociation (as fractions of unity), then from c mols, of H_2SO_4, the dissociation $H_2SO_4 \rightleftharpoons HSO_4^- + H^+$ furnishes $c.a_1$ g ions each of HSO_4^- and H^+; and the dissociation $HSO_4^- \rightleftharpoons SO_4^= + H^+$ furnishes $c.a_1.a_2$ g ions of $SO_4^=$, and an additional $c.a_1.a_2$ g ions of H^+. $c.a_1.a_2$ g ions of HSO_4^- are thereby used up. If c is the total quantity of sulfuric acid, in g mol per liter, we have that $[SO_4^-] = c.a_1.a_2$, $[H^+] = c.a_1 (1 + a_2)$, and $[HSO_4^-] = c.a_1.(1 — a_2)$. Substituting these values in eqn. (3), for $a_1 = 1$ and $c = 0.5$ we find $a_2 = 0.025$, or if $c = 1$, $a_2 = 0.013$.

Statistics. The production of sulfuric acid in the United States in 1940 was equivalent to about seven million tons of the 100% acid.

(*iii*) *Sulfates.* As a dibasic acid, sulfuric acid forms two series of salts—*acid sulfates* (hydrogen sulfates), $M^I HSO_4$, and *normal* (neutral) sulfates, $M^I_2 SO_4$.

Acid sulfates are known only for the alkali metals. They crystallize from solutions of the normal sulfates which have been treated with an excess of sulfuric acid, and are also formed by decomposing the alkali salts of other acids with sulfuric acid at a moderately high temperature. E.g.,

$$Na_2SO_4 + H_2SO_4 = 2NaHSO_4 \text{ (sodium hydrogen sulfate)}$$

$$NaCl + H_2SO_4 = NaHSO_4 + HCl$$

The acid sulfates are very soluble in water, and are also readily fusible. When heated above their melting points, they are first transformed into *pyrosulfates*, $M^I_2S_2O_7$, and these, on stronger heating, decompose into the normal sulfates and sulfur trioxide.

$$2NaHSO_4 = H_2O + Na_2S_2O_7 \text{ (sodium pyrosulfate)}$$

$$Na_2S_2O_7 = Na_2SO_4 + SO_3$$

The normal sulfates are, in some cases, very stable thermally. The alkali sulfates may be melted without decomposition (Na_2SO_4 melts at 884°, K_2SO_4 at 1074°). The alkaline earth sulfates and lead sulfate may be heated to red heat without decomposition, but many sulfates begin to decompose below a red heat, forming the metallic oxide (or the metal, from the sulfates of the noble metals) and sulfur trioxide or its decomposition products, sulfur dioxide and oxygen. Formation of the latter is increasingly favored, the higher the temperature of decomposition. The sulfates of the trivalent metals decompose with particular ease—e.g., the decomposition pressure of iron(III) sulfate, $Fe_2(SO_4)_3$, is perceptible at 500°, and reaches one atmosphere at 721°. The dissociation pressure of aluminum sulfate attains one atmosphere at about 750°. It is therefore possible to obtain sulfur trioxide by heating these substances.

Most sulfates are quite soluble in water. The alkaline earth and lead sulfates are sparingly soluble, and silver sulfate is only moderately soluble. The readily soluble sulfates generally crystallize as hydrates. Most sulfates also have a tendency to form double salts (*sulfato*-salts).

The *vitriols* may be cited as examples of sulfates containing water of crystallization (e.g., green vitriol, $FeSO_4 \cdot 7H_2O$, white vitriol, $ZnSO_4 \cdot 7H_2O$, blue vitriol, $CuSO_4,5H_2O$), and the *alums* as examples of double sulfates. According to Brintzinger (1935), sulfate ions (and also the selenate ions) in solution are firmly bound to two water molecules: $[SO_4(OH_2)_2]^=$.

Sulfates may be prepared:

(1) by dissolving metals in sulfuric acid;

(2) by neutralizing sulfuric acid with metal oxides or hydroxides (or with anhydro-bases such as ammonia);

(3) by decomposing the salts of volatile acids by means of sulfuric acid;

(4) by double decomposition between a sulfate and a salt of the metal of which the sulfate is required;

(5) by oxidation of sulfites or sulfides.

Sulfates find extensive uses. These are discussed for the individual sulfates, under the metals from which they are derived.

(iv) *Polysulfates* are obtained by the action of SO_3 on sulfates. Thus potassium trisulfate, $K_2S_3O_{10}$, is formed from K_2SO_4 and SO_3 (Thilo, 1938). It is probable that less stable compounds still richer in SO_3 could be obtained. $K_2S_3O_{10}$ is stable to about 150°, but decomposes at higher temperatures:

$$K_2S_3O_{10} = K_2S_2O_7 + SO_3$$

This reaction is reversed at about 50°.

(v) *The Constitution of Sulfuric Acid and the Sulfates.* Since neither sulfuric acid nor its anhydride, sulfur trioxide, has the character of a derivative of hydrogen peroxide, there can be no peroxidic bond present in them, and sulfur must be regarded as *hexavalent* in these compounds. The ability of sulfur to function as hexavalent is placed beyond doubt by the existence of sulfur hexafluoride, since the exclusively univalent nature of fluorine makes n other formulation possible. On this basis, formulation of sulfuric acid as a hydroxyl compound leads to the structural formula (I). If, however, it is assumed that the H-atoms are bound to the radical SO_4 as a whole, rather than to particular oxygen atoms, structural formula (II) is obtained. The typical salts of sulfuric acid, in particular, are derived from this (formula (III)). X-ray structure determinations have proved that the SO_4 radical exists in the sulfates as a structural group, of regular tetrahedral configuration, (cf. Fig. 51, p. 218). From the standpoint of the coordination theory, sulfur has the coordination number 4 in the sulfates.

With reference to the bond type and the possibility of an intermediate state between various mesomeric forms of sulphuric acid and the SO_4 radical see p. 291.

(vi) *Peroxysulfuric acid and Peroxysulfates (Persulfates).* If sulfuric acid is electrolyzed at low temperatures (below 30°), using smooth platinum electrodes and high current densities, *peroxysulfuric acid* $H_2S_2O_8$ (more strictly, peroxy*di*sulfuric acid), frequently called *persulfuric acid*, is obtained.

Peroxysulfates (persulfates), $M^I_2S_2O_8$, are similarly obtained by electrolysis of solutions of sulfates under similar conditions, and in better yield than is persulfuric acid itself. In particular the peroxysulfates of ammonium and potassium, which crystallize well, can readily be obtained pure, and are therefore the compounds most frequently prepared technically. To avoid reduction at the cathode, a porous earthenware diaphragm is inserted in the cell, or the cathode is covered with a very thin coating of chromic hydroxide. This is readily achieved by adding chromate, which is reduced at the cathodes and deposited as the hydroxide.

Peroxysulfuric acid and its salts have strong oxidizing properties. They precipitate manganese, lead, nickel and cobalt from solutions of their salts in the form of their higher oxides, convert chromium(III) salts into chromates, etc. The peroxysulfates are quite stable in the dry state. All are readily soluble in water.

The peroxysulfates are used as oxidizing and bleaching agents, as well as for the preparation of hydrogen peroxide. Ammonium peroxysulfate, $(NH_4)_2S_2O_8$, is employed in photography for 'reducing' negatives of too high density. It is also used as an analytical reagent—e.g., for the separation of manganese and chromium.

The correctness of the formula $M^I_2[S_2O_8]$ for the peroxydisulfates, rather than the single formula $M^I[SO_4]$, is confirmed by measurements of the freezing point depressions and conductivities of their solutions. Their constitution follows from their mode of formation:

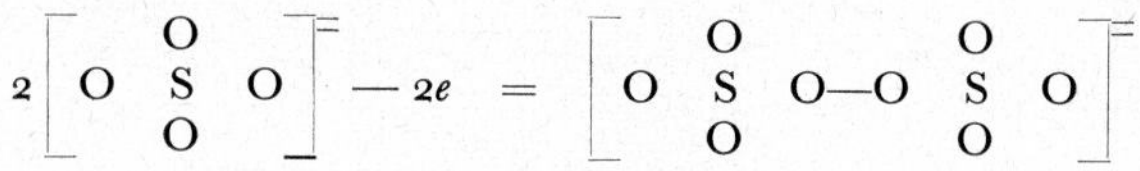

$$2 \begin{bmatrix} & O & \\ O & S & O \\ & O & \end{bmatrix}^{=} - 2e = \begin{bmatrix} & O & & O & \\ O & S & O-O & S & O \\ & O & & O & \end{bmatrix}^{=}$$

The peroxydisulfates are thus derivatives of hydrogen peroxide. Although the salts are relatively stable, the free acid is readily split hydrolytically, forming *peroxymonosulfuric acid* (Caro's acid) also a derivative of hydrogen peroxide:

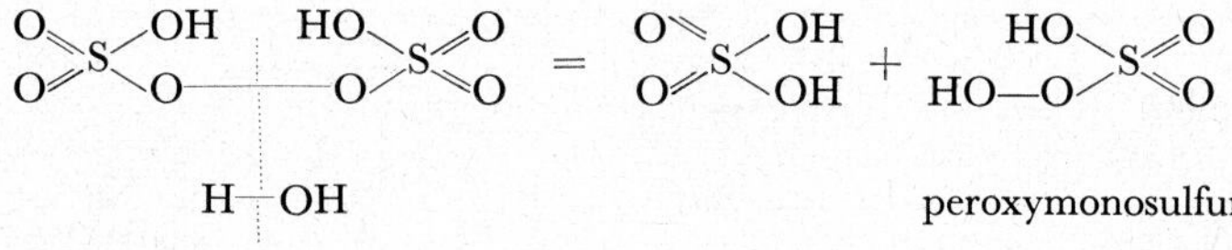

H—OH   peroxymonosulfuric acid

Conversely, ordinary peroxysulfuric acid can be prepared from permonosulfuric acid by the action of chlorsulfonic acid, HSO_3Cl:

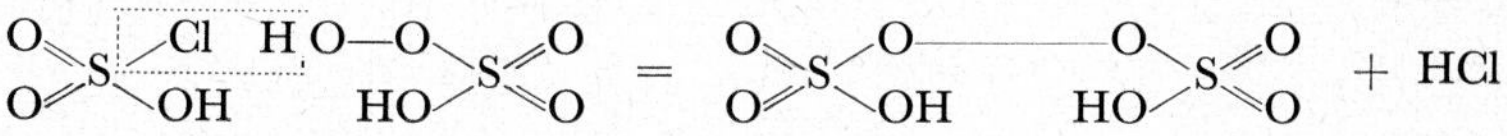

(*vii*) *Peroxymonosulfuric acid* (*Caro's acid*) is formed in the electrolysis of sulfuric acid, as a product of decomposition of the peroxydisulfuric acid which is the primary product. It is also obtained by decomposing peroxydisulfates with concentrated sulfuric acid, and by the action of hydrogen peroxide on cold concentrated sulfuric acid. The latter reaction, however, is reversible:

$$H_2SO_4 + H_2O_2 \rightleftharpoons H_2SO_5 + H_2O$$

The reaction proceeding from right to left—i.e., the hydrolysis of permonosulfuric acid, prepared by electrolyzing concentrated sulfuric acid —is employed for the technical preparation of hydrogen peroxide.

Peroxymonosulfuric acid can be obtained in the pure state, and beautifully crystalline (m.p. $45°$) by the reaction between chlorsulfonic acid and hydrogen peroxide:

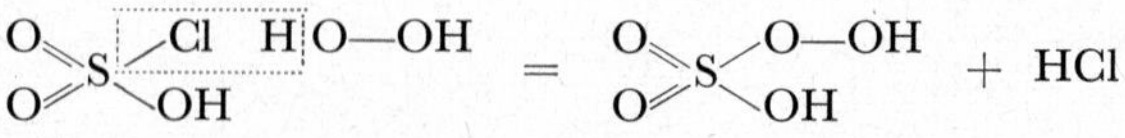

The structure of the acid also follows from this mode formation. It is a strong *mono*basic acid. It is gradually hydrolyzed in solution, the reaction being accelerated by high concentrations of sulfuric acid.

The *salts* of peroxymonosulfuric acid are not very stable. Unlike sulfuric acid, neither peroxymono- nor peroxydisulfuric acid forms an insoluble barium or lead salt. Peroxymonosulfuric acid liberates iodine instantly from potassium iodide solution, even in very great dilution, whereas peroxydisulfuric acid does so much more slowly, and hydrogen peroxide only to a negligible extent. There are also other marked differences in the oxidizing action of these three substances—e.g., towards organic compounds.

(*viii*) *Sulfur heptoxide* (disulfur heptoxide), S_2O_7, was obtained by Berthelot, by the action of the silent electric discharge on a mixture of sulfur dioxide and oxygen. It is said that it can also be prepared by the action of ozone on sulfur dioxide or trioxide. The compound forms oily drops which solidify at $0°$, and readily decomposes, with evolution of oxygen, especially in the presence of water. If the evolution of oxygen is prevented, by suitable experimental methods, the hydrolysis of the compound gives peroxydisulfuric acid. Sulfur heptoxide may accordingly be regarded as the anhydride of this acid.

(*ix*) *Sulfur tetroxide*, SO_4, is formed by the action of fluorine on $SO_4^=$ ions, as discovered by Fichter (1926). It was first obtained pure by Schwarz (1934), by the action of the electric discharge on a mixture of sulfur dioxide and oxygen at a pressure of 0.5 mm., as a white solid, melting with decomposition at $+3°$. It dissolves without decomposition in anhydrous sulfuric acid, and its molecular weight was found cryoscopically to correspond with the formula SO_4. It does not unite with water to form H_2SO_5, but slowly decomposes in aqueous solution, giving off oxygen. It has strong oxidizing properties, converting manganese(II) salts to permanganates, and cuprates(II) to cuprates(III).

(*x*) *Sulfur dioxide*, SO_2, is formed when sulfur is burned in air:

$$S + O_2 = SO_2 + 70.9 \text{ kcal at const. pressure, referred to } 20°;$$

and also when sulfides are heated in air or oxygen. Its preparation by this means has already been discussed in connection with the manufacture of sulfuric acid. It is generally prepared in the laboratory by the reducing action of copper, mercury, carbon or sulfur on hot concentrated sulfuric acid. Sulfur dioxide finds many technical applications in addition to its role in the sulfuric acid industry—e.g., as a bleaching agent, especially for straw, wool and silk, for maize flour and sugar. It is used as a food preservative since it inhibits fermentation. It is toxic if inhaled in large amounts.

Sulfur dioxide is a colorless gas with a pungent smell, which does not burn or support combustion. It will, however, unite with oxygen in the presence of catalysts, as has already been discussed. It has strong reducing properties in aqueous solution, decolorizing many organic dyestuffs by reduction. Its bleaching action depends upon this property.

Sulfur dioxide is also used in refrigeration engineering. It is readily condensed, and has a high latent heat of evaporation (96 cal per g at the boiling point ($-10°$) of the liquefied gas). Its vapor pressure is 1.52 atm at $0°$, 3.3 atm at $20°$; density of the liquid $= 1.46$ at $-10°$; critical temperature $157.2°$, crit. press. 77.7 atm, crit. density 0.51; freezing point $-72.5°$. The gas density under standard conditions, relative to air $= 1$, is 2.2630 (liter weight $= 2.9256$ g at S.T.P.). As accords with its ease of liquefaction, sulfur dioxide deviates markedly from the ideal gas laws.

The SO_2 molecule is bent: $S-O = 1.432$ Å, angle $O-S-O = 119.04°$ [Crable and Smith, *J. Chem. Phys.*, 19 (1951) 502].

SO_2 combines with water to form a solid hydrate, with a composition corresponding approximately to the formula $SO_2 \cdot 6H_2O$. This is most probably a 'gas hydrate' (cf. p. 219), and not a pentahydrate of sulfurous acid.

(xi) *Sulfurous acid and the Sulfites*. Sulfur dioxide is copiously soluble in water (100 g of water at 20° absorb 10.5 g of SO_2, and at 10° 15.4 g. Over 70 volumes of sulfur dioxide are dissolved by 1 volume of water at 0°). The aqueous solution has a distinctly acid reaction, and contains the dibasic acid H_2SO_3, of which sulfur dioxide is therefore the anhydride. Sulfurous acid is not known in the free state, since it readily breaks up into sulfur dioxide and water. For this reason, its solution smells strongly of sulfur dioxide.

As a dibasic acid, sulfurous acid dissociates in 2 stages. The dissociation constants are $K_1 = 1.6 \cdot 10^{-2}$ at 18°, $K_2 = 1.0 \cdot 10^{-7}$ at 18°. The first-stage dissociation constant K_1 gives the ratio of the ion product $[HSO_3^-] \cdot [H^+]$ to the *total concentration* of sulfur dioxide present in the solution in undissociated form ($= [H_2SO_3] + [SO_2]$). The fraction of this which is present in the form of the acid H_2SO_3 is probably only a few per cent, according to spectroscopic measurements.

Two series of salts are derived from sulfurous acid—*acid sulfites*, M^IHSO_3, and *normal sulfites* $M^I_2SO_3$. Both are decomposed by strong acids, liberating sulfur dioxide:

$$2M^IHSO_3 + H_2SO_4 = M^I_2SO_4 + 2H_2O + 2SO_2$$

$$M^I_2SO_3 + H_2SO_4 = M^I_2SO_4 + H_2O + SO_2$$

The acid sulfites (formerly known as *bisulfites*) are all very soluble in water. Many are known only in aqueous solution—e.g., acid calcium sulfite (calcium hydrogen sulfite) $Ca(HSO_3)_2$, a solution of which—*sulfite liquor*—is used in the production of cellulose. Acid sodium sulfite (sodium bisulfite, $NaHSO_3$,) also finds many industrial applications; it is obtained by passing sulfur dioxide into cold saturated sodium carbonate solution, and crystallizes in the form of small, colorless sparkling crystals. It is used as a reducing agent—e.g., in dyeing and in calico printing—, as a bleaching agent, and as the starting material for the preparation of sodium dithionite (sodium hydrosulfite), $Na_2S_2O_4$.

If water is eliminated from sodium hydrogen sulfite:

$$2NaHSO_3 = Na_2S_2O_5 + H_2O$$

or if sodium hydrogen sulfite solution is supersaturated with sulfur dioxide, *sodium pyrosulfite*, $Na_2S_2O_5$, usually known in commerce as 'sodium metabisulfite' is obtained. *Potassium pyrosulfite*, which is obtained similarly, is more widely used; it forms lustrous, rather hard crystals, which dissolve only slowly in water. It is used in photography for the preparation of acid baths, and also in dyeing and printing.

With the exceptions of the alkali and ammonium sulfites, the normal sulfites are sparingly soluble in water, but dissolve in aqueous sulfurous acid. Aqueous solutions of the normal sulfites are sufficiently hydrolyzed to give a basic reaction towards methyl orange or litmus, but not with phenolphthalein.

The sulfites are quite reactive substances. In solution, they are readily oxidized to sulfates. They can, on the other hand, easily be reduced—e.g., to dithionites, $M_2S_2O_4$, by zinc dust in aqueous solution, or to sulfides by heating the dry salts with carbon, zinc, or other metals. Thiosulfates, $M_2S_2O_3$, are obtained by treating solutions of the alkali sulfites with sulfur. Reaction of the dry salts with phosphorus oxychloride yields thionyl chloride $SOCl_2$. The strong tendency of the sufites— especially those of the bivalent heavy metals—to form complex compounds (*sulfito-salts*) should also be noted.

Ammonium sulfite is slowly oxidized in solution, by atmospheric oxygen, forming the sulfate. The reaction is so greatly accelerated by the presence of cobalt ammonia complex salts that it might be considered for the technical preparation of nitrogen and ammonium sulfate.

Sulfites are generally prepared by passing sulfur dioxide into solutions or sus- pensions of metallic hydroxides or carbonates, or by dissolving the corresponding metal in aqueous sulfurous acid. The sparingly soluble sulfites can also be obtained by double decomposition between alkali sulfites and the soluble salts of other metals.

The ions $\begin{bmatrix} & O & \\ O & S & O \end{bmatrix}^=$ are present in the solutions of sulfites; in these ions the sulfur may be considered as formally quadrivalent. It may be assumed that they persist as structural groups in the crystallization of the salts. Two constitutional formulas have to be considered for the free acid:

$$\begin{array}{c} HO \\ HO \end{array} \!\! S\!=\!O \quad \text{(I)} \qquad\qquad \begin{array}{c} H \\ HO \end{array} \!\! S \!\! \begin{array}{c} O \\ O \end{array} \quad \text{(II)}$$

symmetrical formula   unsymmetrical formula

Sulfur is structurally quadrivalent in the former, and hexavalent in the latter. Organic derivatives are known corresponding to both forms, with hydrogen replaced by alkyl groups R; they are termed sulfurous acid esters (alkyl sulfites), and alkyl sulfonic acid esters, respectively:

$$\begin{array}{c} RO \\ RO \end{array} \!\! S\!=\!O \qquad\qquad \begin{array}{c} R \\ RO \end{array} \!\! S \!\! \begin{array}{c} O \\ O \end{array}$$

dialkyl sulfite   alkylsulfonic acid ester

That one alkyl group in the latter compounds is directly bound to sulfur is proved by the fact that only one of the two alkyl groups (i.e., that bound to oxygen) can be hydrolyzed off ('saponified'). Schwarzenbach (1936) considers that the dissociation constants indicate the symmetrical formula (I) for sulfurous acid, whereas both the tautomeric forms $H—SO_3^-$ and $HO—SO_2^-$ come into consideration for the bi-sulfite ion.

Sodium sulfite, Na_2SO_3, is generally prepared by passing sulfur dioxide into sodium carbon- ate solution, thereby first converting it to sodium hydrogen sulfite; a further, equivalent quantity of sodium carbonate is then added. When warm solutions are evaporated, the salt crystallizes anhydrous. Below 37°, it crystallizes with $7H_2O$. Its solubility in water is: at 0°, 14.2 and at 18.2°, 25.3 g of Na_2SO_3 in 100 g of water. It is sparingly soluble in alcohol. Sodium sulfite is used in photography for preserving developing and fixing baths. It is used in medicine as an antiseptic. It is also used to remove chlorine from fabrics after bleaching, and also as a preservative. Its use as a preservative for meat is forbidden in certain countries, since it can restore the color of bad meat. It is also considered to have some injurious effects on the health.

Potassium sulfite, K_2SO_3, is prepared technically by passing sulfur dioxide into concentrated potassium hydroxide until a (previously diluted) sample of the solution no longer reddens phenolphthalein. The completion of neutralization may also be controlled by measurement of the specific gravity. Potassium sulfite crystallizes as the hydrate $K_2SO_3 \cdot 2H_2O$, in colorless, rather deliquescent crystals with a bitter taste. It is insoluble in alcohol, and very soluble in water. Its solution is used in color printing as a reducing agent.

(xii) *Thiosulfuric acid and the Thiosulfates*. When sulfites are boiled with finely divided sulfur, they combine with the latter, being converted into *thiosulfates*, compounds of the general formula $M^I_2S_2O_3$.

$$Na_2SO_3 + S = Na_2S_2O_3; \quad \text{quite generally} \quad SO_3^= + S = S_2O_3^=$$

Alkali thiosulfates are also formed by the action of iodine on mixed solutions of alkali sulfides and alkali sulfites:

$$S^= + SO_3^= + I_2 = S_2O_3^= + 2I^-.$$

The oxidation of polysulfides to thiosulfates by means of atmospheric oxygen is also of technical importance—e.g.,

$$Na_2S_5 + \tfrac{3}{2}O_2 = Na_2S_2O_3 + 3S, \qquad CaS_2 + \tfrac{3}{2}O_2 = CaS_2O_3$$

Most thiosulfates are readily soluble in water. The thiosulfates of lead, silver and thallium(I) are but slightly soluble, and barium thiosulfate is moderately insoluble. The soluble thiosulfates crystallize very well, usually in hydrated form. The thiosulfates of the heavy metals have a strong tendency to form complex salts. Thus silver thiosulfate dissolves very readily in sodium thiosulfate solution, forming trithiosulfato complex ions, $[Ag(S_2O_3)_3]^{5-}$ (cf. Vol. II, Chap. 8). Thiosulfates are decomposed by acids, with deposition of sulfur and evolution of sulfur dioxide. In aqueous solution, the deposition of sulfur does not occur instantaneously, but only after an interval. It is therefore assumed that thiosulfuric acid is first formed in the solution, but that this is stable only for a short time:

$$S_2O_3^= + 2H^+ = H_2S_2O_3 \rightarrow H_2O + SO_2 + S$$

Thiosulfuric acid is too unstable to be isolated. Only one series of salts is known—i.e., the neutral salts in which both the hydrogen atoms of the acid are replaced by metals. Unlike the very unstable acid, the salts are mostly very stable, even in solution.

It follows from the mode of formation and the properties of the thiosulfates that the ion $[S_2O_3]^=$ is derived from the sulfate ion $[SO_4]^=$ by exchange of one oxygen atom for a sulfur atom, or from the sulfite ion $[SO_3]^=$ by the addition of a sulfur atom:

$$\begin{bmatrix} & O & \\ O & S & O \end{bmatrix}^= \qquad \begin{bmatrix} & O & \\ O & S & O \\ & O & \end{bmatrix}^= \qquad \begin{bmatrix} & O & \\ O & S & O \\ & S & \end{bmatrix}^=$$

sulfite ion   sulfate ion   thiosulfate ion

At one time thiosulfuric acid was considered to be a product of the reduction of sulfurous acid, and was therefore termed 'hyposulfurous acid'. According to the foregoing, however, the conversion of the sulfite ion into the thiosulfate ion is not a reduction process, but can in a certain sense be regarded as an *oxidation* (cf. p. 756). The name 'hyposulfites' for the salts of thiosulfuric acid is therefore wrong. Since the replacement of oxygen in any compound by sulfur is generally denoted by the prefix 'thio', the name thiosulfuric acid denotes the relationship of the compound to sulfuric acid. This constitution may be taken as now firmly established.

Sodium thiosulfate (the 'sodium hyposulfite' of commerce) was formerly obtained largely from the calcium sulfide which was a waste product of the Leblanc soda process. On long standing in air, the sulfide is transformed largely into calcium thiosulfate, from which the sodium salt can be obtained by double decomposition with sodium sulfate:

$$CaS_2O_3 + Na_2SO_4 = Na_2S_2O_3 + CaSO_4$$

Large quantities of sodium thiosulfate are formed as by-products of the manufacture of sulfur dyes (cf. p. 735). The salt crystallizes from solution generally as the pentahydrate, $Na_2S_2O_3 \cdot 5H_2O$ ('hypo'). This forms clear, transparent monoclinic crystals, which are easily soluble in water, with considerable absorption of heat (11 kcal per mol). They melt at 48.5° in their water of crystallization. The melt supercools very readily and can be preserved indefinitely in this state; only when it is shaken, scratched with a glass rod, or inoculated with a minute crystal of the solid salt does crystallization suddenly set in, with the evolution of a considerable amount of heat.

Sodium thiosulfate is widely used in photography as 'fixing salt', to dissolve unchanged silver bromide out of the emulsion after development, and in bleaching as an 'anti-chlor'—i.e., to remove chlorine from bleached fabrics.

$$S_2O_3^= + 4Cl_2 + 5H_2O = 2SO_4^= + 8Cl^- + 10H^+$$

Sodium thiosulfate reacts with iodine, forming sodium tetrathionate

$$2S_2O_3^= + I_2 = S_4O_6^= + 2I^-$$

This reaction is quantitative, and is the basis of the use of sodium thiosulfate in iodometry.

The reaction with bromine also takes place quantitatively in acid solution, but follows the same course as that with chlorine. It can be utilized in bromometry (D'Ans, *Angew. Chem.*, 63 (1951) 45).

(xiii) *Thiosulfurous acid*, $H_2S_2O_2$ or $S=S(OH)_2$ probably exists in solution in equilibrium with H_2S and H_2SO_3:

$$SH_2 + OS(OH)_2 \rightleftharpoons H_2O + S=S(OH)_2.$$

It cannot be obtained in the free state. Compounds of the general formula $S_2O_2R_2$, colorless, pungent smelling liquids first obtained by Lengfeld in 1895 by the reaction of sodium ethylate or methylate with disulfur dichloride (sulfur monochloride), S_2Cl_2, dissolved do petroleum ether, are generally referred to as esters of thiosulfurous acid. However, these in not have the structure $S=S(OR)_2$ formerly ascribed to them, but $R\diagdown_O\diagup^S\diagdown_S\diagup^O\diagdown_R$,
as is proved by their parachor, dipole moment and Raman spectrum (Stamm, 1937; Scheibe, 1938).

(xiv) *Polythionic acids and the Polythionates.* When hydrogen sulfide is passed into a solution of sulfur dioxide, the so called *Wackenroder's* solution is obtained. This is a solution containing a mixture of *polythionic acids*, that is, of compounds with the general formula $H_2S_xO_6$, in which x may have the values 3, 4, 5, 6 and possibly greater.* The principal constituents of Wackenroder's solution are *tetrathionic acid*, $H_2S_4O_6$, and *pentathionic acid*, $H_2S_5O_6$. *Thiosulfuric acid*, $H_2S_2O_3$ and *trithionic acid*, $H_2S_3O_6$, are formed chiefly as intermediates in the action of sulfurous acid on hydrogen sulfide, the latter at once combining with more sulfur, if thiosulfuric acid is present, to form tetra- and pentathionic acids.

The reactions which are of importance in the formation of the polythionic acids have been studied chiefly by Debus, Foerster, Riesenfeld, Bassett and Stamm. For a time it was widely assumed that *sulfur monoxide* or *sulfoxylic acid* $S(OH)_2$ was formed as the first intermediate in the reaction between H_2S and H_2SO_3. However, after Schenk succeeded in preparing SO, he was able to show that none of this substance was formed by the action of H_2S with SO_2, and also that SO did not react with H_2SO_3 to form polythionic acids. On the other hand, Noack (1925) and Stamm (1938 and later) proved that the substance obtained by hydrolyzing S_2Cl_2 and $S_2O_2R_2$—i.e., the compound $H_2S_2O_2$**—did react with both H_2SO_3 and H_2S, to form the same substances which are found to be present in Wackenroder's solution. It is now therefore generally considered (as Foerster supposed in 1922) that $H_2S_2O_2$ is formed as the primary product of reaction between H_2S and H_2SO_3, and that the polythionic acids are the products of the following set of reactions:

$$H_2S \quad + \quad H_2SO_3 \quad \rightleftharpoons H_2S_2O_2 + \quad H_2O \qquad (1)$$

$$H_2S_2O_2 + 2H_2SO_3 \quad = H_2S_4O_6 + 2H_2O \qquad (2)$$

$$H_2S_4O_6 + \quad H_2SO_3 \quad \rightleftharpoons H_2S_3O_6 + H_2S_2O_3 \qquad (3)$$

$$H_2S_2O_3 \quad = \quad S + \quad H_2SO_3 \qquad (4)$$

$$H_2S_2O_2 + \quad H_2S \quad = 3S + 2H_2O \qquad (5)$$

$$H_2S_4O_6 + S \quad = H_2S_5O_6 \qquad (6)$$

The crude Wackenroder's solution contains large amounts of sulfur in colloidal dispersion. When it is freed from this—e.g., by adding a little lanthanum chloride, —a colorless, odorless solution is obtained, with a strongly acid reaction, which can be concentrated to a considerable degree on a water bath. However, the polythionic acids contained in this solution cannot be isolated. Their salts, the polythionates, $M^1_2[S_xO_6]$, can be prepared, the alkali salts in particular being notably stable; the alkaline earth and heavy metal polythionates are less stable. The trithionates are most readily decomposed, sulfuric acid, sulfurous acid and sulfur appearing as the end products of decomposition. Most of the polythionates are very soluble in water, but insoluble in alcohol.

Kurtenacker (1933, 1938) gives the following figures for the solubility of the alkali polythionates at 20° (expressed as g of anhydrous salt per 100 g of solution):

$Na_2S_3O_6 \cdot 3H_2O$	$Na_2S_4O_6 \cdot 2H_2O$	$Na_2S_5O_6 \cdot 2H_2O$
52.9	50.4	52.0

* *Dithionic acid*, $H_2S_2O_6$, does not belong to the polythionic acids. Cf. p. 724.
** It is not known whether this compound has the structure HO—S—S—OH or $S=S(OH)_2$.

$(NH_4)_2S_3O_6$
56.9

$(NH_4)_2S_4O_6$
54.4

$K_2S_3O_6$
18.4

$K_2S_4O_6$
23.2

$K_2S_5O_6 \cdot 1\frac{1}{2}H_2O$
24.8

The polythionates of the heavy metals can be obtained from the corresponding perchlorates, by double decomposition with alkali polythionates. They are not very stable (von Deines, 1933, Meuwsen, 1952).

1. *Trithionates. Potassium trithionate*, $K_2S_3O_6$, is obtained by the action of sulfur dioxide on potassium thiosulfate solution.* *Sodium trithionate* can be prepared by adding hydrogen peroxide to a well-cooled, saturated solution of sodium thiosulfate:

$$2Na_2S_2O_3 + 4H_2O_2 = Na_2S_3O_6 + Na_2SO_4 + 4H_2O$$

2. *Trithionic acid*, $H_2S_3O_6$, may be liberated from the potassium salt by double decomposition with fluorosilicic, perchloric or tartaric acid, but is obtainable only in solution.

3. *Tetrathionates.* As has been mentioned, sodium tetrathionate is formed by the action of iodine on sodium thiosulfate. The potassium salt may be prepared similarly.

4. *Tetrathionic acid*, $H_2S_4O_6$, may be liberated from the latter by adding the calculated quantity of tartaric acid. Its solution is appreciably more stable than that of trithionic acid, but the acid cannot be isolated.

5. *Pentathionates. Potassium pentathionate*, $K_2S_5O_6$, can be prepared from Wackenroder's solution by adding potassium acetate. Potassium tetrathionate crystallizes out first, in prisms, and tabular crystals of potassium pentathionate are obtained from the mother liquors. Pentathionic acid, $H_2S_5O_6$, which is also stable only in solution, can be liberated by adding tartaric acid to the potassium salt.

6. *Hexathionates.* Debus (1888) suspected that Wackenroder's solution contained a *hexathionic acid*, $H_2S_6O_6$, but the existence of this acid was proved only in 1928, when Weitz succeeded in preparing its pure, crystalline salts. *Potassium hexathionate*, $K_2S_6O_6$, which is quite stable in the solid state, although readily decomposed in solution, especially in the presence of alkali, may be obtained by the action of potassium nitrite on potassium thiosulfate, in concentrated hydrochloric acid solution at low temperatures (—10 to —15°). It separates out in the form of white, crystalline crusts. The benzidinium salt, $[C_{12}H_{12}N_2H_2][S_6O_6]$, crystallizes particularly well, in fine colorless needles.

7. *Constitution of the Polythionic Acids.* The structure of the tetrathionate ion follows from its mode of formation, by the oxidation of thiosulfate ion by means of iodine:

Since the iodine atoms remove an electron from each of the two sulfur atoms, these can then combine with one another. The reaction is exactly analogous to the formation of the peroxydisulfate ion from two partly discharged sulfate ions (p. 716).

It may be assumed that the trithionate and pentathionate ions have structures similar to the tetrathionate ion—i.e.,

trithionate ion

pentathionate ion

* The mechanism of this reaction is not yet fully explained. It is curious that sodium trithionate cannot be obtained by an analogous method.

From the standpoint of formal valence theory, the polythionic acids always contain two *hexavalent* sulfur atoms, linked through 1, 2 or more *bivalent* sulfur atoms. In terms of the coordination theory, the polythionic acids are binuclear compounds, with two coordination centers of 4-coordinated sulfur atoms, coordinatively bivalent, neutral sulfur atoms functioning as bridge groups between them.

8. *High-molecular Polythionates.* Formation of polythionates with more than six S-atoms was first noticed by Weitz (1928). These compounds are invariably formed as a complex mixture, probably because the high-molecular polythionate ions have a tendency to break up into mixtures of ions of even higher molecular weight and ions of lower molecular weight. It is possible to resolve the mixture, however, by fractional precipitation (e.g., with benzidine salts). By this method, Weitz was able later to prepare crystalline benzidinium salts with up to 10 S atoms. Especially high-molecular polythionates (with 20 to 40 S atoms) are formed in the decomposition of potassium thiosulfate with concentrated hydrochloric acid, in the presence of S_2Cl_2. These are present in aqueous solution in colloidal form, and it would seem that their long-chain anions, like those of the soaps, tend to come together to form micelles (cf. Vol. II). Weitz (1952) considers that the so-called hydrophilic sulfur sols (Odén, 1913) do not really consist of colloidal sulfur, but involve colloidal dispersions of extremely high-molecular polythionates (with 40 to 60 or more S atoms).

(*xv*) *Dithionic acid and the Dithionates. Dithionic acid,* $H_2S_2O_6$, was formerly classified with the polythionic acids but is not really related to them. It forms rather an intermediate step between sulfurous acid and sulfuric acid. It is obtained, in the form of its manganese salt, by the action of sulfur dioxide on a suspension of manganese dioxide hydrate:

$$MnO_2 + 2SO_2 = MnS_2O_6$$

Part of the manganese dioxide always enters into a side reaction, forming sulfate:

$$MnO_2 + SO_2 = MnSO_4$$

Other higher metallic oxides with mild oxidizing properties may be used (in reactive form— i.e., as their oxide hydrates) in place of manganese dioxide—e.g., ferric oxide hydrate. In this instance iron(III) sulfite is first formed, and this decomposes into iron(II) sulfite and iron(II) dithionate:

$$Fe_2O_3 + 3SO_2 = Fe_2(SO_3)_3, \qquad Fe^{III}_2(SO_3)_3 = Fe^{II}SO_3 + Fe^{II}S_2O_6$$

The sulfite ion may also be converted into the dithionate ion by electrolytic oxidation at anodically polarized platinized platinum electrodes.

$$2SO_3^= - 2e = S_2O_6^=$$

However, the reaction

$$SO_3^= - 2e + H_2O = SO_4^= + 2H^+,$$

whereby sulfate ion is formed, generally occurs preferentially.

The well crystallized barium dithionate, $BaS_2O_6 \cdot 2H_2O$, is obtained by double decomposition of the manganese salt with barium hydroxide, and dithionic acid, $H_2S_2O_6$, may be liberated from this by reaction with the calculated quantity of sulfuric acid. It is only stable in solution, and decomposes into sulfuric acid and sulfur dioxide when the solution is much concentrated:

$$H_2S_2O_6 = H_2SO_4 + SO_2$$

Its salts decompose in an analogous manner when heated,—e.g., potassium dithionate at 258°:

$$K_2S_2O_6 = K_2SO_4 + SO_2 - 5.0 \text{ kcal}$$

(Tammann, 1932; Roth, 1935).

All the dithionates are readily soluble in water, and crystallize well. Except for the alkali and alkaline earth salts, they decompose when their solutions are boiled, or in some cases slowly even at ordinary temperature.

The formula proposed by Mendeléeff, $(HO)O_2S—SO_2(OH)$, was formerly generally accepted as representing the structure of dithionic acid, This derives dithionic acid from sulfuric acid, by the union of two sulfonic radicals $—SO_2(OH)$, in the same way as oxalic acid can be considered as built up by the combination of two carboxyl radicals. This formula for dithionic acid (or the corresponding structure $[O_3S—SO_3]^{2-}$ for its ion) has now been fully substantiated by the X-ray structure determination of the cesium salt, $Cs_2[S_2O_6]$, (Hagg, 1932), and by magnetochemical studies (cf. Chap. 9).

(*xvi*) *Dithionites* (*Hyposulfites*). Just as the dithionates are derived from the sulfites by oxidation, so the *dithionites*, $M^I_2[S_2O_4]$ are derived from them by reduction:

$$2SO_3^= + 2e + 4H^+ = S_2O_4^= + 2H_2O$$

In accordance with this equation, they may be obtained by the electrolytic reduction of hydrogen sulfite solutions, but are better prepared by zinc reduction of a solution of sodium hydrogen sulfite, containing excess sulfurous acid:

$$2NaHSO_3 + H_2SO_3 + Zn = ZnSO_3 + Na_2S_2O_4 + 2H_2O$$

Sodium dithionite (sodium hyposulfite, sodium hydrosulfite) may be precipitated from the solution by addition of sodium chloride or alcohol. It crystallizes as the hydrate, $Na_2S_2O_4 \cdot 2H_2O$, but as it is not very stable in this form it is dehydrated, to give the anhydrous salt which keeps better. It is used as a reducing agent in vat dyeing and in discharge printing, and for the preparation of *rongalite* (see below), which is also used in vat dyeing and printing. It is also used as an absorbent for oxygen in gas analysis.

The dithionites are very unstable in solution, and as hydrates, because of their great tendency to be oxidized. Even in the absence of air, they gradually decompose in solution, or if moist, breaking up into thiosulfate and pyrosulfite:

$$2Na_2S_2O_4 = Na_2S_2O_3 + Na_2S_2O_5.$$

When they are heated, decomposition takes place according to the equation*

$$2Na_2S_2O_4 = Na_2S_2O_3 + Na_2SO_3 + SO_2 + 20.7 \text{ kcal}$$

The zinc salt, ZnS_2O_4 is relatively stable, as also are the double salts which it forms with the alkali dithionites. Except for calcium dithionite, all the dithionites are readily soluble. The solubility of the sodium salt is 21.8 g in 100 g of water at 20°. The corresponding acid, *dithionous acid*, is not capable of existence.

The formation of dithionites by the reduction of hydrogen sulfite solutions was first observed by Schönbein, in 1852. Schützenberger (1869) first succeeded in isolating a salt (the sodium salt) in the solid state, but incorrectly assigned to it the composition $NaHSO_2$, and accordingly named it sodium *hydrosulfite*. The correct composition, $NaSO_2$ or $Na_2S_2O_4$, was determined by Bernthsen in 1880, but the incorrect name 'hydrosufite' has persisted to some extent even today, especially in the technical literature. The introduction of the name 'hyposulfite', proposed by Bernthsen for the salts $M^I_2S_2O_4$, conflicted with the former usage whereby that name was applied to the thiosulfates. The use of this name for the salts

---

* This decomposition sets in violently at 190°, in the case of the sodium salt. As some evolution of heat takes place simply through absorption of water, the accumulation of heat can give rise to 'spontaneous combustion' when very large quantities are stored.

$M^I_2S_2O_4$ is also open to the objection that, according to the usual system of nomenclature of oxy-acids, this name should be strictly applied to the salts $M^I_2SO_2$ (sulfoxylates). It was pointed out by Noyes (1929) that, according to the generally accepted rules of nomenclature the salts $M^I_2S_2O_4$ must be called *dithionites*, since the sulfur in them is formally one double oxidation stage lower than the sulfur in dithionates. The name *dithionites* is therefore recommended in the internationally agreed rules for the nomenclature of inorganic compounds.* This name involves no assumptions as to the constitution of the compounds, which was until recently open to doubt. The chemical reactions of the dithionites do not conform to such an unambiguous mechanism as to afford direct evidence on their constitution. Evidence regarding structure has been obtained, however, from the Raman spectra (Simon, *Z. anorg. Chem.*, 260 (1949) 161). It has been shown that in the $[S_2O_4]^{2-}$ ion, there is an S—S single bond (without free rotation), and that the ion is probably planar and completely symmetrical in structure. Hence the constitution of the dithionite ion is

$$\left[\begin{array}{cc} :\ddot{O} & \ddot{O}: \\ & :S\!-\!S: \\ :\ddot{O} & \ddot{O}: \end{array} \right]^{2-}$$

Each of the sulfur atoms thus forms three bonds. The mode of formation and the reactions of the dithionites are in harmony with this structure.

(xvii) *Sulfoxylates. Rongalite.* The action of formaldehyde, CH_2O, on sodium dithionite yields the substance *rongalite*, which is used as a reducing agent in textile dyeing:

$$Na_2S_2O_4 + 2CH_2O + H_2O = CH_2(OH).SO_2Na + CH_2(OH).SO_3Na$$

sodium hydroxy-   sodium hydroxy-
methane sulfinate   methane sulfonate
(rongalite)

Rongalite was formerly generally regarded as a derivative of *sulfoxylic acid*, $S(OH)_2$, which is itself unstable. It is, however, probably not derived directly from this, but from the isomeric acid $HSO(OH)$ (which is also unstable), so that it may be regarded as the sodium salt of *hydroxymethane sulfinic acid*. For a long time, the constitution of rongalite has been a matter of discussion. Many reasons lead to the conclusion that it is a derivative of sulfurous acid. The problem is not yet finally solved; in 1943, Faessler and Goehring found that the wave length of the sulfur K_a doublet was almost the same as in p-toluene sulfinic acid, but differed from that expected for any derivative of sulfurous acid. *Sulfoxylic acid*, $S(OH)_2$, has a transitory existence in the hydrolysis of SCl_2, $S(OC_2H_5)_2$ and similar compounds. Whether any salts of sulfoxylic acid (*sulfoxylates*) exist is still uncertain. Scholder (1935) considered that he had obtained *cobalt sulfoxylate*, $CoSO_2 \cdot 3H_2O$, by carefully acidifying an aqueous-ammoniacal solution of $CoCl_2$ and $Na_2S_2O_4$. The dithionite ion is supposed to be split up hydrolytically in the first place, with the formation of sulfite and sulfoxylate ions:

$$[S_2O_4]^= + H_2O = [SO_3]^= + [SO_2]^= + 2H^+$$

Although the sulfoxylate ion is not itself stable in solution, it is then captured by the cobalt ion, with which it forms an insoluble salt which is thrown out of solution as a dark brown precipitate. However, Bode (1952) finds that the magnetic properties of preparations obtained in this way do not correspond to those which would be expected for a compound $Co^{II}SO_2$. If sulfoxylic acid has the structure $S(OH)_2$ generally assigned to it, the compound $S(OC_2H_5)_2$ prepared by Meuwsen (1936) may be considered to be its ester. This was obtained by the catalytic decomposition of diethyl thiosulfite,

$$S\!=\!S(OC_2H_5)_2 = S + S(OC_2H_5)_2$$

as a pungent, colorless, inflammable liquid, not miscible with water (b.p. 117° at 733 mm pressure).

* See appendix to Vol. II.

(*xviii*) *Disulfur trioxide*, S_2O_3, may be obtained in the form of blue crystalline crusts by the action of sulfur on completely anhydrous sulfur trioxide. The excess of sulfur trioxide, in which the compound is only very sparingly soluble, may be poured off, although a deep blue solution is obtained if any moisture has access. Disulfur trioxide dissolves in fuming sulfuric acid with a blue to brown color, according to the sulfur trioxide content. The brown solutions contain colloidal sulfur, formed from the decomposition of disulfur trioxide, which is extremely sensitive to traces of moisture. The blue solution was also formerly considered by many to be a colloidal dispersion of sulfur, but the well-defined stoichiometric composition of the substance was determined by Weber in 1875, and has more recently been confirmed by Vogel and Partington.

From its composition, disulfur trioxide might be regarded as the anhydride of dithionous acid, $H_2S_2O_4$, but no chemical relationship with this acid or its salts has as yet been established.

(*xix*) *Sulfur monoxide*, SO, was first obtained in the pure state by Schenk, in 1933, by the action of a glow discharge on sulfur dioxide, or on a mixture of sulfur dioxide with sulfur vapor. It is a colorless gas, which decomposes (with deposition of sulfur) only slowly at ordinary temperature in the absence of catalysts. When once condensed by cooling, however, it cannot be vaporized again without undergoing decomposition. It does not react with oxygen at ordinary temperature. Its primary reaction with water is represented by the equation

$$3SO + 3H_2O = H_2S + 2H_2SO_3$$

The deposition of sulfur, and formation of polythionic acids, then take place as secondary processes.

It had already been concluded by Henri, in 1929, from spectroscopic observations, that SO was formed from SO_2 by the action of the glow discharge. The intermediate formation of SO in various chemical reactions had also already been suspected—e.g., in the formation and decomposition of the polythionic acids. Schenk was able to confirm spectroscopically that SO is formed in the reaction of $SOCl_2$ with various metals (e.g., Sn and Sb), and also in the combustion of sulfur if an excess of sulfur vapor is present. It was also found that $SOCl_2$ at 900°, and $SOBr_2$ at 520°, begin to decompose to a perceptible extent into SO and free halogen, whereas the decomposition of SO_2 into SO and $\frac{1}{2}O_2$ does not occur significantly except at a considerably higher temperature.

(b) Halogen Compounds of Sulfur

With fluorine, sulfur forms the compounds S_2F_2, SF_2, SF_4, S_2F_{10} and SF_6, and with chlorine the compounds S_2Cl_2, SCl_2 and SCl_4. It forms only the one, rather unstable bromide S_2Br_2. It has not hitherto proved possible to isolate any compound of sulfur and iodine, although disulfur diiodide, S_2I_2, is formed transiently, according to Rao (1940), by the reaction of KI on S_2Cl_2 dissolved in CCl_4. The same is true of *thionyl iodide*, SOI_2 (see below).

The replacement of hydroxyl groups by halogen atoms in the oxyacids of sulfur, leads to compounds among which *thionyl chloride*, $SOCl_2$, *sulfuryl chloride*, SO_2Cl_2, and *chlorosulfonic acid*, $SO_2(OH)Cl$, are especially worthy of note.

Compounds containing the bivalent radical $SO<$ are termed *thionyl* compounds, whereas *sulfuryl* compounds contain the radical $SO_2<$. Compounds derived from the symmetrical form of sulfurous acid, $OS(OH)_2$, by replacement of one OH-group by some other univalent atom or radical, are known as *sulfinic acids*. Those which are related in a similar way to sulfuric acid, $O_2S(OH)_2$, are *sulfonic acids* or *sulfo-acids*.

Table 101 gives a summary of the simplest halogen compounds of sulfur.

TABLE 101

HALIDES OF SULFUR AND ITS OXYACIDS

Fluorides	Chlorides	Bromides
Disulfur difluoride (sulfur monofluoride) $S=S<^F_F$ Colorless m.p. —120.5° b.p. — 38°	**Disulfur dichloride** (sulfur monochloride) $S=S<^{Cl}_{Cl}$ orange yellow m.p. — 76.5° b.p. +137.1° density 1.709 (at 0°)	**Disulfur dibromide** (sulfur monobromide) $S=S<^{Br}_{Br}$ Deep red m.p. —46° Vaporizes with decomp. density 2.635 (at 20°)
Sulfur difluoride Colorless $F—S—F$ b.p. —35°	**Sulfur dichloride** bright red $Cl—S—Cl$ b.p. 59° density 1.622 (at 15°)	—
Sulfur tetrafluoride $F\backslash S / F \atop F / \backslash F$ Colorless m.p. —124° b.p. — 40°	**Sulfur tetrachloride** $Cl\backslash S / Cl \atop Cl / \backslash Cl$ pale yellow melts about —30° with decomp.	—
Sulfur hexafluoride $F \; F \atop F—S—F \atop F \; F$ colorless m.p. —50.5° b.p. —63.8° density 1.88 (liq., at m.p.)	—	—
Disulfur decafluoride $F \; F \atop F—S—S—F \atop F \; F$ colorless m.p. —92° b.p. +29° density 2.08 (at 0°)	—	—
Thionyl fluoride $O=S<^F_F$ colorless m.p. —110° b.p. — 30°	**Thionyl chloride** $O=S<^{Cl}_{Cl}$ colorless b.p. 76° density 1.677 (at 0°)	**Thionyl bromide** $O=S<^{Br}_{Br}$ Red yellow liquid decomposes when heated. Density 2.61 (at 0°)
Sulfuryl fluoride $O{=}S{<}^F_F$ colorless m.p. —120° b.p. — 52°	**Sulfuryl chloride** $O{=}S{<}^{Cl}_{Cl}$ colorless b.p. 69.3° density 1.708 (at 0°)	—
Fluorosulfonic acid $O{=}S{<}^{OH}_F$ colorless b.p. 162.6°	**Chlorosulfonic acid** $O{=}S{<}^{OH}_{Cl}$ colorless b.p. 152° density 1.784 (at 0°)	—
—	**Pyrosulfuryl chloride** $O{=}S{<}^{O{-}O}_{Cl} \; Cl{-}S{<}^O_O$ colorless b.p. 153° density 1.872 (at 0°)	—

All the halides of sulfur, except the fluorides, are hydrolyzed by water. With the lower halides—e.g., S_2Cl_2 and SCl_2—sulfur is deposited.

(*i*) *Sulfur hexafluoride*, SF_6, is of particular interest, since its existence conclusively proves that sulfur may be hexavalent. It is formed by the direct combination of the elements, with the evolution of a considerable amount of heat. Sulfur hexafluoride is a colorless, odorless, and very dense gas (relative density 5.107 when air $= 1$; liter weight $= 6.602$ g). It is sparingly soluble in water, and rather more soluble in alcohol. It does not burn, but remains unchanged when heated either in oxygen or in hydrogen. It is also very unreactive in other respects, but is decomposed by hydrogen sulfide when heated:

$$SF_6 + 3H_2S = 6HF + 4S$$

S_2F_{10} was obtained by Denbigh and Whytlaw-Gray (1929), in the fractional distillation of the products of reaction of fluorine with sulfur. It is as inert chemically as SF_6. The lower fluorides listed in Table 101 have been described, but not closely characterized.

(*ii*) *Disulfur dichloride* ('sulphur monochloride'), S_2Cl_2 is an orange yellow liquid, which has a repulsive, suffocating smell and fumes in moist air. It is prepared by passing dry chlorine over molten sulfur, and is used as a solvent for sulfur in the vulcanization of rubber. It is also used in the preparation of carbon disulfide and other compounds, and especially of chlorides, since it readily gives up chlorine.

Disulfur dichloride is an excellent solvent for many substances, and is therefore frequently used in cryoscopic or ebullioscopic determinations of molecular weights. The molecular freezing point depression, per 1000 g of disulfur dichloride, is 5.36°, and the boiling point elevation 5.02°. The solubility of sulfur in S_2Cl_2 is about 22 g of S per 100 g of solution at 20°. Disulfur dichloride is slowly hydrolyzed by water, forming hydrogen chloride, sulfur dioxide and sulfur. Thiosulfuric acid and polythionic acids are also formed in the process, together with small amounts of hydrogen sulfide and sulfuric acid. The experiments of Noack (1925) and Stamm (1942) lead to the following reaction mechanism. The primary product of hydrolysis ($S_2Cl_2 + 2HOH = 2HCl + S_2(OH)_2$) would be *thiosulfurous acid*, or a compound isomeric with it. The initial product partly decomposes according to:

$$H_2S_2O_2 = H_2S + SO_2$$

and a further portion may then react with the decomposition products, H_2S and SO_2, according to the equations given on p. 722. The formation of the substances mentioned, other than sulfuric acid, is thereby explained. Formation of the latter can be explained if it may be assumed that S_2Cl_2 in the solution is partially dissociated into chlorine and sulfur. This dissociation has been proved to occur in the vapor state (above 300°) (Yost, 1935).

The structure $S{=}S{\diagup\atop\diagdown}{Cl\atop Cl}$ was formerly regarded as probable for disulfur dichloride, on the basis of its formation by the treatment of thionyl chloride with phosphorus pentasulfide (Carius):

$$P_2S_5 + 5O{=}SCl_2 = P_2O_5 + 5S{=}SCl_2,$$

and of other observations, such as the close similarity in absorption spectra of S_2Cl_2 and $SOCl_2$. However, the Raman spectrum (Venkateswaran, 1931, Mizushima, 1937) points to a *single bond* between the two S-atoms. Electron diffraction experiments by Palmer (1938) also led to the structure $Cl{\diagup}^S{\diagdown}S{\diagup}^{Cl}$ (distances $Cl \leftrightarrow S = 1.99$ Å, $S \leftrightarrow S = 2.05$ Å, angle $Cl{-}S{-}S = 103°$). The dipole moment is equally compatible with the asymmetric formula and the symmetrical formula, when account is taken of the free rotation about the single bonds ($\mu = 1.06 \cdot 10^{-18}$ e.s.u., Scheibe, 1938). G. Giacomello (1935) concluded from the chemical evidence that the two forms were present in equilibrium with one another.

(*iii*) *Polysulfur Chlorides*, S_xCl_2. Disulfur dichloride can not merely dissolve sulfur, but also build it into its molecule. The species so formed are principally S_3Cl_2 and S_4Cl_2. Ruff (1924) was able to isolate these in the form of the addition compounds $AlCl_3 \cdot 2S_3Cl_2$ and

$AlCl_3 \cdot 2S_4Cl_2$ (wine red, unstable liquids). Polysulfur chlorides of extremely high sulfur content have more recently been obtained.

When disulfur dichloride is 'cracked' at about 900° in a hot-and-cold tube arrangement, a viscous liquid can be condensed out, having a composition corresponding to the formula $S_2Cl_2 + 20\text{-}25$ S. Since the solubility of rhombic sulfur in disulfur dichloride is limited to about 1.5 atoms of S per mole of S_2Cl_2, Fehér and Baudler (1952) have suggested that the cracking reaction gives rise to 'poly-sulfur chlorides', with a chain structure—e.g., $Cl\text{—}S_x\text{—}Cl$ (x = about 20–30). The process can be compared with the formation of diradical chains, $\text{—}S_x\text{—}$, from sulfur in the liquid phase (cf. p. 703).

(iv) *Sulfur dichloride.* Disulfur dichloride absorbs chlorine, and becomes increasingly darker in color. It was for long uncertain whether a new compound was thereby formed, or whether chlorine was merely dissolved. In fact, both processes occur. The chlorine is at first only dissolved, and subsequently reacts with the disulfur dichloride, forming sulfur di-chloride. When pure, this is a ruby-red liquid, density 1.62 at 15°, b.p. 59°. It is ordinarily dissociated to a considerable extent into disulfur dichloride and chlorine, and therefore behaves chemically as a mixture of these two.

(v) *Sulfur tetrachloride*, SCl_4, is stable only at low temperatures. Ruff obtained it, in the form of yellow-white crystals melting at about —30°, by the action of liquid chlorine on disulfur dichloride. When it is warmed to ordinary temperature, it dissociates completely into S_2Cl_2 and chlorine. It is hydrolyzed by water to HCl and SO_2. If forms double com-pounds with other chlorides, which are much more stable than free sulfur tetrachloride itself—e.g., $SCl_4 \cdot SnCl_4$, $SCl_4 \cdot SbCl_5$, $SCl_4 \cdot AlCl_3$. These compounds are also vigorously decomposed by water.

(vi) *Thionyl chloride*, $SOCl_2$, is formed by the action of phosphorus pentachloride on sulfur dioxide:

$$PCl_5 + SO_2 = POCl_3 + SOCl_2,$$

or on the sulfites:

$$2PCl_5 + CaSO_3 = 2POCl_3 + CaCl_2 + SOCl_2$$

It is usually prepared industrially by the reaction of sulfur trioxide with sulfur dichloride:

$$SCl_2 + SO_3 = SOCl_2 + SO_2$$

It may be separated by fractional distillation from the *pyrosulfuryl chloride*, $S_2O_5Cl_2$, which is formed simultaneously. Disulfur dichloride may also be used, in place of sulfur dichloride:

$$S_2Cl_2 + SO_3 = SOCl_2 + SO_2 + S$$

In this case the sulfur which is deposited may be reconverted to disulfur dichloride by passing in chlorine.

Thionyl chloride is a colorless, strongly smelling liquid. It boils at 78°, and begins to decompose into sulfur dioxide, chlorine, and disulfur dichloride at slightly higher temperatures. It is hydrolyzed by water to sulfur dioxide and hydrogen chloride:

$$SOCl_2 + H_2O = SO_2 + 2HCl$$

By the reaction of $SOCl_2$ with NH_3, it is possible to prepare the compound $HN(SONH_2)_2$ (imidodisulfinamide). When this is treated with liquid hydrogen chloride, it is converted into the yellow *sulfinimide*, OSNH. This can undergo transformation into a deep red, iso-meric substance, which may be regarded as an *oxime of sulfur monoxide*, S=N—O—H (Goehring, 1951–52). The isomerism between these two compounds corresponds to that between isocyanic and fulminic acids.

Thionyl chloride finds its principal uses in organic chemistry as a chlorinating agent—i.e., for the introduction of chlorine into organic compounds in the place of other atoms or groups.

(*vii*) *Sulfuryl chloride*, SO_2Cl_2, is formed by the direct union of sulfur dioxide and chlorine in sunlight, or in the presence of suitable catalysts (camphor or active charcoal):

$$SO_2 + Cl_2 = SO_2Cl_2$$

It may also be obtained by heating chlorosulfonic acid, $SO_2(OH)Cl$:

$$2SO_2(OH)Cl = H_2SO_4 + SO_2Cl_2$$

When pure, sulfuryl chloride is a colorless liquid, with a suffocating smell. It turns yellowish on standing, as a result of partial dissociation into sulfur dioxide and chlorine. The products of dissociation may be removed by bubbling an indifferent gas through the liquid. Sulfuryl chloride is hydrolyzed by much water into sulfuric acid and hydrogen chloride:

$$SO_2Cl_2 + 2H_2O = H_2SO_4 + 2HCl$$

Sulfuryl chloride is a good solvent for many organic and inorganic substances, and has some ionizing properties (cf. Vol. II). It is used in organic chemistry as an energetic chlorinating agent.

(*vii*) *Sulfuryl chlorofluoride*, SO_2ClF, was prepared in 1936 by Booth, by the reaction of SO_2Cl_2 with SbF_3 in the presence of $SbCl_5$. It is a colorless gas (m.p. —124.7°; b.p. +7.1°s latent heat at the b.p. 6.34 kcal per mol.; density 1.623 at 0°, surface tension 17.2 dyne; per cm).

(*ix*) *Chlorosulfonic acid*, $SO_2(OH)Cl$, may be obtained by the direct combination of sulfur trioxide and hydrogen chloride:

$$HCl + SO_3 = SO_2(OH)Cl$$

or by reaction between sulfuric acid and phosphorus trichloride, oxychloride, or pentachloride:

$$SO_2(OH)_2 + PCl_5 = SO_2(OH)Cl + HCl + POCl_3$$

It is a colorless liquid which fumes strongly in moist air and possesses a suffocating smell. It reacts with water with a hissing sound, being immediately hydrolyzed to sulfuric acid and hydrogen chloride.

Depending upon the experimental conditions, chlorosulfonic acid reacts with organic substances to form either sulfonic acids or their chlorides:

$$O_2N.C_6H_5 + ClSO_3H = O_2N.C_6H_4.SO_3H + HCl$$
nitrobenzene   nitrobenzene sulfonic acid

$$O_2N.C_6H_5 + 2ClSO_3H = O_2N.C_6H_4.SO_2Cl + HCl + H_2SO_4$$
        nitrobenzene sulfonyl chloride

It is therefore occasionally used in organic chemistry to prepare sulfonyl chlorides, but less frequently for the preparation of sulfonic acids, which can, as a rule, be more conveniently prepared by treating the compound to be sulfonated with concentrated or fuming sulfuric acid.

(*x*) *Fluorosulfonic acid*, $SO_2(OH)F$, may be obtained by the combination of SO_3 with HF; by warming CaF_2 with fuming sulfuric acid (60% SO_3)—the best method, according to Ruff (1914)—; or by the action of fuming sulfuric acid on KHF_2 (Meyer, 1932). For many preparative purposes it is more convenient than chlorosulfonic acid, since it is much more stable than the latter, and is not decomposed even at 900° although it decomposes into $SO_3 + HF$ at much lower temperatures in the presence of sulfur. It reacts with NaCl, forming *sodium fluorosulfonate*, $Na[SO_3F]$, and is slowly hydrolyzed by water to H_2SO_4 and HF.

(c) Hydrogen Sulfide and the Sulfides

(*i*) *Hydrogen sulfide*, H_2S, is formed by the direct combination of the elements (see below). It is generally prepared by decomposing sulfides by means of acids—usually ferrous sulfide, with hydrochloric acid:

$$FeS + 2HCl = FeCl_2 + H_2S$$

The gas is washed with water to free it from hydrochloric acid vapors that may be carried over. Dilute sulfuric acid may be employed instead of hydrochloric acid for the decomposition of the sulfide, but in this case the hydrogen sulfide is apt to be contaminated with sulfur dioxide. Iron sulfide, prepared technically by melting iron with sulfur, usually contains some uncombined iron, so that hydrogen sulfide prepared from it contains admixed hydrogen. Oxygen, nitrogen, and water vapor are usually present also, in smaller amount. For most purposes these impurities are unimportant, but they may be removed by liquefaction of the gas with solid carbon dioxide, and fractional distillation.

The usual drying agents cannot be employed with hydrogen sulfide since it reacts with them.

The decomposition of CaS or BaS with dilute hydrochloric acid is often used for the preparation of hydrogen sulfide in a state of high purity (although generally not quite free from CO_2). Completely pure H_2S can most reliably be obtained by synthesizing it from its elements. The best procedure is to pass sulfur vapor, mixed with hydrogen, through a glass tube heated to 600° (Klemenc, 1932).

Hydrogen sulfide is frequently found in nature in volcanic gases and in the gases escaping from the ground in many volcanic regions ('solfatari', near Naples). It is found in solution in the waters of the sulfur springs. The 'sulfur bacteria' found in such waters do not produce hydrogen sulfide (e.g., by the reduction of sulfates), but live on it, in so far as they utilize the energy liberated by oxidizing hydrogen sulfide to sulfur or sulfates. There are, however, microorganisms which can perform the converse, reducing sulfates to sulfides —i.e., salts of hydrogen sulfide.

Hydrogen sulfide is a colorless gas, which has a very unpleasant smell, like rotten eggs, even at great dilutions. *It is extremely poisonous.* Inhaled in even minute quantities it produces headache, giddyness, and sickness. At higher concentrations it can lead to sudden death, in the same way as hydrogen cyanide. Hydrogen sulfide is heavier than air (density 1.1906 relative to air = 1; liter weight = 1.5392 g). It is readily condensed to a colorless fluid (cf. Table 99, p. 687).

The vapor pressure of liquefied hydrogen sulfide is 10.3 atm at 0°. The crit. temperature is 100.4°, crit. pressure 89 atm, crit. density 0.31. Liquid hydrogen sulfide is an excellent solvent for many organic compounds, but a poor solvent for inorganic compounds. It has only a slight ionizing power, in accordance with its rather small dielectric constant (about 10 at —60°, and 6 at 0°).

Solid hydrogen sulfide crystallizes with a face-centered cubic molecular crystal lattice. It undergoes transitions at —169.6° and —145.0°. These do not correspond to changes of crystal structure, but to the onset of free rotation of the H_2S molecules about 2 axes. The heats of transformation are 0.365 and 0.109 kcal per mol (Giauque and Clusius, 1929).

Hydrogen sulfide decomposes into its components at high temperatures. The reaction

$$S + H_2 \rightleftharpoons H_2S + 4.8 \text{ kcal}$$

is therefore readily reversible. Hydrogen sulfide burns with a bluish flame when it is ignited in the air, forming water and sulfur dioxide (1) or—if access of air is restricted—water and sulfur (2)

$$H_2S + \tfrac{3}{2}O_2 = H_2O + SO_2 \tag{1}$$

$$H_2S + \tfrac{1}{2}O_2 = H_2O + S \tag{2}$$

Hydrogen sulfide reacts readily with most metals, forming the corresponding sulfides, especially when heated or in the presence of moisture. It also reacts with many oxides, when heated, forming sulfides. It reacts vigorously with fluorine, chlorine and bromine, and less energetically with iodine. It is decomposed by concentrated sulfuric acid, depositing sulfur, whereas the acid is reduced to sulfur dioxide:

$$H_2SO_4 + H_2S = S + SO_2 + 2H_2O$$

In general, hydrogen sulfide is a powerful reducing agent.

Hydrogen sulfide has a fairly high solubility in water—about twice that of carbon dioxide: 1 volume of water can absorb 4.65 volumes of H_2S at 0°, 3.44 volumes at 10°, 2.75 volumes at 18° and 2.61 volumes at 20°. The heat of solution is 4.52 kcal per mol at 20° (Roth, 1934). A crystalline hydrate, $H_2S \cdot 5\tfrac{3}{4}H_2O$, (cf. p. 220) is formed at low temperatures. Hydrogen sulfide also forms addition compounds with other substances—e.g., with numerous halides. It is much more soluble in alcohol than in water (11.8 volumes in 1 vol of alcohol at 10°).

The saturated aqueous solution of hydrogen sulfide is used in the laboratory as a reagent. It does not keep well, since hydrogen sulfide is slowly oxidized in solution by the oxygen of the air with the deposition of sulfur.

Hydrogen sulfide is a very weak dibasic acid; as such, it dissociates electrolytically in two stages. The extent of the second stage dissociation is so small that its effect is not detectable in the conductivity of the solutions, even at the greatest dilutions. The extent of the first stage dissociation is 0.13% in 0.1 normal solution at 18°.

(ii) *Sulfides.* The compounds of sulfur with the more electropositive elements are called *sulfides*. Most of the sulfides—in particular those of the metals—can be regarded, from their formation and properties, as salts of hydrogen sulfide.

Since hydrogen sulfide is a dibasic acid, two series of salts are derived from it— the *acid sulfides* or *hydrogen sulfides*, M^IHS, and the normal *sulfides*, M^I_2S. The acid sulfides are all soluble in water. Of the normal sulfides, those of the alkali metals are also soluble in water. They are strongly hydrolyzed in solution (to the extent of about 90% in 1-normal solution), according to the equation

$$Na_2S + HOH \rightleftharpoons NaOH + NaHS, \quad \text{or} \quad S^= + HOH \rightleftharpoons OH^- + HS^-$$

Their solutions are therefore strongly basic in reaction. The normal alkaline earth sulfides are themselves insoluble in water, but they undergo hydrolysis when they are treated with water:

$$2CaS + 2HOH = Ca(HS)_2 + Ca(OH)_2$$

and the acid sulfide thereby formed goes into solution. This is in turn decomposed when the solution is boiled:

$$Ca(HS)_2 + 2HOH = Ca(OH)_2 + 2H_2S$$

The sulfides of certain elements of higher valence are yet more readily hydrolyzed —e.g., aluminum sulfide, Al_2S_3, chromium sulfide, Cr_2S_3, and silicon disulfide, SiS_2. These are all decomposed by water, with the evolution of hydrogen sulfide.

Most of the sulfides of the heavy metals are so insoluble in water that they do not undergo hydrolysis, and many are not decomposed even by dilute acids. The solubility products of these compounds are so small that even after repressing the sulfide ion concentration in the solution by the addition of hydrogen ions, only an infinitesimal concentration of metal ions can be present in the solution in equilibrium with solid sulfide. Such sulfides are therefore precipitated by hydrogen sulfide even from quite strongly acid solutions.

The fact that some of the heavy metals are precipitated from acid solution by means of hydrogen sulfide, whereas others are precipitated only from ammoniacal solutions, by means of 'ammonium sulfide' is the basis of the use of these two reagents in the systematic separation of the cations in analysis (cf. p. 736).

The following elements are precipitated by hydrogen sulfide from acid solutions:

(a) arsenic, antimony and tin;

(b) silver, mercury, lead, bismuth, copper and cadmium.

The following are precipitated by ammonium sulfide: zinc, manganese, cobalt, nickel, iron, chromium and aluminum. Since the sulfides of the last two elements are hydrolyzed by water, they are precipitated in the form of their oxide hydrates.

The sulfides grouped under (a) have the property of dissolving in ammonium polysulfide, forming thiosalts, whereas those named under (b) are insoluble in this reagent.

The foregoing summary is restricted to those elements commonly encountered. It should be noted that in the course of the analytical separation, silver, univalent mercury, and part of the lead usually precipitate on addition of hydrochloric acid, before use of hydrogen sulfide. The sulfide precipitate obtained in the course of the analysis therefore contains no silver sulfide.

The following are the principal methods for the preparation of the sulfides.

[1] Double decomposition of hydroxides with hydrogen sulfide.

This method is suitable, in particular, for the sulfides which are soluble in water—i.e., for the alkali sulfides. The procedure is first to saturate the alkali hydroxide solution with hydrogen sulfide, thereby first obtaining the acid sulfide (1). This is then converted into the normal sulfide by adding an equivalent quantity of alkali (2):

$$NaOH + H_2S \quad = NaHS + H_2O \tag{1}$$

$$NaHS + NaOH = Na_2S \quad + H_2O \tag{2}$$

[2] Double decomposition of salts in aqueous solution, with hydrogen sulfide or ammonium sulfide.

This method is suitable, in particular, for the insoluble sulfides. Its significance for analytical chemistry has already been mentioned.

[3] Reduction of sulfates by heating them with carbon.

Sodium sulfide is generally prepared technically by this method

$$Na_2SO_4 + 4C = Na_2S + 4CO \tag{3}$$

This is also the usual method for preparation of the alkaline earth sulfides.

(4) Direct combination of the elements.

As has already been mentioned, iron sulfide is made in this manner:

$$Fe + S = FeS$$

Combination of metals with sulfur usually takes place very readily, and often with the evolution of much heat. However, it rarely leads to absolutely pure products.

The alkali and alkaline earth sulfides are colorless. Those of the heavy metals, however, are mostly very deeply colored, and often black.

Although many sulfides can be heated in the absence of air without undergoing decomposition, others lose sulfur. Thus pyrite, FeS_2, decomposes into iron(II) sulfide and sulfur even on moderate heating. Tin(IV) sulfide similarly decomposes into the tin(II) compound when heated. The sulfides which are stable towards heat can usually also be heated in hydrogen without decomposition. When heated in oxygen or air ('roasted'), however, most sulfides are converted into oxides, and often, to some extent, into sulfates. Sulfides precipitated from aqueous solution may undergo oxidation to a considerable extent even at ordinary temperature, if they remain in contact with atmospheric oxygen for a long time in the moist state. Either the deposition of sulfur (4) or the formation of sulfate (5) can occur in this process:

$$2FeS + 3H_2O + \tfrac{3}{2}O_2 = 2Fe(OH)_3 + 2S \qquad (4)$$

$$CuS + 2O_2 = CuSO_4 \qquad (5)$$

Dissolved sulfides are also readily oxidized, and therefore have strongly reducing properties.

The powerful reducing properties of hydrogen sulfide, and of the sulfides, in solution, depend upon the small free energy of formation of the $S^=$ ion. In a galvanic cell, made up of a normal hydrogen electrode and a platinum electrode dipping into a sulfide solution, the hydrogen electrode forms the positive pole, and the 'sulfur electrode' the negative pole, because of the tendency of the $S^=$ ion to be discharged (cf. p. 685).

(*iii*) *Sodium sulfide*, Na_2S, is generally prepared technically by the reduction of sodium sulfate (eqn. (3) above). It crystallizes from aqueous solution below 48° with $9H_2O$, in hygroscopic, quadratic prisms. It is colorless when pure, but is generally somewhat yellowish, due to the presence of admixed polysulfide (see below).

The solubility of sodium sulfide is 18.06 g in 100 g of water at 18°. The solution is strongly basic in reaction, as a result of hydrolysis. It readily becomes oxidized by atmospheric oxygen, with the formation of thiosulfate:

$$2Na_2S + 2O_2 + H_2O = Na_2S_2O_3 + 2NaOH$$

Sodium sulfide finds industrial uses, especially as a reducing agent for organic nitro-compounds—e.g., in the preparation of sulfur dyes. In this process also, it is oxidized to sodium thiosulfate. Sodium sulfide is used in tanning, for de-hairing hides.

(*iv*) *Potassium sulfide*, K_2S, is generally prepared as shown on p. 734, by mixing equivalent amounts of potassium hydroxide and potassium hydrosulfide solutions.

It is colorless when pure, and generally crystallizes as the pentahydrate $K_2S \cdot 5H_2O$.

The yellow-green to brown potassium sulfide of industry, 'liver of sulfur', is a mixture of potassium polysulfides (see below) and potassium thiosulfate. It is prepared by fusing potassium carbonate with sulfur, with access of air, and is widely used for medicinal purposes—e.g., in lotions and baths for skin diseases.

(v) *Ammonium sulfide.* If a dilute solution of ammonia is saturated with hydrogen sulfide, *ammonium hydrogen sulfide* is formed:

$$NH_3 + H_2S = NH_4HS$$

If an equivalent quantity of ammonia is then added, it remains uncombined. The so called 'ammonium sulfide' solution used for many purposes in the laboratory, especially in analytical work, which is prepared in this manner, is thus not a solution of ammonium sulfide $[NH_4]_2S$, but one of equimolecular amounts of ammonium hydrogen sulfide and ammonia.

The true normal ammonium sulfide, $[NH_4]_2S$, is capable of existence only at low temperatures, and in the absence of water. The acid (primary) salt, ammonium hydrogen sulfide, NH_4HS, which can be obtained by cooling a mixture of gaseous ammonia and hydrogen sulfide to $0°$, volatilizes very readily, by dissociating into its constituents, NH_3 and H_2S. Its dissociation pressure is already 335 mm at $20°$. It is also extensively dissociated in solution. In a 1-normal solution of ammonium hydrogen sulfide, about 4% of the salt is present in the form of ammonia and undissociated hydrogen sulfide:

$$NH_4HS \rightleftharpoons NH_3 + H_2S$$

At the same concentration, the degree of hydrolysis of potassium- or sodium hydrogen sulfide is about one hundred times smaller.

The so called 'yellow ammonium sulfide', which is made by dissolving sulfur in colorless ammonium sulfide solution, is more extensively used as an analytical reagent than the colorless ammonium sulfide; it is a solution of *ammonium polysulfides*.

(vi) *Alkaline Earth Sulfides.* 1. *Calcium sulfide,* CaS, (m.p. about $2400°$) was at one time obtained as a by-product of the Leblanc soda process. It is now generally prepared by igniting calcium sulfate with carbon, and is extensively used in tanning and in cosmetics as a depilatory. The polysulfide, CaS_x, prepared by igniting calcium oxide with sulfur, is used for the same purpose, and also has medicinal applications. The 'lime-sulfur spray' which is produced by boiling sulfur with milk of lime, and which also contains calcium polysulfides, is used as a pesticide.

2. *Barium sulfide,* BaS, is prepared by heating an intimate mixture of finely ground barytes with carbon to 600–$800°$:*

$$BaSO_4 + 2C = BaS + 2CO_2$$

The barium sulfide of commerce is a loose, grey powder. It is used in the manufacture of lithopone, and also, like calcium sulfide, for de-hairing hides. Barium polysulfide, BaS_x, a light brown powder (*barium sulphuratum flavum*) finds a use as an insecticide against plant pests, and also as an efficient depilatory.

The alkaline earth sulfides have the same crystal structure as the oxides (rock salt type)

	MgS	CaS	SrS	BaS
a =	5.19	5.68	6.01	6.37 Å

* If the temperature is raised higher, CO is formed as well as CO_2, thereby requiring a higher consumption of carbon.

3. *Phosphors* [*8–11*]. The addition of traces of a salt of a heavy metal, and strong ignition in the presence of certains fluxes, confers on the sulfides of the alkaline earths the property of glowing for a prolonged period after exposure to light (*phosphorescing*). Preparations possessing this property are called *phosphors*. They are now generally prepared by igniting alkaline earth oxides or carbonates, mixed with sulfur and alkali salts, with suitable additions of very minute amounts of heavy metal salts. Thus 40.0 g of strontium carbonate, 6.0 g of sulfur, 1.0 g of lithium carbonate, 1.0 g of arsenic trisulfide are treated with 2 cc of a solution of thallium nitrate (1 part in 200), and ignited at 1200°. The resulting mass has a bright green phosphorescence. Preparations with blue, bluish green, yellow, orange-red, etc. after-glows can be made in analogous ways. The discovery of phosphors dates back to an accidental observation made by the cobbler Vincentius Casciarolus, of Bologna, a student of alchemy who lived at the turn of the 16th century. In the course of a walk he found a mineral which was unknown to him (heavy spar), and taking it home he ignited it with coal. He was astonished to find that the resulting preparation glowed in the dark. Preparations which had some technical uses as phosphors were first produced around 1870 by Balmain, in England. It had, in the intervening period, been established that strongly ignited alkaline earth sulfides owed their phosphorescent properties to the presence of heavy metal salts as impurities. Lenard considered that three materials must be present in order to have a phosphor: the sulfide matrix, the flux, and the heavy metal. The flux may be present in relatively large amounts, but the heavy metal must be present only in traces (e.g., concentrations around 10^{-4}; certain heavy metals 'quench' the phosphorescence, and must be reduced to much smaller concentrations). Such substances as sodium chloride, calcium fluoride, sodium phosphate, borax, etc. may be used as fluxes. Tiede and Schleede showed that the flux is not essential if the sulfide itself is fused by heating to a sufficiently high temperature under pressure. The sulfides of the alkaline earths do not constitute the only type of phosphors, which include such substances as natural and artificial willemite, Zn_2SiO_4, and zinc sulfide. Numerous substances have been prepared with the property of emitting visible light while they are irradiated with ultraviolet light. The sulfides of the alkaline earths are notable for their long *after-glow*—i.e., for storing up energy by the excitation of the crystal lattice, and subsequently re-emitting it as visible light over a period of time.

(*vii*) *Thiosalts*. Many sulfides dissolve in alkali sulfide solutions, forming *thiosalts*,* e.g.,

$$As_2S_3 + 3K_2S = 2K_3[AsS_3]$$
potassium
thioarsenite.

This process is quite analogous to the formation of oxy-acid salts by the union of an acidic and a basic oxide:

$$As_2O_3 + 3K_2O = 2K_3[AsO_3]$$

The sulfides which are capable of giving rise to thio-salts, and which therefore dissolve in alkali sulfide solutions, have in the past been called the 'acidic sulfides', as distinct from the 'basic sulfides'. However, according to the systematic rules of nomenclature the name 'acid sulfides' must be used to designate the acid salts of hydrogen sulfide—i.e., the compounds M^IHS. The name should, therefore, not be used for the thio-salt forming sulfides, to avoid confusion.

By analogy with the oxy-salts, the formation of the ions of the thio-salts may be represented by the following ionic equations.

---

* According to the internationally agreed rules of nomenclature, acids which are derived from oxy-acids by the replacement of O-atoms by S-atoms are known as *thio-acids*, and their salts as *thio-salts*. They were formerly called *sulfo-acids* also, but this name is also applied to quite a different class of compound—namely to the derivatives of sulfuric acid with the general formula $R.SO_3H$, for which the older name *sulfonic acids* is also used.

$$As_2O_3 + 3O^= \rightleftharpoons 2[AsO_3]^{\equiv} \quad \text{or} \quad As_2O_3 + 6OH^- \rightleftharpoons 2[AsO_3]^{\equiv} + 3H_2O \quad (1)$$

$$As_2S_3 + 3S^= \rightleftharpoons 2[AsS_3]^{\equiv} \quad \text{or} \quad As_2S_3 + 3SH^- + 3OH^- \rightleftharpoons 2[AsS_3]^{\equiv} + 3H_2O \quad (2)$$

$$As_2S_3 + 3O^= \rightleftharpoons [AsS_3]^{\equiv} + [AsO_3]^{\equiv} \quad \text{or} \quad As_2S_3 + 6OH^- \rightleftharpoons [AsS_3]^{\equiv} + [AsO_3]^{\equiv} + 3H_2O \quad (3)$$

The last equation shows that the ions of thio-salts and of oxyacid salts may be formed simultaneously, when sulfides soluble in alkali sulfides are treated with caustic alkalis (cf. p. 534 and 660).

The free thio-acids are unstable, and the thio-salts are therefore decomposed when their solutions are acidified, hydrogen sulfide being liberated and the sulfide re-precipitated.

In this respect also there is no fundamental difference between thio-salts and oxy-salts. The acids corresponding to many of the latter are unstable, and decompose to form the oxide, with loss of water. The decomposition of the thio-salts is simply the result of displacing the equilibria of equations (2) and (3) from right to left, by increasing the hydrogen ion concentration.

The ability to form thio-salts is of considerable analytical importance, especially for the sulfides of arsenic, antimony and tin.

In addition to these, thio-salts are formed by platinum, gold, germanium, tellurium, molybdenum, tungsten, vanadium and carbon. All are formed by treating the corresponding sulfide with alkali sulfide solutions. A further series of thio-salts may be obtained by fusion processes, but it is often uncertain whether compounds so prepared are true thio-salts, or merely double sulfides in the narrower sense (double compounds between sulfides, lacking salt-like character, and not based upon a definite thio-anion as a structural unit).

(*viii*) *Polysulfides and Hydrogen polysulfides*. Solutions of the alkali sulfides can take up considerable amounts of sulfur, forming yellow to brownish-red *polysulfides*— i.e., compounds with the general formula $M^I_2S_x$, where x is usually 2 to 5, but may in some cases be larger.

Polysulfides of the alkalis are also formed when alkali sulfide solutions stand in air, as a consequence of the slow oxidation of the hydrogen sulfide ion by atmospheric oxygen:

$$2HS^- + \tfrac{1}{2}O_2 = H_2O + S_2^=$$

They are also obtained when alkali sulfides, hydroxides or carbonates are fused with sulfur. When hydroxides and carbonates are used, however, the polysulfides are contaminated by thiosulfate, formed simultaneously, and also by sulfate if there is access of air. 'Liver of sulfur' (hepar sulphuris, from ἧπαρ, liver), which has long been known, is an impure mixture of polysulfides of this kind.

The polysulfides of the alkaline earth metals are also well known. The compounds with 4 sulfur atoms appear to be the most stable type.

Table 102 gives a summary of the alkali polysulfides which have been prepared. In the polysulfides, the depth of color increases as the sulfur content rises, whereas the *polyselenides* and *polytellurides*, which are also listed in the Table, are all grey-black in color. On the magnetochemistry and molar volumes of the compounds see Klemm, *Z. anorg. Chem.*, 241 (1939) 281.

TABLE 102

ALKALI POLYSULFIDES, POLYSELENIDES AND POLYTELLURIDES

Polysulfides				Polyselenides		Polytellurides		
Na_2S_2	K_2S_2	Rb_2S_2	Cs_2S_2	Na_2Se_2	K_2Se_2	Na_2Te_2	K_2Te_2	Rb_2Te_2
—	K_2S_3	Rb_2S_3	Cs_2S_3	Na_2Se_3	K_2Se_3	—	K_2Te_3	Rb_2Te_3
Na_2S_4	K_2S_4	Rb_2S_4	Cs_2S_4	Na_2Se_4	K_2Se_4	—		
Na_2S_5	K_2S_5	Rb_2S_5	Cs_2S_5	—	K_2Se_5	—		
—	K_2S_6	Rb_2S_6	Cs_2S_6	Na_2Se_6	?	Na_2Te_6		

The polysulfides are hydrolyzed in solution to a much smaller extent than the ordinary sulfides. Ammonium polysulfides such as $(NH_4)_2S_4$, $(NH_4)_2S_5 \cdot H_2O$, $(NH_4)_2S_9 \cdot \frac{1}{2}H_2O$ are therefore stable at ordinary temperature, in contrast to the normal ammonium sulfide, $(NH_4)_2S$.

Polysulfides are decomposed by acids, usually with the deposition of sulfur:

$$Na_2S_2 + 2HCl = 2NaCl + H_2S + S$$

However, the corresponding acids, the *hydrogen polysulfides*, H_2S_x, may be liberated under suitable conditions, and may be purified by high vacuum distillation.

It was formerly considered that the polysulfides should be represented by coordination formulas, about a central $S^=$ ion. On this basis H_2S_5 would be the analogue of $H_2[SO_4]$ (i.e., $H_2[SS_4]$), and might be characterized by particular stability. This is, however, not the case, and the investigations of Fehér (1941–48) make it almost certain that the hydrogen polysulfides contain chains of sulfur atoms, terminated by hydrogen atoms:

$$H—S—S—H, \qquad H—S—S—S—H, \qquad H—S—S—S—S—S—H \quad etc.$$

The Raman spectrum of H_2S_2 shows its structure to be essentially the same as that of H_2O_2, with two structurally equivalent sulfur atoms as shown above. The Raman spectra, molecular volumes and refractive indices of the higher members of the group accord with the chain structure. The existence of compounds with a much higher sulfur content than H_2S_5—e.g., H_2S_7, H_2S_8, and the salts derived from them, such as $[NH_4]_2S_9$—also accords better with the chain formula than with any centric formula.

Formation of hydrogen polysulfides with chains of linked sulfur atoms is not necessarily limited to the substances of relatively low molecular complexity just cited. There is evidence that compounds H_2S_x, where x is very large, may also be formed. Thus, Fanelli (1949) showed that when hydrogen sulfide is passed through molten sulfur at 120–190°, it may reduce the viscosity of the melt by as much as 1000-fold. From the previous discussion of the constitution of liquid sulfur, it may readily be understood that such an effect would arise from the reaction

$$H_2S + —S_x— = HS—S_{x-1}—SH.$$

Each molecule of hydrogen sulfide would thereby provide two chain-terminating groups, and by combining with the $—S_x—$ diradical would lower the average degree of polymerization. Fehér and Berthold (1952) have shown that the clear, mobile, oily form of sulfur, which is precipitated from sodium thiosulfate solution by means of concentrated acids, does not deposit solid sulfur even when it is kept indefinitely. It contains one molecule of hydrogen sulfide for about 350 atoms of sulfur, and this hydrogen sulfide is not evolved until the material is heated above 100°. Fehér and Berthold therefore regard the material as a mixture of high-molecular polysulfides, H_2S_x (with x > 300).

The molecular heats of formation of the liquid hydrogen polysulfides, from H_2 and S_{rhomb}, are: H_2S_2 4.2, H_2S_3 3.5, and H_2S_4 3.4 kcal. The heat of formation of H_2S_2 from

H_2S and gaseous S_2 is also exothermic ($+6.4$ kcal per g mol), whereas the corresponding reaction for hydrogen peroxide is strongly endothermic ($H_2O_{gas} + \frac{1}{2}O_2 = H_2O_{2gas} - 24.2$ kcal).

For the analytical determination of polysulfide sulfur, see Wintersberger (*Z. anorg. Chem.*, 236 (1938) 369) and Fehér (*ibid.*, 255 (1948) 316 and 260 (1949) 273).

(ix) *Alkyl sulfides and Sulfonium Compounds.* The *mercaptans* or *thioalcohols* $\dfrac{R}{H}{>}S$ and the *alkyl sulfides* or *thioethers* $\dfrac{R}{R}{>}S$ are derived from hydrogen sulfide by exchange of one or both hydrogen atoms for organic radicals R. The thioethers are able to add on alkyl halides (especially alkyl iodides) to form *sulfonium compounds*:

$$R_2S + RI = [R_3S]I$$

When alkyl sulfonium iodides are treated with moist silver oxide, alkyl sulfonium hydroxides are obtained, with the properties of true bases. *Sulfonium salts* may be prepared from these, by neutralizing them with acids. The radical R_3S- present in the sulfonium salts thus enters solution as the univalent positive ion $[R_3S]^+$. The sulfonium compounds thus probably correspond in constitution to the oxonium compounds. They may be regarded as compounds of coordinatively trivalent sulfur.

7. Analytical (Sulfur) [*17*]

The so-called 'silver coin' test serves for the general detection of sulfur compounds; it depends upon the reduction of sulfur compounds to sodium sulfide, by heating them with sodium carbonate in the reducing flame of the blowpipe or bunsen burner. When the cooled mass is moistened and placed upon a silver object it forms a black stain of silver sulfide. Sulfide ores may be recognized because they evolve sulfur dioxide when they are heated in an ignition tube open at both ends, and held obliquely in the flame to set up a current of air.

Hydrogen sulfide is easily recognizable by its unpleasant smell and its reactions with solutions of heavy metal salts. Traces of hydrogen sulfide—e.g., in air— are frequently detected by means of paper soaked in lead acetate, which becomes blackened through formation of lead sulfide. Sodium nitroprusside is a very sensitive reagent for hydrogen sulfide in solution, with which it gives a violet coloration. The reaction of sodium nitroprusside with *sulfite* (red coloration, especially in the presence of much zinc sulfate or nitrate) is much less sensitive. On the addition of strong acids, sulfites evolve sulfur dioxide (recognizable by its smell), as also do thiosulfates. The latter deposit sulfur simultaneously (cf. p. 720). Neutral solutions of sulfites react with iodine:

$$SO_3^= + I_2 + H_2O = SO_4^= + 2I^- + 2H^+.$$

Iodine solution is thus decolorized, and the solution becomes acid. Thiosulfates also decolorize iodine, but without the development of any acidity.

With barium ions, sulfate ions give a white precipitate of barium sulfate, insoluble in dilute strong acids. This may be distinguished from barium fluoride, BaF_2, and fluorosilicate, $BaSiF_6$, which are also sparingly soluble in acids, by means of the silver coin test.

Sulfur is generally determined quantitatively by conversion to barium sulfate, and weighing in that form. It is generally determined in organic compounds in the same manner, after converting it to sulfuric acid by heating it with concentrated nitric acid in a sealed tube (Carius method), or burning it to sulfur dioxide, which is absorbed in sodium hydroxide and hydrogen peroxide (micro-method of Pregl).

The uses of sodium thiosulfate in volumetric analysis (iodometry and bromometry) have already been mentioned.

8. Selenium (Se) and Tellurium (Te)

(a) Occurrence

Selenium occurs in traces in many native sulfides—e.g., pyrite, chalcopyrite, zinc blende, and is considerably enriched in the flue dusts which are formed when such seleniferous ores are roasted. The lead chamber sludge of the sulfuric acid works is also often rich in selenium.

Since the selenides are almost invariably isomorphous with the corresponding sulfides, they are usually present in admixture with the latter. Pure selenium minerals are very rare. *Berzelianite*, Cu_2Se, *tiemannite*, HgSe, and *naumannite*, Ag_2Se, may be cited. Pure tellurium minerals are rather less rare—e.g., *hessite*, Ag_2Te, which is isomorphous with naumannite, *altaite*, PbTe, *coloradoite*, HgTe, *sylvanite* $AgAuTe_4$, and other silver-gold tellurides (petzite, muthmannite, krennerite). Oxygen compounds of tellurium are also found—e.g., *tellurium ocher*, TeO_2. Tellurium also occurs native, both alone and mixed with sulfur and selenium. The reddish Japanese tellurium-sulfur contains 0.17% Te and 0.06% Se.

Tellurides are less commonly found as minor constituents of sulfides than are the selenides. Tellurium is therefore, on the whole, less widely distributed than selenium, although it is more often segregated in individual minerals. One of the most important tellurium ores is *nagyagite*, an isomorphous mixture of sulfides and tellurides, chiefly of lead, gold, copper, silver, and antimony.

(b) History

Tellurium was discovered in 1782, in auriferous ores from the Siebenbürgen, and was named after the earth (*tellus*) by Klaproth, who determined its most important properties. Berzelius, in 1817, discovered in the lead chamber sludge of a sulfuric acid factory an element very similar to tellurium, which he therefore named after the moon (σελήνη).

(c) Preparation

Selenium is recovered from the flue dusts obtained in roasting ores containing selenium, or from lead chamber sludge. When selenium is present, the sludge is reddish-grey, instead of a pure grey. The selenium contained in it is brought into solution by treatment with sulfuric acid and sodium nitrate. It is thereby converted largely into selenious acid, H_2SeO_3, with some selenic acid, H_2SeO_4. The latter is reduced to selenious acid when it is warmed with hydrochloric acid, and elementary selenium is deposited, as a red amorphous precipitate, when sulfur dioxide is then passed through the solution.

It is purified by burning it in oxygen, charged with the vapors of fuming nitric acid, whereby pure selenium dioxide sublimes. Selenium is precipitated once more from the aqueous solution of this oxide, by reduction with sulfur dioxide or hydrazine hydrate, after the addition of hydrochloric acid.

Selenium may also be leached out of the lead chamber sludge by digestion with warm potassium cyanide solution. In this case, potassium selenocyanate, K[SeCN], is formed. This is decomposed by hydrochloric acid with the deposition of selenium.

Tellurium is only rarely present in important amounts in lead chamber sludge, but may be recovered from the anode slimes of copper refining. It is generally prepared, however, from nagyagite.

The mineral may be attacked by various methods. Thus the sulfides present may first be removed by boiling with concentrated hydrochloric acid, and the residue of gold telluride decomposed by nitric acid. The solution is then evaporated, the residue taken up in hydrochloric acid, and the tellurium precipitated by means of sulfur dioxide.

If selenium and tellurium are present together, they may be separated by utilizing the difference in behavior of aqueous solutions of the selenites and tellurites towards sulfuric acid. On addition of this acid, tellurium dioxide, TeO_2 is precipitated, whereas selenious acid remains in solution.

(d) Properties

Like sulfur, *selenium* exists in several modifications. It is obtained as a loose powder, of a fine red color, by the reduction of selenious acid with sulfurous acid or other reducing agents, or by the rapid condensation of selenium vapor. This form is distinguished only by its degree of subdivision from *vitreous selenium*, which can be obtained by suddenly chilling molten selenium—e.g., by pouring it into water. Vitreous selenium forms a red-brown to blue-grey, very brittle, lustrous mass (of density 4.28–4.30), which can readily be reduced to a red powder. Vitreous selenium is a non-conductor of electricity. It is completely insoluble in cold water, but is somewhat soluble in concentrated sulfuric acid and in carbon disulfide, forming a ruby-red solution in the latter and a green solution in the former. Vitreous selenium begins to soften at about 50°, and changes into *grey crystalline selenium*, with a considerable evolution of heat, when more strongly heated. This form also results from the slow cooling of molten selenium. It is much less soluble in carbon disulfide than the red selenium. It forms a grey crystalline mass, which is a very poor electrical conductor in the dark. When illuminated, however, the conductivity rises about a thousandfold, and falls again to its original value in the dark.

This phenomenon is due to a 'loosening' or excitation of electrons. It is well known that when metals are irradiated with light of sufficiently short wave length they emit electrons (photoelectric effect). Selenium also shows this effect when irradiated with ultraviolet light. Since light of long wave length imparts a smaller quantum of energy to an atom or electron (cf. equation (10), p. 85) than light of short wave length, it may be assumed that on irradiation with visible light, electrons are, indeed, removed from the occupied levels of the selenium atoms, but cannot be ejected through the surface of the crystal into free space; they are excited into the *conduction levels*. This effect should, indeed, occur in ordinary metals, as well as in selenium, but it cannot be observed since it is masked by their much greater intrinsic conductivity.

Use is made of the photosensitivity of selenium for the construction of 'selenium cells' (more strictly 'selenium bridges') and of selenium photocells (selenium barrier layer photocells). Both are appliances for translating variations in the intensity of light into variations in an electric current. These are utilized for the operation of alarm apparatus (the 'electric eye') and various devices for the automatic operation of machinery (escalators, etc.), and especially for sound films, telegraphic transmission of pictures, and television [*12–14*]. The selenium bridge gave the first impetus to these inventions, but had too low an efficiency and too slow a response for full scale practical application. This proved possible only when photocells depending upon the photoelectric effect had been developed in place of the selenium bridge. The alkali metals were at first invariably used for this purpose. For many types of application, however, the more recently developed *barrier layer photocell* (Schottky and Lange, 1930) is considerably superior to the alkali metal cell, and for this purpose selenium has again proved particularly useful, as also have certain compounds (e.g., Cu_2O) which are classed as semiconductors [15].

A *selenium bridge* consists of a pair of parallel wires, wound on some insulating base (e.g., porcelain), with a thin coating of fused selenium between them. The selenium, which at first solidifies in the vitreous form is transformed into the crystalline modification ('sensitized') by prolonged annealing. Selenium bridges may be 'hard' or 'soft'. When the former are irradiated, the conductivity gradually rises, during several minutes, to a limiting value, and gradually falls off again in the dark. The conductivity of 'soft' selenium bridges reaches its maximum within the first seconds of irradiation, and then diminishes slowly on continued irradiation, or instantly in the dark. Even without irradiation, the conductivity of metallic selenium increases slowly with time, if too high a voltage is applied. It is also greater for large potential differences than for small. The thermal conductivity of metallic selenium is increased by irradiation in the same way as the electrical conductivity.

An ordinary *photocell* consists essentially of a thin layer of an alkali metal, deposited on the inner side of a glass vessel, and connected by a suitable current lead to the negative pole of a source of current. A second electrode (usually in the form of a ring or gauze) is joined to the positive pole. The tube is either evacuated, or filled with an inert gas. On illumination, electrons are emitted from the surface of the alkali metal, and are attracted by the positive electrode, thereby bringing about the passage of a current. If the space is gas-filled, the current is amplified by ionization. In the barrier layer photocell, the space between the electrodes is replaced by a thin layer of some material within which electrons are liberated upon irradiation. This is the case especially with selenium, and to a lesser extent with such substances as Cu_2O. Whereas a relatively large amount of energy is necessary for an electron to pass from any metal (including the alkali metals) into a vacuum or a gas space, the passage of an electron between two solid conductors, in contact with one another, takes place very easily. This is the reason for the high efficiency of barrier layer photocells. According to Lange (1931), a cuprous oxide barrier layer photocell without bias-potential furnishes a current from 6 to 125 times as great as that obtained from the usual alkali photocell with a bias-potential (120–200 volts). A selenium barrier layer photocell may be 10 to 20 times yet more sensitive. Since barrier layer photocells act as independent sources of current when they are illuminated ('photoelements' [15]), they are particularly suitable for photoelectric devices which must operate without external sources of current—e.g., for photographic exposure meters. They are also used for photoelectric switching and signalling devices (e.g., Lichtschranken, sorting machines, twilight switches, optical train safety devices), and for a wide variety of photometric apparatus—spectrophotometers, colorimeters, etc.

Grey selenium crystallizes in the hexagonal-rhombohedral system (see p. 684 for physical constants). Molten selenium is brownish red, and the vapor is yellow. Above 900°, it consists of Se_2 molecules, but more extensive polymerization takes place at lower temperatures. The cryoscopic measurements of Beckmann and Pfeiffer indicated that selenium dissolved in carbon disulfide has a molecular weight corresponding to the formula Se_8.

Red selenium can be obtained crystalline from carbon disulfide solutions in two distinct monoclinic modifications with different axial ratios—Se_α, density 4.48, and Se_β, density 4.40. α-Selenium can form mixed crystals with sulfur. Crystalline red selenium is metastable, like the vitreous form, and changes into the metallic modification when heated. By heating it rapidly, however, it can be melted directly, at about 180°, without undergoing transformation.

(e) Crystal Structure

(*i*) *Selenium*. The structure of grey selenium can be thought of as derived from a simple cubic lattice by displacing the atoms alternately in opposite directions. Instead of 6 equidistant neighbors, each Se atom then has 2 nearest neighbors (at 2.32 Å), and 4 more distant neighbors, (at 3.46 Å), so as to build up a parallel array of spiral chains. *Tellurium* has a similar structure, with interatomic distances of 2.86 Å and 3.74 Å, respectively. Thus the stable crystalline modifications of selenium and tellurium display the same 'infinite linear polymer' structure that is found in liquid sulfur (S_μ) and the metastable plastic sulfur.

Because of its strong tendency to supercool, grey selenium separates from the melt only in the form of minute crystals. It can, however, be obtained in needles several centimeters long by sublimation at low pressures (Straumanis, 1940). Whereas the prism faces of these are well developed, the ends generally show no regular development of faces at all. This is apparently because the individual chains grow independently in the direction of the *c*-axis, but the binding forces between them are too weak to couple their development in such a way as to build up planar end faces (De Boer, 1943).

Molten selenium is viscous (cf. sulfur); it solidifies to the vitreous form or, if it is stretched while in the plastic state, it gives the X-ray diffraction pattern characteristic of a fibrous structure. In *vitreous selenium* the atoms are linked together in chains, in the same ways as they are in the crystal lattice of hexagonal selenium. These chains are also oriented parallel to one another over small regions or *micelles* (in contrast to the completely random arrangement of flexed chains in the melt), and it is the increased alignment produced by plastic flow when the glass is stretched that produces the fiber structure. The micelles are small, however, and the chains are not regularly stacked as they are in the crystalline state (Prins 1937, Glocker 1942, Richter 1952; see also Krebs, *Angew. Chem.*, 65 (1953) 293). The behavior of vitreous selenium is therefore similar to that of other linear high polymers. Recrystallization is sluggish, since it involves the breaking and re-formation of Se—Se bonds; it takes place slowly above 70°, but is accelerated by impurities which can give rise to ionic and free radical mechanisms for breaking down larger units and opening rings. Such impurities are metals (e.g., Na), halogens, S^{2-}, Se^{2-}, and organic bases, which can break the long chains by processes such as

$$Se_x + 2e \quad \rightleftharpoons \ ^-Se—(Se_y)—Se^-$$

$$Se_x + I_2 \quad \rightleftharpoons I—Se—(Se_y)—Se—I$$

$$Se_x + R_3N \rightleftharpoons —Se—(Se_y)—Se^- + R_3N^+$$

Amorphous and crystalline *arsenic* differ in the same way as do vitreous and crystalline selenium. According to Richter, vitreous-amorphous modifications are most likely to be found for those elements such as S, Se, P, As or Sb, which have a chain or sheet structure in at least one of their crystalline modifications. Since the binding forces between chains or sheets are only weak, it is understandable that they do not always suffice to impose the requisite high degree of order, and that the amorphous state should be produced. Where the binding forces are the same in all spatial directions, it is only in exceptional cases that the conditions exist for the formation of amorphous modifications,—namely when, as with Si and Ge, whole groups of atoms (regular tetrahedra) constitute the structural units, and can link up quite at random (amorphous state) instead of in regular array (crystal). Impurities generally favor the development of the amorphous state, since they hinder the regular arrangement of the atoms or structural units, and serve as chain terminations, or saturate free valences. The two forms of red selenium have structures built up from Se_8 molecules, corresponding to the crystalline forms of sulfur.

(*ii*) *Tellurium* crystallizes in the rhombohedral system, isomorphous with grey selenium. It is silver white, with a metallic luster, its hardness is low (2.5 on Mohs' scale), but it is brittle, and it can therefore be easily powdered. Its electrical conductivity is small, being about 10^{-5} that of silver. It is increased to some extent by illumination, but not nearly as much as that of selenium. Crystalline tellurium has a density of 6.25. Tellurium deposited from solution, by reduction of tellurous acid with sulfurous acid, is obtained as a voluminous brown powder, density 6.0. It has not been definitely established whether this 'amorphous tellurium' is a distinct modification, or merely a very finely divided form of tellurium.

The so-called amorphous tellurium is also obtained, mixed with crystalline tellurium, when molten tellurium is rapidly cooled. Conversely if a solution of potassium telluride is left to be oxidized spontaneously in the air, tellurium is deposited in distinctly crystalline form as thin rhombohedral needles. Tellurium melts at 452° and boils at 1390°, forming a golden yellow vapor which consists substantially of Te_2 molecules up to about 2000°. At higher temperatures it dissociates into atoms to a considerable extent.

Klemm (1952) has shown that molten tellurium has a maximum density at a temperature a little above its melting point, just as is found for water (cf. p. 526).

Selenium and tellurium burn when they are heated in air, the former with a pure blue flame, the latter with a blue flame sheathed with green, forming the dioxides, SeO_2 and TeO_2. A characteristic smell (like decayed radishes) accompanies the combustion of selenium, whereas burning tellurium only smells faintly sour. Selenium and tellurium also combine energetically with other electronegative elements, especially the halogens, and also with many metals. Only selenium unites directly with hydrogen, tellurium reacting only to a minute extent, if at all. Avid combination apparently takes place with sulfur, but it does not lead to the formation of compounds, but to solutions which solidify to form mixed crystals.

Neither selenium nor tellurium is attacked by non-oxidizing acids. Both dissolve in concentrated sulfuric acid, however, and also in nitric acid and caustic alkalis.

(f) Uses

Selenium is used for photoelectric apparatus, such as is employed in sound films, picture telegraphy and television, and also for such purposes as automatically switching light buoys on and off, photometry, etc. Its sensitivity towards small intensities of light, and variations of intensity, greatly exceeds that of the human eye. Selenium is also used for coloring glass, giving pink, red or reddish yellow tones. The uses of selenium and tellurium, other than those mentioned, are very limited. Selenium has been used in place of gold salts for photographic toning baths.

Selenium compounds, like arsenic compounds, are *highly toxic*. Thus even at very great dilutions, hydrogen selenide produces headache and sickness. At higher concentrations, even the smallest amount causes acute irritation of the mucous membrane. When brought in contact with the skin, selenium compounds cause exczema and painful inflammations.

Tellurium compounds are far less toxic than selenium compounds, since they are rapidly reduced within the organism, to harmless elementary tellurium. This is then gradually excreted from the body, in the form of unpleasant smelling organic tellurium compounds.

9. Compounds of Selenium and Tellurium

Selenium and tellurium combine with most elements, but not with each other or with sulfur. They do, however, form mixed crystals with each other and with sulfur. The most stable compounds of selenium and tellurium are those with oxygen and the lighter halogens (chlorine and fluorine), and also those with strongly electropositive elements such as the alkali metals. Thus potassium selenide, K_2Se, is formed from its elements with explosive violence.

Selenium and tellurium resemble sulfur in behaving as electronegatively bivalent towards hydrogen and the metals. Except in combination with fluorine, towards which they preferentially exert a valence of six, the quadrivalent state is the preferred valence in compounds in which they are the more electropositive component. Among the oxygen compounds, the dioxides, and the acids and salts derived from them, are the most stable, in contrast with the preferred hexavalence of sulfur in its oxygen compounds (sulfur trioxide, sulfuric acid, sulfates).

In general, the formal analogy between the compounds of sulfur, selenium and tellurium is close. It is shown in the analogous compositions of the following compounds, as well as in the compounds with oxygen and hydrogen (discussed further below), but is less clearly developed in the halogen compounds,

As_2S_3	Sb_2S_3	CS_2	$(SCN)_2$	$[SCN]^-$	$[S_5O_6]^=$
As_2Se_3	Sb_2Se_3	$CSSe$	$(SeCN)_2$	$[SeCN]^-$	$[SeS_4O_6]^=$
As_2Te_3	Sb_2Te_3	$CSTe$	—	$[TeCN]^-$	$[TeS_4O_6]^=$

The principal difference between the selenium and tellurium compounds and the sulfur compounds listed above (which are formally related by replacement of sulfur by selenium or tellurium), lies in the smaller stability of the former.

Of the compounds with nitrogen, it appears that only selenium nitride, Se_4N_4, is the analogue of sulfur nitride, S_4N_4. According to Strecker (1934), tellurium nitride, which has not yet been obtained in a pure state, has the composition Te_3N_4.

(a) Oxides and Oxyacids

The most stable oxides of selenium and tellurium are the dioxides SeO_2 and TeO_2. The *trioxide* of tellurium, TeO_3, can also readily be prepared, whereas selenium trioxide, SeO_3, can be obtained only with difficulty (cf. p. 749). It has been inferred from spectroscopic observations that a *monoxide* of tellurium, TeO, exists at high temperatures in the gaseous state in equilibrium with its decomposition products, but it is apparently not possible to isolate the compound (Glemser, 1948).

Selenium dioxide combines with water to form *selenious acid*, H_2SeO_3. The salts of this acid are the *selenites*.

Tellurium dioxide is but slightly soluble in water. *Tellurous acid*, which is derived from it, has a variable water content. The corresponding salts, the *tellurites*, have the general formula $M^I_2TeO_3$ in the simplest cases, and are thus similar in composition to the sulfites, as are the selenites. Sulfurous acid, selenious acid and tellurous acid are also corresponding compounds, although the water content of the last is variable, as a result of its strong tendency to undergo 'polymerization' or, more strictly, condensation.

Just as oxidation of sulfurous acid, H_2SO_3 yields sulfuric acid, H_2SO_4, so selenious acid, H_2SeO_3, is oxidized to *selenic acid*, H_2SeO_4, although much stronger oxidizing agents are necessary in the latter case. *Telluric acid* is similarly obtained from tellurous acid by the action of powerful oxidizing agents. The anhydride of this acid corresponds to sulfuric anhydride, but the acid itself differs in water content from sulfuric acid, the ordinary telluric acid being the *orthotelluric acid*, H_6TeO_6. The salts of selenic acid are the *selenates*, and of telluric acid the *tellurates*.

(*i*) *Selenium dioxide* is formed when selenium burns in oxygen or air. The oxygen may advantageously be passed first through fuming nitric acid, as the presence of nitrogen oxides strongly accelerates the combustion. Pure selenium dioxide forms brilliant white needles (density 3.954 at 16°). It sublimes readily, the vapor pressure reaching 1 atm at 315°.

Selenium dioxide can be melted without decomposition when it is heated under pressure, the melt being orange yellow. It also has a distinct, yellow green color in the vapor state. Selenium dioxide dissolves readily in water, and in concentrated sulfuric acid and in alcohol. Large transparent plates of the monoethyl ester of selenious acid are obtained when the alcohol solution is evaporated. Dry selenium dioxide combines with hydrogen chloride, forming $SeO_2 \cdot 2HCl$. It forms similar double compounds with other substances. In contrast with sulfur dioxide, selenium dioxide is easily reduced to elementary selenium. Partial reduction may be brought about even by particles of dust, conferring a reddish coloration on the product.

(*ii*) *Selenious acid and the Selenites*. Selenium dioxide absorbs water when it is exposed to moist air, combining with it to form *selenious acid*, H_2SeO_3. The latter is usually prepared by dissolving powdered selenium in dilute nitric acid. The compound crystallizes in deliquescent, colorless, hexagonal prisms (density 3.007 at 15°), which effloresce in dry air through the elimination of water.

Selenious acid is a weak dibasic acid. Its two dissociation constants are $K_1 = 4 \cdot 10^{-3}$ and $K_2 = 0.9 \cdot 10^{-8}$ at 25°. It accordingly forms both *acid* and *neutral* selenites. The selenites are mostly colorless. Aqueous solutions of the neutral selenites have a basic reaction, as a result of hydrolysis. According to Brintzinger (1935) both the selenite ion and tellurite ion are present in solution as hexahydroxo ions, $[Se(OH)_6]^=$ and $[Te(OH)_6]^=$.

Unlike sulfurous acid, selenious acid exists in only one form, the symmetrical $O=Se(OH))_2$. This is shown, for example, by the existence of only one series of alkyl compounds SeO_3R_2, which, as shown by their saponification, must be represented by the constitutional formula $O=Se(OR)_2$.

Selenious acid is readily reduced to selenium—e.g., by sulfur dioxide. The presence of hydrochloric acid is usually necessary to ensure quantitative reduction, as selenopolythionic acids may otherwise be formed—i.e., polythionic acids in which sulfur is partly replaced by selenium. These are decomposed by hydrochloric acid, with the deposition of selenium. Selenium is also precipitated by other reducing agents—e.g., by sodium dithionite (hydrosulfite), hydrazinium or hydroxylammonium salts. Hydrogen iodide reacts quantitatively, according to:

$$H_2SeO_3 + 4HI = Se + 2I_2 + 3H_2O.$$

Hydrogen sulfide produces a reddish yellow precipitate, consisting of a mixture of sulfur and selenium:

$$H_2SeO_3 + 2H_2S = Se + 2S + 3H_2O.$$

Selenium dioxide has often been used with success in preparative organic chemistry to bring about specific oxidation reactions. Thus it will convert a CH_2 group, adjacent to a ketonic group, into an aldehyde group (Riley, 1929).

(*iii*) *Tellurium dioxide*, TeO_2, is formed by the combustion of tellurium. It is better prepared by oxidizing tellurium with well cooled concentrated nitric acid.

When the solution is evaporated or diluted, tellurium dioxide separates in color-less, tetragonal, octahedron-like crystals, of density 5.7–5.9.

The crystal structure of TeO_2 resembles that of SnO_2, PbO_2 and MgF_2, (Fig. 63); $a = 4.79$, $c = 3.77$ Å. This structural type (rutile type) is widely found among the dioxides and difluorides of the elements of the Sub-groups of the Periodic Table.

Tellurium dioxide turns yellow when it is heated. It melts at an incipient red heat, and begins to vaporize at the same time. It solidifies from the melt in rhombic (or monoclinic?) needles (density 5.78 at 14°). It is very sparingly soluble in water (1 in 150,000). The solu-tion is not sour, but has a bitter metallic taste, and does not noticeably redden litmus paper. Tellurium dioxide dissolves much more readily in concentrated strong acids, with which it forms salts. It also dissolves readily in concentrated alkali hydroxides. The resulting salts of *tellurous acid* are the *tellurites*.

(iv) *Tellurites and Tellurous acid*. The simplest tellurites correspond in composition to the sulfites and normal selenites, and have the formula $M^I_2TeO_3$. The alkali tellurites may be obtained by fusing tellurium dioxide with alkali hydroxide or carbonate, as well as from solutions of TeO_2 in caustic alkalis. They are colorless and soluble in water. The tellurites of other metals are insoluble. The corresponding acid, tellurous acid, is very weak, so that the alkali tellurites are partially decomposed even by carbon dioxide. Tellurous acid has not been isolated in the pure state, since it has a great tendency to condense by loss of water into high-molecular complexes. At higher temperatures all the water is driven off, and tellurium dioxide is formed.

Certain tellurites solidify from the melt in a vitreous state. The tellurite glasses have unusual physical properties in some cases, and according to Stanworth (1952), many of them have a particularly high transparency in the infrared.

(v) *Selenic acid and the Selenates*. Selenic acid, H_2SeO_4, is obtained by treating selenious acid with strong oxidants—e.g., with chloric acid. Its salts, the *selenates*, $M^I_2SeO_4$, are most readily prepared by fusing selenium, selenides, or selenium dioxide with potassium nitrate, or from solutions of selenites by treatment with chlorine (1) or by electrolytic oxidation (2):

$$SeO_3^= + Cl_2 + H_2O = SeO_4^= + 2H^+ + 2Cl^- \tag{1}$$

$$SeO_3^= + 2OH^- - 2e = SeO_4^= + H_2O \tag{2}$$

The insoluble barium and lead selenates may be precipitated from selenate solutions by the addition of barium or lead nitrate. By treating these with sulfuric acid or hydrogen sulfide it is possible to obtain solutions of pure selenic acid.

$$BaSeO_4 + H_2SO_4 = BaSO_4 + H_2SeO_4* \tag{3}$$

$$PbSeO_4 + H_2S \quad = PbS + H_2SeO_4 \tag{4}$$

The anhydrous acid is obtained by evaporation in a vacuum at about 200°. It solidifies at ordinary temperature (m.p. 57°) to colorless hexagonal prisms. Selenic acid very readily loses oxygen when it is heated (above 210°) and it can therefore be evaporated without decomposition only under the highest vacuum. Selenic acid combines very avidly with water, forming two hydrates, $H_2SeO_4 \cdot H_2O$ (m.p. +26°) and $H_2SeO_4 \cdot 4H_2O$ (m.p. —51.7°).

Selenic acid is extensively dissociated electrolytically, even in moderately concentrated solutions, like sulfuric acid, which it closely resembles in many respects. Thus concentrated selenic acid carbonizes organic substances in the same

---

* Barium selenate is considerably more soluble than barium sulfate, so that reaction (3) goes to completion. The solubility of $BaSeO_4$ is 8.25 mg in 100 g of water at 25°.

way as sulfuric acid. The oxidizing properties of selenic acid are, however, much stronger than those of sulfuric acid. Thus a mixture of concentrated selenic acid and concentrated hydrochloric acid evolves chlorine, and can therefore dissolve gold and platinum in the same way as aqua regia. The process represented by eqn (1) is therefore reversible.

The salts of selenic acid, the *selenates*, also resemble the sulfates very closely—e.g., in respect of their solubilities and crystalline forms. They also form double salts corresponding to the alums. They are less stable towards heat than the sulfates. They lose oxygen rather easily, and deflagrate when they are heated on charcoal, forming selenides.

Potassium selenate, K_2SeO_4, is best prepared by double decomposition of K_2CO_3 with $BaSeO_4$:

$$K_2CO_3 + BaSeO_4 = BaCO_3 + K_2SeO_4$$

$BaSeO_4$ can be prepared by adding a solution of H_2SeO_4 (obtained by oxidizing SeO_2 with HNO_3 and $KBrO_3$) to a $Ba(NO_3)_2$ solution (Blumenthal, 1913). K_2SeO_4 is isomorphous with K_2SO_4; like the latter, it will add on SO_3, to form $K_2SeS_2O_{10}$ which (like $K_2S_3O_{10}$) loses SO_3 when it is heated to about 150°, and forms K_2SeSO_7. This, in turn, decomposes at about 440°, according to the equation

$$K_2SeSO_7 = K_2SO_4 + SeO_2 + \tfrac{1}{2}O_2$$

(Lehmann, 1952).

Attempts to prepare *selenium trioxide*, SeO_3, by dehydrating selenic acid have been quite unsuccessful. According to Rheinboldt (1930), however, SeO_3 is obtained from SeO_2 by the action of oxygen, activated by the high frequency glow discharge. Lehmann and Krüger (1952) have prepared selenium trioxide by warming potassium selenate with sulfur trioxide. Potassium polysulfates are thereby formed, and selenium trioxide is liberated

$$K_2SeO_4 + nSO_3 = K_2[S_nO_{3n+1}] + SeO_3$$

Selenium trioxide is a white solid, which combines vigorously with water to form H_2SeO_4. It exists in two crystalline modifications, corresponding closely with the cubic and the asbestos-like forms of sulfur trioxide. As with the latter, the asbestos-like form is obtained if traces of water are present, and it is probably a highly condensed polyselenic acid,

$$\begin{array}{ccc} O & O & O \\ HO.Se.O- & \left[-Se-O- \right]_n & -Se-OH \\ O & O & O \end{array}$$

Selenium trioxide melts at 118°, and decomposes to SeO_2, with loss of oxygen, above 180°.

(vi) Telluric acid and the Tellurates. *Telluric acid* is obtained by treating tellurous acid with strong oxidizing agents. Thus Staudenmeier's method involves dissolving finely divided tellurium in the minimal amount of hot dilute nitric acid, and subsequently adding chromic acid slowly. According to Meyer (1930), it is more convenient to oxidize tellurium directly to telluric acid by means of chloric acid. *Orthotelluric acid*, with the formula H_6TeO_6, crystallizes from the solution upon evaporation. It can also be prepared conveniently by oxidizing tellurium dioxide with hydrogen peroxide in alkaline solution (Horner, 1952). Ammonium tellurate is first prepared, and the free acid is liberated from this by means of concentrated nitric acid. Orthotelluric acid exists in two distinct forms—a cubic form with density 3.053, and a monoclinic form with density 3.071. The former changes gradually into the latter, in contact with the solution. Below 10°, the

hydrate $H_6TeO_6 \cdot 4H_2O$ crystallizes from solution. The solubility of telluric acid in water is rather high.

Unlike sulfuric and selenic acids, telluric acid is a very weak acid—so weak that it cannot be titrated with alkali hydroxide. Its solutions have an extremely small electrical conductivity accordingly. Its taste is also sweetish and metallic, and not acid. Blanc gives the value $6 \cdot 10^{-7}$ for the first stage dissociation constant.

As a rule therefore, only a portion of the hydrogen atoms of telluric acid can be replaced by metals in forming the tellurates. Tellurates are known, however, in which all the hydrogen atoms are replaced—e.g., Ag_6TeO_6 and Hg_3TeO_6. Although the neutral sodium salt cannot be obtained from solution, Na_6TeO_6 was obtained by Zintl (1938) by heating Na_2TeO_4 with Na_2O. Na_2TeO_4 may be prepared by dehydrating $Na_2H_4TeO_6$, which crystallizes from aqueous solution.

*Heteropolyacids** are formed by the linkage of molecules of other acid anhydrides to the oxygen atoms of telluric acid—e.g., $H_6[Te(O.MoO_3)_6]$ and $H_6[Te(O.WO_3)_6]$. Telluric acid also has a tendency to form auto-complexes. Such a great increase in particle size occurs in the solution at the boiling point that it takes on the character of a colloidal dispersion. Dispersion into smaller species, and ultimately into individual molecules, takes place once more when the solutions are cooled.

Orthotelluric acid loses water when it is heated, and above 300° passes ultimately into yellow tellurium trioxide, TeO_3. This is practically insoluble in water, and very indifferent to chemical reagents. It loses oxygen at an incipient red heat, and is transformed into tellurium dioxide. Montignie (1945–47) found that a second modification, β-TeO_3 (density 6.21) existed, in addition to the ordinary α-TeO_3 ($d = 5.075$). β-TeO_3, which is grey and crystalline, is formed by prolonged heating of α-TeO_3 at 310° in a sealed tube.

When orthotelluric acid is heated to its melting point in a sealed tube, *allotelluric acid* is formed, as a colorless, sticky, syrupy mass, which is miscible with water in all proportions. Its solutions differ in characteristic ways from orthotelluric acid. Thus, unlike the latter, solutions of allotelluric acid are distinctly acid in reaction. Allotelluric acid slowly reverts in solution to orthotelluric acid.

(b) Halogen Compounds

The halogen compounds of selenium and tellurium are, in general, more stable than those of sulfur. Towards fluorine, selenium and tellurium can function as both quadrivalent and hexavalent. Tellurium is bivalent and quadrivalent, towards the other halogens, whereas selenium gives compounds of the type Se_2X_2, the analogues of disulfur dichloride, in addition to the tetrahalides.

SUMMARY

Fluorides		Chlorides	
SeF_4	colorless liquid	Se_2Cl_2	brown-yellow liquid
SeF_6	colorless gas	$SeCl_4$	colorless crystals
TeF_4	colorless solid	$TeCl_2$	black green solid
TeF_6	colorless gas	$TeCl_4$	colorless crystals
Te_2F_{10}			
Bromides		Iodides	
Se_2Br_2	dark red liquid	Se_2I_2?	black solid
$SeBr_4$	yellow powder		
$TeBr_2$	black green solid	TeI_2?	black solid
$TeBr_4$	red yellow crystals	TeI_4	black crystals

$SeCl_4$ and $SeBr_4$ are completely dissociated in the vapor state into Cl_2 or Br_2 and $SeCl_2$ or $SeBr_2$, respectively. The dihalides exist only in the vapor state, as dissociation products of the tetrahalides (Yost, 1930).

* See Vol. II further upon this subject.

Several of the halogen compounds listed above, are able to form complex acids, by adding on hydrogen halide. Thus selenium tetrabromide combines with hydrogen bromide, to form hexabromoselenic acid, $H_2[SeBr_6]$, and with alkali bromides to form the salts of this acid. Hexachloroselenates, $M^I_2[SeCl_6]$, have been prepared by Petzold (1932). For tellurium, the complex acids $H[TeCl_5]$, $H[TeBr_5]$, and $H[TeI_5]$ (all crystallizing with water of hydration) are known, and also complex salts of the general formulas $M^I_2[TeCl_6]$, $M^I_2[TeBr_6]$, and $M^I_2[TeI_6]$.

Selenium tetrafluoride, SeF_4, (m.p. —9.5°, b.p. 106°) can dissolve the fluorides of the alkali metals, forming complex salts of the type $M^I[SeF_5]$. The corresponding selenium(IV) oxyfluoride, $SeOF_2$ (m.p. 15°, b.p. 126°) may be prepared by the action of fluorine on dry selenium dioxide. It is very vigorously hydrolyzed by water.

(c) Hydrogen Compounds

Hydrogen selenide, H_2Se, and *hydrogen telluride*, H_2Te, correspond in composition to hydrogen sulfide. They are also very similar to it in their properties, but are far less stable. Their heats of formation are *negative* (cf. Table 99).

Selenium, like sulfur, combines directly with hydrogen. Tellurium will not do so, but hydrogen telluride can be obtained in good yield when hydrogen is liberated electrolytically at a tellurium electrode at low temperatures. It is also obtained in high dilution by decomposing metallic tellurides—e.g., Al_2Te_3—with water or acids. Hydrogen selenide is obtained almost in a pure state by this method, but it may also be prepared by passing hydrogen over elementary selenium at 400°.

(i) *Hydrogen selenide* is a colorless gas, which is easily liquefied and solidified (cf. Table 99, p. 687). One liter weighs 3.672 g at 0°. It is less stable than hydrogen sulfide, and gradually deposits selenium through the action of dust in the presence of moisture, or by reaction with atmospheric oxygen. It burns with a cornflower blue flame when it is ignited, forming a white smoke of selenium dioxide, or depositing red selenium if there is access of insufficient air. At ordinary temperature, hydrogen selenide is metastable, but begins to decompose with measurable velocity only above 300° in the absence of catalysts. The equilibrium between H_2Se and its components is displaced increasingly in favor of the compound at higher temperatures, and it is therefore possible to prepare the compound by passing hydrogen over heated selenium; some decomposition occurs once more as the gas cools, so that a ring of selenium is deposited just beyond the heated zone of the apparatus.

Hydrogen selenide is even more poisonous than hydrogen sulfide. It has a similar smell, but has at the same time a very painful and protracted effect on the mucous membranes of the respiratory organs and eyes.

Hydrogen selenide is rather more soluble in water than is hydrogen sulfide. It can combine with water to form hydrates, although these are very unstable. A solution saturated with hydrogen selenide at atmospheric pressure and at ordinary temperature is almost exactly 0.1 molar. The aqueous solution, from which a deposit of red selenium is soon formed on exposure to air, has a distinctly acid reaction. A 0.1-molar solution at 25° is dissociated to the extent of about 4% (see Table 99 for dissociation constants). Hydrogen selenide is thus a considerably stronger acid than hydrogen sulfide, being about as strong as formic acid.

Two series of salts are derived from hydrogen selenide: the *acid selenides*, M^IHSe, and the *neutral selenides*, M^I_2Se. The selenides of the alkali and alkaline earth metals readily take up additional selenium, forming *polyselenides*, $M^I_2Se_x$, in the same way as the sulfides of these elements form polysulfides. The alkali selenides generally have a reddish color, due to the presence of some polyselenide, but are probably colorless in the pure state. The heavy metal selenides, however, are mostly more or less strongly colored, like the sulfides, and are insoluble in water, and often insoluble in acids. They can usually be prepared by melting their components together, and also by the action of hydrogen selenide on solutions of heavy metal salts.

(ii) *Hydrogen telluride* is a colorless gas, with an unpleasant smell reminiscent of arsine. It is more condensable than hydrogen selenide (cf. Table 99), giving a liquid which is probably colorless when it is absolutely pure, but which is always somewhat yellow as a result of partial decomposition. It is very easily decomposed, and is at once attacked even by weak oxidants. It reacts with atmospheric oxygen even at ordinary temperature, depositing tellurium. It burns in air with a bluish flame, forming tellurium dioxide. It is very soluble in water, but the solutions are rather unstable—thus decomposition occurs almost instantly if air is admitted, and tellurium is deposited. The acid strength of hydrogen telluride approaches that of phosphoric acid.

The *tellurides*, M^I_2Te, of the alkali metals are soluble in water, and colorless when pure. Their solutions at once turn red when exposed to the air, polytellurides being formed by oxidation. The tellurides of the heavy metals are insoluble in water and dark in color. Some of them are decomposed by water, as also is aluminum telluride, Al_2Te_3; others are decomposed only by acids.

(*iii*) *Alkyl Compounds*. When neutral or acid alkali selenides are distilled with potassium alkyl sulfates, *alkyl selenides*, SeR_2, (R = alkyl group) and *alkyl selenomercaptans*, SeRH, are formed—i.e., compounds corresponding to the thioethers (alkyl sulfides) and mercaptans:

$$K_2Se + 2KO.SO_2.OC_2H_5 = (C_2H_5)_2Se + 2K_2SO_4$$
potassium ethyl sulfate   diethyl selenide

$$KHSe + KO.SO_2.OC_2H_5 = C_2H_5.SeH + K_2SO_4$$
ethyl selenomercaptan

The *alkyl tellurides*, R_2Te, can be prepared in a similar manner. The alkyl selenides and tellurides are volatile liquids, with repulsive odors. They readily add on halogens or oxygen, to form compounds such as $(C_2H_5)_2SeCl_2$, $(C_2H_5)_2TeCl_2$, $(C_2H_5)_2SeO$, etc. The last mentioned compound is basic in character, and forms salts with acids—e.g., $(C_2H_5)_2Se(NO_3)_2$. The analogous tellurium compound is similar. The capacity of the alkyl selenides and tellurides to add on alkyl iodides is also noteworthy. They thereby form *alkyl selenonium* and *alkyl telluronium* compounds:

$$(C_2H_5)_2Se + C_2H_5I = [(C_2H_5)_3Se]I$$

$$(C_2H_5)_2Te + C_2H_5I = [(C_2H_5)_3Te]I.$$

The corresponding hydroxides are formed when these compounds are treated with moist silver oxide. Like the sulfonium hydroxides, they are strong bases.

10. Analytical (Selenium and Tellurium)

Selenium compounds are readily detected by converting them to elementary selenium, recognizable by its characteristic red color. Selenium is usually converted to the elementary state for quantitative analysis also. In this case, the red modification which is at first precipitated from the solution is converted into the granular grey form, which filters better, by warming it under the solution for a long time.

Selenium compounds give off a characteristic smell ('decaying radish') when they are heated on a charcoal block before the blowpipe. A cold object placed in the reducing zone of the flame becomes covered with a *red* deposit, which dissolves in concentrated sulfuric acid with a *green* color, as a result of the reaction

$$Se + H_2SO_4 \rightleftharpoons SeSO_3 + H_2O$$

Red selenium is reprecipitated upon dilution with water, since the reaction is reversible.

Tellurium compounds under similar conditions yield no definite smell, but a brown to black deposit which dissolves in sulfuric acid to give a *red* solution. A *black* precipitate is thrown down by water, through the formation of elementary tellurium once more.

$$Te + H_2SO_4 \rightleftharpoons TeSO_3 + H_2O.$$

Tellurium is determined in gravimetric analysis in the same way as selenium. It is converted first to tellurous acid, which is reduced to tellurium by a suitable reducing agent, such as sulfurous acid or hydrazine, and weighed as such.

In the course of systematic analysis, selenium and tellurium are precipitated by hydrogen sulfide, mixed with sulfur. They pass into solution, like sulfur, upon treatment with ammonium polysulfide, and are reprecipitated when the solution is acidified.

11. Polonium (Po)

Polonium [*16*], the heaviest element of the oxygen-sulfur group, is one of the radioactive elements. It was, moreover, the first to be discovered, of the elements which were brought to our knowledge through the study of radioactivity. Shortly after the phenomenon of radioactivity had been discovered for uranium, and had been shown to be possessed by thorium also, polonium was discovered (in 1898) by Pierre and Marie Curie. The discovery was made in the course of an analysis of pitchblende, in which every constituent was examined for radioactive properties, in order to account for the abnormally intense radioactivity of the mineral, as compared with uranium compounds.

Polonium is very much more intensely radioactive than radium. Its life, however, is correspondingly very much shorter. The activity of a polonium preparation diminishes by about $\frac{1}{2}\%$ per day, and falls to a half of the original value in 140 days.

Since it is a disintegration product of radium (Radium-F), polonium is always present in old radium preparations, but can never accumulate much because of its short half-life. The quantity of polonium in equilibrium with 1 g of radium is only about 0.2 mg., so that the polonium content of pitchblende is about 5000 times smaller still than its radium content. It would therefore be necessary to work up over a thousand tons of pitchblende or over 300 g of radium bromide* which had been kept for at least 30 years in the solid state, in order to isolate 40 mg of polonium. The best natural source for the preparation of polonium would probably be Radium-D, which must, however, first be isolated from radium or radon. Very recently it has become possible to prepare polonium by the 'nuclear reaction' of bismuth with neutrons. (For a discussion of nuclear transmutation and 'artificial radioactivity' see Vol. II.)

In the past, the usual starting materials for polonium preparations have been the residues from pitchblende which have been worked for radium. These are dissolved in hydrochloric acid, and polonium is precipitated with hydrogen sulfide, together with other precipitable metals. The precipitate is then worked up for bismuth. This element, the neighbor of

---

* To be precise, $\dfrac{385.8 \cdot 10^4}{226.0 \cdot 25 \cdot 2.14} = 319$ g $RaBr_2$.

polonium in the Periodic System, and rather similar to it in its analytical chemistry, is present in pitchblende in enormously greater concentration than polonium. Since, however, polonium sulfide is less soluble than bismuth sulfide, the latter not only serves as a 'carrier', but can be enriched in polonium sulfide by fractional precipitation. Polonium may also be enriched by fractionally precipitating bismuthyl nitrate, by cautious dilution of the concentrated nitric acid solution; by fractional precipitation with tin(II) chloride from hydrochloric acid solution; or by deposition on a strip of copper, silver, or bismuth, dipped into the solution.

Polonium may often conveniently be deposited from solution by electrolysis. It is deposited at the cathode, which is best made of gold foil, since polonium deposited on platinum foil cannot afterwards be completely removed by boiling with acids. The polonium may, however, be volatilized completely by heating the platinum foil for a time at about $1000°$.

The normal potential of polonium, relative to the normal hydrogen electrode, is —0.9 volts (Joliot 1929; Erbacher 1933). Polonium is thus more noble than silver, and is therefore deposited upon silver from solution. Use is often made of this fact in radiochemical work. It has been stated that polonium is present in solution as a doubly charged positive ion.

In its chemical behavior, polonium resembles the neighboring element, bismuth, and also its higher homologue tellurium. Polonium compounds are hydrolyzed very readily, and therefore tend to form colloidal dispensions of the hydroxide by hydrolysis.

The homology of polonium with tellurium was recognized by Marckwald in 1903. By dipping a polished sheet of bismuth into a hydrochloric acid solution of pitchblende residues, he obtained what he at first supposed to be a new element, and which he named 'radiotellurium' because of the similarity of its reactions to those of tellurium. From a measurement of its disintegration constant he later satisfied himself that it was identical with polonium, already discovered. Paneth showed in 1918 that polonium was able to form a volatile hydride, in the same way as its lighter homologues. This was proved in the same way as for thorium-C hydride (i.e., bismuth hydride, cf. p. 682). Polonium hydride is very similar in properties to bismuth hydride, but is even less stable. It resembles tellurium hydride—e.g., in its sensitivity towards water containing dissolved oxygen.

As Tammann showed (1932), polonium has very little ability to form mixed crystals with metals such as Ag, Cu, Zn, Cd, Pb, Sn, Sb, Bi, or even with tellurium. In the latter case, this is because the crystal structures of polonium and tellurium are very different. The structure of polonium was determined by Rollier in 1936 from electron diffraction photographs taken with 0.1 microgram of the pure metal. Polonium has a unique monoclinic structure, in which each Po atom is surrounded by 6 others which form a highly distorted octahedron in which all the Po—Po distances are different; they vary from 2.81 to 4.13 Å.

The chemistry of polonium has been largely investigated by using tellurium as a 'carrier', and showing that polonium follows tellurium during chemical change, and that its compounds form mixed crystals with corresponding tellurium compounds. The occurrence of mixed crystal formation, rather than adsorption is proved by constancy of the ratio

$$\frac{\text{polonium concentration in solid compound}}{\text{polonium concentration in solution}}$$

as crystallization proceeds. In this way it has been proved that Na_2Po forms mixed crystals with Na_2Te, and the dibenzyl compound $(C_6H_5CH_2)_2Po$ crystallizes with $(C_6H_5CH_2)_2Te$.

Polonium solutions react with sodium dithiocarbamates, $NaS.CS.NR_2$ ($R = C_2H_5$, etc.), forming unionized complex salts, which are soluble in chloroform and can be crystallized together with the similar compounds of Bi(III), Co(III) and Ni(II). Polonium also forms an acetylacetonate which forms mixed crystals with the thorium compound, $Th(C_5H_7O_2)_4$, but which can be separated from the corresponding aluminum compound. The acetylacetonate is therefore thought to be an inner complex salt (cf. Vol. II) of quadrivalent polonium.

References

1 M. LASCHIN, *Der Sauerstoff, seine Gewinnung und seine Anwendung in der Industrie*, 3rd Ed., Halle 1943, 105 pp.

2 O. KAUSCH, *Die Herstellung, Verwendung und Aufbewahrung von flüssiger Luft*, 6th Ed., Weimar 1938, 708 pp.

3 D. M. YOST and H. RUSSELL, *Systematic Inorganic Chemistry of the Fifth and Sixth Group Nonmetallic Elements*, New York 1946, 423 pp.

4 E. THIELER, *Schwefel*, Dresden 1936, 132 pp.

5 G. LUNGE, *Schwefelsäurefabrikation* (Vol. I of the *Handbuch der Sodaindustrie etc.*), Braunschweig 1903, 1117 pp.

6 A. M. FAIRLIE, *Sulfuric Acid Manufacture*, New York 1936, 669 pp.

7 A. SMITS, *The Theory of Allotropy*, London 1922, 397 pp.

8 L. VANINO, *Die Leuchtfarben*, 2nd. Ed., Stuttgart 1935, 168 pp.

9 H. RUPP, *Die Leuchtmassen und ihre Verwendung*, Berlin 1937, 163 pp.

10 P. LENARD, F. SCHMIDT and R. TOMASCHEK, *Phosphoreszenz und Fluoreszenz* (Vol. 23 of the *Handbuch der Experimental-Physik*, Leipzig 1928, 2 parts, 741 and 799 pp.

11 P. PRINGSHEIM, *Fluorescence and Phosphorescence*, New York 1949, 794 pp.

12 G. P. BARNARD, *The Selenium Cell, Its Properties and Applications*, London 1930, 332 pp.

13 J. S. ANDERSON (Editor), *Photoelectric Cells and Their Applications*, London 1930, 236 pp.

14 H. THIRRING and O. P. FUCHS, *Photowiderstände*, Leipzig 1939, 186 pp.

15 B. LANGE, *Die Photoelemente und ihre Anwendung*. Vol. I, *Entwicklung und physikalische Eigenschaften*, 2nd Ed., Leipzig 1940, 144 pp.; Vol II, *Technische Anwendung*, 2nd Ed., Leipzig 1940, 110 pp.

16 M. HAISSINSKY, *Le Polonium*, Paris 1937, 44 pp.

17 A. KURTENACKER, *Analytische Chemie der Sauerstoffsäuren des Schwefels*, Stuttgart 1938, 216 pp.

CHAPTER 16

OXIDATION AND REDUCTION

1. Oxidation

As was stated in the foregoing chapter, the process whereby any substance combines with oxygen is called *oxidation*. A variety of other processes are in many respects analogous to combination with oxygen—e.g., combination with chlorine, bromine, sulfur and other non-metallic elements. The analogy is often obvious in external characteristics. For example, antimony burns in chlorine, in the same way as it does in air or oxygen, and indeed most metals can be burned in an atmosphere of chlorine, bromine vapor or sulfur vapor, just as in oxygen. In some instances, combination with these elements takes place more vigorously than with oxygen, and this is generally true as regards union with fluorine. The products of these processes can be converted, by reactions completely different from those of typical oxidation processes, into the same substances as are obtained by direct combination with oxygen. Thus the product of combustion of tin in chlorine, tin tetrachloride, $SnCl_4$, can be decomposed (hydrolyzed) by water, and after drying and ignition yields the same end product, tin dioxide, SnO_2, as is formed directly by burning tin in air. It has therefore come about that the term 'oxidation' has been extended, to include not only combination with oxygen, but also related processes, and in particular the combination of metals or hydrogen with fluorine, chlorine, bromine, iodine, sulfur and similar elements—i.e., with *elements of electronegative character* generally.

It was first clearly stated by Ostwald that in the case of substances which exist as charged ions in solution or in their compounds, the essential nature of any oxidation process consists in an increase of their positive charge, or in giving up negative charges. More precisely expressed, an atom or radical, in undergoing oxidation, *gives up electrons*. We are equally justified in speaking of the process as oxidation when lead combines with chlorine to form lead chloride, $PbCl_2$, as when it is changed to lead oxide, PbO, by combination with oxygen, in that, ultimately, the lead gives up electrons in both processes, and passes from the elementary electroneutral state into the electropositive bivalent state:

$$Pb + Cl_2 = Pb^{2+}(Cl^{1-})_2 \qquad \text{and} \qquad Pb + \tfrac{1}{2}O_2 = Pb^{2+}O^{2-}$$

We can, perhaps, not assert that lead in lead oxide bears a double positive charge with quite the same certainty as in the case of lead chloride, which is largely dissociated into ions, both in the fused state and in solution. However, the properties of lead oxide leave no real doubt that this compound is also ionic in structure, and that the lead is the electropositive constituent. The same applies to other metallic oxides. It is, however, open to question whether it is justifiable to speak of opposite charges on the atoms of oxygen and the other constituents in the oxygen compounds of non-metals. This is especially true of carbon compounds. In such cases, in which the essential nature of oxidation does not lie in acquirement of positive charge, the term oxidation is applied simply to combination with oxygen (or, in some cases, to the loss of hydrogen), and it is obviulsy not possible to include the addition of chlorine bromine, sulfur, etc. in the concept of oxidation; the terms chlorination, etc. are used for these processes.

We therefore arrive at two definitions of oxidation, which do not exactly coincide, covering respectively the purely chemical and the electrochemical aspects.

Oxidation in the purely chemical sense signifies, primarily, the combination of any substance with oxygen. In addition, many reactions in which hydrogen is eliminated are considered as oxidations in the chemical sense. (It is permissible to term this process an oxidation only when the elimination of hydrogen can be regarded as arising, in principle, from the action of oxygen.)

Oxidation in the electrochemical sense comprises all those chemical and electrochemical processes in which electrons are abstracted from one element or radical by the action of another.

An example of oxidation in the sense of the first definition is provided by the oxidation of nitrous acid to nitric acid: $HNO_2 + O = HNO_3$

The oxidation of hydroxylamine to hyponitrous acid can be regarded as an example of oxidation as defined in terms of the abstraction of hydrogen from a compound.

$$
\begin{array}{ccc}
HO{-}NH_2 & O & HO{-}N \\
& +\ \| & = & \| + 2H_2O \\
HO{-}NH_2 & O & HO{-}N
\end{array}
$$

On the other hand, the loss of hydrogen from calcium hydride, through thermal decomposition:

$$CaH_2 = Ca + H_2$$

is not an oxidation process, but in the light of the second definition must, indeed, be considered the exact converse—a *reduction* process.

The concept of oxidation in the electrochemical sense may be elaborated by a few examples.

Conversion of an element which is already charged, into a state of *higher charge*—e.g., conversion of PbO into PbO_2, of $PbCl_2$ into $PbCl_4$, or of $PbSO_4$ into $Pb(SO_4)_2$—must be considered to be an oxidation, in just the same sense as the conversion of a substance from the elementary state into some compound in which it is electropositively charged (e.g., conversion of Pb into PbO or $PbCl_2$). Conversion of an element from the negatively charged state into the elementary (electroneutral) state, or beyond that to the electropositive state, is also an oxidation—e.g., conversion of S^{2-} to S or to S^{6+} when H_2S is oxidized to S or H_2SO_4.

Oxidation (loss of electrons) may take place electrolytically—e.g.,

$$Pb^{2+}SO_4 + H_2SO_4 - 2e = Pb^{4+}(SO_4)_2 + 2H^+$$

In this case also the electrons are taken up (indirectly) by another substance since the electrolysis, in which oxidation of lead sulfate occurs at the anode, can only take place if some other substance is simultaneously reduced at the cathode (e.g., $PbSO_4$ to Pb, or $2H^+$ to H_2). Oxidation processes in the electrochemical sense also include the dissolution of metals in acids, including the so-called 'non-oxidizing' acids,* e.g., of zinc in hydrochloric acid

$$Zn + 2HCl = ZnCl_2 + H_2 \quad \text{or} \quad Zn + 2H^+ = Zn^{2+} + H_2;$$

* It is usual to use the term 'non-oxidizing' acids to denote those acids which owe their solvent action for metals simply to the tendency of hydrogen ions to become discharged by acquiring electrons—e.g., hydrochloric acid, dilute sulfuric acid, acetic acid.

also the dissolution of metals by the displacement of a nobler metal—e.g., the dissolution of zinc in copper sulfate solution:

$$Zn + CuSO_4 = ZnSO_4 + Cu, \qquad \text{or} \qquad Zn + Cu^{++} = Zn^{++} + Cu.$$

The conversion of the alkali and alkaline earth metals into their hydrides, in which these metals bear a positive ionic charge, is also an oxidation in the electrochemical sense:

$$Li + \tfrac{1}{2}H_2 = Li^+H^-.$$

2. Reduction

As the name implies, reduction signifies the restoration of a (previously oxidized) substance to its original state. It is the converse process to oxidation, and one must distinguish in a similar manner between reduction in the narrower chemical sense, namely abstraction of oxygen, or sometimes the addition of hydrogen, and reduction processes in the electrochemical sense, i.e., processes in which the essential feature is the acquirement of additional electrons.

Thus the conversion of lead oxide into lead by heating it with charcoal is a reduction process:

$$PbO + C = Pb + CO \tag{1}$$

the lead oxide is said to be reduced to lead. In the electrochemical sense the reduction of *lead chloride* to lead is also a reduction, and this is true in general of every process in which the positive charge on an element is reduced or the negative charge increased. Reduction can also take place electrolytically, and then takes place at the cathode, whereas electrolytic oxidations occur at the anode.

Reduction can be briefly defined as follows: reduction processes are chemical or electrochemical processes in which oxygen is abstracted from some substance, or electrons imparted to some substance. Except in so far as it may be coupled with an increase in the negative charge borne by the element with which it combines, the addition of hydrogen to any substance does not fall within this definition. This will be discussed later.

Whenever electrons are acquired by one atom or radical, through some chemical process, some other substance must furnish the electrons. Conversely, electrons can be abstracted from any substance by chemical means only if another substance serves as an electron acceptor. *Every oxidation process is therefore coupled with a reduction process*, and vice versa. As applied to electrochemical oxidations and reductions, this is subject to the qualification that the two complementary processes can occur at different sites. As has been stated, oxidation takes place at the anode and reduction at the cathode, but cathodic reduction can take place only to exactly the same extent as oxidation occurs at the anode; the same quantity of electrons must enter the solution at the cathode, as are carried out of it at the anode.

3. Deoxygenation

In dealing with ionic compounds, the abstraction of oxygen is always equivalent in meaning to a decrease in the positive charge borne by the other element, which must act as an acceptor of electrons since oxygen is invariably the negatively charged constituent of heteropolar compounds. In covalent compounds, however, it need not follow that loss of oxygen is coupled with any decrease in charge on the component originally bound to oxygen. In such instances, which are found especially in the field of organic chemistry, the process of removal of oxygen is frequently referred to as *deoxygenation*.

4. Hydrogenation and Dehydrogenation

Reduction of compounds is often brought about by hydrogen. This can take place either in that hydrogen abstracts oxygen from the substance with which it reacts, or it may itself add on to the substance undergoing reduction, giving up electrons to it. Thus, the reaction of hydrogen with sulfur may be represented:*

$$H_2 + S = \begin{matrix} H^{1+} \\ H^{1+} \end{matrix} S^{2-}$$

Both processes may occur simultaneously—i.e., oxygen may be removed, and be replaced by hydrogen.

Not every addition of hydrogen can be regarded as a reduction. On the contrary, as we have seen, the combination of the alkali and alkaline earth metals with hydrogen represents an *oxidation* process in the electrochemical sense, the hydrogen removing electrons from these metals when it combines with them.

There are many cases in which combination with hydrogen takes place, without it being possible to state that any opposite charges are set up between hydrogen and the other atoms involved. This is especially true of organic compounds. It is, indeed, customary to regard the hydrogen as the more electropositive constituent of the compounds, and in this sense there is some justification for speaking of the addition of hydrogen as a *reduction*. However, where the addition of hydrogen takes place without any evidence that this involves the transfer of electrons from the hydrogen to other atoms—i.e., especially in the addition of hydrogen to unsaturated organic compounds—it is preferable to speak, not of reduction, but of *hydrogenation*. In the same way, the elimination of hydrogen, where it does not involve electron transfer, is conveniently referred to as *dehydrogenation*.

The term dehydrogenation is used, in particular, in order to bring out the fact that a process, which could be regarded as an oxidation in the wider sense, takes place through the elimination of hydrogen, and not through combination with oxygen. Thus the oxidation of alcohols to aldehydes is a dehydrogenation:

$$CH_3.CH_2OH + O = CH_3.CHO + H_2O,$$

as also, according to Wieland [5] is the oxidation of acetaldehyde in aqueous solution to acetic acid. This is so since, as he showed, the reaction proceeds in the presence of water by the removal of hydrogen from the compound $CH_3.CH(OH)_2$, which although not stable in the free state, is formed by addition of water to the aldehyde:

$$CH_3-C\overset{\displaystyle OH}{\underset{\displaystyle OH}{\big|}}H + O \; = \; CH_3-C\overset{\displaystyle OH}{\underset{\displaystyle O}{\diagdown}} + H_2O$$

* The sulfur atom in hydrogen sulfide is of course not ionic since the molecule is essentially covalent. However, in terms of relative electronegativities the sulfur atom is more electronegative than hydrogen and thus may arbitrarily be considered to be in a negative valence state.

5. Oxidizing and Reducing Agents

Substances which can react chemically to give up oxygen, or which accept electrons, are known as *oxidizing agents*. Those which take up oxygen, or which donate electrons, are *reducing agents*. These expressions are usually used in a more restricted sense. Substances are generally considered to be real oxidizing agents only when the tendency to give up oxygen or accept electrons is strongly developed, so that it is exercised towards numerous other materials. In the same way, the real reducing agents are the compounds with a strongly developed tendency to abstract oxygen from, or furnish electrons to, other substances.

For example, in the process shown in eqn. (1), p. 758, litharge oxidizes the carbon, just as much as the carbon reduces the lead oxide. One might say that lead oxide is an oxidizing agent in this process. It is not usual, however, to regard lead oxide as an oxidizing agent in the general sense, since, apart from organic compounds, the number of substances to which lead oxide can give up its oxygen is relatively small.* Carbon, however, is quite generally considered to be a reducing agent, since at high temperatures it has a very marked capacity for abstracting oxygen from compounds of other elements.

Oxidizing agents which have a very marked tendency to give up oxygen, or to accept electrons, are spoken of as *strong oxidants*. In *weak* oxidizing agents this tendency, although not developed to the same extent, is none the less evident. A similar distinction is drawn between strong and weak reducing agents.

The most important oxidizing agents, other than the oxygen of the air, are: hydrogen peroxide and other peroxides, chlorine, bromine, hypochlorites (especially sodium hypochlorite and bleaching powder), chromic acid and chromium trioxide, potassium dichromate, potassium permanganate, concentrated nitric acid. Concentrated sulfuric acid and dilute nitric acid are weaker than the last mentioned. Silver oxide and (at high temperatures) copper oxide are examples of weak oxidizing agents. Molten nitrates and chlorates are powerful oxidizing agents.

The most important reducing agents at high temperatures are: carbon (generally used in the form of coke or coal), carbon monoxide, and hydrogen. The alkali metals and alkaline earth metals, including magnesium, are extremely powerful reducing agents at high temperatures, as also is aluminum. Iron is another reducing agent often used in industry. The reagents principally used to bring about reduction in aqueous solutions are: tin(II) chloride, iron(II) sulfate, sulfurous acid, oxalic acid, formic acid, formaldehyde, hydroxylamine, hydrazine, hydriodic acid, hydrogen sulfide, phosphorous acid, elementary hydrogen in the presence of a catalyst (platinum or palladium black), and 'nascent' hydrogen, i.e., hydrogen evolved by the reaction of metals with acids or alkalis (cf. p. 41).

On the basis of the electrochemical definition of oxidation and reduction, the mechanism of reaction of these oxidizing and reducing agents may be represented by the following schematic equations:

* If the behavior of lead oxide towards organic compounds is being particularly considered, lead oxide could be regarded as an oxidizing agent in this field.

Oxidizing agents:

Atmospheric oxygen: $O_2 + 4e = 2O^{2-}$

Hydrogen peroxide: $H_2O_2 + 2e = H_2O + O^{2-}$

Chlorine or bromine: $Cl_2 + 2e = 2Cl^-$, $Br_2 + 2e = 2Br^-$

Hypochlorites: $ClO^- + 2e = Cl^- + O^{2-}$

Chromium trioxide: $CrO_3 + 3e = Cr^{3+} + 3O^{2-}$

Dichromate: $Cr_2O_7{}^{2-} + 6e = 2Cr^{3+} + 7O^{2-}$

Permanganate: $MnO_4{}^- + 5e = Mn^{2+} + 4O^{2-}$

Nitric acid: $HNO_3 + 3e = NO + H^+ + 2O^{2-}$

Sulfuric acid: $H_2SO_4 + 2e = H_2SO_3 + O^{2-}$

Silver oxide: $Ag_2O + 2e = 2Ag + O^{2-}$

Copper oxide: $CuO + 2e = Cu + O^{2-}$

Potassium nitrate: $KNO_3 + 2e = KNO_2 + O^{2-}$ $NO_3{}^- + 2e = NO_2{}^- + O^{2-}$

Potassium chlorate: $KClO_3 + 6e = KCl + 3O^{2-}$ $ClO_3{}^- + 6e = Cl^- + 3O^{2-}$

Reducing agents:

Carbon: $C + O^{2-} = CO + 2e$

Carbon monoxide: $CO + O^{2-} = CO_2 + 2e$

Hydrogen: $H_2 = 2H^+ + 2e$ $H_2 + O^{2-} = H_2O + 2e$

Metals—e.g., iron: $M = M^{x+} + xe$ $Fe = Fe^{2+} + 2e$

Tin(II) chloride: $Sn^{2+} = Sn^{4+} + 2e$

Iron(II) sulfate: $Fe^{2+} = Fe^{3+} + e$

Sulfurous acid: $H_2SO_3 + O^{2-} = H_2SO_4 + 2e$ $SO_3{}^{2-} + O^{2-} = SO_4{}^{2-} + 2e$

Oxalic acid: $H_2C_2O_4 + O^{2-} = H_2O + 2CO_2 + 2e$ $C_2O_4{}^{2-} = 2CO_2 + 2e$

Formic acid: $HCOOH + O^{2-} = H_2O + CO_2 + 2e$

Formaldehyde: $HCHO + O^{2-} = \underset{\text{formic acid}}{HCOOH} + 2e$

Hydroxylamine: $2NH_2OH + O^{2-} = N_2 + 3H_2O + 2e$

Hydrazine: $N_2H_4 + 2O^{2-} = N_2 + 2H_2O + 4e$

Hydriodic acid: $2HI = I_2 + 2H^+ + 2e$ $2I^- = I_2 + 2e$

Hydrogen sulfide: $H_2S = S + 2H^+ + 2e$ $H_2S + O^{2-} = S + H_2O + 2e$

Phosphorous acid: $H_3PO_3 + O^{2-} = H_3PO_4 + 2e$

Nascent hydrogen: $H = H^+ + e$ $2H + O^{2-} = H_2O + 2e$

The foregoing equations bring out the essential characteristics of oxidizing agents and reducing agents respectively—namely the property common to all of the former, of being electron acceptors, and that common to the latter, of acting as electron donors.

Since chemical processes are associated only with an *exchange* of electrons,—i.e., free electrons never appear*—it is necessary, in order to obtain the over-all equation for any reaction, to combine two equations of the kind given above, in such a way that when they are added up no free electrons are left. E.g.,

$$\left. \begin{array}{rcl} H_2O_2 + 2e &=& H_2O + O^{2-} \\ H_2C_2O_4 + O^{2-} &=& H_2O + 2CO_2 + 2e \end{array} \right\} +$$
$$\overline{H_2C_2O_4 + H_2O_2 = 2H_2O + 2CO_2}$$

$$2 \times \left| \begin{array}{rcl} Cl_2 + 2e &=& 2Cl^- \\ Fe^{2+} &=& Fe^{3+} + e \end{array} \right\} +$$
$$\overline{Cl_2 + 2Fe^{2+} = 2Cl^- + 2Fe^{3+}}$$

Quantities of oxidizing agents and reducing agents which accept or furnish the same numbers of electrons are said to be *equivalent* to one another. Thus one molecule of hydrogen peroxide is equivalent to one molecule of oxalic acid or one molecule of chlorine (it is, of course, possible to compare the oxidizing capacity of oxidizing agents with each other). One molecule of chlorine or hydrogen peroxide is equivalent to two iron(II) ions; two iron(II) ions are equivalent to one molecule of oxalic acid, etc. *One equivalent of any oxidizing or*

---

* For the applicability of this statement to electrolytic processes see p. 757.

reducing agent is defined as that quantity which accepts or furnishes one electron. Since a neutral oxygen atom can accept two electrons, one equivalent of an oxidizing agent or reducing agent may also be described as the quantity corresponding to the transfer of half an atom of oxygen. One *gram equivalent* is the corresponding quantity in grams—i.e., which implies the transference of 8 grams of oxygen or of 1 electrochemical unit of electric charge.*
An *a*-normal solution of an oxidizing agent or reducing agent is one which contains *a* gram equivalents of the oxidant or reducing agent per liter.

When schematic partial equations of the type given above are used to set up the equations for oxidation and reduction processes, it should be noted that the negative oxygen ion O^{2-} never appears as such in significant concentrations. It immediately unites with the electropositive constituents of the reaction mixture—i.e., in aqueous solution, it unites with hydrogen ions to form water, if the solution is acid, or else with water to form hydroxyl ions:

$$O^{2-} + 2H^+ = H_2O \qquad \text{or} \qquad O^{2-} + H_2O = 2OH^-$$

This must be allowed for by adding a corresponding number of H^+ ions or of H_2O molecules, or in the absence of water, by introducing some other electropositive constituent of the reactants. It is likewise necessary to insert in the equations substances to furnish O^{2-} ions needed in any partial reaction. This may be made clear by a few examples.**

First example. Oxidation of hydriodic acid by potassium dichromate in sulfuric acid solution.

$$\left. \begin{array}{l} Cr_2O_7{}^{2-} + \quad 6e \ = 2Cr^{3+} + 7O^{2-} \\ 3 \times | \qquad\qquad 2I^- = I_2 + 2e \end{array} \right\} +$$

$$\left. \begin{array}{l} Cr_2O_7{}^{2-} + \ 6I^- = 2Cr^{3+} + 3I_2 + 7O^{2-} \\ 7O^{2-} \quad + 14H^+ = 7H_2O \end{array} \right\} +$$

$$\overline{Cr_2O_7{}^{2-} + 6I^- + 14H^+ = 2Cr^{3+} + 3I_2 + 7H_2O}$$

Net equation:

$$K_2Cr_2O_7 + 6HI + 4H_2SO_4 = Cr_2(SO_4)_3 + K_2SO_4 + 3I_2 + 7H_2O$$

Second example. Oxidation of oxalic acid by potassium permanganate in sulfuric acid solution.

$$\left. \begin{array}{l} 2 \times | \ MnO_4{}^- + 5e \quad = Mn^{2+} + 4O^{2-} \\ 5 \times | \qquad\qquad C_2O_4{}^{2-} = 2CO_2 + 2e \end{array} \right\} +$$

$$\left. \begin{array}{l} 2MnO_4{}^- + 5C_2O_4{}^{2-} = 2Mn^{2+} + 10CO_2 + 8O^{2-} \\ 8O^{2-} \qquad + 16H^+ = 8H_2O \end{array} \right\} +$$

$$\overline{2MnO_4{}^- + 5C_2O_4{}^{2-} + 16H^+ = 2Mn^{2+} + 10CO_2 + 8H_2O}$$

Net equation:

$$2KMnO_4 + 5H_2C_2O_4 + 3H_2SO_4 = 2MnSO_4 + K_2SO_4 + 10CO_2 + 8H_2O$$

* The electrochemical unit of charge is the *Faraday*, i.e., the quantity of electricity released in discharging one gram atom of a univalent element. 1 Faraday = 96493 coulombs = N_A electrons, where N_A = Avogadro's number.

** No difficulty arises in setting up oxidation and reduction equations, even in complicated cases, when the method shown in the following examples is employed. When the partial equations are written in ionic form, care must be taken not only to balance the kind and number of atoms on both sides (as in any other chemical equation), but also to balance the *sum of ionic charges* on both sides of the equation. Any unbalance of charge must be made up by writing free electrons.

In setting up the equation for an oxidation or reduction, the *oxidation process* is first written down, and then the *reduction process* to give the appropriate final products. These are made equivalent by multiplying the partial equations by such factors as will make the number of electrons used in the oxidation process equal to the number furnished by the reduction process; the partial equations are then added up. If any O^{2-} ions remain on either side of the equation, some additional process must be formulated and added in, whereby these are furnished or removed. With practice, this step may be included in the writing of the first partial equations.

Third example. Oxidation of manganese(II) sulfate by potassium permanganate in neutral solution.

$$
\begin{array}{rl}
2 \times \mid & MnO_4^- + 3e = MnO_2 + 2O^{2-} \\
3 \times \mid & Mn^{2+} + 2O^{2-} = MnO_2 + 2e
\end{array} \Big\} +
$$

$$
\begin{array}{l}
2MnO_4^- + 3Mn^{2+} + 2O^{2-} = 5MnO_2 \\
\phantom{2MnO_4^- + 3Mn^{2+}} 2H_2O = 2O^{2-} + 4H^+
\end{array} \Big\} +
$$

$$
2MnO_4^- + 3Mn^{2+} + 2H_2O = 5MnO_2 + 4H^+
$$

Net equation:

$$
2KMnO_4 + 3MnSO_4 + 2H_2O = 5MnO_2 + 2KHSO_4 + H_2SO_4
$$

Fourth example. Oxidation of chromium sulfate to sodium chromate, by fusing it with soda and potassium nitrate.

$$
\begin{array}{rl}
3 \times \mid & NO_3^- + 2e = NO_2^- + O^{2-} \\
2 \times \mid & Cr^{3+} + 4O^{2-} = CrO_4^{2-} + 3e
\end{array} \Big\} +
$$

$$
\begin{array}{l}
3NO_3^- + 2Cr^{3+} + 5O^{2-} = 3NO_2^- + 2CrO_4^{2-} \\
\phantom{3NO_3^- + 2Cr^{3+}} 5CO_3^{2-} = 5CO_2 + 5O^{2-}
\end{array} \Big\} +
$$

$$
3NO_3^- + 2Cr^{3+} + 5CO_3^{2-} = 3NO_2^- + 2CrO_4^{2-} + 5CO_2
$$

Net equation:

$$
3KNO_3 + Cr_2(SO_4)_3 + 5Na_2CO_3 = 3KNO_2 + 2Na_2CrO_4 + 3Na_2SO_4 + 5CO_2
$$

The two examples of oxidation with potassium permanganate given above show that the decomposition of an oxidizing agent does not necessarily always take place in the same way. In the case of potassium permanganate, it proceeds differently in neutral or weakly acid solution and in strongly acid solution. Other oxidizing agents also often react in different ways, according to the experimental conditions. Indeed, the same substance may function either as an oxidizing agent, or as a reducing agent, depending upon the conditions. Hydrogen peroxide provides an interesting example of this. Its reactivity depends on its tendency to form water, which may occur either according to the reaction

$$
H_2O_2 + 2e = H_2O + O^{2-} \tag{2}
$$

or by the reaction

$$
H_2O_2 + O^{2-} = H_2O + O_2 + 2e \tag{3}
$$

I.e., hydrogen peroxide can react either as an acceptor of electrons or as a donor of electrons —that is, both as an oxidizing agent and as a reducing agent. It acts as a reducing agent towards powerful oxidants, such as potassium permanganate. It reacts with this in acid solution, forming a manganese(II) salt and liberating molecular oxygen:

$$
\begin{array}{rl}
2 \times \mid & MnO_4^- + 5e = Mn^{2+} + 4O^{2-} \\
5 \times \mid & H_2O_2 + O^{2-} = H_2O + O_2 + 2e \\
& 3O^{2-} + 6H^+ = 3H_2O
\end{array} \Big\} +
$$

$$
2MnO_4^- + 5H_2O_2 + 6H^+ = 2Mn^{2+} + 5O_2 + 8H_2O
$$

By combining equations (2) and (3) we obtain

$$
2H_2O_2 = 2H_2O + O_2
$$

This is the reaction previously given in Chap. 2 for the spontaneous decomposition of hydrogen peroxide in the presence of a catalyst.

The peculiar feature of the decomposition of hydrogen peroxide by oxidation or by reduction is that in both cases the same reaction product is formed—namely water, although in the one case molecular oxygen is formed as well. It is common to find that one and the same substance can act as a reducing agent and as an oxidizing agent by yielding different reaction products. In principle this is true of every substance which can exist in more than one valence state. Thus lead oxide, which acts as an oxidizing agent towards carbon at a red heat, behaves as a reducing agent towards bleaching powder, which oxidizes it to lead dioxide.

6. Electrolytic Oxidation and Reduction [1, 7]

The essence of electrolysis lies in the decomposition of substances by the electric current—i.e., their chemical transformation by addition of electrons at the cathode, and removal of electrons at the anode. Thus every electrolytic process connotes an anodic oxidation and a cathodic reduction. For example, the essential processes of the electrolysis of hydrochloric acid consist of the discharge of chloride ions at the anode—i.e., their oxidation to free chlorine—and the discharge of hydrogen ions at the cathode—i.e., their reduction to free hydrogen.

$$\text{Anode} \qquad 2Cl^- = Cl_2 + 2e$$
$$\text{Cathode} \qquad 2H^+ + 2e = H_2$$

What is true of processes brought about by the passage of current applies also to spontaneously occurring processes which furnish current, such as take place in a galvanic cell. Thus the process generating current in the familiar Daniell cell consists in the dissolution of zinc in ionic form, thereby furnishing electrons, while copper is deposited at the other electrode because its ions are discharged by accepting electrons. In this case also, one process is an oxidation, the other a reduction.

$$\text{Oxidation process} \qquad Zn = Zn^{++} + 2e$$
$$\text{Reduction process} \qquad Cu^{++} + 2e = Cu$$

The discharge (or changes of charge) of the ions of oxidizing agents and reducing agents in the narrower sense—i.e., of ions such as MnO_4^-, ClO_3^-, NO_3^-, etc.,—can also be made into current-furnishing processes, by combining them with suitable electrodes. This is important because the potential differences which are thereby set up enable us to determine quantitatively the strengths of oxidizing and reducing agents.

7. Oxidation Potentials [2, 6]

If a platinum foil is dipped into a solution containing, e.g., equal concentrations of both tin(II) and tin(IV) ions, the electrode so produced may be coupled with a normal hydrogen electrode in the manner shown in Fig. 6, p. 29. In this instance, the positive current flows in the external circuit from the electrode dipping into the solution of the tin salts to the normal hydrogen electrode. The current-furnishing processes are:

$$Sn^{++++} + 2e = Sn^{++} \qquad \text{and} \qquad H_2 = 2H^+ + 2e$$

The tendency of the tin to pass from the quadrivalent state to the bivalent state is therefore strong enough to convert hydrogen from the elementary state to the ionic form, under the conditions specified, (i.e., to oxidize it) and at the same time to perform external work. The potential difference between the electrodes—which could, in principle be used to perform external work—amounts to 0.2 volts on open circuit, as was stated in Chap. 12. The potential difference set up between a platinum electrode dipping into an equimolecular solution of Pb^{++} and Pb^{++++} ions, and a normal hydrogen electrode, is as great as 1.8 volts. In this case also the Pb^{++}-Pb^{++++} electrode constitutes the positive pole. The Pb^{++++} ion thus possesses a very much greater oxidizing power than the Sn^{4+} ion. If a platinum foil dipping into a solution of the ions Sn^{++} and Sn^{++++} were combined with another, dipping into a solution of Pb^{++} and Pb^{++++}, the positive current would therefore flow from the latter to the former. Pb^{++++} ions would be reduced, and Sn^{++} ions oxidized. The potential, relative to the normal hydrogen electrode, taken up by a platinum foil dipping into an equimolar mixture of the ions of two valence states of an element is known as the *oxidation potential* of the higher valence state, or as the *redox potential* of the corresponding ions. A selection of oxidation potentials is given in Table 103. In addition to the potentials relating to the simple changes of ionic charge, the table gives also a series of oxidation potentials referring to reactions in which the solvent, water, participates. Thus, in the reduction of the nitrate ion:

$$\left. \begin{array}{l} NO_3^- + 3e = NO + 2O^{2-} \\ 2O^{2-} + 4H^+ = 2H_2O \end{array} \right\} \quad NO_3^- + 3e + 4H^+ = NO + 2H_2O$$

or the reduction of PbO_2, instead of Pb^{++++} ions, to Pb^{++}:

$$PbO_2 + 2e + 4H^+ = Pb^{++} + 2H_2O$$

TABLE 103

NORMAL OXIDATION POTENTIALS RELATIVE TO THE NORMAL HYDROGEN ELECTRODE

Cu^+	$\to Cu^{2+}$	$+2e$	—0.17 volt	H_2	$+2OH^- \to 2H_2O$	$+2e$	$+0.82$ volt
Sn^{2+}	$\to Sn^{4+}$	$+2e$	—0.2 ,,	NO	$+2H_2O \to NO_3^-$	$+4H^+ + 3e$	—0.95 ,,
$Fe(CN)_6]^{4-}$	$\to [Fe(CN)_6]^{3-}$	$+e$	—0.44 ,,		$2H_2O \to O_2$	$+4H^+ + 4e$	—1.23 ,,
Fe^{2+}	$\to Fe^{3+}$	$+e$	—0.75 ,,	Cr^{3+}	$+4H_2O \to HCrO_4^-$	$+7H^+ + 6e$	—1.3 ,,
Hg_2^{2+}	$\to 2Hg^{2+}$	$+2e$	—0.91 ,,	Mn^{2+}	$+2H_2O \to MnO_2$	$+4H^+ + 2e$	—1.35 ,,
Tl^+	$\to Tl^{3+}$	$+2e$	—1.21 ,,	Pb^{2+}	$+2H_2O \to PbO_2$	$+4H^+ + 2e$	—1.44 ,,
Co^{2+}	$\to Co^{3+}$	$+e$	—1.8 ,,	Cl^-	$+3H_2O \to ClO_3^-$	$+6H^+ + 6e$	—1.44 ,,
Pb^{2+}	$\to Pb^{4+}$	$+2e$	—1.8 ,,	MnO_2	$+2H_2O \to MnO_4^-$	$+4H^+ + 3e$	—1.59 ,,
				O_2	$+H_2O \to O_3$	$+2H^+ + 2e$	—1.9 ,,

Since, in the light of the foregoing discussion, the potentials listed in Table 4, p. 30 may also be regarded as oxidation potentials (in the wider sense), the values of Table 103 can be compared with them directly. Thus it follows from the numerical values that if a silver electrode in 1-normal silver ion solution is combined with a tin(II)-tin(IV) electrode, a potential difference of 0.6 volts is set up, in such a direction that the positive current flows externally from the former to the latter—i.e., Sn^{2+} ions are oxidized to Sn^{4+} ions, and Ag^+ ions discharged as metallic silver.

Just as the potential difference between a metal and a solution of its ions depends on the concentration of the latter, so the oxidation potentials depend upon the concentration of the ions in question in the solution. The redox potentials depend upon the ratio of the concentrations of the ions in the two valence states (e.g., $[Sn^{2+}] : [Sn^{4+}]$). Those oxidiation potentials which relate to reactions involving water will also depend upon the hydrogen ion concentration or hydroxyl ion concentration of the solution. The oxidation potentials listed in Table 103 relate to solutions in which all the ions participating in the electrode reaction are present at 1-molar concentration, and represent the potential, relative to the normal hydrogen electrode, assumed by a platinum electrode immersed in such a solution. The potentials shown in the table represent, at the same time, the work in volts which must be *expended* in order to oxidize one gram equivalent of the substance concerned, at the same time discharging 1 gram equivalent of hydrogen ion—i.e., in order to reverse the reactions represented by the arrows. Thus, where the potentials have negative values, the spontaneously occurring processes, which would furnish current, proceed in the direction *opposite* to the arrows.

8. **Oxido-Reduction and Disproportionation**

It is often observed that from a compound containing some element in an intermediate valence state, or in the electroneutral condition, two new substances may be formed—the one derived from a higher valence state, and the other from a lower valence state. For example, potassium chlorate is transformed into a mixture of potassium perchlorate and potassium chloride when it is heated. Free chlorine, with caustic soda, forms hypochlorite and chloride. Phosphorus similarly forms hypophosphite and phosphine.

$$\overset{+5}{4KClO_3} = \overset{+7}{3KClO_4} + \overset{-1}{KCl} \tag{4}$$

$$\overset{0}{Cl_2} + 2NaOH = \overset{+1}{Na[ClO]} + \overset{-1}{NaCl} + H_2O \tag{5}$$

$$\overset{0}{4P} + 3NaOH + 3H_2O = \overset{+1}{3Na[H_2PO_2]} + \overset{-3}{PH_3} \tag{6}$$

Processes of this sort are known as *oxido-reductions*, since they can be regarded as the oxidation of one portion of the element by another portion of the same element, which is thereby reduced. This is at once evident if the net equations are built up from partial equations in the manner considered previously.

$$
\begin{array}{lll}
3 \times \mid & ClO_3^- + O^{2-} = ClO_4^- + 2e & \text{(oxidation)} \\
& ClO_3^- - 6e = Cl^- + 3O^{2-} & \text{(reduction)} \\
\hline
& 4ClO_3^- = 3ClO_4^- + Cl^- &
\end{array}
$$

$$
\begin{array}{lll}
& \tfrac{1}{2}Cl_2 + O^{2-} = ClO^- + e & \text{(oxidation)} \\
& \tfrac{1}{2}Cl_2 + e = Cl^- & \text{(reduction)} \\
& 2OH^- = O^{2-} + H_2O & \\
\hline
& Cl_2 + 2OH^- = ClO^- + Cl^- + H_2O &
\end{array}
$$

$$
\begin{array}{lll}
3 \times \mid & P + 2OH^- = H_2PO_2^- + e & \text{(oxidation)} \\
& P + 3e + 3H^+ = PH_3 & \text{(reduction)} \\
\hline
& 4P + 3OH^- + 3H_2O = 3H_2PO_2^- + PH_3 &
\end{array}
$$

These reactions can be thought of as combinations of oxidation and reduction processes in the electrochemical sense, regardless of whether the atoms of the elements concerned are present in the compounds in the electrically charged state, or whether the compounds are considered to be purely covalent.

The range of oxido-reductions overlaps to some extent with that of *disproportionations*. Disproportionation is a process in which two or more molecules of the same kind react to give products which are less complex than the original,—e.g.,

$$6B_2H_5Cl = 5B_2H_6 + 2BCl_3 \qquad \text{(cf. p. 329)} \tag{7}$$

$$3BCl_2(OCH_3) = 2BCl_3 + B(OCH_3)_3 \qquad \text{(cf. p. 335)} \tag{8}$$

$$2CH_3 \cdot CHO + KOH = [CH_3 \cdot CO_2]K + CH_3 \cdot CH_2OH \tag{9}$$
$$\text{acetaldehyde} \qquad\qquad \text{potassium}$$
$$\text{acetate}$$

Reaction (9) is both a disproportionation and an oxido-reduction. Reaction (8) is a disproportionation, but not an oxido-reduction. The disproportionation represented by (7) can only somewhat artificially be considered as an oxido-reduction—namely by assuming that boron in chlorodiborane is more electropositive or less electronegative than in diborane, an assumption not supported by any experimental evidence. In this and similar cases it is therefore better to speak of disproportionations than of oxido-reductions. The reactions shown in equations (4), (5) and (6), on the other hand, are oxido-reductions but not disproportionations.

9. Autoxidation [3–5]

The term *autoxidation* is applied to oxidation processes brought about by the oxygen of the air at ordinary or slightly elevated temperature. It is not uncommonly observed in these that during the reaction a part of the oxygen is converted into a more reactive form, capable of oxidizing substances which would not ordinarily be attacked by atmospheric oxygen under the experimental conditions used. For example, sodium arsenite in solution by itself is unaffected by oxygen; sodium sulfite in solution is slowly oxidized by the oxygen of the air. If a current of air is passed through a solution containing both sodium sulfite and sodium arsenite, the oxidation of sulfite to sulfate is accompanied by the oxidation of arsenite to arsenate. Indeed, under suitable conditions, one molecule of arsenate is formed for every molecule of sulfate formed from sulfite. The oxygen in such a case is said to be equally shared. The process can be explained on the view that a whole molecule of oxygen is taken up primarily by the sulfite ion, and that the peroxy-compound thus formed—in this instance the ion $[O_2.SO_3]^{2-}$ —promptly gives up one oxygen atom to an arsenite ion:

$$[SO_3]^{2-} + O_2 = [O_2.SO_3]^{2-}$$

$$[O_2.SO_3]^{2-} + [AsO_3]^{3-} = [SO_4]^{2-} + [AsO_4]^{3-}$$

The theory that autoxidations occur by way of the intermediate formation of peroxide-like compounds was first sketched out by Traube, and developed further by Engler. It was at that time assumed that peroxide-like intermediates were involved in *all* autoxidations. It is now known that many take place by a different mechanism—namely through the intermediate formation of free radicals (see below). Substances such as sodium sulfite, which can be directly oxidized by the oxygen of the air, were termed *autoxidants* by Engler. Those to which oxygen can be transferred by autoxidants are called *acceptors*. In the presence of suitable autoxidants—e.g., hydrogen dissolved in palladium—, hydrogen iodide, carbon monoxide, oxalic acid, benzene, indigo white, and a whole series of other organic compounds may act as acceptors. All these are substances which are unattacked by oxygen at ordinary temperature, or attacked imperceptibly slowly. In the presence of hydrogen dissolved in palladium they are rapidly oxidized—hydrogen iodide to iodine, carbon monoxide to the dioxide, ammonia to nitrous acid, etc.* Experimental conditions can be so adjusted, in these cases also, that the oxygen is halved between autoxidant and acceptor.

It is not necessary, however, that the oxygen consumed in forming the labile intermediate peroxide should be halved with the acceptor. The labile peroxide (called the 'moloxide' by Engler)—the ion $[O_2.SO_3]^{2-}$ in the case cited, which is not identical with the ion $[HO.O.SO_3]^-$ of peroxymonosulfuric acid—can also react with the autoxidant itself:

$$[O_2.SO_3]^{2-} + [SO_3]^{2-} = 2[SO_4]^{2-},$$

or two such molecules can react with each other, regenerating a molecule of molecular oxygen:

$$2[O_2.SO_3]^{2-} = 2[SO_4]^{2-} + O_2.$$

Finally, if water is present, the 'moloxide' can react hydrolytically, to form hydrogen peroxide. The formation of hydrogen peroxide is frequently observed when autoxidations take place in aqueous solution.

Hydrogen peroxide may also be formed in autoxidations in another way, and not by the hydrolysis of a primarily formed peroxide of the autoxidant. In this case, the primary mechanism consists of the addition of water to the autoxidant, whereby the hydrogen atoms of the water are so activated that they can react with molecular oxygen. The autoxidation

* The peroxide which is thereby formed as an intermediate is not fully identical with ordinary hydrogen peroxide, but is more reactive; some have assumed that it is a solution of hydrogen peroxide in palladium or a loose addition compound between palladium and hydrogen peroxide. Since Harteck has observed the formation of an isomer of hydrogen peroxide, stable only at low temperatures, by the action of atomic H on O_2, it is plausible to assume the intermediate formation of this substance in these reactions.

of zinc in contact with water containing dissolved oxygen, studied by Traube, follows this mechanism:

$$\underset{\text{Zn}}{\overset{}{}} + \begin{matrix} \text{H} \\ | \\ \text{O—H} \\ \\ \text{O—H} \\ | \\ \text{H} \end{matrix} + \begin{matrix} \text{O} \\ \| \\ \text{O} \end{matrix} = \text{Zn} \begin{matrix} \diagup \text{OH} \\ \diagdown \text{OH} \end{matrix} + \begin{matrix} \text{H—O} \\ | \\ \text{H—O} \end{matrix}$$

Such cases are referred to as 'indirect autoxidations'. Hydrogen peroxide is then the 'indirect autoxidant', and the substance which, instead of activating oxygen, reacts with water and thus activates its hydrogen, is the 'pseudo-autoxidant'. Heavy metals in aqueous solution often act as pseudo-autoxidants. Thus the formation of hydrogen peroxide by the action of the air upon lead, in contact with dilute sulfuric acid—a reaction first observed by Schön-bein—is of this type. One molecule of hydrogen peroxide is formed for every molecule of lead sulfate:

$$Pb + H_2SO_4 + O_2 = PbSO_4 + H_2O_2.$$

If an acceptor is present, the hydrogen peroxide can pass oxygen on to it. In any event, however, both with direct and indirect autoxidations, *not more than half* (and often less) of the oxygen originally taken up by the autoxidant can be transferred to the acceptor. Autoxidants are thereby distinguished from catalysts, which can, in theory at least, pass on indefinite quantities of oxygen. A further distinction is that whereas catalysts can only bring about reactions which do not absorb energy (unless external energy is supplied), autoxidations can also bring about oxidations in which energy is absorbed. The necessary energy is supplied, in such cases, by the oxidation of the autoxidant itself. The only restriction is that the *total* process—*oxidation of autoxidant + oxidation of acceptor*—must be one which furnishes energy.

It appears that autoxidation processes which take place in this way are very important in the living organism. This may perhaps explain the extraordinary oxidative powers exhibited by living organisms. For example, the human and animal organism can oxidize benzene to phenol, or even burn hydrocarbons of the paraffin series (vaseline). Certain bacteria can oxidize hydrogen to water, ammonia to nitrogen, and to nitrate, methane to carbon dioxide. At the temperatures involved, such oxidations can be carried out only by our strongest oxidants, and in some cases not at all under laboratory conditions. Engler assumed that peroxides formed as primary products of autoxidations were the effective oxidizing agents in the organism. This conclusion would be modified today. It is now known that biological oxidation processes take place in a number of stages, with not too high a redox potential at any one stage. Furthermore, the examples cited illustrate the fact that exothermic processes may be inhibited because the reactivity of the system is too low; processes which are not brought about by the usual oxidants may, however, occur readily through the high reactivity of free radicals.

10. Free Radical Mechanisms in Oxidation and Autoxidation Processes

It is now considered that a great many oxidation-reduction processes in solution, as well as in the gaseous phase (cf. Vol. II) involve the transient formation and high reactivity of free radicals. This was first explicitly recognized by Franck and Haber (1931), who suggested a chain mechanism (see Vol. II) involving free radicals , for the catalytic oxidation of sulfite ions in the presence of copper(II) salts. The initial step is considered to be oxidation of the $SO_3^=$ ion to the SO_3^- radical, by the transfer of one electron to the Cu^{++} ion: in general, it is considered that the most important reactions are those involving *transfer of a single electron or breakage of a single chemcial bond.*

$$Cu^{++} + SO_3^{=} = Cu^{+} + SO_3^{-} \tag{1}$$

$$SO_3^{-} + H_3O^{+} = H_2O + HSO_3 \tag{2}$$

In the absence of oxygen, the HSO_3 radicals are destroyed by combining with each other, to form dithionic acid

$$2HSO_3 = H_2S_2O_6 \tag{3}$$

but if oxygen is present, the reactions (4) and (5) may take place:

$$\longrightarrow HSO_3 + O_2 = HSO_5 \tag{4}$$

$$HSO_5 + SO_3^{=} + H_2O = 2HSO_4^{-} + OH \tag{5}$$

$$OH + SO_3^{=} = OH^{-} + SO_3^{-} \tag{6}$$
$$SO_3^{-} + H_3O^{+} = HSO_3 + H_2O$$

The reactive OH radicals thus regenerate the original HSO_3 radical. Reactions (4) to (6) may take place continuously, until the chain of events is terminated by the disappearance of the radicals HSO_3 and OH by adventitious side reactions such as (3).

Since, moreover, the Cu^{+} ion is reoxidized to Cu^{++} ion by atmospheric oxygen, the addition of a very small amount of copper(II) salt can bring about the autoxidation of an indefinite amount of sulfite.

There is now little doubt that analogous radical mechanisms afford the best explanation of a great many other oxidation processes, and notably those in which hydrogen peroxide is consumed or formed. Hydrogen peroxide is decomposed catalytically by iron(II) ions, and a mixture of hydrogen peroxide and an iron (II) salt (known as Fenton's reagent) is one of the most powerful of oxidants. Haber and Weiss (*Proc. Roy. Soc. London*, A 147 (1934) 332) based their interpretation of the decomposition of H_2O_2 on the formation of OH and HO_2 radicals in the solution.

$$Fe^{++} + H_2O_2 = Fe^{+++} + OH^{-} + OH \tag{7}$$

$$\longrightarrow OH + H_2O_2 = HO_2 + H_2O \tag{8}$$

$$HO_2 + H_2O_2 = O_2 + H_2O + OH \tag{9}$$

$$Fe^{++} + OH = Fe^{+++} + OH^{-} \tag{10}$$

Although the Haber-Weiss mechanism has been modified by later work, it expresses the essential features of the catalytic decomposition of H_2O_2. The oxidizing properties of Fenton's reagent are accordingly due to the high reactivity displayed by OH radicals.

The existence of the HO_2 (and OH) radicals was first postulated to interpret the mechanism of gaseous oxidation reactions; it is believed to be also an active intermediate in processes occurring in solution. It can be regarded formally as the parent acid of which potassium hyperoxide, KO_2 (p. 176) is a derivative. It is an important radical in oxidations with molecular oxygen, since its anion O_2^- can be formed by the transfer of one electron from some suitable substrate to the O_2 molecule. It may therefore play a part in reactions formerly interpreted in terms of 'moloxides'.

It seems probable, therefore, that many of the phenomena summarized above as characteristic of autoxidations can be understood in terms of the set of reactions (11) to (14):

$$O_2 + \text{electron (from donor)} \rightarrow O_2^- \tag{11}$$

$$O_2^- + H_3O^+ \longrightarrow HO_2 + H_2O \tag{12}$$

$$HO_2 + \text{electron (from donor)} \rightarrow HO_2^- \tag{13}$$

$$HO_2^- + H_3O^+ \longrightarrow H_2O_2 + H_2O \tag{14}$$

In this set of reactions, two equivalents of the electron donor are oxidized (in the unit steps referred to above), for the consumption of one molecule of O_2 and the simultaneous production of one molecule of H_2O_2. In place of (13) and (14), it is possible for HO_2 radicals to be destroyed by mutual reaction to form ozone: $2HO_2 = H_2O + O_3$ (which has often been reported as formed in traces during autoxidations) or hydrogen peroxide: $2HO_2 = H_2O_2 + O_2$; the latter is the more favored process.

Rather direct proof of the participation of free radicals in inorganic reactions has been furnished by the observation (Baxendale, Evans and Parks, *Trans. Faraday Soc.*, 42 (1946) 155) that the polymerization of vinyl compounds such as styrene or acrylonitrile, which is known to be initiated by free organic radicals,

$$R\!\!-\!\! + CH_2\!\!=\!\!CHX \rightarrow R\!\!-\!\!CH_2\!\!-\!\!CHX\!\!-$$

$$R\!\!-\!\!CH_2\!\!-\!\!CHX\!\!- + CH_2\!\!=\!\!CHX \rightarrow R\!\!-\!\!CH_2\!\!-\!\!CHX\!\!-\!\!CH_2\!\!-\!\!CHX\!\!-, \quad \text{etc.}$$

is also initiated by a variety of inorganic oxidation-reduction processes. These include oxidations by persulfate, the ferrous ion-hydrogen peroxide system, cobalt (III) salts and numerous others. If there is sufficient of the vinyl compound to 'trap' almost all the radicals produced, the oxidation of ferrous ion by hydrogen peroxide follows the course:

$$Fe^{2+} + H_2O_2 \rightarrow Fe^{3+} + OH^- + OH \tag{15}$$

$$OH + \text{vinyl compound} \rightarrow HO\!\!-\!\!\text{vinyl polymer} \tag{16}$$

In this case the ratio of Fe^{3+} to H_2O_2 destroyed is 1 : 1, whereas in the absence of vinyl compounds, following equations (7) and (10) above, the Fe^{3+} : H_2O_2 ratio can rise to 2 : 1.

Free radicals, initiating polymerization, have also been detected in reactions such as the oxidation of iron(II) salts with free halogens or hypobromite:

$$Fe^{2+} + Br_2 \quad \rightarrow \quad Fe^{3+} + Br^- + Br$$

$$Fe^{2+} + HOBr \rightarrow Fe^{3+} + Br + OH^- \qquad \text{(Evans, 1947)}$$

Free radicals may be produced not only as the result of a primary chemical act, but also by the absorption of light, which may ionize off an electron from a neutral molecule or negative ion. Thus the autoxidation of sulfite may also be accelerated by irradiation with light. For every quantum $h\nu$ of light absorbed, since it produces one SO_3^- or HSO_3 radical by ionizing off one electron from a $SO_3^=$ ion, several tens of thousands of $SO_3^=$ ions may be oxidized by oxygen to $SO_4^=$ ions on the basis of the chain reaction shown above (Bäckström, 1927). This observation explains why substances which are completely stable to air when kept in the dark may, nevertheless, be attacked by oxygen if exposed to light.

Autoxidations which proceed under the action of light no longer involve *spontaneously* occurring processes (i.e., requiring no addition of energy) like those discussed previously. Although in some cases, such as that cited, the amount of light energy supplied is extremely minute in comparison with the effect produced, this energy is none the less used up. The process is no longer a catalytic one. The term autoxidation, as its name implies (Greek αὐτός, by itself), was originally understood to include only those oxidations by atmospheric oxygen which proceed spontaneously. However, this restriction has been dropped since, as the example of the oxidation of sulfite shows, the means by which a process is triggered is of minor importance for the essential mechanism of the process.

How far the practically important autoxidation processes, and especially those which are important biologically, proceed by unstable peroxides, and how far they are chain reactions involving free radicals, is as yet uncertain. Recent researches, such as the very exhausitve studies made by Evans and his coworkers, and of Weiss, and others, have made it probable that the chain reaction mechanism is very generally applicable. Wieland [5] showed that it provided an explanation of some biologically important autoxidations, and the work of Calvin, Rabinowitch, Gaffron, and others has extended its use to an understanding of the photosynthetic reduction of CO_2 to carbohydrate in the green leaf. Predominant importance is therefore now ascribed to radical mechanisms in both biological and inorganic oxidation-reduction processes.

Reference should be made to the reviews by Uri (*Chem. Rev.*, 5 (1952) 375) and Weiss (*Chem. Soc. Ann. Rep.*, 44 (1947) 60) for more detailed discussion.

References

1 S. GLASSTONE and A. HICKLING, *Electrolytic Oxidation and Reduction, Inorganic and Organic*, London 1935, 420 pp.

2 R. ABEGG, F. AUERBACH, R. LUTHER and C. DRUCKER, *Messungen elektromotorischer Kräfte galvanischer Ketten mit wässerigen Elektrolyten*, Halle 1911 and 1915, Berlin 1929, *Abhandlungen der Deutschen Bunsen-Gesellschaft*, No. 5, 213 pp., No. 8, 60 pp., and No. 10, 234 pp.

3 R. WURMSER, *Oxydations et Réductions*, Paris 1930, 400 pp.

4 Institut Solvay, *L'Oxygène, ses Réactions chimiques et biologiques*, Paris 1935, 353 pp.

5 H. WIELAND, *Über den Verlauf der Oxydationsvorgänge*, Stuttgart 1933, 96 pp.

6 W. M. LATIMER, *The Oxidation States of the Elements and Their Potentials in Aqueous Solutions*, 2nd. Ed., New York 1952, 392 pp.

7 G. KORTÜM and J. O'M. BOCKRIS, *Textbook of Electrochemistry*, Amsterdam 1951, Vol. I, 365 pp.; Vol, II, 544 pp.

CHAPTER 17

SEVENTH MAIN GROUP OF THE PERIODIC SYSTEM: THE HALOGENS

Ato-mic num-bers	Ele-ments	Sym-bols	Atomic weights	Densi-ties	Melting points	Boiling points	Heats of fusion	Heats of evapo-ration	Valences
							kcal/g-atom		
9	Fluorine	F	19.00	1.108^1	$-223°$	$-187.9°$	0.17	0.80	I
17	Chlorine	Cl	35.457	1.57^1	$-102.4°$	$-34.0°$	0.81	2.2	I, III, IV V, VII
35	Bromine	Br	79.916	3.14	$-7.3°$	$+58.8°$	1.76	3.82	I, III, V
53	Iodine	I	126.91	4.942	$+113.7°$	$+184.5°$	2.0	7.45	I, III, V, VII
85	Astatine	At	210^2	—	—	—	—	—	Probably as for iodine

[1] For liquid fluorine and chlorine at their boiling point
[2] Atomic weight of longest lived isotope (half-life, 8.3 hours)

1. Introduction

(a) General

Main Group VII of the Periodic System comprises the elements fluorine, chlorine, bromine, iodine and the very unstable radio-element astatine; all are typical non-metals. Fluorine and chlorine are gaseous at ordinary temperature, bromine is liquid, and iodine solid. They all form diatomic molecules in the gaseous state. All these elements are very reactive. This is especially true of the lighter members; fluorine, the first of the group, is the most reactive of all the known elements. This high reactivity is associated with the tendency of their atoms to become singly charged negative ions. As a result of this tendency, they combine with the light metals, from which they can abstract electrons, to form compounds of typically *salt-like character*. The elements of Group VIIB are grouped together under the name of the *halogens** (from ἅλς, salt, and γεννᾶν, to produce). The observation that these substances could unite directly with metals to form salts was very striking at the time the halogens were discovered, since it was then assumed that oxygen was an essential constituent of acids (or of strong acids, at least) and of salts. The halogens, however, are able to

* The name 'halogen' was first suggested by Schweigger in 1811 for chlorine, because of its ability to form salts.

form oxygen-free salts and oxygen-free acids, namely the hydrohalogen acids of the type HX (X = halogen). With the exception of hydrofluoric acid, these acids are very strong acids, not only in aqueous solution but in other solvents—e.g., in alcoholic or ethereal solution.

The binary compounds of the halogens with strongly electropositive elements are termed the halides. In these compounds the halogens are always univalent.

The great reactivity of the halogens is due not only to their tendency to form negative ions, but also to the comparatively small amount of energy which must be expended in order to dissociate their molecules into atoms (cf. Table 104).

All the halogens have *positive values* for the *electron affinity*—i.e., for the quantity of energy made available for the performance of work when an electron is added on to a free *atom*. The electron affinity decreases from fluorine to iodine, but probably has its maximum value for chlorine. The electron affinity of fluorine is at present uncertain, owing to the disagreement in the energy of dissociation of F_2 as determined by different methods. As the data of Table 104 show, the *normal (oxidation) potentials* increase much more strongly in the same sequence. All the normal potentials are negative, relative to the normal hydrogen electrode. Thus a 'chlorine electrode' dipping in a solution 1-normal in Cl^- ions has a potential 1.36 volts lower than that of a normal hydrogen electrode; on completing the circuit, positive electricity would flow through the external circuit from the chlorine electrode to the hydrogen electrode, so that hydrogen would go into solution as positive ions and chlorine into solution as negative ions, respectively, at the two electrodes. The quantity of energy liberated in forming dilute hydrochloric acid in this way is found from eqn. (3), p. 142, to be 31.3 kcal per mol. It is equal to the sum of the free energy of formation of HCl and the free energy of solution of HCl in water. If the latter (8.6 kcal per mol) is deducted, the free energy of formation of HCl is found to be 22.7 kcal per mol, as compared with 22.76 kcal per mol found spectroscopically. The normal potentials given in the table have been measured directly. They can, however, be calculated from spectroscopically determined dissociation energies and electron affinities, using the cyclic process discussed on p. 146. Makishima (1935) has shown that when the temperature variation of the energy quantities is taken into

TABLE 104

PROPERTIES OF THE HALOGENS (see also table on p. 772)

	Fluorine	Chlorine	Bromine	Iodine
Color in gaseous state	Pale green-yellow	Yellow green	Red brown	Violet
Molecular volume (liquid, at b.p.), cc	34	45	54	68
Atomic radius in crystals (Å)	—	1.07	1.19	1.36
Ionic radius in crystals (Å)	1.33	1.81	1.96	2.20
Moment of inertia of gas molecule (g cm^2 × 10^{40})	25.3	113.7	340	742
Internuclear distance in gas molecule (Å)	1.44	1.98	2.28	2.67
Degree of thermal dissociation {at 1000° K	4.3	0.035	0.23	2.8
(per cent) {at 2000° K	99	52	72.4	89.5
Dissociation energy ($X_2 = 2X$), kcal/mol	38 ± 3	57.2	45.4	35.5
Electron affinity ($X + e = X^-$), kcal/g atom	83 ± 3	86.5	81.5	74.2
Free energy for the reaction $X_2 + 2e \rightleftharpoons 2 X^-$ (kcal/mol)	—128	—115.8	—117.6	112.9
Normal potential in volts (at 18°, relative to the normal hydrogen electrode)	—2.8	—1.36	—1.08	—0.58
Heat of hydration of the negative ion (kcal g ion)	128	96.9	92.2	85.8
Electrolytic mobility of the negative ion at infinite dilution, 18°	47.6	66.3	68.2	66.8

account, there is good agreement between the calculated and observed values. It turns out, as in the cases considered in Chap. 6 and Chap. 8, that the values of the normal potentials are considerably influenced by the *energies of hydration*. The values given for these in Table 104 are based on the assumption that the heat of hydration of the H^+ ion is 250 kcal per g ion (see p. 157).

The halogens can combine with both electronegative and electropositive elements. In their compounds with electronegative elements, the halogens (other than fluorine, which is invariably univalent) can exhibit valences of seven, five and three, as well as unity. The maximum valence state of chlorine and iodine is seven, but no compounds are known in which bromine has a valence state higher than five. Chlorine also forms an oxide (chlorine dioxide, ClO_2) in which it is quadrivalent. From the chemical standpoint, the reactions associated with the variable valence of the halogens can most simply be grasped if one regards the halogens as being formally electropositive in their compounds with the strongly electronegative elements. This is a very crude approximation, and the compounds of the halogens with other electronegative elements are almost certainly covalent. It leads, however, to the useful rule that the most extreme valence states are generally the most stable—i.e., negative univalent and positive seven- (or for bromine, five-) valent. Iodine also has a marked preference for the five-valent state.

On the Kossel-Lewis theory, the position of the halogens in relation to the inert gases, and the fact that thier atoms contain a group of seven electrons with the same principal quantum number, provide an explanation for their valence properties on the assumption that the outer shells of the atoms tend to assume an inert gas configuration. The electron affinities of the halogen atoms have since been determined, even if not with very great accuracy (Mayer, 1932), and their positive values show that, for the halogens, the addition of an electron to the atom is, in fact, exothermic. The especial stability of compounds in which the halogens are present as negative ions is thereby explained.

The conversion of the atoms into positive ions would require the expenditure of energy, as with all atoms. The ionization energy for the eighth electron of the halogen atoms is far greater than that of the first seven stages, as is shown by the spectroscopic measurements (cf. Table 23, p. 118). The maximum valence of the halogens is explained in this way on Kossel's views, the invariable electronegative univalence of fluorine being attributed to the high value of even the first stage ionization energy, as compared with the energy recovered by the Coulomb attraction between singly charged ions.

This view, as has been stated, is very much over simplified, and it may be taken that, with a few exceptions, the compounds of the halogens are covalent, except when the elements are present as univalent negative ions. The number of covalent bonds which may be formed by an atom with an incompletely filled valence shell is subject, however, to restrictions on energetic grounds, closely allied to those set out in the Kossel theory. That fluorine never exerts a valence higher than one, even in covalent compounds, accords, with the Heitler-London theory and the Pauli principle, since the fluorine atom has one 'unpaired' electron. The homologues of fluorine can exhibit valences higher than unity in their covalent compounds, since the higher principal quantum numbers of their outer electrons enables them to accommodate more than 8 electrons in the outer shell (cf. the maximum valences of oxygen and its homologues).

Iodine is relatively the most electropositive of the halogens, and can function as a positive ion—both electropositively univalent, as in iodine perchlorate, $I[ClO_4]$, and trivalent, in the *iodyl radical* $[IO]^+$, e.g., in iodyl sulfate, $[IO]_2SO_4$. It appears that iodine may be present in very small amount as the positive electrolytic ion I^+ in aqueous solution. Its ability to form electropositive ions is greatly augmented by combination with organic radicals (formation of *iodonium* compounds p. 794) or

addition of neutral molecules (p. 784). It has recently been possible to stabilize
bromine as an electropositive univalent ion in the same way (p. 785). In general,
however, the halogens do not exist as elementary, electropositive, electrolytic ions
but combine with other elements—generally oxygen—to form negative radicals,
which often exist in solution as very stable ions.

The energy which must be expended in order to convert the gas molecule into free singly
charged positive ions, according to: $\frac{1}{2}X_2 = X^+ + e$ is actually much smaller for the
halogens (other than fluorine) than for hydrogen. It may be calculated from the dissociation energies and primary ionization energies (Tables 7, 104 and 22):

for	H^+	F^+	Cl^+	Br^+	I^+
	364	418	327	298	258 kcal per g atom

The fact that chlorine, bromine and iodine, in contrast with hydrogen, nevertheless do
not exist to any appreciable extent in solution as electropositive ions is explained by the
presumably quite small energies of hydration of the positive ions, which would probably
not be very different from those of the negative halogen ions (Table 104). The heat of
hydration of the hydrogen ion, however, is very high, and compensates to a considerable
extent for the energy required to dissociate the H_2 molecule and ionize the H atom when
compounds are formed in aqueous solution. If a corresponding amount of energy is made
available to the halogen atoms, by adding other substances which are more strongly bound
to the positive ion than the H_2O molecules (i.e., which afford a greater liberation of energy),
then compounds of bromine and iodine can be obtained in which these elements are present
in the electropositive univalent state, as the central atoms of complexes. The stabilizing
action of complex formation is interpreted in this way, and can be observed in numerous
other instances. Thus gold cannot exist in solution as the univalent elementary ion Au^+, but
only as a complex ion (cf. Vol. II). In the case of gold, the expenditure of energy of form
the free, singly charged (gaseous) ion from the solid element (305 kcal per g atom) is
actually greater than for bromine and iodine. The energy necessary to stabilize the positive
ion can most readily be obtained by the addition of negative ions to it (e.g., addition of
CN^- ions in the case of gold). In the case of the halogens, we may perceive in this one
reason why the positive ions should be relatively much less stable than the oxy-acids, which
are formally derived from multivalent positive halogen atoms by addition of O^{2-} ions. It is,
however, likely that the stability of the ions derived from higher valence states—e.g.,
ClO_3^- and ClO_4^- —is raised by the fact that the binding in them is largely due to quantum
mechanical resonance forces.

A comparison of the halogens among themselves shows, in general, a quite
regular gradation in properties. The negative electroaffinity—i.e., the tendency to
assume a negative charge—does not decrease quite regularly from fluorine to
iodine, but reaches a maximum with chlorine. The hydration is greatest for the
F^- ion, however. The same is true respecting the influence of the negative ions
on the lattice energy. It also holds for the free energy of formation of the negative
ions from the diatomic gas molecules. One consequence of this is that fluorine
displaces all the other halogens from their compounds with metals. Chlorine
decomposes the metal compounds of bromine and iodine, and bromine only those
of iodine. The process underlying all these reactions is the acquirement of a charge
in order to form the ion affording the largest free energy of formation and the
largest contribution to the crystal- or hydration energy:

$$\frac{1}{2}F_2 + Cl^- = F^- + \frac{1}{2}Cl_2$$

$$\frac{1}{2}Cl_2 + Br^- = Cl^- + \frac{1}{2}Br_2 \quad \text{etc.}$$

Conversely, from the halogen oxyacids and their salts—compounds in which the halogens may be regarded as positively charged—, elementary iodine can liberate chlorine and bromine, and bromine can liberate chlorine. The greater negative electroaffinity corresponds to the smaller tendency to exist in the positively charged state. The trend in electronegativity thus shows itself also in the electropositive state, although to a less marked degree than in the transition from the neutral state to the negative ion. Under suitable conditions, the reaction between elementary iodine and the chlorate ion proceeds smoothly:

$$\tfrac{1}{2}I_2 + [ClO_3]^- = [IO_3]^- + \tfrac{1}{2}Cl_2$$

This is not so true of the neighboring halogens. Nevertheless, the increase in electropositive character with increase in atomic weight, which is typical of the Main Groups of the Periodic System, is unmistakably shown by the halogens.

There is also a regular increase in molecular volumes and in atomic and ionic radii from fluorine to iodine, as the data of Table 104 show. In the same sequence there is a regular rise in the melting points and boiling points of the halogens. This arises not only from the increase in molecular volume, but also from a decrease in the mutual saturation of the atoms within the molecule from fluorine to iodine. The color of the halogens is also connected with the incomplete saturation of the atoms in the halogen molecules, since the property of absorbing visible light—to which the color is due—is generally coupled with the presence of relatively loosely bound—i.e., readily excited—electrons. The intensity of the color increases from fluorine to iodine (Table 104).

At high temperatures, I_2 molecules are dissociated to a perceptible extent into atoms. The molecules of bromine and chlorine undergo thermal dissociation into atoms with much greater difficulty, but even with these the thermal dissociation reaches considerable proportions at very high temperatures. The comparatively large thermal dissociation of the F_2 molecule is noteworthy (see Table 104).

It might be expected that the increase in apparent radius of the negative elementary halogen ions in crystals would imply a decrease in the mobility of the electrolytic ions in solution. In reality, the mobility *increases* from fluoride ion to bromide ion, and then decreases somewhat with the iodide ion (Table 104). The reason for the increase in electrolytic mobility may lie in a decrease in the hydration sheath around the ions, as appears from the increase in transport of water towards the cathode (Table 17, p. 75) in going from the chloride to the bromide of salts containing the same cation. The fact that two factors enter into the electrolytic mobility in opposite senses (namely an increase in ionic volume which diminishes the mobility, and a decrease in ion hydration which augments it) probably explains why the electrolytic mobilities of the halogens have a maximum value for bromine.

As is usual in the Main Groups of the Periodic System, the first element, fluorine, occupies rather a special position as compared with the others. It has been pointed out that fluorine never functions as electropositive. Furthermore, when a comparison is made between compounds of analogous composition, the peculiar position of fluorine is evident. Thus hydrogen fluoride differs from the other hydrohalogen acids in that it is much less dissociated in solution, and has a tendency to form 'acid salts', $M^I HF_2$. The fluorides often differ markedly from the other halides in the their solubilities. The chlorides, bromides and iodides of the alkaline earth metals are all very soluble in water, or even deliquescent, whereas the fluorides are insoluble. Silver chloride, bromide, and iodide are practically insoluble, but silver fluoride is deliquescent.

It should be noticed, however, that there is already some increase in solubility from silver iodide to silver chloride, and some decrease in solubility from the alkaline earth iodides to the alkaline earth chlorides even although these trends are small. Thus in passing from chlorine to fluorine we find in each case only the great increase in a trend which, even if much less clearly marked, could already be detected in the homologues of fluorine.

The halogens can also combine with each other. The further apart their components stand in the group, the more numerous are the compounds formed.

On the basis of the Periodic Table, it would be expected that the there should be a homologue of iodine with the atomic number 85. This has long been sought, but has only recently been discovered. It was formerly thought that this element should accompany iodine in nature, but it has emerged from the rules governing the stability of atomic nuclei (see Vol. II, Chap. 11) that eka-iodine must be an *unstable element*, i.e., is radioactive. If such an element has a short half-life, it cannot exist in nature unless it is a member of one of the natural radioactive distintegration series (cf. Vol. II). Karlik and Bernert (1943) have recently been able to show that an element of atomic number 85 does, indeed, occur in the disintegration products both of thorium and uranium. It has been given the name *astatine* (symbol At), to denote its instability (Gk. τὸ ἄστατον, the unstable). For the isolation of astatine, see Johnson, Leininger and Segré, *J. Chem. Physics*, 17 (1949) 1.

(b) Crystal Structure

In the solid state the halogens form molecular crystal lattices.* The interatomic distances in the (diatomic) molecules of these structures are: $Cl \leftrightarrow Cl = 1.82$ Å (at $-185°$), $Br \leftrightarrow Br = 2.27$ Å (at $-150°$), $I \leftrightarrow I = 2.70$ Å (at room temp.). Fig. 117 shows the I_2

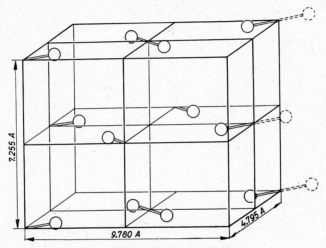

Fig. 117. Unit cell of iodine lattice.

structure. It constitutes a distinct layer lattice, and thereby accounts for the flaky habit of the crystals. Br_2 forms a crystal lattice of similar type. Whereas Br_2 and I_2 crystallize orthorhombic, Cl_2 forms a unique tetragonal structure.

(c) Occurrence

The halogens are almost universally distributed in Nature in the form of their compounds, although in very different amounts. The compounds of chlorine are by far the most abundant, but fluorine compounds are also found in large deposits.

Fluorine is found especially in the form of *fluorspar*, CaF_2, *cryolite*, Na_3AlF_6, and *apatite*, $Ca_5(PO_4)_3(F, Cl)$. Chlorine is present in solution in the water of the oceans, in the form of *sodium chloride* and other alkali and alkaline earth chlorides (cf. p. 243). It is found in the same forms in the salt deposits laid down by the drying up of arms of the sea. Heavy metal chlorides are also found occasionally, especially in the form of double compounds such as *atacamite*, $CuCl_2 \cdot 3Cu(OH)_2$. Chlorine makes up about 0.19% of the composition of the earth's crust, inclusive of the oceans. Bromine is generally found together with chlorine, in the form of the same

* The crystal structure of solid fluorine has not yet been determined.

compounds, but in considerably smaller quantity. For each 200 parts by weight of chlorine in sea water, there is about 1 part of bromine and 0.1 part of iodine. These elements are present in a similar ratio in the solid crust of the earth. Iodine is present in sea water not so much in the form of iodides, but principally in the form of organic compounds. It is found as iodate, and sometimes as periodate, in the nitrate deposits of Chile and Bolivia. Elementary iodine is occasionally present in the products of volcanic activity.

2. The Free Halogens

(a) Historical

(*i*) *Chlorine.* The first of the halogens to be prepared in the free state was *chlorine*, which was first isolated by Scheele, in 1774, by the oxidation of hydrochloric acid with pyrolusite.

In terms of the Phlogiston Theory, Scheele rightly called chlorine 'dephlogisticated muriatic acid'. Berthollet later believed that chlorine must be a substance containing *oxygen*, on the grounds that chlorine water exposed to sunlight forms hydrochloric acid and liberates oxygen. He therefore proposed that it should be named 'acide muriatique oxygéné' (= oxidized hydrochloric acid).* However, after the experiments of Gay-Lussac and Thenard, and of Davy, had proved unsuccessful in removing oxygen from chlorine—e.g., by passing chlorine over white hot charcoal—Davy maintained that chlorine was an element, and proposed the name chloric gas, or chlorine, after its characteristic color (Gk. χλωρός, yellow green). The present shorter name was due to Gay-Lussac, who originally opposed Davy's conclusions, but accepted them in 1813 after his investigation of iodine.

(*ii*) *Iodine* was discovered in soda prepared from the ashes of sea weeds in 1811 by a Paris saltpeter boiler named Courtois. It was more closely studied in 1813 by Clément and Désormes, by Gay-Lussac and by Davy. Gay-Lussac at once perceived its analogy with chlorine. He named it after the violet color of its vapor (ἰωειδής, violet colored).

(*iii*) *Bromine* was discovered by Balard in 1826 in the mother liquors from the evaporation of sea salt. It was thoroughly investigated, its analogy with chlorine and iodine was recognised, and it was named after its pungent smell (βρῶμος, a stench).

(*iv*) *Fluorine.* As soon as the chloristic theory was put forward by Davy, in 1811, Ampère at once suspected that an element analogous to chlorine must be present in hydrofluoric acid (i.e., fluorine). However, the attempts of numerous investigators to isolate this hypothetical element remained fruitless, because of the reaction of fluorine with the walls of the vessels, with water, etc. It was first successfully prepared in 1886 by Moissan, by the electrolysis of potassium fluoride dissolved in anhydrous liquid hydrogen fluoride. The element was named after its naturally occurring calcium compound, fluorspar, which is used in metallurgical processes as a flux (from Latin *fluo*, I flow).

---

* In the sense of the extended definition of oxidation given in the previous chapter, chlorine is indeed to be regarded as an oxidation product of hydrochloric acid. To Berthollet, however, oxidation was equivalent to addition of oxygen.

(b) Preparation

The halogens are prepared generally by the process

$$2X^- = X_2 + 2e$$

i.e., by the discharge of halogen ions, either electrolytically or by means of oxidizing agents. Iodine may also be prepared by the reduction of oxygen compounds of iodine.

(*i*) *Fluorine.* The fluorine ion can be discharged only by electrolytic processes, because of the exceptionally high affinity of the element for a negative charge. The electrolysis of potassium fluoride dissolved in liquefied hydrogen fluoride was employed by Moissan when he first successfully isolated the element, but fluorine is now prepared on a considerable scale, and in many laboratories, by the electrolysis of molten potassium hydrogen fluoride or potassium dihydrogen fluoride.

High temperature processes for the preparation of fluorine involve the electrolysis of molten KHF_2, at 240–250°, as was first carried out by Mather and Argo, about 1919. The salt is contained in an electrically heated vessel, which may be made of copper, and which serves as the cathode, and a rod of Acheson graphite serves as the anode. In Argo's cell this was surrounded by a 'diaphragm', in the form of a copper tube, closed at one end and slotted at the sides, which served to hinder the hydrogen evolved at the cathode from reaching the anode, as explosive recombination would otherwise occur. The electrode processes in such a cell are:

at the anode $F^- = \frac{1}{2}F_2 + e$,   at the cathode $H^+ + e = \frac{1}{2}H_2$

Monel metal and magnesium are resistant to corrosion by fluorine, and have been used in place of copper for the construction of high temperature fluorine cells.

Lebeau and Damiens (1925) and Fredenhagen (1928) introduced the use of $KF \cdot 3HF$ and $KF \cdot 2HF$ as electrolytes, in place of KHF_2, in order to lower the melting point of the electrolyte. Electrolysis of $KF \cdot 3HF$ can be carried out at 50–100°, with nickel serving as anode.

During the war of 1939–45, considerable progress was made in the technical production of fluorine. Medium-temperature cells of 2000 amperes capacity, with graphite anodes, are now regularly operated industrially, and the use of small fluorine cells has become a normal laboratory operation. Complete elimination of water is essential for satisfactory performance. In the presence of traces of water, the fluorine is contaminated by oxygen, which gives rise to strong anodic polarization and to the rapid disintegration of graphite anodes.

(*ii*) *Chlorine.* Chlorine is obtained in large quantities from the electrolysis of alkali chlorides (cf. p. 182).

In the laboratory, if it is not convenient to take chlorine from a cylinder, it is usually prepared by gently warming concentrated hydrochloric acid (or a mixture of common salt and moderately concentrated sulfuric acid) with manganese dioxide:

$$4HCl + MnO_2 = MnCl_2 + 2H_2O + Cl_2$$

In principle, this amounts to an oxidation of hydrogen chloride by the manganese dioxide. In detail, the reaction takes place through the initial formation of manganese tetrachloride, which then decomposes into chlorine and manganese dichloride:

$$MnO_2 + 4HCl = MnCl_4 + 2H_2O, \qquad MnCl_4 = MnCl_2 + Cl_2$$

Chlorine was formerly produced industrially on a large scale by the Weldon process (invented in 1866). This also started with manganese dioxide as oxidizing agent, which was continuously recovered from the manganese(II) chloride solutions, in the form of double compounds with calcium oxide (e.g., $CaO \cdot MnO_2$), by treatment with milk of lime and oxidation with air. The 'manganese mud' can then once more be reacted with hydrochloric acid:

$$MnCl_2 + 2Ca(OH)_2 + \tfrac{1}{2}O_2 = CaO \cdot MnO_2 + CaCl_2 + 2H_2O$$

$$CaO \cdot MnO_2 + 6HCl = CaCl_2 + MnCl_2 + 3H_2O + Cl_2$$

Thus the oxygen of the air is, in the last analysis, the oxidizing agent in the Weldon process, as is indeed almost always the case in industrial oxidation processes, except where these are carried out electrolytically.

The oxygen of the air may also be used *directly* for the oxidation of hydrogen chloride, as in the Deacon process (invented 1868) in which a mixture of air and hydrochloric acid vapor is passed at about 450° over balls of clay, impregnated with copper chloride, $CuCl_2$. The copper chloride serves as a catalyst for the process

$$2HCl + \tfrac{1}{2}O_2 = H_2O + Cl_2.$$

Although the Deacon process produced chlorine only in a diluted state, it proved superior to the Weldon process for the manufacture of bleaching powder, etc. Both of these processes are now used only very occasionally for the production of chlorine, since the enormous expansion of the use of electrolysis of alkali chlorides has, from time to time, even resulted in the over-production of chlorine.

(*iii*) *Bromine* is prepared from naturally occurring bromides by oxidation of Br^- ion to free Br_2 by means of chlorine:

$$2Br^- + Cl_2 = 2Cl^- + Br_2$$

The bromine liberated is always blown out of the solution by some suitable gas, and is enriched by absorption in a solution from which pure bromine can be subsequently liberated.

Where the source is relatively rich in bromine, as are the Stassfurt mother liquors (up to 0.5% Br^-), Dead Sea and Searles Lake (United States) brines, the bromine is blown out of solution by means of steam, and is collected in sodium carbonate:

$$3Na_2CO_3 + 3Br_2 = 5NaBr + NaBrO_3 + 3CO_2$$

In the United States (at Wilmington, North Carolina, and Freeport, Texas), bromine is manufactured directly from sea water. This is treated with chlorine at pH = 3.5, and the bromine is removed by a current of air, and is then reduced to HBr by means of SO_2. The HBr is absorbed, giving about a 4000 fold enrichment. The bromine is finally liberated once more by treating the resulting solution with Cl_2.

(*iv*) *Iodine* is produced to some extent from the ashes of kelp, in which it is present as iodide, by methods similar to those used for chlorine—i.e., either electrolytically (1) or by heating with manganese dioxide and sulfuric acid (2):

$$2I^- = I_2 + 2e \tag{1}$$

$$2NaI + 3H_2SO_4 + MnO_2 = 2NaHSO_4 + MnSO_4 + 2H_2O + I_2 \tag{2}$$

Iodine can also be liberated from the liquors by passing in chlorine:

$$2I^- + Cl_2 = 2Cl^- + I_2$$

but the process is attended with losses, since part of the iodine is oxidized to iodate, which remains in the solution, by any excess of chlorine.

Most of the iodine produced now comes from the mother liquors of Chile saltpeter, in which it is present as *sodium iodate*, $NaIO_3$.

The iodate may be reduced by sodium hydrogen sulfite:

$$2IO_3^- + 5HSO_3^- = 3HSO_4^- + 2SO_4^= + H_2O + I_2$$

A more rational procedure is first to precipitate copper(I) iodide by adding sodium hydrogen sulfite and copper sulfate solution:

$$IO_3^- + 3HSO_3^- = 3HSO_4^- + I^-$$

$$2I^- + 2Cu^{++} + HSO_3^- + H_2O = 2CuI + HSO_4^- + 2H^+$$

The iodine is then liberated by heating the copper(I) iodide with sulfuric acid and pyrolusite or iron(III) oxide:

$$2CuI + 6H_2SO_4 + 2Fe_2O_3 = 2CuSO_4 + 4FeSO_4 + 6H_2O + I_2.$$

After filtering off the iodine which is precipitated, the mixture of copper and iron(II) sulfates can be used once more to precipitate copper(I) iodide.

(c) Physical Properties

(*i*) *Fluorine* is a gas which has a weak greenish yellow color in thick layers. Its smell is penetrating and extremely pungent, but rather reminiscent of hypochlorous acid. Its density is 1.32, relative to air = 1; 1 liter weighs 1.71 g. Fluorine condenses below —187.9° to a pale yellow liquid (density 1.108), having a rather high refractive index. It solidifies at —223°.

The *energy of dissociation* of the fluorine molecule is not yet known with absolute certainty. If has been deduced from the band spectrum of the ClF molecule that this has a dissociation energy of 58.9 kcal per mol at 0° K. When this is combined with the calorimetrically measured heat of formation of ClF (11.6 kcal per mol) and the dissociation energy of the Cl_2 molecule (Table 104), it gives a value of 37.4 kcal per mol (at 0° K) for the dissociation energy of the F_2 molecule. This value is very much smaller than that which was formerly adopted (64.4 kcal per mol). It is in accord, however, with Wicke's recent (1953) measurements of the dissociation energy of fluorine by the explosion method (determination of the maximum pressure occurring in the explosion of $H_2 + F_2$ mixtures). For the heat of formation of ClF, Wicke obtained the same value by the explosion method as he had found calorimetrically. Measurements of the temperature dependence of the thermal conductivity of fluorine (Wicke, 1951) had yielded a higher value for the dissociation energy of the F_2 molecule. Wicke has considered that these measurements may have been attended with a systematic error, in that the attainment of dissociation equilibrium may have been hindered by the high activation energy of the process.

(*ii*) *Chlorine* is a yellow-green, pungent smelling gas, which strongly attacks the mucous membranes and can lead to fatal inflammations of the lungs if it is inhaled even in small amounts. Its liter-weight is 3.220 g, so that it is about $2\frac{1}{2}$ times as dense as air. It is easily liquefied by pressure (vapor pressure 6.6 atm at 20°). It is therefore handled industrially in steel bottles and cylinders, and can even be transported by water in tankers. The boiling point of chlorine at atmospheric pressure is —34.0°, and its melting point —102.4°; crit. temp. 143.5°, crit. press. 76.1 atm, crit. density 0.57.

(*iii*) *Bromine* is a heavy, deep-brown liquid (density = 3.187 at 0°), which evolves red-brown vapors and has a biting unpleasant smell. Liquid bromine produces painful wounds in contact with the skin. The vapor strongly attacks the mucous membranes, and is dangerous even in high dilution. Bromine boils at $+ 58.8°$ and freezes at $—7.3°$ to a brown crystalline mass with a weakly metallic luster.

(*iv*) *Iodine* is solid even at ordinary temperature, forming grey-black flakes with a metallic luster (density 4.942). Although it boils only at 184.5°, it is already perceptibly volatile at ordinary temperature and has a peculiar smell. Iodine vapor is also poisonous, and produces violent catarrhal inflammations of the mucous membranes of the nose and eyes. Iodine melts at 113.7°, but vaporizes so readily below this temperature, that it often sublimes without melting if it is not too rapidly heated.

Villard (1896) and Dewar (1898) made the curious observation that the vapor pressure of bromine and iodine is higher in the presence of foreign gases (e.g., N_2, O_2, CO_2) than in a vacuum. This observation was confirmed by Braune and Strassmann (1929).

(d) Chemical Properties

(*i*) *Fluorine* [*1, 2*] is extraordinarily reactive chemically. When mixed with hydrogen it usually ignites spontaneously* (even in the dark), and usually with a violent explosion. It also combines in the cold with bromine, iodine, sulfur, phosphorus, arsenic, antimony, boron, silicon, wood charcoal,** and with many metals, producing a flame or vigorous incandescence. Many metals, such as copper, are only superficially attacked at ordinary or slightly elevated temperatures, since the coating which is formed hinders further reaction. However, the reaction with such metals (e.g., zinc, tin, aluminum) goes to completion in some cases with lively incandescence, when they are more strongly heated. Even gold and platinum are strongly attacked by fluorine at a red heat. Most chemical compounds are also decomposed by fluorine, as also are glass and quartz, unless their surfaces are rigorously freed from adsorbed water. The reaction of fluorine with amorphous silica may be attended with incandescence; silicon tetrafluoride is formed and oxygen is displaced. In many of these reactions water exerts a strong catalytic effect, and perfectly dry fluorine can, indeed, be safely manipulated in dry glass or silica apparatus. Stopcocks and ground glass joints in such experimentation may be lubricated with suitable fluorocarbon greases. Hydrogen sulfide and ammonia inflame in fluorine, and the hydrogen halides (other than hydrogen fluoride) are also decomposed. Fluorine liberates oxygen from water, some appearing in the form of ozone.

All these reactions arise from the strong tendency of fluorine to pass into the negatively charged state, whereby fluorine is able to abstract electrons from almost all other elements. Only oxygen and nitrogen do not react directly with fluorine; chlorine does so only upon heating. However, if these elements are present in the negatively charged state, fluorine abstracts from them their excess electrons, and liberates then in the elementary state:

$$Na^+Cl^- + \tfrac{1}{2}F_2 = Na^+F^- + \tfrac{1}{2}Cl_2$$

$$Si^{4+}(O^{2-})_2 + 2F_2 = Si^{4+}(F^-)_4 + O_2$$

(*ii*) *Chlorine* is also a highly reactive element, but is much less reactive than

---

* See footnote 1, p. 40.

** Graphite reacts with fluorine only at a red heat, and diamond is not attacked up to 700°.

fluorine. Like the latter, it reacts directly with most of the elements, but less vigorously. It does not react directly with carbon, or with oxygen or nitrogen, but can be brought into combination with all these elements by indirect means. It forms several oxides, although all are very unstable. It substitutes chlorine (partially or completely) for hydrogen in hydrocarbons, whereby one atom of each Cl_2 molecule unites with the hydrogen atom, and the other combines with the free valence formed on the carbon atom—e.g.:

$$CH_4 + Cl_2 = CH_3Cl + HCl$$

It adds on directly to unsaturated organic compounds—e.g.:

$$H_2C=CH_2 + Cl_2 = ClH_2C—CH_2Cl$$

Chlorine has a bleaching action on most organic dyestuffs, since hypochlorous acid, HClO, formed as an intermediate product, destroys the dyestuff by oxidation.

The solubility of chlorine in water is considerable (2.3 volumes in 1 volume at 20°). It forms an addition compound with water, having the theoretical composition $Cl_2 \cdot 5\frac{3}{4}H_2O$ (observed $5.9 \pm 0.3\ H_2O$, von Stackelberg, 1949; for structure cf. p. 220), which separates in greenish yellow crystals when chlorine is passed into ice cold water. It is stable only up to $+9.6°$ in an open vessel.

Chlorine enters to some extent into reversible reaction with water, forming hypochlorous and hydrochloric acids:

$$Cl_2 + H_2O \rightleftharpoons HOCl + HCl \qquad or \qquad Cl_2 + OH^- \rightleftharpoons HOCl + Cl^-$$

(see Morris, *J. Am. Chem. Soc.*, 68 (1946) 1692, for the mechanism of this reaction). The proportion of the chlorine entering into this reaction is quite considerable. At 0° in a solution containing 25 millimols of chlorine per liter, 49% is present as hypochlorous acid and hydrochloric acid, and 89% is so present in a 5-millimolar solution.

The hypochlorous acid gradually decomposes into hydrochloric acid and oxygen:

$$HOCl = HCl + \tfrac{1}{2}O_2$$

This reaction is considerably accelerated by light, and the evolution of oxygen from chlorine water in bright sunlight is readily detectable. The phenomenon was accordingly observed at an early stage in chemical history, and contributed largely to the view, at first held by most chemists, that chlorine was a compound of oxygen.

Chlorine may be obtained in the atomic state by the action of the electric discharge at low pressures (Rodebush 1933, Schwab 1933). Since chlorine is already a highly reactive substance, the reactions of atomic chlorine introduce few special features.

(*iii*) *Bromine* is very similar to chlorine in its chemical behavior, but is less reactive. In both the vapor and liquid states it unites directly with most elements, and often with inflammation—e.g., with phosphorus, arsenic, antimony, bismuth and tin (as tin foil). Aluminum is also vigorously attacked. Gold and platinum behave very differently towards bromine. Whereas gold is readily transformed into the tribromide, $[AuBr_3]_2$, platinum is quite resistant to attack. Bromine also corrodes metallic sodium only weakly, even at 200° (in strong contrast to its behavior toward potassium—cf. p. 154). Like chlorine, bromine undergoes substitution reactions with organic compounds. Its bleaching action is weaker than that of chlorine. It forms an orange red adsorption product with starch paste.

Bromine is rather soluble in water (3.55 g in 100 g of water at 20°). Below 6.2°

it forms a hydrate, which has the composition $Br_2 \cdot 7\frac{2}{3}$ H_2O, according to von Stackelberg (1949–52; see p. 220). It is more soluble in many organic solvents than in water—e.g., in chloroform and carbon disulfide.

Bromine water, like chlorine water, evolves oxygen when exposed to sunlight:

$$Br_2 + H_2O = HOBr + HBr, \qquad HOBr = HBr + \tfrac{1}{2}O_2$$

Atomic bromine, which was obtained by Schwab (1934) in the same way as atomic chlorine, has a particularly high efficiency of recombination. It was found that recombination of Br-atoms, to form Br_2 molecules, took place at every collision with the wall of the container, irrespective of the material of the wall.

(*iv*) *Iodine* is much less reactive than chlorine or bromine. It combines directly and vigorously with a number of the elements (e.g., with sulfur, phosphorus, iron, and mercury), but has only a small tendency to combine with hydrogen.

At ordinary temperature, hydrogen and iodine do not combine with measurable velocity. At higher temperatures, an equilibrium is established between hydrogen iodide and its components (cf. Table 105):

$$H_2 + I_2 \rightleftharpoons 2HI.$$

Like the other halogens, iodine does not unite directly with oxygen. It forms the oxide I_2O_5, however, which may be obtained by other means, and which—unlike the oxides of chlorine and bromine—is an exothermic compound.

The solubility of iodine in water is but small (1 part in 5500 at 10°). The solution has a weak brownish yellow color. Iodine is readily soluble in many organic solvents, forming *violet* solutions in carbon disulfide, chloroform and carbon tetrachloride, a *red* solution in benzene, and *brownish* solutions in many organic compounds containing oxygen, such as alcohol, ether and acetone. The violet solutions contain I_2 molecules. It is considered by some that the brown solutions contain compounds of iodine with the molecules of solvent (solvates), but others think it more likely that the differences in color of iodine solution arise from differences in the forces of interaction between I_2 molecules and solvent molecules, and the resultant perturbations of molecular energy levels, without involving the formation of any stoichiometric addition compounds (Kortüm, 1947; Rees 1951). Iodine forms deep blue adsorption complexes with starch paste and also with the gelatinous precipitates of basic lanthanum or praseodymium acetates.

The solutions of iodine in many organic solvents, such as benzophenone in which it forms a violet solution, possess electrical conductivity. It may be inferred from this that the iodine is partially dissociated into ions:

$$I_2 \rightleftharpoons I^+ + I^-.$$

The reactions undergone by such solutions also indicate that such dissociation occurs. Thus, in benzene or chloroform solution, iodine reacts with silver perchlorate:

$$I_2 + AgClO_4 = AgI + I[ClO_4].$$

The iodine(I) salt can be stabilized by addition of neutral, complex-forming substances such as pyridine, so that it may be separated as the crystalline dipyridine iodine(I) perchlorate, $[I(pyr)_2][ClO_4]$. The nitrate and other salts of such complexes of univalent, positive iodine can also be prepared. Measurements of conductivity and ionic migration have proved the salt-like character of the compounds formed. It has been found that analogous compounds may also be formed by bromine, which may be stabilized as the positive electrolytic ion by complex formation with suitable ligands. The salts are stable in dry air, but are decomposed by water. By the action of sodium hydroxide solution the iodine(I) salts are converted into addition compounds of iodine(I) oxide, which is not known in the free state—e.g., the orange-red $I_2O(pyr)_4$ is formed from the colorless $[I(pyr)_2][NO_3]$ (Carlsohn 1924, 1932; Birckenbach, 1932 and later; Uschakow 1931, 1935).

(e) Uses

(*i*) *Fluorine* has hitherto found but limited application, although very recent technological advances have made it quite a practicable industrial reagent. The 'fluorocarbons', compounds of the general formula C_nF_{2n+2} derived from the paraffin hydrocarbons by complete replacement of hydrogen by fluorine, have physical properties very similar to those of the hydrocarbons, but are not inflammable and cannot be oxidized. They have uses as lubricants and solvents for special purposes (cf. p. 782). In the past, fluorine has been of use only in the laboratory, either for the preparation of fluorides not readily obtainable by other means (Ruff), or as an extremely powerful oxidant (Fichter) [*1*]. Chlorine, bromine and iodine, however, have many uses in the free state.

(*ii*) *Chlorine* is used for the preparation of many chlorine compounds, and especially of *bleaching powder* ('chloride of lime'), used in large quantity in bleaching. Chlorine is being used to an increasing extent for the synthetic preparation of hydrochloric acid, and large quantities are used for the sterilization of drinking water. Sewage is also chlorinated to remove products of decay and odorous substances. Chlorine is also being consumed in increasing quantities for de-tinning tin plate, and for the preparation of chlorinated organic compounds used as solvents for lacquers and synthetic resins. Its use for the manufacture of bromine has already been discussed.

(*iii*) *Bromine* is used for the preparation of inorganic bromides and many organic bromine compounds. It is often used in the laboratory for analytical and preparative purposes, in place of chlorine, since it is more conveniently manipulated. It can also be used as a disinfectant. Potassium bromide and bromate are used as reagents in volumetric analysis (bromometry).

(*iv*) *Iodine* is used in analytical chemistry (in iodometry), and also for the preparation of numerous inorganic and organic compounds of iodine. Many of these, as also iodine itself, are used in medicine, since iodine has excellent antiseptic properties and is also a coagulant for the blood. It has an irritant action on the skin, however. Large amounts are poisonous when taken internally.

3. Compounds of the Halogens

(a) Hydrogen Halides

(*i*) *Properties*. The hydrogen halides are colorless gases which are relatively easily liquefied. Hydrogen fluoride is already liquid below $+19°$. The hydrogen halides are very avidly absorbed by water, and their aqueous solutions are good conductors of the electric current. Hydrogen chloride, -bromide and -iodide are practically completely dissociated in solution:

$$HX \rightleftharpoons H^+ + X^- \tag{1}$$

Hydrofluoric acid is dissociated to a considerably smaller extent, as is evident from the conductivity quotients $\alpha = \Lambda_c/\Lambda_0$, shown in Table 105, although these represent only the apparent degrees of dissociation.

Since hydrofluoric acid is a relatively weak acid, its electrolytic dissociation obeys the law of mass action. However, the F^- ions produced by the dissociation associate to some extent with undissociated HF molecules:

$$F^- + HF \rightleftharpoons HF_2{}^-,$$

especially in concentrated solutions. Hence the ionic concentrations in solutions of hydrofluoric acid are determined by both the latter equation and eqn. (1). The corresponding equilibrium constants are not very reliably known. According to the measurements of Pick (1912), Hudleston (1924) and Roth (1937–39), the following approximate values may be assumed:

$$\frac{[H^+] \cdot [F^-]}{[HF]} = 7 \cdot 10^{-4} \qquad\qquad \frac{[HF_2{}^-]}{[F^-] \cdot [HF]} = 5$$

In a 1-normal solution, about 10% of the hydrogen fluoride is present in the form of $HF_2{}^-$ ions, and 1% in the form of F^- ions. Thus in concentrated solution the most important equilibrium is

$$2HF \rightleftharpoons H^+ + HF_2{}^- \tag{2}$$

It was formerly assumed that, in not too dilute solution, hydrogen fluoride must be present chiefly in the form of double molecules $(HF)_2$, and that these dissociated like a dibasic acid. This assumption has not been confirmed, however. Thus the molecular conductivity of the neutral fluorides, and its dependence upon concentration, shows that these are the salts of a monobasic acid, not of a dibasic acid. The so-called *acid fluorides* or *hydrogen fluorides*, of the general formula $M^I HF_2$, can be quite satisfactorily interpreted as coordination compounds $M^I[HF_2]$, with hydrogen as the central atom of the complex. Fluoride ions interact with the hydrogen ion to a much greater extent than do the other halide ions, because the small radius of the fluoride ion, as compared with the other halide ions, confers on the resulting complex a much greater energy of formation.

The processes represented by equations (1) and (2) are more precisely formulated as

$$HX + H_2O + aq \rightleftharpoons [H_3O]^+ + X^-.aq \tag{3}$$

and

$$2HF + H_2O \qquad \rightleftharpoons [H_3O]^+ + [HF_2]^- \tag{4}$$

The electrolytic dissociation of the hydrogen halides in solution thus depends upon the fact that although the water molecule as a whole is electrically neutral, its attraction for a proton is not very much weaker than that of the singly charged negative halogen ion. Since additional energy is recovered from the hydration of the halogen ions, the reaction (3) as a whole is exothermic.* Similar considerations apply to other strong acids. Dissociation in non-aqueous solutions, where it occurs, is also due to the addition of a proton to the molecule

---

* Thus, from the data given in Tables 104 and 105 and p. 156, we find:

$$
\begin{aligned}
HCl &= H^+ + Cl^- \quad - 328.6 \text{ kcal} \\
H^+ + H_2O + aq &= [H_3O]^+.aq + 250 \quad \text{kcal} \\
Cl^- + aq &= Cl^-.aq \qquad + 96.9 \text{ kcal} \\
\hline
HCl + H_2O + aq &= [H_3O]^+.aq + Cl^-.aq + 18.3 \text{ kcal}
\end{aligned}
$$

The heats of hydration of ions are obtained experimentally by the converse process, from the heats of solution of the compounds and the work required to separate their ions.

TABLE 105

PROPERTIES OF THE HYDROGEN HALIDES

Compound	HF	HCl	HBr	HI
Melting point	$-83.1°$	$-114.8°$	$-86.9°$	$-50.7°$
Boiling point	$+19.54°$	$-84.9°$	$-66.8°$	$-35.4°$
Critical temperature	$+230.2°$	$+51.3°$	$+91.0°$	$+150.5°$
Heat of fusion at the m.p. (in kcal/mol HX)	1.094	0.505	0.575	0.686
Latent heat of vaporization at the b.p. (kcal/mol HX)	1.850	3.85	4.210	4.72
Trouton's constant	6.3	20.4	20.4	19.9
Density at the b.p.	0.991	1.187	2.160	2.799
Surface tension at the b.p. (dynes per cm)	—	23.18	25.40	26.69
Liter weight in g (at 0°)	—	1.6391	3.6443	5.7888
Heat of formation H (at 20° and const. press.), kcal/mol HX	$+64.4$	$+21.9$	$+7.3$	$+1.32$[1]
Free energy of formation $F°$ at 25° (kcal mol)	-65.0	-22.75	-12.86	-1.85[1]
Degree of thermal dissociation at 1 atm. pressure {at 300° C / at 1000° C	— / —	$0.3 \cdot 10^{-6}\%$ / 0.014%	0.003% / 0.5%	19% / 33%
Heat of solution at 20°, kcal/mol	$+12.4$	$+17.4$	$+20.5$	$+19.6$
Apparent degree of electrolytic dissociation, in 0.1-normal solution at 18°	10%	92.6%	93.5%	95%
Work of separation of ions (HX $\rightarrow$ H$^+$ + X$^-$ in a vacuum) kcal/mol	$+364.2$	$+328.6$	$+327.7$	$+316.3$
Dipole moment of gas molecule (in c.g.s. units $\times 10^{18}$)	1.91	1.04	0.79	0.38
Moment of inertia of gas molecule (g cm^2 $\times 10^{40}$)	1.35	2.65	3.26	4.31
Internuclear distance H—X in gas molecule (Å)	0.92	1.276	1.410	1.62
Molar magnetic susceptibility at 0°, $\times 10^6$	-8.6	-22.1	-32.9	-47.7

[1] This value is for the formation of HI from *gaseous* iodine. For formation from *solid* iodine $H_{293°} = -6.23$ and $F_{298°} = +0.46$ kcal per mol HI (at constant pressure)

of the solvent. The *pure hydrogen halides do not conduct the electric current*, and they are not very reactive at ordinary temperature if water is absent.

Doubts have therefore been raised as to an ionic structure for the hydrogen halides, especially as other evidence appears to support a covalent structure. In the first place, the energy which must be expended to dissociate the crystalline compounds into free (i.e., widely separated) ions—that is the sum of the sublimation energies and the work of separation of the ions (Table 105)—is much greater in the case of the hydrogen halides than it is for the ionic lattices present in the alkali halides (cf. lattice energies, p. 156). This is so even though the molecular volumes of the alkali halides and of the crystalline hydrogen halides are about the same. Secondly, the small value of the sublimation energy, as compared with the energy of separation of the ions, appears to indicate a covalent structure. In the case of the alkali halides, the sublimation energies are of the order of 30% of the lattice energies, whereas for the hydrogen halides they are only 1–2%. This considerable difference expresses itself in the much greater volatility of the hydrogen halides. Thirdly, the great difference in melting point between hydrogen halides and alkali halides suggests that the former are covalent compounds. In themselves these properties, and others related to them, do not afford completely unambiguous evidence that the binding forces are covalent —i.e., that wave mechanical resonance forces completely outweigh those due to the electrostatic attraction between separated charges. They might also be explained as arising from the established fact that in the hydrogen halides the proton is drawn right into the electron cloud of the negative ion. That this is the case is shown by the values found by various methods (e.g., by band spectroscopy) for the moments of inertia of the molecules. The internuclear distance between the proton and the nucleus of the halogen atom is given in Table 105, and it may be seen that the values are significantly less than the ionic radii of the halogens (Table 104). The penetration of the proton into the interior of the ions confers on the hydrogen halide molecules a quasi-inert gas structure, but unlike the inert gases, the hydrogen halides have fairly large dipole moments. These two view points represent alternative extremes of approximation. In the wave mechanical treatment of valence, account must be taken not only of the sharing of electrons (Heitler-London-Pauling-Slater), but also of the probability of finding both bonding electrons close to same nucleus. Especially in molecules such as the hydrogen halides, where the atoms differ considerably in electronegativity, these 'ionic terms' in the wave equation are important. Smyth (1946) concluded from the dipole moments of the gaseous hydrogen halides that in HF the binding was 43% ionic, in HCl 17%, in HBr 12% and in HI 5% ionic.

The solid hydrogen halides form molecular crystal lattices. HCl and HBr crystallize above 98° K and 111 °K, respectively, as face-centered cubic solids (like the inert gases), and at lower temperatures they are rhombic. HBr forms another, rather complex, structure above 119° K. HI crystallizes face-centered tetragonal (Simon 1924; Natta 1931). In addition to the transformation points associated with changes in structure, the crystalline hydrogen halides show other transitions corresponding to the onset of free molecular rotation as the temperature is raised.

The trend of the melting points, boiling points, critical temperatures, etc., in the hydrogen halides shows the same anomaly as in the hydrides of Group VI (cf. p. 687). Just as, in that case, the anomaly arose from the polymerization of water, so in this case it is due to polymerization in liquid hydrogen fluoride. Instead of being abnormally high, as for water however, the Trouton constant for hydrogen fluoride is abnormally low. This is because hydrogen fluoride is also polymerized to a considerable extent in the *gaseous* state at ordinary pressures. Dissociation into single HF molecules takes place when the pressure is reduced. The latent heat of vaporization of hydrogen fluoride accordingly increases markedly as the pressure is lowered. At 20 mm pressure it is 390 cal per g, or 7.80 kcal per mol of HF (Fredenhagen, 1934). From this, $\lambda/T_s = 26.6$—i.e., an abnormally high value for the Trouton constant, as would be expected for polymerization in the liquid state.

The *thermal dissociation* of the hydrogen halides, according to the general equation

$$2HX \rightleftharpoons H_2 + X_2,$$

increases from hydrogen chloride to hydrogen iodide. Whereas, for hydrogen chloride and hydrogen bromide, no decomposition into the components is directly detectable at 300°,* hydrogen iodide is decomposed to a considerable degree at this temperature. The affinity of the halogens for hydrogen thus diminishes greatly with increase in atomic weight, and this expresses itself also in the large decrease in the *heat of formation* from hydrogen fluoride to hydrogen iodide. The heat of formation of these compounds differs but little from the free energy of formation.

(*ii*) Preparation. The following methods are available for the *preparation of the hydrogen halides.*

(a) Direct combination of the constituents—e.g.:

$$Cl_2 + H_2 = 2HCl \tag{1}$$

This method, which is the simplest in principle, is of considerable technical importance for the manufacture of hydrochloric acid.

* The values cited in Table 105 are calculated from those determined at higher temperatures, and the temperature variation of the equilibrium constant.

(b) Action of acids on halides.

This is the most important method for the preparation of hydrofluoric acid, and is still the most widely used process for hydrochloric acid:

$$CaF_2 + H_2SO_4 = CaSO_4 \quad + 2HF \tag{2}$$

$$NaCl + H_2SO_4 = NaHSO_4 + HCl \tag{3}$$

(c) Hydrolytic decomposition of halides:

$$PBr_3 + 3H_2O = H_3PO_3 + 3HBr \tag{4}$$

(d) Action of hydrogen compounds, such as H_2S, on the free halogens in aqueous solution:

$$I_2 + H_2S \cdot aq = 2HI \cdot aq + S + 17.5 \text{ kcal} \tag{5}$$

The last two methods are used principally for the preparation of hydrobromic and hydriodic acids. Reaction (5) depends on the fact that the heat of solution of HI is considerably larger than that of H_2S. In the gaseous state the reaction occurs in the opposite direction:

$$2HI + S = I_2 + H_2S + 5.0 \text{ kcal}$$

This can be utilized for the preparation of pure H_2S.

(iii) *Hydrogen fluoride* is generally prepared by warming calcium fluoride (fluorspar) with concentrated sulfuric acid. Since the compound attacks glass, it must be prepared in platinum or lead vessels. Anhydrous hydrogen fluoride can be prepared by heating potassium or sodium hydrogen fluoride:

$$KHF_2 = KF + HF$$

Below 19.54°, hydrogen fluoride forms a colorless, very mobile, fuming liquid. Just above this temperature it is a colorless gas, which is highly polymerized at atmospheric pressure. Although the vapor of hydrogen fluoride departs widely from an ideal gas in its behavior, the vapor pressure of liquid hydrogen fluoride can be represented by a single law:

$$\log_{10} p \text{ (mm)} = -1.91173 - \frac{918.24}{T} + 3.21542 \log_{10} T.$$

Hildebrand (1924, 1943) found that the apparent molecular complexity of hydrogen fluoride vapor could be represented in terms of an equilibrium involving only HF and $(HF)_6$ (presumably cyclic) molecules. However, this model is almost certainly incorrect, although it is compatible with the gas density data, since no evidence for $(HF)_6$ cyclic molecules could be found by electron diffraction (Bauer, Beach and Simon, 1939). Fredenhagen (1932) adduced some evidence for the existence of $(HF)_2$ and larger chain polymer molecules. According to Oriani and Smyth (1948), the dipole moment of hydrogen fluoride vapor is compatible only with the existence of chain polymers, and it is likely that in the vapor there is an equilibirium between a series of polymers, $(HF)_2$, $(HF)_3$, ..., linked by hydrogen bonds.

Diminution of the pressure, or elevation of the temperature, leads to a rapid dissociation into simple HF molecules, and at 90° the gas consists almost exclusively of single molecules.

Hydrogen fluoride is also strongly associated in the liquid state. The liquid is a nonconductor of electricity. The dielectric constant of hydrogen fluoride is almost as great as that of water, according to Fredenhagen (1929). Liquid hydrogen fluoride can be stored in silver, copper or steel vessels.

Hydrogen fluoride is very soluble in water, and miscibility is complete at ordinary temperature. The solution, known as *hydrofluoric acid* [3], dissolves most metals with the evolution of hydrogen, and also attacks glass, since moist hydrogen fluoride reacts with silicon dioxide and silicates to form volatile silicon tetrafluoride, SiF_4. Absolutely dry hydrogen fluoride does not attack glass, even when liquefied, and reacts with most metals only when heated. The use of hydrofluoric acid to attack silicates and etch glass is based on its ability to decompose silicon dioxide and its compounds. The etching test for the detection of fluorides is also based on the same reaction. Gold and platinum are not attacked by hydrofluoric acid, and lead only superficially. Vessels of platinum, lead and gutta percha are therefore used for the storage of and work with hydrofluoric acid. For many purposes, paraffin-coated glass vessels are adequate, and the use of chemically inert plastics has recently been introduced.

Hydrofluoric acid finds industrial applications for etching glass, for removing silica from pipes, and for removing adhering sand from castings in iron founding. It is also used for the preparation of hydrogen peroxide from sodium peroxide. It has disinfectant properties, and is used in brewing for the preservation of yeast. For the same reason it is also used to preserve anatomical preparations.

Commercial hydrofluoric acid generally contains 40% HF (density 1.130 at 20°). It gives off extremely pungent vapors, which are *highly toxic* if inhaled. The vapors, and more especially the solution, corrode the skin strongly, and can cause very painful wounds which heal badly.

If concentrated hydrofluoric acid is heated to boiling, hydrogen fluoride is first evolved. At the same time the boiling point rises, and at 120° a liquid of constant composition distils, containing 35.4% HF (density 1.118). A more dilute solution ultimately reaches the same concentration, since water distils off first.

(*iv*) *Hydrogen chloride*, or its aqueous solution, hydrochloric acid, was first prepared by Glauber (17th century), by warming common salt with sulfuric acid. The same method of preparation is generally used in the laboratory, and is still the most widely used industrial process. However, the synthetic preparation of hydrochloric acid by burning hydrogen with chlorine is coming increasingly into use.

The reaction of sulfuric acid with common salt proceeds in two stages. In the cold, sodium hydrogen sulfate is first formed according to eqn. (3), p. 789, with evolution of hydrogen chloride. When heated (to a red heat) the reaction proceeds further, hydrogen chloride being formed afresh by the reaction of the sodium acid sulfate with sodium chloride:

$$NaCl + NaHSO_4 = Na_2SO_4 + HCl.$$

Hydrogen chloride is obtained synthetically, from the gaseous hydrogen and chlorine which are the by-products of the electrolysis of alkali-chlorides. The gases are burned in a quartz flue, and furnish an especially pure hydrochloric acid.

Hydrochloric acid is a colorless gas, with a pungent smell and taste (see Table 105 for physical constants). From its density relative to oxygen ($= 1.1471$), the molecular weight is 36.71, whereas the formula HCl requires 36.47. Hydrogen chloride accordingly consists at ordinary temperature of single molecules, HCl. It is relatively easily liquefied. Its gas density is also approximately normal in the neighborhood of its liquefaction temperature.

Hydrogen chloride is avidly absorbed by water, in large amounts and with the evolution of much heat (Table 105). At ordinary temperature, 1 volume of water absorbs 450 volumes of hydrogen chloride at atmospheric pressure.

When the solution is strongly cooled, hydrates separate—namely $HCl \cdot 3H_2O$ (m.p. —24.9°), $HCl \cdot 2H_2O$ (m.p. —17.6°), and $HCl \cdot H_2O$ (m.p. —15.3°), according to the composition of the solution. A solution saturated at 0° under atmospheric pressure furnishes only the trihydrate with a content of 40.3% HCl. The other hydrates are produced from solutions standing under an excess pressure of hydrogen chloride.

A solution saturated at atmospheric pressure with hydrogen chloride contains 45.4% at 0°*, or 42.7% HCl at 15°. If such a solution is heated, it first loses hydrogen chloride, until at about 110° a mixture of constant composition distils over, containing 20.24% HCl. A mixture of the same composition is ultimately obtained if one starts with a more dilute solution. However, as Roscoe showed in 1859, the composition of the constant boiling mixture is dependent upon the pressure at which the distillation is carried out. It is therefore not a chemical compound, as had previously been believed (cf. p. 596).

The name *hydrochloric acid* is generally applied to aqueous solutions of hydrogen chloride. Their composition is customarily determined by means of a hydrometer. A hydrochloric acid of

sp.gr.	1.060	1.124	1.16	1.19 at 15°
contains	12.2	24.8	31.5	37.2 % HCl

Solutions containing, e.g., 12.2% HCl, are known commercially as 'dilute acid', and those with 24% HCl and upwards as 'concentrated hydrochloric acid'. The dilute hydrochloric acid of the laboratory is generally 2-normal (7%, sp. gr. 1.035). As the figures given above serve to show, the percentage composition of hydrochloric acid solutions is obtained with some accuracy by the expression:

$$\% \ HCl = 200 \times (sp.gr. - 1)$$

Pure hydrochloric acid is quite colorless, but the crude hydrochloric acid of commerce is colored yellow by impurities, principally ferric chloride.

Hydrochloric acid has a very high electrolytic conductivity. Its specific conductivity at 18° has the following values at various concentrations:

Percentage HCl	5	10	20	30	40
$\varkappa$	0.395	0.630	0.7615	0.662	0.515

The specific conductivity thus displays a maximum in the middle concentration range, as it does for sulfuric acid.

The chemical properties of hydrochloric acid are essentially those of a *typical strong acid*. As such it finds extensive industrial uses in the preparation of metallic chlorides, for the production of ammonium chloride, and for many other purposes. Hydrochloric acid is one of the most important laboratory reagents.

Gaseous hydrogen chloride is often used as a dehydrating agent, or as a condensation agent, because of its strong affinity for water.

☞

* If, as is usual in the laboratory, very concentrated hydrochloric acid is prepared by passing HCl gas into an open vessel cooled in ice water, this value will not be quite attained, but a content of about 42.5% HCl is usually reached.

Hydrogen chloride is copiously soluble in alcohol and ether, as well as in water, and also in many other liquids. Conversely, liquid hydrogen chloride can serve as a solvent for alcohols, ethers and many other substances. Most metals are unattacked by liquid hydrogen chloride, and it is in general without action on oxides, sulfides, and carbonates. Gaseous hydrogen chloride reacts with metals at a red heat, liberating hydrogen. Metals such as copper and silver, which are not attacked by aqueous hydrochloric acid in the absence of air, also react under these conditions. Whereas hydrogen chloride reacts immediately with fluorine at ordinary temperature, and inflames, it reacts with atmospheric oxygen only in the presence of catalysts, and then only at elevated temperatures. The equilibrium

$$4HCl + O_2 \rightleftharpoons 2H_2O + 2Cl_2$$

is displaced to the left, however, with rise of temperature, in such a manner that the equilibrium constant of the mass action law:

$$\frac{p^4_{HCl} \cdot p_{O_2}}{p^2_{H_2O} \cdot p_{Cl_2}} = K$$

reaches a value of unity at about 600°. Thus chlorine is the stronger oxidant above this temperature, and oxygen at lower temperatures.

(v) *Hydrogen bromide* can be prepared by the action of potassium bromide and sulfuric acid (cf. hydrogen chloride), but the latter reagent must be only moderately concentrated (sp. gr. 1.4) as the hydrogen bromide evolved is otherwise partially oxidized to bromine. The aqueous solution of hydrobromic acid obtained by this means can be concentrated by fractional distillation to a content of 48% HBr.

More concentrated acid or hydrogen bromide gas, can be obtained by hydrolysis of phosphorus tribromide (eqn. (4) p. 789). This reaction is usually carried out by placing red phosphorus in a flask and covering it with water. Bromine is added drop by drop, and the mixture is cautiously warmed, whereupon HBr passes over. A convenient method for preparing aqueous hydrobromic acid consists of treating barium sulfide with a mixture of bromine and water:

$$4Br_2 + 4H_2O + BaS = BaSO_4 + 8HBr$$

The reaction is attended with a considerable evolution of heat. When it is complete, the aqueous acid is distilled off.

Hydrogen bromide is a colorless gas, which is very similar to hydrogen chloride in its properties. It is more readily oxidized, however, so that the aqueous solution gradually turns brown when exposed to light.

The solubility of hydrogen bromide in water is very great (600 volumes in 1 volume of water at 0°). Several crystalline hydrates are formed at low temperatures: $HBr \cdot 4H_2O$ (m.p. —56°), $HBr \cdot 3H_2O$ (m.p. —48°), $HBr \cdot 2H_2O$ (m.p. —11.3°) and $HBr \cdot H_2O$. The latter decomposes at —28.5° under 1 atm pressure.

(vi) *Hydrogen iodide* is yet more readily oxidized than hydrogen bromide, and may be considered as a definite reducing agent. In other respects it is very similar to hydrogen chloride and hydrogen bromide. Like them, it is a colorless pungent gas, and is a strong acid in aqueous solution.

1 volume of water will absorb about 425 volumes of hydrogen iodide gas at 10° under atmospheric pressure. Three hydrates exist: $HI \cdot 4H_2O$ (m.p. —36.5°), $HI \cdot 3H_2O$ (m.p. —48°) and $HI \cdot 2H_2O$ (m.p. about —42°). A solution saturated with hydrogen iodide at ordinary temperature under atmospheric pressure contains about 70% HI, and has a

specific gravity of 2.00. A constant boiling mixture, boiling at 127° and containing 57.0% HI (sp. gr. 1.70) is obtained by distillation. Concentrated hydriodic acid dissolves silver with the evolution of hydrogen. The occurrence of this reaction, even though silver lies below hydrogen in the electrode potential series, is due to the formation of a fairly strongly complex H[AgI$_2$], so that the silver ion concentration in the solution is vanishingly small. Copper is not dissolved by hydriodic acid.

Both as a gas and in aqueous solution, hydrogen iodide is completely stable at ordinary temperature if oxygen is excluded. Slow oxidation to iodine takes place, however, if oxygen is admitted. Hydriodic acid solutions, and especially those of high concentration, soon turn brown in air. The process is accelerated by light. The brown coloration can be inhibited by adding copper turnings, which combine with the liberated iodine, forming insoluble copper iodide, CuI.

Hydrogen iodide exerts a reducing action on alkyl iodides and alcohols when heated, and a hydrogenating action on unsaturated compounds:

$$C_2H_5I \quad + \quad HI = C_2H_6 + I_2$$

$$C_2H_5OH + 2HI = C_2H_6 + H_2O + I_2$$

$$C_5H_{10} \quad + \quad 2HI = C_5H_{12} + I_2$$

Hydrogen iodide is generally prepared by the method used for hydrogen bromide—i.e., by hydrolysis of phosphorus iodides. The usual procedure is to drop a suspension of red phosphorus (5 parts) in water (10 parts) into iodine (100 parts) moistened with water (10 parts). Phosphorus pentaiodide is formed primarily, but this is at once hydrolyzed into phosphoric acid and hydrogen iodide:

$$2P + 5I_2 = 2 PI_5, \qquad PI_5 + 4H_2O = H_3PO_4 + 5HI.$$

Moderately concentrated solutions of hydriodic acid can be conveniently obtained by passing hydrogen sulfide into a suspension of iodine in water (cf. eqn. (5) p. 789). The excess of hydrogen sulfide is removed by sweeping out with carbon dioxide, and the precipitated sulfur is filtered off; the solution may first be boiled, if necessary, to coagulate the sulfur.

(b) Halides

(i) *General*. The compounds formed by the halogens with more strongly electropositive elements are known as *halides*. The halides of the metals have, for the most part, a distinctly salt-like character, and are the salts of the hydrohalogen acids. The halides of the light metals, in particular, are typical salts. The halides of the non-metals—e.g., CCl$_4$, SiCl$_4$, S$_2$Cl$_2$ —are quite different in character. Unlike the metallic halides, they are usually volatile, and are either insoluble in water or are immediately hydrolyzed. The halides of the metals are mostly soluble in water, and (apart from the exceptions such as HgCl$_2$, HgBr$_2$, HgI$_2$) are almost completely dissociated* in solution. As is usual for the salts of weak bases with strong acids, the dissociation of halides derived from weak bases is accompanied by a partial hydrolysis, but this is not very extensive in moderately dilute solutions.

With respect to their solubilities, the fluorides behave rather differently from the

---

* The apparent degrees of dissociation determined from the conductivity quotients Λ_c/Λ_0 are considerably smaller than the true degrees of dissociation, especially in the case of compounds of the multivalent elements, as a result of interionic forces. In addition to this, autocomplexes are not infrequently formed (cf. Vol. II).

other halides. Whereas *silver* chloride, bromide and iodide are especially insoluble, and the alkaline earth salts are very soluble and even deliquescent, the alkaline earth *fluorides* are very sparingly soluble (especially calcium fluoride), and silver fluoride is extremely soluble and deliquescent.

Sparingly soluble salts are the halides of univalent mercury, univalent copper, and univalent gold (in so far as they are known), and also lead fluoride. Lead chloride, lead bromide and lead iodide are also sparingly soluble in the cold, but much more soluble in hot solutions.

The metallic halides may be obtained by the general methods for the formation of salts, discussed in the next chapter. The preparation of the non-metallic halides has already been discussed in relation to the individual components.

The halides are among the most important compounds of the individual elements. This is especially true of the chlorides. The chlorides of the elements so far dealt with have already been discussed, and most of the other halides have also been dealt with. It is therefore necessary to describe only the fluorides, bromides, and iodides of the alkali metals and of ammonium.

The iodine compounds occupy a special position among the organic halides, in that they give rise to *iodonium compounds*. These are compounds of the type $[R_2I]X$ (where R = organic radical, especially aryl, and X = univalent acidic or hydroxyl group). These stand in the same relation to the simple alkyl or aryl halides as the ammonium compounds to ammonia. They are salt-like in character, and the iodonium radical is present in the form of the free positive ion $[R_2I]^+$ in their solutions. The iodonium hydroxides, $[R_2I]OH$ are strong bases.

(ii) Fluorides.

1. *Sodium fluoride* (density 2.78, m.p. 988°, b.p. 1700°) is thrown down as a white granular precipitate, consisting of cubes or octahedra, when hydrogen fluoride is passed into a not too dilute solution of sodium hydroxide or carbonate.

Sodium fluoride is prepared technically by fusing cryolite with caustic soda:

$$Na_3AlF_6 + 6NaOH = Na_3[Al(OH)_6] + 6NaF.$$

When the melt is leached, sodium aluminate goes into solution and the fluoride remains undissolved.

The solubility of sodium fluoride in water is rather small (4 parts in 100 parts of water at 15°). It is practically insoluble in alcohol. The aqueous solution has a basic reaction, owing to the occurrence of the reaction

$$2F^- + HOH = [HF_2]^- + OH^-$$

to some extent.

Sodium acid fluoride, $NaHF_2$, crystallizes in colorless rhombohedra when solutions of sodium fluoride containing excess hydrofluoric acid are evaporated. This compound loses hydrogen fluoride when it is heated, and can be used for the preparation of HF (cf. p. 789).

Sodium fluoride is used principally to impregnate timber against rot, and as a disinfectant in the fermentation industry. It has occasional uses in medicine.

2. *Potassium fluoride*, unlike sodium fluoride, is very soluble in water and is deliquescent in air. It can be thrown out of solution by adding alcohol, and is then obtained as the dihydrate, $KF \cdot 2H_2O$. This melts in its water of crystallization at

about 46°. The ahydrous salt melts at 848° and boils at about 1500°. It crystallizes cubic (density 2.48). The aqueous solution is basic in reaction, like that of sodium fluoride. Potassium acid fluoride, KHF_2, (density 2.37) crystallizes from solutions containing an excess of hydrofluoric acid. Moissan was able to prepare other addition compounds, $KF \cdot 2HF$ and $KF \cdot 3HF$, from anhydrous hydrogen fluoride. According to Cady (1934), $KF \cdot 2\frac{1}{2}HF$ and $KF \cdot 4HF$ also exist.

3. *Ammonium fluoride* can be prepared by a method first described by Berzelius, by heating a mixture of 1 part of ammonium chloride with $2\frac{1}{4}$ parts of sodium fluoride. Ammonium fluoride sublimes on gentle heating. It is very soluble in water, and is deliquescent in moist air. The solution is hydrolyzed to a considerable extent, to ammonia and the acid fluoride:

$$2[NH_4]F = [NH_4]^+ + [HF_2]^- + NH_3.$$

For this reason the salt cannot be prepared by evaporating mixed solutions of ammonia and hydrofluoric acid, since ammonia escapes and ammonium hydrogen fluoride, $[NH_4][HF_2]$ (density 1.21, m.p. 124°), crystallizes out. Ammonium hydrogen fluoride can be used for etching glass in the same manner as hydrofluoric acid, but gives coarser etches. It is used as a disinfectant in the brewing and spirit industries, especially for hoses and pipe lines. The neutral salt has medicinal uses.

The addition compounds $NH_4F \cdot 2HF$, $NH_4F \cdot 3HF$ and $NH_4F \cdot 5HF$ may be obtained from anhydrous hydrogen fluoride. They have very low melting points (close to 30°) (Ruff, 1933). $NH_4F \cdot 2HF$ may also be isolated from aqueous solution in the form of a hemihydrate, $NH_4F \cdot 2HF \cdot \frac{1}{2}H_2O$ (Hassel, 1932).

(iii) Bromides.

1. *Sodium bromide*, is generally prepared by double decomposition of soda with iron bromide, Fe_3Br_8:

$$4Na_2CO_3 + Fe_3Br_8 = 8NaBr + Fe_3O_4 + 4CO_2.$$

The process is usually carried out by covering iron turnings and other iron scrap with water. This iron is converted to soluble iron(II) bromide by the gradual addition of bromine. The green solution is decanted off, and treated with one third of the amount of bromine previously added:

$$Fe + Br_2 = FeBr_2, \qquad 3FeBr_2 + Br_2 = Fe_3Br_8.$$

A concentrated solution of the latter compound, obtained by evaporation, is run into a well stirred, boiling solution of soda. The precipitate of hydrated iron(II,III) oxide is filtered off and, after evaporation, anhydrous sodium bromide (density 3.20) crystallizes from the solution above 50.7°. The dihydrate, $NaBr \cdot 2H_2O$ (density 2.18) crystallizes at ordinary temperature. A pentahydrate is stable in contact with the solution between —24° and —28°.

Sodium bromide may also be prepared by saturating sodium hydroxide solution with bromine. The solution is evaporated to dryness, and the residue, consisting of a mixture of sodium bromide and sodium bromate, is heated with powdered wood charcoal:

$$3Br_2 + 6NaOH = 5NaBr + NaBrO_3 + 3H_2O$$

$$2NaBrO_3 + 3C = 2NaBr + 3CO_2$$

Pure sodium bromide forms colorless crystals (m.p. 760°, b.p. 1393°). It is very soluble in water and fairly soluble in alcohol. 100 g of water dissolve 79.5 g of NaBr at 0°, 90.3 g at 20° and 120.5 g at 100°. The boiling point of the saturated solution is 121°. Sodium bromide is used in medicine as a nerve sedative (natrium bromatum).

According to Mattauch (1943), the compound Na_2Br has a transient existence in gas discharges, as shown by its mass spectrographic detection.

2. *Potassium bromide* (colorless cubic crystals, density 2.75; m.p. 748°, b.p. 1380°) is prepared by methods similar to those used for sodium bromide, which it closely resembles in properties. It forms no hydrates. 100 g of water dissolve 54 g of KBr at 0°, 65 g at 20° and 105 g at 100°. 100 g of alcohol dissolve only 0.13 g of KBr at at 25°. Potassium bromide is widely used in medicine as a sedative, and also in the photographic industry (cf. Vol. II).

3. *Ammonium bromide* can be prepared like sodium bromide, by reacting aqueous ammonia with iron bromide, Fe_3Br_8, and also by the careful addition of bromine to ice cold concentrated ammonia solution. The latter reaction is accompanied by the vigorous evolution of heat:

$$\tfrac{3}{2}Br_2 + 4NH_3 = 3NH_4Br + \tfrac{1}{2}N_2 + 105 \text{ kcal.}$$

An excess of bromine must be avoided, as the highly explosive nitrogen tribromide may otherwise be formed:

$$NH_4Br + 3Br_2 = 4HBr + NBr_3$$

Ammonium bromide (density 2.39) is colorless and crystallizes cubic, like potassium bromide, with which it forms mixed crystals. It is readily sublimed and very soluble in water (69.7 g in 100 g of water at 15°). Uses are similar to those for potassium bromide.

(iv) Iodides.

1. *Sodium iodide* is prepared by methods similar to those used for sodium bromide. Like the latter, it crystallizes anhydrous at elevated temperatures (above 65°), but as the dihydrate at ordinary temperature, and as pentahydrate at low temperatures (between —13.5° and —31.5°). Sodium iodide has uses in medicine (natrium iodatum). The anhydrous salt forms colorless cubes, of density 3.66 (density of dihydrate 2.45). It melts at 661° and boils at 1300°. Its solubility is: at 0° 159, at 20° 179 and at 100° 302 g of NaI in 100 g of water. It is also very soluble in absolute alcohol (43 g in 100 g of ethyl alcohol and 78 g in 100 g of methyl alcohol at 23°).

2. *Potassium iodide* is prepared by the methods given for sodium bromide. It crystallizes in colorless cubes (density 3.12, m.p. 677°, b.p. about 1325°). It dissolves in water with considerable absorption of heat. No hydrates are formed.

	0°	20°	100° C	
Solubility:	128	144	209	g of KI in 100 g of water

At 20° 1.75 g of KI dissolve in 100 g of absolute ethyl alochol and 16.5 g in 100 g of methyl alcohol.

Potassium iodide is widely used in medicine (kalium iodatum), although it has an undesirable action on the heart. Sodium iodide is not open to this objection. Potassium iodide is also used in the photographic industry and for the preparation of other iodine compounds.

3. *Ammonium iodide* may be obtained by neutralizing hydriodic acid with ammonia or ammonium carbonate. It forms a white, hygroscopic powder (density 2.44). It is very soluble in water and alcohol (167 g in 100 g of water at 15°). It finds pharmaceutical uses.

(c) Polyhalides

Many of the metal compounds of the heavier halogens are able to add on additional halogen atoms. Thus potassium iodide readily adds on 1 molecule of iodine per gram ion of iodide ion, to form the deep red-brown *potassium triiodide*, KI_3. The addition of yet more iodine leads ultimately to the formation of *potassium enneaiodide*, KI_9:

$$KI + I_2 = KI_3; \qquad KI + 4I_2 = KI_9$$

Many other iodides combine with iodine in a similar manner, giving compounds of the general formula $M^I I_n$, where n = 2 to 9. These are called *polyiodides*, and the term *polyhalides* is applied generally to the compounds $M^I X_n$ (X = halogen).

In addition to the compounds of the alkali metals (cesium and rubidium, in particular, tend to form polyhalides) polyiodides are formed especially by the organic ammonium bases. The triiodides and heptaiodides of these bases are generally reddish-brown in color and the penta- and enneaiodides are dark green. Di-, tetra- and hexaiodides are also known, especially of the heavy metals (Cu and Hg) and of the alkaloids. Of *polybromides*, $RbBr_3$ (red crystals), $CsBr_3$ (yellow-red) and $CsBr_5$ (dark red, very unstable) may be cited.

The polyhalides may be divided into those containing only one species of halogen (only iodine or only bromine), and those containing two or more different halogens. The former are in general much less stable than the latter, and lose the extra halogen rather readily. The polyiodides and polybromides were therefore formerly regarded as loose addition compounds of the halogens with the simple halides—e.g., $KI_3 = KI...I_2$. However, more recent work does not indicate any fundamental distinction between the two classes of polyhalides (see below). Cremer and Duncan (1931 and later) showed that mixed polyhalides of the types $M^I IBr_2$, $M^I ICl_2$, $M^I IBrCl$, $M^I ICl_4$ exist, and the salts $M^I ICl_3 F$ (M^I = Cs, Rb, K) have also been obtained by the reaction

$$M^I F + ICl_3 = M^I ICl_3 F$$

All these salts dissociate reversibly into a simple halide and an interhalogen compound. This reaction is characteristic of the polyhalides.

The compound $KF \cdot IF_5$, prepared by Emeléus (1949) is also a halogeno-salt—a hexafluoroiodate, $K[IF_6]$.

X-ray crystal structure determinations (Wyckoff, 1920; Pauling and Bozorth, 1925; Mooney, 1935 and later) have shown that the compounds contain typical complex anionic groups, and that there is no structural difference between the 'mixed' polyhalides and those containing a single species of halogen. In general, the heaviest halogen atom forms the central atom of the complex, and is multivalent, with the other atoms disposed symmetrically around it. The $[ICl_4]^-$ group is planar, and the ions $[XYZ]^-$ (where X, Y and Z may be the same or different halogen atoms) are linear, as also is the $[I_5]^-$ ion.

The compounds most easily prepared are the *polyiodides*, of which potassium triiodide is the most familiar. This was the first polyiodide obtained, prepared by Johnson in 1877 and hence also known as Johnson's salt. It dissociates in solution into the ions K^+ and $[I_3]^-$. The other polyiodides dissociate similarly. Many of them do not dissolve in water without decomposition, but their electrolytic dissociation can be studied in other suitable solvents.

Solutions of polyiodides, and especially of potassium polyiodide, are frequently used when it is desired to have iodine present at fairly high concentration in an aqueous medium. Potassium polyiodide solution may be prepared simply by treating a not too dilute solution of potassium iodide with iodine. This dissolves, but since the ease of decomposition of the polyiodide leads to an equilibrium in the solution

$$I^- + I_2 \rightleftharpoons [I_3]^-$$

iodine is readily given up again. Such a solution is practically equivalent in its properties to a true solution of iodine.

(d) Halogeno-salts and -acids

Most of the halides of the multivalent elements are capable of combining additively with the halides of strongly electropositive elements of low valence (especially the alkali halides) and often with the hydrogen halides. The resulting double- or complex compounds are known in Werner's nomenclature as *halogeno-salts* and *acids*.

Numerous examples of this class of compound have already been met with. The first to be considered in detail were the fluoroborates and fluoroboric acid discussed in Chap. 10.

As was set out in Chap. 11, the formation of compounds of this type is ultimately to be ascribed to the fact that energy many in may cases be liberated by the addition of negative ions to a structure which, although electrically neutral as a whole, is built up by the association of equivalent numbers of multivalent positive and singly charged negative ions. This is especially true when the negative ion has a small radius. For this reason the *fluoro-compounds* are generally the most stable of the halogeno-salts, whereas the *iodo-compounds* rarely exist, and are generally appreciably less stable than the other compounds when they do.

The halogeno-double salts play an important part in inorganic chemistry. They are frequently useful in organic chemistry also, since they often crystallize excellently and are therefore suitable for purifying materials.

The halogeno-salts and acids are generally prepared simply by mixing solutions of their constituents, the double compounds crystallizing out upon evaporation. The strengths of the complexes giving rise to these compounds vary widely. Some are decomposed completely into their components when they are dissolved in water, but with others it is impossible to detect the presence of free halide ions by means of the usual reagents, even in very dilute solutions; only complex anions are present. Strongly complexed halogeno-salts of this kind are formed especially by the metals of the sub-groups of the Periodic System. The best known examples are chloroplatinic acid, $H_2[PtCl_6]$, and its salts, the chloroplatinates.

The halogeno-salts can be *formally* derived from salts of the oxy-acid, by replacing each bivalent negative oxygen atom by two univalent negative halogen atoms. Thus metaboric acid, $H[BO_2]$, leads to tetrafluoroboric acid, $H[BF_4]$, and metasilicic acid, $H_2[SiO_3]$, to hexafluorosilicic acid $H_2[SiF_6]$, etc. If only a part of the oxygens is replaced by halogen, oxo-halogeno-compounds result. Replacement of a part by oxygen and part by other negative atoms or radicals, such as OH^- or NO_2^- yields hydroxo-halogeno- or nitro-halogeno-compounds, etc.

(e) Analytical (Halides)

In the absence of silicic acid, *fluorides* are generally detected by the etching test (cf. p. 790). The finely powdered substance is treated with concentrated sulfuric acid, and warmed cautiously in a lead or platinum crucible, covered with a watch glass. The liberation of hydrogen fluoride is shown by etching of the glass. If silicic acid is present, however (quartz or silicates), silicon fluoride is evolved, and this does not etch glass. In this case, a cover glass with a hole in it is placed on the crucible, and the hole covered with a moistened black filter paper. If fluorides are present in the sample examined, a white spot soon appears on the filter, arising from the formation of hydrated silica by hydrolysis of silicon tetrafluoride.

It should be noted that fluorides give a precipitate with barium chloride, as do sulfates.*

Chlorides, bromides, and *iodides* give precipitates, insoluble in dilute nitric acid, when treated with silver nitrate. If a soluble chloride is mixed with potassium dichromate and warmed with concentrated sulfuric acid, chromyl chloride, CrO_2Cl_2, is evolved. When this is absorbed in sodium hydroxide, it forms chromate, which is readily detected (cf. Vol. II). When a soluble bromide is treated with chlorine water, bromine is liberated, and an excess of chlorine water forms wine yellow bromine monochloride, BrCl. Iodine is even more easily displaced from iodides by means of chlorine, and dissolves in chloroform or carbon disulfide with a violet color. Excess chlorine oxidizes iodine in aqueous solution up to colorless iodic acid. Hence if both iodine and bromine are present, the characteristic color of the latter may be detected by adding a sufficient excess of chlorine water.

Fluorine may be quantitatively determined by precipitation and weighing as calcium fluoride, CaF_2.

If other ions which give precipitates with Ca^{++} ions are present, as well as F^- ion, the fluorine is volatilized as SiF_4, by heating with quartz sand and sulfuric acid. The SiF_4 is absorbed in dilute alcoholic potassium chloride solution, whereby 4 molecules of HCl are set free for every 3 molecules of SiF_4, by the following reactions:

$$3SiF_4 + 2H_2O = SiO_2 + 2H_2SiF_6; \qquad H_2SiF_6 + 2KCl = K_2SiF_6 + 2HCl.$$

The hydrogen chloride liberated is determined by titration with caustic soda, and each gram equivalent of hydrogen ion represents 3 gram atoms of fluorine (Penfield's method). Since this method gives accurate results only if moisture is completely excluded from the distillation apparatus, Willard (1933) recommended the use of aqueous perchloric acid in place of sulfuric acid. In this case the fluorine distils over in the form of aqueous fluorosilicic acid (p. 482), which can be titrated with thorium nitrate solution, using a mixture of zirconium nitrate and alizarin as indicator (cf. Frers, *Z. anal. Chem.*, 110 (1937) 251). A convenient apparatus for Willard's method has been described by Ehrlich (*Angew. Chem.*, 1953). In technical practice, the methods of Starck and Hawley are generally used. These involve precipitation of F^- ions as lead chlorofluoride, PbClF. The precipitate is dissolved in dilute nitric acid, and the Cl^- content is determined volumetrically (cf. Specht, *Z. anorg. Chem.*, 231 (1937) 181). Ehrlich (*Z. anal. Chem.*, 133 (1951) 84) recommends precipitation as PbBrF, lead bromofluoride. This has the advantage that it can be precipitated from more strongly acid solutions.

The other halogens are generally determined by precipitation as the silver halides. These may either be weighed as such, or the quantity of silver nitrate necessary for their precipitation may be found by titration.

Small quantities of F^- ions (between 20 mg and 20 γ) can conveniently be determined by *conductometric titration* with aluminum chloride solution, even in the presence of Cl^-, NO_3^-, $SO_4^=$ and $SiO_3^=$ ions (Jander, 1936). Small amounts of Cl^- ions (as low as 1 γ, using

---

* Since barium fluoride is considerably more soluble than barium sulfate, no precipitate is obtained in very dilute solution. Calcium fluoride is less soluble than barium fluoride, but its solubility is still 7 times that of barium sulfate.

suitable methods) may also be determined conductometrically by titration with silver nitrate (Jander, 1937). The *nephelometric* procedures worked out by Richards (1904) and Lamb (1920) are very suitable for the determination of very minute amounts of Cl^- ions (see Kolthoff, *J. Am. Chem. Soc.*, 55 (1933) 1915).

(f) Oxides and Oxyacids of the Halogens; General

The halogens display a greater diversity of behavior in their oxygen compounds than in other compounds. Whereas the compounds of fluorine with oxygen have been only recently discovered, chlorine forms a number of oxides, some of which have long been known: i.e., Cl_2O, ClO_2, Cl_2O_6 and Cl_2O_7* and also several oxyacids—$HClO$, $HClO_2$, $HClO_3$ and $HClO_4$. Not all these acids are derived from the same valence states as the oxides. Oxides of bromine (Br_2O, BrO_2 and Br_3O_8, all very unstable) have only recently been successfully prepared, and only two oxyacids of bromine are known with certainty— $HBrO$ and $HBrO_3$. Iodine forms the very stable oxide I_2O_5, which is the anhydride of *iodic acid*, HIO_3. This and *periodic acid*, H_5IO_6, are the most stable of the halogen oxyacids.

Periodic acid exists in several forms which differ in their water content. In solutions of these acids and their salts, equilibria are set up between these forms or their ions:

$$[IO_4]^- + 2H_2O \rightleftharpoons [IO_5]^{3-} + 2H^+ + H_2O \rightleftharpoons [IO_6]^{5-} + 4H^+$$

$$2[IO_4]^- + H_2O \rightleftharpoons [I_2O_9]^{4-} + 2H^+, \qquad \text{etc.}$$

Hypoiodous acid, IOH, which may be regarded as a derivative of electropositively univalent iodine, has only a transient existence in solution.

Oxides of iodine are also known with the composition IO_2 (or I_2O_4) and I_4O_9. It may be inferred from the properties of these substances, however, that they are to be considered not as oxides in the ordinary sense, but as the iodates of trivalent iodine, or of the iodyl radical $[IO]^+$ derived from it:

$$I_4O_9 = I^{III}[I^VO_3]_3, \qquad I_2O_4 = [I^{III}O][I^VO_3]$$

The compounds of fluorine, chlorine, and bromine with oxygen are *endothermic*. They can be formed, therefore, only if energy is supplied. This may be achieved by the action of an electric discharge upon the appropriate gas mixture (cf. the preparation of O_2F_2 and BrO_2). The necessary energy may also be furnished by some other chemical change coupled with the formation of the oxide. Thus in the reaction of Cl_2 with HgO (eqn. (1), p. 802), the energy required for the formation of Cl_2O is made available by the simultaneous combination of Cl_2 with Hg.

* As well as these normal oxides, chlorine appears to form another oxide which can probably be regarded as a peroxide. This was obtained in 1923 by Gomberg, by the action of iodine on silver perchlorate in ether solution. It is known only in solution, but it was shown to contain 4 O atoms per Cl atom. If—as Gomberg considered probable from his experiments—its molecular weight corresponds to the formula Cl_2O_8, its formation can probably be represented by the equation

(g) Oxygen Compounds of Fluorine

(i) *Oxygen difluoride*. Lebeau and Damiens found in 1927 that an oxygen compound of fluorine, with the composition OF_2, was formed as a by product in the electrolysis of molten KHF_2, as long as the salt contained any water. They isolated the compound from the accompanying elementary fluorine by passing the gases through water. The compound is only sparingly soluble, whereas fluorine reacts with water, liberating oxygen. Lebeau and Damiens later found that oxygen difluoride may be obtained in better yield by the action of fluorine on dilute sodium hydroxide solution:

$$2F_2 + 2NaOH = 2NaF + H_2O + OF_2$$

Ruff (1930) succeeded in preparing a quantity of the material in a state of high purity, and definitely established the formula OF_2 by analysis and density determinations. Its heat of formation is about —8 kcal per mol. Oxygen difluoride is a colorless gas with a characteristic smell and powerfully irritant action on the respiratory organs. (Liter weight 2.421 g, b.p. —144.8°, m.p. —223.8°, density at the boiling point 1.53). The liquid has an intense yellow color. It does not attack glass in the cold, and is altogether less reactive than free fluorine. It is not very soluble in water (6.8 cc in 100 cc of water at 0°). It reacts only slowly with water in the cold, but since the reaction

$$OF_2 + H_2O_{vap} = O_2 + 2HF + 74.8 \text{ kcal}$$

is strongly exothermic, it occurs explosively if the mixture is ignited (e.g., by a spark). When dry OF_2 is warmed or exposed to light it gradually decomposes into O_2 and F_2, the latter then attacking glass or quartz vessels.

The OF_2 molecule is bent, like the OH_2 molecule. Electron diffraction measurements (Boersch, 1935) and infrared spectroscopy (Pohlmann, 1935) agree in fixing the angle F—O—F as 100.6°. The nuclear distance $O \leftrightarrow F$ is 1.41 Å.

(ii) *Dioxygen difluoride*, O_2F_2, was prepared by Ruff and Menzel (1933), by the direct combination of O_2 and F_2 under the action of a glow discharge. It is pale brown in the gaseous state, condensing to a cherry red liquid and a orange solid (b.p. —57°, m.p. —163.5°, density at the b.p. 1.45, heat of vaporization 4.57 kcal per mol). O_2F_2 decomposes into its constituents at temperatures a little above its boiling point (Schumacher, 1936).

In addition to the oxygen fluorides, a few other compounds are known in which fluorine is bound to oxygen—viz., NO_3F, ClO_4F and CF_3OF (Cady 1934, 1948; Ruff 1936; Yost 1935). For the constitution of NO_3F see Pauling and Brockway (*J. Am. Chem. Soc.*, 59 (1937) 13) and Hill and Bigelow (*J. Am. Chem. Soc.*, 59 (1937) 2127).

(h) Oxygen Compounds of Chlorine

SUMMARY

Oxides		Acids	Salts
Cl_2O chlorine monoxide	yellow brown gas	$HClO$ hypochlorous acid	$M^I[ClO]$ hypochlorites
ClO_2 chlorine dioxide	greenish yellow gas	$HClO_2$ chlorous acid	$M^I[ClO_2]$ chlorites
Cl_2O_6 chlorine hexoxide	dark red liquid	$HClO_3$ chloric acid	$M^I[ClO_3]$ chlorates
Cl_2O_7 chlorine heptoxide	colorless liquid	$HClO_4$ perchloric acid	$M^I[ClO_4]$ perchlorates

The *oxides* of chlorine are explosive, without exception. Of the *oxyacids*, only perchloric acid can be isolated. This also explodes when it is heated, about 16 kcal per mol being liberated by its decomposition into chlorine, oxygen, and water. The other oxyacids of chlorine are known only in solution. *Salts* of all four oxyacids are known in the crystalline state. Their stability increases markedly from the hypo-chlorites to the perchlorates.

(*i*) *Constitution.* The increase in both the stability and the strength of the oxyacids of chlorine as the number of oxygen atoms increases, apart from any other considerations, would discount the 'chain formulas' once ascribed to them (e.g., the constitution H—O—O—O—O—Cl for perchloric acid). Since perhaps the beginning of the present century it has been generally assumed that all the oxygen atoms must be directly bound to the chlorine, and that the chlorine must accordingly exhibit the successive odd valences 1, 3, 5, 7 in the series of oxyacids. The acids may therefore be formally represented by the formulas

$$
\begin{array}{cccc}
\text{I} & \text{III} & \text{V} & \overset{\displaystyle O}{\overset{\displaystyle \|}{\text{VII}}} \\
\text{H—O—Cl} & \text{H—O—Cl}=\text{O} & \text{H—O—Cl}=\text{O} & \text{H—O—Cl}=\text{O} \\
 & & \overset{\displaystyle \|}{\underset{\displaystyle O}{}} & \overset{\displaystyle \|}{\underset{\displaystyle O}{}}
\end{array}
$$

In the salts, the metal ions are not bound to one particular oxygen atom, but to the corresponding ions as a whole. These 'central formulas' deduced from the chemical evidence have been confirmed by determinations of the crystal structures of their salts.

Such formulas can be satisfactorily represented in terms of modern valence theory, whereby the oxy-acid anions may be related to the Cl⁻ ion, with its inert gas like structure, by the operation of 'semipolar' (coordinate covalent) bonds:

$$
[:\overset{..}{\underset{..}{Cl}}:]^- \quad [:\overset{..}{\underset{..}{O}}:\overset{..}{\underset{..}{Cl}}:]^- \quad [:\overset{..}{\underset{..}{O}}:\overset{..}{\underset{..}{Cl}}:\overset{..}{\underset{..}{O}}:]^- \quad [:\overset{..}{\underset{..}{O}}:\overset{..}{\underset{..}{Cl}}:\overset{..}{\underset{..}{O}}:]^- \quad [:\overset{..}{\underset{..}{O}}:\overset{..}{\underset{..}{Cl}}:\overset{..}{\underset{..}{O}}:]^-
$$

These formulas express the equivalence of the oxygen atoms in the anions. It is not possible to state how far the Cl—O binding is ionic in character, involving at least a partial positive charge upon the chlorine, and to what extent it is covalent. The oxides of chlorine are probably purely covalent compounds. It is at least probable that the difference in behavior of fluorine and chlorine toward oxygen lies in the capacity of chlorine to enlarge its outer electron shell (3rd quantum orbits) beyond the argon configuration—a possibility which is bound up intimately with the relatively greater ease with which electrons may be completely removed (formation of ionic bonds) or excited to higher levels, as compared with the fluorine atom.

(*ii*) *Chlorine monoxide, Hypochlorous acid and the Hypochlorites.*

1. *Chlorine monoxide*, Cl_2O, is formed when chlorine is passed at $0°$ over dry mercuric oxide:

$$ \text{HgO} + 2Cl_2 = HgCl_2 + Cl_2O \tag{1} $$

It is a yellow brown gas, with an unpleasant smell, which strongly attacks the respiratory organs. It is much heavier than air (density $= 3.02$, relative to air $= 1$), and can readily be condensed to a red-brown liquid (b.p. $+3.8°$).

Chlorine monoxide is an endothermic compound (heat of formation $—18.6$ kcal at constant pressure). It decomposes into its components explosively—e.g., if it is heated, or poured in the liquid state from one vessel to another, or when it is brought into contact with combustible materials. It is readily absorbed by water, with which it combines to form *hypochlorous acid*:

$$ Cl_2O + H_2O = 2HOCl $$

It is thus the anhydride of hypochlorous acid.

2. *Hypochlorous acid*, HOCl, is known only in aqueous solution. It is formed, together with the ions of hydrochloric acid, by the reaction of chlorine with water (see p. 783):

$$Cl_2 + H_2O \rightleftharpoons HOCl + H^+ + Cl^- \tag{2}$$

If the H^+ and Cl^- ions are removed* by adding mercuric oxide:

$$HgO + 2H^+ + 2Cl^- = H_2O + HgCl_2$$

process (2) proceeds completely from left to right. It is therefore possible to prepare fairly concentrated solutions of hypochlorous acid by passing chlorine into a suspension of mercuric oxide in water:

$$2Cl_2 + HgO + H_2O = HgCl_2 + 2HOCl$$

Solutions of hypochlorous acid are faintly greenish yellow in color, and have a peculiar smell resembling that of bleaching powder. They gradually decompose, especially in light, evolving oxygen and forming chloric acid, $HClO_3$. The reaction probably takes place by the initial splitting off of oxygen:

$$HClO = HCl + O$$

part of which reacts in the 'nascent' state with undecomposed hypochlorous acid:

$$\times \quad HClO + 2O = HClO_3.$$

Hypochlorous acid is a very weak acid (dissociation constant $K = 3.6 \cdot 10^{-8}$ at 25°). Its salts are the *hypochlorites*.

3. *Hypochlorites*, with the general formula M^IClO, are formed together with chlorides by the action of chlorine on strongly basic hydroxides—e.g.,

$$2NaOH + Cl_2 = NaCl + NaOCl + H_2O$$

Alkali hypochlorite solutions can also be prepared directly, by the electrolysis of alkali chloride solutions. For this purpose, the devices described on p. 183 *et seq.*, which hinder the mixing of anodically evolved chlorine with cathodically formed hydroxide, are omitted.

Potassium hypochlorite solutions were first produced industrially at Berthollet's instigation at Javelle (now Javel, near Paris) in 1792, by passing chlorine into potassium carbonate solution, and were known as 'eau de Javelle'. Labarraque in 1820 prepared *sodium hypochlorite solutions* in a similar way (eau de Labarraque). The bleaching solutions so prepared do not usually keep well, since they generally contain free hypochlorous acid:

$$2Na_2CO_3 + Cl_2 + H_2O = NaCl + 2NaHCO_3 + NaOCl;$$

$$NaOCl + H_2O + Cl_2 = NaCl + 2HOCl.$$

Chloride-free solutions of the alkali and alkaline earth hypochlorites can be prepared by neutralizing solutions or suspensions of the corresponding hydroxides with solutions of hypochlorous acid. When these are evaporated at a low temperature, the corresponding hypochlorites are obtained in the crystalline state. They generally crystallize as hydrates.

The hypochlorites have powerful oxidizing properties, and their solutions are used especially as bleaching solutions [4]. The chloride content of solutions prepared in the usual manner does not interfere with this use.

* Mercuric chloride, unlike other salts, is ionized only to a very small extent.

Bleaching Powder. When chlorine is passed over slaked lime, the resulting product can be regarded as a mixed salt of hypochlorous and hydrochloric acids:

$$Cl_2 + Ca(OH)_2 = CaCl(OCl) + H_2O \qquad (1)$$

This mixed salt is the essential constituent of 'chloride of lime', or 'bleaching powder', which is prepared industrially on a large scale.

Bleaching powder forms a white, crumbly powder, with a peculiar smell, which is generally attributed to hypochlorous acid liberated by reaction with the carbon dioxide of the air. Like all compounds of hypochlorous acid, it loses oxygen readily and has strong oxidizing properties. Thus it oxidizes the monoxides of lead and manganese, or their salts in alkaline solution, up to the dioxides PbO_2 and MnO_2. When hydrochloric or sulfuric acid is added to it, chlorine is evolved:

$$CaOCl_2 + 2HCl = CaCl_2 + H_2O + Cl_2 \qquad (2)$$

$$CaOCl_2 + H_2SO_4 = CaSO_4 + H_2O + Cl_2 \qquad (3)$$

It may be assumed that hypochlorous acid is initially formed, but reacts instantly with hydrochloric acid, evolving chlorine: e.g.,

$$CaOCl_2 + H_2SO_4 = CaSO_4 + HCl + HClO; \qquad HCl + HClO = H_2O + Cl_2$$

In direct sunlight, bleaching powder loses oxygen to the air, especially if carbon dioxide is present. The same decomposition occurs when solutions of bleaching powder are warmed with certain oxides and hydroxides which act as catalysts— e.g., copper oxide, ferric oxide, nickel hydroxide, cobalt hydroxide. This reaction can be used for the preparation of oxygen in the laboratory.

Unlike the very hygroscopic calcium chloride, pure bleaching powder absorbs no water from the air. Neither is any calcium chloride leached out when it is treated for a short time with alcohol. This is the basis of the hypothesis that the compound contained in bleaching powder is the mixed salt $CaCl(OCl)$, a view first put forward by Odling. Neumann (1926) showed that calcium chloride hypochlorite, $CaCl(OCl)$, is present in bleaching powder, in a complex compound with $Ca(OH)_2$. According to the conditions of preparation, any one of the compounds

$$[3CaCl(OCl) \cdot Ca(OH)_2] \cdot 5H_2O$$

$$[3CaCl(OCl) \cdot Ca(OH)_2] \cdot 3H_2O$$

or $$[CaCl(OCl) \cdot Ca(OH)_2] \cdot H_2O$$

may be obtained. The water of crystallization is lost when the compounds are exposed in a vacuum. The first mentioned compound is formed under the conditions employed in the technical preparation. It is stable in closed vessels in the absence of light. In sunlight, in a closed vessel, decomposition occurs:

$$6CaOCl_2 = 5CaCl_2 + Ca(ClO_3)_2.$$

Decomposition, with loss of oxygen, gradually takes place in air, even in the dark, but is promoted to a remarkable degree by carbon dioxide. For this reason, bleaching powder prepared from incompletely burned lime, or by the use of chlorine containing carbon dioxide, cannot be kept, even in closed, opaque containers.

Bleaching powder is used in large amounts as a bleaching agent for cellulose, paper, and textiles, and to some extent in laundering and for household purposes. In these spheres of use, it has been replaced to an increasing extent by peroxidic bleaching agents, which are usually less deleterious to the fiber. Bleaching powder is also widely used as a disinfectant. Considerable amounts are consumed in the refining of crude petroleum.

The bleaching power of technical 'chloride of lime' is measured by its content of hypochlorite, and is tested by determining the amount of chlorine liberated on addition of hydrochloric acid (eqn. (2) p. 804). This is expressed as a percentage of the total weight of the material, and is known as the content of 'available chlorine'. The available chlorine of good technical bleaching powder varies between 35 and 39%. For the pure compound [$3CaCl(OCl) \cdot Ca(OH)_2] \cdot 5H_2O$ it would amount to 39.05%. The content of available chlorine is sometimes expressed in either English or French (or Gay-Lussac) degrees. 'English degrees' represent directly the percentage content of available chlorine. 'French degrees' represent the number of liters of chlorine gas (at 0° and 760 mm) evolved from 1 kg of bleaching powder.

Bleaching powder was first manufactured in England, by Tennant, in 1799.

At the present day, the transportation of liquid chlorine has replaced to a considerable extent the shipment of bleaching powder as such: bleaching powder solution can then be prepared where and when needed, by passing the chlorine into milk of lime. More recently, technical *calcium hypochlorite* has increasingly replaced chloride of lime, especially for export purposes. The commercial material contains up to 80% available chlorine (pure $Ca(OCl)_2$ has 99.19% available chlorine).

(iii) Chlorous acid and the Chlorites.

Chlorous acid, $HClO_2$, is formed as an intermediate product in the hydrolysis of chlorine dioxide, but decomposes rapidly. Its salts, the *chlorites*, $M^I[ClO_2]$, are more stable in solution. They are obtained, together with chlorates, by the action of chlorine dioxide on alkali hydroxides or carbonates:

$$2ClO_2 + 2OH^- = ClO_2^- + ClO_3^- + H_2O,$$

and may be obtained free from chlorates by mixing aqueous solutions of chlorine dioxide and sodium peroxide (or sodium hydroxide + hydrogen peroxide):

$$2ClO_2 + Na_2O_2 = 2Na[ClO_2] + O_2$$

Solutions of chlorites have strong oxidizing properties. The solid chlorites, especially those of the heavy metals, decompose explosively when they are heated or struck. Latimer (1947) gives the free energy of formation of the ClO_2^- ion and $HClO_2 \cdot _{aq}$ as $\Delta F_{298} = +2.9$ kcal and $+0.23$ kcal per mol respectively; ΔH_{298} for the same compounds are $+17.2$ kcal and $+13.68$ kcal per mol

(iv) Chloric acid and the Chlorates.

1. *Chloric acid*, $HClO_3$, is formed by the decomposition of hypochlorous acid. It is generally prepared from its salts, the chlorates,—e.g., by the reaction of barium chlorate with dilute sulfuric acid:

$$Ba[ClO_3]_2 + H_2SO_4 = BaSO_4 + 2HClO_3.$$

The colorless solution can be concentrated in a vacuum without decomposition, to a content of 40% of $HClO_3$. Slow decomposition sets in if evaporation is carried further, and becomes violent if the concentration exceeds 50%. Dilute solutions can be heated almost to boiling without decomposition. Chloric acid readily gives up its oxygen to oxidizable substances, especially in concentrated solution.

Chloric acid is a strong monobasic acid, almost completely dissociated in aqueous solution. The apparent degree of dissociation (determined from the conduct-ivity quotient) is 79% in 1-normal solution at 18°.

2. *Chlorates*. The salts of chloric acid, $M^I[ClO_3]$, are formed by the decomposi-tion of hypochlorites in solution on warming:

$$3[ClO]^- = [ClO_3]^- + 2Cl^-$$

They are therefore formed, together with much chloride, when chlorine is passed into warm alkali or alkaline earth hydroxide solutions, or when alkali chlorides are electrolyzed in cells without a diaphragm in hot solution:

$$3Cl_2 + 6OH^- = [ClO_3]^- + 5Cl^- + 3H_2O$$

The chlorates are colorless, quite stable in the solid state at ordinary tempera-ture, and freely soluble in water. Their solutions are less strongly oxidizing than those of the hypochlorites. When heated, the chlorates lose oxygen. Decomposition into chloride and perchlorate often takes place first:

$$4M^I[ClO_3] = M^ICl + 3M^I[ClO_4]$$

and the perchlorate subsequently decomposes into chloride and oxygen on stronger heating:

$$M^I[ClO_4] = M^ICl + 2O_2.$$

Potassium chlorate, $K[ClO_3]$, is prepared technically by various processes, of which that of Liebig is best known. Chlorine is passed into warm milk of lime, and the solution treated with potassium chloride. Potassium chlorate crystallizes out on cooling, while calcium chloride remains in solution:

$$6Ca(OH)_2 + 6Cl_2 = Ca[ClO_3]_2 + 5CaCl_2 + 6H_2O$$

$$Ca[ClO_3]_2 + 2KCl = 2K[ClO_3] + CaCl_2$$

Potassium chlorate forms colorless, lustrous, monoclinic plates, with a cool taste (density 2.33, m.p. 370°). It is very soluble in hot water, but much less soluble in cold (100 g of water dissolve 3.3 g of $KClO_3$ at 0°, 7.3 g at 20° and 56.0 g at 100°). It is insoluble in absolute alcohol. Potassium chlorate begins to decompose at about 400°, forming potassium perchlorate. This in turn decomposes when more strongly heated, with the loss of oxygen (see equation above). Admixture with manganese dioxide, ferric oxide, and other substances lowers the decomposition temperature of potassium chlorate markedly, and formation of the perchlorate is suppressed. Potassium chlorate can explode violently when rapidly heated above its decompo-sition temperature. It is often used as an oxidant in melts, etc., because of the ease with which it gives up oxygen when heated. It is used in large amounts in the safety match industry.

'Safety matches', which can only be struck on prepared surfaces, were first produced in Sweden. The match head contains a mixture of potassium chlorate, with sulfur, or anti-mony sulfide, dextrin solution, glass powder, and coloring matter. The striking surface is coated with a mixture of red phosphorus, antimony sulfide, and dextrin. A trace of red phosphorus is converted to white by frictional heat; this ignites, combustion spreads to the mixture in the match head, and thence to the match stick.

Potassium chlorate is also used in pyrotechnics and in the production of explosives. It is also used in medicine—e.g., the dilute solution is employed as a gargle. Like all chlorates, it is poisonous in larger quantities (over 1 g).

Sodium chlorate, $NaClO_3$, is extremely soluble in water (101 g in 100 g of water at 20°), and therefore less readily prepared pure than the potassium salt. It is deliquescent in air. It crystallizes anhydrous, however, generally in the cubic system; metastable monoclinic and rhombohedral varieties are also known. It is used as an oxidant, as a starting material for the preparation of perchlorates, and in large quantities as weed killer.

Sodium chlorate is now manufactured—almost entirely by the electrolytic process—in much larger quantities than potassium chlorate. Production, in Germany, for example, is around 30,000 tons per year, and consumes about 200 million kWh, as compared with 600 million kWh for electrolytic chlorine and alkali hydroxide.

(v) Chlorine dioxide and Chlorine hexoxide.

1. *Chlorine dioxide*, ClO_2, is a greenish yellow explosive gas, with a sharp, penetrating smell. It is formed by the action of concentrated sulfuric acid on potassium chlorate, since the chloric acid liberated decomposes under the action of sulfuric acid:

$$KClO_3 + H_2SO_4 = HClO_3 + KHSO_4; \qquad 3HClO_3 = HClO_4 + H_2O + 2ClO_2$$

It is also—and more safely—obtained by the action of dilute sulfuric acid on a mixture of potassium chlorate and oxalic acid:

$$2KClO_3 + H_2C_2O_4 + H_2SO_4 = K_2SO_4 + 2H_2O + 2CO_2 + 2ClO_2$$

It is readily condensed to a red-brown liquid, which solidifies to a hard, orange red brittle crystalline mass when more strongly cooled (m.p. —76°, b.p. +9.9°). The vapor density corresponds to the formula ClO_2.

Chlorine dioxide is a strongly endothermic compound (heat of formation —26.3 kcal per mol), and is very unstable. It decomposes with a violent explosion into chlorine and oxygen merely on gentle warming, or on contact with oxidizable materials.

Explosive decomposition also takes place on contact with fluorine. By careful experimentation, however, it is possible to isolate the compound ClO_2F (colorless gas, b.p. —6°, m.p. —115°). This is probably to be regarded as the fluoride of chloric acid (Schumacher, 1942).

Chlorine dioxide dissolves readily in water, with which it forms a hydrate. This is difficult to isolate in a state of purity. In the dark, and at ordinary temperature, solutions of chlorine dioxide remain stable almost indefinitely. Reaction with water, forming chloric acid and hydrochloric acid, takes place only very gradually, hypochlorous acid being thereby produced as an intermediate. The hydrolysis of chlorine dioxide is greatly accelerated by higher concentrations of chloride ions. In alkaline solution decomposition takes place more rapidly. Under these conditions it does not go beyond the stage of chlorite:

$$2ClO_2 + 2OH^- = ClO_2^- + ClO_3^- + H_2O.$$

2. *Chlorine hexoxide*, Cl_2O_6, was obtained in 1925 by Bodenstein, in the course of a study of the photochemical decomposition of chlorine dioxide. It is also formed by the action of ozone on chlorine dioxide, and can comparatively easily be prepared in quantity by this means in a suitable form of apparatus (all greased taps must be eliminated).

Chlorine hexoxide forms a deep red liquid (density 1.65, and freezing point —1°) which fumes in air. It is fairly stable at ordinary temperature when it is pure, but explodes with great violence on contact with organic materials. It may also react explosively with water under some conditions. The formation of a substance having the properties of chlorine hexoxide was noticed by Millon, as early as 1843, although he was unable to determine its composition correctly. Unlike the other oxides of chlorine, chlorine hexoxide has only a small vapor pressure (about 1 mm) at ordinary temperature. It can therefore be separated fairly easily from the other oxides by fractional distillation in a vacuum.

(vi) Chlorine heptoxide (Perchloric anhydride), Perchloric acid and the Perchlorates

1. *Chlorine heptoxide*, Cl_2O_7, is the anhydride of perchloric acid, $HClO_4$, and can be prepared from the latter by abstracting the elements of water by means of phosphorus pentoxide:

$$2HClO_4 + P_2O_5 = Cl_2O_7 + 2HPO_3$$

Chlorine heptoxide is a colorless, very volatile (b.p. 83°), oily liquid. It explodes violently on shock, or if brought into a flame, and reacts explosively with iodine. It can be poured on to paper, wood and similar materials, however, without entering into reaction. It does not attack sulfur or phosphorus at ordinary temperature. It slowly dissolves in cold water, forming perchloric acid:

$$Cl_2O_7 + H_2O = 2HClO_4$$

2. *Perchloric acid*, $HClO_4$, is much the most stable of the oxyacids of chlorine. It can be isolated by the action of concentrated sulfuric acid on potassium perchlorate:

$$KClO_4 + H_2SO_4 = KHSO_4 + HClO_4.$$

The reaction mixture is cautiously heated in an oil bath, the temperature being slowly raised from 90° to 160°. The anhydrous acid distils over in a water pump vacuum and can be purified by repeated vacuum distillation.

Anhydrous perchloric acid is a colorless, very mobile liquid, which fumes strongly in air and solidifies only when strongly cooled (m.p. —112°). It cannot be distilled without decomposition except under reduced pressure. When it is heated, it becomes red-brown in color as a result of decomposition, and ultimately explodes. Decomposition slowly occurs even at ordinary temperature, especially if impurities are present. Perchloric acid readily gives up oxygen to oxidizable substances. If brought into contact with combustible materials, such as paper, wood, or wood charcoal, inflammation takes place explosively. On the skin it produces painful wounds, which heal badly.

Perchloric acid is miscible in all proportions with water. It forms several hydrates, of which the monohydrate, $HClO_4 \cdot H_2O$ is characterized by a relatively high melting point (+50°). Concentrated solutions of perchloric acid are oily in consistency, like sulfuric acid, in contrast with the anhydrous acid. Perchloric acid is much more stable in solution than in the anhydrous state. This is true also of the very concentrated solutions—e.g., the 72% solution which has a constant boiling point of about 203°, although boiling is accompanied by some decomposition.

Perchloric acid is a monobasic acid, and is one of the strongest acids known. Its salts, $M^I[ClO_4]$, are the *perchlorates*.

Perchloric acid also forms salts with certain positive radicals that are otherwise seldom found—e.g., the nitronium radical $[NO_2]^+$ (cf. p. 600). The monohydrate of perchloric acid is also to be considered as a salt-like compound, hydronium perchlorate, $[H_3O]^+[ClO_4]^-$. This follows, in particular from the X-ray work of Volmer (1924), the diffraction pattern being practically identical with that of the ammonium salt $[NH_4]^+[ClO_4]^-$.

The substance ClO_3F is derived from perchloric acid, by exchange of OH for fluorine. This compound was obtained by Engelbrecht (1952) by electrolysis of a solution of sodium perchlorate in liquid hydrogen fluoride. It is a colorless gas (b.p. —48°, m.p. —152°), which is decomposed by strong caustic alkali to form perchlorate and fluoride, although it is practically unattacked by water. The *chloryl oxyfluoride*, $ClO_2 \cdot OF$ described by Bode (1951) would appear to be isomeric with this substance.

3. *Perchlorates*. The salts of perchloric acid, the perchlorates, are the most stable oxygen compounds of chlorine. Many of them are completely stable both in the solid state and in solution. They are decomposed when heated, forming chloride and losing oxygen. For potassium perchlorate, $KClO_4$, this takes place only above 400°. Perchlorates which crystallize as hydrates, and those containing organic constituents, are less stable. Most perchlorates are readily soluble in water. The perchlorates of potassium, rubidium and cesium are fairly sparingly soluble in cold water, but more soluble in hot water (cf. Table 36, p. 185). They can therefore be crystallized well and prepared readily. This is also true of ammonium perchlorate, which is not very soluble.

The alkali perchlorates are generally prepared from the corresponding chlorates, which may either be decomposed thermally (1) or oxidized electrolytically in solution (2):

$$4KClO_3 = KCl + 3KClO_4 \tag{1}$$

$$ClO_3^- + H_2O - 2e = ClO_4^- + 2H^+ \tag{2}$$

Other perchlorates are usually prepared by dissolving the appropriate metal, oxide, or carbonate in aqueous perchloric acid. This may be obtained, for example, by decomposing potassium perchlorate with fluorosilicic acid and filtering off the precipitated, sparingly soluble potassium fluorosilicate:

$$2KClO_4 + H_2SiF_6 = K_2SiF_6 + 2HClO_4.$$

Aqueous perchloric acid is frequently used as a reagent for the gravimetric determination of potassium. Perchlorates are used in the explosives industry in place of the more unstable, and therefore more dangerous, chlorates. Ammonium perchlorate, NH_4ClO_4, in particular, is not sensitive to shock but explodes if detonated.

Potassium perchlorate occurs in nature in small quantities in *caliche*, the raw material of Chile saltpeter. Since it is very toxic to plants, it must be removed if the nitrate is to be used as fertilizer.

It is frequently stated that the readily soluble perchlorates are deliquescent. It has been shown in a number of cases, however, (e.g., $NaClO_4$, $LiClO_4$, $Ba(ClO_4)_2$) that this is true only of salts contaminated with $HClO_4$. The perchlorates of the alkaline earths are finding increasing technical applications—e.g,. in pyrotechnics. The conditions for the preparation of these compounds by dry methods have been given by G. F. Smith (1935); by this means they are obtained directly in the anhydrous state.

A noteworthy property of the perchlorates is their very slight tendency to undergo hydrolysis. As far as is known, the ClO_4^- ion is not capable of functioning as a constituent of complex ions.

(i) Analytical (Oxygen Compounds of Chlorine)

The presence of hypochlorites is readily shown by the smell of chlorine obtained when their solutions are acidified. They give a precipitate of silver chloride when treated with silver nitrate, since the silver hypochlorite initially formed rapidly undergoes decomposition:

$$3AgClO = AgClO_3 + 2AgCl.$$

The hypochlorite content of 'chloride of lime', bleach liquors, etc. is usually determined by the method of Penot, in which the 'available chlorine' is titrated with arsenic trioxide solution, the latter being oxidized by hypochlorite to arsenic acid. The end point is found by spotting on potassium iodide-starch paper.

Chlorates and perchlorates give no precipitate with silver nitrate. Chlorates, but not perchlorates, can be reduced by sulfur dioxide or by 'nascent' hydrogen in alkaline solution (evolved from Devarda's alloy), and silver chloride may then be precipitated. Characteristic of perchlorates is the formation of a precipitate on addition of potassium chloride; the precipitate redissolves when the solution is warmed.

(j) Oxygen Compounds of Bromine

(i) *General.* The *oxides* of bromine can be prepared only with difficulty, and are extremely unstable. They have therefore only become known in recent years.

The first to be discovered was the oxide Br_3O_8, obtained in 1929 by Schumacher, by the action of ozone on bromine. It forms colorless needles which are not stable above —80°. Schwarz, in 1937, was able to prepare *bromine dioxide*, BrO_2, by the action of a glow discharge upon a mixture of Br_2 and O_2. The compound separated in the central section of the discharge tube, which was cooled in liquid air, as a yellow solid, which decomposed, without melting, into Br_2 and O_2 at 0°. *Bromine monoxide*, Br_2O, was also formed during the decomposition. The formation of this oxide had been previously observed by Zintl, in 1930, by passing bromine vapor over HgO at 50–100°. It was first prepared in a pure state by Schwarz (1938), by decomposing bromine dioxide in a high vacuum (see *J. prakt. Chem.*, 152, (1939) 157). Bromine monoxide is not stable above —40°. It is dark brown, dissolves in CCl_4 with an intense green color, and reacts smoothly with sodium hydroxide to form hypobromite. Schwarz also noticed that a very unstable white oxide of bromine was formed, but it was not possible to isolate this substance.

Only two oxyacids of bromine are known with certainty—*hypobromous acid*, HBrO, and *bromic acid*, $HBrO_3$. These compounds and their salts are closely analogous to the corresponding compounds of chlorine.

Bromous acid, $HBrO_2$, or *bromite* ions, BrO_2^-, although very unstable even in solution, are apparently formed as intermediates in many reactions,—e.g., in the decomposition of hypobromite ions:

$$2BrO^- = BrO_2^- + Br^-; \qquad BrO_2^- + BrO^- = BrO_3^- + Br^-$$

(Clarens 1913). Very little is known of this reaction, however.

In view of the fact that perchloric acid and the perchlorates are the most stable of the oxygen compounds of chlorine, it is striking that there is no evidence for the existence of any analogous compounds of bromine.

(ii) *Hypobromous acid*, HBrO, and the *hypobromites*, $M^I[BrO]$, are obtained by methods similar to those used for hypochlorites. Like the latter, they are strong bleaching agents and oxidants. A compound analogous to 'chloride of lime' can be prepared by passing bromine vapor over slaked lime. Solutions of alkali hypobromites are frequently used as oxidants in quantitative analysis, since they are conveniently prepared. They are straw yellow in color, and have a characteristic aromatic smell. When they are warmed or acidified they are immediately decomposed, forming bromide (or hydrobromic acid) and bromate. Decomposition takes place slowly at ordinary temperature, partly by the reaction of BrO^- ion cited above, and partly because hypobromous acid formed by hydrolysis is unstable. Its electrolytic dissociation constant is about $2 \cdot 10^{-11}$.

Crystalline hypobromites were first isolated by Scholder (1952)—viz., the compounds $NaBrO \cdot 5H_2O$, $NaBrO \cdot 7H_2O$ and $KBrO \cdot 3H_2O$. These compounds (yellow needles, extremely soluble in water) are very unstable—especially the potassium salt, which rapidly decomposes even at $0°$.

(*iii*) *Bromic acid*, $HBrO_3$, may be formed either by passing chlorine into bromine water:

$$Br_2 + 5Cl_2 + 6H_2O = 2HBrO_3 + 10HCl$$

or by decomposing barium bromate with dilute sulfuric acid:

$$Ba[BrO_3]_2 + H_2SO_4 = BaSO_4 + 2HBrO_3.$$

It is stable only in solution, and decomposes when its colorless solutions are evaporated.

(*iv*) *Bromates*. The salts of bromic acid, $M^I[BrO_3]$, are more stable than the free acid. They are obtained like the chlorates, by adding bromine to warm alkali hydroxide solutions, or by the electrolysis of bromide solutions. The bromates, like the chlorates, readily lose oxygen. They deflagrate when they are heated with oxidizable substances, and decompose, evolving oxygen, when heated alone. *Potassium bromate* (colorless, trigonal crystals, $d = 3.27$, solubility at $20°$ 6.9 g at $100°$ 49.7 g in 100 g of solution) is used for bromatometric titrations. It decomposes, with loss of oxygen, above $434°$.

The use of bromate titrations in volumetric analysis depends on the ease with which BrO_3^- ions in hydrochloric acid solution can be reduced quantitatively to Br^- ions. As soon as all the reducing agent is used up, free bromine is formed:

$$BrO_3^- + 5Br^- + 6H^+ = 3Br_2 + 3H_2O;$$

this bleaches an indicator dyestuff)methyl orange or methyl red), thereby showing the endpoint of the titration.

Solutions containing *free bromine* are also used in the volumetric analysis of reducing agents. KBr solutions are also used in determining strong oxidants. This process is known as *bromometry*; indigo carmine is commonly used as color indicator for the endpoint. Errors arising from the high vapor pressure of bromine can be avoided by the use of the apparatus described by D'Ans (*Angew. Chem.*, 63 (1951) 45).

(k) Oxygen Compounds of Iodine

(*i*) *General*. The reaction of iodine with solutions of alkali or alkaline earth hydroxides, or with alkali carbonates, resembles that of chlorine or bromine, in that *hypoiodites*, of the general formula $M^I[IO]$ are first formed. These are so unstable, however, that they are rapidly transformed into *iodates*, $M^I[IO_3]$, even at ordinary temperature. *Hypoiodous acid* itself is even more unstable, but may be obtained transiently in very dilute solution by shaking aqueous iodine solution with mercuric oxide. It is a very weak acid, and is therefore present, as a result of hydrolysis, in solutions of hypoiodites. It can, indeed, be regarded as an amphoteric substance, *iodine hydroxide*. Skrabal (1942) gives the dissociation constant for the acid dissociation, $IOH \rightleftharpoons IO^- + H^+$, as about $4 \cdot 10^{-13}$, whereas for its basic dissociation, $IOH \rightleftharpoons I^+ + OH^-$, the equilibrium constant is $3 \cdot 10^{-10}$ (Murray, 1925). For the hydrolysis of iodine, $I_2 + HOH \rightleftharpoons IOH + I^- + H^+$, $K = 3 \cdot 10^{-13}$, and for the dissociation of iodine itself, $I_2 \rightleftharpoons I^+ + I^-$, $K = 9.6 \cdot 10^{-9}$.

Hypoiodites have not been isolated in a pure state, because of the ease with which they decompose. Their solutions have a characteristic smell, and give a blue color with potassium iodide-starch paper in the absence of excess alkali. They are even more powerful bleaching agents than hypochlorite or hypobromite solutions.

According to Skrabal (1909), hypoiodous acid can be stabilized by the association of a neutral iodine molecule with the electropositive iodine atoms. The compound so formed, $I_2 \cdot IOH$, is stable for a long time in acid solution, whereas IOH under the same conditions decomposes immediately into iodine and iodic acid.

(*ii*) *Iodine pentoxide, Iodic acid and the Iodates. Iodic acid*, HIO_3, is more stable than chloric or bromic acids, and can readily be obtained crystalline. It can be liberated from its salts—e.g., from sodium iodate—by heating them with sulfuric acid (1). It can also be obtained by oxidizing iodine with concentrated nitric acid (2), or by means of chlorine in aqueous solution (3). The hydrochloric acid formed simultaneously in the last process is removed by addition of silver oxide (4).

$$NaIO_3 + H_2SO_4 = HIO_3 + NaHSO_4 \tag{1}$$

$$5HNO_3 + \tfrac{5}{2}I_2 = 3HIO_3 + 5NO + H_2O \tag{2}$$

$$I_2 + 5Cl_2 + 6H_2O = 2HIO_3 + 10HCl \tag{3}$$

$$10HCl + 5Ag_2O = 10AgCl + 5H_2O \tag{4}$$

Iodic acid forms colorless crystals with a glassy luster and bitter taste. It crystallizes in two different modifications which are not interconvertible, but both belong to the bisphenoidal class of the rhombic system. It is very soluble in water (310 g of HIO_3 in 100 g of water at $16°$). When it is heated, it partially melts at $110°$, and is converted into a form poorer in water, the so-called *anhydroiodic acid*, HI_3O_8, which also crystallizes from the solution above $110°$. The same product is formed when iodic acid is allowed to stand in dry air at $30-40°$. All the water is lost when iodic acid is heated to $195°$, and *iodine pentoxide*, I_2O_5, remains as a white powder. In the absence of light, this decomposes only above $300°$. Unlike the oxides of the other halogens, it is an *exothermic compound* (heat of formation 43 kcal per mol of I_2O_5). Iodine pentoxide dissolves freely in water, reforming iodic acid, of which it is therefore the anhydride.

The salts of iodic acid, the *iodates*, mostly correspond in composition to the type M^IIO_3, but hydrogen diiodates, $M^IH[IO_3]_2$, and dihydrogen triiodates, $M^IH_2[IO_3]_3$, are known. These are addition compounds of iodic acid with the normal iodates; they exist in the solid state only, the acid molecules being linked to the oxy-anions by hydrogen bonds.

In concentrated solutions, iodic acid shows some tendency to polymerize or condense, but the conductivities of solutions of its neutral salts show that it is a monobasic acid. These all accord with the conductivity of salts of the type MX, and not with those of the type M_2X (cf. p. 396). The dissociation constant of iodic acid is $K = 0.19$ at $18°$ (Abel, 1934).

Iodic acid and the iodates tend to form addition compounds with alkali halides—e.g., $NaBr \cdot NaIO_3 \cdot 6H_2O$ — and with other acid anhydrides such as SO_3, SeO_3, TeO_3, MoO_3, etc. (Formation of heteropolyacids, cf. Vol. II.)

The iodates are much more stable than the chlorates and bromates, but they are also oxidizing agents. They deflagrate weakly on red hot charcoal, and can be detonated by shock if they are mixed with combustible substances.

Calcium iodate, $Ca[IO_3]_2$, occurs in nature as *lautarite*. Sodium iodate, $NaIO_3$, occurs in minute amounts as an impurity in Chile saltpeter.

Iodates can be prepared by oxidizing iodides in alkaline solution; this can be effected electrolytically. They may also be obtained by dissolving iodine in hot solutions of alkali or alkaline earth hydroxides. An interesting mode of formation is by the action of chlorates on elementary iodine in solution.*

Iodic acid, or an acidified iodate solution, reacts with hydrogen iodide, liberating iodine:

$$HIO_3 + 5HI = 3I_2 + 3H_2O, \quad \text{or} \quad IO_3^- + 5I^- + 6H^+ = 3I_2 + 3H_2O$$

The liberation of iodine from iodic acid takes place much more rapidly than with chloric or bromic acid, even though these are much more unstable than iodic acid. If a mixture of potassium iodate and iodide is added to a solution which has an acid reaction as a result of partial hydrolysis (e.g., aluminum sulfate solution), hydrogen ions are used up in accordance with the reaction shown above. The hydrolysis equilibrium will be disturbed thereby, and if steps are taken to remove the liberated iodine, by adding thiosulfate, the hydrolysis goes to completion. Use is often made of this in analytical chemistry, following a suggestion by Stock, to effect a clean precipitation of oxide hydrates, which under other conditions, have a strong tendency to carry down other substances from solution. Iodate solutions can also be employed in oxidimetric titrations (iodatometric titrations).

(*iii*) *Periodic acid and the Periodates.* Many iodates (e.g., barium iodate) are converted into *periodates* when they are heated:

$$5Ba[IO_3]_2 = Ba_5[IO_6]_2 + 4I_2 + 9O_2$$

Periodates are also obtained by oxidizing iodates with chlorine, in alkaline solution:

$$[IO_3]^- + 6OH^- + Cl_2 = [IO_6]^{5-} + 2Cl^- + 3H_2O.$$

---

* This reaction has a somewhat complex mechanism, since the primary reaction of iodine with the chlorate is to liberate chlorine, and this reacts further, reforming chlorate. E.g.,

$$2KClO_3 + I_2 + H_2O = Cl_2 + KOH + KH[IO_3]_2 \text{ (potassium hydrogen diiodate)}$$

$$Cl_2 + KOH = KCl + HOCl$$

$$3HOCl = HClO_3 + 2HCl \text{ (on warming)}$$

$$3KOH + HClO_3 + 2HCl = KClO_3 + 2KCl + 3H_2O$$

Without taking account of other side reactions, the net equation obtained is

$$11KClO_3 + 6I_2 + 3H_2O = 6KH[IO_3]_2 + 3Cl_2 + 5KCl.$$

Oxidation can also be carried out electrolytically:

$$[IO_3]^- + 6OH^- - 2e = [IO_6]^{5-} + 3H_2O.$$

The free acid may be obtained from the barium salt, by double decomposition with sulfuric acid:

$$Ba_5[IO_6]_2 + 5H_2SO_4 = 5BaSO_4 + 2H_5[IO_6].$$

It can also be prepared directly—e.g., at the anode by the electrolysis of a solution of iodic acid.

Periodic acid, H_5IO_6, forms colorless prismatic crystals, which rapidly deliquesce in moist air. They melt a little above 130°, and begin to decompose at slightly higher temperatures. The acid then loses oxygen and water, and is converted into iodine pentoxide, the anhydride of iodic acid; it is not possible to obtain periodic anhydride. Periodic acid is freely soluble in water. It is a quite weak pentabasic acid. Its normal salts, of the type $M^I_5[IO_6]$, are hydrolyzed to a considerable extent in solution.

Neutral and acid ortho periodates are known. There are also periodates derived from forms of periodic acid poorer in water—especially the *mesoperiodates*, $M^I_3[IO_5]$, the *metaperiodates*, $M^I[IO_4]$, and the *diperiodates*, $M^I_4[I_2O_9]$. Neutral sodium orthoperiodate, $Na_5[IO_6]$, cannot be prepared from solution, but was obtained by Zintl (1938), by heating sodium metaperiodate, $NaIO_4$, with Na_2O. It is also formed when sodium iodide is heated with sodium peroxide or with Na_2O in air, and is slowly formed when sodium iodide is fused with caustic soda. It is more stable than sodium metaperiodate.

Unlike the perchlorates, and unlike periodic acid itself, the periodates are almost all sparingly soluble in water. They are decomposed when heated.

The orthoperiodates are quite stable, and decompose only slowly when strongly heated, evolving iodine. The metaperiodates, however, in some cases decompose with explosive violence. Ammonium periodate may explode even on gentle friction (e.g., when it is touched with a spatula). It is peculiar, however, that whereas many periodates first lose oxygen when heated, forming iodates, many iodates undergo the converse reaction, and are converted to periodates (e.g., barium iodate, equation above). In solution, the iodates are more stable than the periodates, as shown by the oxidation-reduction potentials. A platinum electrode immersed in a solution containing iodate ions and periodate ions at equal concentrations assumes a potential 1.51 volts below the normal hydrogen electrode (Abel, 1932).

Although orthoperiodic acid and orthoantimonic acid are of the same structural type, the metaperiodates—e.g., $NaIO_4$, KIO_4, $RbIO_4$, $(NH_4)IO_4$, $AgIO_4$ —, unlike the metantimonates (cf. p. 668), form crystal structures of the scheelite type (cf. Vol. II), in which IO_4 radicals exist as discrete structural groups.

(l) Analytical (Bromates and Iodates)

Bromate and iodate ions differ from chlorate ions, in that they give sparingly soluble precipitates with silver nitrate. These are soluble in ammonia. If sulfur dioxide is passed into an ammoniacal solution of either silver bromate or iodate, silver bromide or iodide is precipitated, and may be differentiated either by differing solubility in concentrated ammonia solution, or, better, by decomposing the precipitate either with zinc and sulfuric acid, or with hydrogen sulfide water, and treating the solution with chlorine water and chloroform.

(m) Compounds of the Halogens with Each Other

INTERHALOGEN COMPOUNDS: SUMMARY

ClF gaseous m.p. —155.6° b.p. —100.1°	BrCl Only stable in equilibrium with Br_2 and Cl_2 BrF liquid m.p. about —33° b.p. +20°	IBr solid m.p. ca. 40° b.p. 116°	ICl solid Two modifications α (stable form): ruby red needles m.p. 27.2° density 3.22 β (metastable form): brown red rhombic plates m.p. 13.9° b.p. ca. 100°
ClF_3 gaseous m.p. —82.6° b.p. +12.1°	BrF_3 liquid m.p. +8.8° b.p. 127°	ICl_3 solid density 3.11 sublimes	
	BrF_5 liquid m.p. —61.3° b.p. +40.5°	IF_5 liquid m.p. + 8.5° b.p. +97° IF_7 gaseous m.p. +4.5° b.p. +5.5°	

Halogens which are adjacent to one another usually exert a single valence in their interhalogen compounds, although chlorine can also be trivalent towards fluorine. Bromine has a maximum valence of *five* towards fluorine, and iodine a valence of *seven*.

All the interhalogen compounds are volatile. BrF_3 has the highest boiling point 127°). Most of them are rather unstable, but none is explosive. All the compounds are *exothermic*, usually with small heats of formation—e.g., 2.5 kcal for IBr, 7.9 kcal for ICl, 21.5 kcal for ICl_3, 0.34 kcal for BrCl, 15.0 kcal for ClF and 42.0 kcal for ClF_3.

Chlorine monofluoride, ClF, was obtained by Ruff in 1928, by the direct union of the components at 250°. Its preparation is made more difficult by the extraordinary chemical aggressiveness of the compound, which reacts with glass, forming explosive chlorine oxides, and attacks metals, including arsenic and antimony, more vigorously than does fluorine itself. Chlorine monofluoride is an almost colorless gas, which can be liquefied and solidified only at very low temperatures.

Chlorine trifluoride, ClF_3, is formed by the action of excess fluorine on ClF. It is an easily condensable gas (liter weight 3.57 g), light green in the liquid state, and colorless when solidified. Its great chemical reactivity is noteworthy; glass wool catches fire at once in ClF_3 vapor. MgO, CaO, Al_2O_3, and other substances, which are themselves very stable, also react with incandescence in ClF_3 (Ruff, 1930).

Bromine trifluoride, BrF_3, is also formed by the direct combination of the elements; *bromine monofluoride*, BrF is formed as an intermediate product, but at once disproportionates into BrF_3 and Br_2. Bromine trifluoride is a heavy, colorless liquid, which fumes in air and strongly attacks the skin. It is very reactive. It combines with fluorine at 200°, forming *bromine pentafluoride*, BrF_5 (Ruff, 1931–33).

Iodine pentafluoride, IF_5, a colorless liquid, fuming in air, is prepared by passing fluorine over dry iodine. It is highly reactive, and its vapors strongly attack the respiratory organs.

Electron diffraction measurements have shown that in IF_5 all the fluorine atoms are equidistant from the iodine, the molecule probably having the form of a trigonal bipyramid.

Iodine heptafluoride (Ruff, 1930–31), formed by the combination of the pentafluoride with fluorine, is still more reactive than IF_5. It resembles chlorine trifluoride in its powerful fluorinating action.

Bromine monochloride, BrCl, was discovered by Balard in 1828, who obtained it by passing chlorine into liquid bromine. The existence of the compound was definitely established, however, only by recent investigations. It is only stable in equilibrium with its constituents, Br_2 and Cl_2, being dissociated to the extent of 43% at ordinary temperature. The heat of formation is only 0.34 kcal per mol.

Iodine monochloride, ICl, is obtained by passing chlorine over iodine, or by boiling iodine with aqua regia and extracting the diluted solution with ether. It is a red brown oil, but is converted by distillation into ruby-red needles of the more stable a-form, which is solid at ordinary temperature. The primary reaction in its hydrolysis gives hydrochloric acid and iodine hydroxide (hypoiodous acid):

$$ICl + HOH = HCl + IOH.$$

The latter rapidly decomposes, so that the end products of hydrolysis are hydrochloric acid, iodic acid, and free iodine:

$$5ICl + 3HOH = 5HCl + HIO_3 + 2I_2.$$

Iodine bromide, IBr, is very similar to the chloride. Iodine is the electropositive component in these compounds, as is shown by the mode of hydrolysis. This view is substantiated by the fact that other negative radicals can combine with univalent iodine—e.g., the radicals of hydrocyanic acid (forming iodine cyanide, ICN) and of hydrazoic acid (forming iodazide, IN_3).

Iodine trichloride, ICl_3, is obtained by the action of excess chlorine on iodine, or by treating the monochloride with chlorine. It forms yellow needles which are deliquescent in air, and have a penetrating, pungent smell. It dissolves in organic solvents such as alcohol, ether, or benzene, with a dark orange red color. It is decomposed by water according to the equation:

$$2ICl_3 + 3H_2O = 5HCl + ICl + HIO_3.$$

The hydrolysis may be repressed by addition of hydrochloric acid. A solution of iodine trichloride in hydrochloric acid is a very useful reagent for dissolving sulfide minerals such as pyrite, since the compound has very good oxidizing properties towards sulfur (Birk, 1928, Wilke-Dörfurt 1930).

Iodine trichloride can add on to the alkali chlorides and chlorides of other strongly electropositive metals, forming well crystallized salts of the general type $M^I[ICl_4]$. These salts (tetrachloroiodates) can also be obtained by the action of concentrated hydrochloric acid on iodates:

$$K[IO_3] + 6HCl = K[ICl_4] + Cl_2 + 3H_2O.$$

Although iodine trichloride loses chlorine when it is heated, it must be regarded as a definite compound of trivalent iodine ('compound of the first order'), and not merely as an addition compound of ICl and Cl_2.

There is evidence that certain of the interhalogen compounds undergo ionization to a small degree in the liquid state. Fused iodine monochloride and iodine trichloride have specific conductivities of about 10^{-3} ohm^{-1}cm^{-1}, which is distinctly larger than is usual for non-ionic liquids, and they also yield conducting solutions in organic solvents. It is probable that they ionize according to the equations:

$$2ICl \rightleftharpoons I^+ + ICl_2^-; \qquad 2ICl_3 \rightleftharpoons ICl_2^+ + [ICl_4]^-.$$

Bromine trifluoride and iodine pentafluoride are also ionized similarly to a slight extent (e.g., $2BrF_3 \rightleftharpoons BrF_2^+ + BrF_4^-$). Liquid BrF_3 is of particular interest, since it dissolves and combines with a number of metallic fluorides. These may either form salts of the $[BrF_4]^-$ anion with the cation of the added salt—e.g.,

$$KF + BrF_3 = K[BrF_4],$$

or, alternatively, fluorides which can give rise to complex fluoro-anions may combine to form fluoro-salts of the complex cation $[BrF_2]^+$. This is the case with the fluorides of e.g., antimony and gold:

$$SbF_5 + BrF_3 = SbBrF_8 \qquad (= [BrF_2][SbF_6])$$

$$AuF_3 + BrF_3 = AuBrF_6 \qquad (= [BrF_2][AuF_4]).$$

These reactions in liquid BrF_3 solutions have been shown to bear many analogies to the neutralization reactions of basic and acidic oxides in aqueous solutions. In the latter case, the solvent undergoes an auto-ionization similar to that represented above for BrF_3

$$2H_2O \rightleftharpoons [H_3O]^+ + OH^-.$$

(Eméleus 1948 and later; see also Vol. II, Chap. 18).

(n) Salt-like Compounds with Electropositive Halogens

As has already been mentioned on p. 784, bromine and iodine are capable of forming salt-like compounds, in which they function as the cationic constituents. A considerable number of such compounds is known for iodine in particular, derived both from the univalent and trivalent states of this element.

(i) *Salt-like Iodine(I) Compounds.* Iodine(I) salts of the ordinary strong acids such as $I[NO_3]$ and $I[ClO_4]$ are not themselves stable, but they can be stabilized by complex formation,—e.g., by addition of pyridine. The electropositive univalent iodine can also be stabilized by combining it with highly polarizable acid anions—i.e., with anions which have a marked tendency to act as ligands in complexes. The compounds of this group, such as $I[CN]$, $I[CNO]$, $I[CNS]$, are usually prepared, like the addition compounds of the type of $[I \, pyr_2][NO_3]$, by the reaction of the silver salts with iodine in non-aqueous (ethereal) solution—e.g.,

$$Ag[CNO] + I_2 = AgI + I[CNO].$$

They do not possess such markedly salt-like character as do the compounds with a complex iodine(I) cation.

(ii) *Salt-like Iodine(III) Compounds.* The first salt of electropositive, trivalent iodine to be prepared was the *acetate*, $I[C_2H_3O_2]_3$, obtained by Schützenberger in 1861, as colorless hexagonal crystals, stable in the absence of light. A considerable range of similar salts has subsequently been prepared, especially by Fichter (1915) and by Masson—e.g., the perchlorate, $I[ClO_4]_3 \cdot 2H_2O$ (greenish yellow), the sulfate, $I_2[SO_4]_3$ (bright yellow), and the phosphate (yellow). The compounds I_4O_9 and I_2O_4 are also to be assigned to this group, since they can be considered to be iodates, $I_4O_9 = I[IO_3]_3$ (yellow white powder), $I_2O_4 = [IO][IO_3]$ (iodyl iodate, pale yellow crystals).

(iii) *Iodoso Compounds.* Another group of salt like compounds of trivalent iodine is derived from the *iodoso compounds*. The designation iodoso-compound (by analogy with the nitroso-compounds) is applied to compounds of the type $R—I=O$ (R = organic radical), which are obtained by oxidizing certain aryl iodides with fuming nitric acid. The iodoso compounds, which were investigated chiefly by Willgerodt [5], have a definitely basic character, and form true salts with acids, $R—IX_2$ (X = acid radical). These salts are hydrolyzed by water, the iodoso compounds being reformed. The iodoso-compounds, which are light grey yellow substances with a peculiar, penetrating smell, decompose when they are heated, and form an aryl iodide and an iodoxy compound:

$$2R—I=O = R—I + R—IO_2$$

(*iv*) *The aryl iodoxy compounds* can be derived from iodic acid, by exchanging the hydroxyl group for an aryl radical. Unlike the iodoso-compounds they have no basic properties. Their aqueous solutions are neutral in reaction. Whereas the iodoso compounds have strong oxidizing properties, the aryl iodoxy-compounds are only weak oxidants. If heated in the dry state, however, they decompose with a violent detonation.

References

1. O. Ruff, *Die Chemie des Fluors*, Berlin 1920, 136 pp.

2. R. N. Haszeldine and A. G. Sharpe, *Fluorine and Its Compounds*, London 1951, 153 pp.

3. O. Kausch, *Flussäure, Kieselflussäure und deren Metallsalze*, Stuttgart 1936, 438 pp.

4. E. Abel and V. Engelhardt, *Hypochlorite und elektrische Bleiche*, Vol. I, Halle 1905, 111 pp.; Vol. II, Halle 1903, 275 pp.

5. C. Willgerodt, *Die organischen Verbindungen mit mehrwertigem Jod*, Stuttgart 1914, 265 pp.

CHAPTER 18

SALT FORMATION AND NEUTRALIZATION

1. Neutral, Acid, and Basic Salts

The most important and most abundant compound of the halogens, common salt, known in every day life simply as 'salt' is, in fact, the *prototype of salts in the chemical sense*. A salt, in chemistry, is a substance which can be obtained by the combination of an acid with a base, whereby water is eliminated.

Thus common salt may be formed by the combination of hydrochloric acid with sodium hydroxide:

$$HCl + NaOH = NaCl + H_2O.$$

The reaction depends on the fact that hydrogen, which may be dissociated in ionic form from the hydrochloric acid, unites with the hydroxyl group, which may similarly be dissociated as an ion from the base, sodium hydroxide. The remaining ions, Na^+ and Cl^-, remain as such in the solution. If the solution is evaporated, these ions come together to form the crystal lattice represented in Fig. 44 (p. 209).

The same considerations apply to the formation of other salts. If the salts are sparingly soluble—e.g., CaF_2, $Ba[SO_4]$ —, association of ions to build up the crystal lattice takes place even in the dilute solution. Free ions are left in the solution in amounts corresponding to the solubility of the salt. In rare instances (almost completely limited to mercury salts), the ions composing the salt already associate in solution to form molecules. Almost all other salts are practically completely dissociated into ions in their solutions. This is the basis for an important characteristic of typical soluble salts: *all typical salts are strong electrolytes in aqueous solution.*

If all the dissociable hydrogen atoms of an acid enter into reaction with the hydroxyl groups of bases, the acid is said to be *neutralized* by the bases. In the same way one speaks of the neutralization of bases by acids. Such quantities of acid and base as will neutralize one another are said to be *equivalent*. Salts formed by the mutual neutralization of equivalent amounts of acid and base are called *neutral* or *normal salts*.

As has already been seen in numerous instances, it is often possible to obtain, in crystalline form, salts which result from the reaction of several equivalents of acid with one equivalent of base. These are termed *acid* salts or hydrogen salts.*

Examples:

$$2NaOH + H_2SO_4 = 2H_2O + Na_2SO_4 \quad \text{neutral sodium sulfate}$$
$$NaOH + H_2SO_4 = H_2O + NaHSO_4 \quad \text{acid sodium sulfate}$$
$$\text{(sodium hydrogen sulfate)}$$

* For the nomenclature of salts cf. pp. 192 and 630.

$$3NaOH + H_3PO_4 = 3H_2O + Na_3PO_4 \quad \text{neutral sodium phosphate}$$
$$2NaOH + H_3PO_4 = 2H_2O + Na_2HPO_4 \quad \text{monoacid sodium phosphate}$$
$$\text{(disodium hydrogen phosphate)}$$
$$NaOH + H_3PO_4 = H_2O + NaH_2PO_4 \quad \text{diacid sodium phosphate}$$
$$\text{(sodium dihydrogen phosphate)}$$

In a similar manner, many salts have been obtained which contain more than one equivalent of base for each equivalent of acid. These are termed *basic* salts or hydroxy-(oxy-) salts.

Examples:

$PbCO_3$	neutral lead carbonate
$Pb_3(OH)_2(CO_3)_2$	basic lead carbonate (white lead)
$SbCl_3$	antimony chloride
$SbOCl$	basic antimony chloride, antimony oxychloride, antimonyl chloride

2. General Methods for the Preparation of Salts

The following are the principal methods available:

(1) Combination of acids with bases, water being eliminated.
The reaction underlying this method can be represented by the general equation

$$HX + MOH = MX + HOH.$$

For preparative purposes, the anhydrides are often used in place of the acids or bases themselves.

Thus lead nitrate, $Pb(NO_3)_2$, is prepared technically by dissolving lead oxide (the anhydride of the base $Pb(OH)_2$) in nitric acid. Quartz sand, i.e., silicon dioxide, the anhydride of silicic acid, is almost always used in the preparation of silicates.
If anhydrides of both base and acid are used, the process of salt formation appears formally as an addition reaction—e.g.,

$$CaO + SiO_2 = CaSiO_3,$$

whereas the reaction between base and acid proper is a double decomposition

$$HX + M(OH) = MX + HOH.$$

The following three methods of salt formation are also double decompositions.

(2) Reaction between an acid (or its anhydride) and a salt.

$$HX + MY = MX + HY \quad \text{(X and Y = acid radicals)}$$

This double decomposition has already been met with in Chap 2 as a method for the preparation of acids. It is also much used for the preparation of salts, although rather in analytical chemistry than in preparative work. It is used especially when the salt formed by the reaction is sparingly soluble (e.g., the precipitation of barium sulfate by means of sulfuric acid, or of heavy metal sulfides with hydrogen sulfide), or when the resulting acid (or its anhydride) is volatile (e.g., preparation of chlorides, sulfates, nitrates, etc. by adding the corresponding acids to metal carbonates). In these cases the reaction proceeds to completion from left to right because the equilibrium is displaced.

(3) Double decomposition between a base (or its anhydride) and a salt.

$$M(OH) + RX = R(OH) + MX \qquad \text{(M and R = metals or positive radicals)}$$

This reaction, which is one of the most important for the preparation of bases (cf. p. 61), is used only in exceptional cases in preparing salts. It is occasionally of advantage, if the salt to be prepared is derived from such a weak acid that it is only stable in alkaline solution.

(4) Double decomposition between two salts.

$$MY + RX = RY + MX$$

This method is often used in preparing salts. It can always be used if one or other of the salts formed differs substantially from the other either in its solubility or its volatility (or involatility).

Even if the salts do not differ very much in solubility in aqueous solution, they may do so if a different solvent is employed. Only the *difference* in solubility (or volatility) of the salts is involved. If the salts differ sufficiently in solubility, the method can be used for the preparation of quite soluble salts, although in such a case recrystallization is usually necessary to obtain a pure product. The best known example of the use of this method is the preparation of 'conversion saltpeter' (p. 192).

(5) Dissolution of metals in acids.

In this case, either the hydrogen of the acid may be simply substituted by the metal (as, e.g., in the dissolution of zinc in sulfuric acid, $Zn + H_2SO_4 = ZnSO_4 + H_2$), or a portion of the acid may be decomposed and consumed in the oxidation of the metal. This is the case in the dissolution of lead in nitric acid (preparation of lead nitrate):

$$3Pb + 8HNO_3 = 3Pb(NO_3)_2 + 2NO + 4H_2O.$$

Even in the first case, the reaction is strictly speaking not a simple substitution, but involves an exchange of charge or electron transfer, as is evident when the reaction is written in ionic form:

$$Zn + 2H^+ + SO_4^= = Zn^{++} + SO_4^= + H_2$$

or

$$Zn + 2H^+ = Zn^{++} + H_2$$

The difference between the two cases therefore lies only in the fact that in the former the metal is oxidized (i.e., converted to a cation by loss of electrons) by the *hydrogen ion*, and in the second case by the *acid radical*. The lower in the electrochemical series any metal stands, the stronger must be the oxidizing properties of any acid used to dissolve it. Thus, zinc dissolves in dilute sulfuric acid (oxidizing agent—the hydrogen ion), copper only in concentrated sulfuric acid or nitric acid (oxidizing agent in both cases—the acid radical), and gold only dissolves in aqua regia (oxidizing agent—free chlorine).

Transference of charge between two metals is also of importance in the preparation of salts. It is often employed as a means of purifying salts—e.g., ferrous sulfate can be purified from copper (and from iron(III) salts at the same time) by placing some pure iron in the solution:

$$Cu^{++} + Fe = Fe^{++} + Cu, \qquad 2Fe^{+++} + Fe = 3Fe^{++}$$

Transference of charge between non-metals is also occasionally of use in preparing salts. The preparation of potassium hydrogen diiodate, $KH[IO_3]_2$, by the action of iodine on potassium chlorate, discussed in the previous chapter, is an instance of this.

(6) Salts of *hydroacids* can often be prepared by the *direct combination of the elements*.

Examples:

$$Fe + S = FeS; \qquad Sn + 2Cl_2 = SnCl_4;$$

$$2Sb + 3I_2 = 2SbI_3; \qquad 2Al + 3Se = Al_2Se_3.$$

Acid salts are obtained by incomplete neutralization of acids, or by adding an acid to a neutral salt. Acid salts are commonly formed as the first stage in applying method (2) above.

Basic salts are obtained by the incomplete neutralization of bases, but are generally prepared either by mixing solutions of the neutral salt and the free base, or by partial hydrolysis of the neutral salt (e.g., by pouring its solution into hot water). The sparingly soluble basic salts of antimony and bismuth, in particular, are prepared in this way. Basic salts are frequently obtained instead of the neutral salts, when the salts of heavy metals with weak acids are prepared in aqueous solution.

3. Hydrolysis

The term hydrolysis is applied to the decomposition of a salt by water, to reform the acid and base from which the salt was derived. E.g.,

$$AlCl_3 + 3HOH \rightleftharpoons 3HCl + Al(OH)_3$$

In the direction from right to left, the equation would represent the formation of the salt from the acid and the base. The hydrolysis of a salt is thus the converse of salt formation.

The term *hydrolysis* is more widely, and quite generally, used to describe decompositions brought about by water, which have as a feature common to all that the components of water combine with the products of decomposition. It thus includes the splitting of an ester into an acid and an alcohol. Many compounds, especially the halides of non-metals, furnish two different acids as their products of hydrolysis. Thus phosphorus trichloride is hydrolyzed to hydrochloric and phosphorous acid, and silicon tetrachloride to hydrochloric acid and silicic acid (or hydrated silica):

$$PCl_3 + 3HOH = 3HCl + P(OH)_3$$

$$SiCl_4 + 3HOH = 4HCl + H_2SiO_3$$

Whereas processes of the kind last mentioned almost invariably proceed to completion in aqueous solution, the hydrolytic decomposition of salts (and also of esters) often results in an *equilibrium*. They are said to involve *partial hydrolysis*. The position of equilibrium depends upon the experimental conditions. Hydrolysis is promoted by raising the temperature, by diluting the solution, and above all by displacement of the equilibrium through the deposition or volatilization of one or both hydrolysis products.

A salt necessarily undergoes hydrolysis if the ions of water are able to associate to any appreciable extent, forming undissociated molecules, with the ions formed by electrolytic dissociation of the salt. This is the case if the anion of the salt combines with the hydrogen ion to form a weak acid, or if the cation of the salt combines with the hydroxyl ion, forming a weak base. Hence all salts derived either from a weak acid, or from a weak base (or from both) are hydrolyzed; the weaker the acid (or base), the greater is the extent of hydrolysis.

As an example of the hydrolysis of a salt of a weak acid, we may consider the case of sodium cyanide, NaCN. Like all typical salts, this undergoes complete primary dissociation in solution:

$$NaCN = Na^+ + CN^-.$$

Since, however, hydrocyanic acid is a weak acid, the CN^- ions associate with H^+ ions from the water, to form undissociated hydrocyanic acid (the presence of which can be perceived directly, from the smell):

$$CN^- + H^+ = HCN.$$

This disturbs the dissociation equilibrium of the water, $HOH \rightleftharpoons H^+ + OH^-$, and since the mass action law holds:

$$[H^+] \cdot [OH^-] = const.,$$

the OH^- ion concentration must increase in the same proportion as the H^+ ion concentration is diminished. Since sodium cyanide is the salt of a strong base, the OH^- ions remain free in solution, which therefore acquires a basic reaction. This can at once be seen if the two equations above are combined:

$$CN^- + HOH \rightleftharpoons HCN + OH^-.$$

In a similar manner, the hydrolysis of the salt of a weak base and a strong acid produces free H^+ ions in the solution, and the solution displays an acid reaction. If the salt suffering hydrolysis is derived from a weak base and a weak acid, the reaction of the solution depends upon which of these is the more strongly dissociated, and if both are dissociated to a similar degree, the solution may remain practically neutral. Thus a solution of ammonium acetate has a neutral reaction, although the salt is markedly hydrolyzed. In general:

Solutions of salts derived from weak acids with strong bases have a basic reaction as a result of hydrolysis; solutions of salts of strong acids with weak bases react acid because of hydrolysis. It is possible for hydrolysis to confer a basic reaction on solutions of *acid* salts, as well as of 'neutral' (normal) salts. Thus solutions of the acid alkali sulfides (alkali hydrogen sulfides), M^1HS, have a strongly basic reaction.

The fraction of the dissolved amount of salt which is split up by hydrolysis under a particular set of conditions—i.e., which has entered into reaction with water, forming base and acid—is known as the *degree of hydrolysis*, and may be expressed either as a fraction or as a percentage. The degree of hydrolysis depends upon the temperature and the dilution. It may be calculated from the strengths of the acid and base from which the salt is derived. Conversely, the degree of hydrolysis of salts is frequently used to determine the dissociation constants of weak acids and bases.

4. Measurement and Calculation of Degree of Hydrolysis [1]

Account must be taken of the following processes in the solution of a salt M'X.

$$M'X = M^+ + X^- \qquad \text{Dissociation of salt} \qquad (1)$$

$$\left.\begin{array}{l} M^+ + HOH \rightleftharpoons MOH + H^+ \\[4pt] X^- + HOH \rightleftharpoons HX + OH^- \end{array}\right\} \quad \text{Hydrolysis equilibria} \qquad \begin{array}{l}(2)\\[12pt](3)\end{array}$$

$$MOH \rightleftharpoons M^+ + OH^- \qquad \text{Dissociation equilibrium of base} \qquad (4)$$

$$HX \rightleftharpoons H^+ + X^- \qquad \text{Dissociation equilibrium of acid} \qquad (5)$$

$$HOH \rightleftharpoons H^+ + OH^- \qquad \text{Dissociation equilibrium of water} \qquad (6)$$

In the case of the salt of a strong base with a strong acid, the equilibria, (2), (3), (4) and (5) practically go out of consideration. The hydrogen ion concentration of pure water is not altered by dissolving such a salt.

If the salt is derived from a weak acid with a strong base, equilibria (2) and (4) may be left out of account. The degree of hydrolysis β_1 (as a fraction) is in this case given, by definition, as

$$\beta_1 = \frac{[HX]}{c} \qquad (7)$$

where $[HX]$ is the molar concentration of acid resulting from the hydrolysis. and c is the molar concentration of dissolved salt (total or analytical concentration). The quantity of OH^- ions in the solution is the sum of that formed by processes (3) and (6). Hence

$$[OH^-] = [HX] + [H^+] \qquad \text{or} \qquad [HX] = [OH^-] - [H^+] \qquad (8)$$

Since, as was discussed on p. 52,

$$[H^+] \cdot [OH^-] = k_w \qquad (9)$$

where k_w is the dissociation constant of water, we obtain for the degree of hydrolysis:

$$\beta_1 = \frac{[OH^-] - \dfrac{k_w}{[OH^-]}}{c} \qquad \text{or} \qquad \beta_1 = \frac{[OH^-]^2 - k_w}{[OH^-] \cdot c} = \frac{k_w - [H^+]^2}{[H^+] \cdot c} \qquad (10a)$$

The degree of hydrolysis can thus be determined experimentally, by measuring the hydrogen ion concentration (or more accurately, the hydrogen ion activity).* If the hydrolysis confers a distinctly basic reaction to the solution ($[OH^-] > 10^{-6}$ at room temperature),

---

* As was discussed on p. 67 *et seq.*, the *activities* of the ions, and not their *true concentrations*, enter into their equilibria. If a_{H+} and a_{OH-} denote the activities of the H^+ and OH^- ions, we have that

$$a_{H+} = f_{H+} \cdot [H^+] \qquad \text{and} \qquad a_{OH-} = f_{OH-} \cdot [OH^-],$$

where $[H^+]$, $[OH^-]$ are the true concentrations, and f_{H+}, f_{OH-} are the activity coefficients of the ions. Remembering that k_w is the product of the *activities* of the H^+ and OH^- ions not the product of their true concentrations, it follows that in (10a) and (10b), k_w should be divided by $f_{H+} \cdot f_{OH-}$. Equations (11a) and (11b) are thereby changed into

$$\beta_1 = \frac{k_w}{a_{H+} \cdot f_{OH-} \cdot c} \qquad \text{and} \qquad \beta_2 = \frac{a_{H+}}{f_{H+} \cdot c}$$

k_w ($\sim 10^{-14}$) is negligible, compared with $[OH^-]^2$, and $[H^+]^2$ similarly vanishes in comparison with k_w. Equation (10a) then takes on the simpler form (11a):

$$\beta_1 = \frac{[OH^-]}{c} = \frac{k_w}{[H^+] \cdot c} \qquad (11a)$$

For the hydrolysis of the salt of a weak base with a strong acid we have similarly

$$\beta_2 = \frac{[H^+]^2 - k_w}{[H^+] \cdot c}, \quad \text{and if } [H^+] > 10^{-6}, \quad \beta_2 = \frac{[H^+]}{c} \qquad (10b), (11b)$$

These formulas have been derived for the case of uni-univalent salts. They hold also for multivalent salts if c is taken as the equivalent concentration instead of the molar concentration.

These equations are not valid however, for the hydrolysis of the salt of a weak base with a weak acid. In this case all the equilibria (2) to (6) must be taken into account, and there is no simple relation between the hydrogen ion concentration and the degree of hydrolysis. The latter can, however, still be calculated from the dissociation constants of the acid and base.

We can arrive at general expressions for the calculation of the hydrolysis constant and the degree of hydrolysis, involving the dissociation constants of the acids and bases concerned, by the following considerations. It is assumed that reaction takes place in a homogeneous system (an aqueous solution, with no precipitation or volatilization of the products of reaction). The distinction between true concentrations and activities may be ignored. It will be seen later that the values calculated from the dissociation constants—unlike those based on measurements of hydrogen ion concentration, according to (10a), (11a) or (10b), (11b)—are but little affected by this approximation.

(a) *Hydrolysis of the salt of a weak acid and a strong base*. At a constant temperature, we can apply the mass action law to eqn (3), and obtain

$$\frac{[HX] \cdot [OH^-]}{[X^-] \cdot [H_2O]} = \text{const.}$$

At the concentrations usually considered, the amount of water is not significantly altered by reaction (3), so that the concentration of water can be combined with the constant, thus giving

$$K_{hydrol} = \frac{[HX] \cdot [OH^-]}{[X^-]} \qquad (12)$$

K_{hydrol} is called the *hydrolysis constant*. Application of the mass action law to eqn (5) gives

$$K_{HX} = \frac{[H^+] \cdot [X^-]}{[HX]} \quad \text{(Dissociation const. of HX)} \qquad (13a)$$

Combining (13a) with (9) and (12), we obtain

$$K_{hydrol} = \frac{k_w}{K_{HX}} \qquad (14a)$$

The hydrolysis constant of the salt of a weak acid and a strong base is thus equal to the ratio between the dissociation constant of water and the dissociation constant of the acid.

Introducing (13a) and (9), eqn. (8) becomes

$$[OH^-]^2 = k_w \left(1 + \frac{[X^-]}{K_{HX}} \right)$$

For weak or moderate degrees of hydrolysis ($\beta < 0.01$), we may ignore the diminution in

[X⁻] which results from reaction (3),* and replace [X⁻] by the analytical molar concentration of the salt, c.
We then have

$$[OH^-] = \sqrt{k_w\left(1 + \frac{c}{K_{HX}}\right)} \tag{15a}$$

When substituted in eqn. (10a), this gives

$$\beta_1 = \sqrt{\frac{k_w}{K_{HX}(K_{HX} + c)}} \tag{16a}$$

(b) *Hydrolysis of the salt of a weak base and a strong acid.* For this case we arrive by similar reasoning at the expressions

$$\frac{[M^+] \cdot [OH^-]}{[MOH]} = K_{MOH} \text{ (dissociation constant of the weak base MOH)} \tag{13b}$$

$$K_{hydrol} = \frac{k_w}{K_{MOH}} \tag{14b}$$

The hydrolysis constant of the salt of a weak base and a strong acid is equal to the ratio between the dissociation constant of water and the dissociation constant of the base.

$$[H^+] = \sqrt{k_w\left(1 + \frac{c}{K_{MOH}}\right)} \tag{15b}$$

$$\beta_2 = \sqrt{\frac{k_w}{K_{MOH}(K_{MOH} + c)}} \tag{16b}$$

(c) *Hydrolysis of the salt of a weak base and a weak acid.* All the equilibria (2) to (6) must be taken into account, with equations (13a), (13b) and (9) governing them. Multiplying (13a) by (13b), we find

$$\frac{[MOH] \cdot [HX]}{[M^+] \cdot [X^-]} = \frac{k_w}{K_{HX} \cdot K_{MOH}} \tag{14c}$$

In the hydrolysis of salts of this type, as has already been mentioned, the H⁺-ion concentration of water is often practically unaffected. This is the case when almost equivalent amounts of undissociated acid and undissociated base are formed. Since it has been assumed

---

* If the extent of hydrolysis is considerable, the decrease in [X⁻] must be allowed for: [X⁻] = c — [HX]. However, in this case, [H⁺] in eqn (8) vanishes compared with [HX]. We thus have

$$[OH^-] = [HX] = c - [X^-] = c - (K_{HX} \cdot [OH^-]^2/k_w),$$

$$[OH^-]^2 = (k_w/K_{HX})(c - [OH^-]).$$

From this it follows that

$$\beta = \sqrt{\frac{a}{c}}\left(\sqrt{1 + \frac{a}{4c}} - \sqrt{\frac{a}{4c}}\right), \text{ where } a = \frac{k_w}{K_{HX}}$$

The same equation gives the degree of hydrolysis of the salt of a weak base with a strong acid (for $\beta \geqslant 0.01$), by substituting $a = \dfrac{k_w}{K_{MOH}}$.

that these remain in the solution, $[MOH] = [HX]$, and $[M^+] = [X^-]$. We may often write $[M^+] = [X^-] = c$ without significant error. Then, from (14c),

$$\frac{[MOH]^2}{c^2} = \frac{[HX]^2}{c^2} = \frac{k_w}{K_{HX} \cdot K_{MOH}}$$

or

$$[MOH] = [HX] = c \sqrt{\frac{k_w}{K_{HX} \cdot K_{MOH}}} \qquad (15c)$$

and the resulting value for the degree of hydrolysis is

$$\beta_3 = \frac{[MOH]}{c} = \frac{[HX]}{c} = \sqrt{\frac{k_w}{K_{HX} \cdot K_{MOH}}} \qquad (16c)$$

It follows from equation (16c) that the extent of hydrolysis of salts of weak bases and weak acids is independent of the concentration, on the assumptions made initially. For salts of strong bases and weak acids, or of weak bases and strong acids, however, the degree of hydrolysis increases as the concentration is reduced, in accordance with equations (16a) and (16b). In all cases, the smaller the values of K_{HX} and K_{MOH} (i.e., the weaker the acids or bases concerned), the more extensive is the hydrolysis of their salts. Table 106 gives the degree of hydrolysis of salts of strong bases and weak acids, or of weak bases and strong acids as a function of the dissociation constant of the weak acid or base, and of the analytical molar concentration of the salt MX; these values are calculated from the equations given above. For salts of strong acids and strong bases, these equations give $\beta = 0$, since in this case $K_{HX} = K_{MOH} = \infty$.

TABLE 106

DEGREE OF HYDROLYSIS OF SALTS AT 22°

The numbers represent the fraction hydrolyzed in a solution of molar concentration c, for salts formed by a strong base (or strong acid) with a weak acid (or weak base), of dissociation constant K.

K	$c = 1$	$c = 0.1$	$c = 0.01$	$c = 0.001$	$c = 0.0001$
10^{-1}	$0.30 \cdot 10^{-6}$	$0.71 \cdot 10^{-6}$	$0.95 \cdot 10^{-6}$	$1.00 \cdot 10^{-6}$	$1.00 \cdot 10^{-6}$
10^{-2}	$0.10 \cdot 10^{-5}$	$0.30 \cdot 10^{-5}$	$0.71 \cdot 10^{-5}$	$0.95 \cdot 10^{-5}$	$0.995 \cdot 10^{-5}$
10^{-3}	$0.32 \cdot 10^{-5}$	$0.995 \cdot 10^{-5}$	$3.0 \cdot 10^{-5}$	$7.1 \cdot 10^{-5}$	$9.5 \cdot 10^{-5}$
10^{-4}	$0.10 \cdot 10^{-4}$	$0.32 \cdot 10^{-4}$	$0.995 \cdot 10^{-4}$	$3.0 \cdot 10^{-4}$	$7.1 \cdot 10^{-4}$
10^{-5}	$0.32 \cdot 10^{-4}$	$0.10 \cdot 10^{-3}$	$0.32 \cdot 10^{-3}$	$0.995 \cdot 10^{-3}$	$3.0 \cdot 10^{-3}$
10^{-6}	$0.10 \cdot 10^{-3}$	$0.32 \cdot 10^{-3}$	$1.0 \cdot 10^{-3}$	$3.2 \cdot 10^{-3}$	$9.95 \cdot 10^{-3}$
10^{-8}	$0.10 \cdot 10^{-2}$	$0.32 \cdot 10^{-2}$	$1.0 \cdot 10^{-2}$	$3.1 \cdot 10^{-2}$	$9.5 \cdot 10^{-2}$
10^{-10}	$0.10 \cdot 10^{-1}$	$0.32 \cdot 10^{-1}$	$0.95 \cdot 10^{-1}$	$2.7 \cdot 10^{-1}$	$6.2 \cdot 10^{-1}$
10^{-12}	0.095	0.27	0.62	0.92	1.00

In deducing the above equations for the degree of hydrolysis, and in calculating the values given in Table 106, the distinction between true concentrations and activities was not taken into account. If this is to be considered, the values for β_1 calculated from eqn (16a) must be multiplied by the factor $\sqrt{f_{X^-}/f_{OH^-}}$, the values for β_2 (eqn (16b)) by the factor $\sqrt{f_{M^+}/f_{H^+}}$, and those for β_3 (eqn (16c)) by the factor $\sqrt{f_{X^-} \cdot f_{M^+}}$. f_{X^-}, f_{M^+}, f_{OH^-} and f_{H^+} here stand for the activity coefficients of the corresponding ions in the solution concerned. Equations (14a) and (14b) remain unaffected when the distinction between activities and concentrations is considered.

Examples of the calculation of degree of hydrolysis. — For sodium acetate at 18°, substitution of the values for k_w, K_{HX} given on pp. 52 and 454 in eqn. (16a) gives, for $c = 1$,

$$\beta = \sqrt{\frac{0.74 \cdot 10^{-14}}{1.75 \cdot 10^{-5} \cdot 1.00}} = 2.06 \cdot 10^{-5}$$

and for $c = 0.1$,

$$\beta = 6.5 \cdot 10^{-5}.$$

For $c = 0.0001$,

$$\beta = \sqrt{\frac{0.74 \cdot 10^{-14}}{1.75 \cdot 10^{-5} \cdot 1.175 \cdot 10^{-4}}} = 0.19 \cdot 10^{-2}.$$

In percentages, the degree of hydrolysis at these concentrations is thus 0.0021%, 0.0065% and 0.19%. If the activity coefficients are included in the calculation, these become 0.0022%, 0.0066% and 0.19%.

For the hydrolysis of potassium cyanide, using the value for K_{HCN} given on p. 460, we have in 1-molar solution

$$\beta = \sqrt{\frac{0.74 \cdot 10^{-14}}{4.8 \cdot 10^{-10} \cdot 1.00}} = 0.39 \cdot 10^{-2}$$

and in 0.1 – molar solution, $\beta = 1.24 \cdot 10^{-2}$.*
Correction for the activity coefficients makes these values $0.37 \cdot 10^{-2}$ and $1.245 \cdot 10^{-2}$ (found experimentally $0.38 \cdot 10^{-2}$ and $1.20 \cdot 10^{-2}$).

For ammonium acetate, the degree of hydrolysis calculated from eqn (16c) is

$$\beta = \sqrt{\frac{0.74 \cdot 10^{-14}}{1.75 \cdot 10^{-5} \cdot 1.75 \cdot 10^{-5}}} = 0.49 \cdot 10^{-2}.$$

In this case, inclusion of the activity coefficients yields an appreciably lower value, since

$$\sqrt{f_{NH_4^+} \cdot f_{Ac^-}} = f_{NH_4Ac} = 0.54.$$

This gives for the degree of hydrolysis $\beta = 0.54 \cdot 0.0049 = 0.0027$ or 0.27%. In general, the values calculated from eqn (16c) are significantly altered by taking account of the difference between true concentration and activities, whereas those based on (16a) or (16b) are generally but little affected. This is because in (16c) the correction factor involves the square root of the *product*, and in the other case the square root of the *quotient* of the activity coefficients. Since the activity coefficients of univalent ions differ but little in their values, their quotient is not, as a rule, appreciably different from unity.

5. Neutralization

A solution is said to be neutral when it is neither acid not basic in reaction. Neutralization thus signifies removal of the acidic or basic reaction**. An acid reaction, due to the presence of hydrogen ions in high concentration, is removed by adding hydroxyl ions (i.e., by adding a base). Conversely, the basic reaction due

* Since the degree of hydrolysis is only just greater than 1%, equation (16a) can be used with sufficient accuracy. With the equations given in footnote to p. 826, the value for β is found as $1.233 \cdot 10^{-2}$.

** It should be noticed that neutralization in this sense is quite distinct from the usage of the word on p. 819. In German chemical literature a distinction is drawn between 'Neutralisation', in the sense given above, and 'Absättigung' in the sense of chemical equivalence between acid and base. This distinction in terminology is lacking in English.

to the presence of hydroxyl ions is removed by adding hydrogen ions (addition of acid). In both cases, the essential process is the union of hydrogen ions and hydroxyl ions, to form practically undissociated water:

$$H^+ + OH^- = H_2O \tag{17}$$

If, to the solution of a dilute strong acid, an equivalent amount of a dilute strong base is added, the neutralization process consists solely of the reaction represented by eqn (17), in so far as the salt which is formed remains in solution. That this is indeed the case—the ions of the salt being present in the almost completely dissociated solutions of the acid and base, in the same form as in the solution of the salt itself—is shown by the fact that the heat liberated by the reaction of equivalent amounts of acid and base—the *heat of neutralization*—is almost independent, in dilute solution, of the nature of the acid and base.

This is no longer the case if weak acids and bases are involved in the reaction. If the solution of a weak acid is treated with a base, the dissociation equilibrium of the acid is disturbed because hydrogen ions react with hydroxyl ions of the base. The dissociation of the acid goes on as long as it is able to furnish hydrogen ions. Similar considerations apply to a weak base. Hence, with weak acids and weak bases, the heats of neutralization differ considerably, and depend upon the nature of the compound. In neutralization of weak acids and bases, equilibria are set up, whereby the hydrogen ion or hydroxyl ion concentrations are not given simply by the difference in amounts of acid and base mixed, but involve also the strength of the acid or base. It follows at once that this must be so, since solutions of 'neutral' salts (i.e., salts formed from equivalent amounts of acid and base) do not, as a rule, have a neutral reaction if the salts are derived from weak acids or bases.

Determination of the point at which exactly equivalent amounts of acid and base have entered into reaction—the so called *equivalence point*—is of great importance in the volumetric determination of acids and bases [1]. Color indicators are generally used for this purpose. Indicators are substances whose color depends upon the hydrogen ion concentration of the solution. An example is *litmus* dyestuff (*azolitmin*) which is red in acid solutions and blue in basic solutions. Strips of paper impregnated with this substance ('litmus paper')are often used in qualitative testing for acid or basic reaction, but litmus is rarely employed in the volumetric analytical determination of acids and bases. For this purpose the indicators *phenolphthalein* and *methyl orange* are most commonly used, the latter being replaced for some purposes by *methyl red*, which was introduced by Rupp in 1908. Phenolphthalein is colorless in acid solution and red in basic solution. Methyl orange and methyl red are red in acid solution, and orange yellow or yellow in basic solution.

It is important to note that the color change of the usual indicators does not coincide at all with the true *neutralization point* i.e., at the hydrogen ion concentrantion of pure water ($[H^+] = 10^{-7}$). The change does not, in fact, take place at a definite hydrogen ion concentration, but is spread over a range of concentrations which is usually fairly broad. Thus for litmus the color change from red to blue extends over the range $[H^+] = 10^{-5}$ to $[H^+] = 10^{-8}$ i.e., over three powers of ten in the hydrogen ion concentration. The color change ranges of other indicators are shown in Table 107 [2]. It may be seen that that of phenolphthalein lies in the more strongly basic region, and that of methyl orange and methyl red in the more acidic region.

TABLE 107

END POINT RANGES OF INDICATORS

Indicator	Color change range, in pH	Color change acidid ⇄ basic	Amount of indicator in 10 cc of solution
Methyl violet	0.1– 3.2	yellow ⇄ blue ⇄ violet	2– 8 drops of 0.05% aq. sol.
Metanil yellow	1.2– 2.3	red ⇄ yellow	3– 5 drops of 0.1% aq. sol.
Tropaolin oo	1.3– 3.2	red ⇄ yellow	1– 3 drops of 0.1% aq. sol.
Dimethyl yellow	2.9– 4.0	red ⇄ yellow	2– 5 drops of 0.1% alcoholic sol.
Methyl orange	3.1– 4.4	red ⇄ orange yellow	1– 3 drops of 0.1% aq. sol.
Congo red	3.0– 5.2	blue violet ⇄ red	1– 5 drops of 0.1% aq. sol.
Methyl red	4.2– 6.3	red ⇄ yellow	2– 4 drops of 0.2% alcoholic-aq. sol.
Lackmoid	4.4– 6.4	red ⇄ blue	1– 5 drops of 0.2% alcoholic sol.
p-Nitrophenol	5.0– 7.0	colorless ⇄ yellow	3–20 drops of 0.4% aq. sol.
Neutral red	6.8– 8.0	red ⇄ yellow	2– 4 drops of 0.1% alcoholic-aq. sol.
Litmus (Azolitmin)	5.0– 8.0	red ⇄ blue	10–20 drops of 0.2% aq. sol.
Rosolic acid	6.9– 8.0	brown ⇄ red	1– 4 drops of 0.1% alcoholic-aq. sol.
Curcumin	7.8– 9.2	yellow ⇄ red brown	1– 5 drops of 0.1% aq. sol.
Phenolphthalein	8.2–10.0	colorless ⇄ red	3–10 drops of 0.1% alcoholic-aq. sol.
Alizarin yellow	10.1–12.1	yellow ⇄ lilac	5–10 drops of 0.1% aq. sol.
Nitramine	11.0–13.0	colorless ⇄ orange brown	1– 3 drops of 0.1% alcoholic-aq. sol.

Where titrations of strong acids with strong bases are involved neither the deviation of the change point of these indicators from the true neutral point, nor the considerable range over which the color change extends, is of any practical importance. This is because the hydrogen ion concentration changes sharply by a very large factor in the immediate neighborhood of the neutral point.

Fig. 118 shows how the hydrogen ion exponent depends upon the proportions of acid and alkali—i.e., how the hydrogen ion concentration changes progressively, as a strong base is added in increasing amounts to a strong acid. The abscissa represents the proportions in which acid and base are mixed, so that the proportion 50% acid +50% base corresponds to the formation of the neutral salt. The ordinates are the hydrogen ion exponents, pH,—i.e., the negative logarithms of the hydrogen ion concentrations at each composition of the mixutre. It may be seen that the curve representing the dependence of hydrogen ion concentration upon composition of solution passes very steeply through the neutral point (pH = 7), so as to be almost vertical in passing through the color change ranges of both

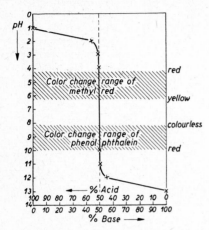

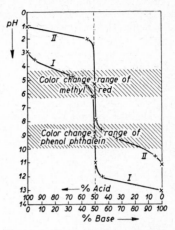

Fig. 118. Neutralization of a strong acid by a strong alkali

Fig. 119. Curve I Neutralization of acetic acid by sodium hydroxide. Curve II Neutralization of hydrochloric acid by aqueous ammonia.

phenolphthalein and methyl orange (shown shaded). Suppose that 50 cc. of 0.1-normal hydrochloric acid has been treated with 49.9 cc of 0.1-normal sodium hydroxide. 99.9 cc of the solution contain 0.00001 g equivalents of excess hydrogen ion, and therefore the hydrogen ion concentration is $10^{-4.0}$ g equivalents per liter (pH = 4.0). It is only necessary to add an additional 0.2 cc of 0.1-normal sodium hydroxide in order to depress the hydrogen ion concentration sharply to $10^{-10.0}$, or pH = 10.0, at which concentration the solution will have a basic reaction towards both indicators. [3]

The curve of the hydrogen exponent in the neutralization of a weak acid with a strong base, or a strong acid with a weak base, takes quite a different course, as is represented in Fig. 119. Curve I, for the neutralization of a weak acid with a strong base, shows the reaction of 0.1-normal acetic acid with 0.1-normal sodium hydroxide. In this case it is evident that the color change range of methyl red is traversed quite slowly, and even when the color change is complete, the solution still contains not quite as much base as is chemically equivalent to the acid. This is still not exactly the case at the neutrality point, pH = 7.0, and chemical equivalence attained only when the hydrogen ion concentration has been reduced to $10^{-8.8}$ (pH = 8.8), as will be shown below. If, in the titration, the acid were added to the sodium hydroxide, the equivalence point would already have been passed before the color change of methyl red commenced. Weak acids, such as acetic acid, can therefore not be titrated with methyl red serving as indicator (or with methyl orange, as is obvious from Table 107), but phenolphthalein can be used, as Curve I shows.

As an example of the conditions during the titration of a weak base with a strong acid, Curve II shows the course of the hydrogen ion exponent in titrating 0.1-normal hydrochloric acid with 0.1-normal ammonia. It is evident that weak bases cannot be titrated with phenolphthalein acting as indicator, but can be titrated with methyl red.

The values of the hydrogen exponents plotted in Figs. 118, 119 can not only be determined experimentally, but can very easily be calculated. For pure 0.1-normal hydrochloric acid, $[H^+] = 10^{-1}$ or $pH = 1$.* If 55 cc of the acid are mixed with 45 cc of 0.1-normal sodium hydroxide, 45 cc of the former are converted to salt, and the remainder (10 cc) is present in 100 cc of solution in all. The hydrogen ion concentration has thus been reduced by a factor of 10, and $pH = 2$. In the same way, for the mixture of 50.5 parts of acid with 49.5 parts of base, $pH = 3$, and for 50.05 parts of acid with 49.95 parts of base, $pH = 4$. The mixture of 50 parts of acid with 50 parts of base should (theoretically) have $pH = 7$,** and will be exactly neutral. A mixture of 49.95 parts of acid with 50.05 parts base contains 0.1 cc of 0.1-normal base in 100 cc of solution; thus $[OH^-] = 10^{-4}$, $pH = 10$. Similarly, for 49.5 parts of acid with 50.5 parts of base, $pH = 11$, etc.

In the case of a weak acid, the hydrogen ion concentration can be derived from the dissociation constant K_{HX} of the acid.

$$[H^+] = \frac{[HX]}{[X^-]} \cdot K_{HX} \tag{18}$$

$$pH = -\log [H^+] = \log [X^-] + \log [HX] - \log K_{HX} \tag{19}$$

For acetic acid at 18°, $K_{HX} = 1.75 \cdot 10^{-5} = 10^{-4.76}$; $-\log K_{HX}$ (now frequently written as pK_{HX}) $= 4.76$. In the pure acid, $[H^+] = [X^-]$. Thus

$$[H^+]^2 = [HX] \cdot K_{HX}$$

$$pH = \tfrac{1}{2} \log [HX] + \tfrac{1}{2} pK_{HX}$$

$[HX]$ can, in practice, be put equal to the total concentration of the acid. Thus, for 0.1-normal acetic acid,

$$pH = -\tfrac{1}{2} \log 0.1 + \tfrac{1}{2} \cdot 4.76 = 0.5 + 2.38 = 2.88$$

A mixture of 95 parts of 0.1-N acetic acic and 5 parts of 0.1-N sodium hydroxide is $\frac{5}{100} \cdot$ 0.1-normal in sodium acetate, and $\frac{90}{100} \cdot$ 0.1-normal in free acetic acid. For this solution, the hydrogen ion exponent is found from (19) as

$$pH = \log \frac{5 \cdot 0.1}{100} - \log \frac{90 \cdot 0.1}{100} + 4.76 = 3.51$$

In a mixture of two parts of acetic acid with one part of sodium hydroxide, the concentration of salt is equal to the concentration of free acid. For this,

$$[X^-] = [HX], \qquad pH = pK_{HX} = 4.76$$

In a similar manner, a mixture of 55 parts of acid with 45 parts of sodium hydroxide has $pH = 5.41$; for 51 parts of acetic acid with 49 parts of sodium hydroxide, $pH = 6.15$. The 'neutral' salt is obtained by mixing 50 parts of 0.1-normal acetic acid with 50 parts of 0.1-normal sodium hydroxide, but this mixture is basic in reaction, through hydrolysis.

* The hydrogen ion activity in 0.1-N HCl has the value $10^{-1.05}$. The slight difference is negligible in this connection.
** Because of inevitable impurities this will not, in practice, be exactly the case—see the last section of this chapter.

The hydrogen ion exponent can be calculated from eqn (15a), p. 826. Since $c/K_{HX} \gg 1$, this becomes

$$[OH^-] = k_w^{\frac{1}{2}} \cdot (c/K_{HX})^{\frac{1}{2}}, \qquad \text{or} \qquad [H^+] = k_w^{\frac{1}{2}} \cdot K_{HX}^{\frac{1}{2}}/c^{\frac{1}{2}}.$$

Hence

$$pH = -\tfrac{1}{2}\log k_w + \tfrac{1}{2}pK_{HX} + \tfrac{1}{2}\log c.$$

The solution is 0.05-normal with respect to the salt, so that

$$pH = 7.06 + 2.38 - 0.65 = 8.79$$

A mixture of 49.5 cc of acetic acid with 50.5 cc of sodium hydroxide contains 1 cc of un-reacted sodium hydroxide in 100 cc and has pH = 11.0. From this point onwards the course of the titration curve is the same as that in the first example.

If aqueous ammonia is mixed with a strong acid, the *hydroxyl ion concentrations* in mixtures with various proportions have practically the same values as the hydrogen ion concentrations in the acetic acid- sodium hydroxide mixtures just discussed, since the dissociation constant of ammonia hydrate is almost the same as that of acetic acid. Curve II, Fig. 119, thus has the same form as Curve I, except that hydroxyl ion exponents, pOH $(= -\log [OH^-])$ are plotted upwards, instead of hydrogen ion exponents being plotted from above, downwards.

6. Colorimetric Determination of Hydrogen Ion Concentration

It is possible to *determine hydrogen ion concentrations* colorimetrically by utilizing the varied position of the color change of different indicators, and the progressive change of color over a range of hydrogen ion concentration [2]. Thus, if it be found that a solution turns red on adding methyl red, but not on adding methyl orange, it follows that the pH must be less than 6, but greater than 3. A more precise value for the hydrogen ion concentration can then be found by comparing the tint of samples, to which methyl orange or methyl red has been added, with that of solutions of known hydrogen ion concentration containing the same indicator.

7. Buffer Mixtures

Over a certain range of hydrogen ion concentrations (depending on the dissociation constant of the weak acid or base), a mixture of a weak acid or a weak base with one of its salts undergoes a relatively small change in hydrogen ion concentration as acid or alkali is added. This phenomenon is utilized when it is required to make up solutions with a low, but accurately specified hydrogen ion concentration. Such solutions cannot be made up simply by diluting solutions of strong acids or strong bases, for if the normality is reduced by dilution below about 0.01-N or 0.001-N the addition of even minimal amounts of any acid or alkali serves to exert a relatively enormous influence on the hydrogen ion concentration. It is, however, difficult to avoid the access of traces of acid (carbon dioxide from the air) or alkali (leached out of glass vessels). Solutions of low, but fixed, hydrogen ion concentration are therefore prepared by mixing weak acids or bases with their salts. Since these solutions resiliently oppose changes in the hydrogen ion concentration, they are termed *buffer mixtures*. The buffering action of such mixtures decreases as they are diluted. At any given concentration, it is greatest for mixtures containing roughly equal amounts of the acid (or base) and its salt. From the previous discussion, it will be evident that at this mid-point of the buffering range, the hydrogen ion exponent is given b

$$pH = pK_{HX}$$

(pK_{HX} = the negative logarithm of the dissociation constant of the weak acid).

Data for the preparation of buffer mixtures covering a wide range of hydrogen ion concentrations are given in the monographs cited below.

References

1 N. BJERRUM, *Die Theorie der alkalimetrischen und azidimetrischen Titrierungen*, Stuttgart 1914, 128 pp.

2. I. M. KOLTHOFF, *Acid-Base Indicators* (translated by Ch. Rosenblum), New York 1937, 414 pp.

3. I. M. KOLTHOFF and H. A. LAITINEN, *pH and Electro Titrations*, 2nd Ed., New York 1941, 190 pp.

APPENDIX

TABLE I

INTERNATIONAL ATOMIC WEIGHTS

Element	Symbol	Atomic Number	Atomic Weight	Element	Symbol	Atomic Number	Atomic Weight
Actinium	Ac	89	227	Mercury	Hg	80	200.61
Aluminum	Al	13	26.98	Molybdenum	Mo	42	95.95
Americium	Am	95	[243]				
Antimony	Sb	51	121.76	Neodymium	Nd	60	144.27
Argon	A	18	39.944	Neon	Ne	10	20.183
Arsenic	As	33	74.91	Neptunium	Np	93	[237]
Astatine	At	85	[210]	Nickel	Ni	28	58.69
				Niobium	Nb	41	92.91
Barium	Ba	56	137.36	Nitrogen	N	7	14.008
Berkelium	Bk	97	[245]				
Beryllium	Be	4	9.013	Osmium	Os	76	190.2
Bismuth	Bi	83	209.00	**Oxygen**	**O**	**8**	**16.0000**
Boron	B	5	10.82				
Bromine	Br	35	79.916	Palladium	Pd	46	106.7
				Phosphorus	P	15	30.975
Cadmium	Cd	48	112.41	Platinum	Pt	78	195.23
Calcium	Ca	20	40.08	Plutonium	Pu	94	[242]
Californium	Cf	98	[248]	Polonium	Po	84	210
Carbon	C	6	12.011	Potassium	K	19	39.100
Cerium	Ce	58	140.13	Praseodymium	Pr	59	140.92
Cesium	Cs	55	132.91	Promethium	Pm	61	[145]
Chlorine	Cl	17	35.457	Protactinium	Pa	91	231
Chromium	Cr	24	52.01				
Cobalt	Co	27	58.94	Radium	Ra	88	226.05
Copper	Cu	29	63.54	Radon	Rn	86	222
Curium	Cm	96	[245]	Rhenium	Re	75	186.31
				Rhodium	Rh	45	102.91
Dysprosium	Dy	66	162.46	Rubidium	Rb	37	85.48
				Ruthenium	Ru	44	101.1
Einsteinium	E	99	253				
Erbium	Er	68	167.2	Samarium	Sm	62	150.43
Europium	Eu	63	152.0	Scandium	Sc	21	44.96
				Selenium	Se	34	78.96
Fermium	Fm	100	254	Silicon	Si	14	28.09
Fluorine	F	9	19.00	Silver	Ag	47	107.880
Francium	Fr	87	[223]	Sodium	Na	11	22.991
				Strontium	Sr	38	87.63
Gadolinium	Gd	64	156.9	Sulfur	S	16	32.066
Gallium	Ga	31	69.72				
Germanium	Ge	32	72.60	Tantalum	Ta	73	180.95
Gold	Au	79	197.0	Technetium	Tc	43	[99]
Hafnium	Hf	72	178.6	Tellurium	Te	52	127.61
Helium	He	2	4.003	Terbium	Tb	65	158.93
Holmium	Ho	67	164.94	Thallium	Tl	81	204.39
Hydrogen	H	1	1.0080	Thorium	Th	90	232.05
				Thulium	Tm	69	168.94
Indium	In	49	114.76	Tin	Sn	50	118.70
Iodine	I	53	126.91	Titanium	Ti	22	47.90
Iridium	Ir	77	192.2	Tungsten	W	74	183.92
Iron	Fe	26	55.85				
				Uranium	U	92	238.07
Krypton	Kr	36	83.80				
				Vanadium	V	23	50.95
Lanthanum	La	57	138.92				
Lead	Pb	82	207.21	Xenon	Xe	54	131.3
Lithium	Li	3	6.940				
Lutetium	Lu	71	174.99	Ytterbium	Yb	70	173.04
				Yttrium	Y	39	88.92
Magnesium	Mg	12	24.32				
Manganese	Mn	25	54.94	Zinc	Zn	30	65.38
Mendelevium	Mv	101	256	Zirconium	Zr	40	91.22

TABLE

THE PERIODIC

The atomic number of each element is given to the left of and below the name, and the atomic weight above. In[...] thanides and for the elements following actinium these are not yet fully established.) Elements marked with an asterisk[...] have been obtained artificially (by nuclear[...]

Periods	Series	I Sub-group	I Main group	II Sub-group	II Main group	III Sub-group	III Main group	IV Sub-group	IV Main group
	1								
1st short Period	2		6.940 **Lithium** 3 2_s		9.013 **Beryllium** 4 2_s^2		10.82 **Boron** 5 $2_s^2 2_p$		12.011 **Carbon** 6 $2_s^2 2_p^2$
2nd short Period	3		22.998 **Sodium** 11 3_s		24.32 **Magnesium** 12 3_s^2		26.978 **Aluminum** 13 $3_s^2 3_p$		28.09 **Silicon** 14 $3_s^2 3_p^2$
1st long Period	4		39.100 **Potassium** 19 4_s		40.08 **Calcium** 20 4_s^2	44.96 **Scandium** 21 $3d4s^2$		47.90 **Titanium** 22 $3d^2 4s^2$	
1st long Period	5	63.54 **Copper** 29 $3d^{10}+$ $4s$		65.38 **Zinc** 30 $4s^2$			69.72 **Gallium** 31 $4s^2 4p$		72.60 **Germanium** 32 $4s^2 4p^2$
2nd long Period — Kr$_{shell}$+	6		85.48 **Rubidium** 37 $5s$		87.63 **Strontium** 38 $5s^2$	88.92 **Yttrium** 39 $4d5s^2$		91.22 **Zirconium** 40 $4d^3 5s$	
2nd long Period — Kr$_{shell}$+	7	107.880 **Silver** 47 $4d^{10}+$ $5s$		112.41 **Cadmium** 48 $5s^2$			114.76 **Indium** 49 $5s^2 5p$		118.70 **Tin** 50 $5s^2 5p^2$
3rd long Period — Xe$_{shell}$+	8		132.91 **Cesium** 55 $6s$		137.36 **Barium** 56 $6s^2$	138.92 **Lanthanum** 57 $5d6s^2$		Lanthanides 58–71 | 178.6 **Hafnium** 72 $5d^2 6s^2$	
3rd long Period — Xe$_{shell}$+	9	197.0 **Gold** 79 $5d^{10}+$ $6s$		200.60 **Mercury** 80 $6s^2$			204.39 **Thallium** 81 $6s^2 6p$		207.21 **Lead** 82 $6s^2 6p^2$
4th long Period	10		223 **Francium*** 87 $7s$		226.05 **Radium*** 88 $7s^2$	227 **Actinium*** 89 $6d7s^2$		232.05 **Thorium*** 90 $6d^2 7s^2$	

Lanthanides:					
140.13 **Cerium** 58 $4f5d6s^2$	140.92 **Praseodymium** 59 $4f^2 5d6s^2$	144.27 **Neodymium** 60 $4f^3 5d6s^2$	145 *Promethium** 61 $4f^4 5d6s^2$	150.43 **Samarium** 62 $4f^5 5d6s^2$	152.0 **Europium** 63 $4f^6 6d6s^2$

Transuranic elements (Actinides; cf. Vol. II, Chap. 14):

	237 **Neptunium*** 93 $5f^5 7s^2$	239 **Plutonium*** 94 $5f^6 6d7s^2$	243 *Americium* 95 $5f^6 6d7s^2$

I

SYSTEM

the lowest lines of each row are given the electronic configurations of the outer shells of the atoms. (For many lan-
are those for which every nuclear species is unstable (i.e. radioactive). The elements with names printed in italics
reactions), but have not been detected in Nature.

V		VI		VII		VIII			Main group (or Group o)
Sub-group	Main group	Sub-group	Main group	Sub-group	Main group	Sub-group			
				1.0080 **Hydrogen** *1* 1_s					4.003 **Helium** *2* 1_s^2
14.008 **Nitrogen** *7* $2_s^2 2_p^3$		16.0000 **Oxygen** *8* $2_s^2 2_p^4$		19.00 **Fluorine** *9* $2_s^2 2_p^5$					20.183 **Neon** *10* $2_s^2 2_p^6$
30.975 **Phosphorus** *15* $3_s^2 3_p^3$		32.066 **Sulphur** *16* $3_s^2 3_p^4$		35.457 **Chlorine** *17* $3_s^2 3_p^5$					39.944 **Argon** *18* $3_s^2 3_p^6$
50.95 **Vanadium** *23* $3_d^3 4_s$		52.01 **Chromium** *24* $3_d^5 4_s$		54.94 **Manganese** *25* $3_d^5 4_s^2$		55.85 **Iron** *26* $3_d^6 4_s^2$	58.94 **Cobalt** *27* $3_d^7 4_s^2$	58.69 **Nickel** *28* $3_d^8 4_s^2$	
74.91 **Arsenic** *33* $4_s^2 4_p^3$		78.96 **Selenium** *34* $4_s^2 4_p^4$		79.916 **Bromine** *35* $4_s^2 4_p^5$					83.80 **Krypton** *36* $4_s^2 4_p^6$
92.91 **Niobium** *41* $4_d^4 5_s$		92.95 **Molybdenum** *42* $4_d^5 5_s$		99 *Technetium** *43* $4_d^5 5_s^2$		101.1 **Ruthenium** *44* $4_d^7 5_s$	102.91 **Rhodium** *45* $4_d^8 5_s$	106.7 **Palladium** *46* 4_d^{10}	
121.76 **Antimony** *51* $5_s^2 5_p^3$		127.61 **Tellurium** *52* $5_s^2 5_p^4$		126.92 **Iodine** *53* $5_s^2 5_p^5$					131.3 **Xenon** *54* $5_s^2 5_p^6$
180.95 **Tantalum** *73* $5_d^3 6_p^2$		183.92 **Tungsten** *74* $5_d^4 6_p^2$		185.31 **Rhenium** *75* $5_d^5 6_p^2$		190.2 **Osmium** *76* $5_d^6 6_s^2$	192.2 **Iridium** *77* $5_d^7 6_p^2$	195.23 **Platinum** *78* $5_d^9 6_s$	
209.00 **Bismuth** *83* $6_s^2 6_p^3$		210 **Polonium** *84* $6_s^2 6_p^4$		211 **Astatine** *85* $6_s^2 6_p^5$					222 **Radon** *86* $6_s^2 6_p^6$
231 **Protactinium** *91* $6_d^3 7_s^2$		238.07 **Uranium** *92* $6_d^4 7_s^2$		Transuranic elements (see below)					

156.9 **Gadolinium** *64* $4_f^7 5_d 6_s^2$	158.93 **Terbium** *65* $4_f^8 5_d 6_s^2$	162.46 **Dysprosium** *66* $4_f^9 5_d 6_s^2$	164.94 **Holmium** *67* $4_f^{10} 5_d 6_s^2$	167.2 **Erbium** *68* $4_f^{11} 5_d 6_s^2$	168.94 **Thulium** *69* $4_f^{13} 6_s^2$	173.04 **Ytterbium** *70* $4_f^{14} 6_s^2$	174.99 **Lutetium** *71* $4_f^{14} 5_d 6_s^2$
245 *Curium** *96* $5_f^7 6_d 7_s^2$	245 *Berkelium** *97* $5_f^8 6_d 7_s^2$	248 *Californium** *98* $5_f^9 6_d 7_s^2$	253 *Einsteinium** *99* $5_f^{10} 6_d 7_s^2$	254 *Fermium** *100* $5_f^{11} 6_d 7_s^2$	256 *Mendelevium** *101* $5_f^{12} 6_d 7_s^2$		

Suggestions for Further Reading

General

W. G. BERL (Editor), *Physical Methods in Chemical Analysis*, 2 vols., New York 1950–51.

J. O' M. BOCKRIS and B. E. CONWAY (Editors), *Modern Aspects of Electrochemistry*, New York 1954.

J. A. V. BUTLER (Editor), *Electrical Phenomena at Interfaces*, New York 1952.

B. E. CONWAY, *Electrochemical Data*, Amsterdam 1952.

R. E. DODD and P. L. ROBINSON, *Experimental Inorganic Chemistry*, Amsterdam 1953.

C. DUVAL, *Inorganic Thermogravimetric Analysis*, Amsterdam 1953.

H. J. EMELEUS and J. S. ANDERSON, *Modern Aspects of Inorganic Chemistry*, 2nd Ed., New York 1952.

A. FINDLAY, *Phase Rule and Its Applications*, 9th Ed., New York 1951.

J. A. N. FRIEND, *Man and the Chemical Elements; From Stone Age Hearth to the Cyclotron*, London 1951.

S. GLASSTONE, *Introduction to Electrochemistry*, New York 1942, London 1947.

S. GLASSTONE, *Textbook of Physical Chemistry*, 2nd Ed., New York 1946.

E. A. GUGGENHEIM, *Thermodynamics*, 2nd Ed., Amsterdam and New York 1950.

E. A. GUGGENHEIM and J. E. PRUE, *Physicochemical Calculations*, Amsterdam 1954, New York 1955.

R. W. GURNEY, *Ionic Processes in Solution*, New York 1953.

H. S. HARNED and B. B. OWEN, *Physical Chemistry of Electrolytic Solutions*, 2nd Ed., New York 1950.

C. D. HODGMAN (Editor), *Handbook of Chemistry and Physics*, 36th Ed., Cleveland (Ohio) 1955.

W. HÜCKEL, *Structural Chemistry of Inorganic Compounds*, 2 vols., Amsterdam, 1950–51.

Inorganic Syntheses, 4 vols., New York 1939–53.

B. JAFFE, *Crucibles; The Story of Chemistry from Ancient Alchemy to Nuclear Fission*, New York 1948, London 1950.

R. E. KIRK and D. F. OTHMER, *Encyclopedia of Chemical Technology*, 15 vols., New York 1947–55.

I. M. KOLTHOFF and E. B. SANDELL, *Textbook of Quantitative Inorganic Analysis*, 3rd Ed., New York 1952.

G. KORTÜM and J. O'M. BOCKRIS, *Textbook of Electrochemistry*, 2 vols., Amsterdam 1951.

W. M. LATIMER and J. H. HILDEBRAND, *Reference Book of Inorganic Chemistry*, 3rd Ed., New York 1951.

J. J. LINGANE, *Electroanalytical Chemistry*, New York 1953.

R. E. LYNN, The Critical Properties of Elements and Compounds, *Chem. Revs.*, 52 (1953) 117–236.

E. B. MAXTED, *Modern Advances in Inorganic Chemistry*, Oxford 1947.

J. W. MELLOR, *A Comprehensive Treatise on Inorganic and Theoretical Chemistry*, 16 vols., London 1922–36.

E. MOLLOY (Editor), *Heavy Chemicals, Manufacture and Use*, Hollywood-by-the-Sea (Fla.) 1955.

W. J. MOORE, *Physical Chemistry*, New York 1950.

O. H. MÜLLER, *The Polarographic Method of Analysis*, 2nd Ed., Easton (Pa.) 1951.

W. G. PALMER, *Experimental Inorganic Chemistry*, Cambridge 1953.

J. R. PARTINGTON, *Advanced Treatise on Physical Chemistry*, 5 vols., London 1949–55.

H. A. J. PIETERS and J. W. CREYGHTON, *Safety in the Chemical Laboratory*, London 1951.

M. J. N. POURBAIX, *Thermodynamics of Dilute Aqueous Solutions*, London 1949.

F. F. PURDON and V. W. SLATER, *Aqueous Solution and the Phase Diagram*, New York 1946.

J. REILLY and W. N. RAE, *Physico-Chemical Methods*, 5th Ed., New York 1954.

J. E. RICCI, *Phase Rule and Heterogeneous Equilibrium*, New York 1951.

F. D. ROSSINI, *Chemical Thermodynamics*, New York 1950.

E. B. SANDELL, *Colorimetric Determination of Traces of Metals*, 2nd Ed., New York 1950.

A. SEIDELL and W. F. LINKE, *Solubilities of Inorganic and Metal Organic Compounds*, 2 vols., 3rd Ed., New York 1940; Supplement New York 1952.

N. V. SIDGWICK, *Chemical Elements and Their Compounds*, 2 vols., Oxford 1950.

M. C. SNEED, J. L. MAYNARD and R. C. BRASTED (Editors), *Comprehensive Inorganic Chemistry*, 11 vols., New York 1953 etc.

F. D. SNELL and C. T. SNELL, *Colorimetric Methods of Analysis*, Vol. II, *Inorganic*, 3rd Ed., New York 1949.
H. S. TAYLOR and S. GLASSTONE (Editors), *Treatise on Physical Chemistry*, 5 vols., 3rd Ed., New York 1951.
Thorpe's Dictionary of Applied Chemistry, 4th Ed., 11 vols., London 1937–54.
A. E. VAN ARKEL, *Molecules and Crystals in Inorganic Chemistry*, New York 1949.
A. F. WELLS, *Structural Inorganic Chemistry*, Oxford 1950.

Chapter 1

A. F. SCOTT and M. BETTMAN, A Comparison of the Chemical and Physical Atomic Weight Values of the Monoisotopic Elements, *Chem. Revs.*, 50 (1952) 363–74.
E. WEEKS, *Discovery of the Elements*, 5th Ed., Easton (Pa.) 1945.

Chapter 2

American Public Health Association, *Standard Methods for the Examination of Water and Sewage*, 9th Ed., New York 1946.
R. G. BATES, Definitions of pH Scales, *Chem. Revs.*, 42 (1948) 1–61.
R. P. BELL, The Use of the Terms "Acid" and "Base", *Quart. Revs.*, 1 (1947) 113–25.
C. S. FOX, *Water: A Study of Its Properties, Its Constitution, Its Circulation on the Earth, and Its Utilization by Man*, London 1952.
N. G. GAYLLORD, *Reduction with Complex Hydrides*, New York 1955.
D. T. HURD, *Introduction to the Chemistry of the Hydrides*, New York 1952.
T. KING, *Water, Miracle of Nature*, New York 1953.
P. H. KUENEN, *Realms of Water: Some Aspects of Its Cycle in Nature*, London 1955.
F. A. LONG and W. F. McDEVIT, Activity Coefficients of Nonelectrolytic Solutes in Aqueous Salt Solutions, *Chem. Revs.*, 51 (1952) 119–69.
W. F. LUDER and S. ZUFFANTI, *Electronic Theory of Acids and Bases*, New York 1946.
J. MITCHELL and D. M. SMITH, *Aquametry*, New York 1948.
P. G. OWSTON, The Structure of Ice, *Quart. Revs.*, 5 (1951) 344–63.
F. C. PHILIPS, Oceanic Salt Deposits, *Quart. Revs.*, 1 (1947) 91–111.
W. C. SCHUMB, C. N. SATTERFIELD and R. L. WENTWORTH, *Hydrogen Peroxide*, New York 1955.
D. P. SMITH, *Hydrogen in Metals*, Chicago 1948.
A. F. WELLS, The Crystal Structure of Salt Hydrates and Complex Halides, *Quart. Revs.*, 8 (1954) 380–403.

Chapter 3

H. EYRING, J. E. WALTER and G. E. KIMBALL, *Quantum Chemistry*, New York 1944.
W. FINKELNBURG, *Atomic Physics*, New York 1950.
H. T. FLINT. *Wave Mechanics*, 8th Ed., London 1953.
S. GLASSTONE, *Theoretical Chemistry; An Introduction to Quantum Mechanics, Statistical Mechanics, and Molecular Spectra for Chemists*, New York 1944.
W. HEITLER, *Elementary Wave Mechanics*, Oxford 1945.
N. F. MOTT and I. N. SNEDDON, *Wave Mechanics and Its Applications*, Oxford 1948.
K. S. PITZER, *Quantum Chemistry*, New York 1953.
G. F. J. TEMPLE, *General Principles of Quantum Theory*, London 1934.

Chapter 5

H. EYRING *et al.*, see *Chapter 3*.
S. GLASSTONE, see *Chapter 3*.
J. A. A. KETELAAR, *Chemical Constitution; An Introduction to the Theory of the Chemical Bond*, Amsterdam 1953.
W. G. PALMER, *Valency, Classical and Modern*, New York 1944.
K. S. PITZER, see *Chapter 3*.
F. O. RICE and E. TELLER, *Structure of Matter*, New York 1949.
J. C. SPEAKMAN, *Introduction to the Electronic Theory of Valence*, 3rd Ed., London 1955.

Chapter 6

T. P. Hou, *Manufacture of Soda*, New York 1942.

W. P. Kelley, *Alkali Soils*, New York 1951.

J. J. Kennedy, The Alkali Metal Cesium and Some of Its Salts, *Chem. Revs.*, 23 (1938) 157–63.

R. Levine and W. Conrad, The Chemistry of the Alkali Amides, *Chem. Revs.*, 54 (1954) 449–66.

G. D. Van Arsdale (Editor), *Hydrometallurgy of Base Metals*, New York 1953.

Chapter 7

R. Beeching, *Electron Diffraction*, 2nd Ed., London 1946.

J. M. Bijvoet, N. H. Kolkmeyer and C. H. McGillavry, *X-Ray Analysis of Crystals*, London 1951.

R. C. Evans, *An Introduction to Crystal Chemistry*, Cambridge 1939.

R. W. James, *X-Ray Crystallography*, 5th Ed., New York 1953.

H. P. Klug and L. E. Alexander, *X-Ray Diffraction Procedures for Polycrystalline and Amorphous Materials*, New York 1954.

P. G. Owston, The Structure of Ice, *Quart. Revs.*, 5 (1951) 344–63.

Z. G. Pinsker, *Electron Diffraction*, London 1953.

A. Taylor, *Introduction to X-Ray Metallography*, New York 1945.

A. F. Wells, The Crystal Structure of Salt Hydrates and Complex Halides, *Quart. Revs.*, 8 (1954) 380–403.

R. W. G. Wyckoff, *Crystal Structures*, New York 1948.

W. H. Zachariasen, *Theory of X-Ray Diffraction in Crystals*, New York 1945.

Chapter 8

W. Bulian and E. Fahrenhorst, *Metallography of Magnesium and Its Alloys*, London 1944.

W. F. Neuman and M. W. Neuman, The Nature of the Mineral Phase of Bone, *Chem. Revs.*, 53 (1953) 1–45.

G. D. Van Arsdale, see *Chapter 6*.

Chapter 9

R. B. Barnes, R. C. Gore, U. Liddel and V. Z. Williams, *Infra-red Spectroscopy; Industrial Applications and Bibliography*, New York 1944.

L. J. Bellamy, *Infra-rad Spectra of Complex Molecules*, New York 1954.

W. R. Brode, *Chemical Spectroscopy*, New York 1943.

T. L. Cottrell, *Strengths of Chemical Bonds*, New York 1954.

W. Gordy, W. V. Smith and R. H. Trambarulo, *Microwave Spectroscopy*, New York 1953.

G. Herzberg, *Molecular Spectra and Molecular Structure*, 2nd Ed., New York 1950.

K. Hoselitz, *Ferromagnetic Properties of Metals and Alloys*, Oxford 1952.

J. A. A. Ketelaar, see *Chapter 5*.

L. H. Long, The Heats of Formation of Simple Inorganic Compounds, *Quart. Revs.*, 7 (1953) 134–74.

R. S. Nyholm, Magnetism and Inorganic Chemistry, *Quart. Revs.*, 7 (1953) 377–406.

Ya. K. Syrkin and M. E. Dyatkina, *Structure of Molecules and the Chemical Bond*, New York 1950.

M. Szwarc, The Determination of Bond Dissociation Energies by Pyrolytic Methods, *Chem. Revs.*, 47 (1950) 75–173.

F. O. Rice and E. Teller, see *Chapter 5*.

Chapter 10

R. P. Bell, The Boron Hydrides and Related Compounds, *Quart. Revs.*, 2 (1948) 132–51.

H. Brown *et al.*, *Aluminum and Its Applications*, New York 1948.

N. N. Greenwood and R. L. Martin, Boron Trifluoride Coordination Compounds, *Quart. Revs.*, 8 (1954) 1–39.

D. R. Martin, Coordination Compounds of Boron Bromide and Boron Iodide, *Chem. Revs.*, 42 (1948) 581–99.

F. G. A. Stone, Chemistry of the Boron Hydrides, *Quart. Revs.*, 9 (1955) 174–201.

C. A. Thomas *et al.*, *Anhydrous Aluminum Chloride in Organic Chemistry*, New York 1941.

Chapter 11

J. S. ANDERSON, Chemistry of the Metal Carbonyls, *Quart. Revs.*, 1 (1947) 331–57.

F. BASOLO, Stereochemistry and Reaction Mechanisms of Hexacovalent Inorganic Complexes, *Chem. Revs.*, 52 (1953) 459–527.

J. BJERRUM, *Metal Ammine Formation in Aqueous Solution; Theory of the Reversible Step Reactions*, Copenhagen 1941.

J. BJERRUM, On the Tendency of the Metal Ions toward Complex Formation, *Chem. Revs.*, 46 (1950) 381–401.

R. E. BURK and O. J. GRUMMITT (Editors), *Chemical Architecture: Stereochemistry*, New York 1948.

A. R. BURKIN, The Stabilities of Complex Compounds, *Quart. Revs.*, 5 (1951) 1–21.

R. W. PARRY, The Molecular Volume as a Criterion of Bond Type in Coordination Compounds, *Chem. Revs.*, 46 (1950) 507–16.

J. V. QUAGLIANO, The Trans-Effect in Complex Inorganic Compounds, *Chem. Revs.*, 50 (1952) 201–60.

Symposium on Complex Inorganic Compounds, *Chem. Revs.*, 21 (1937) 1–128.

H. TAUBE, Rates and Mechanisms of Constitution in Inorganic Complexes in Solution, *Chem. Revs.*, 50 (1952) 69–126.

A. F. WELLS, The Crystal Structures of Salt Hydrates and Complex Halides, *Quart. Revs.*, 8 (1954) 380–403.

Chapter 12

R. H. BOGUE, *Chemistry of Portland Cement*, 2nd Ed., New York 1955.

R. M. GARRELS and F. T. GUCKER, Activity Coefficients and Dissociation of Lead Chloride in Aqueous Solutions, *Chem. Revs.*, 44 (1949) 117–34.

H. GILMAN and G. E. DUNN, Relationship between Analogous Organic Compounds of Silicon and Carbon, *Chem. Revs.*, 52 (1953) 77–115.

O. H. JOHNSON, The Germanes and Their Organo-Derivatives, *Chem. Revs.*, 49 (1951) 259–97.

O. H. JOHNSON, Germanium and Its Inorganic Compounds, *Chem. Revs.*, 51 (1952) 431–69.

F. M. LEA, The Constitution of Portland Cement, *Quart. Revs.*, 3 (1949) 82–93.

Lead Industries Association, *Lead in Modern Industry; Manufacture, Applications and Properties of Lead, Lead Alloys and Lead Compounds*, New York 1952.

R. W. LEEPER, L. SUMMERS and H. GILMAN, Organo-Lead Compounds, *Chem. Revs.*, 54 (1954) 101–67.

G. W. MOREY, *The Properties of Glass*, 2nd Ed., New York 1954.

H. L. RILEY, Amorphous Carbon and Graphite, *Quart. Revs.*, 1 (1947) 59–72.

B. H. SAGE and W. N. LACEY, *Thermodynamic Properties of the Lighter Paraffin Hydrocarbons and Nitrogen*, New York 1950.

O. SAMUELSON, *Ion Exchangers in Analytical Chemistry*, New York 1953.

J. G. VAIL, *Soluble Silicates; Their Properties and Uses*, 2 vols., New York 1952.

Chapter 13

C. E. BEYNON, *Physical Structure of Alloys*, London 1945.

W. BOAS, *Introduction to the Physics of Metals and Alloys*, New York 1947.

R. J. FORBES, *Metallurgy in Antiquity*, Leyden 1950.

W. HUME-ROTHERY, *Electrons, Atoms, Metals and Alloys*, London 1955.

W. HUME-ROTHERY and G. V. RAYNOR, *Structure of Metals and Alloys*, 3rd Ed., London 1954.

O. KUBASCHEWSKI and E. L. EVANS, *Metallurgical Thermochemistry*, New York 1951.

D. M. LIDDELL (Editor), *Handbook of Nonferrous Metallurgy*, 2 vols., 2nd Ed., New York 1945.

J. NEWTON, *Introduction to Metallurgy*, 2nd Ed., New York 1947.

E. SCHMIDT and W. BOAS, *Plasticity of Crystals, with Special Reference to Metals*, London 1950.

F. SEITZ, *Physics of Metals*, New York 1943.

Chapter 14

C. C. ADDISON and J. LEWIS, The Chemistry of the Nitrosyl Group (NO), *Quart. Revs.*, 9 (1955) 115–49.

L. F. Audrieth, Hydrazoic Acid and Its Inorganic Derivatives, *Chem. Revs.*, 15 (1934) 169–224.

L. J. Beckham, W. A. Fessler and M. A. Kise, Nitrosyl Chloride (NOCl), *Chem. Revs.*, 49 (1951) 319–96.

C. F. Callis, J. R. Van Wazer and P. G. Arvan, The Inorganic Phosphates as Polyelectrolytes, *Chem. Revs.*, 54 (1954) 777–96.

B. A. Fry, *The Nitrogen Metabolism of Micro-Organisms*, London 1955.

W. A. Jolly, Heats, Free Energies and Entropies in Liquid Ammonia, *Chem. Revs.*, 50 (1952) 351–61.

B. Topley, The Condensed Phosphates, *Quart. Revs.*, 3 (1949) 345–68.

W. H. Waggaman, *Phosphoric Acid, Phosphates and Phosphatic Fertilizers*, 2nd Ed., New York 1952.

B. Wendrow and K. A. Kobe, The Alkali Orthophosphates – Phase Equilibria in Aqueous Solution, *Chem. Revs.*, 54 (1954) 891–924.

Chapter 15

L. Brewer, The Thermodynamic Properties of the Oxides and Their Vaporization Processes, *Chem. Revs.*, 52 (1953) 1–75.

M. Dole, The Chemistry of the Isotopes of Oxygen, *Chem. Revs.*, 51 (1952) 263–300.

L. Long, The Ozonization Reaction, *Chem. Revs.*, 27 (1940) 437–93.

G. B. L. Smith, Selenium Oxychloride as a Solvent, *Chem. Revs.*, 23 (1938) 165–85.

A. Sommer, *Photoelectric Cells*, London 1946.

R. C. Walker, *Photoelectric Cells in Industry*, London 1948.

Chapter 16

G. Holst, The Chemistry of Bleaching and Oxidizing Agents, *Chem. Revs.*, 54 (1954) 169–94.

O. Kubaschewski and B. E. Hopkins, *Oxidation of Metals and Alloys*, New York 1953.

W. M. Latimer, *Oxidation Potentials*, 2nd Ed., New York 1951.

B. Lewis and G. von Elbe, *Combustion, Flames and Explosions of Gases*, New York 1951.

J. W. Loveland and P. J. Elving, Cathode-Ray Oscilloscopic Investigation of Phenomena at Polarizable Mercury Electrodes, *Chem. Revs.*, 51 (1952) 67–117.

R. J. Marens and H. Eyring, Inorganic Oxidation-Reduction Reactions in Solution, *Chem. Revs.*, 55 (1955) 157–80.

E. W. R. Steacie, *Free Radical Mechanisms*, New York 1946.

E. W. R. Steacie, *Atomic and Free Radical Reactions*, 2 vols., 2nd Ed., New York 1954.

N. Uri, Inorganic Free Radicals in Solution, *Chem. Revs.*, 50 (1952) 375–454.

W. A. Waters, *Chemistry of Free Radicals*, 2nd Ed., Oxford 1948.

Chapter 17

J. Kleinberg and A. W. Davidson, The Nature of Iodine Solutions, *Chem. Revs.*, 42 (1948) 601–09.

H. R. Leech, Laboratory and Technical Production of Fluorine and Its Compounds, *Quart. Revs.*, 3 (1949) 22–35.

A. G. Sharpe, Interhalogen Compounds and Polyhalides, *Quart. Revs.*, 4 (1950) 115–30.

J. H. Simons (Editor), *Fluorine Chemistry*, 2 vols., New York 1950.

C. Slesser and S. R. Schram (Editors), *Preparation, Properties, and Technology of Fluorine and Organic Fluoro Compounds*, New York 1951.

Chapter 18

R. G. Bates, Definitions of pH Scales, *Chem. Revs.*, 42 (1948) 1–61.

R. P. Bell, The Use of the Terms "Acid" and "Base", *Quart. Revs.*, 1 (1947) 113–25.

R. W. Gurney, *Ionic Processes in Solution*, New York 1953.

NAME INDEX

A

Abegg, 127, 128, 386, 764 [2]
Abel, 593, 709, 803 [4], 813, 814
Acheson, 422, 468
Adams, 59
Addison, 594
Aiken, 378
Albrecht, 548 [54]
Alexander-Katz, 492 [38]
Algarotus, 670
Allison, 160
Alten, 366
Aluminium-Zentrale, 345 [5], 348 [5]
Ampère, 303, 778
Anderson, 300, 331, 415, 416, 742 [13]
Andrade, da Costa, 82 [4]
Andress, 641
Angeli, 618
Antropoff, von, 17 [4], 466
Arbusow, 631
Arfvedson, 164
Argo, 779
Aristotle, 163, 651
Arndt, 422, 423 [7], 426 [7]
Arnfelt, 425
Arnold, 334
Arrhenius, 64, 65
Aston, 43, 160
Aten, 701
Audrieth, 616 [11]
Auerbach, 449, 546, 764 [2]
Austerweil, 546
Aynsley, 669

B

Baborowsky, 75
Bacharach, 597
Bacher, 89 [12]
Bäckström, 771
Bacon, 703
Baeyer, von, 603, 697
Bailleul, 425 [8]
Balard, 778, 816
Ball, 528
Ballod, 492
Balmain, 737
Balmer, 80, 358
Balz, 622
Banco, 274 [4]
Barger, 298
Barkla, 213, 224
Barlett, 352

Barnard, 742 [12]
Barnes, 146
Barnik, 642
Barrer, 504
Bassett, H., 722
Bassett, H. N., 542 [53]
Baudler, 631, 730
Bauer, 330, 333, 359, 362, 520, 620, 789
Baumgarten, 333
Baxendale, 770
Bayliss, 296
Beach, 333, 789
Becco, 348 [12]
Beck, 253 [1]
Becker, 307 [17]
Beckmann, 701, 743
Béguyer de Chancourtois, 7
Beintema, 668
Bell, 328, 631
Bellucci, 532, 547
Bémont, 244
Benedict, 688
Benoit, 271, 272
Berg, 284, 370, 680
Berge, 509 [41]
Bergius, 429
Bergman, 675
Bergmann, 644
Berl, 425, 546, 709
Bernal, 50, 207, 265
Bernert, 8, 777
Bernthelot, 429, 717
Berthold, 739
Berthollet, 778, 803
Berzelius, 164, 244, 334, 470, 533, 612, 741, 795
Besson, 644
Bethe, 307
Beyer, 649
Bhatnagar, 302 [14], 304 [14]
Bichowsky, 148 [25], 271
Biesalski, 41
Bigelow, 440, 801
Billig, 295
Billiter, 182 [1], 184
Biltz, 9, 222 [11], 223 [11], 354, 359, 403, 486, 495, 516, 573, 586, 711
Birckenbach, 785
Birk, 816
Birkeland, 599
Bjerrum, 67, 394, 404, 444, 824 [1], 829 [1]
Black, 243
Blanc, 750

Blaser, 642
Bloch, 303 [15], 304 [15]
Bloch-Sée, 435 [19]
Blomstrand, 399
Blumenthal, 749
Bockemüller, 635
Bockris, 27 [26], 764 [7]
Bode, 317, 644, 726, 809
Bodenbender, 506 [37]
Bodenstein, 40, 598, 709, 807
Boersch, 801
Böeseken, 325
Böhm, 354, 372, 664
Bohr, 6, 84 [1], 414
Boisbaudran, Lecoq de, 369, 373
Bonhoeffer, 34, 42, 44, 486
Bonsdorff, von, 356
Booth, 333 [21], 483, 485, 731
Born, 90
Bosch, 608
Böttger, 509
Boudouard, 441
Boutaric, 300
Boyle, 200, 679
Bozorth, 462, 797
Bragg, W. H., 204 [15], 206, 207, 421
Bragg, W. L., 15, 204 [15], 204 [16], 206, 207 [5], 208, 209, 213, 301, 421, 498 [32]
Brand, 625
Brandenberger, 233 [10], 355
Brauer, 176
Braune, 643, 702, 782
Bredig, G., 54
Bredig, M. A., 640
Breuer, 182
Brewer, 159, 523
Bridgman, 628
Briegleb, 130 [7], 288 [6], 290, 685
Brill, 301, 451
Briner, 268, 592, 597, 601
Brintzinger, 299, 349, 453, 522, 535, 622, 655, 658, 714, 747
Briscoe, 323
Britton, 76 [17]
Broadley, 594
Brockway, 333, 360, 643, 801
Brodie, 628
Brodkorb, 175
Broglie, de, 90 [6], 91
Brönsted, 604
Brosset, 356
Brown, 130 [10], 146, 353
Browne, 605

SUBJECT INDEX

A

Absorption limit, 230, 231
Absorption spectrum of sodium, 170
Acceptants, 767
Acceptors, 767
Acetates, 455
Acetic acid, 446, 447, 454
— anhydride, 454
Acetylene, 435, 436, 437
Acetyl fluoroborate, 455
Acetylides, 435, 436, 466
Acids, bases, and salts, 57ff
— dissociation constant, 57
— electrolytic dissociation, 57
— preparation, 58ff
Acid acetates, 455
— amides, 615
— anhydrides, 58, 62, 696
— carbonates, 447
— fluorides, 786
— nitrates, 192
— salts, 822
— strength, 57
Acrylic acid, 438
Acrylic resins, 434
Actinon, 122
Activated oxygen, 694
Active charcoal, 425
Activity, 28, 35, 67
— coefficient, 28, 35, 36, 68, 69, 71
Affinity of formation, 140
Alabaster, 240, 280
Alanates, 362
Albite, 160
Alcoholysis, 335
Alkali alkyl compounds, 159
— carbonates, dissociation pressures, 193
— chlorides, 186
— — boiling points, 186
— — decomposition potentials, 186
— — melting points, 186
— compounds, heats of formation, 155
— hydrides, 173
— hydrogen selenides, 179
— — sulfides, 179
— — — crystal structures, 180
— hydroxides, 178ff
— — dissociation energies, 179
— — melting points, 179
— ions, electrolytic mobilities, 157

Alkali metals, 153ff
— — carbonates, 193
— — compounds, 172, 173
— — compressibilities, 167
— — crystal structure, 211
— — hydrogen carbonates, 193
— — occurrence, 160, 161
— — physical properties, 156
— — preparation, 164, 166
— — properties, 166
— — spectra, 168, 169
— — uses, 172
— metal amides, 168
— — chlorides, electrolysis, 182
— — higher oxides, 176
— — nitrates, 190
— — oxides, 175
— — salts, 157, 184ff
— minerals, 162
— oxides, 175
— polysulfides, polyselenides, polytellurides, 739
— salts, solubilities, 185
— sulfates, 198
Alkaline earth, amide formation, 248
— — ammoniate formation, 248
— — carbonates, 274
— — — decomposition pressures, 277
— — compounds, 255
— — crystal structure, 249
— — decomposition of carbonates, 238
— — decomposition potentials of chlorides, 237
— — flame spectra, lines, 251
— — fluorides, 265, 266
— — group, 234
— — halides, 265
— — heats of formation of compounds, 237
— — hydride formation, 248
— — hydrides, 238, 255, 256
— — hydroxides, 257
— — hyperoxides, 263, 264
— — imide formation, 248
— — melting points of chlorides, 237
— — minerals, 240, 241, 242
— — monohalides, 271
— — nitrates, 272
— — nitride formation, 248

Alkaline earths, normal potentials, 236
— — occurrence, 240
— — oxides and hydroxides, 256
— — pernitrides formation, 248
— — peroxides, 263, 264
— — physical properties, 235
— — preparation, 244
— — properties, 246
— — salts, 265ff
— — solubility of salts, 238
— — sulfates, 278
— — uses, 253
Alkanes, 428
Alkasil, 497
Alkenes, 428
Alkinol synthesis, 437
Alkoxy-aminosilanes, 472
Alkyl selenides, 752
— selenomercaptans, 752
— selenonium compounds, 752
— sulfides, 740
— sulfonic acid esters, 719
— tellurides, 752
— telluronium compounds, 752
Alkynes, 428
Allophanes, 508
Allotelluric acid, 750
Alloys, 561
— mechanical working, 570
— of magnesium, 253
Allylene, 466
Alumina, 350
Aluminates, 355
Aluminite, 363
Aluminothermic reduction, 470
Aluminothermy, 348
Aluminum, 315ff, 344ff
— acetate, 365
— alkyls, 359, 360
— analysis, 366
— borohydride, 330, 361
— bromide, 359
— carbide, 365
— chloride, 357, 358, 359
— compounds, 349ff
— double oxides, 356
— fluoride, 356
— halides, 356
— heats of formation of compounds, 322
— hydrides, 361
— hydroxide, 352, 355
— hydroxyfluoride, 357